Using *The Practice of Statistics*, Fourth Edition, for Advanced Placement (AP) Statistics

(The percents in parentheses reflect coverage on the AP exam.)

		Topic Outline for AP Statistics from the College Board's *AP Statistics Course Description*	*The Practice of Statistics,* 4th ed. Chapter and Section references
		I Exploring data: describing patterns and departures from patterns (20%–30%)	
A		**Constructing and interpreting graphical displays of distributions of univariate data (dotplot, stemplot, histogram, cumulative frequency plot)**	**Dotplot, stemplot, histogram 1.2; Cumulative relative frequency graph 2.1**
	1	Center and spread	1.2
	2	Clusters and gaps	1.2
	3	Outliers and unusual features	1.2
	4	Shape	1.2
B		**Summarizing distributions of univariate data**	**1.3 and 2.1**
	1	Measuring center: median, mean	1.3
	2	Measuring spread: range, interquartile range, standard deviation	1.3
	3	Measuring position: quartiles, percentiles, standardized scores (z-scores)	Quartiles 1.3; percentiles and z-scores 2.1
	4	Using boxplots	1.3
	5	The effect of changing units on summary measures	2.1
C		**Comparing distributions of univariate data (dotplots, back-to-back stemplots, parallel boxplots)**	**Dotplots and stemplots 1.2; boxplots 1.3**
	1	Comparing center and spread	1.2 and 1.3
	2	Comparing clusters and gaps	1.2 and 1.3
	3	Comparing outliers and unusual features	1.2 and 1.3
	4	Comparing shape	1.2 and 1.3
D		**Exploring bivariate data**	**Chapter 3 and Section 12.2**
	1	Analyzing patterns in scatterplots	3.1
	2	Correlation and linearity	3.1
	3	Least-squares regression line	3.2
	4	Residual plots, outliers, and influential points	3.2
	5	Transformations to achieve linearity: logarithmic and power transformations	12.2
E		**Exploring categorical data**	**Sections 1.1, 5.2, 5.3**
	1	Frequency tables and bar charts	1.1 (we call them bar graphs)
	2	Marginal and joint frequencies for two-way tables	Marginal 1.1; joint 5.2
	3	Conditional relative frequencies and association	5.3
	4	Comparing distributions using bar charts	1.1
		II Sampling and experimentation: planning and conducting a study (10%–15%)	
A		**Overview of methods of data collection**	**Sections 4.1 and 4.2**
	1	Census	4.1
	2	Sample survey	4.1
	3	Experiment	4.2
	4	Observational study	4.2
B		**Planning and conducting surveys**	**Section 4.1**
	1	Characteristics of a well-designed and well-conducted survey	4.1
	2	Populations, samples, and random selection	4.1
	3	Sources of bias in sampling and surveys	4.1
	4	Sampling methods, including simple random sampling, stratified random sampling, and cluster sampling	4.1
C		**Planning and conducting experiments**	**Section 4.2**
	1	Characteristics of a a well-designed and well-conducted experiment	4.2
	2	Treatments, control groups, experimental units, random assignments, and replication	4.2
	3	Sources of bias and confounding, including placebo effect and blinding	4.2
	4	Completely randomized design	4.2
	5	Randomized block design, including matched pairs design	4.2
D		**Generalizability of results and types of conclusions that can be drawn from observational studies, experiments, and surveys**	**Section 4.3**

Using *The Practice of Statistics*, Fourth Edition, for Advanced Placement (AP) Statistics

(The percents in parentheses reflect coverage on the AP exam.)

		Topic Outline for AP Statistics from the College Board's *AP Statistics Course Description*	***The Practice of Statistics,* 4th ed.** Chapter and Section references
		III Anticipating patterns: exploring random phenomena using probability and simulation (20%–30%)	
A		**Probability**	**Chapters 5 and 6**
	1	Interpreting probability, including long-run relative frequency interpretation	5.1
	2	"Law of large numbers" concept	5.1
	3	Addition rule, multiplication rule, conditional probability, and independence	Addition rule 5.2; other three topics 5.3
	4	Discrete random variables and their probability distributions, including binomial and geometric	Discrete random variables 6.1; Binomial and geometric 6.3
	5	Simulation of random behavior and probability distributions	5.1
	6	Mean (expected value) and standard deviation of a random variable, and linear transformation of a random variable	Mean and standard deviation 6.1; Linear transformation 6.2
B		**Combining independent random variables**	**Section 6.2**
	1	Notion of independence versus dependence	6.2
	2	Mean and standard deviation for sums and differences of independent random variables	6.2
C		**The Normal distribution**	**Section 2.2**
	1	Properties of the Normal distribution	2.2
	2	Using tables of the Normal distribution	2.2
	3	The Normal distribution as a model for measurements	2.2
D		**Sampling distributions**	**Chapter 7; Sections 8.3, 10.1, 10.2, 11.1**
	1	Sampling distribution of a sample proportion	7.2
	2	Sampling distribution of a sample mean	7.3
	3	Central limit theorem	7.3
	4	Sampling distribution of a difference between two independent sample proportions	10.1
	5	Sampling distribution of a difference between two independent sample means	10.2
	6	Simulation of sampling distributions	7.1
	7	*t* distribution	8.3
	8	Chi-square distribution	11.1
		IV Statistical inference: estimating population parameters and testing hypotheses (30%–40%)	
A		**Estimation (point estimators and confidence intervals)**	**Chapter 8 plus parts of Sections 9.3, 10.1, 10.2, 12.1**
	1	Estimating population parameters and margins of error	8.1
	2	Properties of point estimators, including unbiasedness and variability	8.1
	3	Logic of confidence intervals, meaning of confidence level and confidence intervals, and properties of confidence intervals	8.1
	4	Large-sample confidence interval for a proportion	8.2
	5	Large-sample confidence interval for a difference between two proportions	10.1
	6	Confidence interval for a mean	8.3
	7	Confidence interval for a difference between two means (unpaired and paired)	Paired 8.3; unpaired 10.2
	8	Confidence interval for the slope of a least-squares regression line	12.1
B		**Tests of significance**	**Chapters 9 and 11 plus parts of Sections 10.1, 10.2, 12.1**
	1	Logic of significance testing, null and alternative hypotheses; *P*-values; one-and two-sided tests; concepts of Type I and Type II errors; concept of power	9.1
	2	Large-sample test for a proportion	9.2
	3	Large-sample test for a difference between two proportions	10.1
	4	Test for a mean	9.3
	5	Test for a difference between two means (unpaired and paired)	Paired 9.3; unpaired 10.2
	6	Chi-square test for goodness of fit, homogeneity of proportions, and independence (one- and two-way tables)	Chapter 11
	7	Test for the slope of a least-squares regression line	12.1

The PRACTICE of STATISTICS

FOURTH EDITION

Director, BFW High School: Craig Bleyer
Executive Editor: Ann Heath
Assistant Editor: Dora Figueiredo
Associate Editor: Katrina Wilhelm
Development Editor: Donald Gecewicz
Media Editor: Laura Capuano
Assistant Media Editor: Catriona Kaplan
Executive Marketing Manager: Cindi Weiss
Photo Editor: Cecilia Varas
Photo Researcher: Julie Tesser
Design Manager: Blake Logan
Cover and Text Designer: Rae Grant
Cover Image: James Warwick/The Image Bank/Getty Images
Project Editor: Vivien Weiss
Illustrations: Network Graphics
Illustration Coordinator: Bill Page
Production Manager: Susan Wein
Composition: Preparé
Printing and Binding: RR Donnelley

Library of Congress Control Number: 2010934160

ISBN-13: 978-1-4292-4559-3
ISBN-10: 1-4292-4559-X

Fifth printing

Printed in the United States of America

W. H. Freeman and Company
41 Madison Avenue
New York, NY 10010
Houndmills, Basingstoke RG21 6XS, England
www.whfreeman.com

The PRACTICE of STATISTICS

FOURTH EDITION

Daren S. Starnes
The Lawrenceville School

Daniel S. Yates
Statistics Consultant

David S. Moore
Purdue University

W. H. Freeman and Company
New York

CONTENTS

Additional topics on CD-ROM and at http://www.whfreeman.com/tps4e:

ABOUT THE AUTHORS

Daren S. Starnes is Mathematics Department Chair and holds the endowed Master Teacher chair in Mathematics at The Lawrenceville School near Princeton, New Jersey. He earned his MA in Mathematics from the University of Michigan. Daren received a GTE GIFT Grant to integrate AP Statistics and AP Environmental Science. He was also named a Tandy Technology Scholar. Daren has led numerous one-day and week-long AP Statistics institutes for new and experienced AP teachers, and he has been a reader, table leader, and question leader for the AP Statistics exam since 1998. Daren is a frequent speaker at local, state, regional, national, and international conferences. For two years, he served as coeditor of the Technology Tips column in the NCTM journal *The Mathematics Teacher*. From 2004 to 2009, Daren served on the ASA/NCTM Joint Committee on the Curriculum in Statistics and Probability (which he chaired in 2009). While on the committee, he edited the *Guidelines for Assessment and Instruction in Statistics Education* (GAISE) pre-K–12 report and coauthored (with Roxy Peck) *Making Sense of Statistical Studies*, a capstone module in statistical thinking for high school students. Daren is also coauthor of the popular text *Statistics Through Applications*, First and Second Editions.

Daniel S. Yates has taught Advanced Placement Statistics in the Electronic Classroom (a distance-learning facility) affiliated with Henrico County Public Schools in Richmond, Virginia. Prior to high school teaching, he was on the mathematics faculty at Virginia Tech and Randolph-Macon College. He has a PhD in Mathematics Education from Florida State University. He has served as President of the Greater Richmond Council of Teachers of Mathematics and the Virginia Council of Teachers of Mathematics. Named a Tandy Technology Scholar in 1997, Dan received a College Board/Siemens Foundation Advanced Placement Teaching Award in 2000. Although recently retired from classroom teaching, he stays in step with trends in teaching by frequently conducting College Board workshops for new and experienced AP Statistics teachers and by monitoring and participating in the AP Statistics electronic discussion group. Dan is coauthor of *Statistics Through Applications*, a popular textbook for non–AP Statistics classes.

David S. Moore is Shanti S. Gupta Distinguished Professor of Statistics at Purdue University and 1998 President of the American Statistical Association. David is an elected fellow of the American Statistical Association and of the Institute of Mathematical Statistics and an elected member of the International Statistical Institute. He has served as program director for statistics and probability at the National Science Foundation. David has devoted his attention to the teaching of statistics. He was the content developer for the Annenberg/Corporation for Public Broadcasting college-level telecourse *Against All Odds: Inside Statistics* and for the series of video modules *Statistics: Decisions Through Data*, intended to aid the teaching of statistics in schools. He is the author of influential articles on statistics education and of several leading textbooks, including *Introduction to the Practice of Statistics* (written with George P. McCabe and Bruce Craig), *The Basic Practice of Statistics, Statistics: Concepts and Controversies, The Practice of Business Statistics, The Practice of Statistics in the Life Sciences* (written with Brigitte Baldi), and most recently *Essential Statistics*.

Content Advisory Board and Supplements Team

Josh Tabor
Canyon del Oro High School

Josh is a past member of the AP Statistics Development Committee (2005–2009) and an experienced table leader and question leader at the AP Statistics Reading. He has led many one-week AP Summer Institutes and one-day College Board workshops around the country. Josh is the author of the *Annotated Teacher's Edition* and *Teacher's Resource Binder* for *The Practice of Statistics*.

In addition to teaching AP Statistics, Josh enjoys spending time with his wife and daughter and has combined two of his passions—statistics and sports—to write a textbook for non–AP classes called *Statistical Reasoning in Sports*.

Michael Legacy
Greenhill School

Michael is a past member of the AP Statistics Development Committee (2001–2005) and a former table leader at the Reading. He currently reads the Alternate Exam. He has been a lead teacher at many AP Summer Institutes and student prep sessions over the past twelve years and is a frequent presenter at College Board conferences. Michael is the author of the 2007 *College Board AP Statistics Teacher's Guide* and writes the *Prep for the AP Exam Guide* for *The Practice of Statistics*. In February 2010, the Siemens Corporation named Michael the 2009–2010 AP Math/Science Teacher of the Year for the state of Texas.

In addition to his duties as pre-K–12 Math Chair, Michael teaches AP Statistics and Precalculus, indulges his passionate interest in theater, and does the costume work for the annual Madrigal Dinner at Greenhill School in Addison, Texas.

Tim Brown
The Lawrenceville School

Tim first piloted an AP Statistics course the year before the first exam was administered. He has been an AP reader since 1997 and a table leader since 2004. He earned a BA from Hampshire College and a PhD in ecology from The University of North Carolina at Chapel Hill. Since 1982, he has taught mathematics and statistics at The Lawrenceville School and currently holds the Bruce McClellan Distinguished Teaching Chair.

In addition to his teaching and advising duties, Tim supervises two intramural Ultimate Disk programs at Lawrenceville. Over the past twenty-eight years, he has introduced hundreds of students to the game.

Ann Cannon
Cornell College

Ann has been a reader, table leader, and question leader for the AP Statistics exam. She has taught the equivalent course at the college level for seventeen years at Cornell College in Mount Vernon, Iowa. Ann has also been very active in the Statistics Education Section of the American Statistical Association, serving on the Executive Committee for two three-year terms.

In her spare time, she runs herd on her family, including two sons, and she enjoys music, playing both the French horn and handbells.

Jason Molesky
Lakeville Area Public Schools

Jason has served as an AP Statistics reader for the past four years. He has taught AP Statistics for eight years and is currently the "Assessment and Accountability Coordinator" for Lakeville schools, overseeing his district's testing program and providing professional development. He also maintains the "StatsMonkey" Web site, a clearinghouse for AP Statistics activities, notes, test preparation materials, and other resources.

In his spare time, Jason enjoys spending time with his wife, their two children, and his trusty beagle.

Heather Overstreet
Staunton River High School

Heather has served as an AP Statistics reader for the past three years. She has taught AP Statistics for six years and is currently working with Virginia Advanced Study Strategies, a program for promoting AP math, science, and English courses in Virginia high schools.

In her spare time, Heather enjoys spending time with her husband and chauffeuring her children to their many activities.

Tim Erickson
Lick-Wilmerding High School

Tim is one of the designers of Fathom Dynamic Data software and has been its chief curriculum developer since Fathom began in 1995. He is currently working on the Data Games project and is also teaching a section of high school statistics.

Outside the stats world, Tim is an adult-onset cellist, playing in a community orchestra and in a string quartet.

PREFACE

The Practice of Statistics (*TPS*) was the first book written specifically for the Advanced Placement (AP) Statistics course. *TPS* is organized to closely follow the College Board's AP Statistics Course Description and is designed to prepare students for the AP exam. In the fourth edition (TPS 4e), we have made the AP connection even stronger by:

- Ensuring that every item on the College Board's "Topic Outline" is covered thoroughly in the text. In addition, we have indicated content that goes beyond the AP syllabus, such as Simpson's paradox, data ethics, and Normal approximation to the binomial.
- Continuing to use terms, notation, formulas, and tables that are consistent with those found on the AP Statistics exam.
- Following accepted conventions from AP exam rubrics when presenting model solutions.
- Adding five-option multiple-choice exercises throughout, as well as an AP Statistics Practice Test at the end of each chapter and four Cumulative AP Practice Tests.
- Including AP Exam Tips in the margin where appropriate.
- Consulting a Content Advisory Board and supplements team who are intimately familiar with the AP Statistics course and exam.

TPS 4e is a book for high school students, written by high school teachers. We deliberately chose contexts in the examples and exercises to appeal to a teenage audience. Our writing style is straightforward enough to make ideas clear to students at varied reading levels (including English-language learners) without sacrificing statistical accuracy. Students can learn by reading the narrative, but they will develop deeper understanding by doing Activities and Data Explorations, answering Check Your Understanding questions, and working lots of exercises. The text takes advantage of the simulation, graphing, and computation capabilities of the TI-83/84, TI-89, and TI-Nspire statistical software and online applets to promote active learning.

TPS 4e is aligned with the National Council of Teachers of Mathematics (NCTM) content standard "Data Analysis and Probability," with the American Statistical Association (ASA) *Guidelines for Assessment and Instruction in Statistics Education* (GAISE), and with joint recommendations of the ASA and the Mathematical Association of America concerning first courses in statistics:

- Emphasize statistical thinking
- Present more data and concepts with less theory and fewer formulas
- Foster active learning

The Fourth Edition

In planning the fourth edition of *The Practice of Statistics*, we commissioned numerous written reviews from users of the third edition and competitive books. As a follow-up to the written feedback, we held a focus group and assembled a Content Advisory Board to learn more about the current state of the AP Statistics course and the perceived strengths and weaknesses of prior editions of *TPS*. A number of salient points emerged that helped shape our goals for the new edition:

- **Streamline the text.** *TPS* 3e had 15 chapters plus a "Preliminary Chapter." Dividing the content into so many distinct units created pacing challenges for teachers who wanted to assess students after each chapter and still have time left for a comprehensive review before the AP exam. We have streamlined the table of contents to include 12 focused chapters.
- **Make the link to AP Statistics more explicit.** Even though *TPS* has been the primary text used for AP Statistics since the first edition was published, some reviewers suggested that we were being a bit too subtle about this fact. As you will see, from the front cover to the last page of the book, we have rededicated ourselves to our primary mission: preparing students for success in the course and on the AP exam.
- **Enhance the design.** Today's students are accustomed to operating in a visually stimulating environment. A textbook with an engaging look and feel is more likely to get used regularly. We have attempted to create visual appeal without making the pages too busy or distracting.
- **Focus on readability.** Explanations that are too technical or formal discourage students from reading. Narrative that's excessively casual can lead to sloppy statistics. Finding a comfortable middle ground, where students hear a consistent voice and tone throughout, was a major priority for the new edition. Readability tests and reviewer comments suggest that the fourth edition has improved accessibility without sacrificing accuracy.

Daren Starnes has undertaken a substantial revision of the entire book with these goals in mind. A summary of the important changes to content and features follows.

Content

To streamline the text:

- *TPS* 3e Chapter P was replaced with a brief overview of statistical thinking. Material from Chapter P was relocated as follows: "Data Production: Where Do You Get Good Data?" now appears in Chapter 4; "Data Analysis: Making Sense of Data" is in Chapter 1; "Probability: What Are the Chances?" can be found in Chapter 5; and "Statistical Inference: Drawing Conclusions from Data" appears in Chapters 8 and 9.

- Chapter 3, on exploring relationships between two quantitative variables, was trimmed from three sections to two.
- The content of *TPS* 3e Chapter 4, "More about Relationships between Two Variables," was redistributed as follows: Section 4.1, "Transforming to Achieve Linearity," is now Section 12.2; the material from Section 4.2, "Relationships between Categorical Variables," is included in a new Section 1.1, entitled "Analyzing Categorical Data"; and Section 4.3, "Establishing Causation" is now part of the new Section 4.3, "Using Studies Wisely."
- *TPS* 3e Chapter 7 ("Random Variables") and Chapter 8 ("The Binomial and Geometric Distributions") have been combined in the new Chapter 6, "Random Variables."
- *TPS* 3e Chapter 11 ("Testing a Claim") and Chapter 12 ("Significance Tests in Practice") have been consolidated in the new Chapter 9, "Testing a Claim."

Other content changes, by chapter:

- **Chapter 1** The idea of an association between two variables is introduced. Simpson's paradox is included as an optional topic. Timeplots are no longer included.
- **Chapter 2** Cumulative relative frequency curves (ogives) are now included in Section 2.1 with percentiles. Linear transformations and their effects on shape, center, and spread are introduced here and connected with *z*-scores. Chebyshev's inequality is de-emphasized.
- **Chapter 4** Several new data-based Activities have been added to illustrate key concepts and to reinforce the connection between data production and data analysis. Section 4.1 includes a more thorough discussion of stratified random sampling and cluster sampling. The idea of a lurking variable and the concept of confounding are examined in Section 4.2. A new Section 4.3, "Using Studies Wisely," focuses on the scope of inference that's appropriate for a given study design and includes an optional subsection on data ethics.
- **Chapter 5** Greater emphasis has been placed on using simulation as a tool to estimate probability. Sections 5.2 and 5.3 have been reorganized for smoother presentation of basic rules of probability. More examples with two-way tables have been added.
- **Chapter 6** Section 6.2, "Transforming and Combining Random Variables," has been completely rewritten. Parallels between linear transformations of data and random variables are highlighted. New examples aim to make these topics seem less abstract. In Section 6.3, the emphasis on algebraic properties of geometric random variables has been decreased. The Normal approximation to the binomial distribution is included as an optional topic.
- **Chapter 7** Several new Activities (German tanks, Reese's Pieces, *Rice University CLT* applet) have been added to give students more concrete experience in working with sampling distributions. Section 7.1, "What Is a Sampling Distribution?" has increased focus on the shape, center, and spread of sampling distributions. Bias and variability are examined for a variety of statistics.

Note the reversal of section order from the previous edition in the new Chapters 8, 9, and 10—proportions are discussed first to match the order in Chapter 7 on sampling distributions.

- **Chapter 8** We don't discuss calculation of confidence intervals for μ with σ known.
- **Chapter 9** We don't discuss performing significance tests for μ with σ known. We discuss the concepts of Type I error, Type II error, and power in Section 9.1, "Significance Tests: The Basics."
- **Chapter 10** This chapter includes greater emphasis on the distinction between inference based on random sampling and inference for randomized experiments.
- **Chapter 12** This chapter includes a rewritten treatment of transforming to achieve linearity.

Features

Examples One of the hallmark features of *TPS* is the many interesting examples that permeate the text. However, in response to feedback from both students and teachers, we have redesigned the examples to make them easier to use.

Examples are no longer in a strict narrative format and are not numbered. Now, most examples include a problem followed by a model solution. This format makes the example appear more like an exercise, which will help students see the connection more clearly.

- **Four-Step examples** This special type of example is integrated throughout. Four-step examples using the State, Plan, Do, Conclude (Mantra: Statistics Problems Demand Consistency) framework prepare students to tackle more complex exercises, like those on the AP exam.
- **For Practice Try Exercise . . .** suggestions link examples to similar problems in the Section Exercises to encourage students to practice what they have learned. Icons in the Section Exercises point students back to a relevant example in the text.

Exercises Although *TPS* has always provided a tremendous number and variety of exercises, improving the accessibility and range of problem types was a major goal of the fourth edition. One significant change is that the integrated, intra-section exercises that characterized previous editions of *TPS* have been moved to the end of each section to make them easier to find and use.

- New **Check Your Understanding** boxes are integrated in the body of the text and include questions involving facts, definitions, and procedures. These can be used as in-class exercises or in conjunction with reading assignments.
- **Section Exercises** at the end of each section provide an extensive set of problems.
  - All exercises have descriptive titles.
  - Most exercises are paired so that odd- and even-numbered problems test the same skill or concept.
  - More skill-, definition-, and procedure-based exercises are included.
  - Some multiple-choice questions are included.
  - The four-step process is indicated for some exercises.

- **Recycle and Review** exercises appear at the end of the exercise set and cross-reference the relevant prior section.

- **Chapter Review Exercises** with learning objectives keyed directly to each question provide practice problems that test concepts from throughout the chapter and prepare students for the AP Statistics exam.
- **AP Statistics Practice Tests** consist of approximately ten multiple-choice questions with five distracters and about three free-response questions.
- **Cumulative AP Practice Tests** appear after Chapters 4, 7, 10, and 12. These consist of AP-style multiple-choice and free-response questions developed by Michael Legacy, our *Prep for the AP Exam Guide* author.

Activities Many new activities have been incorporated that aim to involve students directly in the process of learning statistics by doing statistics. More detailed instructions for the Activities and numerous additional Activities will be called out in the *Annotated Teacher's Edition* (*ATE*) and included in the *Teacher's Resource Binder* (*TRB*).

Technology Corners This feature provides step-by-step instructions on how to use the TI-83/84 and TI-89 and how to interpret computer output from statistical software. TI-Nspire instructions are included in Appendix B.

Data Explorations These investigations give students an opportunity to analyze data collected to answer an interesting real-world question.

Case Study Now the chapter-opening case study is resolved in the Case Closed that appears at the end of the chapter.

You'll learn about . . . This is a new feature at the beginning of each section that briefly highlights the main topics in the upcoming section.

Think About It This feature asks students to reflect on newly introduced concepts in greater depth to help them make important connections between ideas.

Margin Notes Several different types of comments have been placed in the margins, including AP Exam Tips, advice to students from the author, and some historical tidbits.

Why Did You Do That?

There is no single best way to organize a first course in statistics. That said, our choices reflect thinking about both content and pedagogy. Here are comments on several "frequently asked questions" about the order and selection of material in *The Practice of Statistics*, Fourth Edition.

Why not begin with data production? It is certainly reasonable to do so—the natural flow of a planned study is from study design to data analysis to inference. In fact, the book is structured so that you can start with Chapter 4 on designing studies if you wish. We put data analysis first because the graphical and numerical techniques introduced there are useful even for data that are not produced according to a careful plan.

Why do Normal distributions appear so early in the book? Density curves such as the Normal curves are just another tool to describe the distribution of a quantitative variable, along with dotplots, stemplots, histograms, and boxplots. Most statistical software offers to make density curves from data just as it offers histograms. We prefer not to suggest that this material is essentially tied to probability, as the traditional order does. And we find it very helpful to break up the indigestible lump of probability that troubles students so much. Encountering Normal distributions early helps to do so and strengthens the "probability distributions are like data distributions" way of approaching probability.

What happened to inference about a population mean μ when σ is known? Previous editions of the text introduced confidence intervals and significance tests about a population mean μ in an "idealized" setting where the population standard deviation σ is known. Doing so allows use of the standard Normal curve and z procedures. Unfortunately, this approach requires considerable time, is rarely used in practice, and sometimes confuses students later when we expect them to use t for means and z for proportions.

Why did you choose this inference sequence? Students find the ideas and methods of estimating an unknown parameter much more straightforward than the logic of testing a claim about a parameter. That's a pedagogical argument for introducing confidence intervals first. Mixing confidence intervals and tests in the opening chapter on inference (from experience) leads to more confusion than clarity. Students also struggle to distinguish settings that involve proportions and means. We attack this issue by including inference methods for both proportions and means in the same chapter. This allows us to say "z is for proportions and t is for means" in Chapter 8 and again in Chapters 9 and 10. Finally, we ask students to put all these ideas together by constructing confidence intervals and performing tests in settings that involve comparing two proportions or two means. For consistency, we tackle inference for proportions first and inference for means second in every case.

For a detailed list of supplements available with *The Practice of Statistics*, Fourth Edition, go to www.whfreeman.com/tps4e.

ACKNOWLEDGMENTS

At a two-day meeting in May 2009, a remarkable team of people gathered in New York to create a vision for the fourth edition of *The Practice of Statistics*. What emerged was an extremely ambitious revision plan. Working under very tight time constraints, team members reviewed manuscript, offered sage advice, and endured countless emails and several lengthy phone calls. Just over a year later, our collective vision became reality. I am forever indebted to the individuals on the team for helping bring this project to a successful conclusion.

My heartfelt thanks go to Ann Heath, Executive Editor at Bedford, Freeman, and Worth (BFW) Publishers, for her unwavering commitment to doing things the right way. Her uncanny ability to push and prod, encourage, commiserate, troubleshoot, and marshal the troops was exceeded only by her limitless patience and good humor.

This was my first time working closely with Josh Tabor, but it certainly won't be my last. Josh's influence on the new edition is writ large from cover to cover. For his thoughtful suggestions, balanced criticisms of my first drafts, and rapid-fire exchanges about issues large and small, I offer Josh my sincere appreciation. He deserves extra credit for reshaping the *Annotated Teacher's Edition* and the *Teacher's Resource Binder*, two critically valuable supplements for AP Statistics teachers.

To Ann Cannon, who played the vital role of "uber content reviewer" throughout the project, thank you for asking tough questions at just the right moments and for reading every line of every chapter several times to prevent embarrassing mistakes from creeping in. I have benefited from Ann's statistical expertise, clear thinking, and incredible talent for meeting deadlines on numerous occasions over the past two years while working on *Statistics Through Applications*, Second Edition, and *The Practice of Statistics*, Fourth Edition. She has also done yeoman's work assembling a thorough *Solutions Manual*.

After collaborating with Michael Legacy on the finishing stages of the *Teacher's Resource Binder* for *Statistics Through Applications*, second edition, I was delighted when he agreed to oversee the development of the new Cumulative AP Practice Tests for the fourth edition of *TPS*. Generating AP-caliber questions is no small feat, and Michael does it well—and often, as you can see from the new installment of his *Prep for the AP Exam Guide*.

Tim Brown, my Lawrenceville colleague, has spent hours and hours creating new quizzes and tests for the *Teacher's Resource Binder* to mirror the changes in the fourth edition. He has also offered frequent counsel around the proverbial water cooler, willingly test-piloted new manuscript with his own classes, and pitched in as needed to give me time to write. Tim has also greatly improved the quality of the Test Bank for the fourth edition.

Heather Overstreet joined our team in the fall to write the very useful TI-Nspire Technology Corners and Projects. Jason Molesky answered our call to write new PowerPoint presentations brillantly, and we are grateful for his outstanding work.

Last and definitely not least, I would like to recognize the contributions of the final member of our team, my loving wife, Judy. For the incredible sacrifices she has made on my behalf these past twelve months, I owe her a very long vacation to an exotic location of her choosing. Judy has done a myriad of behind-the-scenes work, from assembling the Notes and Data Sources, to preparing assorted spreadsheets to track exercises, to reviewing new art. Most importantly, she has endured my long-winded ramblings about statistics for more than two years now—in restaurants; on planes, trains, and in the car; while walking around campus; and pretty much any other time when we were both awake. Without her unconditional love and support, there's simply no way I would have made it through this revision. Thanks, Love.

With such significant changes being made in the fourth edition, it was a relief to have highly skilled development editor Don Gecewicz working with me throughout the project. His keen eye was exceeded only by his ability to find ways to make the book flow more smoothly.

I would also like to offer my thanks to Sarah Street for bringing her careful eye and experience to her accuracy check of all the problems and solutions, to Dora Figueiredo for doing the labor-intensive work of preparing the manuscript for production, and to Katrina Wilhelm, Catriona Kaplan, and Laura Capuano for helping to manage the print supplements and for overseeing the companion Web site, ebook, and media.

I am deeply grateful to Rae Grant, designer extraordinaire, who transformed *TPS* into a book with tremendous eye appeal, and to Blake Logan, who made sure that our aesthetic vision was well executed. My gratitude also goes to Vivien Weiss and Susan Wein at W. H. Freeman for managing the production process with aplomb and to Bill Page for supervising the development of all the new figures for this edition. It was a relief to have our longtime copy editor, Pam Bruton, adding her usual careful touch to the preparation of the final manuscript.

Although Dan Yates and David Moore have now officially retired, *TPS* still reflects their many important contributions over several editions. Both of them have dramatically shaped my views on statistics learning and teaching, and for that, I am sincerely appreciative.

To my 2009–2010 AP Statistics students at Lawrenceville, thanks for your willingness to use the rough-draft manuscript, to live without odd-numbered solutions, and to give me candid feedback on a regular basis during class. I am especially grateful to Kevin Stevick, who served as a student reviewer throughout the development of the fourth edition.

Finally, to colleagues who reviewed chapters of the fourth edition manuscript and offered many constructive suggestions, I give my thanks.

Daren Starnes

Fourth Edition focus group participants and reviewers:

Gloria Barrett, Virginia Advanced Study Strategies, Richmond, VA
David Bernklau, Long Island University, Brookville, NY
Patricia Busso, Shrewsbury High School, Shrewsbury, MA
Lynn Church, Caldwell Academy, Greensboro, NC
Steven Dafilou, Springside High School, Philadelphia, PA
Sandra Daire, Felix Varela High School, Miami, FL
Roger Day, Pontiac High School, Pontiac, IL
Jared Derksen, Rancho Cucamonga High School, Rancho Cucamonga, CA
Michael Drozin, Munroe Falls High School, Stow, OH
Therese Ferrell, I. H. Kempner High School, Sugar Land, TX
Sharon Friedman, Newport High School, Bellevue, WA
Jennifer Gregor, Central High School, Omaha, NE
Julia Guggenheimer, Greenwich Academy, Greenwich, CT
Dorinda Hewitt, Diamond Bar High School, Diamond Bar, CA
Dorothy Klausner, Bronx High School of Science, Bronx, NY
Robert Lochel, Hatboro-Horsham High School, Horsham, PA
Lynn Luton, Duchesne Academy of the Sacred Heart, Houston, TX
Jim Mariani, Woodland Hills High School, Greensburgh, PA
Stephen Miller, Winchester Thurston High School, Pittsburgh, PA
Jason Molesky, Lakeville Area Public Schools, Lakeville, MN
Mary Mortlock, Harker School, San Jose, CA
Heather Nichols, Oak Creek High School, Oak Creek, WI
Jamis Perrett, Texas A&M University, College Station, TX
Heather Pessy, Mount Lebanon High School, Pittsburgh, PA
Kathleen Petko, Palatine High School, Palatine, IL
Todd Phillips, Mills Godwin High School, Richmond, VA
Paula Schute, Mount Notre Dame High School, Cincinnati, OH
Susan Stauffer, Boise High School, Boise, ID
Doug Tyson, Central York High School, York, PA
Bill Van Leer, Flint High School, Oakton, VA
Julie Verne, Pine Crest High School, Fort Lauderdale, FL
Steve Willot, Francis Howell North High School, St. Charles, MO
Jay C. Windley, A. B. Miller High School, Fontana, CA
Scott Fritchman, Liberty High School, Bethlehom, PA

Reviewers of previous editions:

Christopher E. Barat, Villa Julie College, Stevenson, MD
Jason Bell, Canal Winchester High School, Canal Winchester, OH
Zack Bigner, Elkins High School, Missouri City, TX
Naomi Bjork, University High School, Irvine, CA
Robert Blaschke, Lynbrook High School, San Jose, CA
Alla Bogomolnaya, Orange High School, Pepper Pike, OH
Andrew Bowen, Grand Island Central School District, Grand Island, NY
Jacqueline Briant, Bishop Feehan High School, Attleboro, MA
Marlys Jean Brimmer, Ridgeview High School, Bakersfield, CA
Floyd E. Brown, Science Hill High School, Johnson City, TN
James Cannestra, Germantown High School, Germantown, WI
Joseph T. Champine, King High School, Corpus Christi, TX
Jared Derksen, Rancho Cucamonga High School, Rancho Cucamonga, CA
George J. DiMundo, Coast Union High School, Cambria, CA
Jeffrey S. Dinkelmann, Novi High School, Novi, MI
Ronald S. Dirkse, American School in Japan, Tokyo, Japan
Cynthia L. Dishburger, Whitewater High School, Fayetteville, GA
Michael Drake, Springfield High School, Erdenheim, PA
Mark A. Fahling, Gaffney High School, Gaffney, SC
David Ferris, Noblesville High School, Noblesville, IN
David Fong, University High School, Irvine, CA
Terry C. French, Lake Braddock Secondary School, Burke, VA
Glenn Gabanski, Oak Park and River Forest High School, Oak Park, IL
Jason Gould, Eaglecrest High School, Centennial, CO
Dr. Gene Vernon Hair, West Orange High School, Winter Garden, FL
Stephen Hansen, Napa High School, Napa, CA
Katherine Hawks, Meadowcreek High School, Norcross, GA
Gregory D. Henry, Hanford West High School, Hanford, CA
Duane C. Hinders, Foothill College, Los Altos Hills, CA
Beth Howard, Saint Edwards, Sebastian, FL
Michael Irvin, Legacy High School, Broomfield, CO
Beverly A. Johnson, Fort Worth Country Day School, Fort Worth, TX
Matthew L. Knupp, Danville High School, Danville, KY
Kenneth Kravetz, Westford Academy, Westford, MA
Lee E. Kucera, Capistrano Valley High School, Mission Viejo, CA
Christina Lepi, Farmington High School, Farmington, CT
Jean E. Lorenson, Stone Ridge School of the Sacred Heart, Bethesda, MD
Thedora R. Lund, Millard North High School, Omaha, NE
Philip Mallinson, Phillips Exeter Academy, Exeter, NH
Dru Martin, Ramstein American High School, Ramstein, Germany
Richard L. McClintock, Ticonderoga High School, Ticonderoga, NY
Louise McGuire, Pattonville High School, Maryland Heights, MO
Jennifer Michaelis, Collins Hill High School, Suwanee, GA
Dr. Jackie Miller, Ohio State University
Jason M. Molesky, Lakeville South High School, Lakeville, MN
Wayne Nirode, Troy High School, Troy, OH
Heather Pessy, Mt. Lebanon High School, Pittsburgh, PA
Sarah Peterson, University Preparatory Academy, Seattle, WA
Kathleen Petko, Palatine High School, Palatine, IL
German J. Pliego, University of St. Thomas
Stoney Pryor, A&M Consolidated High School, College Station, TX
Judy Quan, Alameda High School, Alameda, CA
Stephanie Ragucci, Andover High School, Andover, MA
James M. Reeder, University School, Hunting Valley, OH
Joseph Reiken, Bishop Garcia Diego High School, Santa Barbara, CA
Roger V. Rioux, Cape Elizabeth High School, Cape Elizabeth, ME
Tom Robinson, Kentridge Senior High School, Kent, WA
Albert Roos, Lexington High School, Lexington, MA
Linda C. Schrader, Cuyahoga Heights High School, Cuyahoga Heights, OH
Daniel R. Shuster, Royal High School, Simi Valley, CA
David Stein, Paint Branch High School, Burtonsville, MD
Vivian Annette Stephens, Dalton High School, Dalton, GA
Charles E. Taylor, Flowing Wells High School, Tucson, AZ
Reba Taylor, Blacksburg High School, Blacksburg, VA
Shelli Temple, Jenks High School, Jenks, OK
David Thiel, Math/Science Institute, Las Vegas, NV
William Thill, Harvard-Westlake School, North Hollywood, CA
Richard Van Gilst, Westminster Christian Academy, St. Louis, MO
Joseph Robert Vignolini, Glen Cove High School, Glen Cove, NY
Ira Wallin, Elmwood Park Memorial High School, Elmwood Park, NJ
Linda L. Wohlever, Hathaway Brown School, Shaker Heights, OH

Preface to the Student:

Getting the most from your text...

READ THE TEXT and use the book's features to help you grasp the big ideas.

Review the outline at the beginning of each section to preview the topics you'll study.

Least-Squares Regression

In Section 3.2, you'll learn about:

- Interpreting a regression line
- Prediction
- Residuals and the least-squares regression line
- Calculating the equation of the least-squares line
- How well the line fits the data: Residual plots
- How well the line fits the data: The role of r^2 in regression
- Interpreting computer

Linear (straight-line) relationships between two quantitative variables are pretty common and easy to understand. In the previous section, we found linear relationships in settings as varied as sparrowhawk colonies, natural-gas consumption, and Florida manatee deaths. Correlation measures the direction and strength of these relationships. When a scatterplot shows a linear relationship, we'd like to summarize the overall pattern by drawing a line on the scatterplot. A **regression line** summarizes the relationship between two variables, but only in a specific setting: when one of the variables helps explain or predict the other. *Regression, unlike correlation, requires that we have an explanatory variable and a response variable.* CAUTION

Watch for the green **DEFINITION** boxes that highlight important vocabulary. Flip back as needed to review the key terms and their definitions.

DEFINITION: Regression line

A **regression line** is a line that describes how a response variable y changes as an explanatory variable x changes. We often use a regression line to predict the value of y for a given value of x.

Pay attention to the **CAUTION ICONS**. They alert you to common mistakes that students make.

Look for the boxes with the blue headings. Some explain how to make graphs or do calculations while others summarize important concepts.

How to Calculate the Correlation r

Suppose that we have data on variables x and y for n individuals. The values for the first individual are x_1 and y_1, the values for the second individual are x_2 and y_2, and so on. The means and standard deviations of the two variables are $\bar{x}$ and s_x for the x-values, and $\bar{y}$ and s_y for the y-values. The correlation r between x and y is

$$r = \frac{1}{n-1}\left[\left(\frac{x_1-\bar{x}}{s_x}\right)\left(\frac{y_1-\bar{y}}{s_y}\right)+\left(\frac{x_2-\bar{x}}{s_x}\right)\left(\frac{y_2-\bar{y}}{s_y}\right)+\cdots+\left(\frac{x_n-\bar{x}}{s_x}\right)\left(\frac{y_n-\bar{y}}{s_y}\right)\right]$$

or, more compactly,

$$r = \frac{1}{n-1}\sum\left(\frac{x_i-\bar{x}}{s_x}\right)\left(\frac{y_i-\bar{y}}{s_y}\right)$$

Read the **AP EXAM TIPS** and other notes in the margin. They give advice on how to be successful in the course and on the **AP Exam**.

AP EXAM TIP Students often have a hard time interpreting the value of r^2 on AP exam questions. They frequently leave out key words in the definition. Our advice: Treat this as a fill-in-the-blank exercise. Write "____% of the variation in [response variable name] is accounted for by the regression line."

Make connections and deepen your understanding of subtle ideas by reflecting on the questions asked in **THINK ABOUT IT** passages.

THINK ABOUT IT

Does a small slope mean that there's no relationship? For the NEA and fat gain regression line, the slope $b = -0.00344$ is a small number. This does *not* mean that change in NEA has little effect on fat gain. The size of the slope depends on the units in which we measure the two variables. In this setting, the slope is the predicted change in fat gain in kilograms when NEA increases by 1 calorie. There are 1000 grams in a kilogram. If we measured fat gain in grams, the slope would be 1000 times larger, $b = 3.44$. *You can't say how important a relationship is by looking at the size of the slope of the regression line.*

LEARN STATISTICS BY *DOING* STATISTICS

ACTIVITY *CSI Stats: The case of the missing cookies*

MATERIALS: Metric ruler, meterstick, handprint, and photo lineup (from Teacher's Resource Binder) for each group of three to four students; one sheet of graph paper per student

Mrs. Hagen keeps a large jar full of cookies on her desk for her students. Over the past few days, a few cookies have disappeared. The only people with access to Mrs. Hagen's desk are the other math teachers at her school. She asks her colleagues whether they have been making withdrawals from the cookie jar. No one confesses to the crime.

But the next day, Mrs. Hagen catches a break—she finds a clear handprint on the cookie jar. The careless culprit has left behind crucial evidence! At this point,

ACTIVITY *Investigating properties of the least-squares regression line*

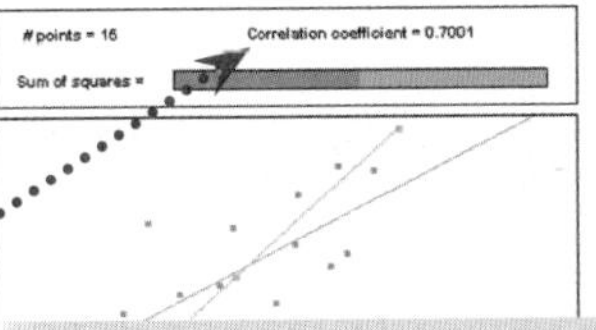

In this Activity, you will use the *Correlation and Regression* applet at the book's Web site, www.whfreeman.com/tps4e, to explore some properties of the least-squares regression line.

1. Click on the scatterplot to create a group of 15 to 20 points from lower left to upper right with a clear positive straight-line pattern (correlation around 0.7). Click the "Draw line" button and use the mouse (right-click and drag) to draw a line

Each chapter begins with a hands-on **ACTIVITY** that introduces the content of the chapter. Many of these activities involve collecting data and drawing conclusions from the data. In other activities, you'll use dynamic applets to explore statistical concepts.

DATA EXPLORATION *The SAT essay: Is longer better?*

Following the debut of the new SAT Writing test in March 2005, Dr. Les Perelman from the Massachusetts Institute of Technology stirred controversy by reporting, "It appeared to me that regardless of what a student wrote, the longer the essay, the higher the score." He went on to say, "I have never found a quantifiable predictor in 25 years of grading that was anywhere as strong as this one. If you just graded them based on length without ever reading them, you'd be right over 90 percent of the time."[15] The table below shows the data that Dr. Perelman used to draw his conclusions.

Length of essay and score for a sample of SAT essays

Words:	460	422	402	365	357	278	236	201	168	156	133
Score:	6	6	5	5	6	5	4	4	4	3	2
Words:	114	108	100	403	401	388	320	258	236	189	128
Score:	2	1	1	5	6	6	5	4	4	3	2
Words:	67	697	387	355	337	325	272	150	135		
Score:	1	6	6	5	5	4	4	2	3		

Source: The College Board, www.collegeboard.com.

Does this mean that if students write a lot, they are guaranteed high scores? Carry out your own analysis of the data. How would you respond to each of Dr. Perelman's claims?

DATA EXPLORATIONS ask you to play the role of data detective. Your goal is to answer a puzzling, real-world question by examining data graphically and numerically.

CHECK YOUR UNDERSTANDING

Refer to the nonexercise activity and fat gain data.

1. Find the residual for the subject who increased NEA by 620 calories. Show your work.
2. Interpret the value of this subject's residual in context.
3. For which subject did the regression line overpredict fat gain by the most? Justify your answer.

CHECK YOUR UNDERSTANDING questions appear throughout the section. They will help you master key definitions and procedures along the way. Be sure to check your answers in the back of the book.

EXAMPLES: Study the model problems and solutions

EXAMPLE *Does Fidgeting Keep You Slim?*

Interpreting the slope and y intercept

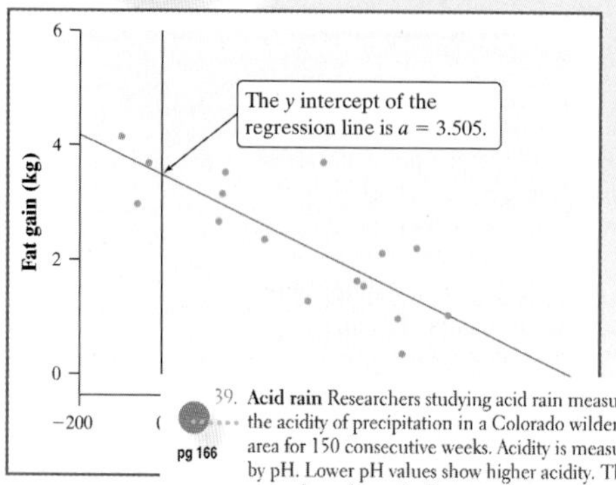

The regression line shown in Figure 3.7 is

$$\widehat{\text{fat gain}} = 3.505 - 0.00344(\text{NEA change})$$

PROBLEM: **Identify the slope and y intercept of the regression line. Interpret each value in context.**

SOLUTION: **The slope $b = -0.00344$ tells us that the amount of fat gained is predicted to go down by 0.00344 kilogram (kg) for each added calorie (cal) of NEA. The y intercept, $a = 3.505$ kg, is the fat gain estimated by this model if NEA does not change when a person overeats.**

For Practice *Try Exercise 39(a) and (b)*

pg 166 39. **Acid rain** Researchers studying acid rain measured the acidity of precipitation in a Colorado wilderness area for 150 consecutive weeks. Acidity is measured by pH. Lower pH values show higher acidity. The researchers observed a linear pattern over time. They reported that the regression line $\widehat{pH} = 5.43 - 0.0053(\text{weeks})$ fit the data well.[16]
(a) Identify the slope of the line and explain what it means in this setting.
(b) Identify the y intercept of the line and explain what it means in this setting.
(c) According to the regression line, what was the pH at the end of this study?

Try out the concept yourself by working the **FOR PRACTICE, *TRY*** exercise that directs you to a similar problem in the Section Exercises.

An icon next to the exercise directs you back to the page where the model **EXAMPLE** appears

EXAMPLE *The Endangered Manatee*

Pulling it all together

STEP 4

FOUR-STEP EXAMPLES: By reading the Four-Step Examples and mastering the special "State, Plan, Do, Conclude" framework, you can develop good problem-solving skills and your ability to tackle more complex problems like those on the AP Exam.

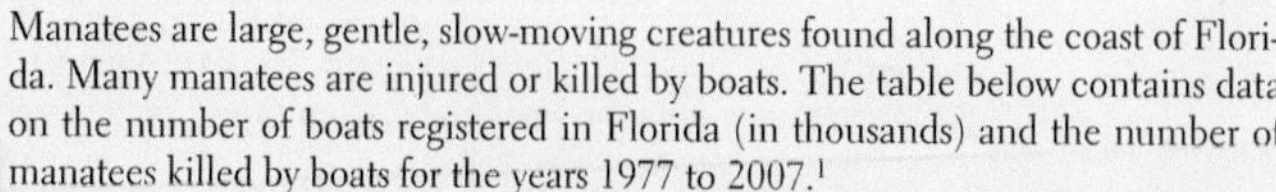

Manatees are large, gentle, slow-moving creatures found along the coast of Florida. Many manatees are injured or killed by boats. The table below contains data on the number of boats registered in Florida (in thousands) and the number of manatees killed by boats for the years 1977 to 2007.[1]

Florida boat registrations (thousands) and manatees killed by boats

Year	Boats	Manatees	Year	Boats	Manatees	Year	Boats	Manatees
1977	447	13	1988	675	43	1999	830	82
1978	460	21	1989	711	50	2000	880	78
1979	481	24	1990	719	47	2001	944	81
1980	498	16	1991	681	53	2002	962	95
1981	513	24	1992	679	38	2003	978	73
1982	512	20	1993	678	35	2004	983	69
1983	526	15	1994	696	49	2005	1010	79
1984	559	34	1995	713	42	2006	1024	92
1985	585	33	1996	732	60	2007	1027	73
1986	614	33	1997	755	54			
1987	645	39	1998	809	66			

FIGURE 3.4 Scatterplot of the number of Florida manatees killed by boats from 1977 to 2007 against the number of boats registered in Florida that year.

Manatees killed: 10, 20, 30, 40, 50, 60, 70, 80, 90, 100
Boats registered in Florida (1000s): 400, 500, 600, 700, 800, 900, 1000, 1100

STATE: **What is the relationship between the number of manatees killed and the number of registered boats?**

PLAN: **First, we'll make a scatterplot with "boats registered" as the explanatory variable and "manatees killed" as the response variable. Then we'll describe the direction, form, and strength of the relationship and identify any outliers.**

DO: **Figure 3.4 is our completed scatterplot. There is a positive association—more boats registered goes with more manatees killed. The form of the relationship is linear. That is, the overall pattern follows a straight line from lower left to upper right. The relationship is strong because the points don't deviate greatly from a line. There are no obvious outliers.**

CONCLUDE: **As more boats are registered, the number of manatees killed by boats goes up linearly.**

For Practice *Try Exercise 13*

EXERCISES: Practice makes perfect!

Start by reading the **SECTION SUMMARY** to be sure that you understand the key concepts.

SECTION 3.2 Summary

- A **regression line** is a straight line that describes how a response variable y changes as an explanatory variable x changes. You can use a regression line to **predict** the value of y for any value of x by substituting this x into the equation of the line.
- The **slope** b of a regression line $\hat{y} = a + bx$ is the rate at which the predicted response $\hat{y}$ changes along the line as the explanatory variable x changes. Specifically, b is the *predicted* change in y when x increases by 1 unit.
- The y **intercept** a of a regression line $\hat{y} = a + bx$ is the predicted response $\hat{y}$ when the explanatory variable $x = 0$. This prediction is of no statistical use unless x can actually take values near 0.

Practice! Work the exercises assigned by your teacher. Compare your answers with those in the Solutions at the back of the book (odd-numbered exercises only).

SECTION 3.2 Exercises

Most of the exercises are paired; odd- and even-numbered problems test the same skill or concept.

35. **What's my line?** You use the same bar of soap to shower each morning. The bar weighs 80 grams when it is new. Its weight goes down by 6 grams per day on the average. What is the equation of the regression line for predicting weight from days of use?

36. **What's my line?** An eccentric professor believes that a child with IQ 100 should have a reading test score of 50, and that reading score should increase by 1 point for every additional point of IQ. What is the equation of the professor's regression line for predicting reading score from IQ?

37. **Gas mileage** We expect a car's highway gas mileage to be related to its city gas mileage. Data for all 1198

40. **How much gas?** In Exercise 4 (page 158), we examined the relationship between the average monthly temperature and the amount of natural gas consumed in Joan's midwestern home. The figure below shows the original scatterplot with the least-squares line added. The equation of the least-squares line is $\hat{y} = 1425 - 19.87x$.

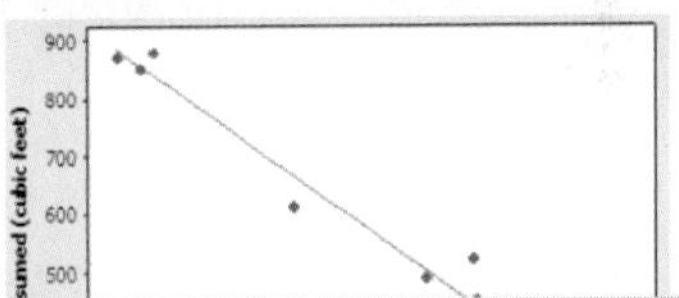

If you answer an assigned problem incorrectly, try to figure out your mistake. Then see if you can solve the paired exercise.

Watch for icons that appear next to selected problems. They will guide you to:
- an **Example** that models the problem.
- the **State, Plan, Do, Conclude** way of solving problems in the four-step framework.
- previous sections on which the problem will draw, here Section 2.2.

47. **Husbands and wives** The mean height of American women in their early twenties is 64.5 inches and the standard deviation is 2.5 inches. The mean height of men the same age is 68.5 inches, with standard devia- tion 2.7 inches. ... of husbands and

pg 173

70. **Beavers and beetles** Do beavers benefit beetles? Researchers laid out 23 circular plots, each 4 meters in diameter, in an area where beavers were cutting down cottonwood trees. In each plot, they counted the number of stumps from trees cut by beavers and the number of clusters of beetle larvae. Ecologists think ... ps are more tender

STEP 4

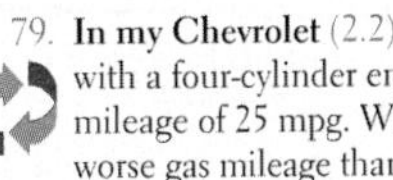

79. **In my Chevrolet** (2.2) The 2008 Chevrolet Malibu with a four-cylinder engine has a combined gas mileage of 25 mpg. What percent of all vehicles have worse gas mileage than the Malibu?

REVIEW and PRACTICE for quizzes, tests, and exams

Keyed to each question, learning objectives describe what you should know and be able to do after studying the section.

Tackle the **CHAPTER REVIEW EXERCISES** for practice in solving problems that test concepts from throughout the chapter.

Chapter Review Exercises

These exercises are designed to help you review the important ideas and methods of the chapter. Relevant learning objectives are provided in bulleted form before each exercise.

- Identify explanatory and response variables in situations where one variable helps explain or influences another.
- Explain why association doesn't imply causation.

R3.1. The risks of obesity A study observes a large group of people over a 10-year period. The goal is to see if overweight and obese people are more likely to die during the study than people who weigh less. Such studies can be misleading, because obese people are more likely to be inactive and poor.

(a) What are the explanatory and response variables in the study?

(b) If the study finds a strong association between these variables, can we conclude that increased weight causes greater risk of dying? Why or why not?

- Describe the direction, form, and strength of the overall pattern of a scatterplot. In particular, recognize positive or negative association and linear (straight-line) patterns. Recognize outliers in a scatterplot.

Each chapter concludes with an **AP STATISTICS PRACTICE TEST,** which includes about 10 AP-style multiple-choice questions and three free-response questions.

Chapter 3 AP Statistics Practice Test

Section I: Multiple Choice *Select the best answer for each question.*

T3.1. A school guidance counselor examines the number of extracurricular activities of students and their grade point average. The guidance counselor says, "The evidence indicates that the correlation between the number of extracurricular activities a student participates in and his or her grade point average is close to zero." A correct interpretation of this statement would be that

(a) active students tend to be students with poor grades, and vice versa.

(c) students involved in many extracurricular activities are just as likely to get good grades as bad grades; the same is true for students involved in few extracurricular activities.

(d) there is no linear relationship between number of activities and grade point average for students at this school.

(e) involvement in many extracurricular activities and good grades go hand in hand.

T3.2 The British government conducts regular surveys of household spending. The average weekly household ... holic ... as re-

Section II: Free Response *Show all your work. Indicate clearly the methods you use, because you will be graded on the correctness of your methods as well as on the accuracy and completeness of your results and explanations.*

T3.11 Sarah's parents are concerned that she seems short for her age. Their doctor has the following record of Sarah's height:

Age (months):	36	48	51	54	57	60
Height (cm):	86	90	91	93	94	95

called diatoms, is a measure of the abundance of life in the lake. A rather complex variable based on the ratio of certain isotopes relative to ocean water gives an indirect measure of moisture, mostly from snow. As we drill down, we look further into the past. Here is a scatterplot of data from 2300 to 12,000 years ago:

Four **CUMULATIVE AP TESTS** simulate the real exam. They are placed after Chapters 4, 7, 10, and 12. The tests expand in length and content coverage from the first through the fourth.

Cumulative AP Practice Test 1

Section I: Multiple Choice *Choose the best answer for Questions AP1.1 to AP1.14.*

AP1.1. You look at real estate ads for houses in Sarasota, Florida. Many houses range from \$200,000 to \$400,000 in price. The few houses on the water, however, have prices up to \$15 million. Which of the following statements best describes the distribution of home prices in Sarasota?

(a) The distribution is most likely skewed to the left, and the mean is greater than the median.

(b) The distribution is most likely skewed to the left, and the mean is less than the median.

(c) The distribution is roughly symmetric with a few high outliers, and the mean is approximately equal to the median.

AP1.4. For a certain experiment, the available experimental units are eight rats, of which four are female (F1, F2, F3, F4) and four are male (M1, M2, M3, M4). There are to be four treatment groups, A, B, C, and D. If a randomized block design is used, with the experimental units blocked by gender, which of the following assignments of treatments is impossible?

(a) A → (F1, M1), B → (F2, M2), C → (F3, M3), D → (F4, M4)

(b) A → (F1, M2), B → (F2, M3), C → (F3, M4), D → (F4, M1)

Use TECHNOLOGY as a tool

TECHNOLOGY CORNERS give step-by-step instructions for using the TI-83/84 and TI-89 calculator. Instructions for the **TI-Nspire** are in Appendix B.

TECHNOLOGY CORNER Residual plots and *s* on the calculator

Let's continue the analysis of the NEA and fat gain study data from the previous Technology Corner (page 170). You should have already made a scatterplot, calculated the equation of the least-squares regression line, and graphed the line on your plot. Now, we want to calculate residuals and make a residual plot. Earlier, we found that $\hat{y} = 3.505 - 0.00344x$.

TI-83/84	TI-89

1. Define L3(list3) as the predicted values from the regression equation.
   - TI-83/84: With L3 highlighted, enter the command `3.505-0.00344*L1` and press ENTER
   - TI-89: With list3 highlighted, enter the command `3.505-0.00344*list1` and press ENTER
2. Define L4/list4 as the observed *y*-value minus the predicted *y*-value.
   - TI-83/84: With L4 highlighted, enter the command `L2-L3` and press ENTER to show the residuals.
   - TI-89: With list4 highlighted, enter the command `list2-list3` and press ENTER to show the residuals.

L2	L3	L4
4.2	3.8286	.37138
3	3.7013	-.7013
3.7	3.6049	.09507
2.7	3.0405	-.3405
3.2	3.013	.18701
3.6	2.9855	.61454
2.4	2.662	-.262

L4(1)=.3713773027...

F1 Tools | F2 Plots | F3 List | F4 Calc | F5 Distr | F6 Tests | F7 Ints

list1	list2	list3	list4
-94	4.2	3.8286	.37138
-57	3.	3.7013	-.7013
-29	3.7	3.6049	.09507
135	2.7	3.0405	-.3405
143	3.2	3.013	.18701
151	3.6	2.9855	.61454

list4[1]=.3713773027852

MAIN RAD AUTO FUNC 4/7

Find the **Technology Corners** easily by consulting the summary table at the end of each Section or the complete table on the back endpaper of the book.

Several types of software displays, including Minitab, Fathom, Excel, and Applet screen captures, appear throughout the book to help you learn to read and interpret many different kinds of output.

63. **Merlins breeding** Exercise 13 (page 160) gives data on the number of breeding pairs of merlins in an isolated area in each of nine years and the percent of males who returned the next year. The data show that the percent returning is lower after successful breeding seasons and that the relationship is roughly linear. The figure below shows Minitab regression output for these data.

pg 183

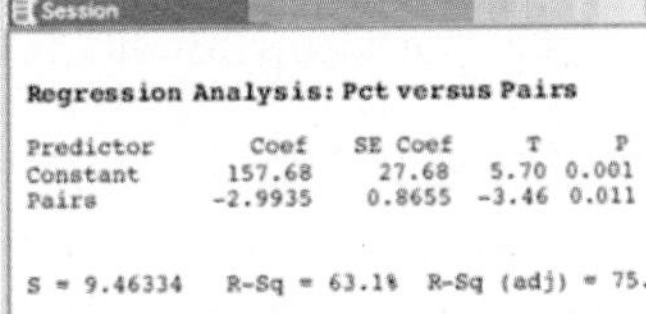

Session

Regression Analysis: Pct versus Pairs

Predictor	Coef	SE Coef	T	P
Constant	157.68	27.68	5.70	0.001
Pairs	-2.9935	0.8655	-3.46	0.011

S = 9.46334 R-Sq = 63.1% R-Sq (adj) = 75.

T3.10 A data set included the number of people per television set and the number of people per physician for 40 countries. The Fathom screen shot below displays a scatterplot of the data with the least-squares regression line added. In Ethiopia, there were 503 people per TV and 36,660 people per doctor. What effect would removing this point have on the regression line?

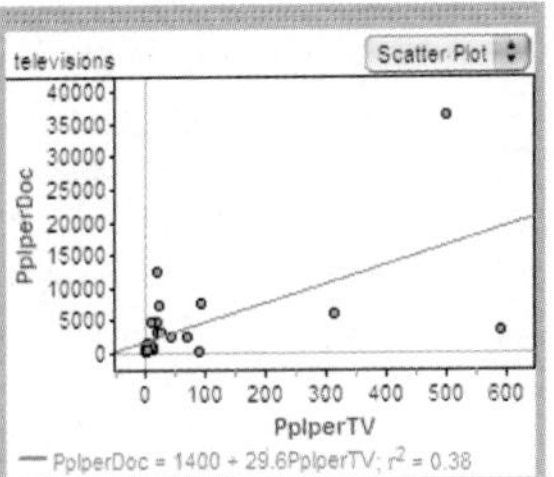

OVERVIEW What Is Statistics?

Does listening to music while studying help or hinder learning? If an athlete fails a drug test, how sure can we be that she took a banned substance? Does having a pet help people live longer? How well do SAT scores predict college success? Do most people recycle? Which of two diets will help obese children lose more weight and keep it off? Should a poker player go "all in" with pocket aces? Can a new drug help people quit smoking? How strong is the evidence for global warming?

These are just a few of the questions that statistics can help answer. But what is statistics? And why should you study it?

Statistics Is the Science of Learning from Data

Data are usually numbers, but they are not "just numbers." *Data are numbers with a context.* The number 10.5, for example, carries no information by itself. But if we hear that a family friend's new baby weighed 10.5 pounds at birth, we congratulate her on the healthy size of the child. The context engages our knowledge about the world and allows us to make judgments. We know that a baby weighing 10.5 pounds is quite large, and that a human baby is unlikely to weigh 10.5 ounces or 10.5 kilograms. The context makes the number meaningful.

In your lifetime, you will be bombarded with data and statistical information. Poll results, television ratings, music sales, gas prices, unemployment rates, medical study outcomes, and standardized test scores are discussed daily in the media. Using data effectively is a large and growing part of most professions. A solid understanding of statistics will enable you to make sound, data-based decisions in your career and everyday life.

Data Beat Personal Experiences

It is tempting to base conclusions on your own experiences or the experiences of those you know. But our experiences may not be typical. In fact, the incidents that stick in our memory are often the unusual ones.

Do cell phones cause brain cancer?

In August 2000, Dr. Chris Newman appeared as a guest on CNN's *Larry King Live*. Dr. Newman had developed brain cancer. He was also a frequent cell phone user. Dr. Newman's physician suggested that the brain tumor may have been caused by cell phone use. So Dr. Newman decided to sue the cell phone maker, Motorola, and the phone company that provided service, Verizon. As people heard Dr. Newman's sad story, they began to worry about whether their own cell phone use might lead to cancer.

Since 2000, several statistical studies have investigated the link between cell phone use and brain cancer. One of the largest was conducted by the Danish Cancer Society. Over 400,000 residents of Denmark who regularly used cell phones were included in the study. Researchers compared the brain cancer rate for the cell phone users with the rate in the general population. The result: no difference.[1] In fact, most studies have produced similar conclusions. In spite of the evidence, many people are still convinced that Dr. Newman's experience is typical.

In the public's mind, the compelling story wins every time. A statistically literate person knows better. *Data are more reliable than personal experiences because they systematically describe an overall picture rather than focus on a few incidents.*

Where the Data Come from Matters

Are you kiddin' me?

The famous advice columnist Ann Landers once asked her readers, "If you had it to do over again, would you have children?" A few weeks later, her column was headlined "70% OF PARENTS SAY KIDS NOT WORTH IT." Indeed, 70% of the nearly 10,000 parents who wrote in said they would not have children if they could make the choice again. Do you believe that 70% of all parents regret having children?

You shouldn't. The people who took the trouble to write Ann Landers are not representative of all parents. Their letters showed that many of them were angry with their children. All we know from these data is that there are some unhappy parents out there. A statistically designed poll, unlike Ann Landers's appeal, targets specific people chosen in a way that gives all parents the same chance to be asked. Such a poll showed that 91% of parents *would* have children again.

Where data come from matters a lot. If you are careless about how you get your data, you may announce 70% "No" when the truth is close to 90% "Yes."

Who talks more—women or men?

According to Louann Brizendine, author of *The Female Brain*, women say nearly three times as many words per day as men. Skeptical researchers devised a study to test this claim. They used electronic devices to record the talking patterns of 396 university students from Texas, Arizona, and Mexico. The device was programmed to record 30 seconds of sound every 12.5 minutes without the carrier's knowledge. What were the results?

According to a published report of the study in *Scientific American*, "Men showed a slightly wider variability in words uttered. . . . But in the end, the sexes came out just about even in the daily averages: women at 16,215 words and men at 15,669."[2]

The most important information about any statistical study is how the data were produced. Only carefully designed studies produce results that can be trusted.

Always Plot Your Data

Yogi Berra, a famous New York Yankees baseball player known for his unusual quotes, had this to say: "You can observe a lot just by watching." That's a motto for learning from data. *A carefully chosen graph is often more instructive than a bunch of numbers.*

Do people live longer in wealthier countries?

The Gapminder Web site, www.gapminder.org, provides loads of data on the health and well-being of the world's inhabitants. The following graph is from Gapminder.[3] The individual points represent all the world's nations for which data are available. Each point shows the income per person and life expectancy in years for one country.

We expect people in richer countries to live longer. The overall pattern of the graph does show this, but the relationship has an interesting shape. Life expectancy rises very quickly as personal income increases and then levels off. People in very rich countries like the United States live no longer than people in poorer but not extremely poor nations. In some less wealthy countries, people live longer than in the United States. Several other nations stand out in the graph. What's special about each of these countries?

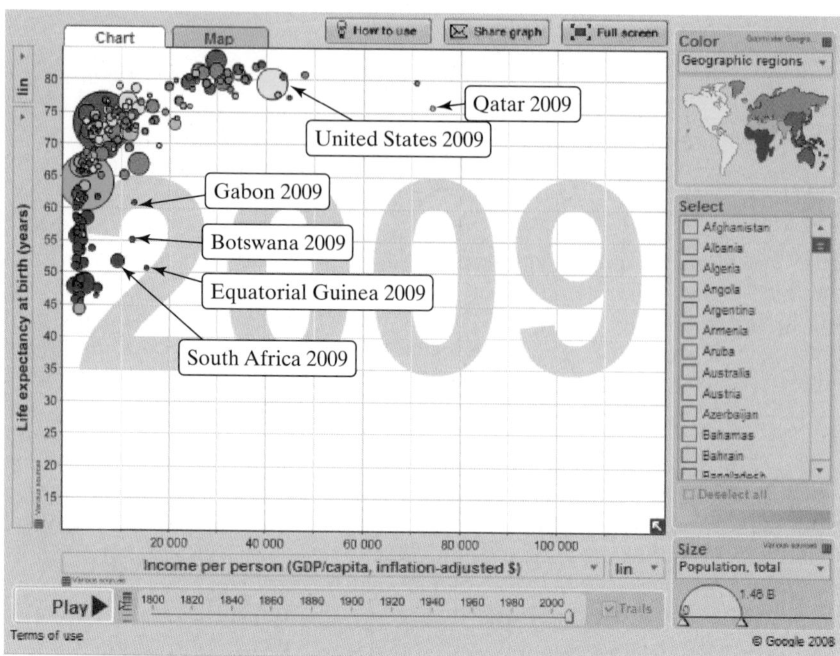

Graph of the life expectancy of people in many nations against each nation's income per person in 2009.

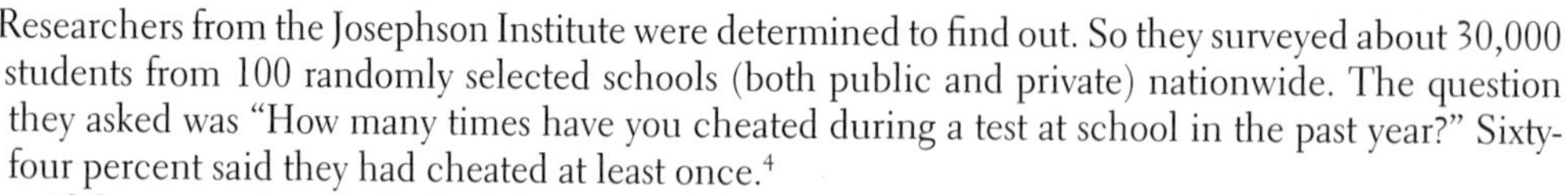

Variation Is Everywhere

Individuals vary. Repeated measurements on the same individual vary. Chance outcomes—like spins of a roulette wheel or tosses of a coin—vary. Almost everything varies over time. Statistics provides tools for understanding variation.

Have most students cheated on a test?

Researchers from the Josephson Institute were determined to find out. So they surveyed about 30,000 students from 100 randomly selected schools (both public and private) nationwide. The question they asked was "How many times have you cheated during a test at school in the past year?" Sixty-four percent said they had cheated at least once.[4]

If the researchers had asked the same question of *all* high school students, would exactly 64% have answered "Yes"? Probably not. If the Josephson Institute had selected a different sample of about 30,000 students to respond to the survey, they would probably have gotten a different estimate. *Variation is everywhere!*

Fortunately, statistics provides a description of how the sample results will vary in relation to the actual population percent. Based on the sampling method that this study used, we can say that the estimate of 64% is very likely to be within 1% of the true population value. That is, we can be quite confident that between 63% and 65% of *all* high school students would say that they have cheated on a test.

Because variation is everywhere, conclusions are uncertain. Statistics gives us a language for talking about uncertainty that is understood by statistically literate people everywhere.

Statistical Thinking and You

The purpose of this book is to give you a working knowledge of the ideas and tools of practical statistics. Because data always come from a real-world context, doing statistics means more than just manipulating data. *The Practice of Statistics*, Fourth Edition, is full of data, and each set of data has some brief background to help you understand what the data say. It is important to form the habit of asking, "What do the data tell me?"

You learn statistics by doing statistical problems. This book offers many different types of problems for you to tackle, arranged to help you learn.

- Short **Check Your Understanding** questions appear from time to time throughout the text. These are straightforward problems that help you solidify the main points before going on.
- **Section Exercises** help you combine all the ideas of a particular section. **Chapter Review Exercises** look back over the entire chapter. The review exercises include a list of specific things you should now be able to do. Go through that list, and be sure you can say "I can do that" to each item. Then prove it by solving some problems.
- The **AP Statistics Practice Test** at the end of each chapter will help you prepare for in-class exams.
- Finally, the **Cumulative AP Practice Tests** provide challenging, cumulative problems like the ones you might find on a midterm, final, or the AP Statistics exam.

The basic principle of learning is persistence. The main ideas of statistics, like the main ideas of any important subject, took a long time to discover and take some time to master. Once you put it all together, statistics will help you make informed decisions based on data.

Chapter 1

case study

Exploring Data

Do Rewards Promote Creativity?

What motivates people to be creative? Is it the possibility of receiving an external reward—like money, praise, fame, or a good grade? Or is the personal satisfaction gained from doing creative work its own reward? Researcher Teresa Amabile designed a study to find out. Her specific research question was: Will competing for a prize improve children's artistic creativity?

Amabile gathered some elementary school students to take part in her study. The children were divided into two groups and instructed to make a "silly" collage using materials that were provided. Before they started, the children in one group were told that their collages would be judged by experts and that the winners would receive prizes. The children in the other group were told that they would share their collages at an art party. In fact, expert judges rated the creativity of all the collages.[1]

Want to know what happened? By the end of this chapter, you'll have your answer.

Introduction

Data Analysis: Making Sense of Data

In the Introduction, you'll learn about:

- Individuals and variables
- From data analysis to inference

Data analysis

Statistics is the science of data. The volume of data available to us is overwhelming. For example, the Census Bureau's American Community Survey collects data from 3,000,000 housing units each year. Astronomers work with data on tens of millions of galaxies. The checkout scanners at Walmart's 6500 stores in 15 countries record hundreds of millions of transactions every week. In all these cases, the data are trying to tell us a story—about U.S. households, objects in space, or Walmart shoppers. To hear what the data are saying, we need to help them speak by organizing, displaying, summarizing, and asking questions. That's **data analysis.**

Individuals and Variables

Any set of data contains information about some group of **individuals.** The characteristics we measure on each individual are called **variables.**

DEFINITION: Individuals and variables

Individuals are the objects described by a set of data. Individuals may be people, animals, or things.

A **variable** is any characteristic of an individual. A variable can take different values for different individuals.

A high school's student data base, for example, includes data about every currently enrolled student. The students are the *individuals* described by the data set. For each individual, the data contain the values of *variables* such as age, gender, grade point average, homeroom, and grade level. In practice, any set of data is accompanied by background information that helps us understand the data. When you first meet a new data set, ask yourself the following questions:

1. *Who* are the individuals described by the data? How many individuals are there?

2. *What* are the variables? In what *units* is each variable recorded? Weights, for example, might be recorded in grams, pounds, thousands of pounds, or kilograms.

We could follow a newspaper reporter's lead and extend our list of questions to include *Why, When, Where,* and *How* were the data produced? For now, we'll focus on the first two questions.

Some variables, like gender and grade level, simply place individuals into categories. Others, like age and grade point average (GPA), take numerical values for which we can do arithmetic. It makes sense to give an average GPA for a group of students, but it doesn't make sense to give an "average" gender.

AP EXAM TIP If you learn to distinguish categorical from quantitative variables now, it will pay big rewards later. The type of data determines what kinds of graphs and which numerical summaries are appropriate. You will be expected to analyze categorical and quantitative data effectively on the AP exam.

DEFINITION: Categorical variable and quantitative variable

A **categorical variable** places an individual into one of several groups or categories.

A **quantitative variable** takes numerical values for which it makes sense to find an average.

CAUTION

Not every variable that takes number values is quantitative. Zip code is one example. Although zip codes are numbers, it doesn't make sense to talk about the average zip code. In fact, zip codes place individuals (people or dwellings) into categories based on location. Some variables—such as gender, race, and occupation—are categorical by nature. Other categorical variables are created by grouping values of a quantitative variable into classes. For instance, we could classify people in a data set by age: 0–9, 10–19, 20–29, and so on.

The proper method of analysis for a variable depends on whether it is categorical or quantitative. As a result, it is important to be able to distinguish these two types of variables.

Census at School

Data, individuals, and variables

CensusAtSchool is an international project that collects data about primary and secondary school students using surveys. Hundreds of thousands of students from Australia, Canada, New Zealand, South Africa, and the United Kingdom have taken part in the project since 2000. Data from the surveys are available at the project's Web site (www.censusatschool.com). We used the site's "Random Data Selector" to choose 10 Canadian students who completed the survey in a recent year. The table below displays the data.

Province	Gender	Languages spoken	Handed	Height (cm)	Wrist circum. (mm)	Preferred communication	Travel to school (min)
Ontario	Male	1	Right	175	175	Internet chat or MSN	25
Alberta	Female	3	Right	147	140	MySpace/Facebook	20
Ontario	Male	1	Right	165	170	Internet chat	4
British Columbia	Female	1	Right	155	145	In person	10
New Brunswick	Male	9	Left	130.5	130	Other	40
Ontario	Male	2	Right	170	165	In person	7
Ontario	Male	3	Left	150	100	Internet chat	10
New Brunswick	Male	2	Both	167.5	220	Internet chat	30
Ontario	Female	1	Right	161	104	Text messaging	10
Ontario	Male	6	Right	190.5	180	Internet chat	10

PROBLEM:

(a) Who are the individuals in this data set?

(b) What variables were measured? Identify each as categorical or quantitative. In what units were the quantitative variables measured?

(c) Describe the individual in the highlighted row.

SOLUTION:

We'll see in Chapter 4 why choosing at random, as we did in this example, is a good idea.

(a) The individuals are the 10 randomly selected Canadian students.

(b) The eight variables measured are province where student lives (categorical), gender (categorical), number of languages spoken (quantitative, in whole numbers), dominant hand (categorical), height (quantitative, in centimeters), wrist circumference (quantitative, in millimeters), preferred communication method (categorical), and travel time to school (quantitative, in minutes).

(c) This student lives in Ontario, is male, speaks three languages, is left-handed, is 150 cm tall (about 59 inches), has a wrist circumference of 100 mm (about 4 inches), prefers to communicate via Internet chat, and travels 10 minutes to get to school.

To make life simpler, we sometimes refer to "categorical data" or "quantitative data" instead of identifying the variable as categorical or quantitative.

Most data tables follow the format shown in the example—each row is an individual, and each column is a variable. Sometimes the individuals are called *cases*.

A variable generally takes values that vary (hence the name "variable"!). Categorical variables sometimes have similar counts in each category and sometimes don't. For instance, we might have expected similar numbers of males and females in the CensusAtSchool data set. But we aren't surprised to see that most students are right-handed. Quantitative variables may take values that are very close together or values that are quite spread out. We call the pattern of variation of a variable its **distribution.**

DEFINITION: Distribution

The **distribution** of a variable tells us what values the variable takes and how often it takes these values.

Section 1.1 begins by looking at how to describe the distribution of a single categorical variable and then examines relationships between categorical variables. Sections 1.2 and 1.3 and all of Chapter 2 focus on describing the distribution of a quantitative variable. Chapter 3 investigates relationships between two quantitative variables. In each case, we begin with graphical displays, then add numerical summaries for a more complete description.

How to Explore Data

- Begin by examining each variable by itself. Then move on to study relationships among the variables.
- Start with a graph or graphs. Then add numerical summaries.

CHECK YOUR UNDERSTANDING

Jake is a car buff who wants to find out more about the vehicles that students at his school drive. He gets permission to go to the student parking lot and record some data. Later, he does some research about each model of car on the Internet. Finally, Jake makes a spreadsheet that includes each car's model, year, color, number of cylinders, gas mileage, weight, and whether it has a navigation system.

1. Who are the individuals in Jake's study?
2. What variables did Jake measure? Identify each as categorical or quantitative.

From Data Analysis to Inference

Inference

Sometimes, we're interested in drawing conclusions that go beyond the data at hand. That's the idea of **inference**. In the CensusAtSchool example, 7 of the 10 randomly selected Canadian students are right-handed. That's 70% of the *sample*. Can we conclude that 70% of the *population* of Canadian students who participated in CensusAtSchool are right-handed? No. If another random sample of 10 students was selected, the percent who are right-handed would probably not be exactly 70%. Can we at least say that the actual population value is "close" to 70%? That depends on what we mean by "close."

The following Activity gives you an idea of how statistical inference works.

ACTIVITY *Hiring discrimination—it just won't fly!*

MATERIALS: Deck of cards for each student or pair of students

An airline has just finished training 25 pilots—15 male and 10 female—to become captains. Unfortunately, only eight captain positions are available right now. Airline managers announce that they will use a lottery to determine which pilots will fill the available positions. The names of all 25 pilots will be written on identical slips of paper, which will be placed in a hat, mixed thoroughly, and drawn out one at a time until all eight captains have been identified.

A day later, managers announce the results of the lottery. Of the 8 captains chosen, 5 are female and 3 are male. Some of the male pilots who weren't selected suspect that the lottery was not carried out fairly. One of these pilots asks your statistics class for advice about whether to file a grievance with the pilots' union.

The key question in this possible discrimination case seems to be: *is it possible that these results happened just by chance?* To find out, you and your classmates will *simulate* the lottery process that airline managers said they used.

1. Separate the cards into piles by suit. Use ten cards from one suit to represent the female pilots. To represent the 15 male pilots, you'll need all 13 cards of another suit plus two extra cards from a third suit. A second student or group can use the leftover cards from the deck to set up their simulation in a similar way.

2. Shuffle your stack of 25 cards thoroughly and deal 8 cards. Count the number of female pilots selected. Record this value in a table like the one shown below.

Trial:	1	2	3	4	5
No. of females:					

3. Return the 8 cards to your stack. Shuffle and deal four more times so that you have a total of five simulated lottery results.

4. Your teacher will draw and label axes for a class dotplot. Each student should plot the number of females obtained in each of the five simulation trials on the graph.

5. Discuss the results with your classmates. Does it seem believable that airline managers carried out a fair lottery? What advice would you give the male pilot who contacted you?

6. Would your advice change if the lottery had chosen 6 female (and 2 male) pilots? Explain.

Our ability to do inference is determined by how the data are produced. Chapter 4 discusses the two primary methods of data production—sampling and experiments—and the types of conclusions that can be drawn from each. As the Activity illustrates, the logic of inference rests on asking, "What are the chances?" *Probability*, the study of chance behavior, is the topic of Chapters 5 through 7. We'll introduce the most common inference techniques in Chapters 8 through 12.

INTRODUCTION Summary

- A data set contains information on a number of **individuals.** Individuals may be people, animals, or things. For each individual, the data give values for one or more **variables.** A variable describes some characteristic of an individual, such as a person's height, gender, or salary.
- Some variables are **categorical** and others are **quantitative.** A categorical variable places each individual into a category, such as male or female. A quantitative variable has numerical values that measure some characteristic of each individual, such as height in centimeters or salary in dollars.
- The **distribution** of a variable describes what values the variable takes and how often it takes them.

INTRODUCTION Exercises

1. **Protecting wood** How can we help wood surfaces resist weathering, especially when restoring historic wooden buildings? In a study of this question, researchers prepared wooden panels and then exposed them to the weather. Here are some of the variables recorded: type of wood (yellow poplar, pine, cedar); type of water repellent (solvent-based, water-based); paint thickness (millimeters); paint color (white, gray, light blue); weathering time (months). Identify each variable as categorical or quantitative.

2. **Medical study variables** Data from a medical study contain values of many variables for each of the people who were the subjects of the study. Here are some of the variables recorded: gender (female or male); age (years); race (Asian, black, white, or other); smoker (yes or no); systolic blood pressure (millimeters of mercury); level of calcium in the blood (micrograms per milliliter). Identify each as categorical or quantitative.

pg 3

3. **A class survey** Here is a small part of the data set that describes the students in an AP Statistics class. The data come from anonymous responses to a questionnaire filled out on the first day of class.

Gender	Hand	Height (in)	Homework time (min)	Favorite music	Pocket change (cents)
F	L	65	200	Hip-hop	50
M	L	72	30	Country	35
M	R	62	95	Rock	35
F	L	64	120	Alternative	0
M	R	63	220	Hip-hop	0
F	R	58	60	Alternative	76
F	R	67	150	Rock	215

(a) What individuals does this data set describe?

(b) Clearly identify each of the variables. Which are quantitative? In what units are they measured?

(c) Describe the individual in the highlighted row.

4. **Coaster craze** Many people like to ride roller coasters. Amusement parks try to increase attendance by building exciting new coasters. The table below displays data on several roller coasters that were opened in 2009.[2]

Roller coaster	Type	Height (ft)	Design	Speed (mph)	Duration (s)
Wild mouse	Steel	49.3	Sit down	28	70
Terminator	Wood	95	Sit down	50.1	180
Manta	Steel	140	Flying	56	155
Prowler	Wood	102.3	Sit down	51.2	150
Diamondback	Steel	230	Sit down	80	180

(a) What individuals does this data set describe?

(b) Clearly identify each of the variables. Which are quantitative? In what units are they measured?

(c) Describe the individual in the highlighted row.

5. **Ranking colleges** Popular magazines rank colleges and universities on their "academic quality" in serving undergraduate students. Describe two categorical variables and two quantitative variables that you might record for each institution. Give the units of measurement for the quantitative variables.

6. **Students and TV** You are preparing to study the television-viewing habits of high school students. Describe two categorical variables and two quantitative variables that you might record for each student. Give the units of measurement for the quantitative variables.

Multiple choice: Select the best answer.

Exercises 7 and 8 refer to the following setting. At the Census Bureau Web site, you can view detailed data collected by the American Community Survey. The table below includes data for 10 people chosen at random from the more than one million people in households contacted by the survey. "School" gives the highest level of education completed.

Weight (lb)	Age (yr)	Travel to work (min)	School	Gender	Income last year ($)
187	66	0	Ninth grade	1	24,000
158	66	n/a	High school grad	2	0
176	54	10	Assoc. degree	2	11,900
339	37	10	Assoc. degree	1	6,000
91	27	10	Some college	2	30,000
155	18	n/a	High school grad	2	0
213	38	15	Master's degree	2	125,000
194	40	0	High school grad	1	800
221	18	20	High school grad	1	2,500
193	11	n/a	Fifth grade	1	0

7. The individuals in this data set are
   (a) households.
   (b) people.
   (c) adults.
   (d) 120 variables.
   (e) columns.

8. This data set contains
   (a) 7 variables, 2 of which are categorical.
   (b) 7 variables, 1 of which is categorical.
   (c) 6 variables, 2 of which are categorical.
   (d) 6 variables, 1 of which is categorical.
   (e) None of these.

Analyzing Categorical Data

In Section 1.1, you'll learn about:

- Bar graphs and pie charts
- Graphs: Good and bad
- Two-way tables and marginal distributions
- Relationships between categorical variables: Conditional distributions
- Organizing a statistical problem
- Simpson's paradox*

The values of a categorical variable are labels for the categories, such as "male" and "female." The distribution of a categorical variable lists the categories and gives either the *count* or the *percent* of individuals who fall in each category. Here's an example.

Radio Station Formats
Distribution of a categorical variable

The radio audience rating service Arbitron places the country's 13,838 radio stations into categories that describe the kinds of programs they broadcast. Here are two different tables showing the distribution of station formats:[3]

Frequency table

Format	Count of stations
Adult contemporary	1,556
Adult standards	1,196
Contemporary hit	569
Country	2,066
News/Talk/Information	2,179
Oldies	1,060
Religious	2,014
Rock	869
Spanish language	750
Other formats	1,579
Total	13,838

Relative frequency table

Format	Percent of stations
Adult contemporary	11.2
Adult standards	8.6
Contemporary hit	4.1
Country	14.9
News/Talk/Information	15.7
Oldies	7.7
Religious	14.6
Rock	6.3
Spanish language	5.4
Other formats	11.4
Total	99.9

Frequency table

Relative frequency table

In this case, the *individuals* are the radio stations and the *variable* being measured is the kind of programming that each station broadcasts. The table on the left, which we call a **frequency table,** displays the counts (*frequencies*) of stations in each format category. On the right, we see a **relative frequency table** of the data that shows the percents (*relative frequencies*) of stations in each format category.

*This is an interesting topic, but it is not required for the AP Statistics exam.

It's a good idea to check data for consistency. The counts should add to 13,838, the total number of stations. They do. The percents should add to 100%. In fact, they add to 99.9%. What happened? Each percent is rounded to the nearest tenth. The exact percents would add to 100, but the rounded percents only come close. This is **roundoff error.** Roundoff errors don't point to mistakes in our work, just to the effect of rounding off results.

Roundoff error

Bar Graphs and Pie Charts

Pie chart
Bar graph

Columns of numbers take time to read. You can use a **pie chart** or a **bar graph** to display the distribution of a categorical variable more vividly. Figure 1.1 illustrates both displays for the distribution of radio stations by format.

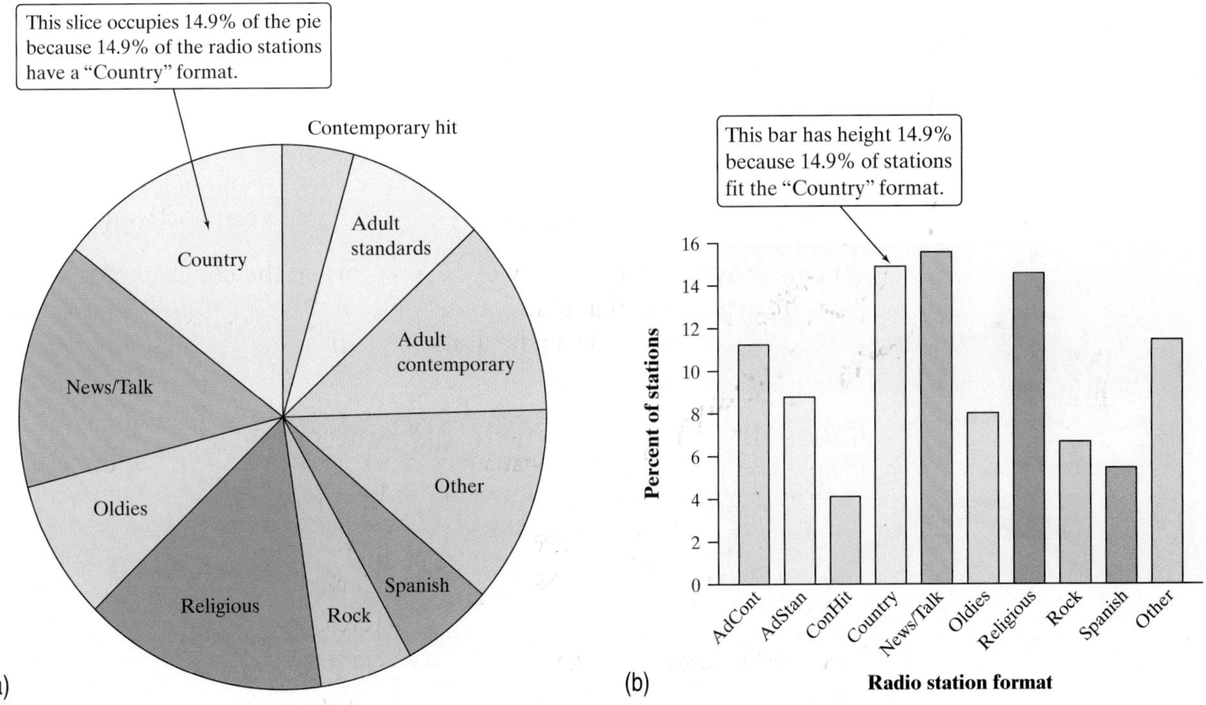

FIGURE 1.1 (a) Pie chart and (b) bar graph of U.S. radio stations by format.

THINK ABOUT IT

Do the data tell you what you want to know? Let's say that you plan to buy radio time to advertise your Web site for downloading MP3 music files. How helpful are the data in Figure 1.1? Not very. You are not interested in counting *stations*, but in counting *listeners*. For example, 14.6% of all stations are religious, but they have only a 5.5% share of the radio audience, according to Arbitron. In fact, you aren't even interested in the entire radio audience, because MP3 users are mostly young people. You really want to know what kinds of radio stations reach the largest numbers of young people. *Always think about whether the data you have help answer your questions.*

Pie charts show the distribution of a categorical variable as a "pie" whose slices are sized by the counts or percents for the categories. A pie chart must include all the categories that make up a whole. In the radio station example, we needed the "Other formats" category to complete the whole (all radio stations) and allow us to make a pie chart. Use a pie chart only when you want to emphasize each category's relation to the whole. Pie charts are awkward to make by hand, but technology will do the job for you.

Bar graphs are also called *bar charts.*

Bar graphs represent each category as a bar. The bar heights show the category counts or percents. Bar graphs are easier to make than pie charts and are also easier to read. To convince yourself, try to use the pie chart in Figure 1.1 to estimate the percent of radio stations that have an "Oldies" format. Now look at the bar graph—it's easy to see that the answer is about 8%.

Bar graphs are also more flexible than pie charts. Both graphs can display the distribution of a categorical variable, but a bar graph can also compare any set of quantities that are measured in the same units.

EXAMPLE

Who Owns an MP3 Player?

Choosing the best graph to display the data

Portable MP3 music players, such as the Apple iPod, are popular—but not equally popular with people of all ages. Here are the percents of people in various age groups who own a portable MP3 player, according to an Arbitron survey of 1112 randomly selected people.[4]

Age group (years)	Percent owning an MP3 player
12 to 17	54
18 to 24	30
25 to 34	30
35 to 54	13
55 and older	5

PROBLEM:

(a) Make a well-labeled bar graph to display the data. Describe what you see.

(b) Would it be appropriate to make a pie chart for these data? Why or why not?

SOLUTION:

(a) We start by labeling the axes: age group goes on the horizontal axis, and percent who own an MP3 player goes on the vertical axis. For the vertical scale, which is measured in percents, we'll start at 0

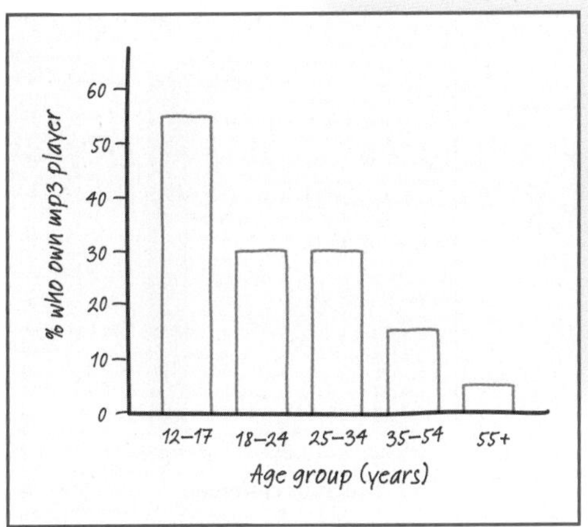

FIGURE 1.2 Bar graph comparing the percents of several age groups who own portable MP3 players.

and go up to 60, with tick marks for every 10. Then for each age category, we draw a bar with height corresponding to the percent of survey respondents who said they have an MP3 player. Figure 1.2 shows the completed bar graph. It appears that MP3 players are more popular among young people and that their popularity generally decreases as the age category increases.

(b) Making a pie chart to display these data is not appropriate because each percent in the table refers to a different age group, not to parts of a single whole.

Graphs: Good and Bad

Bar graphs compare several quantities by comparing the heights of bars that represent the quantities. Our eyes, however, react to the *area* of the bars as well as to their height. When all bars have the same width, the area (width × height) varies in proportion to the height, and our eyes receive the right impression. When you draw a bar graph, make the bars equally wide. Artistically speaking, bar graphs are a bit dull. It is tempting to replace the bars with pictures for greater eye appeal. Don't do it! The following example shows why.

Who Buys iMacs?

Beware the pictograph!

When Apple, Inc., introduced the iMac, the company wanted to know whether this new computer was expanding Apple's market share. Was the iMac mainly being bought by previous Macintosh owners, or was it being purchased by first-time computer buyers and by previous PC users who were switching over? To find out, Apple hired a firm to conduct a survey of 500 iMac customers. Each customer was categorized as a new computer purchaser, a previous PC owner, or a previous Macintosh owner. The table summarizes the survey results.[5]

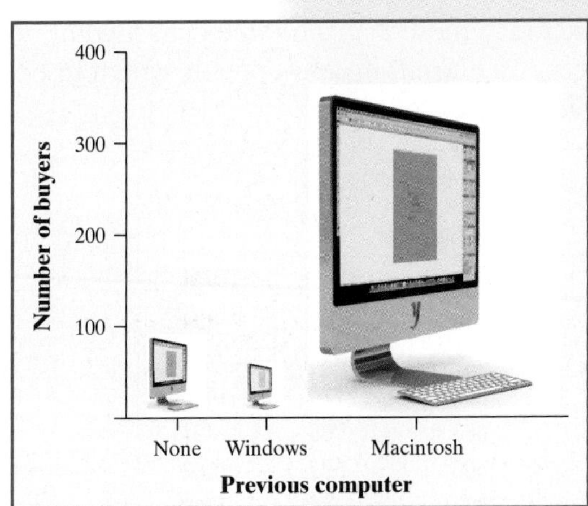

Previous ownership	Count	Percent
None	85	17.0
PC	60	12.0
Macintosh	355	71.0
Total	500	100.0

PROBLEM:

(a) Here's a clever graph of the data that uses pictures instead of the more traditional bars. How is this graph misleading?

(b) Two possible bar graphs of the data are shown on the next page. Which one could be considered deceptive? Why?

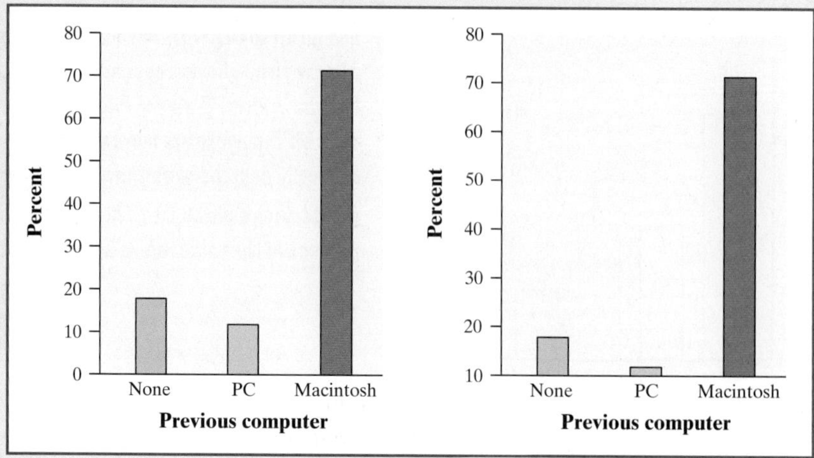

SOLUTION:

(a) Although the heights of the pictures are accurate, our eyes respond to the area of the pictures. The pictograph makes it seem like the percent of iMac buyers who are former Mac owners is at least ten times higher than either of the other two categories, which isn't the case.

(b) The bar graph on the right is misleading. By starting the vertical scale at 10 instead of 0, it looks like the percent of iMac buyers who previously owned a PC is less than half the percent who are first-time computer buyers. We get a distorted impression of the relative percents in the three categories.

For Practice *Try Exercise 17*

There are two important lessons to be learned from this example: (1) beware the pictograph, and (2) watch those scales.

Two-Way Tables and Marginal Distributions

We have learned some techniques for analyzing the distribution of a single categorical variable. What do we do when a data set involves two categorical variables? We begin by examining the counts or percents in various categories for one of the variables. Here's an example to show what we mean.

EXAMPLE

I'm Gonna Be Rich!

Relationship between two categorical variables

A survey of 4826 randomly selected young adults (aged 19 to 25) asked, "What do you think are the chances you will have much more than a middle-class income at age 30?" The table below shows the responses, omitting a few people who refused to respond or who said they were already rich.[6]

Young adults by gender and chance of getting rich

	Gender		
Opinion	**Female**	**Male**	**Total**
Almost no chance	96	98	194
Some chance but probably not	426	286	712
A 50-50 chance	696	720	1416
A good chance	663	758	1421
Almost certain	486	597	1083
Total	2367	2459	4826

Two-way table

This is a **two-way table** because it describes two categorical variables, gender and opinion about becoming rich. Opinion is the *row variable* because each row in the table describes young adults who held one of the five opinions about their chances. Because the opinions have a natural order from "Almost no chance" to "Almost certain," the rows are also in this order. Gender is the *column variable.* The entries in the table are the counts of individuals in each opinion-by-gender class.

How can we best grasp the information contained in the two-way table above? First, *look at the distribution of each variable separately.* The distribution of a categorical variable says how often each outcome occurred. The "Total" column at the right of the table contains the totals for each of the rows. These row totals give the distribution of opinions about becoming rich in the entire group of 4826 young adults: 194 felt that they had almost no chance, 712 thought they had just some chance, and so on. (If the row and column totals are missing, the first thing to do in studying a two-way table is to calculate them.) The distributions of opinion alone and gender alone are called **marginal distributions** because they appear at the right and bottom margins of the two-way table.

DEFINITION: Marginal distribution

The **marginal distribution** of one of the categorical variables in a two-way table of counts is the distribution of values of that variable among all individuals described by the table.

Percents are often more informative than counts, especially when we are comparing groups of different sizes. We can display the marginal distribution of opinions in percents by dividing each row total by the table total and converting to a percent. For instance, the percent of these young adults who think they are almost certain to be rich by age 30 is

$$\frac{\text{almost certain total}}{\text{table total}} = \frac{1083}{4826} = 0.224 = 22.4\%$$

I'm Gonna Be Rich!

Examining a marginal distribution

PROBLEM:

(a) Use the data in the two-way table to calculate the marginal distribution (in percents) of opinions.

(b) Make a graph to display the marginal distribution. Describe what you see.

SOLUTION:

(a) We can do four more calculations like the one shown above to obtain the marginal distribution of opinions in percents. Here is the complete distribution.

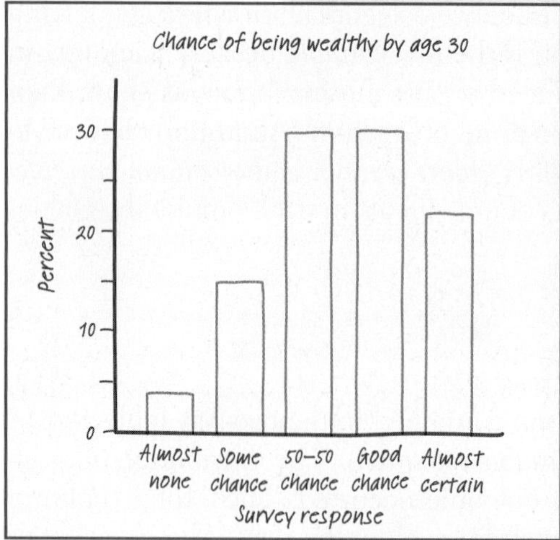

FIGURE 1.3 Bar graph showing the marginal distribution of opinion about chance of being rich by age 30.

Response	Percent
Almost no chance	$\frac{194}{4826} = 4.0\%$
Some chance	$\frac{712}{4826} = 14.8\%$
A 50–50 chance	$\frac{1416}{4826} = 29.3\%$
A good chance	$\frac{1421}{4826} = 29.4\%$
Almost certain	$\frac{1083}{4826} = 22.4\%$

(b) Figure 1.3 is a bar graph of the distribution of opinion among these young adults. It seems that many young adults are optimistic about their future income. Over 50% of those who responded to the survey felt that they had "a good chance" or were "almost certain" to be rich by age 30.

For Practice *Try Exercise 19*

Each marginal distribution from a two-way table is a distribution for a single categorical variable. As we saw earlier, we can use a bar graph or a pie chart to display such a distribution.

CHECK YOUR UNDERSTANDING

1. Use the data in the two-way table on page 12 to calculate the marginal distribution (in percents) of gender.
2. Make a graph to display the marginal distribution. Describe what you see.

Relationships between Categorical Variables: Conditional Distributions

The two-way table contains much more information than the two marginal distributions of opinion alone and gender alone. *Marginal distributions tell us nothing about the relationship between two variables.* To describe a relationship between two categorical variables, we must calculate some well-chosen percents from the counts given in the body of the table.

Young adults by gender and chance of getting rich

	Gender		
Opinion	**Female**	**Male**	**Total**
Almost no chance	96	98	194
Some chance but probably not	426	286	712
A 50-50 chance	696	720	1416
A good chance	663	758	1421
Almost certain	486	597	1083
Total	2367	2459	4826

Conditional distribution of opinion among women	
Response	**Female**
Almost no chance	$\frac{96}{2367} = 4.1\%$
Some chance	$\frac{426}{2367} = 18.0\%$
A 50-50 chance	$\frac{696}{2367} = 29.4\%$
A good chance	$\frac{663}{2367} = 28.0\%$
Almost certain	$\frac{486}{2367} = 20.5\%$

We can study the opinions of women alone by looking only at the "Female" column in the two-way table. To find the percent of *young women* who think they are almost certain to be rich by age 30, divide the count of such women by the total number of women, the column total:

$$\frac{\text{women who are almost certain}}{\text{column total}} = \frac{486}{2367} = 0.205 = 20.5\%$$

Doing this for all five entries in the "Female" column gives the **conditional distribution** of opinion among women. See the table in the margin. We use the term "conditional" because this distribution describes only young adults who satisfy the condition that they are female.

DEFINITION: Conditional distribution

A **conditional distribution** of a variable describes the values of that variable among individuals who have a specific value of another variable. There is a separate conditional distribution for each value of the other variable.

Now let's examine the men's opinions.

I'm Gonna Be Rich!
Calculating a conditional distribution

PROBLEM: Calculate the conditional distribution of opinion among the men.

SOLUTION: To find the percent of *men* who think they are almost certain to be rich by age 30, divide the count of such men by the total number of men, the column total:

$$\frac{\text{men who are almost certain}}{\text{column total}} = \frac{597}{2459} = 24.3\%$$

If we do this for all five entries in the "Male" column, we get the conditional distribution shown in the table.

Conditional distribution of opinion among men	
Response	**Male**
Almost no chance	$\frac{98}{2459} = 4.0\%$
Some chance	$\frac{286}{2459} = 11.6\%$
A 50-50 chance	$\frac{720}{2459} = 29.3\%$
A good chance	$\frac{758}{2459} = 30.8\%$
Almost certain	$\frac{597}{2459} = 24.3\%$

For Practice *Try Exercise 21*

Software will calculate conditional distributions for you. Most programs allow you to choose which conditional distributions you want to compute.

TECHNOLOGY CORNER Analyzing two-way tables

Figure 1.4 presents the two conditional distributions of opinion, for women and for men, and also the marginal distribution of opinion for all of the young adults. The distributions agree (up to rounding) with the results in the last two examples.

Session

	Female	Male	All
A: Almost no chance	96 4.06	98 3.99	194 4.02
B: Some chance but probably not	426 18.00	286 11.63	712 14.75
C: A 50-50 chance	696 29.40	720 29.28	1416 29.34
D: A good chance	663 28.01	758 30.83	1421 29.44
E: Almost certain	486 20.53	597 24.28	1083 22.44
All	2367 100.00	2459 100.00	4826 100.00

Cell Contents: Count
% of Column

FIGURE 1.4 Minitab output for the two-way table of young adults by gender and chance of being rich, along with each entry as a percent of its column total. The "Female" and "Male" columns give the conditional distributions of opinion for women and men, and the "All" column shows the marginal distribution of opinion for all these young adults.

There are *two sets* of conditional distributions for any two-way table. So far, we have looked at the conditional distributions of opinion for the two genders. We could also examine the five conditional distributions of gender, one for each of the five opinions, by looking separately at the rows in the original two-way table. For instance, the conditional distribution of gender among those who responded "Almost certain" is

Female	**Male**
$\frac{486}{1083} = 44.9\%$	$\frac{597}{1083} = 55.1\%$

That is, of the young adults who said they were almost certain to be rich by age 30, 44.9% were female and 55.1% were male.

Because the variable "gender" has only two categories, comparing the five conditional distributions amounts to comparing the percents of women among young adults who hold each opinion. Figure 1.5 makes this comparison in a bar graph. The bar heights do *not* add to 100%, because each bar represents a different group of people.

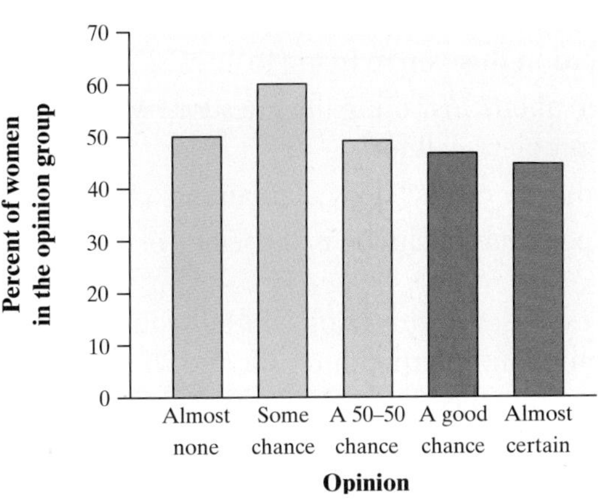

FIGURE 1.5 Bar graph comparing the percents of females among those who hold each opinion about their chance of being rich by age 30.

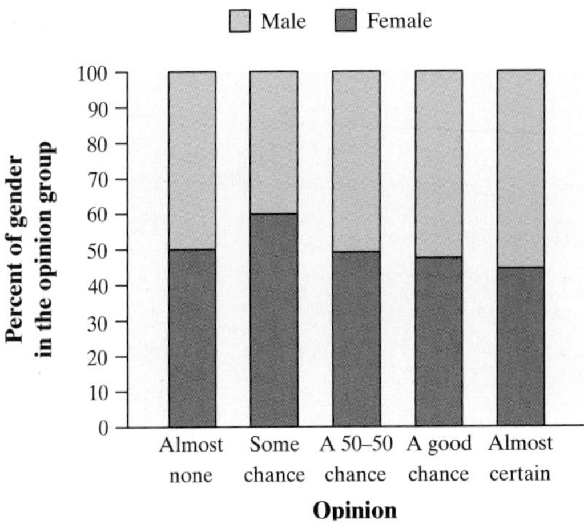

FIGURE 1.6 Segmented bar graph showing the conditional distribution of gender for each opinion category.

Segmented bar graph

An alternative to the bar graph in Figure 1.5 is a **segmented bar graph,** like the one shown in Figure 1.6. For each opinion category, there is a single bar with "segments" that correspond to the different genders. The height of each segment is determined by the percent of young adults having that opinion who were of each gender. We can see the two percents we calculated earlier displayed in the "Almost certain" bar—female 44.9% and male 55.1%. Notice that each bar has a total height of 100%.

Which conditional distributions should we compare? Our goal all along has been to analyze the relationship between gender and opinion about chances of becoming rich for these young adults. We started by examining the conditional distributions of opinion for males and females. Then we looked at the conditional distributions of gender for each of the five opinion categories. Which of these two gives us the information we want? Here's a hint: think about whether changes in one variable might help explain changes in the other. In this case, it seems reasonable to think that gender might influence young adults' opinions about their chances of getting rich. To see whether the data support this idea, we should compare the conditional distributions of opinion for women and men.

CHECK YOUR UNDERSTANDING

1. Find the conditional distributions of gender among each of the other four opinion categories (we did "Almost certain" earlier). Use Figure 1.5 or Figure 1.6 to check that your answers are approximately correct.
2. Make a revised version of Figure 1.4 that includes your results from Question 1.

Organizing a Statistical Problem

As you learn more about statistics, you will be asked to solve more complex problems. Although no single strategy will work on every problem, it might be helpful to have a general framework for organizing your thinking. Here is a four-step process you can follow.

How to Organize a Statistical Problem: A Four-Step Process

State: What's the question that you're trying to answer?

Plan: How will you go about answering the question? What statistical techniques does this problem call for?

Do: Make graphs and carry out needed calculations.

Conclude: Give your practical conclusion in the setting of the real-world problem.

To keep the four steps straight, just remember: **S**tatistics **P**roblems **D**emand **C**onsistency!

Many examples and exercises in this book will tell you what to do—construct a graph, perform a calculation, interpret a result, and so on. Real statistics problems don't come with such detailed instructions, however. From now on, you will encounter some examples and exercises that are more realistic. They are marked with the four-step icon. Use the four-step process as a guide to solving these problems, as the following example illustrates.

EXAMPLE

Women's and Men's Opinions

Conditional distributions and relationships

Based on the survey data, can we conclude that young men and women differ in their opinions about the likelihood of future wealth? Give appropriate evidence to support your answer. Follow the four-step process.

STATE: What is the relationship between gender and responses to the question "What do you think are the chances you will have much more than a middle-class income at age 30?"

PLAN: We suspect that gender might influence a young adult's opinion about the chance of getting rich. So we'll compare the conditional distributions of response for men alone and for women alone.

Response	Female	Male
Almost no chance	$\frac{96}{2367} = 4.1\%$	$\frac{98}{2459} = 4.0\%$
Some chance	$\frac{426}{2367} = 18.0\%$	$\frac{286}{2459} = 11.6\%$
A 50-50 chance	$\frac{696}{2367} = 29.4\%$	$\frac{720}{2459} = 29.3\%$
A good chance	$\frac{663}{2367} = 28.0\%$	$\frac{758}{2459} = 30.8\%$
Almost certain	$\frac{486}{2367} = 20.5\%$	$\frac{597}{2459} = 24.3\%$

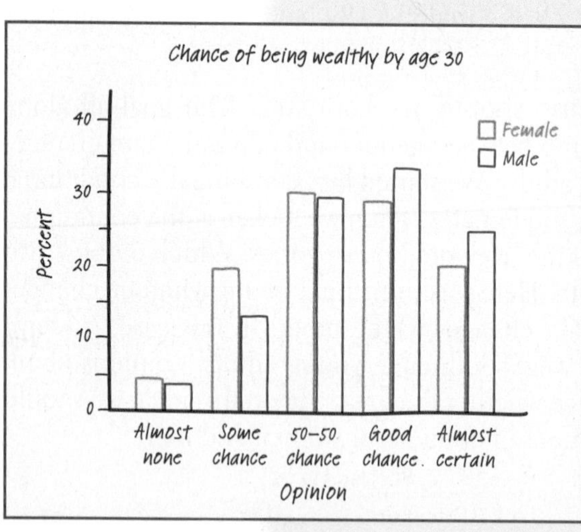

FIGURE 1.7 Side-by-side bar graph comparing the opinions of males and females.

Side-by-side bar graph

DO: We'll make a **side-by-side bar graph** to compare the opinions of males and females. Figure 1.7 displays the completed graph.

CONCLUDE: Based on the sample data, men seem somewhat more optimistic about their future income than women. Men were less likely to say that they have "some chance but probably not" than women (11.6% vs. 18.0%). Men were more likely to say that they have "a good chance" (30.8% vs. 28.0%) or are "almost certain" (24.3% vs. 20.5%) to have much more than a middle-class income by age 30 than women were.

For Practice *Try Exercise 25*

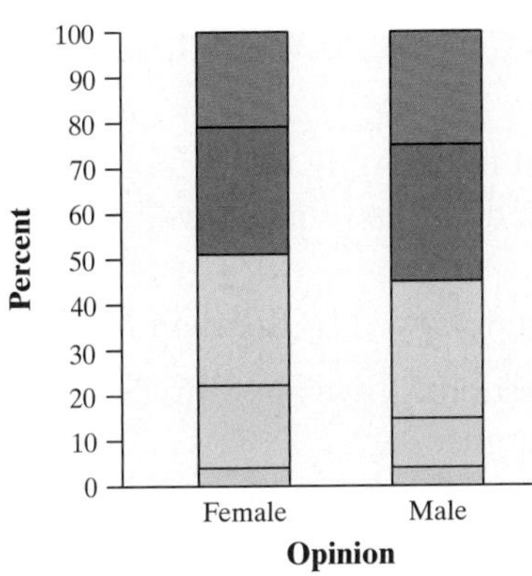

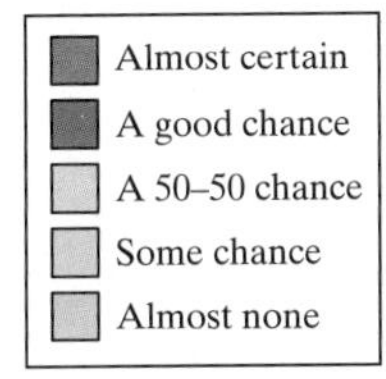

FIGURE 1.8 Segmented bar graph comparing the opinions of males and females.

We could have used a segmented bar graph to compare the distributions of male and female responses in the previous example. Figure 1.8 shows the completed graph. Each bar has five segments—one for each of the opinion categories. It's fairly difficult to compare the percents of males and females in each category because the "middle" segments in the two bars start at different locations on the vertical axis. The side-by-side bar graph in Figure 1.7 makes comparison easier.

Both graphs provide evidence of an **association** between gender and opinion about future wealth in this sample of young adults. That is, the values of one variable (opinion) tend to occur more or less frequently in combination with specific values of the other variable (gender). Men more often rated their chances of becoming rich in the two highest categories; women said "some chance but probably not" much more frequently. Can we say that there is an association between gender and opinion in the *population* of young adults? Making this determination requires formal inference, which will have to wait a few chapters.

DEFINITION: Association

We say that there is an **association** between two variables if specific values of one variable tend to occur in common with specific values of the other.

There's one caution that we need to offer: *even a strong association between two categorical variables can be influenced by other variables lurking in the background.* The Data Exploration that follows gives you a chance to explore this idea using a famous (or infamous) data set.

DATA EXPLORATION *A Titanic disaster*

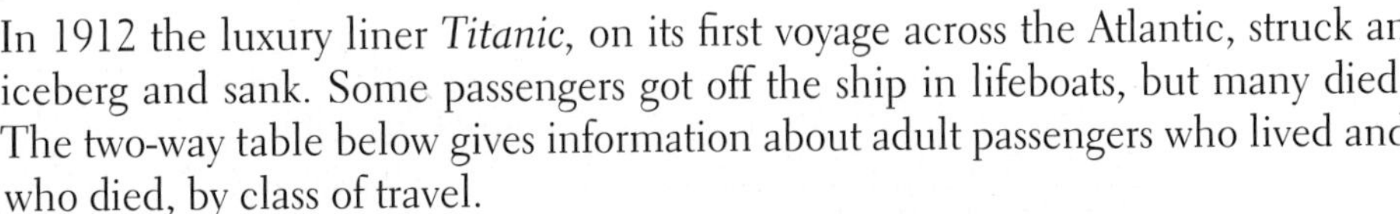
In 1912 the luxury liner *Titanic*, on its first voyage across the Atlantic, struck an iceberg and sank. Some passengers got off the ship in lifeboats, but many died. The two-way table below gives information about adult passengers who lived and who died, by class of travel.

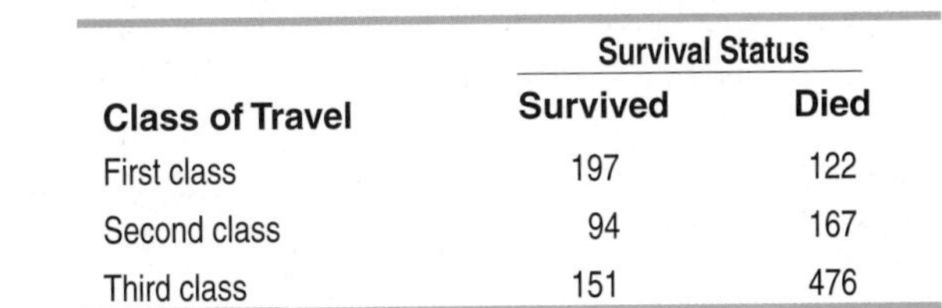

	Survival Status	
Class of Travel	**Survived**	**Died**
First class	197	122
Second class	94	167
Third class	151	476

Here's another table that displays data on survival status by gender and class of travel.

	Gender			
	Female		Male	
Class of Travel	**Survived**	**Died**	**Survived**	**Died**
First class	140	4	57	118
Second class	80	13	14	154
Third class	76	89	75	387

The movie *Titanic*, starring Leonardo DiCaprio and Kate Winslet, suggested the following:

- First-class passengers received special treatment in boarding the lifeboats, while some other passengers were prevented from doing so (especially third-class passengers).
- Women and children boarded the lifeboats first, followed by the men.

1. What do the data tell us about these two suggestions? Give appropriate graphical and numerical evidence to support your answer.
2. How does gender affect the relationship between class of travel and survival status? Explain.

Simpson's Paradox*

In the most extreme cases, it is possible for an association between two categorical variables to be "reversed" when we consider a third variable. Here is an example that demonstrates the surprises that can await the unsuspecting user of data.

Do Medical Helicopters Save Lives?

Reversing an association

Accident victims are sometimes taken by helicopter from the accident scene to a hospital. Helicopters save time. Do they also save lives? Let's compare the percents of accident victims who die with helicopter evacuation and with the usual transport to a hospital by road. Here are hypothetical data that illustrate a practical difficulty:[7]

	Helicopter	Road
Victim died	64	260
Victim survived	136	840
Total	200	1100

We see that 32% (64 out of 200) of helicopter patients died, but only 24% (260 out of 1100) of the others did. That seems discouraging.

The explanation is that the helicopter is sent mostly to serious accidents, so that the victims transported by helicopter are more often seriously injured. They are more likely to die with or without helicopter evacuation. Here are the same data broken down by the seriousness of the accident:

Serious Accidents		
	Helicopter	**Road**
Died	48	60
Survived	52	40
Total	100	100

Less Serious Accidents		
	Helicopter	**Road**
Died	16	200
Survived	84	800
Total	100	1000

Inspect these tables to convince yourself that they describe the same 1300 accident victims as the original two-way table. For example, 200 (100 + 100) were moved by helicopter, and 64 (48 + 16) of these died.

*This is an interesting topic, but it is not required for the AP Statistics exam.

Among victims of serious accidents, the helicopter saves 52% (52 out of 100) compared with 40% for road transport. If we look only at less serious accidents, 84% of those transported by helicopter survive, versus 80% of those transported by road. Both groups of victims have a higher survival rate when evacuated by helicopter.

How can it happen that the helicopter does better for both groups of victims but worse when all victims are lumped together? Examining the data makes the explanation clear. Half the helicopter transport patients are from serious accidents, compared with only 100 of the 1100 road transport patients. So the helicopter carries patients who are more likely to die. The seriousness of the accident was a "lurking variable" that, until we uncovered it, hid the true relationship between survival and mode of transport to a hospital. This example illustrates **Simpson's paradox.**

DEFINITION: Simpson's paradox

An association between two variables that holds for each individual value of a third variable can be changed or even reversed when the data for all values of the third variable are combined. This reversal is called **Simpson's paradox.**

SECTION 1.1 Summary

- The distribution of a categorical variable lists the categories and gives the count (**frequency table**) or percent (**relative frequency table**) of individuals that fall in each category.
- **Pie charts** and **bar graphs** display the distribution of a categorical variable. Bar graphs can also compare any set of quantities measured in the same units. When examining any graph, ask yourself, "What do I see?"
- A **two-way table** of counts organizes data about two categorical variables. Two-way tables are often used to summarize large amounts of information by grouping outcomes into categories.
- The row totals and column totals in a two-way table give the **marginal distributions** of the two individual variables. It is clearer to present these distributions as percents of the table total. Marginal distributions tell us nothing about the relationship between the variables.
- There are two sets of **conditional distributions** for a two-way table: the distributions of the row variable for each value of the column variable, and the distributions of the column variable for each value of the row variable. You may want to use a **side-by-side bar graph** (or possibly a **segmented bar graph**) to display conditional distributions.

- A statistical problem has a real-world setting. You can organize many problems using the four steps **state, plan, do,** and **conclude.**
- To describe the **association** between the row and column variables, compare an appropriate set of conditional distributions. Remember that even a strong association between two categorical variables can be influenced by other variables lurking in the background.
- An association between two variables that holds for each individual value of a third variable can be changed or even reversed when the data for all values of the third variable are combined. This is **Simpson's paradox.**

1.1 TECHNOLOGY CORNER

SECTION 1.1 Exercises

9. **Cool car colors** The most popular colors for cars and light trucks change over time. Silver passed green in 2000 to become the most popular color worldwide, then gave way to shades of white in 2007. Here is the distribution of colors for vehicles sold in North America in 2008.[8]

Color	Percent of vehicles
White	20
Black	17
Silver	17
Blue	13
Gray	12
Red	11
Beige/brown	5
Green	3
Yellow/gold	2

(a) What percent of vehicles had colors other than those listed?

(b) Display these data in a bar graph. Be sure to label your axes and title your graph.

(c) Would it be appropriate to make a pie chart of these data? Explain.

10. **Spam** Email spam is the curse of the Internet. Here is a compilation of the most common types of spam:[9]

Type of spam	Percent
Adult	19
Financial	20
Health	7
Internet	7
Leisure	6
Products	25
Scams	9
Other	??

(a) What percent of spam would fall in the "Other" category?

(b) Display these data in a bar graph. Be sure to label your axes and title your graph.

(c) Would it be appropriate to make a pie chart of these data? Explain.

11. **Birth days** Births are not evenly distributed across the days of the week. Here are the average numbers of babies born on each day of the week in the United States in a recent year:[10]

Day	Births
Sunday	7,374
Monday	11,704
Tuesday	13,169
Wednesday	13,038
Thursday	13,013
Friday	12,664
Saturday	8,459

(a) Present these data in a well-labeled bar graph. Would it also be correct to make a pie chart?

(b) Suggest some possible reasons why there are fewer births on weekends.

12. **Deaths among young people** Among persons aged 15 to 24 years in the United States, the leading causes of death and number of deaths in a recent year were as follows: accidents, 15,567; homicide, 5359; suicide, 4139; cancer, 1717; heart disease, 1067; congenital defects, 483.[11]

(a) Make a bar graph to display these data.

(b) To make a pie chart, you need one additional piece of information. What is it?

13. **Hispanic origins** Below is a pie chart prepared by the Census Bureau to show the origin of the more than 43 million Hispanics in the United States in 2006.[12] About what percent of Hispanics are Mexican? Puerto Rican?

Percent Distribution of Hispanics by Type: 2006

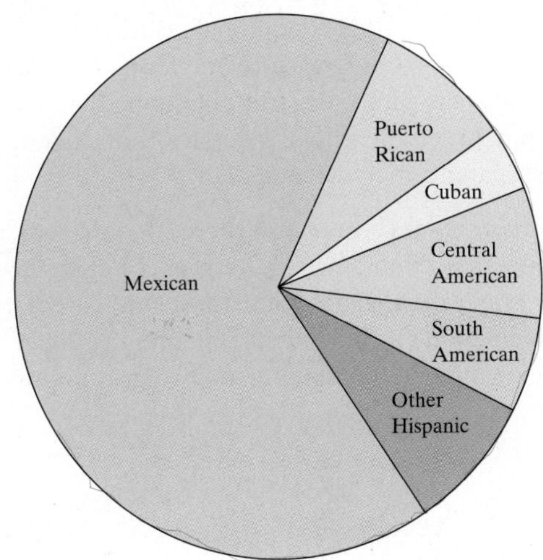

Comment: You see that it is hard to determine numbers from a pie chart. Bar graphs are much easier to use. (The Census Bureau did include the percents in its pie chart.)

14. **Which major?** About 1.6 million first-year students enroll in colleges and universities each year. What do they plan to study? The pie chart displays data on the percents of first-year students who plan to major in several discipline areas.[13] About what percent of first-year students plan to major in business? In social science?

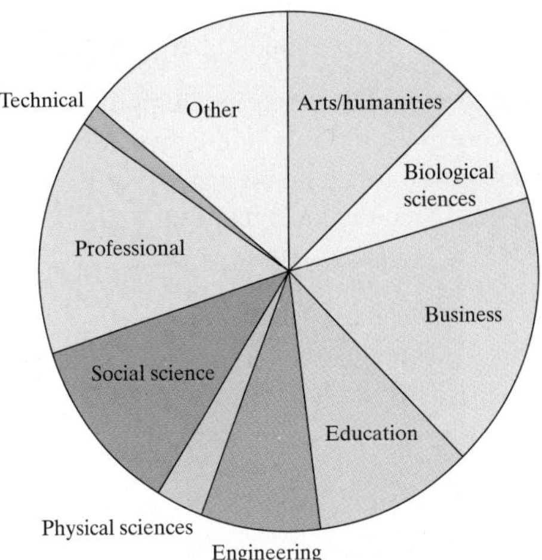

15. **Buying music online** Young people are more likely than older folk to buy music online. Here are the percents of people in several age groups who bought music online in 2006:[14]

pg 10

Age group	Bought music online
12 to 17 years	24%
18 to 24 years	21%
25 to 34 years	20%
35 to 44 years	16%
45 to 54 years	10%
55 to 64 years	3%
65 years and over	1%

(a) Explain why it is *not* correct to use a pie chart to display these data.

(b) Make a bar graph of the data. Be sure to label your axes and title your graph.

16. **The audience for movies** Here are data on the percent of people in several age groups who attended a movie in the past 12 months:[15]

Age group	Movie attendance
18 to 24 years	83%
25 to 34 years	73%
35 to 44 years	68%
45 to 54 years	60%
55 to 64 years	47%
65 to 74 years	32%
75 years and over	20%

(a) Display these data in a bar graph. Describe what you see.

(b) Would it be correct to make a pie chart of these data? Why or why not?
(c) A movie studio wants to know what percent of the total audience for movies is 18 to 24 years old. Explain why these data do not answer this question.

17. **Going to school** Students in a high school statistics class were given data about the primary method of transportation to school for a group of 30 students. They produced the pictograph shown.

pg 11

(a) How is this graph misleading?
(b) Make a new graph that isn't misleading.

18. **Oatmeal and cholesterol** Does eating oatmeal reduce cholesterol? An advertisement included the following graph as evidence that the answer is "Yes."

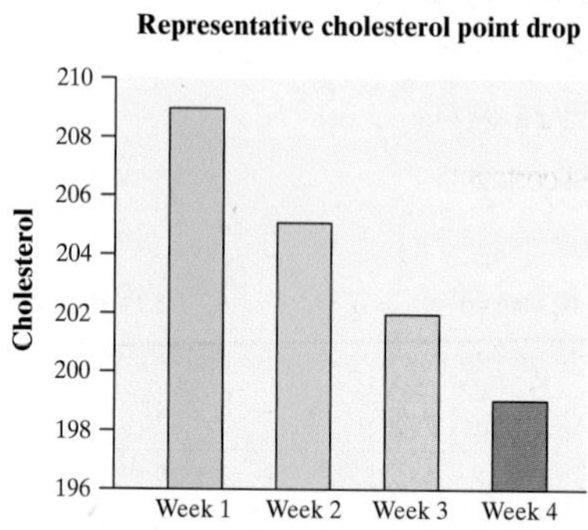

(a) How is this graph misleading?
(b) Make a new graph that isn't misleading. What do you conclude about the effect of eating oatmeal on cholesterol reduction?

19. **Attitudes toward recycled products** Recycling is supposed to save resources. Some people think recycled products are lower in quality than other products, a fact that makes recycling less practical. People who actually use a recycled product may have different opinions from those who don't use it. Here are data on attitudes toward coffee filters made of recycled paper among people who do and don't buy these filters:[16]

pg 13

	Think the quality of the recycled product is:		
	Higher	The same	Lower
Buyers	20	7	9
Nonbuyers	29	25	43

(a) How many people does this table describe? How many of these were buyers of coffee filters made of recycled paper?
(b) Give the marginal distribution of opinion about the quality of recycled filters. What percent think the quality of the recycled product is the same or higher than the quality of other filters?

20. **Smoking by students and parents** Here are data from a survey conducted at eight high schools on smoking among students and their parents:[17]

	Neither parent smokes	One parent smokes	Both parents smoke
Student does not smoke	1168	1823	1380
Student smokes	188	416	400

(a) How many students are described in the two-way table? What percent of these students smoke?
(b) Give the marginal distribution of parents' smoking behavior, both in counts and in percents.

21. **Attitudes toward recycled products** Exercise 19 gives data on the opinions of people who have and have not bought coffee filters made from recycled paper. To see the relationship between opinion and experience with the product, find the conditional distributions of opinion (the response variable) for buyers and nonbuyers. What do you conclude?

pg 15

22. **Smoking by students and parents** Refer to Exercise 20. Calculate three conditional distributions of students' smoking behavior: one for each of the three parental smoking categories. Describe the relationship between the smoking behaviors of students and their parents in a few sentences.

23. **Popular colors—here and there** Favorite vehicle colors may differ among countries. The side-by-side bar graph shows data on the most popular colors of cars in 2008 for the United States and Europe. Write a few sentences comparing the two distributions.

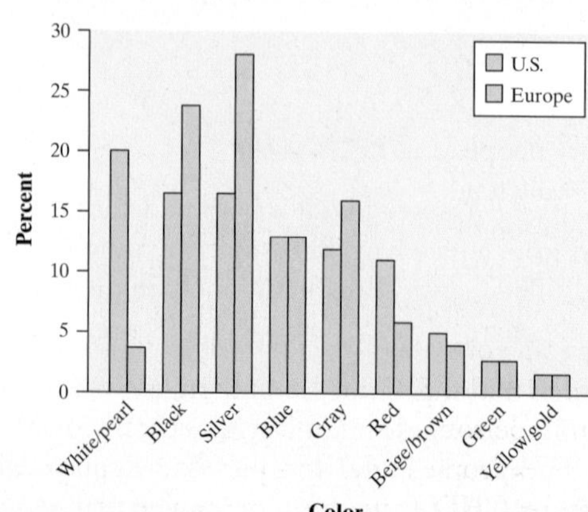

24. **Comparing car colors** Favorite vehicle colors may differ among types of vehicle. Here are data on the most popular colors in 2008 for luxury cars and for SUVs, trucks, and vans.

Color	Luxury cars (%)	SUVs, trucks, vans (%)
Black	22	13
Silver	16	16
White pearl	14	1
Gray	12	13
White	11	25
Blue	7	10
Red	7	11
Yellow/gold	6	1
Green	3	4
Beige/brown	2	6

(a) Make a graph to compare colors by vehicle type.

(b) Write a few sentences describing what you see.

pg 18

25. **Snowmobiles in the park** Yellowstone National Park surveyed a random sample of 1526 winter visitors to the park. They asked each person whether they owned, rented, or had never used a snowmobile. Respondents were also asked whether they belonged to an environmental organization (like the Sierra Club). The two-way table summarizes the survey responses.

	Environmental Clubs		
	No	Yes	Total
Never used	445	212	657
Snowmobile renter	497	77	574
Snowmobile owner	279	16	295
Total	1221	305	1526

Do these data provide convincing evidence of an association between environmental club membership and snowmobile use for the population of visitors to Yellowstone National Park? Follow the four-step process.

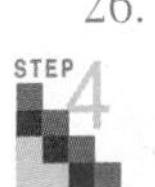

26. **Angry people and heart disease** People who get angry easily tend to have more heart disease. That's the conclusion of a study that followed a random sample of 12,986 people from three locations for about four years. All subjects were free of heart disease at the beginning of the study. The subjects took the Spielberger Trait Anger Scale test, which measures how prone a person is to sudden anger. Here are data for the 8474 people in the sample who had normal blood pressure. CHD stands for "coronary heart disease." This includes people who had heart attacks and those who needed medical treatment for heart disease.

	Low anger	Moderate anger	High anger	Total
CHD	53	110	27	190
No CHD	3057	4621	606	8284
Total	3110	4731	633	8474

Do these data support the study's conclusion about the relationship between anger and heart disease? Follow the four-step process.

Multiple choice: Select the best answer.

Exercises 27 to 32 refer to the following setting. The National Survey of Adolescent Health interviewed several thousand teens (grades 7 to 12). One question asked was "What do you think are the chances you will be married in the next ten years?" Here is a two-way table of the responses by gender:[18]

	Female	Male
Almost no chance	119	103
Some chance, but probably not	150	171
A 50-50 chance	447	512
A good chance	735	710
Almost certain	1174	756

27. The percent of females among the respondents was
(a) 2625. (c) about 46%. (e) None of these.
(b) 4877. (d) about 54%.

28. Your percent from the previous exercise is part of
(a) the marginal distribution of females.
(b) the marginal distribution of gender.
(c) the marginal distribution of opinion about marriage.
(d) the conditional distribution of gender among adolescents with a given opinion.
(e) the conditional distribution of opinion among adolescents of a given gender.

29. What percent of females thought that they were almost certain to be married in the next ten years?
(a) About 16% (c) About 40% (e) About 61%
(b) About 24% (d) About 45%

30. Your percent from the previous exercise is part of
(a) the marginal distribution of gender.
(b) the marginal distribution of opinion about marriage.
(c) the conditional distribution of gender among adolescents with a given opinion.

(d) the conditional distribution of opinion among adolescents of a given gender.

(e) the conditional distribution of "Almost certain" among females.

31. What percent of those who thought they were almost certain to be married were female?

(a) About 16% (c) About 40% (e) About 61%

(b) About 24% (d) About 45%

32. Your percent from the previous exercise is part of

(a) the marginal distribution of gender.

(b) the marginal distribution of opinion about marriage.

(c) the conditional distribution of gender among adolescents with a given opinion.

(d) the conditional distribution of opinion among adolescents of a given gender.

(e) the conditional distribution of females among those who said "Almost certain."

33. **Marginal distributions aren't the whole story** Here are the row and column totals for a two-way table with two rows and two columns:

a	b	50
c	d	50
60	40	100

Find *two different* sets of counts a, b, c, and d for the body of the table that give these same totals. This shows that the relationship between two variables cannot be obtained from the two individual distributions of the variables.

34.* **Baseball paradox** Most baseball hitters perform differently against right-handed and left-handed pitching. Consider two players, Joe and Moe, both of whom bat right-handed. The table below records their performance against right-handed and left-handed pitchers:

Player	Pitcher	Hits	At-bats
Joe	Right	40	100
	Left	80	400
Moe	Right	120	400
	Left	10	100

(a) Use these data to make a two-way table of player (Joe or Moe) versus outcome (hit or no hit).

(b) Show that Simpson's paradox holds: one player has a higher overall batting average, but the other player hits better against both left-handed and right-handed pitching.

(c) The manager doesn't believe that one player can hit better against both left-handers and right-handers yet have a lower overall batting average. Explain in simple language why this happens to Joe and Moe.

35.* **Race and the death penalty** Whether a convicted murderer gets the death penalty seems to be influenced by the race of the victim. Here are data on 326 cases in which the defendant was convicted of murder:[19]

	White Defendant			Black Defendant	
	White victim	Black victim		White victim	Black victim
Death	19	0	Death	11	6
Not	132	9	Not	52	97

(a) Use these data to make a two-way table of defendant's race (white or black) versus death penalty (yes or no).

(b) Show that Simpson's paradox holds: a higher percent of white defendants are sentenced to death overall, but for both black and white victims a higher percent of black defendants are sentenced to death.

(c) Use the data to explain why the paradox holds in language that a judge could understand..

36. **Fuel economy** (Introduction) Here is a small part of a data set that describes the fuel economy (in miles per gallon) of model year 2009 motor vehicles:

Make and model	Vehicle type	Transmission type	Number of cylinders	City mpg	Highway mpg
Aston Martin Vantage	Two-seater	Manual	8	12	19
Honda Civic	Subcompact	Automatic	4	25	36
Toyota Prius	Midsize	Automatic	4	48	45
Chevrolet Impala	Large	Automatic	6	18	29

(a) What are the individuals in this data set?

(b) What variables were measured? Identify each as categorical or quantitative.

*These exercises relate to the optional content on Simpson's paradox.

1.2

Displaying Quantitative Data with Graphs

In Section 1.2, you'll learn about:

- Dotplots
- Describing shape
- Comparing distributions
- Stemplots
- Histograms
- Using histograms wisely

To display the distribution of a categorical variable, use a bar graph or a pie chart. How can we picture the distribution of a quantitative variable? In this section, we present several types of graphs that can be used to display quantitative data.

Dotplots

Dotplot

One of the simplest graphs to construct and interpret is a **dotplot.** Each data value is shown as a dot above its location on a number line. We'll show how to make a dotplot using some sports data.

Gooooaaaaallllll!

How to make a dotplot

How good was the 2004 U.S. women's soccer team? With players like Brandi Chastain, Mia Hamm, and Briana Scurry, the team put on an impressive showing en route to winning the gold medal at the 2004 Olympics in Athens. Here are data on the number of goals scored by the team in 34 games played during the 2004 season:[20]

3	0	2	7	8	2	4	3	5	1	1	4	5	3	1	1	3
3	3	2	1	2	2	2	4	3	5	6	1	5	5	1	1	5

Here are the steps in making a dotplot:

- *Draw a horizontal axis (a number line) and label it with the variable name.* In this case, the variable is number of goals scored.
- *Scale the axis.* Start by looking at the minimum and maximum values of the variable. For these data, the minimum number of goals scored was 0, and the maximum was 8. So we mark our scale from 0 to 8, with tick marks at every whole-number value.
- *Mark a dot above the location on the horizontal axis corresponding to each data value.* Figure 1.9 displays a completed dotplot for the soccer data.

FIGURE 1.9 A dotplot of goals scored by the U.S. women's soccer team in 2004.

Making a graph is not an end in itself. The purpose of a graph is to help us understand the data. After you make a graph, always ask, "What do I see?" Here is a general strategy for interpreting graphs of quantitative data.

How to Examine the Distribution of a Quantitative Variable

In any graph, look for the **overall pattern** and for striking **departures** from that pattern.

- You can describe the overall pattern of a distribution by its **shape, center,** and **spread.**
- An important kind of departure is an **outlier,** an individual value that falls outside the overall pattern.

We'll learn more formal ways of describing shape, center, and spread and identifying outliers shortly. For now, let's use our informal understanding of these ideas to examine the graph in Figure 1.9.

Mode

Shape: The dotplot has a peak at 1. This indicates that the team's most frequent number of goals scored in games that season (known as the **mode**) was 1. In most of its games, the U.S. women's soccer team scored between 1 and 5 goals. However, the distribution has a long tail to the right. (Later, we will describe the shape of Figure 1.9 as *skewed to the right.*)

Center: We can describe the center by finding a value that divides the observations so that about half take larger values and about half take smaller values. This value is called the *median* of the distribution. In Figure 1.9, the median is 3. That is, in a typical game during the 2004 season, the U.S. women's soccer team scored about 3 goals. Of course, we could also summarize the center of the distribution by calculating the average (*mean*) number of goals scored per game. For the 2004 season, the team's mean was 3.06 goals.

Range

Spread: The spread of a distribution tells us how much *variability* there is in the data. One way to describe the variability is to give the smallest and largest values. The spread in Figure 1.9 is from 0 goals to 8 goals scored. Alternatively, we can compute the **range** of the distribution by subtracting the smallest value from the largest value. For these data, the range is $8 - 0 = 8$ goals.

When describing a distribution of quantitative data, don't forget your SOCS (shape, outliers, center, spread)!

Outliers: Was the game in which the women's team scored 8 goals an outlier? How about the team's 7-goal game? These values differ somewhat from the overall pattern. However, they don't clearly stand apart from the rest of the distribution. For now, let's agree to call attention only to potential outliers that suggest something special about an observation. In Section 1.3, we'll establish a procedure for determining whether a particular data value is an outlier.

EXAMPLE

Are You Driving a Gas Guzzler?

Interpreting a dotplot

The Environmental Protection Agency (EPA) is in charge of determining and reporting fuel economy ratings for cars (think of those large window stickers on a new car). For years, consumers complained that their actual gas mileages were noticeably lower than the values reported by the EPA. It seems that the EPA's tests—all of which are done on computerized devices to ensure consistency—did not consider things like outdoor temperature, use of the air conditioner, or realistic acceleration and braking by drivers. In 2008, the EPA changed the method for measuring a vehicle's fuel economy to try to give more accurate estimates.

The table below displays the EPA estimates of highway gas mileage in miles per gallon (mpg) for a sample of 24 model year 2009 midsize cars.

Model	Mpg	Model	Mpg	Model	Mpg
Acura RL	22	Dodge Avenger	30	Mercury Milan	29
Audi A6 Quattro	23	Hyundai Elantra	33	Mitsubishi Galant	27
Bentley Arnage	14	Jaguar XF	25	Nissan Maxima	26
BMW 528I	28	Kia Optima	32	Rolls Royce Phantom	18
Buick Lacrosse	28	Lexus GS 350	26	Saturn Aura	33
Cadillac CTS	25	Lincoln MKZ	28	Toyota Camry	31
Chevrolet Malibu	33	Mazda 6	29	Volkswagen Passat	29
Chrysler Sebring	30	Mercedes-Benz E350	24	Volvo S80	25

WWW

Source: 2009 Fuel Economy Guide, from the U.S. Environmental Protection Agency's Web site at www.fueleconomy.gov.

Here is a dotplot of the data:

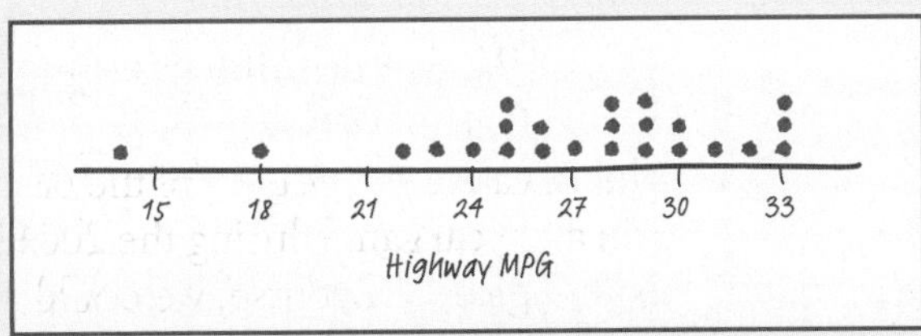

PROBLEM: Describe the shape, center, and spread of the distribution. Are there any outliers?

SOLUTION: Don't forget your SOCS (shape, outliers, center, spread)! **Shape:** In the dotplot, we can see three *clusters* of values: cars that get around 25 mpg, cars that get about 28 to 30 mpg, and cars that get around 33 mpg. We can also see large *gaps* between the Acura RL at 22 mpg, the Rolls Royce Phantom at 18 mpg, and the Bentley Arnage at 14 mpg. **Center:** The median is 28. So a "typical" model year 2009 midsize car got about 28 miles per gallon on the highway. **Spread:** The highest value is 33 mpg and the lowest value is 14 mpg. The range is 33 − 14 = 19 mpg. **Outliers:** We see two midsize cars with unusually low gas mileage ratings—the Bentley Arnage (14 mpg) and the Rolls Royce Phantom (18 mpg). These cars are potential outliers.

For Practice *Try Exercise 39*

Describing Shape

When you describe a distribution's shape, concentrate on the main features. Look for major peaks, not for minor ups and downs in the graph. Look for clusters of values and obvious gaps. Look for potential outliers, not just for the smallest and largest observations. Look for rough **symmetry** or clear **skewness.**

For brevity, we sometimes say "left-skewed" instead of "skewed to the left" and "right-skewed" instead of "skewed to the right." We could also describe a distribution with a long tail to the left as "skewed toward negative values" or "negatively skewed" and a distribution with a long right tail as "positively skewed."

DEFINITION: Symmetric and skewed distributions

A distribution is roughly **symmetric** if the right and left sides of the graph are approximately mirror images of each other.

A distribution is **skewed to the right** if the right side of the graph (containing the half of the observations with larger values) is much longer than the left side. It is **skewed to the left** if the left side of the graph is much longer than the right side.

The direction of skewness is the direction of the long tail, not the direction where most observations are clustered. See the drawing in the margin for a cute but corny way to help you keep this straight.

For his own safety, which way should Mr. Starnes go "skewing"?

EXAMPLE

Die Rolls and Quiz Scores

Describing shape

Figure 1.10 displays dotplots for two different sets of quantitative data. Let's practice describing the shapes of these distributions. Figure 1.10(a) shows the results of rolling a pair of fair, six-sided dice and finding the sum of the up-faces 100 times. This distribution is roughly symmetric. The dotplot in Figure 1.10(b) shows the scores on an AP Statistics class's first quiz. This distribution is skewed to the left.

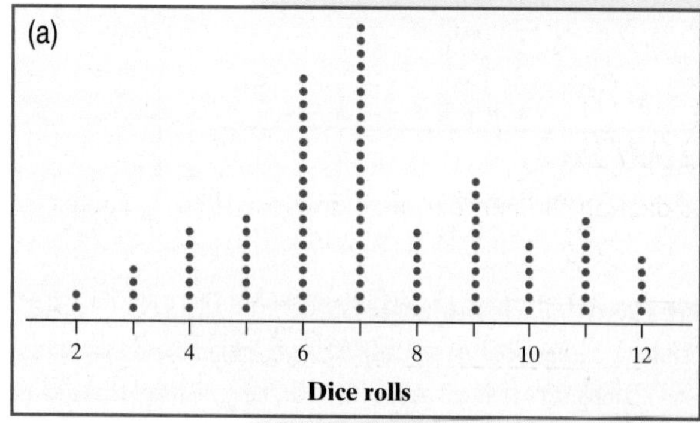

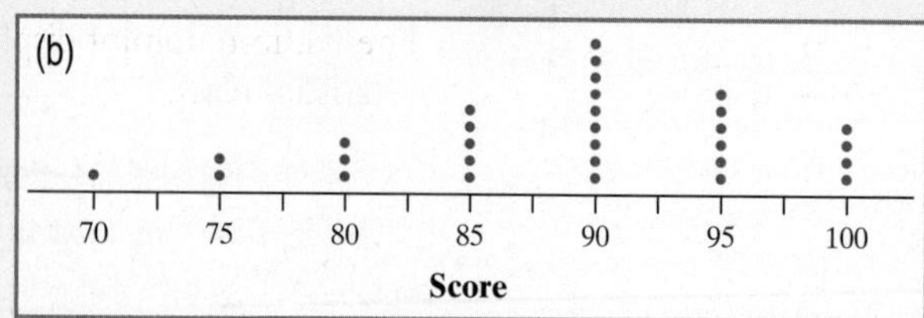

FIGURE 1.10 Dotplots displaying different shapes: (a) roughly symmetric; (b) skewed to the left.

Unimodal

Bimodal

Multimodal

Although the dotplots in the previous example have different shapes, they do have something in common. Both are **unimodal,** that is, they have a single peak: the graph of dice rolls at 7 and the graph of quiz scores at 90. (We don't count minor ups and downs in a graph, like the "bumps" at 9 and 11 in the dice rolls dotplot, as "peaks.") Figure 1.11 is a dotplot of the duration (in minutes) of 220 eruptions of the Old Faithful geyser. We would describe this distribution's shape as **bimodal** since it has two clear peaks: one near 2 minutes and the other near 4.5 minutes. (Although we could continue the pattern with "trimodal" for three peaks and so on, it's more common to refer to distributions with more than two clear peaks as **multimodal.**)

THINK ABOUT IT

What shape will the graph have? Some variables have distributions with predictable shapes. Many biological measurements on individuals from the same species and gender—lengths of bird bills, heights of young women—have symmetric distributions. Salaries and home prices, on the other hand, usually have right-skewed distributions. There are many moderately priced houses, for example, but the few very expensive mansions give the distribution of house prices a

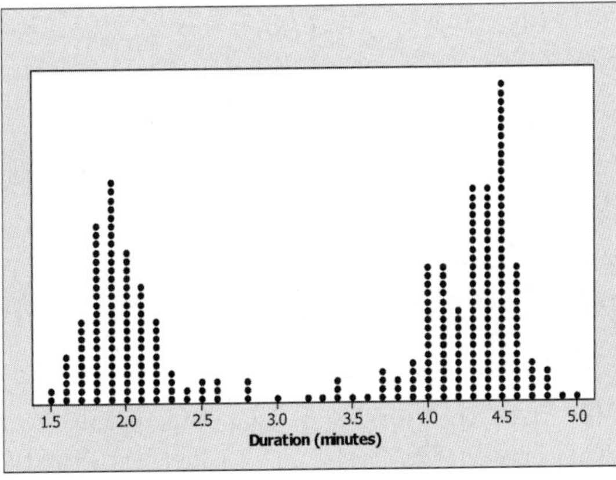

FIGURE 1.11 Dotplot displaying duration (in minutes) of Old Faithful eruptions. This graph has a bimodal shape.

strong right-skew. Many distributions have irregular shapes that are neither symmetric nor skewed. Some data show other patterns, such as the two peaks in Figure 1.11. Use your eyes, describe the pattern you see, and then try to explain the pattern.

CHECK YOUR UNDERSTANDING

The Fathom dotplot displays data on the number of siblings reported by each student in a statistics class.

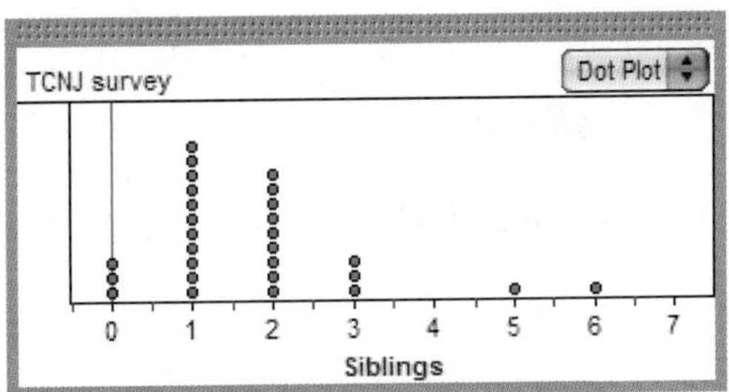

1. Describe the shape of the distribution.
2. Describe the center of the distribution.
3. Describe the spread of the distribution.
4. Identify any potential outliers.

Comparing Distributions

Some of the most interesting statistics questions involve comparing two or more groups. Which of two popular diets leads to greater long-term weight loss? Who texts more—males or females? Does the number of people living in a household differ among countries? As the following example suggests, you should always discuss shape, center, spread, and possible outliers whenever you compare distributions of a quantitative variable.

Household Size: U.K. versus South Africa

Comparing distributions

How do the numbers of people living in households in the United Kingdom (U.K.) and South Africa compare? To help answer this question, we used CensusAtSchool's "Random Data Selector" to choose 50 students from each country. Figure 1.12 is a dotplot of the household sizes reported by the survey respondents.

PROBLEM: **Compare the distributions of household size for these two countries.**

SOLUTION: **Don't forget your SOCS! Shape: The distribution of household size for the U.K. sample is roughly symmetric and unimodal, while the distribution for the South Africa sample is skewed to the right and unimodal. Center: Household sizes for the South African students tended to be larger than for the U.K. students. The median household sizes for the two groups are 6 people and 4 people, respectively. Spread: There is more variability (greater spread) in the household sizes for the South African students than for the U.K. students. The range for the South African data is $26 - 3 = 23$ people, while the range for the U.K. data is $6 - 2 = 4$ people. Outliers: There don't appear to be any potential outliers in the U.K. distribution. The South African distribution has two potential outliers in the right tail of the distribution—students who reported living in households with 15 and 26 people. (The U.K. households with 2 people actually *will* be classified as outliers when we introduce a procedure in the next section.)**

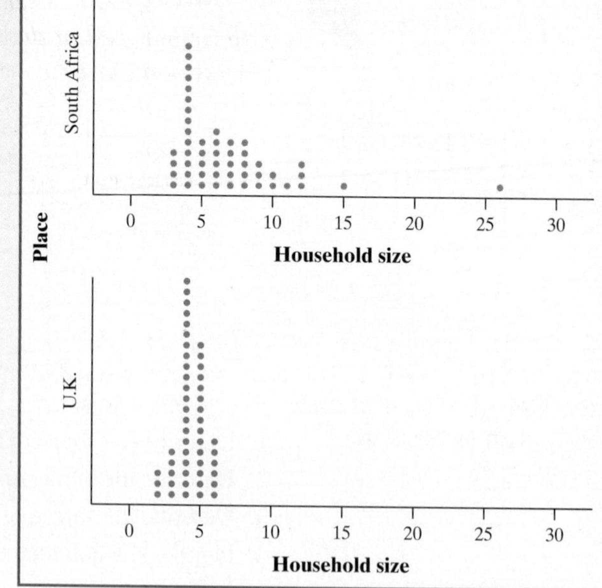

FIGURE 1.12 Dotplot of household size for random samples of 50 students from the United Kingdom and South Africa.

AP EXAM TIP When comparing distributions of quantitative data, it's not enough just to list values for the center and spread of each distribution. You have to explicitly *compare* these values, using words like "greater than," "less than," or "about the same as."

Notice that we discussed the distributions of household size only for the two *samples* of 50 students in the previous example. We might be interested in whether the sample data give us convincing evidence of a difference in the *population* distributions of household size for South Africa and the United Kingdom. We'll have to wait a few chapters to decide whether we can reach such a conclusion, but our ability to make such an inference later will be helped by the fact that the students in our samples were chosen at random.

Stemplots

Another simple graphical display for fairly small data sets is a **stemplot** (also called a stem-and-leaf plot). Stemplots give us a quick picture of the shape of a distribution while including the actual numerical values in the graph. Here's an example that shows how to make a stemplot.

EXAMPLE

How Many Shoes?

Making a stemplot

How many pairs of shoes does a typical teenager have? To find out, a group of AP Statistics students conducted a survey. They selected a random sample of 20 female students from their school. Then they recorded the number of pairs of shoes that each respondent reported having. Here are the data:

50	26	26	31	57	19	24	22	23	38
13	50	13	34	23	30	49	13	15	51

Here are the steps in making a stemplot. Figure 1.13 displays the process.

- *Separate each observation into a **stem**, consisting of all but the final digit, and a **leaf**, the final digit. Write the stems in a vertical column with the smallest at the top, and draw a vertical line at the right of this column. Do not skip any stems, even if there is no data value for a particular stem.* For these data, the tens digits are the stems, and the ones digits are the leaves. The stems run from 1 to 5.
- *Write each leaf in the row to the right of its stem.* For example, the female student with 50 pairs of shoes would have stem 5 and leaf 0, while the student with 31 pairs of shoes would have stem 3 and leaf 1.
- *Arrange the leaves in increasing order out from the stem.*
- *Provide a key that explains in context what the stems and leaves represent.*

Stems	Add leaves		Order leaves		Add a key
1	1	93335	1	33359	Key: 4|9 represents a female student who reported having 49 pairs of shoes.
2	2	664233	2	233466	
3	3	1840	3	0148	
4	4	9	4	9	
5	5	0701	5	0017	

FIGURE 1.13 Making a stemplot of the shoe data. (1) Write the stems. (2) Go through the data and write each leaf on the proper stem. (3) Arrange the leaves on each stem in order out from the stem. (4) Add a key.

The AP Statistics students in the previous example also collected data from a random sample of 20 male students at their school. Here are the numbers of pairs of shoes reported by each male in the sample:

14	7	6	5	12	38	8	7	10	10
10	11	4	5	22	7	5	10	35	7

Splitting stems

What would happen if we tried the same approach as before: using the first digits as stems and the last digits as leaves? The completed stemplot is shown in Figure 1.14(a). What shape does this distribution have? It is difficult to tell with so few stems. We can get a better picture of male shoe ownership by **splitting stems.**

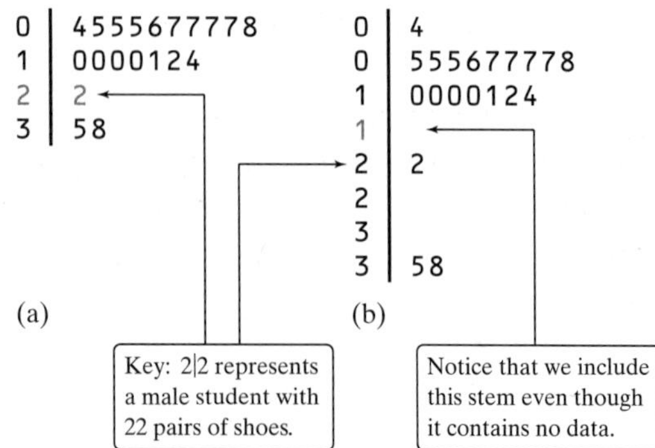

FIGURE 1.14 Two stemplots showing the male shoe data. Figure 1.14(b) improves on the stemplot of Figure 1.14(a) by splitting stems.

Females		Males
	0	4
	0	555677778
333	1	0000124
95	1	
4332	2	2
66	2	
410	3	
8	3	58
	4	
9	4	
100	5	
7	5	

Key: 2|2 represents a male student with 22 pairs of shoes.

FIGURE 1.15 Back-to-back stemplot comparing numbers of pairs of shoes for male and female students at a school.

In Figure 1.14(a), the values from 0 to 9 are placed on the "0" stem. Figure 1.14(b) shows another stemplot of the same data. This time, values having leaves 0 through 4 are placed on one stem, while values ending in 5 through 9 are placed on another stem. Now we can see the single peak, the cluster of values between 4 and 14, and the large gap between 22 and 35 more clearly.

Back-to-back stemplot

What if we want to compare the number of pairs of shoes that males and females have? That calls for a **back-to-back stemplot** with common stems. The leaves on each side are ordered out from the common stem. Figure 1.15 is a back-to-back stemplot for the male and female shoe data. Note that we have used the split stems from Figure 1.14(b) as the common stems. The values on the right are the male data from Figure 1.14(b). The values on the left are the female data, ordered out from the stem from right to left. We'll ask you to compare these two distributions shortly.

Here are a few tips to consider before making a stemplot:

- Stemplots do not work well for large data sets, where each stem must hold a large number of leaves.
- There is no magic number of stems to use, but five is a good minimum. Too few or too many stems will make it difficult to see the distribution's shape.
- If you split stems, be sure that each stem is assigned an equal number of possible leaf digits (two stems, each with five possible leaves; or five stems, each with two possible leaves).
- You can get more flexibility by rounding the data so that the final digit after rounding is suitable as a leaf. Do this when the data have too many digits. For example, in reporting teachers' salaries, using all five digits (for example, \$42,549) would be unreasonable. It would be better to round to the nearest thousand and use 4 as a stem and 3 as a leaf.

Instead of rounding, you can also *truncate* (remove one or more digits) when data have too many digits. The teacher's salary of \$42,549 would truncate to \$42,000.

CHECK YOUR UNDERSTANDING

1. Use the back-to-back stemplot in Figure 1.15 to write a few sentences comparing the number of pairs of shoes owned by males and females. Be sure to address shape, center, spread, and outliers.

Multiple choice: Select the best answer for Questions 2 through 4.

Here is a stemplot of the percents of residents aged 65 and older in the 50 states and the District of Columbia. The stems are whole percents and the leaves are tenths of a percent.

Stem	Leaves
6	8
7	
8	8
9	79
10	08
11	15566
12	0122234444578889999
13	0123333344489 9
14	02666
15	23
16	8

Key: 8|8 represents a state in which 8.8% of residents are 65 and older.

2. The low outlier is Alaska. What percent of Alaska residents are 65 or older?
(a) 0.68 (b) 6.8 (c) 8.8 (d) 16.8 (e) 68
3. Ignoring the outlier, the shape of the distribution is
(a) skewed to the right (c) skewed to the left. (e) skewed to the middle.
(b) roughly symmetric (d) bimodal.
4. The center of the distribution is close to
(a) 13.3%. (b) 12.8%. (c) 12.0%. (d) 11.6%. (e) 6.8% to 16.8%.

Histograms

Histogram

Quantitative variables often take many values. A graph of the distribution is clearer if nearby values are grouped together. The most common graph of the distribution of one quantitative variable is a **histogram.** Let's look at how to make a histogram using data on foreign-born residents in the United States.

Foreign-Born Residents

Making a histogram

What percent of your home state's residents were born outside the United States? The country as a whole has 12.5% foreign-born residents, but the states vary from 1.2% in West Virginia to 27.2% in California. The table below presents the data for all 50 states.[21] The *individuals* in this data set are the states. The *variable* is the percent of a state's residents who are foreign-born. It's much easier to see from a graph than from the table how your state compares with other states.

State	Percent	State	Percent	State	Percent
Alabama	2.8	Louisiana	2.9	Ohio	3.6
Alaska	7.0	Maine	3.2	Oklahoma	4.9
Arizona	15.1	Maryland	12.2	Oregon	9.7
Arkansas	3.8	Massachusetts	14.1	Pennsylvania	5.1
California	27.2	Michigan	5.9	Rhode Island	12.6
Colorado	10.3	Minnesota	6.6	South Carolina	4.1
Connecticut	12.9	Mississippi	1.8	South Dakota	2.2
Delaware	8.1	Missouri	3.3	Tennessee	3.9
Florida	18.9	Montana	1.9	Texas	15.9
Georgia	9.2	Nebraska	5.6	Utah	8.3
Hawaii	16.3	Nevada	19.1	Vermont	3.9
Idaho	5.6	New Hampshire	5.4	Virginia	10.1
Illinois	13.8	New Jersey	20.1	Washington	12.4
Indiana	4.2	New Mexico	10.1	West Virginia	1.2
Iowa	3.8	New York	21.6	Wisconsin	4.4
Kansas	6.3	North Carolina	6.9	Wyoming	2.7
Kentucky	2.7	North Dakota	2.1		

Here are the steps in making a histogram:

• *Divide the range of the data into classes of equal width.* The data in the table vary from 1.2 to 27.2, so we might choose to use classes of width 5, beginning at 0:

0–5 5–10 10–15 15–20 20–25 25–30

But we need to specify the classes so that each individual falls into exactly one class. For instance, what if a state had exactly 5.0% of its residents born outside the United States? Since a value of 0.0% would go in the 0–5 class, we'll agree to place a value of 5.0% in the 5–10 class, a value of 10.0% in the 10–15 class, and so on. In reality, then, our classes for the percent of foreign-born residents in the states are

0 to <5   5 to <10   10 to <15   15 to <20   20 to <25   25 to <30

• *Find the count (frequency) or percent (relative frequency) of individuals in each class.* Here is a frequency table and a relative frequency table for these data:

Notice that the frequencies add to 50, the number of individuals (states) in the data, and that the relative frequencies add to 100%.

Frequency table	
Class	**Count**
0 to < 5	20
5 to < 10	13
10 to < 15	9
15 to < 20	5
20 to < 25	2
25 to < 30	1
Total	50

Relative frequency table	
Class	**Percent**
0 to < 5	40
5 to < 10	26
10 to < 15	18
15 to < 20	10
20 to < 25	4
25 to < 30	2
Total	100

• *Label and scale your axes and draw the histogram.* Label the horizontal axis with the variable whose distribution you are displaying. That's the percent of a state's residents who are foreign-born. The scale on the horizontal axis runs from 0 to 30 because that is the span of the classes we chose. The vertical axis contains the scale of counts or percents. Each bar represents a class. The base of the bar covers the class, and the bar height is the class frequency or relative frequency. Draw the bars with no horizontal space between them unless a class is empty, so that its bar has height zero.

Figure 1.16(a) shows a completed frequency histogram; Figure 1.16(b) shows a completed relative frequency histogram. The two graphs look identical except for the vertical scales.

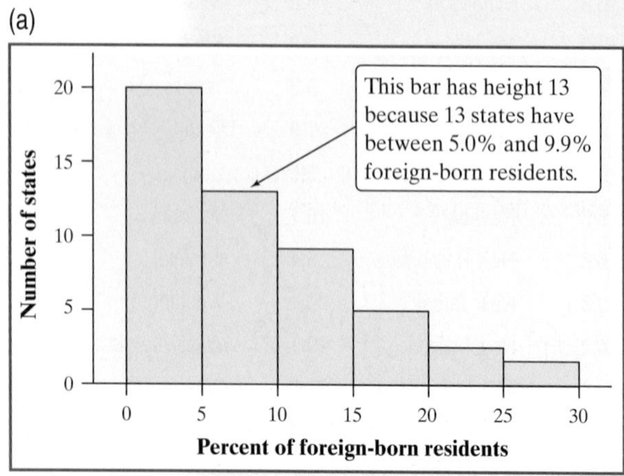

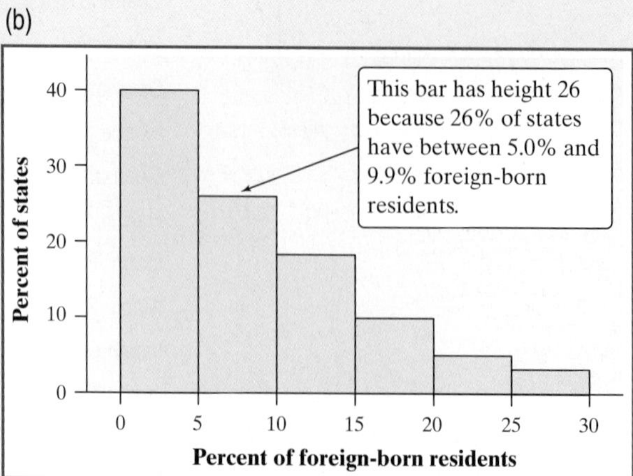

FIGURE 1.16 (a) Frequency histogram and (b) relative frequency histogram of the distribution of the percent of foreign-born residents in the 50 states.

What do the histograms in Figure 1.16 tell us about the percent of foreign-born residents in the states? To find out, we follow our familiar routine: describe the pattern and look for any departures from the pattern.

Shape: The distribution is skewed to the right. A majority of states have fewer than 10% foreign-born residents, but several states have much higher percents, so that the graph extends quite far to the right of its peak. The distribution has a *single peak* at the left, which represents states in which between 0% and 4.9% of residents are foreign-born.

Center: From the graph, we see that the midpoint (median) would fall somewhere in the 5.0% to 9.9% class. Remember that we're looking for the value having 25 states with smaller percents foreign-born and 25 with larger. (Arranging the observations from the table in order of size shows that the median is 6.1%.)

Spread: The histogram shows that the percent of foreign-born residents in the states varies from less than 5% to over 25%. (Using the data in the table, we see that the range is 27.2% − 1.2% = 26.0%.)

Outliers: We don't see any observations outside the overall single-peaked, right-skewed pattern of the distribution.

Figure 1.17 shows (a) a frequency histogram and (b) a relative frequency histogram of the same distribution, with classes half as wide. The new classes are 0–2.4, 2.5–4.9, etc. Now California, at 27.2%, stands out as a potential outlier in the right tail. The choice of classes in a histogram can influence the appearance of a distribution. Histograms with more classes show more detail but may have a less clear pattern.

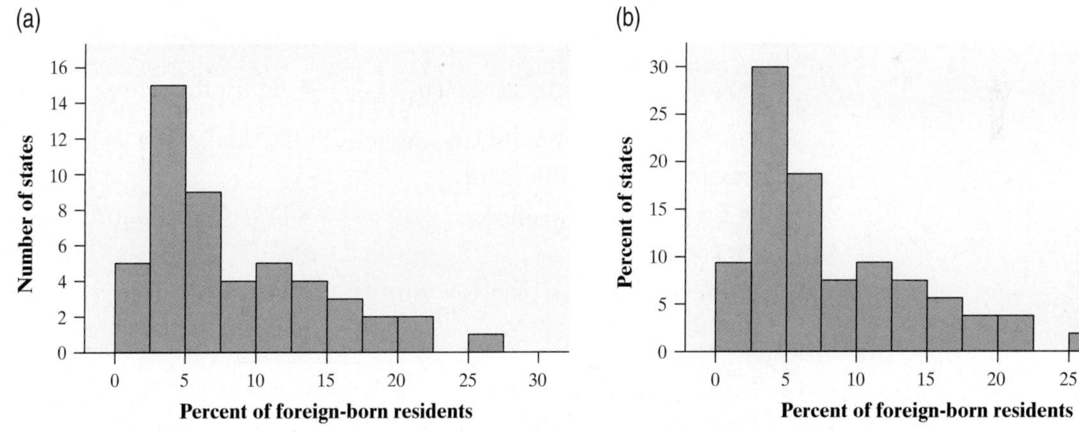

FIGURE 1.17 (a) Frequency histogram and (b) relative frequency histogram of the distribution of the percent of foreign-born residents in the 50 states, with classes half as wide as in Figure 1.16.

Statistical software and graphing calculators will choose the classes for you. The default choice is a good starting point, but you should adjust the classes to suit your needs. To see what we're talking about, launch the *One-Variable Statistical Calculator* applet at the book's Web site, www.whfreeman.com/tps4e. Select the "Percent of foreign-born residents" data set, and then click on the "Histogram" tab. You can change the number of classes by dragging the horizontal axis with your mouse or pointing device. By doing so, it's easy to see how the choice of classes affects the histogram. *Bottom line: Use your judgment in choosing classes to display the shape.*

TECHNOLOGY CORNER Histograms on the calculator

TI-83/84	TI-89

1. Enter the data for the percent of state residents born outside the United States in your Statistics/List Editor.

TI-83/84	TI-89
• Press STAT and choose `1:Edit...`	• Press APPS and select `Stats/List Editor.`
• Type the values into list L1.	• Type the values into list1.

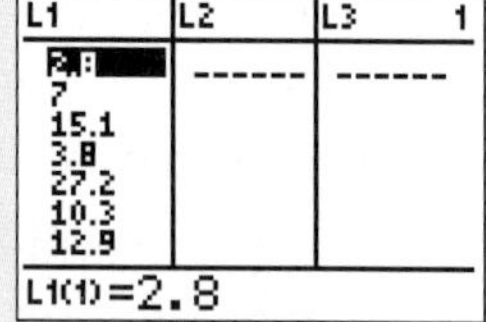

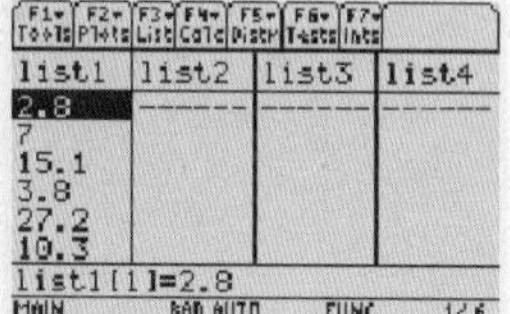

2. Set up a histogram in the Statistics Plots menu.

TI-83/84	TI-89
• Press 2nd Y = (STAT PLOT).	• Press F2 and choose `1:Plot Setup...`
• Press ENTER or 1 to go into Plot1.	• With Plot1 highlighted, press F1 to define.

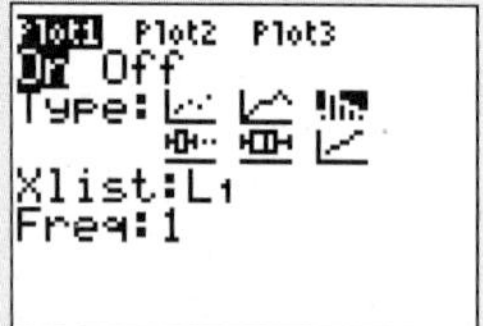

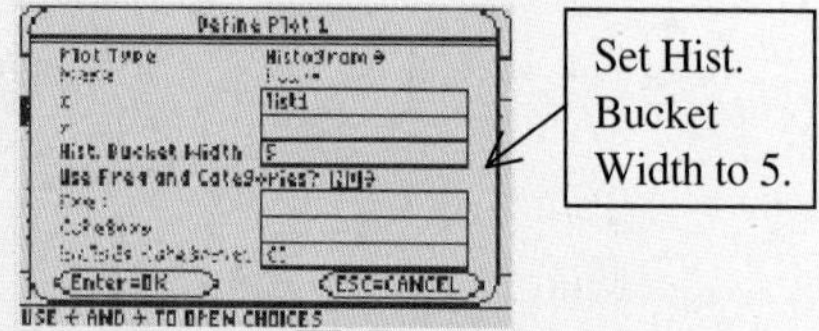

TI-83/84	TI-89
• Adjust the settings as shown.	• Adjust the settings as shown.

3. Use ZoomStat (ZoomData on the TI-89) to let the calculator choose classes and make a histogram.

TI-83/84	TI-89
• Press ZOOM and choose `9:ZoomStat.`	• Press F5 (ZoomData).
• Press TRACE and ◄ ► to examine the classes.	• Press F3 (Trace) and ◄ ► to examine the classes.

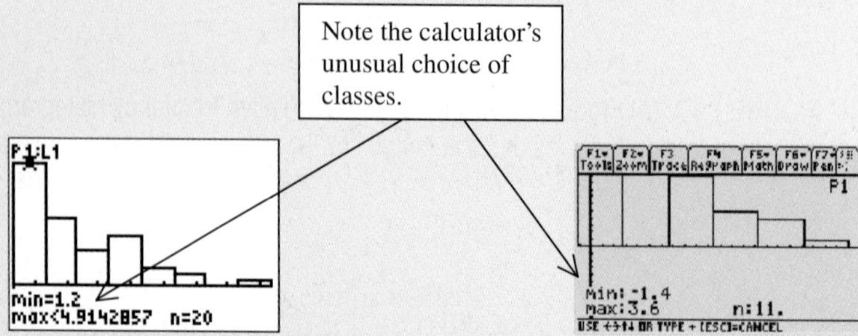

4. Adjust the classes to match those in Figure 1.16, and then graph the histogram.

TI-83/84	TI-89
• Press WINDOW and enter the values shown.	• Press ◆ F2 (WINDOW) and enter the values shown.
• Press GRAPH	• Press ◆ F3 (GRAPH)
• Press TRACE and ◄ ► to examine the classes.	• Press F3 (trace and ◄ ► to examine the classes)

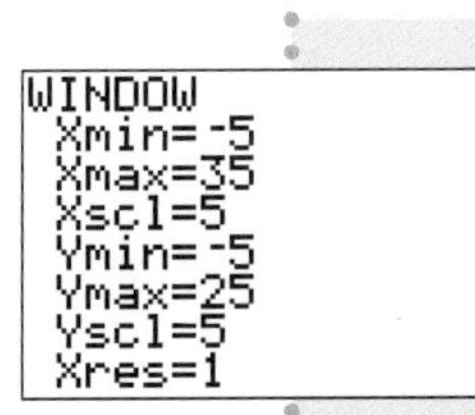

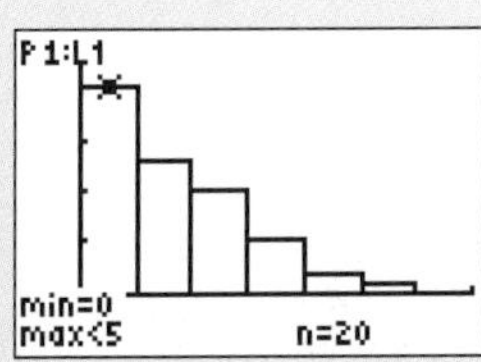

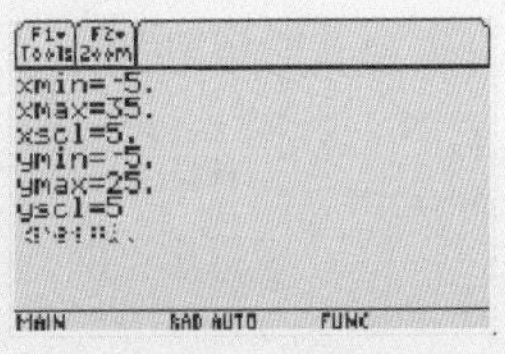

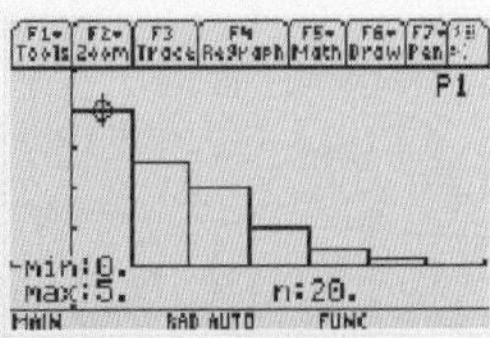

5. See if you can match the histogram in Figure 1.17.

TI-Nspire instructions in Appendix B

AP EXAM TIP If you're asked to make a graph on a free-response question, be sure to label and scale your axes. Unless your calculator shows labels and scaling, don't just transfer a calculator screen shot to your paper.

Here are some important things to consider when you are constructing a histogram:

- Our eyes respond to the area of the bars in a histogram, so *be sure to choose classes that are all the same width.* Then area is determined by height, and all classes are fairly represented.
- There is no one right choice of the classes in a histogram. Too few classes will give a "skyscraper" graph, with all values in a few classes with tall bars. Too many will produce a "pancake" graph, with most classes having one or no observations. Neither choice will give a good picture of the shape of the distribution. Five classes is a good minimum.

CHECK YOUR UNDERSTANDING

Many people believe that the distribution of IQ scores follows a "bell curve," like the one shown in the margin. But is this really how such scores are distributed? The IQ scores of 60 fifth-grade students chosen at random from one school are shown below.[22]

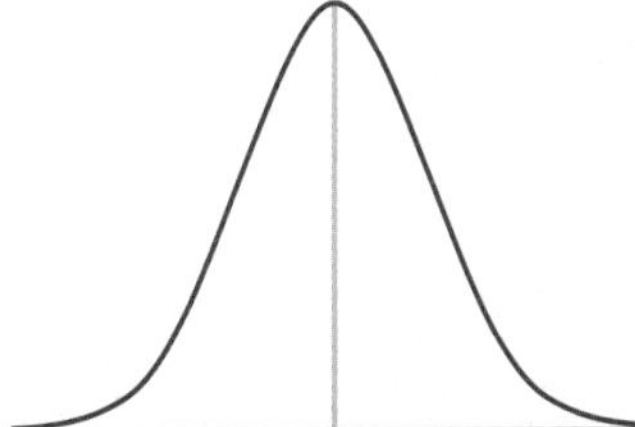

145	139	126	122	125	130	96	110	118	118
101	142	134	124	112	109	134	113	81	113
123	94	100	136	109	131	117	110	127	124
106	124	115	133	116	102	127	117	109	137
117	90	103	114	139	101	122	105	97	89
102	108	110	128	114	112	114	102	82	101

1. Construct a histogram that displays the distribution of IQ scores effectively.
2. Describe what you see. Is the distribution bell-shaped?

Using Histograms Wisely

We offer several cautions based on common mistakes students make when using histograms.

CAUTION

1. *Don't confuse histograms and bar graphs.* Although histograms resemble bar graphs, their details and uses are different. A histogram displays the distribution of a quantitative variable. The horizontal axis of a histogram is marked in the units of measurement for the variable. A bar graph is used to display the distribution of a categorical variable or to compare the sizes of different quantities. The horizontal axis of a bar graph identifies the categories or quantities being compared. Draw bar graphs with blank space between the bars to separate the items being compared. Draw histograms with no space, to show the equal-width classes. For comparison, here is one of each type of graph from previous examples.

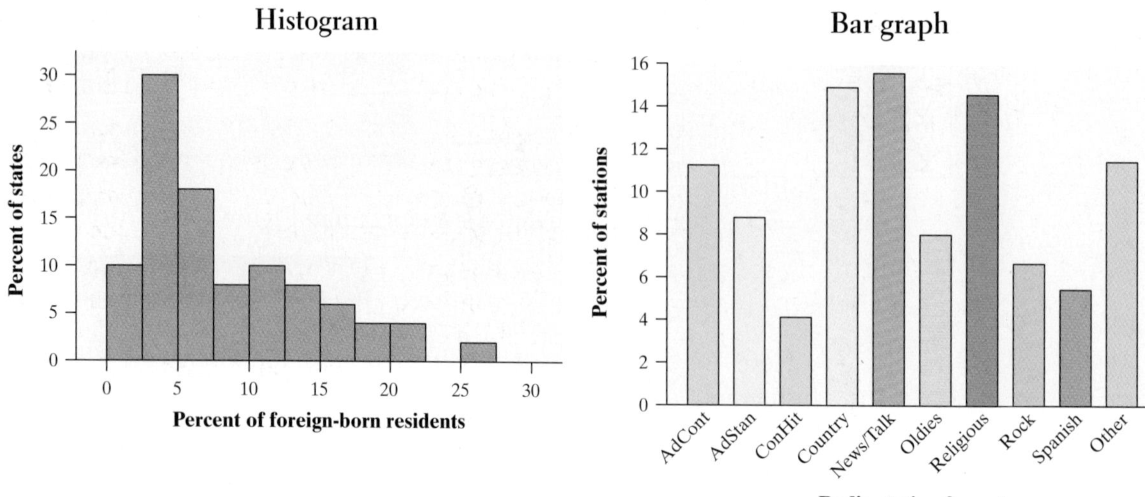

2. *Don't use counts (in a frequency table) or percents (in a relative frequency table) as data.* Below is a frequency table displaying the lengths (number of letters) of the first 100 words in a journal article.

CAUTION

Length:	1	2	3	4	5	6	7	8	9	10	11	12	13
Count:	1	15	25	7	5	7	8	7	7	6	8	3	1

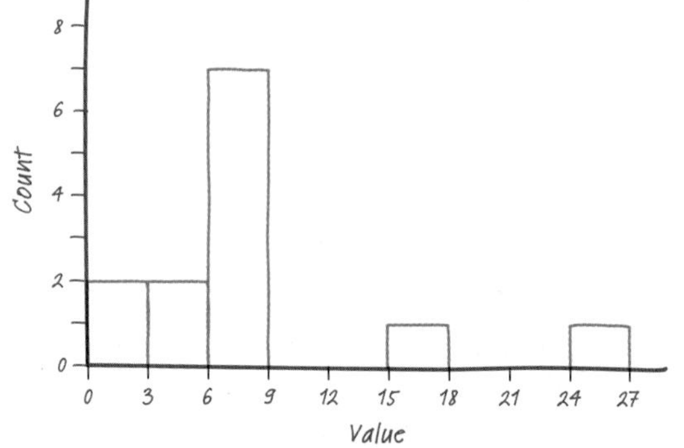

Billy made the histogram shown to display these data. Can you see what Billy did wrong? (He used the counts as data when drawing the histogram—so there were two counts of 1, two counts between 3 and 5, and so on.) Question 1 in the Check Your Understanding below asks you to make a correct graph.

3. *Use percents instead of counts on the vertical axis when comparing distributions with different numbers of observations.* Mary was interested in comparing the reading levels of a medical journal and an airline's in-flight magazine. She counted the number of letters in the first 200 words of an article in the medical journal and of the first 100 words of an article in the airline magazine. Mary then used Minitab statistical software to produce the histograms shown in Figure 1.18(a). This figure is misleading—it compares frequencies, but the two samples were of very different

CAUTION

(a)

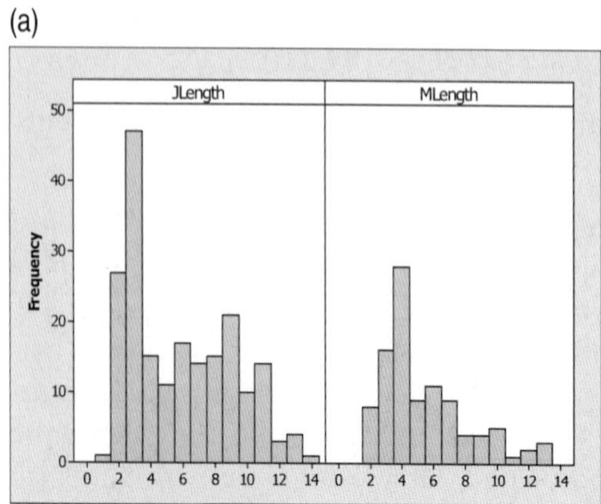

(b)

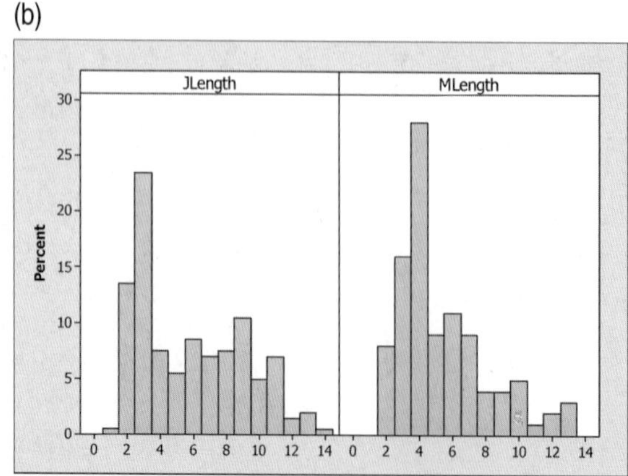

FIGURE 1.18 Two sets of histograms comparing word lengths in articles from a journal and from an airline magazine. In (a), the vertical scale uses frequencies. The graph in (b) fixes this problem by using percents on the vertical scale.

samples were of very different sizes (100 and 200). Using the same data, Mary's teacher produced the histograms in Figure 1.18(b). By using relative frequencies, this figure provides an accurate comparison of word lengths in the two samples.

CAUTION

4. *Just because a graph looks nice, it's not necessarily a meaningful display of data.* The students in a small statistics class recorded the number of letters in their first names. One student entered the data into an Excel spreadsheet and then used Excel's "chart maker" to produce the graph shown. What kind of graph is this? It's neither a bar graph nor a histogram. Both of these types of graphs display the number or percent of individuals in a given category or class. This graph shows the individual data values, in the order that they were entered into the spreadsheet. It is not a very meaningful display of the data.

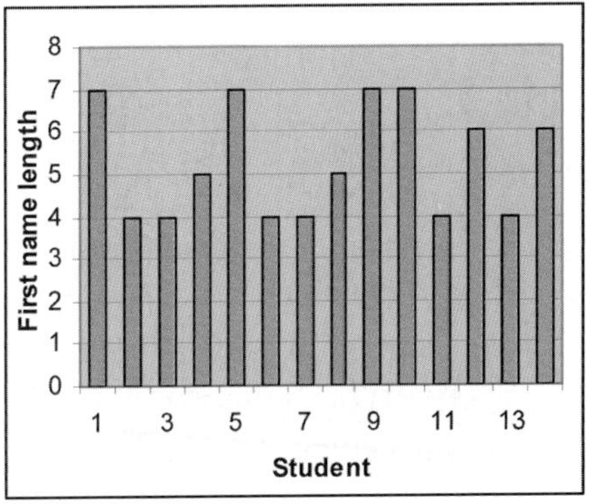

CHECK YOUR UNDERSTANDING

1. Draw a correct histogram to replace Billy's graph of the word length data from Caution 2.
2. Draw a more meaningful graph of the first-name length data from Caution 4.

Questions 3 and 4 relate to the following setting. About 1.6 million first-year students enroll in colleges and universities each year. What do they plan to study? The graph displays data on the percents of first-year students who plan to major in several discipline areas.[23]

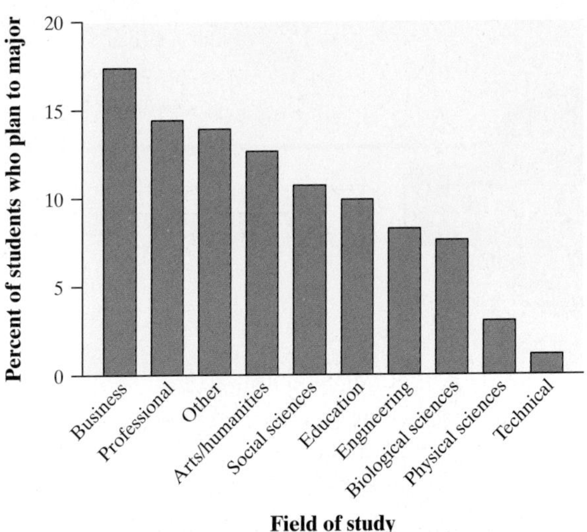

3. Is this a bar graph or a histogram? Explain.
4. Would it be correct to describe this distribution as right-skewed? Why or why not?

SECTION 1.2 Summary

- You can use a **dotplot, stemplot,** or **histogram** to show the distribution of a quantitative variable. A dotplot displays individual values on a number line. Stemplots separate each observation into a stem and a one-digit leaf. Histograms plot the counts (frequencies) or percents (relative frequencies) of values in equal-width classes.
- When examining any graph, look for an **overall pattern** and for notable **departures** from that pattern. **Shape, center, and spread** describe the overall pattern of the distribution of a quantitative variable. **Outliers** are observations that lie outside the overall pattern of a distribution. Always look for outliers and try to explain them. Don't forget your SOCS!
- Some distributions have simple shapes, such as **symmetric** or **skewed.** The number of **modes** (major peaks) is another aspect of overall shape. Not all distributions have a simple overall shape, especially when there are few observations.
- When comparing distributions of quantitative data, be sure to discuss shape, center, spread, and possible outliers.
- Remember: histograms are for quantitative data; bar graphs are for categorical data. Also, be sure to use relative frequency histograms when comparing data sets of different sizes.

1.2 TECHNOLOGY CORNER

TI-Nspire instructions in Appendix B

SECTION 1.2 Exercises

37. **Feeling sleepy?** Students in a college statistics class responded to a survey designed by their teacher. One of the survey questions was "How much sleep did you get last night?" Here are the data (in hours):

9	6	8	6	8	8	6	6.5	6	7	9	4	3	4
5	6	11	6	3	6	6	10	7	8	4.5	9	7	7

(a) Make a dotplot to display the data.

(b) Describe the overall pattern of the distribution and any deviations from that pattern.

38. **Olympic gold!** The following table displays the total number of gold medals won by a sample of countries in the 2008 Summer Olympic Games in China.

Country	Gold medals	Country	Gold medals
Sri Lanka	0	Thailand	2
China	51	Kuwait	0
Vietnam	0	Bahamas	0
Great Britain	19	Kenya	5
Norway	3	Trinidad and Tobago	0
Romania	4	Greece	0
Switzerland	2	Mozambique	0
Armenia	0	Kazakhstan	2
Netherlands	7	Denmark	2
India	0	Latvia	1
Georgia	3	Czech Republic	3
Kyrgyzstan	0	Hungary	3
Costa Rica	0	Sweden	0
Brazil	3	Uruguay	0
Uzbekistan	1	United States	36

(a) Make a dotplot to display these data. Describe the overall pattern of the distribution and any deviations from that pattern.

(b) Overall, 204 countries participated in the 2008 Summer Olympics, of which 55 won at least one gold medal. Do you believe that the sample of countries listed in the table is representative of this larger population? Why or why not?

pg 28

39. **U.S. women's soccer—2004** Earlier, we examined data on the number of goals scored by the U.S. women's soccer team in games during the 2004 season. The dotplot below displays the goal differential for those same games, computed as U.S. score minus opponent's score.

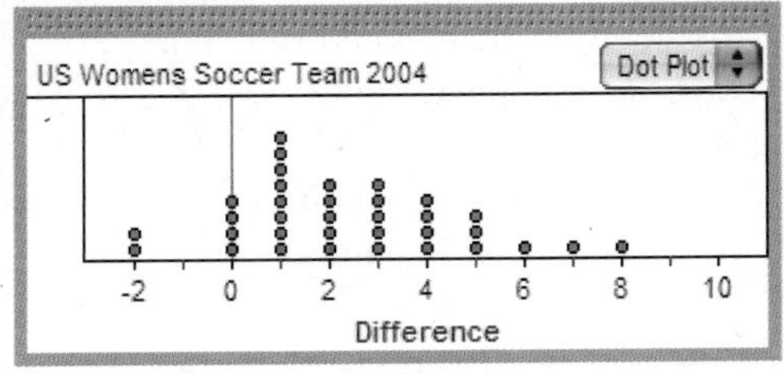

(a) Explain what the two dots above -2 represent.

(b) What does the graph tell us about how well the team did in 2004? Be specific.

40. **Fuel efficiency** In an earlier example, we examined data on highway gas mileages of model year 2009 midsize cars. The dotplot below shows the difference (highway – city) in EPA mileage ratings for each of the 24 car models from the earlier example.

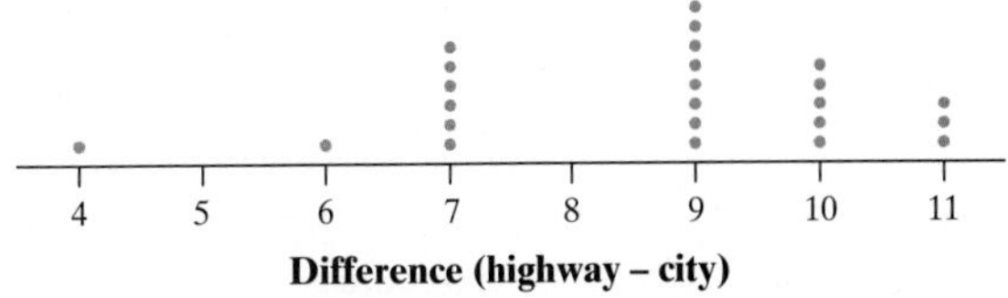

(a) Explain what the dot above 6 represents.

(b) What does the graph tell us about fuel economy in the city versus on the highway for these car models? Be specific.

41. **Dates on coins**

(a) Sketch a dotplot for a distribution that is skewed to the left.

(b) Suppose that you and your friends emptied your pockets of coins and recorded the year marked on each coin. The distribution of dates would be skewed to the left. Explain why.

42. **Phone numbers** The dotplot below displays the last digit of 100 phone numbers chosen at random from a phone book. Describe the shape of the distribution. Does this shape make sense to you? Explain.

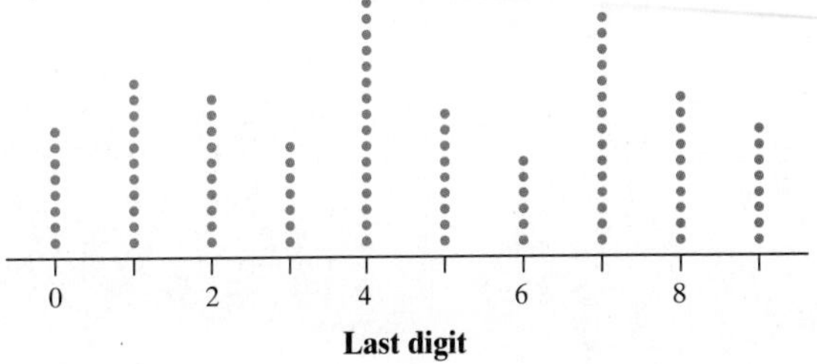

pg 32

43. **Creative writing** The chapter-opening Case Study described research by Teresa Amabile investigating whether external rewards would promote creativity in children's artwork. Dr. Amabile conducted another study involving college students, who were divided into two groups using a chance process (like drawing names from a hat). The students in one group were given a list of statements about external reasons (E) for writing, such as public recognition, making money, or pleasing their parents. Students in the other group were given a list of statements about internal reasons (I) for writing, such as expressing yourself and enjoying playing with words. Both groups were then instructed to write a poem about laughter. Each student's poem was rated separately by 12 different poets using a creativity scale.[24] The 12 poets' ratings of each student's poem were averaged to obtain an overall creativity score.

A dotplot of the two groups' creativity scores is shown below. Compare the two distributions. What do you conclude about whether external rewards promote creativity?

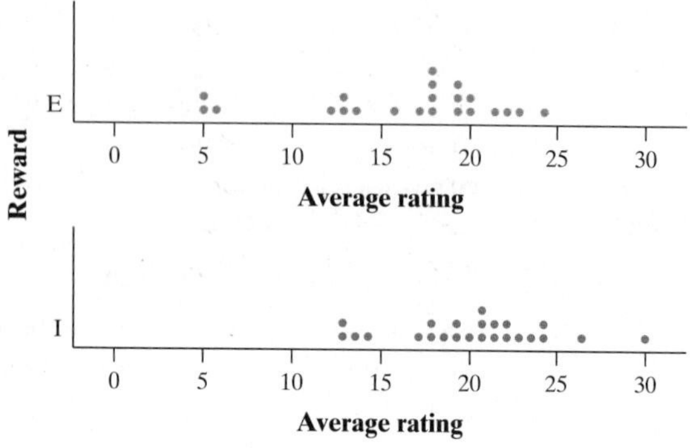

44. **Healthy cereal?** Researchers collected data on 77 brands of cereal at a local supermarket.[25] For each brand, the sugar content (grams per serving) and the shelf in the store on which the cereal was located (1 = bottom, 2 = middle, 3 = top) were recorded. A dotplot of the data is shown below. Compare the three distributions. Critics claim that supermarkets tend to put sugary kids' cereals on lower shelves, where the kids can see them. Do the data from this study support this claim?

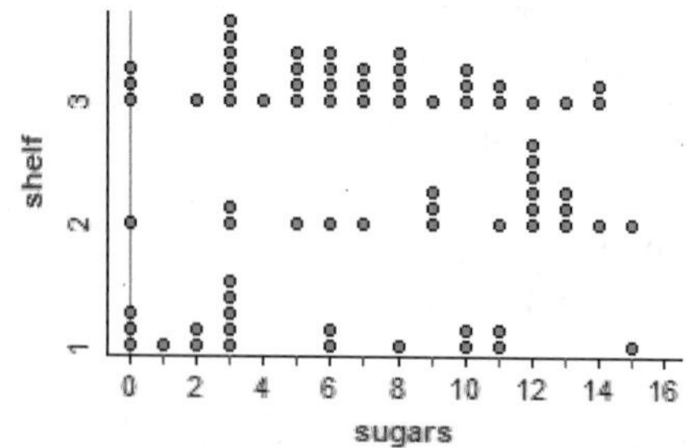

45. **Where do the young live?** Below is a stemplot of the percent of residents aged 25 to 34 in each of the 50 states. As in the stemplot for older residents (page 35), the stems are whole percents, and the leaves are tenths of a percent. This time, each stem has been split in two, with values having leaves 0 through 4 placed on one stem, and values ending in 5 through 9 placed on another stem.

```
11 | 44
11 | 66778
12 | 0134
12 | 666778888
13 | 0000001111444
13 | 7788999
14 | 0044
14 | 567
15 | 11
15 |
16 | 0
```

Key: 12|1 means that 12.1% of that state's residents are aged 25 to 34.

(a) Why did we split stems?

(b) Utah has the highest percent of residents aged 25 to 34. What is that percent? Why do you think Utah has an unusually high percent of residents in this age group?

(c) Describe the shape, center, and spread of the distribution, ignoring Utah.

46. **Watch that caffeine!** The U.S. Food and Drug Administration (USFDA) limits the amount of caffeine in a 12-ounce can of carbonated beverage to 72 milligrams. That translates to a maximum of 48 milligrams of caffeine per 8-ounce serving. Data on the caffeine content of popular soft drinks (in milligrams per 8-ounce serving) are displayed in the stemplot below.

```
1 | 556
2 | 033344
2 | 55667778888899
3 | 113
3 | 55567778
4 | 33
4 | 77
```

(a) Why did we split stems?

(b) Give an appropriate key for this graph.

(c) Describe the shape, center, and spread of the distribution. Compare the caffeine content of these drinks with the USFDA's limit.

47. **El Niño and the monsoon** It appears that El Niño, the periodic warming of the Pacific Ocean west of South America, affects the monsoon rains that are essential for agriculture in India. Here are the monsoon rains (in millimeters) for the 23 strong El Niño years between 1871 and 2004:[26]

628	669	740	651	710	736	717	698	653	604	781	784
790	811	830	858	858	896	806	790	792	957	872	

(a) To make a stemplot of these rainfall amounts, round the data to the nearest 10, so that stems are hundreds of millimeters and leaves are tens of millimeters. Make two stemplots, with and without splitting the stems. Which plot do you prefer?

(b) Describe the shape, center, and spread of the distribution.

(c) The average monsoon rainfall for all years from 1871 to 2004 is about 850 millimeters. What effect does El Niño appear to have on monsoon rains?

48. **Shopping spree** A marketing consultant observed 50 consecutive shoppers at a supermarket. One variable of interest was how much each shopper spent in the store. Here are the data (in dollars), arranged in increasing order:

3.11	8.88	9.26	10.81	12.69	13.78	15.23	15.62	17.00	17.39
18.36	18.43	19.27	19.50	19.54	20.16	20.59	22.22	23.04	24.47
24.58	25.13	26.24	26.26	27.65	28.06	28.08	28.38	32.03	34.98
36.37	38.64	39.16	41.02	42.97	44.08	44.67	45.40	46.69	48.65
50.39	52.75	54.80	59.07	61.22	70.32	82.70	85.76	86.37	93.34

(a) Round each amount to the nearest dollar. Then make a stemplot using tens of dollars as the stems and dollars as the leaves.

(b) Make another stemplot of the data by splitting stems. Which of the plots shows the shape of the distribution better?

(c) Write a few sentences describing the amount of money spent by shoppers at this supermarket.

49. **Do women study more than men?** We asked the students in a large first-year college class how many minutes they studied on a typical weeknight. Here are the responses of random samples of 30 women and 30 men from the class:

Women					Men				
180	120	180	360	240	90	120	30	90	200
120	180	120	240	170	90	45	30	120	75
150	120	180	180	150	150	120	60	240	300
200	150	180	150	180	240	60	120	60	30
120	60	120	180	180	30	230	120	95	150
90	240	180	115	120	0	200	120	120	180

(a) Examine the data. Why are you not surprised that most responses are multiples of 10 minutes? Are there any responses you consider suspicious?

(b) Make a back-to-back stemplot to compare the two samples. Does it appear that women study more than men (or at least claim that they do)? Justify your answer.

50. **Basketball playoffs** Here are the scores of games played in the California Division I-AAA high school basketball playoffs:[27]

71-38	52-47	55-53	76-65	77-63	65-63	68-54	64-62
87-47	64-56	78-64	58-51	91-74	71-41	67-62	106-46

On the same day, the final scores of games in Division V-AA were

98-45	67-44	74-60	96-54	92-72	93-46
98-67	62-37	37-36	69-44	86-66	66-58

(a) Construct a back-to-back stemplot to compare the points scored by the 32 teams in the Division I-AAA playoffs and the 24 teams in the Division V-AA playoffs.

(b) Write a few sentences comparing the two distributions.

51. **Returns on common stocks** The return on a stock is the change in its market price plus any dividend payments made. Total return is usually expressed as a percent of the beginning price. The figure below shows a histogram of the distribution of the monthly returns for all common stocks listed on U.S. markets from January 1985 to September 2007 (273 months).[28] The extreme low outlier represents the market crash of October 1987, when stocks lost 23% of their value in one month.

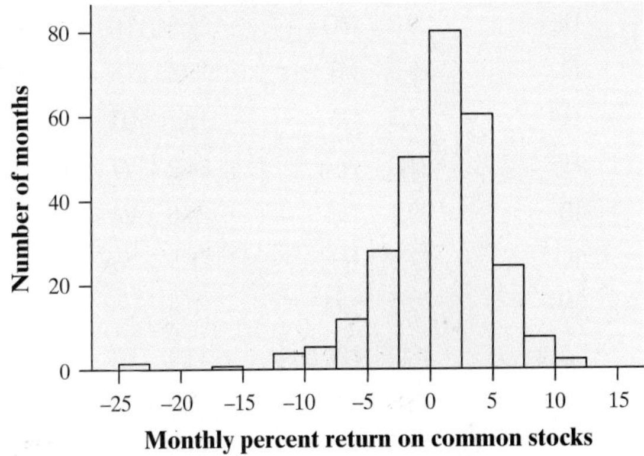

(a) Ignoring the outliers, describe the overall shape of the distribution of monthly returns.

(b) What is the approximate center of this distribution?

(c) Approximately what were the smallest and largest monthly returns, leaving out the outliers?

(d) A return less than zero means that stocks lost value in that month. About what percent of all months had returns less than zero?

52. **Shakespeare** The histogram below shows the distribution of lengths of words used in Shakespeare's plays.[29] Describe the shape, center, and spread of this distribution.

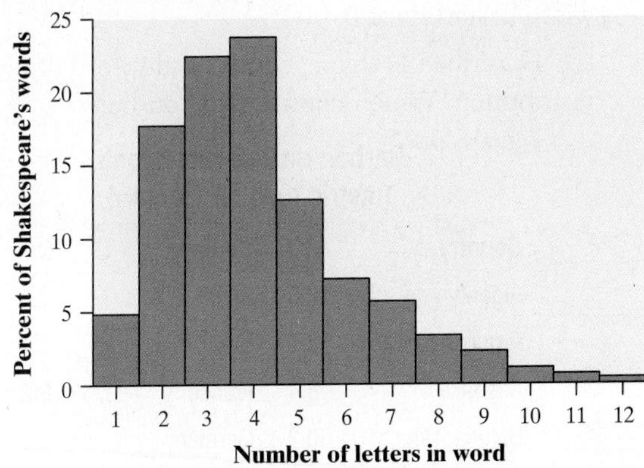

53. **Traveling to work** How long do people travel each day to get to work? The following table gives the average travel times to work (in minutes) for workers in each state and the District of Columbia who are at least 16 years old and don't work at home.[30]

AL	23.6	LA	25.1	OH	22.1
AK	17.7	ME	22.3	OK	20.0
AZ	25.0	MD	30.6	OR	21.8
AR	20.7	MA	26.6	PA	25.0
CA	26.8	MI	23.4	RI	22.3
CO	23.9	MN	22.0	SC	22.9
CT	24.1	MS	24.0	SD	15.9
DE	23.6	MO	22.9	TN	23.5
FL	25.9	MT	17.6	TX	24.6
GA	27.3	NE	17.7	UT	20.8
HI	25.5	NV	24.2	VT	21.2
ID	20.1	NH	24.6	VA	26.9
IL	27.9	NJ	29.1	WA	25.2
IN	22.3	NM	20.9	WV	25.6
IA	18.2	NY	30.9	WI	20.8
KS	18.5	NC	23.4	WY	17.9
KY	22.4	ND	15.5	DC	29.2

(a) Make a histogram of the travel times using classes of width 2 minutes, starting at 14 minutes. That is, the first class is 14 to 16 minutes, the second is 16 to 18 minutes, and so on.

(b) The shape of the distribution is a bit irregular. Is it closer to symmetric or skewed? About where is the center of the data? What is the spread in terms of the smallest and largest values? Are there any outliers?

54. **Carbon dioxide emissions** Burning fuels in power plants and motor vehicles emits carbon dioxide (CO_2), which contributes to global warming. The table below displays CO_2 emissions per person from countries with populations of at least 20 million.[31]

(a) Make a histogram of the data using classes of width 2, starting at 0.

(b) Describe the shape, center, and spread of the distribution. Which countries are outliers?

Carbon dioxide emissions (metric tons per person)

Country	CO_2	Country	CO_2
Algeria	2.6	Egypt	2.0
Argentina	3.6	Ethiopia	0.1
Australia	18.4	France	6.2
Bangladesh	0.3	Germany	9.9
Brazil	1.8	Ghana	0.3
Canada	17.0	India	1.1
China	3.9	Indonesia	1.6
Colombia	1.3	Iran	6.0
Congo	0.2	Iraq	2.9

Country	CO_2	Country	CO_2
Italy	7.8	Romania	4.2
Japan	9.5	Russia	10.8
Kenya	0.3	Saudi Arabia	13.8
Korea, North	3.3	South Africa	7.0
Korea, South	9.3	Spain	7.9
Malaysia	5.5	Sudan	0.3
Mexico	3.7	Tanzania	0.1
Morocco	1.4	Thailand	3.3
Myanmar	0.2	Turkey	3.0
Nepal	0.1	Ukraine	6.3
Nigeria	0.4	United Kingdom	8.8
Pakistan	0.8	United States	19.6
Peru	1.0	Uzbekistan	4.2
Philippines	0.9	Venezuela	5.4
Poland	7.8	Vietnam	1.0

55. **DRP test scores** There are many ways to measure the reading ability of children. One frequently used test is the Degree of Reading Power (DRP). In a research study on third-grade students, the DRP was administered to 44 students.[32] Their scores were:

40	26	39	14	42	18	25	43	46	27	19
47	19	26	35	34	15	44	40	38	31	46
52	25	35	35	33	29	34	41	49	28	52
47	35	48	22	33	41	51	27	14	54	45

Make a histogram to display the data. Write a paragraph describing the distribution of DRP scores.

56. **Drive time** Professor Moore, who lives a few miles outside a college town, records the time he takes to drive to the college each morning. Here are the times (in minutes) for 42 consecutive weekdays:

8.25	7.83	8.30	8.42	8.50	8.67	8.17	9.00	9.00	8.17	7.92
9.00	8.50	9.00	7.75	7.92	8.00	8.08	8.42	8.75	8.08	9.75
8.33	7.83	7.92	8.58	7.83	8.42	7.75	7.42	6.75	7.42	8.50
8.67	10.17	8.75	8.58	8.67	9.17	9.08	8.83	8.67		

Make a histogram of these drive times. Is the distribution roughly symmetric, clearly skewed, or neither? Are there any clear outliers?

57. **The statistics of writing style** Numerical data can distinguish different types of writing, and sometimes even individual authors. Here are data on the percent of words of 1 to 15 letters used in articles in *Popular Science* magazine:[33]

Length:	1	2	3	4	5	6	7	8	9	10	11	12	13	14	15
Percent:	3.6	14.8	18.7	16.0	12.5	8.2	8.1	5.9	4.4	3.6	2.1	0.9	0.6	0.4	0.2

(a) Make a histogram of this distribution. Describe its shape, center, and spread.

(b) How does the distribution of lengths of words used in *Popular Science* compare with the similar distribution for Shakespeare's plays in Exercise 52? Look in particular at short words (2, 3, and 4 letters) and very long words (more than 10 letters).

58. **Chest out, Soldier!** In 1846, a published paper provided chest measurements (in inches) of 5738 Scottish militiamen. The table below summarizes the data.

Chest size	Count	Chest size	Count
33	3	41	934
34	18	42	658
35	81	43	370
36	185	44	92
37	420	45	50
38	749	46	21
39	1073	47	4
40	1079	48	1

Source: Online Data and Story Library (DASL).

(a) Make a histogram.

(b) Describe the shape, center, and spread of the chest measurements distribution. Why might this information be useful?

59. **Paying for championships** Does paying high salaries lead to more victories in professional sports? The New York Yankees have long been known for having Major League Baseball's highest team payroll. And over the years, the team has won many championships. This strategy didn't pay off in 2008, when the Philadelphia Phillies won the World Series. Maybe the Yankees didn't spend enough money that year. The graph below shows histograms of the salary distributions for the two teams during the 2008 season. Why can't you use this graph to effectively compare the team payrolls?

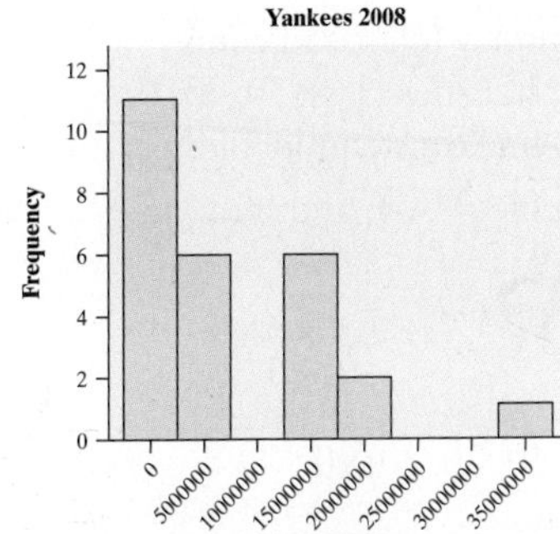

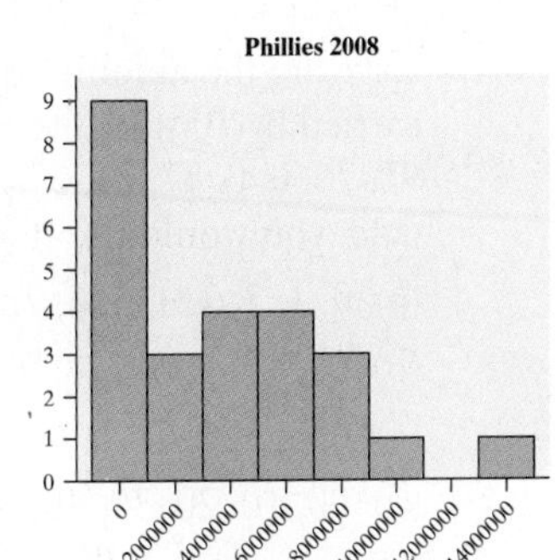

60. **Paying for championships** Refer to Exercise 59. Here is another graph of the 2008 salary distributions for the Yankees and the Phillies. Write a few sentences comparing these two distributions.

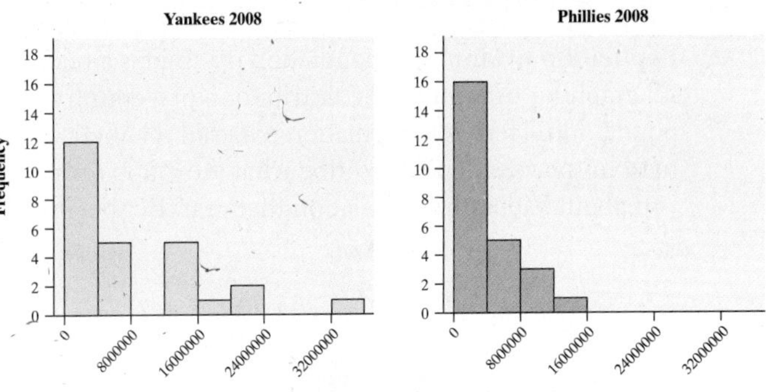

61. **Birth months** Imagine asking a random sample of 60 students from your school about their birth months. Draw a plausible graph of the distribution of birth months. (*Hint:* Should you use a bar graph or a histogram?)

62. **Die rolls** Imagine rolling a fair, six-sided die 60 times. Draw a plausible graph of the distribution of die rolls. (*Hint:* Should you use a bar graph or a histogram?)

63. **Who makes more?** A manufacturing company is reviewing the salaries of its full-time employees below the executive level at a large plant. The clerical staff is almost entirely female, while a majority of the production workers and technical staff are male. As a result, the distributions of salaries for male and female employees may be quite different. The table below gives the frequencies and relative frequencies for women and men.

	Women		Men	
Salary ($1000)	Number	%	Number	%
10–15	89	11.8	26	1.1
15–20	192	25.4	221	9.0
20–25	236	31.2	677	27.6
25–30	111	14.7	823	33.6
30–35	86	11.4	365	14.9
35–40	25	3.3	182	7.4
40–45	11	1.5	91	3.7
45–50	3	0.4	33	1.3
50–55	2	0.3	19	0.8
55–60	0	0.0	11	0.4
60–65	0	0.0	0	0.0
65–70	1	0.1	3	0.1
Total	**756**	**100.1**	**2451**	**99.9**

(a) Explain why the total for women is greater than 100%.

(b) Make histograms for these data, choosing the vertical scale that is most appropriate for comparing the two distributions.

(c) Write a few sentences comparing the salary distributions for men and women.

64. **Population pyramids** A population pyramid is a helpful graph for examining the distribution of a country's population. Here is a population pyramid for Vietnam in the year 2010. Describe what the graph tells you about Vietnam's population that year. Be specific.

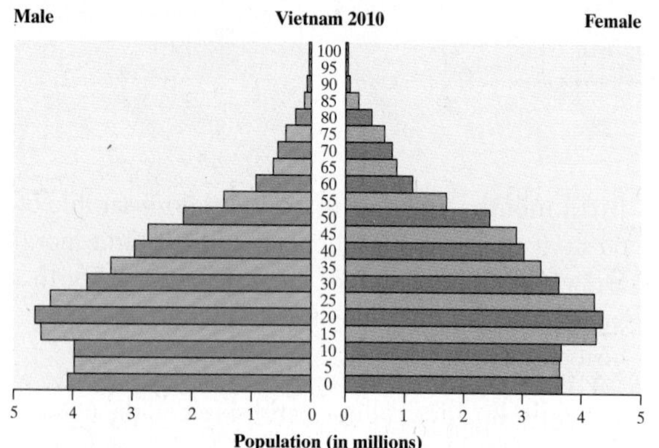

65. **Comparing AP scores** The table below gives the distribution of grades earned by students taking the AP Calculus AB and AP Statistics exams in 2009.[34]

		Grade				
	No. of exams	5	4	3	2	1
Calculus AB	230,588	52,921	43,140	41,204	35,843	57,480
Statistics	116,876	14,353	26,050	28,276	22,283	25,914

(a) Make an appropriate graphical display to compare the grade distributions for AP Calculus AB and AP Statistics.

(b) Write a few sentences comparing the two distributions of exam grades.

66. **Population pyramids** Refer to Exercise 64. Here is a graph of the projected population distribution for China in the year 2050. Describe what the graph suggests about China's future population. Be specific.

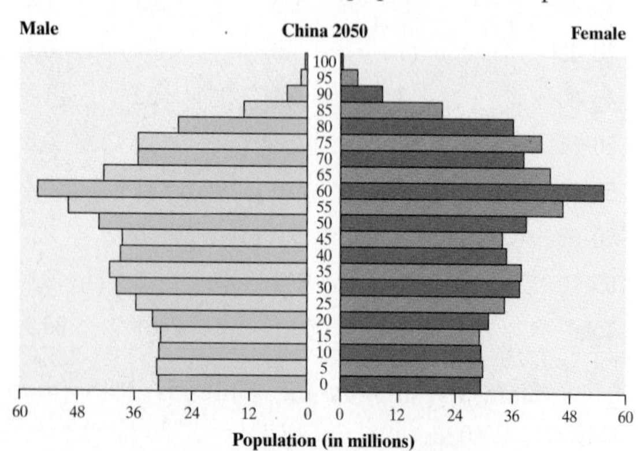

67. **Student survey** A survey of a large high school class asked the following questions:

(i) Are you female or male? (In the data, male = 0, female = 1.)

(ii) Are you right-handed or left-handed? (In the data, right = 0, left = 1.)

(iii) What is your height in inches?

(iv) How many minutes do you study on a typical weeknight?

The figure below shows histograms of the student responses, in scrambled order and without scale markings. Which histogram goes with each variable? Explain your reasoning.

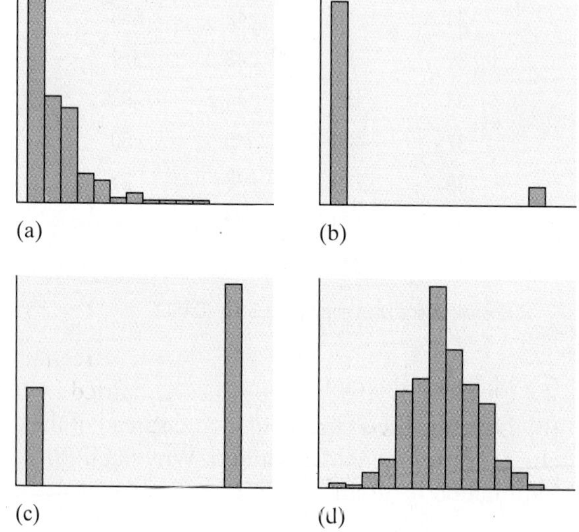

68. **Choose a graph** What type of graph or graphs would you plan to make in a study of each of the following issues at your school? Explain your choices.

(a) Which radio stations are most popular with students?

(b) How many hours per week do students study?

(c) How many calories do students consume per day?

Multiple choice: Select the best answer for Exercises 69 to 74.

69. Here are the amounts of money (cents) in coins carried by 10 students in a statistics class: 50, 35, 0, 97, 76, 0, 0, 87, 23, 65. To make a stemplot of these data, you would use stems

(a) 0, 1, 2, 3, 4, 5, 6, 7, 8, 9.

(b) 0, 2, 3, 5, 6, 7, 8, 9.

(c) 0, 3, 5, 6, 7.

(d) 00, 10, 20, 30, 40, 50, 60, 70, 80, 90.

(e) None of these.

70. One of the following 12 scores was omitted from the stemplot below:

84 76 92 92 88 96 68 80 92 88 76 96

```
6 | 8
7 | 66
8 | 0488
9 | 2266
```

The missing number is

(a) 76. (b) 88. (c) 90. (d) 92. (e) 96.

71. You look at real estate ads for houses in Naples, Florida. There are many houses ranging from $200,000 to $500,000 in price. The few houses on the water, however, have prices up to $15 million. The distribution of house prices will be

(a) skewed to the left.

(b) roughly symmetric.

(c) skewed to the right.

(d) unimodal.

(e) too high.

Exercises 72 to 74 refer to the following setting. The histogram below shows the distribution of the percents of women aged 15 and over who have never married in each of the 50 states and the District of Columbia.

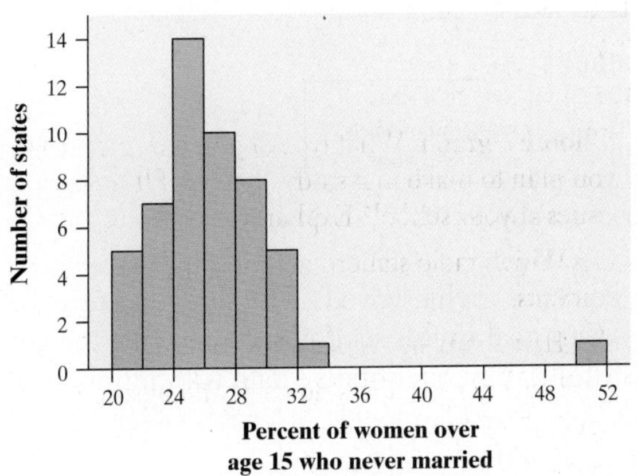

72. The leftmost bar in the histogram covers percents of never-married women ranging from about

(a) 20% to 24%. (d) 0% to 5%.

(b) 20% to 22%. (e) None of these.

(c) 0% to 20%.

73. The center of this distribution is in the interval

(a) 22% to 24%. (d) 28% to 30%.

(b) 24% to 26%. (e) 36% to 38%.

(c) 26% to 28%.

74. In about what percent of states have at least 30% of women aged 15 and over never married?

(a) 4% (b) 7% (c) 10% (d) 14% (e) 32%

75. **Baseball players** (Introduction) Here is a small part of a data set that describes Major League Baseball players as of opening day of the 2009 season:

Player	Team	Position	Age	Height	Weight	Salary
Rodriguez, Alex	Yankees	Infielder	33	6-3	230	33,000,000
Ramirez, Manny	Dodgers	Outfielder	36	6-0	200	23,854,494
Santana, Johan	Mets	Pitcher	30	6-0	210	18,876,139
Zambrano, Carlos	Cubs	Pitcher	27	6-5	255	18,750,000
Suzuki, Ichiro	Mariners	Outfielder	35	5-11	170	18,000,000

(a) What individuals does this data set describe?

(b) In addition to the player's name, how many variables does the data set contain? Which of these variables are categorical and which are quantitative?

(c) What do you think are the units of measurement for each of the quantitative variables?

76. **I love my iPod!** (1.1) The rating service Arbitron asked adults who used several high-tech devices and services whether they "loved" using them. Below is a graph of the percents who said they did.[35]

(a) Summarize what this graph tells you in a sentence or two.

(b) Would it be appropriate to make a pie chart of these data? Why or why not?

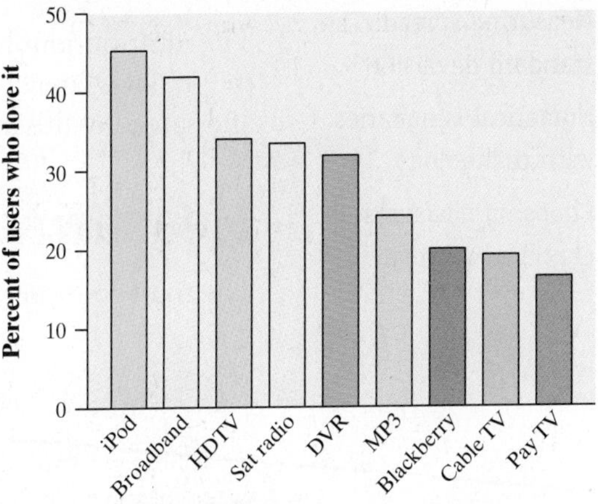

77. **Risks of playing soccer** (1.1) A study in Sweden looked at former elite soccer players, people who had played soccer but not at the elite level, and

people of the same age who did not play soccer. Here is a two-way table that classifies these individuals by whether or not they had arthritis of the hip or knee by their mid-fifties:[36]

	Elite	Non-Elite	Did not play
Arthritis	10	9	24
No arthritis	61	206	548

(a) What percent of the people in this study were elite soccer players? What percent had arthritis?

(b) What percent of the elite soccer players had arthritis? What percent of those who had arthritis were elite soccer players?

78. **Risks of playing soccer** (1.1) Refer to Exercise 77. We suspect that the more serious soccer players have more arthritis later in life. Do the data confirm this suspicion? Give graphical and numerical evidence to support your answer.

1.3

Describing Quantitative Data with Numbers

In Section 1.3, you'll learn about:

- Measuring center: The mean
- Measuring center: The median
- Comparing the mean and the median
- Measuring spread: The interquartile range (*IQR*)
- Identifying outliers
- The five-number summary and boxplots
- Measuring spread: The standard deviation
- Numerical summaries with technology
- Choosing measures of center and spread

How long do people spend traveling to work? The answer may depend on where they live. Here are the travel times in minutes for 15 workers in North Carolina, chosen at random by the Census Bureau:[37]

30 20 10 40 25 20 10 60 15 40 5 30 12 10 10

We aren't surprised that most people estimate their travel time in multiples of 5 minutes. Here is a stemplot of these data:

```
0 | 5
1 | 000025
2 | 005
3 | 00
4 | 00
5 |
6 | 0
```

Key: 2|5 is a NC worker who travels 25 minutes to work.

The distribution is single-peaked and right-skewed. The longest travel time (60 minutes) may be an outlier. Our goal in this section is to describe the center and spread of this and other distributions of quantitative data with numbers.

Measuring Center: The Mean

The most common measure of center is the ordinary arithmetic average, or **mean.**

DEFINITION: The mean $\bar{x}$

To find the **mean** $\bar{x}$ (pronounced "x-bar") of a set of observations, add their values and divide by the number of observations. If the n observations are $x_1, x_2, \ldots, x_n$, their mean is

$$\bar{x} = \frac{\text{sum of observations}}{n} = \frac{x_1 + x_2 + \cdots + x_n}{n}$$

or, in more compact notation,

$$\bar{x} = \frac{\sum x_i}{n}$$

The Σ (capital Greek letter sigma) in the formula for the mean is short for "add them all up." The subscripts on the observations x_i are just a way of keeping the n observations distinct. They do not necessarily indicate order or any other special facts about the data.

Actually, the notation $\bar{x}$ refers to the mean of a *sample*. Most of the time, the data we'll encounter can be thought of as a sample from some larger population. When we need to refer to a *population* mean, we'll use the symbol μ (Greek letter mu, pronounced "mew"). If you have the entire population of data available, then you calculate μ in just the way you'd expect: add the values of all the observations, and divide by the number of observations.

Travel Times to Work in North Carolina

Calculating the mean

Refer to the data on travel times to work for the sample of 15 North Carolinians.

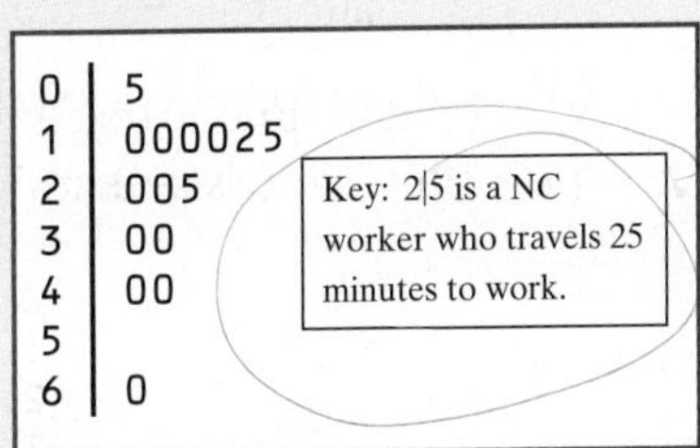

PROBLEM:

(a) Find the mean travel time for all 15 workers.

(b) Calculate the mean again, this time excluding the person who reported a 60-minute travel time to work. What do you notice?

SOLUTION:

(a) The mean travel time for the sample of 15 North Carolina workers is

$$\bar{x} = \frac{\sum x_i}{n} = \frac{x_1 + x_2 + \cdots + x_n}{n} = \frac{30 + 20 + \cdots + 10}{15} = \frac{337}{15} = 22.5 \text{ minutes}$$

(b) Notice that only 6 of the 15 travel times are larger than the mean. If we leave out the longest travel time, 60 minutes, the mean for the remaining 14 people is

$$\bar{x} = \frac{\sum x_i}{n} = \frac{x_1 + x_2 + \cdots + x_n}{n} = \frac{277}{14} = 19.8 \text{ minutes}$$

That one observation raises the mean by 2.7 minutes.

For Practice *Try Exercise 79*

CAUTION

Resistant measure

The previous example illustrates an important weakness of the mean as a measure of center: *the mean is sensitive to the influence of extreme observations*. These may be outliers, but a skewed distribution that has no outliers will also pull the mean toward its long tail. Because the mean cannot resist the influence of extreme observations, we say that it is not a **resistant measure** of center.

THINK ABOUT IT

What does the mean mean? A group of elementary school children was asked how many pets they have. Here are their responses, arranged from lowest to highest:[38]

1 3 4 4 4 5 7 8 9

What's the mean number of pets for this group of children? It's

$$\bar{x} = \frac{\text{sum of observations}}{n} = \frac{1+3+4+4+4+5+7+8+9}{9} = 5 \text{ pets}$$

But what does that number tell us? Here's one way to look at it: if every child in the group had the same number of pets, each would have 5 pets. In other words, the mean is the "fair share" value.

The mean tells us how large each observation in the data set would be if the total were split equally among all the observations. In the language of young children, the mean is the "fair share" value. The mean of a distribution also has a physical interpretation, as the following Activity shows.

ACTIVITY *Mean as a "balance point"*

MATERIALS: Foot-long ruler, pencil, and 5 pennies per group of 3 to 4 students

In this Activity, you'll investigate an interesting property of the mean.

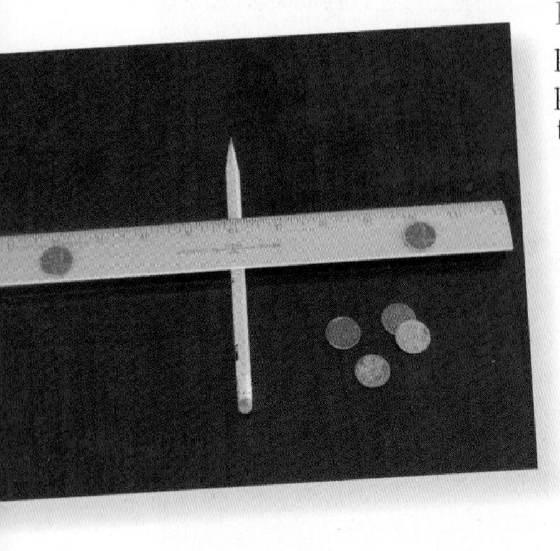

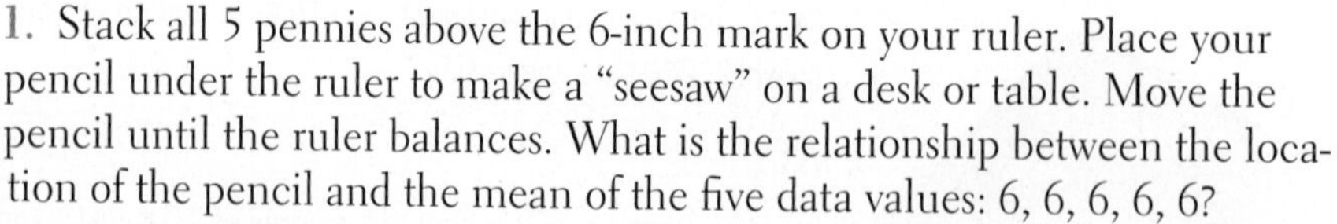

1. Stack all 5 pennies above the 6-inch mark on your ruler. Place your pencil under the ruler to make a "seesaw" on a desk or table. Move the pencil until the ruler balances. What is the relationship between the location of the pencil and the mean of the five data values: 6, 6, 6, 6, 6?

2. Move one penny off the stack to the 8-inch mark on your ruler. Now move one other penny so that the ruler balances again without moving the pencil. Where did you put the other penny? What is the mean of the five data values represented by the pennies now?

3. Move one more penny off the stack to the 2-inch mark on your ruler. Now move both remaining pennies from the 6-inch mark so that the ruler still balances with the pencil in the same location. Is the mean of the data values still 6?

4. Do you see why the mean is sometimes called the "balance point" of a distribution?

Measuring Center: The Median

In Section 1.2, we introduced the median as an informal measure of center that describes the "midpoint" of a distribution. Now it's time to offer an official "rule" for calculating the median.

DEFINITION: The median *M*

The **median *M*** is the midpoint of a distribution, the number such that half the observations are smaller and the other half are larger. To find the median of a distribution:

1. Arrange all observations in order of size, from smallest to largest.
2. If the number of observations n is odd, the median M is the center observation in the ordered list.
3. If the number of observations n is even, the median M is the average of the two center observations in the ordered list.

Medians require little arithmetic, so they are easy to find by hand for small sets of data. Arranging even a moderate number of values in order is tedious, however, so finding the median by hand for larger sets of data is unpleasant.

Travel Times to Work in North Carolina

Finding the median when n is odd

What is the median travel time for our 15 North Carolina workers? Here are the data arranged in order:

5 10 10 10 10 12 15 **20** 20 25 30 30 40 40 60

The count of observations $n = 15$ is odd. The bold **20** is the center observation in the ordered list, with 7 observations to its left and 7 to its right. This is the median, $M = 20$ minutes.

Stuck in Traffic

Finding the median when n is even

People say that it takes a long time to get to work in New York State due to the heavy traffic near big cities. What do the data say? Here are the travel times in minutes of 20 randomly chosen New York workers:

10 30 5 25 40 20 10 15 30 20 15 20 85 15 65 15 60 60 40 45

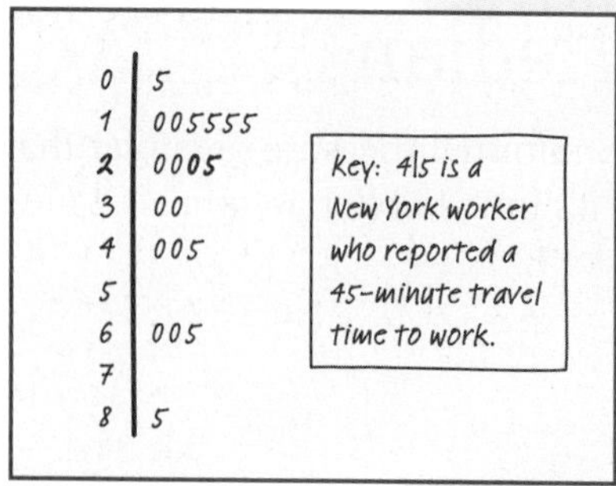

PROBLEM:

(a) Make a stemplot of the data. Be sure to include a key.

(b) Find and interpret the median.

SOLUTION:

(a) Here is a stemplot of the data. The stems are in 10 minutes and the leaves are in minutes.

(b) Since there is an even number of data values, there is no center observation. There is a center pair—the bold 20 and 25 in the stemplot—which have 9 observations before them and 9 after them in the ordered list. The median is the average of these two observations:

$$M = \frac{20 + 25}{2} = 22.5 \text{ minutes}$$

A stemplot makes finding the median easy because it arranges the observations in order.

Interpretation: In the sample of New York workers, about half of the people reported traveling less than 22.5 minutes to work, and about half reported traveling more.

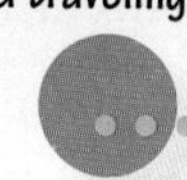

For Practice *Try Exercise 81*

Comparing the Mean and the Median

0 | 5
1 | 000025
2 | 005
3 | 00
4 | 00
5 |
6 | 0

Key: 2|5 is a NC worker who travels 25 minutes to work.

Our discussion of travel times to work in North Carolina illustrates an important difference between the mean and the median. The median travel time (the midpoint of the distribution) is 20 minutes. The mean travel time is higher, 22.5 minutes. The mean is pulled toward the right tail of this right-skewed distribution. The median, unlike the mean, is *resistant.* If the longest travel time were 600 minutes rather than 60 minutes, the mean would increase to more than 58 minutes but the median would not change at all. The outlier just counts as one observation above the center, no matter how far above the center it lies. The mean uses the actual value of each observation and so will chase a single large observation upward.

You can compare the behavior of the mean and median by using the *Mean and Median* applet at the book's Web site, www.whfreeman.com/tps4e.

Comparing the Mean and Median

The mean and median of a roughly symmetric distribution are close together. If the distribution is exactly symmetric, the mean and median are exactly the same. In a skewed distribution, the mean is usually farther out in the long tail than is the median.[39]

The mean and median measure center in different ways, and both are useful. *Don't confuse the "average" value of a variable (the mean) with its "typical" value, which we might describe by the median.*

THINK ABOUT IT

Should we choose the mean or the median? Many economic variables have distributions that are skewed to the right. College tuitions, home prices, and personal incomes are all right-skewed. In Major League Baseball (MLB), for instance, most players earn close to the minimum salary (which was \$400,000 in 2009), while a few earn more than \$10 million. The median salary for MLB players in 2009 was about \$1.15 million—but the mean salary was over \$3.24 million. Alex

Rodriguez, Manny Ramirez, Derek Jeter, and several other highly paid superstars pull the mean up but do not affect the median. Reports about incomes and other strongly skewed distributions usually give the median ("midpoint") rather than the mean ("arithmetic average"). However, a county that is about to impose a tax of 1% on the incomes of its residents cares about the mean income, not the median. The tax revenue will be 1% of total income, and the total is the mean times the number of residents.

CHECK YOUR UNDERSTANDING

Questions 1 through 4 refer to the following setting. Here, once again, is the stemplot of travel times to work for 20 randomly selected New Yorkers. Earlier, we found that the median was 22.5 minutes.

```
0 | 5
1 | 005555
2 | 0005
3 | 00
4 | 005
5 |
6 | 005
7 |
8 | 5
```

Key: 4|5 is a New York worker who reported a 45-minute travel time to work.

1. Based only on the stemplot, would you expect the mean travel time to be less than, about the same as, or larger than the median? Why?

2. Use your calculator to find the mean travel time. Was your answer to Question 1 correct?

3. Interpret your result from Question 2 in context without using the words "mean" or "average."

4. Would the mean or the median be a more appropriate summary of the center of this distribution of drive times? Justify your answer.

Measuring Spread: The Interquartile Range (*IQR*)

A measure of center alone can be misleading. The mean annual temperature in San Francisco, California, is 57°F—the same as in Springfield, Missouri. But the wardrobe needed to live in these two cities is very different! That's because daily temperatures vary a lot more in Springfield than in San Francisco. *A useful numerical description of a distribution requires both a measure of center and a measure of spread.*

The simplest measure of variability is the *range*, which we defined earlier as the difference between the largest and smallest observations. The range shows the full spread of the data. But it depends on only the maximum and minimum values, which may be outliers.

We can improve our description of spread by also looking at the spread of the middle half of the data. Here's the idea. Count up the ordered list of observations, starting from the minimum. The **first quartile Q_1** lies one-quarter of the way up the list. The second quartile is the median, which is halfway up the list. The **third quartile Q_3** lies three-quarters of the way up the list. These **quartiles** mark out the middle half of the distribution. The **interquartile range** (*IQR*) measures the range of the middle 50% of the data. We need a rule to make this idea exact. The process for calculating the quartiles and the *IQR* uses the rule for finding the median.

How to Calculate the Quartiles Q_1 and Q_3 and the Interquartile Range (*IQR*)

To calculate the **quartiles:**

1. Arrange the observations in increasing order and locate the median M in the ordered list of observations.
2. The **first quartile Q_1** is the median of the observations whose position in the ordered list is to the left of the median.
3. The **third quartile Q_3** is the median of the observations whose position in the ordered list is to the right of the median.

The **interquartile range** (*IQR*) is defined as

$$IQR = Q_3 - Q_1$$

Be careful in locating the quartiles when several observations take the same numerical value. Write down all the observations, arrange them in order, and apply the rules just as if they all had distinct values.

Let's look at how this process works using a familiar set of data.

Travel Times to Work in North Carolina

Calculating quartiles

Our North Carolina sample of 15 workers' travel times, arranged in increasing order, is

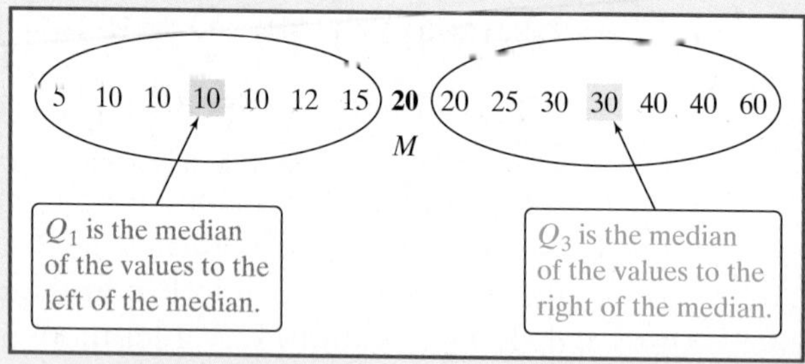

There is an odd number of observations, so the median is the middle one, the bold **20** in the list. The first quartile is the median of the 7 observations to the left of the median. This is the 4th of these 7 observations, so $Q_1 = 10$ minutes (shown in blue). The third quartile is the median of the 7 observations to the right of the median, $Q_3 = 30$ minutes (shown in green). So the spread of the middle 50% of the travel times is $IQR = Q_3 - Q_1 = 30 - 10 = 20$ minutes. *Be sure to leave out the overall median M when you locate the quartiles.*

CAUTION

The quartiles and the interquartile range are *resistant* because they are not affected by a few extreme observations. For example, Q_3 would still be 30 and the *IQR* would still be 20 if the maximum were 600 rather than 60.

Stuck in Traffic Again

Finding and interpreting the *IQR*

In an earlier example, we looked at data on travel times to work for 20 randomly selected New Yorkers. Here is the stemplot once again:

```
0 | 5
1 | 005555
2 | 0005
3 | 00
4 | 005
5 |
6 | 005
7 |
8 | 5
```

Key: 4|5 is a New York worker who reported a 45-minute travel time to work.

PROBLEM: Find and interpret the interquartile range (*IQR*).

SOLUTION: We begin by writing the travel times arranged in increasing order:

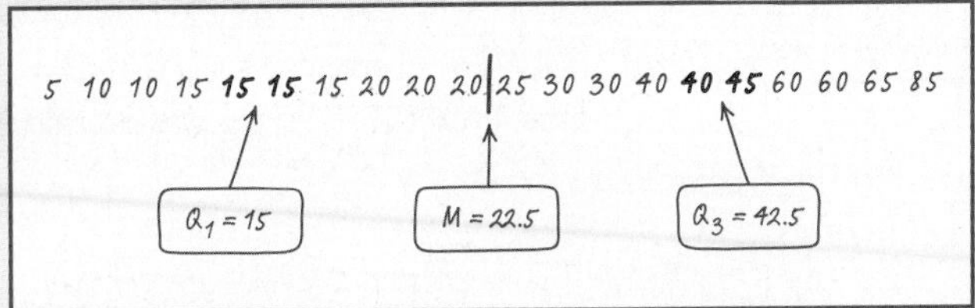

There is an even number of observations, so the median lies halfway between the middle pair. Its value is $M = 22.5$ minutes. (We marked the location of the median by |.) The first quartile is the median of the 10 observations to the left of $M = 22.5$. So it's the average of the two bold 15s: $Q_1 = 15$ minutes. The third quartile is the median of the 10 observations to the right of $M = 22.5$. It's the average of the bold numbers 40 and 45: $Q_3 = 42.5$ minutes. The interquartile range is

$$IQR = Q_3 - Q_1 = 42.5 - 15 = 27.5 \text{ minutes}$$

Interpretation: The range of the middle half of travel times for the New Yorkers in the sample is 27.5 minutes.

For Practice *Try Exercise 89(a)*

Identifying Outliers

In addition to serving as a measure of spread, the interquartile range (*IQR*) is used as part of a rule of thumb for identifying outliers.

DEFINITION: The 1.5 × *IQR* rule for outliers

Call an observation an outlier if it falls more than 1.5 × *IQR* above the third quartile or below the first quartile.

```
0 | 5
1 | 005555
2 | 0005
3 | 00
4 | 005
5 |
6 | 005
7 |
8 | 5
```

Key: 4|5 is a New York worker who reported a 45-minute travel time to work.

Does the 1.5 × *IQR* rule identify any outliers for the New York travel time data? In the previous example, we found that $Q_1 = 15$ minutes, $Q_3 = 42.5$ minutes, and $IQR = 27.5$ minutes. For these data,

$$1.5 \times IQR = 1.5(27.5) = 41.25$$

Any values not falling between

$$Q_1 - 1.5 \times IQR = 15 - 41.25 = -26.25 \quad \text{and}$$
$$Q_3 + 1.5 \times IQR = 42.5 + 41.25 = 83.75$$

are flagged as outliers. Look again at the stemplot: the only outlier is the longest travel time, 85 minutes. The $1.5 \times IQR$ rule suggests that the three next-longest travel times (60 and 65 minutes) are just part of the long right tail of this skewed distribution.

Travel Times to Work in North Carolina

Identifying outliers

```
0 | 5
1 | 000025
2 | 005
3 | 00
4 | 00
5 |
6 | 0
```

Key: 2|5 is a NC worker who travels 25 minutes to work.

Earlier, we noted the influence of one long travel time of 60 minutes in our sample of 15 North Carolina workers.

PROBLEM: Determine whether this value is an outlier.

SOLUTION: Earlier, we found that $Q_1 = 10$ minutes, $Q_3 = 30$ minutes, and $IQR = 20$ minutes. To check for outliers, we first calculate

$$1.5 \times IQR = 1.5(20) = 30$$

By the $1.5 \times IQR$ rule, any value *greater than*

$$Q_3 + 1.5 \times IQR = 30 + 30 = 60$$

or *less than*

$$Q_1 - 1.5 \times IQR = 10 - 30 = -20$$

would be classified as an outlier. The maximum value of 60 minutes is not quite large enough to be flagged as an outlier.

For Practice *Try Exercise 89(b)*

AP EXAM TIP You may be asked to determine whether a quantitative data set has any outliers. Be prepared to state and use the rule for identifying outliers.

Whenever you find outliers in your data, try to find an explanation for them. Sometimes the explanation is as simple as a typing error, like typing 10.1 as 101. Sometimes a measuring device broke down or someone gave a silly response, like the student in a class survey who claimed to study 30,000 minutes per night. (Yes, that really happened.) In all these cases, you can simply remove the outlier from your data. When outliers are "real data," like the long travel times of some New York workers, you should choose statistical methods that are not greatly affected by the outliers.

The Five-Number Summary and Boxplots

The smallest and largest observations tell us little about the distribution as a whole, but they give information about the tails of the distribution that is missing if we know only the median and the quartiles. To get a quick summary of both center and spread, combine all five numbers.

DEFINITION: The five-number summary

The **five-number summary** of a distribution consists of the smallest observation, the first quartile, the median, the third quartile, and the largest observation, written in order from smallest to largest. In symbols, the five-number summary is

$$\text{Minimum} \quad Q_1 \quad M \quad Q_3 \quad \text{Maximum}$$

These five numbers divide each distribution roughly into quarters. About 25% of the data values fall between the minimum and Q_1, about 25% are between Q_1 and the median, about 25% are between the median and Q_3, and about 25% are between Q_3 and the maximum.

Boxplot

The five-number summary of a distribution leads to a new graph, the **boxplot** (sometimes called a box and whisker plot).

How to Make a Boxplot

- A central box is drawn from the first quartile (Q_1) to the third quartile (Q_3).
- A line in the box marks the median.
- Lines (called whiskers) extend from the box out to the smallest and largest observations that are not outliers.

Here's an example that shows how to make a boxplot.

Home Run King

Making a boxplot

Barry Bonds set the major league record by hitting 73 home runs in a single season in 2001. On August 7, 2007, Bonds hit his 756th career home run, which broke Hank Aaron's longstanding record of 755. By the end of the 2007 season when Bonds retired, he had increased the total to 762. Here are data on the number of home runs that Bonds hit in each of his 21 complete seasons:

16 25 24 19 33 25 34 46 37 33 42 40 37 34 49 73 46 45 45 26 28

PROBLEM: We want to make a boxplot for these data.

SOLUTION: Let's start by sorting the data values so that we can find the five-number summary.

16 19 24 25 (25 26) 28 33 33 34 34 37 37 40 42 (45 45) 46 46 49 73

Min 16, $Q_1 = 25.5$, *M* 34, $Q_3 = 45$, *Max* 73

Now we check for outliers. Since $IQR = 45 - 25.5 = 19.5$, by the $1.5 \times IQR$ rule, any value *greater than* $Q_3 + 1.5 \times IQR = 45 + 1.5 \times 19.5 = 74.25$ or *less than* $Q_1 - 1.5 \times IQR = 25.5 - 1.5 \times 19.5 = -3.75$ would be classified as an outlier. So there are no outliers in this data set. Now we are ready to draw the boxplot. See the finished graph in the margin.

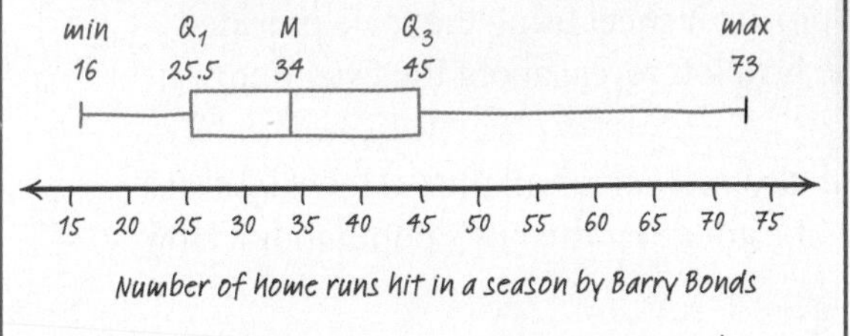

For Practice *Try Exercise 91*

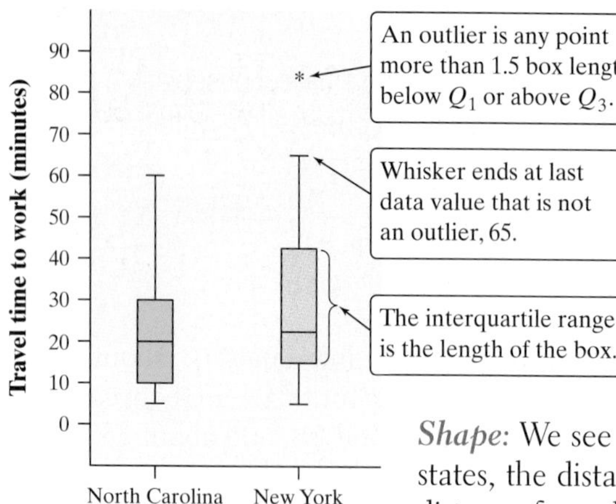

FIGURE 1.19 Boxplots comparing the travel times to work of samples of workers in North Carolina and New York.

Figure 1.19 shows boxplots (this time, they are oriented vertically) comparing travel times to work for the samples of workers from North Carolina and New York. We will identify outliers as isolated points in the boxplot (like the * for the maximum value in the New York data set).

Boxplots show less detail than histograms or stemplots, so they are best used for side-by-side comparison of more than one distribution, as in Figure 1.19. As always, be sure to discuss shape, center, spread, and outliers as part of your comparison. For the travel time to work data:

Shape: We see from the graph that both distributions are right-skewed. For both states, the distance from the minimum to the median is much smaller than the distance from the median to the maximum.

Center: It appears that travel times to work are generally a bit longer in New York than in North Carolina. The median, both quartiles, and the maximum are all larger in New York.

Spread: Travel times are also more variable in New York, as shown by the heights of the boxes (the *IQR*) and the spread from smallest to largest time.

Outliers: Earlier, we showed that the maximum travel time of 85 minutes is an outlier for the New York data. There are no outliers in the North Carolina sample.

The following Activity reinforces the important ideas of working with boxplots.[40]

ACTIVITY *Do you know your geography?*

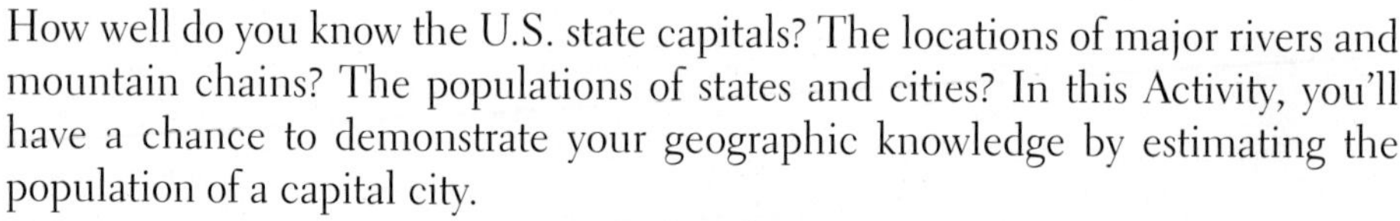

How well do you know the U.S. state capitals? The locations of major rivers and mountain chains? The populations of states and cities? In this Activity, you'll have a chance to demonstrate your geographic knowledge by estimating the population of a capital city.

1. Your teacher will give each student a slip of paper with two questions about the population of a U.S. capital city. Look only at your paper. Answer the questions to the best of your ability.
2. Now your teacher will share the details of what was on the slips of paper. About half the students in the class received one version of the first question, and the other half received a different version. The second question was the same on all the slips.
3. Students in the class should divide into two groups based on which version of the first question they received. Within each group, line up in order of your population estimates, from smallest to largest. Then find the five-number summary for your estimates.
4. With the other members of your group, determine whether there are any outliers.
5. Have a representative from each group draw a boxplot of the group's population estimates on the board or overhead-projector sheet using the scale provided by your teacher. What do the side-by-side boxplots reveal about the two groups' estimates?
6. Your teacher will provide the capital city's actual population. How close was your estimate? Which group did better at estimating the city's population? How can you tell?

CHECK YOUR UNDERSTANDING

The 2009 roster of the Dallas Cowboys professional football team included 10 offensive linemen. Their weights (in pounds) were

338 318 353 313 318 326 307 317 311 311

1. Find the five-number summary for these data by hand. Show your work.
2. Calculate the *IQR*. Interpret this value in context.
3. Determine whether there are any outliers using the $1.5 \times IQR$ rule.
4. Draw a boxplot of the data.

TECHNOLOGY CORNER Making calculator boxplots

The TI-83/84 and TI-89 can plot up to three boxplots in the same viewing window. Let's use the calculator to make side-by-side boxplots of the travel time to work data for the samples from North Carolina and New York.

1. Enter the travel time data for North Carolina in L1/list1 and for New York in L2/list2.

2. Set up two statistics plots: Plot1 to show a boxplot of the North Carolina data and Plot2 to show a boxplot of the New York data.

TI-83/84

TI-89

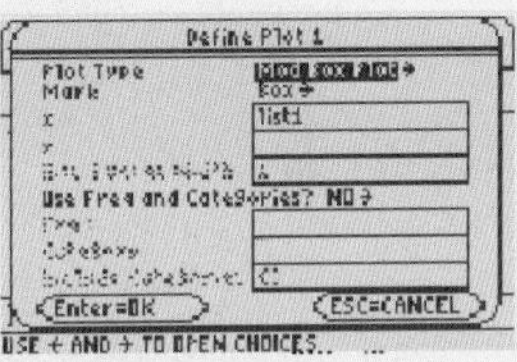

Note: The calculator offers two types of boxplots: a "modified" boxplot that shows outliers and a standard boxplot that doesn't. We'll always use the modified boxplot.

3. Use the calculator's Zoom feature to display the side-by-side boxplots. Then Trace to view the five-number summary.

TI-83/84

- Press ZOOM and select `9 : ZoomStat`.
- Press TRACE.

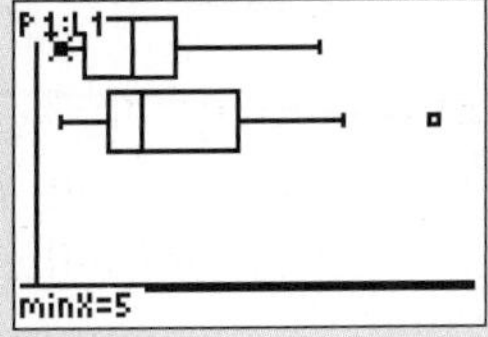

TI-89

- Press F5 `(ZoomData)`.
- Press F3 `(Trace)`.

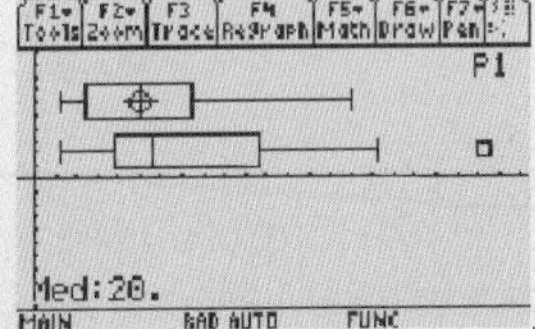

TI-Nspire instructions in Appendix B

Measuring Spread: The Standard Deviation

The five-number summary is not the most common numerical description of a distribution. That distinction belongs to the combination of the mean to measure center and the *standard deviation* to measure spread. The standard deviation and its close relative, the *variance*, measure spread by looking at how far the observations are from their mean. Let's explore this idea using a simple set of data.

How Many Pets?

Investigating spread around the mean

In the Think About It on page 52, we examined data on the number of pets owned by a group of 9 children. Here are the data again, arranged from lowest to highest:

1 3 4 4 4 5 7 8 9

Earlier, we found the mean number of pets to be $\bar{x} = 5$. Let's look at where the observations in the data set are relative to the mean.

Figure 1.20 displays the data in a dotplot, with the mean clearly marked. The data value 1 is 4 units below the mean. We say that its *deviation* from the mean is −4. What about the data value 7? Its deviation is $7 - 5 = 2$ (it is 2 units above the mean). The arrows in the figure mark these two deviations from the mean. The deviations show how much the data vary about their mean. They are the starting point for calculating the variance and standard deviation.

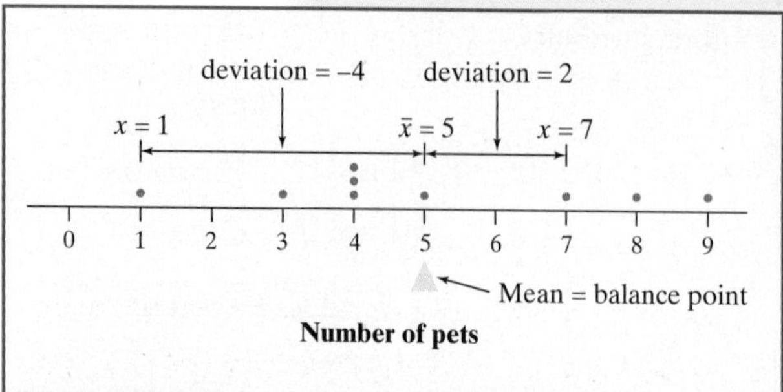

FIGURE 1.20 Dotplot of the pet data with the mean and two of the deviations marked.

The table at top right shows the deviation from the mean $(x_i - \bar{x})$ for each value in the data set. Sum the deviations from the mean. You should get 0, because the mean is the balance point of the distribution. Since the sum of the deviations from the mean will be 0 for *any* set of data, we need another way to calculate spread around the mean. How can we fix the problem of the positive and negative deviations canceling out? We could take the absolute value of each deviation. Or we could square the deviations. For mathematical reasons beyond the scope of this book, statisticians choose to square rather than to use absolute values.

Observations x_i	Deviations $x_i - \bar{x}$	Squared deviations $(x_i - \bar{x})^2$
1	$1 - 5 = -4$	$(-4)^2 = 16$
3	$3 - 5 = -2$	$(-2)^2 = 4$
4	$4 - 5 = -1$	$(-1)^2 = 1$
4	$4 - 5 = -1$	$(-1)^2 = 1$
4	$4 - 5 = -1$	$(-1)^2 = 1$
5	$5 - 5 = 0$	$0^2 = 0$
7	$7 - 5 = 2$	$2^2 = 4$
8	$8 - 5 = 3$	$3^2 = 9$
9	$9 - 5 = 4$	$4^2 = 16$
	sum = ??	sum = ??

We have added a column to the table that shows the square of each deviation $(x_i - \bar{x})^2$. Add up the squared deviations. Did you get 52? Now we compute the average squared deviation—sort of. Instead of dividing by the number of observations n, we divide by $n - 1$:

$$\text{“average” squared deviation} = \frac{16+4+1+1+1+0+4+9+16}{9-1} = \frac{52}{8} = 6.5$$

This value, 6.5, is called the **variance.**

Because we squared all the deviations, our units are in "squared pets." That's no good. We'll take the square root to get back to the correct units—pets. The resulting value is the **standard deviation:**

$$\text{standard deviation} = \sqrt{\text{“average” squared deviation}} = \sqrt{6.5} = 2.55$$

This 2.55 is roughly the average distance of the values in the data set from the mean.

As you can see, the average distance in the standard deviation is found in a rather unexpected way. Why do we divide by $n - 1$ instead of n when calculating the variance and standard deviation? The answer is complicated but will be revealed in Chapter 7.

DEFINITION: The standard deviation s_x and variance s_x^2

The **standard deviation** s_x measures the average distance of the observations from their mean. It is calculated by finding an average of the squared distances and then taking the square root. This average squared distance is called the **variance.** In symbols, the variance s_x^2 is given by

$$s_x^2 = \frac{(x_1 - \bar{x})^2 + (x_2 - \bar{x})^2 + \cdots + (x_n - \bar{x})^2}{n-1} = \frac{1}{n-1}\sum (x_i - \bar{x})^2$$

Here's a brief summary of the process for calculating the standard deviation.

How to Find the Standard Deviation

To find the standard deviation of n observations:

1. Find the distance of each observation from the mean and square each of these distances.
2. Average the distances by dividing their sum by $n - 1$.
3. The standard deviation s_x is the square root of this average squared distance:

$$s_x = \sqrt{\frac{1}{n-1}\sum (x_i - \bar{x})^2}$$

Many calculators report two standard deviations, giving you a choice of dividing by n or by $n - 1$. The former is usually labeled σ_x, the symbol for the standard deviation of a population. If your data set consists of the entire population, then it's appropriate to use σ_x. More often, the data we're examining come from a sample. In that case, we should use s_x.

More important than the details of calculating s_x are the properties that determine the usefulness of the standard deviation:

- s_x measures *spread about the mean* and should be used only when the mean is chosen as the measure of center.
- s_x is *always greater than or equal to 0.* $s_x = 0$ only when there is no variability. This happens only when all observations have the same value. Otherwise, $s_x > 0$. As the observations become more spread out about their mean, s_x gets larger.
- s_x has the *same units of measurement as the original observations.* For example, if you measure metabolic rates in calories, both the mean $\bar{x}$ and the standard deviation s_x are also in calories. This is one reason to prefer s_x to the variance s_x^2, which is in squared calories.
- Like the mean $\bar{x}$, s_x is *not resistant.* A few outliers can make s_x very large.

CAUTION

The use of squared deviations makes s_x even more sensitive than $\bar{x}$ to a few extreme observations. For example, the standard deviation of the travel times for the 15 North Carolina workers is 15.23 minutes. If we omit the maximum value of 60 minutes, the standard deviation drops to 11.56 minutes.

CHECK YOUR UNDERSTANDING

The heights (in inches) of the five starters on a basketball team are 67, 72, 76, 76, and 84.

1. Find and interpret the mean.
2. Make a table that shows, for each value, its deviation from the mean and its squared deviation from the mean.
3. Show how to calculate the variance and standard deviation from the values in your table.
4. Interpret the meaning of the standard deviation in this setting.

Numerical Summaries with Technology

Graphing calculators and computer software will calculate numerical summaries for you. That will free you up to concentrate on choosing the right methods and interpreting your results.

TECHNOLOGY CORNER Computing numerical summaries with technology

Let's find numerical summaries for the travel times of North Carolina and New York workers from the previous Technology Corner (page 61). We'll start by showing you the necessary calculator techniques and then look at output from computer software.

I. One-variable statistics on the calculator If you haven't done so already, enter the North Carolina data in L1/list1 and the New York data in L2/list2.

1. Find the summary statistics for the North Carolina travel times.

TI-83/84

- Press STAT ▶ (CALC); choose `1:1-VarStats`.
- Press ENTER. Now press 2nd 1 (L1) and ENTER.

TI-89

- Press F4 (Calc); choose `1:1-Var Stats`.
- Type list1 in the list box. Press ENTER

Press ▼ to see the rest of the one-variable statistics for North Carolina.

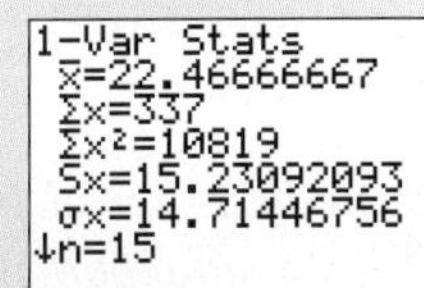

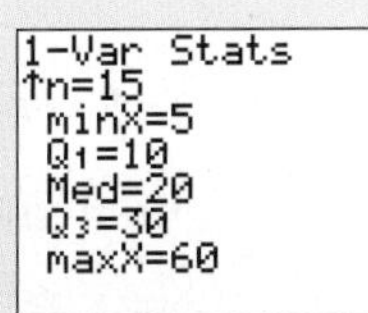

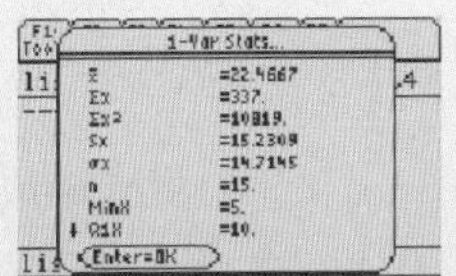

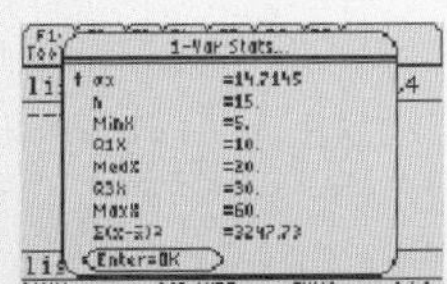

2. Repeat Step 1 using L2/list2 to find the summary statistics for the New York travel times.

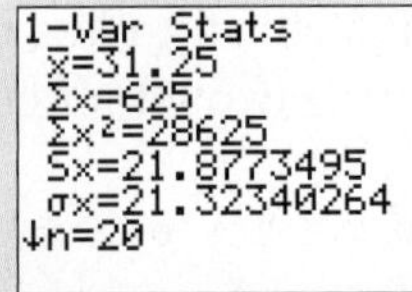

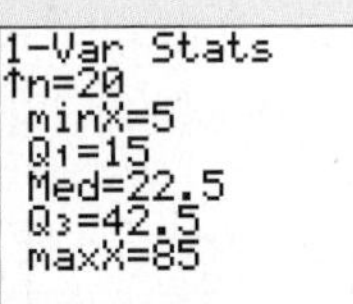

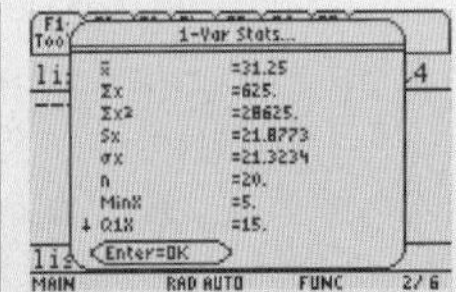

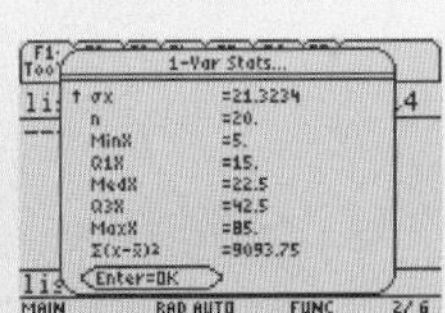

II. Output from statistical software We used Minitab statistical software to produce descriptive statistics for the New York and North Carolina travel time data. Minitab allows you to choose which numerical summaries are included in the output.

Descriptive Statistics: Travel time to work

Variable	N	Mean	StDev	Minimum	Q_1	Median	Q_3	Maximum
NY Time	20	31.25	21.88	5.00	15.00	22.50	43.75	85.00
NC Time	15	22.47	15.23	5.00	10.00	20.00	30.00	60.00

TI-Nspire instructions in Appendix B

THINK ABOUT IT

What's with that third quartile? Earlier, we saw that the quartiles of the New York travel times are $Q_1 = 15$ and $Q_3 = 42.5$. Look at the Minitab output in the Technology Corner. Minitab says that $Q_3 = 43.75$. What happened? Minitab and some other software use different rules for locating quartiles. Results from the various rules are always close to each other, so the differences are rarely important in practice. But because of the slight difference, Minitab wouldn't identify the maximum value of 85 as an outlier by the $1.5 \times IQR$ rule.

Choosing Measures of Center and Spread

We now have a choice between two descriptions of the center and spread of a distribution: the median and IQR, or $\bar{x}$ and s_x. Because $\bar{x}$ and s_x are sensitive to extreme observations, they can be misleading when a distribution is strongly skewed or has outliers. In these cases, the median and IQR, which are both resistant to extreme values, provide a better summary. We'll see in the next chapter that the mean and standard deviation are the natural measures of center and spread for a very important class of symmetric distributions, the Normal distributions.

Choosing Measures of Center and Spread

The median and IQR are usually better than the mean and standard deviation for describing a skewed distribution or a distribution with strong outliers. Use $\bar{x}$ and s_x only for reasonably symmetric distributions that don't have outliers.

CAUTION

Remember that a graph gives the best overall picture of a distribution. Numerical measures of center and spread report specific facts about a distribution, but they do not describe its entire shape. Numerical summaries do not highlight the presence of multiple peaks or clusters, for example. *Always plot your data.*

Here's a final example that shows the thinking involved in choosing measures of center and spread when comparing two sets of quantitative data.

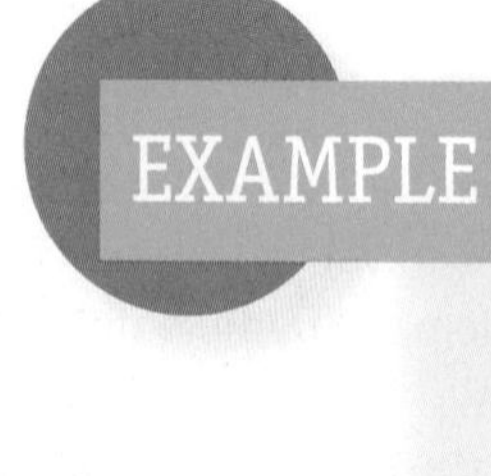

Who Texts More—Males or Females?

Pulling it all together

STEP 4

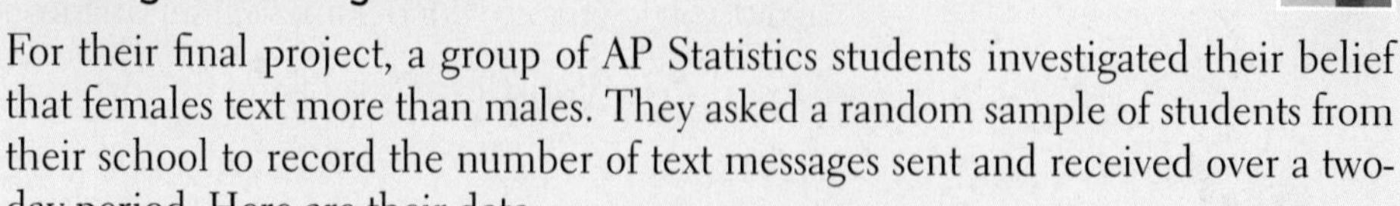

For their final project, a group of AP Statistics students investigated their belief that females text more than males. They asked a random sample of students from their school to record the number of text messages sent and received over a two-day period. Here are their data:

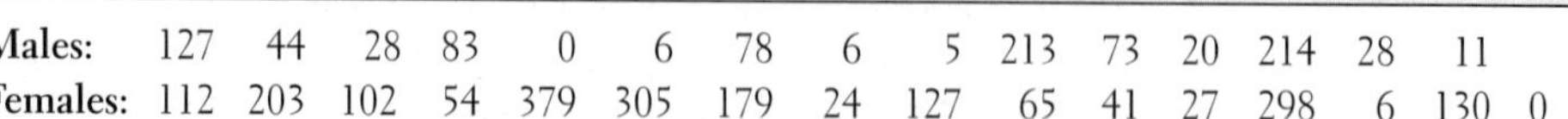

Males:	127	44	28	83	0	6	78	6	5	213	73	20	214	28	11	
Females:	112	203	102	54	379	305	179	24	127	65	41	27	298	6	130	0

What conclusion should the students draw? Give appropriate evidence to support your answer.

STATE: **Do the data give convincing evidence that females text more than males?**

PLAN: We'll begin by making side-by-side boxplots. Then we'll calculate one-variable statistics. Finally, we'll compare shape, center, spread, and outliers for the two distributions.

DO: Figure 1.21 is a sketch of the boxplots we obtained from our calculator. The table below shows numerical summaries for males and females.

	$\bar{x}$	s_x	Min	Q_1	M	Q_3	Max	IQR
Male	62.4	71.4	0	6	28	83	214	77
Female	128.3	116.0	0	34	107	191	379	157

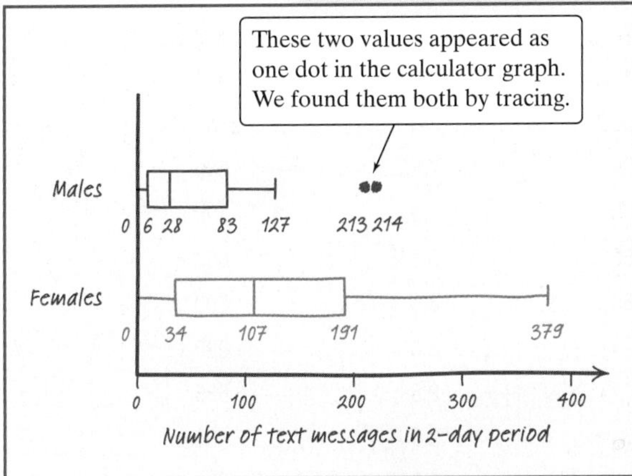

FIGURE 1.21 Side-by-side boxplots of the texting data.

Due to the strong skewness and outliers, we'll use the median and *IQR* instead of the mean and standard deviation when discussing center and spread. Shape: Both distributions are heavily right-skewed. Center: On average, females text more than males. The median number of texts for females (107) is about four times as high as for males (28). In fact, the median for the females is above the third quartile for the males. This indicates that over 75% of the males texted less than the "typical" (median) female. Spread: There is much more variability in texting among the females than the males. The *IQR* for females (157) is about twice the *IQR* for males (77). Outliers: There are two outliers in the male distribution: students who reported 213 and 214 texts in two days. The female distribution has no outliers.

CONCLUDE: The data from this survey project give very strong evidence to support the students' belief that females text more than males. Females sent and received a median of 107 texts over the two-day period, which exceeded the number of texts reported by over 75% of the males.

For Practice *Try Exercise 105*

AP EXAM TIP Use statistical terms carefully and correctly on the AP exam. Don't say "mean" if you really mean "median." Range is a single number; so are Q_1, Q_3, and *IQR*. Avoid colloquial use of language, like "the outlier *skews* the mean." Skewed is a shape. If you misuse a term, expect to lose some credit.

Now it's time for you to put what you have learned into practice in the following Data Exploration.

DATA EXPLORATION *Did Mr. Starnes stack his class?*

Mr. Starnes teaches AP Statistics, but he also does the class scheduling for the high school. There are two AP Statistics classes—one taught by Mr. Starnes and one taught by Ms. McGrail. The two teachers give the same first test to their classes and grade the test together. Mr. Starnes's students earned an average score that was 8 points higher than the average for Ms. McGrail's class. Ms. McGrail wonders whether Mr. Starnes might have "adjusted" the class rosters from the computer scheduling program. In other words, she thinks he might have "stacked" his class. He denies this, of course.

To help resolve the dispute, the teachers collect data on the cumulative grade point averages and SAT Math scores of their students. Mr. Starnes provides the GPA data from his computer. The students report their SAT Math scores. The tables below show the data for each student in the two classes. Note that the two data values in each row come from a single student.

Starnes GPA	Starnes SAT-M	McGrail GPA	McGrail SAT-M
2.9	670	2.9	620
2.86	520	3.3	590
2.6	570	3.98	650
3.6	710	2.9	600
3.2	600	3.2	620
2.7	590	3.5	680
3.1	640	2.8	500
3.085	570	2.9	502.5
3.75	710	3.95	640
3.4	630	3.1	630
3.338	630	2.85	580
3.56	670	2.9	590
3.8	650	3.245	600
3.2	660	3.0	600
3.1	510	3.0	620
		2.8	580
		2.9	600
		3.2	600

Did Mr. Starnes stack his class? Give appropriate graphical and numerical evidence to support your conclusion.

At the end of each chapter, we'll return to the chapter-opening Case Study to let you know what happened. We call this feature "Case Closed!"

case closed

Do Rewards Promote Creativity?

Refer to the chapter-opening Case Study (page 1). In Dr. Amabile's experiment, the children who believed they were competing for a prize produced *less creative* collages than the children who were told that they would share their finished collages at an art party. As a result of this study, Dr. Amabile argued that external rewards—things like money, grades, and fame—actually reduce creativity. She went on to conduct several more experiments, like the one investigating the effects of internal and external motivation on college students' creative writing (Exercise 43, page 43). Her findings were remarkably consistent: for most people, creativity is its own reward.

SECTION 1.3 Summary

- A numerical summary of a distribution should report at least its **center** and its **spread, or variability.**
- The **mean** $\bar{x}$ and the **median** M describe the center of a distribution in different ways. The mean is the average of the observations, and the median is the midpoint of the values.
- When you use the median to indicate the center of a distribution, describe its spread using the **quartiles.** The **first quartile** Q_1 has about one-fourth of the observations below it, and the **third quartile** Q_3 has about three-fourths of the observations below it. The **interquartile range (*IQR*)** is the range of the middle 50% of the observations and is found by $IQR = Q_3 - Q_1$. An extreme observation is an **outlier** if it is smaller than $Q_1 - (1.5 \times IQR)$ or larger than $Q_3 + (1.5 \times IQR)$.
- The **five-number summary** consisting of the median, the quartiles, and the maximum and minimum values provides a quick overall description of a distribution. The median describes the center, and the quartiles and extremes show the spread.
- **Boxplots** based on the five-number summary are useful for comparing distributions. The box spans the quartiles and shows the spread of the central half of the distribution. The median is marked within the box. Lines extend from the box to the smallest and the largest observations that are not outliers. Outliers are plotted as isolated points.
- The **variance** s_x^2 and especially its square root, the **standard deviation** s_x, are common measures of spread about the mean as center. The standard deviation s_x is zero when there is no variability and gets larger as the spread increases.
- The median is a **resistant** measure of center because it is relatively unaffected by extreme observations. The mean is nonresistant. Among measures of spread, the *IQR* is resistant, but the standard deviation is not.
- The mean and standard deviation are good descriptions for symmetric distributions without outliers. They are most useful for the Normal distributions introduced in the next chapter. The median and *IQR* are a better description for skewed distributions.
- Numerical summaries do not fully describe the shape of a distribution. *Always plot your data.*

1.3 TECHNOLOGY CORNERS

TI-Nspire instructions in Appendix B

SECTION 1.3 Exercises

79. **Quiz grades** Joey's first 14 quiz grades in a marking period were

pg 51

86	84	91	75	78	80	74
87	76	96	82	90	98	93

Calculate the mean. Show your work. Interpret your result in context.

80. **Cowboys** The 2009 roster of the Dallas Cowboys professional football team included 7 defensive linemen. Their weights (in pounds) were 306, 305, 315, 303, 318, 309, and 285. Calculate the mean. Show your work. Interpret your result in context.

81. **Quiz grades** Refer to Exercise 79.

pg 53

(a) Find the median by hand. Show your work. Interpret your result in context.

(b) Suppose Joey has an unexcused absence for the 15th quiz, and he receives a score of zero. Recalculate the mean and the median. What property of measures of center does this illustrate?

82. **Cowboys** Refer to Exercise 80.

(a) Find the median by hand. Show your work. Interpret your result in context.

(b) Suppose the lightest lineman had weighed 265 pounds instead of 285 pounds. How would this change affect the mean and the median? What property of measures of center does this illustrate?

83. **Incomes of college grads** According to the Census Bureau, the mean and median 2008 income of people at least 25 years old who had a bachelor's degree but no higher degree were \$48,097 and \$60,954. Which of these numbers is the mean and which is the median? Explain your reasoning.

84. **House prices** The mean and median selling prices of existing single-family homes sold in November 2009 were \$216,400 and \$172,600.[41] Which of these numbers is the mean and which is the median? Explain how you know.

85. **Baseball salaries** Suppose that a Major League Baseball team's mean yearly salary for its players is \$1.2 million and that the team has 25 players on its active roster. What is the team's total annual payroll? If you knew only the median salary, would you be able to answer this question? Why or why not?

86. **Mean salary?** Last year a small accounting firm paid each of its five clerks \$22,000, two junior accountants \$50,000 each, and the firm's owner \$270,000. What is the mean salary paid at this firm? How many of the employees earn less than the mean? What is the median salary? Write a sentence to describe how an unethical recruiter could use statistics to mislead prospective employees.

87. **Domain names** When it comes to Internet domain names, is shorter better? According to one ranking of Web sites in 2008, the top 8 sites (by number of "hits") were yahoo.com, google.com, youtube.com, live.com, msn.com, myspace.com, wikipedia.org, and facebook.com. These familiar sites certainly have short domain names. The histogram below shows the domain name lengths (in number of letters in the name, not including the extensions .com and .org) for the 500 most popular Web sites.

www

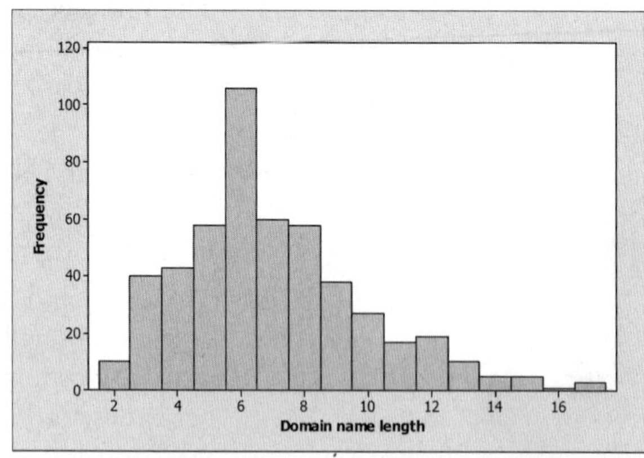

(a) Estimate the mean and median of the distribution. Explain your method clearly.

(b) If you wanted to argue that shorter domain names were more popular, which measure of center would you choose—the mean or the median? Justify your answer.

88. **Do adolescent girls eat fruit?** We all know that fruit is good for us. Below is a histogram of the number of servings of fruit per day claimed by 74 seventeen-year-old girls in a study in Pennsylvania.[42]

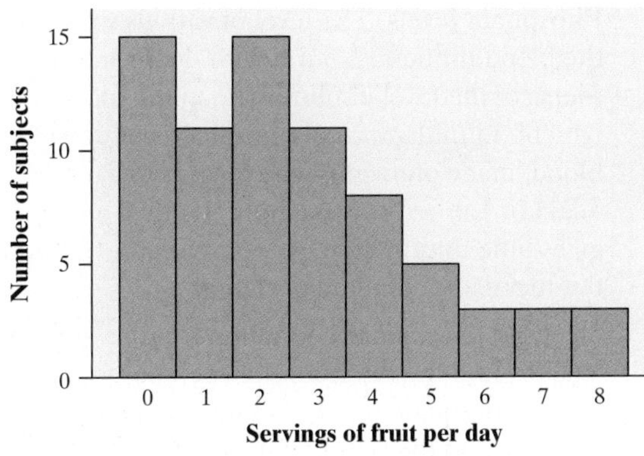

(a) With a little care, you can find the median and the quartiles from the histogram. What are these numbers? How did you find them?

(b) Estimate the mean of the distribution. Explain your method clearly.

89. **Quiz grades** Refer to Exercise 79.

pg 57, 58

(a) Find and interpret the interquartile range (IQR).

(b) Determine whether there are any outliers. Show your work.

90. **Cowboys** Refer to Exercise 80.

(a) Find and interpret the interquartile range (IQR).

(b) Determine whether there are any outliers. Show your work.

91. **Don't call me** In a September 28, 2008, article titled "Letting Our Fingers Do the Talking," the *New York Times* reported that Americans now send more text messages than they make phone calls. According to a study by Nielsen Mobile, "Teenagers ages 13 to 17 are by far the most prolific texters, sending or receiving 1,742 messages a month." Mr. Williams, a high school statistics teacher, was skeptical about the claims in the article. So he collected data from his first-period statistics class on the number of text messages and calls they had sent or received in the past 24 hours. Here are the texting data:

pg 59

0	7	1	29	25	8	5	1	25	98	9	0	26
8	118	72	0	92	52	14	3	3	44	5	42	

(a) Make a boxplot of these data by hand. Be sure to check for outliers.

(b) Do these data support the claim in the article about the number of texts sent by teens? Justify your answer with appropriate evidence.

92. **Acing the first test** Here are the scores of Mrs. Liao's students on their first statistics test:

93	93	87.5	91	94.5	72	96	95	93.5	93.5	73
82	45	88	80	86	85.5	87.5	81	78	86	89
92	91	98	85	82.5	88	94.5	43			

(a) Make a boxplot of the test score data by hand. Be sure to check for outliers.

(b) How did the students do on Mrs. Liao's first test? Justify your answer.

93. **Texts or calls?** Refer to Exercise 91. A boxplot of the difference (texts – calls) in the number of texts and calls for each student is shown below.

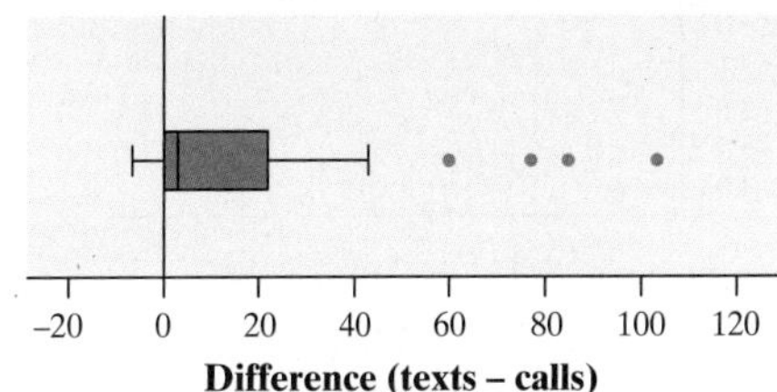

(a) Do these data support the claim in the article about texting versus calling? Justify your answer with appropriate evidence.

(b) Can we draw any conclusion about the preferences of all students in the school based on the data from Mr. Williams's statistics class? Why or why not?

94. **Electoral votes** To become president of the United States, a candidate does not have to receive a majority of the popular vote. The candidate does have to win a majority of the 538 electoral votes that are cast in the Electoral College. Here is a stemplot of the number of electoral votes for each of the 50 states and the District of Columbia.

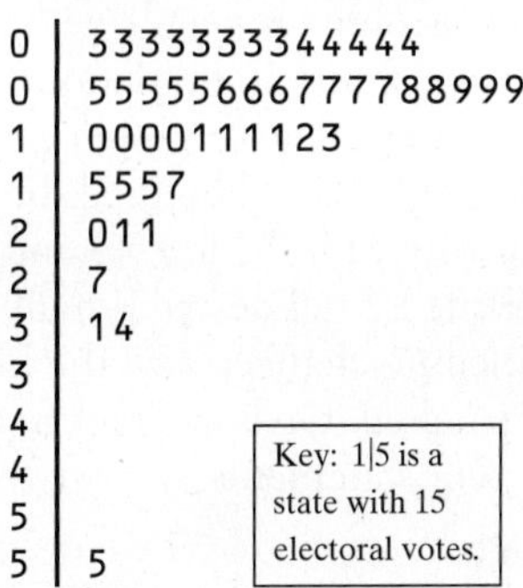

(a) Make a boxplot of these data by hand. Be sure to check for outliers.

(b) Which measure of center and spread would you use to summarize the distribution—the mean and standard deviation or the median and IQR? Justify your answer.

95. **Comparing investments** Should you put your money into a fund that buys stocks or a fund that invests in real estate? The boxplots compare the daily returns (in percent) on a "total stock market" fund and a real estate fund over a year ending in November 2007.[43]

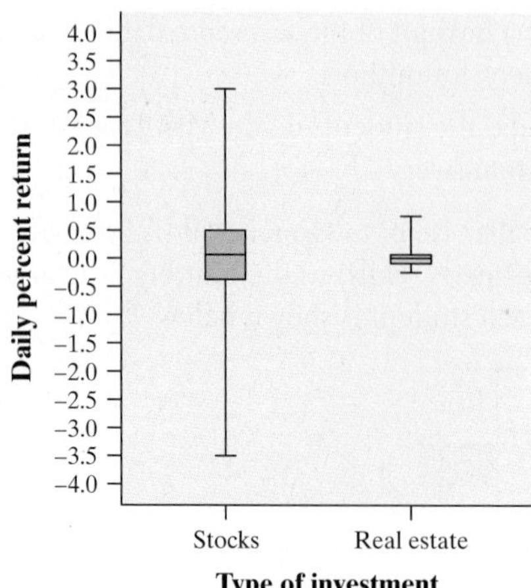

(a) Read the graph: about what were the highest and lowest daily returns on the stock fund?

(b) Read the graph: the median return was about the same on both investments. About what was the median return?

(c) What is the most important difference between the two distributions?

96. **Income and education level** Each March, the Bureau of Labor Statistics compiles an Annual Demographic Supplement to its monthly Current Population Survey.[44] Data on about 71,067 individuals between the ages of 25 and 64 who were employed full-time were collected in one of these surveys. The boxplots below compare the distributions of income for people with five levels of education. This figure is a variation of the boxplot idea: because large data sets often contain very extreme observations, we omitted the individuals in each category with the top 5% and bottom 5% of incomes. Write a brief description of how the distribution of income changes with the highest level of education reached. Give specifics from the graphs to support your statements.

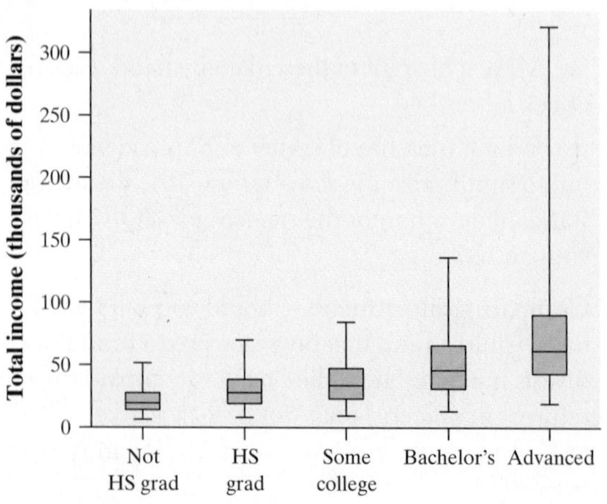

97. **Phosphate levels** The level of various substances in the blood influences our health. Here are measurements of the level of phosphate in the blood of a patient, in milligrams of phosphate per deciliter of blood, made on 6 consecutive visits to a clinic: 5.6, 5.2, 4.6, 4.9, 5.7, 6.4. A graph of only 6 observations gives little information, so we proceed to compute the mean and standard deviation.

(a) Find the standard deviation from its definition. That is, find the deviations of each observation from the mean, square the deviations, then obtain the variance and the standard deviation.

(b) Interpret the value of s_x you obtained in (a).

98. **Feeling sleepy?** The first four students to arrive for a first-period statistics class were asked how much sleep (to the nearest hour) they got last night. Their responses were 7, 7, 9, and 9.

(a) Find the standard deviation from its definition. That is, find the deviations of each observation from the mean, square the deviations, then obtain the variance and the standard deviation.

(b) Interpret the value of s_x you obtained in (a).

(c) Do you think it's safe to conclude that the mean amount of sleep for all 30 students in this class is close to 8 hours? Why or why not?

99. **Shopping spree** The figure displays computer output from Data Desk for data on the amount spent by 50 grocery shoppers.

DataDesk

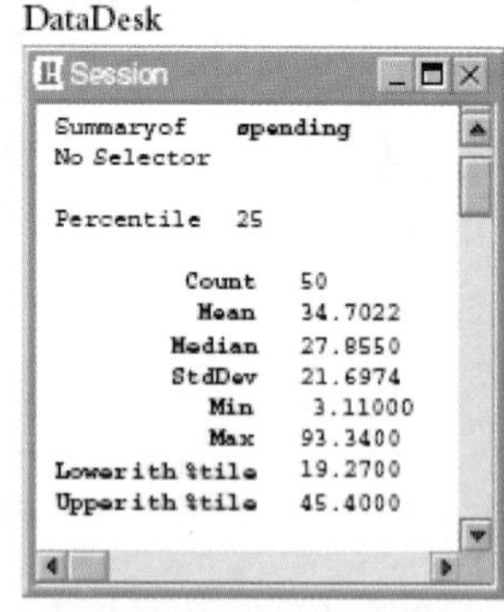

(a) What would you guess is the shape of the distribution based only on the computer output? Explain.

(b) Interpret the value of the standard deviation.

(c) Are there any outliers? Justify your answer.

100. **C-sections** Do male doctors perform more cesarean sections (C-sections) than female doctors? A study in Switzerland examined the number of cesarean sections (surgical deliveries of babies) performed in a

year by samples of male and female doctors. Here are summary statistics for the two distributions:

	$\bar{x}$	s_x	Min.	Q_1	M	Q_3	Max.	IQR
Male doctors	41.333	20.607	20	27	34	50	86	23
Female doctors	19.1	10.126	5	10	18.5	29	33	19

(a) Based on the computer output, which distribution would you guess has a more symmetrical shape? Explain.

(b) Explain how the *IQR*s of these two distributions can be so similar even though the standard deviations are quite different.

(c) Does it appear that males perform more C-sections? Justify your answer.

101. **The *IQR*** Is the interquartile range a resistant measure of spread? Give an example of a small data set that supports your answer.

102. **Measuring spread** Which of the distributions shown has a larger standard deviation? Justify your answer.

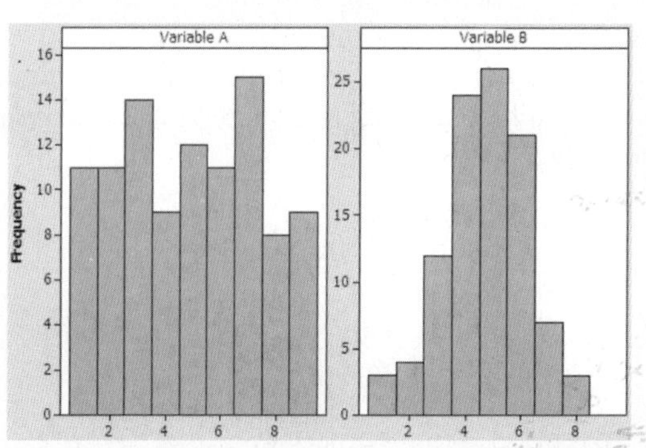

103. **SD contest** This is a standard deviation contest. You must choose four numbers from the whole numbers 0 to 10, with repeats allowed.

(a) Choose four numbers that have the smallest possible standard deviation.

(b) Choose four numbers that have the largest possible standard deviation.

(c) Is more than one choice possible in either (a) or (b)? Explain.

104. **What do they measure?** For each of the following summary statistics, decide (i) whether it could be used to measure center or spread and (ii) whether it is resistant.

(a) $\dfrac{Q_1 + Q_3}{2}$ (b) $\dfrac{\text{Max} - \text{Min}}{2}$

105. **SSHA scores** Here are the scores on the Survey of Study Habits and Attitudes (SSHA) for 18 first-year college women:

pg 66

STEP 4

154	109	137	115	152	140	154	178	101
103	126	126	137	165	165	129	200	148

and for 20 first-year college men:

108	140	114	91	180	115	126
92	169	146	109	132	75	88
113	151	70	115	187	104	

Do these data support the belief that women have better study habits and attitudes toward learning than men? (Note that high scores indicate good study habits and attitudes toward learning.) Follow the four-step process.

106. **Hummingbirds and tropical flowers** Researchers from Amherst College studied the relationship between varieties of the tropical flower *Heliconia* on the island of Dominica and the different species of hummingbirds that fertilize the flowers.[45] Over time, the researchers believe, the lengths of the flowers and the forms of the hummingbirds' beaks have evolved to match each other. If that is true, flower varieties fertilized by different hummingbird species should have distinct distributions of length.

STEP 4

The table below gives length measurements (in millimeters) for samples of three varieties of *Heliconia*, each fertilized by a different species of hummingbird. Do these data support the researchers' belief? Follow the four-step process.

H. bihai

47.12	46.75	46.80	47.12	46.67	47.43	46.44	46.64
48.07	48.34	48.15	50.26	50.12	46.34	46.94	48.36

***H. caribaea* red**

41.90	42.01	41.93	43.09	41.47	41.69	39.78	40.57
39.63	42.18	40.66	37.87	39.16	37.40	38.20	38.07
38.10	37.97	38.79	38.23	38.87	37.78	38.01	

***H. caribaea* yellow**

36.78	37.02	36.52	36.11	36.03	35.45	38.13	37.10
35.17	36.82	36.66	35.68	36.03	34.57	34.63	

Multiple choice: Select the best answer for Exercises 107 to 110.

107. If a distribution is skewed to the right with no outliers,
(a) mean < median.
(b) mean ≈ median.
(c) mean = median.
(d) mean > median.
(e) We can't tell without examining the data.

108. You have data on the weights in grams of 5 baby pythons. The mean weight is 31.8 and the standard deviation of the weights is 2.39. The correct units for the standard deviation are
(a) no units—it's just a number.
(b) grams.
(c) grams squared.
(d) pythons.
(e) pythons squared.

109. Which of the following is least affected if an extreme high outlier is added to your data?
(a) Median
(b) Mean
(c) Standard deviation
(d) Range
(e) Maximum

110. What are all the values that a standard deviation s_x can possibly take?
(a) $s_x \geq 0$
(b) $s_x > 0$
(c) $0 \leq s_x \leq 1$
(d) $-1 \leq s_x \leq 1$
(e) Any number

Exercises 111 and 112 refer to the following setting. We used CensusAtSchool's "Random Data Selector" to choose a sample of 50 Canadian students who completed a survey in 2007–2008.

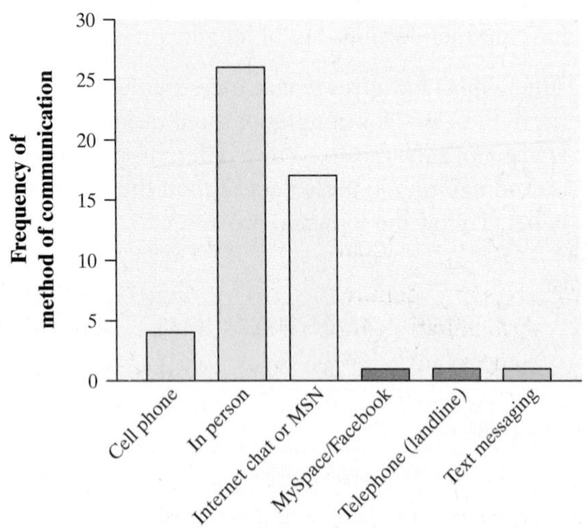

111. **Let's chat** (1.1) The bar graph displays data on students' responses to the question "Which of these methods do you most often use to communicate with your friends?"

(a) Would it be appropriate to make a pie chart for these data? Why or why not?

(b) Jerry says that he would describe this bar graph as skewed to the right. Explain why Jerry is wrong.

112. **How tall are you?** (1.2) Here are the heights (in centimeters) of 50 randomly selected Canadian students who participated in CensusAtSchool in 2007–2008:

166.5	170	178	163	150.5	169	173	169	171	166
190	183	178	161	171	170	191	168.5	178.5	173
175	160.5	166	164	163	174	160	174	182	167
166	170	170	181	171.5	160	178	157	165	187
168	157.5	145.5	156	182	168.5	177	162.5	160.5	185.5

Make an appropriate graph to display these data. Describe the shape, center, and spread of the distribution. Are there any outliers?

113. **Success in college** (1.1) The 2007 Freshman Survey asked first-year college students about their "habits of mind"—specific behaviors that college faculty have identified as being important for student success. One question asked students, "How often in the past year did you revise your papers to improve your writing?" Another asked, "How often in the past year did you seek feedback on your academic work?" The figure is a bar graph comparing male and female responses to these two questions.[46]

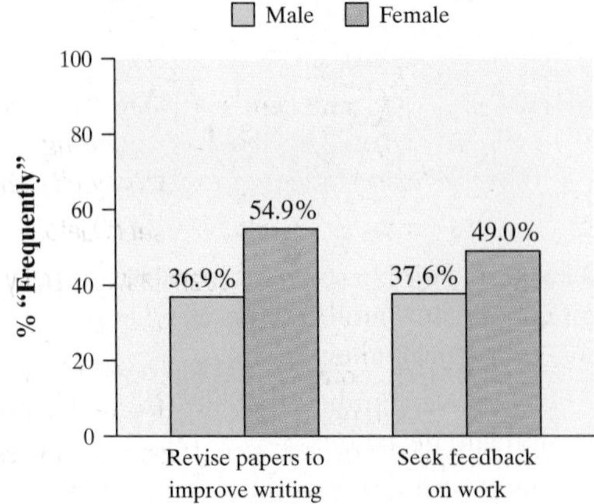

What does the graph tell us about the habits of mind of male and female college freshmen?

Chapter 1 Chapter Review

Statistics is the art and science of data. Data come from measuring one or more variables on some group of individuals. These variables can be categorical or quantitative. Data analysis is a process of describing data using graphs and numerical summaries. This chapter introduced the basic tools of data analysis for categorical and quantitative variables.

Bar graphs and sometimes pie charts are used to display distributions of categorical variables. Counts and percents help summarize the data. To describe the relationship between two categorical variables, start with a two-way table. Then compare appropriate percents.

For quantitative variables, the figure below outlines the data analysis process.

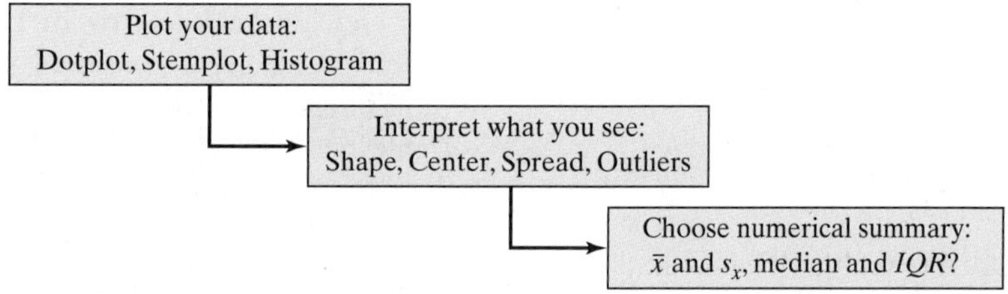

Chapter Review Exercises

These exercises are designed to help you review the important ideas and methods of the chapter. Relevant learning objectives are provided in bulleted form before each exercise.

- **Identify the individuals and variables in a set of data.**
- **Classify variables as categorical or quantitative. Identify units of measurement for a quantitative variable.**

R1.1. Hit movies According to the Internet Movie Database, *Avatar* is tops based on box office sales worldwide. The following table displays data on several popular movies.[47]

Movie	Year	Rating	Time	Genre	Box office
Avatar	2009	PG-13	162	Action	2,714,767,458
Titanic	1997	PG-13	194	Drama	1,835,300,000
The Lord of the Rings: The Return of the King	2003	PG -13	201	Action	1,129,219,252
Pirates of the Caribbean: Dead Man's Chest	2006	PG-13	151	Action	1,060,332,628
The Dark Knight	2008	PG-13	152	Action	1,001,921,825

(a) What individuals does this data set describe?

(b) Clearly identify each of the variables. Which are quantitative? In what units do you think they are measured?

(c) Describe the individual in the highlighted row.

- **Make a bar graph of the distribution of a categorical variable or, in general, to compare related quantities.**
- **Recognize when a pie chart can and cannot be used.**

R1.2. Movie ratings The movie rating system we use today was first established on November 1, 1968. Back then, the possible ratings were G, PG, R, and X. In 1984, the PG-13 rating was created. And in 1990, NC-17 replaced the X rating. Here is a summary of the ratings assigned to movies between 1968 and 2000: 8% rated G, 24% rated PG, 10% rated PG-13, 55% rated R, and 3% rated NC-17.[48] Make an appropriate graph for displaying these data.

- **Identify what makes some graphs deceptive.**

R1.3. I'd die without my phone! In a July 2008 survey of over 2000 U.S. teenagers by Harris Interactive, 47% said that "their social life would end or be worsened without their cell phone."[49] One survey question asked the teens how important it is for their phone to have certain features. The figure below displays data on the percent who indicated that a particular feature is vital.

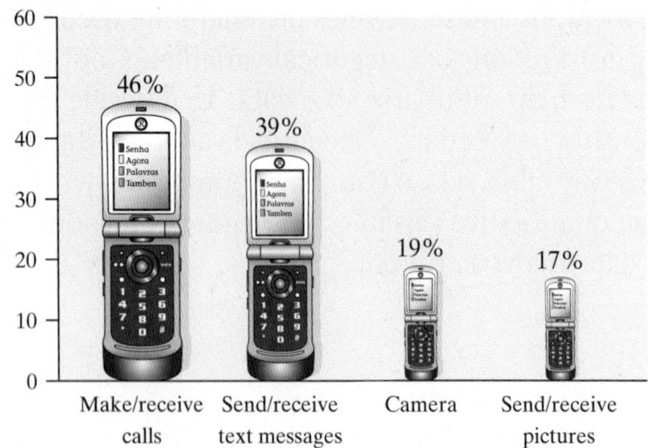

(a) Explain how the graph gives a misleading impression.

(b) Would it be appropriate to make a pie chart to display these data? Why or why not?

(c) Make a graph of the data that isn't misleading.

- From a two-way table of counts, answer questions involving marginal and conditional distributions.

R1.4. Facebook and age Is there a relationship between Facebook use and age among college students? The following two-way table displays data for the 219 students who responded to the survey.[50]

	Facebook User?	
Age	Yes	No
Younger (18–22)	78	4
Middle (23–27)	49	21
Older (28 and up)	21	46

(a) What percent of the students who responded were Facebook users? Is this percent part of a marginal distribution or a conditional distribution? Explain.

(b) What percent of the younger students in the sample were Facebook users? What percent of the Facebook users in the sample were younger students?

- Describe the relationship between two categorical variables by computing appropriate conditional distributions.
- Construct bar graphs to display the relationship between two categorical variables.

R1.5. Facebook and age Use the data in the previous exercise to determine whether there is an association between Facebook use and age. Give appropriate graphical and numerical evidence to support your answer.

- Make a dotplot or stemplot to display small sets of data.
- Describe the overall pattern (shape, center, spread) of a distribution, and identify any major departures from the pattern (like outliers).

R1.6. Density of the earth In 1798, the English scientist Henry Cavendish measured the density of the earth several times by careful work with a torsion balance. The variable recorded was the density of the earth as a multiple of the density of water. Here are Cavendish's 29 measurements:[51]

5.50	5.61	4.88	5.07	5.26	5.55	5.36	5.29	5.58	5.65
5.57	5.53	5.62	5.29	5.44	5.34	5.79	5.10	5.27	5.39
5.42	5.47	5.63	5.34	5.46	5.30	5.75	5.68	5.85	

(a) Present these measurements graphically in a stemplot.

(b) Discuss the shape, center, and spread of the distribution. Are there any outliers?

(c) What is your estimate of the density of the earth based on these measurements? Explain.

- Make a histogram with a reasonable choice of classes.
- Identify the shape of a distribution from a dotplot, stemplot, or histogram as roughly symmetric or skewed.
- Make a boxplot.

R1.7. Guinea pig survival times Here are the survival times in days of 72 guinea pigs after they were injected with infectious bacteria in a medical experiment.[52] Survival times, whether of machines under stress or cancer patients after treatment, usually have distributions that are skewed to the right.

43	45	53	56	56	57	58	66	67	73	74	79
80	80	81	81	81	82	83	83	84	88	89	91
91	92	92	97	99	99	100	100	101	102	102	102
103	104	107	108	109	113	114	118	121	123	126	128
137	138	139	144	145	147	156	162	174	178	179	184
191	198	211	214	243	249	329	380	403	511	522	598

(a) Make a histogram of the data and describe its main features. Does it show the expected right-skew?

(b) Now make a boxplot of the data. Be sure to check for outliers.

- Interpret histograms.

R1.8. Household incomes Rich and poor households differ in ways that go beyond income. Following are histograms that compare the distributions of household size (number of people) for low-income and high-income households.[53] Low-income households had annual incomes less than $15,000, and high-income households had annual incomes of at least $100,000.

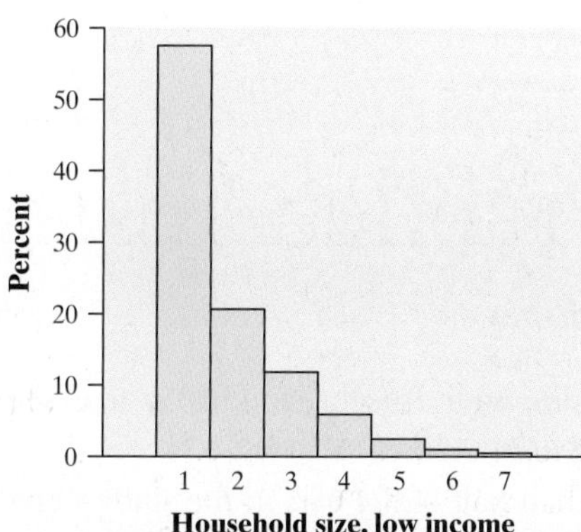

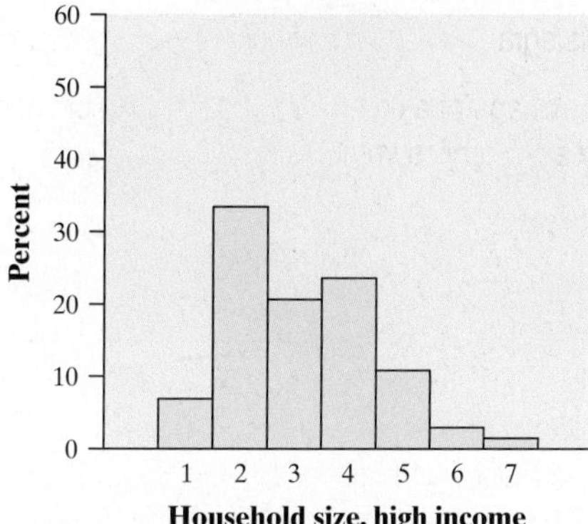

(a) About what percent of each group of households consisted of two people?

(b) What are the important differences between these two distributions? What do you think explains these differences?

Exercises R1.9 and R1.10 refer to the following setting. Do you like to eat tuna? Many people do. Unfortunately, some of the tuna that people eat may contain high levels of mercury. Exposure to mercury can be especially hazardous for pregnant women and small children. How much mercury is safe to consume? The Food and Drug Administration will take action (like removing the product from store shelves) if the mercury concentration in a six-ounce can of tuna is 1.00 ppm (parts per million) or higher.

What is the typical mercury concentration in cans of tuna sold in stores? A study conducted by Defenders of Wildlife set out to answer this question. Defenders collected a sample of 164 cans of tuna from stores across the United States. They sent the selected cans to a laboratory that is often used by the Environmental Protection Agency for mercury testing.[54]

- Interpret the value of a numerical summary in context: mean, median, *IQR*, standard deviation.
- Identify outliers using the 1.5 × *IQR* rule.
- Select the more appropriate measures of center and spread: $\bar{x}$ and s_x or the median and *IQR*.

R1.9. Mercury in tuna A histogram and some computer output provide information about the mercury concentration in the sampled cans (in parts per million, ppm).

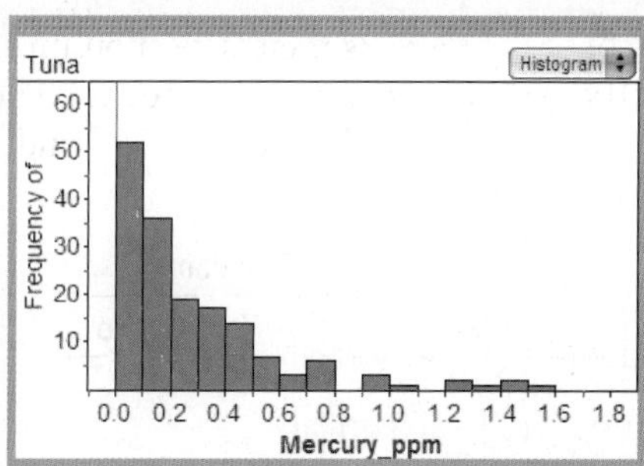

Descriptive Statistics: Mercury_ppm

Variable	N	Mean	StDev	Min
Mercury	164	0.285	0.300	0.012
Variable	**Q_1**	**Med**	**Q_3**	**Max**
Mercury	0.071	0.180	0.380	1.500

(a) Interpret the standard deviation in context.

(b) Determine whether there are any outliers.

(c) Describe the shape, center, and spread of the distribution.

- Use appropriate graphs and numerical summaries to compare distributions of quantitative variables.

R1.10. Mercury in tuna Is there a difference in the mercury concentration of light tuna and albacore tuna? Use the side-by-side boxplots and the computer output to write a few sentences comparing the two distributions.

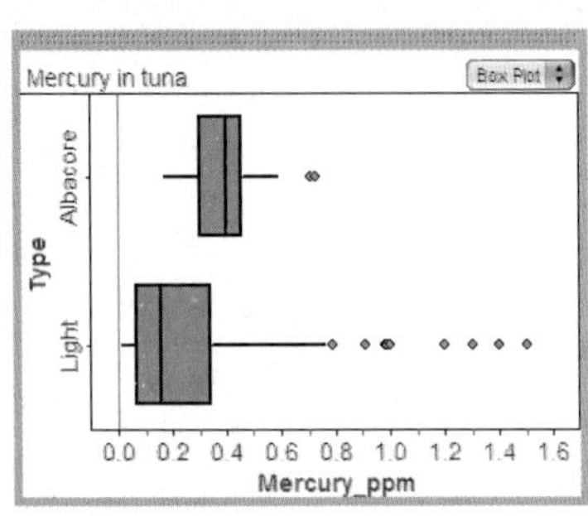

Descriptive Statistics: Mercury_ppm

Type	N	Mean	StDev	Min
Albacore	20	0.401	0.152	0.170
Light	144	0.269	0.312	0.012
Type	**Q_1**	**Med**	**Q_3**	**Max**
Albacore	0.293	0.400	0.460	0.730
Light	0.059	0.160	0.347	1.500

Chapter 1 AP Statistics Practice Test

Section I: Multiple Choice *Select the best answer for each question.*

T1.1. You record the age, marital status, and earned income of a sample of 1463 women. The number and type of variables you have recorded is

(a) 3 quantitative, 0 categorical

(b) 4 quantitative, 0 categorical

(c) 3 quantitative, 1 categorical

(d) 2 quantitative, 1 categorical

(e) 2 quantitative, 2 categorical

T1.2. Consumers Union measured the gas mileage in miles per gallon of 38 vehicles from the same model year on a special test track. The pie chart provides information about the country of manufacture of the model cars tested by Consumers Union. Based on the pie chart, we conclude

(a) Japanese cars get significantly lower gas mileage than cars from other countries.

(b) U.S. cars get significantly higher gas mileage than cars from other countries.

(c) Swedish cars get gas mileages that are between those of Japanese and U.S. cars.

(d) Mercedes, Audi, Porsche, and BMW represent approximately a quarter of the cars tested.

(e) More than half of the cars in the study were from the United States.

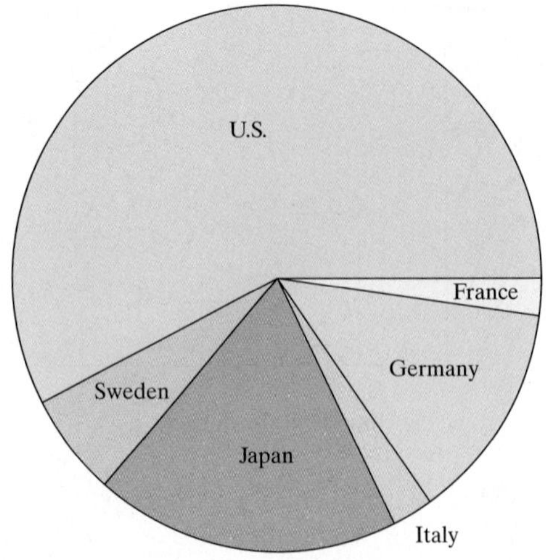

T1.3. Which of the following bar graphs is equivalent to the pie chart in Question T1.2?

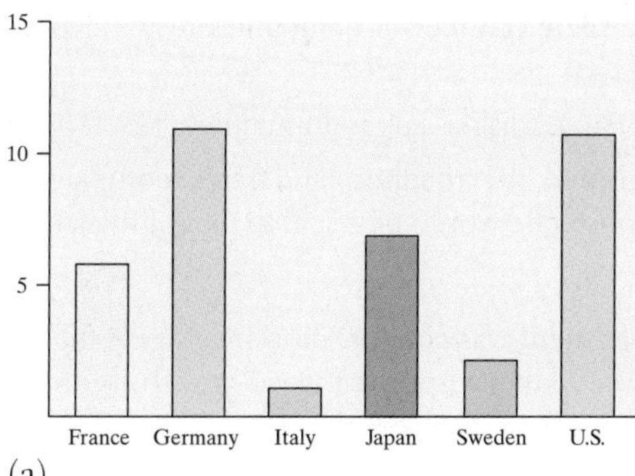

(a)

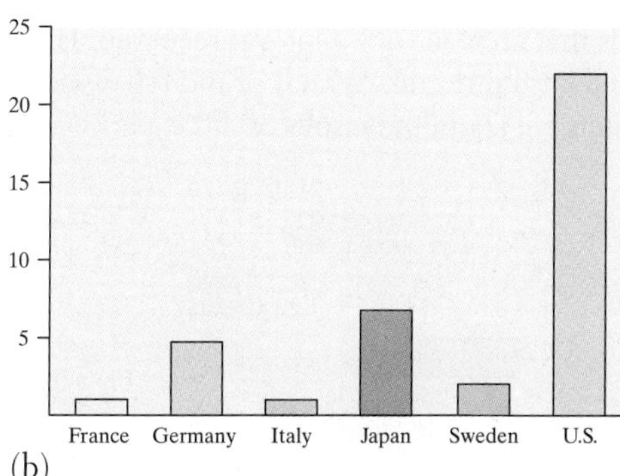

(b)

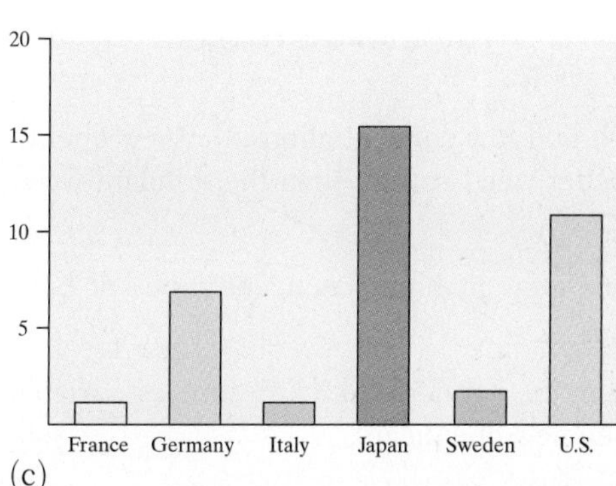

(c)

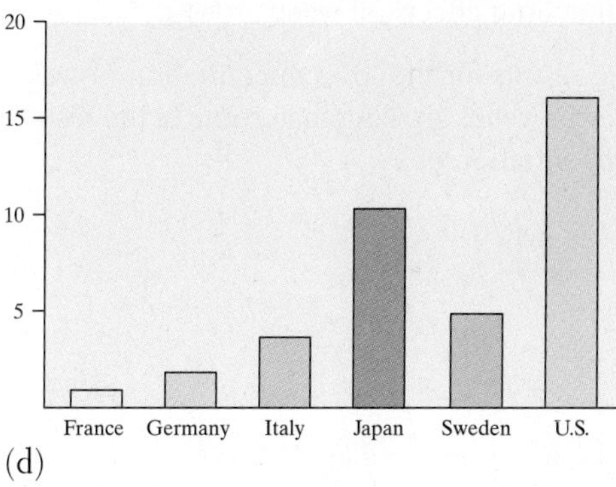

(d)

(e) None of these.

T1.4. Earthquake intensities are measured using a device called a seismograph, which is designed to be most sensitive to earthquakes with intensities between 4.0 and 9.0 on the Richter scale. Measurements of nine earthquakes gave the following readings:

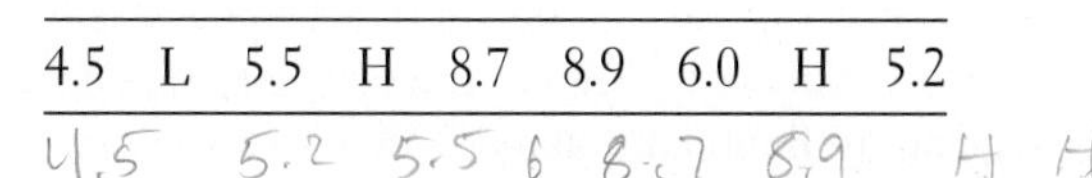

4.5	L	5.5	H	8.7	8.9	6.0	H	5.2

where L indicates that the earthquake had an intensity below 4.0 and an H indicates that the earthquake had an intensity above 9.0. The median earthquake intensity of the sample is

(a) 5.75. (c) 6.47. (e) Cannot be determined.
(b) 6.00. (d) 8.70.

Questions T1.5 and T1.6 refer to the following setting. In a statistics class with 136 students, the professor records how much money (in dollars) each student has in his or her possession during the first class of the semester. The histogram shows the data that were collected.

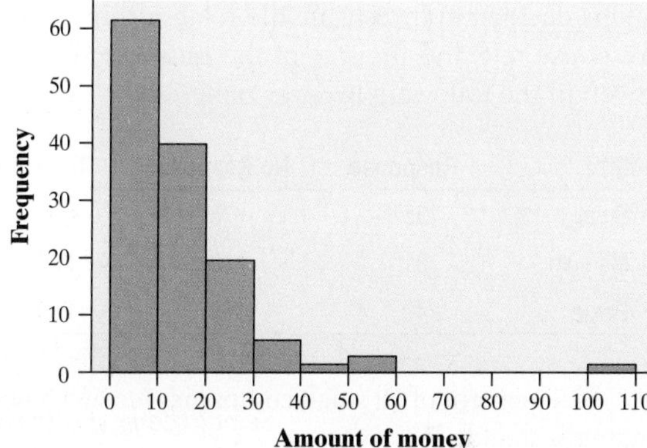

T1.5. The percentage of students with less than $10 in their possession is closest to

(a) 30%. (b) 35%. (c) 50%. (d) 60%. (e) 70%.

T1.6. Which of the following statements about this distribution is *not* correct?

(a) The histogram is right-skewed.

(b) The median is less than $20.

(c) The *IQR* is $35.

(d) The mean is greater than the median.

(e) The histogram is unimodal.

T1.7. Forty students took a statistics examination having a maximum of 50 points. The score distribution is given in the following stem-and-leaf plot:

Stem	Leaves
0	28
1	2245
2	01333358889
3	001356679
4	22444466788
5	000

The third quartile of the score distribution is equal to

(a) 45. (b) 44. (c) 43. (d) 32. (e) 23.

T1.8. The mean salary of all female workers is $35,000. The mean salary of all male workers is $41,000. What must be true about the mean salary of all workers?

(a) It must be $38,000.

(b) It must be larger than the median salary.

(c) It could be any number between $35,000 and $41,000.

(d) It must be larger than $38,000.

(e) It cannot be larger than $40,000.

Questions T1.9 and T1.10 refer to the following setting. A survey was designed to study how business operations vary according to their size. Companies were classified as small, medium, or large. Questionnaires were sent to 200 randomly selected businesses of each size. Since not all questionnaires in a survey of this type are returned, researchers decided to investigate the relationship between the response rate and the size of the business. The data are given in the following two-way table:

Size	Response	No Response	Total
Small	125	75	200
Medium	81	119	200
Large	40	160	200

T1.9. What percent of all small companies receiving questionnaires responded?

(a) 12.5% (b) 20.8% (c) 33.3% (d) 50.8% (e) 62.5%

T1.10. Which of the following conclusions seems to be supported by the data?

(a) There are more small companies than large companies in the survey.

(b) Small companies appear to have a higher response rate than medium or big companies.

(c) Exactly the same number of companies responded as didn't respond.

(d) Small companies dislike larger companies.

(e) If we combined the medium and large companies, then their response rate would be equal to that of the small companies.

T1.11. An experiment was conducted to investigate the effect of a new weed killer to prevent weed growth in onion crops. Two chemicals were used: the standard weed killer (C) and the new chemical (W). Both chemicals were tested at high and low concentrations on a total of 50 test plots. The percent of weeds that grew in each plot was recorded. Here are some boxplots of the results. Which of the following is *not* a correct statement about the results of this experiment?

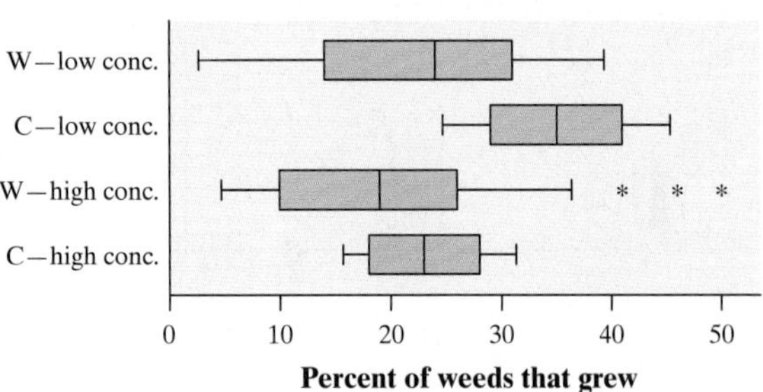

(a) At both high and low concentrations, the new chemical (W) gives better weed control than the standard weed killer (C).

(b) Fewer weeds grew at higher concentrations of both chemicals.

(c) The results for the standard weed killer are less variable than those for the new chemical.

(d) High and low concentrations of either chemical have approximately the same effects on weed growth.

(e) Some of the results for the low concentration of weed killer W show fewer weeds growing than some of the results for the high concentration of W.

Section II: Free Response *Show all your work. Indicate clearly the methods you use, because you will be graded on the correctness of your methods as well as on the accuracy and completeness of your results and explanations.*

T1.12. You are interested in how much time students spend on the Internet each day. Here are data on the time spent on the Internet (in minutes) for a particular day reported by a random sample of 30 students at a large high school:

7	20	24	25	25	28	28	30	32	35
42	43	44	45	46	47	48	48	50	51
72	75	77	78	79	83	87	88	135	151

(a) Construct a histogram of these data.

(b) Are there any outliers? Justify your answer.

(c) Would it be better to use the mean and standard deviation or the median and *IQR* to describe the center and spread of this distribution? Why?

T1.13. A study among the Pima Indians of Arizona investigated the relationship between a mother's diabetic status and the appearance of birth defects in her children. The results appear in the two-way table below.

	Diabetic Status			
Birth Defects	**Nondiabetic**	**Prediabetic**	**Diabetic**	**Total**
None	754	362	38	
One or more	31	13	9	
Total				

(a) Fill in the row and column totals in the margins of the table.

(b) Compute (in percents) the conditional distributions of birth defects for each diabetic status.

(c) Display the conditional distributions in a graph. Don't forget to label your graph completely.

(d) Comment on any clear associations you see.

T1.14. The back-to-back stemplot shows the lifetimes of several Brand X and Brand Y batteries.

Brand X		Brand Y
	1	
	1	7
	2	2
	2	6
2110	3	
99775	3	
3221	4	223334
	4	56889
4	5	0
5	5	

Key: 4|2 represents 420–429 hours.

(a) What is the longest that any battery lasted?

(b) Give a reason someone might prefer a Brand X battery.

(c) Give a reason someone might prefer a Brand Y battery.

T1.15. During the early part of the 1994 baseball season, many fans and players noticed that the number of home runs being hit seemed unusually large. Here are the data on the number of home runs hit by American League and National League teams in the early part of the 1994 season:

American League: 35 40 43 49 51 54 57 58 58 64 68 68 75 77

National League: 29 31 42 46 47 48 48 53 55 55 55 63 63 67

Compare the distributions of home runs for the two leagues graphically and numerically. Write a few sentences summarizing your findings.

Chapter 2

case study

Modeling Distributions of Data

Do You Sudoku?

The sudoku craze has officially swept the globe. Here's what Will Shortz, crossword puzzle editor for the *New York Times*, said about sudoku in August 2005:

> *As humans we seem to have an innate desire to fill up empty spaces. This might explain part of the appeal of sudoku, the new international craze, with its empty squares to be filled with digits. Since April, when sudoku was introduced to the United States in The New York Post, more than half the leading American newspapers have begun printing one or more sudoku a day. No puzzle has had such a fast introduction in newspapers since the crossword craze of 1924–25.*[1]

Since then, millions of people have made sudoku part of their daily routines.

WWW One of the authors played an online game of sudoku at www.websudoku.com. The graph provides information about how well he did. (His time is marked with an arrow.)

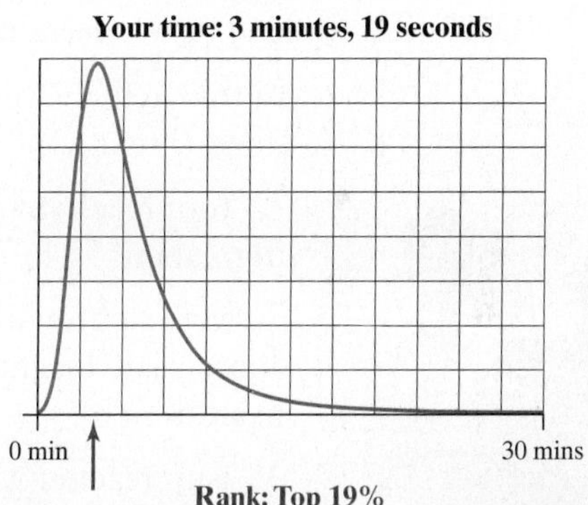

> In this chapter, you'll learn more about how to describe the location of an individual observation—like the author's sudoku time—within a distribution.

Introduction

Suppose Jenny earns an 86 (out of 100) on her next statistics test. Should she be satisfied or disappointed with her performance? That depends on how her score compares with the scores of the other students who took the test. If 86 is the highest score, Jenny might be very pleased. Maybe her teacher will "curve" the grades so that Jenny's 86 becomes an "A." But if Jenny's 86 falls below the "average" in the class, she may not be so happy.

Section 2.1 focuses on describing the location of an individual within a distribution. We begin by discussing a familiar measure of position: *percentiles*. Next, we introduce a new type of graph that is useful for displaying percentiles. Then we consider another way to describe an individual's position that is based on the mean and standard deviation. In the process, we examine the effects of transforming data on the shape, center, and spread of a distribution.

Sometimes it is helpful to use graphical models called *density curves* to describe the location of individuals within a distribution, rather than relying on actual data values. Such models are especially helpful when data fall in a bell-shaped pattern called a *Normal distribution*. Section 2.2 examines the properties of Normal distributions and shows you how to perform useful calculations with them.

ACTIVITY *Where do I stand?*

MATERIALS: Masking tape to mark number line scale

In this Activity, you and your classmates will explore ways to describe where you stand (literally!) within a distribution.

1. Your teacher will mark out a number line on the floor with a scale running from about 58 to 78 inches.

2. Make a human dotplot. Each member of the class should stand at the appropriate location along the number line scale based on height (to the nearest inch).

3. Your teacher will make a copy of the dotplot on the board for your reference.

4. What percent of the students in the class have heights less than yours? This is your *percentile* in the distribution of heights.

5. Work with a partner to calculate the mean and standard deviation of the class's height distribution from the dotplot. Confirm these values with your classmates.

6. Where does your height fall in relation to the mean: above or below? How far above or below the mean is it? How many standard deviations above or below the mean is it? This last number is the *z-score* corresponding to your height.

7. *Class discussion*: What would happen to the class's height distribution if you converted each data value from inches to centimeters? (There are 2.54 centimeters in 1 inch.) How would this change of units affect the measures of center, spread, and location (percentile and *z*-score) that you calculated?

WWW Want to know more about where you stand—in terms of height, weight, or even body mass index? Do a Web search for "Clinical Growth Charts" at the National Center for Health Statistics site, www.cdc.gov/nchs.

Describing Location in a Distribution

In Section 2.1, you'll learn about:

- Measuring position: Percentiles
- Cumulative relative frequency graphs
- Measuring position: z-scores
- Transforming data
- Density curves
- Describing density curves

Here are the scores of all 25 students in Mr. Pryor's statistics class on their first test:

79 81 80 77 73 83 74 93 78 80 75 67 73
77 83 **86** 90 79 85 83 89 84 82 77 72

The bold score is Jenny's **86**. How did she perform on this test relative to her classmates?

The stemplot displays this distribution of test scores. Notice that the distribution is roughly symmetric with no apparent outliers. From the stemplot, we can see that Jenny did better than all but three students in the class.

6	7
7	2334
7	5777899
8	00123334
8	5**6**9
9	03

Measuring Position: Percentiles

One way to describe Jenny's location in the distribution of test scores is to tell what percent of students in the class earned scores that were below Jenny's score. That is, we can calculate Jenny's **percentile.**

DEFINITION: Percentile

The ***p*th percentile** of a distribution is the value with p percent of the observations less than it.

Using the stemplot, we see that Jenny's 86 places her fourth from the top of the class. Since 21 of the 25 observations (84%) are below her score, Jenny is at the 84th percentile in the class's test score distribution.

Mr. Pryor's First Test

Finding percentiles

PROBLEM: Use the scores on Mr. Pryor's first statistics test to find the percentiles for the following students:

(a) Norman, who earned a 72.

(b) Katie, who scored 93.

(c) The two students who earned scores of 80.

SOLUTION:

(a) Only 1 of the 25 scores in the class is below Norman's 72. His percentile is computed as follows: $1/25 = 0.04$, or 4%. So Norman scored at the 4th percentile on this test.

(b) Katie's 93 puts her by definition at the 96th percentile, since 24 out of 25 test scores fall below her result.

(c) Two students scored an 80 on Mr. Pryor's first test. By our definition of percentile, 12 of the 25 scores in the class were less than 80, so these two students are at the 48th percentile.

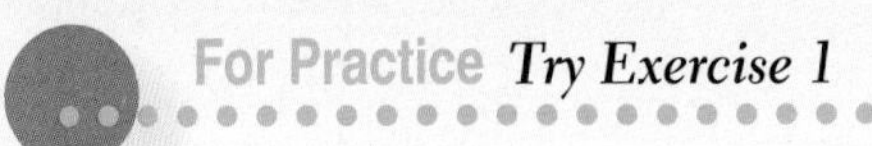

For Practice Try Exercise 1

Note: Some people define the pth percentile of a distribution as the value with p percent of observations *less than or equal to* it. Using this alternative definition of percentile, it is possible for an individual to fall at the 100th percentile. If we used this definition, the two students in part (c) of the example would fall at the 56th percentile (14 of 25 scores were less than or equal to 80). Of course, since 80 is actually the median score, it is also possible to think of it as being the 50th percentile. Calculating percentiles is not an exact science, especially with small data sets! We'll stick with the definition of percentile we gave earlier for consistency.

Cumulative Relative Frequency Graphs

There are some interesting graphs that can be made with percentiles. One of the most common graphs starts with a frequency table for a quantitative variable. For instance, here is a frequency table that summarizes the ages of the first 44 U.S. presidents when they were inaugurated:

Age	Frequency
40–44	2
45–49	7
50–54	13
55–59	12
60–64	7
65–69	3

Let's expand this table to include columns for relative frequency, cumulative frequency, and cumulative relative frequency.

- To get the values in the *relative frequency* column, divide the count in each class by 44, the total number of presidents. Multiply by 100 to convert to a percent.
- To fill in the *cumulative frequency* column, add the counts in the frequency column for the current class and all classes with smaller values of the variable.
- For the *cumulative relative frequency* column, divide the entries in the cumulative frequency column by 44, the total number of individuals. Multiply by 100 to convert to a percent.

Here is the original frequency table with the relative frequency, cumulative frequency, and cumulative relative frequency columns added.

Age	Frequency	Relative frequency	Cumulative frequency	Cumulative relative frequency
40–44	2	2/44 = 0.045, or 4.5%	2	2/44 = 0.045, or 4.5%
45–49	7	7/44 = 0.159, or 15.9%	9	9/44 = 0.205, or 20.5%
50–54	13	13/44 = 0.295, or 29.5%	22	22/44 = 0.500, or 50.0%
55–59	12	12/44 = 0.273, or 27.3%	34	34/44 = 0.773, or 77.3%
60–64	7	7/44 = 0.159, or 15.9%	41	41/44 = 0.932, or 93.2%
65–69	3	3/44 = 0.068, or 6.8%	44	44/44 = 1.000, or 100%

Cumulative relative frequency graph

To make a **cumulative relative frequency graph,** we plot a point corresponding to the cumulative relative frequency in each class at the smallest value of the *next* class. For example, for the 40 to 44 class, we plot a point at a height of 4.5% above the age value of 45. This means that 4.5% of presidents were inaugurated *before* they were 45 years old. (In other words, age 45 is the 4.5th percentile of the inauguration age distribution.) It is customary to start a cumulative relative frequency graph with a point at a height of 0% at the smallest value of the first class (in this case, 40). The last point we plot should be at a height of 100%. We connect consecutive points with a line segment to form the graph. Figure 2.1 shows the completed cumulative relative frequency graph.

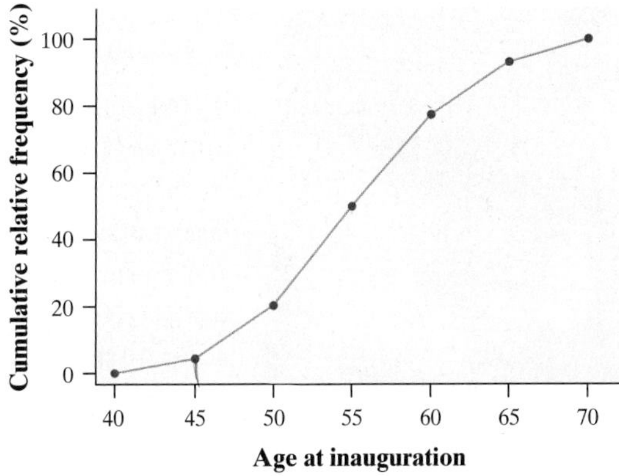

FIGURE 2.1 Cumulative relative frequency graph for the ages of U.S. presidents at inauguration.

Some people refer to cumulative relative frequency graphs as "ogives" (pronounced "o-jives").

Here's an example that shows how to interpret a cumulative relative frequency graph.

EXAMPLE

Age at Inauguration

Interpreting a cumulative relative frequency graph

What can we learn from Figure 2.1? The graph grows very gradually at first because few presidents were inaugurated when they were in their 40s. Then the graph gets very steep beginning at age 50. Why? Because most U.S. presidents were in their 50s when they were inaugurated. The rapid growth in the graph slows at age 60.

Suppose we had started with only the graph in Figure 2.1, without any of the information in our original frequency table. Could we figure out what percent of presidents were between 55 and 59 years old at their inaugurations? Sure. Since the point at age 60 has a cumulative relative frequency of about 77%, we know that about 77% of presidents were inaugurated before they were 60 years old. Similarly, the point at age 55 tells us that about 50% of presidents were younger than 55 at inauguration. As a result, we'd estimate that about 77% − 50% = 27% of U.S. presidents were between 55 and 59 when they were inaugurated.

A cumulative relative frequency graph can be used to describe the position of an individual within a distribution or to locate a specified percentile of the distribution, as the following example illustrates.

Ages of U.S. Presidents

Interpreting cumulative relative frequency graphs

PROBLEM: Use the graph in Figure 2.1 to help you answer each question.

(a) Was Barack Obama, who was inaugurated at age 47, unusually young?

(b) Estimate and interpret the 65th percentile of the distribution.

SOLUTION:

(a) To find President Obama's location in the distribution, we draw a vertical line up from his age (47) on the horizontal axis until it meets the graphed line. Then we draw a horizontal line from this point of intersection to the vertical axis. Based on Figure 2.2(a), we would estimate that Barack Obama's inauguration age places him at the 11% cumulative relative frequency mark. That is, he's at the 11th percentile of the distribution. In other words, about 11% of all U.S. presidents were younger than Barack Obama when they were inaugurated. Put another way, President Obama was inaugurated at a younger age than about 89% of all U.S. presidents.

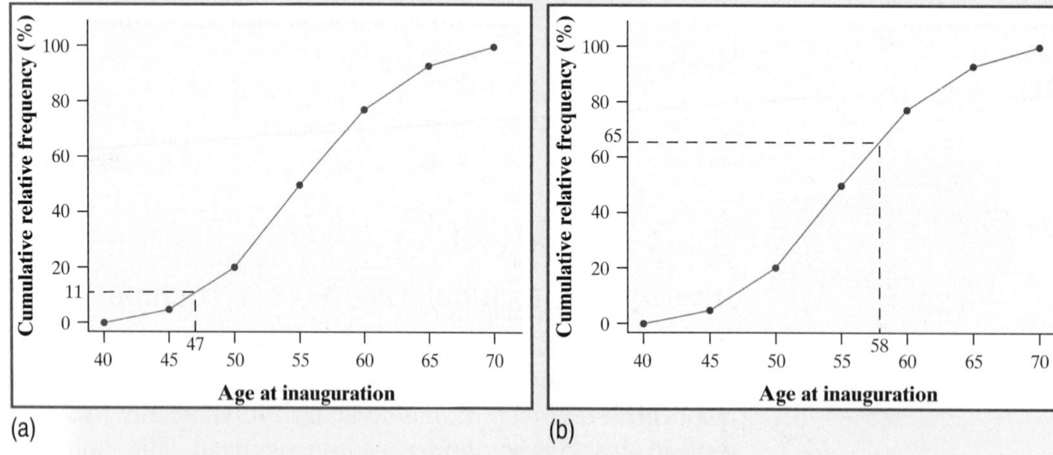

FIGURE 2.2 The cumulative relative frequency graph of presidents' ages at inauguration is used to (a) locate President Obama within the distribution and (b) determine the 65th percentile, which is about 58 years.

(b) The 65th percentile of the distribution is the age with cumulative relative frequency 65%. To find this value, draw a horizontal line across from the vertical axis at a height of 65% until it meets the graphed line. From the point of intersection, draw a vertical line down to the horizontal axis. In Figure 2.2(b), the value on the horizontal axis is about 58. So about 65% of all U.S. presidents were younger than 58 when they took office.

THINK ABOUT IT

Percentiles and quartiles Have you made the connection between percentiles and the quartiles from Chapter 1? Earlier, we noted that the median (second quartile) corresponds to the 50th percentile. What about the first quartile, Q_1? It's at the median of the lower half of the ordered data, which puts it about one-fourth of the way through the distribution. In other words, Q_1 is roughly the 25th percentile. By similar reasoning, Q_3 is approximately the 75th percentile of the distribution.

CHECK YOUR UNDERSTANDING

1. *Multiple choice: Select the best answer.* Mark receives a score report detailing his performance on a statewide test. On the math section, Mark earned a raw score of 39, which placed him at the 68th percentile. This means that

(a) Mark did better than about 39% of the students who took the test.
(b) Mark did worse than about 39% of the students who took the test.
(c) Mark did better than about 68% of the students who took the test.
(d) Mark did worse than about 68% of the students who took the test.
(e) Mark got fewer than half of the questions correct on this test.

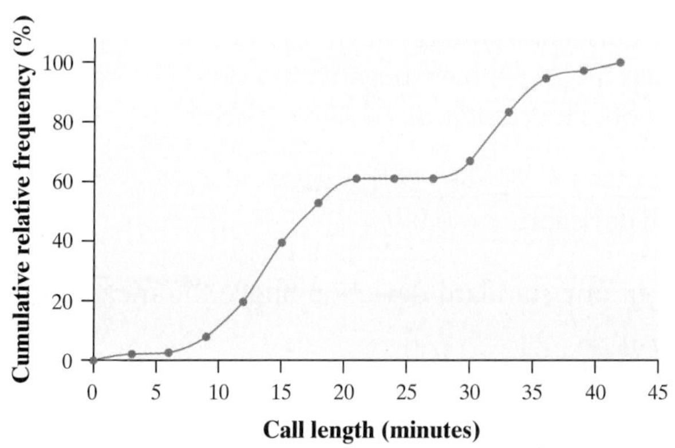

2. Mrs. Munson is concerned about how her daughter's height and weight compare with those of other girls of the same age. She uses an online calculator to determine that her daughter is at the 87th percentile for weight and the 67th percentile for height. Explain to Mrs. Munson what this means.

Questions 3 and 4 relate to the following setting. The graph displays the cumulative relative frequency of the lengths of phone calls made from the mathematics department office at Gabalot High last month.

3. About what percent of calls lasted less than 30 minutes? 30 minutes or more?

4. Estimate Q_1, Q_3, and the IQR of the distribution.

Measuring Position: *z*-Scores

```
6 | 7
7 | 2334
7 | 5777899
8 | 00123334
8 | 569
9 | 03
```

Let's return to the data from Mr. Pryor's first statistics test, which are shown in the stemplot. Figure 2.3 provides numerical summaries from Minitab for these data. Where does Jenny's score of 86 fall relative to the mean of this distribution? Since the mean score for the class is 80, we can see that Jenny's score is "above average." But how much above average is it?

The relationship between the mean and the median is about what you'd expect in this fairly symmetric distribution.

Minitab

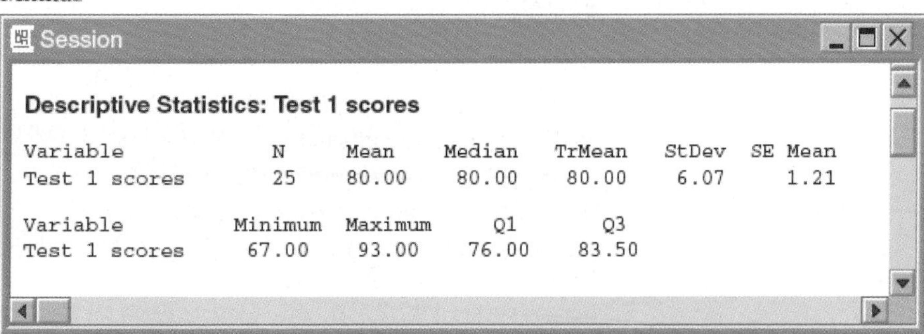

Session

Descriptive Statistics: Test 1 scores

Variable	N	Mean	Median	TrMean	StDev	SE Mean
Test 1 scores	25	80.00	80.00	80.00	6.07	1.21

Variable	Minimum	Maximum	Q1	Q3
Test 1 scores	67.00	93.00	76.00	83.50

FIGURE 2.3 Minitab output for the scores of Mr. Pryor's students on their first statistics test.

We can describe Jenny's location in the class's test score distribution by telling how many standard deviations above or below the mean her score is. Since the mean is 80 and the standard deviation is about 6, Jenny's score of 86 is about one standard deviation above the mean. Converting observations like this from original values to standard deviation units is known as **standardizing.** To standardize a value, subtract the mean of the distribution and then divide by the standard deviation.

DEFINITION: Standardized value (*z*-score)

If *x* is an observation from a distribution that has known mean and standard deviation, the **standardized value** of *x* is

$$z = \frac{x - \text{mean}}{\text{standard deviation}}$$

A standardized value is often called a ***z*-score.**

A *z*-score tells us how many standard deviations from the mean an observation falls, and in what direction. Observations larger than the mean have positive *z*-scores; observations smaller than the mean have negative *z*-scores. For example, Jenny's score on the test was $x = 86$. Her *standardized* score (*z*-score) is

$$z = \frac{x - \text{mean}}{\text{standard deviation}} = \frac{86 - 80}{6.07} = 0.99$$

That is, Jenny's test score is about one standard deviation above the mean score of the class.

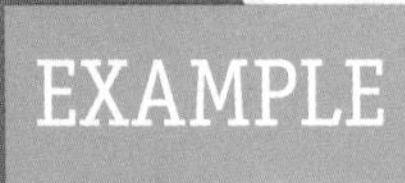

Mr. Pryor's First Test, Again

Finding and interpreting *z*-scores

PROBLEM: **Use the information in Figure 2.3 to find the standardized scores (*z*-scores) for each of the following students in Mr. Pryor's class. Interpret each value in context.**

(a) Katie, who scored 93.

(b) Norman, who earned a 72.

SOLUTION:

(a) Katie's 93 was the highest score in the class. Her corresponding *z*-score is

$$z = \frac{93 - 80}{6.07} = 2.14$$

In other words, Katie's result is 2.14 standard deviations above the mean score for this test.

(b) For Norman's 72, his standardized score is

$$z = \frac{72 - 80}{6.07} = -1.32$$

Norman's score is 1.32 standard deviations below the class mean of 80.

For Practice *Try Exercise 15(b)*

We can also use z-scores to compare the position of individuals in different distributions, as the following example illustrates.

Jenny Takes Another Test

Using z-scores for comparisons

The day after receiving her statistics test result of 86 from Mr. Pryor, Jenny earned an 82 on Mr. Goldstone's chemistry test. At first, she was disappointed. Then Mr. Goldstone told the class that the distribution of scores was fairly symmetric with a mean of 76 and a standard deviation of 4.

PROBLEM: **On which test did Jenny perform better relative to the class? Justify your answer.**

SOLUTION: **Jenny's z-score for her chemistry test result is**

$$z = \frac{82 - 76}{4} = 1.50$$

Her 82 in chemistry was 1.5 standard deviations above the mean score for the class. Since she scored only 1 standard deviation above the mean on the statistics test, Jenny actually did better relative to the class in chemistry.

We often standardize observations to express them on a common scale. We might, for example, compare the heights of two children of different ages by calculating their z-scores. At age 2, Jordan is 89 centimeters (cm) tall. Her height puts her at a z-score of 0.5; that is, she is one-half standard deviation above the mean height of 2-year-old girls. Zayne's height at age 3 is 101 cm, which yields a z-score of 1. In other words, he is one standard deviation above the mean height of 3-year-old boys. So Zayne is taller relative to boys his age than Jordan is relative to girls her age. The standardized heights tell us where each child stands (pun intended!) in the distribution for his or her age group.

CHECK YOUR UNDERSTANDING

Mrs. Navard's statistics class has just completed the first three steps of the "Where Do I Stand?" Activity (page 84). The figure below shows a dotplot of the class's height distribution, along with summary statistics from computer output.

1. Lynette, a student in the class, is 65 inches tall. Find and interpret her z-score.

2. Another student in the class, Brent, is 74 inches tall. How tall is Brent compared with the rest of the class? Give appropriate numerical evidence to support your answer.

3. Brent is a member of the school's basketball team. The mean height of the players on the team is 76 inches. Brent's height translates to a z-score of −0.85 in the team's height distribution. What is the standard deviation of the team members' heights?

60 62 64 66 68 70 72 74

Height (inches)

Variable	n	$\bar{x}$	s_x	Min	Q_1	M	Q_3	Max
Height	25	67	4.29	60	63	66	69	75

Transforming Data

To find the standardized score (z-score) for an individual observation, we transform this data value by subtracting the mean and dividing by the standard deviation. Transforming converts the observation from the original units of measurement (inches, for example) to a standardized scale. What effect do these kinds of transformations—adding or subtracting; multiplying or dividing—have on the shape, center, and spread of the entire distribution? Let's investigate using an interesting data set from "down under."

Soon after the metric system was introduced in Australia, a group of students was asked to guess the width of their classroom to the nearest meter. Here are their guesses in order from lowest to highest:[2]

8	9	10	10	10	10	10	10	11	11	11	11	12
12	13	13	13	14	14	14	15	15	15	15	15	15
15	15	16	16	16	17	17	17	17	18	18	20	22
25	27	35	38	40								

Figure 2.4 includes a dotplot of the data and some numerical summaries.

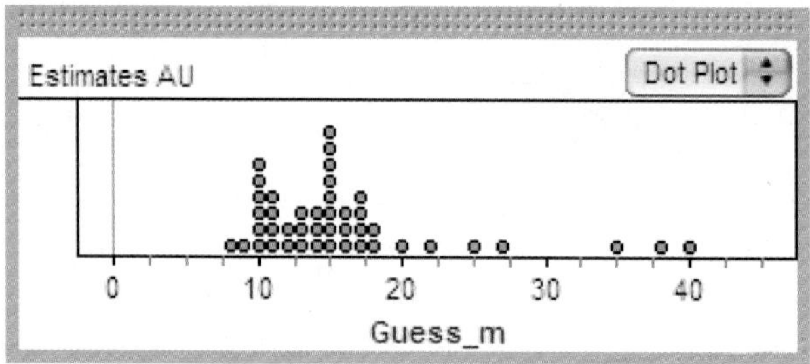

	n	$\bar{x}$	s_x	Min	Q_1	M	Q_3	Max	IQR	Range
Guess	44	16.02	7.14	8	11	15	17	40	6	32

FIGURE 2.4 Fathom dotplot and summary statistics for Australian students' guesses of the classroom width.

Let's practice what we learned in Chapter 1 and describe what we see.

Shape: The distribution of guesses appears bimodal, with peaks at 10 and 15 meters. It is also skewed to the right.

Center: The median guess was 15 meters and the mean guess was about 16 meters. Due to the clear skewness and potential outliers, the median is a better choice for summarizing the "typical" guess.

Spread: Since $Q_1 = 11$, about 25% of the students estimated the width of the room at 11 meters or less. The 75th percentile of the distribution is at about $Q_3 = 17$. The IQR of 6 meters describes the spread of the middle 50% of students' guesses. The standard deviation tells us that the average distance of students' guesses from the mean was about 7 meters. Since s_x is not resistant to extreme values, we prefer the five-number summary and IQR to describe the variability of this distribution.

Outliers: By the $1.5 \times IQR$ rule, values greater than $17 + 9 = 26$ meters or less than $11 - 9 = 2$ meters are identified as outliers. So the four highest guesses—27, 35, 38, and 40 meters—are outliers.

Effect of adding or subtracting a constant By now, you're probably wondering what the actual width of the room was. In fact, it was 13 meters wide. How close were students' guesses? The student who guessed 8 meters was too low by 5 meters. The student who guessed 40 meters was too high by 27 meters (and probably needs to study the metric system more carefully). We can examine the distribution of students' guessing errors by defining a new variable as follows:

$$\text{error} = \text{guess} - 13$$

That is, we'll subtract 13 from each observation in the data set. Try to predict what the shape, center, and spread of this new distribution will be. Refer to Figure 2.4 above as needed.

Estimating Room Width

Effect of subtracting a constant

Let's see how accurate your predictions were (you did make predictions, right?). Figure 2.5 shows dotplots of students' original guesses and their errors on the same scale. We can see that the original distribution of guesses has been shifted to the left. By how much? Since the peak at 15 meters in the original graph is located at 2 meters in the error distribution, the original data values have been translated 13 units to the left. That should make sense: we calculated the errors by subtracting the actual room width, 13 meters, from each student's guess.

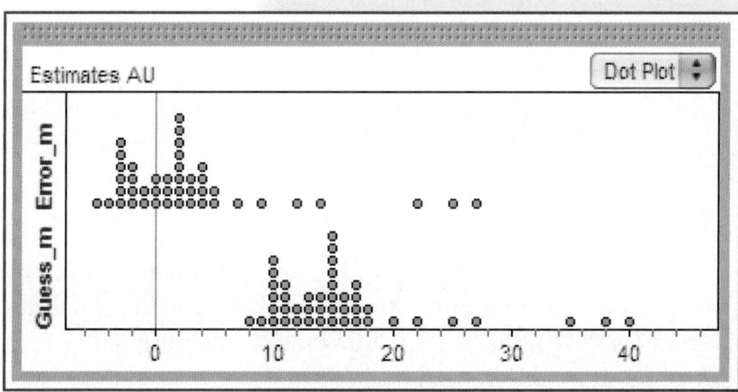

FIGURE 2.5 Fathom dotplots of students' original guesses of classroom width and the errors in their guesses.

From Figure 2.5, it seems clear that subtracting 13 from each observation did not affect the shape or spread of the distribution. But this transformation appears to have decreased the center of the distribution by 13 meters. The summary statistics in the table below confirm our beliefs.

	n	$\bar{x}$	s_x	Min	Q_1	M	Q_3	Max	IQR	Range
Guess (m)	44	16.02	7.14	8	11	15	17	40	6	32
Error (m)	44	3.02	7.14	−5	−2	2	4	27	6	32

The error distribution is centered at a value that is clearly positive—the median error is 2 meters and the mean error is 3 meters. So the students generally tended to overestimate the width of the room.

Let's summarize what we've learned so far about transforming data.

Effect of Adding (or Subtracting) a Constant

Adding the same number *a* (either positive, zero, or negative) to each observation

- adds *a* to measures of center and location (mean, median, quartiles, percentiles), but
- does not change the shape of the distribution or measures of spread (range, *IQR*, standard deviation).

Effect of multiplying or dividing by a constant Since our group of Australian students is having some difficulty with the metric system, it may not be very helpful to tell them that their guesses tended to be about 2 to 3 meters too high. Let's convert the error data to feet before we report back to them. There are roughly 3.28 feet in a meter. So for the student whose error was −5 meters, that translates to

$$-5\text{ meters} \times \frac{3.28\text{ feet}}{1\text{ meter}} = -16.4\text{ feet}$$

To change the units of measurement from meters to feet, we need to multiply each of the error values by 3.28. What effect will this have on the shape, center, and spread of the distribution? (Go ahead, make some predictions!)

Estimating Room Width

Effect of multiplying by a constant

Figure 2.6 includes dotplots of the students' guessing errors in meters and feet, along with summary statistics from computer software. The shape of the two distributions is the same—bimodal and right-skewed. However, the centers and spreads of the two distributions are quite different. The bottom dotplot is centered at a value that is to the right of the top dotplot's center. In addition, the bottom dotplot shows much greater spread than the top dotplot.

When the errors were measured in meters, the median was 2 and the mean was 3.02. For the transformed error data in feet, the median is 6.56 and the mean is 9.91. Can you see that the measures of center were multiplied by 3.28? That makes sense—if we multiply all of the observations by 3.28, then the mean and median should also be multiplied by 3.28.

What about the spread? Multiplying each observation by 3.28 increases the variability of the distribution. By how much? You guessed it—by a factor of 3.28. The numerical summaries in Figure 2.6 show that the standard deviation s_x, the interquartile range, and the range have been multiplied by 3.28.

We can safely tell our group of Australian students that their estimates of the classroom's width tended to be too high by about 6.5 feet. (Notice that we choose not to report the mean error, which is affected by the strong skewness and the three high outliers.)

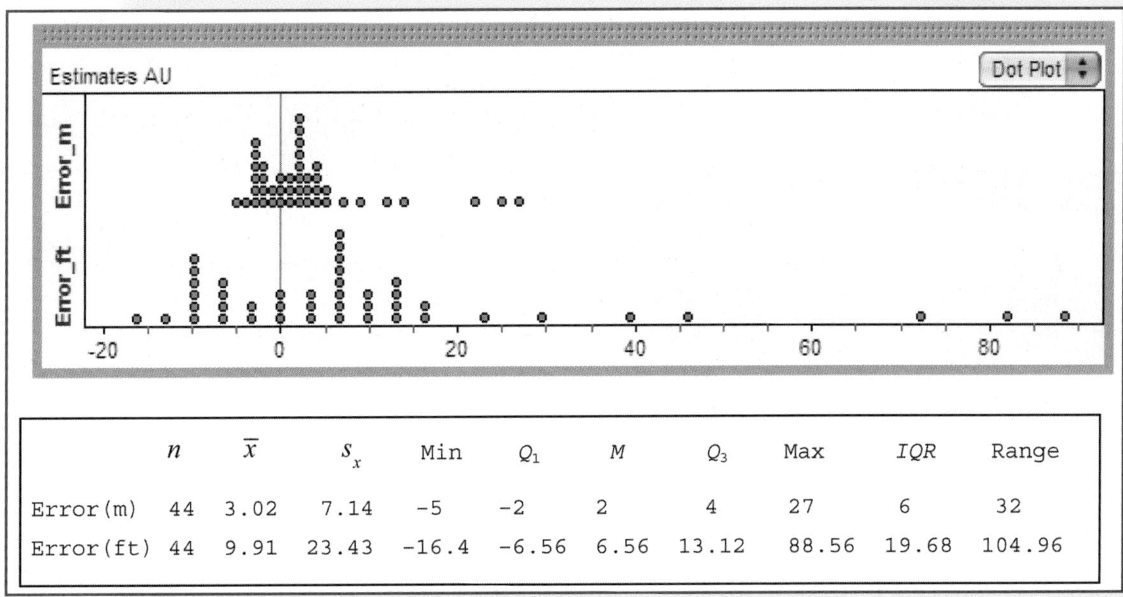

	n	$\bar{x}$	s_x	Min	Q_1	M	Q_3	Max	IQR	Range
Error(m)	44	3.02	7.14	-5	-2	2	4	27	6	32
Error(ft)	44	9.91	23.43	-16.4	-6.56	6.56	13.12	88.56	19.68	104.96

FIGURE 2.6 Fathom dotplots and numerical summaries of students' errors guessing the width of their classroom in meters and feet.

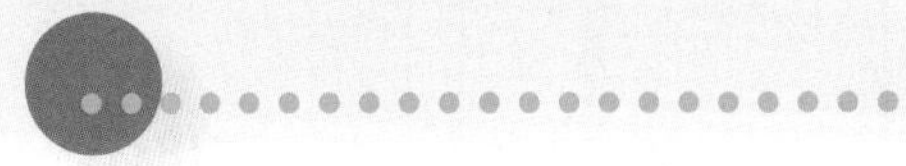

As before, let's recap what we discovered about the effects of transforming data.

Effect of Multiplying (or Dividing) by a Constant

Multiplying (or dividing) each observation by the same number b (positive, negative, or zero)

- multiplies (divides) measures of center and location (mean, median, quartiles, percentiles) by b,
- multiplies (divides) measures of spread (range, *IQR*, standard deviation) by $|b|$, but
- does not change the shape of the distribution.

Note that multiplying all the values in a data set by a negative number multiplies the measures of spread by the *absolute value* of that number. We can't have a negative amount of variability!

Putting it all together: Adding/subtracting and multiplying/dividing What happens if we transform a data set by both adding or subtracting a constant and multiplying or dividing by a constant? For instance, if we need to convert temperature data from Celsius to Fahrenheit, we have to use the formula °F = 9/5(°C) + 32. That is, we would multiply each of the observations by 9/5 and then add 32. As the following example shows, we just use the facts about transforming data that we've already established.

Too Cool at the Cabin?

Analyzing the effects of transformations

During the winter months, the temperatures at the Starnes's Colorado cabin can stay well below freezing (32°F or 0°C) for weeks at a time. To prevent the pipes from freezing, Mrs. Starnes sets the thermostat at 50°F. She also buys a digital thermometer that records the indoor temperature each night at midnight. Unfortunately, the thermometer is programmed to measure the temperature in degrees Celsius. A dotplot and numerical summaries of the midnight temperature readings for a 30-day period are shown below.

PROBLEM: Use the fact that °F = (9/5)°C + 32 to help you answer the following questions.

(a) Find the mean temperature in degrees Fahrenheit. Does the thermostat setting seem accurate?

(b) Calculate the standard deviation of the temperature readings in degrees Fahrenheit. Interpret this value in context.

(c) The 90th percentile of the temperature readings was 11°C. What is the 90th percentile temperature in degrees Fahrenheit?

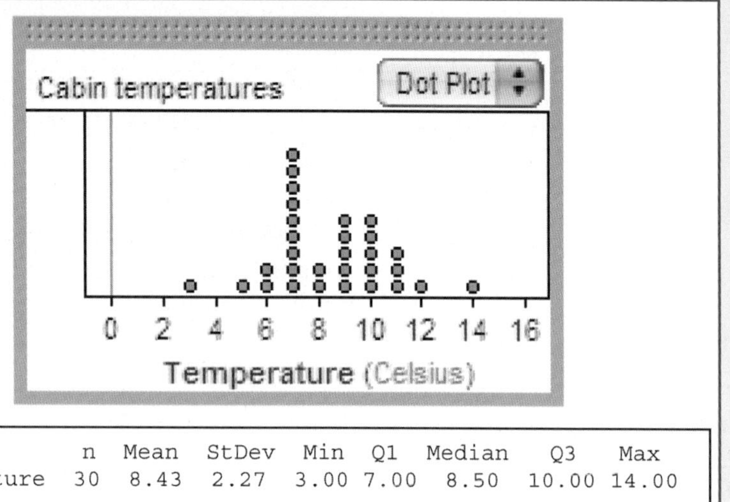

	n	Mean	StDev	Min	Q1	Median	Q3	Max
Temperature	30	8.43	2.27	3.00	7.00	8.50	10.00	14.00

SOLUTION:

(a) To convert the temperature measurements from Celsius to Fahrenheit, we multiply each value by 9/5 and then add 32. Multiplying the observations by 9/5 also multiplies the mean by 9/5. Adding 32 to each observation increases the mean by 32. So the mean temperature in degrees Fahrenheit is (9/5)(8.43) + 32 = 47.17°F. The thermostat doesn't seem to be very accurate. It is set at 50°F, but the mean temperature over the 30-day period is about 47°F.

(b) Multiplying each observation by 9/5 multiplies the standard deviation by 9/5. However, adding 32 to each observation doesn't affect the spread. So the standard deviation of the temperature measurements in degrees Fahrenheit is (9/5)(2.27) = 4.09°F. This means that the average distance of the temperature readings from the mean is about 4°F. That's a lot of variation!

(c) Both multiplying by a constant and adding a constant affect the value of the 90th percentile. To find the 90th percentile in degrees Fahrenheit, we need to multiply the 90th percentile in degrees Celsius by 9/5 and then add 32: (9/5)(11) + 32 = 51.8°F.

As part (c) of the previous example suggests, adding (or subtracting) a constant does not change an individual data value's location within a distribution. Neither does multiplying or dividing by a constant. Consider this: if you're at the 90th percentile for height in inches, then you're also at the 90th percentile for height in meters.

Connecting transformations and z-scores What does all this transformation business have to do with z-scores? To standardize an observation, you subtract the mean of the distribution and then divide by the standard deviation. What if we standardized *every* observation in a distribution?

Returning to Mr. Pryor's statistics test scores, we recall that the distribution was roughly symmetric with a mean of 80 and a standard deviation of 6.07. To convert the entire class's test results to z-scores, we would subtract 80 from each observation and then divide by 6.07. What effect would these transformations have on the shape, center, and spread of the distribution?

Mr. Pryor's test scores Dot Plot

65 70 75 80 85 90 95

Test_scores

Mr. Pryor's test scores Dot Plot

-2 -1 0 1 2

zscores

- **Shape:** The shape of the distribution of z-scores would be the same as the shape of the original distribution of test scores. Neither subtracting a constant nor dividing by a constant would change the shape of the graph. The dotplots confirm that the combination of these two transformations does not affect the shape.
- **Center:** Subtracting 80 from each data value would also reduce the mean by 80. Since the mean of the original distribution was 80, the mean of the transformed data would be 0. Dividing each of these new data values by 6.07 would also divide the mean by 6.07. But since the mean is now 0, dividing by 6.07 would leave the mean at 0. That is, the mean of the z-score distribution would be 0.
- **Spread:** The spread of the distribution would not be affected by subtracting 80 from each observation. However, dividing each data value by 6.07 would also divide our common measures of spread by 6.07. The standard deviation of the distribution of z-scores would therefore be $6.07/6.07 = 1$.

The Minitab computer output below confirms the result: *If we standardize every observation in a distribution, the resulting set of z-scores has mean 0 and standard deviation 1.*

Descriptive Statistics: Test scores, z-scores

Variable	N	Mean	StDev	Minimum	Q1	Median	Q3	Maximum
Test scores	25	80.00	6.07	67.00	76.00	80.00	83.50	93.00
z-scores	25	0.00	1.00	-2.14	-0.66	0.00	0.58	2.14

There are many other types of transformations that can be very useful in analyzing data. We have only studied what happens when you transform data through addition, subtraction, multiplication, or division.

CHECK YOUR UNDERSTANDING

The figure below shows a dotplot of the height distribution for Mrs. Navard's class, along with summary statistics from computer output.

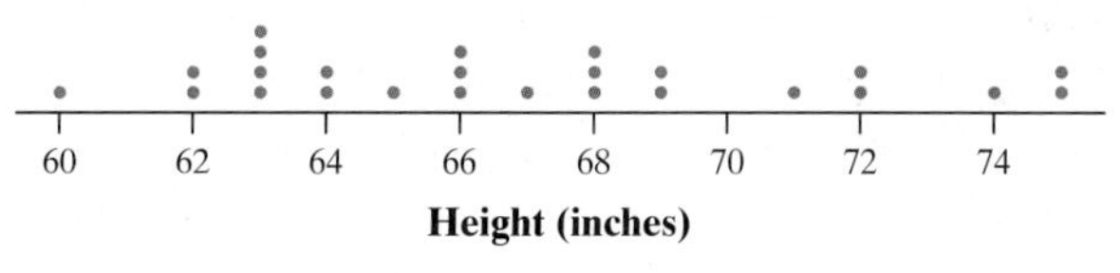

Variable	n	$\bar{x}$	s_x	Min	Q_1	M	Q_3	Max
Height	25	67	4.29	60	63	66	69	75

1. Suppose that you convert the class's heights from inches to centimeters (1 inch = 2.54 cm). Describe the effect this will have on the shape, center, and spread of the distribution.

2. If Mrs. Navard had the entire class stand on a 6-inch-high platform and then had the students measure the distance from

the top of their heads to the ground, how would the shape, center, and spread of this distribution compare with the original height distribution?

3. Now suppose that you convert the class's heights to z-scores. What would be the shape, center, and spread of this distribution? Explain.

DATA EXPLORATION *The speed of light*

Light travels fast, but it is not transmitted instantly. Light takes over a second to reach us from the moon and over 10 billion years to reach us from the most distant objects in the universe. Because radio waves and radar also travel at the speed of light, having an accurate value for that speed is important in communicating with astronauts and orbiting satellites. An accurate value for the speed of light is also important to computer designers because electrical signals travel at light speed. The first reasonably accurate measurements of the speed of light were made more than a hundred years ago by A. A. Michelson and Simon Newcomb. The table below contains 66 measurements made by Newcomb between July and September 1882.[3]

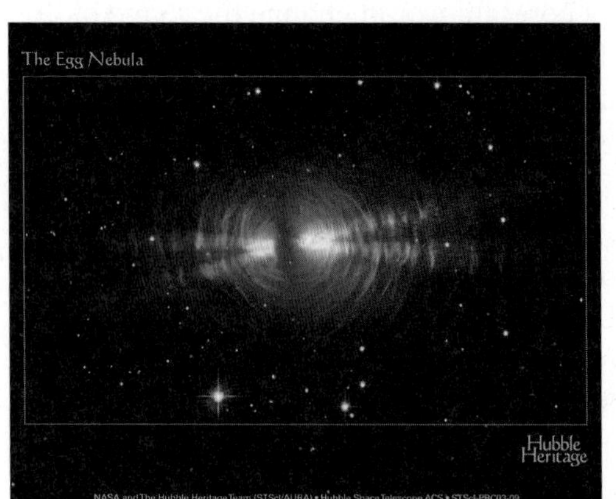

Newcomb measured the time in seconds that a light signal took to pass from his laboratory on the Potomac River to a mirror at the base of the Washington Monument and back, a total distance of about 7400 meters. Newcomb's first measurement of the passage time of light was 0.000024828 second, or 24,828 nanoseconds. (There are 10^9 nanoseconds in a second.) The entries in the table record only the deviation from 24,800 nanoseconds.

28	26	33	24	34	−44	27	16	40	−2	29	22	24	21
25	30	23	29	31	19	24	20	36	32	36	28	25	21
28	29	37	25	28	26	30	32	36	26	30	22	36	23
27	27	28	27	31	27	26	33	26	32	32	24	39	28
24	25	32	25	29	27	28	29	16	23				

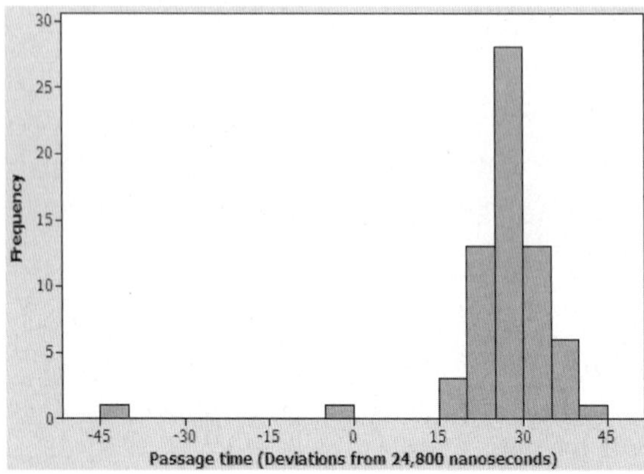

Descriptive Statistics: Passage time

Variable	N	Mean	StDev	Min	Q1	Med	Q3	Max
P. Time	66	26.21	10.75	-44	24	27	31	40
P. Time*	64	27.75	5.08	16	24.5	27	31	40

The figure provides a histogram and numerical summaries (computed with and without the two outliers) from Minitab for these data.

1. We could convert the passage time measurements to nanoseconds by adding 24,800 to each of the data values in the table. What effect would this have on the shape, center, and spread of the distribution? Be specific.

2. After performing the transformation to nanoseconds, we could convert the measurements from nanoseconds to seconds by dividing each value by 10^9. What effect would this have on the shape, center, and spread of the distribution? Be specific.

3. Use the information provided to estimate the speed of light in meters per second. Be prepared to justify the method you used.

Density Curves

In Chapter 1, we developed a kit of graphical and numerical tools for describing distributions. Our work gave us a clear strategy for exploring data from a single quantitative variable.

Exploring Quantitative Data

1. Always plot your data: make a graph, usually a dotplot, stemplot, or histogram.
2. Look for the overall pattern (shape, center, spread) and for striking departures such as outliers.
3. Calculate a numerical summary to briefly describe center and spread.

In this section, we add one more step to this strategy.

4. Sometimes the overall pattern of a large number of observations is so regular that we can describe it by a smooth curve.

Figure 2.7 is a histogram of the scores of all 947 seventh-grade students in Gary, Indiana, on the vocabulary part of the Iowa Test of Basic Skills (ITBS).[4] Scores on this national test have a very regular distribution. The histogram is symmetric, and both tails fall off smoothly from a single center peak. There are no large gaps or obvious outliers. The smooth curve drawn through the tops of the histogram bars in Figure 2.7 is a good description of the overall pattern of the data.

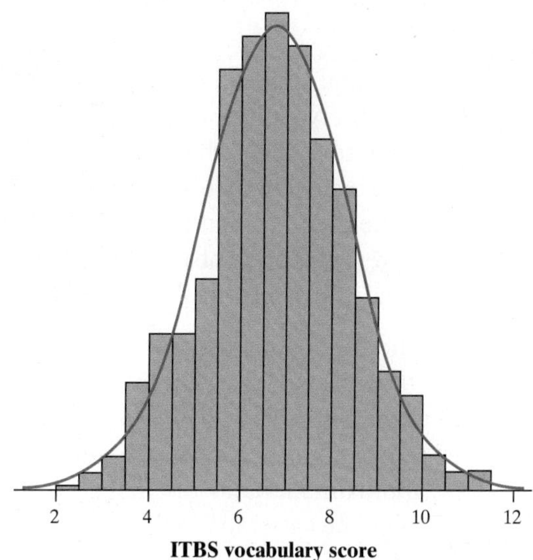

FIGURE 2.7 Histogram of the Iowa Test of Basic Skills (ITBS) vocabulary scores of all seventh-grade students in Gary, Indiana. The smooth curve shows the overall shape of the distribution.

EXAMPLE Seventh-Grade Vocabulary Scores

From histogram to density curve

Our eyes respond to the areas of the bars in a histogram. The bar areas represent relative frequencies (proportions) of the observations. Figure 2.8(a) is a copy of Figure 2.7 with the leftmost bars shaded. The area of the shaded bars in Figure 2.8(a) represents the proportion of students with vocabulary scores less than 6.0. There are 287 such students, who make up the proportion 287/947 = 0.303 of all Gary seventh-graders. In other words, a score of 6.0 corresponds to about the 30th percentile.

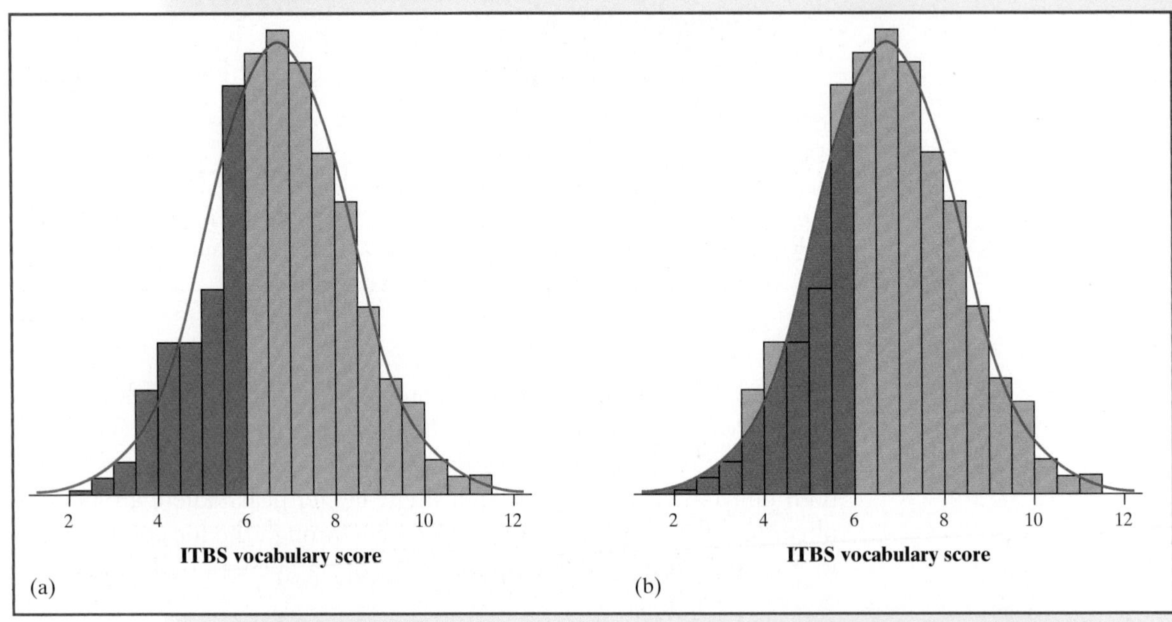

FIGURE 2.8 (a) The proportion of scores less than or equal to 6.0 in the actual data is 0.303. (b) The proportion of scores less than or equal to 6.0 from the density curve is 0.293.

The total area of the bars in the histogram is 100% (a proportion of 1), since all of the observations are represented. Now look at the curve drawn through the bars. In Figure 2.8(b), the area under the curve to the left of 6.0 is shaded. In moving from histogram bars to a smooth curve, we make a specific choice: adjust the scale of the graph so that *the total area under the curve is exactly 1*. Now the total area represents all the observations, just like with the histogram. We can then interpret areas under the curve as proportions of the observations.

The shaded area under the curve in Figure 2.8(b) represents the proportion of students with scores lower than 6.0. This area is 0.293, only 0.010 away from the actual proportion 0.303. So our estimate based on the curve is that a score of 6.0 falls at about the 29th percentile. You can see that areas under the curve give good approximations to the actual distribution of the 947 test scores. In practice, it might be easier to use this curve to estimate relative frequencies than to determine the actual proportion of students by counting data values.

A curve like the one in the previous example is called a **density curve.**

DEFINITION: Density curve

A **density curve** is a curve that

- is always on or above the horizontal axis, and
- has area exactly 1 underneath it.

A density curve describes the overall pattern of a distribution. The area under the curve and above any interval of values on the horizontal axis is the proportion of all observations that fall in that interval.

Density curves, like distributions, come in many shapes. A density curve is often a good description of the overall pattern of a distribution. Outliers, which are departures from the overall pattern, are not described by the curve. *No set of real data is exactly described by a density curve. The curve is an approximation that is easy to use and accurate enough for practical use.* CAUTION

Describing Density Curves

Median of a density curve

Our measures of center and spread apply to density curves as well as to actual sets of observations. Areas under a density curve represent proportions of the total number of observations. The median of a data set is the point with half the observations on either side. So the **median of a density curve** is the "equal-areas point," the point with half the area under the curve to its left and the remaining half of the area to its right.

Because density curves are idealized patterns, a symmetric density curve is exactly symmetric. The median of a symmetric density curve is therefore at its center. Figure 2.9(a) shows a symmetric density curve with the median marked. It isn't so easy to spot the equal-areas point on a skewed curve. There are mathematical ways of finding the median for any density curve. That's how we marked the median on the skewed curve in Figure 2.9(b).

What about the mean? The mean of a set of observations is their arithmetic average. As we saw in Chapter 1, the mean is also the "balance point" of a distribution. That is, if we think of the observations as weights strung out along a thin rod, the mean is the point at which the rod would balance. This fact is also true of density curves. The **mean of a density curve** is the point at which the curve would balance if made of solid material. Figure 2.10 illustrates this fact about the mean.

Mean of a density curve

A symmetric curve balances at its center because the two sides are identical. *The mean and median of a symmetric density curve are equal,* as in Figure 2.9(a).

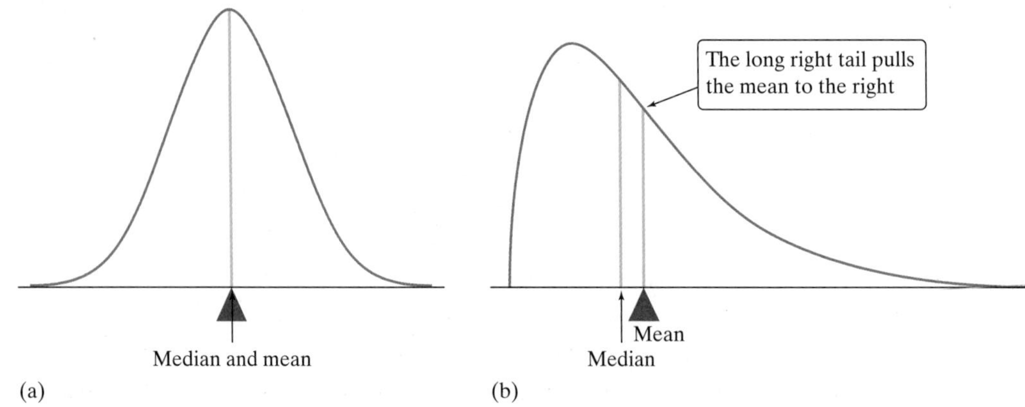

FIGURE 2.9 (a) The median and mean of a symmetric density curve both lie at the center of symmetry. (b) The median and mean of a right-skewed density curve. The mean is pulled away from the median toward the long tail.

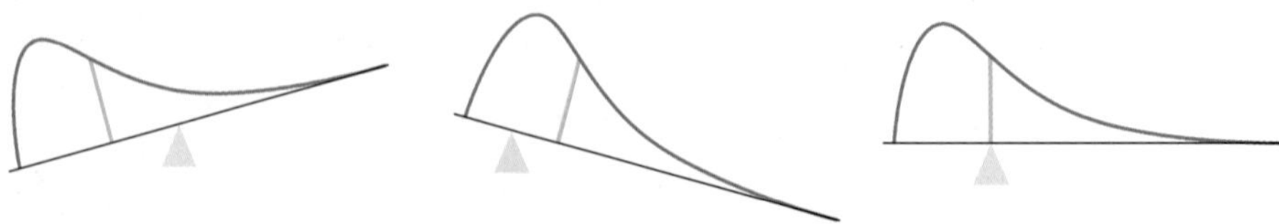

FIGURE 2.10 The mean is the balance point of a density curve.

We know that the mean of a skewed distribution is pulled toward the long tail. Figure 2.9(b) shows how the mean of a skewed density curve is pulled toward the long tail more than is the median.

Distinguishing the Median and Mean of a Density Curve

The **median** of a density curve is the equal-areas point, the point that divides the area under the curve in half.

The **mean** of a density curve is the balance point, at which the curve would balance if made of solid material.

The median and mean are the same for a symmetric density curve. They both lie at the center of the curve. The mean of a skewed curve is pulled away from the median in the direction of the long tail.

You probably noticed that we used the same notation for the mean and standard deviation of a population in Chapter 1, μ and σ, as we do here for the mean and standard deviation of a density curve.

Because a density curve is an idealized description of a distribution of data, we need to distinguish between the mean and standard deviation of the density curve and the mean $\bar{x}$ and standard deviation s_x computed from the actual observations. The usual notation for the mean of a density curve is μ (the Greek letter mu). We write the standard deviation of a density curve as σ (the Greek letter sigma). We can roughly locate the mean μ of any density curve by eye, as the balance point. There is no easy way to locate the standard deviation σ by eye for density curves in general.

CHECK YOUR UNDERSTANDING

Use the figure shown to answer the following questions.

1. Explain why this is a legitimate density curve.
2. About what proportion of observations lie between 7 and 8?
3. Trace the density curve onto your paper. Mark the approximate location of the median.
4. Now mark the approximate location of the mean. Explain why the mean and median have the relationship that they do in this case.

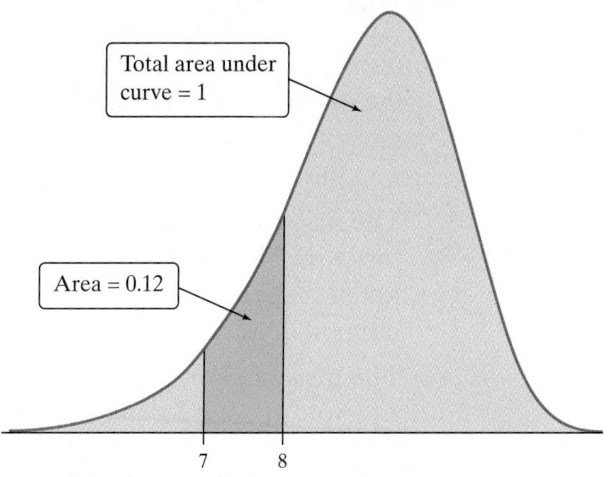

SECTION 2.1 Summary

- Two ways of describing an individual's location within a distribution are **percentiles** and **z-scores.** An observation's percentile is the percent of the distribution that is below the value of that observation. To standardize any observation x, subtract the mean of the distribution and then divide by the standard deviation. The resulting z-score

$$z = \frac{x - \text{mean}}{\text{standard deviation}}$$

says how many standard deviations x lies from the distribution mean. We can also use percentiles and z-scores to compare the relative standing of individuals in different distributions.

- A **cumulative relative frequency graph** allows us to examine location within a distribution. Cumulative relative frequency graphs begin by grouping the observations into equal-width classes (much like the process of making a histogram). The completed graph shows the accumulating percent of observations as you move through the classes in increasing order.
- It is common to **transform data**, especially when changing units of measurement. When you add a constant a to all the values in a data set, measures of center (median and mean) and location (quartiles and percentiles) increase by a. Measures of spread do not change. When you multiply all the values in a data set by a constant b, measures of center and location are multiplied by b, while measures of spread are multiplied by $|b|$. Neither of these transformations changes the shape of the distribution.
- We can sometimes describe the overall pattern of a distribution by a **density curve.** A density curve always remains on or above the horizontal axis and has total area 1 underneath it. An area under a density curve gives the proportion of observations that fall in a range of values.
- A density curve is an idealized description of the overall pattern of a distribution that smooths out the irregularities in the actual data. We write the **mean of a density curve** as μ and the **standard deviation of a density curve** as σ to distinguish them from the mean $\bar{x}$ and the standard deviation s_x of the actual data.
- The mean and the median of a density curve can be located by eye. The mean μ is the balance point of the curve. The median divides the area under the curve in half. The standard deviation σ cannot be located by eye on most density curves.
- The mean and median are equal for symmetric density curves. The mean of a skewed curve is located farther toward the long tail than the median is.

SECTION 2.1 Exercises

1. **Shoes** How many pairs of shoes do students have? Do girls have more shoes than boys? Here are data from a random sample of 20 female and 20 male students at a large high school:

pg 85

Female:	50	26	26	31	57	19	24	22	23	38
	13	50	13	34	23	30	49	13	15	51
Male:	14	7	6	5	12	38	8	7	10	10
	10	11	4	5	22	7	5	10	35	7

(a) Find and interpret the percentile in the female distribution for the girl with 22 pairs of shoes.

(b) Find and interpret the percentile in the male distribution for the boy with 22 pairs of shoes.

(c) Who is more unusual: the girl with 22 pairs of shoes or the boy with 22 pairs of shoes? Explain.

2. **Old folks** Here is a stemplot of the percents of residents aged 65 and older in the 50 states:

Stem	Leaves
7	0
8	8
9	8
10	019
11	16777
12	01122456778999
13	0001223344455689
14	023568
15	24
16	9

Key: 15|2 means 15.2% of this state's residents are 65 or older

(a) Find and interpret the percentile for Colorado, which has 10.1% of its residents aged 65 or older.

(b) Find and interpret the percentile for Rhode Island, with 13.9% of residents aged 65 or older.

(c) Which of these two states is more unusual? Explain.

3. **Speed limits** According to the *Los Angeles Times*, speed limits on California highways are set at the 85th percentile of vehicle speeds on those stretches of road. Explain what that means to someone who knows little statistics.

4. **Blood pressure** Larry came home very excited after a visit to his doctor. He announced proudly to his wife, "My doctor says my blood pressure is at the 90th percentile among men like me. That means I'm better off than about 90% of similar men." How should his wife, who is a statistician, respond to Larry's statement?

5. **Growth charts** We used an online growth chart to find percentiles for the height and weight of a 16-year-old girl who is 66 inches tall and weighs 118 pounds. According to the chart, this girl is at the 48th percentile for weight and the 78th percentile for height. Explain what these values mean in plain English.

6. **Run fast** Peter is a star runner on the track team. In the league championship meet, Peter records a time that would fall at the 80th percentile of all his race times that season. But his performance places him at the 50th percentile in the league championship meet. Explain how this is possible. (Remember that lower times are better in this case!)

Exercises 7 and 8 involve a new type of graph called a ***percentile plot.*** *Each point gives the value of the variable being measured and the corresponding percentile for one individual in the data set.*

7. **Text me** The *percentile plot* below shows the distribution of text messages sent and received in a two-day period by a random sample of 16 females from a large high school.

(a) Describe the student represented by the highlighted point.

(b) Use the graph to estimate the median number of texts. Explain your method.

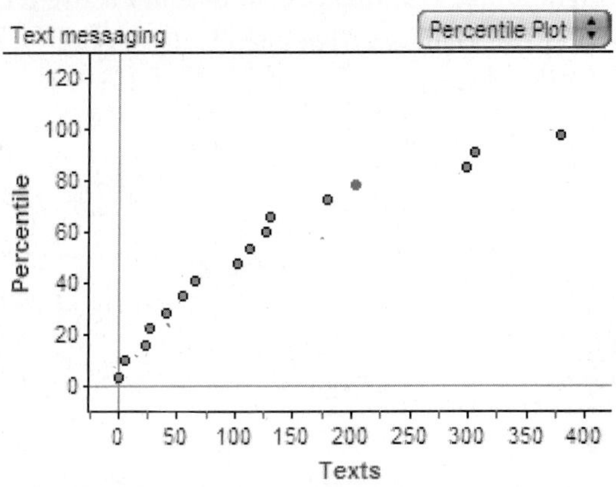

8. **Foreign-born residents** The *percentile plot* below shows the distribution of the percent of foreign-born residents in the 50 states.

(a) The highlighted point is for Maryland. Describe what the graph tells you about this state.

(b) Use the graph to estimate the 30th percentile of the distribution. Explain your method.

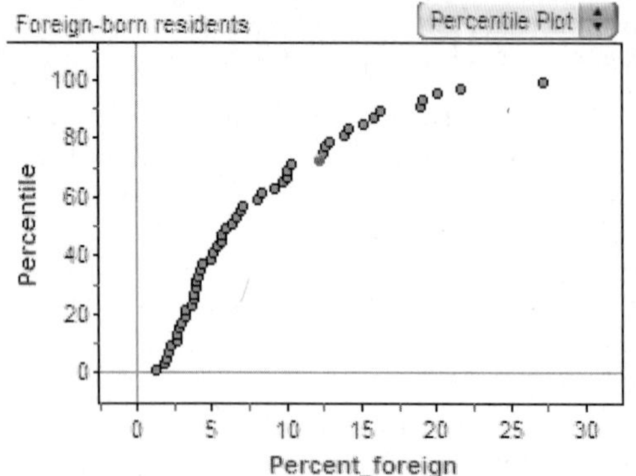

9. **Shopping spree** The figure below is a cumulative relative frequency graph of the amount spent by 50 consecutive grocery shoppers at a store.

pg 88

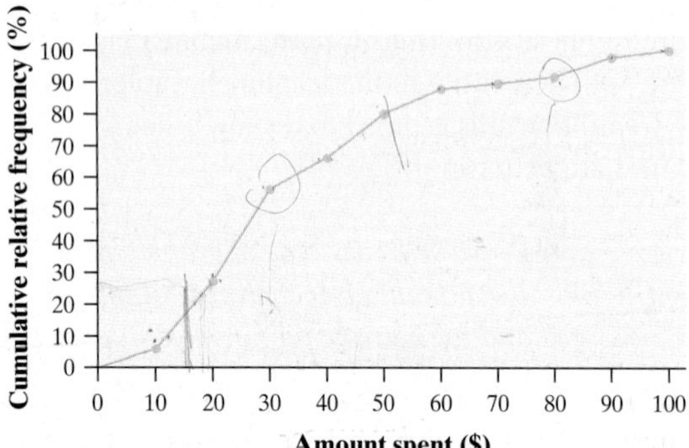

(a) Estimate the interquartile range of this distribution. Show your method.

(b) What is the percentile for the shopper who spent \$19.50?

(c) *Challenge:* Draw the histogram that corresponds to this graph.

10. **Light it up!** The graph below is a cumulative relative frequency graph showing the lifetimes (in hours) of 200 lamps.[5]

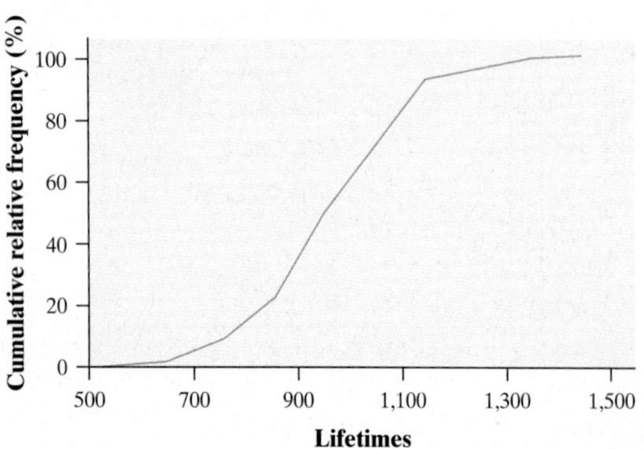

(a) Estimate the 60th percentile of this distribution. Show your method.

(b) What is the percentile for a lamp that lasted 900 hours?

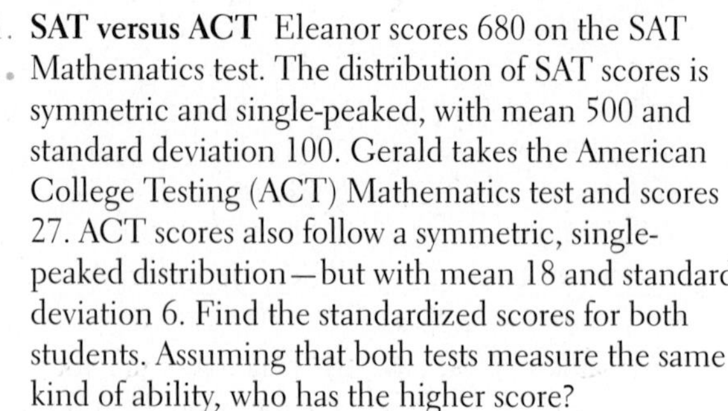

11. **SAT versus ACT** Eleanor scores 680 on the SAT Mathematics test. The distribution of SAT scores is symmetric and single-peaked, with mean 500 and standard deviation 100. Gerald takes the American College Testing (ACT) Mathematics test and scores 27. ACT scores also follow a symmetric, single-peaked distribution—but with mean 18 and standard deviation 6. Find the standardized scores for both students. Assuming that both tests measure the same kind of ability, who has the higher score?

pg 91

12. **Comparing batting averages** Three landmarks of baseball achievement are Ty Cobb's batting average of .420 in 1911, Ted Williams's .406 in 1941, and George Brett's .390 in 1980. These batting averages cannot be compared directly because the distribution of major league batting averages has changed over the years. The distributions are quite symmetric, except for outliers such as Cobb, Williams, and Brett. While the mean batting average has been held roughly constant by rule changes and the balance between hitting and pitching, the standard deviation has dropped over time. Here are the facts:

Decade	Mean	Standard deviation
1910s	0.266	0.0371
1940s	0.267	0.0326
1970s	0.261	0.0317

Compute the standardized batting averages for Cobb, Williams, and Brett to compare how far each stood above his peers.[6]

13. **Measuring bone density** Individuals with low bone density have a high risk of broken bones (fractures). Physicians who are concerned about low bone density (osteoporosis) in patients can refer them for specialized testing. Currently, the most common method for testing bone density is dual-energy X-ray absorptiometry (DEXA). A patient who undergoes a DEXA test usually gets bone density results in grams per square centimeter (g/cm^2) and in standardized units.

Judy, who is 25 years old, has her bone density measured using DEXA. Her results indicate a bone density in the hip of 948 g/cm^2 and a standardized score of $z = -1.45$. In the reference population of 25-year-old women like Judy, the mean bone density in the hip is 956 g/cm^2.[7]

(a) Judy has not taken a statistics class in a few years. Explain to her in simple language what the standardized score tells her about her bone density.

(b) Use the information provided to calculate the standard deviation of bone density in the reference population.

14. **Comparing bone density** Refer to the previous exercise. One of Judy's friends, Mary, has the bone density in her hip measured using DEXA. Mary is 35 years old. Her bone density is also reported as 948 g/cm^2, but her standardized score is $z = 0.50$. The mean bone density in the hip for the reference population of 35-year-old women is 944 grams/cm^2.

(a) Whose bones are healthier—Judy's or Mary's? Justify your answer.

(b) Calculate the standard deviation of the bone density in Mary's reference population. How does this compare with your answer to Exercise 13(b)? Are you surprised?

Exercises 15 and 16 refer to the dotplot and summary statistics of salaries for players on the World Champion 2008 Philadelphia Phillies baseball team.[8]

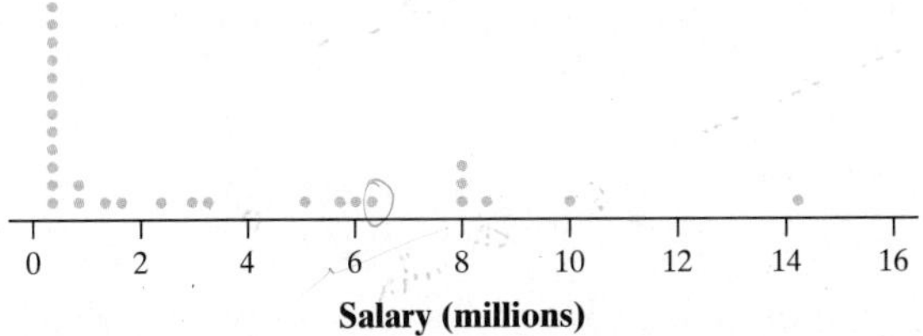

Variable	n	Mean	Std. dev.	Min	Q_1	M	Q_3	Max
Salary	29	3388617	3767484	390000	440000	1400000	6000000	14250000

pg 90

15. **Baseball salaries** Brad Lidge played a crucial role as the Phillies' "closer," pitching the end of many games throughout the season. Lidge's salary for the 2008 season was $6,350,000.

(a) Find the percentile corresponding to Lidge's salary. Explain what this value means.

(b) Find the z-score corresponding to Lidge's salary. Explain what this value means.

16. **Baseball salaries** Did Ryan Madson, who was paid $1,400,000, have a high salary or a low salary compared with the rest of the team? Justify your answer by calculating and interpreting Madson's percentile and z-score.

Exercises 17 and 18 refer to the following setting. Each year, about 1.5 million college-bound high school juniors take the PSAT. In a recent year, the mean score on the Critical Reading test was 46.9 and the standard deviation was 10.9. Nationally, 5.2% of test takers earned a score of 65 or higher on the Critical Reading test's 20 to 80 scale.[9]

17. **PSAT scores** Scott was one of 50 junior boys to take the PSAT at his school. He scored 64 on the Critical Reading test. This placed Scott at the 68th percentile within the group of boys. Looking at all 50 boys' Critical Reading scores, the mean was 58.2 and the standard deviation was 9.4.

(a) Write a sentence or two comparing Scott's percentile among the national group of test takers and among the 50 boys at his school.

(b) Calculate and compare Scott's z-score among these same two groups of test takers.

18. **PSAT scores** How well did the boys at Scott's school perform on the PSAT? Give appropriate evidence to support your answer.

pg 96

19. **Tall or short?** Mr. Walker measures the heights (in inches) of the students in one of his classes. He uses a computer to calculate the following numerical summaries:

Mean	Std.dev.	Min	Q_1	M	Q_3	Max
69.188	3.20	61.5	67.75	69.5	71	74.5

Next, Mr. Walker has his entire class stand on their chairs, which are 18 inches off the ground. Then he measures the distance from the top of each student's head to the floor.

(a) Find the mean and median of these measurements. Show your work.

(b) Find the standard deviation and *IQR* of these measurements. Show your work.

20. **Teacher raises** A school system employs teachers at salaries between $28,000 and $60,000. The teachers' union and the school board are negotiating the form of next year's increase in the salary schedule.

(a) If every teacher is given a flat $1000 raise, what will this do to the mean salary? To the median salary? Explain your answers.

(b) What would a flat $1000 raise do to the extremes and quartiles of the salary distribution? To the standard deviation of teachers' salaries? Explain your answers.

21. **Tall or short?** Refer to Exercise 19. Mr. Walker converts his students' original heights from inches to feet.

(a) Find the mean and median of the students' heights in feet. Show your work.

(b) Find the standard deviation and *IQR* of the students' heights in feet. Show your work.

22. **Teacher raises** Refer to Exercise 20. If each teacher receives a 5% raise instead of a flat $1000 raise, the amount of the raise will vary from $1400 to $3000, depending on the present salary.

(a) What will this do to the mean salary? To the median salary? Explain your answers.

(b) Will a 5% raise increase the *IQR*? Will it increase the standard deviation? Explain your answers.

23. **Cool pool?** Coach Ferguson uses a thermometer to measure the temperature (in degrees Celsius) at 20 different locations in the school swimming pool. An analysis of the data yields a mean of 25°C and a standard deviation of 2°C. Find the mean and standard deviation of the temperature readings in degrees Fahrenheit (recall that °F = (9/5)°C + 32).

24. **Measure up** Clarence measures the diameter of each tennis ball in a bag with a standard ruler. Unfortunately, he uses the ruler incorrectly so that each of his measurements is 0.2 inches too large. Clarence's data had a mean of 3.2 inches and a standard deviation of 0.1 inches. Find the mean and standard deviation of the corrected measurements in centimeters (recall that 1 inch = 2.54 cm).

25. **Density curves** Sketch a density curve that might describe a distribution that is symmetric but has two peaks.

26. **Density curves** Sketch a density curve that might describe a distribution that has a single peak and is skewed to the left.

Exercises 27 to 30 involve a special type of density curve—one that takes constant height (looks like a horizontal line) over some interval of values. This density curve describes a variable whose values are distributed evenly (uniformly) over some interval of values. We say that such a variable has a ***uniform distribution.***

27. **Biking accidents** Accidents on a level, 3-mile bike path occur uniformly along the length of the path. The figure below displays the density curve that describes the uniform distribution of accidents.

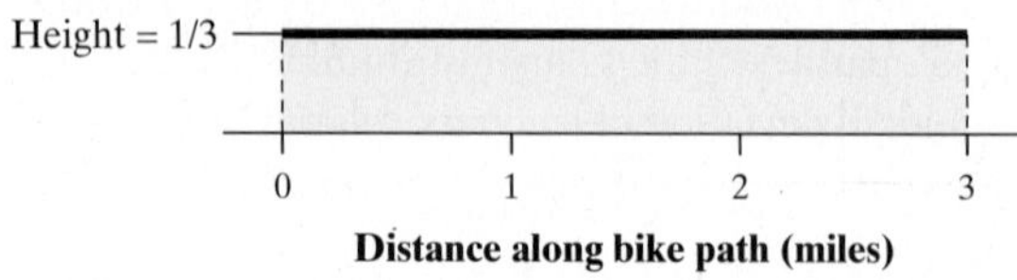

(a) Explain why this curve satisfies the two requirements for a density curve.

(b) The proportion of accidents that occur in the first mile of the path is the area under the density curve between 0 miles and 1 mile. What is this area?

(c) Sue's property adjoins the bike path between the 0.8 mile mark and the 1.1 mile mark. What proportion of accidents happen in front of Sue's property? Explain.

28. **A uniform distribution** The figure below displays the density curve of a uniform distribution. The curve takes the constant value 1 over the interval from 0 to 1 and is 0 outside the range of values. This means that data described by this distribution take values that are uniformly spread between 0 and 1.

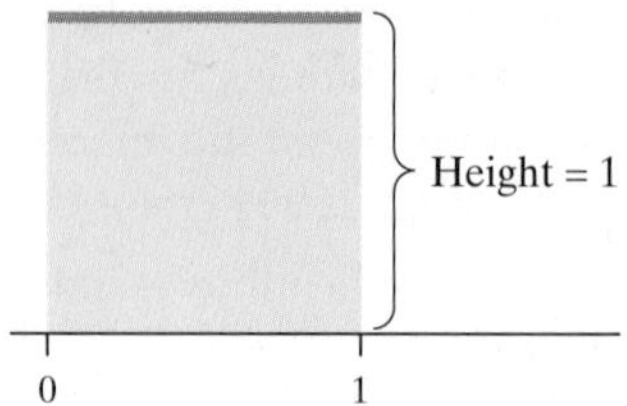

(a) Explain why this curve satisfies the two requirements for a density curve.

(b) What percent of the observations are greater than 0.8?

(c) What percent of the observations lie between 0.25 and 0.75?

29. **Biking accidents** What is the mean μ of the density curve pictured in Exercise 27? (That is, where would the curve balance?) What is the median? (That is, where is the point with area 0.5 on either side?)

30. **A uniform distribution** What is the mean μ of the density curve pictured in Exercise 28? What is the median?

31. **Mean and median** The figure below displays two density curves, each with three points marked. At which of these points on each curve do the mean and the median fall?

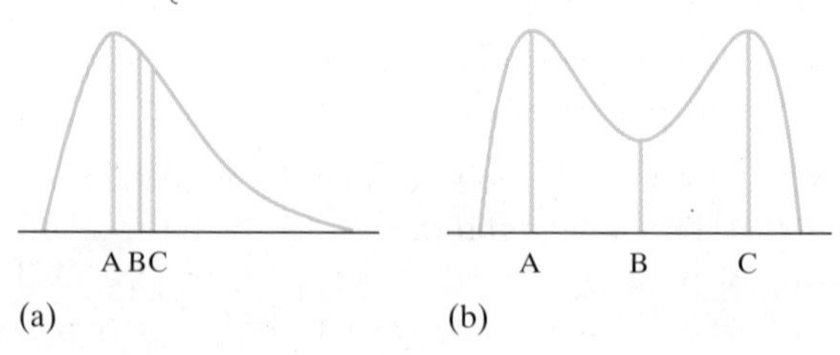

32. **Mean and median** The figure below displays two density curves, each with three points marked. At which of these points on each curve do the mean and the median fall?

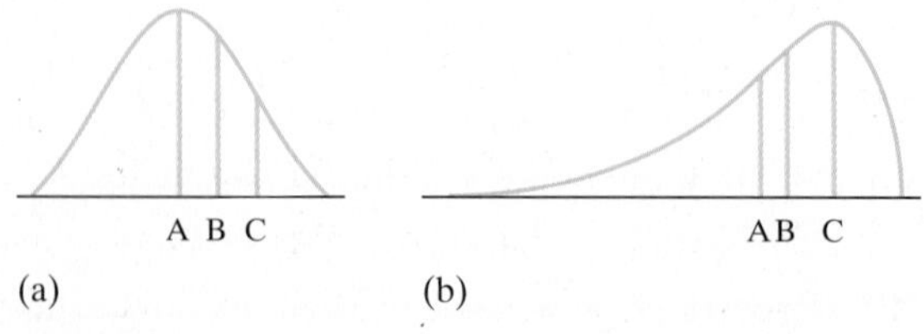

Multiple choice: Select the best answer for Exercises 33 to 38.

33. Jorge's score on Exam 1 in his statistics class was at the 64th percentile of the scores for all students. His score falls
(a) between the minimum and the first quartile.
(b) between the first quartile and the median.
(c) between the median and the third quartile.
(d) between the third quartile and the maximum.
(e) at the mean score for all students.

34. Two measures of center are marked on the density curve shown.

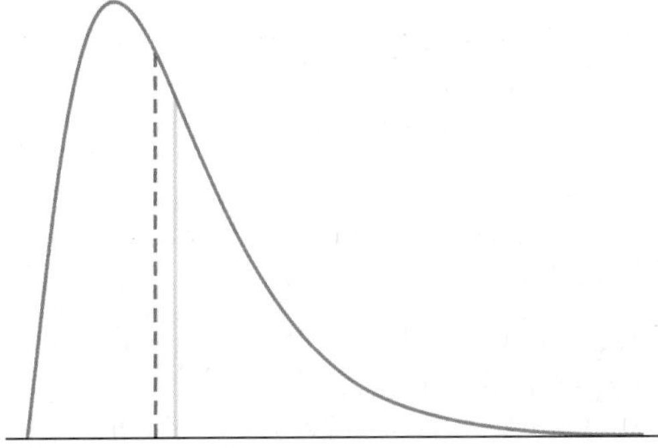

(a) The median is at the yellow line and the mean is at the red line.
(b) The median is at the red line and the mean is at the yellow line.
(c) The mode is at the red line and the median is at the yellow line.
(d) The mode is at the yellow line and the median is at the red line.
(e) The mode is at the red line and the mean is at the yellow line.

35. Scores on the ACT college entrance exam follow a bell-shaped distribution with mean 18 and standard deviation 6. Wayne's standardized score on the ACT was −0.7. What was Wayne's actual ACT score?
(a) 4.2 (c) 13.8 (e) 22.2
(b) −4.2 (d) 17.3

36. George has an average bowling score of 180 and bowls in a league where the average for all bowlers is 150 and the standard deviation is 20. Bill has an average bowling score of 190 and bowls in a league where the average is 160 and the standard deviation is 15. Who ranks higher in his own league, George or Bill?
(a) Bill, because his 190 is higher than George's 180.
(b) Bill, because his standardized score is higher than George's.
(c) Bill and George have the same rank in their leagues, because both are 30 pins above the mean.
(d) George, because his standardized score is higher than Bill's.
(e) George, because the standard deviation of bowling scores is higher in his league.

37. If 30 is added to every observation in a data set, the only one of the following that is *not* changed is
(a) the mean. (d) the standard deviation.
(b) the 75th percentile. (e) the minimum.
(c) the median.

38. If every observation in a data set is multiplied by 10, the only one of the following that is *not* multiplied by 10 is
(a) the mean. (d) the standard deviation.
(b) the median. (e) the variance.
(c) the *IQR*.

Exercises 39 and 40 refer to the following setting. We used CensusAtSchool's Random Data Selector to choose a sample of 50 Canadian students who completed a survey in 2007–2008.

39. **Travel time** (1.2) The dotplot below displays data on students' responses to the question "How long does it usually take you to travel to school?" Describe the shape, center, and spread of the distribution. Are there any outliers?

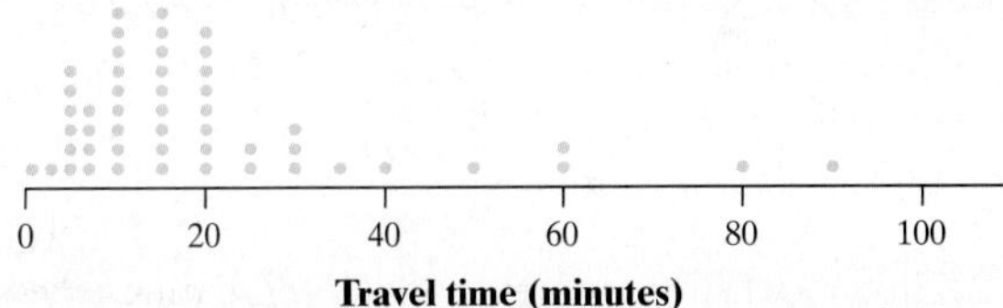

40. **Lefties** (1.1) Students were asked, "Are you right-handed, left-handed, or ambidextrous?" The responses are shown below (R = right-handed; L = left-handed; A = ambidextrous).

R	R	R	R	R	R	R	R	R	R	R	L	R	R
R	R	R	R	R	R	R	R	R	R	R	R	R	A
R	R	R	R	A	R	R	L	R	R	R	R	L	A
R	R	R	R	R	R	R	R						

(a) Make an appropriate graph to display these data.
(b) Over 10,000 Canadian high school students took the CensusAtSchool survey in 2007–2008. What percent of this population would you estimate is left-handed? Justify your answer.

Normal Distributions

In Section 2.2, you'll learn about:

- The 68–95–99.7 rule
- The standard Normal distribution
- Normal distribution calculations
- Assessing Normality

One particularly important class of density curves has already appeared in Figures 2.7, 2.8, and 2.9(a). They are called **Normal curves.** The distributions they describe are called **Normal distributions.** Normal distributions play a large role in statistics, but they are rather special and not at all "normal" in the sense of being usual or typical. We capitalize Normal to remind you that these curves are special.

Look at the two Normal curves in Figure 2.11. They illustrate several important facts:

- All Normal curves have the same overall shape: symmetric, single-peaked, bell-shaped.
- Any specific Normal curve is completely described by giving its mean μ and its standard deviation σ.

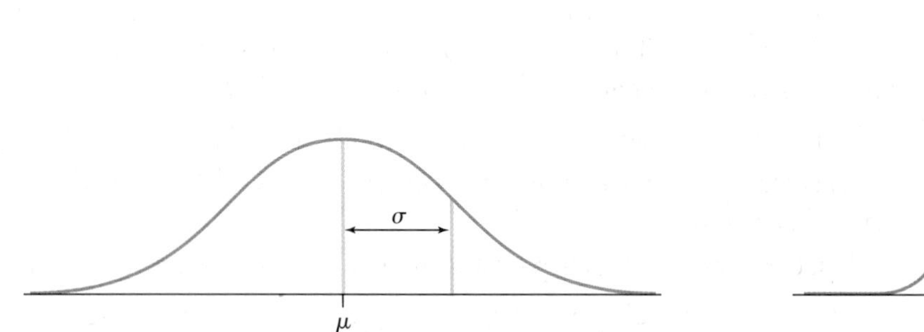

FIGURE 2.11 Two Normal curves, showing the mean μ and standard deviation σ.

- The mean is located at the center of the symmetric curve and is the same as the median. Changing μ without changing σ moves the Normal curve along the horizontal axis without changing its spread.
- The standard deviation σ controls the spread of a Normal curve. Curves with larger standard deviations are more spread out.

The standard deviation σ is the natural measure of spread for Normal distributions. Not only do μ and σ completely determine the shape of a Normal curve, but we can locate σ by eye on a Normal curve. Here's how. Imagine that you are skiing down a mountain that has the shape of a Normal curve. At first, you descend at an ever-steeper angle as you go out from the peak:

Fortunately, before you find yourself going straight down, the slope begins to grow flatter rather than steeper as you go out and down:

The points at which this change of curvature takes place are located at a distance σ on either side of the mean μ. (Advanced math students know these points as "inflection points.") You can feel the change as you run a pencil along a Normal

curve and so find the standard deviation. *Remember that μ and σ alone do not specify the shape of most distributions.* The shape of density curves in general does not reveal σ. These are special properties of Normal distributions.

DEFINITION: Normal distribution and Normal curve

A **Normal distribution** is described by a Normal density curve. Any particular Normal distribution is completely specified by two numbers: its mean μ and standard deviation σ. The mean of a Normal distribution is at the center of the symmetric **Normal curve.** The standard deviation is the distance from the center to the change-of-curvature points on either side.

We abbreviate the Normal distribution with mean μ and standard deviation σ as $N(\mu, \sigma)$.

Normal curves were first applied to data by the great mathematician Carl Friedrich Gauss (1777–1855). He used them to describe the small errors made by astronomers and surveyors in repeated careful measurements of the same quantity. You will sometimes see Normal distributions labeled "Gaussian" in honor of Gauss.

Why are the Normal distributions important in statistics? Here are three reasons. First, Normal distributions are good descriptions for some distributions of *real data*. Distributions that are often close to Normal include

- scores on tests taken by many people (such as SAT exams and IQ tests),
- repeated careful measurements of the same quantity (like the diameter of a tennis ball), and
- characteristics of biological populations (such as lengths of crickets and yields of corn).

Second, Normal distributions are good approximations to the results of many kinds of *chance outcomes*, like the number of heads in many tosses of a fair coin. Third, and most important, we will see that many *statistical inference* procedures are based on Normal distributions.

CAUTION

Even though many sets of data follow a Normal distribution, many do not. Most income distributions, for example, are skewed to the right and so are not Normal. Some distributions are symmetric but not Normal or even close to Normal. The uniform distribution of Exercise 27 (page 108) is one such example. Non-Normal data, like non-normal people, not only are common but are sometimes more interesting than their Normal counterparts.

The 68–95–99.7 Rule

ACTIVITY *The* Normal Curve *applet*

MATERIALS: Computer with Internet access

In this Activity, you will use the *Normal Curve* applet at the book's Web site to explore some interesting properties of Normal distributions. A graph similar to what you will see when you launch the applet is shown below. The applet finds the area under the curve in the region indicated by the green flags. Complete Steps 1 to 3 before you go to the Web site.

1. If you were to put one flag at the extreme left of the curve and the second flag exactly in the middle, what would be the proportion reported by the applet?

2. If you were to place the two flags exactly one standard deviation on either side of the mean, what do you think the applet would say is the area between them?

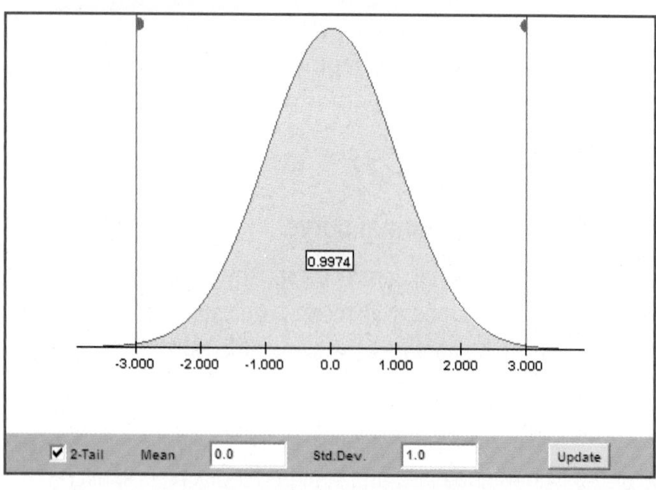

3. In the figure, about 99.7% of the area under the Normal density curve lies within three standard deviations of the mean. Does this mean that about 99.7%/2 = 49.85% will lie within one and a half standard deviations? Explain.

Now go to www.whfreeman.com/tps4e and launch the *Normal Curve* applet. Drag the green flags to have the applet compute the part of the Normal curve that is in the bright-yellow shaded region you create. If you drag one flag past the other, the applet will show the area under the curve between the two flags. When the "2-Tail" box is checked, the applet calculates symmetric areas around the mean.

4. Use the applet to test your answers to Questions 1 to 3 above. Did the proportion of shaded area under the curve behave as you thought it would in each case?

5. What percent of the area under the Normal curve lies within 2 standard deviations of the mean?

6. Change the mean to 100 and the standard deviation to 15. Then click "Update." What percent of the area under this Normal density curve lies within one, two, and three standard deviations of the mean?

7. Change the mean to 500 and the standard deviation to 100. Answer the question from Step 6.

8. *Summarize:* Complete the following sentence: "For any Normal density curve, the area under the curve within one, two, and three standard deviations of the mean is about ___%, ___%, and ___%.

Although there are many Normal curves, they all have properties in common. In particular, all Normal distributions obey the following rule.

DEFINITION: The 68–95–99.7 rule

In the Normal distribution with mean μ and standard deviation σ:

- Approximately **68%** of the observations fall within σ of the mean μ.
- Approximately **95%** of the observations fall within 2σ of μ.
- Approximately **99.7%** of the observations fall within 3σ of μ.

Figure 2.12 illustrates the 68–95–99.7 rule. (Some people refer to this result as the "empirical rule.") By remembering these three numbers, you can think about Normal distributions without constantly making detailed calculations.

Here's an example that shows how we can use the 68–95–99.7 rule to estimate the percent of observations in a specified interval.

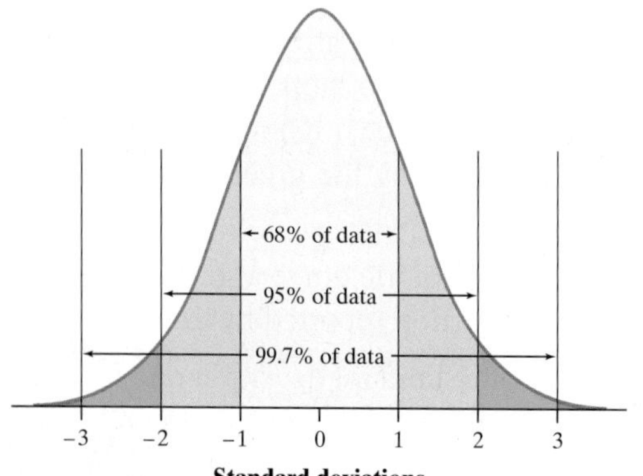

FIGURE 2.12 The 68–95–99.7 rule for Normal distributions.

ITBS Vocabulary Scores

Using the 68–95–99.7 rule

Earlier, we saw that the distribution of Iowa Test of Basic Skills (ITBS) vocabulary scores for seventh-grade students in Gary, Indiana, is close to Normal. Suppose that the distribution is exactly Normal with mean $\mu = 6.84$ and standard deviation $\sigma = 1.55$. (These are the mean and standard deviation of the 947 actual scores.)

PROBLEM:

(a) Sketch a Normal density curve for this distribution of test scores. Label the points that are one, two, and three standard deviations from the mean.

(b) What percent of the ITBS vocabulary scores are less than 3.74? Show your work.

(c) What percent of the scores are between 5.29 and 9.94? Show your work.

SOLUTION:

(a) The distribution of scores is approximately $N(6.84, 1.55)$. Figure 2.13(a) is a sketch of the Normal density curve with the desired points labeled.

(b) Notice that a score of 3.74 is exactly two standard deviations below the mean. By the 68–95–99.7 rule, about 95% of all scores are between

$$\mu - 2\sigma = 6.84 - (2)(1.55) = 6.84 - 3.10 = 3.74$$

and

$$\mu + 2\sigma = 6.84 + (2)(1.55) = 6.84 + 3.10 = 9.94$$

The other 5% of scores are outside this range. Because Normal distributions are symmetric, half of these scores are lower than 3.74 and half are higher than 9.94. That is, about 2.5% of the ITBS scores are below 3.74. Figure 2.13(b) shows this reasoning in picture form.

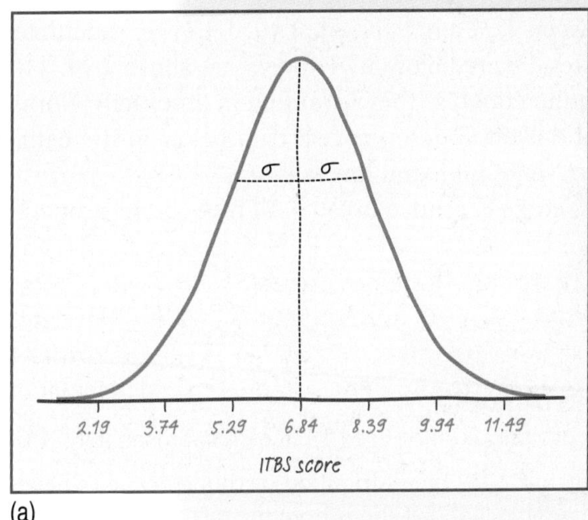

(a)

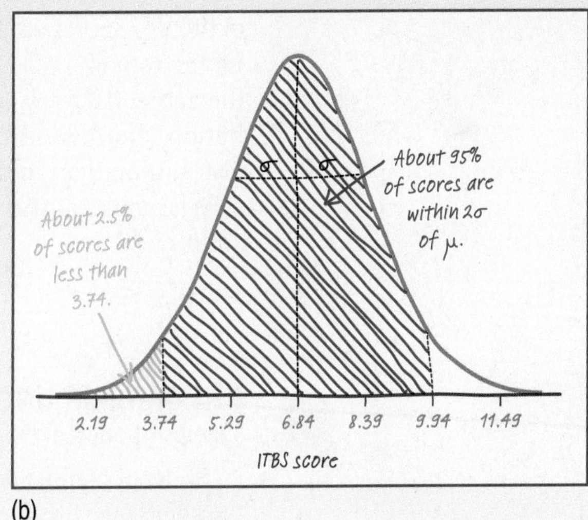

(b)

FIGURE 2.13 (a) Normal density curve for the Iowa Test scores. (b) Finding the percent of Iowa Test scores less than 3.74.

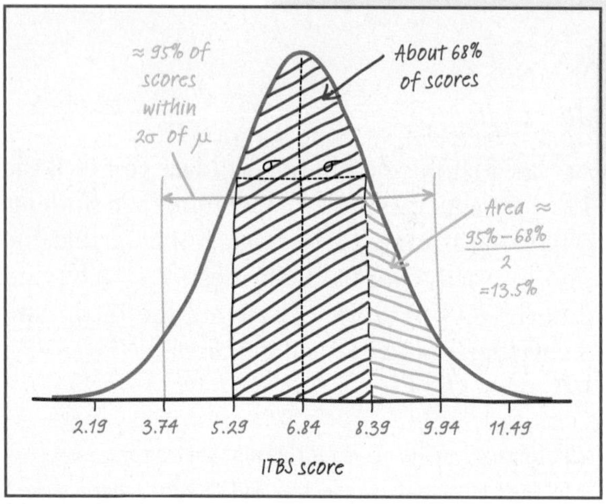

(c) Let's start with a picture. Figure 2.13(c) shows the area under the Normal density curve between 5.29 and 9.94. We can see that about 68% + 13.5% = 81.5% of ITBS scores are between 5.29 and 9.94.

For Practice *Try Exercises 41 and 43*

(c)

FIGURE 2.13 (c) Finding the percent of Iowa Test scores between 5.29 and 9.94.

The 68–95–99.7 rule applies *only* to Normal distributions. Is there a similar rule that would apply to *any* distribution? Sort of. A result known as **Chebyshev's inequality** says that in any distribution, the proportion of observations falling within k standard deviations of the mean is *at least* $1 - \frac{1}{k^2}$. If $k = 2$, for example, Chebyshev's inequality tells us that at least $1 - \frac{1}{2^2} = 0.75$ of the observations in *any* distribution are within 2 standard deviations of the mean. For Normal distributions, we know that this proportion is much higher than 0.75. In fact, it's approximately 0.95.

Chebyshev's inequality

THINK ABOUT IT

All models are wrong, but some are useful! The 68–95–99.7 rule describes distributions that are exactly Normal. Real data such as the actual ITBS scores are never exactly Normal. For one thing, ITBS scores are reported only to the nearest tenth. A score can be 9.9 or 10.0 but not 9.94. We use a Normal distribution because it's a good approximation, and because we think of the knowledge that the test measures as continuous rather than stopping at tenths.

How well does the 68–95–99.7 rule describe the actual ITBS scores? Well, 900 of the 947 scores are between 3.74 and 9.94. That's 95.04%, very accurate indeed. Of the remaining 47 scores, 20 are below 3.74 and 27 are above 9.94. The tails of the actual data are not quite equal, as they would be in an exactly Normal distribution. Normal distributions often describe real data better in the center of the distribution than in the extreme high and low tails.

As famous statistician George Box once noted, "All models are wrong, but some are useful!"

CHECK YOUR UNDERSTANDING

The distribution of heights of young women aged 18 to 24 is approximately N(64.5, 2.5).

1. Sketch a Normal density curve for the distribution of young women's heights. Label the points one, two, and three standard deviations from the mean.
2. What percent of young women have heights greater than 67 inches? Show your work.
3. What percent of young women have heights between 62 and 72 inches? Show your work.

The Standard Normal Distribution

As the 68–95–99.7 rule suggests, all Normal distributions share many properties. In fact, all Normal distributions are the same if we measure in units of size σ from the mean μ as center. Changing to these units requires us to standardize, just as we did in Section 2.1:

$$z = \frac{x - \mu}{\sigma}$$

If the variable we standardize has a Normal distribution, then so does the new variable z. (Recall that subtracting a constant and dividing by a constant don't change the shape of a distribution.) This new distribution is called the **standard Normal distribution.**

DEFINITION: Standard Normal distribution

The **standard Normal distribution** is the Normal distribution with mean 0 and standard deviation 1 (Figure 2.14).

If a variable x has any Normal distribution $N(\mu, \sigma)$ with mean μ and standard deviation σ, then the standardized variable

$$z = \frac{x - \mu}{\sigma}$$

has the standard Normal distribution.

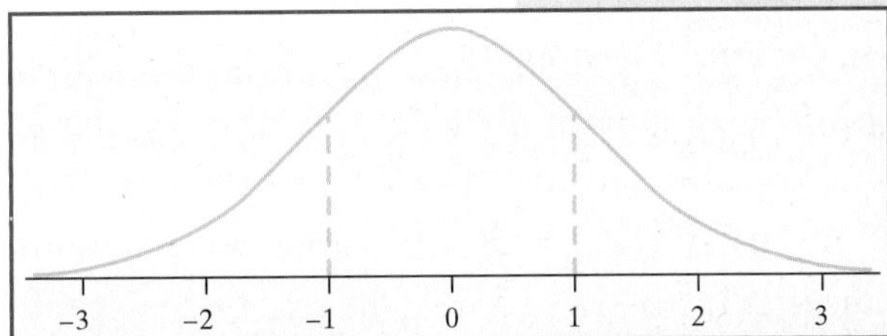

FIGURE 2.14 The standard Normal distribution.

An area under a density curve is a proportion of the observations in a distribution. Any question about what proportion of observations lie in some range of values can be answered by finding an area under the curve. In a standard Normal distribution, the 68–95–99.7 rule tells us that about 68% of the observations fall between $z = -1$ and $z = 1$ (that is, within one standard deviation of the mean). What if we want to find the percent of observations that fall between $z = -1.25$ and $z = 1.25$? The 68–95–99.7 rule can't help us.

Because all Normal distributions are the same when we standardize, we can find areas under any Normal curve from a single table, a table that gives areas under the curve for the standard Normal distribution. Table A, the **standard Normal table,** gives areas under the standard Normal curve. You can find Table A in the back of the book.

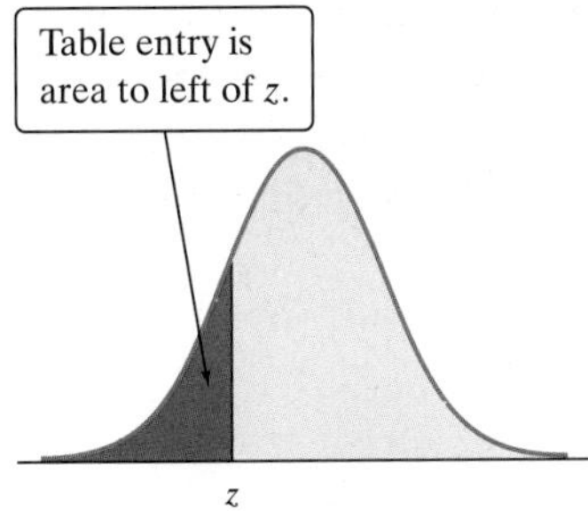

DEFINITION: The standard Normal table

Table A is a table of areas under the standard Normal curve. The table entry for each value z is the area under the curve to the left of z.

z	.00	.01	.02
0.7	.7580	.7611	.7642
0.8	.7881	.7910	.7939
0.9	.8159	.8186	.8212

For instance, suppose we wanted to find the proportion of observations from the standard Normal distribution that are less than 0.81. To find the area to the left of 0.81, locate 0.8 in the left-hand column of Table A, then locate the remaining digit 1 as .01 in the top row. The entry opposite 0.8 and under .01 is .7910. This is the area we seek. A reproduction of the relevant portion of Table A is shown in the margin.

Figure 2.15 illustrates the relationship between the value $z = 0.81$ and the area 0.7910.

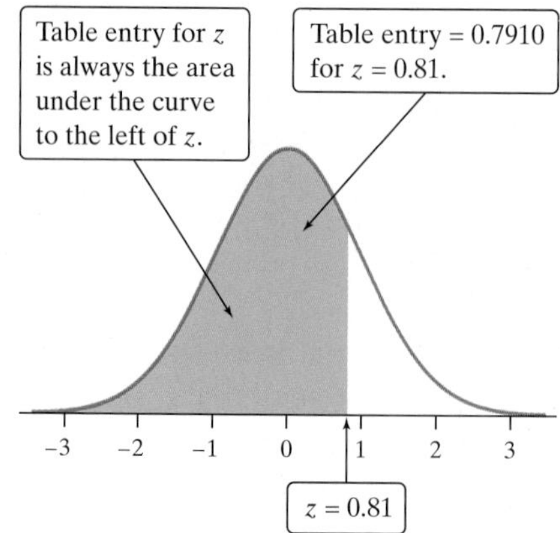

FIGURE 2.15 The area under a standard Normal curve to the left of the point $z = 0.81$ is 0.7910.

EXAMPLE

Standard Normal Distribution

Finding area to the right

What if we wanted to find the proportion of observations from the standard Normal distribution that are *greater than* -1.78? To find the area to the right of $z = -1.78$, locate -1.7 in the left-hand column of Table A, then locate the remaining digit 8 as .08 in the top row. The corresponding entry is .0375. (See the excerpt from Table A in the margin.)

z	.07	.08	.09
−1.8	.0307	.0301	.0294
−1.7	.0384	.0375	.0367
−1.6	.0475	.0465	.0455

This is the area *to the left* of $z = -1.78$. To find the area *to the right* of $z = -1.78$, we use the fact that the total area under the standard Normal density curve is 1. So the desired proportion is $1 - 0.0375 = 0.9625$.

Figure 2.16 illustrates the relationship between the value $z = -1.78$ and the area 0.9625.

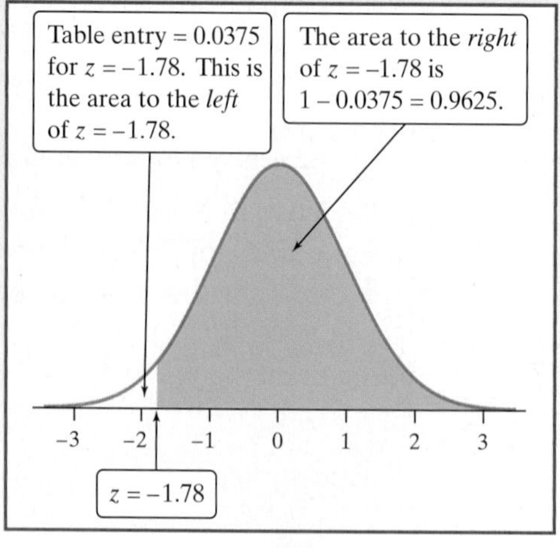

FIGURE 2.16 The area under a standard Normal curve to the right of the point $z = -1.78$ is 0.9625.

CAUTION

A common student mistake is to look up a z-value in Table A and report the entry corresponding to that z-value, regardless of whether the problem asks for the area to the left or to the right of that z-value. To prevent making this mistake, always sketch the standard Normal curve, mark the *z*-value, and shade the area of interest. And before you finish, make sure your answer is reasonable in the context of the problem.

B.C. by johnny hart

NORMALIZE

Catching Some "z"s

Finding areas under the standard Normal curve

PROBLEM: Find the proportion of observations from the standard Normal distribution that are between −1.25 and 0.81.

SOLUTION: From Table A, the area to the left of $z = 0.81$ is 0.7910 and the area to the left of $z = -1.25$ is 0.1056. So the area under the standard Normal curve between these two z-scores is $0.7910 - 0.1056 = 0.6854$. Figure 2.17 shows why this approach works.

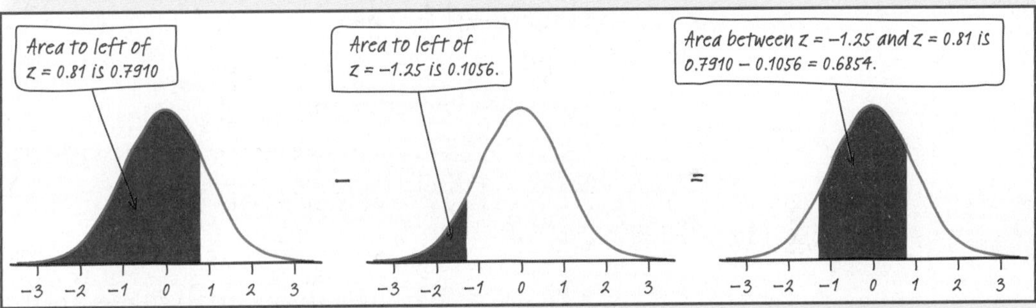

FIGURE 2.17 One way to find the area between $z = -1.25$ and $z = 0.81$ under the standard Normal curve.

Here's another way to find the desired area. The area to the left of $z = -1.25$ under the standard Normal curve is 0.1056. The area to the *right* of $z = 0.81$ is $1 - 0.7910 = 0.2090$. So the area *between* these two z-scores is

$$1 - (0.1056 + 0.2090) = 1 - 0.3146 = 0.6854$$

Figure 2.18 shows this approach in picture form.

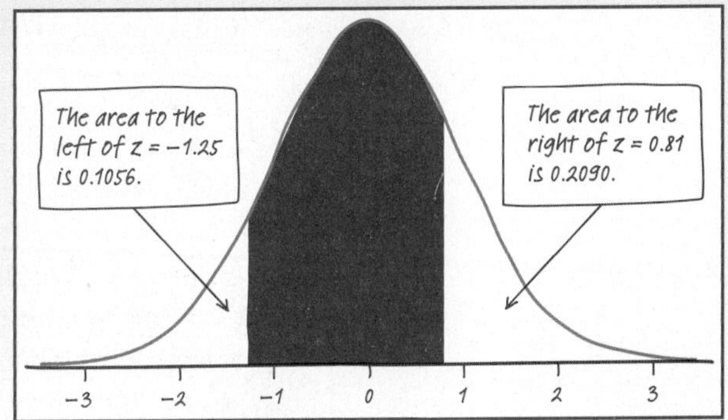

FIGURE 2.18 The area under the standard Normal curve between $z = -1.25$ and $z = 0.81$ is 0.6854.

You can use the *Normal curve* applet at www.whfreeman.com/tps4e to confirm areas under the standard Normal curve. Just enter mean 0 and standard deviation 1, and then drag the flags to the appropriate locations. Of course, you can also sketch a standard Normal curve and find areas under it with your calculator.

TECHNOLOGY CORNER The standard Normal curve

The TI-83/84 and TI-89 can be used to draw and find areas under a standard Normal curve.

To draw a standard Normal curve: First, turn off all statistics plots.

1. Enter the formula for the standard Normal density curve in Y1.

TI-83/84

Define `Y1=normalpdf(X,0,1)`.

- Press `Y=`. With the cursor next to `Y1=`, press [2nd] [VARS] `(DISTR)` and choose `1:normalpdf(`.
- Press [X,T,Θ,n] [,] [0] [,] [1] [)] to complete the formula.

TI-89

Define `Y1=tistat.normPdf(x,0,1)`.

- Press [♦] [F1] `(Y=)`. With the cursor next to `y1=`, press [CATALOG] and choose `F3 Flash Apps`.
- Press [6](N), arrow to `normPdf` (. . . and press [ENTER]. Type [X] [)] to complete the formula.

2. Adjust window settings and graph.

TI-83/84

- Press [WINDOW]. Enter Xmin = −4, Xmax = 4, Xscl = 1, Ymin = −.1, Ymax = .5, Yscl = 1.
- Press [GRAPH] to display the curve.

TI-89

- Press [♦] [F2] `(WINDOW)`. Enter Xmin = −4, Xmax = 4, Xscl = 1, Ymin = −1, Ymax = .5, Yscl = .1.
- Press [♦] [F3] `(GRAPH)` to display the curve.

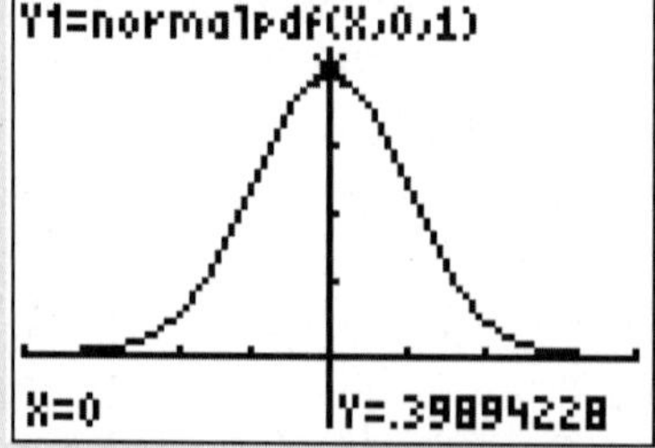

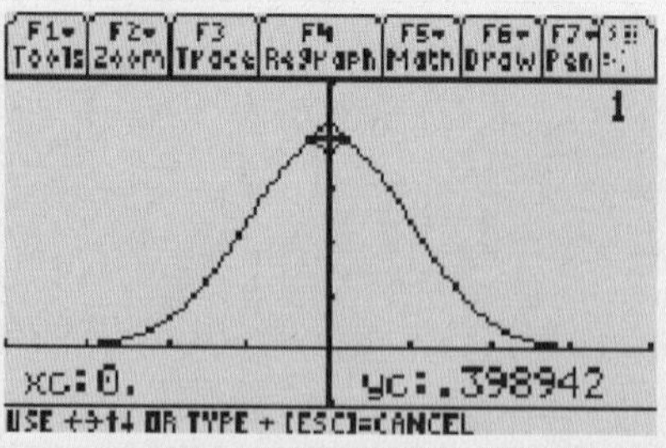

To find areas under the standard Normal curve: Go to the home screen. Use the shadeNorm command to find the desired area.

TI-83/84: Press [2nd] [VARS] `(DISTR)`, arrow right to `DRAW`, and choose `1:ShadeNorm(`. Complete the command `shadeNorm(lower bound,upper bound, μ, σ)`.

TI-89: Press [CATALOG] and choose `F3 Flash Apps`. Press [3](S), arrow to `shadNorm` (. . . and press [ENTER]. Complete the command `shadNorm(lower bound,upper bound, μ, σ)`.

Note: After *each* time you use shadeNorm, execute the command `ClrDraw` to remove the shading. (`ClrDraw` can be found in the CATALOG or in the DRAW menu on the TI-83/84.)

To find

- the area to the left of $z = 2.22$, use -100 for the lower bound: `shadeNorm(−100,2.22,0,1)`.
- the area to the right of $z = -1.78$, use 100 for the upper bound: `shadeNorm(−1.78,100,0,1)`.
- the area between $z = -1.25$ and $z = 0.81$: `shadeNorm(−1.25,0.81,0,1)`.

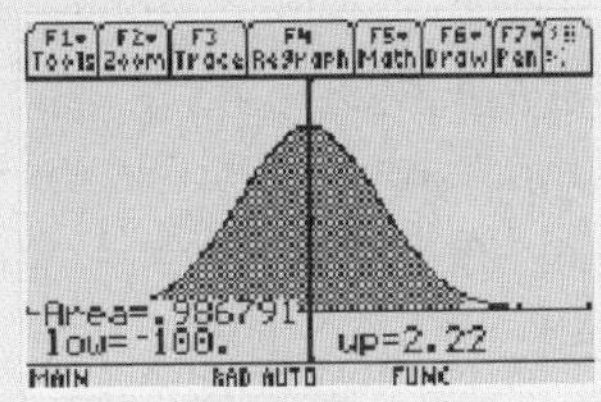

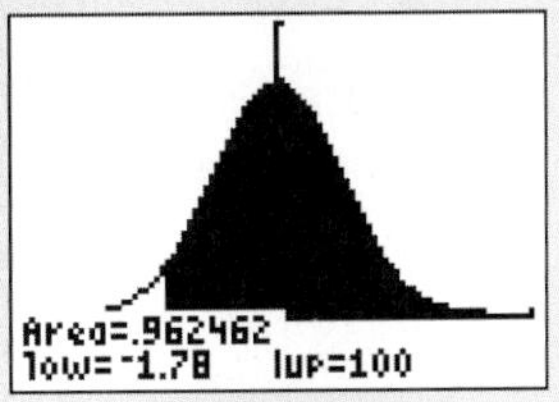

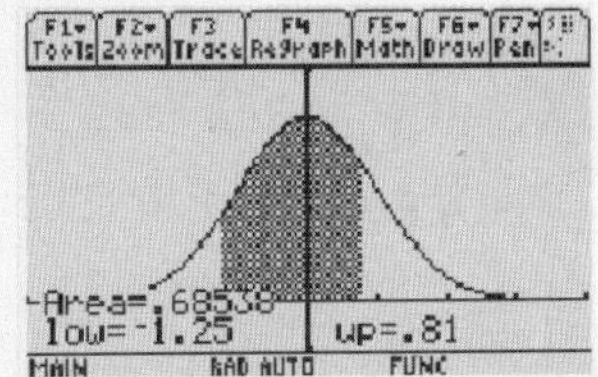

TI-Nspire instructions in Appendix B

So far, we have used Table A to find areas under the standard Normal curve from z-scores. What if we want to find the z-score that corresponds to a particular area? For example, let's find the 90th percentile of the standard Normal curve. We're looking for the z-score that has 90 percent of the area to its left, as shown in Figure 2.19.

The area to the *left* of z is 0.90. What's z?

z = ???

FIGURE 2.19 The z-score with area 0.90 to its left under the standard Normal curve.

Since Table A gives areas to the left of a specified z-score, all we need to do is find the value closest to 0.90 in the middle of the table. From the reproduced portion of Table A, you can see that the desired z-score is $z = 1.28$. That is, the area to the left of $z = 1.28$ is approximately 0.90.

z	.07	.08	.09
1.1	.8790	.8810	.8830
1.2	.8980	.8997	.9015
1.3	.9147	.9162	.9177

CHECK YOUR UNDERSTANDING

Use Table A in the back of the book to find the proportion of observations from a standard Normal distribution that fall in each of the following regions. In each case, sketch a standard Normal curve and shade the area representing the region.

1. $z < 1.39$
2. $z > -2.15$
3. $-0.56 < z < 1.81$

Use Table A to find the value z from the standard Normal distribution that satisfies each of the following conditions. In each case, sketch a standard Normal curve with your value of z marked on the axis.

4. The 20th percentile
5. 45% of all observations are greater than z

Normal Distribution Calculations

We can answer a question about proportions of observations in *any* Normal distribution by standardizing and then using the standard Normal table. Here is an outline of the method for finding the proportion of the distribution in any region. You should recognize the familiar four-step process. (Remember, **S**tatistics **P**roblems **D**emand **C**onsistency!)

How to Solve Problems Involving Normal Distributions

State: Express the problem in terms of the observed variable x.

Plan: Draw a picture of the distribution and shade the area of interest under the curve.

Do: Perform calculations.

- *Standardize* x to restate the problem in terms of a standard Normal variable z.
- *Use Table* A and the fact that the total area under the curve is 1 to find the required area under the standard Normal curve.

Conclude: Write your conclusion in the context of the problem.

Here's an example of the method at work.

Tiger on the Range

Normal calculations

On the driving range, Tiger Woods practices his swing with a particular club by hitting many, many balls. When Tiger hits his driver, the distance the ball travels follows a Normal distribution with mean 304 yards and standard deviation 8 yards. What percent of Tiger's drives travel at least 290 yards?

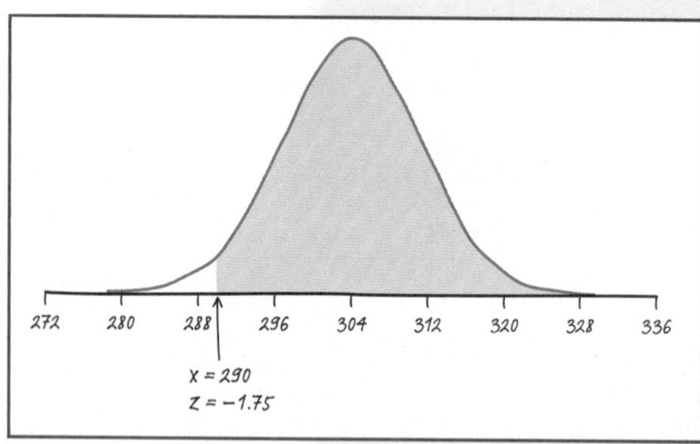

FIGURE 2.20 Distance traveled by Tiger Woods's drives on the range.

STATE: Let x = the distance that Tiger's ball travels. The variable x has a Normal distribution with $\mu = 304$ and $\sigma = 8$. We want the proportion of Tiger's drives with $x \geq 290$.

PLAN: Figure 2.20 is a picture of the distribution with the area of interest shaded.

DO:

- *Standardize.* For $x = 290$, we have

$$z = \frac{x - \mu}{\sigma} = \frac{290 - 304}{8} = -1.75$$

So $x \geq 290$ corresponds to $z \geq -1.75$ under the standard Normal curve.

- *Use Table A.* We see that the proportion of observations less than -1.75 is 0.0401. About 4% of Tiger's drives travel less than 290 yards. The area to the right of -1.75 is therefore $1 - 0.0401 = 0.9599$. This is about 0.96, or 96%.

CONCLUDE: About 96% of Tiger Woods's drives on the range travel at least 290 yards.

For Practice *Try Exercise 53(a)*

In a Normal distribution, the proportion of observations with $x \geq 290$ is the same as the proportion with $x > 290$. There is no *area* under the curve exactly above the point 290 on the horizontal axis, so the areas under the curve with $x \geq 290$ and $x > 290$ are the same. This isn't true of the actual data.

Tiger may hit a drive exactly 290 yards. The Normal distribution is just an easy-to-use approximation, not a description of every detail in the actual data.

The key to doing a Normal calculation is to sketch the area you want, then match that area with the area that the table gives you. Here's another example.

Tiger on the Range (Continued)

More complicated calculations

STEP 4

What percent of Tiger's drives travel between 305 and 325 yards?

STATE: As in the previous example, we let x = the distance that Tiger's ball travels. We want the proportion of Tiger's drives with $305 \leq x \leq 325$.

PLAN: Figure 2.21 is a picture of the distribution with the area of interest shaded.

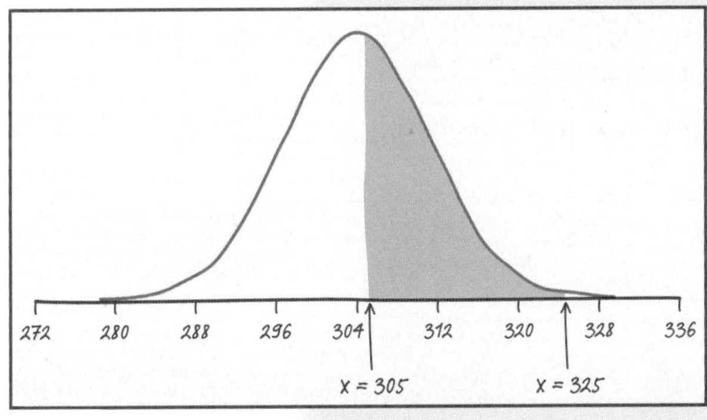

FIGURE 2.21 Distance traveled by Tiger Woods's drives on the range.

DO:

- *Standardize.* When $x = 305$, $z = \frac{305 - 304}{8} = 0.13$.

The standardized score for a 325-yard drive ($x = 325$) is

$$z = \frac{325 - 304}{8} = 2.63.$$

- *Use Table A.* The area between $z = 0.13$ and $z = 2.63$ is the area to the left of 2.63 minus the area to the left of 0.13. Look at the picture below to check this. From Table A, area between 0.13 and 2.63 = area to the left of 2.63 − area to the left of 0.13 = 0.9957 − 0.5517 = 0.4440.

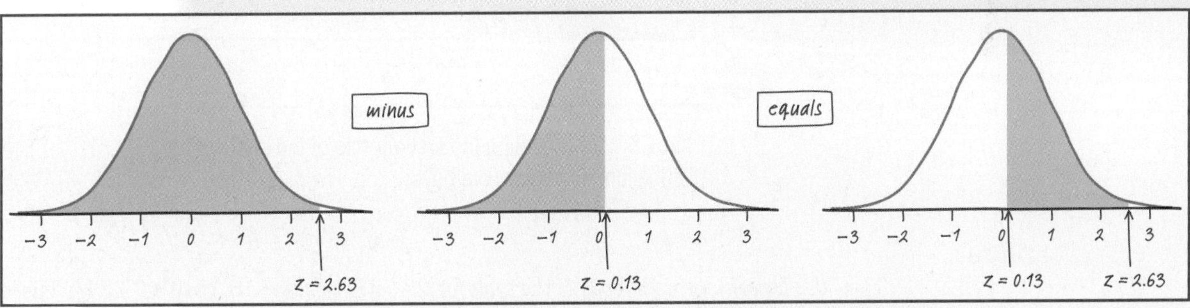

CONCLUDE: About 44% of Tiger's drives travel between 305 and 325 yards.

For Practice *Try Exercise 53(b)*

Sometimes we encounter a value of z more extreme than those appearing in Table A. For example, the area to the left of $z = -4$ is not given directly in the table. The z-values in Table A leave only area 0.0002 in each tail unaccounted for. For practical purposes, we can act as if there is zero area outside the range of Table A.

Finding a value when given a proportion The previous two examples illustrated the use of Table A to find what proportion of the observations satisfies some condition, such as "Tiger's drive travels between 305 and 325 yards." Sometimes, we may want to find the observed value that corresponds to a given percentile.

To do this, use Table A backward. Find the given proportion in the body of the table, read the corresponding z from the left column and top row, then "unstandardize" to get the observed value. Here's an example.

Cholesterol in Young Boys

Using Table A in reverse

High levels of cholesterol in the blood increase the risk of heart disease. For 14-year-old boys, the distribution of blood cholesterol is approximately Normal with mean $\mu = 170$ milligrams of cholesterol per deciliter of blood (mg/dl) and standard deviation $\sigma = 30$ mg/dl.[10] What is the first quartile of the distribution of blood cholesterol?

STATE: Call the cholesterol level x. This variable has the $N(170, 30)$ distribution. The first quartile is the value of x with 25% of the distribution to its left.

PLAN: Figure 2.22 shows a picture of what we are trying to find.

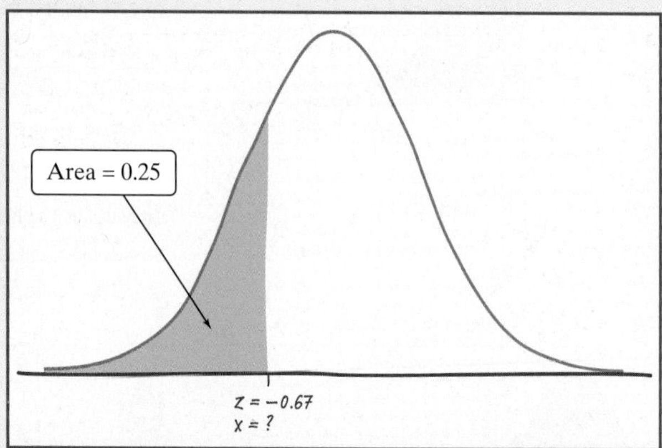

FIGURE 2.22 Locating the first quartile of the cholesterol distribution for 14-year-old boys.

DO:

- *Use Table A.* Look in the body of the table for the entry closest to 0.25. It's 0.2514. This is the entry corresponding to $z = -0.67$. So $z = -0.67$ is the standardized value with area 0.25 to its left.
- *Unstandardize.* We know that the standardized value of the unknown cholesterol level x is $z = -0.67$. So x satisfies the equation $\frac{x - 170}{30} = -0.67$. Solving for x gives

$$x = 170 + (-0.67)(30) = 149.9$$

CONCLUDE: The first quartile of blood cholesterol levels in 14-year-old boys is about 150 mg/dl.

For Practice *Try Exercise 53(c)*

As you've probably guessed, your calculator will help you perform Normal distribution calculations.

TECHNOLOGY CORNER From *z*-scores to areas, and vice versa

Finding areas: The `normalcdf` command on the TI-83/84 (`normCdf` on the TI-89) can be used to find areas under a Normal curve. This method is quicker than `shadeNorm` (see the Technology Corner on page 118) but has the disadvantage of not providing a picture of the area it is finding. The syntax is familiar: `normalcdf(lower bound, upper bound,` μ, σ`)`. Let's confirm our earlier answers in the two Tiger Woods examples. Recall that $\mu = 304$ yards and $\sigma = 8$ yards.

1. What proportion of Tiger's drives on the range travel at least 290 yards?

TI-83/84

- Press [2nd][VARS] `(DISTR)` and choose `2:normalcdf(`.
- Complete the command `normalcdf(290,400,304,8)` and press [ENTER].

```
normalcdf(290,40
0,304,8)
        .9599408865
```

TI-89

- Press [CATALOG][F3] `(Flash Apps)` and choose `normCdf(`.
- Complete the command `tistat.normCdf(290,400,304,8)` and press [ENTER].

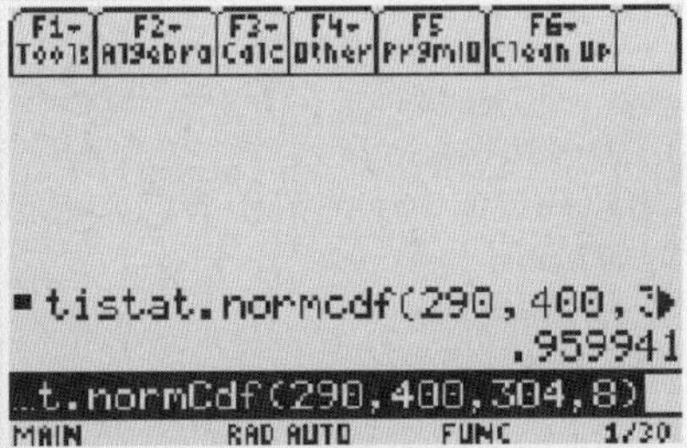

Note: We chose 400 as the upper bound because it's many, many standard deviations above the mean.

These results agree with our previous answer using Table A: 0.9599.

2. What percent of Tiger's drives travel between 305 and 325 yards? The screen shots below indicate that our earlier result of 0.4440 using Table A was a little off. This discrepancy was caused by the fact that we rounded our *z*-scores to two decimal places in order to use the table.

```
normalcdf(290,40
0,304,8)
        .9599408865
normalcdf(305,32
5,304,8)
        .4459292478
```

```
F1- Tools | F2- Algebra | F3- Calc | F4- Other | F5 PrgmIO | F6- Clean Up
■ tistat.normcdf(290, 400, 3▸
                     .959941
■ tistat.normcdf(305, 325, 3▸
                     .445929
..t.normCdf(305,325,304,8)
MAIN      RAD AUTO      FUNC      2/30
```

Working backward: The TI-83/84 and TI-89 `invNorm` function calculates the value corresponding to a given percentile in a Normal distribution. For this command, the syntax is `invNorm(percentile,` μ, σ`)`. Let's confirm our result from the example involving cholesterol levels in 14-year-old boys. Recall that $\mu = 170$ mg/dl and $\sigma = 30$ mg/dl.

3. What is the first quartile of the distribution of blood cholesterol?

AP EXAM TIP Don't use "calculator speak" when showing your work on free-response questions. Writing `normalcdf (305,325,304,8)` will not earn you credit for a Normal calculation. At the very least, you must indicate what each of those calculator inputs represents. For example, "I used normalcdf on my calculator with lower bound 305, upper bound 325, mean 304, and standard deviation 8." Better yet, sketch and label a Normal curve to show what you're finding.

TI-83/84

- Press [2nd][VARS] `(DISTR)` and choose `3:invNorm(`.
- Complete the command `invNorm(.25,170,30)` and press [ENTER].
- Compare this with the result of `invNorm(.25)`.

```
invNorm(.25,170,
30)
        149.7653075
invNorm(.25)
       -.6744897495
```

TI-89

- Press [CATALOG][F3] `(Flash Apps)` and choose `normCdf(`.
- Complete the command `tistat.invnorm(.25,170,30)` and press [ENTER].
- Compare this with the result of `tistat.invnorm(.25)`.

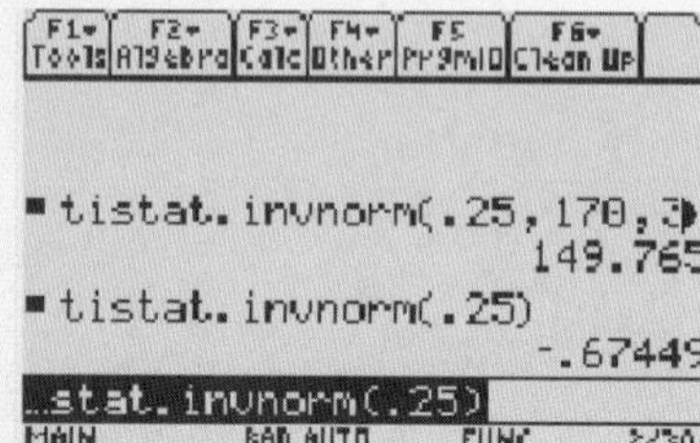

TECHNOLOGY TIP: For `normpdf`, `shadeNorm`, `normcdf`, and `invNorm`, the default values are $\mu = 0$ and $\sigma = 1$.

The first command shows that the first quartile (25th percentile) of the cholesterol distribution for 14-year-old males is 149.8 mg/dl. (Our earlier answer using Table A was 149.9 mg/dl.) The second command shows that in the standard Normal distribution, 25% of the observations fall below $z = -0.67449$. (We used $z = -0.67$ for our previous calculations, which explains the small discrepancy.)

TI-Nspire instructions in Appendix B

CHECK YOUR UNDERSTANDING

Follow the method shown in the examples to answer each of the following questions. Use your calculator or the *Normal Curve* applet to check your answers.

1. Cholesterol levels above 240 mg/dl may require medical attention. What percent of 14-year-old boys have more than 240 mg/dl of cholesterol?
2. People with cholesterol levels between 200 and 240 mg/dl are at considerable risk for heart disease. What percent of 14-year-old boys have blood cholesterol between 200 and 240 mg/dl?
3. What distance would a ball have to travel to be at the 80th percentile of Tiger Woods's drive lengths?

Assessing Normality

The Normal distributions provide good models for some distributions of real data. Examples include SAT and IQ test scores, the highway gas mileage of 2009 Corvette convertibles, state unemployment rates, and weights of 9-ounce bags of potato chips. The distributions of some other common variables are usually skewed and therefore distinctly non-Normal. Examples include economic variables such as personal income and total sales of business firms, the survival times of cancer patients after treatment, and the lifetime of electronic devices. While experience can suggest whether or not a Normal distribution is a reasonable model in a particular case, it is risky to assume that a distribution is Normal without actually inspecting the data.

In the latter part of this course, we will use various statistical inference procedures to try to answer questions that are important to us. These tests involve sampling individuals and recording data to gain insights about the populations from which they come. Many of these procedures are based on the assumption that the population is approximately Normally distributed. Consequently, we need to develop a strategy for assessing Normality.

Unemployment in the States

Are the data close to Normal?

Let's start by examining data on unemployment rates in the 50 states in November 2009. Here are the data arranged from lowest (North Dakota's 4.1%) to highest (Michigan's 14.7%).[11]

4.1	4.5	5.0	6.3	6.3	6.4	6.4	6.6	6.7	6.7	6.7	6.9	7.0
7.0	7.2	7.4	7.4	7.4	7.8	8.0	8.0	8.2	8.2	8.4	8.5	8.5
8.6	8.7	8.8	8.9	9.1	9.2	9.5	9.6	9.6	9.7	10.2	10.3	10.5
10.6	10.6	10.8	10.9	11.1	11.5	12.3	12.3	12.3	12.7	14.7		

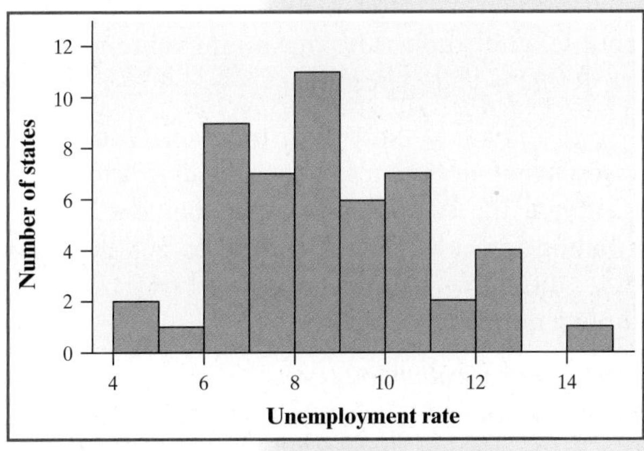

FIGURE 2.23 Histogram of September 2009 state unemployment rates.

- *Plot the data.* Make a dotplot, stemplot, or histogram. See if the graph is approximately symmetric and bell-shaped.

Figure 2.23 is a histogram of the state unemployment rates. The graph is roughly symmetric, single-peaked, and somewhat bell-shaped.

- *Check whether the data follow the 68–95–99.7 rule.*

We entered the unemployment rates into computer software and requested summary statistics. Here's what we got:

```
Mean = 8.682 Standard deviation = 2.225.
```

Now we can count the number of observations within one, two, and three standard deviations of the mean.

Mean ± 1std. dev.:	6.457 to 10.907	36 out of 50 = 72%
Mean ± 2std. dev.:	4.232 to 13.132	48 out of 50 = 96%
Mean ± 3std. dev.:	2.007 to 15.357	50 out of 50 = 100%

These percents are quite close to the 68%, 95%, and 99.7% targets for a Normal distribution.

If a graph of the data is clearly skewed, has multiple peaks, or isn't bell-shaped, that's evidence that the distribution is *not* Normal. However, just because a plot of the data *looks* Normal, we can't say that the distribution *is* Normal. The 68–95–99.7 rule can give additional evidence in favor of or against Normality. A **Normal probability plot** also provides a good assessment of whether a data set follows a Normal distribution.

Normal probability plot

EXAMPLE

Unemployment in the States

Making a Normal probability plot

Most software packages, including Minitab, Fathom, and JMP, can construct Normal probability plots (sometimes called Normal quantile plots) from entered data. The TI-83/84 and TI-89 will also make these graphs. Here's how a Normal probability plot is constructed.

1. *Arrange the observed data values from smallest to largest.* Record the percentile corresponding to each observation (but remember that there are several definitions of "percentile"). For example, the smallest observation in a set of 50 values is at either the 0th percentile (since 0 out of 50 values are below this observation) or the 2nd percentile (since 1 out of 50 values are at or below this observation). Technology usually "splits the difference," declaring this minimum value to be at the $(0 + 2)/2 = 1$st percentile. By similar reasoning, the second-smallest value is at the 3rd percentile, the third-smallest value is at the 5th percentile, and so on. The maximum value is at the $(98 + 100)/2 = 99$th percentile.

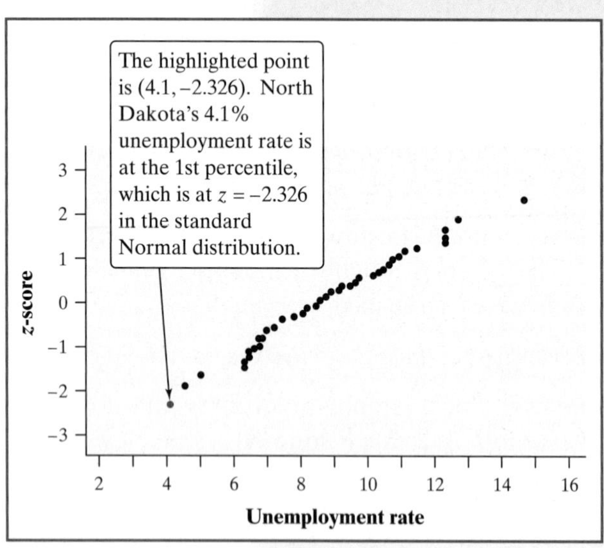

FIGURE 2.24 Normal probability plot of the percent of unemployed individuals in each of the 50 states.

2. *Use the standard Normal distribution (Table A or* `invNorm`*) to find the z-scores at these same percentiles.* For example, the 1st percentile of the standard Normal distribution is $z = -2.326$. The third percentile is $z = -1.881$; the fifth percentile is $z = -1.645$; . . . ; the 99th percentile is $z = 2.326$.

3. *Plot each observation x against the corresponding z.* If the data distribution is close to Normal, the plotted points will lie close to some straight line. Figure 2.24 shows a Normal probability plot for the state unemployment data. There is a strong linear pattern, which suggests that the distribution of unemployment rates is close to Normal.

Some software plots the x-values on the horizontal axis and the z-scores on the vertical axis, while other software does just the reverse. The TI-83/84 and TI-89 give you both options.

CAUTION

As Figure 2.24 indicates, real data almost always show some departure from Normality. *When you examine a Normal probability plot, look for shapes that show clear departures from Normality. Don't overreact to minor wiggles in the plot.* When we discuss statistical methods that are based on the Normal model, we will pay attention to the sensitivity of each method to depar-

AP EXAM TIP Normal probability plots are not included on the AP Statistics course outline. However, these graphs are very useful tools for assessing Normality. You may use them on the AP exam if you wish—just be sure that you know what you're looking for (linear pattern).

tures from Normality. Many common methods work well as long as the data are approximately Normal and outliers are not present.

Interpreting Normal Probability Plots

If the points on a **Normal probability plot** lie close to a straight line, the plot indicates that the data are Normal. Systematic deviations from a straight line indicate a non-Normal distribution. Outliers appear as points that are far away from the overall pattern of the plot.

Let's look at an example of some data that are *not* Normally distributed.

Guinea Pig Survival

Assessing Normality

In Chapter 1 Review Exercise R1.7 (page 77), we introduced data on the survival times in days of 72 guinea pigs after they were injected with infectious bacteria in a medical experiment.

PROBLEM: **Determine whether these data are approximately Normally distributed.**

SOLUTION: **Let's follow the first step in our strategy for assessing Normality: plot the data! Figure 2.25(a) shows a histogram of the guinea pig survival times. We can see that the distribution is heavily right-skewed. Figure 2.25(b) is a Normal probability plot of the data. The clear curvature in this graph confirms that these data do not follow a Normal distribution.**

We won't bother checking the 68–95–99.7 rule for these data since the graphs in Figure 2.25 indicate serious departures from Normality.

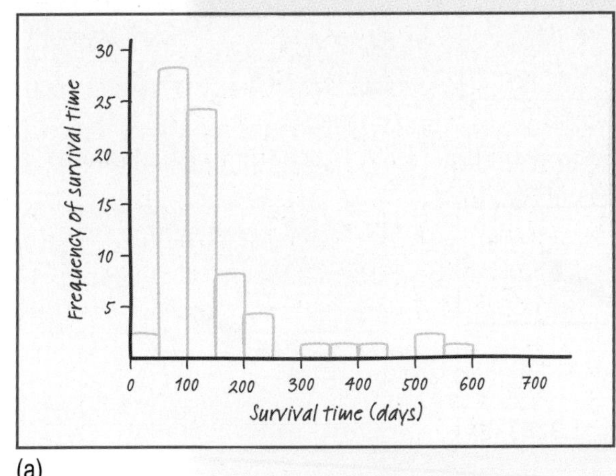

(a)

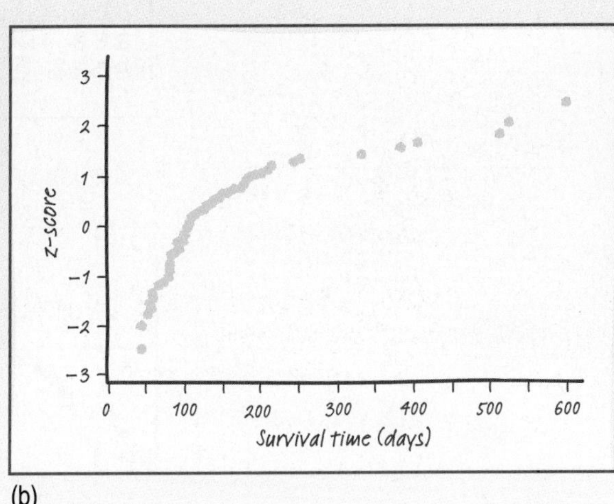

(b)

FIGURE 2.25 (a) Histogram and (b) Normal probability plot of the guinea pig survival data.

For Practice *Try Exercise 63*

How can we determine shape from a Normal probability plot? Look at the Normal probability plot of the guinea pig survival data in Figure 2.25(b). Imagine drawing a line through the leftmost points, which correspond to the smaller observations. The larger observations fall systematically to the right of this line. That is, the right-of-center observations have much larger values than expected based on their percentiles and the corresponding *z*-scores from the standard Normal distribution.

This Normal probability plot indicates that the guinea pig survival data are strongly right-skewed. *In a right-skewed distribution, the largest observations fall distinctly to the right of a line drawn through the main body of points.* Similarly, left-skewness is evident when the smallest observations fall to the left the line.

If you're wondering how to make a Normal probability plot on your calculator, the following Technology Corner shows you the process.

TECHNOLOGY CORNER Normal probability plots

To make a Normal probability plot for a set of quantitative data:

- Enter the data values in L1/list1. We'll use the state unemployment rates data from page 125.
- Define Plot1 as shown.

TI-83/84 TI-89

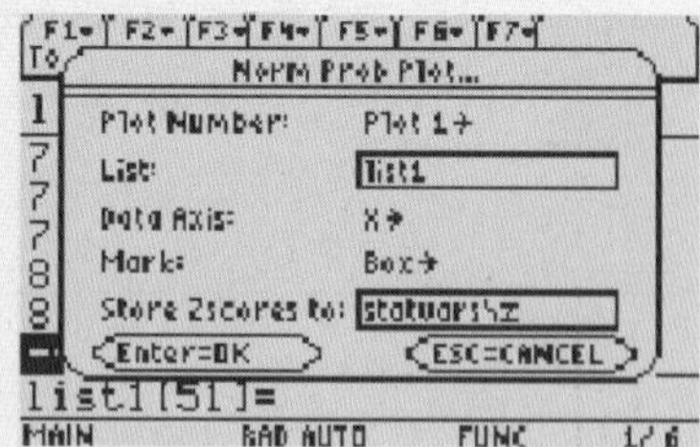

- Use `ZoomStat` (`ZoomData` on the TI-89) to see the finished graph.

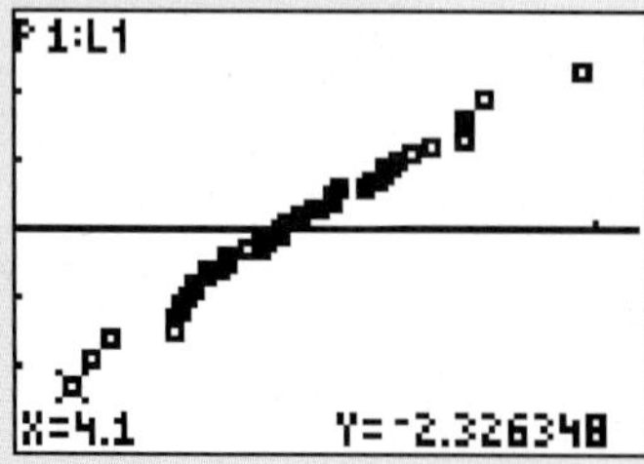

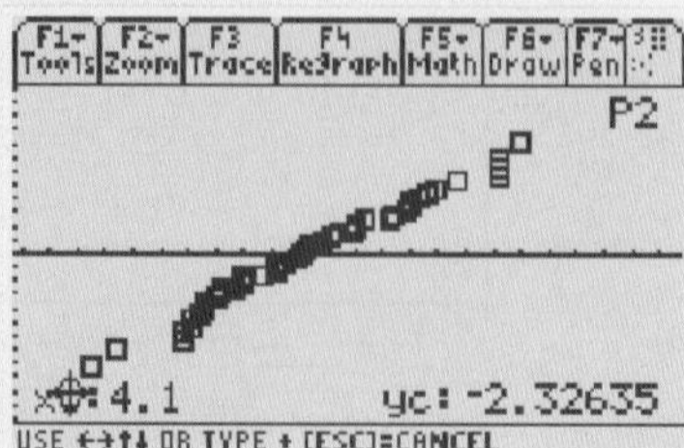

Interpretation: The Normal probability plot is quite linear, so it is reasonable to believe that the data follow a Normal distribution.

TI-Nspire instructions in Appendix B

DATA EXPLORATION *The vending machine problem*

Have you ever purchased a hot drink from a vending machine? The intended sequence of events runs something like this. You insert your money into the machine and select your preferred beverage. A cup falls out of the machine, landing upright. Liquid pours out until the cup is nearly full. You reach in, grab the piping-hot cup, and drink happily.

Sometimes, things go wrong. The machine might swipe your money. Or the cup might fall over. More frequently, everything goes smoothly until the liquid begins to flow. It might stop flowing when the cup is only half full. Or the liquid might keep coming until your cup overflows. Neither of these results leaves you satisfied.

The vending machine company wants to keep customers happy. So they have decided to hire you as a statistical consultant. They provide you with the following summary of important facts about the vending machine:

- Cups will hold 8 ounces.
- The amount of liquid dispensed varies according to a Normal distribution centered at the mean μ that is set in the machine.
- $\sigma = 0.2$ ounces.

If a cup contains too much liquid, a customer may get burned from a spill. This could result in an expensive lawsuit for the company. On the other hand, customers may be irritated if they get a cup with too little liquid from the machine. Given these issues, what mean setting for the machine would you recommend? Write a brief report to the vending machine company president that explains your answer.

case closed

Do You Sudoku?

In the chapter-opening Case Study (page 83), one of the authors played an online game of sudoku. At the end of his game, this graph was displayed. The density curve shown was constructed from a histogram of times from 4,000,000 games played in one week at this Web site. What can we learn from the graph?

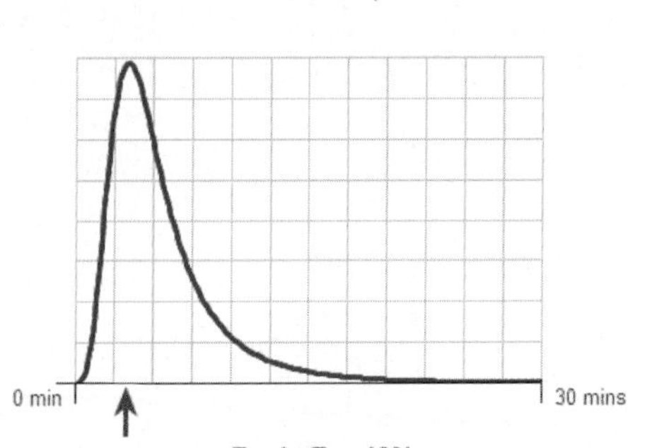

The author's time of 3 minutes and 19 seconds was well below the mean time of 5 minutes and 6 seconds to solve the easy-level puzzles. In fact, his performance placed him at the 19th percentile of sudoku solution times. The density curve's right-skewed shape reveals that most players were able to solve the easy puzzles in a fairly short time, but a few people took much longer.

SECTION 2.2 Summary

- The **Normal distributions** are described by a special family of bell-shaped, symmetric density curves, called **Normal curves.** The mean μ and standard deviation σ completely specify a Normal distribution $N(\mu,\sigma)$. The mean is the center of the curve, and σ is the distance from μ to the change-of-curvature points on either side.
- All Normal distributions obey the **68–95–99.7 rule,** which describes what percent of observations lie within one, two, and three standard deviations of the mean.
- All Normal distributions are the same when measurements are standardized. If x follows a Normal distribution with mean μ and standard deviation σ, we can standardize using

$$z = \frac{x - \mu}{\sigma}.$$

  The variable z has the **standard Normal distribution** with mean 0 and standard deviation 1.
- **Table A** at the back of the book gives percentiles for the standard Normal curve. By standardizing, we can use Table A to determine the percentile for a given z-score or the z-score corresponding to a given percentile in any Normal distribution. You can use your calculator or the *Normal Curve* applet to perform Normal calculations quickly.
- To perform certain inference procedures in later chapters, we will need to know that the data come from populations that are approximately Normally distributed. To assess Normality for a given set of data, we first observe the shape of a dotplot, stemplot, or histogram. Then we can check how well the data fit the 68–95–99.7 rule for Normal distributions. Another good method for assessing Normality is to construct a **Normal probability plot.**

2.2 TECHNOLOGY CORNERS

TI-Nspire instructions in Appendix B

SECTION 2.2 Exercises

41. **Men's heights** The distribution of heights of adult American men is approximately Normal with mean 69 inches and standard deviation 2.5 inches. Draw a Normal curve on which this mean and standard deviation are correctly located. (*Hint:* Draw the curve first, locate the points where the curvature changes, then mark the horizontal axis.)

pg 113

42. **Potato chips** The distribution of weights of 9-ounce bags of a particular brand of potato chips is approximately Normal with mean $\mu = 9.12$ ounces and standard deviation $\sigma = 0.05$ ounce. Draw an accurate sketch of the distribution of potato chip bag weights. Be sure to label the mean, as well as the points one, two, and three standard deviations away from the mean on the horizontal axis.

43. **Men's heights** Refer to Exercise 41. Use the 68–95–99.7 rule to answer the following questions. Show your work!

pg 113

(a) What percent of men are taller than 74 inches?

(b) Between what heights do the middle 95% of men fall?

(c) What percent of men are between 64 and 66.5 inches tall?

(d) A height of 71.5 inches corresponds to what percentile of adult male American heights?

44. **Potato chips** Refer to Exercise 42. Use the 68–95–99.7 rule to answer the following questions. Show your work!

(a) What percent of bags weigh less than 9.02 ounces?

(b) Between what weights do the middle 68% of bags fall?

(c) What percent of 9-ounce bags of this brand of potato chips weigh between 8.97 and 9.17 ounces?

(d) A bag that weighs 9.07 ounces is at what percentile in this distribution?

45. **Estimating SD** The figure at top right shows two Normal curves, both with mean 0. Approximately what is the standard deviation of each of these curves?

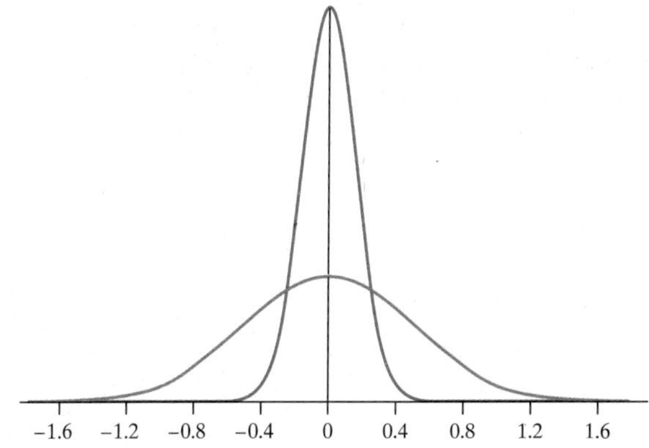

46. **A Normal curve** Estimate the mean and standard deviation of the Normal density curve in the figure below.

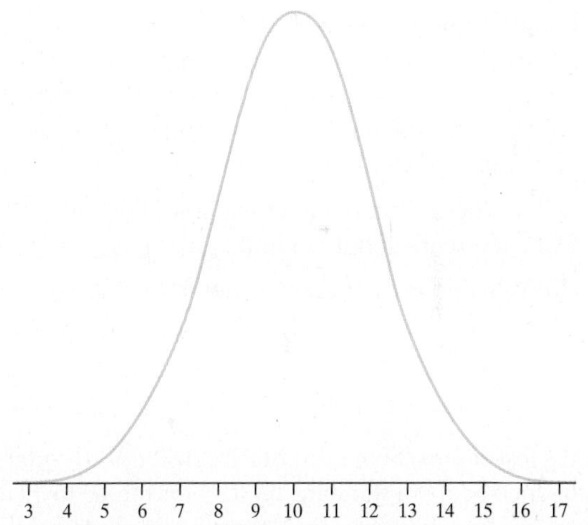

For Exercises 47 to 50, use Table A to find the proportion of observations from the standard Normal distribution that satisfies each of the following statements. In each case, sketch a standard Normal curve and shade the area under the curve that is the answer to the question. Use your calculator or the *Normal Curve* applet to check your answers.

47. **Table A practice**

(a) $z < 2.85$
(b) $z > 2.85$
(c) $z > -1.66$
(d) $-1.66 < z < 2.85$

48. **Table A practice**

(a) $z < -2.46$
(b) $z > 2.46$
(c) $0.89 < z < 2.46$
(d) $-2.95 < z < -1.27$

49. **More Table A practice**
(a) z is between -1.33 and 1.65
pg 117 (b) z is between 0.50 and 1.79

50. **More Table A practice**
(a) z is between -2.05 and 0.78
(b) z is between -1.11 and -0.32

For Exercises 51 and 52, use Table A to find the value z from the standard Normal distribution that satisfies each of the following conditions. In each case, sketch a standard Normal curve with your value of z marked on the axis. Use your calculator or the *Normal Curve* applet to check your answers.

51. **Working backward**
(a) The 10th percentile.
(b) 34% of all observations are greater than z.

52. **Working backward**
(a) The 63rd percentile.
(b) 75% of all observations are greater than z.

53. **Length of pregnancies** The length of human pregnancies from conception to birth varies according to a distribution that is approximately Normal with mean 266 days and standard deviation 16 days. For each part, follow the four-step process.
pg 120–122

(a) At what percentile is a pregnancy that lasts 240 days (that's about 8 months)?
(b) What percent of pregnancies last between 240 and 270 days (roughly between 8 months and 9 months)?
(c) How long do the longest 20% of pregnancies last?

54. **IQ test scores** Scores on the Wechsler Adult Intelligence Scale (a standard IQ test) for the 20 to 34 age group are approximately Normally distributed with $\mu = 110$ and $\sigma = 25$. For each part, follow the four-step process.
(a) At what percentile is an IQ score of 150?
(b) What percent of people aged 20 to 34 have IQs between 125 and 150?
(c) MENSA is an elite organization that admits as members people who score in the top 2% on IQ tests. What score on the Wechsler Adult Intelligence Scale would an individual have to earn to qualify for MENSA membership?

55. **I think I can!** An important measure of the performance of a locomotive is its "adhesion," which is the locomotive's pulling force as a multiple of its weight. The adhesion of one 4400-horsepower diesel locomotive varies in actual use according to a Normal distribution with mean $\mu = 0.37$ and standard deviation $\sigma = 0.04$. For each part that follows, sketch and shade an appropriate Normal distribution. Then show your work.
(a) For a certain small train's daily route, the locomotive needs to have an adhesion of at least 0.30 for the train to arrive at its destination on time. On what proportion of days will this happen? Show your method.
(b) An adhesion greater than 0.50 for the locomotive will result in a problem because the train will arrive too early at a switch point along the route. On what proportion of days will this happen? Show your method.
(c) Compare your answers to (a) and (b). Does it make sense to try to make one of these values larger than the other? Why or why not?

56. **Put a lid on it!** At some fast-food restaurants, customers who want a lid for their drinks get them from a large stack left near straws, napkins, and condiments. The lids are made with a small amount of flexibility so they can be stretched across the mouth of the cup and then snuggly secured. When lids are too small or too large, customers can get very frustrated, especially if they end up spilling their drinks. At one particular restaurant, large drink cups require lids with a "diameter" of between 3.95 and 4.05 inches. The restaurant's lid supplier claims that the diameter of their large lids follows a Normal distribution with mean 3.98 inches and standard deviation 0.02 inches. Assume that the supplier's claim is true.
(a) What percent of large lids are too small to fit? Show your method.
(b) What percent of large lids are too big to fit? Show your method.
(c) Compare your answers to (a) and (b). Does it make sense for the lid manufacturer to try to make one of these values larger than the other? Why or why not?

57. **I think I can!** Refer to Exercise 55. The locomotive's manufacturer is considering two changes that could reduce the percent of times that the train arrives late. One option is to increase the mean adhesion of the locomotive. The other possibility is to decrease the variability in adhesion from trip to trip, that is, to reduce the standard deviation.
(a) If the standard deviation remains at $\sigma = 0.04$, to what value must the manufacturer change the mean adhesion of the locomotive to reduce its proportion of late arrivals to less than 2% of days? Show your work.
(b) If the mean adhesion stays at $\mu = 0.37$, how much must the standard deviation be decreased to ensure that the train will arrive late less than 2% of the time? Show your work.

(c) Which of the two options in (a) and (b) do you think is preferable? Justify your answer. (Be sure to consider the effect of these changes on the percent of days that the train arrives early to the switch point.)

58. **Put a lid on it!** Refer to Exercise 56. The supplier is considering two changes to reduce the percent of its large-cup lids that are too small to less than 1%. One strategy is to adjust the mean diameter of its lids. Another option is to alter the production process, thereby decreasing the standard deviation of the lid diameters.

(a) If the standard deviation remains at $\sigma = 0.02$ inches, at what value should the supplier set the mean diameter of its large-cup lids to ensure that less than 1% are too small to fit? Show your method.

(b) If the mean diameter stays at $\mu = 3.98$ inches, what value of the standard deviation will result in less than 1% of lids that are too small to fit? Show your method.

(c) Which of the two options in (a) and (b) do you think is preferable? Justify your answer. (Be sure to consider the effect of these changes on the percent of lids that are too large to fit.)

59. **Deciles** The deciles of any distribution are the points that mark off the lowest 10% and the highest 10%. The deciles of a density curve are therefore the points with area 0.1 and 0.9 to their left under the curve.

(a) What are the deciles of the standard Normal distribution?

(b) The heights of young women are approximately Normal with mean 64.5 inches and standard deviation 2.5 inches. What are the deciles of this distribution? Show your work.

60. **Outliers** The percent of the observations that are classified as outliers by the $1.5 \times IQR$ rule is the same in any Normal distribution. What is this percent? Show your method clearly.

61. **Flight times** An airline flies the same route at the same time each day. The flight time varies according to a Normal distribution with unknown mean and standard deviation. On 15% of days, the flight takes more than an hour. On 3% of days, the flight lasts 75 minutes or more. Use this information to determine the mean and standard deviation of the flight time distribution.

62. **Brush your teeth** The amount of time Ricardo spends brushing his teeth follows a Normal distribution with unknown mean and standard deviation. Ricardo spends less than one minute brushing his teeth about 40% of the time. He spends more than two minutes brushing his teeth 2% of the time. Use this information to determine the mean and standard deviation of this distribution.

63. **Sharks** Here are the lengths in feet of 44 great white sharks:[12]

pg 127

18.7	12.3	18.6	16.4	15.7	18.3	14.6	15.8	14.9	17.6	12.1
16.4	16.7	17.8	16.2	12.6	17.8	13.8	12.2	15.2	14.7	12.4
13.2	15.8	14.3	16.6	9.4	18.2	13.2	13.6	15.3	16.1	13.5
19.1	16.2	22.8	16.8	13.6	13.2	15.7	19.7	18.7	13.2	16.8

(a) Enter these data into your calculator and make a histogram. Then calculate one-variable statistics. Describe the shape, center, and spread of the distribution of shark lengths.

(b) Calculate the percent of observations that fall within one, two, and three standard deviations of the mean. How do these results compare with the 68–95–99.7 rule?

(c) Use your calculator to construct a Normal probability plot. Interpret this plot.

(d) Having inspected the data from several different perspectives, do you think these data are approximately Normal? Write a brief summary of your assessment that combines your findings from (a) through (c).

64. **Density of the earth** In 1798, the English scientist Henry Cavendish measured the density of the earth several times by careful work with a torsion balance. The variable recorded was the density of the earth as a multiple of the density of water. Here are Cavendish's 29 measurements:[13]

5.50	5.61	4.88	5.07	5.26	5.55	5.36	5.29	5.58	5.65
5.57	5.53	5.62	5.29	5.44	5.34	5.79	5.10	5.27	5.39
5.42	5.47	5.63	5.34	5.46	5.30	5.75	5.68	5.85	

(a) Enter these data into your calculator and make a histogram. Then calculate one-variable statistics. Describe the shape, center, and spread of the distribution of density measurements.

(b) Calculate the percent of observations that fall within one, two, and three standard deviations of the mean. How do these results compare with the 68–95–99.7 rule?

(c) Use your calculator to construct a Normal probability plot. Interpret this plot.

(d) Having inspected the data from several different perspectives, do you think these data are approximately Normal? Write a brief summary of your assessment that combines your findings from (a) through (c).

65. **Runners' heart rates** The figure below is a Normal probability plot of the heart rates of 200 male runners after six minutes of exercise on a treadmill.[14] The distribution is close to Normal. How can you see this? Describe the nature of the small deviations from Normality that are visible in the plot.

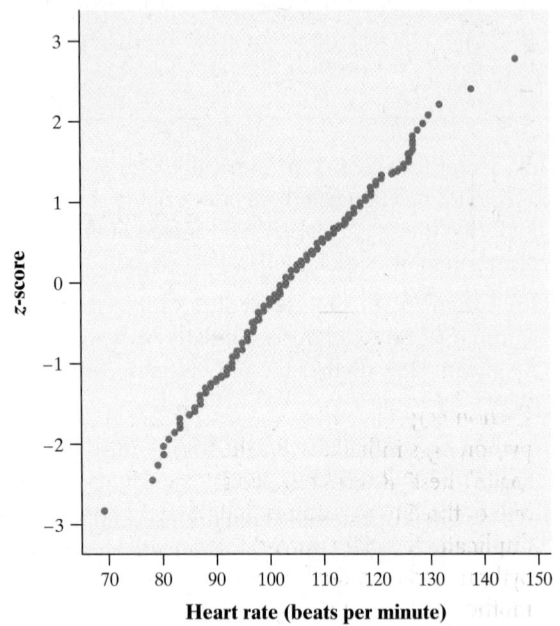

66. **Carbon dioxide emissions** The figure below is a Normal probability plot of the emissions of carbon dioxide per person in 48 countries.[15] In what ways is this distribution non-Normal?

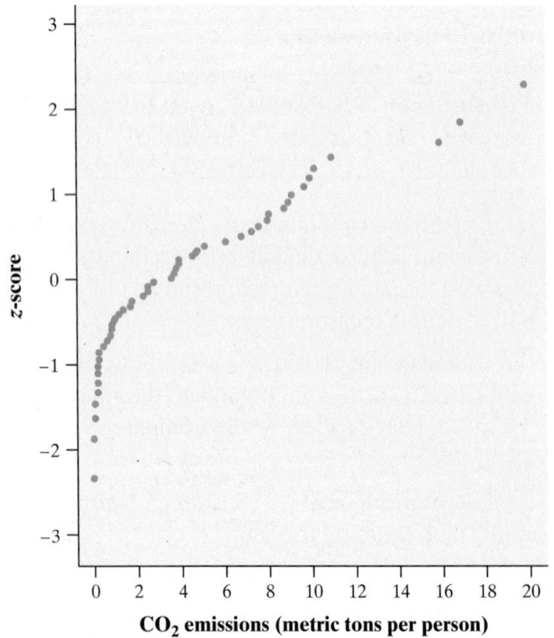

67. **Normal is only approximate: ACT scores** Scores on the ACT test for the 2007 high school graduating class had mean 21.2 and standard deviation 5.0. In all, 1,300,599 students in this class took the test. Of these, 149,164 had scores higher than 27 and another 50,310 had scores exactly 27. ACT scores are always whole numbers. The exactly Normal $N(21.2, 5.0)$ distribution can include any value, not just whole numbers. What's more, there is *no* area exactly above 27 under the smooth Normal curve. So ACT scores can be only approximately Normal. To illustrate this fact, find

(a) the percent of 2007 ACT scores greater than 27.

(b) the percent of 2007 ACT scores greater than or equal to 27.

(c) the percent of observations from the $N(21.2, 5.0)$ distribution that are greater than 27. (The percent greater than or equal to 27 is the same, because there is no area exactly over 27.)

68. **Weights aren't Normal** The heights of people of the same gender and similar ages follow Normal distributions reasonably closely. Weights, on the other hand, are not Normally distributed. The weights of women aged 20 to 29 have mean 141.7 pounds and median 133.2 pounds. The first and third quartiles are 118.3 pounds and 157.3 pounds. What can you say about the shape of the weight distribution? Why?

Multiple choice: Select the best answer for Exercises 69 to 74.

69. Which of the following is *least* likely to have a nearly Normal distribution?

(a) Heights of all female students taking STAT 001 at State Tech.

(b) IQ scores of all students taking STAT 001 at State Tech.

(c) SAT Math scores of all students taking STAT 001 at State Tech.

(d) Family incomes of all students taking STAT 001 at State Tech.

(e) All of (a)–(d) will be approximately Normal.

Exercises 70 to 72 refer to the following setting. The weights of laboratory cockroaches follow a Normal distribution with mean 80 grams and standard deviation 2 grams. The figure below is the Normal curve for this distribution of weights.

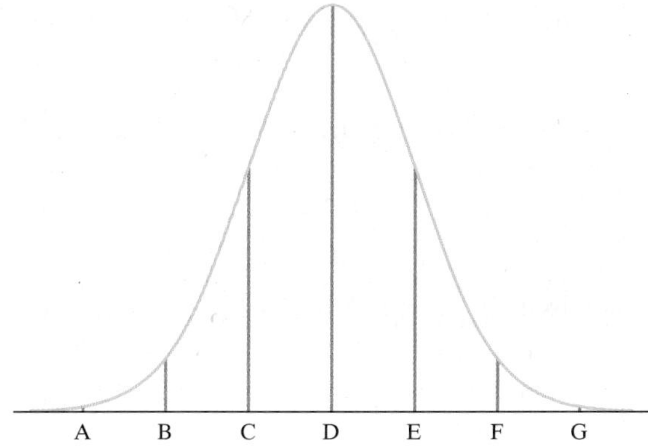

70. Point C on this Normal curve corresponds to

(a) 84 grams. (c) 78 grams. (e) 74 grams.
(b) 82 grams. (d) 76 grams.

71. About what percent of the cockroaches have weights between 76 and 84 grams?

(a) 99.7% (c) 68% (e) 34%
(b) 95% (d) 47.5%

72. About what percent of the cockroaches have weights less than 78 grams?

(a) 34% (c) 16% (e) none of these
(b) 32% (d) 2.5%

73. The proportion of observations from a standard Normal distribution with values less than 1.15 is

(a) 0.1251. (c) 0.8749. (e) none of these.
(b) 0.8531. (d) 0.8944

74. The proportion of observations from a standard Normal distribution with values larger than −0.75 is

(a) 0.2266. (c) 0.7734. (e) none of these
(b) 0.7422. (d) 0.8023.

75. **Gas it up!** (1.3) Interested in a sporty car? Worried that it might use too much gas? The Environmental Protection Agency lists most such vehicles in its "two-seater" or "minicompact" categories. The figure shows boxplots for both city and highway gas mileages for our two groups of cars. Write a few sentences comparing these distributions.

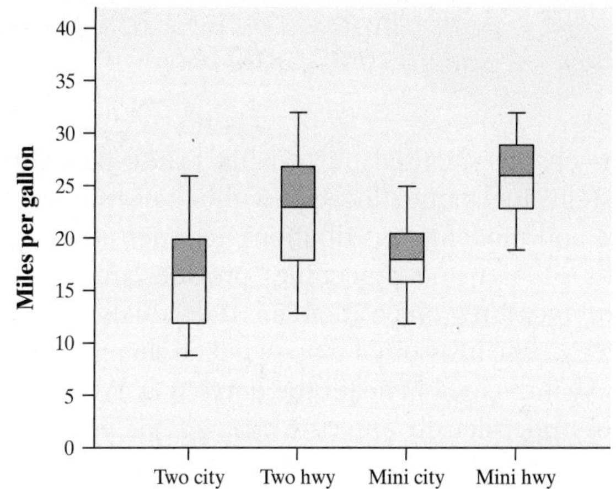

76. **Python eggs** (1.1) How is the hatching of water python eggs influenced by the temperature of the snake's nest? Researchers assigned newly laid eggs to one of three temperatures: hot, neutral, or cold. Hot duplicates the extra warmth provided by the mother python, and cold duplicates the absence of the mother. Here are the data on the number of eggs and the number that hatched:[16]

	Cold	Neutral	Hot
Number of eggs	27	56	104
Number hatched	16	38	75

(a) Make a two-way table of temperature by outcome (hatched or not).

(b) Calculate the percent of eggs in each group that hatched. The researchers believed that eggs would not hatch in cold water. Do the data support that belief?

Chapter 2 Chapter Review

This chapter focused on two big issues: describing an individual value's location within a distribution of data and modeling distributions with density curves. Both z-scores and percentiles provide easily calculated measures of position for individuals. Density curves come in assorted shapes, but all share the property that the area beneath the curve is 1. We can use areas under density curves to estimate the proportion of individuals in a distribution whose values fall in a specified range. In the special case of Normally distributed data, we can use the standard Normal curve and Table A to calculate such areas. There are many real-world examples of Normal distributions. When you meet a new set of data, you can use the graphical and numerical tools discussed in this chapter to assess the Normality of the data.

Chapter Review Exercises

These exercises are designed to help you review the important ideas and methods of the chapter. Relevant learning objectives are provided in bulleted form before each exercise.

- Find the standardized value (z-score) of an observation. Interpret z-scores in context.
- Use percentiles to locate individual values within distributions of data.

Exercises R2.1 and R2.2 refer to the following setting. According to the National Center for Health Statistics, the distribution of heights for 15-year-old males is symmetric, single-peaked, and bell-shaped. For this distribution, a z-score of 0 corresponds to a height of 170 centimeters (cm) and a z-score of 1 corresponds to a height of 177.5 cm.

R2.1 Male heights Consider the height distribution for 15-year-old males.

(a) Find its mean and standard deviation. Show your method clearly.

(b) What height would correspond to a z-score of 2.5? Show your work.

R2.2 Is Paul tall? Paul is 15 years old and 179 cm tall.

(a) Find the z-score corresponding to Paul's height. Explain what this value means.

(b) Paul's height puts him at the 85th percentile among 15-year-old males. Explain what this means to someone who knows no statistics.

- Interpret a cumulative relative frequency graph.

R2.3 Computer use Mrs. Causey asked her students how much time they had spent using a computer during the previous week. The following figure shows a cumulative relative frequency graph of her students' responses.

(a) At what percentile does a student who used her computer for 10 hours last week fall?

(b) Estimate the median, Q_1, and Q_3 from the graph. Are there any outliers?

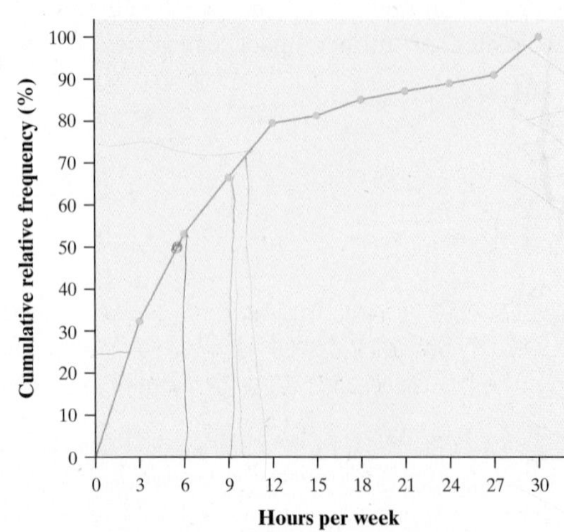

- Describe the effect of adding, subtracting, multiplying by, or dividing by a constant on the shape, center, and spread of a distribution of data.

R2.4 Aussie, Aussie, Aussie A group of Australian students were asked to estimate the width of their classroom in feet. Use the dotplot and summary statistics below to answer the following questions.

(a) Suppose we converted each student's guess from feet to meters (3.28 ft = 1 m). How would the shape of the distribution be affected? Find the mean, median, standard deviation, and *IQR* for the transformed data.

(b) The actual width of the room was 42.6 feet. Suppose we calculated the error in each student's guess as follows: guess − 42.6. Find the mean and standard deviation of the errors. How good were the students' guesses? Justify your answer.

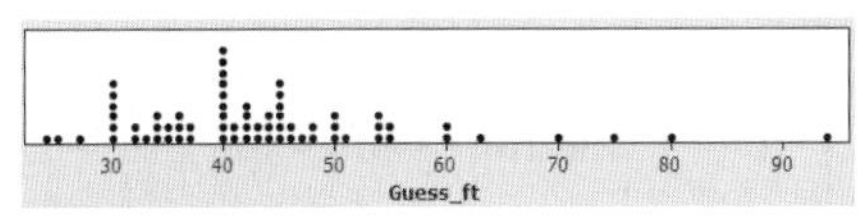

Variable	N	Mean	StDev	Minimum	Q1	Median	Q3	Maximum
Guess_ft	69	43.70	12.50	24.00	35.50	42.00	48.00	94.00

- Approximately locate the median (equal-areas point) and the mean (balance point) on a density curve.

R2.5 What the mean means The figure below is a density curve. Trace the curve onto your paper.

(a) Mark the approximate location of the median. Justify your choice of location.

(b) Mark the approximate location of the mean. Justify your choice of location.

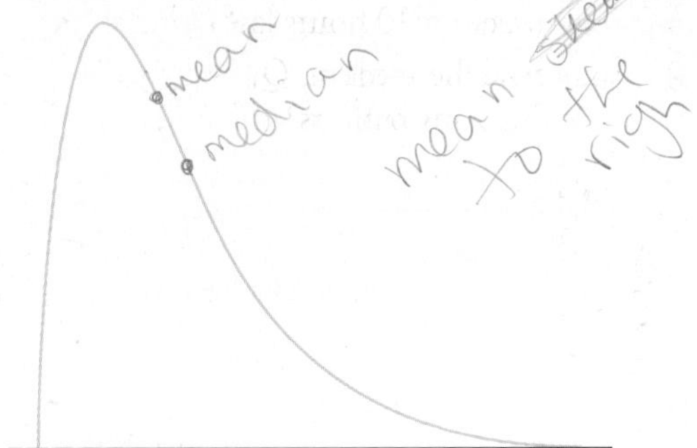

- Use the 68–95–99.7 rule to estimate the percent of observations from a Normal distribution that fall in an interval involving points one, two, or three standard deviations on either side of the mean.

R2.6 Horse pregnancies Bigger animals tend to carry their young longer before birth. The length of horse pregnancies from conception to birth varies according to a roughly Normal distribution with mean 336 days and standard deviation 3 days. Use the 68–95–99.7 rule to answer the following questions.

(a) Almost all (99.7%) horse pregnancies fall in what range of lengths?

(b) What percent of horse pregnancies are longer than 339 days? Show your work.

- Use the standard Normal distribution to calculate the proportion of values in a specified interval.

R2.7 Standard Normal areas Use Table A to find the proportion of observations from a standard Normal distribution that falls in each of the following regions. In each case, sketch a standard Normal curve and shade the area representing the region.

(a) $z \leq -2.25$ (c) $z > 1.77$

(b) $z \geq -2.25$ (d) $-2.25 < z < 1.77$

- Use the standard Normal distribution to determine a *z*-score from a percentile.

R2.8 Working backward

(a) Find the number z at the 80th percentile of a standard Normal distribution.

(b) Find the number z such that 35% of all observations from a standard Normal distribution are greater than z.

- Use Table A to find the percentile of a value from any Normal distribution and the value that corresponds to a given percentile.

R2.9 Low-birth-weight babies Researchers in Norway analyzed data on the birth weights of 400,000 newborns over a 6-year period. The distribution of birth weights is Normal with a mean of 3668 grams and a standard deviation of 511 grams.[17] Babies that weigh less than 2500 grams at birth are classified as "low birth weight."

(a) What percent of babies will be identified as low birth weight? Show your work.

(b) Find the quartiles of the birth weight distribution. Show your work.

R2.10 Grading managers Many companies "grade on a bell curve" to compare the performance of their managers and professional workers. This forces the use of some low performance ratings, so that not all workers are listed as "above average." Ford Motor Company's "performance management process" for a time assigned 10% A grades, 80% B grades, and 10% C grades to the company's 18,000 managers. Suppose that Ford's performance scores really are Normally distributed. This year, managers with scores less than 25 received C's, and those with scores above 475 received A's. What are the mean and standard deviation of the scores? Show your work.

- Make an appropriate graph to determine if a distribution is bell-shaped.
- Use the 68–95–99.7 rule to assess Normality of a data set.

R2.11 Fruit fly thorax lengths Here are the lengths in millimeters of the thorax for 49 male fruit flies:[18]

0.64	0.64	0.64	0.68	0.68	0.68	0.72	0.72	0.72	0.72	0.74	0.76	0.76
0.76	0.76	0.76	0.76	0.76	0.76	0.78	0.80	0.80	0.80	0.80	0.80	0.82
0.82	0.84	0.84	0.84	0.84	0.84	0.84	0.84	0.84	0.84	0.84	0.88	0.88
0.88	0.88	0.88	0.88	0.88	0.88	0.92	0.92	0.92	0.94			

Are these data approximately Normally distributed? Give appropriate graphical and numerical evidence to support your answer.

- Interpret a Normal probability plot.

R2.12 Assessing Normality A Normal probability plot of a set of data is shown here. Would you say that these measurements are approximately Normally distributed? Why or why not?

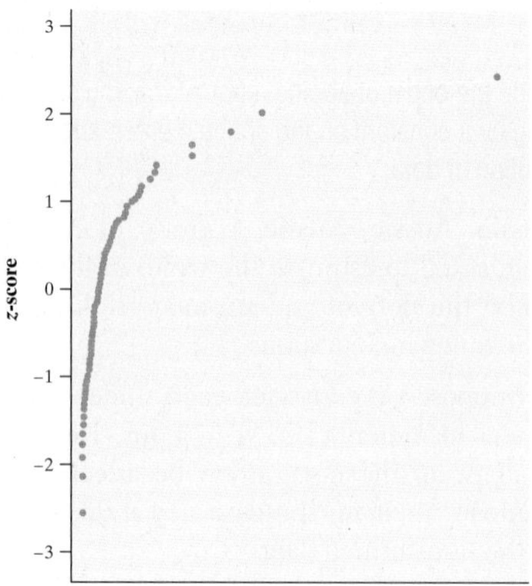

Chapter 2 AP Statistics Practice Test

Section I: Multiple Choice *Select the best answer for each question.*

T2.1. Many professional schools require applicants to take a standardized test. Suppose that 1000 students take such a test. Several weeks after the test, Pete receives his score report: he got a 63, which placed him at the 73rd percentile. This means that

(a) Pete's score was below the median.
(b) Pete did worse than about 63% of the test takers.
(c) Pete did worse than about 73% of the test takers.
(d) Pete did better than about 63% of the test takers.
(e) Pete did better than about 73% of the test takers.

T2.2. For the Normal distribution shown, the standard deviation is closest to

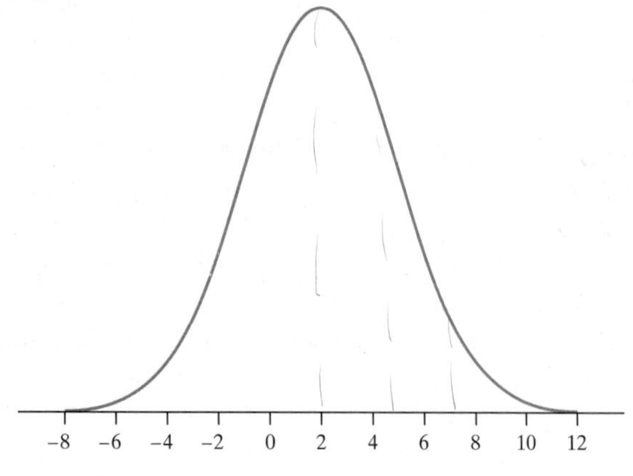

(a) 0 (b) 1 (c) 2 (d) 3 (e) 5

T2.3. Rainwater was collected in water collectors at 30 different sites near an industrial complex, and the amount of acidity (pH level) was measured. The mean and standard deviation of the values are 4.60 and 1.10, respectively. When the pH meter was recalibrated back at the laboratory, it was found to be in error. The error can be corrected by adding 0.1 pH units to all of the values and then multiplying the result by 1.2. The mean and standard deviation of the corrected pH measurements are

(a) 5.64, 1.44 (b) 5.64, 1.32 (c) 5.40, 1.44 (d) 5.40, 1.32 (e) 5.64, 1.20

T2.4. The figure shows a cumulative relative frequency graph of the number of ounces of alcohol consumed per week in a sample of 150 adults. About what percent of these adults consume between 4 and 8 ounces per week?

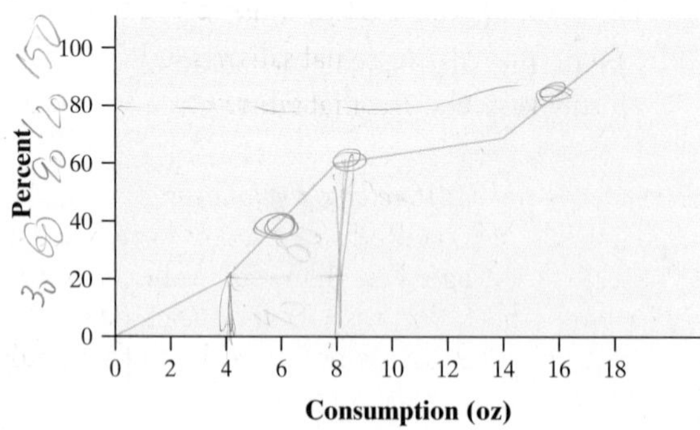

(a) 20% (b) 40% (c) 50% (d) 60% (e) 80%

T2.5. The average yearly snowfall in Chillyville is Normally distributed with a mean of 55 inches. If the snowfall in Chillyville exceeds 60 inches in 15% of the years, what is the standard deviation?

(a) 4.83 inches
(b) 5.18 inches
(c) 6.04 inches
(d) 8.93 inches
(e) The standard deviation cannot be computed from the given information.

T2.6. The figure shown is the density curve of a distribution. Five of the seven points marked on the density curve make up the five-number summary for this distribution. Which two points are *not* part of the five-number summary?

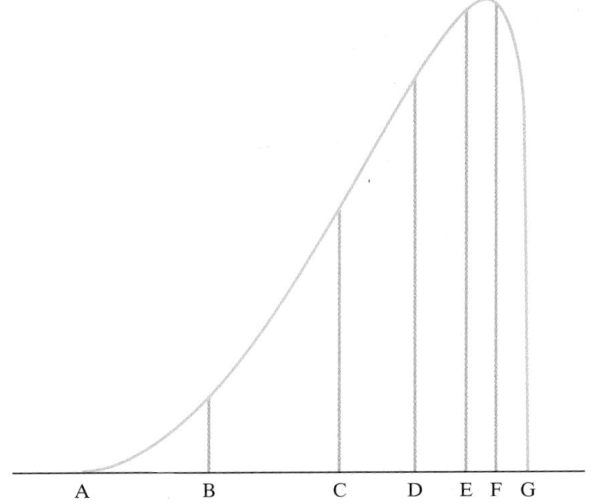

(a) B and E (c) C and E (e) A and G
(b) C and F (d) B and F

T2.7. If the heights of American men follow a Normal distribution, and 99.7% have heights between 5′0″ and 7′0″, what is your estimate of the standard deviation of the height of American men?

(a) 1″ (b) 3″ (c) 4″ (d) 6″ (e) 12″

T2.8. Which of the following is *not* correct about a standard Normal distribution?

(a) The proportion of scores that satisfy $0 < z < 1.5$ is 0.4332.

(b) The proportion of scores that satisfy $z < -1.0$ is 0.1587.

(c) The proportion of scores that satisfy $z > 2.0$ is 0.0228.

(d) The proportion of scores that satisfy $z < 1.5$ is 0.9332.

(e) The proportion of scores that satisfy $z > -3.0$ is 0.9938.

Questions T2.9 and T2.10 refer to the following setting. Until the scale was changed in 1995, SAT scores were based on a scale set many years ago. For Math scores, the mean under the old scale in the 1990s was 470 and the standard deviation was 110. In 2009, the mean was 515 and the standard deviation was 116.

T2.9. What is the standardized score (z-score) for a student who scored 500 on the old SAT scale?

(a) −30 (b) −0.27 (c) −0.13 (d) 0.13 (e) 0.27

T2.10. Jane took the SAT in 1994 and scored 500. Her sister Colleen took the SAT in 2009 and scored 530. Who did better on the exam, and how can you tell?

(a) Colleen—she scored 30 points higher than Jane.

(b) Colleen—her standardized score is higher than Jane's.

(c) Jane—her standardized score is higher than Colleen's.

(d) Jane—the standard deviation was bigger in 2009.

(e) The two sisters did equally well—their z-scores are the same.

Section II: Free Response *Show all your work. Indicate clearly the methods you use, because you will be graded on the correctness of your methods as well as on the accuracy and completeness of your results and explanations.*

T2.11. As part of the President's Challenge, students can attempt to earn the Presidential Physical Fitness Award or the National Physical Fitness Award by meeting qualifying standards in five events: curl-ups, shuttle run, sit and reach, one-mile run, and pull-ups. The qualifying standards are based on the 1985 School Population Fitness Survey. For the Presidential award, the standard for each event is the 85th percentile of the results for a specific age group and gender among students who participated in the 1985 survey. For the National award, the standard is the 50th percentile. To win either award, a student must meet the qualifying standard for all five events.

Jane, who is 9 years old, did 40 curl-ups in one minute. Matt, who is 12 years old, also did 40 curl-ups in one minute. The qualifying standard for the Presidential award is 39 curl-ups for Jane and 50 curl-ups for Matt. For the National award, the standards are 30 and 40, respectively.

(a) Compare Jane's and Matt's performances using percentiles. Explain in language simple enough for someone who knows little statistics to understand.

(b) Who has the higher standardized value (z-score), Jane or Matt? Justify your answer.

T2.12. The army reports that the distribution of head circumference among male soldiers is approximately Normal with mean 22.8 inches and standard deviation 1.1 inches.

(a) A male soldier whose head circumference is 23.9 inches would be at what percentile? Show your method clearly.

(b) The army's helmet supplier regularly stocks helmets that fit male soldiers with head circumferences between 20 and 26 inches. Anyone with a head circumference outside that interval requires a customized helmet order. What percent of male soldiers require custom helmets? Show your work, including a well-labeled sketch of a Normal curve.

(c) Find the interquartile range for the distribution of head circumference among male soldiers. Show your method clearly.

T2.13. A study recorded the amount of oil recovered from the 64 wells in an oil field. Here are descriptive statistics for that set of data from Minitab.

Descriptive Statistics: Oilprod

Variable	N	Mean	Median	StDev	Min	Max	Q_1	Q_3
Oilprod	64	48.25	37.80	40.24	2.00	204.90	21.40	60.75

Does the amount of oil recovered from all wells in this field seem to follow a Normal distribution? Give appropriate statistical evidence to support your answer.

Chapter 3

case study

Introduction

Describing Relationships

How Faithful Is Old Faithful?

The Starnes family visited Yellowstone National Park in hopes of seeing the Old Faithful geyser erupt. They had only about four hours to spend in the park. When they pulled into the parking lot near Old Faithful, a large crowd of people was headed back to their cars from the geyser. Old Faithful had just finished erupting. How long would the Starnes family have to wait until the next eruption?

Let's look at some data. Figure 3.1 shows a histogram of times (in minutes) between consecutive eruptions of Old Faithful in the month before the Starnes family's visit. The shortest interval was 47 minutes, and the longest was 113 minutes. That's a lot of variability! The distribution has two clear peaks—one at about 60 minutes and the other at about 90 minutes.

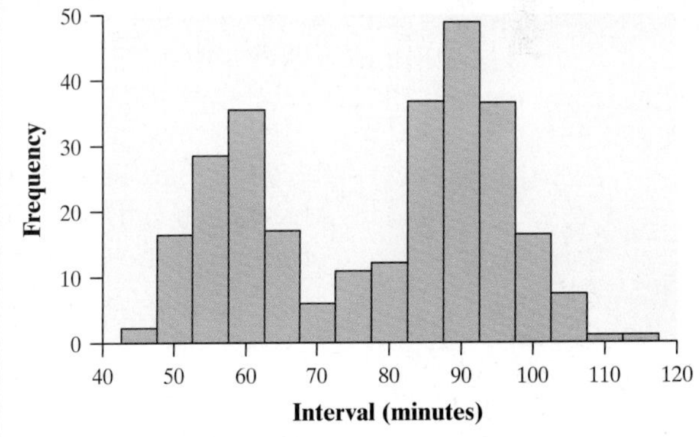

FIGURE 3.1 Histogram of the interval (in minutes) between eruptions of the Old Faithful geyser in the month prior to the Starnes family's visit.

If the Starnes family hopes for a 60-minute gap between eruptions, but the actual interval is closer to 90 minutes, the kids will get impatient. If they plan for a 90-minute interval and go somewhere else in the park, they won't get back in time to see the next eruption if the gap is only about 60 minutes.

What should the Starnes family do?

Later in the chapter, we'll answer this question. For now, keep this in mind: to understand one variable (like eruption interval), you often have to look at how it is related to other variables.

Introduction

A medical study finds that short women are more likely to have heart attacks than women of average height, while tall women have the fewest heart attacks. An insurance group reports that heavier cars have fewer deaths per 100,000 vehicles than lighter cars do. These and many other statistical studies look at the relationship between two variables. However, statistical relationships are overall tendencies, not absolute rules. Although smokers on average die younger than nonsmokers, there are exceptions, such as a person who lives to 90 while smoking three packs a day.

In Section 1.1, we studied relationships between categorical variables. This chapter focuses on relationships between two quantitative variables. To understand a statistical relationship between two variables, we measure both variables on the same group of individuals. Often, we must examine other variables as well. To conclude that shorter women have higher risk from heart attacks, for example, the researchers had to eliminate the effect of other variables like weight and exercise habits. One of our main themes is that *the relationship between two variables can be strongly influenced by other variables that are lurking in the background.*

CAUTION

ACTIVITY *CSI Stats: The case of the missing cookies*

MATERIALS: Metric ruler, meterstick, handprint, and photo lineup (from *Teacher's Resource Binder*) for each group of three to four students; one sheet of graph paper per student

Mrs. Hagen keeps a large jar full of cookies on her desk for her students. Over the past few days, a few cookies have disappeared. The only people with access to Mrs. Hagen's desk are the other math teachers at her school. She asks her colleagues whether they have been making withdrawals from the cookie jar. No one confesses to the crime.

But the next day, Mrs. Hagen catches a break—she finds a clear handprint on the cookie jar. The careless culprit has left behind crucial evidence! At this point, Mrs. Hagen calls in the CSI Stats team (your class) to help her identify the prime suspect in "The Case of the Missing Cookies."

1. Measure the height and hand span of each member of your group to the nearest centimeter (cm). (Hand span is the maximum distance from the tip of the thumb to the tip of the pinkie finger on a person's fully stretched-out hand.)

2. Your teacher will make a data table on the board with two columns, labeled as follows:

Hand span (cm)	Height (cm)

Send a representative to record the data for each member of your group in the table.

3. Copy the data table onto your graph paper very near the left margin of the page. Next, you will make a graph of these data. Begin by constructing a set of coordinate axes. Allow plenty of space on the page for your graph. Label the horizontal axis "Hand span (cm)" and the vertical axis "Height (cm)."

4. Since neither hand span nor height can be close to 0 cm, we want to start our horizontal and vertical scales at larger numbers. Scale the horizontal axis in 0.5 cm increments starting with 15 cm. Scale the vertical axis in 5 cm incre-

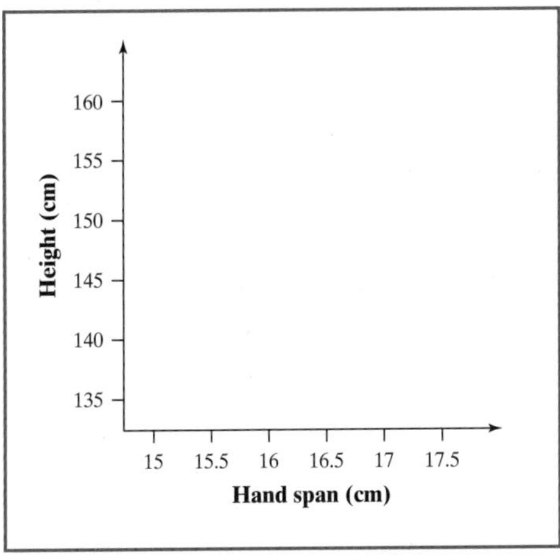

ments starting with 135 cm. Refer to the sketch in the margin for comparison.

5. Plot each point from your class data table as accurately as you can on the graph. Compare your graph with those of your group members.

6. As a group, discuss what the graph tells you about the relationship between hand span and height. Summarize your observations in a sentence or two.

7. Ask your teacher for a copy of the handprint found at the scene and the photo lineup of the math department. Which math teacher does your group believe is the "prime suspect"? Justify your answer with appropriate statistical evidence.

Scatterplots and Correlation

In Section 3.1, you'll learn about:

- Explanatory and response variables
- Displaying relationships: Scatterplots
- Interpreting scatterplots
- Measuring linear association: Correlation
- Facts about correlation

Most statistical studies examine data on more than one variable. Fortunately, analysis of several-variable data builds on the tools we used to examine individual variables. The principles that guide our work also remain the same:

- Plot the data, then add numerical summaries.
- Look for overall patterns and deviations from those patterns.
- When there's a regular overall pattern, use a simplified model to describe it.

Explanatory and Response Variables

We think that car weight helps explain accident deaths and that smoking influences life expectancy. In these relationships, the two variables play different roles. Accident death rate and life expectancy are the **response variables** of interest. Car weight and number of cigarettes smoked are the **explanatory variables.**

DEFINITION: Response variable, explanatory variable

A **response variable** measures an outcome of a study. An **explanatory variable** may help explain or influence changes in a response variable.

It is easiest to identify explanatory and response variables when we actually specify values of one variable to see how it affects another variable. For instance, to study the effect of alcohol on body temperature, researchers gave several different amounts of alcohol to mice. Then they measured the change in each mouse's body temperature 15 minutes later. In this case, amount of alcohol is the explanatory variable, and change in body temperature is the response variable. When we don't specify the values of either variable but just observe both variables, there may or may not be explanatory and response variables. Whether there are depends on how you plan to use the data.

You will often see explanatory variables called *independent variables*, and response variables called *dependent variables*. Because the words "independent" and "dependent" have other meanings in statistics, we won't use them here.

Linking SAT Math and Critical Reading Scores

Explanatory or response?

Julie asks, "Can I predict a state's mean SAT Math score if I know its mean SAT Critical Reading score?" Jim wants to know how the mean SAT Math and Critical Reading scores this year in the 50 states are related to each other.

PROBLEM: For each student, identify the explanatory variable and the response variable if possible.

SOLUTION: Julie is treating the mean SAT Critical Reading score as the explanatory variable and the mean SAT Math score as the response variable. Jim is simply interested in exploring the relationship between the two variables. For him, there is no clear explanatory or response variable.

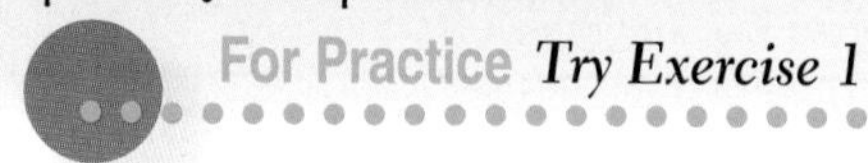

In many studies, the goal is to show that changes in one or more explanatory variables actually *cause* changes in a response variable. However, other explanatory-response relationships don't involve direct causation. In the alcohol and mice study, alcohol actually *causes* a change in body temperature. But there is no cause-and-effect relationship between SAT Math and Critical Reading scores. Because the scores are closely related, we can still use a state's mean SAT Critical Reading score to predict its mean Math score. We will learn how to do such predictions in Section 3.2.

CHECK YOUR UNDERSTANDING

Identify the explanatory and response variables in each setting.

1. How does drinking beer affect the level of alcohol in our blood? The legal limit for driving in all states is 0.08%. In a study, adult volunteers drank different numbers of cans of beer. Thirty minutes later, a police officer measured their blood alcohol levels.

2. The National Student Loan Survey provides data on the amount of debt for recent college graduates, their current income, and how stressed they feel about college debt. A sociologist looks at the data with the goal of using amount of debt and income to explain the stress caused by college debt.

Displaying Relationships: Scatterplots

The most useful graph for displaying the relationship between two quantitative variables is a **scatterplot.** Figure 3.2 shows a scatterplot of the percent of high school graduates in each state who took the SAT and the state's mean SAT Math score in a recent year. We think that "percent taking" will help explain "mean score." So "percent taking" is the explanatory variable and "mean score" is the response variable. We want to see how mean score changes when percent taking changes, so we put percent taking (the explanatory variable) on the horizontal axis. Each point represents a single state. In Colorado, for example, 21% took the SAT, and their mean SAT Math score was 570. Find 21 on the x (horizontal) axis and 570 on the y (vertical) axis. Colorado appears as the point (21, 570).

FIGURE 3.2 Scatterplot of the mean SAT Math score in each state against the percent of that state's high school graduates who took the SAT. The dotted lines intersect at the point (21, 570), the data for Colorado.

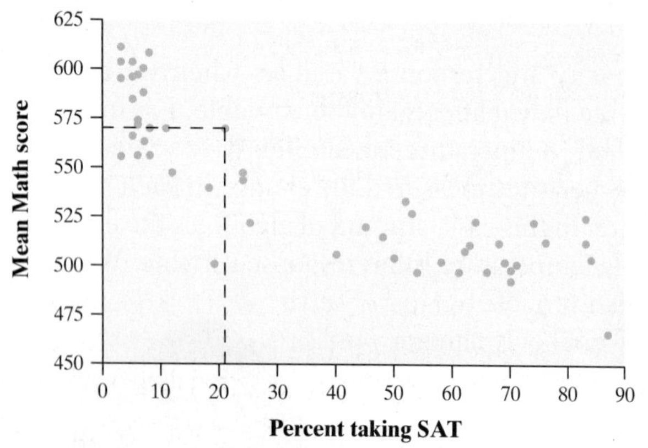

DEFINITION: Scatterplot

A **scatterplot** shows the relationship between two quantitative variables measured on the same individuals. The values of one variable appear on the horizontal axis, and the values of the other variable appear on the vertical axis. Each individual in the data appears as a point in the graph.

Here's a helpful way to remember: the e**X**planatory variable goes on the *x* axis.

Always plot the explanatory variable, if there is one, on the horizontal axis (the *x* axis) of a scatterplot. As a reminder, we usually call the explanatory variable *x* and the response variable *y*. If there is no explanatory-response distinction, either variable can go on the horizontal axis.

We used computer software to produce Figure 3.2. For some problems, you'll be expected to make scatterplots by hand. Here's how to do it.

How to Make a Scatterplot

1. Decide which variable should go on each axis.
2. Label and scale your axes.
3. Plot individual data values.

The following example illustrates the process of constructing a scatterplot.

Heavy Backpacks

Making a scatterplot

Ninth-grade students at the Webb Schools go on a backpacking trip each fall. Students are divided into hiking groups of size 8 by selecting names from a hat. Before leaving, students and their backpacks are weighed. Here are data from one hiking group in a recent year:

Body weight (lb):	120	187	109	103	131	165	158	116
Backpack weight (lb):	26	30	26	24	29	35	31	28

PROBLEM: Make a scatterplot of the relationship between body weight and pack weight.

SOLUTION: We follow the steps described earlier to make the scatterplot.

1. *Decide which variable should go on each axis.* The weight a student can carry depends on his body weight. So we'll use body weight as the explanatory variable (*x* axis) and backpack weight as the response variable (*y* axis).

2. *Label and scale your axes.* We labeled the *x* axis "Body weight (lb)" and the *y* axis "Pack weight (lb)." Since the hikers' body weights range from 103 to 187 pounds, we chose a horizontal scale starting at 100 pounds, with tick marks every 10 pounds. The hikers' pack weights vary from 24 to 35 pounds, so we chose a vertical scale starting at 20 pounds, with tick marks every 2 pounds.

3. *Plot individual data values.* The first student in the group weighs 120 pounds and his pack weighs 26 pounds. We plot this point directly above 120 on the horizontal axis and to the right of 26 on the vertical axis, as shown in Figure 3.3. For the second student in the group, we add the point (187, 30) to the graph. By adding the points for the remaining six students in the group, we get the completed scatterplot in Figure 3.3.

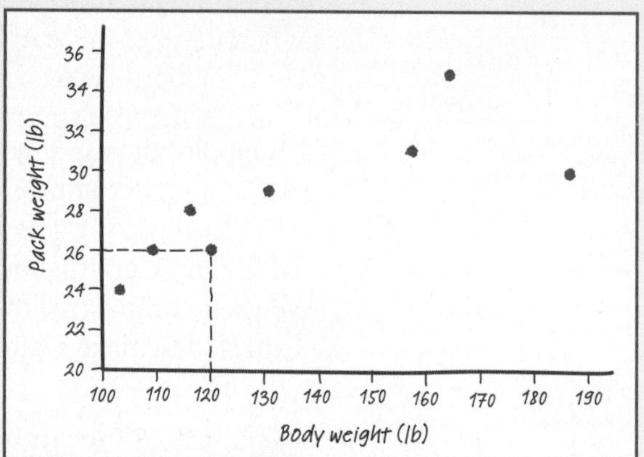

FIGURE 3.3 Completed scatterplot of body weight and backpack weight for the 8 students.

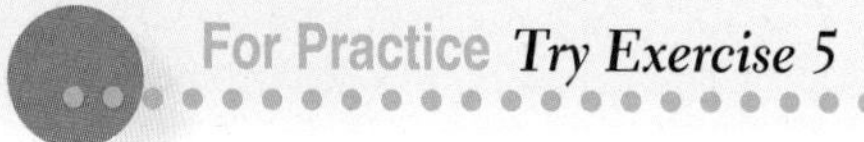

Interpreting Scatterplots

To interpret a scatterplot, follow the basic strategy of data analysis from Chapters 1 and 2: look for patterns and important deviations from those patterns. Let's take a closer look at the scatterplot from Figure 3.2. What do we see?

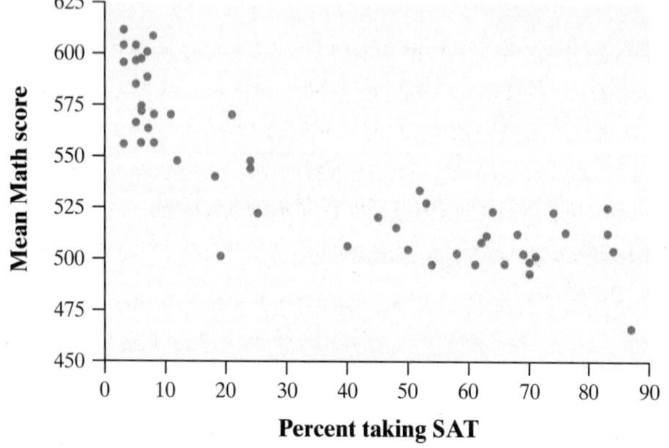

- The graph shows a clear **direction:** the overall pattern moves from upper left to lower right. That is, states in which higher percents of high school graduates take the SAT tend to have lower mean SAT Math scores. We call this a *negative association* between the two variables.
- The **form** of the relationship is slightly curved. More important, most states fall into one of two distinct *clusters.* In about half of the states, 25% or fewer graduates took the SAT. In the other half, more than 40% took the SAT.
- The **strength** of a relationship in a scatterplot is determined by how closely the points follow a clear form. The overall relationship in Figure 3.2 is moderately strong: states with similar percents taking the SAT tend to have roughly similar mean SAT Math scores.

• Two states stand out in the scatterplot: West Virginia at (19, 501) and Maine at (87, 466). These points can be described as **outliers** since they fall outside the overall pattern.

THINK ABOUT IT

What explains the clusters? There are two widely used college entrance exams, the SAT and the American College Testing (ACT) exam. Each state favors one or the other. The ACT states cluster at the left of Figure 3.2 and the SAT states at the right. In ACT states, most students who take the SAT are applying to a selective college that prefers SAT scores. This select group of students has a higher mean score than the much larger group of students who take the SAT in SAT states.

How to Examine a Scatterplot

As in any graph of data, look for the *overall pattern* and for striking *departures* from that pattern.

• You can describe the overall pattern of a scatterplot by the **direction, form,** and **strength** of the relationship.

• An important kind of departure is an **outlier,** an individual value that falls outside the overall pattern of the relationship.

Let's practice examining scatterplots using the hiker data from the previous example.

Heavy Backpacks

Interpreting a scatterplot

In the last example, we constructed the scatterplot shown that displays the body weights and backpack weights of a group of hikers.

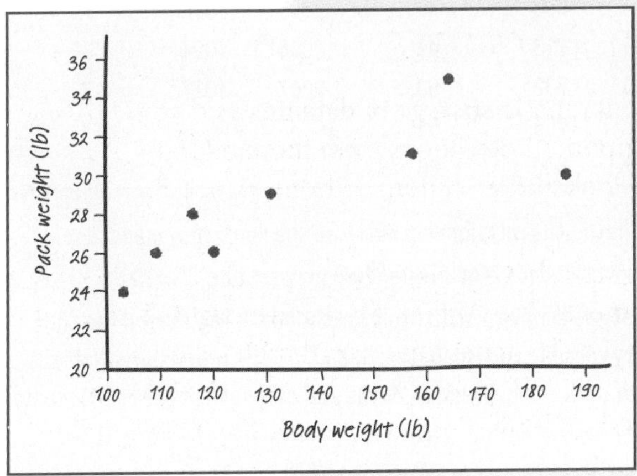

PROBLEM: Describe what the scatterplot reveals about the relationship between body weight and pack weight.

SOLUTION: First, we'll describe the general pattern. Then, we'll identify any deviations from the pattern. *Direction:* In general, it appears that lighter students are carrying lighter backpacks and heavier students are carrying heavier packs. We say that there is a *positive association* between the variables. *Form:* There seems to be a somewhat linear pattern in the graph (that is, the overall pattern follows a straight line). *Strength:* Since the points vary somewhat from the linear pattern, the relationship is only moderately strong. We see one possible *outlier* in the graph—the hiker with body weight 187 pounds and pack weight 30 pounds. He seems to be carrying less relative to his weight than are the other group members.

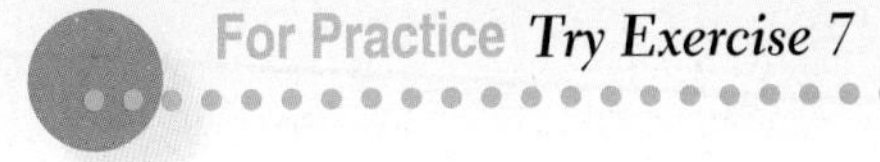

You could also say that the form of the scatterplot is slightly curved or concave down.

So far, we've seen relationships with two different directions. Backpack weight generally increases as body weight increases (**positive association**). The mean SAT score goes down as the percent of graduates taking the test increases (**negative association**). Let's give a careful definition for these terms.

DEFINITION: Positive association, negative association

Two variables have a **positive association** when above-average values of one tend to accompany above-average values of the other, and when below-average values also tend to occur together.

Two variables have a **negative association** when above-average values of one tend to accompany below-average values of the other.

Of course, not all relationships have a clear direction that we can describe as a positive association or a negative association. Exercise 9 involves a relationship that doesn't have a single direction. This next example, however, illustrates a strong positive association with a simple and important form.

The Endangered Manatee

Pulling it all together

STEP 4

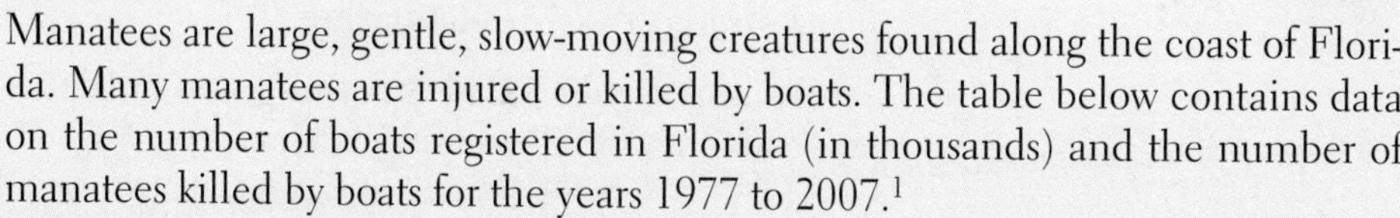

Manatees are large, gentle, slow-moving creatures found along the coast of Florida. Many manatees are injured or killed by boats. The table below contains data on the number of boats registered in Florida (in thousands) and the number of manatees killed by boats for the years 1977 to 2007.[1]

Florida boat registrations (thousands) and manatees killed by boats

Year	Boats	Manatees	Year	Boats	Manatees	Year	Boats	Manatees
1977	447	13	1988	675	43	1999	830	82
1978	460	21	1989	711	50	2000	880	78
1979	481	24	1990	719	47	2001	944	81
1980	498	16	1991	681	53	2002	962	95
1981	513	24	1992	679	38	2003	978	73
1982	512	20	1993	678	35	2004	983	69
1983	526	15	1994	696	49	2005	1010	79
1984	559	34	1995	713	42	2006	1024	92
1985	585	33	1996	732	60	2007	1027	73
1986	614	33	1997	755	54			
1987	645	39	1998	809	66			

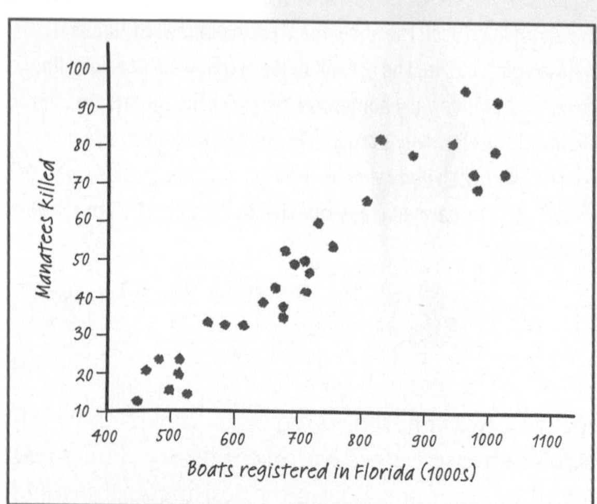

FIGURE 3.4 Scatterplot of the number of Florida manatees killed by boats from 1977 to 2007 against the number of boats registered in Florida that year.

STATE: What is the relationship between the number of manatees killed and the number of registered boats?

PLAN: First, we'll make a scatterplot with "boats registered" as the explanatory variable and "manatees killed" as the response variable. Then we'll describe the direction, form, and strength of the relationship and identify any outliers.

DO: Figure 3.4 is our completed scatterplot. There is a positive association—more boats registered goes with more manatees killed. The form of the relationship is linear. That is, the overall pattern follows a straight line from lower left to upper right. The relationship is strong because the points don't deviate greatly from a line. There are no obvious outliers.

CONCLUDE: As more boats are registered, the number of manatees killed by boats goes up linearly.

For Practice *Try Exercise 13*

The previous example deserves an important caution: *association does not imply causation.* Although the scatterplot shows a strong linear relationship between the variables, we can't conclude that the increase in manatee deaths was caused by the change in boat registrations. Always ask what other variables lurking in the background might contribute to the relationship between two variables. Because both boats registered and manatees killed are recorded year by year, any change in conditions over time might affect the relationship. For example, if boats in Florida have tended to go faster over the years, that might result in more manatees killed by the same number of boats.

CAUTION

CHECK YOUR UNDERSTANDING

In the chapter-opening Case Study (page 141), the Starnes family arrived at Old Faithful after it had erupted. They wondered how long it would be until the next eruption. Here is a scatterplot that plots the interval between consecutive eruptions of Old Faithful against the duration of the previous eruption, for the month prior to their visit.

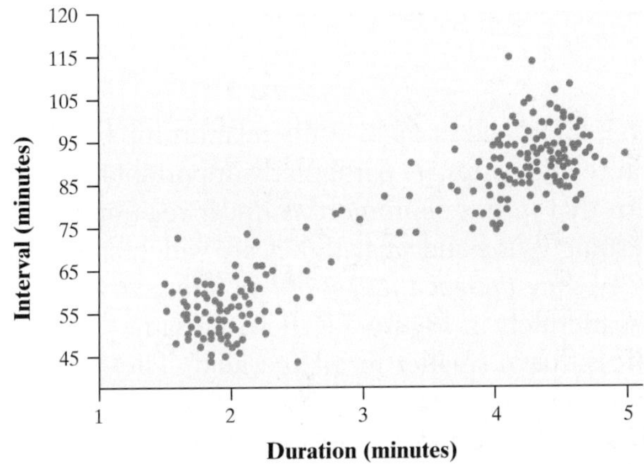

1. Describe the direction of the relationship. Explain why this makes sense.
2. What form does the relationship take? Why are there two clusters of points?
3. How strong is the relationship? Justify your answer.
4. Are there any outliers?
5. What information does the Starnes family need to predict when the next eruption will occur?

TECHNOLOGY CORNER Scatterplots on the calculator

Making scatterplots with technology is much easier than constructing them by hand. We'll use the hiker data from page 145 to show how to construct a scatterplot on a TI-83/84 or TI-89.

- Enter the data values into your lists. Clear lists L1(list1) and L2(list2). Put the body weights in L1(list1) and the backpack weights in L2(list2).
- Define a scatterplot in the statistics plot menu (press F2 on the TI-89). Specify the settings shown.

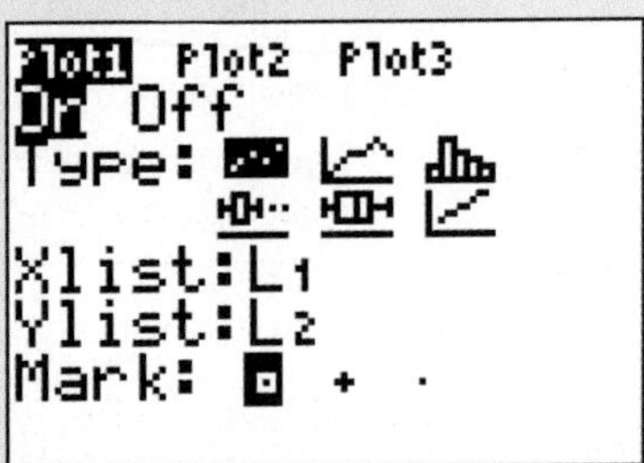

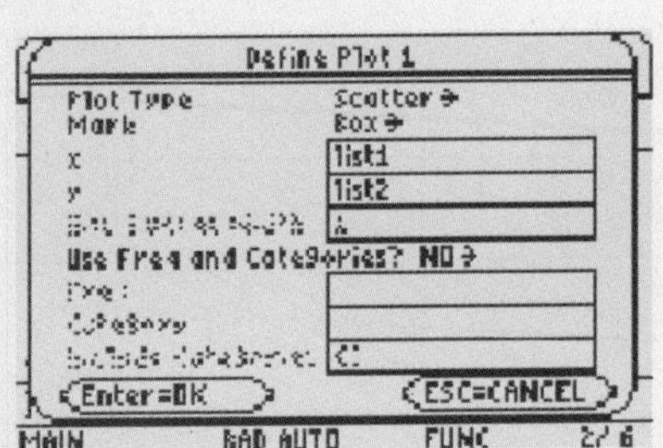

• Use ZoomStat (ZoomData on the TI-89) to obtain a graph. The calculator will set the window dimensions automatically by looking at the values in L1/list1 and L2/list2.

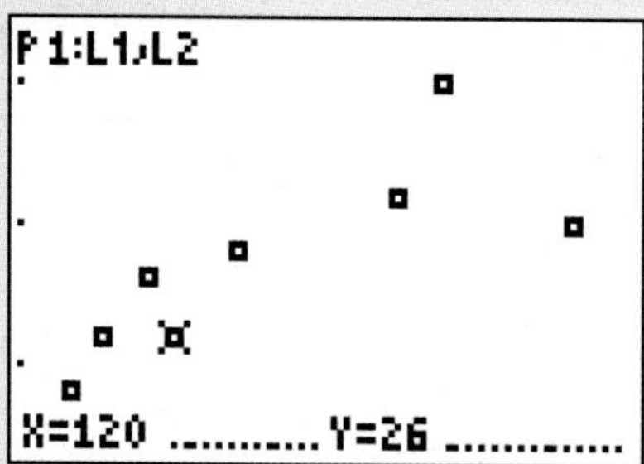

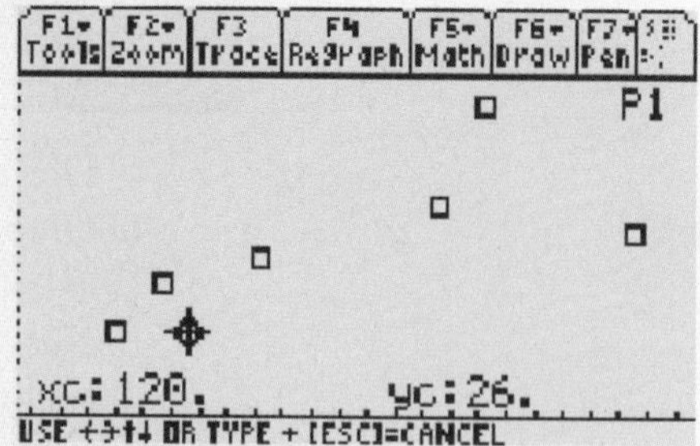

AP EXAM TIP If you are asked to make a scatterplot on a free-response question, be sure to label and scale both axes. *Don't* copy an unlabeled calculator graph directly onto your paper.

Notice that there are no scales on the axes and that the axes are not labeled. If you copy a scatterplot from your calculator onto your paper, make sure that you scale and label the axes. You can use TRACE to help you get started (like we did).

TI-Nspire instructions in Appendix B

Measuring Linear Association: Correlation

A scatterplot displays the direction, form, and strength of the relationship between two quantitative variables. Linear relationships are particularly important because a straight line is a simple pattern that is quite common. A linear relationship is strong if the points lie close to a straight line and weak if they are widely scattered about a line. Unfortunately, *our eyes are not good judges of how strong a linear relationship is.* The two scatterplots in Figure 3.5 show the same data, but the graph on the right is drawn smaller in a large field. The right-hand graph seems to show a stronger linear relationship.

CAUTION

Since it's easy to be fooled by different scales or by the amount of space around the cloud of points in a scatterplot, we need to use a numerical measure to supplement the graph. **Correlation** is the measure we use.

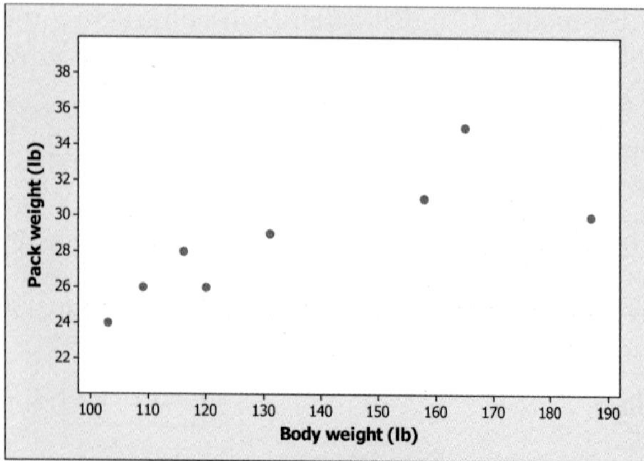

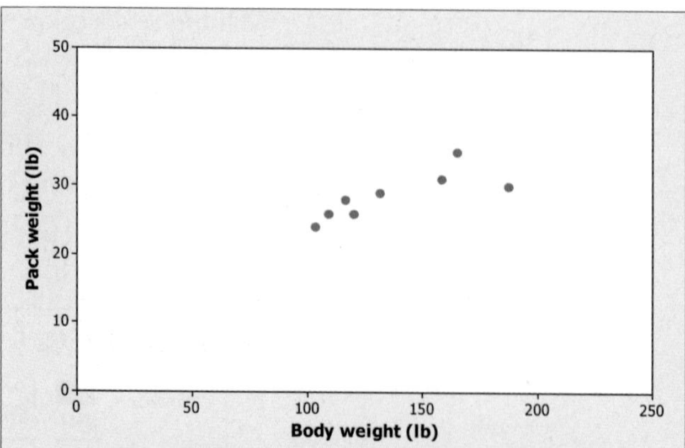

FIGURE 3.5 Two Minitab scatterplots of the same data. The straight-line pattern in the graph on the right appears stronger because of the surrounding space.

Some people refer to r as the "correlation coefficient."

DEFINITION: Correlation *r*

The **correlation *r*** measures the direction and strength of the linear relationship between two quantitative variables.

How good are you at estimating the correlation by eye from a scatterplot? To find out, try one of the interactive correlation matching games on the Web. Just search for "correlation game."

The correlation r is always a number between -1 *and* 1. Correlation indicates the direction of a linear relationship by its sign: $r > 0$ for a positive association and $r < 0$ for a negative association. Values of r near 0 indicate a very weak linear relationship. The strength of the linear relationship increases as r moves away from 0 toward either -1 or 1. The extreme values $r = -1$ and $r = 1$ occur *only* in the case of a perfect linear relationship, when the points lie exactly along a straight line.

Figure 3.6 shows scatterplots that correspond to various values of r. To make the meaning of r clearer, the standard deviations of both variables in these plots are equal, and the horizontal and vertical scales are the same. The correlation describes the direction and strength of the linear relationship in each graph.

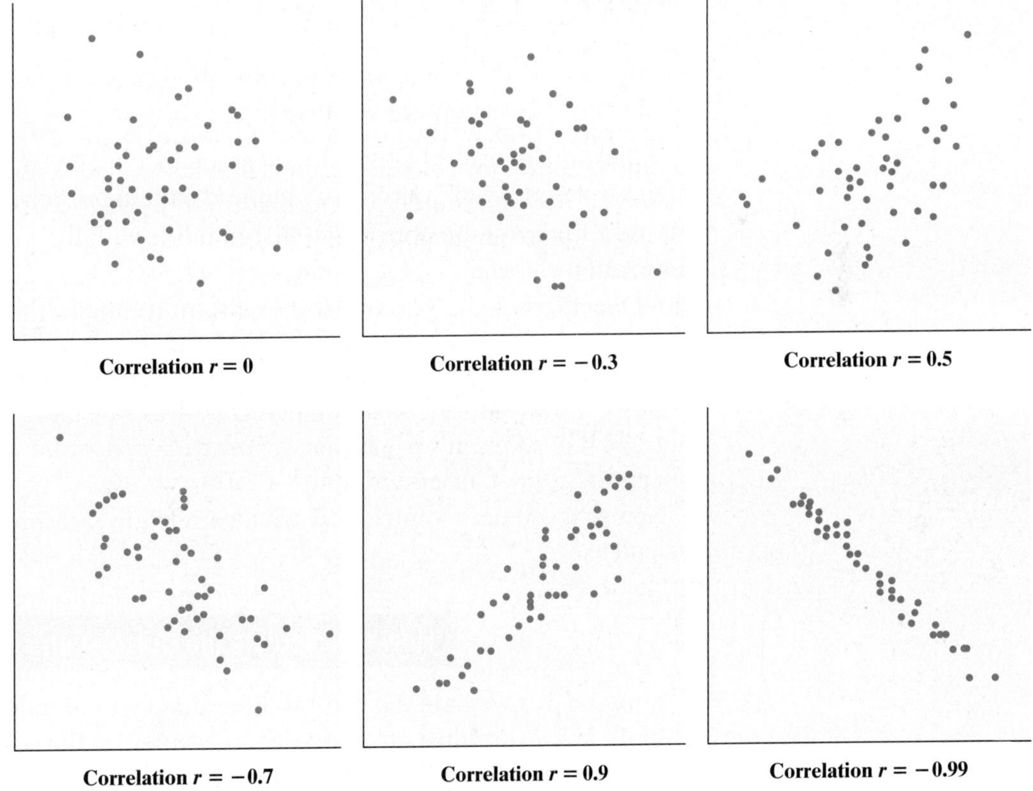

FIGURE 3.6 How correlation measures the strength of a linear relationship. Patterns closer to a straight line have correlations closer to 1 or -1.

The following Activity lets you explore some important properties of the correlation.

ACTIVITY Correlation and regression *applet*

MATERIALS: Computer with Internet connection

Go to the book's Web site, www.whfreeman.com/tps4e, and launch the *Correlation and Regression* applet.

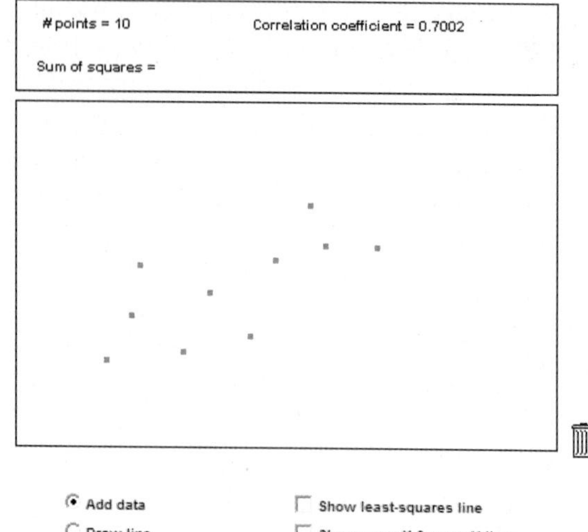

1. You are going to use the *Correlation and Regression* applet to make several scatterplots with 10 points that have correlation close to 0.7.

(a) Start by putting two points on the graph. What's the value of the correlation? Why does this make sense?

(b) Make a lower-left to upper-right pattern of 10 points with correlation about $r = 0.7$. (You can drag points up or down to adjust r after you have 10 points.)

(c) Make another scatterplot: this one should have 9 points in a vertical stack at the left of the plot. Add 1 point far to the right and move it until the correlation is close to 0.7.

(d) Make a third scatterplot: make this one with 10 points in a curved pattern that starts at the lower left, rises to the right, then falls again at the far right. Adjust the points up or down until you have a very smooth curve with correlation close to 0.7.

Summarize: If you know that the correlation between two variables is $r = 0.7$, what can you say about the relationship?

2. Click on the scatterplot to create a group of 10 points in the lower-left corner of the scatterplot with a strong straight-line pattern (correlation about 0.9).

(a) Add 1 point at the upper right that is in line with the first 10. How does the correlation change?

(b) Drag this last point down. How small can you make the correlation? Can you make the correlation negative?

Summarize: What did you learn from (b) about the effect of a single point on the correlation?

Now that you have some idea of what correlation measures, let's look at how it's calculated.

How to Calculate the Correlation *r*

Suppose that we have data on variables x and y for n individuals. The values for the first individual are x_1 and y_1, the values for the second individual are x_2 and y_2, and so on. The means and standard deviations of the two variables are $\bar{x}$ and s_x for the x-values, and $\bar{y}$ and s_y for the y-values. The correlation r between x and y is

$$r = \frac{1}{n-1}\left[\left(\frac{x_1-\bar{x}}{s_x}\right)\left(\frac{y_1-\bar{y}}{s_y}\right)+\left(\frac{x_2-\bar{x}}{s_x}\right)\left(\frac{y_2-\bar{y}}{s_y}\right)+\cdots+\left(\frac{x_n-\bar{x}}{s_x}\right)\left(\frac{y_n-\bar{y}}{s_y}\right)\right]$$

or, more compactly,

$$r = \frac{1}{n-1}\sum\left(\frac{x_i-\bar{x}}{s_x}\right)\left(\frac{y_i-\bar{y}}{s_y}\right)$$

The formula for the correlation r is a bit complex. It helps us see what correlation is, but in practice, you should use your calculator or software to find r. Exercises 19 and 20 ask you to calculate a correlation step-by-step from the definition to solidify its meaning.

The formula for r begins by standardizing the observations. Let's use the familiar hiker data to perform the required calculations. The table below shows the values of body weight x (in pounds) and backpack weight y (also in pounds) for the 8 hikers. For these data, $\bar{x} = 136.125$ and $s_x = 30.296$ are the mean and standard deviation of the 8 body weights, both in pounds.

Body weight (lb):	120	187	109	103	131	165	158	116
Backpack weight (lb):	26	30	26	24	29	35	31	28

The value

$$\frac{x_i - \bar{x}}{s_x}$$

in the correlation formula is the standardized weight (z-score) of the ith person. For the first hiker in the table, the corresponding z-score is

$$z_x = \frac{120 - 136.125}{30.296} = -0.53$$

That is, the first hiker's body weight (120 lb) is about one-half standard deviation below the mean body weight for the group of hikers. Standardized values have no units—in this example, they are no longer measured in pounds.

To standardize the pack weights, we use $\bar{y} = 28.625$ and $s_y = 3.462$. For this same hiker, $z_y = \frac{26 - 28.625}{3.462} = -0.76$. His pack weight (26 lb) is about three-fourths of a standard deviation below the mean backpack weight for the group. When we multiply this hiker's two negative z-scores, we get a positive product. The correlation r is an "average" of the products of the standardized scores for all the hikers. Just as in the case of the standard deviation s_x, the average here divides by one fewer than the number of individuals. Finishing the calculation reveals that $r = 0.795$ for our happy hikers.

Some people like to write the correlation formula as

$$r = \frac{1}{n-1}\sum z_x z_y$$

to emphasize the product of standardized scores in the calculation.

EXAMPLE

Back to the Backpackers

Interpreting correlation

PROBLEM: Our earlier scatterplot of the body weight and pack weight data for the 8 hikers is repeated in the margin. For these data, $r = 0.795$.

(a) Interpret the value of r in context.

(b) What effect would removing the hiker with the heaviest body weight from the data have on the correlation? Justify your answer.

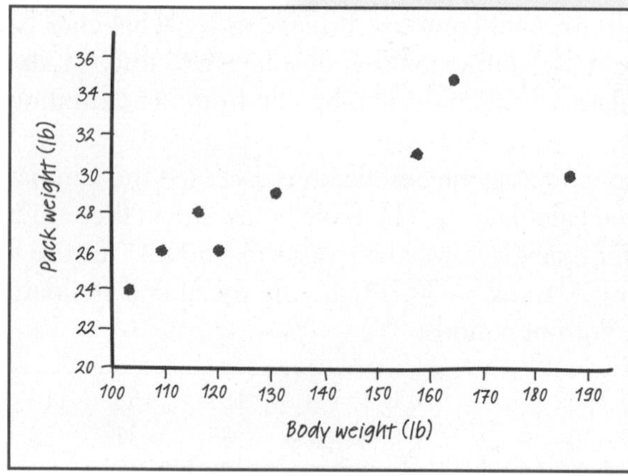

SOLUTION:

(a) The correlation of 0.795 confirms what we see in the scatterplot: there is a moderately strong, positive linear relationship between body weight and backpack weight for these 8 hikers.

(b) Removing the outlier would increase the correlation since the remaining 7 points are more tightly clustered in a linear pattern.

For Practice *Try Exercise 21*

AP EXAM TIP If you're asked to interpret a correlation, start by looking at a scatterplot of the data. Then be sure to address direction, form, strength, and outliers (sound familiar?) and to put your answer in context.

Be careful: a value of r close to 1 or −1 does not guarantee a linear relationship between two variables. A scatterplot with a clear curved form can have a correlation that's near 1 or −1. *Always plot your data!*

What does correlation measure? The calculator screen shots below provide more detail. At the left is a scatterplot of the hiker data with two lines added—a vertical line at the group's mean body weight and a horizontal line at the mean pack weight of the group. Most of the points fall in the upper-right or lower-left "quadrants" of the graph. That is, hikers with above-average body weights in the group tend to have above-average pack weights, and hikers with below-average body weights tend to have pack weights that are below average. This confirms the positive association between the variables.

Below on the right is a scatterplot of the standardized scores. To get this graph, we transformed both the x- and the y-values by subtracting their mean and dividing by their standard deviation. As we saw in Chapter 2, standardizing a data set converts the mean to 0 and the standard deviation to 1. That's why the vertical and horizontal lines in the right-hand graph are both at 0.

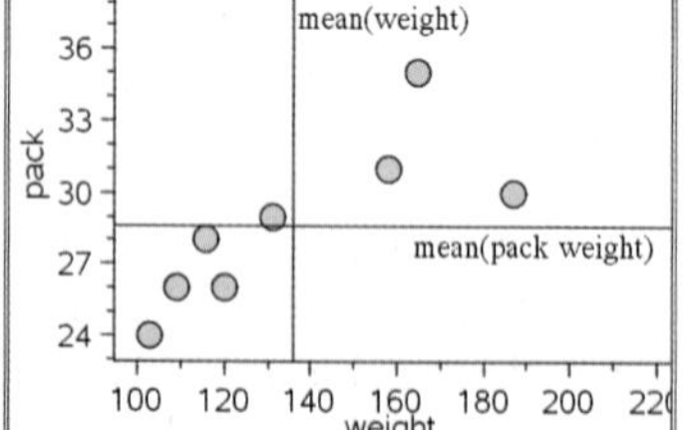

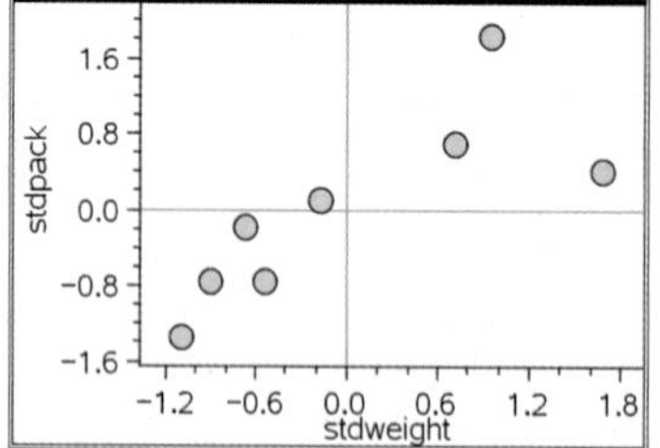

Notice that all the products of the standardized values will be positive except for one—the hiker who had slightly below-average weight in the group but a pack weight that was slightly above average. His point appears in the upper-left "quadrant" of both graphs.

CHECK YOUR UNDERSTANDING

The scatterplots below show four sets of real data: (a) repeats the manatee plot in Figure 3.4 (page 148); (b) shows the number of named tropical storms and the number predicted before the start of hurricane season each year between 1984 and 2007 by William Gray of Colorado State University; (c) plots the healing rate in micrometers (millionths of a meter) per hour for the two front limbs of several newts in an experiment; and (d) shows stock market performance in consecutive years over a 56-year period.

1. For each graph, estimate the correlation r. Then interpret the value of r in context.

2. The scatterplot in (b) contains an outlier: the disastrous 2005 season, which had 27 named storms, including Hurricane Katrina. What effect would removing this point have on the correlation? Explain.

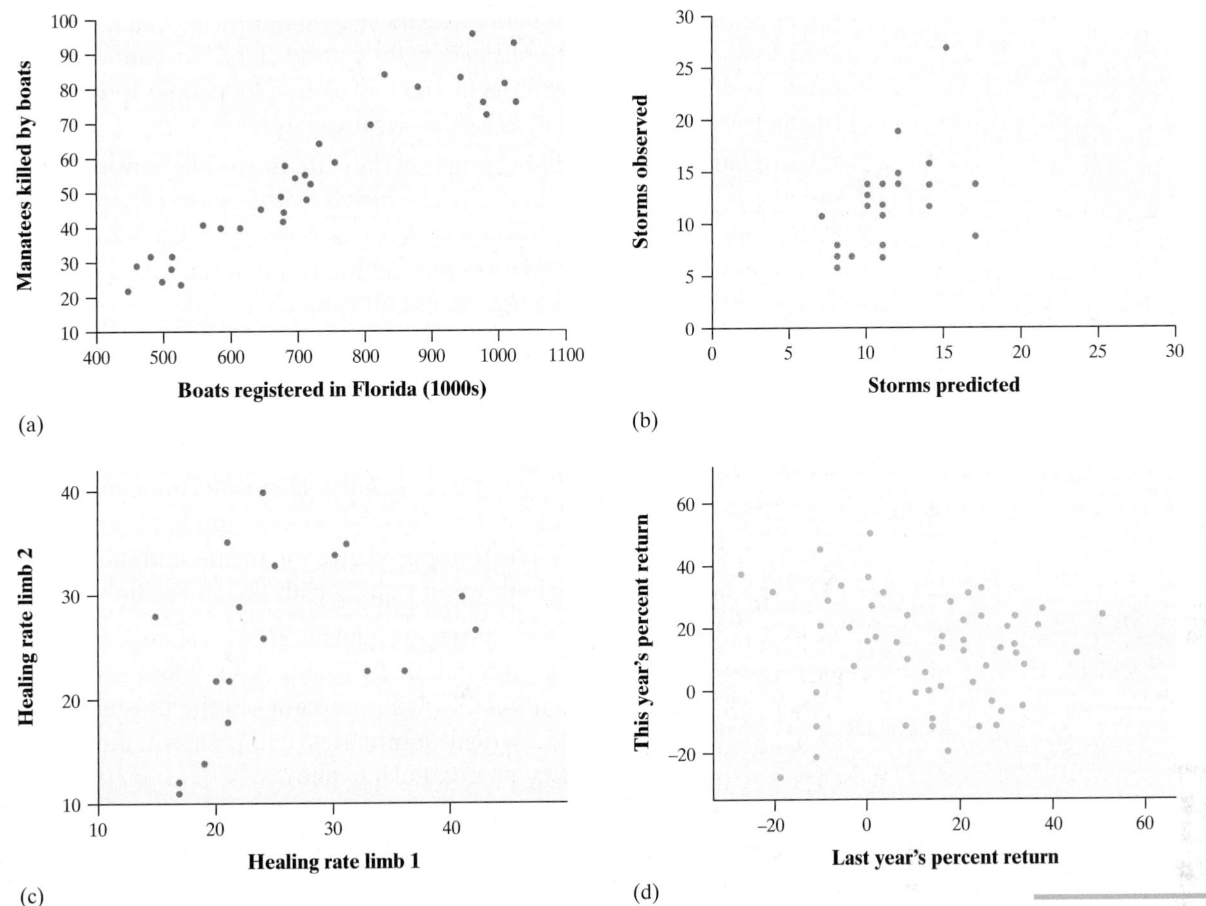

Facts about Correlation

How correlation behaves is more important than the details of the formula. Here's what you need to know in order to interpret correlation.

1. *Correlation makes no distinction between explanatory and response variables.* It makes no difference which variable you call x and which you call y in calculating the correlation. Can you see why from the formula?

$$r = \frac{1}{n-1}\sum\left(\frac{x_i - \bar{x}}{s_x}\right)\left(\frac{y_i - \bar{y}}{s_y}\right)$$

"He says we've ruined his positive correlation between height and weight."

2. Because r uses the standardized values of the observations, *r does not change when we change the units of measurement of x, y, or both.* Measuring height in centimeters rather than inches and weight in kilograms rather than pounds does not change the correlation between height and weight.

3. *The correlation r itself has no unit of measurement.* It is just a number.

Describing the relationship between two variables is more complex than describing the distribution of one variable. Here are some cautions to keep in mind when you use correlation.

- *Correlation requires that both variables be quantitative*, so that it makes sense to do the arithmetic indicated by the formula for r. We cannot calculate a correlation between the incomes of a group of people and what city they live in, because city is a categorical variable.
- Correlation measures the strength of only the linear relationship between two variables. *Correlation does not describe curved relationships between variables, no matter how strong the relationship is.* A correlation of 0 doesn't guarantee that there's *no* relationship between two variables, just that there's no *linear* relationship. Exercise 26 illustrates this important fact.

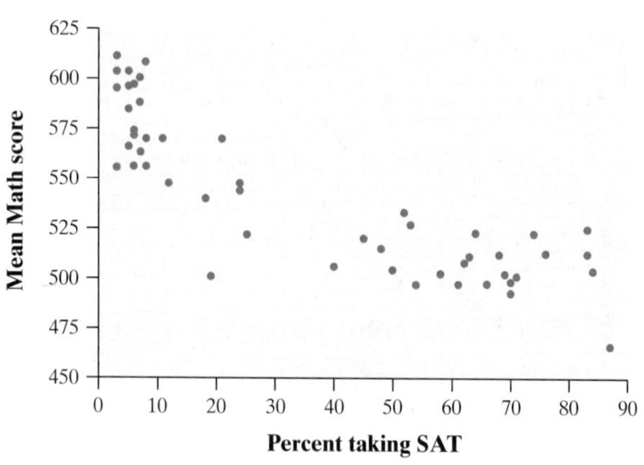

- *Like the mean and standard deviation, the correlation is not resistant: r is strongly affected by a few outlying observations.* Use r with caution when outliers appear in the scatterplot.
- *Correlation is not a complete summary of two-variable data*, even when the relationship between the variables is linear. You should give the means and standard deviations of both x and y along with the correlation.

Of course, even giving means, standard deviations, and the correlation for "state SAT Math scores" and "percent taking" will not point out the clusters in Figure 3.2. Numerical summaries complement plots of data, but they do not replace them.

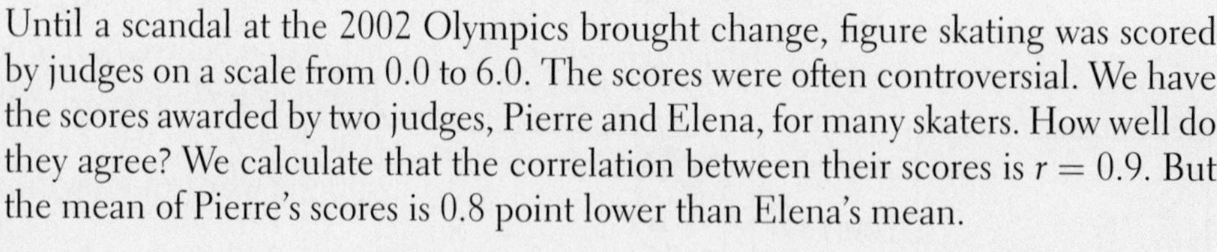
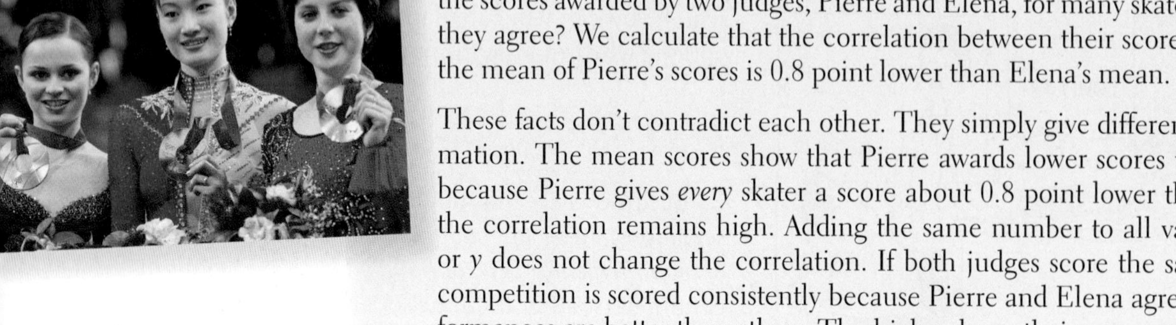

EXAMPLE

Scoring Figure Skaters

Why correlation doesn't tell the whole story

Until a scandal at the 2002 Olympics brought change, figure skating was scored by judges on a scale from 0.0 to 6.0. The scores were often controversial. We have the scores awarded by two judges, Pierre and Elena, for many skaters. How well do they agree? We calculate that the correlation between their scores is $r = 0.9$. But the mean of Pierre's scores is 0.8 point lower than Elena's mean.

These facts don't contradict each other. They simply give different kinds of information. The mean scores show that Pierre awards lower scores than Elena. But because Pierre gives *every* skater a score about 0.8 point lower than Elena does, the correlation remains high. Adding the same number to all values of either x or y does not change the correlation. If both judges score the same skaters, the competition is scored consistently because Pierre and Elena agree on which performances are better than others. The high r shows their agreement. But if Pierre scores some skaters and Elena others, we should add 0.8 point to Pierre's scores to arrive at a fair comparison.

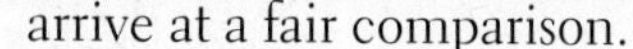

SECTION 3.1 Summary

- A **scatterplot** displays the relationship between two quantitative variables measured on the same individuals. Mark values of one variable on the horizontal axis (x axis) and values of the other variable on the vertical axis (y axis). Plot each individual's data as a point on the graph.
- If we think that a variable x may help explain, predict, or even cause changes in another variable y, we call x an **explanatory variable** and y a **response variable.** Always plot the explanatory variable, if there is one, on the x axis of a scatterplot. Plot the response variable on the y axis.
- In examining a scatterplot, look for an overall pattern showing the **direction, form,** and **strength** of the relationship and then look for **outliers** or other departures from this pattern.
- **Direction:** If the relationship has a clear direction, we speak of either **positive association** (high values of the two variables tend to occur together) or **negative association** (high values of one variable tend to occur with low values of the other variable).
- **Form:** Linear relationships, where the points show a straight-line pattern, are an important form of relationship between two variables. Curved relationships and clusters are other forms to watch for.
- **Strength:** The strength of a relationship is determined by how close the points in the scatterplot lie to a simple form such as a line.
- The **correlation** r measures the strength and direction of the linear association between two quantitative variables x and y. Although you can calculate a correlation for any scatterplot, r measures only straight-line relationships.
- Correlation indicates the direction of a linear relationship by its sign: $r > 0$ for a positive association and $r < 0$ for a negative association. Correlation always satisfies $-1 \leq r \leq 1$ and indicates the strength of a relationship by how close it is to -1 or 1. Perfect correlation, $r = \pm 1$, occurs only when the points on a scatterplot lie exactly on a straight line.
- Remember these important facts about r: Correlation ignores the distinction between explanatory and response variables. The value of r is not affected by changes in the unit of measurement of either variable. Correlation is not resistant, so outliers can greatly change the value of r.

3.1 TECHNOLOGY CORNER

TI-Nspire instructions in Appendix B

SECTION 3.1 Exercises

1. **Coral reefs** How sensitive to changes in water temperature are coral reefs? To find out, measure the growth of corals in aquariums where the water temperature is controlled at different levels. Growth is measured by weighing the coral before and after the experiment. What are the explanatory and response variables? Are they categorical or quantitative?

pg 144

2. **Treating breast cancer** Early on, the most common treatment for breast cancer was removal of the breast. It is now usual to remove only the tumor and nearby lymph nodes, followed by radiation. The change in policy was due to a large medical experiment that compared the two treatments. Some breast cancer patients, chosen at random, were given one or the other treatment. The patients were closely followed to see how long they lived following surgery. What are the explanatory and response variables? Are they categorical or quantitative?

3. **IQ and grades** Do students with higher IQ test scores tend to do better in school? The figure below shows a scatterplot of IQ and school grade point average (GPA) for all 78 seventh-grade students in a rural midwestern school. (GPA was recorded on a 12-point scale with A+ = 12, A = 11, A− = 10, B+ = 9, . . ., D− = 1, and F = 0.)[2]

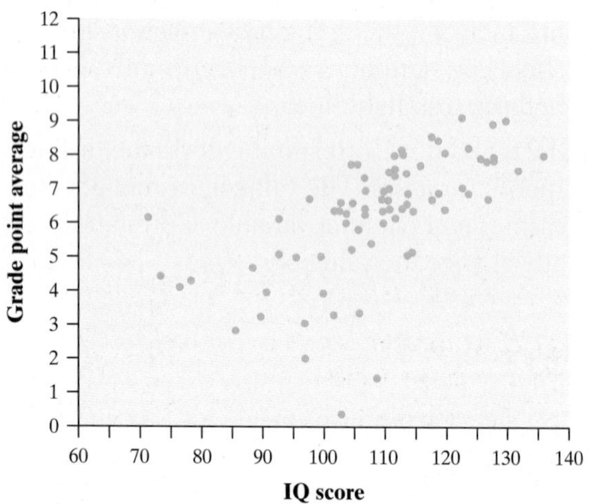

(a) Say in words what a positive association between IQ and GPA would mean. Does the plot show a positive association?

(b) What is the form of the relationship? Is it very strong? Explain your answers.

(c) At the bottom of the plot are several points that we might call outliers. One student in particular has a very low GPA despite an average IQ score. What are the approximate IQ and GPA for this student?

4. **How much gas?** Joan is concerned about the amount of energy she uses to heat her home. The graph below plots the mean number of cubic feet of gas per day that Joan used each month against the average temperature that month (in degrees Fahrenheit) for one heating season.

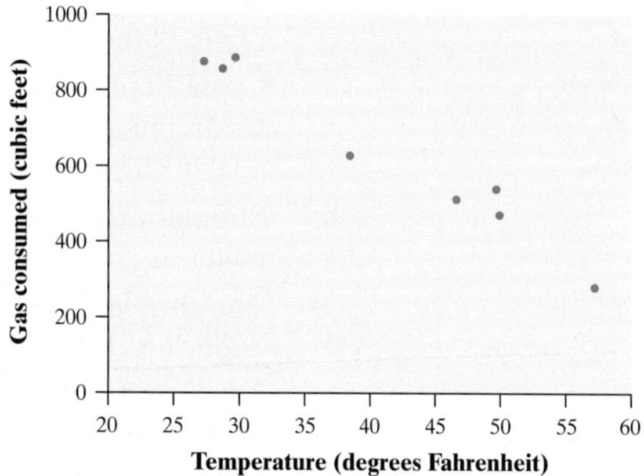

(a) Does the plot show a positive or negative association between the variables? Why does this make sense?

(b) What is the form of the relationship? Is it very strong? Explain your answers.

(c) Explain what the point at the bottom right of the plot represents.

5. **Outsourcing by airlines** Airlines have increasingly outsourced the maintenance of their planes to other companies. Critics say that the maintenance may be less carefully done, so that outsourcing creates a safety hazard. As evidence, they point to government data on percent of major maintenance outsourced and percent of flight delays blamed on the airline (often due to maintenance problems): Make a scatterplot by hand that shows how delays relate to outsourcing.[3]

pg 145

Airline	Outsource percent	Delay percent
AirTran	66	14
Alaska	92	42
American	46	26
America West	76	39
ATA	18	19
Continental	69	20
Delta	48	26
Frontier	65	31
Hawaiian	80	70
JetBlue	68	18
Northwest	76	43
Southwest	68	20
United	63	27
US Airways	77	24

6. **Bird colonies** One of nature's patterns connects the percent of adult birds in a colony that return from the previous year and the number of new adults that join the colony. Here are data for 13 colonies of sparrowhawks:[4]

Percent return:	74	66	81	52	73	62	52	45	62	46	60	46	38
New adults:	5	6	8	11	12	15	16	17	18	18	19	20	20

Make a scatterplot by hand that shows how the number of new adults relates to the percent of returning birds.

7. **Outsourcing by airlines** Refer to your graph from Exercise 5.

pg 147

(a) Describe the direction, form, and strength of the relationship between maintenance outsourcing and delays blamed on the airline.

(b) One airline is a high outlier in delay percent. Which airline is this? Aside from the outlier, does the plot show a roughly linear form? Is the relationship very strong?

8. **Bird colonies** Refer to your graph from Exercise 6.

(a) Describe the direction, form, and strength of the relationship between number of new sparrowhawks in a colony and percent of returning adults.

(b) For short-lived birds, the association between these variables is positive: changes in weather and food supply drive the populations of new and returning birds up or down together. For long-lived territorial birds, on the other hand, the association is negative because returning birds claim their territories in the colony and don't leave room for new recruits. Which type of species is the sparrowhawk? Explain.

9. **Does fast driving waste fuel?** How does the fuel consumption of a car change as its speed increases? Here are data for a British Ford Escort. Speed is measured in kilometers per hour, and fuel consumption is measured in liters of gasoline used per 100 kilometers traveled.[5]

Speed (km/h)	Fuel used (liters/100 km)	Speed (km/h)	Fuel used (liters/100 km)
10	21.00	90	7.57
20	13.00	100	8.27
30	10.00	110	9.03
40	8.00	120	9.87
50	7.00	130	10.79
60	5.90	140	11.77
70	6.30	150	12.83
80	6.95		

(a) Make a scatterplot on your calculator.

(b) Describe the form of the relationship. Why is it not linear? Explain why the form of the relationship makes sense.

(c) It does not make sense to describe the variables as either positively associated or negatively associated. Why?

(d) Is the relationship reasonably strong or quite weak? Explain your answer.

10. **Do heavier people burn more energy?** Metabolic rate, the rate at which the body consumes energy, is important in studies of weight gain, dieting, and exercise. We have data on the lean body mass and resting metabolic rate for 12 women who are subjects in a study of dieting. Lean body mass, given in kilograms, is a person's weight leaving out all fat. Metabolic rate is measured in calories burned per 24 hours. The researchers believe that lean body mass is an important influence on metabolic rate.

Mass:	36.1	54.6	48.5	42.0	50.6	42.0	40.3	33.1	42.4	34.5	51.1	41.2
Rate:	995	1425	1396	1418	1502	1256	1189	913	1124	1052	1347	1204

(a) Make a scatterplot on your calculator to examine the researchers' belief.

(b) Describe the direction, form, and strength of the relationship.

11. **Southern education** For a long time, the South has lagged behind the rest of the United States in the performance of its schools. Efforts to improve education have reduced the gap. We wonder if the South stands out in our study of state average SAT Math scores.

The figure below enhances the scatterplot in Figure 3.2 (page 144) by plotting 12 southern states in red.

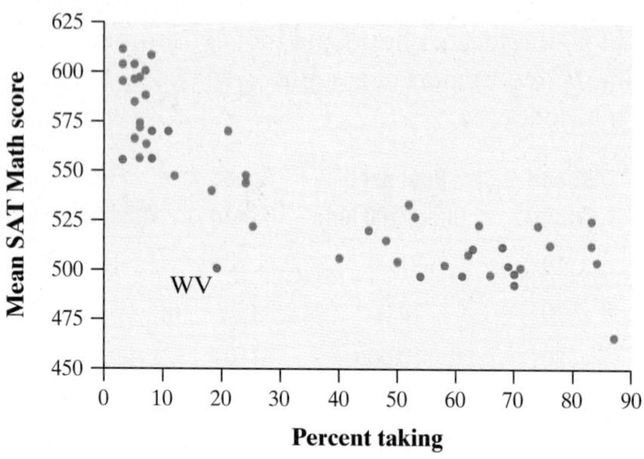

(a) What does the graph suggest about the southern states?

(b) The point for West Virginia is labeled in the graph. Explain how this state is an outlier.

12. **Do heavier people burn more energy?** The study of dieting described in Exercise 10 collected data on the lean body mass (in kilograms) and metabolic rate (in calories) for 12 female and 7 male subjects. The figure below is a scatterplot of the data for all 19 subjects, with separate symbols for males and females.

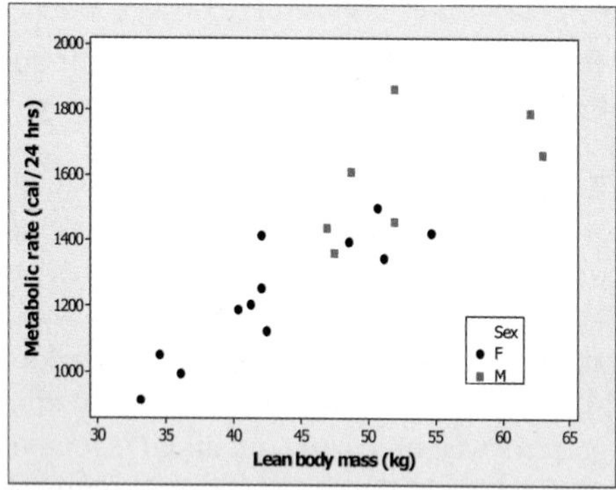

Does the same overall pattern hold for both women and men? What is the most important difference between the sexes?

13. **Merlins breeding** The percent of an animal species in the wild that survives to breed again is often lower following a successful breeding season. A study of merlins (small falcons) in northern Sweden observed the number of breeding pairs in an isolated area and the percent of males (banded for identification) that returned the next breeding season. Here are data for nine years:[6]

pg 148

Breeding pairs:	28	29	29	29	30	32	33	38	38
Percent return:	82	83	70	61	69	58	43	50	47

Investigate the relationship between breeding pairs and percent return. Follow the four-step process.

14. **Does social rejection hurt?** We often describe our emotional reaction to social rejection as "pain." Does social rejection cause activity in areas of the brain that are known to be activated by physical pain? If it does, we really do experience social and physical pain in similar ways. Psychologists first included and then deliberately excluded individuals from a social activity while they measured changes in brain activity. After each activity, the subjects filled out questionnaires that assessed how excluded they felt. The table below shows data for 13 subjects.[7] "Social distress" is measured by each subject's questionnaire score after exclusion relative to the score after inclusion. (So values greater than 1 show the degree of distress caused by exclusion.) "Brain activity" is the change in activity in a region of the brain that is activated by physical pain. (So positive values show more pain.)

Subject	Social distress	Brain activity
1	1.26	−0.055
2	1.85	−0.040
3	1.10	−0.026
4	2.50	−0.017
5	2.17	−0.017
6	2.67	0.017
7	2.01	0.021
8	2.18	0.025
9	2.58	0.027
10	2.75	0.033
11	2.75	0.064
12	3.33	0.077
13	3.65	0.124

Discuss what the data show. Follow the four-step process.

15. **Matching correlations** Five scatterplots are shown below. Match each graph to the r below that best describes it. (Some r's will be left over.)

$r = -0.9 \quad r = -0.7 \quad r = -0.3 \quad r = 0$

$r = 0.3 \quad r = 0.7 \quad r = 0.9$

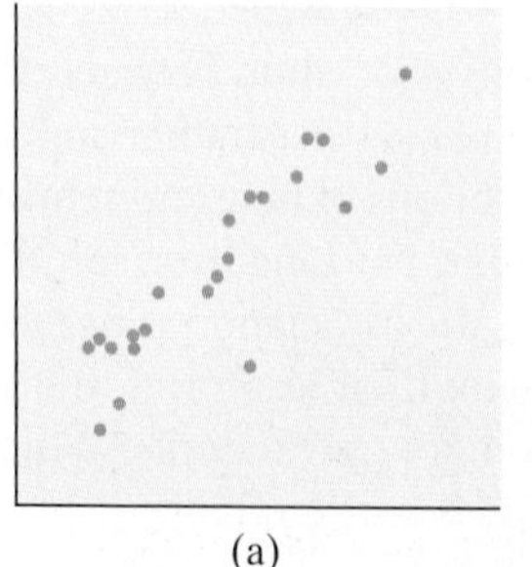

(a)

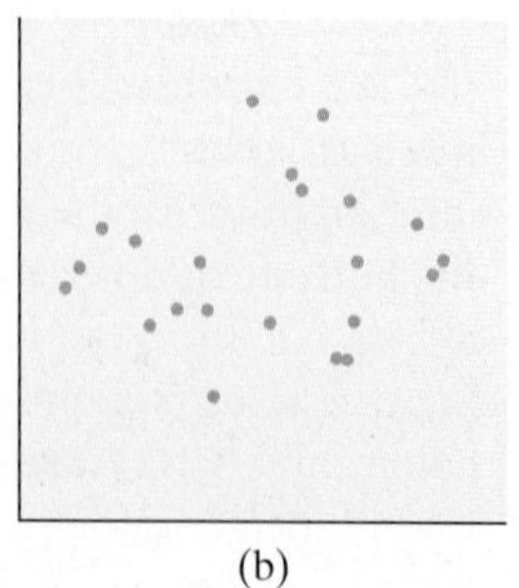

(b)

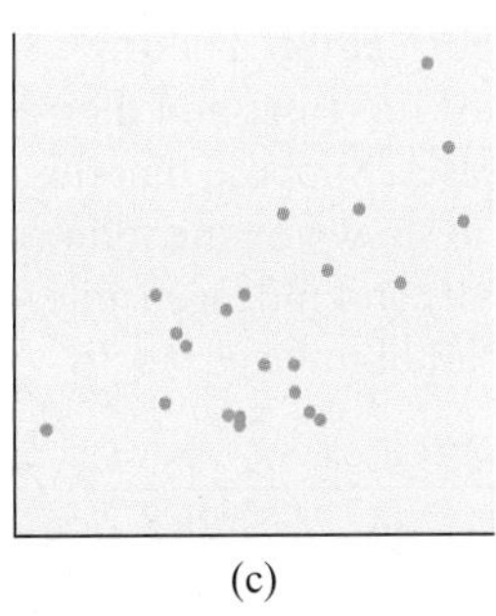
(c)

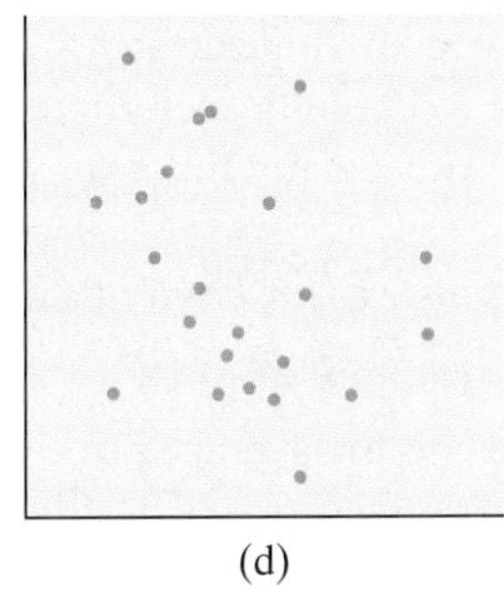
(d)

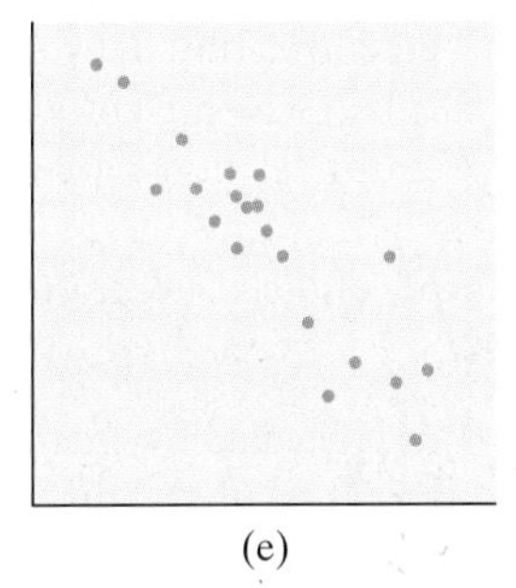
(e)

16. **Rank the correlations** Consider each of the following relationships: the heights of fathers and the heights of their adult sons, the heights of husbands and the heights of their wives, and the heights of women at age 4 and their heights at age 18. Rank the correlations between these pairs of variables from highest to lowest. Explain your reasoning.

17. **Correlation blunders** Each of the following statements contains an error. Explain what's wrong in each case.

(a) "There is a high correlation between the gender of American workers and their income."

(b) "We found a high correlation ($r = 1.09$) between students' ratings of faculty teaching and ratings made by other faculty members."

(c) "The correlation between planting rate and yield of corn was found to be $r = 0.23$ bushel."

pg 153

18. **Teaching and research** A college newspaper interviews a psychologist about student ratings of the teaching of faculty members. The psychologist says, "The evidence indicates that the correlation between the research productivity and teaching rating of faculty members is close to zero." The paper reports this as "Professor McDaniel said that good researchers tend to be poor teachers, and vice versa." Explain why the paper's report is wrong. Write a statement in plain language (don't use the word "correlation") to explain the psychologist's meaning.

19. **Dem bones** Archaeopteryx is an extinct beast having feathers like a bird but teeth and a long bony tail like a reptile. Only six fossil specimens are known. Because these specimens differ greatly in size, some scientists think they are different species rather than individuals from the same species. We will examine some data. If the specimens belong to the same species and differ in size because some are younger than others, there should be a positive linear relationship between the lengths of a pair of bones from all individuals. An outlier from this relationship would suggest a different species. Here are data on the lengths in centimeters of the femur (a leg bone) and the humerus (a bone in the upper arm) for the five specimens that preserve both bones:[8]

Femur (x):	38	56	59	64	74
Humerus (y):	41	63	70	72	84

(a) Make a scatterplot. Do you think that all five specimens come from the same species? Explain.

(b) Find the correlation r step-by-step. First, find the mean and standard deviation of each variable. Then find the six standardized values for each variable. Finally, use the formula for r. Explain how your value for r matches your graph in (a).

20. **Data on dating** A student wonders if tall women tend to date taller men than do short women. She measures herself, her dormitory roommate, and the women in the adjoining rooms. Then she measures the next man each woman dates. Here are the data (heights in inches):

Women (x):	66	64	66	65	70	65
Men (y):	72	68	70	68	71	65

(a) Make a scatterplot of these data. Based on the scatterplot, do you expect the correlation to be positive or negative? Near ±1 or not?

(b) Find the correlation r step-by-step. First, find the mean and standard deviation of each variable. Then find the six standardized values for each variable. Finally, use the formula for r. Do the data show that taller women tend to date taller men?

21. **Hot dogs** Are hot dogs that are high in calories also high in salt? The figure below is a scatterplot of the calories and salt content (measured as milligrams of sodium) in 17 brands of meat hot dogs.[9]

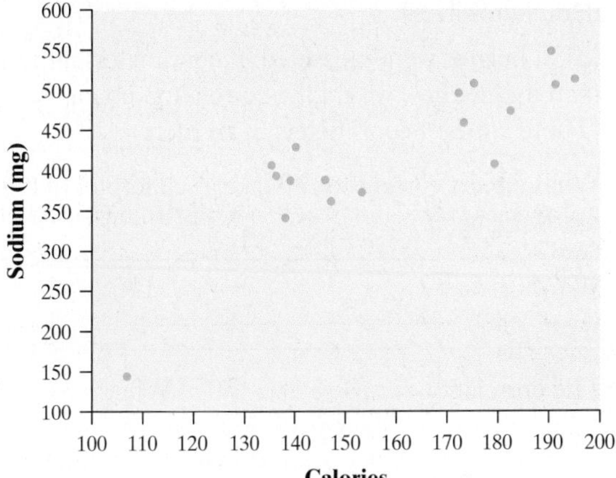

(a) The correlation for these data is $r = 0.87$. Explain what this value means.

(b) What effect would removing the hot dog brand with the lowest calorie content have on the correlation? Justify your answer.

22. **All brawn?** The figure below plots the average brain weight in grams versus average body weight in kilograms for 96 species of mammals.[10] There are many small mammals whose points at the lower left overlap.

(a) The correlation between body weight and brain weight is $r = 0.86$. Explain what this value means.

(b) What effect would removing the elephant have on the correlation? Justify your answer.

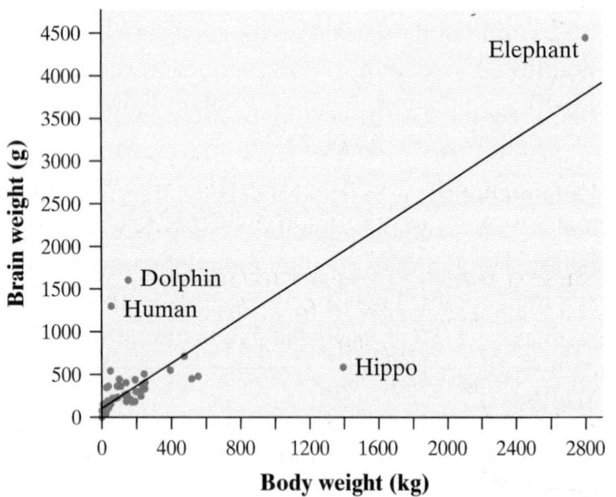

23. **Dem bones** Refer to Exercise 19.

(a) How would r change if the bones had been measured in millimeters instead of centimeters? (There are 10 millimeters in a centimeter.)

(b) If the x and y variables are reversed, how would the correlation change? Explain.

24. **Data on dating** Refer to Exercise 20.

(a) How would r change if all the men were 6 inches shorter than the heights given in the table? Does the correlation tell us if women tend to date men taller than themselves?

(b) If heights were measured in centimeters rather than inches, how would the correlation change? (There are 2.54 centimeters in an inch.)

25. **What affects correlation?** Make a scatterplot of the following data:

x:	1	2	3	4	10	10
y:	1	3	3	5	1	11

The correlation for these data is 0.5. What is responsible for reducing the correlation to this value despite a strong straight-line relationship between x and y in most of the observations?

26. **Strong association but no correlation** The gas mileage of an automobile first increases and then decreases as the speed increases. Suppose that this relationship is very regular, as shown by the following data on speed (miles per hour) and mileage (miles per gallon). Make a scatterplot of mileage versus speed.

Speed:	20	30	40	50	60
Mileage:	24	28	30	28	24

The correlation between speed and mileage is $r = 0$. Explain why the correlation is 0 even though there is a strong relationship between speed and mileage.

Multiple choice: Select the best answer for Exercises 27 to 32.

27. You have data for many years on the average price of a barrel of oil and the average retail price of a gallon of unleaded regular gasoline. If you want to see how well the price of oil predicts the price of gas, then you should make a scatterplot with ______ as the explanatory variable.

(a) the price of oil
(b) the price of gas
(c) the year
(d) either oil price or gas price
(e) time

28. In a scatterplot of the average price of a barrel of oil and the average retail price of a gallon of gas, you expect to see

(a) very little association.
(b) a weak negative association.
(c) a strong negative association.
(d) a weak positive association.
(e) a strong positive association.

29. The graph at top right plots the gas mileage (miles per gallon) of various cars from the same model year versus the weight of these cars in thousands of pounds. The points marked with red dots correspond to cars made in Japan. From this plot, we may conclude that

(a) there is a positive association between weight and gas mileage for Japanese cars.

(b) the correlation between weight and gas mileage for all the cars is close to 1.

(c) there is little difference between Japanese cars and cars made in other countries.

(d) Japanese cars tend to be lighter in weight than other cars.

(e) Japanese cars tend to get worse gas mileage than other cars.

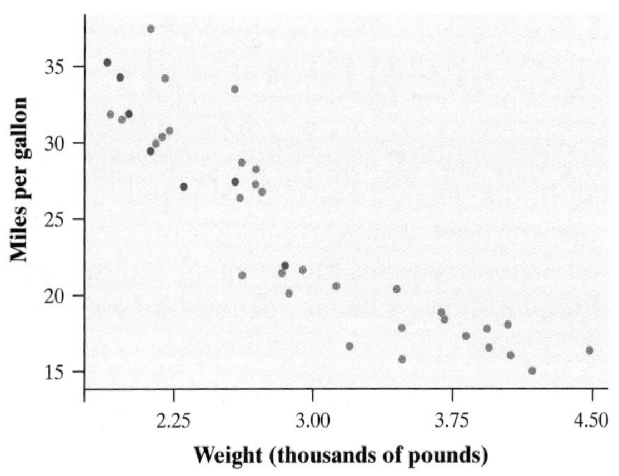

30. If women always married men who were 2 years older than themselves, what would the correlation between the ages of husband and wife be?
(a) 2
(b) 1
(c) 0.5
(d) 0
(e) Can't tell without seeing the data

31. The figure below is a scatterplot of reading test scores against IQ test scores for 14 fifth-grade children. There is one low outlier in the plot. The IQ and reading scores for this child are
(a) IQ = 10, reading = 124.
(b) IQ = 96, reading = 49.
(c) IQ = 124, reading = 10.
(d) IQ = 145, reading = 100.
(e) IQ = 125, reading = 54.

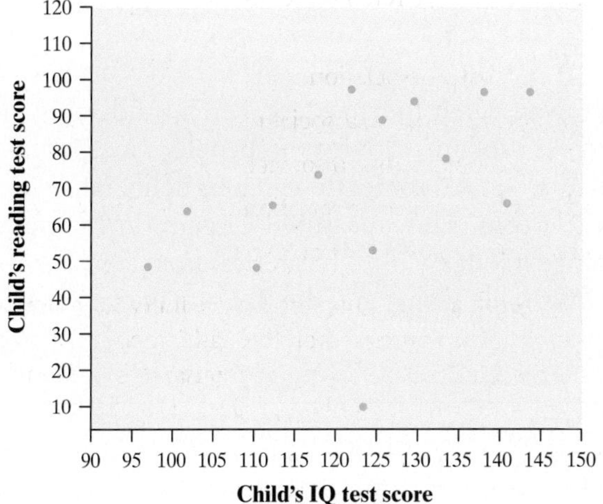

32. If we leave out the low outlier, the correlation for the remaining 13 points in the figure above is closest to
(a) −0.95.
(b) −0.5.
(c) 0.
(d) 0.5.
(e) 0.95.

33. **Big diamonds** (1.2, 1.3) Here are the weights (in milligrams) of 58 diamonds from a nodule carried up to the earth's surface in surrounding rock. These data represent a single population of diamonds formed in a single event deep in the earth.[11]

13.8	3.7	33.8	11.8	27.0	18.9	19.3	20.8	25.4	23.1	7.8
10.9	9.0	9.0	14.4	6.5	7.3	5.6	18.5	1.1	11.2	7.0
7.6	9.0	9.5	7.7	7.6	3.2	6.5	5.4	7.2	7.8	3.5
5.4	5.1	5.3	3.8	2.1	2.1	4.7	3.7	3.8	4.9	2.4
1.4	0.1	4.7	1.5	2.0	0.1	0.1	1.6	3.5	3.7	2.6
4.0	2.3	4.5								

Make a graph that shows the distribution of weights of these diamonds. Describe the shape of the distribution and any outliers. Use numerical measures appropriate for the shape to describe the center and spread.

34. **Student loans** (2.2) A government report looked at the amount borrowed for college by students who graduated in 2000 and had taken out student loans.[12] The mean amount was $\bar{x} = \$17,776$ and the standard deviation was $s_x = \$12,034$. The median was \$15,532 and the quartiles were $Q_1 = \$9900$ and $Q_3 = \$22,500$.

(a) Compare the mean and the median. Also compare the distances of Q_1 and Q_3 from the median. Explain why both comparisons suggest that the distribution is right-skewed.

(b) The right-skew pulls the standard deviation up. So a Normal distribution with the same mean and standard deviation would have a third quartile larger than the actual Q_3. Find the third quartile of the Normal distribution with $\mu = \$17,776$ and $\sigma = \$12,034$ and compare it with $Q_3 = \$22,500$.

3.2 Least-Squares Regression

In Section 3.2, you'll learn about:

- Interpreting a regression line
- Prediction
- Residuals and the least-squares regression line
- Calculating the equation of the least-squares line
- How well the line fits the data: Residual plots
- How well the line fits the data: The role of r^2 in regression
- Interpreting computer regression output
- Correlation and regression wisdom

Linear (straight-line) relationships between two quantitative variables are pretty common and easy to understand. In the previous section, we found linear relationships in settings as varied as sparrowhawk colonies, natural-gas consumption, and Florida manatee deaths. Correlation measures the direction and strength of these relationships. When a scatterplot shows a linear relationship, we'd like to summarize the overall pattern by drawing a line on the scatterplot. A **regression line** summarizes the relationship between two variables, but only in a specific setting: when one of the variables helps explain or predict the other. CAUTION *Regression, unlike correlation, requires that we have an explanatory variable and a response variable.*

DEFINITION: Regression line

A **regression line** is a line that describes how a response variable y changes as an explanatory variable x changes. We often use a regression line to predict the value of y for a given value of x.

Let's look at a situation where a regression line provides a useful model.

EXAMPLE

Does Fidgeting Keep You Slim?

Regression lines as models

Some people don't gain weight even when they overeat. Perhaps fidgeting and other "nonexercise activity" (NEA) explains why—some people may spontaneously increase nonexercise activity when fed more. Researchers deliberately overfed 16 healthy young adults for 8 weeks. They measured fat gain (in kilograms) as the response variable and change in energy use (in calories) from activity other than deliberate exercise—fidgeting, daily living, and the like—as the explanatory variable. Here are the data:[13]

NEA change (cal):	−94	−57	−29	135	143	151	245	355
Fat gain (kg):	4.2	3.0	3.7	2.7	3.2	3.6	2.4	1.3
NEA change (cal):	392	473	486	535	571	580	620	690
Fat gain (kg):	3.8	1.7	1.6	2.2	1.0	0.4	2.3	1.1

Do people with larger increases in NEA tend to gain less fat?

Figure 3.7 is a scatterplot of these data. The plot shows a moderately strong, negative linear association between NEA change and fat gain with no outliers. The correlation is $r = -0.7786$. The line on the plot is a regression line for predicting fat gain from change in NEA.

To "regress" means to go backward. Why are statistical methods for predicting a response from an explanatory variable called "regression"? Sir Francis Galton (1822–1911) looked at data on the heights of children versus the heights of their parents. He found that the taller-than-average parents tended to have children who were also taller than average but not as tall as their parents. Galton called this fact "regression toward the mean," and the name came to be applied to the statistical method.

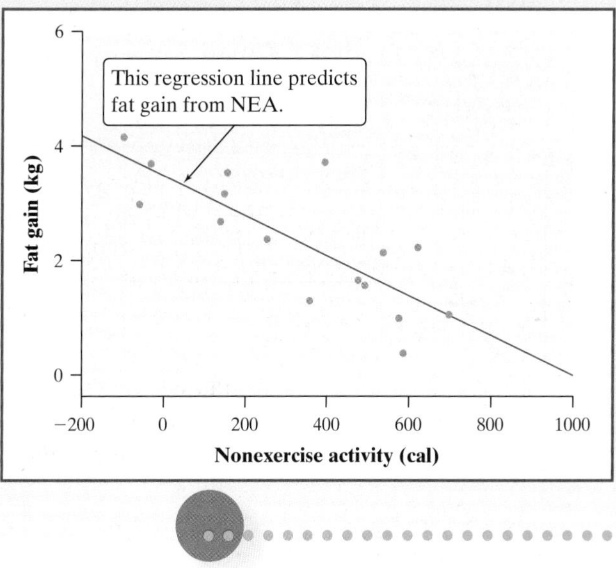

FIGURE 3.7 Scatterplot showing fat gain after 8 weeks of overeating against increase in nonexercise activity over the same period with a regression line added.

Interpreting a Regression Line

A regression line is a *model* for the data, much like the density curves of Chapter 2. The equation of a regression line gives a compact mathematical description of what this model tells us about the relationship between the response variable y and the explanatory variable x.

DEFINITION: Regression line, predicted value, slope, *y* intercept

Suppose that y is a response variable (plotted on the vertical axis) and x is an explanatory variable (plotted on the horizontal axis). A **regression line** relating y to x has an equation of the form

$$\hat{y} = a + bx$$

In this equation,

- $\hat{y}$ (read "y hat") is the **predicted value** of the response variable y for a given value of the explanatory variable x.
- b is the **slope,** the amount by which y is predicted to change when x increases by one unit.
- a is the ***y* intercept,** the predicted value of y when $x = 0$.

Although you are probably used to the form $y = mx + b$ for the equation of a line from algebra, statisticians have adopted a different form for the equation of a regression line. Some use $\hat{y} = b_0 + b_1x$. We prefer $\hat{y} = a + bx$ for two reasons: (1) it's simpler and (2) your calculator uses this form. Don't get so caught up in the symbols that you lose sight of what they mean! The coefficient of x is always the slope, no matter what symbol is used.

Many calculators and software programs will give you the equation of a regression line from keyed-in data. Understanding and using the line is more important than the details of where the equation comes from.

Does Fidgeting Keep You Slim?

Interpreting the slope and y intercept

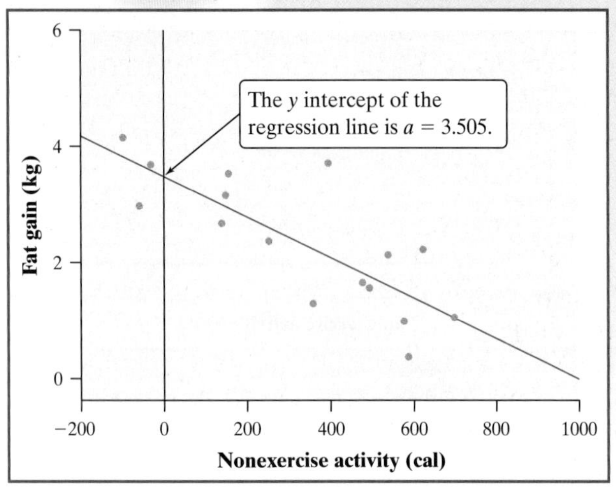

The regression line shown in Figure 3.7 is

$$\widehat{\text{fat gain}} = 3.505 - 0.00344(\text{NEA change})$$

PROBLEM: Identify the slope and y intercept of the regression line. Interpret each value in context.

SOLUTION: The slope $b = -0.00344$ tells us that the amount of fat gained is predicted to go down by 0.00344 kilogram (kg) for each added calorie (cal) of NEA. The y intercept, $a = 3.505$ kg, is the fat gain estimated by this model if NEA does not change when a person overeats.

For Practice *Try Exercise 39(a) and (b)*

The slope of a regression line is an important numerical description of the relationship between the two variables. Although we need the value of the y intercept to draw the line, it is statistically meaningful only when the explanatory variable can actually take values close to zero, as in this setting.

THINK ABOUT IT

Does a small slope mean that there's no relationship? For the NEA and fat gain regression line, the slope $b = -0.00344$ is a small number. This does *not* mean that change in NEA has little effect on fat gain. The size of the slope depends on the units in which we measure the two variables. In this setting, the slope is the predicted change in fat gain in kilograms when NEA increases by 1 calorie. There are 1000 grams in a kilogram. If we measured fat gain in grams, the slope would be 1000 times larger, $b = 3.44$. *You can't say how important a relationship is by looking at the size of the slope of the regression line.*

Prediction

We can use a regression line to predict the response $\hat{y}$ for a specific value of the explanatory variable x. Here's how we do it.

Does Fidgeting Keep You Slim?

Predicting with a regression line

For the NEA and fat gain data, the equation of the regression line is

$$\widehat{\text{fat gain}} = 3.505 - 0.00344(\text{NEA change})$$

If a person's NEA increases by 400 calories when she overeats, substitute $x = 400$ in the equation. The predicted fat gain is

$$\widehat{\text{fat gain}} = 3.505 - 0.00344(400) = 2.13 \text{ kilograms}$$

This prediction is illustrated in Figure 3.8.

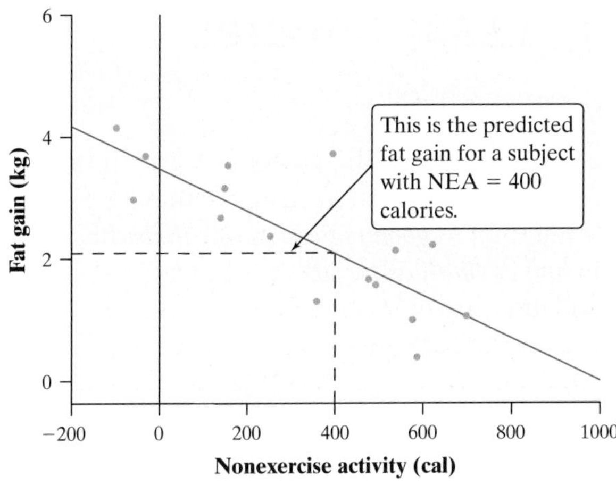

FIGURE 3.8 Using the least-squares regression line to predict fat gain for a person with a 400-calorie increase in NEA.

The accuracy of predictions from a regression line depends on how much the data scatter about the line. In this case, fat gains for similar changes in NEA show a spread of 1 or 2 kilograms. The regression line summarizes the pattern but gives only roughly accurate predictions.

Can we predict the fat gain for someone whose NEA increases by 1500 calories when she overeats? We can certainly substitute 1500 calories into the equation of the line. The prediction is

$$\widehat{\text{fat gain}} = 3.505 - 0.00344(1500) = -1.66 \text{ kilograms}$$

That is, we predict that this individual *loses* 1.66 kilograms of fat when she overeats.

Look again at Figure 3.8. An NEA increase of 1500 calories is far outside the set of *x*-values for our data. We can't say whether increases this large ever occur, or whether the relationship remains linear at such extreme values. Predicting fat gain when NEA increases by 1500 calories is an **extrapolation** of the relationship beyond what the data show.

Often, using the regression line to make a prediction for $x = 0$ is an extrapolation. That's why the *y* intercept isn't always statistically meaningful.

DEFINITION: Extrapolation

Extrapolation is the use of a regression line for prediction far outside the interval of values of the explanatory variable *x* used to obtain the line. Such predictions are often not accurate.

Few relationships are linear for all values of the explanatory variable. *Don't make predictions using values of x that are much larger or much smaller than those that actually appear in your data.*

CHECK YOUR UNDERSTANDING

Some data were collected on the weight of a male white laboratory rat for the first 25 weeks after its birth. A scatterplot of the weight (in grams) and time since birth (in weeks) shows a fairly strong, positive linear relationship. The linear regression equation $\widehat{\text{weight}} = 100 + 40(\text{time})$ models the data fairly well.

1. What is the slope of the regression line? Explain what it means in context.
2. What's the *y* intercept? Explain what it means in context.
3. Predict the rat's weight after 16 weeks. Show your work.
4. Should you use this line to predict the rat's weight at age 2 years? Use the equation to make the prediction and think about the reasonableness of the result. (There are 454 grams in a pound.)

Residuals and the Least-Squares Regression Line

In most cases, no line will pass exactly through all the points in a scatterplot. Because we use the line to predict y from x, the prediction errors we make are errors in y, the vertical direction in the scatterplot. *A good regression line makes the vertical distances of the points from the line as small as possible.*

In Section 3.1, we explored the relationship between body weight and backpack weight for a group of 8 hikers. Here are the data once again:

Body weight (lb):	120	187	109	103	131	165	158	116
Backpack weight (lb):	26	30	26	24	29	35	31	28

Figure 3.9 shows a scatterplot of the data with a regression line added. The prediction errors are marked as bold segments in the graph. These vertical deviations represent "leftover" variation in the response variable after fitting the regression line. For that reason, they are called **residuals.**

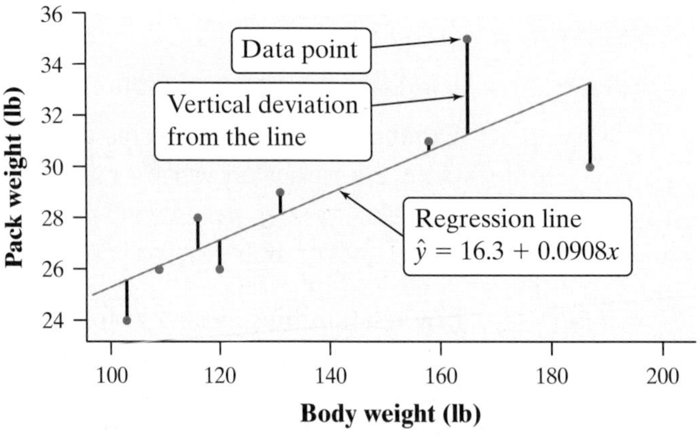

FIGURE 3.9 Scatterplot of the hiker data with a regression line added. A good regression line should make the prediction errors (shown as bold vertical segments) as small as possible.

DEFINITION: Residual

A **residual** is the difference between an observed value of the response variable and the value predicted by the regression line. That is,

$$\begin{aligned}\text{residual} &= \text{observed } y - \text{predicted } y \\ &= y - \hat{y}\end{aligned}$$

The following example shows you how to calculate a residual.

Back to the Backpackers

Finding a residual

PROBLEM: Find and interpret the residual for the hiker who weighed 187 pounds.

SOLUTION: The regression line predicts a pack weight of

$$\widehat{\text{pack weight}} = 16.3 + 0.0908(187) = 33.28 \text{ pounds}$$

AP EXAM TIP There's no firm rule for how many decimal places to show for answers on the AP exam. *Our advice:* Give your answer correct to two or three nonzero decimal places. *Exception:* If you're using one of the tables in the back of the book, give the value shown in the table.

for this hiker, but his actual pack weight was 30 pounds. This hiker's residual is

$$\text{residual} = \text{observed response} - \text{predicted response}$$
$$= y - \hat{y} = 30 - 33.28 = -3.28 \text{ pounds}$$

That is, the regression line *overpredicts* this hiker's pack weight by 3.28 pounds.

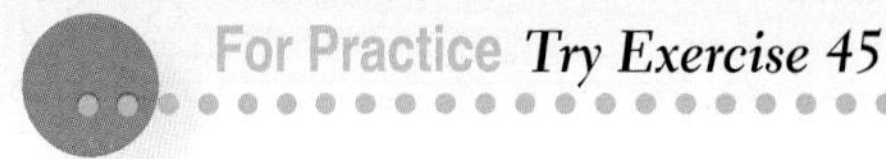

For Practice *Try Exercise 45*

The line shown in Figure 3.9 makes the residuals for the 8 hikers "as small as possible." But what does that mean? Maybe this line minimizes the *sum* of the residuals. Actually, if we add up the prediction errors for all 8 hikers, the positive and negative residuals cancel out. That's the same issue we faced when we tried to measure deviation around the mean in Chapter 1. We'll solve the current problem in much the same way: by squaring the residuals. The regression line we want is the one that minimizes the sum of the squared residuals. That's what the line shown in Figure 3.9 does for the hiker data, which is why we call it the **least-squares regression line.**

DEFINITION: Least-squares regression line

The **least-squares regression line** of y on x is the line that makes the sum of the squared residuals as small as possible.

Figure 3.10 gives a geometric interpretation of the least-squares idea for the hiker data. The least-squares regression line shown minimizes the sum of the squared prediction errors, 30.90. No other regression line would give a smaller sum of squared residuals.

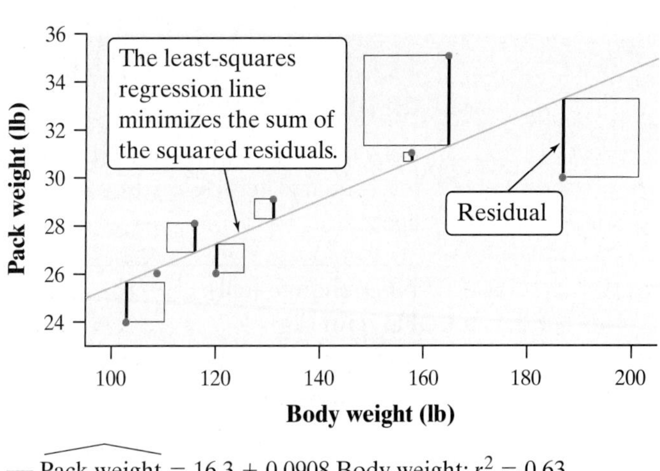

FIGURE 3.10 The least-squares idea: make the errors in predicting y as small as possible by minimizing the sum of the squares of the residuals (in this case, 30.9). Fathom software allows you to explore this fact dynamically.

ACTIVITY *Investigating properties of the least-squares regression line*

MATERIALS: Computer with Internet connection

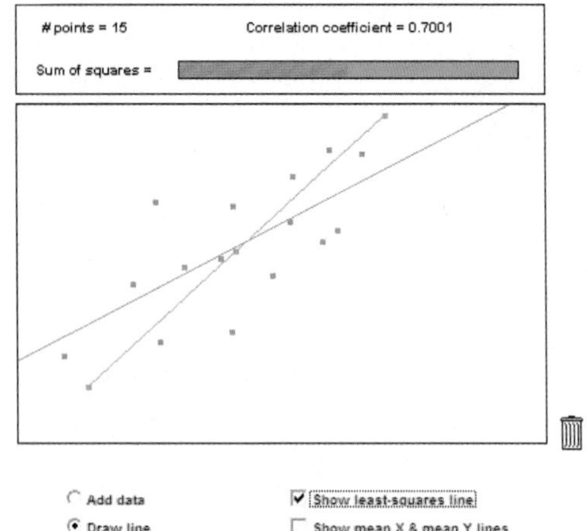

In this Activity, you will use the *Correlation and Regression* applet at the book's Web site, www.whfreeman.com/tps4e, to explore some properties of the least-squares regression line.

1. Click on the scatterplot to create a group of 15 to 20 points from lower left to upper right with a clear positive straight-line pattern (correlation around 0.7). Click the "Draw line" button and use the mouse (right-click and drag) to draw a line through the middle of the cloud of points from lower left to upper right. Use your mouse to adjust the slope and *y* intercept of your line until you are confident that your line fits the data well.

2. Now click the "Show least-squares line" box. How close did you get? The "thermometer" above the plot helps answer this question. The blue portion is the sum of the squared vertical distances from the points in the plot to the least-squares line. The green portion shows by how much your line misses the smallest possible "sum of squares."

3. Click the "Draw line" button again to remove your line. Then click the "Show mean *x* and mean *y* lines" button. What do you notice? Move or add points, one at a time, in your scatterplot to see if this result continues to hold true.

4. *Investigate*: How does an outlier affect the slope and *y* intercept of the least-squares line? Does it depend on whether the outlier has an *x*-value close to the center of the plot or toward the far edges of the plot?

Your calculator or statistical software will give the equation of the least-squares line from data that you enter. Then you can concentrate on understanding and using the regression line.

TECHNOLOGY CORNER Least-squares regression lines on the calculator

Let's use the fat gain and NEA data to show how to find the equation of the least-squares regression line on the TI-83/84 and TI-89. Here are the data again:

NEA change (cal):	−94	−57	−29	135	143	151	245	355
Fat gain (kg):	4.2	3.0	3.7	2.7	3.2	3.6	2.4	1.3
NEA change (cal):	392	473	486	535	571	580	620	690
Fat gain (kg):	3.8	1.7	1.6	2.2	1.0	0.4	2.3	1.1

1. Enter the NEA change data into L1(list1) and the fat gain data into L2(list2). Then make a scatterplot. Refer to the Technology Corner on page 149.

The TI-83/84 command tells the calculator to compute the equation of the least-squares regression line using L1 as the explanatory variable and L2 as the response variable and then to store the result in slot Y1. This method is useful if you want to graph the regression line or use its equation to make predictions. If you're interested in only the equation of the line, `LinReg(a+bx)L1,L2` will do.

If r^2 and r do not appear on the TI-83/84 screen, do this one-time series of keystrokes: Press [2nd] [0] (CATALOG), scroll down to `DiagnosticOn`, and press [ENTER]. Press [ENTER] again to execute the command. The screen should say "Done." Then press [2nd] [ENTER] (ENTRY) to recall the regression command and [ENTER] again to calculate the least-squares line. The r^2- and r-values should now appear.

2. To determine the least-squares regression line:

TI-83/84

- Press [STAT]; choose CALC and then `8:LinReg(a+bx)`. Finish the command to read `LinReg(a+bx) L1,L2,Y1` and press [ENTER]. (Y1 is found under VARS/Y-VARS/1: Function.)

```
LinReg
 y=a+bx
 a=3.505122916
 b=-.003441487
 r²=.6061492049
 r=-.7785558457
```

TI-89

- In the Statistics/List Editor, press [F4] (CALC); choose `3:Regressions` and then `1:LinReg(a+bx)`.
- Enter list1 for the Xlist, list2 for the Ylist; choose to store the RegEqn to y1(x); and press [ENTER]

3. Graph the regression line. Turn off all other equations in the Y= screen and press [GRAPH] ([♦] [F3] on the TI-89) to add the least-squares line to the scatterplot.

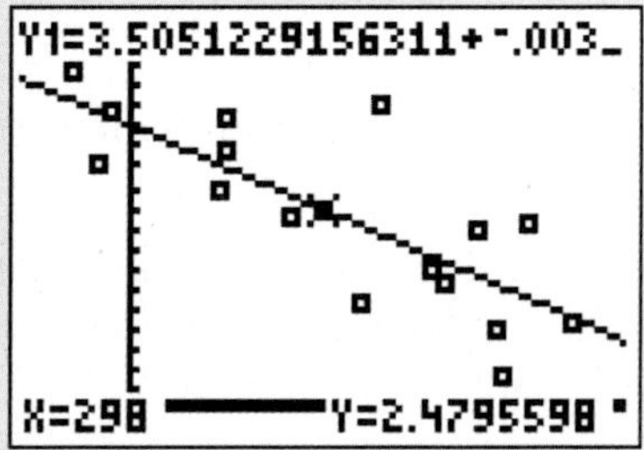

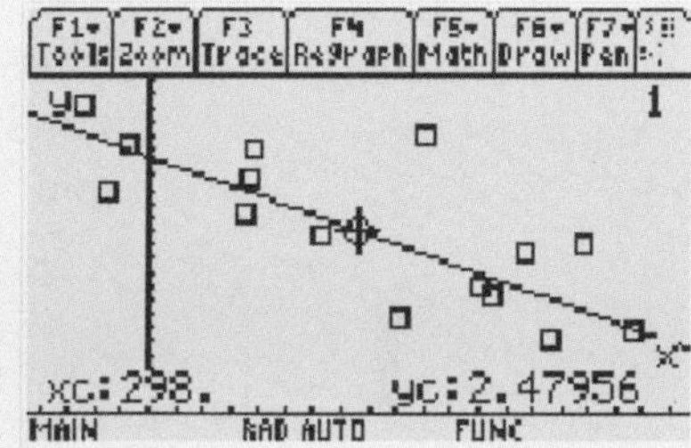

4. Save these lists for later use. On the home screen, execute the command `L1→NEA:L2→FAT` (`list1→NEA:list2→FAT` on the TI-89).

Although the calculator will report the values for a and b to nine decimal places, we usually round off to fewer decimal places. You would write the equation as $\hat{y} = 3.505 - 0.00344x$.

TI-Nspire instructions in Appendix B

CHECK YOUR UNDERSTANDING

It's time to practice your calculator regression skills. Using the familiar hiker data in the table below, repeat the steps in the previous Technology Corner. You should get $\hat{y} = 16.3 + 0.0908x$ as the equation of the regression line.

Body weight (lb):	120	187	109	103	131	165	158	116
Backpack weight (lb):	26	30	26	24	29	35	31	28

Calculating the Equation of the Least-Squares Line

Another reason for studying the least-squares regression line is that the problem of finding its equation has a simple answer. We can give the **equation of the least-squares regression line** in terms of the means and standard deviations of the two variables and their correlation.

> **AP EXAM TIP** The formula sheet for the AP exam uses different notation for these equations: $b_1 = r\frac{s_y}{s_x}$ and $b_0 = \bar{y} - b_1\bar{x}$. That's because the least-squares line is written as $\hat{y} = b_0 + b_1x$. We prefer our simpler versions without the subscripts!

DEFINITION: Equation of the least-squares regression line

We have data on an explanatory variable x and a response variable y for n individuals. From the data, calculate the means $\bar{x}$ and $\bar{y}$ and the standard deviations s_x and s_y of the two variables and their correlation r. The least-squares regression line is the line $\hat{y} = a + bx$ with **slope**

$$b = r\frac{s_y}{s_x}$$

and ***y* intercept**

$$a = \bar{y} - b\bar{x}$$

We know from Step 3 of the Activity on page 170 that the least-squares regression line for any data set passes through the point $(\bar{x}, \bar{y})$. If we start with the equation $\hat{y} = a + bx$ and plug in $(\bar{x}, \bar{y})$, we get

$$\bar{y} = a + b\bar{x}$$

Now solve for the y intercept (a) by subtracting $b\bar{x}$ from both sides of the equation:

$$\bar{y} - b\bar{x} = a$$

That's the formula for the y intercept given in the definition box.

What does the slope of the least-squares line tell us? The figure below shows the regression line in black for the hiker data.

We have added four more lines to the graph: a vertical line at the mean body weight $\bar{x}$, a vertical line at $\bar{x} + s_x$ (one standard deviation above the mean body weight), a horizontal line at the mean pack weight $\bar{y}$, and a horizontal line at $\bar{y} + s_y$ (one standard deviation above the mean pack weight). Note that the regression line passes through $(\bar{x}, \bar{y})$ as expected. From the graph, the slope of the line is

$$b = \text{slope} = \frac{\text{change in } y}{\text{change in } x} = \frac{??}{s_x}$$

From the definition box, we know that the slope is

$$b = r\frac{s_y}{s_x}$$

Setting the two formulas equal to each other, we have

$$b = r\frac{s_y}{s_x} = \frac{??}{s_x} \Rightarrow ?? = r \cdot s_y$$

Pack weight = 16.3 + 0.0908 Body weight; $r^2 = 0.83$

So the unknown distance ?? in the figure must be equal to $r \cdot s_y$. In other words, for an increase of one standard deviation in the value of the explanatory variable x, the least-squares regression line predicts an increase of r standard deviations in the response variable y.

There is a close connection between correlation and the slope of the least-squares line. The slope is

$$b = r\frac{s_y}{s_x}$$

This equation says that along the regression line, *a change of one standard deviation in x corresponds to a change of r standard deviations in y*. When the variables are perfectly correlated ($r = 1$ or $r = -1$), the change in the predicted response $\hat{y}$ is the same (in standard deviation units) as the change in x. Otherwise, because $-1 \leq r \leq 1$, the change in $\hat{y}$ is less than the change in x. As the correlation grows less strong, the prediction $\hat{y}$ moves less in response to changes in x.

Fat Gain and NEA

Calculating the least-squares regression line

Refer to the data from the study of nonexercise activity and fat gain on page 164. The mean and standard deviation of the 16 changes in NEA are $\bar{x} = 324.8$ calories (cal) and $s_x = 257.66$ cal. For the 16 fat gains, the mean and standard deviation are $\bar{y} = 2.388$ kg and $s_y = 1.1389$ kg. The correlation between fat gain and NEA change is $r = -0.7786$.

PROBLEM: (a) Find the equation of the least-squares regression line for predicting fat gain from NEA change. Show your work.

(b) What change in fat gain does the regression line predict for each additional 257.66 cal of NEA? Explain.

SOLUTION: (a) The least-squares regression line of fat gain y on NEA change x has slope

$$b = r\frac{s_y}{s_x} = (-0.7786)\frac{1.1389}{257.66} = -0.00344 \text{ kg/cal}$$

To find the y intercept, we use the fact that the least-squares line passes through $(\bar{x}, \bar{y})$:

$$\bar{y} = a + b\bar{x}$$
$$2.388 = a + (-0.00344)(324.8)$$
$$a = 3.505 \text{ kg}$$

So the equation of the least-squares line is $\hat{y} = 3.505 - 0.00344x$.

(b) An NEA increase of 257.66 cal is a one standard deviation increase. Since

$$b = r\frac{s_y}{s_x}$$

the least-squares line predicts an r standard deviation increase in fat gain; that is,

$$-0.7786(1.1389) = -0.89 \text{ kg}$$

The regression line predicts a decrease of 0.89 kg in fat gain for an additional 257.66 cal of NEA.

When doing calculations like this, don't round until the end of the problem. Use as many decimal places as your calculator stores to get accurate values of the slope and y intercept.

For Practice *Try Exercise 47*

What happens if we standardize both variables? Standardizing a variable converts its mean to 0 and its standard deviation to 1. Doing this to both x and y will transform the point $(\bar{x}, \bar{y})$ to (0, 0). So the least-squares line for the standardized values will pass through (0, 0). What about the slope of this line? From the formula, it's $b = rs_y/s_x$. Since we standardized, $s_x = s_y = 1$. That means $b = r$. In other words, the slope is equal to the correlation. The Fathom screen shot confirms these results. It shows that $r^2 = 0.63$, so $r = \sqrt{0.63} \approx 0.795$.

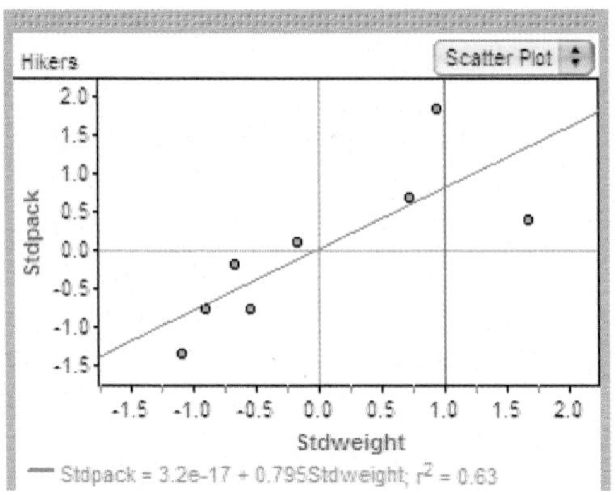

How Well the Line Fits the Data: Residual Plots

One of the first principles of data analysis is to look for an overall pattern and for striking departures from the pattern. A regression line describes the overall pattern of a linear relationship between an explanatory variable and a response variable. We see departures from this pattern by looking at the residuals.

Does Fidgeting Keep You Slim?

Examining residuals

Let's return to the fat gain and NEA study involving 16 young people who volunteered to overeat for 8 weeks. Those whose NEA rose substantially gained less fat than others. In the Technology Corner, we confirmed that the least-squares regression line for these data is $\widehat{\text{fat gain}} = 3.505 - 0.00344(\text{NEA change})$. The calculator screen shot in the margin shows a scatterplot of the data with the least-squares line added.

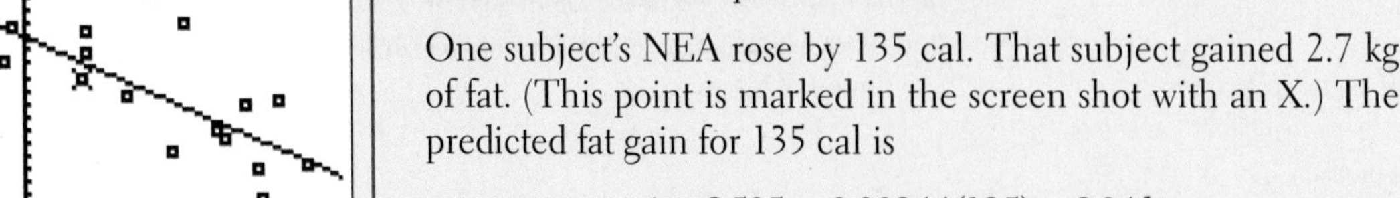

One subject's NEA rose by 135 cal. That subject gained 2.7 kg of fat. (This point is marked in the screen shot with an X.) The predicted fat gain for 135 cal is

$$\hat{y} = 3.505 - 0.00344(135) = 3.04 \text{ kg}$$

The residual for this subject is therefore

$$\text{residual} = \text{observed } y - \text{predicted } y = y - \hat{y} = 2.7 - 3.04 = -0.34 \text{ kg}$$

This residual is negative because the data point lies below the line.

The 16 data points used in calculating the least-squares line produce 16 residuals. Rounded to two decimal places, they are

0.37	−0.70	0.10	−0.34	0.19	0.61	−0.26	−0.98
1.64	−0.18	−0.23	0.54	−0.54	−1.11	0.93	−0.03

Most graphing calculators and statistical software will calculate and store residuals for you.

Because the residuals show how far the data fall from our regression line, examining the residuals helps assess how well the line describes the data. Although residuals can be calculated from any model that is fitted to the data, the residuals from the least-squares line have a special property: *the mean of the least-squares residuals is always zero*. You can check that the sum of the residuals in the above example is 0.01. The sum is not exactly 0 because we rounded to two decimal places.

You can see the residuals in the scatterplot of Figure 3.11(a) by looking at the vertical deviations of the points from the line. The **residual plot** in Figure 3.11(b) makes it easier to study the residuals by plotting them against the explanatory variable, change in NEA. Because the mean of the residuals is always zero, the horizontal line at zero in Figure 3.11(b) helps orient us. This "residual = 0" line corresponds to the regression line in Figure 3.11(a).

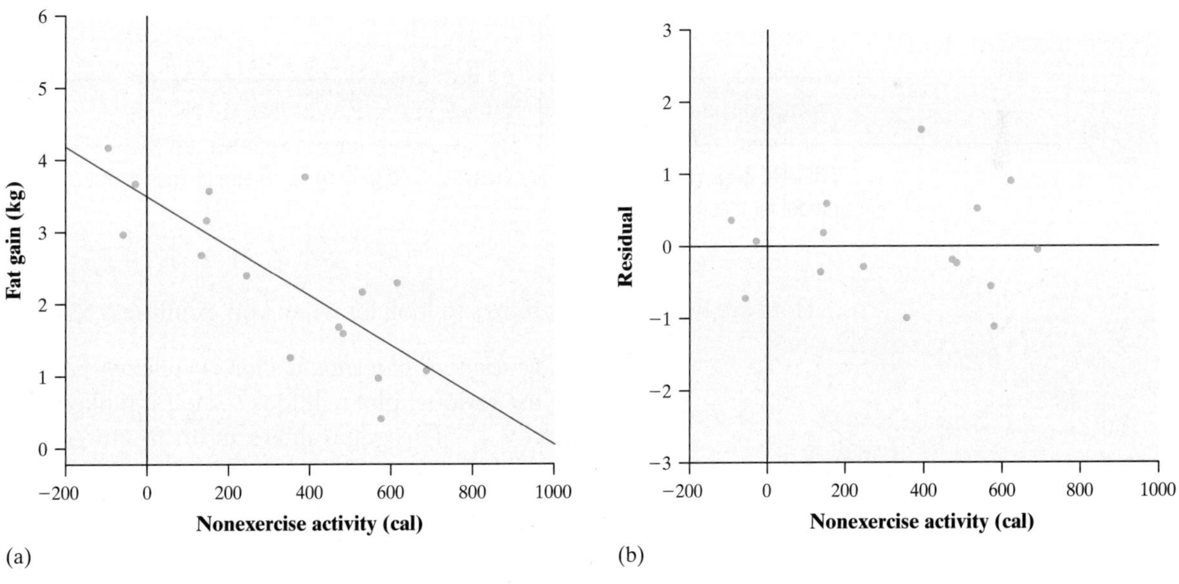

FIGURE 3.11 (a) Scatterplot of fat gain versus nonexercise activity, with the least-squares line. (b) Residual plot for the regression line displayed in Figure 3.11(a). The line at $y = 0$ marks the sum (and mean) of the residuals.

Some software packages prefer to plot the residuals against the predicted values $\hat{y}$ instead of against the values of the explanatory variable. The basic shape of the two plots is the same because $\hat{y}$ is linearly related to x.

DEFINITION: Residual plot

A **residual plot** is a scatterplot of the residuals against the explanatory variable. Residual plots help us assess how well a regression line fits the data.

CHECK YOUR UNDERSTANDING

Refer to the nonexercise activity and fat gain data.

1. Find the residual for the subject who increased NEA by 620 calories. Show your work.
2. Interpret the value of this subject's residual in context.
3. For which subject did the regression line overpredict fat gain by the most? Justify your answer.

Examining residual plots A residual plot in effect turns the regression line horizontal. It magnifies the deviations of the points from the line, making it easier to see unusual observations and patterns. If the regression line captures the overall pattern of the data, there should be *no pattern* in the residuals. Figure 3.12(a) shows a residual plot with a clear curved pattern. A straight line is not an appropriate model for these data, as Figure 3.12(b) confirms.

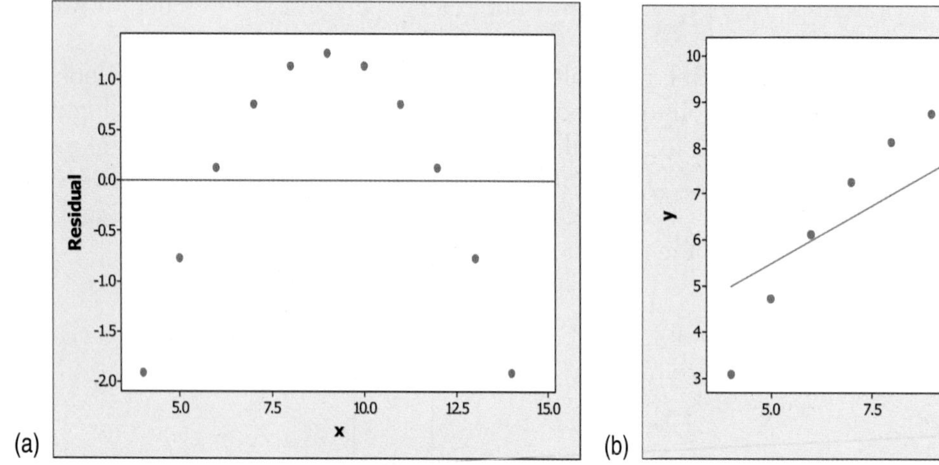

FIGURE 3.12 (a) The residual plot has a curved pattern. (b) A straight line is not a good model for these data.

Here are two important things to look for when you examine a residual plot.

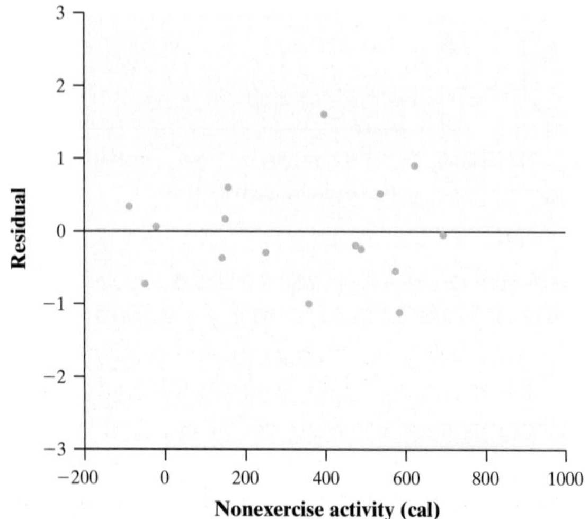

FIGURE 3.13 The unstructured (random) scatter of points indicates that the regression line isn't missing any patterns in the data, so the line is a good model.

1. *The residual plot should show no obvious pattern.* Ideally, the residual plot will look something like the one in Figure 3.13. This graph shows an unstructured (random) scatter of points in a horizontal band centered at zero.

A curved pattern in a residual plot shows that the relationship is not linear. We'll look at how to deal with curved relationships in Chapter 12. Another type of pattern is shown in Figure 3.14. This residual plot reveals increasing spread about the regression line as x increases. Predictions of y using this line will be less accurate for larger values of x.

2. *The residuals should be relatively small in size.* A regression line that fits the data well should come "close" to most of the points. That is, the residuals should be fairly small. How do we decide whether the residuals are "small enough"? We consider the size of a "typical" prediction error.

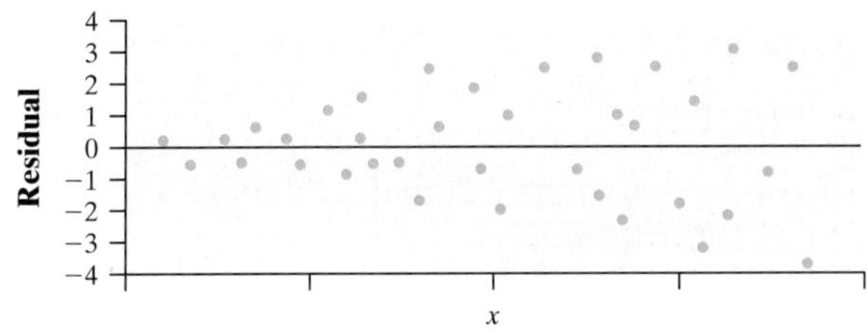

FIGURE 3.14 The response variable y has more spread for larger values of the explanatory variable x, so prediction will be less accurate when x is large.

In Figure 3.13, for example, most of the residuals are between −0.7 and 0.7. For these individuals, the predicted fat gain from the least-squares line is within 0.7 kilogram (kg) of their actual fat gain during the study. That sounds pretty good. But the subjects gained only between 0.4 and 4.2 kg, so a prediction error of 0.7 kg is relatively large compared with the actual fat gain for an individual. The largest residual, 1.64, corresponds to a prediction error of 1.64 kg. This subject's actual fat gain was 3.8 kg, but the regression line predicted a fat gain of only 2.16 kg. That's a pretty large error, especially from the subject's perspective!

Standard deviation of the residuals We have already seen that the average prediction error (that is, the mean of the residuals) is 0 whenever we use a least-squares regression line. That's because the positive and negative residuals "balance out." But that doesn't tell us how far off the predictions are, on average. Instead, we use the **standard deviation of the residuals:**

$$s = \sqrt{\frac{\sum \text{residuals}^2}{n-2}}$$

For the NEA and fat gain data, the sum of the squared residuals is 7.663. So the standard deviation of the residuals is

$$s = \sqrt{\frac{7.663}{14}} = 0.740 \text{ kg}$$

Researchers would need to decide whether they would feel comfortable using the regression line to make predictions that are "off" by an average of 0.74 kg.

DEFINITION: Standard deviation of the residuals (s)

If we use a least-squares line to predict the values of a response variable y from an explanatory variable x, the **standard deviation of the residuals (s)** is given by

$$s = \sqrt{\frac{\sum \text{residuals}^2}{n-2}} = \sqrt{\frac{\sum (y_i - \hat{y})^2}{n-2}}$$

This value gives the approximate size of a "typical" or "average" prediction error (residual).

Does the formula for s look slightly familiar? It should. In Chapter 1, we defined the standard deviation of a set of quantitative data as

$$s_x = \sqrt{\frac{\sum (x_i - \bar{x})^2}{n-1}}$$

We interpreted the resulting value as (roughly) the average distance of the data points from the mean. In the case of two-variable data, we're interested in the average (vertical) distance of the data points from the regression line. We find this value in much the same way: by adding up the squared deviations, then averaging (again in a funny way), and taking the square root to get back to the original units of measurement. Why do we divide by $n - 2$ this time instead of $n - 1$? You'll have to wait until Chapter 12 to find out.

TECHNOLOGY CORNER Residual plots and *s* on the calculator

Let's continue the analysis of the NEA and fat gain study data from the previous Technology Corner (page 170). You should have already made a scatterplot, calculated the equation of the least-squares regression line, and graphed the line on your plot. Now, we want to calculate residuals and make a residual plot. Earlier, we found that $\hat{y} = 3.505 - 0.00344x$.

TI-83/84	TI-89

1. Define L3(list3) as the predicted values from the regression equation.
   - TI-83/84: With L3 highlighted, enter the command `3.505-0.00344*L1` and press ENTER
   - TI-89: With list3 highlighted, enter the command `3.505-0.00344*list1` and press ENTER
2. Define L4/list4 as the observed *y*-value minus the predicted *y*-value.
   - TI-83/84: With L4 highlighted, enter the command `L2-L3` and press ENTER to show the residuals.
   - TI-89: With list4 highlighted, enter the command `list2-list3` and press ENTER to show the residuals.

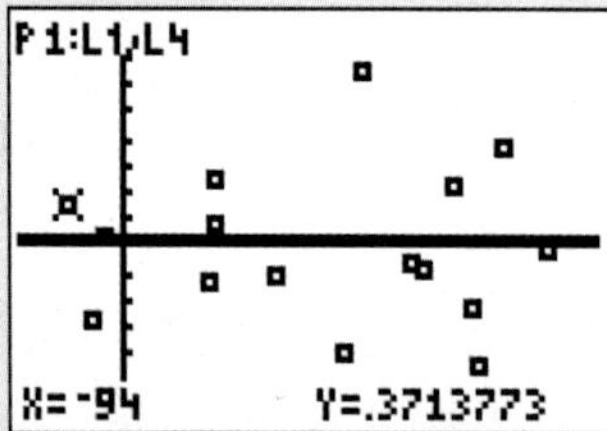
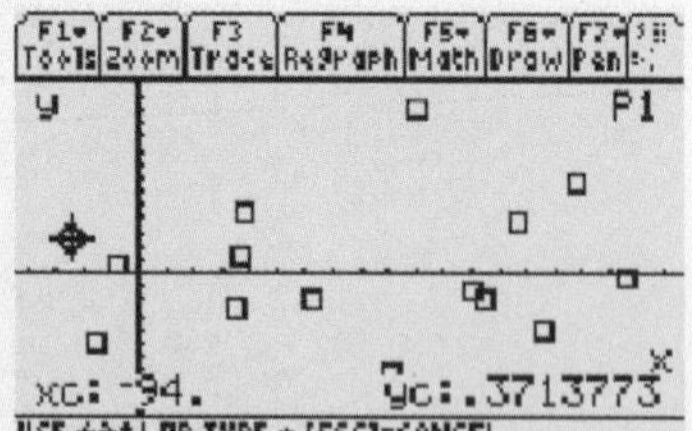

3. Turn off Plot1 and the regression equation. Specify Plot2 with L1/list1 as the *x* variable and L4/list4 as the *y* variable. Use ZoomStat (ZoomData) to see the residual plot.

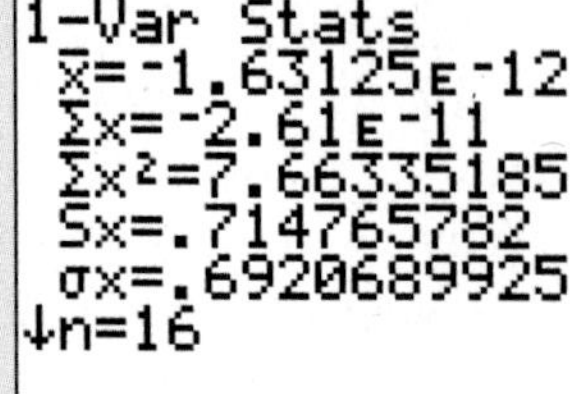
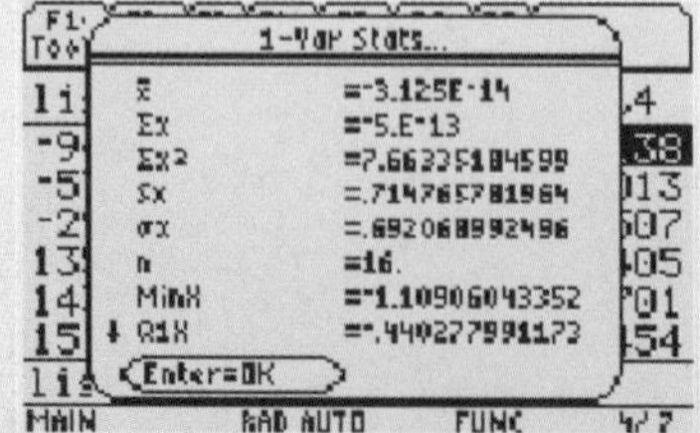

The *x* axis in the residual plot serves as a reference line: points above this line correspond to positive residuals and points below the line correspond to negative residuals. We used TRACE to see the residual for the individual with an NEA change of −94 calories.

4. Finally, we want to compute the standard deviation *s* of the residuals. Calculate one-variable statistics on the residuals list (L4 or list4). The mean of the residuals is 0 (up to roundoff error). The sum of the squared residuals is $\sum x^2 = 7.663$. To find *s*, use the formula

$$s = \sqrt{\frac{\sum \text{residuals}^2}{n-2}} = \sqrt{\frac{7.663}{14}} = 0.74 \text{ kg}$$

Picky point: the calculator's value for the standard deviation of the residuals, "$s_x = 0.7147\ldots$," doesn't quite match our value of $s = 0.74$. That's because the calculator divides by $n - 1$ while we divide by $n - 2$.

```
1-Var Stats
 x̄=-1.63125E-12
 Σx=-2.61E-11
 Σx²=7.66335185
 Sx=.714765782
 σx=.6920689925
↓n=16
```

TI-Nspire instructions in Appendix B

CHECK YOUR UNDERSTANDING

In the Check Your Understanding on page 171, we asked you to perform least-squares regression on the familiar hiker data shown in the table below. The graph shown is a residual plot for the least-squares regression of pack weight on body weight for the 8 hikers.

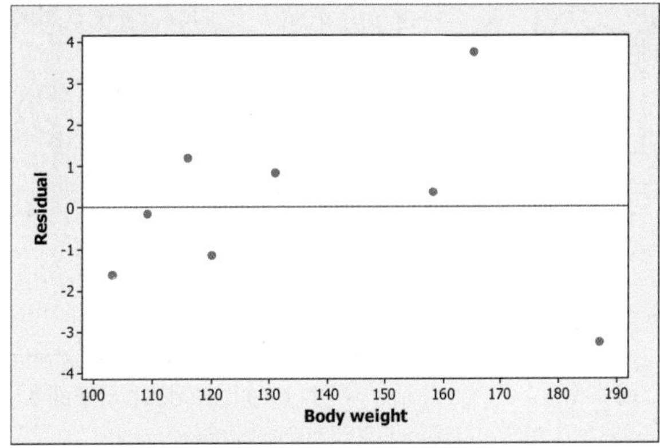

Body weight (lb):	120	187	109	103	131	165	158	116
Backpack weight (lb):	26	30	26	24	29	35	31	28

1. The residual plot does not show a random scatter. Describe the pattern you see.
2. For this regression, $s = 2.27$. Interpret this value in context.

How Well the Line Fits the Data: The Role of r^2 in Regression

A residual plot is a graphical tool for evaluating how well a regression line fits the data. The standard deviation of the residuals, s, gives us a numerical estimate of the average size of our prediction errors from the regression line. There is another numerical quantity that tells us how well the least-squares line predicts values of the response variable y. It is r^2, the **coefficient of determination.** Some computer packages call it "R-sq." You may have noticed this value in some of the calculator and computer regression output that we showed earlier. Although it's true that r^2 is equal to the square of r, there is much more to this story.

Pack weight and body weight

How can we predict y if we don't know x?

Suppose a new student is assigned at the last minute to our group of 8 hikers. What would we predict for his pack weight? Figure 3.15 shows a scatterplot of the hiker data that we have studied throughout this chapter. The least-squares line is drawn on the plot in green. Another line has been added in blue: a horizontal line at the mean y-value, $\bar{y} = 28.625$ lb. If we don't know this new student's body weight, then we can't use the regression line to make a prediction. What should we do? Our best strategy is to use the mean pack weight of the other 8 hikers as our prediction.

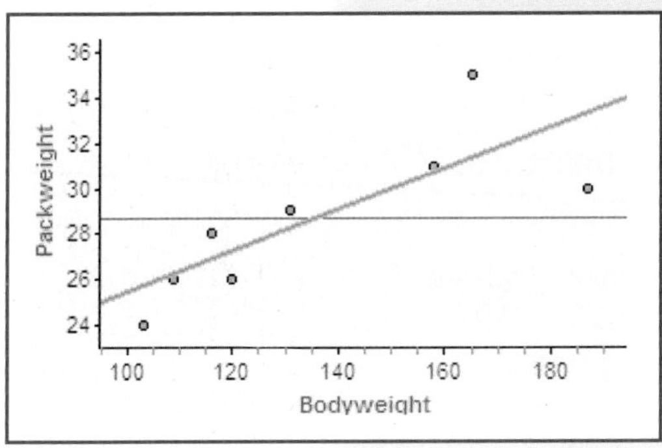

FIGURE 3.15 Scatterplot and least-squares line for the hiker data with a horizontal line added at the mean backpack weight.

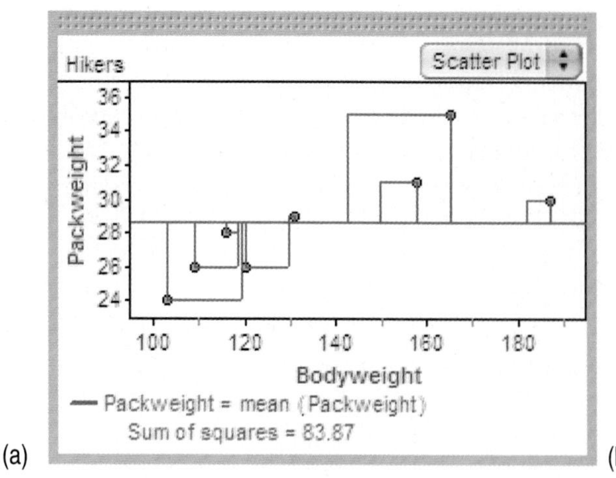

(a)

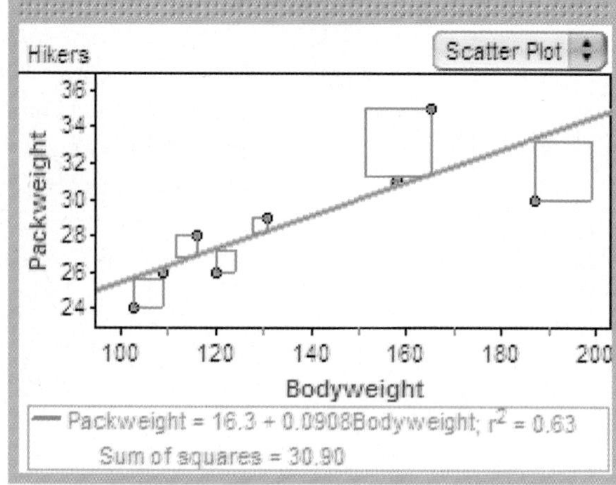

(b)

FIGURE 3.16 (a) The sum of the squared residuals would be 83.87 if we used the mean backpack weight as our prediction for all 8 hikers. (b) The sum of the squared residuals from the least-squares line is 30.90.

SST is closely related to the variance of y:

$$s_y^2 = \frac{\sum (y_i - \bar{y})^2}{n-1}$$

$$\Rightarrow (n-1)s_y^2 = \text{SST}.$$

In other words, SST is just a constant multiple of the variance.

Figure 3.16(a) shows the prediction errors if we use the average pack weight $\bar{y}$ as our prediction for the original group of 8 hikers. We can see that the sum of the squared residuals for this line is $\text{SST} = \sum (y_i - \bar{y})^2 = 83.87$. SST measures the total variation in the y-values.

If we learn our new hiker's body weight, then we could use the least-squares line to predict his pack weight. How much better does the regression line do at predicting pack weights than simply using the average pack weight $\bar{y}$ of all 8 hikers? Figure 3.16(b) reminds us that the sum of squared residuals for the least-squares line is $\sum \text{residual}^2 = 30.90$. We'll call this SSE, for sum of squared errors.

The ratio SSE/SST tells us what proportion of the total variation in y still remains after using the regression line to predict the values of the response variable. In this case,

$$\frac{\text{SSE}}{\text{SST}} = \frac{30.90}{83.87} = 0.368$$

This means that 36.8% of the variation in pack weight is *unaccounted for* by the least-squares regression line. Taking this one step further, the proportion of the total variation in y that *is accounted for* by the regression line is

$$1 - \frac{\text{SSE}}{\text{SST}} = 1 - 0.368 = 0.632$$

We interpret this by saying that "*63.2% of the variation in backpack weight is accounted for by the linear model relating pack weight to body weight.*" For this reason, we define

$$r^2 = 1 - \frac{\text{SSE}}{\text{SST}}$$

DEFINITION: The coefficient of determination: r^2 in regression

The **coefficient of determination** r^2 is the fraction of the variation in the values of y that is accounted for by the least-squares regression line of y on x. We can calculate r^2 using the following formula:

$$r^2 = 1 - \frac{\text{SSE}}{\text{SST}}$$

where $\text{SSE} = \sum \text{residual}^2$ and $\text{SST} = \sum (y_i - \bar{y})^2$.

If all of the points fall directly on the least-squares line, SSE = 0 and $r^2 = 1$. Then all of the variation in y is accounted for by the linear relationship with x. Since the least-squares line yields the smallest possible sum of squared prediction errors, SSE can never be more than SST, which is based on the line $y = \bar{y}$. In the worst-case scenario, the least-squares line does no better at predicting y than $y = \bar{y}$ does. Then SSE = SST and $r^2 = 0$. If $r^2 = 0.632$ (as in the hiker example), then about 63% of the variation in y among the individual hikers is due to the straight-line relationship between y and x. The other 37% is individual variation among hikers that is not explained by the linear relationship. What other factors besides body weight might help explain a hiker's pack weight? Physical fitness, desire to carry "creature comforts," and willingness to share the group's equipment load are just a few possible sources of variation in pack weight.

It seems pretty remarkable that the coefficient of determination is actually the correlation squared. This fact provides an important connection between correlation and regression. When you report a regression, give r^2 as a measure of how successful the regression was in explaining the response. When you see a correlation, square it to get a better feel for the strength of the linear relationship.

CHECK YOUR UNDERSTANDING

Multiple choice: Select the best answer.

1. For the least-squares regression of fat gain on NEA, $r^2 = 0.606$. Which of the following gives a correct interpretation of this value in context?

(a) 60.6% of the points lie on the least-squares regression line.

(b) 60.6% of the fat gain values are accounted for by the least-squares line.

(c) 60.6% of the variation in fat gain is accounted for by the least-squares line.

(d) 77.8% of the variation in fat gain is accounted for by the least-squares line.

2. A recent study discovered that the correlation between the age at which an infant first speaks and the child's score on an IQ test given upon entering elementary school is −0.68. A scatterplot of the data shows a linear form. Which of the following statements about this finding is correct?

(a) Infants who speak at very early ages will have higher IQ scores by the beginning of elementary school than those who begin to speak later.

(b) 68% of the variation in IQ test scores is explained by the least-squares regression of age at first spoken word and IQ score.

(c) Encouraging infants to speak before they are ready can have a detrimental effect later in life, as evidenced by their lower IQ scores.

(d) There is a moderately strong, negative linear relationship between age at first spoken word and later IQ test score for the individuals in this study.

Interpreting Computer Regression Output

Figure 3.17 (on the next page) displays the basic regression output for the NEA data from two statistical software packages: Minitab and JMP. Other software produces very similar output. Each output records the slope and y intercept of the least-squares line. The software also provides information that we don't yet need (or understand!), although we will use much of it later. Be sure that you can locate the slope, the y intercept, and the values of s and r^2 on both computer outputs. *Once you understand the statistical ideas, you can read and work with almost any software output.*

AP EXAM TIP Students often have a hard time interpreting the value of r^2 on AP exam questions. They frequently leave out key words in the definition. Our advice: Treat this as a fill-in-the-blank exercise. Write "____% of the variation in [response variable name] is accounted for by the regression line."

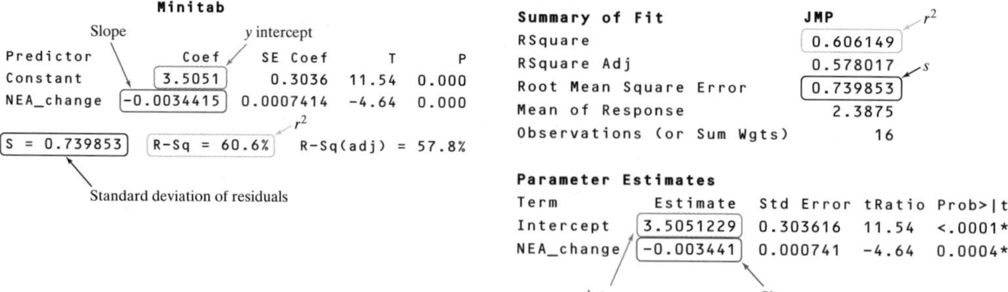

Minitab

Predictor	Coef	SE Coef	T	P
Constant	3.5051	0.3036	11.54	0.000
NEA_change	-0.0034415	0.0007414	-4.64	0.000

S = 0.739853 R-Sq = 60.6% R-Sq(adj) = 57.8%

JMP

Summary of Fit

RSquare	0.606149
RSquare Adj	0.578017
Root Mean Square Error	0.739853
Mean of Response	2.3875
Observations (or Sum Wgts)	16

Parameter Estimates

Term	Estimate	Std Error	tRatio	Prob>|t|
Intercept	3.5051229	0.303616	11.54	<.0001*
NEA_change	-0.003441	0.000741	-4.64	0.0004*

FIGURE 3.17 Least-squares regression results for the nonexercise activity data from two statistical software packages. Other software produces similar output.

Beer and Blood Alcohol

Interpreting regression output

How well does the number of beers a person drinks predict his or her blood alcohol content (BAC)? Sixteen volunteers with an initial BAC of 0 drank a randomly assigned number of cans of beer. Thirty minutes later, a police officer measured their BAC. Least-squares regression was performed on the data. A scatterplot with the regression line added, a residual plot, and some computer output from the regression are shown below.

```
Dependent variable is:  BAC
No Selector
R squared = 80%
R squared (adjusted) = 78.6%
s = 0.0204 with
16 - 2 = 14 degrees of freedom
```

Variable	Coefficient	s.e. of Coeff	t-ratio	prob
Constant	-0.012701	0.0126	-1.00	0.3320
Beers	0.017964	0.0024	7.48	≤0.0001

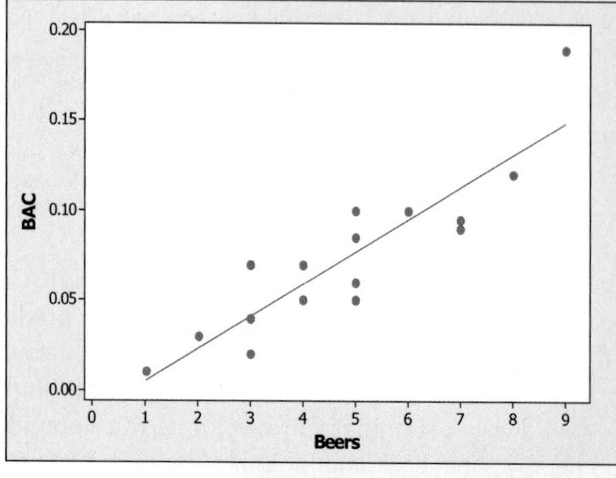

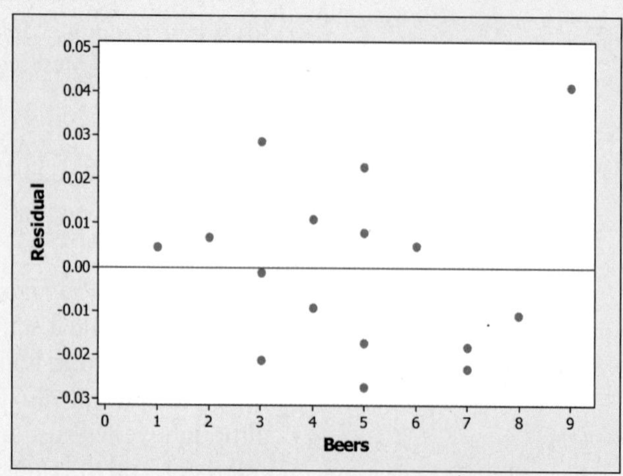

PROBLEM: (a) What is the equation of the least-squares regression line that describes the relationship between beers consumed and blood alcohol content? Define any variables you use.

(b) Interpret the slope of the regression line in context.

(c) Find the correlation.

(d) Is a line an appropriate model to use for these data? What information tells you this?

(e) What was the BAC reading for the person who consumed 9 beers? Show your work.

SOLUTION: (a) The equation is $\hat{y} = -0.0127 + 0.017964x$, where x is the number of beers consumed and y is the blood alcohol content (BAC). We could also write

$$\text{predicted BAC} = -0.0127 + 0.017964(\text{\# of beers}).$$

(b) The slope, 0.017964, says that for every additional beer consumed, the least-squares line predicts an increase of about 0.018 in BAC.

(c) To find the correlation, we take the square root of r^2: $r = \pm\sqrt{0.80} = \pm 0.894$. Since the scatterplot shows a positive direction, $r = 0.894$.

(d) A line does seem to be an appropriate model for these data. The residual plot shows a fairly random scatter of points about the horizontal line representing prediction error = 0. The residuals are between −0.03 and 0.03 except for the subject who drank 9 beers. On average, predictions of BAC using the regression line would be off by about $s = 0.02$ for the 16 people in the study. That may be too much prediction error, because the legal limit for BAC is 0.08. Based on the value of r^2, 80% of the variation in subjects' BAC readings is accounted for by the least-squares regression line of beers consumed on BAC.

(e) For the person who drank 9 beers, the least-squares line predicts a BAC of

$$\widehat{\text{BAC}} = -0.0127 + 0.017964(9) = 0.149$$

From the residual plot, this subject's residual is about 0.045. Since

$$\text{residual} = \text{observed } y - \text{predicted } y$$
$$0.045 = y - 0.149 \Rightarrow y = 0.194$$

This person's BAC reading was about 0.194.

For Practice *Try Exercise 63*

Correlation and Regression Wisdom

Correlation and regression are powerful tools for describing the relationship between two variables. When you use these tools, you should be aware of their limitations.

CAUTION

1. *The distinction between explanatory and response variables is important in regression.* This isn't true for correlation: switching x and y doesn't affect the value of r. Least-squares regression makes the distances of the data points from the line small only in the y direction. If we reverse the roles of the two variables, we get a different least-squares regression line.

Predicting Fat Gain, Predicting NEA

Two different regression lines

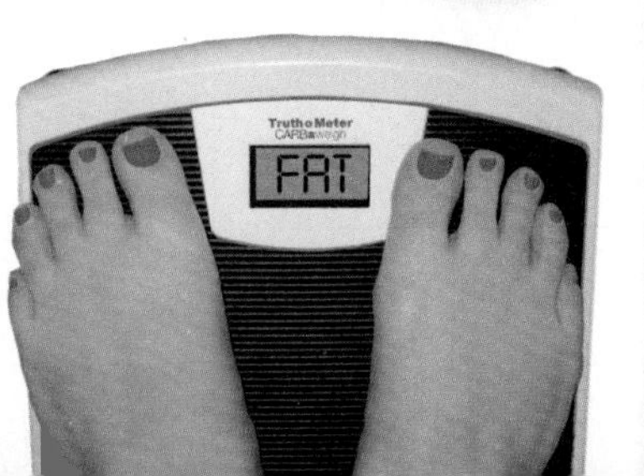

Figure 3.18(a) on the next page repeats the scatterplot of the NEA data with the least-squares regression line for predicting fat gain from change in NEA added. We might also use the data on these 16 subjects to predict the NEA change for another subject from that subject's fat gain when overfed for 8 weeks. Now the roles of the variables are reversed: fat gain is the explanatory variable and change in

NEA is the response variable. Figure 3.18(b) shows a scatterplot of these data with the least-squares line for predicting NEA change from fat gain. The two regression lines are very different. However, no matter which variable we put on the x axis, $r^2 = 0.606$ and the correlation is $r = -0.778$.

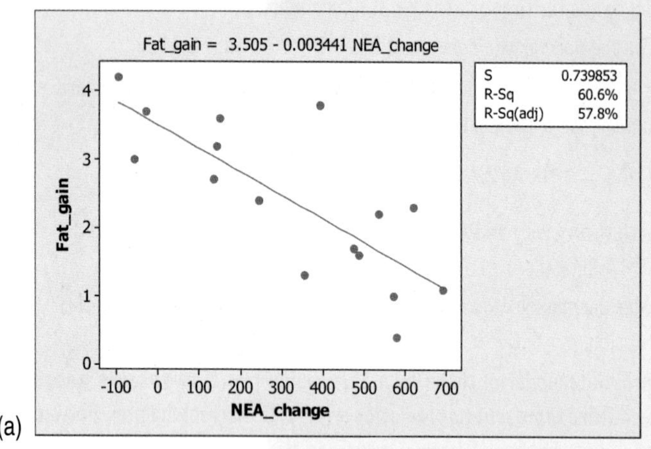

(a)

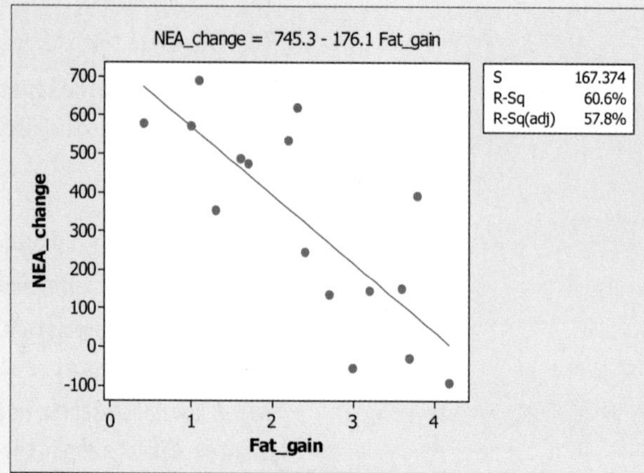

(b)

FIGURE 3.18 (a) Scatterplot with least-squares line for predicting fat gain from change in NEA activity. (b) Scatterplot with least-squares line for predicting NEA change from fat gain.

2. *Correlation and regression lines describe only linear relationships.* You can calculate the correlation and the least-squares line for any relationship between two quantitative variables, but the results are useful only if the scatterplot shows a linear pattern. *Always plot your data!* The following Data Exploration makes this point clear.

DATA EXPLORATION *Anscombe's data*

The table below presents four sets of data prepared by the statistician Frank Anscombe to illustrate the dangers of calculating without first plotting the data.

Four data sets for exploring correlation and regression

Data Set A											
x:	10	8	13	9	11	14	6	4	12	7	5
y:	8.04	6.95	7.58	8.81	8.33	9.96	7.24	4.26	10.84	4.82	5.68
Data Set B											
x:	10	8	13	9	11	14	6	4	12	7	5
y:	9.14	8.14	8.74	8.77	9.26	8.10	6.13	3.10	9.13	7.26	4.74
Data Set C											
x:	10	8	13	9	11	14	6	4	12	7	5
y:	7.46	6.77	12.74	7.11	7.81	8.84	6.08	5.39	8.15	6.42	5.73
Data Set D											
x:	8	8	8	8	8	8	8	8	8	8	19
y:	6.58	5.76	7.71	8.84	8.47	7.04	5.25	5.56	7.91	6.89	12.50

Source: Frank J. Anscombe, "Graphs in statistical analysis," *American Statistician*, 27 (1973), pp. 17–21.

1. Without making scatterplots, find the correlation and the least-squares regression line for all four data sets. What do you notice? Use the regression line to predict y for $x = 10$ for all four data sets.
2. Make a scatterplot for each of the data sets and add the regression line to each plot.
3. Now make a sketch of the residual plot for each of the four data sets.
4. For each of the four cases, discuss whether you would be willing to use the regression line to describe the relationship between x and y.

CAUTION

3. *Correlation and least-squares regression lines are not resistant.* You already know that the correlation r is not resistant. One unusual point in a scatterplot can greatly change the value of r. Is the least-squares line resistant? Not surprisingly, the answer is no. The following example sheds some light on this issue.

STEP 4

Gesell Scores

Dealing with unusual points in regression

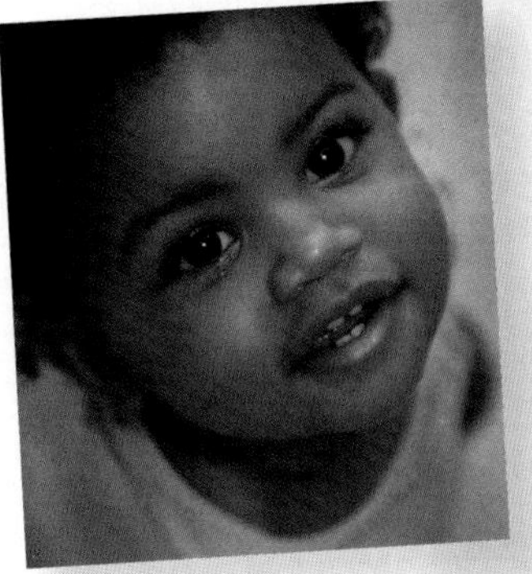

Does the age at which a child begins to talk predict a later score on a test of mental ability? A study of the development of young children recorded the age in months at which each of 21 children spoke their first word and their Gesell Adaptive Score, the result of an aptitude test taken much later. The data appear in the table below.

Age (months) at first word and Gesell score								
Child	Age	Score	Child	Age	Score	Child	Age	Score
1	15	95	8	11	100	15	11	102
2	26	71	9	8	104	16	10	100
3	10	83	10	20	94	17	12	105
4	9	91	11	7	113	18	42	57
5	15	102	12	9	96	19	17	121
6	20	87	13	10	83	20	11	86
7	18	93	14	11	84	21	10	100

Source: N. R. Draper and J. A. John, "Influential observations and outliers in regression," *Technometrics*, 23 (1981), pp. 21–26.

STATE: Can we use a child's age at first word to predict his or her Gesell score? How accurate will our predictions be?

PLAN: Let's start by making a scatterplot with age at first word as the explanatory variable and Gesell score as the response variable. If the graph shows a linear form, we'll fit a least-squares line to the data. Then we should make a residual plot. The residuals, r^2, and s will tell us how well the line fits the data and how large our prediction errors will be.

DO: Figure 3.19(a) on the next page shows a scatterplot of the data. Children 3 and 13, and also Children 16 and 21, have identical values of both variables. We used a different plotting symbol to show that one point stands for two individuals. The scatterplot shows a negative association. That is, children who begin to speak later tend to have lower test scores than early talkers. The overall pattern is moderately linear (a calculator gives $r = -0.640$). There are two outliers on the scatterplot: Child 18 and Child 19. These two children are unusual in different ways. Child 19 is an *outlier in the y direction*, with a Gesell score so high that we should check for a mistake in recording it. (In fact, the score is

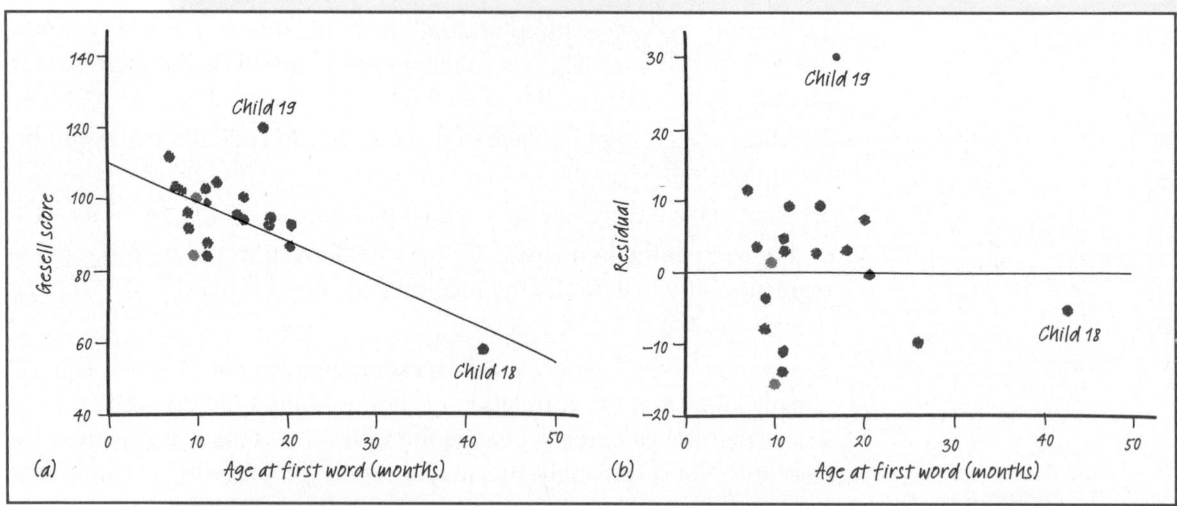

FIGURE 3.19 (a) Scatterplot of Gesell Adaptive Scores versus the age at first word for 21 children. The line is the least-squares regression line for predicting Gesell score from age at first word. (b) Residual plot for the regression. Child 18 and 19 are outliers.

correct.) Child 18 is an *outlier in the x direction.* This child began to speak much later than any of the other children.

We used a calculator to perform least-squares regression. The equation of the least-squares line is $\hat{y} = 109.8738 - 1.1270x$. We added this line to the scatterplot in Figure 3.19(a). The slope suggests that for every month older a child is when she first speaks, her Gesell score is predicted to decrease by 1.127 points. Since a child isn't going to speak her first word at age 0 months, the y intercept of this line has no statistical meaning.

How well does the least-squares line fit the data? Figure 3.19(b) shows a residual plot. The graph shows a fairly "random" scatter of points around the "residual = 0" line with one very large positive residual (Child 19). Most of the prediction errors (residuals) are 10 points or fewer on the Gesell score. We calculated the standard error of the residuals to be $s = 11.023$. This is roughly the size of an average prediction error using the regression line. Since $r^2 = 0.41$, 41% of the variation in Gesell scores is accounted for by the least-squares regression of Gesell score on age at first spoken word. That leaves 59% of the variation in Gesell scores unaccounted for by the linear relationship for these data.

CONCLUDE: We can use the equation $\widehat{\text{score}} = 109.8738 - 1.1270\,(\text{age})$ to predict a child's score on the Gesell test from the age at which the child first speaks. Our predictions may not be very accurate, though. On average, we'll be off by about 11 points on the Gesell score. Also, most of the variation in Gesell score from child to child is not accounted for by this linear model. We should hesitate to use this model to make predictions, especially until we better understand the effect of the two outliers on the regression results.

For Practice *Try Exercise 69*

In the previous example, Child 18 and Child 19 were identified as outliers in the scatterplot of Figure 3.19(a). These points are also marked in the residual plot of Figure 3.19(b). Child 19 has a very large residual because this point lies far from the regression line. However, Child 18 has a pretty small residual. That's because Child 18's point is close to the line. How do these two outliers affect the regression?

Figure 3.20 shows the results of removing each of these points on the correlation and the regression line. The graph adds two more regression lines, one calculated after leaving out Child 18 and the other after leaving out Child 19.

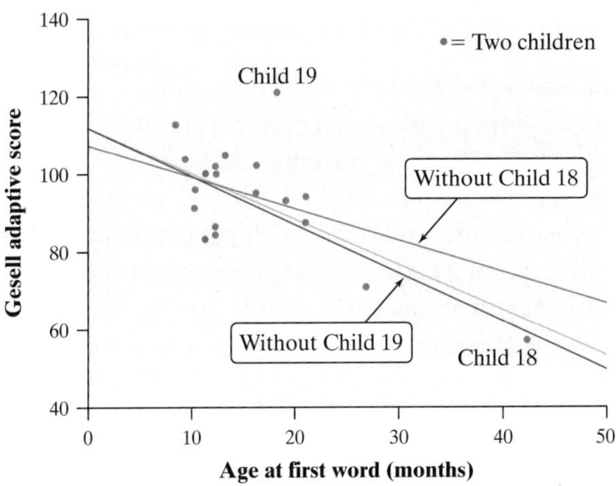

FIGURE 3.20 Three least-squares regression lines of Gesell score on age at first word. The green line is calculated from all the data. The dark blue line is calculated leaving out Child 18. Child 18 is an influential observation because leaving out this point moves the regression line quite a bit. The red line is calculated leaving out Child 19.

With all 19 children:
$r = -0.64$
$\hat{y} = 109.874 - 1.127x$
Without Child 19:
$r = -0.76$
$\hat{y} = 109.305 - 1.193x$
Without Child 18:
$r = -0.33$
$\hat{y} = 105.630 - 0.779x$

You can see that removing the point for Child 18 moves the line quite a bit. (In fact, the equation of the new least-squares line is $\hat{y} = 105.630 - 0.779x$). Because of Child 18's extreme position on the age scale, this point has a strong *influence* on the position of the regression line. However, removing Child 19 has little effect on the regression line.

Least-squares lines make the sum of the squares of the vertical distances to the points as small as possible. A point that is extreme in the x direction with no other points near it pulls the line toward itself. We call such points **influential.**

DEFINITION: Outliers and influential observations in regression

An **outlier** is an observation that lies outside the overall pattern of the other observations. Points that are outliers in the y direction but not the x direction of a scatterplot have large residuals. Other outliers may not have large residuals.

An observation is **influential** for a statistical calculation if removing it would markedly change the result of the calculation. Points that are outliers in the x direction of a scatterplot are often influential for the least-squares regression line.

We did not need the distinction between outliers and influential observations in Chapter 1. A single large salary that pulls up the mean salary $\bar{x}$ for a group of workers is an outlier because it lies far above the other salaries. It is also influential, because the mean changes when it is removed. In the regression setting, however, not all outliers are influential. The least-squares line is most likely to be heavily influenced by observations that are outliers in the x direction. The scatterplot will alert you to such observations. Influential points often have small residuals, because they pull the regression line toward themselves. If you look at just a residual plot, you may miss influential points.

The best way to verify that a point is influential is to find the regression line both with and without the unusual point, as in Figure 3.20. If the line moves more than a small amount when the point is deleted, the point is influential.

THINK ABOUT IT

How much difference can one point make? The strong influence of Child 18 makes the original regression of Gesell score on age at first word misleading. The original data have $r^2 = 0.41$. That is, the least-squares line relating age at which a child begins to talk with Gesell score explains 41% of the variation on this later test of mental ability. This relationship is strong enough to be interesting to parents. If we leave out Child 18, r^2 drops to only 11%. The apparent strength of the association was largely due to a single influential observation.

What should the child development researcher do? She must decide whether Child 18 is so slow to speak that this individual should not be allowed to influence the analysis. If she excludes Child 18, much of the evidence for a connection between the age at which a child begins to talk and later ability score vanishes. If she keeps Child 18, she needs data on other children who were also slow to begin talking, so that the analysis no longer depends so heavily on just one child.

We finish with our most important caution about correlation and regression.

4. *Association does not imply causation.* When we study the relationship between two variables, we often hope to show that changes in the explanatory variable *cause* changes in the response variable. A *strong association between two variables is not enough to draw conclusions about cause and effect.* Sometimes an observed association really does reflect cause and effect. A household that heats with natural gas uses more gas in colder months because cold weather requires burning more gas to stay warm. In other cases, an association is explained by lurking variables, and the conclusion that x causes y is not valid.

CAUTION

Does Having More Cars Make You Live Longer?

Association, not causation

A serious study once found that people with two cars live longer than people who own only one car.[14] Owning three cars is even better, and so on. There is a substantial positive correlation between number of cars x and length of life y.

The basic meaning of causation is that by changing x we can bring about a change in y. Could we lengthen our lives by buying more cars? No. The study used number of cars as a quick indicator of wealth. Well-off people tend to have more cars. They also tend to live longer, probably because they are better educated, take better care of themselves, and get better medical care. The cars have nothing to do with it. There is no cause-and-effect tie between number of cars and length of life.

Remember: It only makes sense to talk about the *correlation* between two *quantitative* variables. If one or both variables are categorical, you should refer to the *association* between the two variables. To be safe, you can use the more general term "association" when describing the relationship between any two variables.

Correlations such as those in the previous example are sometimes called "nonsense correlations." The correlation is real. What is nonsense is the conclusion that changing one of the variables causes changes in the other. A "lurking variable"—such as personal wealth in this example—that influences both x and y can create a high correlation even though there is no direct connection between x and y.

Association Does Not Imply Causation

An association between an explanatory variable x and a response variable y, even if it is very strong, is not by itself good evidence that changes in x actually cause changes in y.

The following Data Exploration gives you a chance to put what you have learned in this chapter into practice in a very relevant setting: scores on the SAT Writing test.

DATA EXPLORATION *The SAT essay: Is longer better?*

Following the debut of the new SAT Writing test in March 2005, Dr. Les Perelman from the Massachusetts Institute of Technology stirred controversy by reporting, "It appeared to me that regardless of what a student wrote, the longer the essay, the higher the score." He went on to say, "I have never found a quantifiable predictor in 25 years of grading that was anywhere as strong as this one. If you just graded them based on length without ever reading them, you'd be right over 90 percent of the time."[15] The table below shows the data that Dr. Perelman used to draw his conclusions.

Length of essay and score for a sample of SAT essays

Words:	460	422	402	365	357	278	236	201	168	156	133
Score:	6	6	5	5	6	5	4	4	4	3	2
Words:	114	108	100	403	401	388	320	258	236	189	128
Score:	2	1	1	5	6	6	5	4	4	3	2
Words:	67	697	387	355	337	325	272	150	135		
Score:	1	6	6	5	5	4	4	2	3		

Source: The College Board, www.collegeboard.com.

Does this mean that if students write a lot, they are guaranteed high scores? Carry out your own analysis of the data. How would you respond to each of Dr. Perelman's claims?

How Faithful Is Old Faithful?

In the chapter-opening Case Study (page 141), the Starnes family had just missed seeing Old Faithful erupt. They wondered how long it would be until the next eruption. The Fathom screen shot at right shows results from a least-squares regression on the Old Faithful eruption data from the previous month. Both variables were measured in minutes.

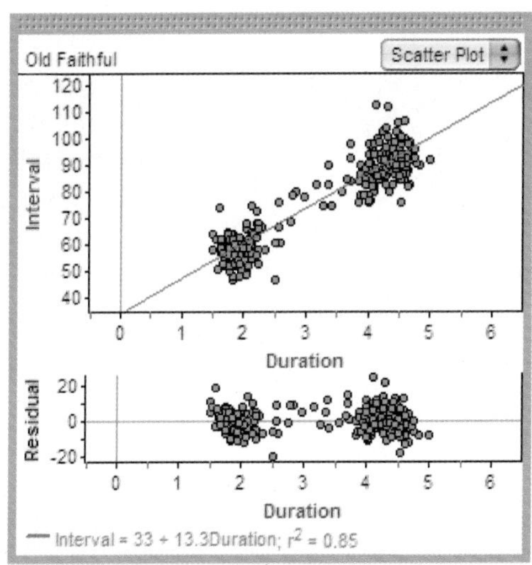

Even though there were two distinct clusters in the distribution of eruption intervals, the graph shows a strong, positive linear relationship between the duration of the previous eruption and the time until the next eruption. The least-squares regression line fits the data well, as evidenced by the random scatter of the residuals and the high r^2-value. If the Starnes family can find out how long the previous eruption lasted, they can use the regression equation to get a fairly accurate prediction of how long it will be before Old Faithful erupts again.

SECTION 3.2 Summary

- A **regression line** is a straight line that describes how a response variable y changes as an explanatory variable x changes. You can use a regression line to **predict** the value of y for any value of x by substituting this x into the equation of the line.
- The **slope** b of a regression line $\hat{y} = a + bx$ is the rate at which the predicted response $\hat{y}$ changes along the line as the explanatory variable x changes. Specifically, b is the *predicted* change in y when x increases by 1 unit.
- The **y intercept a** of a regression line $\hat{y} = a + bx$ is the predicted response $\hat{y}$ when the explanatory variable $x = 0$. This prediction is of no statistical use unless x can actually take values near 0.
- Avoid **extrapolation,** the use of a regression line for prediction using values of the explanatory variable outside the range of the data from which the line was calculated.
- The most common method of fitting a line to a scatterplot is least squares. The **least-squares regression line** is the straight line $\hat{y} = a + bx$ that minimizes the sum of the squares of the vertical distances of the observed points from the line.
- The least-squares regression line of y on x is the line with slope $b = r\,(s_y/s_x)$ and intercept $a = \bar{y} - b\bar{x}$. This line always passes through the point $(\bar{x}, \bar{y})$.
- You can examine the fit of a regression line by studying the **residuals,** which are the differences between the observed and predicted values of y. Be on the lookout for points with unusually large residuals and also for nonlinear patterns and uneven variation about the residual = 0 line in the **residual plot.**
- The **standard deviation of the residuals s** measures the average size of the prediction errors (residuals) when using the regression line.
- The **coefficient of determination r^2** is the fraction of the variation in one variable that is accounted for by least-squares regression on the other variable.
- Correlation and regression must be interpreted with caution. Plot the data to be sure that the relationship is roughly linear and to detect **outliers.** Also look for **influential observations,** individual points that substantially change the correlation or the regression line. Outliers in x are often influential for the regression line.
- Most of all, be careful not to conclude that there is a cause-and-effect relationship between two variables just because they are strongly associated.

3.2 TECHNOLOGY CORNERS

TI-Nspire instructions in Appendix B

SECTION 3.2 Exercises

35. **What's my line?** You use the same bar of soap to shower each morning. The bar weighs 80 grams when it is new. Its weight goes down by 6 grams per day on the average. What is the equation of the regression line for predicting weight from days of use?

36. **What's my line?** An eccentric professor believes that a child with IQ 100 should have a reading test score of 50, and that reading score should increase by 1 point for every additional point of IQ. What is the equation of the professor's regression line for predicting reading score from IQ?

37. **Gas mileage** We expect a car's highway gas mileage to be related to its city gas mileage. Data for all 1198 vehicles in the government's *2008 Fuel Economy Guide* give the regression line predicted highway mpg = 4.62 + 1.109 (city mpg).
(a) What's the slope of this line? Interpret this value in context.
(b) What's the intercept? Explain why the value of the intercept is not statistically meaningful.
(c) Find the predicted highway mileage for a car that gets 16 miles per gallon in the city. Do the same for a car with city mileage 28 mpg.

38. **IQ and reading scores** Data on the IQ test scores and reading test scores for a group of fifth-grade children give the following regression line: predicted reading score = −33.4 + 0.882(IQ score).
(a) What's the slope of this line? Interpret this value in context.
(b) What's the intercept? Explain why the value of the intercept is not statistically meaningful.
(c) Find the predicted reading scores for two children with IQ scores of 90 and 130, respectively.

39. **Acid rain** Researchers studying acid rain measured the acidity of precipitation in a Colorado wilderness area for 150 consecutive weeks. Acidity is measured by pH. Lower pH values show higher acidity. The researchers observed a linear pattern over time. They reported that the regression line $\widehat{pH} = 5.43 - 0.0053(\text{weeks})$ fit the data well.[16]
pg 166
(a) Identify the slope of the line and explain what it means in this setting.
(b) Identify the y intercept of the line and explain what it means in this setting.
(c) According to the regression line, what was the pH at the end of this study?

40. **How much gas?** In Exercise 4 (page 158), we examined the relationship between the average monthly temperature and the amount of natural gas consumed in Joan's midwestern home. The figure below shows the original scatterplot with the least-squares line added. The equation of the least-squares line is $\hat{y} = 1425 - 19.87x$.

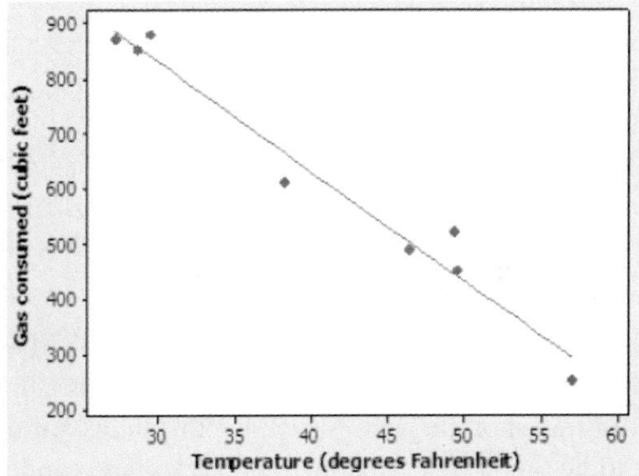

(a) Identify the slope of the line and explain what it means in this setting.
(b) Identify the y intercept of the line. Explain why it's risky to use this value as a prediction.
(c) Use the regression line to predict the amount of natural gas Joan will use in a month with an average temperature of 30°F.

41. **Acid rain** Refer to Exercise 39. Would it be appropriate to use the regression line to predict pH after 1000 months? Justify your answer.

42. **How much gas?** Refer to Exercise 40. Would it be appropriate to use the regression line to predict Joan's natural-gas consumption in a future month with an average temperature of 65°F? Justify your answer.

43. **Least-squares idea** The table below gives a small set of data. Which of the following two lines fits the data better: $\hat{y} = 1 - x$ or $\hat{y} = 3 - 2x$? Make a graph of the data and use it to help justify your answer. (*Note:* Neither of these two lines is the least-squares regression line for these data.)

x:	−1	1	1	3	5
y:	2	0	1	−1	−5

44. **Least-squares idea** Trace the graph from Exercise 40 on your paper. Show why the line drawn on the plot is called the least-squares line.

45. **Acid rain** In the acid rain study of Exercise 39, the actual pH measurement for Week 50 was 5.08. Find and interpret the residual for this week.

pg 168

46. **How much gas?** Refer to Exercise 40. During March, the average temperature was 46.4°F and Joan used 490 cubic feet of gas per day. Find and interpret the residual for this month.

47. **Husbands and wives** The mean height of American women in their early twenties is 64.5 inches and the standard deviation is 2.5 inches. The mean height of men the same age is 68.5 inches, with standard deviation 2.7 inches. The correlation between the heights of husbands and wives is about $r = 0.5$.

pg 173

(a) Find the equation of the least-squares regression line for predicting husband's height from wife's height. Show your work.

(b) Use your regression line to predict the height of the husband of a woman who is 67 inches tall. Explain why you could have given this result without doing the calculation.

48. **The stock market** Some people think that the behavior of the stock market in January predicts its behavior for the rest of the year. Take the explanatory variable x to be the percent change in a stock market index in January and the response variable y to be the change in the index for the entire year. We expect a positive correlation between x and y because the change during January contributes to the full year's change. Calculation from data for an 18-year period gives

$$\bar{x} = 1.75\% \quad s_x = 5.36\% \quad \bar{y} = 9.07\%$$
$$s_y = 15.35\% \quad r = 0.596$$

(a) Find the equation of the least-squares line for predicting full-year change from January change. Show your work.

(b) The mean change in January is $\bar{x} = 1.75\%$. Use your regression line to predict the change in the index in a year in which the index rises 1.75% in January. Why could you have given this result (up to roundoff error) without doing the calculation?

49. **Husbands and wives** Refer to Exercise 47.

(a) Find r^2 and interpret this value in context.

(b) For these data, $s = 1.2$. Explain what this value means.

50. **The stock market** Refer to Exercise 48.

(a) What percent of the observed variation in yearly changes in the index is explained by a straight-line relationship with the change during January?

(b) For these data, $s = 8.3$. Explain what this value means.

51. **IQ and grades** Exercise 3 (page 158) included the plot shown below of school grade point average (GPA) against IQ test score for 78 seventh-grade students. (GPA was recorded on a 12-point scale with A+ = 12, A = 11, A− = 10, B+ = 9, . . ., D− = 1, and F = 0.) Calculation shows that the mean and standard deviation of the IQ scores are $\bar{x} = 108.9$ and $s_x = 13.17$. For the GPAs, these values are $\bar{y} = 7.447$ and $s_y = 2.10$. The correlation between IQ and GPA is $r = 0.6337$.

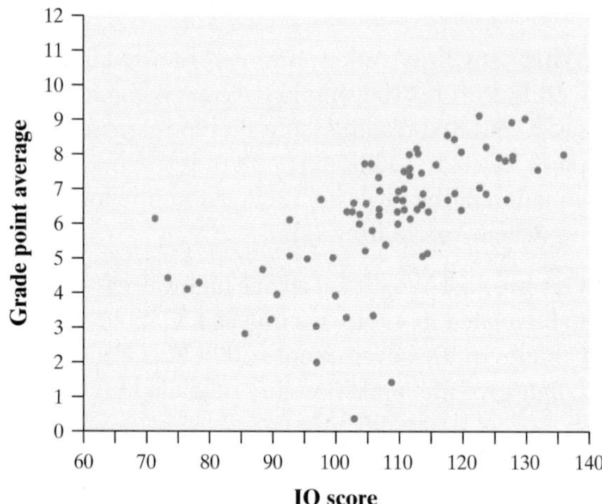

(a) Find the equation of the least-squares line for predicting GPA from IQ. Show your work.

(b) What percent of the observed variation in these students' GPAs can be explained by the linear relationship between GPA and IQ?

(c) One student has an IQ of 103 but a very low GPA of 0.53. Find and interpret the residual for this student.

52. **Will I bomb the final?** We expect that students who do well on the midterm exam in a course will usually also do well on the final exam. Gary Smith of Pomona College looked at the exam scores of all 346 students who took his statistics class over a 10-year period.[17] The least-squares line for predicting final-exam score from midterm-exam score was $\hat{y} = 46.6 + 0.41x$. Octavio scores 10 points above the class mean on the midterm. How many points above the class mean do you predict that he will score on the final? (This is an example of the phenomenon that gave "regression" its name: students who do well on the midterm will on the average do less well, but still above average, on the final.)

53. **Bird colonies** Exercise 6 (page 159) examined the relationship between the number of new birds y and percent of returning birds x for 13 sparrowhawk colonies. Here are the data once again.

Percent return:	74	66	81	52	73	62	52	45	62	46	60	46	38
New adults:	5	6	8	11	12	15	16	17	18	18	19	20	20

(a) Enter the data into your calculator and make a scatterplot.

(b) Use your calculator's regression function to find the equation of the least-squares regression line. Add this line to your scatterplot from (a).

(c) Explain in words what the slope and y intercept of the regression line tell us.

(d) An ecologist uses the line to predict how many birds will join another colony of sparrowhawks, to which 60% of the adults from the previous year return. What's the prediction?

54. **Do heavier people burn more energy?** Exercise 10 (page 159) presented data on the lean body mass and resting metabolic rate for 12 women who were subjects in a study of dieting. Lean body mass, given in kilograms, is a person's weight leaving out all fat. Metabolic rate, in calories burned per 24 hours, is the rate at which the body consumes energy. Here are the data again.

Mass:	36.1	54.6	48.5	42.0	50.6	42.0	40.3	33.1	42.4	34.5	51.1	41.2
Rate:	995	1425	1396	1418	1502	1256	1189	913	1124	1052	1347	1204

(a) Enter the data into your calculator and make a scatterplot.

(b) Use your calculator's regression function to find the equation of the least-squares regression line. Add this line to your scatterplot from (a).

(c) Explain in words what the slope of the regression line tells us.

(d) Another woman has a lean body mass of 45 kilograms. What is her predicted metabolic rate?

55. **Bird colonies** Refer to Exercise 53.

(a) Use your calculator to make a residual plot. Describe what this graph tells you about how well the line fits the data.

(b) Which point has the largest size residual? Explain what this residual means in context.

56. **Do heavier people burn more energy?** Refer to Exercise 54.

(a) Use your calculator to make a residual plot. Describe what this graph tells you about how well the line fits the data.

(b) Which point has the largest residual? Explain what the value of that residual means in context.

57. **Bird colonies** Refer to Exercises 53 and 55. For the regression you performed earlier, $r^2 = 0.56$ and $s = 3.67$. Explain what each of these values means in this setting.

58. **Do heavier people burn more energy?** Refer to Exercises 54 and 56. For the regression you performed earlier, $r^2 = 0.768$ and $s = 95.08$. Explain what each of these values means in this setting.

59. **Oil and residuals** The Trans-Alaska Oil Pipeline is a tube that is formed from 1/2-inch-thick steel and that carries oil across 800 miles of sensitive arctic and subarctic terrain. The pipe segments and the welds that join them were carefully examined before installation. How accurate are field measurements of the depth of small defects? The figure below compares the results of measurements on 100 defects made in the field with measurements of the same defects made in the laboratory.[18] The line $y = x$ is drawn on the scatterplot.

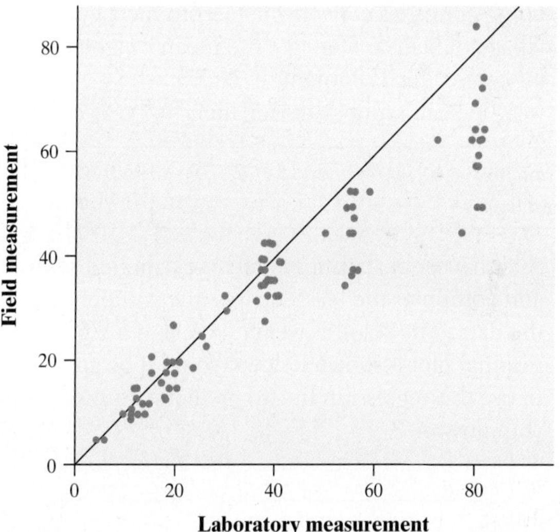

(a) Describe the overall pattern you see in the scatterplot, as well as any deviations from that pattern.

(b) If field and laboratory measurements all agree, then the points should fall on the $y = x$ line drawn on the plot, except for small variations in the measurements. Is this the case? Explain.

(c) The line drawn on the scatterplot ($y = x$) is *not* the least-squares regression line. How would the slope and y intercept of the least-squares line compare? Justify your answer.

60. **Oil and residuals** Refer to Exercise 59. The following figure shows a residual plot for the least-squares regression line. Discuss what the residual plot tells

you about how well the least-squares regression line fits the data.

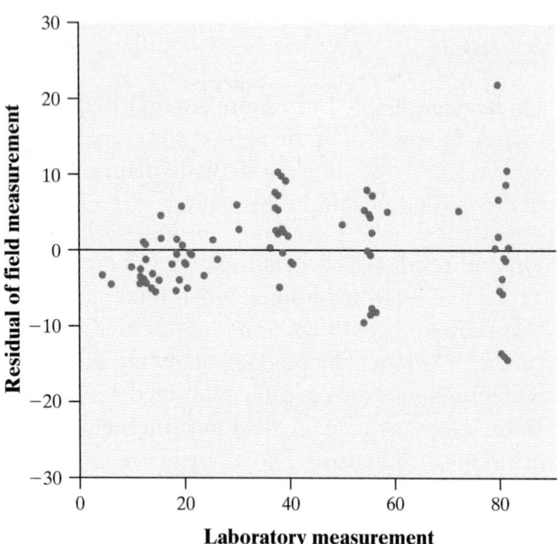

61. **Nahya infant weights** A study of nutrition in developing countries collected data from the Egyptian village of Nahya. Here are the mean weights (in kilograms) for 170 infants in Nahya who were weighed each month during their first year of life:

Age (months):	1	2	3	4	5	6	7	8	9	10	11	12
Weight (kg):	4.3	5.1	5.7	6.3	6.8	7.1	7.2	7.2	7.2	7.2	7.5	7.8

A hasty user of statistics enters the data into software and computes the least-squares line without plotting the data. The result is $\widehat{\text{weight}} = 4.88 + 0.267\,(\text{age})$. A residual plot is shown below. Would it be appropriate to use this regression line to predict y from x? Justify your answer.

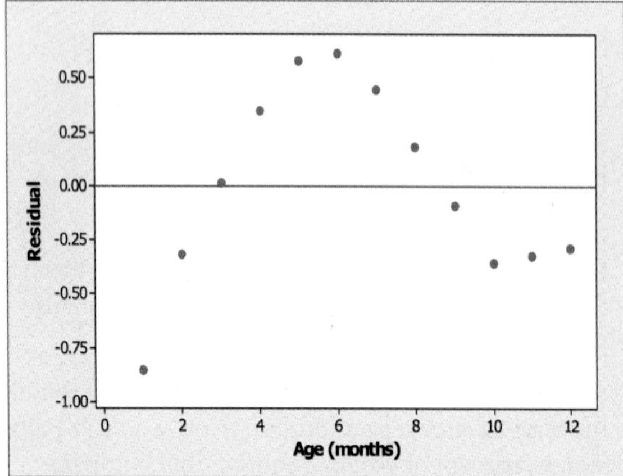

62. **Driving speed and fuel consumption** Exercise 9 (page 159) gives data on the fuel consumption y of a car at various speeds x. Fuel consumption is measured in liters of gasoline per 100 kilometers driven and speed is measured in kilometers per hour. A statistical software package gives the least-squares regression line and the residual plot shown below. The regression line is $\hat{y} = 11.058 - 0.01466x$. Would it be appropriate to use the regression line to predict y from x? Justify your answer.

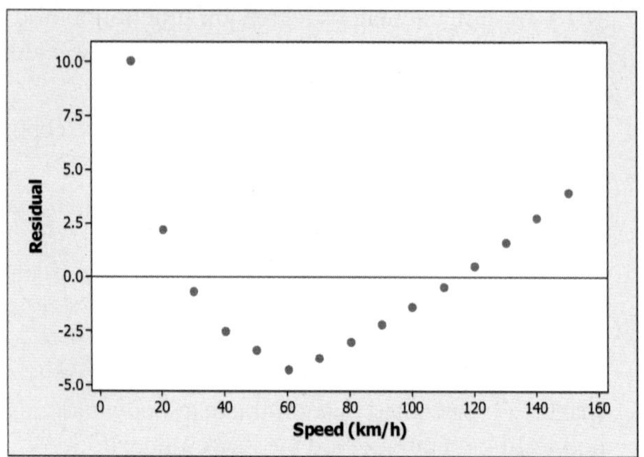

63. **Merlins breeding** Exercise 13 (page 160) gives data on the number of breeding pairs of merlins in an isolated area in each of nine years and the percent of males who returned the next year. The data show that the percent returning is lower after successful breeding seasons and that the relationship is roughly linear. The figure below shows Minitab regression output for these data.

pg 182

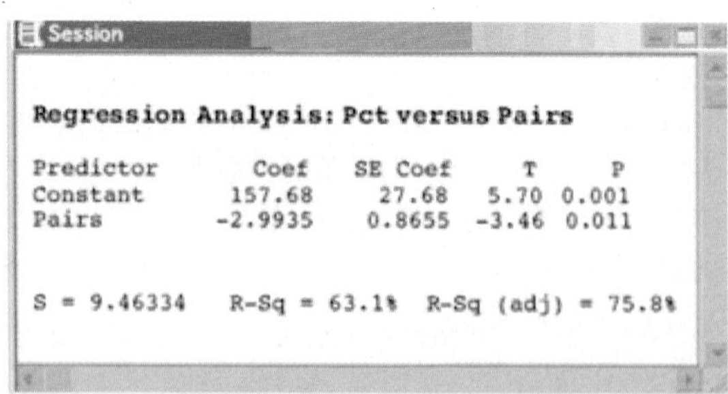
Session

Regression Analysis: Pct versus Pairs

Predictor	Coef	SE Coef	T	P
Constant	157.68	27.68	5.70	0.001
Pairs	-2.9935	0.8655	-3.46	0.011

S = 9.46334 R-Sq = 63.1% R-Sq (adj) = 75.8%

(a) What is the equation of the least-squares regression line for predicting the percent of males that return from the number of breeding pairs? Use the equation to predict the percent of returning males after a season with 30 breeding pairs.

(b) What percent of the year-to-year variation in percent of returning males is explained by the straight-line relationship with number of breeding pairs the previous year?

(c) Use the information in the figure to find the correlation r between percent of males that return and number of breeding pairs. How do you know whether the sign of r is + or −?

(d) Interpret the value of s in this setting.

64. **Does social rejection hurt?** Exercise 14 (page 160) gives data from a study that shows that social exclusion causes "real pain." That is, activity in an area of the brain that responds to physical pain goes up as distress from social exclusion goes up. A scatterplot shows a moderately strong, linear relationship. The figure below shows Minitab regression output for these data.

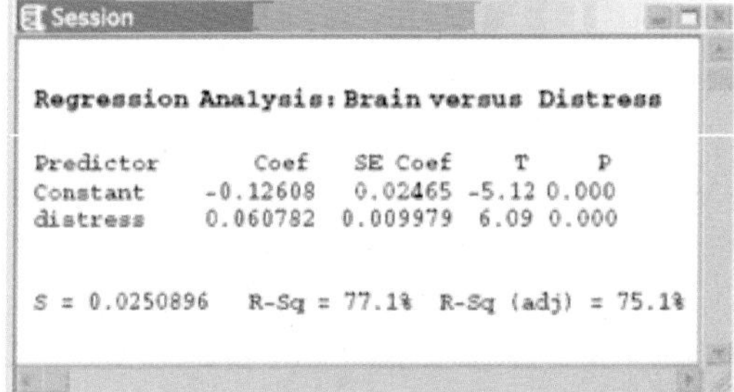

```
Session

Regression Analysis: Brain versus Distress

Predictor      Coef    SE Coef      T      P
Constant   -0.12608    0.02465  -5.12  0.000
distress   0.060782   0.009979   6.09  0.000

S = 0.0250896   R-Sq = 77.1%   R-Sq (adj) = 75.1%
```

(a) What is the equation of the least-squares regression line for predicting brain activity from social distress score? Use the equation to predict brain activity for social distress score 2.0.

(b) What percent of the variation in brain activity among these subjects is explained by the straight-line relationship with social distress score?

(c) Use the information in the figure to find the correlation r between social distress score and brain activity. How do you know whether the sign of r is $+$ or $-$?

(d) Interpret the value of s in this setting.

65. **Outsourcing by airlines** Exercise 5 (page 158) gives data for 14 airlines on the percent of major maintenance outsourced and the percent of flight delays blamed on the airline.

Airline	Outsource percent	Delay percent
AirTran	66	14
Alaska	92	42
American	46	26
America West	76	39
ATA	18	19
Continental	69	20
Delta	48	26
Frontier	65	31
Hawaiian	80	70
JetBlue	68	18
Northwest	76	43
Southwest	68	20
United	63	27
US Airways	77	24

(a) Make a scatterplot with outsourcing percent as x and delay percent as y. Hawaiian Airlines is a high outlier in the y direction. Because several other airlines have similar values of x, the influence of this outlier is unclear without actual calculation.

(b) Find the correlation r with and without Hawaiian Airlines. How influential is the outlier for correlation?

(c) Find the least-squares line for predicting y from x with and without Hawaiian Airlines. Draw both lines on your scatterplot. Use both lines to predict the percent of delays blamed on an airline that has outsourced 76% of its major maintenance. How influential is the outlier for the least-squares line?

66. **Managing diabetes** People with diabetes measure their fasting plasma glucose (FPG; measured in units of milligrams per milliliter) after fasting for at least 8 hours. Another measurement, made at regular medical checkups, is called HbA. This is roughly the percent of red blood cells that have a glucose molecule attached. It measures average exposure to glucose over a period of several months. The table below gives data on both HbA and FPG for 18 diabetics five months after they had completed a diabetes education class.[19]

Subject	HbA (%)	FPG (mg/mL)	Subject	HbA (%)	FPG (mg/mL)
1	6.1	141	10	8.7	172
2	6.3	158	11	9.4	200
3	6.4	112	12	10.4	271
4	6.8	153	13	10.6	103
5	7.0	134	14	10.7	172
6	7.1	95	15	10.7	359
7	7.5	96	16	11.2	145
8	7.7	78	17	13.7	147
9	7.9	148	18	19.3	255

(a) Make a scatterplot with HbA as the explanatory variable. There is a positive linear relationship, but it is surprisingly weak.

(b) Subject 15 is an outlier in the y direction. Subject 18 is an outlier in the x direction. Find the correlation for all 18 subjects, for all except Subject 15, and for all except Subject 18. Are either or both of these subjects influential for the correlation? Explain in simple language why r changes in opposite directions when we remove each of these points.

(c) Add three regression lines for predicting FPG from HbA to your scatterplot: for all 18 subjects, for all except Subject 15, and for all except Subject 18. Is either Subject 15 or Subject 18 strongly influential for the least-squares line? Explain in simple language what features of the scatterplot explain the degree of influence.

67. **Bird colonies** Return to the data of Exercise 53 on sparrowhawk colonies. We'll use these data to illustrate influence.

(a) Make a scatterplot of the data suitable for predicting new adults from percent of returning adults. Then add two new points. Point A: 10% return, 15 new adults. Point B: 60% return, 28 new adults. In which direction is each new point an outlier?

(b) Add three least-squares regression lines to your plot: for the original 13 colonies, for the original colonies plus Point A, and for the original colonies plus Point B. Which new point is more influential for the regression line? Explain in simple language why each new point moves the line in the way your graph shows.

68. **Beer and blood alcohol** The example on page 182 describes a study in which adults drank different amounts of beer. The response variable was their blood alcohol content (BAC). BAC for the same amount of beer might depend on other facts about the subjects. Name two other variables that could account for the fact that $r^2 = 0.80$.

pg 185

69. **Predicting tropical storms** William Gray heads the Tropical Meteorology Project at Colorado State University. His forecasts before each year's hurricane season attract lots of attention. Here are data on the number of named Atlantic tropical storms predicted by Dr. Gray and the actual number of storms for the years 1984 to 2008:[20]

Year	Forecast	Actual	Year	Forecast	Actual
1984	10	12	1997	11	7
1985	11	11	1998	10	14
1986	8	6	1999	14	12
1987	8	7	2000	12	14
1988	11	12	2001	12	15
1989	7	11	2002	11	12
1990	11	14	2003	14	16
1991	8	8	2004	14	14
1992	8	6	2005	15	27
1993	11	8	2006	17	9
1994	9	7	2007	17	14
1995	12	19	2008	15	16
1996	10	13			

Analyze these data. How accurate are Dr. Gray's forecasts? How many tropical storms would you expect in a year when his preseason forecast calls for 16 storms? What is the effect of the disastrous 2005 season on your answers? Follow the four-step process.

70. **Beavers and beetles** Do beavers benefit beetles? Researchers laid out 23 circular plots, each 4 meters in diameter, in an area where beavers were cutting down cottonwood trees. In each plot, they counted the number of stumps from trees cut by beavers and the number of clusters of beetle larvae. Ecologists think that the new sprouts from stumps are more tender than other cottonwood growth, so that beetles prefer them. If so, more stumps should produce more beetle larvae. Here are the data:[21]

Stumps:	2	2	1	3	3	4	3	1	2	5	1	3
Beetle larvae:	10	30	12	24	36	40	43	11	27	56	18	40
Stumps:	2	1	2	2	1	1	4	1	2	1	4	
Beetle larvae:	25	8	21	14	16	6	54	9	13	14	50	

Analyze these data to see if they support the "beavers benefit beetles" idea. Follow the four-step process.

Multiple choice: Select the best answer for Exercises 71 to 78.

71. The figure below is a scatterplot of reading test scores against IQ test scores for 14 fifth-grade children. The line is the least-squares regression line for predicting reading score from IQ score. If another child in this class has IQ score 110, you predict the reading score to be close to

(a) 50. (b) 60. (c) 70. (d) 80. (e) 90.

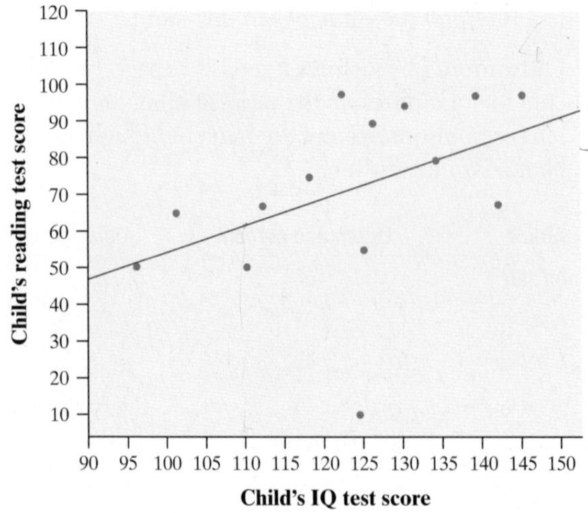

72. The slope of the line in the figure above is closest to

(a) −1. (b) 0. (c) 1. (d) 2. (e) 46.

73. Smokers don't live as long (on average) as nonsmokers, and heavy smokers don't live as long as light smokers. You perform least-squares regression on the age at death y of a large group of male smokers and the number of packs per day x they smoked. The slope of your regression line

(a) will be greater than 0.

(b) will be less than 0.

(c) will be equal to 0.

(d) You can't perform regression on these data.

(e) You can't tell without seeing the data.

Exercises 74 to 78 refer to the following setting. Measurements on young children in Mumbai, India, found this least-squares line for predicting height y from arm span x:[22]

$$\hat{y} = 6.4 + 0.93x$$

Measurements are in centimeters (cm).

74. How much does height increase on average for each additional centimeter of arm span?
    (a) 0.93 cm (b) 1.08 cm (c) 5.81 cm (d) 6.4 cm (e) 7.33 cm

75. According to the regression line, the predicted height of a child with an arm span of 100 cm is about
    (a) 106.4 cm. (b) 99.4 cm. (c) 93 cm. (d) 15.7 cm. (e) 7.33 cm.

76. By looking at the equation of the least-squares regression line, you can see that the correlation between height and arm span is
    (a) greater than zero.
    (b) less than zero.
    (c) 0.93.
    (d) 6.4.
    (e) Can't tell without seeing the data.

77. In addition to the regression line, the report on the Mumbai measurements says that $r^2 = 0.95$. This suggests that
    (a) although arm span and height are correlated, arm span does not predict height very accurately.
    (b) height increases by $\sqrt{0.95} = 0.97$ cm for each additional centimeter of arm span.
    (c) 95% of the relationship between height and arm span is accounted for by the regression line.
    (d) 95% of the variation in height is accounted for by the regression line.
    (e) 95% of the height measurements are accounted for by the regression line.

78. One child in the Mumbai study had height 59 cm and arm span 60 cm. This child's residual is
    (a) −3.2 cm. (b) −2.2 cm. (c) −1.3 cm. (d) 3.2 cm. (e) 62.2 cm.

Exercises 79 and 80 refer to the following setting. In its *Fuel Economy Guide* for 2008 model vehicles, the Environmental Protection Agency gives data on 1152 vehicles. There are a number of outliers, mainly vehicles with very poor gas mileage. If we ignore the outliers, however, the combined city and highway gas mileage of the other 1120 or so vehicles is approximately Normal with mean 18.7 miles per gallon (mpg) and standard deviation 4.3 mpg.

79. **In my Chevrolet** (2.2) The 2008 Chevrolet Malibu with a four-cylinder engine has a combined gas mileage of 25 mpg. What percent of all vehicles have worse gas mileage than the Malibu?

80. **The top 10%** (2.2) How high must a 2008 vehicle's gas mileage be in order to fall in the top 10% of all vehicles? (The distribution omits a few high outliers, mainly hybrid gas-electric vehicles.)

81. **Marijuana and traffic accidents** (1.1) Researchers in New Zealand interviewed 907 drivers at age 21. They had data on traffic accidents and they asked the drivers about marijuana use. Here are data on the numbers of accidents caused by these drivers at age 19, broken down by marijuana use at the same age:[23]

	Marijuana Use per Year			
	Never	1–10 times	11–50 times	51 + times
Drivers	452	229	70	156
Accidents caused	59	36	15	50

(a) Make a graph that displays the accident rate for each class. Is there evidence of an association between marijuana use and traffic accidents?

(b) Explain why we can't conclude that marijuana use *causes* accidents.

Chapter 3 Chapter Review

Chapters 1 and 2 dealt with data analysis for a single variable and relationships between categorical variables. In this chapter, we have studied relationships between two quantitative variables.

Data analysis begins with graphs, then adds numerical summaries of specific aspects of the data. When the data show a regular pattern, we can use a simplified model to summarize the pattern. We should also examine any deviations from the pattern.

Scatterplots show the relationship. Correlation describes the strength of a linear relationship. Least-squares regression fits a line to data that have an explanatory-response relationship. Residual plots, r^2, and s help us assess how well the linear model fits the data.

Correlation and regression are powerful tools, but they have their limitations. Outliers and influential points can drastically affect our interpretations of correlation and regression results. Always remember: even a strong association doesn't imply causation!

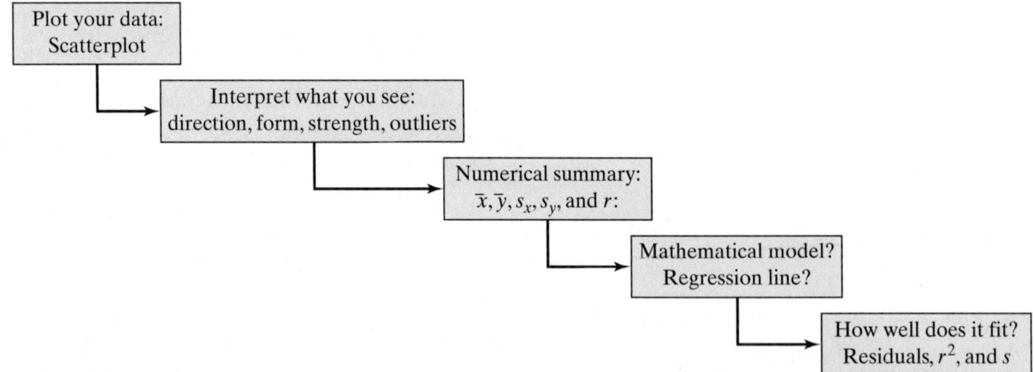

Chapter Review Exercises

These exercises are designed to help you review the important ideas and methods of the chapter. Relevant learning objectives are provided in bulleted form before each exercise.

- Identify explanatory and response variables in situations where one variable helps explain or influences another.
- Explain why association doesn't imply causation.

R3.1. The risks of obesity A study observes a large group of people over a 10-year period. The goal is to see if overweight and obese people are more likely to die during the study than people who weigh less. Such studies can be misleading, because obese people are more likely to be inactive and poor.

(a) What are the explanatory and response variables in the study?

(b) If the study finds a strong association between these variables, can we conclude that increased weight causes greater risk of dying? Why or why not?

- Describe the direction, form, and strength of the overall pattern of a scatterplot. In particular, recognize positive or negative association and linear (straight-line) patterns. Recognize outliers in a scatterplot.

R3.2 Born to be old? Is there a relationship between the gestational period (time from conception to birth) of an animal and its average life span? The figure shows a scatterplot of the gestational period and average life span for 43 species of animals.[24]

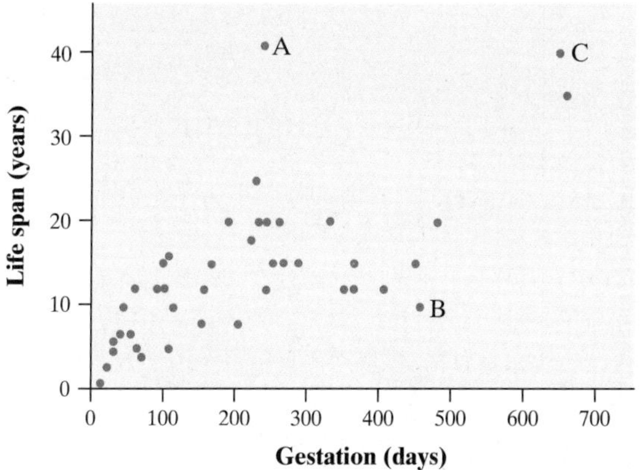

(a) Describe the direction, form, and strength of the scatterplot.

(b) Three "unusual" points are labeled on the graph: Point A is for the hippopotamus, Point B is for the giraffe, and Point C is for the Asian elephant. In what way is each of these animals "unusual"?

- Explain what the slope b and the y intercept a mean in the equation $\hat{y} = a + bx$ of a regression line.
- Use the regression line to predict y for a given x. Beware of extrapolation.

R3.3 Penguins diving A study of king penguins looked for a relationship between how deep the penguins dive to seek food and how long they stay under water.[25] For all but the shallowest dives, there is a linear relationship that is different for different penguins. The study gives a scatterplot for one penguin titled "The Relation of Dive Duration (y) to Depth (x)." Duration y is measured in minutes and depth x is in meters. The report then says, "The regression equation for this bird is: $\hat{y} = 2.69 + 0.0138x$."

(a) What is the slope of the regression line? Explain in specific language what this value says about this penguin's dives.

(b) According to the regression line, how long does a typical dive to a depth of 200 meters last?

(c) Does the y intercept of the regression line make any sense? If so, interpret it. If not, explain why not.

- Interpret computer regression output.

R3.4 Stats teachers' cars A random sample of AP Statistics teachers was asked to report the age (in years) and mileage of their primary vehicles. A scatterplot of the data, a least-squares regression printout, and a residual plot are provided below.

R squared = 82.0% R squared (adjusted) = 81.1%
s = 19280 with 21 - 2 = 19 degrees of freedom

Variable	Coefficient	s.e. of Coeff	t-ratio	prob
Constant	7288.54	6591	1.11	<0.2826
Car age	11630.6	1249	9.31	<0.0001

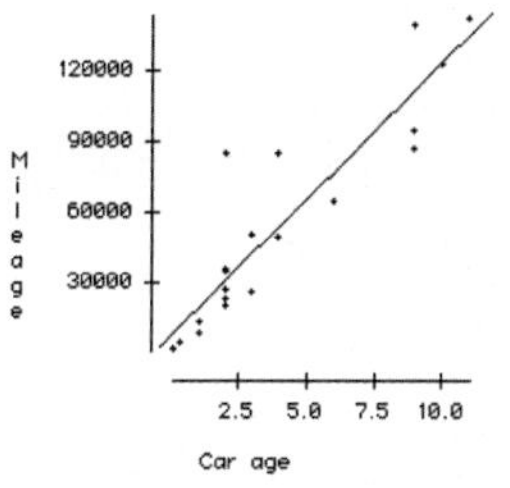

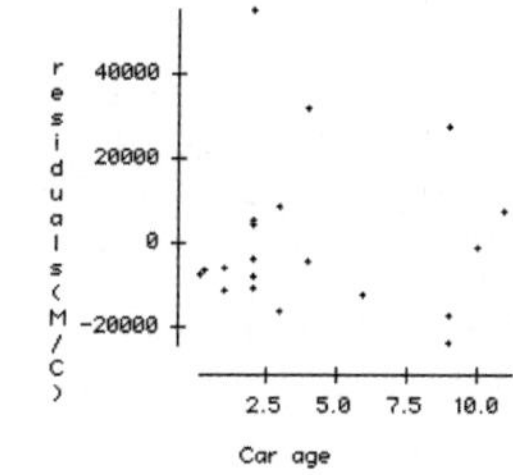

(a) Give the equation of the least-squares regression line for these data. Identify any variables you use.

(b) One teacher reported that her 6-year-old car had 65,000 miles on it. Find its residual.

(c) Interpret the slope of the line in context.

(d) What's the correlation between car age and mileage? Interpret this value in context.

(e) How well does the regression line fit the data? Justify your answer using the residual plot and s.

- Make a scatterplot to display the relationship between two quantitative variables.
- Use your calculator to find a least-squares regression line and make a residual plot.
- Calculate residuals and interpret residual plots.
- Explain what r^2 and s tell you about how well a regression line fits a set of data.

R3.5 Late bloomers? Japanese cherry trees tend to blossom early when spring weather is warm and later when spring weather is cool. Here are some data on the average March temperature (in °C) and the day in April when the first cherry blossom appeared over a 24-year period:[26]

Temperature (°C)	4.0	5.4	3.2	2.6	4.2	4.7	4.9	4.0	4.9	3.8	4.0	5.1
Days in April to first bloom:	14	8	11	19	14	14	14	21	9	14	13	11

Temperature (°C)	4.3	1.5	3.7	3.8	4.5	4.1	6.1	6.2	5.1	5.0	4.6	4.0
Days in April to first bloom:	13	28	17	19	10	17	3	3	11	6	9	11

(a) Make a well-labeled scatterplot that's suitable for predicting when the cherry trees will bloom from the temperature. Describe the direction, form, and strength of the relationship.

(b) Use technology to find the equation of the least-squares regression line. Interpret the slope and y intercept of the line in this setting.

(c) The average March temperature this year was 3.5°C. When would you predict that the first cherry blossom would appear? Show your method clearly.

(d) Find the residual for the year when the average March temperature was 4.5°C. Show your work.

(e) Use technology to construct a residual plot. Describe what you see.

(f) Find and interpret the value of r^2 and s in this setting.

- Find the slope and y intercept of the least-squares regression line from the means and standard deviations of x and y and their correlation.

R3.6 What's my grade? In Professor Friedman's economics course, the correlation between the students' total scores prior to the final examination and their final-examination scores is $r = 0.6$. The pre-exam totals for all students in the course have mean 280 and standard deviation 30. The final-exam scores have mean 75 and standard deviation 8. Professor Friedman has lost Julie's final exam but knows that her total before the exam was 300. He decides to predict her final-exam score from her pre-exam total.

(a) Find the equation for the appropriate least-squares regression line for Professor Friedman's prediction. Interpret the slope of this line in context.

(b) Use the regression line to predict Julie's final-exam score.

(c) Julie doesn't think this method accurately predicts how well she did on the final exam. Determine r^2. Use this result to argue that her actual score could have been much higher (or much lower) than the predicted value.

- Recognize outliers and potentially influential observations from a scatterplot.
- Explain how the correlation r and the least-squares regression line are influenced by extreme observations.
- Know the basic properties of correlation.

R3.7 When it rains, it pours The figure below plots the record-high yearly precipitation in each state against that state's record-high 24-hour precipitation. Hawaii is a high outlier, with a record-high yearly record of 704.83 inches of rain recorded at Kukui in 1982.

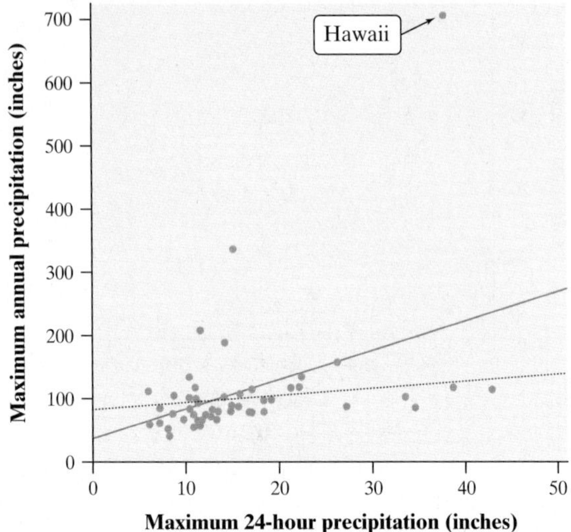

(a) The correlation for all 50 states in the figure is 0.408. If we leave out Hawaii, would the correlation increase, decrease, or stay the same? Explain.

(b) Two least-squares lines are shown on the graph. One was calculated using all 50 states, and the other omits Hawaii. Which line is which? Explain.

(c) Explain how each of the following would affect the correlation, s, and the least-squares line:
- Measuring record precipitation in feet instead of inches for both variables
- Switching the explanatory and response variables

Chapter 3 AP Statistics Practice Test

Section I: Multiple Choice *Select the best answer for each question.*

T3.1. A school guidance counselor examines the number of extracurricular activities that students do and their grade point average. The guidance counselor says, "The evidence indicates that the correlation between the number of extracurricular activities a student participates in and his or her grade point average is close to zero." A correct interpretation of this statement would be that

(a) active students tend to be students with poor grades, and vice versa.

(b) students with good grades tend to be students who are not involved in many extracurricular activities, and vice versa.

(c) students involved in many extracurricular activities are just as likely to get good grades as bad grades; the same is true for students involved in few extracurricular activities.

(d) there is no linear relationship between number of activities and grade point average for students at this school.

(e) involvement in many extracurricular activities and good grades go hand in hand.

T3.2 The British government conducts regular surveys of household spending. The average weekly household spending (in pounds) on tobacco products and alcoholic beverages for each of 11 regions in Great Britain was recorded. A scatterplot of spending on alcohol versus spending on tobacco is shown below. Which of the following statements is true?

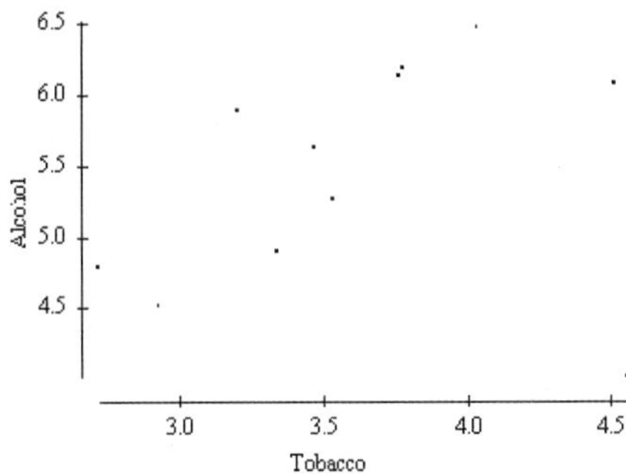

(a) The observation (4.5, 6.0) is an outlier.

(b) There is clear evidence of a negative association between spending on alcohol and tobacco.

(c) The equation of the least-squares line for this plot would be approximately $\hat{y} = 10 - 2x$.

(d) The correlation for these data is $r = 0.99$.

(e) The observation in the lower-right corner of the plot is influential for the least-squares line.

T3.3 The fraction of the variation in the values of y that is explained by the least-squares regression of y on x is

(a) the correlation.

(b) the slope of the least-squares regression line.

(c) the square of the correlation coefficient.

(d) the intercept of the least-squares regression line.

(e) the residual.

T3.4 An AP Statistics student designs an experiment to see whether today's high school students are becoming too calculator dependent. She prepares two quizzes, both of which contain 40 questions that are best done using paper-and-pencil methods. A random sample of 30 students participates in the experiment. Each student takes both quizzes—one with a calculator and one without—in a random order. To analyze the data, the student constructs a scatterplot that displays the number of correct answers with and without a calculator for each of the 30 students. A least-squares regression yields the equation

$$\widehat{\text{Calculator}} = -1.2 + 0.865(\text{Pencil}) \quad r = 0.79$$

Which of the following statements is/are true?

I. If the student had used Calculator as the explanatory variable, the correlation would remain the same.

II. If the student had used Calculator as the explanatory variable, the slope of the least-squares line would remain the same.

III. The standard deviation of the number of correct answers on the paper-and-pencil quizzes was larger than the standard deviation on the calculator quizzes.

(a) I only (b) II only (c) III only (d) I and III only (e) I, II, and III

Questions T3.5 and T3.6 refer to the following setting. Scientists examined the activity level of fish at 7 different temperatures. Fish activity was rated on a scale of 0 (no activity) to 100 (maximal activity). The temperature was measured in degrees Celsius. A computer regression printout and a residual plot are given below. Notice that the horizontal axis on the residual plot is labeled "predicted (F/T)."

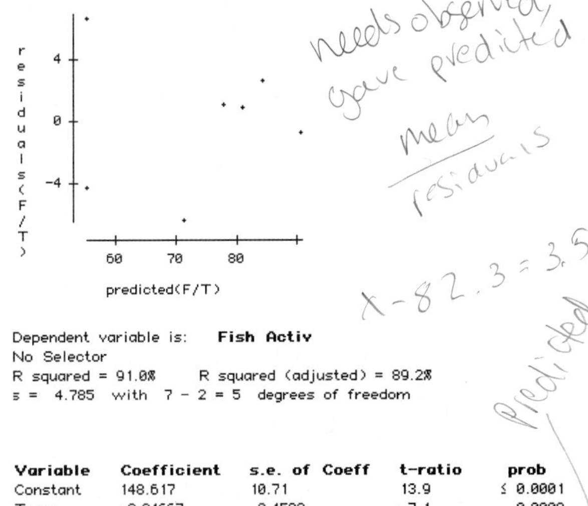

Dependent variable is: **Fish Activ**
No Selector
R squared = 91.0% R squared (adjusted) = 89.2%
s = 4.785 with 7 - 2 = 5 degrees of freedom

Variable	Coefficient	s.e. of Coeff	t-ratio	prob
Constant	148.617	10.71	13.9	≤ 0.0001
Temp	-3.21667	0.4533	-7.1	0.0009

T3.5 What was the activity level rating for the fish at a temperature of 20.4°C?

(a) 86 (b) 83 (c) 80 (d) 66 (e) 3

T3.6 Which of the following gives a correct interpretation of s in this setting?

(a) For every 1°C increase in temperature, fish activity is predicted to increase by 4.785 units.

(b) The average distance of the temperature readings from their mean is about 4.785°C.

(c) The average distance of the activity level ratings from the least-squares line is about 4.785 units.

(d) The average distance of the activity level readings from their mean is about 4.785.

(e) At a temperature of 0°C, this model predicts an activity level of 4.785.

T3.7 Which of these is *not* true of the correlation r between the lengths in inches and weights in pounds of a sample of brook trout?

(a) r must take a value between -1 and 1.

(b) r is measured in inches.

(c) if longer trout tend to also be heavier, then $r > 0$.

(d) r would not change if we measured the lengths of the trout in centimeters instead of inches.

(e) r would not change if we measured the weights of the trout in kilograms instead of pounds.

T3.8 When we standardize the values of a variable, the distribution of standardized values has mean 0 and standard deviation 1. Suppose we measure two variables X and Y on each of several subjects. We standardize both variables and then compute the least-squares regression line. Suppose the slope of the least-squares regression line is −0.44. We may conclude that

(a) the correlation will be 1/−0.44.
(b) the intercept will also be −0.44.
(c) the intercept will be 1.0.
(d) the correlation will be 1.0.
(e) the correlation will also be −0.44.

T3.9 There is a linear relationship between the number of chirps made by the striped ground cricket and the air temperature. A least-squares fit of some data collected by a biologist gives the model $\hat{y} = 25.2 + 3.3x$, where x is the number of chirps per minute and $\hat{y}$ is the estimated temperature in degrees Fahrenheit. What is the predicted increase in temperature for an increase of 5 chirps per minute?

(a) 3.3°F
(b) 16.5°F
(c) 25.2°F
(d) 28.5°F
(e) 41.7°F

T3.10 A data set included the number of people per television set and the number of people per physician for 40 countries. The Fathom screen shot below displays a scatterplot of the data with the least-squares regression line added. In Ethiopia, there were 503 people per TV and 36,660 people per doctor. What effect would removing this point have on the regression line?

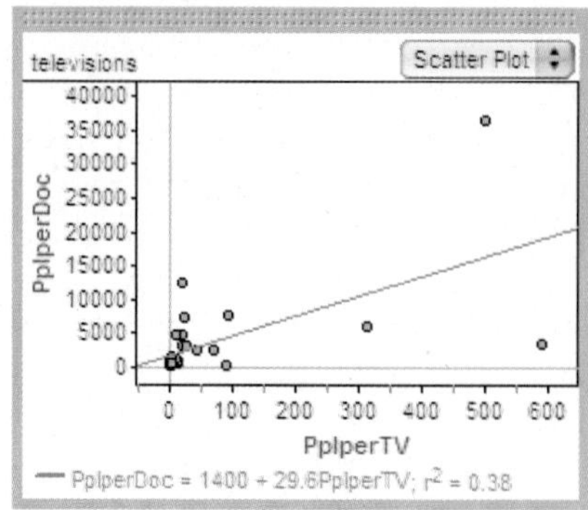

(a) Slope would increase; y intercept would increase.
(b) Slope would increase; y intercept would decrease.
(c) Slope would decrease; y intercept would increase.
(d) Slope would decrease; y intercept would decrease.
(e) Slope and y intercept would stay the same.

Section II: Free Response *Show all your work. Indicate clearly the methods you use, because you will be graded on the correctness of your methods as well as on the accuracy and completeness of your results and explanations.*

T3.11 Sarah's parents are concerned that she seems short for her age. Their doctor has the following record of Sarah's height:

Age (months):	36	48	51	54	57	60
Height (cm):	86	90	91	93	94	95

(a) Make a scatterplot of these data.
(b) Using your calculator, find the equation of the least-squares regression line of height on age.
(c) Use your regression line to predict Sarah's height at age 40 years (480 months). Convert your prediction to inches (2.54 cm = 1 inch).
(d) The prediction is impossibly large. Explain why this happened.

T3.12 Drilling down beneath a lake in Alaska yields chemical evidence of past changes in climate. Biological silicon, left by the skeletons of single-celled creatures called diatoms, is a measure of the abundance of life in the lake. A rather complex variable based on the ratio of certain isotopes relative to ocean water gives an indirect measure of moisture, mostly from snow. As we drill down, we look further into the past. Here is a scatterplot of data from 2300 to 12,000 years ago:

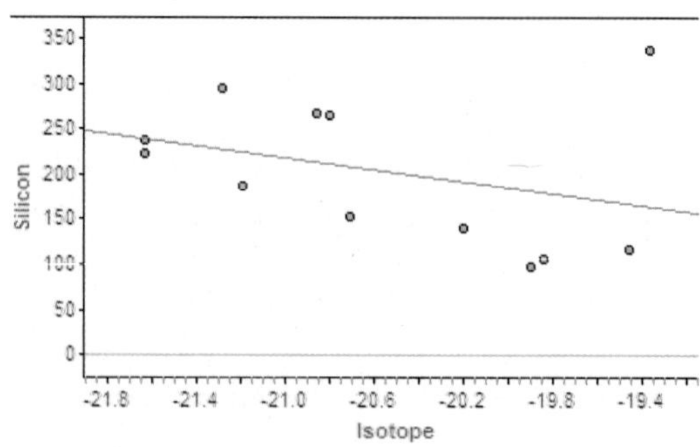

(a) Identify the unusual point in the scatterplot. Explain what's unusual about this point.

(b) If this point was removed, describe the effect on

i. the correlation.

ii. the slope and y intercept of the least-squares line.

T3.13 Long-term records from the Serengeti National Park in Tanzania show interesting ecological relationships. When wildebeest are more abundant, they graze the grass more heavily, so there are fewer fires and more trees grow. Lions feed more successfully when there are more trees, so the lion population increases. Researchers collected data on one part of this cycle, wildebeest abundance (in thousands of animals) and the percent of the grass area burned in the same year. The results of a least-squares regression on the data are shown here.[27]

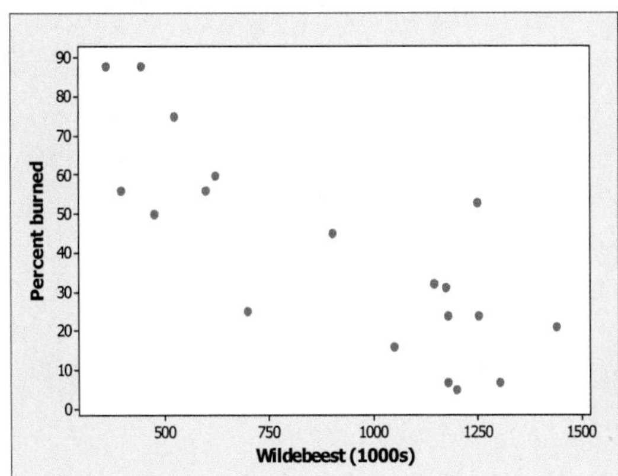

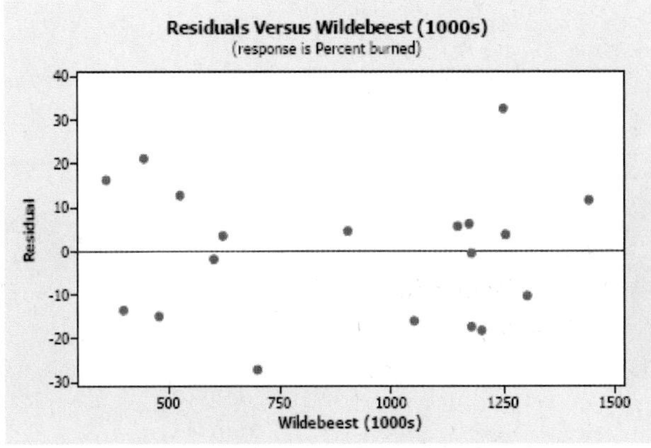

Predictor	Coef	SE Coef	T	P
Constant	92.29	10.06	9.17	0.000
Wildebeest (1000s)	-0.05762	0.01035	-5.56	0.000

S = 15.9880 R-Sq = 64.6% R-Sq(adj) = 62.5%

(a) Give the equation of the least-squares regression line. Be sure to define any variables you use.

(b) Explain what the slope of the regression line means in this setting.

(c) Find the correlation. Interpret this value in context.

(d) Is a linear model appropriate for describing the relationship between wildebeest abundance and percent of grass area burned? Support your answer with appropriate evidence.

Chapter 4

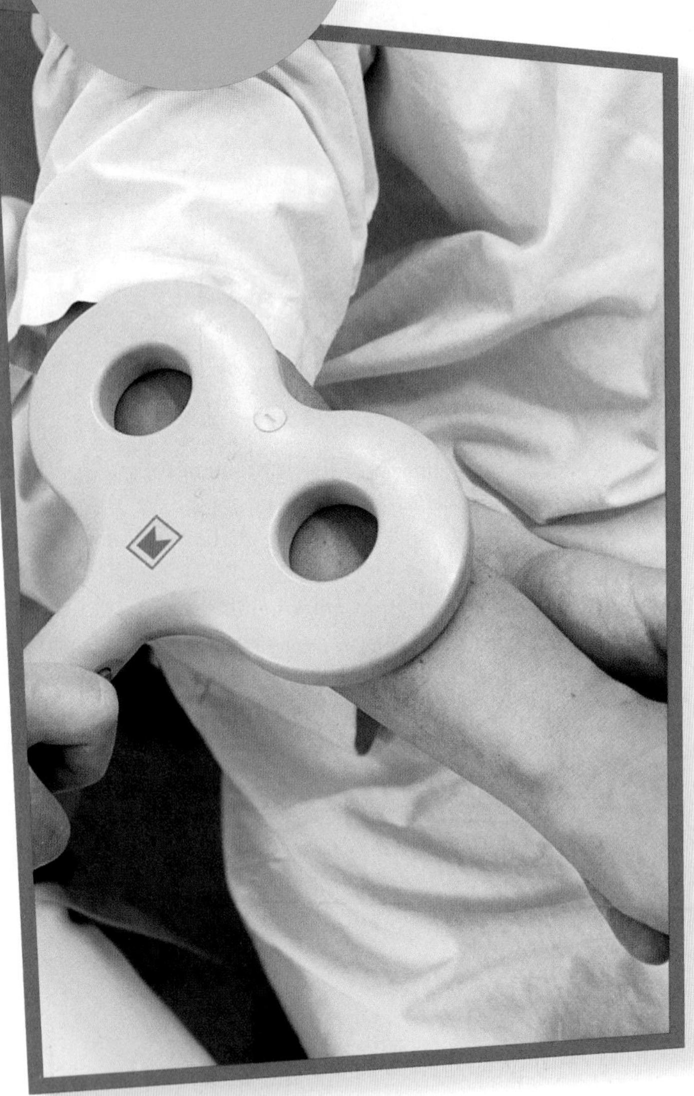

case study

Designing Studies

Can Magnets Help Reduce Pain?

Early research showed that magnetic fields affected living tissue in humans. Some doctors have begun to use magnets to treat patients with chronic pain. Scientists wondered whether this type of therapy really worked. They designed a study to find out.

Fifty patients with chronic pain were recruited for the study. A doctor identified a painful site on each patient and asked him or her to rate the pain on a scale from 0 (mild pain) to 10 (severe pain). Then, the doctor selected a sealed envelope containing a magnet from a box that contained both active and inactive magnets. That way, neither the doctor nor the patient knew which type of magnet was being used. The chosen magnet was applied to the site of the pain for 45 minutes. After "treatment," each patient was again asked to rate the level of pain from 0 to 10.

In all, 29 patients were given active magnets and 21 patients received inactive magnets. All but one of the patients rated their initial pain as an 8, 9, or 10. So scientists decided to focus on patients' final pain ratings. Here they are, grouped by the type of magnet used:[1]

Active:	0, 4, 7, 0, 4, 2, 5, 5, 3, 2, 2, 2, 3, 4, 6, 4, 3, 0, 2, 0, 4, 4, 5, 9, 10, 10, 10, 10, 7
Inactive:	4, 7, 5, 8, 8, 6, 8, 10, 10, 6, 10, 8, 10, 10, 10, 10, 9, 9, 10, 10, 9

What do the data tell us about whether the active magnets helped reduce pain? By the end of the chapter, you'll be ready to interpret the results of this study.

Introduction

You can hardly go a day without hearing the results of a statistical study. Here are some examples:

- The National Highway Traffic Safety Administration (NHTSA) reports that seat belt use in passenger vehicles increased from 83% in 2008 to 84% in 2009.[2]

- According to a recent survey, U.S. teens aged 13 to 18 spend an average of 26.8 hours per week online. Although 59% of the teens said that posting personal information or photos online is unsafe, 62% said they had posted photos of themselves.[3]
- A recent study suggests that lack of sleep increases the risk of catching a cold.[4]
- For their final project, two AP Statistics students showed that listening to music while studying decreased subjects' performance on a memory task.[5]

Can we trust these results? As you'll learn in this chapter, that depends on how the data were produced. Let's take a closer look at where the data came from in each of these studies.

Each year, the NHTSA conducts an *observational study* of seat belt use in vehicles. The NHTSA sends trained observers to record the actual behavior of people in vehicles at randomly selected locations across the country. The idea of an observational study is simple: you can learn a lot just by watching. Or by asking a few questions, as in the survey of teens' online habits. Harris Interactive conducted this survey using a "representative sample" of 655 U.S. 13- to 18-year-olds. Both of these studies use information from a *sample* to draw conclusions about some larger *population*. Section 4.1 examines the issues involved in sampling and surveys.

In the sleep and catching a cold study, 153 volunteers took part. They answered questions about their sleep habits over a two-week period. Then, researchers gave them a virus and waited to see who developed a cold. This was a complicated observational study. Compare this with the *experiment* performed by the AP Statistics students. They recruited 30 students and divided them into two groups of 15 by drawing names from a hat. Students in one group tried to memorize a list of words while listening to music. Students in the other group tried to memorize the same list of words while sitting in silence. Section 4.2 focuses on designing experiments.

The goal of many statistical studies is to show that changes in one variable *cause* changes in another variable. In Section 4.3, we'll look at why establishing causation is so difficult, especially in observational studies. We'll also consider some of the ethical issues involved in planning and conducting a study.

Here's an Activity that gives you a preview of what lies ahead.

ACTIVITY *See no evil, hear no evil?*

MATERIALS: Two index cards, each with 10 pairs of digits written on it (prepared by your teacher); clock, watch, or stopwatch to measure 30 seconds; and a coin for each pair of students

Confucius said, "I hear and I forget. I see and I remember. I do and I understand." Do people really remember what they see better than what they hear?[6] In this Activity, you will perform an experiment to try to find out.

1. Divide the class into pairs of students by drawing names from a hat.

2. Your teacher will give each pair two index cards with 10 sets of numbers on them. *Do not look at the numbers* until it is time for you to do the experiment.

3. Flip a coin to decide which of you is Student A and which is Student B. Shuffle the index cards and deal one face down to each partner.

4. Student A will attempt a memory task while Student B keeps time.

Directions: Study the pairs of numbers on the index card for 30 seconds. Then turn the card over. Recite the alphabet aloud (A, B, C, etc.). Then tell your partner (Student B) what you think the numbers on the card are. Student B will record how many pairs of numbers you recalled correctly.

5. Now it's Student B's turn to do a memory task while Student A records the data.

Directions: Your partner will read the pairs of numbers on your index card aloud three times slowly. Next, you will recite the alphabet aloud (A, B, C, etc.) and then tell your partner what you think the numbers on the card are. Student A will record how many pairs of numbers you recalled correctly.

6. Your teacher will scale and label axes on the board for parallel dotplots of the results. Plot the number of pairs you remembered correctly on the appropriate graph.

7. Did students in your class remember numbers better when they saw them or when they heard them? Give appropriate evidence to support your answer.

8. Based on the results of this experiment, can we conclude that people in general remember better when they see than when they hear? Why or why not?

Sampling and Surveys

In Section 4.1, you'll learn about:

- The idea of a sample survey
- How to sample badly
- How to sample well: Random sampling
- Other sampling methods
- Inference for sampling
- Sample surveys: What can go wrong?

Suppose we want to find out what percent of young drivers in the United States text while driving. To answer the question, we will survey 16- to 20-year-olds who live in the United States and drive. Ideally, we would ask them all (take a *census*). But contacting every driver in this age group wouldn't be practical: it would take too much time and cost too much money. Instead, we put the question to a **sample** chosen to represent the entire **population** of young drivers.

DEFINITION: Population and sample

The **population** in a statistical study is the entire group of individuals about which we want information.

A **sample** is the part of the population from which we actually collect information. We use information from a sample to draw conclusions about the entire population.

The distinction between population and sample is basic to statistics. To make sense of any sample result, you must know what population the sample represents. Here's an example that illustrates this distinction and also introduces some major uses of sampling.

EXAMPLE *Sampling Hardwood and Humans*

Populations and samples

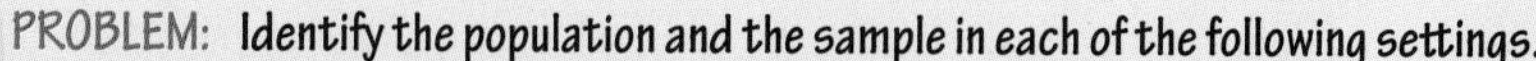

PROBLEM: Identify the population and the sample in each of the following settings.

(a) A furniture maker buys hardwood in large batches. The supplier is supposed to dry the wood before shipping (wood that isn't dry won't hold its size and shape). The furniture maker chooses five pieces of wood from each batch and tests their moisture content. If any piece exceeds 12% moisture content, the entire batch is sent back.

(b) Each week, the Gallup Poll questions a sample of about 1500 adult U.S. residents to determine national opinion on a wide variety of issues.

SOLUTION:

(a) The population is all the pieces of hardwood in a batch. The sample is the five pieces of wood that are selected from that batch and tested for moisture content.

(b) Gallup's population is all adult U.S. residents. Their sample is the 1500 adults who actually respond to the survey questions.

For Practice *Try Exercise 1*

The Idea of a Sample Survey

We often draw conclusions about a whole population on the basis of a sample. Have you ever tasted a sample of ice cream and ordered a cone if the sample tastes good? Since ice cream is fairly uniform, the single taste represents the whole. Choosing a representative sample from a large and varied population (like all young U.S. drivers) is not so easy. The first step in planning a **sample survey** is to say exactly *what population* we want to describe. The second step is to say exactly *what we want to measure*, that is, to give exact definitions of our variables.

Sample survey

We reserve the term "sample survey" for studies that use an organized plan to choose a sample that represents some specific population, like the pieces of hardwood and the U.S. adults in the previous example. By our definition, the population in a sample survey can consist of people, animals, or things. Some people use the terms "survey" or "sample survey" to refer only to studies in which people are asked one or more questions, like the Gallup Poll of the last example. We'll avoid this more restrictive terminology.

EXAMPLE *How Does the Current Population Survey Work?*

A sample survey

One of the most important government sample surveys in the United States is the monthly Current Population Survey (CPS). The CPS contacts about 60,000 households each month. It produces the monthly unemployment rate and much other economic and social information. To measure unemployment, we must first specify the population we want to describe. The CPS defines its population as all U.S. residents (legal or not) 16 years of age and over who are civilians and are not in an institution such as a prison. The unemployment rate announced in the news refers to this specific population.

What does it mean to be "unemployed"? Someone who is not looking for work—for example, a full-time student—should not be called unemployed just because she is not working for pay. If you are chosen for the CPS sample, the interviewer first asks whether you are available to work and whether you actually looked for work in the past four weeks. If not, you are neither employed nor unemployed—you are not in the labor force.

If you are in the labor force, the interviewer goes on to ask about employment. If you did any work for pay or in your own business during the week of the survey, you are employed. If you worked at least 15 hours in a family business without pay, you are employed. You are also employed if you have a job but didn't work because of vacation, being on strike, or other good reason. An unemployment rate of 9.7% means that 9.7% of the sample was unemployed, using the exact CPS definitions of both "labor force" and "unemployed."

The final step in planning a sample survey is to decide how to choose a sample from the population. Let's take a closer look at some good and not-so-good sampling methods.

How to Sample Badly

How can we choose a sample that we can trust to represent the population? The easiest—but not the best—sampling method just chooses individuals who are close by. Suppose we're interested in finding out how long students at a large high school spent on homework last week, for example. We might go to the school library and ask the first 30 students we see about their homework time. A sample selected by taking the members of the population that are easiest to reach is called a **convenience sample.** Convenience samples often produce unrepresentative data.

DEFINITION: Convenience sample

Choosing individuals who are easiest to reach results in a **convenience sample.**

What's wrong with our convenience sample of students in the library? It's unlikely that this sample represents the homework habits of all students at the school well. In fact, we'd expect the sample to overestimate the average homework time in the population since students who hang out in the library might tend to be more studious. This is **bias:** using a method that will consistently overestimate or underestimate the value you want to know.

DEFINITION: Bias

The design of a statistical study shows **bias** if it systematically favors certain outcomes.

AP EXAM TIP If you're asked to describe how the design of a study leads to bias, you're expected to identify the *direction* of the bias. Suppose you were asked, "Explain how using a convenience sample of students in your statistics class to estimate the proportion of all high school students who own a graphing calculator could result in bias." You might respond, "This sample would probably include a much higher proportion of students with a graphing calculator than in the population at large because a graphing calculator is required for the statistics class. That is, this method would probably lead to an overestimate of the actual population proportion."

Convenience samples are almost guaranteed to show bias. So are **voluntary response samples,** in which people decide whether to join the sample in response to an open invitation. Call-in, write-in, and many Internet polls rely on voluntary response samples. Unfortunately, people who choose to participate are usually not representative of any clearly defined population. Voluntary response samples attract people who feel strongly about the issue in question, often in the same direction. That leads to bias.

The Internet brings voluntary response samples to the computer nearest you. Visit www.misterpoll.com to become part of the sample in any of dozens of online polls. As the site says, "None of these polls are 'scientific,' but do represent the collective opinion of everyone who participates." Unfortunately, such polls don't tell you anything about the views of the population at large.

DEFINITION: Voluntary response sample

A **voluntary response sample** consists of people who choose themselves by responding to a general appeal. Voluntary response samples show bias because people with strong opinions (often in the same direction) are most likely to respond.

Write-in and call-in opinion polls are almost sure to lead to strong bias. In fact, only about 15% of the public has ever responded to a call-in poll, and these tend to be the same people who call radio talk shows. That's not a representative sample of the population as a whole.

Illegal Immigrants and Driver's Licenses

Online polls

Former CNN commentator Lou Dobbs doesn't like illegal immigration. One of his shows was largely devoted to attacking a proposal to offer driver's licenses to illegal immigrants. During the show, Mr. Dobbs invited his viewers to go to loudobbs.com to vote on the question "Would you be more or less likely to vote for a presidential candidate who supports giving drivers' licenses to illegal aliens? The result: 97% of the 7350 people who voted by the end of the show said "Less likely."

PROBLEM: **What type of sample did Mr. Dobbs use in his poll? Explain how this sampling method could lead to bias in the poll results.**

SOLUTION: **Mr. Dobbs used a voluntary response sample: people chose to go online and respond. Those who voted were viewers of Mr. Dobbs's program, which means that they are likely to support his views. The 97% poll result is probably an extreme overestimate of the percent of people in the population who would be less likely to support a presidential candidate with this position.**

For Practice *Try Exercise 5*

CHECK YOUR UNDERSTANDING

For each of the following situations, identify the sampling method used. Then explain how the sampling method could lead to bias.

1. A farmer brings a juice company several crates of oranges each week. A company inspector looks at 10 oranges from the top of each crate before deciding whether to buy all the oranges.

2. The ABC program *Nightline* once asked whether the United Nations should continue to have its headquarters in the United States. Viewers were invited to call one telephone number to respond "Yes" and another for "No." There was a charge for calling either number. More than 186,000 callers responded, and 67% said "No."

How to Sample Well: Random Sampling

In a voluntary response sample, people choose whether to respond. In a convenience sample, the interviewer makes the choice. In both cases, personal choice produces bias. The statistician's remedy is to allow impersonal chance to choose the sample. A sample chosen by chance rules out both favoritism by the sampler and self-selection by respondents. **Random sampling,** the use of chance to select a sample, is the central principle of statistical sampling.

Random sampling

In everyday life, some people use the word "random" to mean haphazard, as in, "That's so random." In statistics, random means "due to chance." Don't say that a sample was chosen at random if a chance process wasn't used to select the individuals.

The simplest way to use chance to select a sample is to place names in a hat (the population) and draw out a handful (the sample). This is the idea of a **simple random sample.**

DEFINITION: Simple random sample

A **simple random sample (SRS)** of size *n* consists of *n* individuals from the population chosen in such a way that every set of *n* individuals has an equal chance to be the sample actually selected.

Statisticians fall asleep faster by taking a random sample of sheep.

An SRS not only gives each individual an equal chance to be chosen but also gives every possible sample an equal chance to be chosen. There are other random sampling methods that give each individual, but not each sample, an equal chance. Exercise 26 describes one such method. For now, let's think about how to actually select an SRS.

When you think of an SRS, picture drawing names from a hat to remind yourself that an SRS doesn't favor any part of the population. That's why an SRS is a better method of choosing samples than convenience or voluntary response samples. But writing names on slips of paper and drawing them from a hat doesn't work as well if the population is large. Think about the Current Population Survey, for example, which must draw a sample of size 60,000 every month from the population of over 117 million U.S. households.

In practice, people use random numbers generated by a computer or calculator to choose samples. If you don't have technology handy, you can use a **table of random digits.**

DEFINITION: Table of random digits

A **table of random digits** is a long string of the digits 0, 1, 2, 3, 4, 5, 6, 7, 8, 9 with these properties:

- Each entry in the table is equally likely to be any of the 10 digits 0 through 9.
- The entries are independent of each other. That is, knowledge of one part of the table gives no information about any other part.

Table D at the back of the book is a table of random digits. It begins with the digits 19223950340575628713. To make the table easier to read, the digits appear in groups of five and in numbered rows. The groups and rows have no meaning—the table is just a long list of randomly chosen digits. There are two steps in using the table to choose a simple random sample.

How to Choose an SRS Using Table D

STEP 1: LABEL. Give each member of the population a numerical label of the *same length.*

STEP 2: TABLE. Read consecutive groups of digits of the appropriate length from Table D.

Your sample contains the individuals whose labels you find.

Always use the shortest labels that will cover your population. For instance, you can label up to 100 individuals with two digits: 01, 02, . . . , 99, 00. As standard practice, we recommend that you begin with label 1 (or 01 or 001, as needed). Reading groups of digits from the table gives all individuals the same chance to be chosen because all labels of the same length have the same chance to be found in the table. For example, any pair of digits in the table is equally likely to be any of the 100 possible labels 01, 02, . . . , 99, 00. Ignore any group of digits that was not used as a label or that duplicates a label already in the sample. You can read digits from Table D in any order—across a row, down a column, and so on—because the table has no order. We recommend reading across rows from left to right.

Spring Break!
Choosing an SRS with Table D

The school newspaper is planning an article on family-friendly places to stay over spring break at a nearby beach town. The editors intend to call 4 randomly chosen hotels to ask about their amenities for families with children. They have an alphabetized list of all 28 hotels in the town.

PROBLEM: Use Table D at line 130 to choose an SRS of 4 hotels for the editors to call.

SOLUTION: We'll use the two-step process for selecting an SRS using Table D.

Step 1: **Label.** Two digits are needed to label the 28 resorts. We have added labels 01 to 28 to the alphabetized list of resorts below.

01 Aloha Kai	08 Captiva	15 Palm Tree	22 Sea Shell
02 Anchor Down	09 Casa del Mar	16 Radisson	23 Silver Beach
03 Banana Bay	10 Coconuts	17 Ramada	24 Sunset Beach
04 Banyan Tree	11 Diplomat	18 Sandpiper	25 Tradewinds
05 Beach Castle	12 Holiday Inn	19 Sea Castle	26 Tropical Breeze
06 Best Western	13 Lime Tree	20 Sea Club	27 Tropical Shores
07 Cabana	14 Outrigger	21 Sea Grape	28 Veranda

Step 2: **Table.** To use Table D, start at the left-hand side of line 130 and read two-digit groups. Skip any groups that aren't between 01 and 28, as well as any repeated groups. Continue until you have chosen four resorts. Here is the beginning of line 130:

69051 64817 87174 09517 84534 06489 87201 97245

The first 10 two-digit groups are

69	05	16	48	17	87	17	40	95	17
Skip Too big	✓	✓	Skip Too big	✓	Skip Too big	Skip Repeat	Skip Too big	Skip Too big	Skip Repeat

We skip 5 of these 10 groups because they are too high (over 28) and 2 because they are repeats (both 17s). The hotels labeled 05, 16, and 17 go into the sample. We need one more hotel to complete the sample. The remaining 10 two-digit groups in line 130 are

84	53	40	64	89	87	20	19	72	45
Skip Too big →	Skip	Skip	Skip	Skip	Skip	✓			

Our SRS of 4 hotels for the editors to contact is: 05 Beach Castle, 16 Radisson, 17 Ramada, and 20 Sea Club.

For Practice *Try Exercise 11*

As you saw in the previous example, using Table D to select an SRS can be time-consuming. The *Simple Random Sample* applet can help you quickly choose an SRS for populations of up to 500 individuals. Your calculator can do even better.

TECHNOLOGY CORNER Choosing an SRS

APPLET

You can use the *Simple Random Sample* applet at www.whfreeman.com/tps4e or your graphing calculator to select the SRS of 4 hotels in the previous example.

***Simple Random Sample* applet:** Enter 28 in the "Population = 1 to" box and 4 in the "Select a sample of size" box, click "Reset," and click "Sample." Figure 4.1 shows the result of one sample.

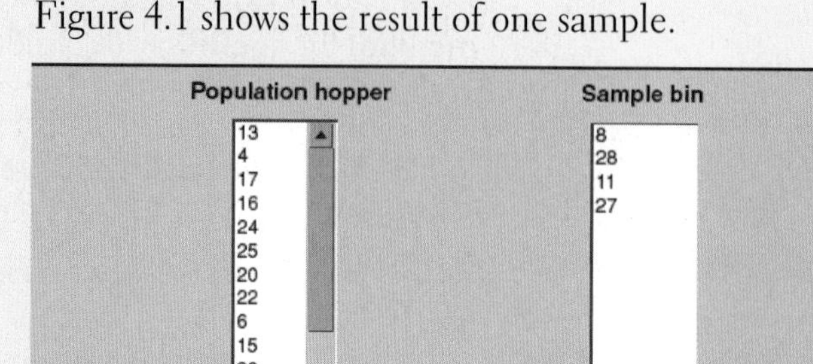

FIGURE 4.1 The *Simple Random Sample* applet used to choose an SRS of size $n = 4$ from a population of size 28.

Graphing calculator:

1. Check that your calculator's random number generator is working properly.

TI-83/84: Press MATH, then select PRB and `5:randInt(`. Complete the command `randInt(1,28)` and press ENTER.

TI-89: Press CATALOG, then F3 (Flash Apps), and choose `randInt(`. Complete the command `TIStat.randInt(1,28)` and press ENTER.

Compare results with your classmates. If several students got the same number, you'll need to seed your calculator's random number generator with different numbers before you proceed. Directions for doing this are given in the *Annotated Teachers Edition*.

2. Now do `randInt(1,28)` again. Keep pressing ENTER until you have chosen four different labels. (Notice the repeated label "25" in the TI-84 screen shot.)

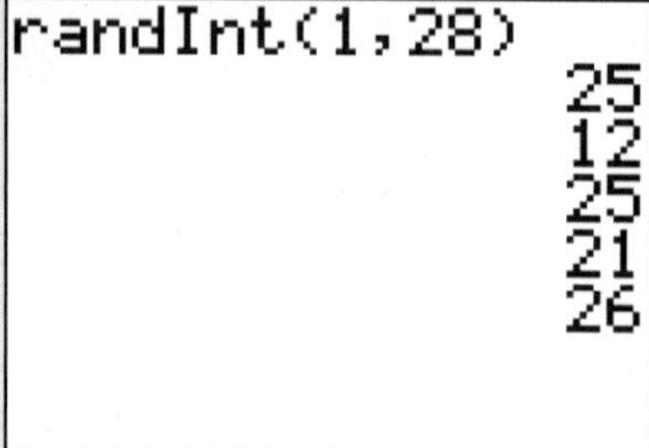

TI-Nspire instructions in Appendix B

To see how software speeds up choosing an SRS, try the random number generator at www.random.org. The Research Randomizer at www.randomizer.org is another option. Click on "Randomize" and fill in the boxes. You can even ask the Randomizer to arrange your sample in order.

There's an important difference between the samples produced by the *Simple Random Sample* applet and the calculator's `randInt` command in the Technology Corner. The applet sampled *without replacement* from the "Population hopper," but the calculator sampled *with replacement* from the specified population. As a result, the calculator sometimes selects the same number more than once in a given sample. To deal with this problem, you can generate additional random numbers as needed to replace any repeats. Or you can use a method other than `randInt` (like `randSamp` on the TI-Nspire) to sample without replacement. Refer to your device's reference manual or ask your teacher.

We can trust results from an SRS, as well as from other types of random samples that we will meet later, because the use of impersonal chance avoids bias. Online and call-in polls also produce samples. But we can't trust results from these samples because they are chosen in ways that invite bias. *The first question to ask about any sample is whether it was chosen at random.*

ACTIVITY *How large is a typical U.S. state?*

In this Activity, you and your classmates will use two different sampling methods to estimate the average area (in square miles) of the 50 states.

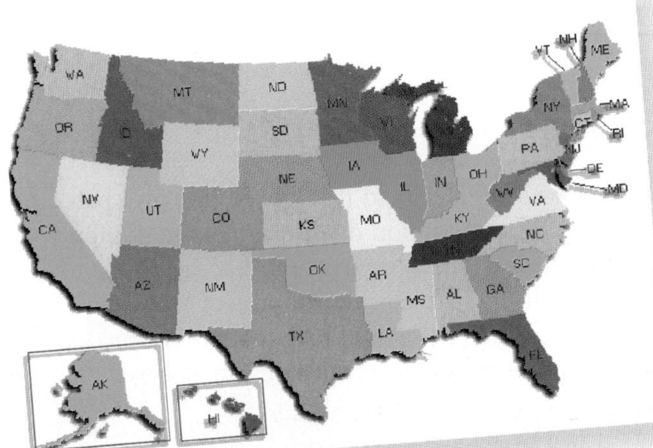

1. Use the map shown to choose a sample of 5 states to estimate the average (mean) land area. You have 15 seconds.
2. Refer to the table of land areas on page N/DS-5. Find the mean land area for your sample.
3. Make a class dotplot of the mean land areas from Step 2.
4. Use a line of Table D assigned by your teacher to choose an SRS of 5 states from the table on page N/DS-5. Find the mean land area for this sample.
5. Make a class dotplot of the mean land areas from Step 4 right above your dotplot from Step 3.
6. How do the class's estimates using the two methods compare? What advantage(s) does random sampling provide?

Other Sampling Methods

The basic idea of sampling is straightforward: take an SRS from the population and use your sample results to gain information about the population. Unfortunately, it's usually very difficult to actually get an SRS from the population of interest. Imagine trying to get a simple random sample of all the batteries produced in one day at a factory. Or an SRS of all U.S. high school students. In either case, it's just not practical to choose an SRS. For starters, it would be difficult to obtain an accurate list of the population from which to draw the sample. It would also be very time-consuming to collect data from each individual that's randomly selected. Sometimes, there are also statistical advantages to using more complex sampling methods.

One of the most common alternatives to an SRS involves sampling important groups (called **strata**) within the population separately. Then these separate "subsamples" are combined to form one **stratified random sample**. This method works best when the individuals within each stratum are similar to one another in a way that affects the variable being measured but there are large differences between strata.

DEFINITION: Stratified random sample and strata

To select a **stratified random sample,** first classify the population into groups of similar individuals, called **strata.** Then choose a separate SRS in each stratum and combine these SRSs to form the full sample.

Unfortunately, "stratified random sample" has the same initials as "simple random sample." Only a simple random sample gets abbreviated SRS, however.

Choose the strata based on facts known before the sample is taken. For example, in a study of sleep habits on school nights, the population of students in a large high school might be divided into freshman, sophomore, junior, and senior strata. In a preelection poll, a population of election districts might be divided into urban, suburban, and rural strata. *If the individuals in each stratum are less varied than the population as a whole, a stratified random sample can produce better information about the population than an SRS of the same size.* Not convinced? Try the following Activity.

ACTIVITY *Sampling sunflowers*

A British farmer grows sunflowers for making sunflower oil. Her field is arranged in a grid pattern, with 10 rows and 10 columns as shown in the figure. Irrigation ditches run along the top and bottom of the field, as shown. The farmer would like to estimate the number of healthy plants in the field so she can project how much money she'll make from selling them. It would take too much time to count the plants in all 100 squares, so she'll accept an estimate based on a sample of 10 squares.

1. Use Table D or technology to take a simple random sample of 10 grid squares. Record the location (for example, B6) of each square you select.

2. This time, you'll take a stratified random sample using the *rows* as strata. Use Table D or technology to randomly select one square from each (horizontal) row. Record the location of each square.

3. Now, take a stratified random sample using the *columns* as strata. Use Table D or technology to randomly select one square from each (vertical) column. Record the location of each square.

4. The table on page N/DS-6 gives the actual number of sunflowers in each grid square. Use the information provided to calculate your estimate of the mean number of sunflowers per square for each of your samples in Steps 1, 2, and 3.

	A	B	C	D	E	F	G	H	I	J
1										
2										
3										
4										
5										
6										
7										
8										
9										
10										

5. Make comparative dotplots showing the mean number of sunflowers obtained using the three different sampling methods for all members of the class. Describe any similarities and differences you see.

6. Your teacher will provide you with the mean number of sunflowers in the population of all 100 grid squares in the field. How did the three sampling methods do?

The following example describes an interesting use of stratified random samples in the music business.

Who Wrote That Song?

Using a stratified random sample

A radio station that broadcasts a piece of music owes a royalty to the composer. The organization of composers (called ASCAP) collects these royalties for all its members by charging stations a license fee for the right to play members' songs. ASCAP has four million songs in its catalog and collects $435 million in fees each year. How should ASCAP distribute this income among its members? By sampling: ASCAP tapes about 60,000 hours from the 53 million hours of local radio programs across the country each year.

Radio stations are stratified by type of community (metropolitan, rural), geographic location (New England, Pacific, etc.), and the size of the license fee paid to ASCAP, which reflects the size of the audience. In all, there are 432 strata. Tapes are made at random hours of randomly selected members of each stratum. The tapes are reviewed by experts who can recognize almost every piece of music ever written, and the composers are then paid according to their popularity.[7]

Although a stratified random sample can sometimes give more precise information about a population than an SRS, both sampling methods are hard to use when populations are large and spread out over a wide area. In that situation, we'd prefer a method that selects groups of individuals that are "near" one another. That's the idea of a **cluster sample.**

DEFINITION: Cluster sample and clusters

To take a **cluster sample,** first divide the population into smaller groups. Ideally, these **clusters** should mirror the characteristics of the population. Then choose an SRS of the clusters. All individuals in the chosen clusters are included in the sample.

In a cluster sample, some people take an SRS from each cluster rather than including all members of the cluster.

Imagine a large high school that assigns its students to homerooms alphabetically by last name. The school administration is considering a new schedule and would like student input. Administrators decide to survey 200 randomly selected students. It would be hard to track down an SRS of 200 students, so the administration opts for a cluster sample of homerooms. The principal (who knows some statistics) takes a simple random sample of 8 homerooms and gives the survey to all 25 students in each homeroom.

You might say that strata are ideally "similar within, but different between," while clusters are ideally "different within, but similar between."

Cluster samples are often used for practical reasons, as in the school survey example. They don't offer the statistical advantage of better information about the population that stratified random samples do. That's because clusters are often chosen for ease or convenience, so they may have as much variability as the population itself. Be sure you understand the difference between strata and clusters. We want each stratum to contain similar individuals, and for there to be large differences between strata. For a cluster sample, we'd *like* each cluster to look just like the population, but on a smaller scale.

EXAMPLE *Sampling at a School Assembly*

Strata or clusters?

The student council wants to conduct a survey during the first five minutes of an all-school assembly in the auditorium about use of the school library. They would like to announce the results of the survey at the end of the assembly. The student council president asks your statistics class to help carry out the survey.

PROBLEM: There are 800 students present at the assembly. A map of the auditorium is shown below. Note that students are seated by grade level and that the seats are numbered from 1 to 800.

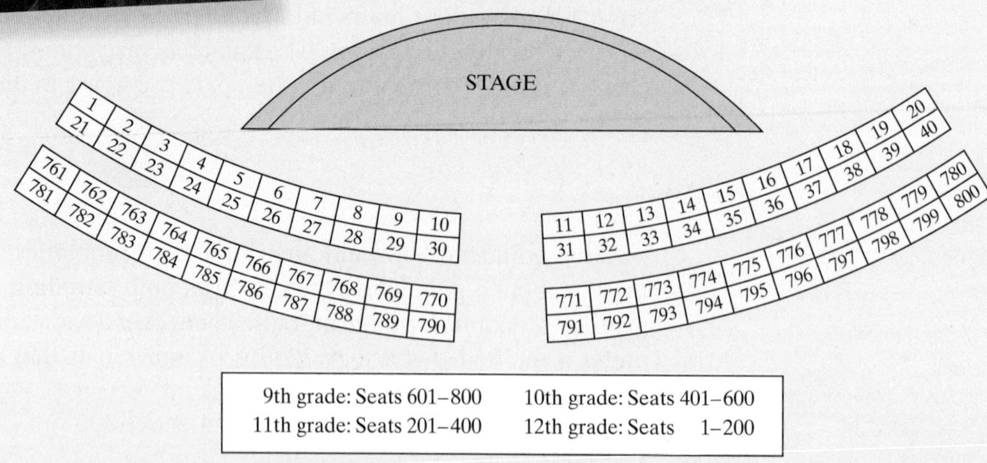

Describe how you would use each of the following sampling methods to select 80 students to complete the survey.

(a) Simple random sample

(b) Stratified random sample

(c) Cluster sample

SOLUTION:

(a) To take an SRS, we need to choose 80 of the seat numbers at random. Use randInt(1,800) on your calculator until 80 different seats are selected. Then give the survey to the students in those seats.

Can you give an advantage and a disadvantage of each sampling method used in the example?

(b) The students in the assembly are seated by grade level. Since students' library use might be similar within grade levels but different across grade levels, we'll use the grade level seating areas as our strata. Within each grade's seating area, we'll select 20 seats at random. For the 9th grade, use randInt(601,800) to select 20 seats. Use randInt(401,600) to pick 20 sophomore seats, randInt(201,400) to get 20 junior seats, and randInt(1,200) to choose 20 senior seats. Give the survey to the students in the selected seats.

(c) With the way students are seated, each column of seats from the stage to the back of the auditorium could be used as a cluster. Note that each cluster contains students from all four grade levels, so each should represent the population well. Since there are 20 clusters, each with 40 seats, we need to choose 2 clusters at random to get 80 students for the survey. Use randInt(1,20) to select two clusters, and then give the survey to all 40 students in each column of seats.

For Practice *Try Exercise 21*

Most large-scale sample surveys use *multistage samples* that combine two or more sampling methods. For example, the Current Population Survey's monthly survey of 60,000 households uses the following design:

Stage 1: Divide the United States into 2007 geographical areas called Primary Sampling Units, or PSUs. Select a sample of 756 PSUs. This sample includes the 428 PSUs with the largest population and a stratified random sample of 328 of the others.

Stage 2: Divide each PSU selected into smaller areas called "neighborhoods." Stratify the neighborhoods using ethnic and other information, and take a stratified random sample of the neighborhoods in each PSU.

Stage 3: Sort the housing units in each neighborhood into clusters of four nearby units. Interview the households in a random sample of these clusters.

Analyzing data from sampling methods more complex than an SRS takes us beyond basic statistics. But the SRS is the building block of more elaborate methods, and the principles of analysis remain much the same for these other methods.

CHECK YOUR UNDERSTANDING

The manager of a sports arena wants to learn more about the financial status of the people who are attending an NBA basketball game. He would like to give a survey to a representative sample of the more than 20,000 fans in attendance. Ticket prices for the game vary a great deal: seats near the court cost over $100 each, while seats in the top rows of the arena cost $25 each. The arena is divided into 30 numbered sections, from 101 to 130. Each section has rows of seats labeled with letters from A (nearest the court) to ZZ (top row of the arena).

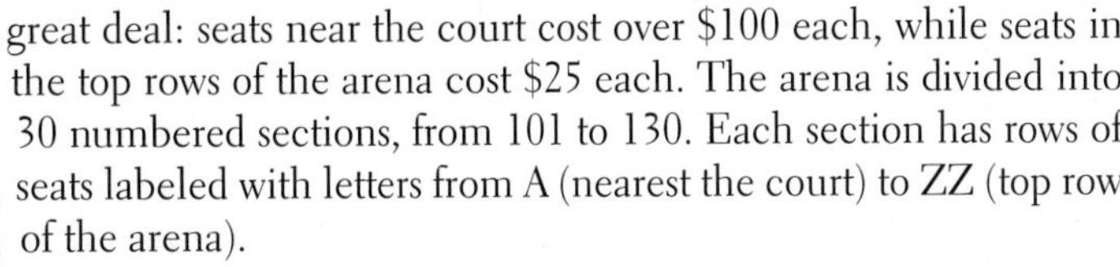

1. Explain why it might be difficult to give the survey to an SRS of 200 fans.
2. Which would be a better way to take a stratified random sample of fans: using the lettered rows or the numbered sections as strata? Explain.
3. Which would be a better way to take a cluster sample of fans: using the lettered rows or the numbered sections as clusters? Explain.

Inference for Sampling

The purpose of a sample is to give us information about a larger population. The process of drawing conclusions about a population on the basis of sample data is called **inference** because we *infer* information about the population from what we *know* about the sample. Inference from convenience samples or voluntary response samples would be misleading because these methods of choosing a sample are biased. We are almost certain that the sample does *not* fairly represent the population. *The first reason to rely on random sampling is to eliminate bias in selecting samples from the list of available individuals.*

Inference

Still, it is unlikely that results from a random sample are exactly the same as for the entire population. Sample results, like the unemployment rate obtained from the monthly Current Population Survey, are only estimates of the truth about the population. If we select two samples at random from the same population, we will almost certainly choose different individuals. So the sample results will differ somewhat, just by chance. Properly designed samples avoid systematic bias. But their results are rarely exactly correct, and we expect them to vary from sample to sample.

ACTIVITY *Results may vary . . .*

MATERIALS: At least 200 colored chips with exactly 60% of a particular color (say, red); large bag or other container that students can't see through

In this Activity, you will explore how the results of repeated random samples vary in relation to the population truth.

1. Your teacher will prepare a large population of colored chips with 60% of a particular color.

2. One at a time, each student in the class will take an SRS of 20 chips and record the proportion of chips obtained that are red. The chips should be returned to the container before the next student chooses a sample.

3. Make a class dotplot of the sample proportions for this color of chip. Where is the graph centered? How much do the sample proportions vary around the center? Is there a clear shape?

Why can we trust random samples? As the previous Activity illustrates, the results of random sampling don't change haphazardly from sample to sample. Because we deliberately use chance, the results obey the laws of probability that govern chance behavior. These laws allow us to say how likely it is that sample results are close to the truth about the population. *The second reason to use random sampling is that the laws of probability allow trustworthy inference about the population.* Results from random samples come with a **margin of error** that sets bounds on the size of the likely error. We will discuss the details of inference for sampling later.

Margin of error

One point is worth making now: *larger random samples give better information about the population than smaller samples.* For instance, in the Activity, you would get a better estimate of the proportion of red chips in the population using random samples of 40 chips than 20 chips. By taking a very large sample, you can be confident that the sample result is very close to the truth about the population.

The Current Population Survey contacts about 60,000 households, so we'd expect its estimate of the national unemployment rate to be within about 0.1% of the actual population value. Opinion polls that contact 1000 or 1500 people give less precise results—we expect the sample result to be within about 3% of the actual population percent with a given opinion. Of course, only samples chosen by chance carry this guarantee. Lou Dobbs's online sample tells us little about overall American public opinion even though 7350 people clicked a response.

Sample Surveys: What Can Go Wrong?

Random sampling eliminates bias in choosing a sample. But even a large random sample will give a result that differs from the truth about the population. This "sampling variability" is described by the margin of error that comes with most poll results. So once we see the magic words *randomly selected*, do we know we have trustworthy information? It certainly beats voluntary response, but not always by as much as we might hope. Sampling in the real world is more complex and less reliable than choosing an SRS from a list of names in a textbook exercise.

Most sample surveys are affected by errors in addition to sampling variability. These errors can introduce bias that makes a survey result meaningless. Good sampling technique includes the art of reducing all sources of error. Let's look at the two main sources of errors in sample surveys: sampling errors and nonsampling errors.

Sampling errors The margin of error tells us how much sampling variability to expect, and we can control it by choosing the size of our random sample. It doesn't tell us about sampling errors, mistakes made in the process of taking a sample that could lead to inaccurate information about the population. One source of sampling error is the use of *bad sampling methods*, such as voluntary response. We can avoid bad methods. Other sampling errors are not so easy to handle. Sampling often begins with a list of individuals from which we will draw our sample. This list is called the **sampling frame.** Ideally, the sampling frame should list every individual in the population. Because a list of the entire population is rarely available, most samples suffer from some degree of **undercoverage.**

Sampling frame

DEFINITION: Undercoverage

Undercoverage occurs when some groups in the population are left out of the process of choosing the sample.

A sample survey of households, for example, will miss not only homeless people but prison inmates and students in dormitories. An opinion poll conducted by calling landline telephone numbers will miss households that have only cell phones as well as households without a phone. The results of national sample surveys therefore have some bias due to undercoverage if the people not covered differ systematically from the rest of the population.

Sampling errors in careful sample surveys are usually quite small. The real problems start when someone picks up (or doesn't pick up) the phone. Now nonsampling errors take over.

Nonsampling errors Nonsampling errors are those that can plague even a census. One of the most serious sources of bias in sample surveys is **nonresponse**, which occurs when a selected individual cannot be contacted or refuses to cooperate. Nonresponse to sample surveys often exceeds 50%, even with careful planning and several callbacks. Because nonresponse is higher in urban areas, most sample surveys substitute other people in the same area to avoid favoring rural areas in the final sample. If the people contacted differ from those who are rarely at home or who refuse to answer questions, some bias remains. For example, retired people may be more likely to respond, which would give their opinions more weight. In a poll about Social Security reform, this could give a misleading impression of the population's views.

DEFINITION: Nonresponse

Nonresponse occurs when an individual chosen for the sample can't be contacted or refuses to participate.

CAUTION

Some students misuse the term "voluntary response" to explain why certain individuals don't respond in a sample survey. Their idea is that participation in the survey is optional (voluntary), so anyone can refuse to take part. What the students are actually describing is nonresponse. Think about it this way: nonresponse can occur only after a sample has been selected. In a voluntary response sample, every individual has opted to take part, so there won't be any nonresponse.

The ACS, GSS, and Opinion Polls

How bad is nonresponse?

The Census Bureau's American Community Survey (ACS) has the lowest nonresponse rate of any poll we know: only about 1% of the households in the sample refuse to respond. The overall nonresponse rate, including "never at home" and other causes, is just 2.5%.[8] This monthly survey of about 250,000 households replaces the "long form" that in the past was sent to some households in the every-ten-years national census. Participation in the ACS is mandatory, and the Census Bureau follows up by telephone and then in person if a household doesn't return the mail questionnaire.

The University of Chicago's General Social Survey (GSS) is the nation's most important social science survey (see Figure 4.2). The GSS contacts its sample in person, and it is run by a university. Despite these advantages, its most recent survey had a 30% rate of nonresponse.

FIGURE 4.2 The home page of the General Social Survey at the University of Chicago's National Opinion Research Center. The GSS has tracked opinions about a wide variety of issues since 1972.

What about opinion polls by news media and opinion-polling firms? We don't know their rates of nonresponse because they won't say. That's a bad sign. The Pew Research Center for the People and the Press imitated a careful random digit dialing survey and published the results: over 5 days, the survey reached 76% of the households in its chosen sample, but "because of busy schedules, skepticism and outright refusals, interviews were completed in just 38% of households that were reached." Combining households that could not be contacted with those who did not complete the interview gave a nonresponse rate of 73%.[9]

Response bias

Another type of nonsampling error occurs when someone gives an incorrect response. A systematic pattern of incorrect responses in a sample survey leads to **response bias.** People know that they should take the trouble to vote, for example, so many who didn't vote in the last election will tell an interviewer that they did. The race or gender of the interviewer can influence responses to questions about race relations or attitudes toward feminism. Answers to questions that ask people to recall past events are often inaccurate because of faulty memory. "Have you visited a dentist in the last 6 months?" will often draw a "Yes" from someone who last visited a dentist 8 months ago.[10] Careful training of interviewers and careful supervision to avoid variation among the interviewers can reduce response bias. Good interviewing technique is another aspect of a well-done sample survey.

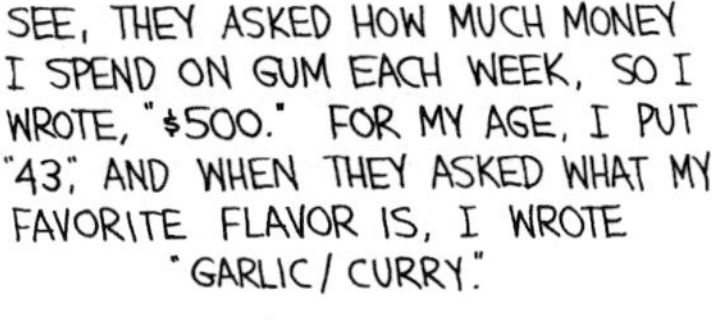

Wording of questions

The **wording of questions** is the most important influence on the answers given to a sample survey. Confusing or leading questions can introduce strong bias, and changes in wording can greatly change a survey's outcome. Even the order in which questions are asked matters.

EXAMPLE

How Do Americans Feel about Illegal Immigrants?

Question wording matters

"Should illegal immigrants be prosecuted and deported for being in the U.S. illegally, or shouldn't they?" Asked this question in an opinion poll, 69% favored deportation. But when the very same sample was asked whether illegal immigrants who have worked in the United States for two years "should be given a chance to keep their jobs and eventually apply for legal status," 62% said that they should. Different questions give quite different impressions of attitudes toward illegal immigrants.

Don't trust the results of a sample survey until you have read the exact questions asked. The amount of nonresponse and the date of the survey are also important. Good statistical design is a part, but only a part, of a trustworthy survey.

THINK ABOUT IT

Does the order matter? Ask a sample of college students these two questions:

"How happy are you with your life in general?" (Answers on a scale of 1 to 5)
"How many dates did you have last month?"

There is almost no association between responses to the two questions when asked in this order. It appears that dating has little to do with happiness. Reverse the order of the questions, however, and a much stronger association appears: college students who say they had more dates tend to give higher ratings of happiness about life. Asking a question that brings dating to mind makes dating success a big factor in happiness.

CHECK YOUR UNDERSTANDING

1. Each of the following is a source of error in a sample survey. Label each as *sampling error* or *nonsampling error,* and explain your answers.

(a) The telephone directory is used as a sampling frame.

(b) The person cannot be contacted in five calls.

(c) Interviewers choose people walking by on the sidewalk to interview.

2. A survey paid for by makers of disposable diapers found that 84% of the sample opposed banning disposable diapers. Here is the actual question:

> It is estimated that disposable diapers account for less than 2% of the trash in today's landfills. In contrast, beverage containers, third-class mail and yard wastes are estimated to account for about 21% of the trash in landfills. Given this, in your opinion, would it be fair to ban disposable diapers?[11]

Explain how the wording of the question could result in bias. Be sure to specify the direction of the bias.

SECTION 4.1 Summary

- A **sample survey** selects a **sample** from the **population** of all individuals about which we desire information. We base conclusions about the population on data from the sample. It is important to specify exactly what population you are interested in and what variables you will measure.
- **Random sampling** uses chance to select a sample.
- The basic random sampling method is a **simple random sample (SRS).** An SRS gives every possible sample of a given size the same chance to be chosen. Choose an SRS by labeling the members of the population and using **random digits** to select the sample. Technology can automate this process.
- To choose a **stratified random sample,** divide the population into **strata,** groups of individuals that are similar in some way that might affect their responses. Then choose a separate SRS from each stratum.
- To choose a **cluster sample,** divide the population into groups, or **clusters.** Randomly select some of these clusters. All the individuals in the chosen clusters are included in the sample.
- Failure to use random sampling often results in **bias,** or systematic errors in the way the sample represents the population. **Voluntary response samples,** in which the respondents choose themselves, and **convenience samples,** in which individuals close by are included in the sample, are particularly prone to large bias.
- **Sampling errors** come from the act of choosing a sample. Random sampling error and **undercoverage** are common types of sampling error. Undercoverage occurs when some members of the population are left out of the **sampling frame,** the list from which the sample is actually chosen.
- The most serious errors in most careful surveys, however, are **nonsampling errors.** These have nothing to do with choosing a sample—they are present even in a census. The single biggest problem for sample surveys is **nonresponse:** people can't be contacted or refuse to answer. Incorrect answers by respondents can lead to **response bias.** Finally, the exact **wording of questions** has a big influence on the answers.

4.1 TECHNOLOGY CORNER

TI-Nspire instructions in Appendix B

SECTION 4.1 Exercises

1. **Students as customers** A high school's student newspaper plans to survey local businesses about the importance of students as customers. From telephone book listings, the newspaper staff chooses 150 businesses at random. Of these, 73 return the questionnaire mailed by the staff. Identify the population and the sample.

pg 208

2. **Student archaeologists** An archaeological dig turns up large numbers of pottery shards, broken stone tools, and other artifacts. Students working on the project classify each artifact and assign it a number. The counts in different categories are important for understanding the site, so the project director chooses 2% of the artifacts at random and checks the students' work. Identify the population and the sample.

3. **Sampling stuffed envelopes** A large retailer prepares its customers' monthly credit card bills using an automatic machine that folds the bills, stuffs them into envelopes, and seals the envelopes for mailing. Are the envelopes completely sealed? Inspectors choose 40 envelopes from the 1000 stuffed each hour for visual inspection. Identify the population and the sample.

4. **Customer satisfaction** A department store mails a customer satisfaction survey to people who make credit card purchases at the store. This month, 45,000 people made credit card purchases. Surveys are mailed to 1000 of these people, chosen at random, and 137 people return the survey form. Identify the population and the sample.

5. **Call the shots** A newspaper advertisement for an upcoming TV show said: "Should handgun control be tougher? You call the shots in a special call-in poll tonight. If yes, call 1-900-720-6181. If no, call 1-900-720-6182. Charge is 50 cents for the first minute." Explain why this opinion poll is almost certainly biased.

pg 210

6. **Explain it to the congresswoman** You are on the staff of a member of Congress who is considering a bill that would provide government-sponsored insurance for nursing-home care. You report that 1128 letters have been received on the issue, of which 871 oppose the legislation. "I'm surprised that most of my constituents oppose the bill. I thought it would be quite popular," says the congresswoman. Are you convinced that a majority of the voters oppose the bill? How would you explain the statistical issue to the congresswoman?

7. **Instant opinion** A recent online poll posed the question "Should female athletes be paid the same as men for the work they do?" In all, 13,147 (44%) said "Yes," 15,182 (50%) said "No," and the remaining 1448 said "Don't know." In spite of the large sample size for this survey, we can't trust the result. Why not?

8. **Online polls** In June 2008, *Parade* magazine posed the following question: "Should drivers be banned from using all cell phones?" Readers were encouraged to vote online at parade.com. The July 13, 2008, issue of *Parade* reported the results: 2407 (85%) said "Yes" and 410 (15%) said "No."

WWW

(a) What type of sample did the *Parade* survey obtain?

(b) Explain why this sampling method is biased. Is 85% probably higher or lower than the true percent of all adults who believe that cell phone use while driving should be banned? Why?

9. **Sleepless nights** How much sleep do high school students get on a typical school night? An interested student designed a survey to find out. To make data collection easier, the student surveyed the first 100 students to arrive at school on a particular morning. These students reported an average of 7.2 hours of sleep on the previous night.

(a) What type of sample did the student obtain?

(b) Explain why this sampling method is biased. Is 7.2 hours probably higher or lower than the true average amount of sleep last night for all students at the school? Why?

10. **Sampling at the mall** You have probably seen the mall interviewer, approaching people passing by with clipboard in hand. Explain why even a large sample of mall shoppers would not provide a trustworthy estimate of the current unemployment rate.

11. **Do you trust the Internet?** You want to ask a sample of high school students the question "How much do you trust information about health that you find on the Internet—a great deal, somewhat, not much, or not at all?" You try out this and other questions on a pilot group of 5 students chosen from your class. The class members are listed at top right.

pg 213

(a) Explain how you would use a line of Table D to choose an SRS of 5 students from the following list. Explain your method clearly enough for a classmate to obtain your results.

(b) Use line 107 to select the sample. Show how you use each of the digits.

Anderson	Deng	Glaus	Nguyen	Samuels
Arroyo	De Ramos	Helling	Palmiero	Shen
Batista	Drasin	Husain	Percival	Tse
Bell	Eckstein	Johnson	Prince	Velasco
Burke	Fernandez	Kim	Puri	Wallace
Cabrera	Fullmer	Molina	Richards	Washburn
Calloway	Gandhi	Morgan	Rider	Zabidi
Delluci	Garcia	Murphy	Rodriguez	Zhao

12. **Apartment living** You are planning a report on apartment living in a college town. You decide to select three apartment complexes at random for in-depth interviews with residents.

(a) Explain how you would use a line of Table D to choose an SRS of 3 complexes from the list below. Explain your method clearly enough for a classmate to obtain your results.

(b) Use line 117 to select the sample. Show how you use each of the digits.

Ashley Oaks	Chauncey Village	Franklin Park	Richfield
Bay Pointe	Country Squire	Georgetown	Sagamore Ridge
Beau Jardin	Country View	Greenacres	Salem Courthouse
Bluffs	Country Villa	Lahr House	Village Manor
Brandon Place	Crestview	Mayfair Village	Waterford Court
Briarwood	Del-Lynn	Nobb Hill	Williamsburg
Brownstone	Fairington	Pemberly Courts	
Burberry	Fairway Knolls	Peppermill	
Cambridge	Fowler	Pheasant Run	

13. **Sampling the forest** To gather data on a 1200-acre pine forest in Louisiana, the U.S. Forest Service laid a grid of 1410 equally spaced circular plots over a map of the forest. A ground survey visited a sample of 10% of these plots.[12]

(a) Explain how you would use technology or Table D to choose an SRS of 141 plots. Your description should be clear enough for a classmate to obtain your results.

(b) Use your method from (a) to choose the first 3 plots.

14. **Sampling gravestones** The local genealogical society in Coles County, Illinois, has compiled records on all 55,914 gravestones in cemeteries in the county for the years 1825 to 1985. Historians plan to use these records to learn about African Americans in Coles County's history. They first choose an SRS of 395 records to check their accuracy by visiting the actual gravestones.[13]

(a) Explain how you would use technology or Table D to choose the SRS. Your description should be clear enough for a classmate to obtain your results.

(b) Use your method from (a) to choose the first 3 gravestones.

15. **Random digits** In using Table D repeatedly to choose random samples, you should not always begin at the same place, such as line 101. Why not?

16. **Random digits** Which of the following statements are true of a table of random digits, and which are false? Briefly explain your answers.

(a) There are exactly four 0s in each row of 40 digits.

(b) Each pair of digits has chance 1/100 of being 00.

(c) The digits 0000 can never appear as a group, because this pattern is not random.

17. **iPhones** Suppose 1000 iPhones are produced at a factory today. Management would like to ensure that the phones' display screens meet their quality standards before shipping them to retail stores. Since it takes about 10 minutes to inspect an individual phone's display screen, managers decide to inspect a sample of 20 phones from the day's production.

(a) Explain why it would be difficult for managers to inspect an SRS of 20 iPhones that are produced today.

(b) An eager employee suggests that it would be easy to inspect the last 20 iPhones that were produced today. Why isn't this a good idea?

(c) Another employee recommends inspecting every fiftieth iPhone that is produced. Explain carefully why this sampling method is *not* an SRS.

18. **Dead trees** On the west side of Rocky Mountain National Park, many mature pine trees are dying due to infestation by pine beetles. Scientists would like to use sampling to estimate the proportion of all pine trees in the area that have been infected.

(a) Explain why it wouldn't be practical for scientists to obtain an SRS in this setting.

(b) A possible alternative would be to use every pine tree along the park's main road as a sample. Why is this sampling method biased?

(c) Suppose that a more complicated random sampling plan is carried out, and that 35% of the pine trees in the sample are infested by the pine beetle. Can scientists conclude that 35% of *all* the pine trees on the west side of the park are infested? Why or why not?

19. **Who goes to the convention?** A club has 30 student members and 10 faculty members. The students are

Abel	Fisher	Huber	Miranda	Reinmann
Carson	Ghosh	Jimenez	Moskowitz	Santos
Chen	Griswold	Jones	Neyman	Shaw
David	Hein	Kim	O'Brien	Thompson
Deming	Hernandez	Klotz	Pearl	Utts
Elashoff	Holland	Liu	Potter	Varga

The faculty members are

Andrews	Fernandez	Kim	Moore	West
Besicovitch	Gupta	Lightman	Phillips	Yang

The club can send 4 students and 2 faculty members to a convention. It decides to choose those who will go by random selection. How will you label the two strata? Use Table D, beginning at line 123, to choose a stratified random sample of 4 students and 2 faculty members.

20. **Sampling by accountants** Accountants often use stratified samples during audits to verify a company's records of such things as accounts receivable. The stratification is based on the dollar amount of the item and often includes 100% sampling of the largest items. One company reports 5000 accounts receivable. Of these, 100 are in amounts over $50,000; 500 are in amounts between $1000 and $50,000; and the remaining 4400 are in amounts under $1000. Using these groups as strata, you decide to verify all the largest accounts and to sample 5% of the midsize accounts and 1% of the small accounts. How would you label the two strata from which you will sample? Use Table D, starting at line 115, to select only the first 3 accounts from each of these strata.

21. **Go Blue!** Michigan Stadium, also known as "The Big House," seats over 100,000 fans for a football game. The University of Michigan athletic department plans to conduct a survey about concessions that are sold during games. Tickets are most expensive for seats near the field and on the sideline. The cheapest seats are high up in the end zones (where one of the authors sat as a student). A map of the stadium is shown.

pg 218

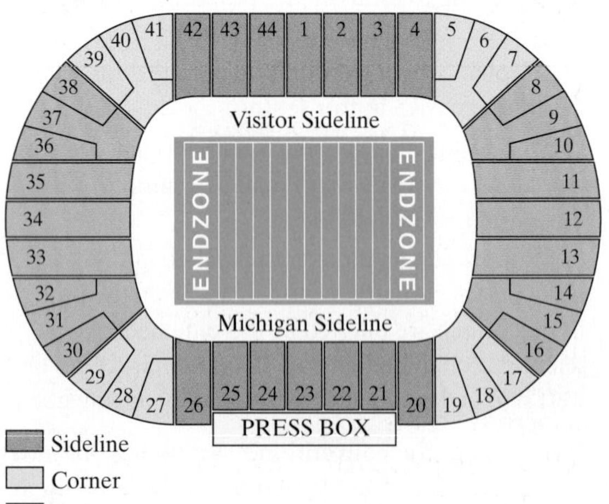

(a) The athletic department is considering a stratified random sample. What would you recommend as the strata? Why?

(b) Explain why a cluster sample might be easier to obtain. What would you recommend for the clusters? Why?

22. **How was your stay?** A hotel has 30 floors with 40 rooms per floor. The rooms on one side of the hotel face the water, while rooms on the other side face a golf course. There is an extra charge for the rooms with a water view. The hotel manager wants to survey 120 guests who stayed at the hotel during a convention about their overall satisfaction with the property.

(a) Explain why choosing a stratified random sample might be preferable to an SRS in this case. What would you use as strata?

(b) Why might a cluster sample be a simpler option? What would you use as clusters?

23. **Is it an SRS?** A corporation employs 2000 male and 500 female engineers. A stratified random sample of 200 male and 50 female engineers gives each engineer 1 chance in 10 to be chosen. This sample design gives every individual in the population the same chance to be chosen for the sample. Is it an SRS? Explain your answer.

24. **Attitudes toward alcohol** At a party there are 30 students over age 21 and 20 students under age 21. You choose at random 3 of those over 21 and separately choose at random 2 of those under 21 to interview about attitudes toward alcohol. You have given every student at the party the same chance to be interviewed: what is the chance? Why is your sample not an SRS?

25. **High-speed Internet** Laying fiber-optic cable is expensive. Cable companies want to make sure that, if they extend their lines out to less dense suburban or rural areas, there will be sufficient demand and the work will be cost-effective. They decide to conduct a survey to determine the proportion of households in a rural subdivision that would buy the service. They select a simple random sample of 5 blocks in the subdivision and survey each family that lives on one of those blocks.

(a) What is the name for this kind of sampling method?

(b) Suppose there are 65 blocks in the subdivision. Use technology or Table D to select 5 blocks to be sampled. Explain your method clearly.

26. **Systematic random sample** Sample surveys often use a systematic random sample to choose a sample of apartments in a large building or housing units in a block at the last stage of a multistage sample. Here is a description of how to choose a systematic random sample.

Suppose that we must choose 4 addresses out of 100. Because 100/4 = 25, we can think of the list as four lists of 25 addresses. Choose 1 of the first 25 addresses at random using Table D. The sample contains this address and the addresses 25, 50, and 75 places down the list from it. If the table gives 13, for example, then the systematic random sample consists of the addresses numbered 13, 38, 63, and 88.

(a) Use Table D to choose a systematic random sample of 5 addresses from a list of 200. Enter the table at line 120.

(b) Like an SRS, a systematic random sample gives all individuals the same chance to be chosen. Explain why this is true. Then explain carefully why a systematic sample is not an SRS.

27. **Sampling frame** Ideally, the sampling frame in a sample survey should list every individual in the population, but in practice, this is often difficult. Suppose that a sample of households in a community is selected at random from the telephone directory. Explain how this sampling method results in undercoverage that could lead to bias.

28. **Sampling frame** Refer to the previous exercise. It is more common in telephone surveys to use random digit dialing equipment that selects the last four digits of a telephone number at random after being given the exchange (the first three digits). Explain how this sampling method results in undercoverage that could lead to bias.

29. **Baseball tickets** Suppose you want to know the average amount of money spent by the fans attending opening day for the Cleveland Indians baseball season. You get permission from the team's management to conduct a survey at the stadium, but they will not allow you to bother the fans in the club seating or box seats (the most expensive seating). Using a computer, you randomly select 500 seats from the rest of the stadium. During the game, you ask the fans in those seats how much they spent that day.

(a) Provide a reason why this survey might yield a biased result.

(b) Explain whether the reason you provided in (a) is a sampling error or a nonsampling error.

30. **What kind of error?** Which of the following are sources of *sampling error* and which are sources of *nonsampling error?* Explain your answers.

(a) The subject lies about past drug use.

(b) A typing error is made in recording the data.

(c) Data are gathered by asking people to mail in a coupon printed in a newspaper.

31. **Nonresponse** A survey of drivers began by randomly sampling all listed residential telephone numbers in the United States. Of 45,956 calls to these numbers, 5029 were completed. The goal of the survey was to estimate how far people drive, on average, per day.[14]

(a) What was the rate of nonresponse for this sample?

(b) Explain how nonresponse can lead to bias in this survey. Be sure to give the direction of the bias.

32. **Ring-no-answer** A common form of nonresponse in telephone surveys is "ring-no-answer." That is, a call is made to an active number but no one answers. The Italian National Statistical Institute looked at nonresponse to a government survey of households in Italy during the periods January 1 to Easter and July 1 to August 31. All calls were made between 7 and 10 P.M., but 21.4% gave "ring-no-answer" in one period versus 41.5% "ring-no-answer" in the other period.[15] Which period do you think had the higher rate of no answers? Why? Explain why a high rate of nonresponse makes sample results less reliable.

33. **Running red lights** The sample described in Exercise 31 produced a list of 5024 licensed drivers. The investigators then chose an SRS of 880 of these drivers to answer questions about their driving habits. One question asked was: "Recalling the last ten traffic lights you drove through, how many of them were red when you entered the intersections?" Of the 880 respondents, 171 admitted that at least one light had been red. A practical problem with this survey is that people may not give truthful answers. What is the likely direction of the bias: do you think more or fewer than 171 of the 880 respondents really ran a red light? Why?

34. **Seat belt use** A study in El Paso, Texas, looked at seat belt use by drivers. Drivers were observed at randomly chosen convenience stores. After they left their cars, they were invited to answer questions that included questions about seat belt use. In all, 75% said they always used seat belts, yet only 61.5% were wearing seat belts when they pulled into the store parking lots.[16] Explain the reason for the bias observed in responses to the survey. Do you expect bias in the same direction in most surveys about seat belt use?

35. **Wording bias** Comment on each of the following as a potential sample survey question. Is the question clear? Is it slanted toward a desired response?

(a) "Some cell phone users have developed brain cancer. Should all cell phones come with a warning label explaining the danger of using cell phones?"

(b) "Do you agree that a national system of health insurance should be favored because it would provide health insurance for everyone and would reduce administrative costs?"

(c) "In view of escalating environmental degradation and incipient resource depletion, would you favor economic incentives for recycling of resource-intensive consumer goods?"

36. **Checking for bias** Comment on each of the following as a potential sample survey question. Is the question clear? Is it slanted toward a desired response?

(a) Which of the following best represents your opinion on gun control?

1. The government should confiscate our guns.
2. We have the right to keep and bear arms.

(b) A freeze in nuclear weapons should be favored because it would begin a much-needed process to stop everyone in the world from building nuclear weapons now and reduce the possibility of nuclear war in the future. Do you agree or disagree?

Multiple choice: Select the best answer for Exercises 37 to 42.

37. The Web portal AOL places opinion poll questions next to many of its news stories. Simply click your response to join the sample. One of the questions in January 2008 was "Do you plan to diet this year?" More than 30,000 people responded, with 68% saying "Yes." You can conclude that

(a) about 68% of Americans planned to diet in 2008.

(b) the poll used a convenience sample, so the results tell us little about the population of all adults.

(c) the poll uses voluntary response, so the results tell us little about the population of all adults.

(d) the sample is too small to draw any conclusion.

(e) None of these.

38. Archaeologists plan to examine a sample of 2-meter-square plots near an ancient Greek city for artifacts visible in the ground. They choose separate random samples of plots from floodplain, coast, foothills, and high hills. What kind of sample is this?

(a) A cluster sample

(b) A convenience sample

(c) A simple random sample

(d) A stratified random sample

(e) A voluntary response sample

39. Your statistics class has 30 students. You want to call an SRS of 5 students from your class to ask where they use a computer for the online exercises. You label the students 01, 02, . . . , 30. You enter the table of random digits at this line:

14459 26056 31424 80371 65103 62253 22490 61181

Your SRS contains the students labeled

(a) 14, 45, 92, 60, 56.

(b) 14, 31, 03, 10, 22.

(c) 14, 03, 10, 22, 22.

(d) 14, 03, 10, 22, 06.

(e) 14, 03, 10, 22, 11.

40. When we take a census, we attempt to collect data from

(a) a stratified random sample.

(b) every individual selected in an SRS.

(c) every individual in the population.

(d) a voluntary response sample.

(e) a convenience sample.

41. An example of a nonsampling error that can reduce the accuracy of a sample survey is

(a) using voluntary response to choose the sample.

(b) using the telephone directory as the sampling frame.

(c) interviewing people at shopping malls to obtain a sample.

(d) variation due to chance in choosing a sample at random.

(e) inability to contact many members of the sample.

42. A simple random sample of 1200 adult Americans is selected, and each person is asked the following question: "In light of the huge national deficit, should the government at this time spend additional money to establish a national system of health insurance?" Only 39% of those responding answered "Yes." This survey

(a) is reasonably accurate since it used a large simple random sample.

(b) needs to be larger since only about 24 people were drawn from each state.

(c) probably understates the percent of people who favor a system of national health insurance.

(d) is very inaccurate but neither understates nor overstates the percent of people who favor a system of national health insurance. Since simple random sampling was used, it is unbiased.

(e) probably overstates the percent of people who favor a system of national health insurance.

43. **Sleep debt** (3.2) A researcher reported that the average teenager needs 9.3 hours of sleep per night but gets only 6.3 hours.[17] By the end of a 5-day school week, a teenager would accumulate about 15 hours of "sleep debt." Students in a high school statistics class were skeptical, so they gathered data on the amount of sleep debt (in hours) accumulated over time (in days) by a random sample of 25 high school students. The resulting least-squares regression equation for their data is $\widehat{\text{Sleep debt}} = 2.23 + 3.17(\text{days})$. Do the students have reason to be skeptical of the research study's reported results? Explain.

44. **Internet charges** (2.1) Some Internet service providers (ISPs) charge companies based on how much bandwidth they use in a month. One method that ISPs use for calculating bandwidth is to find the 95th percentile of a company's usage based on samples of hundreds of 5-minute intervals during a month.

(a) Explain what "95th percentile" means in this setting.

(b) Which would cost a company more: the 95th percentile method or a similar approach using the 98th percentile? Justify your answer.

4.2 Experiments

In Section 4.2, you'll learn about:

- Observational study versus experiment
- The language of experiments
- How to experiment badly
- How to experiment well: The randomized comparative experiment
- Three principles of experimental design
- Experiments: What can go wrong?
- Inference for experiments
- Blocking
- Matched pairs design

A sample survey aims to gather information about a population without disturbing the population in the process. Sample surveys are one kind of **observational study.** Other observational studies watch the behavior of animals in the wild or the interactions between teacher and students in the classroom. This section is about statistical designs for **experiments,** a very different way to produce data.

Observational Study versus Experiment

In contrast to observational studies, experiments don't just observe individuals or ask them questions. They actively impose some *treatment* in order to measure the response. Experiments can answer questions like "Does aspirin reduce the chance of a heart attack?" and "Can yoga help dogs live longer?"

DEFINITION: Observational study and experiment

An **observational study** observes individuals and measures variables of interest but does not attempt to influence the responses.

An **experiment** deliberately imposes some treatment on individuals to measure their responses.

The goal of an observational study can be to describe some group or situation, to compare groups, or to examine relationships between variables. The purpose of an experiment is to determine whether the treatment causes a change in the response. An observational study, even one based on a random sample, is a poor way to gauge the effect that changes in one variable have on another variable. To see the response to a change, we must actually impose the change. *When our goal is to understand cause and effect, experiments are the only source of fully convincing data.* For this reason, the distinction between observational study and experiment is one of the most important in statistics.

EXAMPLE

Does Taking Hormones Reduce Heart Attack Risk after Menopause?

Observation versus experiment

Should women take hormones such as estrogen after menopause, when natural production of these hormones ends? In 1992, several major medical organizations said "Yes." Women who took hormones seemed to reduce their risk of a heart attack by 35% to 50%. The risks of taking hormones appeared small compared with the benefits.

The evidence in favor of hormone replacement came from a number of *observational studies* that compared women who were taking hormones with others who were not. But the women who chose to take hormones were richer and better educated and saw doctors more often than women who didn't take hormones. Because the women who took hormones did many other things to maintain their health, it isn't surprising that they had fewer heart attacks.

To get convincing data on the link between hormone replacement and heart attacks, we should do an *experiment.* Experiments don't let women decide what to do. They assign women to either hormone replacement or to dummy pills that look and taste the same as the hormone pills. The assignment is done by a coin toss, so that all kinds of women are equally likely to get either treatment. By 2002, several experiments with women of different ages agreed that hormone replacement does *not* reduce the risk of heart attacks. The National Institutes of Health, after reviewing the evidence, concluded that the first studies were wrong. Taking hormones after menopause quickly fell out of favor.[18]

From Chapter 3: A **response variable** measures an outcome of a study. An **explanatory variable** may help explain or influence changes in a response variable.

For each of these studies, the *explanatory variable* was whether a woman took hormones, and the *response variable* was whether the woman had a heart attack. Researchers wanted to argue that changes in the explanatory variable (taking hormones or not) actually caused changes in the response variable (having a heart attack or not). In the early observational studies, however, the effect of taking hormones was mixed up with the characteristics of women who chose to take them. These characteristics are **lurking variables** that make it hard to see the true relationship between the explanatory and response variables.

DEFINITION: Lurking variable

A **lurking variable** is a variable that is not among the explanatory or response variables in a study but that may influence the response variable.

Let's consider two lurking variables from the observational studies of hormone replacement: number of doctor visits per year and age. The women who chose to take hormones visited their doctors more often than the women who didn't take hormones. Did the women in the hormone group have fewer heart attacks because they got better health care or because they took hormones? We can't be sure. A situation like this, in which the effects of two variables on a response variable cannot be separated from each other, is called **confounding.**

Some people call a lurking variable that results in confounding, like the number of doctor visits per year in this case, a *confounding variable.*

What about age? Older women are at greater risk of having a heart attack than younger women. If the women who took hormones were generally younger than those who didn't, we'd have more confounding. That wasn't the case, however. There was no link between age and group membership (hormones or not) in the observational studies. *With no association between the lurking variable and the explanatory variable, there can be no confounding.*

DEFINITION: Confounding

Confounding occurs when two variables are associated in such a way that their effects on a response variable cannot be distinguished from each other.

AP EXAM TIP If you are asked to identify a possible confounding variable in a given setting, you are expected to explain how the variable you choose (1) is associated with the explanatory variable and (2) affects the response variable.

CAUTION

Observational studies of the effect of one variable on another often fail because of confounding between the explanatory variable and one or more lurking variables. Well-designed experiments take steps to prevent confounding. The later hormone therapy experiments avoided confounding by letting chance decide who took hormones and who didn't. That way, women who took better care of themselves were split about evenly between the two groups. So were older women and younger women. When these experiments found no reduction in heart attack risk for women taking hormones, researchers began to doubt the results of the earlier observational studies. The moral of the story is simple: *beware the lurking variable!*

CHECK YOUR UNDERSTANDING

1. Does reducing screen brightness increase battery life in laptop computers? To find out, researchers obtained 30 new laptops of the same brand. They chose 15 of the computers at random and adjusted their screens to the brightest setting. The other 15 laptop screens were left at the default setting—moderate brightness. Researchers then measured how long each machine's battery lasted. Was this an observational study or an experiment? Justify your answer.

Questions 2 to 4 refer to the following setting. Does eating dinner with their families improve students' academic performance? According to an ABC News article, "Teenagers who eat with their families at least five times a week are more likely to get better grades in school."[19] This finding was based on a sample survey conducted by researchers at Columbia University.

2. Was this an observational study or an experiment? Justify your answer.

3. What are the explanatory and response variables?

4. Explain clearly why such a study cannot establish a cause-and-effect relationship. Suggest a lurking variable that may be confounded with whether families eat dinner together.

The Language of Experiments

An experiment is a statistical study in which we actually do something (a **treatment**) to people, animals, or objects (the **experimental units**) to observe the response. Here is the basic vocabulary of experiments.

DEFINITION: Treatment, experimental units, subjects

A specific condition applied to the individuals in an experiment is called a **treatment.** If an experiment has several explanatory variables, a treatment is a combination of specific values of these variables.

The **experimental units** are the smallest collection of individuals to which treatments are applied. When the units are human beings, they often are called **subjects.**

The best way to learn the language of experiments is to practice using it.

EXAMPLE

When Will I Ever Use This Stuff?

Vocabulary of experiments

Researchers at the University of North Carolina were concerned about the increasing dropout rate in the state's high schools, especially for low-income students. Surveys of recent dropouts revealed that many of these students had started to lose interest during middle school. They said they saw little connection between what they were studying in school and their future plans. To change this perception, researchers developed a program called CareerStart. The big idea of the program is that teachers show students how the topics they learn get used in specific careers.

To test the effectiveness of CareerStart, the researchers recruited 14 middle schools in Forsyth County to participate in an experiment. Seven of the schools, chosen at random, used CareerStart along with the district's standard curriculum. The other seven schools just followed the standard curriculum. Researchers followed both groups of students for several years, collecting data on students' attendance, behavior, standardized test scores, level of engagement in school, and whether the students graduated from high school. *Results:* Students at schools that used CareerStart generally had better attendance and fewer discipline problems, earned higher test scores, reported greater engagement in their classes, and were more likely to graduate.[20]

Note that the experimental units in the CareerStart example are the schools, not individual students. Experimental units are the smallest collection of individuals to which treatments are applied. The curricular treatments were administered to entire schools, so those are the experimental units.

PROBLEM: **Identify the experimental units, explanatory and response variables, and the treatments in the CareerStart experiment.**

SOLUTION: **The *experimental units* are 14 middle schools in Forsyth County, NC. The *explanatory variable* is whether the school used the CareerStart program with its students. This experiment compares two *treatments:* (1) the standard middle school curriculum and (2) the standard curriculum plus CareerStart. Several *response variables* were measured, including test scores, attendance, behavior, student engagement, and graduation rates.**

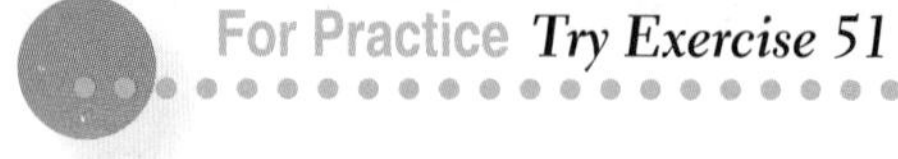

For Practice *Try Exercise 51*

The previous example illustrates the big advantage of experiments over observational studies: *experiments can give good evidence for causation.* In an experiment, we study the effects of the specific treatments we are interested in, while trying to control for the effect of lurking variables. For instance, the students in all 14 schools followed the standard curriculum. To ensure that the two groups of schools were as similar as possible before the treatments were administered, researchers let chance decide which 7 schools would use CareerStart. The only systematic difference between the schools was the educational treatment. When students from the CareerStart schools did better, we can be confident that the program made the difference.

Factors

Level

Sometimes, the explanatory variables in an experiment are called **factors.** Many experiments study the joint effects of several factors. In such an experiment, each treatment is formed by combining a specific value (often called a **level**) of each of the factors. Here's an example of a multifactor experiment.

TV Advertising

Experiments with multiple explanatory variables

What are the effects of repeated exposure to an advertising message? The answer may depend on both the length of the ad and on how often it is repeated. An experiment investigated this question using undergraduate students as subjects. All subjects viewed a 40-minute television program that included ads for a digital camera. Some subjects saw a 30-second commercial; others, a 90-second version. The same commercial was shown either 1, 3, or 5 times during the program. After viewing, all the subjects answered questions about their recall of the ad, their attitude toward the camera, and their intention to purchase it.[21]

PROBLEM: **For the advertising study,**

(a) identify the explanatory and response variables, and

(b) list all the treatments.

SOLUTION:

(a) This experiment has 2 explanatory variables (factors): length of the commercial and number of repetitions. The response variables include measures of subjects' recall of the ad, their attitudes about the digital camera, and whether they intend to purchase it.

(b) There are 2 different lengths of commercial (30 and 90 seconds) and three different numbers of repetitions (1, 3, and 5). The 6 combinations consisting of one level of each factor form the 6 treatments shown in Figure 4.3: (1) 30 seconds, 1 time; (2) 30 seconds, 3 times; (3) 30 seconds, 5 times; (4) 90 seconds, 1 time; (5) 90 seconds, 3 times; (6) 90 seconds, 5 times.

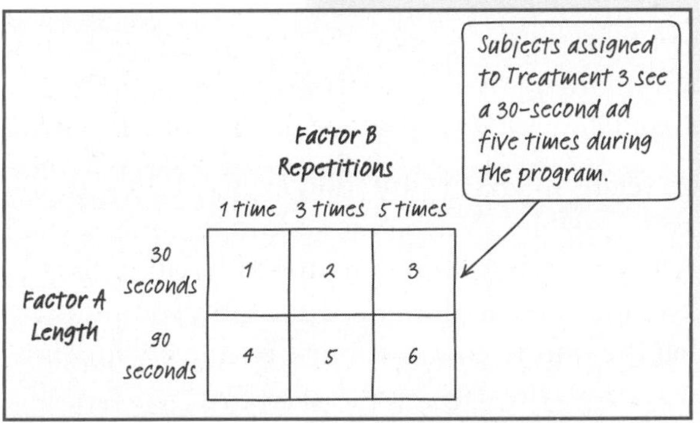

FIGURE 4.3 The six treatments in the TV ad experiment. Combinations of values of the two explanatory variables (factors) form six treatments.

This example shows how experiments allow us to study the combined effect of several factors. The interaction of several factors can produce effects that could not be predicted from looking at the effect of each factor alone. Perhaps longer commercials increase interest in a product, and more commercials also increase interest. But if we both make a commercial longer and show it more often, viewers get annoyed and their interest in the product drops. The two-factor experiment in the TV advertising example will help us find out.

How to Experiment Badly

Experiments are the preferred method for examining the effect of one variable on another. By imposing the specific treatment of interest and controlling other influences, we can pin down cause and effect. Good designs are essential for effective experiments, just as they are for sampling. To see why, let's start with an example of a bad experimental design.

Which Works Better: Online or In-Class SAT Preparation?

A bad experiment

A high school regularly offers a review course to prepare students for the SAT. This year, budget cuts will allow the school to offer only an online version of the course. Over the past 10 years, the average SAT score of students in the classroom course was 1620. The online group gets an average score of 1780. That's roughly 10% higher than the long-time average for those who took the classroom review course. Is the online course more effective?

This experiment has a very simple design. A group of subjects (the students) were exposed to a treatment (the online course), and the outcome (SAT scores) was observed. Here is the design:

Students → Online course → SAT scores

A closer look showed that the students in the online review course were quite different from the students who took the classroom course in past years. They had higher GPAs and were taking more AP classes.

The effect of online versus in-class instruction is mixed up with the effect of these lurking variables. Maybe the online students earned higher SAT scores because they were smarter to begin with, not because the online course prepared them better. This confounding prevents us from concluding that the online course is more effective than classroom instruction.

Many laboratory experiments use a design like the one in the online SAT course example:

Experimental units → Treatment → Measure response

In the lab environment, simple designs often work well. Field experiments and experiments with animals or people deal with more variable conditions. *Outside the lab, badly designed experiments often yield worthless results because of confounding.*

How to Experiment Well: The Randomized Comparative Experiment

The remedy for the confounding in the SAT prep course example is to do a *comparative experiment* in which some students are taught in the classroom and other, similar students take the course online. Most well-designed experiments compare two or more treatments.

Comparison alone isn't enough to produce results we can trust. If the treatments are given to groups that differ greatly when the experiment begins, *bias* will result. For example, if we allow students to select online or classroom instruction, more self-motivated students are likely to sign up for the online course. Allowing personal choice will bias our results in the same way that volunteers bias the results of online opinion polls. The solution to the problem of bias in sampling is random selection. In experiments, the solution is **random assignment.**

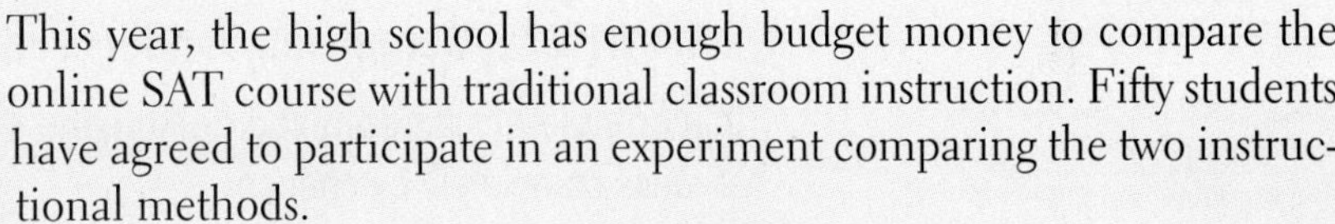

DEFINITION: Random assignment

In an experiment, **random assignment** means that experimental units are assigned to treatments at random, that is, using some sort of chance process.

Let's look at how random assignment can be used to improve the design of the SAT prep course experiment.

EXAMPLE

SAT Prep: Online versus Classroom

How random assignment works

This year, the high school has enough budget money to compare the online SAT course with traditional classroom instruction. Fifty students have agreed to participate in an experiment comparing the two instructional methods.

PROBLEM: Describe how you would randomly assign 25 students to each of the two methods:

(a) Using 50 identical slips of paper

(b) Using Table D

(c) Using technology

SOLUTION:

(a) The simplest way would be to use the "hat method." Write each subject's name on one of the slips. Put all the slips in a hat and mix them thoroughly. Draw them out one at a time until you have 25 slips. These 25 students will take the online course. The remaining 25 students will take the classroom course. Alternatively, you could write "online" on 25 of the slips and "classroom" on the other 25 slips. Then put the slips in a hat and mix them well. Have students come up one by one and (without looking) pick a slip from the hat. This guarantees 25 students per group, with the treatments assigned by chance.

(b) We can use the two-step process from random sampling to do the random assignment.

Step 1: **Label.** Give labels 01, 02, 03, . . . , 49, 50 to the subjects in alphabetical order by last name.

Step 2: **Table.** Go to a line of Table D and read two-digit groups moving from left to right. The first 25 distinct labels between 01 and 50 that you encounter select the online group. The remaining 25 students will take the classroom course. Ignore repeated labels and groups of digits from 51 to 00.

(c) Give numbers 1, 2, 3, . . . , 49, 50 to the subjects in alphabetical order by last name. Then use your calculator's randInt command or a computer's random number generator to produce numbers between 1 and 50. The first 25 different numbers chosen select the students for the online course. The remaining 25 subjects will take the classroom course.

For Practice *Try Exercise 59*

The experimental design in this example is *comparative* because it compares two treatments (the two instructional settings). It is *randomized* because the subjects are assigned to the treatments by chance. The diagram in Figure 4.4 presents the details: random assignment, the sizes of the groups and which treatment they receive, and the response variable. There are, as we will see later, statistical reasons for using treatment groups that are about equal in size. This type of design is called a **completely randomized design.**

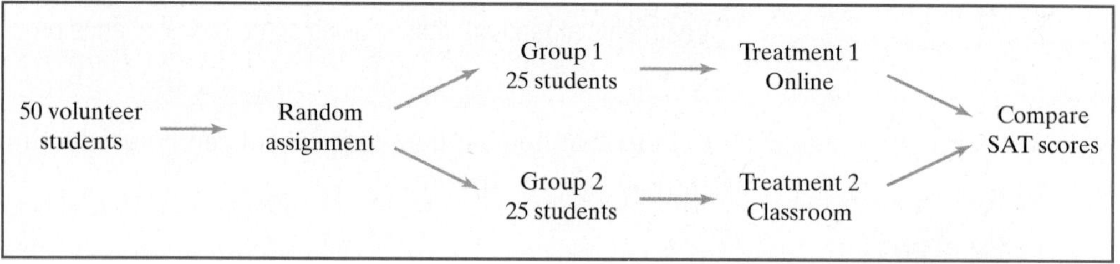

FIGURE 4.4 Outline of a completely randomized design to compare online and classroom instruction.

DEFINITION: Completely randomized design

In a **completely randomized design,** the treatments are assigned to all the experimental units completely by chance.

Notice that the definition of a completely randomized design does not require that each treatment be assigned to an equal number of experimental units. It does specify that the assignment of treatments must occur completely at random.

THINK ABOUT IT

Does using chance to assign treatments in an experiment guarantee a completely randomized design? Actually, no. Let's return to the SAT prep course experiment involving 50 students. Another way to randomly assign the subjects to the two treatments is by tossing a coin. Have each student come forward and toss a coin. If it's heads, then the student will take the course online. If it's tails, then the student will take the classroom course.

As long as all 50 students toss a coin, this is still a completely randomized design. Of course, the two experimental groups are unlikely to contain exactly 25 students each due to the chance variation in coin tosses.

The problem comes if we try to force the two groups to have equal sizes. Suppose we let the coin tossing continue until one of the groups has 25 students and then place the remaining students in the other group. This is no longer a completely randomized design, because the last few students aren't being assigned to one of the treatments by chance. In fact, these students will all end up in the same group, which could lead to bias if these individuals share some characteristic that would systematically affect the response variable. For example, if the students came to toss the coin last because they're lazier than the other students who volunteered, then the SAT prep class that they're in will seem less effective than it really is.

Completely randomized designs can compare any number of treatments. Here is an experiment that compares three treatments.

Conserving Energy

A completely randomized design

Many utility companies have introduced programs to encourage energy conservation among their customers. An electric company considers placing small digital displays in households to show current electricity use and what the cost would be if this use continued for a month. Will the displays reduce electricity use? One cheaper approach is to give customers a chart and information about monitoring their electricity use from their outside meter. Would this method work almost as well? The company decides to conduct an experiment to compare these two approaches (display, chart) with a *control group* of customers who receive information about energy consumption but no help in monitoring electricity use.

PROBLEM: Outline a completely randomized design involving 60 single-family residences in the same city who are willing to participate in such an experiment. Write a few sentences describing how you would implement your design.

SOLUTION: Figure 4.5 outlines the design. We'll randomly assign 20 houses to each of three treatments: digital display, chart plus information, and control. Our response variable is the total amount of electricity used in a year.

AP EXAM TIP If you are asked to describe the design of an experiment on the AP exam, you won't get full credit for a diagram like Figure 4.5. You are expected to describe how the treatments are assigned to the experimental units and to clearly state what will be measured or compared. Some students prefer to start with a diagram and then add a few sentences. Others choose to skip the diagram and put their entire response in narrative form.

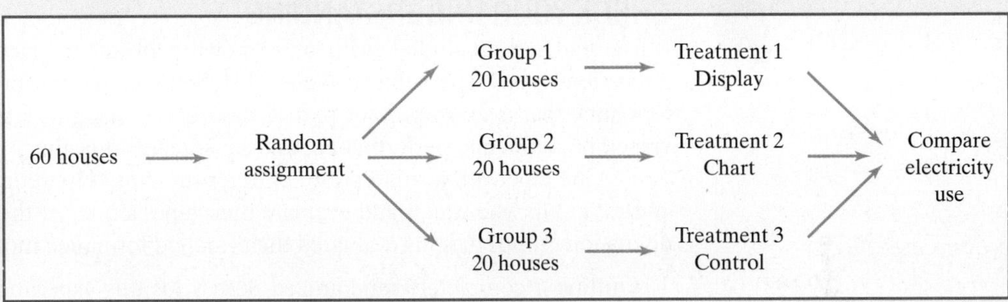

FIGURE 4.5 Outline of a completely randomized design to compare three energy-saving programs.

To implement the design, start by labeling each household with a distinct number from 1 to 60. Write the labels on 60 identical slips of paper, put them in a hat, and mix them well. Draw out 20 slips. The corresponding households will be given digital displays showing current electricity use. Now draw out 20 more slips. Those houses will use a chart. The remaining 20 houses will be given information about energy consumption but no way to monitor their usage. At the end of the year, compare how much electricity was used by the homes in the three groups.

Control group

Why did we include a **control group** of 20 houses in the energy conservation experiment? *The primary purpose of a control group is to provide a baseline for comparing the effects of the other treatments.* Without such a comparison group, we wouldn't be able to tell whether homes with digital displays or charts used less electricity than homes without such aids.

Was a control group really necessary? You might be thinking that the change in electricity use from last year to this year in the houses with displays and charts would tell us whether these treatments helped. Unfortunately, it's not that simple. Suppose last year's temperatures were more extreme than this year's. Then many households might show a decrease in electricity use, but we couldn't be sure whether this change was due to the weather or to the treatments. (Can you say confounding?!)

Many experiments (like the one in the previous example) include a control group that receives an inactive treatment. However, a control group can be given an active treatment. Suppose we want to compare the effectiveness of a newly developed drug for treating a serious disease with a drug that's already known to work. In that case, the experimental units that receive the existing drug form the control group.

Some experimental designs don't include a control group. That's appropriate if researchers simply want to compare the effects of several treatments, and not to determine whether any of them works better than an inactive treatment. For instance, maybe a state's highway department wants to see which of three brands of paint will last longest when marking lane lines on the freeway.

CHECK YOUR UNDERSTANDING

Music students often don't evaluate their own performances accurately. Can small-group discussions help? The subjects were 29 students preparing for the end-of-semester performance that is an important part of their grade. Assign 15 students to the treatment: videotape a practice performance, ask the student to evaluate it, then have the student discuss the tape with a small group of other students. The remaining 14 students form a control group who watch and evaluate their tapes alone. At the end of the semester, the discussion-group students evaluated their final performance more accurately.[22]

1. Outline a completely randomized design for this experiment. Follow the model of Figure 4.4.

2. Describe how you would carry out the random assignment. Provide enough detail that a classmate could implement your procedure.

3. What is the purpose of the control group in this experiment?

Three Principles of Experimental Design

Randomized comparative experiments are designed to give good evidence that differences in the treatments actually *cause* the differences we see in the response. The logic is as follows:

- A proper comparative design ensures that influences other than the experimental treatments operate equally on all groups. This helps control for the effects of lurking variables.
- Random assignment helps balance out the effects of lurking variables that we can't control, or don't think of, on the treatment groups. That is, random assignment forms groups of experimental units that should be similar in all respects before the treatments are applied.

• Therefore, since the groups are roughly equivalent except for the treatments, any differences in average response must be due either to the treatments or to the play of chance in the random assignment of experimental units to the treatments.

The first two bullets give us two basic principles of experimental design: **control** and **random assignment.** That "either-or" in the last bullet deserves some more thought. In the SAT prep course example, we can't say that *any* difference between the average SAT scores of students enrolled online and in the classroom must be caused by a difference in the effectiveness of the two types of instruction. There would be *some* difference even if both groups received the same instruction, because of variation among students in background and study habits. Chance assigns students to one group or the other, and this creates a chance difference between the groups.

We would not trust an experiment with just one student in each group, for example. The results would depend too much on which group got lucky and received the stronger student. If we assign many subjects to each group, however, the effects of chance will average out, and there will be little difference in the average responses in the two groups unless the treatments themselves cause a difference. Our third design principle is **replication:** use enough experimental units to distinguish a difference in the effects of the treatments from chance variation.

Replication

In statistics, replication means "use enough subjects." In other fields, the term "replication" has a different meaning. If one experiment is conducted and then the same or a similar experiment is independently conducted in a different location by different investigators, this is known as replication. That is, in this case, replication means repeatability.

Summary: Principles of Experimental Design

The basic principles for designing experiments are as follows:

1. **Control** for lurking variables that might affect the response: Use a comparative design and ensure that the only systematic difference between the groups is the treatment administered.
2. **Random assignment:** Use impersonal chance to assign experimental units to treatments. This helps create roughly equivalent groups of experimental units by balancing the effects of lurking variables that aren't controlled on the treatment groups.
3. **Replication:** Use enough experimental units in each group so that any differences in the effects of the treatments can be distinguished from chance differences between the groups.

Let's see how these principles were used in designing a famous medical experiment.

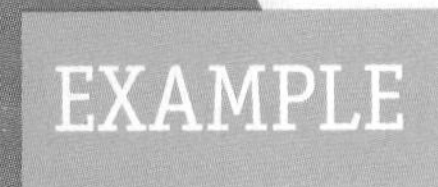

EXAMPLE

The Physicians' Health Study

A well-designed experiment

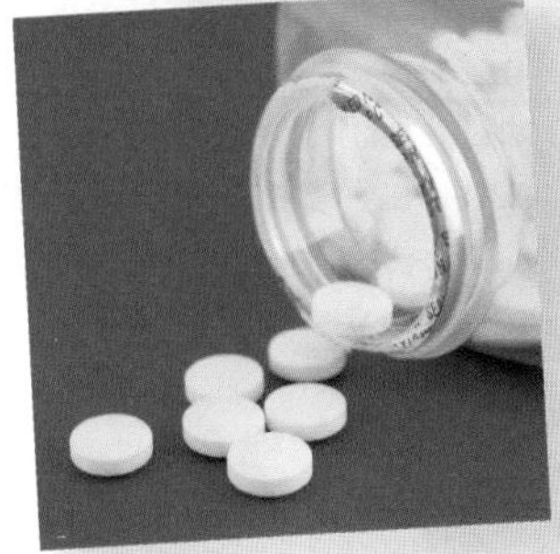

Does regularly taking aspirin help protect people against heart attacks? The Physicians' Health Study was a medical experiment that helped answer this question. In fact, the Physicians' Health Study looked at the effects of two drugs: aspirin and beta-carotene. Researchers wondered whether beta-carotene would help prevent some forms of cancer. The subjects in this experiment were 21,996 male physicians. There were two explanatory variables (factors), each having two levels: aspirin (yes or no) and beta-carotene (yes or no). Combinations of the levels of these factors form the four treatments shown in Figure 4.6. One-fourth of the subjects were assigned at random to each of these treatments.

Placebo

On odd-numbered days, the subjects took either a tablet that contained aspirin or a dummy pill that looked and tasted like the aspirin but had no active ingredient (a **placebo**). On even-numbered days, they took either a capsule containing beta-carotene or a placebo. There were several response variables—the study looked for heart attacks, several kinds of cancer, and other medical outcomes. After several years, 239 of the placebo group but only 139 of the aspirin group had suffered heart attacks. This difference is large enough to give good evidence that taking aspirin does reduce heart attacks.[23] It did not appear, however, that beta-carotene had any effect on preventing cancer.

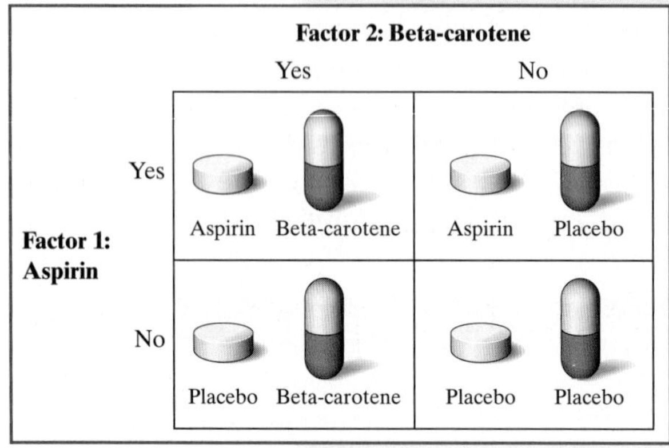

FIGURE 4.6 The treatments in the Physicians' Health Study.

PROBLEM: Explain how each of the three principles of experimental design was used in the Physicians' Health Study.

SOLUTION: Researchers attempted to *control* for the effects of lurking variables by using a design that compared both of the active treatments to a placebo and by having all subjects follow the same schedule of pill taking. *Random assignment* was used to determine which subjects received each of the four treatment combinations. This helped ensure that the treatment groups were roughly equivalent to begin with. There were over 5000 subjects per treatment group. This *replication* helped ensure that any sizable difference in response among the groups was due to the treatments and not to chance variation in the random assignment.

For Practice *Try Exercise 67*

The control group in the Physicians' Health Study received inactive versions (placebos) of both aspirin and beta-carotene. This group served as a comparison for the effects of aspirin alone, beta-carotene alone, and aspirin plus beta-carotene.

Why did researchers decide to do the Physicians' Health Study (PHS)? The interesting history that led to this experiment is detailed at the PHS Web site. You can also find out about the Physicians' Health Study II, which ended in December 2007.

Experiments: What Can Go Wrong?

The logic of a randomized comparative experiment depends on our ability to treat all the subjects the same in every way except for the actual treatments being compared. Good experiments, therefore, require careful attention to details to ensure that all subjects really are treated identically.

If some subjects in a medical experiment take a pill each day and a control group takes no pill, the subjects are not treated identically. Many medical experiments are therefore "placebo-controlled," like the Physicians' Health Study. All the subjects received the same medical attention during the several years of the experiment. On odd-numbered days, all of them took an aspirin or a placebo. On even-numbered days, all subjects took either a beta-carotene pill or a placebo. Many patients respond favorably to any treatment, even a placebo, perhaps because they trust the doctor. The response to a dummy treatment is called the

Placebo effect

placebo effect. If some subjects did not take any pills, the effect of aspirin or beta-carotene would be confounded with the placebo effect, the effect of simply taking pills.

Curing Baldness and Soothing Pain

Do placebos work?

Want to help balding men keep their hair? Give them a placebo—one study found that 42% of balding men maintained or increased the amount of hair on their heads when they took a placebo. In another study, researchers zapped the wrists of 24 test subjects with a painful jolt of electricity. Then they rubbed a cream with no active medicine on subjects' wrists and told them the cream should help soothe the pain. When researchers shocked them again, 8 subjects said they experienced significantly less pain.[24]

When the ailment is vague and psychological, like depression, some experts think that about three-quarters of the effect of the most widely used drugs is just the placebo effect.[25] Others disagree. The strength of the placebo effect in medical treatments is hard to pin down because it depends on the exact environment. How enthusiastic the doctor is seems to matter a lot. But "placebos work" is a good place to start when you think about planning medical experiments.

The strength of the placebo effect is a strong argument for randomized comparative experiments. In the baldness study, 42% of the placebo group kept or increased their hair, but 86% of the men getting a new drug to fight baldness did so. The drug beats the placebo, so it has something besides the placebo effect going for it. Of course, the placebo effect is still part of the reason this and other treatments work.

Because the placebo effect is so strong, it would be foolish to tell subjects in a medical experiment whether they are receiving a new drug or a placebo. Knowing that they are getting "just a placebo" might weaken the placebo effect and bias the experiment in favor of the other treatments. It is also foolish to tell doctors and other medical personnel what treatment each subject received. If they know that a subject is getting "just a placebo," they may expect less than if they know the subject is receiving a promising experimental drug. Doctors' expectations change how they interact with patients and even the way they diagnose a patient's condition. Whenever possible, experiments with human subjects should be **double-blind.**

Reports in medical journals regularly begin with words like these, from a study of a flu vaccine given as a nose spray: "This study was a randomized, double-blind, placebo-controlled trial. Participants were enrolled from 13 sites across the continental United States between mid-September and mid-November."[26] Doctors are supposed to know what this means. Now you know, too.

DEFINITION: Double-blind

In a **double-blind** experiment, neither the subjects nor those who interact with them and measure the response variable know which treatment a subject received.

The idea of a double-blind design is simple. Until the experiment ends and the results are in, only the study's statistician knows for sure which treatment a subject is receiving. However, some experiments cannot be carried out in a double-blind manner. If researchers are comparing the effects of exercise and dieting on weight loss, then subjects will know which treatment they are receiving. Such an experiment can still be **single-blind** if the individuals who are interacting with the subjects and measuring the response variable don't know who is dieting and who is exercising. In other single-blind experiments, the subjects are unaware of which treatment they are receiving, but the people interacting with them and measuring the response variable do know.

Single-blind

CHECK YOUR UNDERSTANDING

In an interesting experiment, researchers examined the effect of ultrasound on birth weight. Pregnant women participating in the study were randomly assigned to one of two groups. The first group of women received an ultrasound; the second group did not. When the subjects' babies were born, their birth weights were recorded. The women who received the ultrasounds had heavier babies.[27]

1. Did the experimental design take the placebo effect into account? Why is this important?
2. Was the experiment double-blind? Why is this important?
3. Based on your answers to Questions 1 and 2, describe an improved design for this experiment.

Inference for Experiments

In an experiment, researchers usually hope to see a difference in the responses so large that it is unlikely to happen just because of chance variation. We can use the laws of probability, which describe chance behavior, to learn whether the treatment effects are larger than we would expect to see if only chance were operating. If they are, we call them **statistically significant.**

DEFINITION: Statistically significant

An observed effect so large that it would rarely occur by chance is called **statistically significant.**

If we observe statistically significant differences among the groups in a randomized comparative experiment, we have good evidence that the treatments actually caused these differences. You will often see the phrase "statistically significant" in published research reports in many fields. The great advantage of randomized comparative experiments is that they can produce data that give good evidence for a cause-and-effect relationship between the explanatory and response variables. We know that in general a strong association does not imply causation. A statistically significant association in data from a well-designed experiment *does* imply causation.

ACTIVITY *Distracted driving*

MATERIALS: Deck of cards for each team of 3 to 4 students

Is talking on a cell phone while driving more distracting than talking to a passenger? David Strayer and his colleagues at the University of Utah designed an experiment to help answer this question. They used 48 undergraduate students as subjects. The researchers randomly assigned half of the subjects to drive in a simulator while talking on a cell phone, and the other half to drive in the simulator while talking to a passenger. One response variable was whether the driver stopped at a rest area that was specified by researchers before the simulation started. The table below shows the results:[28]

	Distraction	
Stopped at rest area?	**Cell phone**	**Passenger**
Yes	12	21
No	12	3

Are these results statistically significant? To find out, let's see what would happen just by chance if we randomly reassign the 48 people in this experiment to the two groups many times, *assuming the treatment received doesn't affect whether a driver stops at the rest area.*

1. We need 48 cards from the deck to represent the drivers in this study. In the original experiment, 33 drivers stopped at the rest area and 15 didn't. Since we're assuming that the treatment received won't change whether each driver stops at the rest area, we use 33 cards to represent drivers who stop and 15 cards to represent those who don't. Remove the ace of spades and any three of the 2s from the deck.
   - Stop: All cards with denominations 2 through 10 (36 – 3 missing 2s = 33)
   - Don't stop: All jacks, queens, kings, and aces (16 – 1 missing ace = 15)
2. Shuffle and deal two piles of 24 cards each—the first pile represents the cell phone group and the second pile represents the passenger group. The shuffling reflects our assumption that the outcome for each subject is not affected by the treatment. Record the number of drivers who failed to stop at the rest area in each group.
3. Repeat this process 9 more times so that you have a total of 10 trials. Record your results in a table like this:

Trial	Number who didn't stop in cell phone group	Number who didn't stop in passenger group
1		

In how many of your 10 trials did 12 or more drivers in the cell phone group fail to stop?

4. Make a class dotplot of the number of drivers in the cell phone group who failed to stop at the rest area in each trial. In what percent of the class's trials did 12 or more people in the cell phone group fail to stop at the rest area?

5. In the original experiment, 12 of the 24 drivers using cell phones didn't stop at the rest area. Based on the class's simulation results, how surprising would it be to get a result this large or larger simply due to the chance involved in the random assignment? Is the result statistically significant?

6. What conclusion would you draw about whether talking on a cell phone is more distracting than talking to a passenger?

THINK ABOUT IT

Can an "unlucky" random assignment lead to confounding? Let's return to the distracted-driver Activity. Some people are more forgetful than others. Suppose that the random assignment happens to put most of the forgetful subjects in one group. If more drivers in that group fail to stop at the rest area, we don't know if it's because of the treatment they received (cell phone or passenger) or their forgetfulness. Is this confounding?

You might be surprised that the answer is "No!" Although people's memory is a lurking variable that might affect whether they stop at the rest area (the response variable), the design of the experiment takes care of this by randomly assigning subjects to the two treatment groups. The "unlucky" random assignments are taken into account in determining statistical significance. In an experiment, confounding occurs when the design doesn't account for existing differences in the experimental units that might systematically affect their response to the treatments.

Blocking

Completely randomized designs are the simplest statistical designs for experiments. They illustrate clearly the principles of control, random assignment, and replication. But just as with sampling, there are times when the simplest method doesn't yield the most precise results. When a population consists of groups of individuals that are "similar within but different between," a stratified random sample gives a better estimate than a simple random sample. This same logic applies in experiments.

EXAMPLE

Doing the Laundry

The idea of blocking

Suppose researchers want to test whether a new detergent for clothes that require hand-washing cleans better in warm or in cold water. They decide to perform an experiment using numerous pieces of dirty laundry as the experimental units. The response variable is a cleanliness rating on a scale of 0 (very dirty) to 10 (very clean). How should researchers deal with the fact that light-colored clothing tends to come cleaner in warm water? They could use a completely randomized design and hope that the random assignment distributes the light-colored clothing about evenly between the cold- and warm-water treatment groups. Even so, there might be a lot of variability in cleanliness ratings in both groups due to the color of the clothing. This

Color is a lurking variable in the completely randomized design. The random assignment prevents confounding due to this variable, however.

lurking variable might make it difficult to detect a difference in the effects of the treatments. What should the researchers do?

Since the researchers know that color may affect cleanliness in a systematic way, they could start by separating the laundry into two piles—one for lighter colors and one for darker colors. Each of these piles of similar experimental units is known as a **block.** Within each block, the researchers could then randomly assign about half the pieces of laundry to be hand-washed in cold water and the other half to be hand-washed in warm water. The same person should do all the washing as a form of control. This **randomized block design** reduces the effect of variation among the experimental units due to color on the response variable.

Figure 4.7 outlines the randomized block design for our laundry experiment. The experimental units are first separated into blocks to deal with the potential confounding variable, color. Then, the two treatments are randomly assigned within each block.

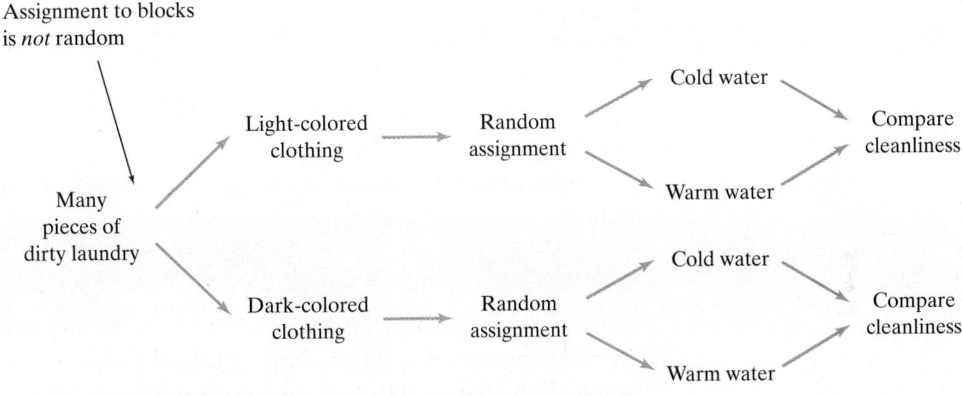

FIGURE 4.7 Outline of a randomized block design for the laundry experiment. The blocks consist of light-colored and dark-colored pieces of laundry. The treatments are warm and cold water.

DEFINITION: Block and randomized block design

A **block** is a group of experimental units that are known before the experiment to be similar in some way that is expected to affect the response to the treatments.

In a **randomized block design,** the random assignment of experimental units to treatments is carried out separately within each block.

Blocks are another form of *control.* They control the effects of some outside variables by bringing those variables into the experiment to form the blocks. A randomized block design allows us to draw separate conclusions about each block, for example, about light-colored and dark-colored clothes in the laundry experiment. Blocking also allows more precise overall conclusions, because the

AP EXAM TIP Don't mix the language of experiments and the language of sample surveys or other observational studies. You will lose credit for saying things like "use a randomized block design to select the sample for this survey" or "this experiment suffers from nonresponse since some subjects dropped out during the study."

systematic differences between light- and dark-colored clothes can be removed when we study the overall effects of using this detergent in warm and cold water.

The idea of blocking is an important additional principle of experimental design. A wise experimenter will form blocks based on the most important unavoidable sources of variability (lurking variables) among the experimental units. Randomization will then average out the effects of the remaining lurking variables and allow an unbiased comparison of the treatments. The moral of the story is: *control what you can, block on what you can't control, and randomize to create comparable groups.*

Men, Women, and Advertising

Blocking in an experiment

Women and men respond differently to advertising. Researchers would like to design an experiment to compare the effectiveness of three advertisements for the same product.

PROBLEM:

(a) Explain why a randomized block design might be preferable to a completely randomized design for this experiment.

(b) Outline a randomized block design using 300 volunteers (180 men and 120 women) as subjects. Describe how you would carry out the random assignment required by your design.

SOLUTION:

(a) A completely randomized design considers all subjects, both men and women, as a single pool. The random assignment would send subjects to three treatment groups without regard to their gender. This ignores the differences between men and women, which would probably result in a great deal of variability in responses to the advertising in all three groups. For example, if an ad appealed

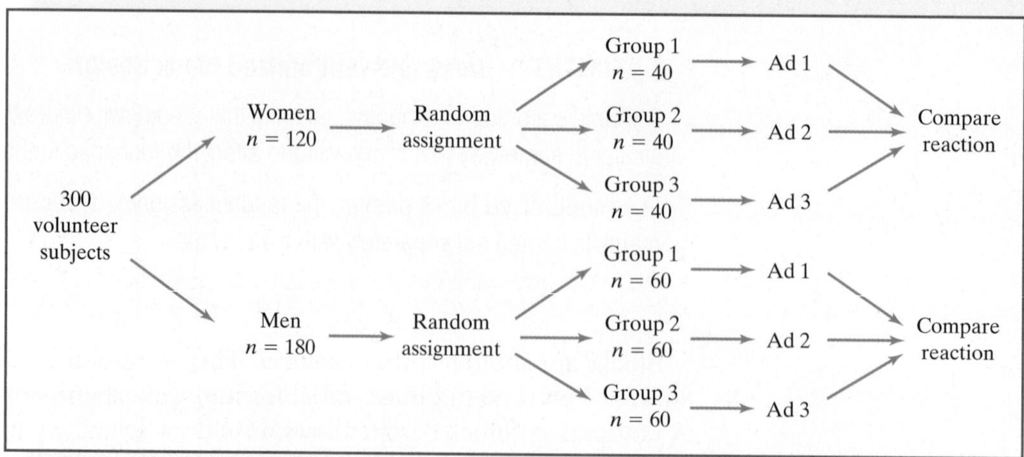

FIGURE 4.8 Randomized block design for comparing responses to three advertisements. The blocks consist of male and female subjects.

much more to men, you would get a wide range of reactions to that ad from the two genders. That would make it harder to compare the effectiveness of the ads.

A randomized block design would consider women and men separately. In this case, the random assignment would occur separately in each block. Blocking will control for the variability in responses to advertising due to gender. This will allow researchers to look separately at the reactions of men and women, as well as to more effectively assess the overall response to the ads.

(b) Figure 4.8 outlines the randomized block design. We randomly assign the 120 women into three groups of 40, one for each of the advertising treatments. Give each woman a distinct label between 1 and 120. Use a computer's random number generator to sort the numbers from 1 to 120 in a random order. The women with labels corresponding to the first 40 numbers in the list will view Ad 1, the next 40 will view Ad 2, and the last 40 will view Ad 3. Randomly assign the 180 men into three groups of 60 using a similar process. After each subject has viewed the assigned ad, compare reactions to the three ads within the gender blocks.

For Practice *Try Exercise 79*

Matched Pairs Design

Matched pairs design

A common type of randomized block design for comparing two treatments is a **matched pairs design.** The idea is to create blocks by matching pairs of similar experimental units. Then we can use chance to decide which member of a pair gets the first treatment. The other subject in that pair gets the other treatment. That is, the random assignment of subjects to treatments is done within each matched pair. Just as with other forms of blocking, matching helps reduce the effect of variation among the experimental units.

Sometimes each "pair" in a matched pairs design consists of just one experimental unit that gets both treatments one after the other. In that case, each experimental unit serves as its own control. The *order* of the treatments can influence the response, so we randomize the order for each experimental unit.

ACTIVITY *Get your heart beating*

MATERIALS: Clock or stopwatch

Are standing pulse rates generally higher than sitting pulse rates? In this Activity, you will perform two experiments to try to answer this question.

1. *Completely randomized design* For the first experiment, you'll randomly assign half of the students in your class to stand and the other half to sit. You can use the hat method, Table D, or technology to carry out the random assignment. Once the two treatment groups have been formed, students should stand or sit as required. Then they should measure their pulses for one minute. Have the subjects in each group record their data on the board.

2. *Matched pairs design* In a matched pairs design, each student should receive both treatments in a random order. Since you already sat or stood in Step 1, you just need to do the opposite now. As before, everyone should measure their pulses for one minute after completing the treatment (that is, once they are standing or sitting). Have all the subjects record their data (both measurements) in a chart on the board.

3. Analyze the data for the completely randomized design. Make a dotplot and calculate the mean pulse rate for each group. Is there evidence that standing pulse rates are higher? Explain.

4. Analyze the data for the matched pairs design. Since the data are paired by student, your first step should be to calculate the difference in pulse rate (standing – sitting) for each subject. Make a dotplot of these differences and calculate their mean. Is there evidence that standing pulse rates are higher? Explain.

5. What advantage does the matched pairs design have over the completely randomized design?

An AP Statistics class with 24 students performed the "Get Your Heart Beating" Activity. We'll analyze the results of their experiment in the following example.

Standing and Sitting Pulse Rate

Design determines analysis

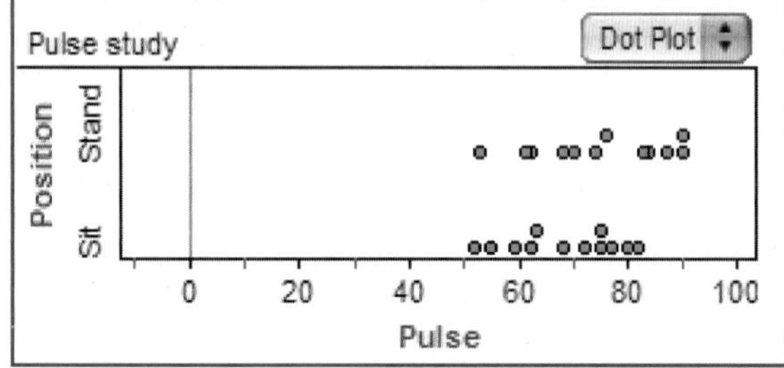

A Fathom dotplot of the pulse rates for their completely randomized design is shown. The mean pulse rate for the standing group is 74.83; the mean for the sitting group is 68.33. So the average pulse rate is 6.5 beats per minute higher in the standing group. However, the variability in pulse rates for the two groups creates a lot of overlap in the graph. These data don't provide convincing evidence that standing pulse rates tend to be higher.

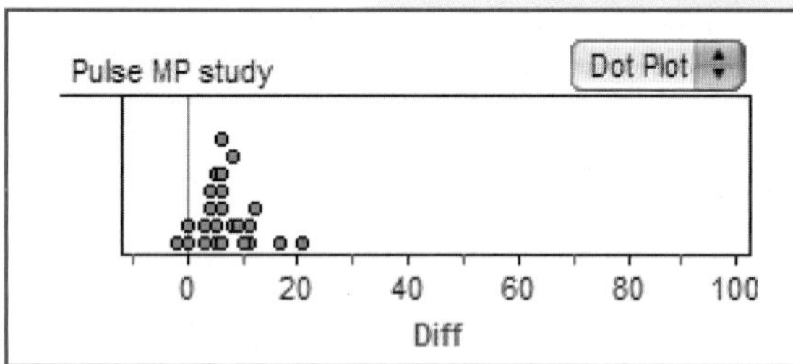

What about the class's matched pairs experiment? The Fathom dotplot shows their data on the difference in pulse rates (standing – sitting). For these 24 students, the mean difference was 6.8 beats per minute. In addition, 21 of the 24 students recorded a positive difference (meaning the standing pulse rate was higher). These data provide strong evidence that people's standing pulse rates tend to be higher than their sitting pulse rates.

Let's take one more look at the two Fathom dotplots in the example. Notice that we used the same scale for both graphs. This is to help you visually compare the amount of variability in the response variable for each of the two experimental designs. Blocking by subject in the matched pairs design greatly reduced the variability in the response variable. That made it easier to detect the fact that standing causes an increase in pulse rate. With the large amount of variability in the completely randomized design, we were unable to draw such a conclusion.

Another important lesson to take away from the example is this: *the design of the study determines the appropriate method of analysis*. For the completely randomized design, it makes sense to compare pulse rates for the two groups with parallel dotplots and means. In the matched pairs design, each student is a block. We compare the effects of the treatments within each block by examining the differences in standing and sitting pulse rates for each student.

The following Data Exploration asks you to apply what you have learned about analyzing data from an experiment.

DATA EXPLORATION *Nitrogen in tires—a lot of hot air?*

Most automobile tires are inflated with compressed air, which consists of about 78% nitrogen. Aircraft tires are filled with pure nitrogen, which is safer than air in case of fire. Could filling automobile tires with nitrogen improve safety, performance, or both?

Consumers Union designed a study to test whether nitrogen-filled tires would maintain pressure better than air-filled tires. They obtained two tires from each of several brands and then filled one tire in each pair with air and one with nitrogen. All tires were inflated to a pressure of 30 pounds per square inch and then placed outside for a year. At the end of the year, Consumers Union measured the pressure in each tire. The amount of pressure lost (in pounds per square inch) during the year for the air-filled and nitrogen-filled tires of each brand is shown in the table below.[29]

Brand	Air	Nitrogen	Brand	Air	Nitrogen
BF Goodrich Traction T/A HR	7.6	7.2	Pirelli P6 Four Seasons	4.4	4.2
Bridgestone HP50 (Sears)	3.8	2.5	Sumitomo HTR H4	1.4	2.1
Bridgestone Potenza G009	3.7	1.6	Yokohama Avid H4S	4.3	3.0
Bridgestone Potenza RE950	4.7	1.5	BF Goodrich Traction T/A V	5.5	3.4
Bridgestone Potenza EL400	2.1	1.0	Bridgestone Potenza RE950	4.1	2.8
Continental Premier Contact H	4.9	3.1	Continental ContiExtreme Contact	5.0	3.4
Cooper Lifeliner Touring SLE	5.2	3.5	Continental ContiProContact	4.8	3.3
Dayton Daytona HR	3.4	3.2	Cooper Lifeliner Touring SLE	3.2	2.5
Falken Ziex ZE-512	4.1	3.3	General Exclaim UHP	6.8	2.7
Fuzion Hrl	2.7	2.2	Hankook Ventus V4 H105	3.1	1.4
General Exclaim	3.1	3.4	Michelin Energy MXV4 Plus	2.5	1.5
Goodyear Assurance Tripletred	3.8	3.2	Michelin Pilot Exalto A/S	6.6	2.2
Hankook Optimo H418	3.0	0.9	Michelin Pilot HX MXM4	2.2	2.0
Kumho Solus KH16	6.2	3.4	Pirelli P6 Four Seasons	2.5	2.7
Michelin Energy MXV4 Plus	2.0	1.8	Sumitomo HTR+	4.4	3.7
Michelin Pilot XGT H4	1.1	0.7			

Does filling tires with nitrogen instead of compressed air reduce pressure loss? Give appropriate graphical and numerical evidence to support your answer.

251

CTION 4.2 Summary

- We can produce data intended to answer specific questions by **observational studies** or **experiments.** An observational study gathers data on individuals as they are. Experiments actively do something to people, animals, or objects in order to measure their response.
- Statistical studies often try to show that changing one variable (the **explanatory variable**) causes changes in another variable (the **response variable**). Variables are **confounded** when their effects on a response can't be distinguished from each other. Observational studies and uncontrolled experiments often fail to show that changes in an explanatory variable actually cause changes in a response variable because the explanatory variable is confounded with **lurking variables.**
- In an experiment, we impose one or more **treatments** on a group of **experimental units** (sometimes called **subjects** if they are human). Each treatment is a combination of values of the explanatory variables (also called **factors**).
- The **design** of an experiment describes the choice of treatments and the manner in which the subjects are assigned to the treatments. The basic principles of experimental design are **control** for lurking variables, **random assignment** of treatments, and **replication** (using enough experimental units).
  - The simplest form of *control* is comparison. Experiments should compare two or more treatments to avoid confounding of the effect of a treatment with other influences, such as lurking variables.
  - *Random assignment* uses chance to assign subjects to the treatments. Random assignment creates treatment groups that are similar (except for chance variation) before the treatments are applied. Randomization and comparison together prevent **bias,** or systematic favoritism, in experiments.
  - Applying each treatment to many experimental units (*replication*) reduces the role of chance variation and makes the experiment more likely to detect differences in the effects of the treatments.
- Many behavioral and medical experiments are **double-blind.** That is, neither the subjects nor those interacting with them and measuring their responses know who is receiving which treatment. If one party knows and the other doesn't, then the experiment is **single-blind.**
- Some experiments give a **placebo** (fake treatment) to a control group. That helps prevent confounding due to the **placebo effect,** in which some patients get better because they expect the treatment to work even though they have received an inactive treatment.
- In addition to comparison, a second form of control is to form **blocks of individuals that are similar in some way that is important to the response.** Randomization is then carried out separately within each block. Blocking helps reduce unwanted variability among experimental units.
- **Matched pairs** are a common form of blocking for comparing just two treatments. In some matched pairs designs, each subject receives both treatments in a random order. In others, the subjects are matched in pairs as closely as possible, and each subject in a pair receives one of the treatments.

SECTION 4.2 Exercises

45. **Chocolate and happy babies** A University of Helsinki (Finland) study wanted to determine if chocolate consumption during pregnancy had an effect on infant temperament at age 6 months. Researchers began by asking 305 healthy pregnant women to report their chocolate consumption. Six months after birth, the researchers asked mothers to rate their infants' temperament, including smiling, laughter, and fear. The babies born to women who had been eating chocolate daily during pregnancy were found to be more active and "positively reactive"—a measure that the investigators said encompasses traits like smiling and laughter.[30]

   (a) Was this an observational study or an experiment? Justify your answer.

   (b) What are the explanatory and response variables?

   (c) Does this study show that eating chocolate regularly during pregnancy helps produce infants with good temperament? Explain.

46. **Child care and aggression** A study of child care enrolled 1364 infants and followed them through their sixth year in school. Later, the researchers published an article in which they stated that "the more time children spent in child care from birth to age four-and-a-half, the more adults tended to rate them, both at age four-and-a-half and at kindergarten, as less likely to get along with others, as more assertive, as disobedient, and as aggressive."[31]

   (a) Is this an observational study or an experiment? Justify your answer.

   (b) What are the explanatory and response variables?

   (c) Does this study show that child care causes children to be more aggressive? Explain.

47. **Learning biology with computers** An educator wants to compare the effectiveness of computer software for teaching biology with that of a textbook presentation. She gives a biology pretest to each of a group of high school juniors, then randomly divides them into two groups. One group uses the computer, and the other studies the text. At the end of the year, she tests all the students again and compares the increase in biology test scores in the two groups.

   (a) Is this an observational study or an experiment? Justify your answer.

   (b) If the group using the computer has a much higher average increase in test scores than the group using the textbook, what conclusions, if any, could the educator draw?

48. **Cell phones and brain cancer** One study of cell phones and the risk of brain cancer looked at a group of 469 people who have brain cancer. The investigators matched each cancer patient with a person of the same age, gender, and race who did not have brain cancer, then asked about the use of cell phones. Result: "Our data suggest that the use of handheld cellular phones is not associated with risk of brain cancer."[32]

   (a) Is this an observational study or an experiment? Justify your answer.

   (b) Based on this study, would you conclude that cell phones do not increase the risk of brain cancer? Why or why not?

49. **Effects of class size** Do smaller classes in elementary school really benefit students in areas such as scores on standardized tests, staying in school, and going on to college? We might do an observational study that compares students who happened to be in smaller and larger classes in their early school years. Identify a lurking variable that may lead to confounding with the effects of small classes. Explain how confounding might occur.

50. **Effects of binge drinking** A common definition of "binge drinking" is 5 or more drinks at one sitting for men and 4 or more for women. An observational study finds that students who binge drink have lower average GPA than those who don't. Identify a lurking variable that may be confounded with the effects of binge drinking. Explain how confounding might occur.

For the experiments described in Exercises 51 to 56, identify the experimental units or subjects, the explanatory variables (factors), the treatments, and the response variables.

51. **Growing in the shade** Ability to grow in shade may help pines found in the dry forests of Arizona to resist drought. How well do these pines grow in shade? Investigators planted pine seedlings in a greenhouse in either full light, light reduced to 25% of normal by shade cloth, or light reduced to 5% of normal. At the end of the study, they dried the young trees and weighed them.

pg 234

52. **Internet telephone calls** You can use Voice over Internet Protocol (VoIP) to make long-distance telephone calls over the Internet. How will the cost affect the use of this service? A university plans an experiment to find out. It will offer the service to all

350 students in one of its dormitories. Some students will pay a low flat rate. Others will pay higher rates at peak periods and very low rates off-peak. The university is interested in the amount and time of use and in the effect on the congestion of the network.

53. **Improving response rate** How can we reduce the rate of refusals in telephone surveys? Most people who answer at all listen to the interviewer's introductory remarks and then decide whether to continue. One study made telephone calls to randomly selected households to ask opinions about the next election. In some calls, the interviewer gave her name, in others she identified the university she was representing, and in still others she identified both herself and the university. For each type of call, the interviewer either did or did not offer to send a copy of the final survey results to the person interviewed. Do these differences in the introduction affect whether the interview is completed?

54. **Eat well and exercise** Most American adolescents don't eat well and don't exercise enough. Can middle schools increase physical activity among their students? Can they persuade students to eat better? Investigators designed a "physical activity intervention" to increase activity in physical education classes and during leisure periods throughout the school day. They also designed a "nutrition intervention" that improved school lunches and offered ideas for healthy home-packed lunches. Each participating school was randomly assigned to one of the interventions, both interventions, or no intervention. The investigators observed physical activity and lunchtime consumption of fat.

pg 235

55. **Fabric science** A maker of fabric for clothing is setting up a new line to "finish" the raw fabric. The line will use either metal rollers or natural-bristle rollers to raise the surface of the fabric; a dyeing-cycle time of either 30 or 40 minutes; and a temperature of either 150° or 175° Celsius. An experiment will compare all combinations of these choices. Three specimens of fabric will be subjected to each treatment and scored for quality.

56. **Exercise and heart rate** A student project measured the increase in the heart rates of fellow students when they stepped up and down for three minutes to the beat of a metronome. The step was either 5.75 or 11.5 inches high and the metronome beat was 14, 21, or 28 steps per minute. Five students stepped at each combination of height and speed.

57. **Cocoa and blood flow** A study conducted by Norman Hollenberg, professor of medicine at Brigham and Women's Hospital and Harvard Medical School, involved 27 healthy people aged 18 to 72. Each subject consumed a cocoa beverage containing 900 milligrams of flavonols (a class of flavonoids) daily for five days. Using a finger cuff, blood flow was measured on the first and fifth days of the study. After five days, researchers measured what they called "significant improvement" in blood flow and the function of the cells that line the blood vessels.[33] What flaw in the design of this experiment makes it impossible to say whether the cocoa really caused the improved blood flow? Explain.

58. **Reducing unemployment** Will cash bonuses speed the return to work of unemployed people? A state department of labor notes that last year 68% of people who filed claims for unemployment insurance found a new job within 15 weeks. As an experiment, this year the state offers $500 to people filing unemployment claims if they find a job within 15 weeks. The percent who do so increases to 77%. What flaw in the design of this experiment makes it impossible to say whether the bonus really caused the increase? Explain.

pg 237

59. **Layoffs and "survivor guilt"** Workers who survive a layoff of other employees at their location may suffer from "survivor guilt." A study of survivor guilt and its effects used as subjects 120 students who were offered an opportunity to earn extra course credit by doing proofreading. Each subject worked in the same cubicle as another student, who was an accomplice of the experimenters. At a break midway through the work, one of three things happened:

Treatment 1: The accomplice was told to leave; it was explained that this was because she performed poorly.
Treatment 2: It was explained that unforeseen circumstances meant there was only enough work for one person. By "chance," the accomplice was chosen to be laid off.
Treatment 3: Both students continued to work after the break.

The subjects' work performance after the break was compared with performance before the break.[34] Describe how you would randomly assign the subjects to the treatments

(a) using slips of paper.
(b) using Table D.
(c) using technology.

60. **Effects of TV advertising** Figure 4.3 (page 235) displays the 6 treatments for a two-factor experiment on TV advertising. Suppose we have 150 students who are willing to serve as subjects. Describe how you would randomly assign the subjects to the treatments

(a) using slips of paper.
(b) using Table D.
(c) using technology.

pg 239

61. **Headache relief** Doctors identify "chronic tension-type headaches" as headaches that occur almost daily for at least six months. Can antidepressant medications or stress management training reduce the number and severity of these headaches? Are both together more effective than either alone? Investigators compared four treatments: antidepressant alone, placebo alone, antidepressant plus stress management, and placebo plus stress management. The headache sufferers named in the following table have agreed to participate in the study.

Acosta	Duncan	Han	Liang	Padilla	Valasco
Asihiro	Durr	Howard	Maldonado	Plochman	Vaughn
Bennett	Edwards	Hruska	Marsden	Rosen	Wei
Bikalis	Farouk	Imrani	Montoya	Solomon	Wilder
Chen	Fleming	James	O'Brian	Trujillo	Willis
Clemente	George	Kaplan	Ogle	Tullock	Zhang

(a) Outline the design of the experiment. What is this type of design called?

(b) Explain how you would randomly assign the subjects to the four treatment groups. Then carry out your random assignment.

62. **More rain for California?** The changing climate will probably bring more rain to California, but we don't know whether the additional rain will come during the winter wet season or extend into the long dry season in spring and summer. Kenwyn Suttle of the University of California at Berkeley and his coworkers carried out an experiment to study the effects of more rain in either season. They randomly assigned plots of open grassland to 3 treatments: added water equal to 20% of annual rainfall either during January to March (winter) or during April to June (spring), and no added water (control). Thirty-six circular plots of area 70 square meters were available (see the photo), of which 18 were used for this study. One response variable was total plant biomass, in grams per square meter, produced in a plot over a year.[35]

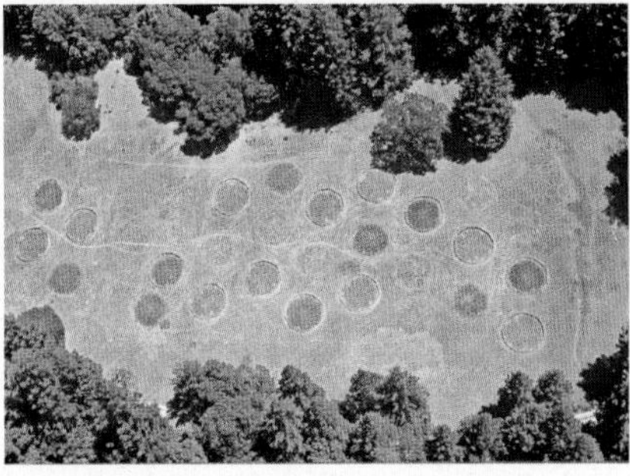

(a) Outline the design of the experiment. What is this type of design called?

(b) Explain how you would randomly assign the experimental units to the three treatments. Then carry out your random assignment.

63. **Treating prostate disease** A large study used records from Canada's national health care system to compare the effectiveness of two ways to treat prostate disease. The two treatments are traditional surgery and a new method that does not require surgery. The records described many patients whose doctors had chosen each method. The study found that patients treated by the new method were significantly more likely to die within 8 years.[36]

(a) Further study of the data showed that this conclusion was wrong. The extra deaths among patients who got the new method could be explained by lurking variables. What lurking variables might be confounded with a doctor's choice of surgical or nonsurgical treatment?

(b) You have 300 prostate patients who are willing to serve as subjects in an experiment to compare the two methods. Write a few sentences describing how you would design this experiment.

64. **Getting teachers to come to school** Elementary schools in rural India are usually small, with a single teacher. The teachers often fail to show up for work. Here is an idea for improving attendance: give the teacher a digital camera with a tamperproof time and date stamp and ask a student to take a photo of the teacher and class at the beginning and end of the day. Offer the teacher better pay for good attendance, verified by the photos. Will this work? Researchers obtained permission to use 120 rural schools in Rajasthan for an experiment to find out.[37]

(a) Explain why it would not be a good idea to offer better pay for good attendance to the teachers in all 120 schools and then to compare this year's attendance with last year's.

(b) Write a few sentences describing how you would design this experiment.

65. **Stronger players** A football coach hears that a new exercise program will increase upper-body strength better than lifting weights. He is eager to test this new program in the off-season with the players on his high school team. The coach decides to let his players choose which of the two treatments they will undergo for 3 weeks—exercise or weight lifting. He will use the number of push-ups a player can do at the end of the experiment as the response variable.

(a) Which principle of experimental design does the coach's plan violate? Explain how this violation could lead to confounding.

(b) Comment on the coach's choice of response variable.

66. **Prayer and meditation** You read in a magazine that "nonphysical treatments such as meditation and prayer have been shown to be effective in controlled scientific studies for such ailments as high blood

pressure, insomnia, ulcers, and asthma." Explain in simple language what the article means by "controlled scientific studies." Why can such studies provide good evidence that meditation is an effective treatment for high blood pressure?

67. **The effects of day care** Does day care help low-income children stay in school and hold good jobs later in life? The Carolina Abecedarian Project (the name suggests the ABCs) has followed a group of 111 children since 1972. Back then, these individuals were all healthy but low-income black infants in Chapel Hill, North Carolina. All the infants received nutritional supplements and help from social workers. Half were also assigned at random to an intensive preschool program.[38]

pg 241

(a) Explain the purpose of each of the three experimental design principles.

(b) Describe how each of these principles was used in this study.

68. **Killing weeds** A biologist would like to determine which of two brands of weed killer is less likely to harm the plants in a garden at the university. Before spraying near the plants, the biologist decides to conduct an experiment using 24 individual plants. Which of the following two plans for randomly assigning the treatments should the biologist use? Why?

Plan A: Choose the 12 healthiest-looking plants. Apply Brand X weed killer to all 12 of those plants. Apply Brand Y weed killer to the remaining 12 plants.

Plan B: Choose 12 of the 24 plants at random. Apply Brand X weed killer to those 12 plants and Brand Y weed killer to the remaining 12 plants.

69. **Do placebos really work?** Researchers in Japan conducted an experiment on 13 individuals who were extremely allergic to poison ivy. On one arm, each subject was rubbed with a poison ivy leaf and told the leaf was harmless. On the other arm, each subject was rubbed with a harmless leaf and told it was poison ivy. All the subjects developed a rash on the arm where the harmless leaf was rubbed. Of the 13 subjects, 11 did not have any reaction to the real poison ivy leaf.[39]

(a) What was the placebo in this experiment?

(b) Explain how the results of this study support the idea of a placebo effect.

70. **Pain relief study** Fizz Laboratories, a pharmaceutical company, has developed a new drug for relieving chronic pain. Sixty patients suffering from arthritis and needing pain relief are available. Each patient will be treated and asked an hour later, "About what percent of pain relief did you experience?"

(a) Why should Fizz not simply administer the new drug and record the patients' responses?

(b) Should the patients be told whether they are getting the new drug or a placebo? How would this knowledge probably affect their reactions?

71. **Meditation for anxiety** An experiment that claimed to show that meditation lowers anxiety proceeded as follows. The experimenter interviewed the subjects and rated their level of anxiety. Then the subjects were randomly assigned to two groups. The experimenter taught one group how to meditate and they meditated daily for a month. The other group was simply told to relax more. At the end of the month, the experimenter interviewed all the subjects again and rated their anxiety level. The meditation group now had less anxiety. Psychologists said that the results were suspect because the ratings were not blind. Explain what this means and how lack of blindness could bias the reported results.

72. **Testosterone for older men** As men age, their testosterone levels gradually decrease. This may cause a reduction in lean body mass, an increase in fat, and other undesirable changes. Do testosterone supplements reverse some of these effects? A study in the Netherlands assigned 237 men aged 60 to 80 with low or low-normal testosterone levels to either a testosterone supplement or a placebo. The report in the *Journal of the American Medical Association* described the study as a "double-blind, randomized, placebo-controlled trial."[40] Explain each of these terms to someone who knows no statistics.

73. **Acupuncture and pregnancy** A study sought to determine whether the ancient Chinese art of acupuncture could help infertile women become pregnant.[41] One hundred sixty healthy women undergoing treatment with artificial insemination were recruited for the study. Half of the subjects were randomly assigned to receive acupuncture treatment 25 minutes before embryo transfer and again 25 minutes after the transfer. The remaining 80 subjects were instructed to lie still for 25 minutes after the embryo transfer. *Results:* In the acupuncture group, 34 women became pregnant. In the control group, 21 women became pregnant.

(a) Describe how the three principles of experimental design were addressed in this study.

(b) The difference in the percent of women who became pregnant in the two groups is statistically significant. Explain what this means to someone who knows little statistics.

(c) Explain why the placebo effect prevents us from concluding that acupuncture caused the difference in pregnancy rates.

74. **Do reducing diets work?** Dr. Linda Stern and her colleagues recruited 132 obese adults at the Philadelphia Veterans Affairs Medical Center in Pennsylvania. Half the participants were randomly assigned to a low-carbohydrate diet and the other half to a low-fat diet. Researchers measured each participant's change in weight and cholesterol level after six months and again after one year. Subjects in the low-carb diet group lost significantly more weight than subjects in the low-fat diet group during the first six months. At the end of a year, however, the average weight loss for subjects in the two groups was not significantly different.[42]
(a) Why did researchers randomly assign the subjects to the diet treatments?
(b) Explain to someone who knows little statistics what "lost significantly more weight" means.
(c) The subjects in the low-carb diet group lost an average of 5.1 kg in a year. The subjects in the low-fat diet group lost an average of 3.1 kg. Explain how this information could be consistent with the fact that weight loss in the two groups was not significantly different.

75. **Is yawning contagious?** According to the popular TV show *Mythbusters*, the answer is "Yes." In the March 9, 2005, episode, the *Mythbusters* team presented the results of an experiment involving 50 subjects. All the subjects were placed in a booth for an extended period of time and monitored by hidden camera. Two-thirds of the subjects were given a "yawn seed" by one of the experimenters; that is, the experimenter yawned in the subject's presence prior to leaving the room. The remaining subjects were given no yawn seed. What were the results? Of the 16 subjects who had no yawn seed, 4 yawned. Of the 34 subjects given a yawn seed, 10 yawned. Adam Savage and Jamie Hyneman, the cohosts of *Mythbusters*, used these results to conclude that yawning is contagious.
(a) Explain how you could use slips of paper to randomly *reassign* the subjects to the treatment groups, assuming that the treatment received doesn't affect the response.
(b) Suppose we used your method in (a) to redo the random assignment 50 times. The Fathom dotplot displays the number of subjects in the yawn seed group who yawned in each of these random assignments. What conclusion would you draw about whether yawning is contagious? Explain.

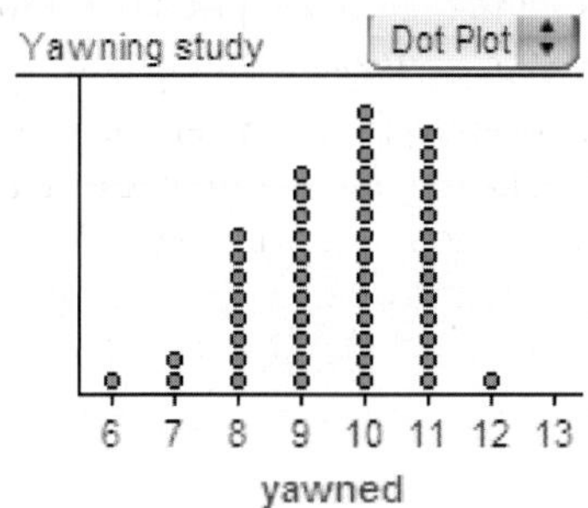

76. **Magnets and pain** Refer to the chapter-opening Case Study (page 205). The researchers decided to analyze the patients' final pain ratings. It also makes sense to examine the *difference* between patients' initial pain ratings and their final pain ratings in both groups. Here are the data:

Active:	10, 6, 1, 10, 6, 8, 5, 5, 6, 8, 7, 8, 7, 6, 4, 4, 7, 10, 6, 10, 6, 5, 5, 1, 0, 0, 0, 0, 1
Inactive:	4, 3, 5, 2, 1, 4, 1, 0, 0, 1, 0, 0, 0, 0, 0, 0, 0, 1, 0, 0, 1

(a) Construct a comparative dotplot of the data. Describe what you see.
(b) Calculate the mean (average) change in pain rating for each group. Find the difference in the average changes for the two groups (Active – Inactive).
(c) Describe how you could use index cards to randomly reassign the subjects to the treatment groups, assuming that the treatment received doesn't affect the response.
(d) Suppose we used your method in (c) to redo the random assignment 50 times. The Fathom dotplot displays the difference (Active – Inactive) for the average change in pain rating for each of these random assignments. What conclusion would you draw about the effect of magnets on pain relief? Explain.

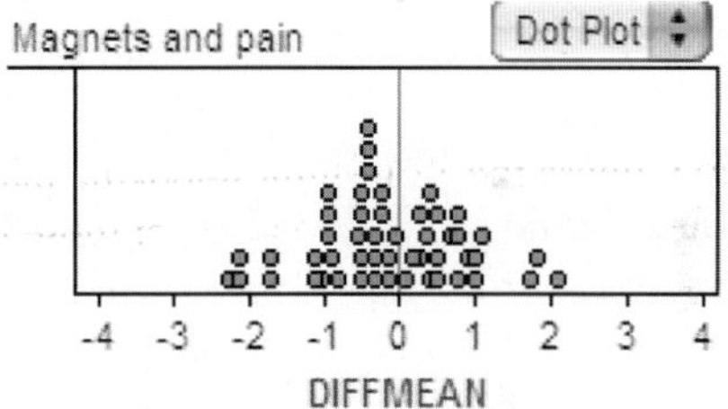

77. **Doctors and nurses** Nurse-practitioners are nurses with advanced qualifications who often act much like primary-care physicians. Are they as effective as doctors at treating patients with chronic conditions? An experiment was conducted with 1316 patients who had been diagnosed with asthma, diabetes, or high blood pressure. Within each condition, patients were randomly assigned to either a doctor or a nurse-practitioner. The response variables included measures of the patients' health and of their satisfaction with their medical care after 6 months.[43]
(a) Which are the blocks in this experiment: the different diagnoses (asthma, etc.) or the type of care (nurse or doctor)? Why?
(b) Explain why a randomized block design is preferable to a completely randomized design here.

78. **Comparing cancer treatments** The progress of a type of cancer differs in women and men. Researchers want to design an experiment to compare three therapies for this cancer. They recruit 500 male and 300 female patients who are willing to serve as subjects.

2

(a) Which are the blocks in this experiment: the cancer therapies or the two sexes? Why?

(b) What are the advantages of a randomized block design over a completely randomized design using these 800 subjects?

(c) Suppose the researchers had 800 male and no female subjects available for the study. What advantage would this offer? What disadvantage?

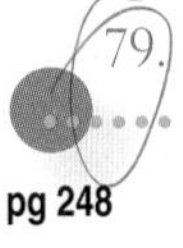
pg 248

79. **In the cornfield** An agriculture researcher wants to compare the yield of 5 corn varieties: A, B, C, D, and E. The field in which the experiment will be carried out increases in fertility from north to south. The researcher therefore divides the field into 25 plots of equal size, arranged in 5 east–west rows of 5 plots each, as shown in the diagram.

North

(a) Explain why a randomized block design would be better than a completely randomized design in this setting.

(b) Should the researcher use the rows or the columns of the field as blocks? Justify your answer.

(c) Use technology or Table D to carry out the random assignment required by your design. Explain your method clearly.

80. **Comparing weight-loss treatments** Twenty overweight females have agreed to participate in a study of the effectiveness of four weight-loss treatments: A, B, C, and D. The researcher first calculates how overweight each subject is by comparing the subject's actual weight with her "ideal" weight. The subjects and their excess weights in pounds are as follows:

Birnbaum	35	Hernandez	25	Moses	25	Smith	29
Brown	34	Jackson	33	Nevesky	39	Stall	33
Brunk	30	Kendall	28	Obrach	30	Tran	35
Cruz	34	Loren	32	Rodriguez	30	Wilansky	42
Deng	24	Mann	28	Santiago	27	Williams	22

The response variable is the weight lost after 8 weeks of treatment. Previous studies have shown that the effects of a diet may vary based on a subject's initial weight.

(a) Explain why a randomized block design would be better than a completely randomized design in this setting.

(b) Should researchers form blocks of size 4 based on subjects' last names in alphabetical order or by how overweight the subjects are? Explain.

(c) Use technology or Table D to carry out the random assignment required by your design. Explain your method clearly.

81. **Aw, rats!** A nutrition experimenter intends to compare the weight gain of newly weaned male rats fed Diet A with that of rats fed Diet B. To do this, she will feed each diet to 10 rats. She has available 10 rats from one litter and 10 rats from a second litter. Rats in the first litter appear to be slightly healthier.

(a) If the 10 rats from Litter 1 were fed Diet A, the effects of genetics and diet would be confounded, and the experiment would be biased. Explain this statement carefully.

(b) Describe a better design for this experiment.

82. **Technology for teaching statistics** The Brigham Young University (BYU) statistics department is performing experiments to compare teaching methods. Response variables include students' final-exam scores and a measure of their attitude toward statistics. One study compares two levels of technology for large lectures: standard (overhead projectors and chalk) and multimedia. There are 8 lecture sections of a basic statistics course at BYU, each with about 200 students. There are four instructors, each of whom teaches two sections.[44] Suppose the sections and lecturers are as follows:

Section	Lecturer	Section	Lecturer
1	Hilton	5	Tolley
2	Christensen	6	Hilton
3	Hadfield	7	Tolley
4	Hadfield	8	Christensen

(a) Suppose we randomly assign two lecturers to use standard technology in their sections and the other two lecturers to use multimedia technology. Explain how this could lead to confounding.

(b) Describe a better design for this experiment.

83. **Look, Ma, no hands!** Does talking on a hands-free cell phone distract drivers? Researchers recruit 40 student subjects for an experiment to investigate this question. They have a driving simulator equipped with a hands-free phone for use in the study.

(a) Researchers are considering the design shown in the figure below. What type of design is this?

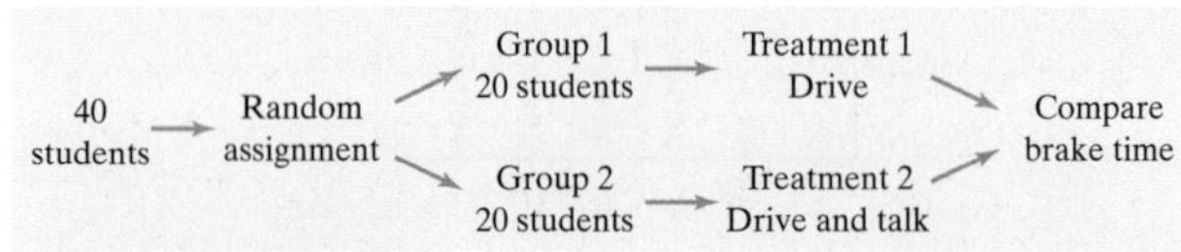

(b) Explain how blocking could be used to improve the design in (a).

(c) Why is it important to randomly assign the treatments within each block?

84. **Chocolate gets my heart pumping** Cardiologists at Athens Medical School in Greece wanted to test whether chocolate affected blood flow in the blood vessels. The researchers recruited 17 healthy young volunteers, who were each given a 3.5-ounce bar of dark chocolate, either bittersweet or fake chocolate. On another day, the volunteers were switched. The subjects had no chocolate outside the study, and investigators didn't know whether a subject had eaten the real or the fake chocolate. An ultrasound was taken of each volunteer's upper arm to see the functioning of the cells in the walls of the main artery. The investigators found that blood vessel function was improved when the subjects ate bittersweet chocolate, and that there were no such changes when they ate the placebo (fake chocolate).[45]

(a) What type of design did the investigators use in their study?

(b) Explain why the investigators chose this design instead of a completely randomized design.

(c) Why is it important to randomly assign the order of the treatments for the subjects?

85. **Room temperature and dexterity** An expert on worker performance is interested in the effect of room temperature on the performance of tasks requiring manual dexterity. She chooses temperatures of 70°F and 90°F as treatments. The response variable is the number of correct insertions, during a 30-minute period, in a peg-and-hole apparatus that requires the use of both hands simultaneously. Each subject is trained on the apparatus and then asked to make as many insertions as possible in 30 minutes of continuous effort.

(a) Outline a completely randomized design to compare dexterity at 70° and 90°. Twenty subjects are available.

(b) Because individuals differ greatly in dexterity, the wide variation in individual scores may hide the systematic effect of temperature unless there are many subjects in each group. Describe in detail the design of a matched pairs experiment in which each subject serves as his or her own control.

86. **Carbon dioxide and tree growth** The concentration of carbon dioxide (CO_2) in the atmosphere is increasing rapidly due to our use of fossil fuels. Because plants use CO_2 to fuel photosynthesis, more CO_2 may cause trees and other plants to grow faster. An elaborate apparatus allows researchers to pipe extra CO_2 to a 30-meter circle of forest. We want to compare the growth in base area of trees in treated and untreated areas to see if extra CO_2 does in fact increase growth. We can afford to treat three circular areas.[46]

(a) Describe the design of a completely randomized experiment using 6 well-separated 30-meter circular areas in a pine forest. Sketch the circles and carry out the randomization your design calls for.

(b) Areas within the forest may differ in soil fertility. Describe a matched pairs design using three pairs of circles that will reduce the extra variation due to different fertility. Sketch the circles and carry out the randomization your design calls for.

87. **Got deodorant?** A group of students wants to perform an experiment to determine whether Brand A or Brand B deodorant lasts longer. One group member suggests the following design: Recruit 40 student volunteers—20 male and 20 female. Separate by gender, since male and female bodies might respond differently to deodorant. Give all the males Brand A deodorant and all the females Brand B. Have each student rate how well the deodorant is still working at the end of the school day on a 0 to 10 scale. Then compare ratings for the two treatments.

(a) Identify any flaws you see in the proposed design for this experiment.

(b) Describe how you would design the experiment. Explain how your design addresses each of the problems you identified in (a).

88. **Wake-up call** Do people naturally wake up earlier when they set an alarm before going to sleep? Justin decides to conduct his own experiment to find out. On Friday and Saturday nights, he doesn't set the alarm before going to bed. On Monday and Tuesday, he sets the alarm for 7 A.M. Justin records the time when he wakes up each day and then compares his average wake-up time with and without the alarm.

(a) Identify any flaws you see in the proposed design for this experiment.

(b) Describe how you would design the experiment. Explain how your design addresses each of the problems you identified in (a).

89. **Close shave** Which of two brands of electric razor shaves closer? Describe how you would design and carry out an experiment to answer this question using 50 adult male volunteers.

90. **Music and studying** Does listening to music while reading a story help or hinder recall of factual details? Describe how you would design and carry out an experiment to answer this question using 30 students at your school who have agreed to participate.

Multiple choice: Select the best answer for Exercises 91 to 98.

91. Can changing diet reduce high blood pressure? Vegetarian diets and low-salt diets are both promising. Men with high blood pressure are assigned at random to four diets: (1) normal diet with unrestricted salt; (2) vegetarian with unrestricted salt; (3) normal with restricted salt; and (4) vegetarian with restricted salt. This experiment has
(a) one factor, the type of diet.
(b) two factors, high blood pressure and type of diet.
(c) two factors, normal/vegetarian diet and unrestricted/restricted salt.
(d) three factors, men, high blood pressure, and type of diet.
(e) four factors, the four diets being compared.

92. In the experiment of the previous exercise, the 240 subjects are labeled 001 to 240. Software randomly assigns 60 subjects to each of the four diets. This is
(a) a completely randomized design.
(b) a randomized block design.
(c) a matched pairs design.
(d) an observational study.
(e) an SRS.

93. The Community Intervention Trial for Smoking Cessation asked whether a community-wide advertising campaign would reduce smoking. The researchers located 11 pairs of communities, each pair similar in location, size, economic status, and so on. One community in each pair participated in the advertising campaign and the other did not. This is
(a) an observational study.
(b) a matched pairs experiment.
(c) a completely randomized experiment.
(d) a randomized block design, but not matched pairs.
(e) a stratified random sample.

94. The drug manufacturer Merck recently stopped testing a promising new drug to treat depression. It turned out that in a randomized, double-blind trial, a dummy pill did almost as well as the new drug. The fact that many people respond to a dummy treatment is called
(a) confounding.
(b) nonresponse.
(c) comparison.
(d) the placebo effect.
(e) bias.

95. Corn variety 1 yielded 140 bushels per acre last year at a research farm. This year, corn variety 2, planted in the same location, yielded only 110 bushels per acre. Unfortunately, we don't know whether the difference is due to the superiority of variety 1 or to the effect of this year's drought. This is an example of
(a) bias.
(b) matched pairs design.
(c) confounding.
(d) the placebo effect.
(e) replication.

96. A report in a medical journal notes that the risk of developing Alzheimer's disease among subjects who (voluntarily) regularly took the anti-inflammatory drug ibuprofen (the active ingredient in Advil) was about half the risk among those who did not. Is this good evidence that ibuprofen is effective in preventing Alzheimer's disease?
(a) Yes, because the study was a randomized, comparative experiment.
(b) No, because the effect of ibuprofen is confounded with the placebo effect.
(c) Yes, because the results were published in a reputable professional journal.
(d) No, because this is an observational study. An experiment would be needed to confirm (or not confirm) the observed effect.
(e) Yes, because a 50% reduction can't happen just by chance.

97. A double-blind experiment was conducted to evaluate the effectiveness of the Salk polio vaccine. The purpose of keeping the diagnosing physicians unaware of the treatment status of the experimental subjects was to
(a) eliminate grounds for malpractice suits.
(b) ensure that subjects were randomly assigned to treatments.
(c) eliminate a possible source of bias.
(d) make sure nobody is harmed.
(e) avoid the placebo effect.

98. Two essential features of all statistically designed experiments are
(a) compare several treatments; use the double-blind method.
(b) compare several treatments; use chance to assign subjects to treatments.
(c) always have a placebo group; use the double-blind method.
(d) use a block design; use chance to assign subjects to treatments.
(e) use enough subjects; always have a control group.

99. **Seed weights** (2.2) Biological measurements on the same species often follow a Normal distribution quite closely. The weights of seeds of a variety of winged bean are approximately Normal with mean 525 milligrams (mg) and standard deviation 110 mg.

(a) What percent of seeds weigh more than 500 mg? Show your method.

(b) If we discard the lightest 10% of these seeds, what is the smallest weight among the remaining seeds? Show your method.

Exercises 100 and 101 refer to the following setting. A researcher studied a group of identical twins who had been separated and adopted at birth. In each case, one twin (Twin A) was adopted by a low-income family and the other (Twin B) by a high-income family. Both twins were given an IQ test as adults. Here are their scores:[47]

Twin A:	120	99	99	94	111	97	99	94	104	114	113	100
Twin B:	128	104	108	100	116	105	100	100	103	124	114	112

100. **IQ and twins** (3.1) How well does one twin's IQ predict the other's? Give appropriate evidence to support your answer.

101. **IQ and income** (1.3) Do identical twins living in low-income homes tend to have lower IQs later in life than their twins who live in high-income homes? Give appropriate evidence to support your answer.

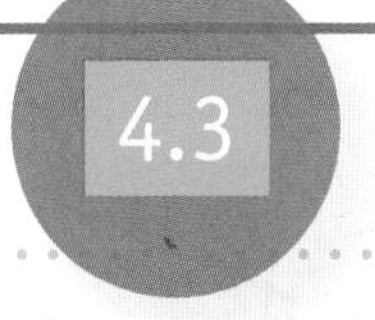

4.3

Using Studies Wisely

In Section 4.3, you'll learn about:

- Scope of inference
- The challenges of establishing causation
- Data ethics

Researchers who conduct statistical studies often want to draw conclusions (make inferences) that go beyond the data they produce. Here are two examples.

- The U.S. Census Bureau carries out a monthly Current Population Survey of about 60,000 households. Their goal is to use data from these randomly selected households to estimate the percent of unemployed individuals in the population.
- Scientists performed an experiment that randomly assigned 21 volunteer subjects to one of two treatments: sleep deprivation for one night or unrestricted sleep. The experimenters hoped to show that sleep deprivation causes a decrease in performance two days later.[48]

What type of inference can be made from a particular study? The answer depends on the design of the study.

Scope of Inference

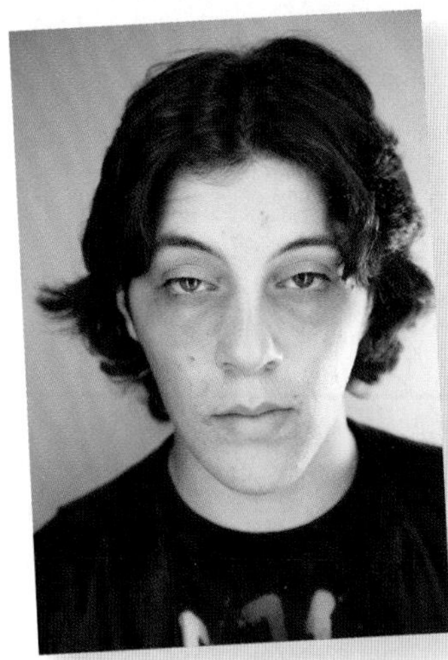

In the Census Bureau's sample survey, the individuals who responded were chosen at random from the population of interest. Random sampling avoids bias and produces trustworthy estimates of the truth about the population. The Census Bureau should be safe making an *inference about the population* based on the results of the sample.

In the sleep deprivation experiment, subjects were randomly assigned to the sleep deprivation and unrestricted sleep treatments. Random assignment helps ensure that the two groups of subjects are as similar as possible before the treatments are imposed. If the unrestricted sleep group performs much better than the sleep deprivation group, and the difference is too large to be explained by chance variation in the random assignment, it must be due to the treatments. In that case, the scientists could safely conclude that sleep deprivation caused the decrease in performance. That is, they can make an *inference about cause and effect*. However, since the experiment used volunteer subjects, this limits scientists' ability to generalize their findings to some larger population of individuals.

Let's recap what we've learned about the scope of inference in a statistical study. Random selection of individuals allows inference about the population. Random assignment of individuals to groups permits inference about cause and effect. The chart below summarizes the possibilities.[49]

Both random sampling and random assignment introduce chance variation into a statistical study. When performing inference, statisticians use the laws of probability to describe this chance variation. You'll learn how this works later in the book.

		Were individuals randomly assigned to groups?	
		Yes	**No**
Were individuals randomly selected?	**Yes**	Inference about the population: YES Inference about cause and effect: YES	Inference about the population: YES Inference about cause and effect: NO
	No	Inference about the population: NO Inference about cause and effect: YES	Inference about the population: NO Inference about cause and effect: NO

Well-designed experiments randomly assign individuals to treatment groups. However, most experiments don't select experimental units at random from the larger population. That limits such experiments to inference about cause and effect. Observational studies don't randomly assign individuals to groups, which rules out inference about cause and effect. An observational study that uses random sampling can make an inference about the population. The following example illustrates all four cases from the table above in a single setting.

EXAMPLE

Vitamin C and Canker Sores

Determining scope of inference

A small-town dentist wants to know if a daily dose of 500 milligrams (mg) of vitamin C will result in fewer canker sores in the mouth than taking no vitamin C.[50]

The dentist is considering the following four study designs.

Design 1: Get all dental patients in town with appointments in the next two weeks to take part in a study. Give each patient a survey with two questions: (1) Do you take at least 500 of vitamin C each day? (2) Do you frequently have canker sores? Based on patients' answers to Question 1, divide them into two groups: those who take at least 500 mg of vitamin C daily and those who don't.

Design 2: Get all dental patients in town with appointments in the next two weeks to take part in a study. Randomly assign half of them to take 500 mg of vitamin C each day and the other half to abstain from taking vitamin C for three months.

Design 3: Select a random sample of dental patients in town and get them to take part in a study. Divide the patients into two groups as in Design 1.

Design 4: Select a random sample of dental patients in town and get them to take part in a study. Randomly assign half of them to take 500 mg of vitamin C each day and the other half to abstain from taking vitamin C for three months.

For whichever design the dentist chooses, suppose she compares the proportion of patients in each group who complain of canker sores. Also suppose that she finds a statistically significant difference, with a smaller proportion of those taking vitamin C having canker sores.

PROBLEM: **What can the dentist conclude for each design?**

SOLUTION:

Design 1: **Since the patients were not randomly selected, the dentist cannot infer that this result holds for a larger population of dental patients. This was an observational study since no treatments**

were deliberately imposed on the patients. With no random assignment to the two groups, no inference about cause and effect can be made. The dentist just knows that for these patients, those who took vitamin C had fewer canker sores than those who didn't.

Design 2: As in Design 1, the dentist can't make any inference about this result holding for a larger population of dental patients. However, the treatments were randomly assigned to the subjects. Assuming proper control in the experiment, she can conclude that taking vitamin C reduced the chance of getting canker sores in her subjects.

Design 3: Since the patients were randomly selected from the population of dental patients in the town, the dentist can generalize the results of this study to the population. Because this was an observational study, no inference about cause and effect can be made. The dentist would conclude that for the population of dental patients in this town, those taking vitamin C have fewer canker sores than those who don't. She can't say whether the vitamin C causes this reduction or some other confounding variable. (Note that the dentist can't draw a conclusion about whether those people in the general population of the town who take vitamin C will have fewer canker sores, since the sample was only of dental patients.)

Design 4: As in Design 3, the random sampling allows the dentist to generalize the results of this study to the population of dental patients in the town. As in Design 2, the random assignment would allow her to conclude (assuming proper control in the experiment) that taking vitamin C reduced the chance of getting canker sores. So the dentist would conclude that for the population of dental patients in this town, those taking vitamin C will tend to have fewer canker sores than those who don't *due to the vitamin C.* As with Design 3, she can't extend this result to the general population of the town.

For Practice *Try Exercise 105*

The Challenges of Establishing Causation

A well-designed experiment tells us that changes in the explanatory variable cause changes in the response variable. More precisely, it tells us that this happened for specific individuals in the specific environment of this specific experiment. The serious threat is that the treatments, the subjects, or the environment of our experiment may not be realistic. *Lack of realism* can limit our ability to apply the conclusions of an experiment to the settings of greatest interest.

Do Center Brake Lights Reduce Rear-End Crashes?

Lack of realism

Do those high center brake lights, required on all cars sold in the United States since 1986, really reduce rear-end collisions? Randomized comparative experiments with fleets of rental and business cars, done before the lights were required, showed that the third brake light reduced rear-end collisions by as much as 50%. But requiring the third light in all cars led to only a 5% drop.

What happened? Most cars did not have the extra brake light when the experiments were carried out, so it caught the eye of following drivers. Now that almost all cars have the third light, they no longer capture attention.

In some cases, it isn't practical or even ethical to do an experiment. Consider these important questions:

- Does texting while driving increase the risk of having an accident?
- Does going to church regularly help people live longer?
- Does smoking cause lung cancer?

To answer these cause-and-effect questions, we just need to perform a randomized comparative experiment. Unfortunately, we can't randomly assign people to text while driving or to attend church or to smoke cigarettes. The best data we have about these and many other cause-and-effect questions come from observational studies.

It is sometimes possible to build a strong case for causation in the absence of experiments. The evidence that smoking causes lung cancer is about as strong as nonexperimental evidence can be.

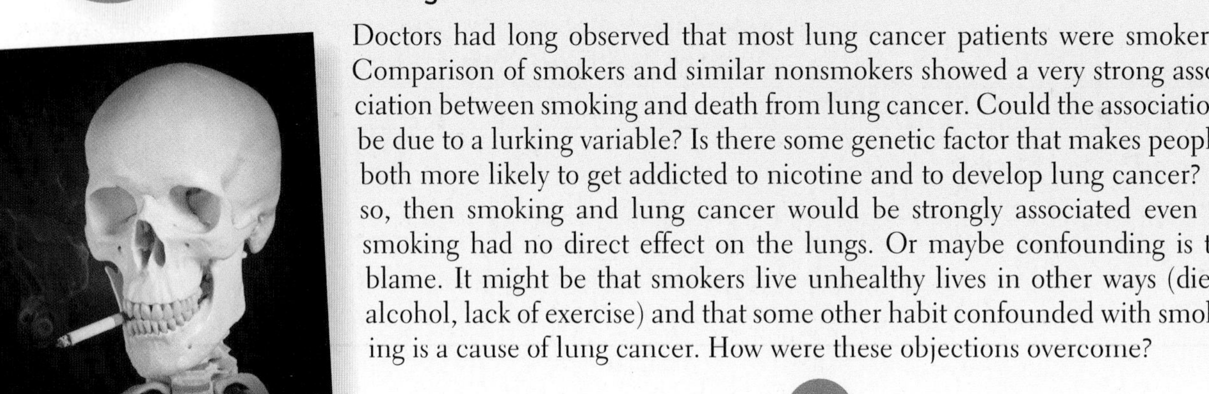

Does Smoking Cause Lung Cancer?

Living with observational studies

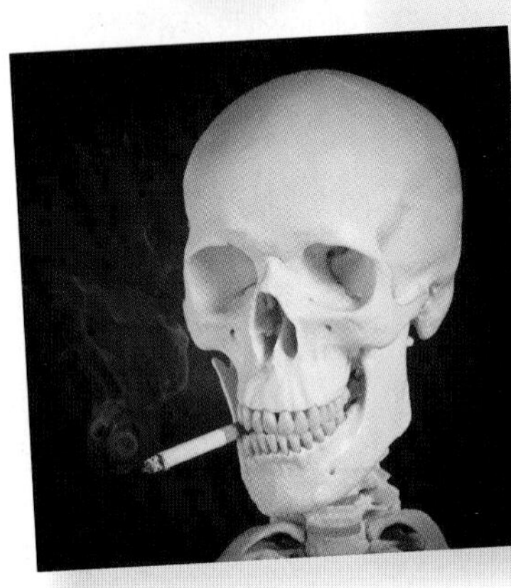

Doctors had long observed that most lung cancer patients were smokers. Comparison of smokers and similar nonsmokers showed a very strong association between smoking and death from lung cancer. Could the association be due to a lurking variable? Is there some genetic factor that makes people both more likely to get addicted to nicotine and to develop lung cancer? If so, then smoking and lung cancer would be strongly associated even if smoking had no direct effect on the lungs. Or maybe confounding is to blame. It might be that smokers live unhealthy lives in other ways (diet, alcohol, lack of exercise) and that some other habit confounded with smoking is a cause of lung cancer. How were these objections overcome?

What are the criteria for establishing causation when we can't do an experiment?

- *The association is strong.* The association between smoking and lung cancer is very strong.
- *The association is consistent.* Many studies of different kinds of people in many countries link smoking to lung cancer. That reduces the chance that a lurking variable specific to one group or one study explains the association.
- *Larger values of the explanatory variable are associated with stronger responses.* People who smoke more cigarettes per day or who smoke over a longer period get lung cancer more often. People who stop smoking reduce their risk.
- *The alleged cause precedes the effect in time.* Lung cancer develops after years of smoking. The number of men dying of lung cancer rose as smoking became more common, with a lag of about 30 years. Lung cancer kills more men than any other form of cancer. Lung cancer was rare among women until women began to smoke. Lung cancer in women rose along with smoking, again with a lag of about 30 years, and has now passed breast cancer as the leading cause of cancer death among women.

• *The alleged cause is plausible.* Experiments with animals show that tars from cigarette smoke do cause cancer.

Medical authorities do not hesitate to say that smoking causes lung cancer. The U.S. Surgeon General states that cigarette smoking is "the largest avoidable cause of death and disability in the United States."[51] The evidence for causation is overwhelming—but it is not as strong as the evidence provided by well-designed experiments. Conducting an experiment in which some subjects were forced to smoke and others were not allowed to would be unethical. In cases like this, observational studies are our best source of reliable information.

Data Ethics*

Medical professionals are taught to follow the basic principle "First, do no harm." Shouldn't those who carry out statistical studies follow the same principle? Most reasonable people think so. But this may not always be as simple as it sounds. Decide whether you think each of the following studies is ethical or unethical.

• A promising new drug has been developed for treating cancer in humans. Before giving the drug to human subjects, researchers want to administer the drug to animals to see if there are any potentially serious side effects.

• Are companies discriminating against some individuals in the hiring process? To find out, researchers prepare several equivalent résumés for fictitious job applicants, with the only difference being the gender of the applicant. They send the fake résumés to companies advertising positions and keep track of the number of males and females who are contacted for interviews.

• In a medical study of a new drug for migraine sufferers, volunteer subjects are randomly assigned to two groups. Members of the first group are given a placebo pill. Subjects in the second group are given the new drug. None of the subjects knows whether they are taking a placebo or the active drug. Neither do any of the physicians who are interacting with the subjects.

• Will people try to stop someone from driving drunk? A television news program hires an actor to play a drunk driver and uses a hidden camera to record the behavior of individuals who encounter the driver.

The most complex issues of data ethics arise when we collect data from people. The ethical difficulties are more severe for experiments that impose some treatment on people than for sample surveys that simply gather information. Trials of new medical treatments, for example, can do harm as well as good to their subjects. Here are some basic standards of data ethics that must be obeyed by all studies that gather data from human subjects, both observational studies and experiments.

Basic Data Ethics

All planned studies must be reviewed in advance by an **institutional review board** charged with protecting the safety and well-being of the subjects.

All individuals who are subjects in a study must give their **informed consent** before data are collected.

All individual data must be kept **confidential.** Only statistical summaries for groups of subjects may be made public.

*This is an important topic, but it is not required for the AP Statistics exam.

The law requires that studies carried out or funded by the federal government obey these principles.[52] But neither the law nor the consensus of experts is completely clear about the details of their application.

Institutional review boards The purpose of an institutional review board is not to decide whether a proposed study will produce valuable information or whether it is statistically sound. The board's purpose is, in the words of one university's board, "to protect the rights and welfare of human subjects (including patients) recruited to participate in research activities." The board reviews the plan of the study and can require changes. It reviews the consent form to be sure that subjects are informed about the nature of the study and about any potential risks. Once research begins, the board monitors its progress at least once a year.

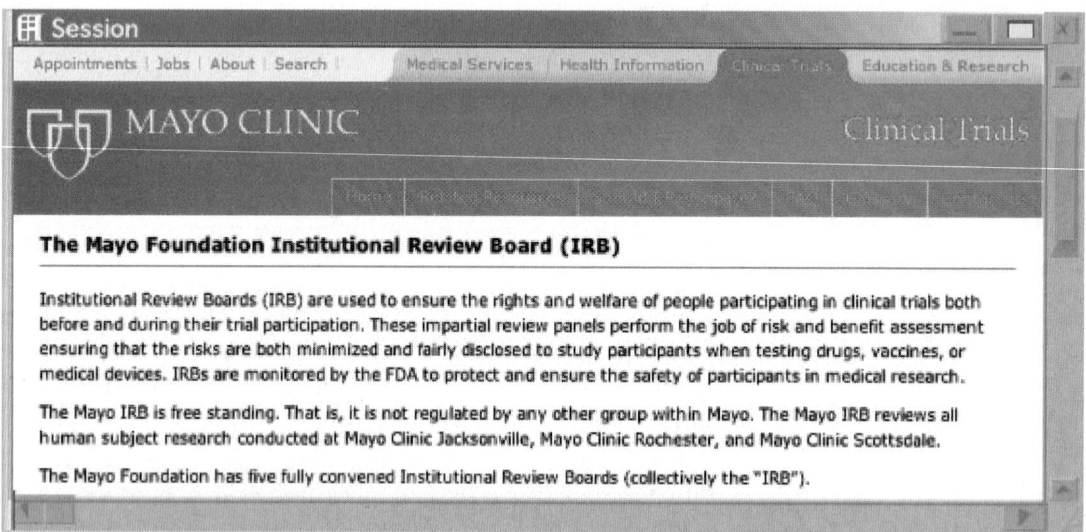

The Web page of the Mayo Clinic's institutional review board. It begins by describing the job of such boards.

Informed consent Both words in the phrase "informed consent" are important, and both can be controversial. Subjects must be *informed* in advance about the nature of a study and any risk of harm it may bring. In the case of a sample survey, physical harm is not possible. But a survey on sensitive issues could result in emotional harm. The participants should be told what kinds of questions the survey will ask and about how much of their time it will take. Experimenters must tell subjects the nature and purpose of the study and outline possible risks. Subjects must then *consent* in writing.

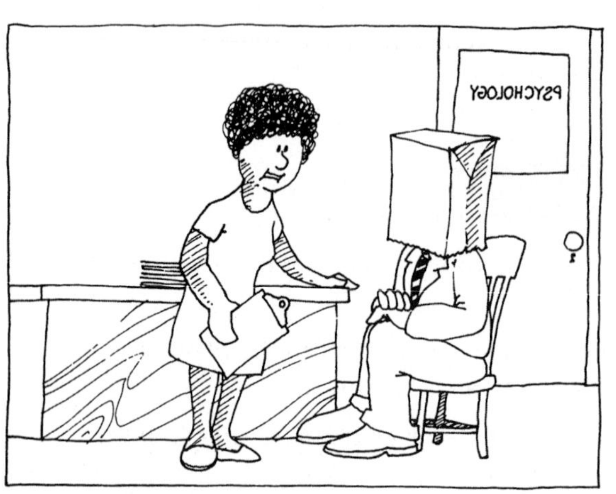

"I realize the participants in this study are to be anonymous, but you're going to have to expose your eyes."

Confidentiality Ethical problems do not disappear once a study has been cleared by the review board, has obtained consent from its participants, and has actually collected data about them. It is important to protect individuals' privacy by keeping all data about them **confidential.** The report of an opinion poll may say what percent of the 1200 respondents felt that legal immigration should be reduced. It may not report what *you* said about this or any other issue.

Confidentiality is not the same as **anonymity.** Anonymity means that individuals are anonymous—their names are not known even to the director of the study. Anonymity is rare in statistical studies. Even where anonymity is possible (mainly in surveys conducted by mail), it prevents any follow-up to improve nonresponse or inform individuals of results.

Any breach of confidentiality is a serious violation of data ethics. The best practice is to separate the identity of the study's participants from the rest of the data at once. A clever computer search of several data bases might be able, by combining information, to identify you and learn a great deal about you even if your name and other identification have been removed from the data available for search. Privacy and confidentiality of data are hot issues among statisticians in the computer age.

ACTIVITY *Response bias*

In this Activity, your team will design and conduct an experiment to investigate the effects of response bias in surveys.[53] You may choose the topic for your surveys, but you must design your experiment so that it can answer at least one of the following questions:

- Can the wording of a question create response bias?
- Do the characteristics of the interviewer create response bias?
- Does anonymity change the responses to sensitive questions?
- Does manipulating the answer choices change the response?

1. Write a proposal describing the design of your experiment. Be sure to include

   (a) your chosen topic and which of the above questions you'll try to answer.

   (b) a detailed description of how you will obtain your subjects (minimum of 50). Your plan must be practical!

   (c) what your questions will be and how they will be asked.

   (d) a clear explanation of how you will implement your design.

   (e) precautions you will take to collect data ethically.

   Here are two examples of successful student projects:

   "Make-Up," by Caryn S. and Trisha T. (all questions asked to males)

   1. "Do you find females who wear makeup attractive?" (questioner wearing makeup: 75% answered yes)
   2. "Do you find females who wear makeup attractive?" (questioner not wearing makeup: 30% answered yes)

   "Cartoons" by Sean W. and Brian H.

   1. "Do you watch cartoons?" (90% answered yes)
   2. "Do you *still* watch cartoons?" (60% answered yes)

2. Once your teacher has approved your design, carry out the experiment. Record your data in tabular form.

3. Analyze your data. What conclusion do you draw? Provide appropriate graphical and numerical evidence to support your answer.

4. Prepare a report that includes the data you collected, your analysis from Step 3, and a discussion of any problems you encountered and how you dealt with them.

Magnets and Pain

Let's return to the chapter-opening Case Study (page 205). Researchers carried out an experiment using as subjects 50 volunteer patients with chronic pain. A completely randomized design was used, with 29 subjects receiving the active-magnet treatment and 21 subjects getting the fake-magnet treatment. The Fathom dotplot shows the final pain ratings for both groups (A = active magnet, I = inactive magnet).

The average pain rating after the treatment was 8.43 for the inactive-magnet group and 4.38 for the active-magnet group. This difference is statistically significant. What conclusion can we draw?

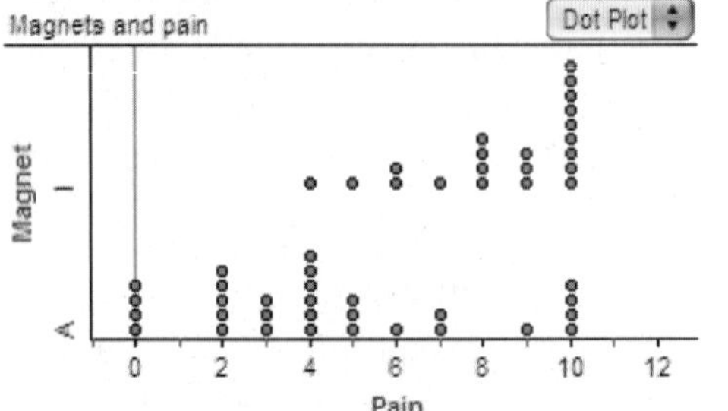

The difference in the average responses for the two groups was too large to be explained by chance variation in the random assignment. So we can infer that the magnet treatment caused a reduction in pain. But since the subjects were volunteers, we can't safely generalize the findings of this experiment to any larger population.

SECTION 4.3 Summary

- Most statistical studies aim to make inferences that go beyond the data actually produced. **Inference about the population** requires that the individuals taking part in a study be randomly selected from the larger population. A well-designed experiment that randomly assigns treatments to experimental units allows **inference about cause and effect.**
- **Lack of realism** in an experiment can prevent us from generalizing its results.
- In the absence of an experiment, good evidence of causation requires a strong association that appears consistently in many studies, a clear explanation for the alleged causal link, and careful examination of possible lurking variables.
- Studies involving humans must be screened in advance by an **institutional review board.** All participants must give their **informed consent** before taking part. Any information about the individuals in the study must be kept **confidential.**
- Remember that randomized comparative experiments can answer questions that can't be answered without them. Also remember that "the interests of the subject must always prevail over the interests of science and society."[54]

SECTION 4.3 Exercises

102. **Random sampling versus random assignment** Explain the difference between the types of inference that can be made as a result of random sampling and random assignment.

103. **Foster care versus orphanages** Do abandoned children placed in foster homes do better than similar children placed in an institution? The Bucharest Early Intervention Project found that the answer is a clear "Yes." The subjects were 136 young children abandoned at birth and living in orphanages in Bucharest, Romania. Half of the children, chosen at random, were placed in foster homes. The other half remained in the orphanages.[55] (Foster care was not easily available in Romania at the time and so was paid for by the study.) What conclusion can we draw from this study? Explain.

104. **Frozen batteries** Will storing batteries in a freezer make them last longer? To find out, a company that produces batteries takes a random sample of 100 AA batteries from its warehouse. The company statistician randomly assigns 50 batteries to be stored in the freezer and the other 50 to be stored at room temperature for 3 years. At the end of that time period, each battery's charge is tested. *Result:* Batteries stored in the freezer had a higher average charge, and the difference between the groups was statistically significant. What conclusion can we draw from this study? Explain.

105. **Who talks more—women or men?** According to Louann Brizendine, author of *The Female Brain*, women say nearly three times as many words per day as men. Skeptical researchers devised a study to test this claim. They used electronic devices to record the talking patterns of 396 university students who volunteered to participate in the study. The device was programmed to record 30 seconds of sound every 12.5 minutes without the carrier's knowledge. According to a published report of the study in *Scientific American*, "Men showed a slightly wider variability in words uttered.... But in the end, the sexes came out just about even in the daily averages: women at 16,215 words and men at 15,669."[56] This difference was not statistically significant. What conclusion can we draw from this study? Explain.

pg 262

106. **Attend church, live longer?** One of the better studies of the effect of regular attendance at religious services gathered data from a random sample of 3617 adults. The researchers then measured lots of variables, not just the explanatory variable (religious activities) and the response variable (length of life). A news article said: "Churchgoers were more likely to be nonsmokers, physically active, and at their right weight. But even after health behaviors were taken into account, those not attending religious services regularly still were about 25% more likely to have died."[57] What conclusion can we draw from this study? Explain.

107. **Daytime running lights** Canada requires that cars be equipped with "daytime running lights," headlights that automatically come on at a low level when the car is started. Some manufacturers are now equipping cars sold in the United States with running lights. Will running lights reduce accidents by making cars more visible? An experiment conducted in a driving simulator suggests that the answer may be "Yes." What concerns would you have about generalizing the results of such an experiment?

108. **Studying frustration** A psychologist wants to study the effects of failure and frustration on the relationships among members of a work team. She forms a team of students, brings them to the psychology lab, and has them play a game that requires teamwork. The game is rigged so that they lose regularly. The psychologist observes the students through a one-way window and notes the changes in their behavior during an evening of game playing. Can the psychologist generalize the results of her study to a team of employees that spends months developing a new product that never works right and is finally abandoned by their company? Explain.

109.* **Minimal risk?** You have been invited to serve on a college's institutional review board. You must decide whether several research proposals qualify for lighter review because they involve only minimal risk to subjects. Federal regulations say that "minimal risk" means the risks are no greater than "those ordinarily encountered in daily life or during the performance of routine physical or psychological examinations or tests." That's vague. Which of these do you think qualifies as "minimal risk"?

*Exercises 109 to 116: This is an important topic, but it is not required for the AP Statistics exam.

(a) Draw a drop of blood by pricking a finger to measure blood sugar.

(b) Draw blood from the arm for a full set of blood tests.

(c) Insert a tube that remains in the arm, so that blood can be drawn regularly.

110.* **Who reviews?** Government regulations require that institutional review boards consist of at least five people, including at least one scientist, one nonscientist, and one person from outside the institution. Most boards are larger, but many contain just one outsider.

(a) Why should review boards contain people who are not scientists?

(b) Do you think that one outside member is enough? How would you choose that member? (For example, would you prefer a medical doctor? A member of the clergy? An activist for patients' rights?)

111.* **No consent needed?** In which of the circumstances below would you allow collecting personal information without the subjects' consent?

(a) A government agency takes a random sample of income tax returns to obtain information on the average income of people in different occupations. Only the incomes and occupations are recorded from the returns, not the names.

(b) A social psychologist attends public meetings of a religious group to study the behavior patterns of members.

(c) A social psychologist pretends to be converted to membership in a religious group and attends private meetings to study the behavior patterns of members.

112.* **Surveys of youth** A survey asked teenagers whether they had ever consumed an alcoholic beverage. Those who said "Yes" were then asked, "How old were you when you first consumed an alcoholic beverage?" Should consent of parents be required to ask minors about alcohol, drugs, and other such issues, or is consent of the minors themselves enough? Give reasons for your opinion.

113.* **Anonymous? Confidential?** One of the most important nongovernment surveys in the United States is the National Opinion Research Center's General Social Survey. The GSS regularly monitors public opinion on a wide variety of political and social issues. Interviews are conducted in person in the subject's home. Are a subject's responses to GSS questions anonymous, confidential, or both? Explain your answer.

114.* **Anonymous? Confidential?** Texas A&M, like many universities, offers screening for HIV, the virus that causes AIDS. Students may choose either anonymous or confidential screening. An announcement says, "Persons who sign up for screening will be assigned a number so that they do not have to give their name." They can learn the results of the test by telephone, still without giving their name. Does this describe the *anonymous* or the *confidential* screening? Why?

115.* **The Willowbrook hepatitis studies** In the 1960s, children entering the Willowbrook State School, an institution for the mentally retarded, were deliberately infected with hepatitis. The researchers argued that almost all children in the institution quickly became infected anyway. The studies showed for the first time that two strains of hepatitis existed. This finding contributed to the development of effective vaccines. Despite these valuable results, the Willowbrook studies are now considered an example of unethical research. Explain why, according to current ethical standards, useful results are not enough to allow a study.

116.* **Unequal benefits** Researchers on aging proposed to investigate the effect of supplemental health services on the quality of life of older people. Eligible patients on the rolls of a large medical clinic were to be randomly assigned to treatment and control groups. The treatment group would be offered hearing aids, dentures, transportation, and other services not available without charge to the control group. The review board felt that providing these services to some but not other persons in the same institution raised ethical questions. Do you agree?

117. **Animal testing** (1.1) "It is right to use animals for medical testing if it might save human lives." The General Social Survey asked 1152 adults to react to this statement. Here is the two-way table of their responses:

	Male	Female
Strongly agree	76	59
Agree	270	247
Neither agree nor disagree	87	139
Disagree	61	123
Strongly disagree	22	68

How do the distributions of opinion differ between men and women? Give appropriate graphical and numerical evidence to support your answer.

118. **Initial public offerings** (1.3) The business magazine *Forbes* reports that 4567 companies sold their first stock to the public between 1990 and 2000. The *mean* change in the stock price of these companies since the first stock was issued was +111%. The *median* change was −31%.[58] Explain how this could happen. (*Hint:* Start with the fact that Cisco Systems stock went up 60,600%.)

*Exercises 109 to 116: This is an important topic, but it is not required for the AP Statistics exam.

Chapter 4 Chapter Summary

This chapter has explained good techniques for producing data and has also explained why bad techniques often produce worthless data. If we want to draw conclusions that go beyond the data at hand, then we have to design the data production appropriately. The figures below display the big ideas visually.

Random sampling and randomized comparative experiments are perhaps the two most important statistical inventions of the twentieth century. Unfortunately, you will still see many voluntary response samples and uncontrolled experiments.

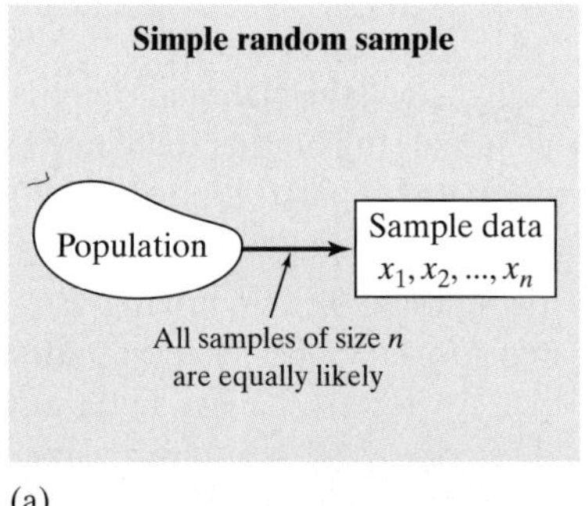

(a)

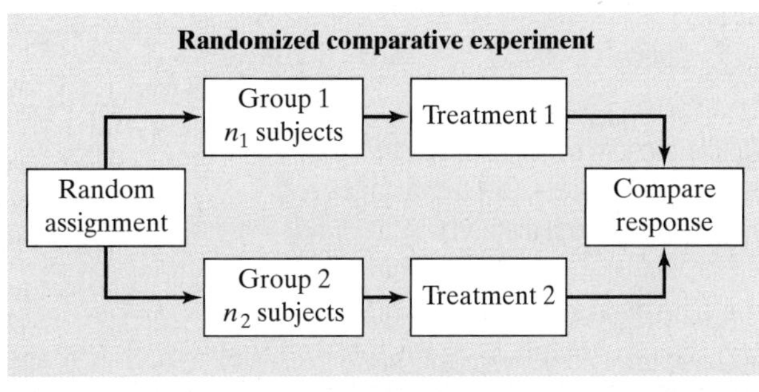

(b)

Chapter Review Exercises

These exercises are designed to help you review the important ideas and methods of the chapter. Relevant learning objectives are provided in bulleted form before each exercise.

- Identify the population and sample in a sample survey.
- Describe the scope of inference that is appropriate in an observational study.

R4.1. Ontario Health Survey The Ministry of Health in the province of Ontario, Canada, wants to know whether the national health care system is achieving its goals in the province. Much information about health care comes from patient records, but that source doesn't allow us to compare people who use health services with those who don't. So the Ministry of Health conducted the Ontario Health Survey, which interviewed a random sample of 61,239 people who live in the province of Ontario.[59]

(a) What is the population for this sample survey? What is the sample?

(b) The survey found that 76% of males and 86% of females in the sample had visited a general practitioner at least once in the past year. Do you think these estimates are close to the truth about the entire population? Why?

- Identify voluntary response samples and convenience samples. Explain how these bad sampling methods can lead to bias.

R4.2. Bad sampling A large high school wants to gather student opinion about parking for students on campus. It isn't practical to contact all students.

(a) Give an example of a way to choose a voluntary response sample of students. Explain how this method could lead to bias.

(b) Give an example of a way to choose a convenience sample of students. Explain how this method could lead to bias.

- **Describe how to use Table D to select a simple random sample (SRS).**

R4.3. Drug testing A baseball team regularly conducts random drug tests on its players. The 25 members of the team are listed below.

Agarwal	Chen	Healy	Moser	Roberts
Andrews	Frank	Hixson	Musselman	Shen
Baer	Fuest	Lee	Pavnica	Smith
Berger	Fuhrmann	Lynch	Petrucelli	Sundheim
Brockman	Garcia	Milhalko	Reda	Wilson

(a) Explain how you would use the line of random digits below to select an SRS of 3 team members for a random drug test.

(b) Use your method from (a) to choose the SRS. Show how you are using the digits below.

17521 78009 46239 84569 03316

- **Distinguish a simple random sample from a stratified random sample or cluster sample. Give advantages and disadvantages of each sampling method.**

R4.4. Polling the faculty A researcher wants to study the attitudes of college faculty members about the work habits of entering freshmen. These attitudes appear to differ depending on the type of college. The American Association of University Professors classifies colleges as follows:

Class I: Offer doctorate degrees and award at least 15 per year.

Class IIA: Award degrees above the bachelor's but are not in Class I.

Class IIB: Award no degrees beyond the bachelor's.

Class III: Two-year colleges.

The researcher would like to survey about 200 faculty members. Would you recommend a simple random sample, stratified random sample, or cluster sample? Justify your answer.

- **Explain how undercoverage, nonresponse, and question wording can lead to bias in a sample survey.**

R4.5. Been to the movies? An opinion poll calls 2000 randomly chosen residential telephone numbers, then asks to speak with an adult member of the household. The interviewer asks, "How many movies have you watched in a movie theater in the past 12 months?" In all, 1131 people respond.

(a) Identify a potential source of bias related to the question being asked. Suggest a change that would help fix this problem.

(b) Identify a potential source of bias in this survey that is not related to question wording. Suggest a change that would help fix this problem.

- **Distinguish between an observational study and an experiment.**

- **Explain how a lurking variable in an observational study can lead to confounding.**

R4.6. Are anesthetics safe? The National Halothane Study was a major investigation of the safety of anesthetics used in surgery. Records of over 850,000 operations performed in 34 major hospitals showed the following death rates for four common anesthetics:[60]

Anesthetic:	A	B	C	D
Death rate:	1.7%	1.7%	3.4%	1.9%

There seems to be a clear association between the anesthetic used and the death rate of patients. Anesthetic C appears to be dangerous.

(a) Explain why we call the National Halothane Study an observational study rather than an experiment, even though it compared the results of using different anesthetics in actual surgery.

(b) When the study looked at lurking variables that are related to a doctor's choice of anesthetic, it found that Anesthetic C was not causing extra deaths. Describe how a lurking variable could lead to confounding in this setting.

- **Identify the experimental units or subjects, explanatory variables (factors), treatments, and response variables in an experiment.**

R4.7. Ugly fries Few people want to eat discolored french fries. Potatoes are kept refrigerated before being cut for french fries to prevent spoiling and preserve flavor. But immediate processing of cold potatoes causes discoloring due to complex chemical reactions. The potatoes must therefore be brought to room temperature before processing. Researchers want to design an experiment in which tasters will rate the color and flavor of french fries prepared from several groups of potatoes. The potatoes will be freshly picked or stored for a month at room temperature or stored for a month refrigerated. They will then be sliced and cooked either immediately or after an hour at room temperature. Identify the experimental units, explanatory variable(s), the treatments, and the response variable(s).

- **Describe a completely randomized design for an experiment.**

R4.8. Attitudes toward homeless people Negative attitudes toward poor people are common. Are attitudes more negative when a person is homeless? To find out, read to subjects a description of a poor person. There are two versions. One begins

Jim is a 30-year-old single man. He is currently living in a small single-room apartment.

The other description begins

Jim is a 30-year-old single man. He is currently homeless and lives in a shelter for homeless people.

After reading the description, ask subjects what they believe about Jim and what they think should be done to help him. The subjects are 544 adults interviewed by telephone.[61] Outline a completely randomized design for this experiment. Write a few sentences describing how you would implement your design.

- Describe how to avoid the placebo effect in an experiment.
- Explain the meaning and the purpose of blinding in an experiment.

R4.9. An herb for depression? Does the herb Saint-John's-wort relieve major depression? Here is an excerpt from the report of a study of this issue: "Design: Randomized, double-blind, placebo-controlled clinical trial."[62] The study concluded that the herb is no more effective than a placebo.

(a) How did the design of this experiment account for the placebo effect?

(b) Why is a double-blind design a good idea in this setting?

- Distinguish between a completely randomized design and a randomized block design.
- Explain why random assignment is an important experimental design principle.
- Explain in context what "statistically significant" means.

R4.10. Vitamin C for marathon runners An ultramarathon, as you might guess, is a footrace longer than the 26.2 miles of a marathon. Runners commonly develop respiratory infections after an ultramarathon. Will taking 600 milligrams of vitamin C daily reduce those infections? Researchers randomly assigned ultramarathon runners to receive either vitamin C or a placebo. Separately, they also randomly assigned these treatments to a group of nonrunners the same age as the runners. All subjects were watched for 14 days after the big race to see if infections developed.

(a) What type of experimental design did the researchers use? Why is this design preferable in this setting? Give one reason.

(b) Explain the purpose of the random assignment.

(c) The report of the study said: "Sixty-eight percent of the runners in the placebo group reported the development of symptoms of upper respiratory tract infection after the race; this was significantly more than that reported by the vitamin C–supplemented group (33%)."[63] Explain to someone who knows no statistics why "significantly more" means there is good reason to think that vitamin C works.

- Know when a matched pairs experimental design is appropriate and how to implement such a design.

R4.11. How long did I work? A psychologist wants to know if the difficulty of a task influences our estimate of how long we spend working at it. She designs two sets of mazes that subjects can work through on a computer. One set has easy mazes and the other has hard mazes. Subjects work until told to stop (after 6 minutes, but subjects do not know this). They are then asked to estimate how long they worked. The psychologist has 30 students available to serve as subjects.

(a) Describe an experiment using a completely randomized design to learn the effect of difficulty on estimated time.

(b) Describe a matched pairs experimental design using the same 30 subjects.

(c) Which design would be more likely to detect a difference in the effects of the treatments? Explain.

- Evaluate whether a statistical study has been carried out in an ethical manner.
- Determine the scope of inference for a statistical study.

R4.12* Deceiving subjects Students sign up to be subjects in a psychology experiment. When they arrive, they are told that interviews are running late and are taken to a waiting room. The experimenters then stage a theft of a valuable object left in the waiting room. Some subjects are alone with the thief, and others are in pairs—these are the treatments being compared. Will the subject report the theft?

(a) The students had agreed to take part in an unspecified study, and the true nature of the experiment is explained to them afterward. Does this meet the requirement of informed consent? Explain.

(b) What two other ethical principles should be followed in this study?

(c) Does this study allow for inference about a population, inference about cause and effect, both, or neither? Justify your answer.

*This is an important topic, but it is not required for the AP Statistics exam.

Chapter 4 AP Statistics Practice Test

Section I: Multiple Choice *Select the best answer for each question.*

T4.1. When we take a census, we attempt to collect data from

(a) a stratified random sample.

(b) every individual chosen in a simple random sample.

(c) every individual in the population.

(d) a voluntary response sample.

(e) a convenience sample.

T4.2. You want to take a simple random sample (SRS) of 50 of the 816 students who live in a dormitory on campus. You label the students 001 to 816 in alphabetical order. In the table of random digits, you read the entries

95592 94007 69769 33547 72450 16632 81194 14873

The first three students in your sample have labels

(a) 955, 929, 400.
(b) 400, 769, 769.
(c) 559, 294, 007.
(d) 929, 400, 769.
(e) 400, 769, 335.

T4.3. A study of treatments for angina (pain due to low blood supply to the heart) compared bypass surgery, angioplasty, and use of drugs. The study looked at the medical records of thousands of angina patients whose doctors had chosen one of these treatments. It found that the average survival time of patients given drugs was the highest. What do you conclude?

(a) This study proves that drugs prolong life and should be the treatment of choice.

(b) We can conclude that drugs prolong life because the study was a comparative experiment.

(c) We can't conclude that drugs prolong life because the patients were volunteers.

(d) We can't conclude that drugs prolong life because this was an observational study.

(e) We can't conclude that drugs prolong life because no placebo was used.

T4.4. A simple random sample (SRS) is

(a) any sample selected by using chance.

(b) any sample that gives every individual the same chance to be selected.

(c) a sample that gives every possible sample of the same size the same chance to be selected.

(d) a sample that doesn't involve strata or clusters.

(e) a sample that is guaranteed to be representative of the population.

T4.5. Consider an experiment to investigate the effectiveness of different insecticides in controlling pests and their impact on the productivity of tomato plants. What is the best reason for randomly assigning treatment levels (spraying or not spraying) to the experimental units (farms)?

(a) Random assignment makes the experiment easier to conduct since we can apply the insecticide in any pattern rather than in a systematic fashion.

(b) Random assignment will tend to average out all other uncontrolled factors such as soil fertility so that they are not confounded with the treatment effects.

(c) Random assignment makes the analysis easier since the data can be collected and entered into the computer in any order.

(d) Random assignment is required by statistical consultants before they will help you analyze the experiment.

(e) Random assignment implies that it is not necessary to be careful during the experiment, during data collection, and during data analysis.

T4.6. The most important advantage of experiments over observational studies is that

(a) experiments are usually easier to carry out.

(b) experiments can give better evidence of causation.

(c) confounding cannot happen in experiments.

(d) an observational study cannot have a response variable.

(e) observational studies cannot use random samples.

T4.7. A TV station wishes to obtain information on the TV viewing habits in its market area. The market area contains one city of population 170,000, another city of 70,000, and four towns of about 5000 inhabitants each. The station suspects that the viewing habits may be different in larger and smaller cities and in the rural areas. Which of the following sampling designs would give the type of information that the station requires?

(a) A cluster sample using the cities and towns as clusters

(b) A convenience sample from the market area

(c) A simple random sample from the whole market area

(d) A stratified sample from the cities and towns in the market area

(e) An online poll that invites all people from the cities and towns in the market area to participate

T4.8. *Bias* in a sampling method is

(a) any error in the sample result, that is, any deviation of the sample result from the truth about the population.

(b) the random error due to using chance to select a sample.

(c) any error due to practical difficulties such as contacting the subjects selected.

(d) any systematic error that tends to occur in the same direction whenever you use this sampling method.

(e) racism or sexism on the part of those who take the sample.

T4.9. You wonder if TV ads are more effective when they are longer or repeated more often or both. So you design an experiment. You prepare 30-second and 60-second ads for a camera. Your subjects all watch the same TV program, but you assign them at random to four groups. One group sees the 30-second ad once during the program; another sees it three times; the third group sees the 60-second ad once; and the last group sees the 60-second ad three times. You ask all subjects how likely they are to buy the camera.

(a) This is a randomized block design, but not a matched pairs design.

(b) This is a matched pairs design.

(c) This is a completely randomized design with one explanatory variable (factor).

(d) This is a completely randomized design with two explanatory variables (factors).

(e) This is a completely randomized design with four explanatory variables (factors).

T4.10. A researcher wishes to compare the effects of 2 fertilizers on the yield of soybeans. She has 20 plots of land available for the experiment, and she decides to use a matched pairs design with 10 pairs of plots. To carry out the random assignment for this design, the researcher should

(a) use a table of random numbers to divide the 20 plots into 10 pairs and then, for each pair, flip a coin to assign the fertilizers to the 2 plots.

(b) subjectively divide the 20 plots into 10 pairs (making the plots within a pair as similar as possible) and then, for each pair, flip a coin to assign the fertilizers to the 2 plots.

(c) use a table of random numbers to divide the 20 plots into 10 pairs and then use the table of random numbers a second time to decide upon the fertilizer to be applied to each member of the pair.

(d) flip a coin to divide the 20 plots into 10 pairs and then, for each pair, use a table of random numbers to assign the fertilizers to the 2 plots.

(e) use a table of random numbers to assign the 2 fertilizers to the 20 plots and then use the table of random numbers a second time to place the plots into 10 pairs.

T4.11. You want to know the opinions of American high school teachers on the issue of establishing a national proficiency test as a prerequisite for graduation from high school. You obtain a list of all high school teachers belonging to the National Education Association (the country's largest teachers' union) and mail a survey to a random sample of 2500 teachers. In all, 1347 of the teachers return the survey. Of those who responded, 32% say that they favor some kind of national proficiency test. Which of the following statements about this situation is true?

(a) Since random sampling was used, we can feel confident that the percent of all American high school teachers who would say they favor a national proficiency test is close to 32%.

(b) We cannot trust these results, because the survey was mailed. Only survey results from face-to-face interviews are considered valid.

(c) Because over half of those who were mailed the survey actually responded, we can feel pretty confident that the actual percent of all American high school teachers who would say they favor a national proficiency test is close to 32%.

(d) The results of this survey may be affected by nonresponse bias.

(e) The results of this survey cannot be trusted due to voluntary response bias.

Section II: Free Response *Show all your work. Indicate clearly the methods you use, because you will be graded on the correctness of your methods as well as on the accuracy and completeness of your results and explanations.*

T4.12. Elephants sometimes damage trees in Africa. It turns out that elephants dislike bees. They recognize beehives in areas where they are common and avoid them. Can this be used to keep elephants away from trees? A group in Kenya placed active beehives in some trees and empty beehives in others. Will elephant damage be less in trees with hives? Will even empty hives keep elephants away? Researchers want to design an experiment to answer these questions using 72 acacia trees.[64]

(a) Identify the experimental units, treatments, and the response variable.

(b) Describe how the researchers could carry out a completely randomized design for this experiment. Include a description of how the treatments should be assigned.

T4.13. A 2008 *New York Times* article on public opinion about steroid use in baseball discussed the results of a sample survey. The survey found that 34% of adults think that at least half of Major League Baseball (MLB) players "use steroids to enhance their athletic performance." Another 36% thought that about a quarter of MLB players use steroids; 8% had no opinion. Here is part of the *Times*'s statement on "How the Poll Was Conducted":

> The latest New York Times/CBS News Poll is based on telephone interviews conducted March 15 through March 18 with 1,067 adults throughout the United States.... The sample of telephone numbers called was randomly selected by a computer from a list of more than 42,000 active residential exchanges across the country. The exchanges were chosen to ensure that each region of the country was represented in proportion to its population. In each exchange, random digits were added to form a complete telephone number, thus permitting access to listed and unlisted numbers. In each household, one adult was designated by a random procedure to be the respondent for the survey.[65]

(a) Explain why the sampling method used in this survey was *not* a simple random sample.

(b) Why was one adult chosen at random in each household to respond to the survey?

(c) Explain how undercoverage could lead to bias in this sample survey.

T4.14. Many people start their day with a jolt of caffeine from coffee or a soft drink. Most experts agree that people who take in large amounts of caffeine each day may suffer from physical withdrawal symptoms if they stop ingesting their usual amounts of caffeine. Researchers recruited 11 volunteers who were caffeine dependent and who were willing to take part in a caffeine withdrawal experiment. The experiment was conducted on two 2-day periods that occurred one week apart. During one of the 2-day periods, each subject was given a capsule containing the amount of caffeine normally ingested by that subject in one day. During the other study period, the subjects were given placebos. The order in which each subject received the two types of capsules was randomized. The subjects' diets were restricted during each of the study periods. At the end of each 2-day study period, subjects were evaluated using a tapping task in which they were instructed to press a button 200 times as fast as they could.[66]

(a) How and why was blocking used in the design of this experiment?

(b) Why did researchers randomize the order in which subjects received the two treatments?

(c) Could this experiment have been carried out in a double-blind manner? Explain.

Cumulative AP Practice Test 1

Section I: Multiple Choice *Choose the best answer for Questions AP1.1 to AP1.14.*

AP1.1. You look at real estate ads for houses in Sarasota, Florida. Many houses range from \$200,000 to \$400,000 in price. The few houses on the water, however, have prices up to \$15 million. Which of the following statements best describes the distribution of home prices in Sarasota?

(a) The distribution is most likely skewed to the left, and the mean is greater than the median.

(b) The distribution is most likely skewed to the left, and the mean is less than the median.

(c) The distribution is roughly symmetric with a few high outliers, and the mean is approximately equal to the median.

(d) The distribution is most likely skewed to the right, and the mean is greater than the median.

(e) The distribution is most likely skewed to the right, and the mean is less than the median.

AP1.2. A child is 40 inches tall, which places her at the 90th percentile of all children of similar age. The heights for children of this age form an approximately Normal distribution with a mean of 38 inches. Based on this information, what is the standard deviation of the heights of all children of this age?

(a) 0.20 inches (c) 0.65 inches (e) 1.56 inches
(b) 0.31 inches (d) 1.21 inches

AP1.3. A large set of test scores has mean 60 and standard deviation 18. If each score is doubled, and then 5 is subtracted from the result, the mean and standard deviation of the new scores are

(a) mean 115; std. dev. 31. (d) mean 120; std. dev. 31.
(b) mean 115; std. dev. 36. (e) mean 120; std. dev. 36.
(c) mean 120; std. dev. 6.

AP1.4. For a certain experiment, the available experimental units are eight rats, of which four are female (F1, F2, F3, F4) and four are male (M1, M2, M3, M4). There are to be four treatment groups, A, B, C, and D. If a randomized block design is used, with the experimental units blocked by gender, which of the following assignments of treatments is impossible?

(a) A → (F1, M1), B → (F2, M2), C → (F3, M3), D → (F4, M4)

(b) A → (F1, M2), B → (F2, M3), C → (F3, M4), D → (F4, M1)

(c) A → (F1, M2), B → (F3, F2), C → (F4, M1), D → (M3, M4)

(d) A → (F4, M1), B → (F2, M3), C → (F3, M2), D → (F1, M4)

(e) A → (F4, M1), B → (F1, M4), C → (F3, M2), D → (F2, M3)

AP1.5. For a biology project, you measure the weight in grams (g) and the tail length in millimeters (mm) of a group of mice. The equation of the least-squares line for predicting tail length from weight is

$$\text{predicted tail length} = 20 + 3 \times \text{weight}$$

Which of the following is *not* correct?

(a) The slope is 3, which indicates that a mouse's weight should increase by about 3 grams for each additional millimeter of tail length.

(b) The predicted tail length of a mouse that weighs 38 grams is 134 millimeters.

(c) By looking at the equation of the least-squares line, you can see that the correlation between weight and tail length is positive.

(d) If you had measured the tail length in centimeters instead of millimeters, the slope of the regression line would have been $3/10 = 0.3$.

(e) One mouse weighed 29 grams and had a tail length of 100 millimeters. The residual for this mouse is -7.

AP1.6. The figure below shows a Normal density curve. Which of the following gives the best estimates for the mean and standard deviation of this Normal distribution?

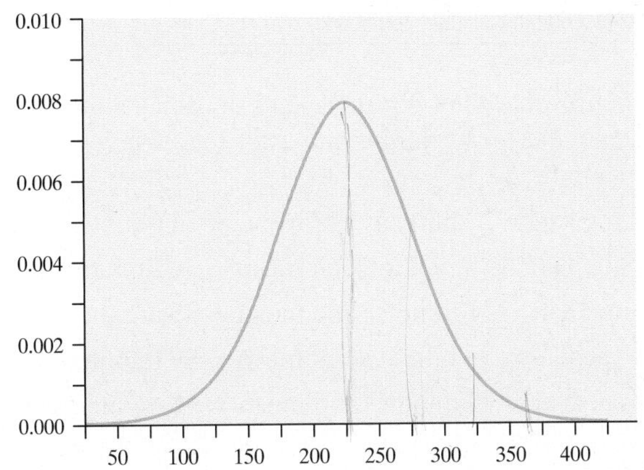

(a) $\mu = 200, \sigma = 50$
(b) $\mu = 200, \sigma = 25$
(c) $\mu = 225, \sigma = 50$
(d) $\mu = 225, \sigma = 25$
(e) $\mu = 225, \sigma = 275$

AP1.7. The owner of a chain of supermarkets notices that there is a positive correlation between the sales of beer and the sales of ice cream over the course of the previous year. During seasons when sales of beer were above average, sales of ice cream also tended to be above average. Likewise, during seasons when sales of beer were below average, sales of ice cream also tended to be below average. Which of the following would be a valid conclusion from these facts?

(a) Sales records must be in error. There should be no association between beer and ice cream sales.

(b) Evidently, for a significant proportion of customers of these supermarkets, drinking beer causes a desire for ice cream or eating ice cream causes a thirst for beer.

(c) A scatterplot of monthly ice cream sales versus monthly beer sales would show that a straight line describes the pattern in the plot, but it would have to be a horizontal line.

(d) There is a clear negative association between beer sales and ice cream sales.

(e) The positive correlation is most likely a result of the lurking variable temperature; that is, as temperatures increase, so do both beer sales and ice cream sales.

AP1.8. Here are the IQ scores of 10 randomly chosen fifth-grade students:

145 139 126 122 125 130 96 110 118 118

Which of the following statements about this data set is *not* true?

(a) The student with an IQ of 96 is considered an outlier by the $1.5 \times IQR$ rule.

(b) The five-number summary of the 10 IQ scores is 96, 118, 123.5, 130, 145.

(c) If the value 96 were removed from the data set, the mean of the remaining 9 IQ scores would be higher than the mean of all 10 IQ scores.

(d) If the value 96 were removed from the data set, the standard deviation of the remaining 9 IQ scores would be lower than the standard deviation of all 10 IQ scores.

(e) If the value 96 were removed from the data set, the *IQR* of the remaining 9 IQ scores would be lower than the *IQR* of all 10 IQ scores.

AP1.9. Before he goes to bed each night, Mr. Kleen pours dishwasher powder into his dishwasher and turns it on. Each morning, Mrs. Kleen weighs the box of dishwasher powder. From an examination of the data, she concludes that Mr. Kleen dispenses a rather consistent amount of powder each night. Which of the following statements is true?

I. There is a high positive correlation between the number of days that have passed since the box of dishwasher powder was opened and the amount of powder left in the box.

II. A scatterplot with days since purchase as the explanatory variable and amount of dishwasher powder used as the response variable would display a strong positive association.

III. The correlation between the amount of powder left in the box and the amount of powder used should be -1.

(a) I only
(b) II only
(c) III only
(d) II and III only
(e) I, II, and III

AP1.10. The General Social Survey (GSS), conducted by the National Opinion Research Center at the University of Chicago, is a major source of data on social attitudes in the United States. Once each year, 1500 adults are interviewed in their homes all across the country. The subjects are asked their opinions about sex and marriage, attitudes toward women, welfare, foreign policy, and many other issues. The GSS begins by selecting a sample of counties from the 3000 counties in the country. The counties are divided into urban, rural, and suburban; a separate sample is chosen at random from each group. This is a

(a) simple random sample.

(b) systematic random sample.

(c) cluster sample.

(d) stratified random sample.

(e) voluntary response sample.

AP1.11. You are planning an experiment to determine the effect of the brand of gasoline and the weight of a car on gas mileage measured in miles per gallon. You will use a single test car, adding weights so that its total weight is 3000, 3500, or 4000 pounds. The car will drive on a test track at each weight using each of Amoco, Marathon, and Speedway gasoline. Which is the best way to organize the study?

(a) Start with 3000 pounds and Amoco and run the car on the test track. Then do 3500 and 4000 pounds. Change to Marathon and go through the three weights in order. Then change to Speedway and do the three weights in order once more.

(b) Start with 3000 pounds and Amoco and run the car on the test track. Then change to Marathon and then to Speedway without changing the weight. Then add weights to get 3500 pounds and go through the three gasolines in the same order. Then change to 4000 pounds and do the three gasolines in order again.

(c) Choose a gasoline at random, and run the car with this gasoline at 3000, 3500, and 4000 pounds in order. Choose one of the two remaining gasolines at random and again run the car at 3000, then 3500, then 4000 pounds. Do the same with the last gasoline.

(d) There are nine combinations of weight and gasoline. Run the car several times using each of these combinations. Make all these runs in random order.

(e) Randomly select an amount of weight and a brand of gasoline, and run the car on the test track. Repeat this process a total of 30 times.

AP1.12. A linear regression was performed using the five following data points: A(2, 22), B(10, 4), C(6, 14), D(14, 2), E(18, −4). The residual for which of the five points has the largest absolute value?

(a) A (b) B (c) C (d) D (e) E

AP1.13. The frequency table below summarizes the times in the last month that patients at the emergency room of a small-city hospital waited to receive medical attention.

Waiting time	Frequency
Less than 10 minutes	5
At least 10 but less than 20 minutes	24
At least 20 but less than 30 minutes	45
At least 30 but less than 40 minutes	38
At least 40 but less than 50 minutes	19
At least 50 but less than 60 minutes	7
At least 60 but less than 70 minutes	2

Which of the following represents possible values for the median and mean waiting times for the emergency room last month?

(a) median = 27 minutes and mean = 24 minutes

(b) median = 28 minutes and mean = 30 minutes

(c) median = 31 minutes and mean = 35 minutes

(d) median = 35 minutes and mean = 39 minutes

(e) median = 45 minutes and mean = 46 minutes

AP1.14. Boxplots of two data sets are shown.

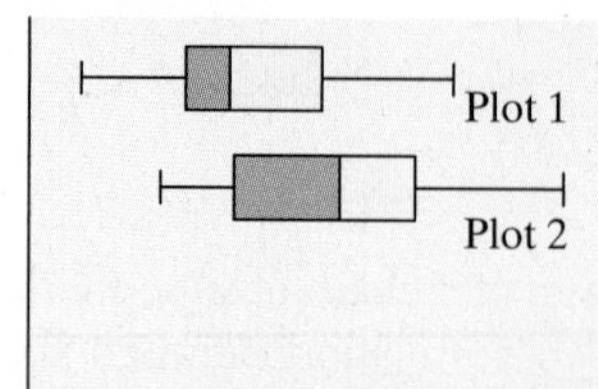

Based on the boxplots, which statement below is true?

(a) The spread of both plots is about the same.

(b) The means of both plots are approximately equal.

(c) Plot 2 contains more data points than Plot 1.

(d) The medians are approximately equal.

(e) Plot 1 is more symmetric than Plot 2.

Section II: Free Response *Show all your work. Indicate clearly the methods you use, because you will be graded on the correctness of your methods as well as on the accuracy and completeness of your results and explanations.*

AP1.15. The manufacturer of exercise machines for fitness centers has designed two new elliptical machines that are meant to increase cardiovascular fitness. The two machines are being tested on 30 volunteers at a fitness center near the company's headquarters. The volunteers are randomly assigned to one of the machines and use it daily for two months. A measure of cardiovascular fitness is administered at the start of the experiment and again at the end. The following table contains the differences in the two scores (After – Before) for the two machines. Note that higher scores indicate larger gains in fitness.

Machine A		Machine B
	5	3, 5, 9
6, 1	4	2, 5, 7
9, 7, 4, 1, 1	3	2, 4, 8, 9
8, 7, 6, 3, 2, 0	2	1, 5, 9
5, 4	1	0
	0	2

(a) Write a few sentences comparing the distributions of cardiovascular fitness gains from the two elliptical machines.

(b) Which machine should be chosen if the company wants to advertise it as achieving the highest overall gain in cardiovascular fitness? Explain your reasoning.

(c) Which machine should be chosen if the company wants to advertise it as achieving the most consistent gain in cardiovascular fitness? Explain your reasoning.

(d) Give one reason why the advertising claims of the company (the scope of inference) for this experiment would be limited. Explain how the company could broaden that scope of inference.

AP1.16. Those who advocate for monetary incentives in a work environment claim that this type of incentive has the greatest appeal because it allows the winners to do what they want with their winnings. Those in favor of tangible incentives argue that money lacks the emotional appeal of, say, a weekend for two at a romantic country inn or elegant hotel or a weeklong trip to Europe.

A few years ago a national tire company, in an effort to improve sales of a new line of tires, decided to test which method—offering cash incentives or offering non-cash prizes such as vacations—was more successful in increasing sales. The company had 60 retail sales districts of various sizes across the country and data on the previous sales volume for each district.

(a) Describe a completely randomized design using the 60 retail sales districts that would help answer this question.

(b) Explain how you would use the table of random digits below to do the randomization that your design requires. Then use your method to assign treatments to the first 3 experimental units. Show your work clearly.

07511 88915 41267 16853 84569 79367 32337 03316
81486 69487 60513 09297 00412 71238 27649 39950

(c) One of the company's officers suggested that it would be better to use a matched pairs design instead of a completely randomized design. Explain how you would change your design to accomplish this.

AP1.17. In retail stores, there is a lot of competition for shelf space. Not only are there national brands for most products, but many stores have their own in-house brands. Since shelf space is not infinite, the question is how many linear feet to allocate to each product and which shelf (top, bottom, or somewhere in the middle) to put it on. The middle shelf is the most popular and lucrative, since many shoppers, if undecided, will simply pick the product that is at eye level.

A local store that sells many upscale goods is trying to determine how much shelf space to allocate to its own brand of men's personal-grooming products. The middle shelf space is randomly varied between three and six linear feet over the next 12 weeks, and weekly sales revenue (in dollars) from the store's brand of personal-grooming products for men is recorded. Below is some computer output from the study, along with a scatterplot.

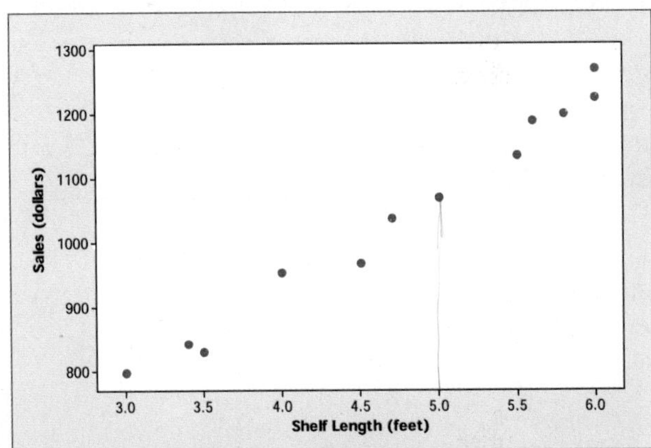


```
Predictor        Coef      SE Coef      T        P
Constant         317.94    31.32        10.15    0.000
Shelf length     152.680   6.445        23.69    0.000
S = 22.9212  R-Sq = 98.2%  R-Sq(adj) = 98.1%
```

(a) Does it appear that the weekly sales revenue is related to the shelf length allocated to the house brand? Justify your answer.

(b) Write the equation of the least-squares regression line. Be sure to define any variables you use.

(c) If the store manager were to decide to allocate five linear feet of shelf space to the store's brand of men's grooming products, what is the best estimate of the weekly sales revenue?

(d) Interpret the value of *s*.

(e) Identify and interpret the coefficient of determination.

(f) The store manager questions the intercept of the regression line: "Am I supposed to believe that this analysis tells me that I can sell these products with no shelf space?" How do you answer her?

Chapter 5

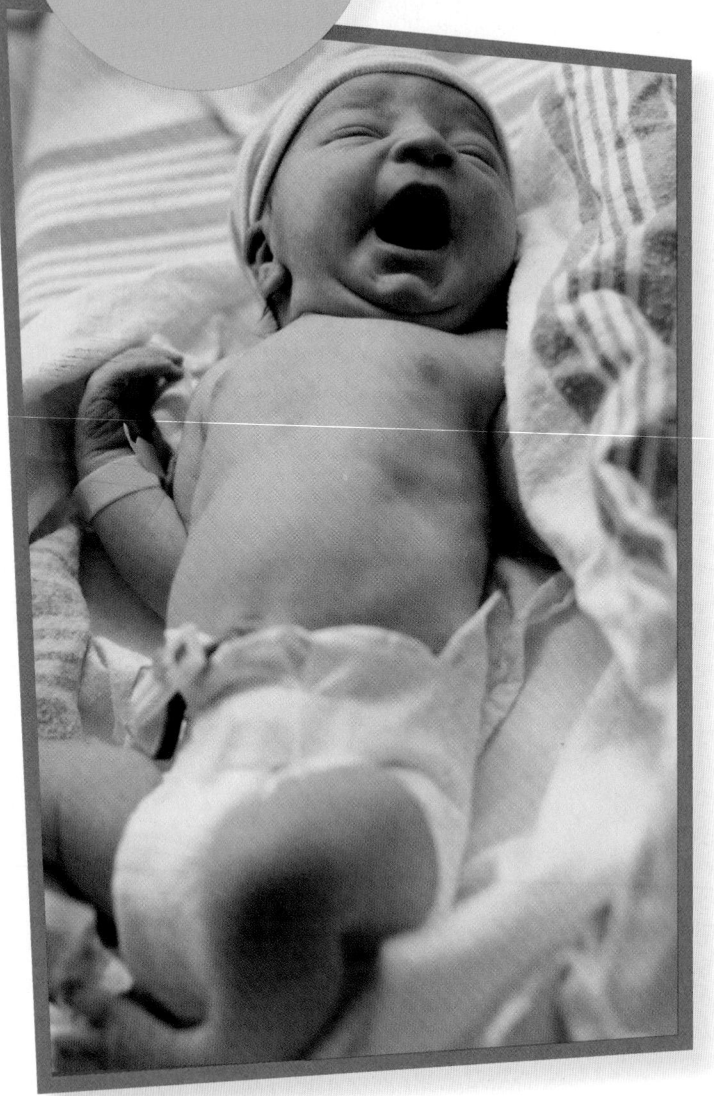

case study

Probability: What Are the Chances?

How Well Can Babies Hear?

According to the National Campaign for Hearing Health, deafness is the most common birth defect. Experts estimate that 3 of every 1000 babies are born with some kind of hearing loss. Early detection and treatment of hearing loss can improve a child's ability to speak and learn. Screening tests are given shortly after birth. A sound is made in a baby's ear and the baby's response is measured. Just because a baby fails the test, however, does not mean there is a hearing problem.

Studies were conducted at the University of Utah to test a new screening device. The accuracy of this new device was judged based on results found using standard equipment. Data from these studies are shown in the following table.[1]

		Baby's actual hearing	
		Loss	**Normal**
Test result	Loss	54	6
	Normal	4	36

If the new test says that a baby has hearing loss, what's the chance that the test is correct? By the end of this chapter, you should be able to answer this question.

Introduction

Chance is all around us. You and your friend play rock-paper-scissors to determine who gets the last slice of pizza. A coin toss decides which team gets to receive the ball first in a football game. Many adults regularly play the lottery, hoping to win a big jackpot with a few lucky numbers. Others head to casinos or racetracks, hoping that some combination of luck and skill will pay off. People young and old play games of chance involving cards or dice or spinners. The traits that children inherit—gender, hair and eye color, blood type, handedness, dimples, whether they can roll their tongues—are determined by the chance involved in which genes get passed along by their parents.

A roll of a die, a simple random sample, and even genetic inheritance represent chance behavior that we can understand and work with. We can roll the die again and again and again. The outcomes are governed by chance, but in many repetitions a pattern emerges. We use mathematics to understand the regular patterns of chance behavior when we repeat the same chance process again and again.

"What kind of childish nonsense are you working on now?"

The mathematics of chance is called *probability*. Probability is the topic of this chapter. Here is an Activity that gives you some idea of what lies ahead.

ACTIVITY *The "1 in 6 wins" game*

MATERIALS: One six-sided die for each student

As a special promotion for its 20-ounce bottles of soda, a soft drink company printed a message on the inside of each bottle cap. Some of the caps said, "Please try again!" while others said, "You're a winner!" The company advertised the promotion with the slogan "1 in 6 wins a prize." Seven friends each buy one 20-ounce bottle at a local convenience store. The store clerk is surprised when three of them win a prize. Is this group of friends just lucky, or is the company's 1-in-6 claim inaccurate? In this Activity, you and your classmates will perform a *simulation* to help answer this question.

For now, let's assume that the company is telling the truth, and that every 20-ounce bottle of soda it fills has a 1-in-6 chance of getting a cap that says, "You're a winner!" We can model the status of an individual bottle with a six-sided die: let 1 through 5 represent "Please try again!" and 6 represent "You're a winner!"

1. Roll your die seven times to imitate the process of the seven friends buying their sodas. How many of them won a prize?

2. Repeat Step 1 four more times. In your five repetitions of this simulation, how many times did three or more of the group win a prize?

3. Combine results with your classmates. What percent of the time did the friends come away with three or more prizes, just by chance?

4. Based on your answer in Step 3, does it seem plausible that the company is telling the truth, but that the seven friends just got lucky? Discuss this as a class.

As the Activity shows, simulation is a powerful method for modeling chance behavior. Section 5.1 begins by examining the idea of probability and then illustrates how simulation can be used to estimate probabilities. In Sections 5.2 and 5.3, we develop the basic rules of probability. Along the way, we introduce some helpful tools for displaying possible outcomes from a chance process: two-way tables, Venn diagrams, and tree diagrams.

Probability calculations are the basis for inference. When we produce data by random sampling or randomized comparative experiments, the laws of probability answer the question "What would happen if we did this many times?" Many of the examples, exercises, and activities in this chapter focus on the connection between probability and inference.

5.1 Randomness, Probability, and Simulation

In Section 5.1, you'll learn about:

- The idea of probability
- Myths about randomness
- Simulation

Toss a coin 10 times. How likely are you to get a run of 3 or more consecutive heads or tails? An airline knows that a certain percent of customers who purchased tickets will not show up for a flight. If the airline overbooks a particular flight, what are the chances that they'll have enough seats for the passengers who show up? A couple plans to have children until they have at least one child of each gender. How many children should they expect to have? To answer these questions, you need a better understanding of how chance behavior operates.

The Idea of Probability

In football, a coin toss helps determine which team gets the ball first. Why do the rules of football require a coin toss? Because tossing a coin seems a "fair" way to decide. That's one reason why statisticians recommend random samples and randomized experiments. They avoid bias by letting chance decide who gets selected or who receives which treatment. A big fact emerges when we watch coin tosses or the results of random sampling and random assignment closely: *chance behavior is unpredictable in the short run but has a regular and predictable pattern in the long run.* This remarkable fact is the basis for the idea of probability.

ACTIVITY Probability *applet*

MATERIALS: Computer with Internet connection

If you toss a fair coin, what's the chance that it shows heads? It's 1/2, or 0.5, right? In this Activity, you'll use the *Probability* applet at www.whfreeman.com/tps4e to investigate what probability really means.

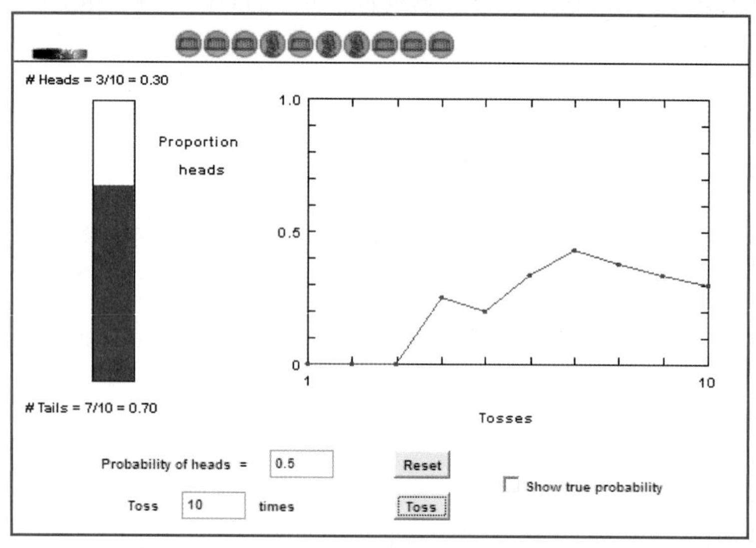

1. If you toss a fair coin 10 times, how many heads will you get? Before you answer, launch the *Probability* applet. Set the number of tosses at 10 and click "Toss." What proportion of the tosses were heads? Click "Reset" and toss the coin 10 more times. What proportion of heads did you get this time? Repeat this process several more times. What do you notice?

2. What if you toss the coin 100 times? Reset the applet and tell it to toss the coin 40 times (that's the maximum number of tosses it will do at one time). Then toss 40 more times, and 20 times after that, without hitting "Reset," for a total of 100. Is the proportion of heads exactly equal to 0.5? Close to 0.5?

3. Keep on tossing without hitting "Reset." What happens to the proportion of heads?

4. As a class, discuss what the following statement means: "If you toss a fair coin, the probability of heads is 0.5."

5. Predict what will happen if you change the probability of heads to 0.3 (an unfair coin). Then use the applet to test your prediction.

6. If you toss a coin, it can land heads or tails. If you "toss" a thumbtack, it can land with the point sticking up or with the point down. Does that mean that the probability of a tossed thumbtack landing point up is 0.5? How could you find out? Discuss with your classmates.

The following example looks at some results from the *Probability* applet Activity.

EXAMPLE *Tossing Coins*

Short-run and long-run behavior

When you toss a coin, there are only two possible outcomes, heads or tails. Figure 5.1(a) shows the results of tossing a coin 20 times. For each number of tosses from 1 to 20, we have plotted the proportion of those tosses that gave a head. You can see that the proportion of heads starts at 0 on the first toss, rises to 0.5 when the second toss gives a head, then falls to 0.33, 0.25, and 0.2 as we get three more tails. After that, the proportion of heads begins to increase, rising well above 0.5 before it decreases again.

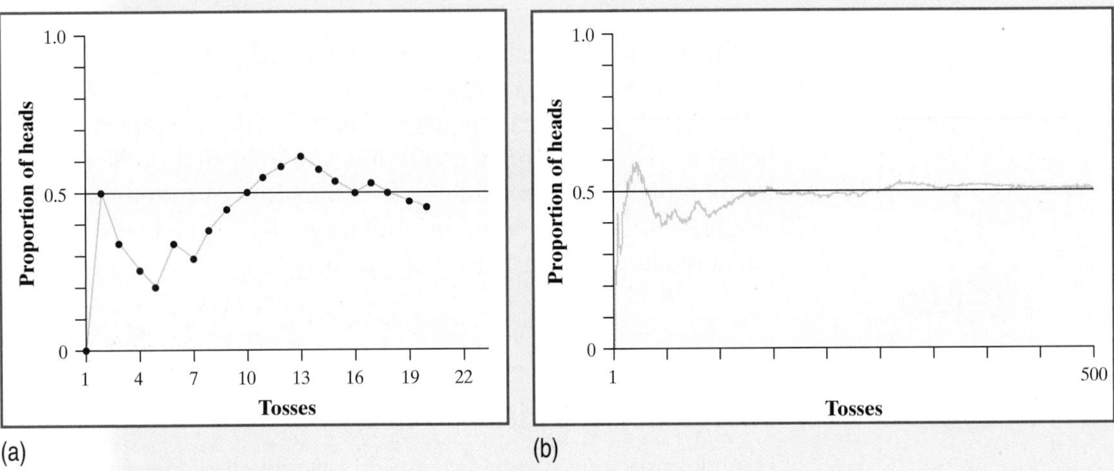

FIGURE 5.1 (a) The proportion of heads in the first 20 tosses of a coin. (b) The proportion of heads in the first 500 tosses of a coin.

Suppose we keep tossing the coin until we have made 500 tosses. Figure 5.1(b) shows the results. The proportion of tosses that produce heads is quite variable at first. As we make more and more tosses, however, the proportion of heads gets close to 0.5 and stays there.

Law of large numbers

The fact that the proportion of heads in many tosses eventually closes in on 0.5 is guaranteed by the **law of large numbers.** This important result says that if we observe more and more repetitions of any chance process, the proportion of times that a specific outcome occurs approaches a single value. We call this value the **probability.** The previous example confirms that the probability of getting a head when we toss a fair coin is 0.5. Probability 0.5 means "occurs half the time in a very large number of trials."

We might suspect that a coin has probability 0.5 of coming up heads just because the coin has two sides. But we can't be sure. In fact, spinning a penny on a flat surface, rather than tossing the coin, gives heads a probability of about 0.45 rather than 0.5.[2] What about thumbtacks? They also have two ways to land—point up or point down—but the chance that a tossed thumbtack lands point up isn't 0.5. How do we know that? From tossing a thumbtack over and over and over again. Probability describes what happens in very many trials, and we must actually observe many tosses of a coin or thumbtack to pin down a probability.

DEFINITION: Probability

The **probability** of any outcome of a chance process is a number between 0 and 1 that describes the proportion of times the outcome would occur in a very long series of repetitions.

Outcomes that never occur have probability 0. An outcome that happens on every repetition has probability 1. An outcome that happens half the time in a very long series of trials has probability 0.5. Of course, we can never observe a probability exactly. We could always continue tossing the coin, for example. The mathematics of probability is based on imagining what would happen in an indefinitely long series of trials.

Recall from Chapter 4 that "random" doesn't mean haphazard. In statistics, random means "by chance."

Probability gives us a language to describe the long-term regularity of random behavior. The outcome of a coin toss and the gender of the next baby born in a local hospital are both random. So is the result of a random sample or a random assignment. Even life insurance, for example, is based on the fact that deaths occur at random among many individuals.

Life Insurance

Probability and risk

How do insurance companies decide how much to charge for life insurance? We can't predict whether a particular person will die in the next year. But the National Center for Health Statistics says that the proportion of men aged 20 to 24 years who die in any one year is 0.0015. This is the *probability* that a randomly selected young man will die next year. For women that age, the probability of death is about 0.0005. If an insurance company sells many policies to people aged 20 to 24, it knows that it will have to pay off next year on about 0.15% of the policies sold to men and on about 0.05% of the policies sold to women. Therefore, the company will charge about three times more to insure a man because the probability of having to pay is three times higher.

We often encounter the unpredictable side of randomness in our everyday lives, but we rarely see enough repetitions of the same chance process to observe the long-run regularity that probability describes. Life insurance companies, casinos, and others who make important decisions based on probability rely on the long-run predictability of random behavior.

CHECK YOUR UNDERSTANDING

1. According to the "Book of Odds," the probability that a randomly selected U.S. adult usually eats breakfast is 0.61.

(a) Explain what probability 0.61 means in this setting.

(b) Why doesn't this probability say that if 100 U.S. adults are chosen at random, exactly 61 of them usually eat breakfast?

2. Probability is a measure of how likely an outcome is to occur. Match one of the probabilities that follow with each statement. Be prepared to defend your answer.

0 0.01 0.3 0.6 0.99 1

(a) This outcome is impossible. It can never occur.

(b) This outcome is certain. It will occur on every trial.

(c) This outcome is very unlikely, but it will occur once in a while in a long sequence of trials.

(d) This outcome will occur more often than not.

Myths about Randomness

The idea of probability seems straightforward. It answers the question "What would happen if we did this many times?" In fact, both the behavior of random phenomena and the idea of probability are a bit subtle. We meet chance behavior constantly, and psychologists tell us that we deal with it poorly.

ACTIVITY *Investigating randomness*

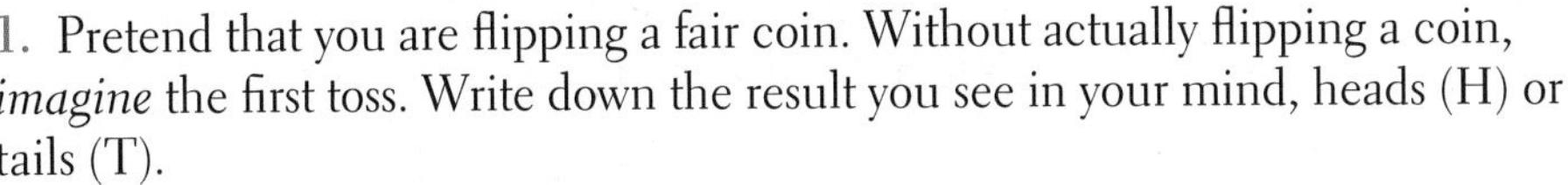

1. Pretend that you are flipping a fair coin. Without actually flipping a coin, *imagine* the first toss. Write down the result you see in your mind, heads (H) or tails (T).
2. Imagine a second coin flip. Write down the result.
3. Keep doing this until you have recorded the results of 50 imaginary flips. Write your results in groups of 5 to make them easier to read, like this: HTHTH TTHHT, etc.
4. A *run* is a repetition of the same result. In the example in Step 3, there is a run of two tails followed by a run of two heads in the first 10 coin flips. Read through your 50 imagined coin flips and count the number of runs of size 2, 3, 4, and so on. Record the number of runs of each size in a table like this:

Run length:	2	3	4	5	6	7	8
Frequency:							

5. Use Table D or technology to generate a similar list of 50 coin flips. Record the number of runs of size 2, 3, 4, and so forth in a table like this:

Run length:	2	3	4	5	6	7	8
Frequency:							

6. Compare the two results. Did you or your calculator have the longest run? How much longer?

The myth of short-run regularity The idea of probability is that randomness is predictable in the long run. Unfortunately, our intuition about randomness tries to tell us that random phenomena should also be predictable in the short run. When they aren't, we look for some explanation other than chance variation.

EXAMPLE *Runs in Coin Tossing*

What looks random?

Toss a coin six times and record heads (H) or tails (T) on each toss. Which of the following outcomes is more probable?

HTHTTH TTTHHH

Almost everyone says that HTHTTH is more probable, because TTTHHH does not "look random." In fact, both are equally likely. That heads and tails are equally probable says only that about half of a very long sequence of tosses will be heads. It doesn't say that heads and tails must come close to alternating in the short run. The coin has no memory. It doesn't know what past outcomes were, and it can't try to create a balanced sequence.

The outcome TTTHHH in tossing six coins looks unusual because of the runs of 3 straight tails and 3 straight heads. Runs seem "not random" to our intuition but are quite common. Here's a more striking example than tossing coins.

EXAMPLE

That Shooter Seems "Hot"

Chance variation or skill?

Is there such a thing as a "hot hand" in basketball? Belief that runs must result from something other than "just chance" influences behavior. If a basketball player makes several consecutive shots, both the fans and her teammates believe that she has a "hot hand" and is more likely to make the next shot. This is wrong. Careful study has shown that runs of baskets made or missed are no more frequent in basketball than would be expected if each shot is independent of the player's previous shots. If a player makes half her shots in the long run, her made shots and misses behave just like tosses of a coin—and that means that runs of makes and misses are more common than our intuition expects.[3]

The myth of the "law of averages" Once, at a convention in Las Vegas, one of the authors roamed the gambling floors, watching money disappear into the drop boxes under the tables. You can see some interesting human behavior in a casino. When the shooter in the dice game, craps, rolls several winners in a row, some gamblers think she has a "hot hand" and bet that she will keep on winning. Others say that "the law of averages" means that she must now lose so that wins and losses will balance out. Believers in the law of averages think that if you toss a coin six times and get TTTTTT, the next toss must be more likely to give a head. It's true that in the long run heads will appear half the time. What is a myth is that future outcomes must make up for an imbalance like six straight tails.

Don't confuse the law of large numbers, which describes the big idea of probability, with the "law of averages" described here.

Coins and dice have no memories. A coin doesn't know that the first six outcomes were tails, and it can't try to get a head on the next toss to even things out. Of course, things do even out *in the long run*. That's the law of large numbers in action. After 10,000 tosses, the results of the first six tosses don't matter. They are overwhelmed by the results of the next 9994 tosses.

Aren't We Due for a Boy?

Don't fall for the "law of averages"

"So the law of averages doesn't guarantee me a girl after seven straight boys, but can't I at least get a group discount on the delivery fee?"

Belief in this phony "law of averages" can lead to serious consequences. A few years ago, an advice columnist published a letter from a distraught mother of eight girls. It seems that she and her husband had planned to limit their family to four children. When all four were girls, they tried again—and again and again. After seven straight girls, even her doctor had assured her that "the law of averages was in our favor 100 to 1." Unfortunately for this couple, having children is like tossing coins. Eight girls in a row is highly unlikely, but once seven girls have been born, it is not at all unlikely that the next child will be a girl—and it was.

Simulation

Simulation

The imitation of chance behavior, based on a model that accurately reflects the situation, is called a **simulation.** You already have some experience with simulations. In Chapter 1's "Hiring Discrimination—It Just Won't Fly!" Activity (page 5), you drew cards to imitate a random lottery to choose which pilots would become captains. Chapter 4's "Distracted Driving" Activity (page 245) asked you to shuffle and deal piles of cards to mimic the random assignment of subjects to treatments. The "1 in 6 wins" game that opened this chapter had you roll a die several times to simulate buying 20-ounce sodas and looking under the cap. These simulations involved different chance "devices"—cards or dice. But the same basic strategy was followed in all three simulations. We can summarize this strategy using our familiar four-step process: State, Plan, Do, Conclude.

Performing a Simulation

State: What is the question of interest about some chance process?

Plan: Describe how to use a chance device to imitate one repetition of the process. Explain clearly how to identify the outcomes of the chance process and what variable to measure.

Do: Perform many repetitions of the simulation.

Conclude: Use the results of your simulation to answer the question of interest.

The following table summarizes this four-step process for each of our previous simulations.

	Hiring Discrimination	Distracted Driving	"1 In 6 Wins" Game
State	How likely is it that a fair lottery would result in 5 female and 3 male pilots being selected from an initial pool of 15 male and 10 female pilots?	Is it plausible that just by the luck of the random assignment, 12 of the subjects who were going to miss the rest area anyway ended up in the cell phone group, while only 3 were assigned to the passenger conversation group?	What's the probability that three of seven people who buy a 20-ounce bottle of soda win a prize if each bottle has a 1/6 chance of saying, "You're a winner!"?
Plan	Make a special deck with all 13 hearts plus any two diamonds and 10 spades. Red card = male pilot Black card = female pilot Shuffle and deal 8 cards. Count the number of female pilots selected.	Remove the ace of spades and three 2s from a deck of 52 cards. J, Q, K, A = miss rest area 2 through 10 = stop at rest area Shuffle well and deal two piles of 24 cards—the cell phone group and the passenger group. Record the number of drivers who miss the rest area in each group.	Use a six-sided die to determine the outcome for each person's bottle of soda. 6 = wins a prize 1 to 5 = no prize Roll the die seven times, once for each person. Record whether 3 or more people win a prize (yes or no).
Do	Have each student in class do this 5 times.	Have each pair of students repeat the process 10 times.	Each student should perform 5 repetitions.
Conclude	In 100 repetitions, 18 yielded 5 or more female pilots. So about 18% of the time, a fair lottery would choose at least 5 female pilots to become captains. We conclude that the company could have carried out a fair lottery.	In 300 repetitions of the random assignment, there were only 2 times when 12 or more drivers who missed the rest area ended up in the cell phone group. That's less than 1% of the time! It doesn't seem plausible that the random assignment is the explanation for the difference between the groups.	Out of 125 total repetitions of the simulation, there were 15 times when three or more of the seven people won a prize. So our estimate of the probability is 15/125, or about 12%.

So far, we have used physical devices for our simulations. Using random numbers from Table D or technology is another option, as the following examples illustrate.

EXAMPLE

STEP 4

Golden Ticket Parking Lottery
Simulations with Table D

At a local high school, 95 students have permission to park on campus. Each month, the student council holds a "golden ticket parking lottery" at a school assembly. The two lucky winners are given reserved parking spots next to the school's main entrance. Last month, the winning tickets were drawn by a student council member from the AP Statistics class. When both golden tickets went to members of that same class, some people thought the lottery had been rigged. There are 28 students in the AP Statistics class, all of whom are eligible to park on campus. Design and carry out a simulation to decide whether it's plausible that the lottery was carried out fairly.

STATE: **What's the probability that a fair lottery would result in two winners from the AP Statistics class?**

PLAN: **We'll use Table D to simulate choosing the golden ticket lottery winners. Since there are 95 eligible students in the lottery, we'll label the students in the AP Statistics class from 01 to 28, and the remaining students from 29 to 95. Numbers from 96 to 00 will be skipped. Moving left to right across the row, we'll look at pairs of digits until we come across two *different* labels from 01 to 95. The two students with these labels will win the prime parking spaces. We will record whether both winners are members of the AP Statistics class (Yes or No).**

DO: Let's perform many repetitions of our simulation. We'll use Table D starting at line 139. The digits from that row are shown below. We have drawn vertical bars to separate the pairs of digits. Underneath each pair, we have marked a √ if the chosen student is in the AP Statistics class, × if the student is not in the class, and "skip" if the number isn't between 01 and 95 or if that student was already selected. (Note that if the consecutive "70" labels had been in the same repetition, we would have skipped the second one.)

55 | 58	89 | 94	04 | 70	70 | 84	10 | 98 | 43	56 | 35	69 | 34	48 | 39	45 | 17
× ×	× ×	√ ×	× ×	√ skip ×	× ×	× ×	× ×	× √
Rep 1	Rep 2	Rep 3	Rep 4	Rep 5	Rep 6	Rep 7	Rep 8	Rep 9
No	No	No	No	No	No	No	No	No

In the first 9 repetitions, the two winners have never both been from the AP Statistics class. But 9 isn't many repetitions of the simulation. Continuing where we left off,

19 | 12	97 | 51 | 32	58 | 13	04 | 84	51 | 44	72 | 32	18 | 19	40 | 00 | 36	00 | 24 | 28
√ √	skip × ×	× √	√ ×	× ×	× ×	√ √	× skip ×	skip √ √
Rep 10	Rep 11	Rep 12	Rep 13	Rep 14	Rep 15	Rep 16	Rep 17	Rep 18
Yes	No	No	No	No	No	**Yes**	No	**Yes**

AP EXAM TIP On the AP exam, you may be asked to describe how you will perform a simulation using rows of random digits. If so, provide a clear enough description of your simulation process for the reader to get the same results you did from *only* your written explanation.

So after 18 repetitions, there have been 3 times when both winners were in the AP Statistics class. If we keep going for 32 more repetitions (to bring our total to 50), we find 30 more "No" and 2 more "Yes" results. All totaled, that's 5 "Yes" and 45 "No" results.

CONCLUDE: In our simulation of a fair lottery, both winners came from the AP Statistics class in 10% of the repetitions. So about 1 in every 10 times the student council holds the golden ticket lottery, this will happen just by chance. It seems plausible that the lottery was conducted fairly.

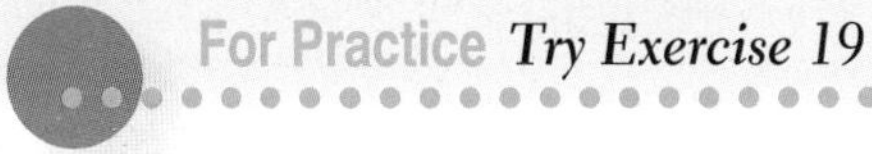

In the previous example, we could have saved a little time by using `randInt(1,95)` repeatedly instead of Table D (so we wouldn't have to worry about numbers 96 to 00). We'll take this alternate approach in the next example.

NASCAR Cards and Cereal Boxes

Simulations with technology

STEP 4

In an attempt to increase sales, a breakfast cereal company decides to offer a NASCAR promotion. Each box of cereal will contain a collectible card featuring one of these NASCAR drivers: Jeff Gordon, Dale Earnhardt, Jr., Tony Stewart, Danica Patrick, or Jimmie Johnson. The company says that each of the 5 cards is equally likely to appear in any box of cereal. A NASCAR fan decides to keep buying boxes of the cereal until she has all 5 drivers' cards. She is surprised when it takes her 23 boxes to get the full set of cards. Should she be surprised? Design and carry out a simulation to help answer this question.

STATE: What is the probability that it will take 23 or more boxes to get a full set of 5 NASCAR collectible cards?

PLAN: We need five numbers to represent the five possible cards. Let's let 1 = Jeff Gordon, 2 = Dale Earnhardt, Jr., 3 = Tony Stewart, 4 = Danica Patrick, and 5 = Jimmie Johnson. We'll use randInt(1,5) to simulate buying one box of cereal and looking at which card is inside.

In the golden ticket lottery example, we ignored repeated numbers from 01 to 95 within a given repetition. That's because the chance process involved sampling students *without* replacement. In the NASCAR example, we allowed repeated numbers from 1 to 5 in a given repetition. That's because the chance process of pretending to buy boxes of cereal and looking inside could have resulted in the same driver's card appearing in more than one box.

Since we want a full set of cards, we'll keep pressing Enter until we get all five of the labels from 1 to 5. We'll record the number of boxes that we had to open.

DO: It's time to perform many repetitions of the simulation. Here are our first few results:

352152354 9 boxes   5125141412224453 16 boxes   5552412153 10 boxes
435351115315452 15 boxes   3322124334223332334225 22 boxes

The Fathom dotplot shows the number of boxes we had to buy in 50 repetitions of the simulation.

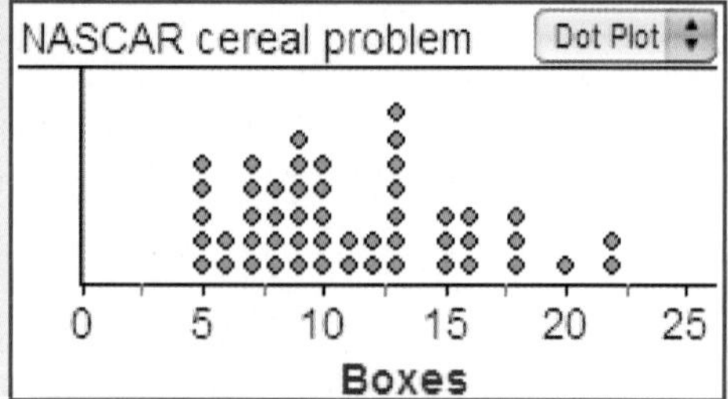

CONCLUDE: We never had to buy more than 22 boxes to get the full set of NASCAR drivers' cards in 50 repetitions of our simulation. So our estimate of the probability that it takes 23 or more boxes to get a full set is roughly 0. The NASCAR fan should be surprised about how many boxes she had to buy.

For Practice *Try Exercise 25*

THINK ABOUT IT

What *don't* these simulations tell us? For the golden ticket parking lottery, we concluded that it's plausible the drawing was done fairly. Does that mean the lottery *was* conducted fairly? Not necessarily. All we did was estimate that the probability of getting two winners from the AP Statistics class was about 10% *if* the drawing was fair. So the result isn't unlikely enough to convince us that the lottery was rigged. What about the cereal box simulation? It took our NASCAR fan 23 boxes to complete the set of 5 cards. Does that mean the company didn't tell the truth about how the cards were distributed? Not necessarily. Our simulation says that it's very unlikely for someone to have to buy 23 boxes to get a full set *if* each card is equally likely to appear in a box of cereal. The evidence suggests that the company's statement is incorrect. It's still possible, however, that the NASCAR fan was just very unlucky.

CHECK YOUR UNDERSTANDING

1. Refer to the golden ticket parking lottery example. At the following month's school assembly, the two lucky winners were once again members of the AP Statistics class. This raised suspicions about how the lottery was being conducted. How would you modify the simulation in the example to estimate the probability of this happening just by chance?

2. Refer to the NASCAR and breakfast cereal example. What if the cereal company decided to make it harder to get some drivers' cards than others? For instance, suppose the chance that each card appears in a box of the cereal is Jeff Gordon, 10%; Dale Earnhardt, Jr., 30%; Tony Stewart, 20%; Danica Patrick, 25%; and Jimmie Johnson, 15%. How would you modify the simulation in the example to estimate the chance that a fan would have to buy 23 or more boxes to get the full set?

SECTION 5.1 Summary

- A chance process has outcomes that we cannot predict but that nonetheless have a regular distribution in very many repetitions. The **law of large numbers** says that the proportion of times that a particular outcome occurs in many repetitions will approach a single number. This long-run relative frequency of a chance outcome is its **probability.** A probability is a number between 0 (never occurs) and 1 (always occurs).
- Probabilities describe only what happens in the long run. Short runs of random phenomena like tossing coins or shooting a basketball often don't look random to us because they do not show the regularity that in fact emerges only in very many repetitions.
- A **simulation** is an imitation of chance behavior, most often carried out with random numbers. To perform a simulation, follow the familiar four-step process:

STATE: Formulate a question of interest about some chance process.

PLAN: Describe how to use a chance device to imitate one repetition of the process. Explain clearly how to identify the outcomes of the chance process and what variable to measure.

DO: Perform many repetitions of the simulation.

CONCLUDE: Use the results of your simulation to answer the question of interest.

SECTION 5.1 Exercises

1. **Liar, liar!** Sometimes police use a lie detector (also known as a polygraph) to help determine whether a suspect is telling the truth. A lie detector test isn't foolproof—sometimes it suggests that a person is lying when they're actually telling the truth (a "false positive"). Other times, the test says that the suspect is being truthful when the person is actually lying (a "false negative"). For one brand of polygraph machine, the probability of a false positive is 0.08.

   (a) Interpret this probability as a long-run relative frequency.

   (b) Which is a more serious error in this case: a false positive or a false negative? Justify your answer.

2. **Drug testing** Athletes are often tested for use of performance-enhancing drugs. Drug tests aren't perfect—they sometimes say that an athlete took a banned substance when that isn't the case (a "false positive"). Other times, the test concludes that the athlete is "clean" when he or she actually took a banned substance (a "false negative"). For one commonly used drug test, the probability of a false negative is 0.03.

   (a) Interpret this probability as a long-run relative frequency.

   (b) Which is a more serious error in this case: a false positive or a false negative? Justify your answer.

3. **Genetics** Suppose a married man and woman both carry a gene for cystic fibrosis but don't have the disease themselves. According to the laws of genetics, the probability that their first child will develop cystic fibrosis is 0.25.

(a) Explain what this probability means.

(b) Why doesn't this probability say that if the couple has 4 children, one of them is guaranteed to get cystic fibrosis?

4. **Texas hold 'em** In the popular Texas hold 'em variety of poker, players make their best five-card poker hand by combining the two cards they are dealt with three of five cards available to all players. You read in a book on poker that if you hold a pair (two cards of the same rank) in your hand, the probability of getting four of a kind is 88/1000.

(a) Explain what this probability means.

(b) Why doesn't this probability say that if you play 1000 such hands, exactly 88 will be four of a kind?

5. **Spinning a quarter** With your forefinger, hold a new quarter (with a state featured on the reverse) upright, on its edge, on a hard surface. Then flick it with your other forefinger so that it spins for some time before it falls and comes to rest. Spin the coin a total of 25 times, and record the results.

(a) What's your estimate for the probability of heads? Why?

(b) Explain how you could get an even better estimate.

6. **Nickels falling over** You may feel it's obvious that the probability of a head in tossing a coin is about 1/2 because the coin has two faces. Such opinions are not always correct. Stand a nickel on edge on a hard, flat surface. Pound the surface with your hand so that the nickel falls over. Do this 25 times, and record the results.

(a) What's your estimate for the probability that the coin falls heads up? Why?

(b) Explain how you could get an even better estimate.

7. **Free throws** The figure below shows the results of a basketball player shooting several free throws. Explain what this graph says about chance behavior in the short run and long run.

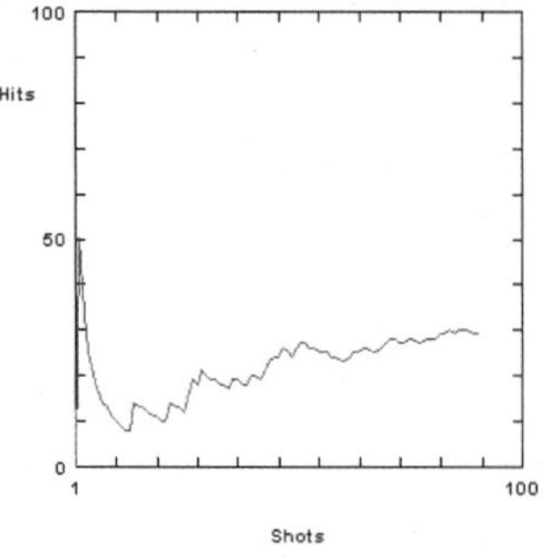

8. **Keep on tossing** The figure below shows the results of two different sets of 5000 coin tosses. Explain what this graph says about chance behavior in the short run and the long run.

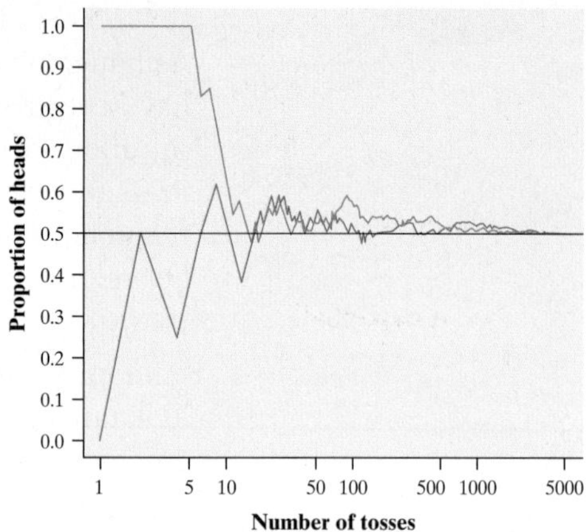

9. **Due for a hit** A very good professional baseball player gets a hit about 35% of the time over an entire season. After the player failed to hit safely in six straight at-bats, a TV commentator said, "He is due for a hit by the law of averages." Is that right? Why?

10. **Cold weather coming** A TV weather man, predicting a colder-than-normal winter, said, "First, in looking at the past few winters, there has been a lack of really cold weather. Even though we are not supposed to use the law of averages, we are due." Do you think that "due by the law of averages" makes sense in talking about the weather? Why or why not?

11. **Playing "Pick 4"** The Pick 4 games in many state lotteries announce a four-digit winning number each day. You can think of the winning number as a four-digit group from a table of random digits. You win (or share) the jackpot if your choice matches the winning number. The winnings are divided among all players who matched the winning number. That suggests a way to get an edge.

(a) The winning number might be, for example, either 2873 or 9999. Explain why these two outcomes have exactly the same probability.

(b) If you asked many people whether 2873 or 9999 is more likely to be the randomly chosen winning number, most would favor one of them. Use the information in this section to say which one and to explain why. How might this affect the four-digit number you would choose?

12. **An unenlightened gambler**

(a) A gambler knows that red and black are equally likely to occur on each spin of a roulette wheel. He observes five consecutive reds occur and bets heavily on black at the next spin. Asked why, he explains that black is "due by the law of averages." Explain to the gambler what is wrong with this reasoning.

(b) After hearing you explain why red and black are still equally likely after five reds on the roulette wheel, the gambler moves to a poker game. He is dealt five straight red cards. He remembers what you said and assumes that the next card dealt in the same hand is equally likely to be red or black. Is the gambler right or wrong, and why?

13. **Free throws** A basketball player has probability 0.75 of making a free throw. Explain how you would use each chance device to simulate one free throw by the player.

(a) A six-sided die

(b) Table D of random digits

(c) A standard deck of playing cards

14. **Stoplight** On her drive to work every day, Ilana passes through an intersection with a traffic light. The light has probability 1/3 of being green when she gets to the intersection. Explain how you would use each chance device to simulate whether the light is red or green on a given day.

(a) A six-sided die

(b) Table D of random digits

(c) A standard deck of playing cards

15. **Simulation blunders** Explain what's wrong with each of the following simulation designs.

(a) A roulette wheel has 38 colored slots—18 red, 18 black, and 2 green. To simulate one spin of the wheel, let numbers 00 to 18 represent red, 19 to 37 represent black, and 38 to 40 represent green.

(b) About 10% of U.S. adults are left-handed. To simulate randomly selecting one adult at a time until you find a left-hander, use two digits. Let 01 to 10 represent being left-handed and 11 to 00 represent being right-handed. Move across a row in Table D, two digits at a time, skipping any numbers that have already appeared, until you find a number between 01 and 10. Record the number of people selected.

16. **Simulation blunders** Explain what's wrong with each of the following simulation designs.

(a) According to the Centers for Disease Control and Prevention, about 26% of U.S. adults were obese in 2008. To simulate choosing 8 adults at random and seeing how many are obese, we could use two digits. Let 01 to 26 represent obese and 27 to 00 represent not obese. Move across a row in Table D, two digits at a time, until you find 8 distinct numbers (no repeats). Record the number of obese people selected.

(b) Assume that the probability of a newborn being a boy is 0.5. To simulate choosing a random sample of 9 babies who were born at a local hospital today and observing their gender, use one digit. Use `randInt(0,9)` on your calculator to determine how many babies in the sample are male.

17. **Is this valid?** Determine whether each of the following simulation designs is valid. Justify your answer.

(a) According to a recent poll, 75% of American adults regularly recycle. To simulate choosing a random sample of 100 U.S. adults and seeing how many of them recycle, roll a 4-sided die 100 times. A result of 1, 2, or 3 means the person recycles; a 4 means that the person doesn't recycle.

(b) An archer hits the center of the target with 60% of her shots. To simulate having her shoot 10 times, use a coin. Flip the coin once for each of the 10 shots. If it lands heads, then she hits the center of the target. If the coin lands tails, she doesn't.

18. **Is this valid?** Determine whether each of the following simulation designs is valid. Justify your answer.

(a) According to a recent survey, 50% of people aged 13 and older in the United States are addicted to email. To simulate choosing a random sample of 20 people in this population and seeing how many of them are addicted to email, use a deck of cards. Shuffle the deck well, and then draw one card at a time. A red card means that person is addicted to email; a black card means he isn't. Continue until you have drawn 20 cards (without replacement) for the sample.

(b) A tennis player gets 95% of his second serves in play during practice (that is, the ball doesn't go out of bounds). To simulate the player hitting 5 second serves, look at pairs of digits going across a row in Table D. If the number is between 00 and 94, the serve is in; numbers between 95 and 99 indicate that the serve is out.

19. **Airport security** The Transportation Security Administration (TSA) is responsible for airport safety. On some flights, TSA officers randomly select passengers for an extra security check prior to boarding. One such flight had 76 passengers—12 in first class and 64 in coach class. Some passengers were surprised when none of the 10 passengers chosen for screening were seated in first class. We can use a simulation to see if this result is likely to happen by chance.

pg 290

(a) State the question of interest using the language of probability.

(b) How would you use random digits to imitate one repetition of the process? What variable would you measure?

(c) Use the line of random digits below to perform one repetition. Copy these digits onto your paper. Mark directly on or above them to show how you determined the outcomes of the chance process.

71487 09984 29077 14863 61683 47052 62224 51025

(d) In 100 repetitions of the simulation, there were 15 times when none of the 10 passengers chosen was seated in first class. What conclusion would you draw?

20. **Scrabble** In the game of Scrabble, each player begins by drawing 7 tiles from a bag containing 100 tiles. There are 42 vowels, 56 consonants, and 2 blank tiles in the bag. Cait chooses her 7 tiles and is surprised to discover that all of them are vowels. We can use a simulation to see if this result is likely to happen by chance.

(a) State the question of interest using the language of probability.

(b) How would you use random digits to imitate one repetition of the process? What variable would you measure?

(c) Use the line of random digits below to perform one repetition. Copy these digits onto your paper. Mark directly on or above them to show how you determined the outcomes of the chance process.

00694 05977 19664 65441 20903 62371 22725 53340

(d) In 1000 repetitions of the simulation, there were 2 times when all 7 tiles were vowels. What conclusion would you draw?

21. **The birthday problem** What's the probability that in a randomly selected group of 30 unrelated people, at least two have the same birthday? Let's make two assumptions to simplify the problem. First, we'll ignore the possibility of a February 29 birthday. Second, we assume that a randomly chosen person is equally likely to be born on each of the remaining 365 days of the year.

(a) How would you use random digits to imitate one repetition of the process? What variable would you measure?

(b) Use technology to perform 5 repetitions. Record the outcome of each repetition.

(c) Would you be surprised to learn that the theoretical probability is 0.71? Why or why not?

22. **Monty Hall problem** In *Parade* magazine, a reader posed the following question to Marilyn vos Savant and the "Ask Marilyn" column:

> Suppose you're on a game show, and you're given the choice of three doors. Behind one door is a car, behind the others, goats. You pick a door, say #1, and the host, who knows what's behind the doors, opens another door, say #3, which has a goat. He says to you, "Do you want to pick door #2?" Is it to your advantage to switch your choice of doors?

The game show in question was *Let's Make a Deal* and the host was Monty Hall. Here's the first part of Marilyn's response: "Yes; you should switch. The first door has a 1/3 chance of winning, but the second door has a 2/3 chance." Thousands of readers wrote to Marilyn to disagree with her answer. But she held her ground.

(a) Use an online *Let's Make a Deal* applet to perform at least 50 repetitions of the simulation. Record whether you stay or switch (try to do each about half the time) and the outcome of each repetition.

(b) Do you agree with Marilyn or her readers? Explain.

23. **Recycling** Do most teens recycle? To find out, an AP Statistics class asked an SRS of 100 students at their school whether they regularly recycle. How many students in the sample would need to say "Yes" to provide convincing evidence that more than half of the students at the school recycle? The Fathom dotplot below shows the results of taking 200 SRSs of 100 students from a population in which the true proportion who recycle is 0.50.

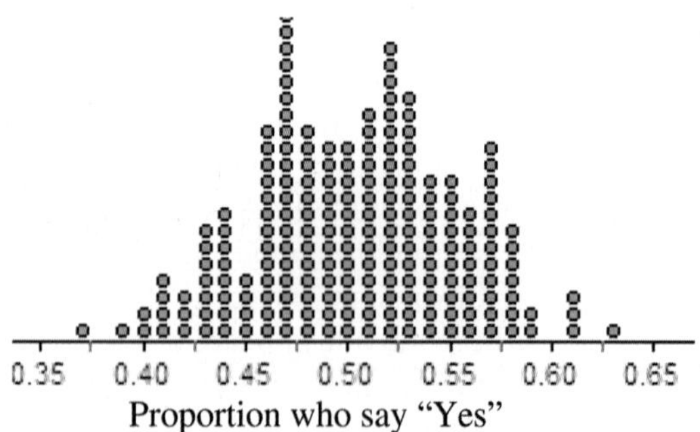

(a) Suppose 55 students in the class's sample say "Yes." Explain why this result does not give convincing evidence that more than half of the school's students recycle.

(b) Suppose 63 students in the class's sample say "Yes." Explain why this result gives strong evidence that a majority of the school's students recycle.

24. **Brushing teeth, wasting water?** A recent study reported that fewer than half of young adults turn off the water while brushing their teeth. Is the same true for teenagers? To find out, a group of statistics students asked an SRS of 60 students at their school if they usually brush with the water off. How many students in the sample would need to say "No" to provide convincing evidence that fewer than half of the students at the school brush with the water off? The Fathom dotplot below shows the results of taking 200 SRSs of 60 students from a population in which the true proportion who brush with the water off is 0.50.

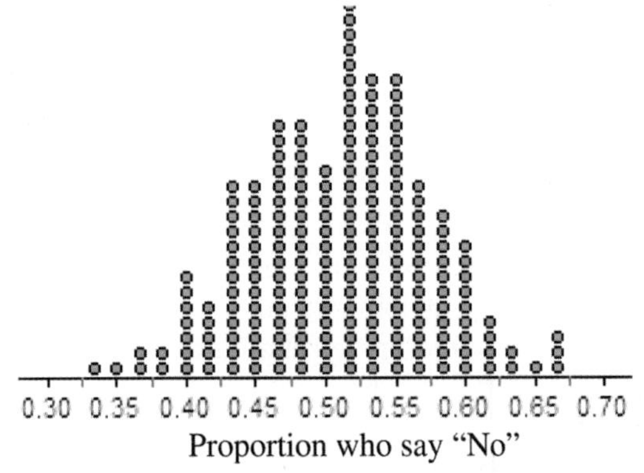

(a) Suppose 27 students in the class's sample say "No." Explain why this result does not give convincing evidence that fewer than half of the school's students brush their teeth with the water off.

(b) Suppose 18 students in the class's sample say "No." Explain why this result gives strong evidence that fewer than 50% of the school's students brush their teeth with the water off.

25. **Color-blind men** About 7% of men in the United States have some form of red-green color blindness. Suppose we randomly select one U.S. adult male at a time until we find one who is red-green color-blind. How many men would we expect to choose, on average? Design and carry out a simulation to answer this question. Follow the four-step process.

pg 291

26. **Scrabble** Refer to Exercise 20. About 3% of the time, the first player in Scrabble can "bingo" by playing all 7 tiles on the first turn. How many games of Scrabble would you expect to have to play, on average, for this to happen? Design and carry out a simulation to answer this question. Follow the four-step process.

27. **Color-blind men** Refer to Exercise 25. Suppose we randomly select 4 U.S. adult males. What's the probability that at least one of them is red-green color-blind? Design and carry out a simulation to answer this question. Follow the four-step process.

28. **Lotto** In the United Kingdom's Lotto game, a player picks six numbers from 1 to 49 for each ticket. Rosemary bought one ticket for herself and one for each of her four adult children. She had the lottery computer randomly select the six numbers on each ticket. When the six winning numbers were drawn, Rosemary was surprised to find that none of these numbers appeared on any of the five Lotto tickets she had bought. Should she be? Design and carry out a simulation to answer this question. Follow the four-step process.

29. **Random assignment** Researchers recruited 20 volunteers—8 men and 12 women—to take part in an experiment. They randomly assigned the subjects into two groups of 10 people each. To their surprise, 6 of the 8 men were randomly assigned to the same treatment. Should they be surprised? Design and carry out a simulation to estimate the probability that the random assignment puts 6 or more men in the same group. Follow the four-step process.

30. **Taking the train** According to New Jersey Transit, the 8:00 A.M. weekday train from Princeton to New York City has a 90% chance of arriving on time. To test this claim, an auditor chooses 6 weekdays at random during a month to ride this train. The train arrives late on 2 of those days. Does the auditor have convincing evidence that the company's claim isn't true? Design and carry out a simulation to answer this question. Follow the four-step process.

Multiple choice: Select the best answer for Exercises 31 to 36.

31. You read in a book about bridge that the probability that each of the four players is dealt exactly one ace is about 0.11. This means that

(a) in every 100 bridge deals, each player has one ace exactly 11 times.

(b) in one million bridge deals, the number of deals on which each player has one ace will be exactly 110,000.

(c) in a very large number of bridge deals, the percent of deals on which each player has one ace will be very close to 11%.

(d) in a very large number of bridge deals, the average number of aces in a hand will be very close to 0.11.

(e) None of these

32. If I toss a fair coin five times and the outcomes are TTTTT, then the probability that tails appears on the next toss is

(a) 0.5. (b) less than 0.5. (c) greater than 0.5. (d) 0. (e) 1.

Exercises 33 to 35 refer to the following setting. A basketball player makes 47% of her shots from the field during the season.

33. To simulate whether a shot hits or misses, you would assign random digits as follows:

(a) One digit simulates one shot; 4 and 7 are a hit; other digits are a miss.

(b) One digit simulates one shot; odd digits are a hit and even digits are a miss.

(c) Two digits simulate one shot; 00 to 47 are a hit and 48 to 99 are a miss.

(d) Two digits simulate one shot; 00 to 46 are a hit and 47 to 99 are a miss.

(e) Two digits simulate one shot; 00 to 45 are a hit and 46 to 99 are a miss.

34. Use the correct choice from the previous question and these random digits to simulate 10 shots:

82734 71490 20467 47511 81676 55300 94383 14893

How many of these 10 shots are hits?

(a) 2 (b) 3 (c) 4 (d) 5 (e) 6

35. You want to estimate the probability that the player makes 5 or more of 10 shots. You simulate 10 shots 25 times and get the following numbers of hits:

5 7 5 4 1 5 3 4 3 4 5 3 4 4 6 3 4 1 7 4 5 5 6 5 7

What is your estimate of the probability?

(a) 5/25, or 0.20 (b) 11/25, or 0.44 (c) 12/25, or 0.48 (d) 16/25, or 0.64 (e) 19/25, or 0.76

36. Ten percent of U.S. households contain 5 or more people. You want to simulate choosing a household at random and recording whether or not it contains 5 or more people. Which of these are correct assignments of digits for this simulation?

(a) Odd = Yes (5 or more people); Even = No (not 5 or more people)

(b) 0 = Yes; 1, 2, 3, 4, 5, 6, 7, 8, 9 = No

(c) 5 = Yes; 0, 1, 2, 3, 4, 6, 7, 8, 9 = No

(d) All three are correct.

(e) Choices (b) and (c) are correct, but (a) is not.

37. **Are you feeling stressed?** (4.1) A Gallup Poll asked whether people experienced stress "a lot of the day yesterday." Forty percent said they did. Gallup's report said, "Results are based on telephone interviews with 178,545 national adults, aged 18 and older, conducted Jan. 2–June 30, 2009."[4]

(a) Identify the population and the sample.

(b) Explain how undercoverage could lead to bias in this survey.

38. **Waiting to park** (1.3) Do drivers take longer to leave their parking spaces when someone is waiting? Researchers hung out in a parking lot and collected some data. The graphs and numerical summaries below display information about how long it took drivers to exit their sp\aces.

(a) Write a few sentences comparing these distributions.

(b) Can we conclude that having someone waiting causes drivers to leave their spaces more slowly? Why or why not?

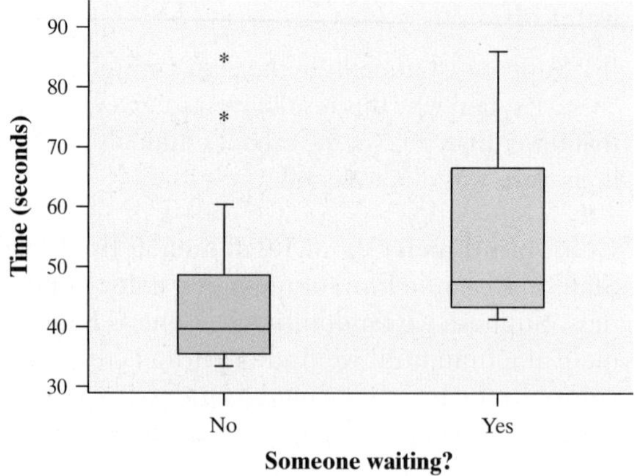

Descriptive Statistics: Time

Variable	Waiting	N	Mean	StDev	Minimum	Q1	Median	Q3	Maximum
Time	No	20	44.42	14.10	33.76	35.61	39.56	48.48	84.92
	Yes	20	54.11	14.39	41.61	43.41	47.14	66.44	85.97

Probability Rules

In Section 5.2, you'll learn about:

- Probability models
- Basic rules of probability
- Two-way tables and probability
- Venn diagrams and probability

The idea of probability rests on the fact that chance behavior is predictable in the long run. In Section 5.1, we used simulation to imitate chance behavior. Do we always need to repeat a chance process many times to determine the probability of a particular outcome? Fortunately, the answer is no.

Probability Models

In Chapter 2, we saw that a Normal density curve could be used to model some distributions of data. In Chapter 3, we modeled linear relationships between two quantitative variables with a least-squares regression line. Now we're ready to develop a model for chance behavior.

Let's start with a very simple chance process: tossing a coin once. When we toss a coin, we can't know the outcome in advance. What *do* we know? We are willing to say that the outcome will be either heads or tails. We believe that each of these outcomes has probability 1/2. This description of coin tossing has two parts:

- A list of possible outcomes (the **sample space** S)
- A probability for each outcome

Such a description is the basis for a **probability model.** Here is the basic vocabulary we use.

DEFINITION: Sample space, probability model

The **sample space *S*** of a chance process is the set of all possible outcomes.

A **probability model** is a description of some chance process that consists of two parts: a sample space *S* and a probability for each outcome.

A sample space S can be very simple or very complex. When we toss a coin once, there are only two possible outcomes, heads and tails. We can write the sample space using set notation as S = {H, T}. When Gallup draws a random sample of 1523 adults and asks a survey question, the sample space contains all possible sets of responses from 1523 of the 235 million adults in the country. This S is extremely large. Each member of S lists the answers from one possible sample, which explains the term *sample space.*

Let's look at how to set up a probability model in a familiar setting—rolling a pair of dice.

Roll the Dice

Building a probability model

Many board games involve rolling dice. Imagine rolling two fair, six-sided dice—one that's red and one that's green.

PROBLEM: *Give a probability model for this chance process.*

SOLUTION: *There are 36 possible outcomes when we roll two dice and record the number of spots showing on the up-faces. Figure 5.2 (on the next page) displays these outcomes. They* make up the sample space S. If the dice are perfectly balanced, all 36 outcomes will be *equally likely*. That is, each of the 36 outcomes will come up on one-thirty-sixth of all rolls in the long run. So each outcome has probability 1/36.

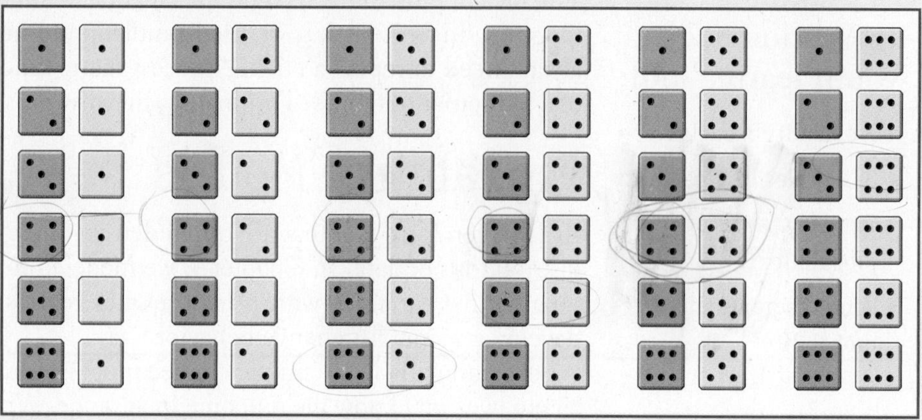

FIGURE 5.2 The 36 possible outcomes in rolling two dice. If the dice are carefully made, all of these outcomes have the same probability.

For Practice *Try Exercise 39*

A probability model does more than just assign a probability to each outcome. It allows us to find the probability of any collection of outcomes, which we call an **event.**

DEFINITION: Event

An **event** is any collection of outcomes from some chance process. That is, an event is a subset of the sample space. Events are usually designated by capital letters, like *A, B, C,* and so on.

If A is any event, we write its probability as $P(A)$. In the dice-rolling example, suppose we define event A as "sum is 5." What's $P(A)$, the probability that event A occurs? There are four outcomes in event A:

Since each of these outcomes has probability 1/36, $P(A) = 4/36$. Now consider event B: sum is not 5. To find $P(B)$, we could list all the outcomes that make up event B, but that would take a while. Fortunately, there's an easier way. Of the 36 equally likely outcomes in Figure 5.2, event A (sum is 5) occurs in 4 of them. So event A does *not* occur in 32 of these outcomes. Then $P(B) = P(\text{sum isn't } 5) = P(\text{not } A) = 32/36$. Notice that $P(A) + P(B) = 1$.

Let's consider one more event, which we'll call C: sum is 6. The outcomes in event C are

So $P(C) = 5/36$. What's the probability that we get a sum of 5 *or* 6, that is, $P(A \text{ or } C)$? Since these two events have no outcomes in common, we can add the probabilities of the individual events:

$$P(\text{sum is 5 or sum is 6}) = P(\text{sum is 5}) + P(\text{sum is 6}) = 4/36 + 5/36 = 9/36$$

In other words, $P(A \text{ or } C) = P(A) + P(C)$.

Basic Rules of Probability

Our dice-rolling example revealed some basic rules that any probability model must obey:

- *The probability of any event is a number between 0 and 1.* The probability of an event is the long-run proportion of repetitions on which that event occurs. Any proportion is a number between 0 and 1, so any probability is also a number between 0 and 1. An event with probability 0 never occurs, and an event with probability 1 occurs on every trial. An event with probability 0.5 occurs in half the trials in the long run.
- *All possible outcomes together must have probabilities whose sum is 1.* Because some outcome must occur on every trial, the sum of the probabilities for all possible outcomes must be exactly 1.
- *If all outcomes in the sample space are equally likely, the probability that event A occurs can be found using the formula*

$$P(A) = \frac{\text{number of outcomes corresponding to event A}}{\text{total number of outcomes in sample space}}$$

- *The probability that an event does not occur is 1 minus the probability that the event does occur.* If an event occurs in (say) 70% of all trials, it fails to occur in the other 30%. The probability that an event occurs and the probability that it does not occur always add to 100%, or 1. (This explains why $P(\text{sum isn't 5}) = 1 - P(\text{sum is 5})$ in the dice-rolling example.) We refer to the event "not A" as the **complement** of A and denote it by A^C.

Complement

- *If two events have no outcomes in common, the probability that one or the other occurs is the sum of their individual probabilities.* If one event occurs in 40% of all trials, a different event occurs in 25% of all trials, and the two can never occur together, then one or the other occurs on 65% of all trials because 40% + 25% = 65%. When two events have no outcomes in common, we refer to them as **mutually exclusive** or **disjoint.**

DEFINITION: Mutually exclusive (disjoint)

Two events are **mutually exclusive (disjoint)** if they have no outcomes in common and so can never occur together.

We can summarize the basic probability rules more concisely in symbolic form.

Basic Probability Rules

- For any event A, $0 \leq P(A) \leq 1$.
- If S is the sample space in a probability model, $P(S) = 1$.
- In the case of equally likely outcomes,

$$P(A) = \frac{\text{number of outcomes corresponding to event A}}{\text{total number of outcomes in sample space}}$$

- **Complement rule:** $P(A^C) = 1 - P(A)$.
- **Addition rule for mutually exclusive events:** If A and B are mutually exclusive, $P(A \text{ or } B) = P(A) + P(B)$.

Here's an example that illustrates use of the basic probability rules.

Distance Learning

Applying probability rules

Distance-learning courses are rapidly gaining popularity among college students. Randomly select an undergraduate student who is taking a distance-learning course for credit, and record the student's age. Here is the probability model:[5]

Age group (yr):	18 to 23	24 to 29	30 to 39	40 or over
Probability:	0.57	0.17	0.14	0.12

PROBLEM:

(a) Show that this is a legitimate probability model.

(b) Find the probability that the chosen student is not in the traditional college age group (18 to 23 years).

SOLUTION:

(a) The probability of each outcome is a number between 0 and 1, and the probabilities of all the possible outcomes add to 1, so this is a legitimate probability model.

(b) There are two ways to find this probability. By the complement rule,

$$\begin{aligned} P(\text{not 18 to 23 years}) &= 1 - P(\text{18 to 23 years}) \\ &= 1 - 0.57 = 0.43 \end{aligned}$$

That is, if 57% of distance learners are 18 to 23 years old, then the remaining 43% are not in this age group.

Using the addition rule for mutually exclusive events,

$$\begin{aligned} P(\text{not 18 to 23 years}) &= P(\text{24 to 29 years}) + P(\text{30 to 39 years}) + P(\text{40 years or over}) \\ &= 0.17 + 0.14 + 0.12 = 0.43 \end{aligned}$$

CHECK YOUR UNDERSTANDING

Choose an American adult at random. Define two events:

A = the person has a cholesterol level of 240 milligrams per deciliter of blood (mg/dl) or above (high cholesterol)
B = the person has a cholesterol level of 200 to 239 mg/dl (borderline high cholesterol)

According to the American Heart Association, $P(A) = 0.16$ and $P(B) = 0.29$.

1. Explain why events A and B are mutually exclusive.
2. Say in plain language what the event "A or B" is. What is $P(A \text{ or } B)$?
3. If C is the event that the person chosen has normal cholesterol (below 200 mg/dl), what's $P(C)$?

Two-Way Tables and Probability

When we're trying to find probabilities involving two events, a two-way table can display the sample space in a way that makes probability calculations easier.

Who Has Pierced Ears?

Two-way tables and probability

Students in a college statistics class wanted to find out how common it is for young adults to have their ears pierced. They recorded data on two variables—gender and whether the student had a pierced ear—for all 178 people in the class. The two-way table below displays the data.

	Pierced Ears?		
Gender	**Yes**	**No**	**Total**
Male	19	71	**90**
Female	84	4	**88**
Total	**103**	**75**	**178**

PROBLEM: Suppose we choose a student from the class at random. Find the probability that the student

(a) has pierced ears.

(b) is a male with pierced ears.

(c) is male or has pierced ears.

SOLUTION: We'll define events A: is male and B: has pierced ears.

(a) Since each of the 178 students in the class is equally likely to be chosen, and there are 103 students with pierced ears, $P(\text{pierced ears}) = P(B) = 103/178$.

(b) We want to find P(male *and* pierced ears), that is, $P(A \text{ and } B)$. Looking at the intersection of the "Male" row and "Yes" column, we see that there are 19 males with pierced ears. So P(male and pierced ears) $= P(A \text{ and } B) = 19/178$.

(c) This time, we're interested in P(male *or* pierced ears), that is, $P(A \text{ or } B)$. (Note the mathematical use of the word *or* here—the person could be a male or have pierced ears or both.) From the two-way table, we see that there are 90 males in the class, so $P(A) = 90/178$. Can we just add $P(A)$

When we found the probability of getting a male with pierced ears in the example, we could have described this as either *P*(*A* and *B*) or *P*(*B* and *A*). Why? Because "*A* and *B*" describes the same event as "*B* and *A*." Likewise, *P*(*A* or *B*) is the same as *P*(*B* or *A*). *Don't get so caught up in the notation that you lose sight of what's really happening!*

to *P*(*B*) to get the correct answer? No! These two events are *not* mutually exclusive, since there are 19 males with pierced ears. (If we did add the two probabilities, we'd get 90/178 + 103/178 = 193/178, which is clearly wrong, since the probability is bigger than 1!) From the two-way table, we see that there are 19 + 71 + 84 = 174 students who are male *or* have pierced ears. So *P*(*A* or *B*) = 174/178.

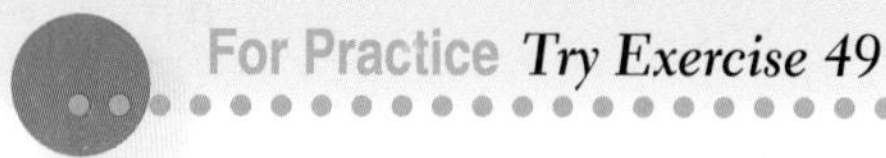

The previous example revealed two important facts about finding the probability P(A or B) when the two events are not mutually exclusive. First, the use of the word "or" in probability questions is different from that in everyday life. If someone says, "I'll either watch a movie or go to the football game," that usually means they'll do one thing or the other, but not both. In statistics, "A or B" could mean one or the other or both. Second, we can't use the addition rule for mutually exclusive events unless two events have no outcomes in common. The **Venn diagram** in Figure 5.3 shows why. If events A and B are *not* mutually exclusive, they can occur together. The probability that one or the other occurs is then *less* than the sum of their probabilities. As Figure 5.3 illustrates, outcomes common to both are counted twice when we add probabilities.

Venn diagram

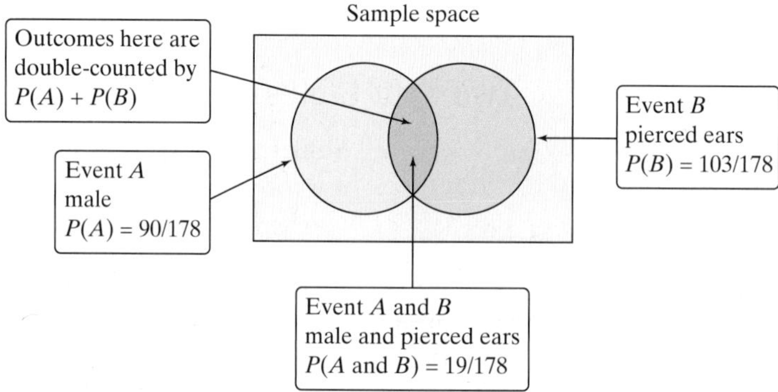

FIGURE 5.3 Venn diagram showing events *A* and *B* from the pierced-ears example. These events are *not* mutually exclusive.

The Venn diagram also suggests how to fix this "double-counting" problem: by subtracting the probability P(A and B) from the sum. That is,

$$P(A \text{ or } B) = P(A) + P(B) - P(A \text{ and } B)$$

This intuitive result is known as the **general addition rule.** Let's check that it works for the pierced-ears example. In that case, P(A) = 90/178, P(B) = 103/178, and P(A and B) = 19/178. The general addition rule says

$$\begin{aligned} P(A \text{ or } B) &= P(A) + P(B) - P(A \text{ and } B) \\ &= 90/178 + 103/178 - 19/178 \\ &= 174/178 \end{aligned}$$

This matches our earlier result.

General Addition Rule for Two Events

If A and B are any two events resulting from some chance process, then

$$P(A \text{ or } B) = P(A) + P(B) - P(A \text{ and } B)$$

What happens if we use the general addition rule for two mutually exclusive events A and B? In that case, P(A and B) = 0, and the formula reduces to P(A or B) = P(A) + P(B). In other words, the addition rule for mutually exclusive events is just a special case of the general addition rule.

You might be wondering whether there is also a rule for finding P(A and B). There is, but it's not quite as intuitive. Stay tuned for that later.

CHECK YOUR UNDERSTANDING

A standard deck of playing cards (with jokers removed) consists of 52 cards in four suits—clubs, diamonds, hearts, and spades. Each suit has 13 cards, with denominations ace, 2, 3, 4, 5, 6, 7, 8, 9, 10, jack, queen, and king. The jack, queen, and king are referred to as "face cards." Imagine that we shuffle the deck thoroughly and deal one card. Let's define events A: getting a face card and B: getting a heart.

1. Make a two-way table that displays the sample space.
2. Find P(A and B).
3. Explain why P(A or B) ≠ P(A) + P(B). Then use the general addition rule to find P(A or B).

Venn Diagrams and Probability

We have already seen that Venn diagrams can be used to illustrate the sample space of a chance process. Because Venn diagrams have uses in other branches of mathematics, some standard vocabulary and notation have been developed.

- We introduced the *complement* of an event earlier. In Figure 5.4(a), the complement A^C contains exactly the outcomes that are not in A.
- The events A and B in Figure 5.4(b) are *mutually exclusive* (*disjoint*) because they do not overlap; that is, they have no outcomes in common.

(a) 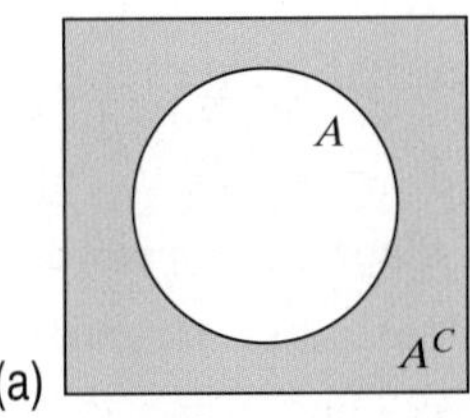

(b) 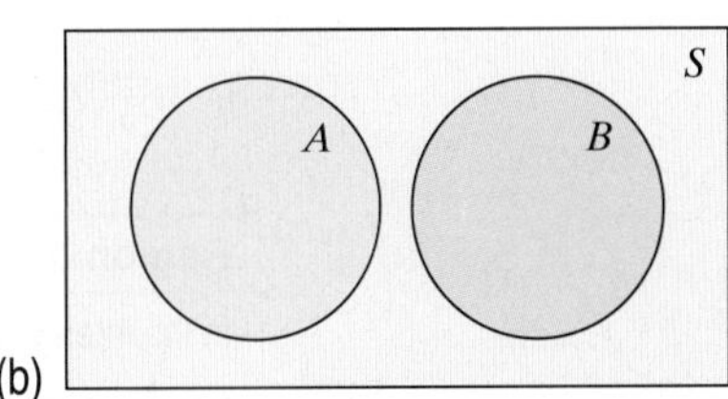

FIGURE 5.4 Venn diagrams showing: (a) event *A* and its complement A^C and (b) mutually exclusive (disjoint) events *A* and *B*.

Here's a way to keep the symbols straight in your head: ∪ for union; ∩ for intersection.

- Figure 5.5(a) shows the event "A *and* B." You can see why this event is also called the **intersection** of A and B. The corresponding notation is $A \cap B$.
- The event "A or B" is shown in Figure 5.5(b). This event is also known as the **union** of A and B. The corresponding notation is $A \cup B$.

(a) 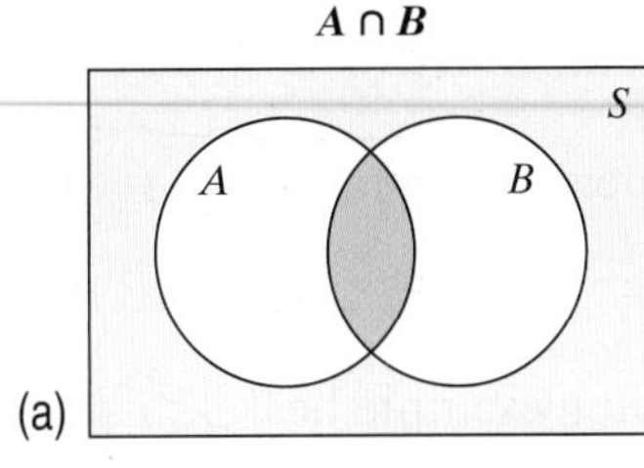

(b) 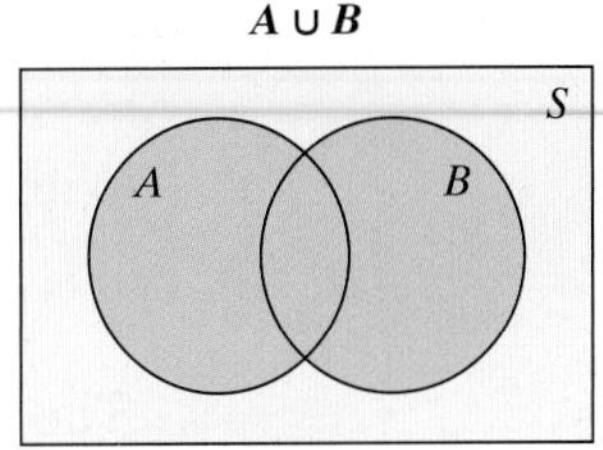

FIGURE 5.5 Venn diagrams showing (a) the intersection and (b) the union of events *A* and *B*.

Who Has Pierced Ears?
Understanding Venn diagrams

In the preceding example, we looked at data from a survey on gender and ear piercings for a large group of college students. The chance process came from selecting a student in the class at random. Our events of interest were A: is male and B: has pierced ears. Here is the two-way table that summarizes the sample space.

	Pierced Ears?	
Gender	**Yes**	**No**
Male	19	71
Female	84	4

How would we construct a Venn diagram that displays the information in the two-way table?

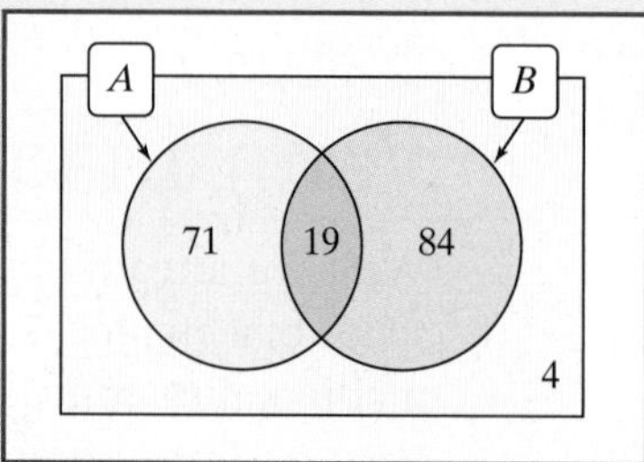

FIGURE 5.6 Completed Venn diagram for the large group of college students. The circles represent the two events *A:* is male and *B:* has pierced ears.

There are four distinct regions in the Venn diagram shown in Figure 5.6. These regions correspond to the four cells in the two-way table. We can describe this correspondence in tabular form as follows:

Region in Venn diagram	In words	In symbols	Count
In the intersection of two circles	Male and pierced ears	$A \cap B$	19
Inside circle *A*, outside circle *B*	Male and no pierced ears	$A \cap B^C$	71
Inside circle *B*, outside circle *A*	Female and pierced ears	$A^C \cap B$	84
Outside both circles	Female and no pierced ears	$A^C \cap B^C$	4

We have added the appropriate counts of students to the four regions in Figure 5.6.

With this new notation, we can rewrite the general addition rule in symbols as

$$P(A \cup B) = P(A) + P(B) - P(A \cap B)$$

The following example ties all this together.

Who Reads the Paper?

Venn diagrams, two-way tables, and probability

In an apartment complex, 40% of residents read *USA Today*. Only 25% read the *New York Times*. Five percent of residents read both papers. Suppose we select a resident of the apartment complex at random and record which of the two papers the person reads.

PROBLEM:

(a) Make a two-way table that displays the sample space of this chance process.

(b) Construct a Venn diagram to represent the outcomes of this chance process.

(c) Find the probability that the person reads at least one of the two papers.

(d) Find the probability that the person doesn't read either paper.

SOLUTION: We'll define events A: reads *USA Today* and B: reads *New York Times*. From the problem statement, we know that $P(A) = 0.40$, $P(B) = 0.25$, and $P(A \cap B) = 0.05$.

	Reads *USA Today?*		
Reads *New York Times*	**Yes**	**No**	**Total**
Yes	5 (5%)	20 (20%)	25 (25%)
No	35 (35%)	40 (40%)	75 (75%)
Total	40 (40%)	60 (60%)	100 (100%)

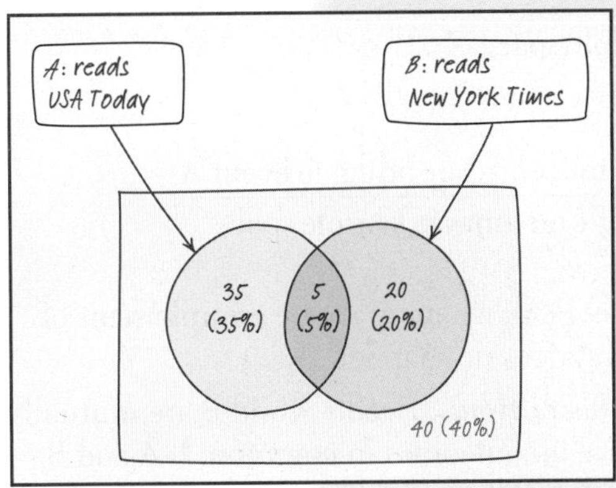

FIGURE 5.7 Venn diagram showing the residents of an apartment complex who A: read *USA Today* and B: read the *New York Times*.

(a) To make the calculations simple, let's suppose there are 100 residents in the apartment complex. Translating the probabilities into counts means that $0.40(100) = 40$ residents read *USA Today*, $0.25(100) = 25$ read the *New York Times*, and $0.05(100) = 5$ read both papers. We can enter the value 40 (40%) as the total for the "Yes" column, 25 (25%) as the total for the "Yes" row, 5 (5%) in the "Yes, Yes" cell, and 100 (100%) as the grand total in the two-way table shown here. This gives us enough information to fill in the empty cells of the table, starting with the missing row total for "No" ($100 - 25 = 75$ (75%)) and the missing column total for "No" ($100 - 40 = 60$ (60%)). In a similar way, we can determine the missing number in the "Yes" row ($25 - 5 = 20$ (20%)) and the "Yes" column ($40 - 5 = 35$). That leaves 40 residents for the "No, No" cell.

(b) Figure 5.7 shows the Venn diagram that corresponds to the completed two-way table from (a).

(c) If the randomly selected person reads *at least one* of the two papers, then he or she reads *USA Today*, the *New York Times*, or both papers. But that's the same as the event $A \cup B$. From the two-way table, the Venn diagram, or the general addition rule, we have

$$P(A \cup B) = P(A) + P(B) - P(A \cap B)$$
$$= 0.40 + 0.25 - 0.05 = 0.60$$

So there's a 60% chance that the randomly selected resident reads at least one of the two papers.

(d) From the two-way table or Venn diagram, $P(\text{reads neither paper}) = P(A^C \cap B^C) = 0.40$.

For Practice Try Exercise 55

AP EXAM TIP Many probability problems involve simple computations that you can do on your calculator. It may be tempting to just write down your final answer without showing the supporting work. Don't do it! A "naked answer," even if it's correct, will usually earn you no credit on a free-response question.

In the previous example, the event "reads neither paper" is the complement of the event "reads at least one of the papers." To solve part (d) of the problem, we could have used our answer from (c) and the complement rule:

$$P(\text{reads neither paper}) = 1 - P(\text{reads at least one paper}) = 1 - 0.60 = 0.40$$

As you'll see in Section 5.3, the fact that "none" is the opposite of "at least one" comes in handy for a variety of probability questions.

SECTION 5.2 Summary

- A **probability model** describes chance behavior by listing the possible outcomes in the **sample space S** and giving the probability that each outcome occurs.
- An **event** is a subset of the possible outcomes in a chance process. To find the probability that an event A happens, we can rely on some basic probability rules:
  - For any event A, $0 \leq P(A) \leq 1$.
  - $P(S) = 1$, where $S =$ the sample space
  - If all outcomes in the sample space are equally likely,

    $$P(A) = \frac{\text{number of outcomes corresponding to event A}}{\text{total number of outcomes in sample space}}$$

  - *Complement rule:* $P(A^C) = 1 - P(A)$, where A^C is the **complement** of event A; that is, the event that A does not happen
  - *Addition rule for mutually exclusive events:* Events A and B are **mutually exclusive** (**disjoint**) if they have no outcomes in common. If A and B are disjoint, $P(A \text{ or } B) = P(A) + P(B)$.
- A **two-way table** or a **Venn diagram** can be used to display the sample space for a chance process. Two-way tables and Venn diagrams can also be used to find probabilities involving events A and B, like the **union** $(A \cup B)$ and **intersection** $(A \cap B)$. The event $A \cup B$ ("A or B") consists of all outcomes in event A, event B, or both. The event $A \cap B$ ("A and B") consists of outcomes in both A and B.
- The **general addition rule** can be used to find $P(A \text{ or } B)$:

  $$P(A \cup B) = P(A) + P(B) - P(A \cap B)$$

SECTION 5.2 Exercises

39. **Role-playing games** Computer games in which the players take the roles of characters are very popular. They go back to earlier tabletop games such as Dungeons & Dragons. These games use many different types of dice. A four-sided die has faces with 1, 2, 3, and 4 spots.

pg 299

(a) List the sample space for rolling the die twice (spots showing on first and second rolls).

(b) What is the assignment of probabilities to outcomes in this sample space? Assume that the die is perfectly balanced.

40. **Tossing coins** Imagine tossing a fair coin 3 times.

(a) What is the sample space for this chance process?

(b) What is the assignment of probabilities to outcomes in this sample space?

41. **Role-playing games** Refer to Exercise 39. Define event A: sum is 5. Find $P(A)$.

42. **Tossing coins** Refer to Exercise 40. Define event B: get more heads than tails. Find $P(B)$.

43. **Probability models?** In each of the following situations, state whether or not the given assignment of probabilities to individual outcomes is legitimate, that is, satisfies the rules of probability. If not, give specific reasons for your answer.

(a) Roll a die and record the count of spots on the up-face: $P(1) = 0$, $P(2) = 1/6$, $P(3) = 1/3$, $P(4) = 1/3$, $P(5) = 1/6$, $P(6) = 0$.

(b) Choose a college student at random and record gender and enrollment status: $P(\text{female full-time}) = 0.56$, $P(\text{male full-time}) = 0.44$, $P(\text{female part-time}) = 0.24$, $P(\text{male part-time}) = 0.17$.

(c) Deal a card from a shuffled deck: $P(\text{clubs}) = 12/52$, $P(\text{diamonds}) = 12/52$, $P(\text{hearts}) = 12/52$, $P(\text{spades}) = 16/52$.

44. **Rolling a die** The following figure displays several possible probability models for rolling a die. We can learn which model is actually *accurate* for a particular die only by rolling the die many times. However, some of the models are not *legitimate*. That is, they do not obey the rules. Which are legitimate and which are not? In the case of the illegitimate models, explain what is wrong.

	Probability			
Outcome	Model 1	Model 2	Model 3	Model 4
⚀	1/7	1/3	1/3	1
⚁	1/7	1/6	1/6	1
⚂	1/7	1/6	1/6	2
⚃	1/7	0	1/6	1
⚄	1/7	1/6	1/6	1
⚅	1/7	1/6	1/6	2

45. **Blood types** All human blood can be typed as one of O, A, B, or AB, but the distribution of the types varies a bit with race. Here is the distribution of the blood type of a randomly chosen black American:

pg 302

Blood type:	O	A	B	AB
Probability:	0.49	0.27	0.20	?

(a) What is the probability of type AB blood? Why?

(b) What is the probability that the person chosen does not have type AB blood?

(c) Maria has type B blood. She can safely receive blood transfusions from people with blood types O and B. What is the probability that a randomly chosen black American can donate blood to Maria?

46. **Languages in Canada** Canada has two official languages, English and French. Choose a Canadian at random and ask, "What is your mother tongue?" Here is the distribution of responses, combining many separate languages from the broad Asia/Pacific region:[6]

Language:	English	French	Asian/Pacific	Other
Probability:	0.63	0.22	0.06	?

(a) What probability should replace "?" in the distribution? Why?

(b) What is the probability that a Canadian's mother tongue is not English?

(c) What is the probability that a Canadian's mother tongue is a language other than English or French?

47. **Education among young adults** Choose a young adult (aged 25 to 29) at random. The probability is 0.13 that the person chosen did not complete high school, 0.29 that the person has a high school diploma but no further education, and 0.30 that the person has at least a bachelor's degree.

(a) What must be the probability that a randomly chosen young adult has some education beyond high school but does not have a bachelor's degree? Why?

(b) What is the probability that a randomly chosen young adult has at least a high school education? Which rule of probability did you use to find the answer?

48. **Preparing for the GMAT** A company that offers courses to prepare students for the Graduate Management Admission Test (GMAT) has the following information about its customers: 20% are currently undergraduate students in business; 15% are undergraduate students in other fields of study; 60% are college graduates who are currently employed; and 5% are college graduates who are not employed. Choose a customer at random.

(a) What's the probability that the customer is currently an undergraduate? Which rule of probability did you use to find the answer?

(b) What's the probability that the customer is not an undergraduate business student? Which rule of probability did you use to find the answer?

49. **Who eats breakfast?** Students in an urban school were curious about how many children regularly eat breakfast. They conducted a survey, asking, "Do you eat breakfast on a regular basis?" All 595 students in the school responded to the survey. The resulting data are shown in the two-way table below.[7]

pg 303

	Male	Female	Total
Eats breakfast regularly	190	110	300
Doesn't eat breakfast regularly	130	165	295
Total	320	275	595

(a) Who are the individuals? What variables are being measured?

(b) If we select a student from the school at random, what is the probability that we choose

- a female?
- someone who eats breakfast regularly?
- a female who eats breakfast regularly?
- a female or someone who eats breakfast regularly?

50. **Sampling senators** The two-way table below describes the members of the U.S Senate in a recent year.

	Male	Female
Democrats	47	13
Republicans	36	4

(a) Who are the individuals? What variables are being measured?

(b) If we select a U.S. senator at random, what's the probability that we choose

- a Democrat?
- a female?
- a female Democrat?
- a female or a Democrat?

51. **Roulette** An American roulette wheel has 38 slots with numbers 1 through 36, 0, and 00, as shown in the figure. Of the numbered slots, 18 are red, 18 are black, and 2—the 0 and 00—are green. When the wheel is spun, a metal ball is dropped onto the middle of the wheel. If the wheel is balanced, the ball is equally likely to settle in any of the numbered slots. Imagine spinning a fair wheel once. Define events B: ball lands in a black slot, and E: ball lands in an even-numbered slot. (Treat 0 and 00 as even numbers.)

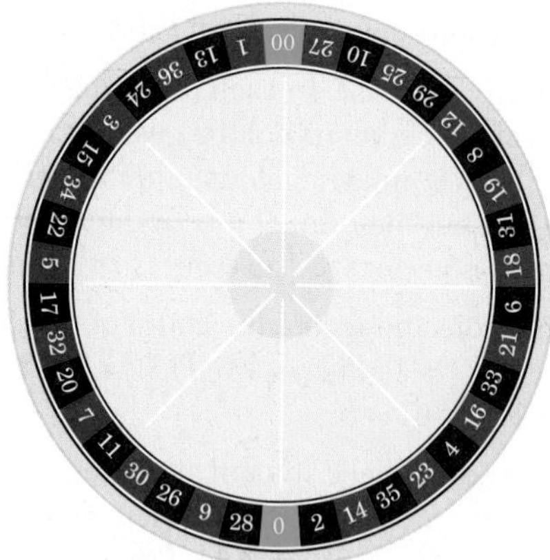

(a) Make a two-way table that displays the sample space in terms of events B and E.

(b) Find $P(B)$ and $P(E)$.

(c) Describe the event "B and E" in words. Then find $P(B \text{ and } E)$.

(d) Explain why $P(B \text{ or } E) \neq P(B) + P(E)$. Then use the general addition rule to compute $P(B \text{ or } E)$.

52. **Playing cards** Shuffle a standard deck of playing cards and deal one card. Define events J: getting a jack, and R: getting a red card.

(a) Construct a two-way table that describes the sample space in terms of events J and R.

(b) Find $P(J)$ and $P(R)$.

(c) Describe the event "J and R" in words. Then find $P(J \text{ and } R)$.

(d) Explain why $P(J \text{ or } R) \neq P(J) + P(R)$. Then use the general addition rule to compute $P(J \text{ or } R)$.

53. **Who eats breakfast?** Refer to Exercise 49.

(a) Construct a Venn diagram that models the chance process using events B: eats breakfast regularly, and M: is male.

(b) Find $P(B \cup M)$. Interpret this value in context.

(c) Find $P(B^C \cap M^C)$. Interpret this value in context.

54. **Sampling senators** Refer to Exercise 50.

(a) Construct a Venn diagram that models the chance process using events R: is a Republican, and F: is female.

(b) Find $P(R \cup F)$. Interpret this value in context.

(c) Find $P(R^C \cap F^C)$. Interpret this value in context.

55. **MySpace versus Facebook** A recent survey suggests that 85% of college students have posted a profile on Facebook, 54% use MySpace regularly, and 42% do both. Suppose we select a college student at random. pg 307

(a) Assuming that there are 20 million college students, make a two-way table for this chance process.

(b) Construct a Venn diagram to represent this setting.

(c) Consider the event that the randomly selected college student has posted a profile on at least one of these two sites. Write this event in symbolic form using the two events of interest that you chose in (b).

(d) Find the probability of the event described in (c). Explain your method.

56. **Mac or PC?** A recent census at a major university revealed that 40% of its students primarily used Macintosh computers (Macs). The rest mainly used PCs. At the time of the census, 67% of the school's students were undergraduates. The rest were graduate students. In the census, 23% of respondents were graduate students who said that they used PCs as their main computers. Suppose we select a student at random from among those who were part of the census.

(a) Assuming that there were 10,000 students in the census, make a two-way table for this chance process.

(b) Construct a Venn diagram to represent this setting.

(c) Consider the event that the randomly selected student is a graduate student who uses a Mac. Write this event in symbolic form using the two events of interest that you chose in (b).

(d) Find the probability of the event described in (c). Explain your method.

Multiple choice: Select the best answer for Exercises 57 to 60.

57. In government data, a household consists of all occupants of a dwelling unit. Choose an American household at random and count the number of people it contains. Here is the assignment of probabilities for your outcome:

Number of persons:	1	2	3	4	5	6	7
Probability:	0.25	0.32	???	???	0.07	0.03	0.01

The probability of finding 3 people in a household is the same as the probability of finding 4 people. These probabilities are marked ??? in the table of the distribution. The probability that a household contains 3 people must be

(a) 0.68. (b) 0.32. (c) 0.16. (d) 0.08.
(e) between 0 and 1, and we can say no more.

58. In a table of random digits such as Table D, each digit is equally likely to be any of 0, 1, 2, 3, 4, 5, 6, 7, 8, or 9. What is the probability that a digit in the table is 7 or greater?

(a) 7/10 (c) 4/10 (e) 1/10
(b) 6/10 (d) 3/10

59. Twenty of a sample of 275 students say they are vegetarians. Of the vegetarians, 9 eat both fish and eggs, 3 eat eggs but not fish, and 8 eat neither. Choose one of the vegetarians at random. What is the probability that the chosen student eats neither fish nor eggs?

(a) 8/275 = 0.03 (c) 8/20 = 0.4 (e) 1
(b) 20/275 = 0.07 (d) 0.5

60. The casino game craps is based on rolling two dice. Here is the assignment of probabilities to the sum of the numbers on the up-faces when two dice are rolled:

Outcome:	2	3	4	5	6	7	8	9	10	11	12
Probability:	1/36	2/36	3/36	4/36	5/36	6/36	5/36	4/36	3/36	2/36	1/36

The most common bet in craps is the "pass line." A pass line bettor wins immediately if either a 7 or an 11 comes up on the first roll. This is called a *natural*. What is the probability of a natural?

(a) 2/36 (c) 8/36 (e) 20/36
(b) 6/36 (d) 12/36

61. **Crawl before you walk** (3.2) At what age do babies learn to crawl? Does it take longer to learn in the winter, when babies are often bundled in clothes that restrict their movement? Perhaps there might even be an association between babies' crawling age and the average temperature during the month they first try to

crawl (around six months after birth). Data were collected from parents who brought their babies to the University of Denver Infant Study Center to participate in one of a number of studies. Parents reported the birth month and the age at which their child was first able to creep or crawl a distance of four feet within one minute. Information was obtained on 414 infants (208 boys and 206 girls). Crawling age is given in weeks, and average temperature (in °F) is given for the month that is six months after the birth month.[8]

Birth month	Average crawling age	Average temperature
January	29.84	66
February	30.52	73
March	29.70	72
April	31.84	63
May	28.58	52
June	31.44	39
July	33.64	33
August	32.82	30
September	33.83	33
October	33.35	37
November	33.38	48
December	32.32	57

Analyze the relationship between average crawling age and average temperature. What do you conclude about when babies learn to crawl?

62. **Treating low bone density** (4.2) Fractures of the spine are common and serious among women with advanced osteoporosis (low mineral density in the bones). Can taking strontium ranelate help? A large medical trial assigned 1649 women to take either strontium ranelate or a placebo each day. All of the subjects had osteoporosis and had had at least one fracture. All were taking calcium supplements and receiving standard medical care. The response variables were measurements of bone density and counts of new fractures over three years. The subjects were treated at 10 medical centers in 10 different countries.[9] Outline an appropriate design for this experiment. Explain why this is the proper design.

5.3

Conditional Probability and Independence

In Section 5.3, you'll learn about:

- What is conditional probability?
- Conditional probability and independence
- Tree diagrams and the general multiplication rule
- Independence: A special multiplication rule
- Calculating conditional probabilities

The probability we assign to an event can change if we know that some other event has occurred. This idea is the key to many applications of probability.

What Is Conditional Probability?

Let's return to the setting of the pierced-ears example (page 303). Earlier, we used the two-way table below to find probabilities involving events A: is male and B: has pierced ears for a randomly selected student. Here is a summary of our previous results:

$P(A) = P(\text{male}) = 90/178$ $\quad$ $P(A \cap B) = P(\text{male and pierced ears}) = 19/178$

$P(B) = P(\text{pierced ears}) = 103/178$ $\quad$ $P(A \cup B) = P(\text{male or pierced ears}) = 174/178$

	Pierced Ears?		
Gender	Yes	No	Total
Male	19	71	**90**
Female	84	4	**88**
Total	**103**	**75**	**178**

Now let's turn our attention to some other interesting probability questions.

Who Has Pierced Ears?

The idea of conditional probability

1. If we know that a randomly selected student has pierced ears, what is the probability that the student is male? There are a total of 103 students in the class with pierced ears. We can restrict our attention to this group, since we are told that the chosen student has pierced ears. Because there are 19 males among the 103 students with pierced ears, the desired probability is

$$P(\text{is male } \textit{given} \text{ has pierced ears}) = 19/103, \text{ or about } 18.4\%$$

2. If we know that a randomly selected student is male, what's the probability that the student has pierced ears? This time, our attention is focused on the males in the class. Since 19 of the 90 males in the class have pierced ears,

$$P(\text{has pierced ears } \textit{given} \text{ is male}) = 19/90, \text{ or about } 21.1\%$$

These two questions sound alike, but they're asking about two very different things.

A probability like "the probability that a randomly selected student is male *given that* the student has pierced ears" is known as a **conditional probability.** The name comes from the fact that we are trying to find the probability that one event will happen under the *condition* that some other event is already known to have occurred. We often use the phrase "given that" to signal the condition. There's even a special notation to indicate a conditional probability. In the example above, we would write $P(\text{is male} \mid \text{has pierced ears})$, where the | means "given that" or "under the condition that." Since we already defined the events A: is male and B: has pierced ears, we could write the conditional probability as $P(A \mid B) = 19/103$.

DEFINITION: Conditional probability

The probability that one event happens given that another event is already known to have happened is called a **conditional probability.** Suppose we know that event A has happened. Then the probability that event B happens *given that* event A has happened is denoted by $P(B \mid A)$.

With this new notation available, we can restate the answers to our two previous questions:

$$P(A \mid B) = 19/103 \quad \text{and} \quad P(B \mid A) = 19/90$$

Notice that the conditional probability $P(A \mid B) = 19/103$ is very different from the *unconditional* probability $P(A) = 90/178$. About half of the students in the class are male, but far fewer than half of the students with pierced ears are male. Also note that $P(B \mid A) = 19/90$ is much smaller than $P(B) = 103/178$. Over half of the students in the class have pierced ears, but only 19 out of 90 males in the class have pierced ears.

THINK ABOUT IT

Is there a connection between conditional probability and the conditional distributions of Chapter 1? Of course! For the college statistics class that we were just discussing, Figure 5.8(a) shows the conditional distribution of ear-piercing status for each gender. Above, we found that $P(\text{pierced ears} \mid \text{male}) = 19/90$, or about 21%. Note that $P(\text{pierced ears} \mid \text{female}) = 84/88$. You can see these values displayed in the "Yes" bars of Figure 5.8(a).

Figure 5.8(b) shows the conditional distribution of gender for each ear-piercing status. We can see that $P(\text{male} \mid \text{pierced ears}) = 19/103$, or about 18%, and that $P(\text{male} \mid \text{no pierced ears}) = 71/75$, or about 95%.

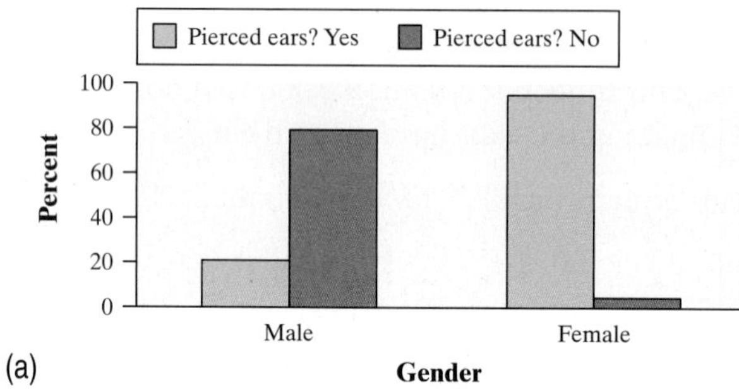

(a)

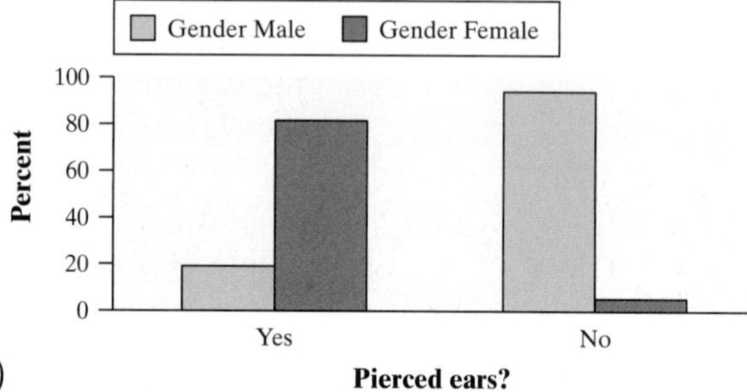

(b)

FIGURE 5.8 (a) Conditional distribution of ear-piercing status given gender in a large college class. (b) Conditional distribution of gender given ear-piercing status in a large college class.

CHECK YOUR UNDERSTANDING

Students at the University of New Harmony received 10,000 course grades last semester. The two-way table below breaks down these grades by which school of the university taught the course. The schools are Liberal Arts, Engineering and Physical Sciences, and Health and Human Services.

	Grade Level		
School	**A**	**B**	**Below B**
Liberal Arts	2,142	1,890	2,268
Engineering and Physical Sciences	368	432	800
Health and Human Services	882	630	588

(This table is based closely on grade distributions at an actual university, simplified a bit for clarity.)[10]

College grades tend to be lower in engineering and the physical sciences (EPS) than in liberal arts and social sciences (which includes Health and Human Services). Consider the two events E: the grade comes from an EPS course, and L: the grade is lower than a B.

1. Find $P(L)$. Interpret this probability in context.
2. Find $P(E \mid L)$ and $P(L \mid E)$. Which of these conditional probabilities tells you whether this college's EPS students tend to earn lower grades than students in liberal arts and social sciences? Explain.

Conditional Probability and Independence

To get a better feel for how conditional probability works, let's return to two familiar settings involving chance behavior—tossing coins and sampling students.

Coins and Pierced Ears

Understanding conditional probability

Suppose you toss a fair coin twice. Define events A: first toss is a head, and *B*: second toss is a head. We have already established that $P(A) = 1/2$ and $P(B) = 1/2$. What's $P(B \mid A)$? It's the conditional probability that the second toss is a head given that the first toss was a head. The coin has no memory, so $P(B \mid A) = 1/2$. In this case, $P(B \mid A) = P(B)$. Knowing that the first toss was a head does not affect the probability that the second toss is a head.

Let's contrast the coin-toss scenario with our earlier pierced-ears example. In that case, the chance process involved randomly selecting a student from a college class. The events of interest were A: is male, and *B*: has pierced ears. We already found that $P(A) = 90/178$, $P(B) = 103/178$, and $P(B \mid A) = 19/90$. If we know that the chosen student is male, the probability that he has pierced ears is 19/90. This conditional probability is very different from the unconditional probability $P(B) = 103/178$ that a randomly selected student from the class has pierced ears.

In the preceding example, we observed that $P(B \mid A) = P(B)$ for events A and *B* in the coin-toss setting. For the pierced-ears scenario, however, $P(B \mid A) \neq P(B)$. When knowledge that one event has happened does not change the likelihood that another event will happen, we say that the two events are **independent**.

DEFINITION: Independent events

Two events *A* and *B* are **independent** if the occurrence of one event has no effect on the chance that the other event will happen. In other words, events *A* and *B* are independent if $P(A \mid B) = P(A)$ and $P(B \mid A) = P(B)$.

The idea of independence is pretty straightforward. Determining whether two events related to the same chance process are independent requires us to compute probabilities. Here's an example that shows what we mean.

EXAMPLE Lefties Down Under

Checking for independence

Is there a relationship between gender and handedness? To find out, we used CensusAtSchool's Random Data Selector to choose an SRS of 50 Australian high school students who completed a survey. The two-way table displays data on the gender and dominant hand of each student.

	Dominant Hand		
Gender	Right	Left	Total
Male	20	3	**23**
Female	23	4	**27**
Total	**43**	**7**	50

PROBLEM: Are the events "male" and "left-handed" independent? Justify your answer.

SOLUTION: To check whether the two events are independent, we need to see whether knowing that one event has happened affects the probability that the other event occurs. Suppose we are told that the chosen student is male. From the two-way table, $P(\text{left-handed} \mid \text{male}) = 3/23 = 0.13$. The unconditional probability $P(\text{left-handed}) = 7/50 = 0.14$. These two probabilities are close, but they're not equal. So the events "male" and "left-handed" are not independent.

For Practice *Try Exercise 69*

You might have thought, "Surely there's no connection between gender and handedness. The events 'male' and 'left-handed' are bound to be independent." As the example shows, you can't use your intuition to check whether events are independent. To be sure, you have to calculate some probabilities.

In the previous example, we could have also compared $P(\text{male} \mid \text{left-handed})$ with $P(\text{male})$. Of the 7 left-handed students in the sample, 3 were male. So $P(\text{male} \mid \text{left-handed}) = 3/7 = 0.43$. We can see from the two-way table that $P(\text{male}) = 23/50 = 0.46$. Once again, the two probabilities are close, but not equal.

THINK ABOUT IT

Is there a connection between independence of events and association between two variables? In the previous example, we found that the events "male" and "left-handed" were *not* independent. Does that mean there actually *is* a relationship between the variables gender and handedness in the larger population? No. The data are much more consistent with there being *no* association between the variables. It would be surprising to choose a random sample of 50 students for which $P(\text{left-handed} \mid \text{male})$ was *exactly* equal to $P(\text{left-handed})$. But these two probabilities should be close to equal (as they are in this case) if there's no association between the variables. How close is close? You'll have to wait a few chapters to find out.

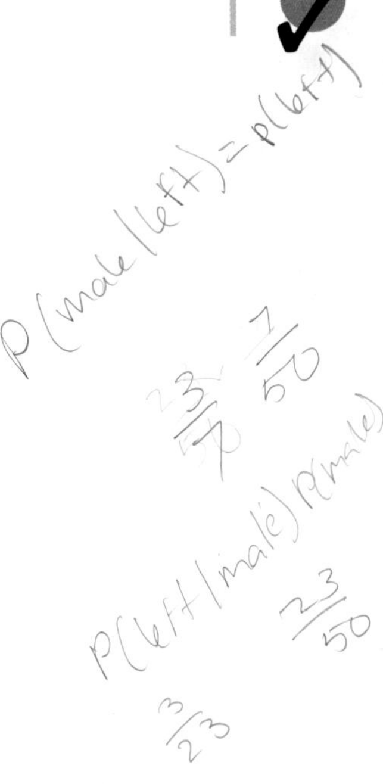

CHECK YOUR UNDERSTANDING

For each chance process below, determine whether the events are independent. Justify your answer.

1. Shuffle a standard deck of cards, and turn over the top card. Put it back in the deck, shuffle again, and turn over the top card. Define events A: first card is a heart, and B: second card is a heart.

2. Shuffle a standard deck of cards, and turn over the top two cards, one at a time. Define events A: first card is a heart, and B: second card is a heart.

3. The 28 students in Mr. Tabor's AP Statistics class completed a brief survey. One of the questions asked whether each student was right- or left-handed. The two-way table summarizes the class data. Choose a student from the class at random. The events of interest are "female" and "right-handed."

	Gender	
Handedness	**Female**	**Male**
Left	3	1
Right	18	6

Tree Diagrams and the General Multiplication Rule

If you toss a fair coin twice, what's the probability that you get two heads? By the reasoning of Section 5.1, it's the proportion of times you would get HH if you tossed the coin two times over and over and over again. We could perform a simulation to estimate the true probability. Using the methods of Section 5.2, we would take a different approach to answering the question. We begin by listing the sample space for this chance process:

HH HT TH TT

Since the result of one coin toss is *independent* from the result of any other toss, the 4 outcomes in the sample space are equally likely. So $P(\text{two heads}) = P(\text{HH}) = 1/4$.

Tree diagram

There's another way to model chance behavior that involves a sequence of outcomes, like tossing coins. It's called a **tree diagram.** Figure 5.9 shows a tree diagram for the coin-tossing scenario. After the first toss, there are only two possible results—heads (H) or tails (T)—each with probability 1/2. These two possibilities are

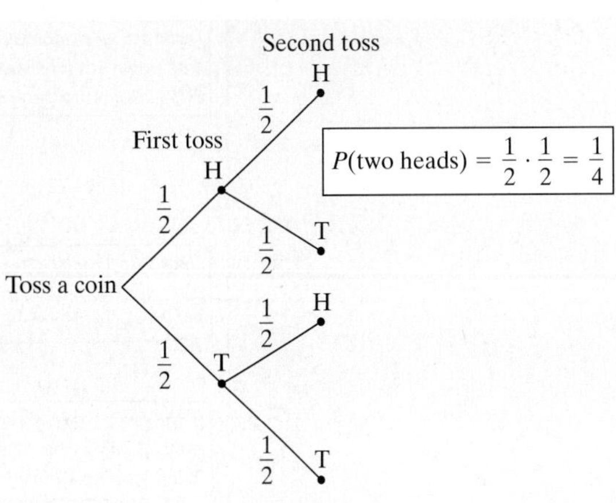

FIGURE 5.9 A tree diagram displaying the sample space of tossing a coin twice.

shown on the leftmost branches of the tree diagram. From there, we toss a second time, which results in either an H or a T, again with equal probability. This puts us on one of the four "nodes" beneath the "Second toss" label on the tree diagram after two tosses.

What's the probability of getting heads, heads? We already know it's 1/4. Do you see how we can obtain that value from the tree diagram? If we multiply probabilities along the top branches, we get

$$P(\text{two heads}) = \frac{1}{2} \cdot \frac{1}{2} = \frac{1}{4}$$

This same general strategy will work even when trials are *not* independent, as the following example illustrates.

EXAMPLE

Picking Two Lefties?

Tree diagrams in action

Let's return to Mr. Tabor's AP Statistics class from the preceding Check Your Understanding (page 317). The two-way table showing the gender and handedness of the students in the class is repeated below.

	Gender	
Handedness	**Female**	**Male**
Left	3	1
Right	18	6

PROBLEM: Suppose we choose two students at random.

(a) Draw a tree diagram that shows the sample space for this chance process.

(b) Find the probability that both students are left-handed.

SOLUTION:

(a)

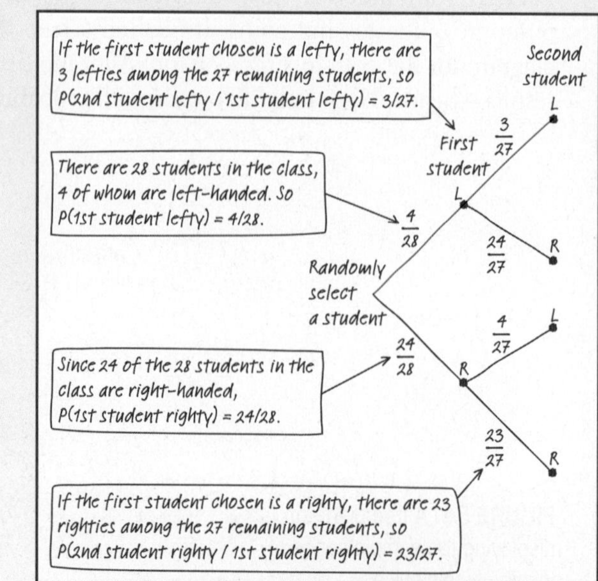

FIGURE 5.10 Tree diagram for the chance process of choosing two students at random from Mr. Tabor's class and recording handedness.

(b) To get two left-handed students, we'd have to get a lefty for the first student chosen *and* a lefty for the second student chosen. Following along the top branches of the tree, we see that the probability is

$$P(\text{two lefties}) = P(\text{1st student lefty } and \text{ 2nd student lefty})$$
$$= P(\text{1st student lefty}) \cdot P(\text{2nd student lefty} \mid \text{1st student lefty})$$
$$= \frac{4}{28} \cdot \frac{3}{27} \approx 0.016$$

There's about a 1.6% chance of randomly selecting two lefties.

This idea of multiplying along the branches in a tree diagram leads to a general method for finding the probability $P(A \cap B)$ that two events happen together. We simply multiply the probability that event A happens by the probability of event B happening given that event A has happened. In symbols, we write

$$P(A \cap B) = P(A) \cdot P(B \mid A)$$

This formula is known as the **general multiplication rule.**

General Multiplication Rule

The probability that events A and B both occur can be found using the **general multiplication rule**

$$P(A \cap B) = P(A) \cdot P(B \mid A)$$

where $P(B \mid A)$ is the conditional probability that event B occurs given that event A has already occurred.

In words, this rule says that for both of two events to occur, first one must occur, and then given that the first event has occurred, the second must occur. This is just common sense expressed in the language of probability, as the following example illustrates.

Teens with Online Profiles

Using the general multiplication rule

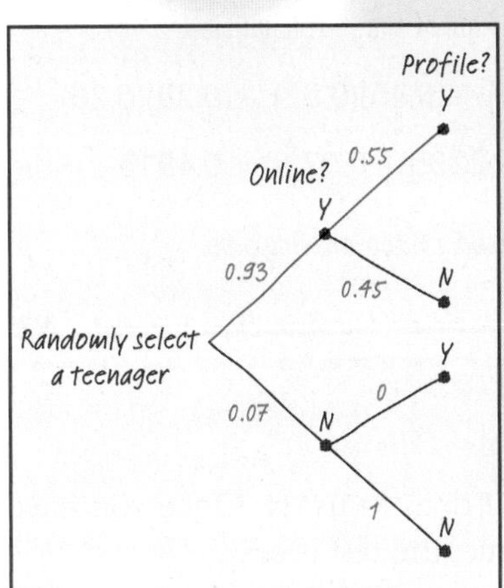

The Pew Internet and American Life Project finds that 93% of teenagers (ages 12 to 17) use the Internet, and that 55% of online teens have posted a profile on a social-networking site.[11]

PROBLEM: What percent of teens are online *and* have posted a profile? Explain why your answer makes sense.

SOLUTION: We know that $P(\text{online}) = 0.93$ and $P(\text{profile} \mid \text{online}) = 0.55$. Use the general multiplication rule:

$$P(\text{online and have profile}) = P(\text{online}) \cdot P(\text{profile} \mid \text{online})$$
$$= (0.93)(0.55) = 0.5115$$

About 51% of all teens use the Internet and have a profile on a social-networking site. This calculation makes sense: just look at the tree diagram in the margin.

For Practice *Try Exercise* 79

Probability models with several stages often require us to combine several rules into a more elaborate calculation. Here's an example.

Who Visits YouTube?

Tree diagrams and probability calculations

STEP 4

Video-sharing sites, led by YouTube, are popular destinations on the Internet. Let's look only at adult Internet users, aged 18 and over. About 27% of adult Internet users are 18 to 29 years old, another 45% are 30 to 49 years old, and the remaining 28% are 50 and over. The Pew Internet and American Life Project finds that 70% of Internet users aged 18 to 29 have visited a video-sharing site, along with 51% of those aged 30 to 49 and 26% of those 50 or older. Do most Internet users visit YouTube and similar sites?

STATE: What percent of all adult Internet users visit video-sharing sites?

PLAN: Let's start by restating the given percents as probabilities. If we choose an online adult at random,

$$P(\text{age 18 to 29}) = 0.27 \quad P(\text{age 30 to 49}) = 0.45 \quad P(\text{age 50 and older}) = 0.28$$

The percents of each group who visit video-sharing sites are *conditional* probabilities.

$$P(\text{video yes} \mid \text{age 18 to 29}) = 0.70 \quad P(\text{video yes} \mid \text{age 30 to 49}) = 0.51$$
$$P(\text{video yes} \mid \text{age 50 and older}) = 0.26$$

We want to find the unconditional probability $P(\text{video yes})$. Making a tree diagram should help.

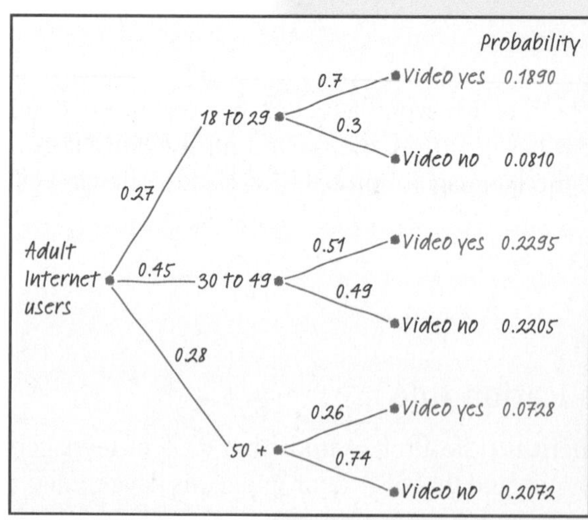

FIGURE 5.11 Tree diagram for use of the Internet and video-sharing sites such as YouTube. The three disjoint paths to the outcome that an adult Internet user visits video-sharing sites are colored red.

DO: The tree diagram in Figure 5.11 organizes the given information. To get the probabilities at the far right, we used the general multiplication rule. For instance, the probability that a randomly chosen Internet user is an 18- to 29-year-old who visits video-sharing sites is

$$P(\text{18 to 29 and video yes}) = P(\text{18 to 29}) \cdot P(\text{video yes} \mid \text{18 to 29})$$
$$= (0.27)(0.70) = 0.1890$$

This probability appears at the end of the topmost branch.

There are three disjoint paths to "video yes," one for each of the three age groups. These paths are colored red in Figure 5.11. Because the three paths are disjoint, the probability that an adult Internet user visits video-sharing sites is the sum of their probabilities:

$$P(\text{video yes}) = (0.27)(0.70) + (0.45)(0.51) + (0.28)(0.26)$$
$$= 0.1890 + 0.2295 + 0.0728 = 0.4913$$

CONCLUDE: About 49% of all adult Internet users have visited a video-sharing site.

It takes longer to explain a tree diagram than it does to use it. Once you have understood a problem well enough to draw the tree, the rest is easy.

CHECK YOUR UNDERSTANDING

A computer company makes desktop and laptop computers at factories in three states—California, Texas, and New York. The California factory produces 40% of the company's computers, the Texas factory makes 25%, and the remaining 35% are manufactured in New York. Of the computers made in California, 75% are laptops. Of those made in Texas and New York, 70% and 50%, respectively, are laptops. All computers are first shipped to a distribution center in Missouri before being sent out to stores. Suppose we select a computer at random from the distribution center.[12]

1. Construct a tree diagram to represent this situation.
2. Find the probability that the computer is a laptop. Show your work.

Independence: A Special Multiplication Rule

What happens to the general multiplication rule in the special case when events A and B are independent? In that case, $P(B \mid A) = P(B)$. We can simplify the general multiplication rule as follows:

$$\begin{aligned} P(A \cap B) &= P(A) \cdot P(B \mid A) \\ &= P(A) \cdot P(B) \end{aligned}$$

This result is known as the **multiplication rule for independent events.**

DEFINITION: Multiplication rule for independent events

If A and B are independent events, then the probability that A and B both occur is

$$P(A \cap B) = P(A) \cdot P(B)$$

Note that this rule *only* applies to independent events. Let's look at an example that uses the multiplication rule for independent events to analyze an important historical event.

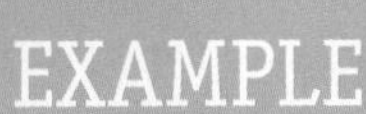

The Challenger *Disaster*

Independence and the multiplication rule

On January 28, 1986, Space Shuttle *Challenger* exploded on takeoff. All seven crew members were killed. Following the disaster, scientists and statisticians helped analyze what went wrong. They determined that the failure of O-ring joints in the shuttle's booster rockets was to blame. Under the cold conditions that day, experts estimated that the probability that an individual O-ring joint would function properly was 0.977. But there were six of these O-ring joints, and all six had to function properly for the shuttle to launch safely.

PROBLEM: Assuming that O-ring joints succeed or fail independently, find the probability that the shuttle would launch safely under similar conditions.

SOLUTION: For the shuttle to launch safely, all six O-ring joints need to function properly. The chance that this happens is given by

$$P(\text{joint 1 OK } and \text{ joint 2 OK } and \text{ joint 3 OK } and \text{ joint 4 OK } and \text{ joint 5 OK } and \text{ joint 6 OK})$$

By the multiplication rule for independent events, this probability is

$$P(\text{joint 1 OK}) \cdot P(\text{joint 2 OK}) \cdot P(\text{joint 3 OK}) \cdot P(\text{joint 4 OK}) \cdot P(\text{joint 5 OK}) \cdot P(\text{joint 6 OK})$$
$$= (0.977)(0.977)(0.977)(0.977)(0.977)(0.977) = 0.87$$

There's an 87% chance that the shuttle would launch safely under similar conditions (and a 13% chance that it wouldn't).

Note: As a result of the statistical analysis following the *Challenger* disaster, NASA made important safety changes to the design of the shuttle's booster rockets.

For Practice *Try Exercise 89*

The next example uses the fact that "at least one" and "none" are opposites.

Rapid HIV Testing

Finding the probability of "at least one"

Many people who come to clinics to be tested for HIV, the virus that causes AIDS, don't come back to learn the test results. Clinics now use "rapid HIV tests" that give a result while the client waits. In a clinic in Malawi, for example, use of rapid tests increased the percent of clients who learned their test results from 69% to 99.7%.

The trade-off for fast results is that rapid tests are less accurate than slower laboratory tests. Applied to people who have no HIV antibodies, one rapid test has probability about 0.004 of producing a false positive (that is, of falsely indicating that antibodies are present).[13]

STATE: If a clinic tests 200 people who are free of HIV antibodies, what is the chance that at least one false positive will occur?

PLAN: It is reasonable to assume that the test results for different individuals are independent. We have 200 independent events, each with probability 0.004. "At least one" combines many possible outcomes. It will be easier to use the fact that

$$P(\text{at least one positive}) = 1 - P(\text{no positives})$$

DO: We'll find $P(\text{no positives})$ first. The probability of a negative result for any one person is $1 - 0.004 = 0.996$. To find the probability that all 200 people tested have negative results, use the multiplication rule for independent events:

$$P(\text{no positives}) = P(\text{all 200 negative})$$
$$= (0.996)(0.996)\ldots(0.996) \quad \{200 \text{ times}\}$$
$$= 0.996^{200} = 0.4486$$

The probability we want is therefore

$$P(\text{at least one positive}) = 1 - 0.4486 = 0.5514$$

CONCLUDE: The probability is greater than 1/2 that at least 1 of the 200 people will test positive for HIV, even though no one has the virus.

CAUTION

The multiplication rule $P(A \text{ and } B) = P(A) \cdot P(B)$ holds if A and B are *independent* but not otherwise. The addition rule $P(A \text{ or } B) = P(A) + P(B)$ holds if A and B are *mutually exclusive* but not otherwise. Resist the temptation to use these simple rules when the conditions that justify them are not present.

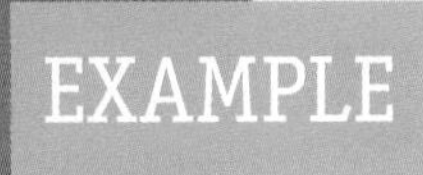

Sudden Infant Death

Condemned by independence

Assuming independence when it isn't true can lead to disaster. Several mothers in England were convicted of murder simply because two of their children had died in their cribs with no visible cause. An "expert witness" for the prosecution said that the probability of an unexplained crib death in a nonsmoking middle-class family is 1/8500. He then multiplied 1/8500 by 1/8500 to claim that there is only a 1-in-72-million chance that two children in the same family could have died naturally. This is nonsense: it assumes that crib deaths are independent, and data suggest that they are not. Some common genetic or environmental cause, not murder, probably explains the deaths.

THINK ABOUT IT

Is there a connection between mutually exclusive and independent? Let's start with a new chance process. Choose a U.S. adult at random. Define event A: the person is male, and event B: the person is pregnant. It's pretty clear that these two events are mutually exclusive (can't happen together)! What about independence? If you know that event A has occurred, does this affect the probability that event B happens? Of course! If we know the person is male, then the chance that the person is pregnant is 0. Since $P(B \mid A) \neq P(B)$, the two events are not independent. *Two mutually exclusive events can never be independent,* because if one event happens, the other event is guaranteed not to happen.

CHECK YOUR UNDERSTANDING

1. During World War II, the British found that the probability that a bomber is lost through enemy action on a mission over occupied Europe was 0.05. Assuming that missions are independent, find the probability that a bomber returned safely from 20 missions.

2. Government data show that 8% of adults are full-time college students and that 30% of adults are age 55 or older. Since $(0.08)(0.30) = 0.024$, can we conclude that about 2.4% of adults are college students 55 or older? Why or why not?

Calculating Conditional Probabilities

If we rearrange the terms in the general multiplication rule, we can get a formula for the conditional probability $P(B \mid A)$. Start with the general multiplication rule

$$P(A \cap B) = P(A) \cdot P(B \mid A)$$

Divide both sides of the equation by $P(A)$. This yields

$$\frac{P(A \cap B)}{P(A)} = P(B \mid A)$$

We can switch sides of the equation to get

$$P(B \mid A) = \frac{P(A \cap B)}{P(A)}$$

Conditional Probability Formula

To find the conditional probability $P(B \mid A)$, use the formula

$$P(B \mid A) = \frac{P(A \cap B)}{P(A)}$$

Here's an example that illustrates the use of this formula in a familiar setting.

Who Reads the Paper?

Conditional probability formula

In Section 5.2 (page 307), we classified the residents of a large apartment complex based on the events A: reads *USA Today*, and B: reads the *New York Times*. The completed Venn diagram is reproduced here.

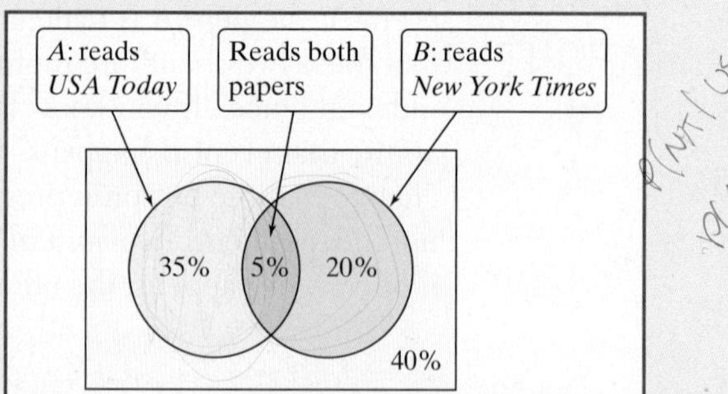

PROBLEM: What's the probability that a randomly selected resident who reads *USA Today* also reads the *New York Times*?

SOLUTION: Because we're given that the randomly chosen resident reads *USA Today*, we want to find *P*(reads *New York Times* | reads *USA Today*), or $P(B \mid A)$. By the conditional probability formula,

$$P(B \mid A) = \frac{P(A \cap B)}{P(A)}$$

AP EXAM TIP You can write statements like $P(A \mid B)$ if events A and B are defined clearly, or you can use a verbal equivalent, such as *P*(reads *New York Times* | reads *USA Today*). Use the approach that makes the most sense to you.

Since $P(A \cap B) = 0.05$ and $P(A) = 0.40$, we have

$$P(B \mid A) = \frac{0.05}{0.40} = 0.125$$

There's a 12.5% chance that a randomly selected resident who reads *USA Today* also reads the *New York Times.*

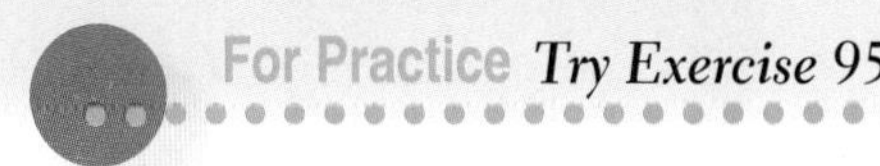

Sometimes we are interested in computing a conditional probability that amounts to "going backward" in a tree diagram. The following example shows how to do that.

Young Adults at Video-Sharing Sites

Conditioning "backward" in a tree diagram

In an earlier example (page 320), we looked at whether adult Internet users of various ages regularly visit video-sharing sites like YouTube. The tree diagram summarizing the probabilities is shown in the margin.

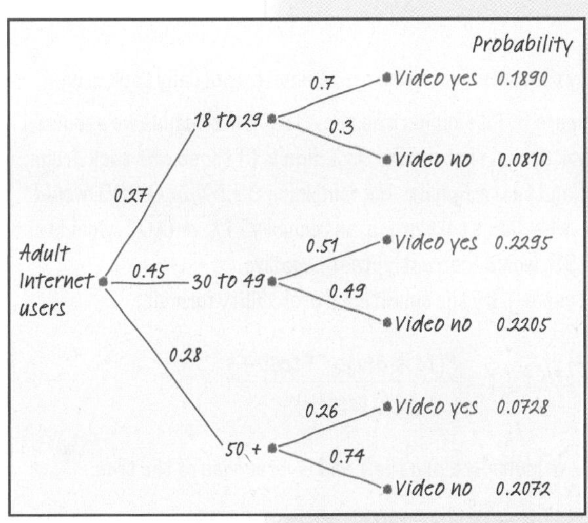

STATE: What percent of adult Internet users who visit video-sharing sites are aged 18 to 29?

PLAN: In probability language, we want the conditional probability *P*(18 to 29 | video yes). Use the tree diagram and the definition of conditional probability:

$$P(\text{18 to 29} \mid \text{video yes}) = \frac{P(\text{18 to 29 and video yes})}{P(\text{video yes})}$$

DO: Look at the tree diagram. *P*(video yes) is the sum of the three red probabilities. *P*(18 to 29 and video yes) is the result of following just the top branch in the tree diagram. So

$$P(\text{18 to 29} \mid \text{video yes}) = \frac{P(\text{18 to 29 and video yes})}{P(\text{video yes})}$$

$$= \frac{0.1890}{0.4913} = 0.3847$$

CONCLUDE: About 38% of adults who visit video-sharing sites are between 18 and 29 years old. Compare this conditional probability with the original information (unconditional) that 27% of adult Internet users are between 18 and 29 years old. Knowing that a person visits video-sharing sites increases the probability that he or she is young (which indicates that these two events are not independent).

For Practice *Try Exercise 97*

Some people use a result known as Bayes's theorem to solve probability questions that require going "backward" in a tree diagram, like the one in this example. To be honest, Bayes's theorem is just a complicated formula for computing conditional probabilities. For that reason, we won't introduce it.

Lately, it seems that we've been hearing more in the news about athletes who have tested positive for performance-enhancing drugs. From Major League Baseball, to the Tour de France, to the Olympics, we have learned about athletes who took banned substances. Sports at all levels now have programs in place to test athletes' urine or blood samples. There's an important question about drug testing that probability can help answer: if an athlete tests positive for a banned substance, did the athlete necessarily attempt to cheat?

EXAMPLE

Athletes and Drug Testing

Conditional probability in real life

Over 10,000 athletes competed in the 2008 Olympic Games in Beijing. The International Olympic Committee wanted to ensure that the competition was as fair as possible. So the committee administered more than 5000 drug tests to athletes. All medal winners were tested, as well as other randomly selected competitors.

Suppose that 2% of athletes had actually taken (banned) drugs. No drug test is perfect. Sometimes the test says that an athlete took drugs, but the athlete actually didn't. We call this a false positive result. Other times, the drug test says an athlete is "clean," but the athlete actually took drugs. This is called a false negative result. Suppose that the testing procedure used at the Olympics has a false positive rate of 1% and a false negative rate of 0.5%.

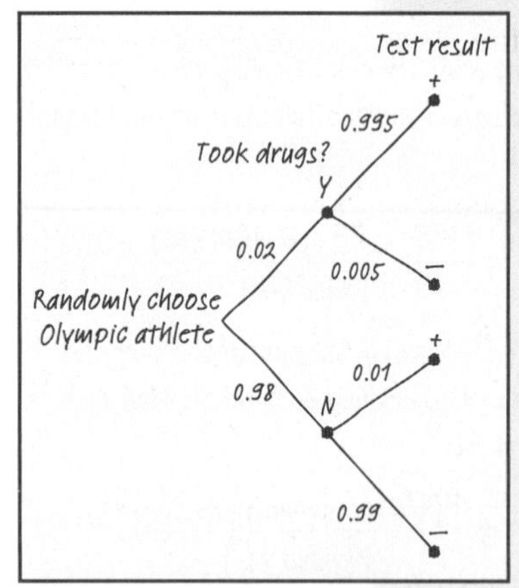

FIGURE 5.12 Tree diagram showing whether Olympic athletes took drugs and the likelihood of getting a positive or a negative test result from a drug test.

PROBLEM: **What's the probability that an athlete who tests positive actually took drugs?**

SOLUTION: **The tree diagram in Figure 5.12 summarizes the situation. Because we assumed that 2% of all Olympic athletes took drugs, that means 98% didn't. Of those who took drugs, 0.5%, or 0.005, would "get lucky" and test negative. The remaining 99.5%, or 0.995, would test positive. Among the athletes who didn't take drugs, an "unlucky" 1%, or 0.01, would test positive. The remaining 99%, or 0.99, would (correctly) test negative.**

We want to find $P(\text{took drugs} \mid \text{test} +)$. By the conditional probability formula,

$$P(\text{took drugs} \mid \text{test}+) = \frac{P(\text{took drugs} \cap \text{test}+)}{P(\text{test}+)}$$

- **To find $P(\text{took drugs} \cap \text{test}+)$, we multiply along the Y and + branches of the tree,**

$$P(\text{took drugs} \cap \text{test}+) = (0.02)(0.995) = 0.0199$$

- **To find $P(\text{test}+)$, we need to calculate the probability that a randomly selected athlete gets a positive test result. There are two ways this can happen: (1) if the athlete took drugs and the test result is positive, and (2) if the athlete didn't take drugs, but the test gives a false positive. From the tree diagram, the desired probability is**

$$P(\text{test}+) = (0.02)(0.995) + (0.98)(0.01) = 0.0297$$

Using these two results, we can find the conditional probability:

$$P(\text{took drugs} \mid \text{test}+) = \frac{P(\text{took drugs} \cap \text{test}+)}{P(\text{test}+)} = \frac{0.0199}{0.0297} = 0.67$$

That is, 67% of athletes who test positive actually took drugs.

For Practice *Try Exercise 99*

Are you surprised by the final result of the example? Most people are. Sometimes a two-way table that includes counts is more convincing. To make calculations simple, we'll suppose that there were exactly 10,000 athletes at the 2008 Olympics, and that exactly 200 took banned substances (that's 2% of the athletes). How many of those 200 would test positive? It would be 99.5% of 200, or 199 of them. That leaves 1 who would test negative. How many of the 9800 athletes who didn't take drugs would get a false positive test result? One percent of them, or $(0.01)(9800) = 98$ athletes. The remaining $9800 - 98 = 9702$ would test negative. In total, $199 + 98 = 297$ athletes would have positive test results and $1 + 9702 = 9703$ athletes would have negative test results. This information is summarized in the two-way table below.

	Took Drugs		
Positive Test?	**Yes**	**No**	**Total**
Yes	199	98	**297**
No	1	9,702	**9,703**
Total	**200**	**9,800**	**10,000**

This example illustrates a fact that is important when considering proposals for widespread testing for illegal drugs or serious diseases: if the condition being tested is uncommon in the population, many positives will be false positives.

Given that a randomly selected athlete has tested positive, we can see from the two-way table that the conditional probability $P(\text{took drugs} \mid \text{test} +) = 199/297 = 0.67$.

Now you can see why most drug-testing programs require an athlete to submit to *two* urine or blood samples at each test. If one sample tests positive, we can't be sure that the athlete took drugs. If both samples test positive, however, we can be pretty sure that the athlete took a banned substance. (That's what happened to Floyd Landis, who was stripped of his victory in the 2006 Tour de France.)

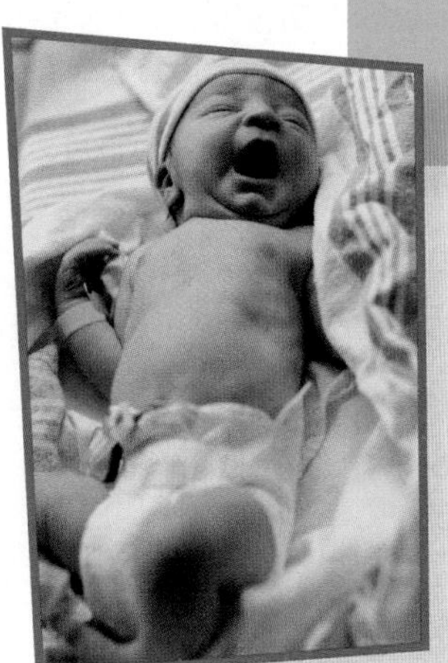

case closed

How Well Can Babies Hear?

Let's return to the chapter-opening Case Study (page 281) about screening babies' hearing.

		Baby's Actual Hearing	
		Loss	**Normal**
Test result	Loss	54	6
	Normal	4	36

Earlier, we asked this question: if the new test says that a baby has hearing loss, what's the chance that the test is correct? By now, you should recognize this as a conditional probability question. To find $P(\text{hearing loss} \mid \text{test says loss})$, we can use the conditional probability formula:

$$P(\text{hearing loss} \mid \text{test says loss}) = \frac{P(\text{hearing loss and test says loss})}{P(\text{test says loss})} = \frac{54/100}{60/100} = 0.90$$

Alternatively, we can see from the two-way table that of the 60 cases in which the new test says there is a hearing loss, the baby actually had hearing loss in 54 of those cases. That means

$$P(\text{hearing loss} \mid \text{test says loss}) = 54/60 = 0.90$$

If the test says that the baby has hearing loss, there's a 90% chance that the baby actually does have hearing loss.

About 4 million babies are born in the United States each year. Suppose all these babies are screened with the new device. If roughly 3 of every 1000 babies has some form of hearing loss, about how many babies with hearing loss will be missed by the new device? We would expect the total number of babies with hearing loss per year to be about 4,000,000(0.003), or about 12,000. According to the results in the two-way table above, 4 out of 58 babies with a hearing loss will be missed by the new screening device. Thus, the estimated total number of babies with hearing loss missed by the screening test per year is

$$12{,}000\left(\frac{4}{58}\right) \approx 827.59$$

or about 828.

SECTION 5.3 Summary

- If one event has happened, the chance that another event will happen is a **conditional probability.** The notation $P(B \mid A)$ represents the probability that event B occurs given that event A has occurred.
- When the chance that event B occurs is not affected by whether event A occurs, we say that events A and B are **independent.** For independent events A and B, $P(B \mid A) = P(B)$ and $P(A \mid B) = P(A)$. *If two events A and B are mutually exclusive (disjoint), they cannot be independent.*
- When chance behavior involves a sequence of outcomes, a **tree diagram** can be used to describe the sample space. Tree diagrams can also help in finding the probability that two or more events occur together. We simply multiply along the branches that correspond to the outcomes of interest.
- The **general multiplication rule** states that the probability of events A and B occurring together is

  $$P(A \cap B) = P(A) \cdot P(B \mid A)$$

  In the special case of *independent* events, the multiplication rule becomes

  $$P(A \cap B) = P(A) \cdot P(B)$$

- Dividing both sides of the equation by $P(A)$, we get the **conditional probability formula**

  $$P(B \mid A) = \frac{P(A \cap B)}{P(A)}$$

SECTION 5.3 Exercises

63. **Get rich** A survey of 4826 randomly selected young adults (aged 19 to 25) asked, "What do you think are the chances you will have much more than a middle-class income at age 30?" The two-way table shows the responses.[14] Choose a survey respondent at random.

	Sex		
Opinion	**Female**	**Male**	**Total**
Almost no chance	96	98	194
Some chance but probably not	426	286	712
A 50-50 chance	696	720	1416
A good chance	663	758	1421
Almost certain	486	597	1083
Total	**2367**	**2459**	**4826**

(a) Given that the person selected is male, what's the probability that he answered "almost certain"?

(b) If the person selected said "some chance but probably not," what's the probability that the person is female?

64. **A *Titanic* disaster** In 1912 the luxury liner *Titanic*, on its first voyage across the Atlantic, struck an iceberg and sank. Some passengers got off the ship in lifeboats, but many died. The two-way table gives information about adult passengers who lived and who died, by class of travel. Suppose we choose an adult passenger at random.

	Survival Status	
Class of Travel	**Survived**	**Died**
First class	197	122
Second class	94	167
Third class	151	476

(a) Given that the person selected was in first class, what's the probability that he or she survived?

(b) If the person selected survived, what's the probability that he or she was a third-class passenger?

65. **Get rich** Refer to Exercise 63.

(a) Find P("a good chance" | female).

(b) Find P("a good chance").

(c) Use your answers to (a) and (b) to determine whether the events "a good chance" and "female" are independent. Explain your reasoning.

66. **A *Titanic* disaster** Refer to Exercise 64.

(a) Find P(survived | second class).

(b) Find P(survived).

(c) Use your answers to (a) and (b) to determine whether the events "survived" and "second class" are independent. Explain your reasoning.

67. **Sampling senators** The two-way table describes the members of the U.S. Senate in a recent year. Suppose we select a senator at random. Consider events D: is a democrat, and F: is female.

	Male	**Female**
Democrats	47	13
Republicans	36	4

(a) Find $P(D \mid F)$. Explain what this value means.

(b) Find $P(F \mid D)$. Explain what this value means.

68. **Who eats breakfast?** The two-way table describes the 595 students who responded to a school survey about eating breakfast. Suppose we select a student at random. Consider events B: eats breakfast regularly, and M: is male.

	Male	**Female**	**Total**
Eats breakfast regularly	190	110	**300**
Doesn't eat breakfast regularly	130	165	**295**
Total	**320**	**275**	**595**

(a) Find $P(B \mid M)$. Explain what this value means.

(b) Find $P(M \mid B)$. Explain what this value means.

69. **Sampling senators** Refer to Exercise 67. Are events D and F independent? Justify your answer.

pg 316 70. **Who eats breakfast?** Refer to Exercise 68. Are events B and M independent? Justify your answer.

71. **Foreign-language study** Choose a student in grades 9 to 12 at random and ask if he or she is studying a language other than English. Here is the distribution of results:

Language:	Spanish	French	German	All others	None
Probability:	0.26	0.09	0.03	0.03	0.59

(a) What's the probability that the student is studying a language other than English?

(b) What is the conditional probability that a student is studying Spanish given that he or she is studying some language other than English?

72. **Income tax returns** Here is the distribution of the adjusted gross income (in thousands of dollars) reported on individual federal income tax returns in a recent year:

Income:	<25	25–49	50–99	100–499	≥500
Probability:	0.431	0.248	0.215	0.100	0.006

(a) What is the probability that a randomly chosen return shows an adjusted gross income of \$50,000 or more?

(b) Given that a return shows an income of at least \$50,000, what is the conditional probability that the income is at least \$100,000?

73. **Tall people and basketball players** Select an adult at random. Define events *T*: person is over 6 feet tall, and *B*: person is a professional basketball player. Rank the following probabilities from smallest to largest. Justify your answer.

$P(T) \quad P(B) \quad P(T \mid B) \quad P(B \mid T)$

74. **Teachers and college degrees** Select an adult at random. Define events A: person has earned a college degree, and *T*: person's career is teaching. Rank the following probabilities from smallest to largest. Justify your answer.

$P(A) \quad P(T) \quad P(A \mid T) \quad P(T \mid A)$

75. **Rolling dice** Suppose you roll two fair, six-sided dice—one red and one green. Are the events "sum is 7" and "green die shows a 4" independent? Justify your answer.

76. **Rolling dice** Suppose you roll two fair, six-sided dice—one red and one green. Are the events "sum is 8" and "green die shows a 4" independent? Justify your answer.

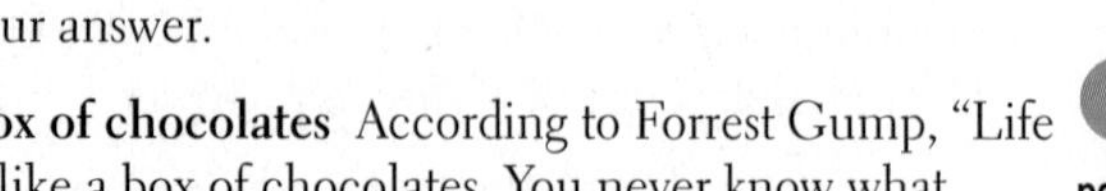

77. **Box of chocolates** According to Forrest Gump, "Life is like a box of chocolates. You never know what you're gonna get." Suppose a candy maker offers a special "Gump box" with 20 chocolate candies that look the same. In fact, 14 of the candies have soft centers and 6 have hard centers. Choose 2 of the candies from a Gump box at random. pg 318

(a) Draw a tree diagram that shows the sample space of this chance process.

(b) Find the probability that one of the chocolates has a soft center and the other one doesn't.

78. **Inspecting switches** A shipment contains 10,000 switches. Of these, 1000 are bad. An inspector draws 2 switches at random, one after the other.

(a) Draw a tree diagram that shows the sample space of this chance process.

(b) Find the probability that both switches are defective.

79. **Free downloads?** Illegal music downloading has become a big problem: 29% of Internet users download music files, and 67% of downloaders say they don't care if the music is copyrighted.[15] What percent of Internet users download music and don't care if it's copyrighted? Write the information given in terms of probabilities, and use the general multiplication rule. pg 319

80. **At the gym** Suppose that 10% of adults belong to health clubs, and 40% of these health club members go to the club at least twice a week. What percent of all adults go to a health club at least twice a week? Write the information given in terms of probabilities, and use the general multiplication rule.

81. **Going pro** Only 5% of male high school basketball, baseball, and football players go on to play at the college level. Of these, only 1.7% enter major league professional sports. About 40% of the athletes who compete in college and then reach the pros have a career of more than 3 years.[16] What is the probability that a high school athlete who plays basketball, baseball, or football competes in college and then goes on to have a pro career of more than 3 years? Show your work.

82. **Teens online** We saw in an earlier example (page 319) that 93% of teenagers are online and that 55% of online teens have posted a profile on a social-networking site. Of online teens with a profile, 76% have placed comments on a friend's blog. What percent of all teens are online, have a profile, and comment on a friend's blog? Show your work.

83. **Fill 'er up!** In a recent month, 88% of automobile drivers filled their vehicles with regular gasoline, 2% purchased midgrade gas, and 10% bought premium gas.[17] Of those who bought regular gas, 28% paid with a credit card; of customers who bought midgrade and premium gas, 34% and 42%, respectively, paid with a credit card. Suppose we select a customer at random. Draw a tree diagram to represent this situation. What's the probability that the customer paid with a credit card? Use the four-step process to guide your work. STEP 4 pg 320

84. **Urban voters** The voters in a large city are 40% white, 40% black, and 20% Hispanic. (Hispanics may be of any race in official statistics, but here we are speaking of political blocks.) A mayoral candidate anticipates attracting 30% of the white vote, 90% of the black vote, and 50% of the Hispanic vote. Draw a tree diagram to represent this situation. What percent of the overall vote does the candidate expect to get? Use the four-step process to guide your work. STEP 4

85. **Fill 'er up** Refer to Exercise 83. Given that the customer paid with a credit card, find the probability that she bought premium gas.

86. **Urban voters** In the election described in Exercise 84, if the candidate's predictions come true, what percent of her votes come from black voters? (Write this as a conditional probability and use the definition of conditional probability.)

87. **Medical risks** Morris's kidneys are failing, and he is awaiting a kidney transplant. His doctor gives him this information for patients in his condition: 90% survive the transplant and 10% die. The transplant succeeds in 60% of those who survive, and the other 40% must return to kidney dialysis. The proportions who survive five years are 70% for those with a new kidney and 50% for those who return to dialysis.

(a) Make a tree diagram to represent this setting.

(b) Find the probability that Morris will survive for five years. Show your work.

88. **Winning at tennis** A player serving in tennis has two chances to get a serve into play. If the first serve goes out of bounds, the player serves again. If the second serve is also out, the player loses the point. Here are probabilities based on four years of the Wimbledon Championship:[18]

$$P(\text{1st serve in}) = 0.59 \quad P(\text{win point} \mid \text{1st serve in}) = 0.73$$

$$P(\text{2nd serve in} \mid \text{1st serve out}) = 0.86$$

$$P(\text{win point} \mid \text{1st serve out and 2nd serve in}) = 0.59$$

(a) Make a tree diagram for the results of the two serves and the outcome (win or lose) of the point.

(b) What is the probability that the serving player wins the point? Show your work.

pg 321

89. **Bright lights?** A string of Christmas lights contains 20 lights. The lights are wired in series, so that if any light fails the whole string will go dark. Each light has probability 0.02 of failing during a 3-year period. The lights fail independently of each other. Find the probability that the string of lights will remain bright for 3 years.

90. **Common names** The Census Bureau says that the 10 most common names in the United States are (in order) Smith, Johnson, Williams, Brown, Jones, Miller, Davis, Garcia, Rodriguez, and Wilson. These names account for 9.6% of all U.S. residents. Out of curiosity, you look at the authors of the textbooks for your current courses. There are 9 authors in all. Would you be surprised if none of the names of these authors were among the 10 most common? (Assume that authors' names are independent and follow the same probability distribution as the names of all residents.)

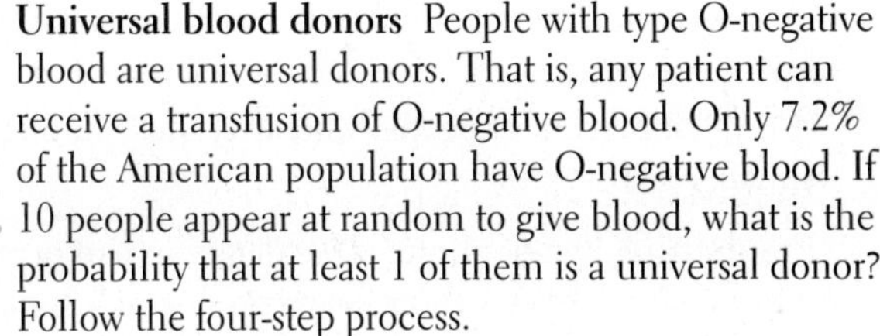

pg 322

91. **Universal blood donors** People with type O-negative blood are universal donors. That is, any patient can receive a transfusion of O-negative blood. Only 7.2% of the American population have O-negative blood. If 10 people appear at random to give blood, what is the probability that at least 1 of them is a universal donor? Follow the four-step process.

92. **Lost Internet sites** Internet sites often vanish or move, so that references to them can't be followed. In fact, 13% of Internet sites referenced in major scientific journals are lost within two years after publication.[19] If a paper contains seven Internet references, what is the probability that at least one of them doesn't work two years later? Follow the four-step process.

93. **Late shows** Some TV shows begin after their scheduled times when earlier programs run late. According to a network's records, about 3% of its shows start late. To find the probability that three consecutive shows on this network start on time, can we multiply (0.97)(0.97)(0.97)? Why or why not?

94. **Late flights** An airline reports that 85% of its flights arrive on time. To find the probability that its next four flights into LaGuardia Airport all arrive on time, can we multiply (0.85)(0.85)(0.85)(0.85)? Why or why not?

pg 324

95. **MySpace versus Facebook** A recent survey suggests that 85% of college students have posted a profile on Facebook, 54% use MySpace regularly, and 42% do both. Suppose we select a college student at random and learn that the student has a profile on Facebook. Find the probability that the student uses MySpace regularly. Show your work.

96. **Mac or PC?** A recent census at a major university revealed that 40% of its students mainly used Macintosh computers (Macs). The rest mainly used PCs. At the time of the census, 67% of the school's students were undergraduates. The rest were graduate students. In the census, 23% of the respondents were graduate students who said that they used PCs as their primary computers. Suppose we select a student at random from among those who were part of the census and learn that the student mainly uses a Mac. Find the probability that this person is a graduate student. Show your work.

pg 325

97. **Lactose intolerance** Lactose intolerance causes difficulty in digesting dairy products that contain lactose (milk sugar). It is particularly common among people of African and Asian ancestry. In the United States (ignoring other groups and people who consider themselves to belong to more than one race), 82% of the population is white, 14% is black, and 4% is Asian. Moreover, 15% of whites, 70% of blacks, and 90% of Asians are lactose intolerant.[20]

(a) What percent of the entire population is lactose intolerant?

(b) What percent of people who are lactose intolerant are Asian?

98. **Fundraising by telephone** Tree diagrams can organize problems having more than two stages. The figure shows probabilities for a charity calling potential donors by telephone.[21] Each person called is either a recent donor, a past donor, or a new prospect. At the next stage, the person called either does or does not pledge to contribute, with conditional probabilities that depend on the donor class the person belongs to. Finally, those who make a pledge either do or don't actually make a contribution.

(a) What percent of calls result in a contribution?

(b) What percent of those who contribute are recent donors?

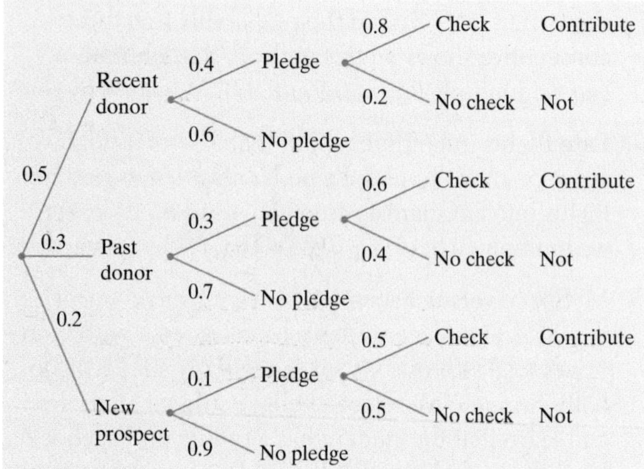

99. **HIV testing** Enzyme immunoassay (EIA) tests are used to screen blood specimens for the presence of antibodies to HIV, the virus that causes AIDS. Antibodies indicate the presence of the virus. The test is quite accurate but is not always correct. Here are approximate probabilities of positive and negative EIA outcomes when the blood tested does and does not actually contain antibodies to HIV:[22]

pg 326

	Test Result	
	+	−
Antibodies present	0.9985	0.0015
Antibodies absent	0.006	0.994

Suppose that 1% of a large population carries antibodies to HIV in their blood.

(a) Draw a tree diagram for selecting a person from this population (outcomes: antibodies present or absent) and for testing his or her blood (outcomes: EIA positive or negative).

(b) What is the probability that the EIA test is positive for a randomly chosen person from this population?

(c) What is the probability that a person has the antibody given that the EIA test is positive?

100. **Testing the test** Are false positives too common in some medical tests? Researchers conducted an experiment involving 250 patients with a medical condition and 750 other patients who did not have the medical condition. The medical technicians who were reading the test results were unaware that they were subjects in an experiment.

(a) Technicians correctly identified 240 of the 250 patients with the condition. They also identified 50 of the healthy patients as having the condition. What were the false positive and false negative rates for the test?

(b) Given that a patient got a positive test result, what is the probability that the patient actually had the medical condition? Show your work.

101. **The probability of a flush** A poker player holds a flush when all 5 cards in the hand belong to the same suit. We will find the probability of a flush when 5 cards are dealt. Remember that a deck contains 52 cards, 13 of each suit, and that when the deck is well shuffled, each card dealt is equally likely to be any of those that remain in the deck.

(a) We will concentrate on spades. What is the probability that the first card dealt is a spade? What is the conditional probability that the second card is a spade given that the first is a spade?

(b) Continue to count the remaining cards to find the conditional probabilities of a spade on the third, the fourth, and the fifth card given in each case that all previous cards are spades.

(c) The probability of being dealt 5 spades is the product of the five probabilities you have found. Why? What is this probability?

(d) The probability of being dealt 5 hearts or 5 diamonds or 5 clubs is the same as the probability of being dealt 5 spades. What is the probability of being dealt a flush?

102. **The geometric distributions** You are tossing a pair of fair, six-sided dice in a board game. Tosses are independent. You land in a danger zone that requires you to roll doubles (both faces showing the same number of spots) before you are allowed to play again. How long will you wait to play again?

(a) What is the probability of rolling doubles on a single toss of the dice? (If you need review, the possible outcomes appear in Figure 5.2 (page 300). All 36 outcomes are equally likely.)

(b) What is the probability that you do not roll doubles on the first toss, but you do on the second toss?

(c) What is the probability that the first two tosses are not doubles and the third toss is doubles? This is the probability that the first doubles occurs on the third toss.

(d) Now you see the pattern. What is the probability that the first doubles occurs on the fourth toss? On the fifth toss? Give the general result: what is the probability that the first doubles occurs on the kth toss?

103. **A probability teaser** Suppose (as is roughly correct) that each child born is equally likely to be a boy or a girl and that the genders of successive children are independent. If we let BG mean that the older child is a boy and the younger child is a girl, then each of the combinations BB, BG, GB, and GG has probability 0.25. Ashley and Brianna each have two children.

(a) You know that at least one of Ashley's children is a boy. What is the conditional probability that she has two boys?

(b) You know that Brianna's older child is a boy. What is the conditional probability that she has two boys?

Multiple choice: Select the best answer for Exercises 104 to 106.

104. An athlete suspected of using steroids is given two tests that operate independently of each other. Test A has probability 0.9 of being positive if steroids have been used. Test B has probability 0.8 of being positive if steroids have been used. What is the probability that neither test is positive if steroids have been used?

(a) 0.72 (c) 0.02 (e) 0.08
(b) 0.38 (d) 0.28

105. In an effort to find the source of an outbreak of food poisoning at a conference, a team of medical detectives carried out a study. They examined all 50 people who had food poisoning and a random sample of 200 people attending the conference who didn't get food poisoning. The detectives found that 40% of the people with food poisoning went to a cocktail party on the second night of the conference, while only 10% of the people in the random sample attended the same party. Which of the following statements is appropriate for describing the 40% of people who went to the party? (Let F = got food poisoning and A = attended party.)

(a) $P(F \mid A) = 0.40$ (d) $P(A^C \mid F) = 0.40$
(b) $P(A \mid F^C) = 0.40$ (e) None of these
(c) $P(F \mid A^C) = 0.40$

106. Suppose a loaded die has the following probability model:

Outcome:	1	2	3	4	5	6
Probability:	0.3	0.1	0.1	0.1	0.1	0.3

If this die is thrown and the top face shows an odd number, what is the probability that the die shows a 1?

(a) 0.1 (d) 0.6
(b) 0.3 (e) None of these
(c) 0.5

Exercises 107 and 108 refer to the following setting. Your body mass index (BMI) is your weight in kilograms divided by the square of your height in meters. Online BMI calculators allow you to enter weight in pounds and height in inches. High BMI is a common but controversial indicator of overweight or obesity. A study by the National Center for Health Statistics found that the BMI of American young women (ages 20 to 29) is approximately Normal with mean 26.8 and standard deviation 7.4.[23]

107. **BMI** (2.2) People with BMI less than 18.5 are often classed as "underweight." What percent of young women are underweight by this criterion? Sketch and shade the area of interest under a Normal curve.

108. **BMI** (5.2) Suppose we select two American young women in this age group at random. Find the probability that at least one of them is classified as underweight. Show your work.

109. **Life at work** (1.1) The University of Chicago's General Social Survey asked a representative sample of adults this question: "Which of the following statements best describes how your daily work is organized? (1) I am free to decide how my daily work is organized. (2) I can decide how my daily work is organized, within certain limits. (3) I am not free to decide how my daily work is organized." Here is a two-way table of the responses for three levels of education:[240]

	Highest Degree Completed		
Response	Less than High School	High School	Bachelor's
1	31	161	81
2	49	269	85
3	47	112	14

Is there an association between level of education and freedom to organize one's work? Give graphical and numerical evidence to support your answer.

Chapter 5 Chapter Review

Probability is used to describe the long-term regularity that results from many repetitions of the same chance process. To calculate the probability of a particular outcome without any formulas, we can use random digits to simulate many repetitions. Section 5.1 discussed randomness and simulation.

Probability models assign probabilities to outcomes. Any such model must obey the rules of probability. Section 5.2 presented these rules, along with two useful tools for describing outcomes of chance behavior—Venn diagrams and two-way tables.

In some cases, we have partial information about the result of a chance process. This leads to the study of conditional probability in Section 5.3. Tree diagrams help organize our thinking about conditional probability in situations that involve a sequence of outcomes.

Chapter Review Exercises

These exercises are designed to help you review the important ideas and methods of the chapter. Relevant learning objectives are provided in bulleted form before the relevant exercise(s).

- Interpret probability as a long-run relative frequency.

R5.1. Rainy days The TV weatherman says, "There's a 30% chance of rain tomorrow." Explain what this statement means.

- Use simulation to model chance behavior.

R5.2. Click it or else From police records, it has been determined that 15% of drivers stopped for routine license checks are not wearing seat belts. If a police officer stops 10 vehicles, how likely is it that two consecutive drivers won't be wearing their seat belts?

(a) Describe the design of a simulation to estimate this probability. *Explain clearly* how you will use the partial table of random digits below to carry out your simulation.

(b) Carry out three repetitions of the simulation. Copy the random digits below onto your paper. Then mark on or directly above the table to show your results.

29077	14863	61683	47052	62224	51025
95052	90908	73592	75186	87136	95761
27102	56027	55892	33063	41842	81868
43367	49497	72719	96758	27611	91596

R5.3. Sandblasters Eight teams of the world's best sand sculptors gathered for an "extreme" competition. At five points during the competition, a randomly selected team's sculpture was blown up. That team was then forced to build a new sculpture from scratch. No team could be selected for destruction more than once. Last year's winning team did not have a sculpture blown up either last year or this year. Other teams were suspicious. Should they be? Design and carry out a simulation to answer this question. (Team Sanding Ovation won the contest with their "zipperhead" sculpture.)

- Describe a probability model for a chance process.

R5.4. Weird dice Nonstandard dice can produce interesting distributions of outcomes. Suppose you have two balanced, six-sided dice. Die A has faces with 2, 2, 2, 2, 6, and 6 spots. Die B has three faces with 5 spots and three faces with 1 spot. Imagine that you roll both dice at the same time.

(a) Find a probability model for the difference (Die A − Die B) in the total number of spots on the up-faces.

(b) Which die is more likely to roll a higher number? Justify your answer.

- Use basic probability rules, including the complement rule and the addition rule for mutually exclusive events.

R5.5. Race and ethnicity The Census Bureau allows each person to choose from a long list of races. That is, in the eyes of the Census Bureau, you belong to whatever race you say you belong to. Hispanic (also called Latino) is a separate category. Hispanics may be of any race. If we choose a resident of the United States at random, the Census Bureau gives these probabilities:[25]

	Hispanic	Not Hispanic
Asian	0.001	0.044
Black	0.006	0.124
White	0.139	0.674
Other	0.003	0.009

(a) Verify that this is a legitimate assignment of probabilities.

(b) What is the probability that a randomly chosen American is Hispanic?

(c) Non-Hispanic whites are the historical majority in the United States. What is the probability that a randomly chosen American is *not* a member of this group?

(d) Explain why $P(\text{white or Hispanic}) \neq P(\text{white}) + P(\text{Hispanic})$. Then find $P(\text{white or Hispanic})$.

- Use a Venn diagram to model a chance process involving two events.
- Use the general addition rule to calculate $P(A \cup B)$.

R5.6. Getting into college Ramon has applied to both Princeton and Stanford. According to his counselor, the probability that Princeton will admit him is 0.4, the probability that Stanford will admit him is 0.5, and the probability that both will admit him is 0.2.

(a) Make a Venn diagram to model this chance process.

(b) What is the probability that neither university admits Ramon?

(c) What's the probability that he gets into at least one of the two schools? Use the general addition rule to confirm that your answer is correct.

- When appropriate, use a tree diagram to describe chance behavior.
- Use the general multiplication rule to solve probability questions.
- Compute conditional probabilities.

R5.7. Steroids A company has developed a drug test to detect steroid use by athletes. The test is accurate 95% of the time when an athlete has taken steroids. It is 97% accurate when an athlete hasn't taken steroids. Suppose that the drug test will be used in a population of athletes in which 10% have actually taken steroids. Let's choose an athlete at random and administer the drug test.

(a) Make a tree diagram showing the sample space of this chance process.

(b) What's the probability that the randomly selected athlete tests positive? Show your work.

(c) Suppose that the chosen athlete tests positive. What's the probability that he or she actually used steroids? Show your work.

- Determine whether two events are independent.

R5.8. Mike's pizza You work at Mike's pizza shop. You have the following information about the 7 pizzas in the oven: 3 of the 7 have thick crust, and of these 1 has only sausage and 2 have only mushrooms. The remaining 4 pizzas have regular crust, and of these 2 have only sausage and 2 have only mushrooms. Choose a pizza at random from the oven.

(a) Are the events {getting a thick-crust pizza} and {getting a pizza with mushrooms} independent? Explain.

(b) You add an eighth pizza to the oven. This pizza has thick crust with only cheese. Now are the events {getting a thick-crust pizza} and {getting a pizza with mushrooms} independent? Explain.

- Find the probability that an event occurs using a two-way table.

R5.9. Deer and pine seedlings As suburban gardeners know, deer will eat almost anything green. In a study of pine seedlings at an environmental center in Ohio, researchers noted how deer damage varied with how much of the seedling was covered by thorny undergrowth:[26]

	Deer Damage	
Thorny Cover	Yes	No
None	60	151
<1/3	76	158
1/3 to 2/3	44	177
>2/3	29	176

(a) What is the probability that a randomly selected seedling was damaged by deer?

(b) What are the conditional probabilities that a randomly selected seedling was damaged, given each level of cover?

(c) Does knowing about the amount of thorny cover on a seedling change the probability of deer damage? Justify your answer.

- When appropriate, use the multiplication rule for independent events to compute probabilities.

R5.10. A random walk on Wall Street? The "random walk" theory of stock prices holds that price movements in disjoint time periods are independent of each other. Suppose that we record only whether the price is up or down each year, and that the probability that our portfolio rises in price in any one year is 0.65. (This probability is approximately correct for a portfolio containing equal dollar amounts of all common stocks listed on the New York Stock Exchange.)

(a) What is the probability that our portfolio goes up for three consecutive years?

(b) What is the probability that the portfolio's value moves in the same direction (either up or down) for three consecutive years?

R5.11. Blood types Each of us has an ABO blood type, which describes whether two characteristics called A and B are present. Every human being has two blood type alleles (gene forms), one inherited from our mother and one from our father. Each of these alleles can be A, B, or O. Which two we inherit determines our blood type. The table shows what our blood type is for each combination of two alleles. We inherit each of a parent's two alleles with probability 0.5. We inherit independently from our mother and father.

Alleles inherited	Blood type
A and A	A
A and B	AB
A and O	A
B and B	B
B and O	B
O and O	O

(a) Hannah and Jacob both have alleles A and B. Diagram the sample space that shows the alleles that their next child could receive. Then give the possible blood types that this child could have, along with the probability for each blood type.

(b) Jennifer has alleles A and O. Jose has alleles A and B. They have two children. What is the probability that at least one of the two children has blood type B? Show your method.

Chapter 5 AP Statistics Practice Test

Section I: Multiple Choice *Select the best answer for each question.*

T5.1. Dr. Stats plans to toss a fair coin 10,000 times in the hope that it will lead him to a deeper understanding of the laws of probability. Which of the following statements is true?

(a) It is unlikely that Dr. Stats will get more than 5000 heads.

(b) Whenever Dr. Stats gets a string of 15 tails in a row, it becomes more likely that the next toss will be a head.

(c) The fraction of tosses resulting in heads should be close to 1/2.

(d) The chance that the 100th toss will be a head depends somewhat on the results of the first 99 tosses.

(e) All of the above statements are true.

T5.2. China has 1.2 billion people. Marketers want to know which international brands they have heard of. A large study showed that 62% of all Chinese adults have heard of Coca-Cola. You want to simulate choosing a Chinese at random and asking if he or she has heard of Coca-Cola. One correct way to assign random digits to simulate the answer is:

(a) One digit simulates one person's answer; odd means "Yes" and even means "No."

(b) One digit simulates one person's answer; 0 to 6 mean "Yes" and 7 to 9 mean "No."

(c) One digit simulates the result; 0 to 9 tells how many in the sample said "Yes."

(d) Two digits simulate one person's answer; 00 to 61 mean "Yes" and 62 to 99 mean "No."

(e) Two digits simulate one person's answer; 00 to 62 mean "Yes" and 63 to 99 mean "No."

T5.3. Choose an American household at random and record the number of vehicles they own. Here is the probability model if we ignore the few households that own more than 5 cars:

Number of cars:	0	1	2	3	4	5
Probability:	0.09	0.36	0.35	0.13	0.05	0.02

A housing company builds houses with two-car garages. What percent of households have more cars than the garage can hold?

(a) 7% (b) 13% (c) 20% (d) 45% (e) 55%

T5.4. Computer voice recognition software is getting better. Some companies claim that their software correctly recognizes 98% of all words spoken by a trained user. To simulate recognizing a single word when the probability of being correct is 0.98, let two digits simulate one word; 00 to 97 mean "correct." The program recognizes words (or not) independently. To simulate the program's performance on 10 words, use these random digits:

60970 70024 17868 29843 61790 90656 87964 18883

The number of words recognized correctly out of the 10 is

(a) 10 (b) 9 (c) 8 (d) 7 (e) 6

Questions T5.5 to T5.7 refer to the following setting. One thousand students at a city high school were classified according to both GPA and whether or not they consistently skipped classes. The two-way table below summarizes the data.

	GPA		
Skipped Classes	**<2.0**	**2.0–3.0**	**>3.0**
Many	80	25	5
Few	175	450	265

T5.5. What is the probability that a student has a GPA under 2.0?

(a) 0.227 (b) 0.255 (c) 0.450 (d) 0.475 (e) 0.506

T5.6. What is the probability that a student has a GPA under 2.0 or has skipped many classes?

(a) 0.080 (b) 0.281 (c) 0.285 (d) 0.365 (e) 0.727

T5.7. What is the probability that a student has a GPA under 2.0 given that he or she has skipped many classes?

(a) 0.080 (b) 0.281 (c) 0.285 (d) 0.314 (e) 0.727

T5.8. For events A and B related to the same chance process, which of the following statements is true?

(a) If A and B are mutually exclusive, then they must be independent.

(b) If A and B are independent, then they must be mutually exclusive.

(c) If A and B are not mutually exclusive, then they must be independent.

(d) If A and B are not independent, then they must be mutually exclusive.

(e) If A and B are independent, then they cannot be mutually exclusive.

T5.9. Choose an American adult at random. The probability that you choose a woman is 0.52. The probability that the person you choose has never married is 0.25. The probability that you choose a woman who has never married is 0.11. The probability that the person you choose is either a woman or has never been married (or both) is therefore about

(a) 0.77. (b) 0.66. (c) 0.44. (d) 0.38. (e) 0.13.

T5.10. A deck of playing cards has 52 cards, of which 12 are face cards. If you shuffle the deck well and turn over the top 3 cards, one after the other, what's the probability that all 3 are face cards?

(a) 0.001 (b) 0.005 (c) 0.010 (d) 0.012 (e) 0.02

Section II: Free Response *Show all your work. Indicate clearly the methods you use, because you will be graded on the correctness of your methods as well as on the accuracy and completeness of your results and explanations.*

T5.11. Your teacher has invented a "fair" dice game to play. Here's how it works. Your teacher will roll one fair *eight-sided* die, and you will roll a fair *six-sided* die. Each player rolls once, and the winner is the person with the higher number. In case of a tie, neither player wins. The table shows the sample space of this chance process.

	Teacher Rolls							
You Roll	**1**	**2**	**3**	**4**	**5**	**6**	**7**	**8**
1								
2								
3								
4								
5								
6								

(a) Let A be the event "your teacher wins." Find $P(A)$.

(b) Let B be the event "you get a 3 on your first roll." Find $P(A \cup B)$.

(c) Are events A and B independent? Justify your answer.

T5.12. Three machines—A, B, and C—are used to produce a large quantity of identical parts at a factory. Machine A produces 60% of the parts, while Machines B and C produce 30% and 10% of the parts, respectively. Historical records indicate that 10% of the parts produced by Machine A are defective, compared with 30% for Machine B and 40% for Machine C.

(a) Draw a tree diagram to represent this chance process.

(b) If we choose a part produced by one of these three machines, what's the probability that it's defective? Show your work.

(c) If a part is inspected and found to be defective, which machine is most likely to have produced it? Give appropriate evidence to support your answer.

T5.13. Researchers are interested in the relationship between cigarette smoking and lung cancer. Suppose an adult male is randomly selected from a particular population. The following table shows the probabilities of some events related to this chance process:

Event	Probability
Smokes	0.25
Smokes and gets cancer	0.08
Does not smoke and does not get cancer	0.71

(a) Find the probability that the individual gets cancer given that he is a smoker. Show your work.

(b) Find the probability that the individual smokes or gets cancer. Show your work.

(c) Two adult males are selected at random. Find the probability that at least one of the two gets cancer. Show your work.

T5.14. Based on previous records, 17% of the vehicles passing through a tollbooth have out-of-state plates. A bored tollbooth worker decides to pass the time by counting how many vehicles pass through until he sees two with out-of-state plates.[27]

(a) Describe the design of a simulation to estimate the average number of vehicles it takes to find two with out-of-state plates. Explain clearly how you will use the partial table of random digits below to carry out your simulation.

(b) Perform three repetitions of the simulation you described in part (a). Copy the random digits below onto your paper. Then mark on or directly above the table to show your results.

```
41050 92031 06449 05059 59884 31880
53115 84469 94868 57967 05811 84514
84177 06757 17613 15582 51506 81435
75011 13006 63395 55041 15866 06589
```

Chapter 6

case study

Random Variables

Does Therapeutic Touch Really Work?

For her fourth-grade science-fair project, Emily Rosa designed an experiment to test whether an alternative healing method known as therapeutic touch actually works. Therapeutic touch requires the "healer" to detect the patient's "human energy field." Emily doubted that this was possible. So she invited 21 people who had practiced therapeutic touch for between 1 and 27 years to participate in her experiment.

Emily's experimental design was fairly simple. She sat at one side of a table and the subject sat at the other side. A tall screen with two holes was placed in the middle of the table, so that the subject could not see Emily. The subject was asked to extend one arm, palm side up, through each of the holes. Unknown to the subject, Emily then flipped a coin to decide which of the subject's two hands she would place her hand above. She then asked the subject to identify the location of the "energy field." This process was repeated either 10 or 20 times with each of the 21 subjects.

Want to know what happened in Emily's experiment? We'll tell you at the end of the chapter, after you have learned some more about probability.

Introduction

Do you drink bottled water or tap water? According to a recent report in *U.S. Mayor Newspaper*, about 75% of people drink bottled water regularly. Some people do so because they believe bottled water is safer than tap water. (There's little evidence to support this belief.) Others say they prefer the taste of bottled water. Can people really tell the difference?

ACTIVITY *Bottled Water versus Tap Water*

MATERIALS: *3 small paper cups per student; enough tap water for two cups per student and enough bottled water for one cup per student; one six-sided die and one index card per student*

This Activity will give you and your classmates a chance to discover whether or not you can taste the difference between bottled water and tap water.

1. Before class begins, your teacher will prepare numbered stations with cups of water. You will be given an index card with a station number on it.
2. Go to the corresponding station. Pick up three cups (labeled A, B, and C) and take them back to your seat.
3. Your task is to determine which one of the three cups contains the bottled water. Drink all the water in Cup A first, then the water in Cup B, and finally the water in Cup C. Write down the letter of the cup that you think held the bottled water. Do not discuss your results with any of your classmates yet!
4. While you taste, your teacher will make a chart on the board like this one:

Station number	Bottled water cup?	Truth

5. When you are told to do so, go to the board and record your station number and the letter of the cup you identified as containing bottled water.
6. Your teacher will now reveal the truth about the cups of drinking water. How many students in the class identified the bottled water correctly? What percent of the class is this?
7. Let's assume that no one in your class can distinguish tap water from bottled water. In that case, students would just be guessing which cup of water tastes different. If so, what's the probability that an individual student would guess correctly?
8. How many correct identifications would you need to see to be convinced that the students in your class aren't just guessing? With your classmates, design and carry out a simulation to answer this question. What do you conclude about your class's ability to distinguish tap water from bottled water?

The ABC News program *20/20* set up a blind taste test in which people were asked to rate four different brands of bottled water and New York City tap water without knowing which they were drinking. Can you guess the result? Tap water came out the clear winner in terms of taste.

When Mr. Bullard's class did the preceding Activity, 13 out of 21 students made correct identifications. If we assume that the students in his class can't tell tap water from bottled water, then each one is basically guessing, with a 1/3 chance of being correct. So we'd expect about one-third of his 21 students, that is, about 7 students, to guess correctly. How likely is it that as many as 13 of his 21 students would guess correctly? To answer this question without a simulation, we need a different kind of probability model than the ones we saw in Chapter 5.

Section 6.1 introduces the concept of a *random variable*, a numerical outcome of some chance process (like the 13 students who guessed correctly in Mr. Bullard's class). Each random variable has a *probability distribution* that gives us information about the likelihood that a specific event happens (like 13 or more correct guesses out of 21) and about what's expected to happen if the chance behavior is repeated many times. Section 6.2 examines the effect of transforming and combining random variables on the shape, center, and spread of their probability distributions. In Section 6.3, we'll look at two random variables with probability distributions that are used enough to have their own names—*binomial* and *geometric*.

6.1 Discrete and Continuous Random Variables

In Section 6.1, you'll learn about:

- Discrete random variables
- Mean (expected value) of a discrete random variable
- Standard deviation (and variance) of a discrete random variable
- Continuous random variables

A probability model describes the possible outcomes of a chance process and the likelihood that those outcomes will occur. For example, suppose we toss a fair coin 3 times. The sample space for this chance process is

HHH HHT HTH THH HTT THT TTH TTT

Since there are 8 equally likely outcomes, the probability is 1/8 for each possible outcome. Define the variable X = the number of heads obtained. The value of X will vary from one set of tosses to another but will always be one of the numbers 0, 1, 2, or 3. How likely is X to take each of those values? It will be easier to answer this question if we group the possible outcomes by the number of heads obtained:

X = 0: TTT
X = 1: HTT THT TTH
X = 2: HHT HTH THH
X = 3: HHH

We can summarize the **probability distribution** of X as follows:

Value:	0	1	2	3
Probability:	1/8	3/8	3/8	1/8

Figure 6.1 on the next page shows the probability distribution of X in graphical form. Notice the symmetric shape.

We can use the probability distribution to answer questions about the variable X. What's the probability that we get at least one head in three tosses of the coin? In symbols, we want to find $P(X \geq 1)$. We could add probabilities to get the answer:

$$P(X \geq 1) = P(X = 1) + P(X = 2) + P(X = 3)$$
$$= 1/8 + 3/8 + 3/8 = 7/8$$

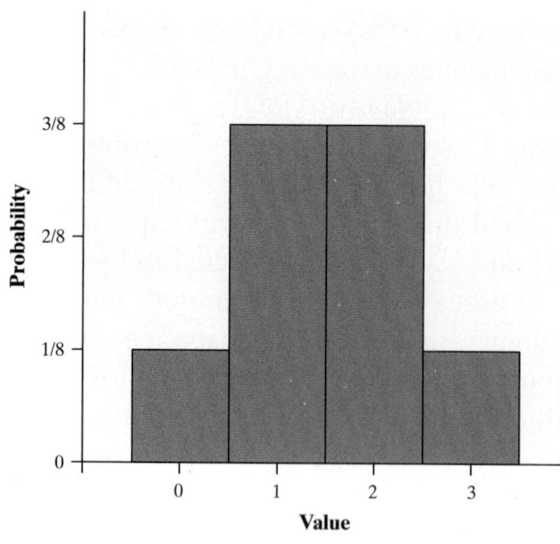

Or we could use the complement rule from Chapter 5:

$$P(X \geq 1) = 1 - P(X < 1) = 1 - P(X = 0)$$
$$= 1 - 1/8 = 7/8$$

A numerical variable that describes the outcomes of a chance process (like X in the coin-tossing scenario) is called a **random variable.** The probability model for a random variable is its probability distribution.

FIGURE 6.1 Histogram of the probability distribution for X = number of heads in three tosses of a fair coin.

DEFINITION: Random variable and probability distribution

A **random variable** takes numerical values that describe the outcomes of some chance process. The **probability distribution** of a random variable gives its possible values and their probabilities.

There are two main types of random variables, corresponding to two types of probability distributions: *discrete* and *continuous.*

Discrete Random Variables

We have learned several rules of probability but only one way of assigning probabilities to events: assign a probability to every individual outcome, then add these probabilities to find the probability of any event. This idea works well if we can find a way to list all possible outcomes. We will call random variables having probability assigned in this way **discrete random variables.**[1] The probability distribution for a discrete random variable must have outcome probabilities that are between 0 and 1 and that add up to 1.

Discrete Random Variables and Their Probability Distributions

A **discrete random variable** X takes a fixed set of possible values with gaps between. The probability distribution of a discrete random variable X lists the values x_i and their probabilities p_i:

Value:	x_1	x_2	x_3	...
Probability:	p_1	p_2	p_3	...

The probabilities p_i must satisfy two requirements:

1. Every probability p_i is a number between 0 and 1.
2. The sum of the probabilities is 1: $p_1 + p_2 + p_3 + \cdots = 1$.

To find the probability of any event, add the probabilities p_i of the particular values x_i that make up the event.

Here's an example of a discrete random variable that involves something a bit more serious than tossing coins.

Apgar Scores: Babies' Health at Birth

Discrete random variables

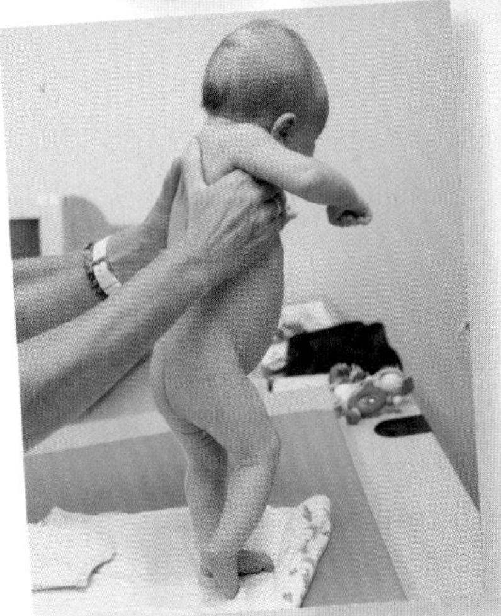

In 1952, Dr. Virginia Apgar suggested five criteria for measuring a baby's health at birth: skin color, heart rate, muscle tone, breathing, and response when stimulated. She developed a 0-1-2 scale to rate a newborn on each of the five criteria. A baby's Apgar score is the sum of the ratings on each of the five scales, which gives a whole-number value from 0 to 10. Apgar scores are still used today to evaluate the health of newborns.

What Apgar scores are typical? To find out, researchers recorded the Apgar scores of over 2 million newborn babies in a single year.[2] Imagine selecting one of these newborns at random. (That's our chance process.) Define the random variable X = Apgar score of a randomly selected baby one minute after birth. The table below gives the probability distribution for X.

Value:	0	1	2	3	4	5	6	7	8	9	10
Probability:	0.001	0.006	0.007	0.008	0.012	0.020	0.038	0.099	0.319	0.437	0.053

PROBLEM:

(a) Show that the probability distribution for X is legitimate.

(b) Make a histogram of the probability distribution. Describe what you see.

(c) Doctors decided that Apgar scores of 7 or higher indicate a healthy baby. What's the probability that a randomly selected baby is healthy?

SOLUTION:

(a) The probabilities are all between 0 and 1, and they add up to 1. So this is a legitimate probability distribution.

(b) Figure 6.2 shows a histogram of the probability distribution of X. The heavily left-skewed shape tells us that a randomly selected newborn will most likely have an Apgar score at the high end of the scale, which means that the baby was pretty healthy at birth. There's a much smaller chance of getting a baby whose Apgar score was 5 or below.

(c) The probability of choosing a healthy baby is $P(X \geq 7)$. We can calculate this probability as follows:

$$P(X \geq 7) = P(X = 7) + P(X = 8) + P(X = 9) + P(X = 10)$$
$$= 0.099 + 0.319 + 0.437 + 0.053 = 0.908$$

That is, we'd have about a 91% chance of randomly choosing a healthy baby.

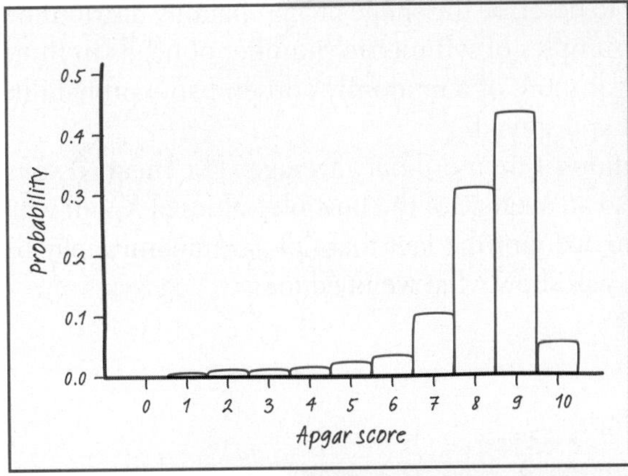

FIGURE 6.2 Histogram showing the probability distribution of the random variable X = Apgar score of a randomly selected newborn at one minute after birth.

For Practice *Try Exercise 5*

Although this procedure was later named for Dr. Apgar, the acronym APGAR also represents the five scales: Appearance, Pulse, Grimace, Activity, and Respiration.

Note that the probability of randomly selecting a newborn whose Apgar score is greater than or equal to 7 is not the same as the probability that the baby's Apgar score is strictly greater than 7. The latter probability is

$$P(X > 7) = P(X = 8) + P(X = 9) + P(X = 10)$$
$$= 0.319 + 0.437 + 0.053 = 0.809$$

The outcome $X = 7$ is included in "greater than or equal to" and is not included in "greater than."

CHECK YOUR UNDERSTANDING

North Carolina State University posts the grade distributions for its courses online.[3] Students in Statistics 101 in a recent semester received 26% As, 42% Bs, 20% Cs, 10% Ds, and 2% Fs. Choose a Statistics 101 student at random. The student's grade on a four-point scale (with A = 4) is a discrete random variable X with this probability distribution:

Value of X:	0	1	2	3	4
Probability:	0.02	0.10	0.20	0.42	0.26

1. Say in words what the meaning of $P(X \geq 3)$ is. What is this probability?
2. Write the event "the student got a grade worse than C" in terms of values of the random variable X. What is the probability of this event?
3. Sketch a graph of the probability distribution. Describe what you see.

Mean (Expected Value) of a Discrete Random Variable

When we analyzed distributions of quantitative data in Chapter 1, we made it a point to discuss their shape, center, and spread, and to identify any outliers. We'll follow the same strategy with probability distributions of random variables. You can use what you learned earlier to describe the shape of a probability distribution histogram. We've already seen examples of symmetric (number of heads in three coin tosses) and left-skewed (Apgar score of a randomly chosen baby) probability distributions. What about center and spread?

The mean $\bar{x}$ of a set of observations is their ordinary average. The **mean μ_X of a discrete random variable** X is also an average of the possible values of X, but with an important change to take into account the fact that not all outcomes may be equally likely. A simple example will show what we need to do.

EXAMPLE

Winning (and Losing) at Roulette

Finding the mean of a discrete random variable

On an American roulette wheel, there are 38 slots numbered 1 through 36, plus 0 and 00. Half of the slots from 1 to 36 are red; the other half are black. Both the 0 and 00 slots are green. Suppose that a player places a simple \$1 bet on red. If the ball lands in a red slot, the player gets the original dollar back, plus an additional dollar for winning the bet. If the ball lands in a different-colored slot, the player loses the dollar bet to the casino.

Let's define the random variable X = net gain from a single $1 bet on red. The possible values of X are −$1 and $1. (The player either gains a dollar or loses a dollar.) What are the corresponding probabilities? The chance that the ball lands in a red slot is 18/38. The chance that the ball lands in a different-colored slot is 20/38. Here is the probability distribution of X:

Value:	−$1	$1
Probability:	20/38	18/38

What is the player's average gain? The ordinary average of the two possible outcomes −$1 and $1 is $0. But $0 isn't the average winnings because the player is less likely to win $1 than to lose $1. In the long run, the player gains a dollar 18 times in every 38 games played and loses a dollar on the remaining 20 of 38 bets. The player's long-run average gain for this simple bet is

$$\mu_X = (-\$1)\left(\frac{20}{38}\right) + (\$1)\left(\frac{18}{38}\right) = -\$0.05$$

You see that in the long run the player loses (and the casino gains) five cents per bet.

If someone actually played several games of roulette, we would call the mean amount the person gained $\bar{x}$. The mean in the previous example is a different quantity—it is the long-run average gain we'd expect if someone played roulette a very large number of times. For this reason, the mean of a random variable is often referred to as its **expected value.** Just as probabilities are an idealized description of long-run proportions, the mean of a discrete random variable describes the long-run average outcome. There are two ways of denoting the mean of a random variable X. We can use the notation μ_X, or we can write $E(X)$, as in the "expected value of X." In the roulette example, $\mu_X = E(X) = -\$0.05$.

The mean of any discrete random variable is found just as in the roulette example. It is an average of the possible outcomes, but a weighted average in which each outcome is weighted by its probability. Here (finally!) is the definition.

DEFINITION: Mean (expected value) of a discrete random variable

Suppose that X is a discrete random variable whose probability distribution is

Value:	x_1	x_2	x_3	...
Probability:	p_1	p_2	p_3	...

To find the **mean (expected value)** of X, multiply each possible value by its probability, then add all the products:

$$\mu_X = E(X) = x_1p_1 + x_2p_2 + x_3p_3 + \cdots$$
$$= \sum x_ip_i$$

Let's put the definition to use in calculating the mean of a familiar random variable.

Apgar Scores: What's Typical?

Mean and expected value as an average

In our earlier example, we defined the random variable X to be the Apgar score of a randomly selected baby. The table below gives the probability distribution for X once again.

Value x_i:	0	1	2	3	4	5	6	7	8	9	10
Probability p_i:	0.001	0.006	0.007	0.008	0.012	0.020	0.038	0.099	0.319	0.437	0.053

PROBLEM: **Compute the mean of the random variable X and interpret this value in context.**

SOLUTION: **From the probability distribution for X, we see that 1 in every 1000 babies would have an Apgar score of 0; 6 in every 1000 babies would have an Apgar score of 1; and so on. So the mean (expected value) of X is**

$$\mu_X = E(X) = \sum x_i p_i$$

$$= (0)(0.001) + (1)(0.006) + (2)(0.007) + \cdots + (10)(0.053) = 8.128$$

The mean Apgar score of a randomly selected newborn is 8.128. This is the long-run average Apgar score of many, many randomly chosen babies.

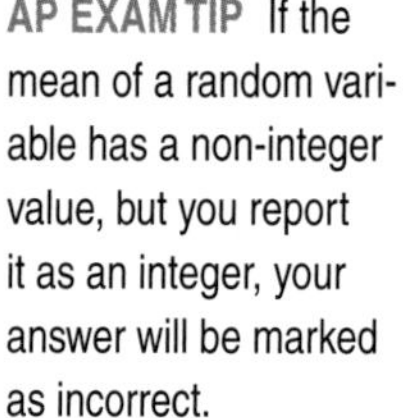
AP EXAM TIP If the mean of a random variable has a non-integer value, but you report it as an integer, your answer will be marked as incorrect.

Notice that the mean Apgar score, 8.128, is not a possible value of the random variable X. It's also not an integer. If you think of the mean as a long-run average over many repetitions, these facts shouldn't bother you.

Standard Deviation (and Variance) of a Discrete Random Variable

With the mean as our measure of center for a discrete random variable, it shouldn't surprise you that we'll use the standard deviation as our measure of spread. In Chapter 1, we first defined the sample variance s_x^2 as the "average squared deviation from the mean" and then took the square root of the variance to get the sample standard deviation s_x. The definition of the **variance of a random variable** σ_X^2 is similar to the definition of the variance for a set of quantitative data. That is, the variance is an average of the squared deviation $(x_i - \mu_X)^2$ of the values of the variable X from its mean μ_X. As with the mean, the average we use is a weighted average in which each outcome is weighted by its probability to take account of

Recall that the formula for the sample variance is

$$s_x^2 = \frac{\sum(x_i - \bar{x})^2}{n-1}$$

outcomes that are not equally likely. To get the **standard deviation of a random variable,** we take the square root of the variance. Here are the details.

DEFINITION: Variance and standard deviation of a discrete random variable

Suppose that X is a discrete random variable whose probability distribution is

Value:	x_1	x_2	x_3	...
Probability:	p_1	p_2	p_3	...

and that μ_X is the mean of X. The **variance** of X is

$$\text{Var}(X) = \sigma_X^2 = (x_1 - \mu_X)^2 p_1 + (x_2 - \mu_X)^2 p_2 + (x_3 - \mu_X)^2 p_3 + \cdots$$
$$= \sum (x_i - \mu_X)^2 p_i$$

The **standard deviation** of X, σ_X, is the square root of the variance.

The standard deviation of a random variable X is a measure of how much the values of the variable tend to vary, on average, from the mean μ_X. Let's compute the variance and standard deviation of a familiar discrete random variable.

Apgar Scores: How Variable Are They?

Calculating measures of spread

In the last example, we calculated the mean Apgar score of a randomly chosen newborn to be $\mu_X = 8.128$. The table below gives the probability distribution for X one more time.

Value x_i:	0	1	2	3	4	5	6	7	8	9	10
Probability p_i:	0.001	0.006	0.007	0.008	0.012	0.020	0.038	0.099	0.319	0.437	0.053

PROBLEM: **Compute and interpret the standard deviation of the random variable X.**

SOLUTION: **The formula for the variance of X is $\sigma_X^2 = \sum(x_i - \mu_X)^2 p_i$. Plugging in values gives**

$$\sigma_X^2 = (0 - 8.128)^2(0.001) + (1 - 8.128)^2(0.006)$$
$$+ (2 - 8.128)^2(0.007) + \cdots + (10 - 8.128)^2(0.053)$$

$$\sigma_X^2 = 2.066$$

The standard deviation of X is $\sigma_X = \sqrt{2.066} = 1.437$. On average, a randomly selected baby's Apgar score will differ from the mean (8.128) by about 1.4 units.

For Practice *Try Exercise 15*

You can use your calculator to graph the probability distribution of a discrete random variable and to calculate measures of center and spread, as the following Technology Corner illustrates.

TECHNOLOGY CORNER Analyzing random variables on the calculator

Let's explore what the calculator can do using the random variable X = Apgar score of a randomly selected newborn.

1. Start by entering the values of the random variable in L1 (list1) and the corresponding probabilities in L2 (list2).

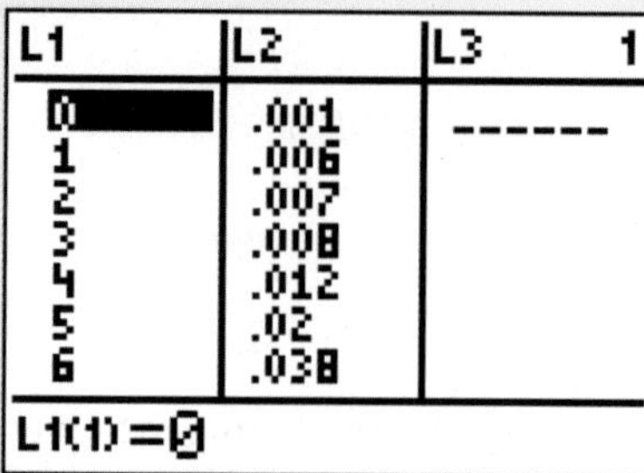

2. To graph a histogram of the probability distribution:
   - Set up a statistics plot with Xlist: L1 (list1) and Freq: L2 (list2).
   - Adjust your window settings as follows: Xmin = −1, Xmax = 11, Xscl = 1, Ymin = −0.1, Ymax = 0.5, Yscl = 0.1.
   - Press GRAPH (◆ F3 on the TI-89).

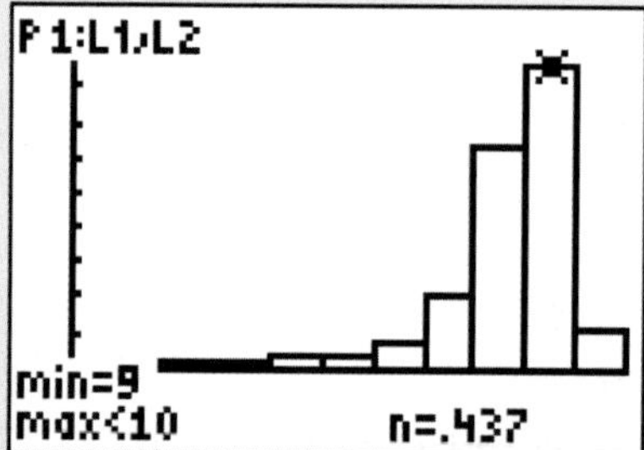

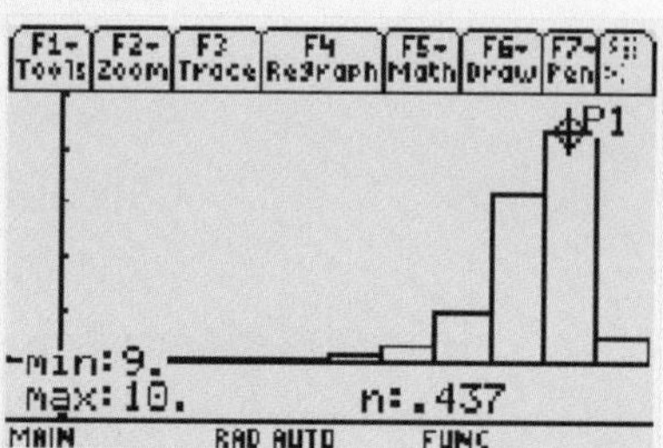

3. To calculate the mean and standard deviation of the random variable, use one-variable statistics with the values in L1 (list1) and the probabilities (relative frequencies) in L2 (list2).

TI-83/84: Execute the command `1-Var Stats L1,L2`.

TI-89: In the Statistics/List Editor, press F4 (Calc) and choose `1: 1-Var Stats...` Use the inputs List: list1 and Freq: list2.

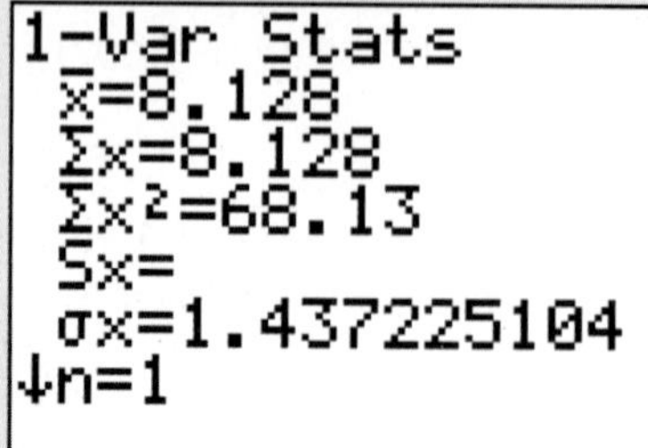

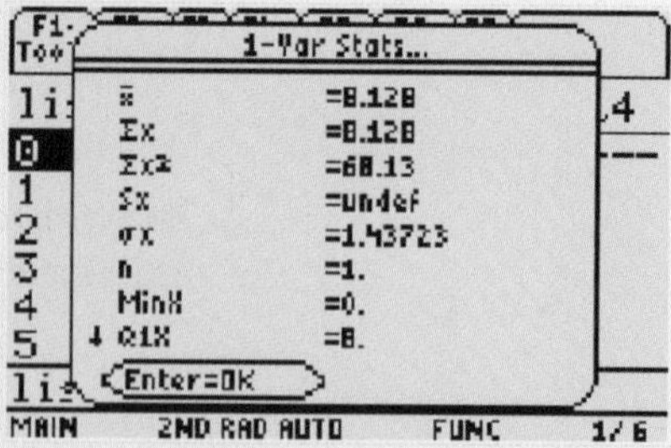

TI-Nspire instructions in Appendix B

CHECK YOUR UNDERSTANDING

A large auto dealership keeps track of sales made during each hour of the day. Let X = the number of cars sold during the first hour of business on a randomly selected Friday. Based on previous records, the probability distribution of X is as follows:

Cars sold:	0	1	2	3
Probability:	0.3	0.4	0.2	0.1

1. Compute and interpret the mean of X.
2. Compute and interpret the standard deviation of X.

Continuous Random Variables

When we use the table of random digits to select a digit between 0 and 9, the result is a discrete random variable (call it X). The probability model assigns probability 1/10 to each of the 10 possible values of X. Figure 6.3 shows the probability distribution for this random variable.

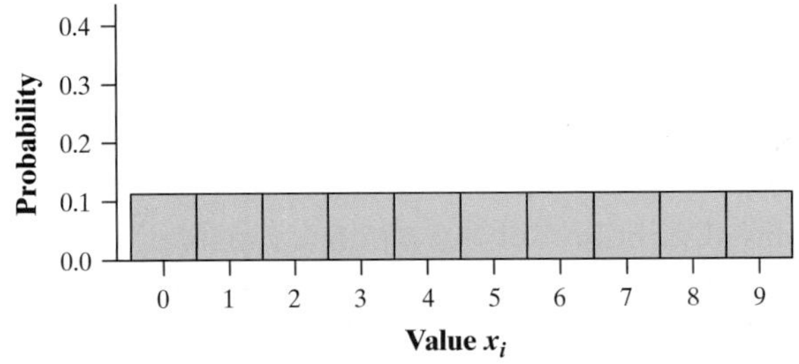

FIGURE 6.3 Probability distribution for the random variable X = random digit from 0 to 9.

Suppose we want to choose a number at random between 0 and 1, allowing *any* number between 0 and 1 as the outcome. Calculator and computer random number generators will do this. The sample space of this chance process is an entire interval of numbers:

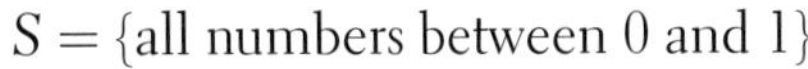

$$S = \{\text{all numbers between } 0 \text{ and } 1\}$$

Call the outcome of the random number generator Y for short. How can we find probabilities of events like $P(0.3 \leq Y \leq 0.7)$? As in the case of selecting a random digit, we would like all possible outcomes to be equally likely. But we cannot assign probabilities to each individual value of Y and then add them, because there are infinitely many possible values.

The calculator command `rand` will generate a random number from 0 to 1. Can you figure out how to modify the command to find a random number between, say, 1 and 3?

In situations like this, we use a different way of assigning probabilities directly to events—as *areas under a density curve*. Recall from Chapter 2 that any density curve has area exactly 1 underneath it, corresponding to total probability 1.

Random Numbers

Density curves and probability distributions

The random number generator will spread its output uniformly across the entire interval from 0 to 1 as we allow it to generate a long sequence of random numbers. The results of many trials are represented by the density curve of a *uniform distribution*. This density curve appears in purple in Figure 6.4 (on the next page). It has height 1 over the interval from 0 to 1. The area under the density curve is 1, and the probability of any event is the area under the density curve and above the event in question.

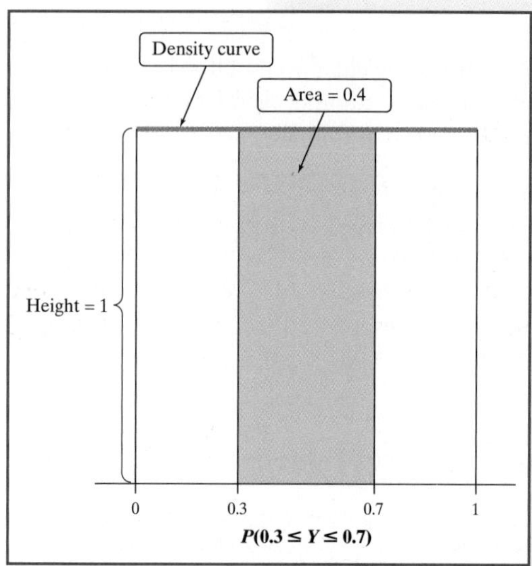

As Figure 6.4 shows, the probability that the random number generator produces a number Y between 0.3 and 0.7 is

$$P(0.3 \le Y \le 0.7) = 0.4$$

That's because the area of the shaded rectangle is

$$\text{length} \times \text{width} = 0.4 \times 1 = 0.4$$

FIGURE 6.4 Assigning probabilities for generating a random number between 0 and 1. The probability of any interval of numbers is the area above the interval and under the density curve.

Figure 6.4 shows the probability distribution of the random variable Y = random number between 0 and 1. We call Y a **continuous random variable** because its values are not isolated numbers but rather an entire interval of numbers.

In many cases, discrete random variables arise from counting something—for instance, the number of siblings that a randomly selected student has. Continuous random variables often arise from measuring something—for instance, height, time to run a mile, or blood pressure of a randomly selected student.

DEFINITION: Continuous random variable

A **continuous random variable** X takes all values in an interval of numbers. The probability distribution of X is described by a density curve. The probability of any event is the area under the density curve and above the values of X that make up the event.

CAUTION

The probability distribution for a continuous random variable assigns probabilities to intervals of outcomes rather than to individual outcomes. In fact, *all continuous probability models assign probability 0 to every individual outcome.* Only intervals of values have positive probability. To see that this is true, consider a specific outcome from the random number generator of the previous example, such as $P(Y = 0.7)$. The probability of this event is the area under the density curve that's above the point 0.7 on the horizontal axis. But this vertical line segment has no width, so the area is 0. For that reason,

$$P(0.3 \le Y \le 0.7) = P(0.3 \le Y < 0.7) = P(0.3 < Y < 0.7) = 0.4$$

We can use any density curve to assign probabilities. The density curves that are most familiar to us are the Normal curves. Normal distributions can be probability distributions as well as descriptions of data. There is a close connection between a Normal distribution as an idealized description for data and a Normal probability model. The following example shows what we mean.

Young Women's Heights

Normal probability distributions

The heights of young women closely follow the Normal distribution with mean $\mu = 64$ inches and standard deviation $\sigma = 2.7$ inches. This is a distribution for a large set of data. Now choose one young woman at random. Call her height Y. If we repeat the random choice very many times, the distribution of values of Y is the same Normal distribution that describes the heights of all young women. Find the probability that the chosen woman is between 68 and 70 inches tall.

STATE: What's the probability that a randomly chosen young woman has height between 68 and 70 inches?

PLAN: The height Y of the woman we choose has the $N(64, 2.7)$ distribution. We want to find $P(68 \le Y \le 70)$. This is the area under the Normal curve in Figure 6.5. We'll standardize the heights and then use Table A to find the shaded area.

DO: The standardized scores for the two heights are

$$z = \frac{68 - 64}{2.7} = 1.48 \quad \text{and} \quad z = \frac{70 - 64}{2.7} = 2.22$$

If we let Z represent the random variable that follows a standard Normal distribution, then the desired probability is $P(1.48 \le Z \le 2.22)$. From Table A, we find that $P(Z \le 2.22) = 0.9868$ and $P(Z \le 1.48) = 0.9306$. So we have

$$\begin{aligned} P(1.48 \le Z \le 2.22) &= P(Z \le 2.22) - P(Z \le 1.48) \\ &= 0.9868 - 0.9306 = 0.0562 \end{aligned}$$

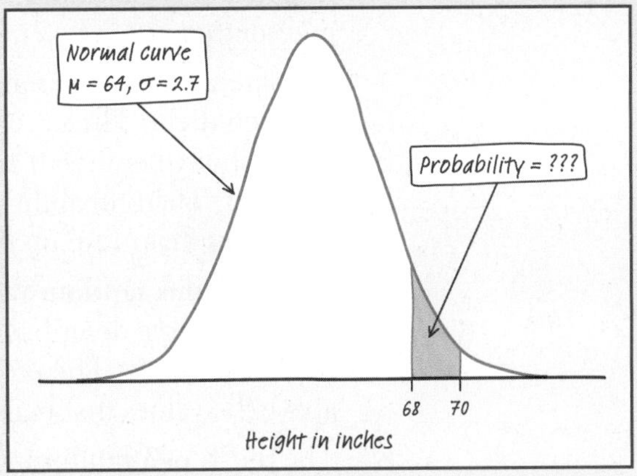

FIGURE 6.5 The probability that a randomly chosen young woman has height between 68 and 70 inches as an area under a Normal curve.

We can check that our answer is correct using the *Normal curve* applet or the normalcdf command on the TI-83/84 (normCdf on the TI-89).

CONCLUDE: There's about a 5.6% chance that a randomly chosen young woman has a height between 68 and 70 inches.

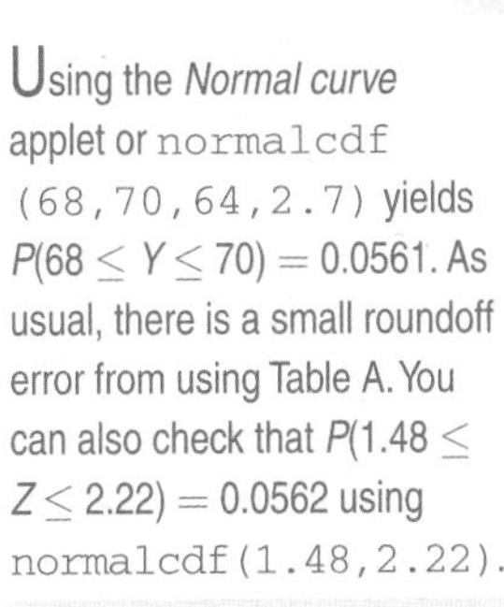

Using the *Normal curve* applet or `normalcdf (68,70,64,2.7)` yields $P(68 \le Y \le 70) = 0.0561$. As usual, there is a small roundoff error from using Table A. You can also check that $P(1.48 \le Z \le 2.22) = 0.0562$ using `normalcdf(1.48,2.22)`.

AP EXAM TIP When you solve problems involving random variables, start by defining the random variable of interest. For example, let $X =$ the Apgar score of a randomly selected baby or let $Y =$ the height of a randomly selected young woman. Then state the probability you're trying to find in terms of the random variable: $P(68 \leq Y \leq 70)$ or $P(X \geq 7)$.

The calculation in the preceding example is the same as those we did in Chapter 2. Only the language of probability is new.

What about the mean and standard deviation for continuous random variables? The probability distribution of a continuous random variable X is described by a density curve. Chapter 2 showed how to find the mean of the distribution: it is the point at which the area under the density curve would balance if it were made out of solid material. The mean lies at the center of symmetric density curves such as the Normal curves. We can locate the standard deviation of a Normal distribution from its inflection points. Exact calculation of the mean and standard deviation for most continuous random variables requires advanced mathematics.[4]

SECTION 6.1 Summary

- A **random variable** is a variable taking numerical values determined by the outcome of a chance process. The **probability distribution** of a random variable X tells us what the possible values of X are and how probabilities are assigned to those values. There are two types of random variables: *discrete* and *continuous.*
- A **discrete random variable** has a fixed set of possible values with gaps between them. The probability distribution assigns each of these values a probability between 0 and 1 such that the sum of all the probabilities is exactly 1. The probability of any event is the sum of the probabilities of all the values that make up the event.
- A **continuous random variable** takes all values in some interval of numbers. A density curve describes the probability distribution of a continuous random variable. The probability of any event is the area under the curve above the values that make up the event.
- The **mean of a random variable** μ_X is the balance point of the probability distribution histogram or density curve. Since the mean is the long-run average value of the variable after many repetitions of the chance process, it is also known as the **expected value** of the random variable.
- If X is a discrete random variable, the mean is the average of the values of X, each weighted by its probability:
$$\mu_X = \sum x_i p_i = x_1p_1 + x_2p_2 + x_3p_3 + \cdots$$
- The **variance of a random variable** σ_X^2 is the average squared deviation of the values of the variable from their mean. The **standard deviation** σ_X is the square root of the variance. The standard deviation measures the variability of the distribution about the mean.

- For a discrete random variable X,
$$\sigma_X^2 = \sum (x_i - \mu_X)^2 p_i = (x_1 - \mu_X)^2 p_1 + (x_2 - \mu_X)^2 p_2 + (x_3 - \mu_X)^2 p_3 + \cdots$$

6.1 TECHNOLOGY CORNER

TI-Nspire instructions in Appendix B

SECTION 6.1 Exercises

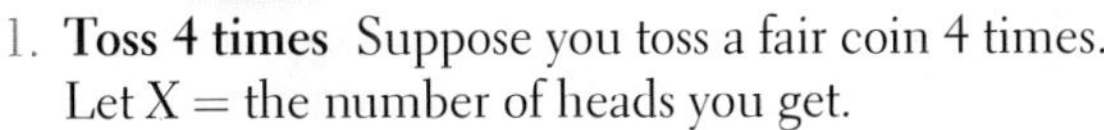

1. **Toss 4 times** Suppose you toss a fair coin 4 times. Let X = the number of heads you get.
   (a) Find the probability distribution of X.
   (b) Make a histogram of the probability distribution. Describe what you see.
   (c) Find $P(X \leq 3)$ and interpret the result.

2. **Pair-a-dice** Suppose you roll a pair of fair, six-sided dice. Let T = the sum of the spots showing on the up-faces.
   (a) Find the probability distribution of T.
   (b) Make a histogram of the probability distribution. Describe what you see.
   (c) Find $P(T \geq 5)$ and interpret the result.

3. **Spell-checking** Spell-checking software catches "nonword errors," which result in a string of letters that is not a word, as when "the" is typed as "teh." When undergraduates are asked to write a 250-word essay (without spell-checking), the number X of nonword errors has the following distribution:

Value of X:	0	1	2	3	4
Probability:	0.1	0.2	0.3	0.3	0.1

   (a) Write the event "at least one nonword error" in terms of X. What is the probability of this event?
   (b) Describe the event $X \leq 2$ in words. What is its probability? What is the probability that $X < 2$?

4. **Kids and toys** In an experiment on the behavior of young children, each subject is placed in an area with five toys. Past experiments have shown that the probability distribution of the number X of toys played with by a randomly selected subject is as follows:

Number of toys x_i:	0	1	2	3	4	5
Probability p_i:	0.03	0.16	0.30	0.23	0.17	0.11

   (a) Write the event "plays with at most two toys" in terms of X. What is the probability of this event?
   (b) Describe the event $X > 3$ in words. What is its probability? What is the probability that $X \geq 3$?

5. **Benford's law** Faked numbers in tax returns, invoices, or expense account claims often display patterns that aren't present in legitimate records. Some patterns, like too many round numbers, are obvious and easily avoided by a clever crook. Others are more subtle. It is a striking fact that the first digits of numbers in legitimate records often follow a model known as Benford's law.[5] Call the first digit of a randomly chosen record X for short. Benford's law gives this probability model for X (note that a first digit can't be 0):

pg 343

First digit X:	1	2	3	4	5	6	7	8	9
Probability:	0.301	0.176	0.125	0.097	0.079	0.067	0.058	0.051	0.046

   (a) Show that this is a legitimate probability distribution.
   (b) Make a histogram of the probability distribution. Describe what you see.
   (c) Describe the event $X \geq 6$ in words. What is $P(X \geq 6)$?
   (d) Express the event "first digit is at most 5" in terms of X. What is the probability of this event?

6. **Working out** Choose a person aged 19 to 25 years at random and ask, "In the past seven days, how many times did you go to an exercise or fitness center or work out?" Call the response Y for short. Based on a large sample survey, here is a probability model for the answer you will get:[6]

Days:	0	1	2	3	4	5	6	7
Probability:	0.68	0.05	0.07	0.08	0.05	0.04	0.01	0.02

(a) Show that this is a legitimate probability distribution.

(b) Make a histogram of the probability distribution. Describe what you see.

(c) Describe the event $Y < 7$ in words. What is $P(Y < 7)$?

(d) Express the event "worked out at least once" in terms of Y. What is the probability of this event?

7. **Benford's law** Refer to Exercise 5. The first digit of a randomly chosen expense account claim follows Benford's law. Consider the events A = first digit is 7 or greater and B = first digit is odd.

(a) What outcomes make up the event A? What is $P(A)$?

(b) What outcomes make up the event B? What is $P(B)$?

(c) What outcomes make up the event "A or B"? What is $P(A \text{ or } B)$? Why is this probability not equal to $P(A) + P(B)$?

8. **Working out** Refer to Exercise 6. Consider the events A = works out at least once and B = works out less than 5 times per week.

(a) What outcomes make up the event A? What is $P(A)$?

(b) What outcomes make up the event B? What is $P(B)$?

(c) What outcomes make up the event "A and B"? What is $P(A \text{ and } B)$? Why is this probability not equal to $P(A) \cdot P(B)$?

pg 346

9. **Keno** Keno is a favorite game in casinos, and similar games are popular with the states that operate lotteries. Balls numbered 1 to 80 are tumbled in a machine as the bets are placed, then 20 of the balls are chosen at random. Players select numbers by marking a card. The simplest of the many wagers available is "Mark 1 Number." Your payoff is \$3 on a \$1 bet if the number you select is one of those chosen. Because 20 of 80 numbers are chosen, your probability of winning is 20/80, or 0.25. Let X = the amount you gain on a single play of the game.

(a) Make a table that shows the probability distribution of X.

(b) Compute the expected value of X. Explain what this result means for the player.

10. **Fire insurance** Suppose a homeowner spends \$300 for a home insurance policy that will pay out \$200,000 if the home is destroyed by fire. Let Y = the profit made by the company on a single policy. From previous data, the probability that a home in this area will be destroyed by fire is 0.0002.

(a) Make a table that shows the probability distribution of Y.

(b) Compute the expected value of Y. Explain what this result means for the insurance company.

11. **Spell-checking** Refer to Exercise 3. Calculate the mean of the random variable X and interpret this result in context.

12. **Kids and toys** Refer to Exercise 4. Calculate the mean of the random variable X and interpret this result in context.

13. **Benford's law and fraud** A not-so-clever employee decided to fake his monthly expense report. He believed that the first digits of his expense amounts should be equally likely to be any of the numbers from 1 to 9. In that case, the first digit Y of a randomly selected expense amount would have the probability distribution shown in the histogram.

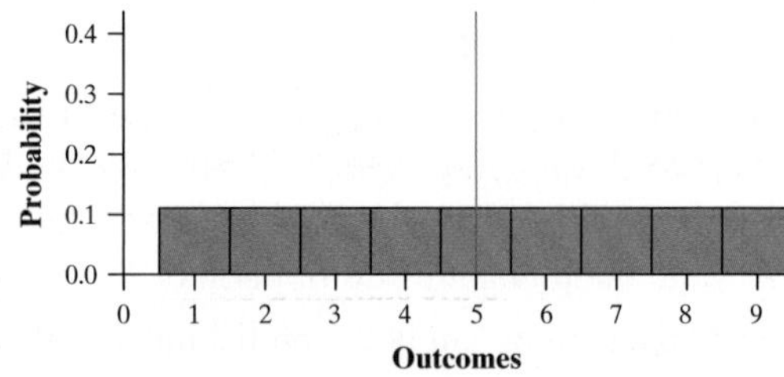

(a) Explain why the mean of the random variable Y is located at the solid red line in the figure.

(b) The first digits of randomly selected expense amounts actually follow Benford's law (Exercise 5). What's the expected value of the first digit? Explain how this information could be used to detect a fake expense report.

(c) What's $P(Y > 6)$? According to Benford's law, what proportion of first digits in the employee's expense amounts should be greater than 6? How could this information be used to detect a fake expense report?

14. **Life insurance** A life insurance company sells a term insurance policy to a 21-year-old male that pays \$100,000 if the insured dies within the next 5 years. The probability that a randomly chosen male will die each year can be found in mortality tables. The company collects a premium of \$250 each year as

payment for the insurance. The amount Y that the company earns on this policy is \$250 per year, less the \$100,000 that it must pay if the insured dies. Here is a partially completed table that shows information about risk of mortality and the values of Y = profit earned by the company:

Age at death:	21	22	23	24	25	26 or more
Profit:	−\$99,750	−\$99,500				
Probability:	0.00183	0.00186	0.00189	0.00191	0.00193	

(a) Copy the table onto your paper. Fill in the missing values of Y.

(b) Find the missing probability. Show your work.

(c) Calculate the mean μ_Y. Interpret this value in context.

15. **Spell-checking** Refer to Exercise 3. Calculate and interpret the standard deviation of the random variable X. Show your work.

pg 347

16. **Kids and toys** Refer to Exercise 4. Calculate and interpret the standard deviation of the random variable X. Show your work.

17. **Benford's law and fraud** Refer to Exercise 13. It might also be possible to detect an employee's fake expense records by looking at the variability in the first digits of those expense amounts.

(a) Calculate the standard deviation σ_Y. This gives us an idea of how much variation we'd expect in the employee's expense records if he assumed that first digits from 1 to 9 were equally likely.

(b) Now calculate the standard deviation of first digits that follow Benford's law (Exercise 5). Would using standard deviations be a good way to detect fraud? Explain.

18. **Life insurance**

(a) It would be quite risky for you to insure the life of a 21-year-old friend under the terms of Exercise 14. There is a high probability that your friend would live and you would gain \$1250 in premiums. But if he were to die, you would lose almost \$100,000. Explain carefully why selling insurance is not risky for an insurance company that insures many thousands of 21-year-old men.

(b) The risk of an investment is often measured by the standard deviation of the return on the investment. The more variable the return is, the riskier the investment. We can measure the great risk of insuring a single person's life in Exercise 14 by computing the standard deviation of the income Y that the insurer will receive. Find σ_Y using the distribution and mean found in Exercise 14.

19. **Housing in San Jose** How do rented housing units differ from units occupied by their owners? Here are the distributions of the number of rooms for owner-occupied units and renter-occupied units in San Jose, California:[7]

	Number of Rooms									
	1	2	3	4	5	6	7	8	9	10
Owned	0.003	0.002	0.023	0.104	0.210	0.224	0.197	0.149	0.053	0.035
Rented	0.008	0.027	0.287	0.363	0.164	0.093	0.039	0.013	0.003	0.003

Let X = the number of rooms in a randomly selected owner-occupied unit and Y = the number of rooms in a randomly chosen renter-occupied unit.

(a) Make histograms suitable for comparing the probability distributions of X and Y. Describe any differences that you observe.

(b) Find the mean number of rooms for both types of housing unit. Explain why this difference makes sense.

(c) Find the standard deviations of both X and Y. Explain why this difference makes sense.

20. **Size of American households** In government data, a household consists of all occupants of a dwelling unit, while a family consists of two or more persons who live together and are related by blood or marriage. So all families form households, but some households are not families. Here are the distributions of household size and family size in the United States:

	Number of Persons						
	1	2	3	4	5	6	7
Household probability	0.25	0.32	0.17	0.15	0.07	0.03	0.01
Family probability	0	0.42	0.23	0.21	0.09	0.03	0.02

Let X = the number of people in a randomly selected U.S. household and Y = the number of people in a randomly chosen U.S. family.

(a) Make histograms suitable for comparing the probability distributions of X and Y. Describe any differences that you observe.

(b) Find the mean for each random variable. Explain why this difference makes sense.

(c) Find the standard deviations of both X and Y. Explain why this difference makes sense.

21. **Random numbers** Let X be a number between 0 and 1 produced by a random number generator. Assuming that the random variable X has a uniform distribution, find the following probabilities:

(a) $P(X > 0.49)$

(b) $P(X \geq 0.49)$

(c) $P(0.19 \leq X < 0.37 \text{ or } 0.84 < X \leq 1.27)$

22. **Random numbers** Let Y be a number between 0 and 1 produced by a random number generator. Assuming that the random variable Y has a uniform distribution, find the following probabilities:
(a) $P(Y \leq 0.4)$
(b) $P(Y < 0.4)$
(c) $P(0.1 < Y \leq 0.15 \text{ or } 0.77 \leq Y < 0.88)$

pg 351

23. **ITBS scores** The Normal distribution with mean $\mu = 6.8$ and standard deviation $\sigma = 1.6$ is a good description of the Iowa Test of Basic Skills (ITBS) vocabulary scores of seventh-grade students in Gary, Indiana. Call the score of a randomly chosen student X for short. Find $P(X \geq 9)$ and interpret the result. Follow the four-step process.

24. **Running a mile** A study of 12,000 able-bodied male students at the University of Illinois found that their times for the mile run were approximately Normal with mean 7.11 minutes and standard deviation 0.74 minute.[8] Choose a student at random from this group and call his time for the mile Y. Find $P(Y < 6)$ and interpret the result. Follow the four-step process.

25. **Did you vote?** A sample survey contacted an SRS of 663 registered voters in Oregon shortly after an election and asked respondents whether they had voted. Voter records show that 56% of registered voters had actually voted. We will see later that in repeated random samples of size 663, the proportion in the sample who voted (call this proportion V) will vary according to the Normal distribution with mean $\mu = 0.56$ and standard deviation $\sigma = 0.019$.
(a) If the respondents answer truthfully, what is $P(0.52 \leq V \leq 0.60)$? This is the probability that the sample proportion V estimates the population proportion 0.56 within ± 0.04.
(b) In fact, 72% of the respondents said they had voted ($V = 0.72$). If respondents answer truthfully, what is $P(V \geq 0.72)$? This probability is so small that it is good evidence that some people who did not vote claimed that they did vote.

26. **Friends** How many close friends do you have? An opinion poll asks this question of an SRS of 1100 adults. Suppose that the number of close friends adults claim to have varies from person to person with mean $\mu = 9$ and standard deviation $\sigma = 2.5$. We will see later that in repeated random samples of size 1100, the mean response $\bar{x}$ will vary according to the Normal distribution with mean 9 and standard deviation 0.075. What is $P(8.9 \leq \bar{x} \leq 9.1)$, the probability that the sample result $\bar{x}$ estimates the population truth $\mu = 9$ to within ± 0.1?

Multiple choice: Select the best answer for Exercises 27 to 30.

Exercises 27 and 28 refer to the following setting. Choose an American household at random and let the random variable X be the number of cars (including SUVs and light trucks) they own. Here is the probability model if we ignore the few households that own more than 5 cars:

Number of cars X:	0	1	2	3	4	5
Probability:	0.09	0.36	0.35	0.13	0.05	0.02

27. A housing company builds houses with two-car garages. What percent of households have more cars than the garage can hold?
(a) 13% (b) 20% (c) 45% (d) 55% (e) 80%

28. What's the expected number of cars in a randomly selected American household?
(a) Between 0 and 5
(b) 1.00
(c) 1.75
(d) 1.84
(e) 2.00

29. A deck of cards contains 52 cards, of which 4 are aces. You are offered the following wager: Draw one card at random from the deck. You win \$10 if the card drawn is an ace. Otherwise, you lose \$1. If you make this wager very many times, what will be the mean amount you win?
(a) About −\$1, because you will lose most of the time.
(b) About \$9, because you win \$10 but lose only \$1.
(c) About −\$0.15; that is, on average you lose about 15 cents.
(d) About \$0.77; that is, on average you win about 77 cents.
(e) About \$0, because the random draw gives you a fair bet.

30. The deck of 52 cards contains 13 hearts. Here is another wager: Draw one card at random from the deck. If the card drawn is a heart, you win \$2. Otherwise, you lose \$1. Compare this wager (call it Wager 2) with that of the previous exercise (call it Wager 1). Which one should you prefer?
(a) Wager 1, because it has a higher expected value.
(b) Wager 2, because it has a higher expected value.
(c) Wager 1, because it has a higher probability of winning.
(d) Wager 2, because it has a higher probability of winning.
(e) Both wagers are equally favorable.

Exercises 31 to 34 refer to the following setting. Many chess masters and chess advocates believe that chess play develops general intelligence, analytical skill, and the ability to concentrate. According to such beliefs, improved reading skills should result from study to improve chess-playing skills. To investigate this belief, researchers conducted a study. All of the subjects in the study participated in a comprehensive chess program, and their reading performances were measured before and after the program. The graphs and numerical summaries below provide information on the subjects' pretest scores, posttest scores, and the difference (post – pre) between these two scores.

Descriptive Statistics: Pretest, Posttest, Post – pre

Variable	N	Mean	Median	StDev	Min	Max	Q_1	Q_3
Pretest	53	57.70	58.00	17.84	23.00	99.00	44.50	70.50
Posttest	53	63.08	64.00	18.70	28.00	99.00	48.00	76.00
Post-pre	53	5.38	3.00	13.02	−19.00	42.00	−3.50	14.00

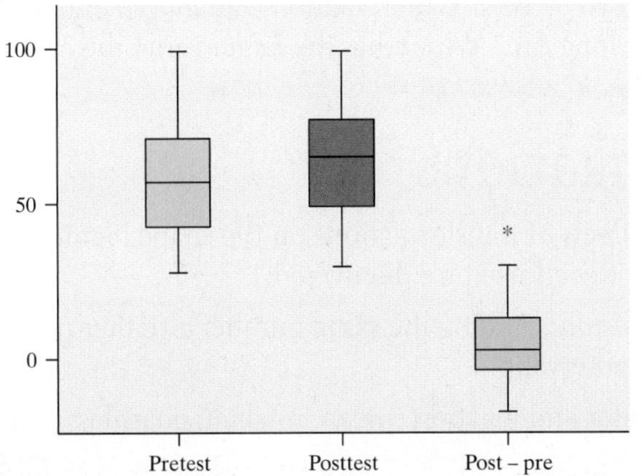

Some graphical and numerical information about the relationship between pretest and posttest scores is provided below.

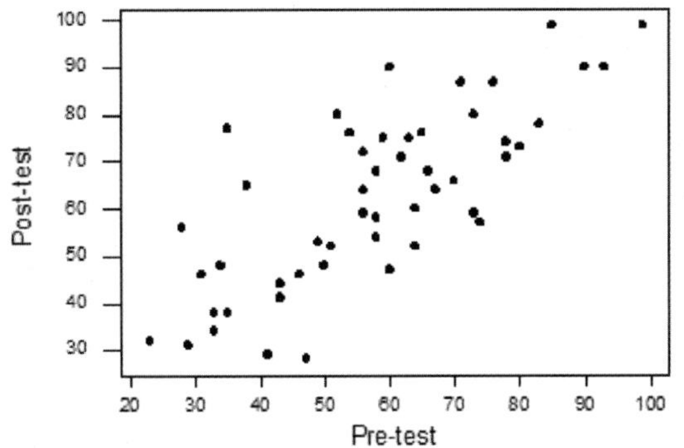

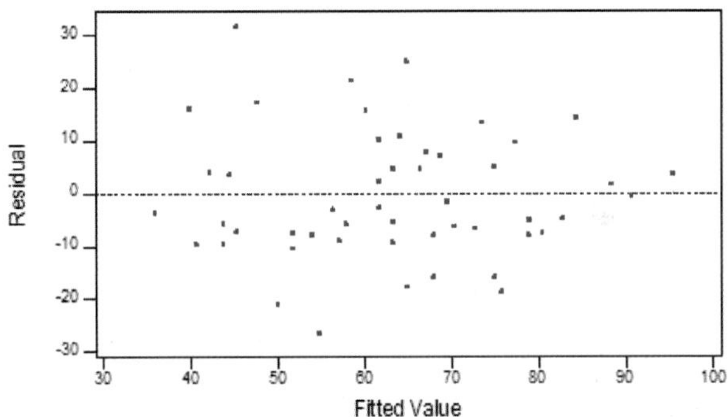

Regression Analysis: Posttest versus Pretest

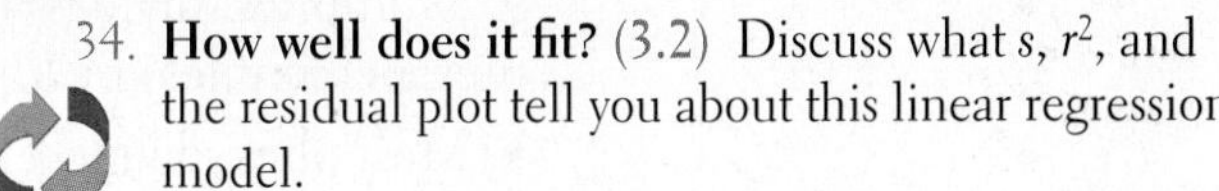

Predictor	Coef	SE Coef	T	P
Constant	17.897	5.889	3.04	0.004
Pretest	0.78301	0.09758	8.02	0.000

S = 12.55 R-Sq = 55.8% R-Sq(adj) = 54.9%

31. **Better readers?** (1.3) Did students have higher reading scores after participating in the chess program? Give appropriate statistical evidence to support your answer.

32. **Chess and reading** (4.3) If the study found a statistically significant improvement in reading scores, could you conclude that playing chess causes an increase in reading skills? Justify your answer.

33. **Predicting posttest scores** (3.2) What is the equation of the linear regression model relating posttest and pretest scores? Define any variables used.

34. **How well does it fit?** (3.2) Discuss what s, r^2, and the residual plot tell you about this linear regression model.

6.2

Transforming and Combining Random Variables

In Section 6.2, you'll learn about:

- Linear transformations
- Combining random variables
- Combining Normal random variables

In Section 6.1, we looked at several examples of random variables and their probability distributions. We also saw that the mean μ_X and standard deviation σ_X give us important information about a random variable. For instance, for X = the amount gained on a single \$1 bet on red in a game of roulette, we already showed that $\mu_X = -\$0.05$. You can verify that the standard deviation is $\sigma_X = \$1.00$. That is, a player can expect to lose an average of 5 cents per \$1 bet if he plays many games. But if he plays only a few games, his actual gain could be much better or worse than this expected value.

Would the player be better off playing one game of roulette with a \$2 bet on red or playing two games and betting \$1 on red each time? To find out, we need to compare the probability distributions of the random variables Y = gain from a \$2 bet and T = total gain from two \$1 bets. Which random variable (if either) has the higher expected gain in the long run? Which has the larger variability? By the end of this section, you'll be able to answer questions like these.

Linear Transformations

In Chapter 2, we studied the effects of transformations on the shape, center, and spread of a distribution of data. Recall what we discovered:

1. *Adding (or subtracting) a constant*: Adding the same number a (either positive, zero, or negative) to each observation:
   - Adds a to measures of center and location (mean, median, quartiles, percentiles).
   - Does not change shape or measures of spread (range, *IQR*, standard deviation).

2. Multiplying (or dividing) each observation by a constant b (positive, negative, or zero):
   - Multiplies (divides) measures of center and location (mean, median, quartiles, percentiles) by b.
   - Multiplies (divides) measures of spread (range, *IQR*, standard deviation) by $|b|$.
   - Does not change the shape of the distribution.

How are the probability distributions of random variables affected by similar transformations to the values of the variable? For reasons that will be clear later, we'll start by considering multiplication (or division) by a constant.

Effect of multiplying or dividing by a constant Let's start with a simple example of a discrete random variable. Pete's Jeep Tours offers a popular half-day trip in a tourist area. There must be at least 2 passengers for the trip to run, and the vehicle will hold up to 6 passengers. The number of passengers X on a randomly selected day has the following probability distribution.

No. of passengers x_i:	2	3	4	5	6
Probability p_i:	0.15	0.25	0.35	0.20	0.05

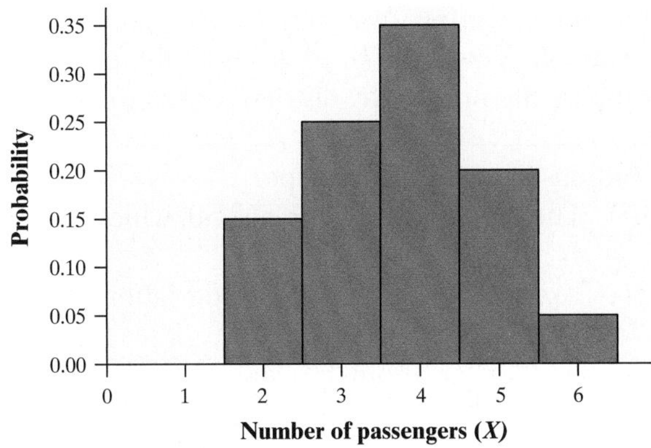

FIGURE 6.6 The probability distribution of the random variable $X =$ the number of passengers on Pete's trip on a randomly chosen day.

Figure 6.6 shows a histogram of the probability distribution.

Using what we learned in Section 6.1, the mean of X is

$$\mu_X = \sum x_i p_i = (2)(0.15) + (3)(0.25) + (4)(0.35) + (5)(0.20) + (6)(0.05) = 3.75$$

That is, Pete expects an average of 3.75 passengers per trip in the long run. The variance of X is given by

$$\sigma_X^2 = \sum (x_i - \mu_X)^2 p_i = (2-3.75)^2(0.15) + (3-3.75)^2(0.25) + \cdots + (6-3.75)^2(0.05) = 1.1875$$

So the standard deviation of X is

$$\sigma_X = \sqrt{1.1875} = 1.090$$

Pete's Jeep Tours

Multiplying a random variable by a constant

Pete charges \$150 per passenger. Let C = the total amount of money that Pete collects on a randomly selected trip. Because the amount of money Pete collects from the trip is just \$150 times the number of passengers, we can write $C = 150X$. From the probability distribution of X, we can see that the chance of having two people ($X = 2$) on the trip is 0.15. In that case, $C = (150)(2) = 300$. So one possible value of C is \$300, and its corresponding probability is 0.15. If $X = 3$, then $C = (150)(3) = 450$, and the corresponding probability is 0.25. Thus, the probability distribution of C is

Total collected c_i:	300	450	600	750	900
Probability p_i:	0.15	0.25	0.35	0.20	0.05

FIGURE 6.7 The probability distribution of the random variable $C =$ the amount of money Pete collects from his trip on a randomly chosen day.

Figure 6.7 is a histogram of this probability distribution.

The mean of C is $\mu_C = \sum c_i p_i = (300)(0.15) + (450)(0.25) + \cdots + (900)(0.05) = 562.50$.

On average, Pete will collect a total of \$562.50 from the half-day trip. The variance of C is

$$\sigma_C^2 = \sum (c_i - \mu_C)^2 p_i = (300 - 562.50)^2(0.15) + (450 - 562.50)^2(0.25) + \cdots + (900 - 562.50)^2(0.05) = 26{,}718.75$$

So the standard deviation of C is $\sigma_C = \sqrt{26{,}718.75} = \$163.50$.

In the previous example, the random variable C was obtained by multiplying the values of our earlier random variable X by 150. To understand the effect of multiplying by a constant, let's compare the probability distributions of these two random variables.

Shape: The two probability distributions have the same shape.

Center: The mean of X is $\mu_X = 3.75$. The mean of C is $\mu_C = 562.50$, which is (150)(3.75). That is, $\mu_C = 150\mu_X$.

Spread: The standard deviation of X is $\sigma_X = 1.090$. The standard deviation of C is $\sigma_C = 163.5$, which is (150)(1.090). That is, $\sigma_C = 150\sigma_X$.

Let's summarize what we've learned so far about transforming a random variable.

Effect on a Random Variable of Multiplying (or Dividing) by a Constant

Multiplying (or dividing) each value of a random variable by a number *b*:

- Multiplies (divides) measures of center and location (mean, median, quartiles, percentiles) by *b*.
- Multiplies (divides) measures of spread (range, *IQR*, standard deviation) by $|b|$.
- Does not change the shape of the distribution.

As with data, if we multiply a random variable by a negative constant *b*, our common measures of spread are multiplied by $|b|$.

THINK ABOUT IT

How does multiplying by a constant affect the variance? For Pete's Jeep Tours, the variance of the number of passengers on a randomly selected trip is $\sigma_X^2 = 1.1875$. The variance of the total amount of money that Pete collects from such a trip is $\sigma_C^2 = 26{,}718.75$. That's (22,500)(1.1875). So $\sigma_C^2 = 22{,}500\sigma_X^2$. Where did 22,500 come from? It's just $(150)^2$. In other words, $\sigma_C^2 = (150)^2\sigma_X^2$. Multiplying a random variable by a constant *b* multiplies the variance by b^2.

Effect of adding or subtracting a constant What happens to the probability distribution of a random variable if we add or subtract a constant? Let's return to Pete's Jeep Tours to find out.

Pete's Jeep Tours

Effect of adding or subtracting a constant

It costs Pete \$100 to buy permits, gas, and a ferry pass for each half-day trip. The amount of profit V that Pete makes from the trip is the total amount of money C that he collects from passengers minus \$100. That is, $V = C - 100$. If Pete has only two passengers on the trip ($X = 2$), then $C = 300$ and $V = 200$. From the probability distribution of C, the chance that this happens is 0.15. So the smallest possible value of V is \$200; its corresponding probability is 0.15. If $X = 3$, then

C = 450 and V = 350, and the corresponding probability is 0.25. The probability distribution of V is

Total collected v_i:	200	350	500	650	800
Probability p_i:	0.15	0.25	0.35	0.20	0.05

Figure 6.8 shows a histogram of this probability distribution.

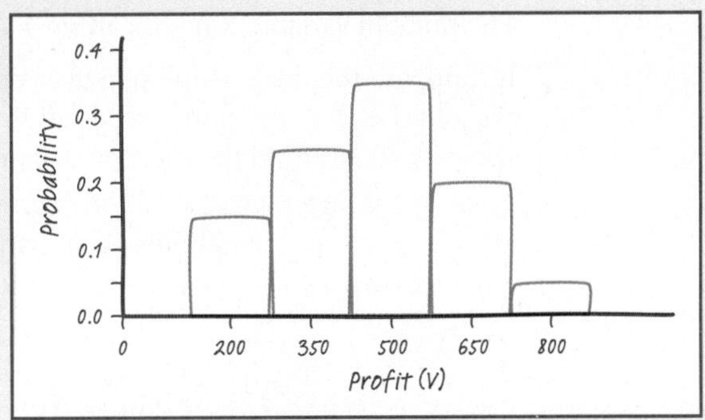

FIGURE 6.8 The probability distribution of the random variable V = the profit that Pete makes from his trip on a randomly chosen day.

The mean of V is $\mu_V = \sum v_i p_i = (200)(0.15) + (350)(0.25) + \cdots + (800)(0.05) = 462.50$. On average, Pete will make a profit of \$462.50 from the trip. The variance of V is

$$\begin{aligned}\sigma_V^2 &= \sum (v_i - \mu_V)^2 p_i \\ &= (200 - 462.50)^2(0.15) + (350 - 462.50)^2(0.25) \\ &\quad + \cdots + (800 - 462.50)^2(0.05) = 26{,}718.75\end{aligned}$$

So the standard deviation of V is

$$\sigma_V = \sqrt{26{,}718.75} = \$163.50.$$

It's pretty clear from the previous example that subtracting 100 from the values of the random variable C just shifts the probability distribution to the left by 100. This transformation decreases the mean by 100 (from \$562.50 to \$462.50) but doesn't change the standard deviation (\$163.50) or the shape. These results can be generalized for any random variable.

Effect on a Random Variable of Adding (or Subtracting) a Constant

Adding the same number *a* (which could be negative) to each value of a random variable:

- Adds *a* to measures of center and location (mean, median, quartiles, percentiles).
- Does not change shape or measures of spread (range, *IQR*, standard deviation).

CHECK YOUR UNDERSTANDING

A large auto dealership keeps track of sales made during each hour of the day. Let X = the number of cars sold during the first hour of business on a randomly selected Friday. Based on previous records, the probability distribution of X is as follows:

Cars sold:	0	1	2	3
Probability:	0.3	0.4	0.2	0.1

The random variable X has mean $\mu_X = 1.1$ and standard deviation $\sigma_X = 0.943$.

1. Suppose the dealership's manager receives a \$500 bonus from the company for each car sold. Let Y = the bonus received from car sales during the first hour on a randomly selected Friday. Find the mean and standard deviation of Y.

2. To encourage customers to buy cars on Friday mornings, the manager spends \$75 to provide coffee and doughnuts. The manager's net profit *T* on a randomly selected Friday is the bonus earned minus this \$75. Find the mean and standard deviation of *T*.

Putting it all together: Adding/subtracting and multiplying/dividing What happens if we transform a random variable by both adding or subtracting a constant and multiplying or dividing by a constant? Let's consider Pete's Jeep Tours again. We could have gone directly from the number of passengers X on a randomly selected jeep tour to Pete's profit *V* with the equation $V = 150X - 100$ or, equivalently, $V = -100 + 150X$. This **linear transformation** of the random variable X includes both of the transformations that we performed earlier: (1) multiplying by 150 and (2) subtracting 100. (In general, a linear transformation can be written in the form $Y = a + bX$, where *a* and *b* are constants.) The net effect of this sequence of transformations is as follows:

Linear transformation

For the linear transformation $V = -100 + 150X$, it would *not* be correct to apply the transformations in the reverse order: subtract 100 and then multiply by 150. Doing so would yield the same standard deviation but a different (wrong) mean.

Shape: Neither transformation changes the shape of the probability distribution.

Center: The mean of X is multiplied by 150 and then decreased by 100; that is, $\mu_V = 150\mu_X - 100 = -100 + 150\mu_X$.

Spread: The standard deviation of X is multiplied by 150 and is unchanged by the subtraction: $\sigma_V = 150\sigma_X$.

This logic generalizes to any linear transformation.

Effects of a Linear Transformation on the Mean and Standard Deviation

If $Y = a + bX$ is a linear transformation of the random variable X, then

- the probability distribution of Y has the same shape as the probability distribution of X.
- $\mu_Y = a + b\mu_X$.
- $\sigma_Y = |b|\sigma_X$ (since *b* could be a negative number).

Can you see why this is called a "linear" transformation? The equation describing the sequence of transformations has the form $Y = a + bX$, which you should recognize as a linear equation.

Whether we're dealing with data or random variables, the effects of a linear transformation are the same. Note that these results apply to both discrete and continuous random variables.

The Baby and the Bathwater
Linear transformations

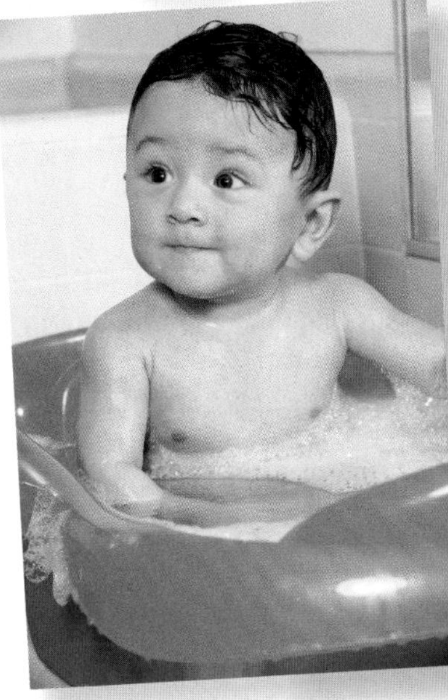

PROBLEM: One brand of bathtub comes with a dial to set the water temperature. When the "babysafe" setting is selected and the tub is filled, the temperature X of the water follows a Normal distribution with a mean of 34°C and a standard deviation of 2°C.

(a) Define the random variable Y to be the water temperature in degrees Fahrenheit (recall that $F = \frac{9}{5}C + 32$) when the dial is set on "babysafe." Find the mean and standard deviation of Y.

(b) According to Babies R Us, the temperature of a baby's bathwater should be between 90°F and 100°F. Find the probability that the water temperature on a randomly selected day when the "babysafe" setting is used meets the Babies R Us recommendation. Show your work.

SOLUTION:

(a) According to the formula for converting Celsius to Fahrenheit, $Y = \frac{9}{5}X + 32$. We could also write this in the form $Y = 32 + \frac{9}{5}X$. The mean of Y is

$$\mu_Y = 32 + \frac{9}{5}\mu_X = 32 + \frac{9}{5}(34) = 93.2°\text{F}$$

The standard deviation of Y is

$$\sigma_Y = \frac{9}{5}\sigma_X = \frac{9}{5}(2) = 3.6°\text{F}$$

(b) The linear transformation doesn't change the shape of the probability distribution, so the random variable Y is Normally distributed with a mean of 93.2 and a standard deviation of 3.6. We want to find $P(90 \le Y \le 100)$. The shaded area in Figure 6.9 shows the desired probability. To find this area, we can standardize and use Table A:

$$z = \frac{90 - 93.2}{3.6} = -0.89 \qquad \text{and} \qquad z = \frac{100 - 93.2}{3.6} = 1.89$$

Then $P(-0.89 \le Z \le 1.89) = 0.9706 - 0.1867 = 0.7839$. There's about a 78% chance that the water temperature meets the recommendation on a randomly selected day.

We can check our answer with `normalcdf(90,100,93.2,3.6)`, which yields 0.7835.

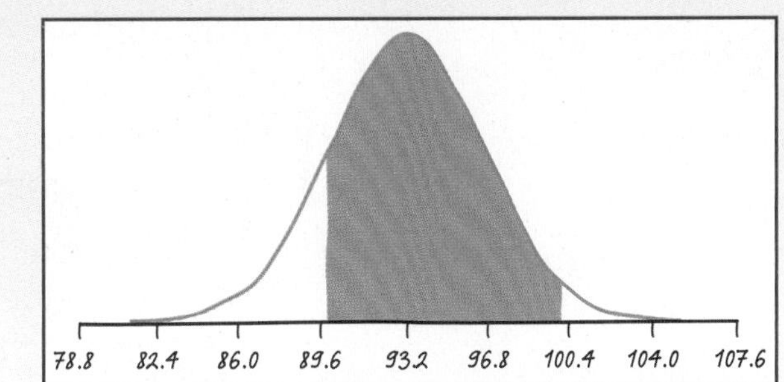

FIGURE 6.9 The Normal probability distribution of the random variable Y = the temperature (in °F) of the bathwater when the dial is set on "babysafe." The shaded area is the probability that the water temperature is between 90°F and 100°F.

For Practice *Try Exercise 45*

Combining Random Variables

So far, we have looked at settings that involved a single random variable. Many interesting statistics problems require us to examine two or more random variables.

Put a Lid on It!

When one random variable isn't enough

If we buy a large drink at a fast-food restaurant and grab a large-sized lid, we assume that the lid will fit snugly on the cup. Usually, it does. Sometimes, though, the lid is a bit too small or too large to fit securely. Why does this happen? Even though the large cups are made to be identical in size, the width actually varies slightly from cup to cup. In fact, the diameter of a randomly selected large drink cup is a random variable (call it C) that has a particular probability distribution. Likewise, the diameter of a randomly selected large lid (call it L) is a random variable with its own probability distribution. For a lid to fit on a cup, the value of L has to be bigger than the value of C, but not too much bigger. To find the probability that the lid fits, we need to know more about what happens when we combine these two random variables.

Before we tackle the cup and lid problem, let's investigate the result of adding and subtracting random variables.

Sums of random variables Earlier, we examined the probability distribution for the random variable X = the number of passengers on a randomly selected half-day trip with Pete's Jeep Tours. Here's a brief recap:

No. of passengers x_i:	2	3	4	5	6
Probability p_i:	0.15	0.25	0.35	0.20	0.05

Mean: $\mu_X = 3.75$ Standard deviation: $\sigma_X = 1.090$

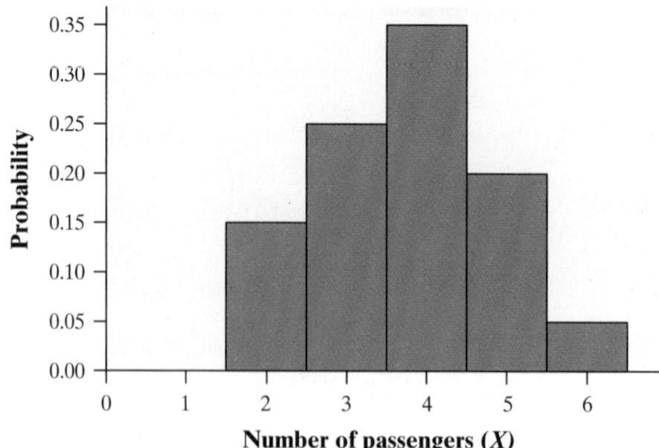

Pete's sister Erin, who lives near a tourist area in another part of the country, is impressed by the success of Pete's business. She decides to join the business, running tours on the same days as Pete in her slightly smaller vehicle, under the name Erin's Adventures. After a year of steady bookings, Erin discovers that the number of passengers Y on her half-day tours has the following probability distribution. Figure 6.10 is a histogram of this probability distribution.

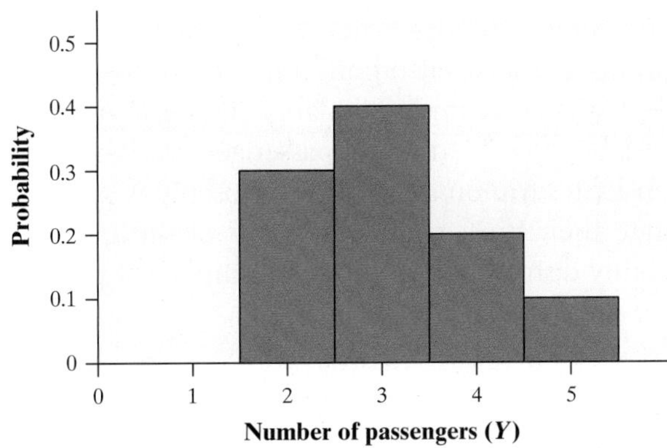

FIGURE 6.10 The probability distribution of the random variable $Y =$ the number of passengers on Erin's trip on a randomly chosen day.

No. of passengers y_i:	2	3	4	5
Probability p_i:	0.3	0.4	0.2	0.1

Mean: $\mu_Y = 3.10$ Standard deviation: $\sigma_Y = 0.943$

How many total passengers T can Pete and Erin expect to have on their tours on a randomly selected day? Since Pete averages $\mu_X = 3.75$ passengers per trip and Erin averages $\mu_Y = 3.10$ passengers per trip, they will average a total of $\mu_T = 3.75 + 3.10 = 6.85$ passengers per day. We can generalize this result for any two random variables as follows: if $T = X + Y$, then $\mu_T = \mu_X + \mu_Y$. In other words, the expected value (mean) of the sum of two random variables is equal to the sum of their expected values (means).

Mean of the Sum of Random Variables

For any two random variables X and Y, if $T = X + Y$, then the expected value of T is

$$E(T) = \mu_T = \mu_X + \mu_Y$$

In general, the mean of the sum of several random variables is the sum of their means.

How much variability is there in the total number of passengers who go on Pete's and Erin's tours on a randomly chosen day? Let's think about the possible values of $T = X + Y$. The number of passengers X on Pete's tour is between 2 and 6, and the number of passengers Y on Erin's tour is between 2 and 5. So the total number of passengers T is between 4 and 11. Thus, the range of T is $11 - 4 = 7$. How is this value related to the ranges of X and Y? The range of X is 4 and the range of Y is 3, so

$$\text{range of } T = \text{range of } X + \text{range of } Y$$

That is, there's more variability in the values of T than in the values of X or Y alone. This makes sense, because the variation in X and the variation in Y both contribute to the variation in T.

What about the standard deviation σ_T? If we had the probability distribution of the random variable T, then we could calculate σ_T. Let's try to construct this probability distribution starting with the smallest possible value, $T = 4$. The only way to get a total of 4 passengers is if Pete has $X = 2$ passengers and Erin has $Y = 2$ passengers. We know that $P(X = 2) = 0.15$ and that $P(Y = 2) = 0.3$. If the two events $X = 2$ and $Y = 2$ are independent, then we can multiply these two probabilities. Otherwise, we're stuck. In fact, we can't calculate the probability for any value of T unless X and Y are **independent random variables.**

DEFINITION: Independent random variables

If knowing whether any event involving X alone has occurred tells us nothing about the occurrence of any event involving Y alone, and vice versa, then X and Y are **independent random variables.**

Probability models often assume independence when the random variables describe outcomes that appear unrelated to each other. You should always ask whether the assumption of independence seems reasonable. For instance, it's reasonable to treat the random variables X = number of passengers on Pete's trip and Y = number of passengers on Erin's trip on a randomly chosen day as independent, since the siblings operate their trips in different parts of the country. Now we can calculate the probability distribution of the total number of passengers that day.

Pete's Jeep Tours and Erin's Adventures

Sum of two random variables

Let $T = X + Y$, as before. Assume that X and Y are independent random variables. We'll begin by considering all the possible combinations of values of X and Y.

x_i	p_i	y_i	p_i	$t_i = x_i + y_i$	p_i
2	0.15	2	0.3	4	(0.15)(0.3) = 0.045
2	0.15	3	0.4	5	(0.15)(0.4) = 0.060
2	0.15	4	0.2	6	(0.15)(0.2) = 0.030
2	0.15	5	0.1	7	(0.15)(0.1) = 0.015
3	0.25	2	0.3	5	(0.25)(0.3) = 0.075
3	0.25	3	0.4	6	(0.25)(0.4) = 0.100
3	0.25	4	0.2	7	(0.25)(0.2) = 0.050
3	0.25	5	0.1	8	(0.25)(0.1) = 0.025
4	0.35	2	0.3	6	(0.35)(0.3) = 0.105
4	0.35	3	0.4	7	(0.35)(0.4) = 0.140
4	0.35	4	0.2	8	(0.35)(0.2) = 0.070
4	0.35	5	0.1	9	(0.35)(0.1) = 0.035
5	0.20	2	0.3	7	(0.20)(0.3) = 0.060
5	0.20	3	0.4	8	(0.20)(0.4) – 0.080
5	0.20	4	0.2	9	(0.20)(0.2) = 0.040
5	0.20	5	0.1	10	(0.20)(0.1) = 0.020
6	0.05	2	0.3	8	(0.05)(0.3) = 0.015
6	0.05	3	0.4	9	(0.05)(0.4) = 0.020
6	0.05	4	0.2	10	(0.05)(0.2) = 0.010
6	0.05	5	0.1	11	(0.05)(0.1) = 0.005

Now we can construct the probability distribution by adding the probabilities for each possible value of T. For instance, $P(T = 6) = 0.030 + 0.100 + 0.105 = 0.235$.

Value t_i:	4	5	6	7	8	9	10	11
Probability p_i:	0.045	0.135	0.235	0.265	0.190	0.095	0.030	0.005

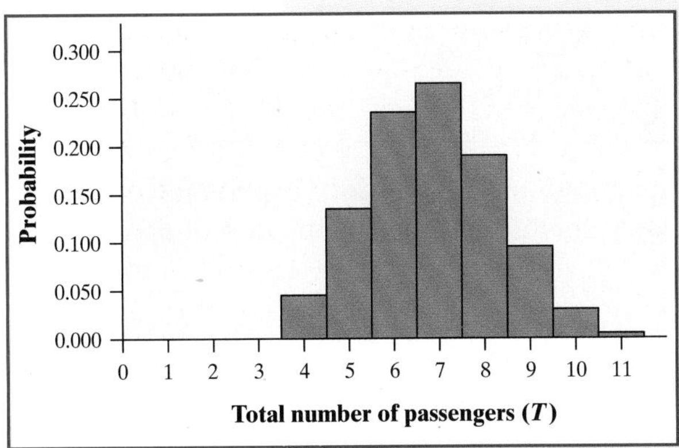

FIGURE 6.11 The probability distribution of the random variable T = the total number of passengers on Pete's and Erin's trips on a randomly chosen day.

You can check that the probabilities add to 1. A histogram of the probability distribution is shown in Figure 6.11.

The mean of T is $\mu_T = \sum t_i p_i = (4)(0.045) + (5)(0.135) + \cdots + (11)(0.005) = 6.85$. Recall that $\mu_X = 3.75$ and $\mu_Y = 3.10$. Our calculation confirms that

$$\mu_T = \mu_X + \mu_Y = 3.75 + 3.10 = 6.85$$

What about the variance of T? It's

$$\sigma_T^2 = \sum (t_i - \mu_T)^2 p_i$$
$$= (4-6.85)^2(0.045) + (5-6.85)^2(0.135) + \cdots + (11-6.85)^2(0.005) = 2.0775$$

Now $\sigma_X^2 = 1.1875$ and $\sigma_Y^2 = 0.89$. Do you see the relationship? Variances add!

$$\sigma_T^2 = \sigma_X^2 + \sigma_Y^2$$

To find the standard deviation of T, take the square root of the variance

$$\sigma_T = \sqrt{2.0775} = 1.441$$

As the preceding example illustrates, when we add two *independent* random variables, their variances add. *Standard deviations do not add.* For Pete's and Erin's passenger totals,

$$\sigma_X + \sigma_Y = 1.090 + 0.943 = 2.033$$

This is very different from $\sigma_T = 1.441$.

Variance of the Sum of Independent Random Variables

For any two *independent* random variables X and Y, if $T = X + Y$, then the variance of T is

$$\sigma_T^2 = \sigma_X^2 + \sigma_Y^2$$

In general, the variance of the sum of several *independent* random variables is the sum of their variances.

You might be wondering whether there's a formula for computing the variance of the sum of two random variables that are *not* independent. There is, but it's beyond the scope of this course. *Just remember that you can add variances only if the two random variables are independent, and that you can never add standard deviations.*

CAUTION

EXAMPLE *SAT Scores*

The role of independence

A college uses SAT scores as one criterion for admission. Experience has shown that the distribution of SAT scores among its entire population of applicants is such that

SAT Math score X:	$\mu_X = 519$	$\sigma_X = 115$
SAT Critical Reading score Y:	$\mu_Y = 507$	$\sigma_Y = 111$

PROBLEM: **What are the mean and standard deviation of the total score $X + Y$ among students applying to this college?**

SOLUTION: The mean overall SAT score is

$$\mu_{X+Y} = \mu_X + \mu_Y = 519 + 507 = 1026$$

The variance and standard deviation of the total *cannot be computed* from the information given. SAT Math and Critical Reading scores are not independent, because students who score high on one exam tend to score high on the other also.

For Practice *Try Exercise 51*

The next example involves two independent random variables and some transformations.

EXAMPLE *Pete's and Erin's Tours*

Rules for adding random variables

Earlier, we defined X = the number of passengers on Pete's trip, Y = the number of passengers on Erin's trip, and C = the amount of money that Pete collects on a randomly selected day. We also found the means and standard deviations of these variables:

$\mu_X = 3.75$	$\mu_Y = 3.10$	$\mu_C = 562.50$
$\sigma_X = 1.090$	$\sigma_Y = 0.943$	$\sigma_C = 163.50$

PROBLEM:

(a) Erin charges \$175 per passenger for her trip. Let $G =$ the amount of money that she collects on a randomly selected day. Find the mean and standard deviation of G.

(b) Calculate the mean and the standard deviation of the total amount that Pete and Erin collect on a randomly chosen day.

SOLUTION:

(a) We can write $G = 175Y$. The mean of G is

$$\mu_G = 175\mu_Y = 175(3.1) = 542.50$$

Erin can expect to make \$542.50 from her trip on a randomly chosen day. The standard deviation of G is

$$\sigma_G = 175\sigma_Y = 175(0.943) = 165.03$$

(b) Let W = the total amount collected. Notice that $W = C + G$. We know that the mean of the sum of two random variables equals the sum of their means: $\mu_W = \mu_C + \mu_G$. So

$$\mu_W = \mu_C + \mu_G = 562.50 + 542.50 = 1105$$

On average, Pete and Erin expect to collect a total of $1105 per day.

Since the number of passengers X and Y are *independent* random variables, so are the amounts of money collected C and G. Therefore, the variance of W is the sum of the variances of C and G.

$$\sigma_W^2 = \sigma_C^2 + \sigma_G^2 = (163.5)^2 + (165.03)^2 = 53{,}967.15$$

To get the standard deviation, we take the square root of the variance:

$$\sigma_W = \sqrt{53{,}967.15} = 232.31$$

The standard deviation of the total amount they collect is $232.31.

For Practice *Try Exercise 53*

We can extend our rules for adding random variables to situations involving repeated observations from the same chance process. For instance, suppose a gambler plays two games of roulette, each time placing a $1 bet on either red or black. What can we say about his total gain (or loss) from playing two games? Earlier, we showed that if X = the amount gained on a single $1 bet on red or black, then $\mu_X = -\$0.05$ and $\sigma_X = \$1.00$. Since we're interested in the player's total gain over two games, we'll define X_1 as the amount he gains from the first game and X_2 as the amount he gains from the second game. Then, his total gain $T = X_1 + X_2$. Both X_1 and X_2 have the same probability distribution as X and, therefore, the same mean (−$0.05) and standard deviation ($1.00). The player's expected gain in two games is

$$\mu_T = \mu_{X_1} + \mu_{X_2} = (-\$0.05) + (-\$0.05) = -\$0.10$$

Since the result of one game has no effect on the result of the other game, X_1 and X_2 are independent random variables. As a result,

$$\sigma_T^2 = \sigma_{X_1}^2 + \sigma_{X_2}^2 = (1)^2 + (1)^2 = 2$$

and the standard deviation of the player's total gain is

$$\sigma_T = \sqrt{2.00} = \$1.41$$

THINK ABOUT IT

$X_1 + X_2$ is not the same as $2X$ At the beginning of the section, we asked whether a roulette player would be better off placing two separate $1 bets on red or a single $2 bet on red. The player's total gain from two $1 bets on red is $T = X_1 + X_2$. This sum of random variables has mean $\mu_T = -\$0.10$ and standard deviation $\sigma_T = \$1.41$. Now think about what happens if the gambler places a $2 bet on red in a single game of roulette. Since the random variable X represents a player's gain from a $1 bet, the random variable $Y = 2X$ represents his gain from a $2 bet.

What's the player's expected gain from a single \$2 bet on red? It's

$$\mu_Y = 2\mu_X = 2(-\$0.05) = -\$0.10$$

That's the same as his expected gain from playing two games of roulette with a \$1 bet each time. But the standard deviation of the player's gain from a single \$2 bet is

$$\sigma_Y = 2\sigma_X = 2(\$1.00) = \$2.00$$

Compare this result to $\sigma_T = \$1.41$. There's more variability in the gain from a single \$2 bet than in the total gain from two \$1 bets.

Let's take this one step further. Would it be better for the player to place a single \$100 bet on red or to play 100 games and bet \$1 each time on red? For the single \$100 bet, the mean and standard deviation of the amount gained would be

$$\text{mean} = 100\mu_X = 100(-\$0.05) = -\$5.00$$
$$\text{standard deviation} = 100\sigma_X = 100(\$1.00) = \$100.00$$

For 100 games with a \$1 bet, the mean and standard deviation of the amount gained would be

$$\text{mean} = \mu_{X_1} + \mu_{X_2} + \cdots + \mu_{X_{100}}$$
$$= (-\$0.05) + (-\$0.05) + \cdots + (-\$0.05) = -\$5.00$$
$$\text{variance} = \sigma^2_{X_1} + \sigma^2_{X_2} + \cdots + \sigma^2_{X_{100}} = (1)^2 + (1)^2 + \cdots + (1)^2 = 100$$
$$\text{standard deviation} = \sqrt{100} = \$10.00$$

The player has a much better chance of winning (or losing) big with a single \$100 bet than with 100 separate \$1 bets. Of course, the casino accepts thousands of bets each day, so it can count on being fairly close to its expected return of 5 cents per dollar bet.

CHECK YOUR UNDERSTANDING

A large auto dealership keeps track of sales and lease agreements made during each hour of the day. Let X = the number of cars sold and Y = the number of cars leased during the first hour of business on a randomly selected Friday. Based on previous records, the probability distributions of X and Y are as follows:

Cars sold x_i:	0	1	2	3
Probability p_i:	0.3	0.4	0.2	0.1

Mean: $\mu_X = 1.1$ Standard deviation: $\sigma_X = 0.943$

Cars leased y_i:	0	1	2
Probability p_i:	0.4	0.5	0.1

Mean: $\mu_Y = 0.7$ Standard deviation: $\sigma_Y = 0.64$

Define $T = X + Y$.

1. Find and interpret μ_T.
2. Compute σ_T assuming that X and Y are independent. Show your work.
3. The dealership's manager receives a \$500 bonus for each car sold and a \$300 bonus for each car leased. Find the mean and standard deviation of the manager's total bonus B. Show your work.

Differences of random variables Now that we've examined sums of random variables, it's time to investigate the difference of two random variables. Let's start by looking at the difference in the number of passengers that Pete and Erin have on their tours on a randomly selected day, $D = X - Y$. Since Pete averages $\mu_X = 3.75$ passengers per trip and Erin averages $\mu_Y = 3.10$ passengers per trip, the average difference is $\mu_D = 3.75 - 3.10 = 0.65$ passengers. That is, Pete averages 0.65 more passengers per day than Erin does. We can generalize this result for any two random variables as follows: if $D = X - Y$, then $\mu_D = \mu_X - \mu_Y$. In other words, the expected value (mean) of the difference of two random variables is equal to the difference of their expected values (means).

Mean of the Difference of Random Variables

For any two random variables X and Y, if $D = X - Y$, then the expected value of D is

$$E(D) = \mu_D = \mu_X - \mu_Y$$

In general, the mean of the difference of several random variables is the difference of their means.

CAUTION

The order of subtraction is important. If we had defined $D = Y - X$, then $\mu_D = \mu_Y - \mu_X = 3.10 - 3.75 = -0.65$. In other words, Erin averages 0.65 fewer passengers than Pete does on a randomly chosen day.

Earlier, we saw that the variance of the sum of two independent random variables is the sum of their variances. Can you guess what the variance of the *difference* of two independent random variables will be? (If you were thinking something like "the difference of their variances," think again!) Let's return to the jeep tours scenario. On a randomly selected day, the number of passengers X on Pete's tour is between 2 and 6, and the number of passengers Y on Erin's tour is between 2 and 5. So the difference in the number of passengers $D = X - Y$ is between -3 and 4. Thus, the range of D is $4 - (-3) = 7$. How is this value related to the ranges of X and Y? The range of X is 4 and the range of Y is 3, so

$$\text{range of } D = \text{range of } X + \text{range of } Y$$

As with sums of random variables, there's more variability in the values of the difference D than in the values of X or Y alone. This should make sense, because the variation in X and the variation in Y both contribute to the variation in D.

If you follow the process we used earlier with the random variable $T = X + Y$, you can build the probability distribution of $D = X - Y$ and confirm that

(1) $\mu_D = \mu_X - \mu_Y = 3.75 - 3.10 = 0.65$

(2) $\sigma_D^2 = \sigma_X^2 + \sigma_Y^2 = 1.1875 + 0.89 = 2.0775$

(3) $\sigma_D = \sqrt{2.0775} = 1.441$

Result 2 shows that, just like with addition, when we subtract two independent random variables, variances add.

Variance of the Difference of Random Variables

For any two *independent* random variables X and Y, if $D = X - Y$, then the variance of D is

$$\sigma_D^2 = \sigma_X^2 + \sigma_Y^2$$

Let's put our new rules for subtracting random variables to use in a familiar setting.

EXAMPLE Pete's Jeep Tours and Erin's Adventures

Difference of random variables

We have defined several random variables related to Pete's and Erin's tour businesses. For a randomly selected day,

X = number of passengers on Pete's trip Y = number of passengers on Erin's trip
C = amount of money that Pete collects G = amount of money that Erin collects

Here are the means and standard deviations of these random variables:

$\mu_X = 3.75$	$\mu_Y = 3.10$	$\mu_C = 562.50$	$\mu_G = 542.50$
$\sigma_X = 1.090$	$\sigma_Y = 0.943$	$\sigma_C = 163.50$	$\sigma_G = 165.03$

PROBLEM: Calculate the mean and the standard deviation of the difference $D = C - G$ in the amounts that Pete and Erin collect on a randomly chosen day. Interpret each value in context.

SOLUTION: We know that the mean of the difference of two random variables is the difference of their means. That is,

$$\mu_D = \mu_C - \mu_G = 562.50 - 542.50 = 20.00$$

On average, Pete collects \$20 more per day than Erin does. Some days the difference will be more than \$20, other days it will be less, but the average difference after lots of days will be \$20.

Since the number of passengers X and Y are *independent* random variables, so are the amounts of money collected C and G. Therefore, the variance of D is the sum of the variances of C and G:

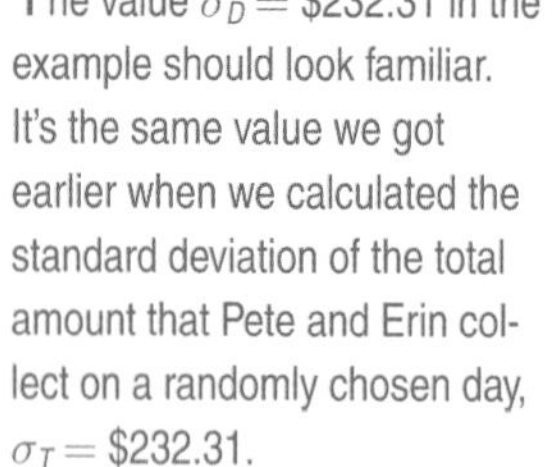

The value $\sigma_D = \$232.31$ in the example should look familiar. It's the same value we got earlier when we calculated the standard deviation of the total amount that Pete and Erin collect on a randomly chosen day, $\sigma_T = \$232.31$.

$$\sigma_D^2 = \sigma_C^2 + \sigma_G^2 = (163.5)^2 + (165.03)^2 = 53{,}967.15$$

$$\sigma_D = \sqrt{53{,}967.15} = 232.31$$

The standard deviation of the difference in the amounts collected by Pete and Erin is \$232.31. Even though the *average* difference in the amounts collected is \$20, the difference on individual days will vary somewhat. How much? By about \$232, on average.

For Practice *Try Exercise 55*

CHECK YOUR UNDERSTANDING

A large auto dealership keeps track of sales and lease agreements made during each hour of the day. Let X = the number of cars sold and Y = the number of cars leased during the first hour of business on a randomly selected Friday. Based on previous records, the probability distributions of X and Y are as follows:

Cars sold x_i:	0	1	2	3
Probability p_i:	0.3	0.4	0.2	0.1

Mean: $\mu_X = 1.1$ Standard deviation: $\sigma_X = 0.943$

Cars leased y_i:	0	1	2
Probability p_i:	0.4	0.5	0.1

Mean: $\mu_Y = 0.7$ Standard deviation: $\sigma_Y = 0.64$

Define $D = X - Y$.

1. Find and interpret μ_D.
2. Compute σ_D assuming that X and Y are independent. Show your work.
3. The dealership's manager receives a $500 bonus for each car sold and a $300 bonus for each car leased. Find the mean and standard deviation of the difference in the manager's bonus for cars sold and leased. Show your work.

Combining Normal Random Variables

So far, we have concentrated on finding rules for means and variances of random variables. If a random variable is Normally distributed, we can use its mean and standard deviation to compute probabilities. The earlier example on young women's heights (page 351) shows the method. What if we combine two Normal random variables?

A Computer Simulation

Sums and differences of Normal random variables

We used Fathom software to simulate taking independent SRSs of 1000 observations from each of two Normally distributed random variables, X and Y. Figure 6.12(a) shows the results. The random variable X is N(3, 0.9) and the random variable Y is N(1, 1.2). What do we know about the sum and difference of these two random variables? The histograms in Figure 6.12(b) came from adding and subtracting the values of X and Y for the 1000 randomly generated observations from each distribution.

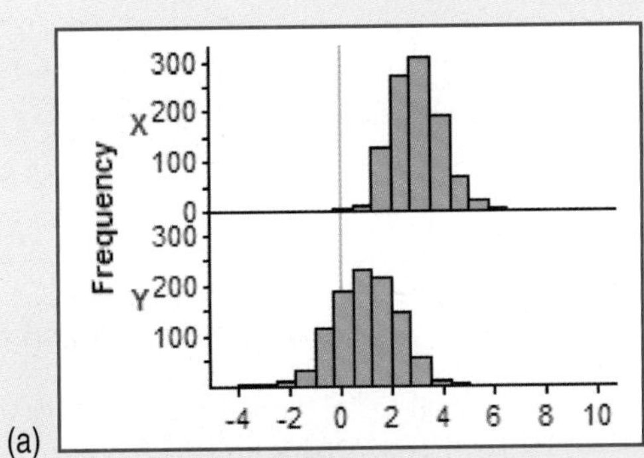

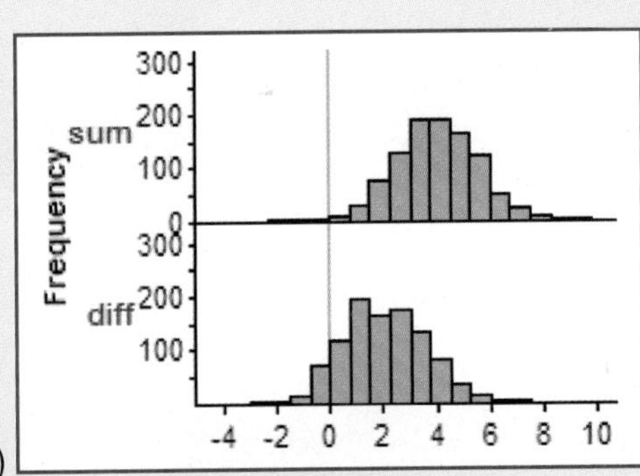

FIGURE 6.12 (a) Histograms showing the results of randomly selecting 1000 observations from two different Normal random variables *X* and *Y*. (b) Histograms of the sum and difference of the 1000 randomly selected values of *X* and *Y*.

Let's summarize what we see:

	Sum X + Y	**Difference X − Y**
Shape:	Looks Normal	Looks Normal
Center:	$\mu_{X+Y} = \mu_X + \mu_Y = 3+1 = 4$	$\mu_{X-Y} = \mu_X - \mu_Y = 3-1 = 2$
Spread:		
	$\sigma^2_{X+Y} = \sigma^2_X + \sigma^2_Y = (0.9)^2 + (1.2)^2 = 2.25$	$\sigma^2_{X-Y} = \sigma^2_X + \sigma^2_Y = (0.9)^2 + (1.2)^2 = 2.25$
	$\sigma_{X+Y} = \sqrt{2.25} = 1.5$	$\sigma_{X-Y} = \sqrt{2.25} = 1.5$

As the previous example illustrates, *any sum or difference of independent Normal random variables is also Normally distributed.* The mean and standard deviation of the resulting Normal distribution can be found using the appropriate rules for means and variances.

EXAMPLE

Give Me Some Sugar!

Sums of Normal random variables

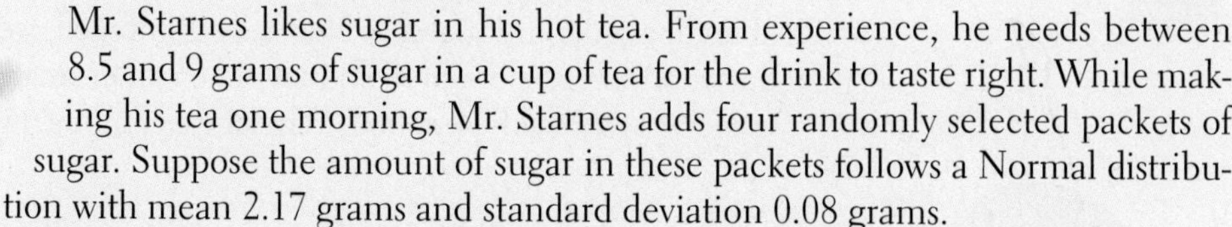

Mr. Starnes likes sugar in his hot tea. From experience, he needs between 8.5 and 9 grams of sugar in a cup of tea for the drink to taste right. While making his tea one morning, Mr. Starnes adds four randomly selected packets of sugar. Suppose the amount of sugar in these packets follows a Normal distribution with mean 2.17 grams and standard deviation 0.08 grams.

STATE: What's the probability that Mr. Starnes's tea tastes right?

PLAN: Let X = the amount of sugar in a randomly selected packet. Then X_1 = amount of sugar in Packet 1, X_2 = amount of sugar in Packet 2, X_3 = amount of sugar in Packet 3, and X_4 = amount of sugar in Packet 4. Each of these random variables has a Normal distribution with mean 2.17 grams and standard deviation 0.08 grams. We're interested in the total amount of sugar that Mr. Starnes puts in his tea, which is given by $T = X_1 + X_2 + X_3 + X_4$. Our goal is to find $P(8.5 \leq T \leq 9)$.

DO: The random variable T is a sum of four independent Normal random variables. So T follows a Normal distribution with mean

$$\mu_T = \mu_{X_1} + \mu_{X_2} + \mu_{X_3} + \mu_{X_4} = 2.17 + 2.17 + 2.17 + 2.17 = 8.68$$

and variance

$$\sigma_T^2 = \sigma_{X_1}^2 + \sigma_{X_2}^2 + \sigma_{X_3}^2 + \sigma_{X_4}^2 = (0.08)^2 + (0.08)^2 + (0.08)^2 + (0.08)^2 = 0.0256$$

The standard deviation of T is

$$\sigma_T = \sqrt{0.0256} = 0.16$$

We want to find the probability that the total amount of sugar in Mr. Starnes's tea is between 8.5 and 9 grams. Figure 6.13 shows this probability as the area under a Normal curve. To find this area, we can standardize and use Table A:

$$z = \frac{8.5 - 8.68}{0.16} = -1.13 \quad \text{and} \quad z = \frac{9 - 8.68}{0.16} = 2.00$$

Then $P(-1.13 \leq Z \leq 2.00) = 0.9772 - 0.1292 = 0.8480$.

CONCLUDE: There's about an 85% chance that Mr. Starnes's tea will taste right.

8.20 8.36 8.52 8.68 8.84 9.00 9.16

FIGURE 6.13 Normal distribution of the total amount of sugar in Mr. Starnes's tea.

Using `normalcdf(8.5,9,8.68,0.16)` or the *Normal curve* applet yields $P(8.5 \leq T \leq 9) = 0.8470$.

Here's an example that involves subtracting two Normal random variables.

EXAMPLE

Put a Lid on It!

Differences of Normal random variables

The diameter C of a randomly selected large drink cup at a fast-food restaurant follows a Normal distribution with a mean of 3.96 inches and a standard deviation of 0.01 inches. The diameter L of a randomly selected large lid at this restaurant follows a Normal distribution with mean 3.98 inches and standard deviation 0.02 inches. For a lid to fit on a cup, the value of L has to be bigger than the value of C, but not by more than 0.06 inches.

STATE: What's the probability that a randomly selected large lid will fit on a randomly chosen large drink cup?

PLAN: We'll define the random variable $D = L - C$ to represent the difference between the lid's diameter and the cup's diameter. Our goal is to find $P(0 < D \le 0.06)$.

DO: The random variable D is the difference of two independent Normal random variables. So D follows a Normal distribution with mean

$$\mu_D = \mu_L - \mu_C = 3.98 - 3.96 = 0.02$$

and variance

$$\sigma_D^2 = \sigma_L^2 + \sigma_C^2 = (0.02)^2 + (0.01)^2 = 0.0005$$

The standard deviation of D is

$$\sigma_D = \sqrt{0.0005} = 0.0224$$

We want to find the probability that the difference D is between 0 and 0.06 inches. Figure 6.14 shows this probability as the area under a Normal curve. To find this area, we can standardize and use Table A:

$$z = \frac{0 - 0.02}{0.0224} = -0.89 \quad \text{and} \quad z = \frac{0.06 - 0.02}{0.0224} = 1.79$$

Then $P(-0.89 \le Z \le 1.79) = 0.9633 - 0.1867 = 0.7766$.

Using `normalcdf(0,0.06,0.02,0.0224)` or the *Normal curve* applet yields $P(0 < D \le 0.06) = 0.7770$.

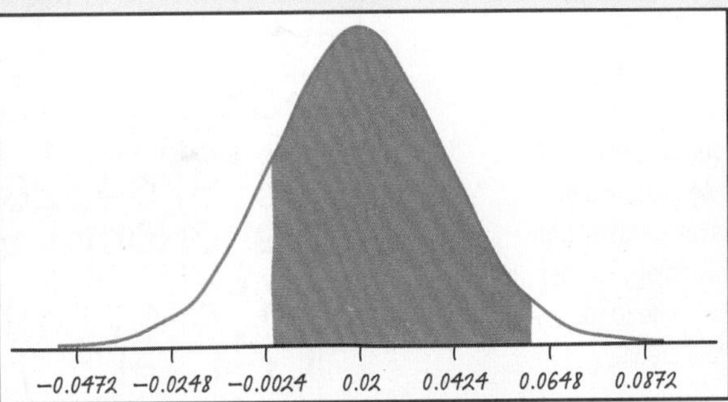

FIGURE 6.14 Normal distribution of the difference in lid diameter and cup diameter at a fast-food restaurant.

CONCLUDE: There's about a 78% chance that a randomly selected large lid will fit on a randomly chosen large drink cup at this fast-food restaurant. Roughly 22% of the time, the lid won't fit. This seems like an unreasonably high chance of getting a lid that doesn't fit. Maybe the restaurant should find a new supplier!

For Practice *Try Exercise 61*

The previous two examples show how to calculate probabilities involving sums and differences of independent Normal random variables. As an alternative, we can use simulation to estimate these probabilities.

TECHNOLOGY CORNER Simulating with randNorm

The `randNorm` command on the TI-83/84 and TI-89 allows you to simulate observations from a Normal distribution with a specified mean and standard deviation. (You can find `randNorm` under MATH/PRB on the TI-83/84 and in the CATALOG under Flash Apps on the TI-89.) For instance, `randNorm(3.98,.02)` will randomly select a value from the Normal distribution with mean 3.98 and standard deviation 0.02. This simulates choosing a large cup lid at random from the fast-food restaurant of the prior example and measuring its diameter (in inches). To simulate choosing a large drink cup and measuring its width, you can use `randNorm(3.96,.01)`.

To estimate the probability that a randomly selected lid will fit on a randomly chosen cup:

1. Simulate choosing 100 large cup lids at random, and store their widths in L1 (list1):

TI-83/84	TI-89
`randNorm(3.98,.02,100)->L1`	`tistat.randnorm(3.98,.02,100)->list1`

2. Simulate choosing 100 large drink cups at random, and store their diameters in L2 (list2):

TI-83/84	TI-89
`randNorm(3.96,.01,100)->L2`	`tistat.randnorm(3.96,.01,100)->list2`

3. Compute the difference between the lid and cup diameters for these 100 pairs of values:

TI-83/84	TI-89
`L1-L2->L3`	`list1-list2->list3`

4. Count the number of values in L3 (list3) that are between 0 and 0.06. (You may want to sort the values in L3 (list3) first!) This number divided by 100 is your estimate of the probability.

```
randNorm(3.98,.0
2,100)→L1
{3.973432644 3.…
randNorm(3.96,.0
1,100)→L2
{3.965754864 3.…
L1-L2→L3
```

TI-Nspire instructions in Appendix B

AP EXAM TIP If you have trouble solving problems involving sums and differences of Normal random variables with the algebraic methods shown earlier in the section, use the simulation strategy from the Technology Corner to earn some (or possibly full) credit.

SECTION 6.2 Summary

- Adding a constant a (which could be negative) to a random variable increases (or decreases) the mean of the random variable by a but does not affect its standard deviation or the shape of its probability distribution.
- Multiplying a random variable by a constant b (which could be negative) multiplies the mean of the random variable by b and the standard deviation by $|b|$ but does not change the shape of its probability distribution.
- A **linear transformation** of a random variable involves adding a constant a, multiplying by a constant b, or both. We can write a linear transformation of the random variable X in the form $Y = a + bX$. The shape, center, and spread of the probability distribution of Y are as follows:
  Shape: Same as the probability distribution of X.
  Center: $\mu_Y = a + b\mu_X$
  Spread: $\sigma_Y = |b|\sigma_X$
- If X and Y are *any* two random variables,
  $\mu_{X+Y} = \mu_X + \mu_Y$ The mean of the sum of two random variables is the sum of their means.
  $\mu_{X-Y} = \mu_X - \mu_Y$ The mean of the difference of two random variables is the difference of their means.
- If X and Y are **independent random variables**, then there is no association between the values of one variable and the values of the other. In that case, variances add:
  $\sigma^2_{X+Y} = \sigma^2_X + \sigma^2_Y$ The variance of the sum of two random variables is the sum of their variances.
  $\sigma^2_{X-Y} = \sigma^2_X + \sigma^2_Y$ The variance of the difference of two random variables is the sum of their variances.
- The sum or difference of independent Normal random variables follows a Normal distribution.

6.2 TECHNOLOGY CORNER

TI-Nspire instructions in Appendix B

SECTION 6.2 Exercises

35. **Crickets** The length in inches of a cricket chosen at random from a field is a random variable X with mean 1.2 inches and standard deviation 0.25 inches. Find the mean and standard deviation of the length Y of a randomly chosen cricket from the field in centimeters. There are 2.54 centimeters in an inch.

36. **Men's heights** A report of the National Center for Health Statistics says that the height of a 20-year-old man chosen at random is a random variable H with mean 5.8 feet (ft) and standard deviation 0.24 ft. Find the mean and standard deviation of the height J of a randomly selected 20-year-old man in inches. There are 12 inches in a foot.

37. **Get on the boat!** A small ferry runs every half hour from one side of a large river to the other. The number of cars X on a randomly chosen ferry trip has the probability distribution shown below. You can check that $\mu_X = 3.87$ and $\sigma_X = 1.29$.

Cars:	0	1	2	3	4	5
Probability:	0.02	0.05	0.08	0.16	0.27	0.42

(a) The cost for the ferry trip is \$5. Make a graph of the probability distribution for the random variable M = money collected on a randomly selected ferry trip. Describe its shape.

(b) Find and interpret μ_M.

(c) Compute and interpret σ_M.

38. **Skee Ball** Ana is a dedicated Skee Ball player (see photo) who always rolls for the 50-point slot. The probability distribution of Ana's score X on a single roll of the ball is shown below. You can check that $\mu_X = 23.8$ and $\sigma_X = 12.63$.

Score:	10	20	30	40	50
Probability:	0.32	0.27	0.19	0.15	0.07

(a) A player receives one ticket from the game for every 10 points scored. Make a graph of the probability distribution for the random variable T = number of tickets Ana gets on a randomly selected throw. Describe its shape.

(b) Find and interpret μ_T.

(c) Compute and interpret σ_T.

Exercises 39 and 40 refer to the following setting. Ms. Hall gave her class a 10-question multiple-choice quiz. Let X = the number of questions that a randomly selected student in the class answered correctly. The computer output below gives information about the probability distribution of X. To determine each student's grade on the quiz (out of 100), Ms. Hall will multiply his or her number of correct answers by 10. Let G = the grade of a randomly chosen student in the class.

N	Mean	Median	StDev	Min	Max	Q1	Q3
30	7.6	8.5	1.32	4	10	8	9

39. **Easy quiz**

(a) Find the mean of G. Show your method.

(b) Find the standard deviation of G. Show your method.

(c) How do the variance of X and the variance of G compare? Justify your answer.

40. **Easy quiz**

(a) Find the median of G. Show your method.

(b) Find the *IQR* of G. Show your method.

(c) What shape would the probability distribution of G have? Justify your answer.

41. **Get on the boat!** Refer to Exercise 37. The ferry company's expenses are \$20 per trip. Define the random variable Y to be the amount of profit (money collected minus expenses) made by the ferry company on a randomly selected trip. That is, $Y = M - 20$.

(a) How does the mean of Y relate to the mean of M? Justify your answer. What is the practical importance of μ_Y?

(b) How does the standard deviation of Y relate to the standard deviation of M? Justify your answer. What is the practical importance of σ_Y?

42. **The Tri-State Pick 3** Most states and Canadian provinces have government-sponsored lotteries. Here is a simple lottery wager, from the Tri-State Pick 3

game that New Hampshire shares with Maine and Vermont. You choose a number with 3 digits from 0 to 9; the state chooses a three-digit winning number at random and pays you \$500 if your number is chosen. Because there are 1000 numbers with three digits, you have probability 1/1000 of winning. Taking X to be the amount your ticket pays you, the probability distribution of X is

Payoff X:	\$0	\$500
Probability:	0.999	0.001

(a) Show that the mean and standard deviation of X are $\mu_X = \$0.50$ and $\sigma_X = \$15.80$.

(b) If you buy a Pick 3 ticket, your winnings are $W = X - 1$, because it costs \$1 to play. Find the mean and standard deviation of W. Interpret each of these values in context.

43. **Get on the boat!** Based on the analysis in Exercise 41, the ferry company decides to increase the cost of a trip to \$6. We can calculate the company's profit Y on a randomly selected trip from the number of cars X. Find the mean and standard deviation of Y. Show your work.

44. **Making a profit** Rotter Partners is planning a major investment. The amount of profit X (in millions of dollars) is uncertain, but an estimate gives the following probability distribution:

Profit:	1	1.5	2	4	10
Probability:	0.1	0.2	0.4	0.2	0.1

Based on this estimate, $\mu_X = 3$ and $\sigma_X = 2.52$. Rotter Partners owes its lender a fee of \$200,000 plus 10% of the profits X. So the firm actually retains $Y = 0.9X - 0.2$ from the investment. Find the mean and standard deviation of Y. Show your work.

pg 363

45. **Too cool at the cabin?** During the winter months, the temperatures at the Starneses' Colorado cabin can stay well below freezing (32°F or 0°C) for weeks at a time. To prevent the pipes from freezing, Mrs. Starnes sets the thermostat at 50°F. She also buys a digital thermometer that records the indoor temperature each night at midnight. Unfortunately, the thermometer is programmed to measure the temperature in degrees Celsius. Based on several years' worth of data, the temperature T in the cabin at midnight on a randomly selected night follows a Normal distribution with mean 8.5°C and standard deviation 2.25°C.

(a) Let Y = the temperature in the cabin at midnight on a randomly selected night in degrees Fahrenheit (recall that $F = (9/5)C + 32$). Find the mean and standard deviation of Y.

(b) Find the probability that the midnight temperature in the cabin is below 40°F. Show your work.

46. **Cereal** A company's single-serving cereal boxes advertise 9.63 ounces of cereal. In fact, the amount of cereal X in a randomly selected box follows a Normal distribution with a mean of 9.70 ounces and a standard deviation of 0.03 ounces.

(a) Let Y = the excess amount of cereal beyond what's advertised in a randomly selected box, measured in grams (1 ounce = 28.35 grams). Find the mean and standard deviation of Y.

(b) Find the probability of getting at least 3 grams more cereal than advertised. Show your work.

Exercises 47 and 48 refer to the following setting. Two independent random variables X and Y have the probability distributions, means, and standard deviations shown.

X:	1	2	5
P(X):	0.2	0.5	0.3

$\mu_X = 2.7, \sigma_X = 1.55$

Y:	2	4
P(Y):	0.7	0.3

$\mu_Y = 2.6, \sigma_Y = 0.917$

47. **Sum** Let the random variable $T = X + Y$.

(a) Find all possible values of T. Compute the probability that T takes each of these values. Summarize the probability distribution of T in a table.

(b) Show that the mean of T is equal to $\mu_X + \mu_Y$.

(c) Confirm that the variance of T is equal to $\sigma_X^2 + \sigma_Y^2$. Show that $\sigma_T \neq \sigma_X + \sigma_Y$.

48. **Difference** Let the random variable $D = X - Y$.

(a) Find all possible values of D. Compute the probability that D takes each of these values. Summarize the probability distribution of D in a table.

(b) Show that the mean of D is equal to $\mu_X - \mu_Y$.

(c) Confirm that the variance of D is equal to $\sigma_X^2 + \sigma_Y^2$.

49. **Checking independence** In which of the following games of chance would you be willing to assume independence of X and Y in making a probability model? Explain your answer in each case.

(a) In blackjack, you are dealt two cards and examine the total points X on the cards (face cards count 10 points). You can choose to be dealt another card and compete based on the total points Y on all three cards.

(b) In craps, the betting is based on successive rolls of two dice. X is the sum of the faces on the first roll, and Y is the sum of the faces on the next roll.

50. **Checking independence** For each of the following situations, would you expect the random variables X and Y to be independent? Explain your answers.

(a) X is the rainfall (in inches) on November 6 of this year, and Y is the rainfall at the same location on November 6 of next year.

(b) X is the amount of rainfall today, and Y is the rainfall at the same location tomorrow.

(c) X is today's rainfall at the airport in Orlando, Florida, and Y is today's rainfall at Disney World just outside Orlando.

pg 368

51. **His and her earnings** A study of working couples measures the income X of the husband and the income Y of the wife in a large number of couples in which both partners are employed. Suppose that you knew the means μ_X and μ_Y and the variances σ_X^2 and σ_Y^2 of both variables in the population.

(a) Is it reasonable to take the mean of the total income X + Y to be $\mu_X + \mu_Y$? Explain your answer.

(b) Is it reasonable to take the variance of the total income to be $\sigma_X^2 + \sigma_Y^2$? Explain your answer.

52. **Study habits** The academic motivation and study habits of female students as a group are better than those of males. The Survey of Study Habits and Attitudes (SSHA) is a psychological test that measures these factors. The distribution of SSHA scores among the women at a college has mean 120 and standard deviation 28, and the distribution of scores among male students has mean 105 and standard deviation 35. You select a single male student and a single female student at random and give them the SSHA test.

(a) Explain why it is reasonable to assume that the scores of the two students are independent.

(b) What are the expected value and standard deviation of the difference (female minus male) between their scores?

(c) From the information given, can you find the probability that the woman chosen scores higher than the man? If so, find this probability. If not, explain why you cannot.

pg 368

53. **The Tri-State Pick 3** Refer to Exercise 42. Suppose you buy a $1 ticket on each of two different days.

(a) Find the mean and standard deviation of the total payoff. Show your work.

(b) Find the mean and standard deviation of your total winnings. Interpret each of these values in context.

54. **The Tri-State Pick 3** Refer to Exercise 42. Suppose you buy a Pick 3 ticket every day for a year (365 days).

(a) Find the mean and standard deviation of your total winnings. Show your work.

(b) Interpret each of the values from (a) in context.

55. **A balanced scale** You have two scales for measuring weights in a chemistry lab. Both scales give answers that vary a bit in repeated weighings of the same item. If the true weight of a compound is 2.00 grams (g), the first scale produces readings X that have mean 2.000 g and standard deviation 0.002 g. The second scale's readings Y have mean 2.001 g and standard deviation 0.001 g. The readings X and Y are independent. Find the mean and standard deviation of the difference Y − X between the readings. Interpret each value in context.

pg 372

56. **Essay errors** Typographical and spelling errors can be either "nonword errors" or "word errors." A nonword error is not a real word, as when "the" is typed as "teh." A word error is a real word, but not the right word, as when "lose" is typed as "loose." When students are asked to write a 250-word essay (without spell-checking), the number of nonword errors X has the following probability distribution:

Value of X:	0	1	2	3	4
Probability:	0.1	0.2	0.3	0.3	0.1

$$\mu_X = 2.1 \qquad \sigma_X = 1.136$$

The number of word errors Y has this probability distribution:

Value of Y:	0	1	2	3
Probability:	0.4	0.3	0.2	0.1

$$\mu_Y = 1.0 \qquad \sigma_Y = 1.0$$

(a) Find the mean and standard deviation of the difference Y − X in the number of errors made by a randomly selected student. Interpret each value in context.

(b) ***Challenge:*** Find the probability that a randomly selected student makes more word errors than nonword errors.

Exercises 57 and 58 refer to the following setting. In Exercises 14 and 18 of Section 6.1, we examined the probability distribution of the random variable X = the amount a life insurance company earns on a 5-year term life policy. Calculations reveal that μ_X = \$303.35 and σ_X = \$9707.57.

57. **Life insurance** The risk of insuring one person's life is reduced if we insure many people. Suppose that we insure two 21-year-old males, and that their ages at death are independent. If X_1 and X_2 are the insurer's income from the two insurance policies, the insurer's average income W on the two policies is

$$W = \frac{X_1 + X_2}{2} = 0.5X_1 + 0.5X_2$$

Find the mean and standard deviation of W. (You see that the mean income is the same as for a single policy but the standard deviation is less.)

58. **Life insurance** If four 21-year-old men are insured, the insurer's average income is

$$V = \frac{X_1 + X_2 + X_3 + X_4}{4} = 0.25X_1 + 0.25X_2 + 0.25X_3 + 0.25X_4$$

where X_i is the income from insuring one man. Assuming that the amount of income earned on individual policies is independent, find the mean and standard deviation of V. (If you compare with the results of Exercise 57, you should see that averaging over more insured individuals reduces risk.)

59. **Time and motion** A time-and-motion study measures the time required for an assembly-line worker to perform a repetitive task. The data show that the time required to bring a part from a bin to its position on an automobile chassis varies from car to car according to a Normal distribution with mean 11 seconds and standard deviation 2 seconds. The time required to attach the part to the chassis follows a Normal distribution with mean 20 seconds and standard deviation 4 seconds. The study finds that the times required for the two steps are independent. A part that takes a long time to position, for example, does not take more or less time to attach than other parts.

(a) Find the mean and standard deviation of the time required for the entire operation of positioning and attaching the part.

(b) Management's goal is for the entire process to take less than 30 seconds. Find the probability that this goal will be met. Show your work.

60. **Electronic circuit** The design of an electronic circuit for a toaster calls for a 100-ohm resistor and a 250-ohm resistor connected in series so that their resistances add. The components used are not perfectly uniform, so that the actual resistances vary independently according to Normal distributions. The resistance of 100-ohm resistors has mean 100 ohms and standard deviation 2.5 ohms, while that of 250-ohm resistors has mean 250 ohms and standard deviation 2.8 ohms.

(a) What is the distribution of the total resistance of the two components in series?

(b) What is the probability that the total resistance lies between 345 and 355 ohms? Show your work.

pg 375

61. **Auto emissions** The amount of nitrogen oxides (NOX) present in the exhaust of a particular type of car varies from car to car according to a Normal distribution with mean 1.4 grams per mile (g/mi) and standard deviation 0.3 g/mi. Two randomly selected cars of this type are tested. One has 1.1 g/mi of NOX; the other has 1.9 g/mi. The test station attendant finds this difference in emissions between two similar cars surprising. If the NOX levels for two randomly chosen cars of this type are independent, find the probability that the difference is at least as large as the value the attendant observed. Follow the four-step process.

62. **Loser buys the pizza** Leona and Fred are friendly competitors in high school. Both are about to take the ACT college entrance examination. They agree that if one of them scores 5 or more points better than the other, the loser will buy the winner a pizza. Suppose that in fact Fred and Leona have equal ability, so that each score varies Normally with mean 24 and standard deviation 2. (The variation is due to luck in guessing and the accident of the specific questions being familiar to the student.) The two scores are independent. What is the probability that the scores differ by 5 or more points in either direction? Follow the four-step process.

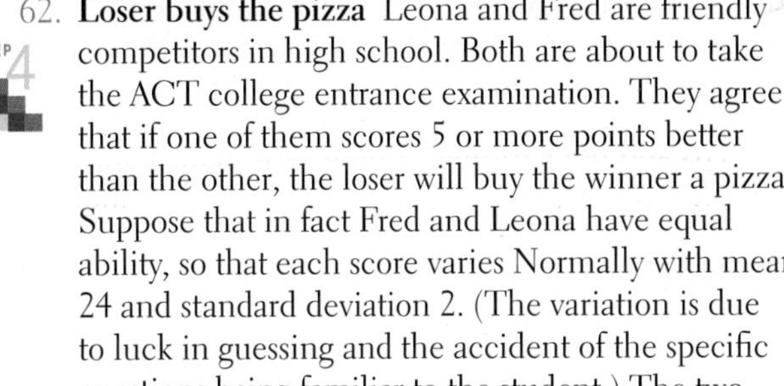

pg 374

63. **Swim team** Hanover High School has the best women's swimming team in the region. The 400-meter freestyle relay team is undefeated this year. In the 400-meter freestyle relay, each swimmer swims 100 meters. The times, in seconds, for the four swimmers this season are approximately Normally distributed with means and standard deviations as shown. Assuming that the swimmer's individual times are independent, find the probability that the total team time in the 400-meter freestyle relay is less than 220 seconds. Follow the four-step process.

Swimmer	Mean	Std. dev.
Wendy	55.2	2.8
Jill	58.0	3.0
Carmen	56.3	2.6
Latrice	54.7	2.7

64. **Toothpaste** Ken is traveling for his business. He has a new 0.85-ounce tube of toothpaste that's supposed to last him the whole trip. The amount of toothpaste Ken squeezes out of the tube each time he brushes varies according to a Normal distribution with mean 0.13 ounces and standard deviation 0.02 ounces. If Ken brushes his teeth six times during the trip, what's the probability that he'll use all the toothpaste in the tube? Follow the four-step process.

Multiple choice: Select the best answer for Exercises 65 and 66, which refer to the following setting. The number of calories in a one-ounce serving of a certain breakfast cereal is a random variable with mean 110 and standard deviation 10. The number of calories in a cup of whole milk is a random variable with mean 140 and standard deviation 12. For breakfast, you eat one ounce of the cereal with 1/2 cup of whole milk. Let T be the random variable that represents the total number of calories in this breakfast.

65. The mean of T is

(a) 110. (b) 140. (c) 180. (d) 195. (e) 250.

66. The standard deviation of T is

(a) 22. (b) 16. (c) 15.62. (d) 11.66. (e) 4.

67. **Statistics for investing** (3.2) Joe's retirement plan invests in stocks through an "index fund" that follows the behavior of the stock market as a whole, as measured by the Standard & Poor's (S&P) 500 stock index. Joe wants to buy a mutual fund that does not track the index closely. He reads that monthly returns from Fidelity Technology Fund have correlation $r = 0.77$ with the S&P 500 index and that Fidelity Real Estate Fund has correlation $r = 0.37$ with the index.

(a) Which of these funds has the closer relationship to returns from the stock market as a whole? How do you know?

(b) Does the information given tell Joe anything about which fund has had higher returns?

68. **Buying stock** (5.3, 6.1) You purchase a hot stock for \$1000. The stock either gains 30% or loses 25% each day, each with probability 0.5. Its returns on consecutive days are independent of each other. You plan to sell the stock after two days.

(a) What are the possible values of the stock after two days, and what is the probability for each value? What is the probability that the stock is worth more after two days than the \$1000 you paid for it?

(b) What is the mean value of the stock after two days?

(*Comment:* You see that these two criteria give different answers to the question "Should I invest?")

6.3

Binomial and Geometric Random Variables

In Section 6.3, you'll learn about:

- Binomial settings and binomial random variables
- Binomial probabilities
- Mean and standard deviation of a binomial distribution
- Binomial distributions in statistical sampling
- Geometric random variables

When the same chance process is repeated several times, we are often interested in whether a particular outcome does or doesn't happen on each repetition. Here are some examples:

- To test whether someone has extrasensory perception (ESP), choose one of four cards at random—a star, wave, cross, or circle. Ask the person to identify the card without seeing it. Do this a total of 50 times and see how many cards the person identifies correctly. *Chance process:* choose a card at random. *Outcome of interest:* person identifies card correctly. *Random variable:* X = number of correct identifications.

- In the game of Pass the Pigs, a player rolls a pair of pig-shaped dice. On each roll, the player earns points according to how the pigs land. If the player gets a "pig out," in which the two pigs land on opposite sides, she loses all points earned in that round and must pass the pigs to the next player. A player can choose to stop rolling at any point during her turn and to keep the points that she has earned before passing the pigs. *Chance process:* roll the pig dice. *Outcome of interest:* pig out. *Random variable:* T = number of rolls until the player pigs out.

• A shipping company claims that 90% of its shipments arrive on time. To test this claim, take a random sample of 100 shipments made by the company last month and see how many arrived on time. *Chance process:* Randomly select a shipment and check when it arrived. *Outcome of interest:* arrived on time. *Random variable:* Y = number of on-time shipments.

Some random variables, like X and Y in the first and third examples above, count the number of times the outcome of interest occurs in a fixed number of repetitions. They are called *binomial random variables.* Other random variables, like T in the Pass the Pigs setting, count the number of repetitions of the chance process it takes for the outcome of interest to occur. They are known as *geometric random variables.* These two special types of discrete random variables are the focus of this section.

Binomial Settings and Binomial Random Variables

What do the following scenarios have in common?

• Toss a coin 5 times. Count the number of heads.

• Spin a roulette wheel 8 times. Record how many times the ball lands in a red slot.

• Take a random sample of 100 babies born in U.S. hospitals today. Count the number of females.

In each case, we're performing repeated *trials* of the same chance process. The number of trials is fixed in advance. In addition, the outcome of one trial has no effect on the outcome of any other trial. That is, the trials are *independent.* We're interested in the number of times that a specific event (we'll call it a "success") occurs. Our chances of getting a "success" are the same on each trial. When these conditions are met, we have a **binomial setting.**

DEFINITION: Binomial setting

A **binomial setting** arises when we perform several independent trials of the same chance process and record the number of times that a particular outcome occurs. The four conditions for a binomial setting are

- **B**inary? The possible outcomes of each trial can be classified as "success" or "failure."
- **I**ndependent? Trials must be independent; that is, knowing the result of one trial must not have any effect on the result of any other trial.
- **N**umber? The number of trials n of the chance process must be fixed in advance.
- **S**uccess? On each trial, the probability p of success must be the same.

The boldface letters in the box give you a helpful way to remember the conditions for a binomial setting: just check the BINS!

When checking the **B**inary condition, note that there can be more than two possible outcomes per trial—a roulette wheel has numbered slots of *three* colors: red, black, and green. If we define "success" as having the ball land in a red slot, then "failure" occurs when the ball lands in a black or a green slot.

Think of tossing a coin n times as an example of the binomial setting. Each toss gives either heads or tails. Knowing the outcome of one toss doesn't change the probability of a head on any other toss, so the tosses are independent. If we call heads a success, then p is the probability of a head and remains the same as long as we toss the same coin. For tossing a coin, p is close to 0.5. The number of heads we count is a **binomial random variable** X. The probability distribution of X is called a **binomial distribution.**

DEFINITION: Binomial random variable and binomial distribution

The count X of successes in a binomial setting is a **binomial random variable.** The probability distribution of X is a **binomial distribution** with parameters n and p, where n is the number of trials of the chance process and p is the probability of a success on any one trial. The possible values of X are the whole numbers from 0 to n.

Later in the section, we'll learn how to assign probabilities to outcomes and how to find the mean and standard deviation of a binomial random variable. For now, it's important to be able to distinguish situations in which a binomial distribution does and doesn't apply.

EXAMPLE

From Blood Types to Aces

Binomial settings and random variables

PROBLEM: Here are three scenarios involving chance behavior. In each case, determine whether the given random variable has a binomial distribution. Justify your answer.

(a) Genetics says that children receive genes from each of their parents independently. Each child of a particular pair of parents has probability 0.25 of having type O blood. Suppose these parents have 5 children. Let X = the number of children with type O blood.

(b) Shuffle a deck of cards. Turn over the first 10 cards, one at a time. Let Y = the number of aces you observe.

(c) Shuffle a deck of cards. Turn over the top card. Put the card back in the deck, and shuffle again. Repeat this process until you get an ace. Let W = the number of cards required.

SOLUTION:

(a) To see if this is a binomial setting, we'll check the BINS:

- Binary? "Success" = has type O blood. "Failure" = doesn't have type O blood.
- Independent? Children inherit genes determining blood type independently from each of their parents.
- Number? There are $n = 5$ trials of this chance process.
- Success? The probability of a success is 0.25 on each trial.

This is a binomial setting. Since X counts the number of successes, it is a binomial random variable with parameters $n = 5$ and $p = 0.25$.

(b) Let's check the BINS:

- Binary? "Success" = get an ace. "Failure" = don't get an ace.
- Independent? No. If the first card you turn over is an ace, then the next card is less likely to be an ace because you're not replacing the top card in the deck. Similarly, if the first card isn't an ace, the second card is more likely to be an ace.

Because the trials are not independent, this is not a binomial setting.

(c) Let's check the BINS:

- Binary? "Success" = get an ace. "Failure" = don't get an ace.
- Independent? Because you are replacing the card in the deck and shuffling each time, the result of one trial does not affect the outcome of any other trial.
- Number? The number of trials is not set in advance. You could get an ace on the first card you turn over, or it may take many cards to get an ace.

Because there is no fixed number of trials, this is not a binomial setting.

For Practice *Try Exercises 71 and 73*

Part (c) of the example raises an important point about binomial random variables. In addition to checking the BINS, make sure that you're being asked to count the number of successes in a certain number of trials. In part (c), you're asked to count the number of *trials* until you get a success. That can't be a binomial random variable. (As you'll see later, W is actually a *geometric* random variable.)

CHECK YOUR UNDERSTANDING

For each of the following situations, determine whether the given random variable has a binomial distribution. Justify your answer.

1. Shuffle a deck of cards. Turn over the top card. Put the card back in the deck, and shuffle again. Repeat this process 10 times. Let X = the number of aces you observe.
2. Choose three students at random from your class. Let Y = the number who are over 6 feet tall.
3. Flip a coin. If it's heads, roll a 6-sided die. If it's tails, roll an 8-sided die. Repeat this process 5 times. Let W = the number of 5s you roll.

Binomial Probabilities

In a binomial setting, we can define a random variable (say, X) as the number of successes in n independent trials. What's the probability distribution of X? Let's see if an example can help shed some light on this question.

Inheriting Blood Type

Calculating binomial probabilities

Each child of a particular pair of parents has probability 0.25 of having type O blood. Genetics says that children receive genes from each of their parents independently. If these parents have 5 children, the count X of children with type O blood is a binomial random variable with $n = 5$ trials and probability $p = 0.25$ of a success on each trial. In this setting, a child with type O blood is a "success" (S) and a child with another blood type is a "failure" (F).

What's $P(X = 0)$? That is, what's the probability that *none* of the 5 children has type O blood? It's the chance that all 5 children *don't* have type O blood. The probability that any one of this couple's children doesn't have type O blood is $1 - 0.25 = 0.75$ (complement rule). By the multiplication rule for independent events (Chapter 5),

$$P(X = 0) = P(\text{FFFFF}) = (0.75)(0.75)(0.75)(0.75)(0.75) = (0.75)^5 = 0.2373$$

How about $P(X = 1)$? There are several different ways in which exactly 1 of the 5 children could have type O blood. For instance, the first child born might have type O blood, while the remaining 4 children don't have type O blood. The probability that this happens is

$$P(\text{SFFFF}) = (0.25)(0.75)(0.75)(0.75)(0.75) = (0.25)(0.75)^4$$

Alternatively, Child 2 could be the one that has type O blood. The corresponding probability is

$$P(\text{FSFFF}) = (0.75)(0.25)(0.75)(0.75)(0.75) = (0.25)(0.75)^4$$

There are three more possibilities to consider:

$$P(\text{FFSFF}) = (0.75)(0.75)(0.25)(0.75)(0.75) = (0.25)(0.75)^4$$

$$P(\text{FFFSF}) = (0.75)(0.75)(0.75)(0.25)(0.75) = (0.25)(0.75)^4$$

$$P(\text{FFFFS}) = (0.75)(0.75)(0.75)(0.75)(0.25) = (0.25)(0.75)^4$$

In all, there are five different ways in which exactly 1 child would have type O blood, each with the same probability of occurring. As a result,

$$\begin{aligned} P(X = 1) &= P(\text{exactly 1 child with type O blood}) \\ &= P(\text{SFFFF}) + P(\text{FSFFF}) + P(\text{FFSFF}) + P(\text{FFFSF}) + P(\text{FFFFS}) \\ &= 5(0.25)(0.75)^4 = 0.39551 \end{aligned}$$

There's about a 40% chance that exactly 1 of the couple's 5 children will have type O blood.

Let's continue with the scenario from the previous example. What if we wanted to find $P(X = 2)$, the probability that exactly 2 of the couple's children have type O blood? Because the method doesn't depend on the specific setting, we use "S" for success and "F" for failure for short. Do the work in two steps, as shown in the example.

Step 1. Find the probability that a specific 2 of the 5 tries—say, the first and the third—give successes. This is the outcome SFSFF. Because tries are independent, the multiplication rule for independent events applies. The probability we want is

$$\begin{aligned} P(\text{SFSFF}) &= P(\text{S})P(\text{F})P(\text{S})P(\text{F})P(\text{F}) \\ &= (0.25)(0.75)(0.25)(0.75)(0.75) \\ &= (0.25)^2(0.75)^3 \end{aligned}$$

Step 2. Observe that *any one arrangement* of 2 S's and 3 F's has this same probability. This is true because we multiply together 0.25 twice and 0.75 three times whenever we have 2 S's and 3 F's. The probability that $X = 2$ is the probability of getting 2 S's and 3 F's in any arrangement whatsoever. Here are all the possible arrangements:

SSFFF SFSFF SFFSF SFFFS FSSFF

FSFSF FSFFS FFSSF FFSFS FFFSS

There are 10 of them, all with the same probability. The overall probability of 2 successes is therefore

$$P(X = 2) = 10(0.25)^2(0.75)^3 = 0.26367$$

The pattern of this calculation works for any binomial probability. That is,

$$\begin{aligned} P(X = k) &= P(\text{exactly } k \text{ successes in } n \text{ trials}) \\ &= \text{number of arrangements} \cdot p^k(1 - p)^{n-k} \end{aligned}$$

To use this formula, we must count the number of arrangements of k successes in n observations. This number is called the **binomial coefficient**. We use the following fact to do the counting without actually listing all the arrangements.

DEFINITION: Binomial coefficient

The number of ways of arranging k successes among n observations is given by the **binomial coefficient**

$$\binom{n}{k} = \frac{n!}{k!(n-k)!}$$

for $k = 0, 1, 2, \ldots, n$ where

$$n! = n(n-1)(n-2) \cdot \ldots \cdot (3)(2)(1)$$

and $0! = 1$.

Factorial

The formula for binomial coefficients uses the **factorial** notation. For any positive whole number n, its factorial $n!$ is

$$n! = n(n-1)(n-2) \cdot \ldots \cdot (3)(2)(1)$$

In addition, we define $0! = 1$.

The larger of the two factorials in the denominator of a binomial coefficient will cancel much of the $n!$ in the numerator. For example, the binomial coefficient we need in order to find the probability that exactly 2 of the couple's 5 children inherit type O blood is

$$\binom{5}{2} = \frac{5!}{2!3!} = \frac{(5)(4)(\cancel{3})(\cancel{2})(\cancel{1})}{(2)(1)(\cancel{3})(\cancel{2})(\cancel{1})} = \frac{(5)(4)}{(2)(1)} = 10$$

CAUTION

The binomial coefficient $\binom{5}{2}$ is not related to the fraction $\frac{5}{2}$. A helpful way to remember its meaning is to read it as "5 choose 2." Binomial coefficients have many uses, but we are interested in them only as an aid to finding binomial probabilities. If you need to compute a binomial coefficient, use your calculator.

Some people prefer the notation ${}_5C_2$ instead of $\binom{5}{2}$ for the binomial coefficient.

TECHNOLOGY CORNER Binomial coefficients on the calculator

To calculate a binomial coefficient like $\binom{5}{2}$ on the TI-83/84 or TI-89, proceed as follows:

TI-83/84: Type 5, press MATH, arrow over to `PRB`, choose `3:nCr`, and press ENTER. Then type 2 and press ENTER again to execute the command `5 nCr 2`.

TI-89: From the home screen, press 2nd 5 (MATH), choose `7:Probability`, `3:nCr(`, and press ENTER. Complete the command `nCr(5,2)` and press ENTER.

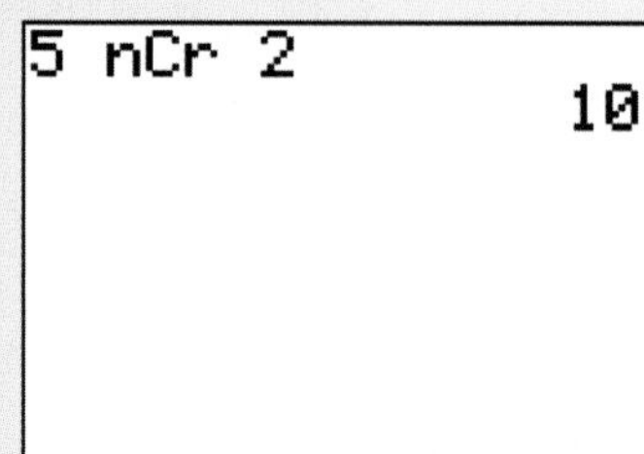

TI-Nspire instructions in Appendix B

The binomial coefficient $\binom{n}{k}$ counts the number of different ways in which k successes can be arranged among n trials. The binomial probability $P(X = k)$ is this count multiplied by the probability of any one specific arrangement of the k successes. Here is the result we seek.

Binomial Probability

If X has the binomial distribution with n trials and probability p of success on each trial, the possible values of X are 0, 1, 2, . . . , n. If k is any one of these values,

$$P(X = k) = \binom{n}{k} p^k (1 - p)^{n-k}$$

With our formula in hand, we can now calculate any binomial probability.

EXAMPLE

Inheriting Blood Type

Using the binomial probability formula

PROBLEM: Each child of a particular pair of parents has probability 0.25 of having type O blood. Suppose the parents have 5 children.

(a) Find the probability that exactly 3 of the children have type O blood.

(b) Should the parents be surprised if more than 3 of their children have type O blood? Justify your answer.

SOLUTION: Let X = the number of children with type O blood. We know that X has a binomial distribution with $n = 5$ and $p = 0.25$.

(a) The probability that exactly 3 of the children have type O blood is

$$P(X = 3) = \binom{5}{3}(0.25)^3(0.75)^2 = 10(0.25)^3(0.75)^2 = 0.08789$$

(b) To answer this question, we need to find $P(X > 3)$.

$$P(X > 3) = P(X = 4) + P(X = 5) = \binom{5}{4}(0.25)^4(0.75)^1 + \binom{5}{5}(0.25)^5(0.75)^0$$

$$= 5(0.25)^4(0.75)^1 + 1(0.25)^5(0.75)^0 = 0.01465 + 0.00098 = 0.01563$$

Since there's only about a 1.5% chance of having more than 3 children with type O blood, the parents should definitely be surprised if this happens.

For Practice *Try Exercise 79*

We could also use the calculator's `binompdf` and `binomcdf` commands to perform the calculations in the previous example. The following Technology Corner shows how to do it.

TECHNOLOGY CORNER Binomial probability on the calculator

There are two handy commands on the TI-83/84 and TI-89 for finding binomial probabilities:

`binompdf(n,p,k)` computes $P(X = k)$
`binomcdf(n,p,k)` computes $P(X \leq k)$

These two commands can be found in the distributions menu (2ND/VARS) on the TI-83/84 and in the CATALOG under Flash Apps on the TI-89.

For the parents having $n = 5$ children, each with probability $p = 0.25$ of type O blood:

$P(X = 3) =$ `binompdf(5,0.25,3)` $= 0.08789$

To find $P(X > 3)$, we used the complement rule:

$P(X > 3) = 1 - P(X \leq 3)$
$=$ `1-binomcdf(5,0.25,3)`
$= 0.01563$

```
binompdf(5,.25,3
)
            .087890625
1-binomcdf(5,.25
,3)
              .015625
```

Of course, we could also have done this as

$P(X > 3) = P(X = 4) + P(X = 5)$
$=$ `binompdf(5,0.25,4)` $+$ `binompdf(5,0.25,5)`
$= 0.01465 + 0.00098 = 0.01563$

TI-Nspire instructions in Appendix B

When n is very large, the binomial coefficient $\binom{n}{k}$ is too large even for a graphing calculator to compute. (Try `500 nCr 100` to see what we mean.) Fortunately, we can still use `binompdf` and `binomcdf` to calculate binomial probabilities for large values of n.

AP EXAM TIP Don't rely on "calculator speak" when showing your work on free-response questions. Writing `binompdf(5,0.25,3)` $= 0.08789$ will *not* earn you full credit for a binomial probability calculation. At the very least, you must indicate what each of those calculator inputs represents. For example, "I used binompdf(5,0.25,3) on my calculator with $n = 5$, $p = 0.25$, and $k = 3$." Better yet, show the binomial probability formula with these numbers plugged in.

CHECK YOUR UNDERSTANDING

To introduce her class to binomial distributions, Mrs. Desai gives a 10-item, multiple-choice quiz. The catch is, students must simply guess an answer (A through E) for each question. Mrs. Desai uses her computer's random number generator to produce the answer key, so that each possible answer has an equal chance to be chosen. Patti is one of the students in this class. Let X = the number of Patti's correct guesses.

1. Show that X is a binomial random variable.
2. Find $P(X = 3)$. Explain what this result means.
3. To get a passing score on the quiz, a student must guess correctly at least 6 times. Would you be surprised if Patti earned a passing score? Compute an appropriate probability to support your answer.

Mean and Standard Deviation of a Binomial Distribution

What does the probability distribution of a binomial random variable look like? The table below shows the possible values and corresponding probabilities for X = the number of children with type O blood. This is a binomial random variable with $n = 5$ and $p = 0.25$. Figure 6.15 shows a histogram of the probability distribution.

Value x_i:	0	1	2	3	4	5
Probability p_i:	0.23730	0.39551	0.26367	0.08789	0.01465	0.00098

Let's describe what we see.

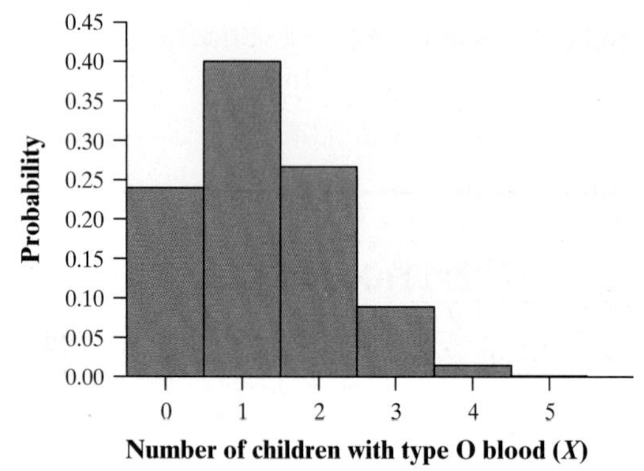

FIGURE 6.15 Histogram showing the probability distribution of the binomial random variable X = number of children with type O blood in a family with 5 children.

Shape: The probability distribution of X is skewed to the right. Since the chance that any one of the couple's children inherits type O blood is 0.25, it's quite likely that 0, 1, or 2 of the children will have type O blood. Larger values of X are much less likely.

Center: The median number of children with type O blood is 1 because that's where the 50th percentile of the distribution falls. How about the mean? It's

$$\mu_X = \sum x_i p_i = (0)(0.23730)+(1)(0.39551)+\cdots+(5)(0.00098) = 1.25$$

So the expected number of children with type O blood is 1.25.

Spread: The variance of X is

$$\sigma_X^2 = \sum (x_i - \mu_X)^2 p_i$$
$$= (0 - 1.25)^2(0.23730) + (1-1.25)^2(0.39551) + \cdots + (5-1.25)^2(0.00098) = 0.9375$$

So the standard deviation of X is

$$\sigma_X = \sqrt{0.9375} = 0.968$$

Did you think about why the mean is $\mu_X = 1.25$? Since each child has a 0.25 chance of inheriting type O blood, we'd expect one-fourth of the 5 children to have this blood type. In other words, $\mu_X = 5(0.25) = 1.25$. This method can be used to find the mean of any binomial random variable with parameters n and p:

$$\mu_X = np$$

There are fairly simple formulas for the variance and standard deviation, too, but they aren't as easy to explain:

$$\sigma_X^2 = np(1-p) \quad \text{and} \quad \sigma_X = \sqrt{np(1-p)}$$

For our family with 5 children,

$$\sigma_X^2 = 5(0.25)(0.75) = 0.9375 \quad \text{and} \quad \sigma_X = \sqrt{0.9375} = 0.968$$

Mean and Standard Deviation of a Binomial Random Variable

If a count X has the binomial distribution with number of trials n and probability of success p, the **mean** and **standard deviation** of X are

$$\mu_X = np$$
$$\sigma_X = \sqrt{np(1-p)}$$

Remember that these formulas work only for binomial distributions. They can't be used for other distributions.

Where do the binomial mean and variance formulas come from? We can derive the formulas for the mean and variance of a binomial distribution using what we learned about combining random variables in Section 6.2. Let's start with the random variable B that's described by the following probability distribution:

Value b_i:	0	1
Probability p_i:	$1-p$	p

You can think of B as representing the result of a single trial of some chance process. If a success occurs (probability p), then $B = 1$. If a failure occurs, then $B = 0$. Notice that the mean of B is

$$\mu_B = \sum b_i p_i = (0)(1-p) + (1)(p) = p$$

and that the variance of B is

$$\begin{aligned}\sigma_B^2 = \sum (b_i - \mu_B)^2 p_i &= (0-p)^2(1-p) + (1-p)^2 p \\ &= p(1-p)[p + (1-p)] \\ &= p(1-p)\end{aligned}$$

Now consider the random variable $X = B_1 + B_2 + \cdots + B_n$. We can think of X as counting the number of successes in n independent trials of this chance process, with each trial having success probability p. In other words, X is a binomial random variable with parameters n and p. By the rules from Section 6.2, the mean of X is

$$\mu_X = \mu_{B_1} + \mu_{B_2} + \cdots + \mu_{B_n} = p + p + \cdots + p = np$$

and the variance of X is

$$\begin{aligned}\sigma_X^2 &= \sigma_{B_1}^2 + \sigma_{B_2}^2 + \cdots + \sigma_{B_n}^2 \\ &= p(1-p) + p(1-p) + \cdots + p(1-p) \\ &= np(1-p)\end{aligned}$$

EXAMPLE

Bottled Water versus Tap Water

Binomial distribution in action

Mr. Bullard's AP Statistics class did the Activity on page 340. There are 21 students in the class. If we assume that the students in his class *cannot* tell tap water from bottled water, then each one is basically guessing, with a 1/3 chance of being correct. Let X = the number of students who correctly identify the cup containing the different type of water.

PROBLEM:

(a) Explain why X is a binomial random variable.

(b) Find the mean and standard deviation of X. Interpret each value in context.

(c) Of the 21 students in the class, 13 made correct identifications. Are you convinced that Mr. Bullard's students can tell bottled water from tap water? Justify your answer.

SOLUTION:

(a) Assuming that students are just guessing, the Activity consists of 21 repetitions of this chance process. Let's check the BINS:

- Binary: On each trial, "success" = correct identification; "failure" = incorrect identification.
- Independent: One student's result should have no effect on any other student's result.
- Number: There are $n = 21$ trials.
- Success: Each student has a $p = 1/3$ chance of guessing correctly.

Since X is counting the number of successes in this binomial setting, it is a binomial random variable.

(b) The mean of X is

$$\mu_X = np = 21(1/3) = 7$$

We'd expect about one-third of his 21 students, that is, about 7 students, to guess correctly. The standard deviation of X is

$$\sigma_X = \sqrt{np(1-p)} = \sqrt{21(1/3)(2/3)} = 2.16$$

If the Activity were repeated many times with groups of 21 students who were just guessing, the number of correct identifications would differ from 7 by an average of 2.16.

(c) The class's result corresponds to $X = 13$, a value that's nearly 3 standard deviations above the mean. How likely is it that 13 or more of Mr. Bullard's students would guess correctly? It's

$$P(X \geq 13) = 1 - P(X \leq 12)$$

Using the calculator's `binomcdf` command with $n = 21$, $p = 1/3$, and $k = 12$:

$$P(X \geq 13) = 1 - 0.9932 = 0.0068$$

The students have less than a 1% chance of getting so many right if they are all just guessing. This is strong evidence that some of the students in the class can tell bottled water from tap water.

For Practice *Try Exercise 85*

Figure 6.16 shows the probability distribution for the number of correct guesses in Mr. Bullard's class if no one can tell bottled water from tap water. As you can see from the graph, the chance of 13 or more guessing correctly is quite small.

Although the histogram is slightly right-skewed (there's a long tail that extends out to $X = 21$), it looks like a Normal density curve might fit the bulk of the distribution pretty well.

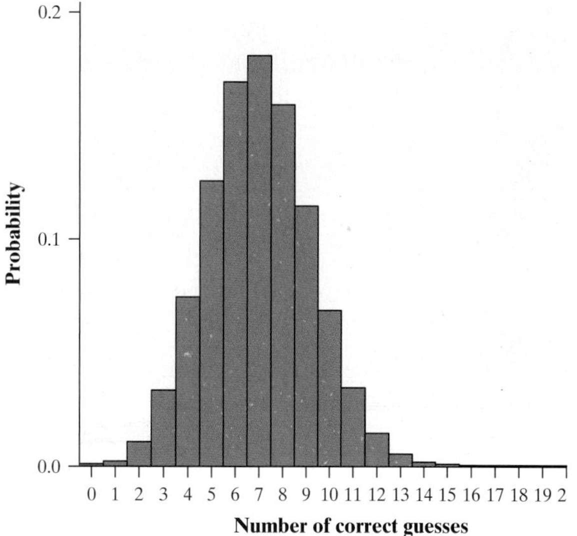

FIGURE 6.16 Histogram of the probability distribution for the binomial random variable X = number of correct guesses in Mr. Bullard's class.

CHECK YOUR UNDERSTANDING

Refer to the previous Check Your Understanding (page 390) about Mrs. Desai's special multiple-choice quiz on binomial distributions. We defined X = the number of Patti's correct guesses.

1. Find μ_X. Interpret this value in context.
2. Find σ_X. Interpret this value in context.
3. What's the probability that the number of Patti's correct guesses is more than 2 standard deviations above the mean? Show your method.

Binomial Distributions in Statistical Sampling

The binomial distributions are important in statistics when we wish to make inferences about the proportion p of successes in a population. Here is an example involving a familiar product.

Choosing an SRS of CDs

Binomial distributions and sampling

A music distributor inspects an SRS of 10 CDs from a shipment of 10,000 music CDs. Suppose that (unknown to the distributor) 10% of the CDs in the shipment have defective copy-protection schemes that will harm personal computers. Count the number X of bad CDs in the sample.

This is not quite a binomial setting. Removing 1 CD changes the proportion of bad CDs remaining in the shipment. So the probability that the second CD chosen is bad changes when we know whether the first is good or bad. But removing 1 CD from a shipment of 10,000 changes the makeup of the remaining 9999 CDs very little. In practice, the distribution of X is very close to the binomial distribution with $n = 10$ and $p = 0.1$. To illustrate this, let's compute the probability that none of the 10 CDs is defective. Using the binomial distribution, it's

$$P(X = 0) = \binom{10}{0}(0.10)^0(0.90)^{10} = 0.3487$$

The actual probability of getting no defective CDs is

$$P(\text{no defectives}) = \frac{9000}{10000} \cdot \frac{8999}{9999} \cdot \frac{8998}{9998} \cdot \ldots \cdot \frac{8991}{9991} = 0.3485$$

Those two probabilities are pretty close!

As the previous example illustrates, sampling without replacement leads to a violation of the independence condition. Almost all real-world sampling, such as taking an SRS from a population of interest, is done without replacement. The CDs example shows how we can use binomial distributions in the statistical setting of selecting an SRS. When the population is much larger than the sample, a count of successes in an SRS of size n has approximately the binomial distribution with n equal to the sample size and p equal to the proportion of successes in the population. What counts as "much larger"? In practice, the binomial distribution gives a good approximation as long as we don't sample more than 10% of the population.

Sampling Without Replacement Condition

When taking an SRS of size n from a population of size N, we can use a binomial distribution to model the count of successes in the sample as long as $n \leq \frac{1}{10}N$.

Here's an example that shows why it's important to check the sampling without replacement condition before calculating a binomial probability. You might recognize the setting from the first activity in the book (page 5).

EXAMPLE

Hiring Discrimination—It Just Won't Fly!

Sampling without replacement

An airline has just finished training 25 first officers—15 male and 10 female—to become captains. Unfortunately, only eight captain positions are available right now. Airline managers decide to use a lottery to determine which pilots will fill the available positions. Of the 8 captains chosen, 5 are female and 3 are male.

PROBLEM: **Explain why the probability that 5 female pilots are chosen in a fair lottery is *not***

$$P(X=5)=\binom{8}{5}(0.40)^5(0.60)^3=0.124$$

(The correct probability is 0.106.)

SOLUTION: **The managers are sampling without replacement when they do the lottery. There's a 0.40 chance that the first pilot selected for a captain position is female. Once that person is chosen, the probability that the next pilot selected will be female is no longer 0.40. The binomial formula assumes that the probability of success stays constant at 0.40 throughout the eight trials of this chance process. This calculation will be approximately correct if the success probability doesn't change too much—as long as we don't sample more than 10% of the population. In this case, managers are sampling 8 out of 25 pilots—almost 1/3 of the population. That explains why the binomial probability is off by about 17% (0.018/0.106) from the correct answer.**

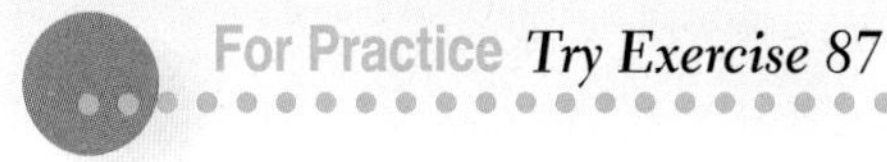

The Normal approximation to binomial distributions* As n gets larger, something interesting happens to the shape of the binomial distribution. Figure 6.17 shows histograms of binomial distributions for different values of n and p. As the number of observations n gets larger, the binomial distribution gets close to a Normal distribution. You can investigate the relationship between n and p yourself using the *Normal Approximation to Binomial* applet at the book's Web site.

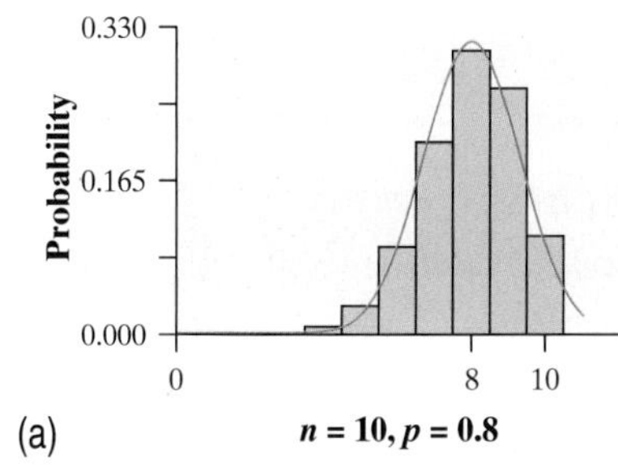

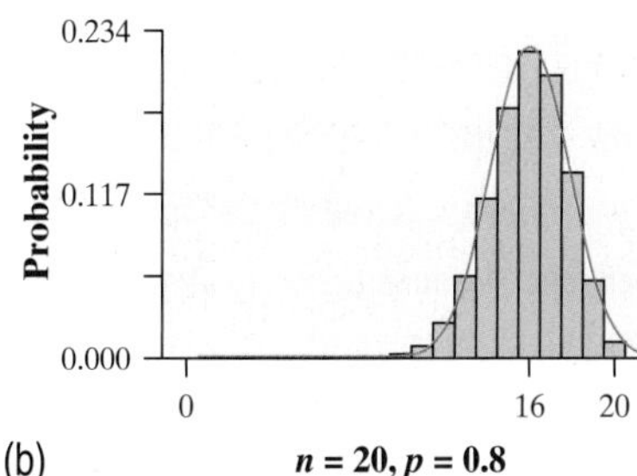

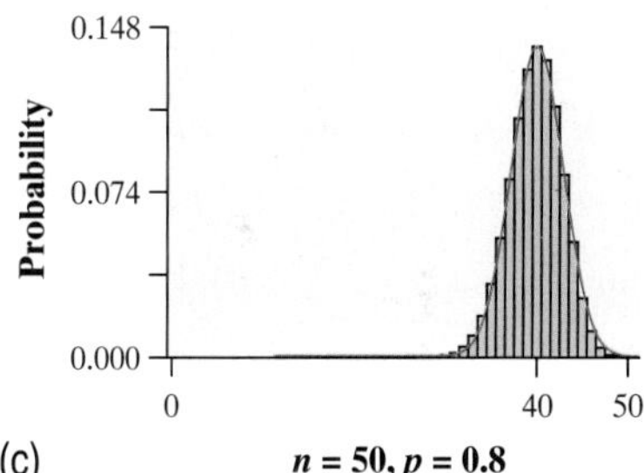

FIGURE 6.17 Histograms of binomial distributions with (a) $n = 10$ and $p = 0.8$, (b) $n = 20$ and $p = 0.8$, and (c) $n = 50$ and $p = 0.8$. As n increases, the shape of the probability distribution gets more and more Normal.

When n is large, we can use Normal probability calculations to approximate binomial probabilities. Here are the facts.

Normal Approximation for Binomial Distributions

Suppose that a count X has the binomial distribution with n trials and success probability p. When n is large, the distribution of X is approximately Normal with

$$\text{mean: } \mu_X = np \quad \text{and} \quad \text{standard deviation: } \sigma_X = \sqrt{np(1-p)}$$

As a rule of thumb, we will use the Normal approximation when n is so large that

$$np \geq 10 \qquad \text{and} \qquad n(1-p) \geq 10$$

That is, the expected number of successes and failures are both at least 10.

---

*This topic is not required for the AP Statistics exam. Some teachers prefer to discuss this topic when presenting the sampling distribution of $\hat{p}$ (Chapter 7).

The Normal approximation is easy to remember because it says to act as if X is Normal with exactly the same mean and standard deviation as the binomial. The accuracy of the Normal approximation improves as the sample size n increases. It is most accurate for any fixed n when p is close to 1/2 and least accurate when p is near 0 or 1. This is why the rule of thumb in the box depends on p as well as n.

EXAMPLE

Attitudes toward Shopping

Normal approximation to a binomial

Sample surveys show that fewer people enjoy shopping than in the past. A survey asked a nationwide random sample of 2500 adults if they agreed or disagreed that "I like buying new clothes, but shopping is often frustrating and time-consuming."[9] The population that the poll wants to draw conclusions about is all U.S. residents aged 18 and over.

PROBLEM: Suppose that exactly 60% of all adult U.S. residents would say "Agree" if asked the same question. Let $X =$ the number in the sample who agree.

(a) Show that X is approximately a binomial random variable.

(b) Check the conditions for using a Normal approximation in this setting.

(c) Use a Normal distribution to estimate the probability that 1520 or more of the sample agree.

SOLUTION:

(a) Let's check the BINS.

- Binary: Success = agree that shopping is frustrating, failure = don't agree.
- Independent: Because there are about 235 million adults in the United States, the responses of 2500 randomly chosen adults are very close to independent.
- Number: There are $n = 2500$ trials of this chance process.
- Success: The probability of selecting an adult who agrees is $p = 0.6$.

So the number in our sample who agree that shopping is frustrating is a random variable X having roughly the binomial distribution with $n = 2500$ and $p = 0.6$.

(b) Since $np = 2500(0.6) = 1500$ and $n(1 - p) = 2500(0.4) = 1000$ are both at least 10, we should be safe using the Normal approximation.

(c) Act as though the count X has the Normal distribution with the same mean and standard deviation as the binomial distribution:

$$\mu = np = 2500(0.6) = 1500 \qquad \text{and} \qquad \sigma = \sqrt{np(1-p)} = \sqrt{2500(0.6)(0.4)} = 24.49$$

Standardizing X gives

$$z = \frac{1520 - 1500}{24.49} = 0.82$$

The probability we want is

$$P(X \geq 1520) = P(Z \geq 0.82) = 1 - 0.7939 = 0.2061$$

For Practice Try Exercise 91

Figure 6.18 is a probability histogram of the binomial distribution from the example. As the Normal approximation suggests, the shape of the distribution looks Normal. The probability we want is the sum of the areas of the shaded bars. We can get this answer using the command `1-binomcdf(2500,0.6,1519)`, which yields 0.2131. The Normal approximation, 0.2061, misses the more accurate binomial probability by about 0.007.

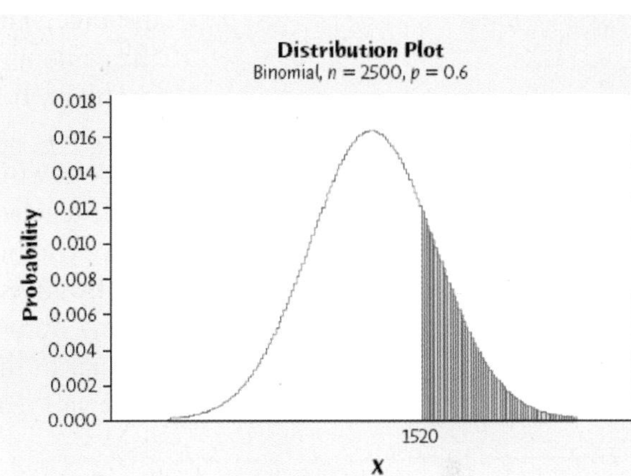

FIGURE 6.18 Probability histogram from Minitab for the binomial distribution with $n = 2500$, $p = 0.6$. The bars at and above 1520 are shaded to highlight the probability of getting at least 1520 successes. The shape of this binomial probability distribution closely resembles a Normal curve.

Geometric Random Variables

In a binomial setting, the number of trials n is fixed in advance, and the binomial random variable X counts the number of successes. The possible values of X are 0, 1, 2, . . . , n. In other situations, the goal is to repeat a chance process *until a success occurs:*

- Roll a pair of dice until you get doubles.
- In basketball, attempt a three-point shot until you make one.
- Keep placing a \$1 bet on the number 15 in roulette until you win.

These are all examples of a **geometric setting.** Although the number of trials isn't fixed in advance, the trials are independent and the probability of success remains constant.

The boldface letters in the box give you a helpful way to remember the conditions for a geometric setting: just check the BITS!

DEFINITION: Geometric setting

A **geometric setting** arises when we perform independent trials of the same chance process and record the number of trials until a particular outcome occurs. The four conditions for a geometric setting are

- **B**inary? The possible outcomes of each trial can be classified as "success" or "failure."
- **I**ndependent? Trials must be independent; that is, knowing the result of one trial must not have any effect on the result of any other trial.
- **T**rials? The goal is to count the number of trials until the first success occurs.
- **S**uccess? On each trial, the probability p of success must be the same.

Here's an Activity your class can try that involves a geometric setting.

ACTIVITY *The Birth Day Game*

Your teacher is planning to give you 10 problems for homework. As an alternative, you can agree to play the Birth Day Game. Here's how it works. A student will be selected at random from your class and asked to guess the day of the week (for instance, Thursday) on which one of your teacher's friends was born. If the student guesses correctly, then the class will have only one homework problem. If the student guesses the wrong day of the week, your teacher will once again select a student from the class at random. The chosen student will try to guess the day of the week on which a *different* one of your teacher's friends was born. If this student gets it right, the class will have two homework problems. The game continues until a student correctly guesses the day on which one of your teacher's (many) friends was born. Your teacher will assign a number of homework problems that is equal to the total number of guesses made by members of your class. Are you ready to play the Birth Day Game?

In a geometric setting, if we define the random variable Y to be the number of trials needed to get the first success, then Y is called a **geometric random variable.** The probability distribution of Y is a **geometric distribution.**

DEFINITION: Geometric random variable and geometric distribution

The number of trials Y that it takes to get a success in a geometric setting is a **geometric random variable.** The probability distribution of Y is a **geometric distribution** with parameter p, the probability of a success on any trial. The possible values of Y are $1, 2, 3, \ldots$.

As with binomial random variables, it's important to be able to distinguish situations in which a geometric distribution does and doesn't apply.

The Birth Day Game

Geometric settings and random variables

The random variable of interest in this game is Y = the number of guesses it takes to correctly match the birth day of one of your teacher's friends. Each guess is one trial of the chance process. Let's check the conditions for a geometric setting:

- **B**inary: "Success" = correct guess; "failure" = incorrect guess.
- **I**ndependent: The result of one student's guess has no effect on the result of any other guess because each student is trying to guess a different person's birth day.

- Trials: We're counting the number of trials up to and including the first correct guess.
- Success: On each trial, the probability of a correct guess is 1/7.

This is a geometric setting. Because Y counts the number of trials to get the first success, it is a geometric random variable with parameter $p = 1/7$.

What is the probability that the first student guesses correctly and wins the Birth Day Game? It's $P(Y = 1) = 1/7$. That's also the class's chance of getting only one homework problem. For the class to have two homework problems, the first student selected must guess an incorrect day of the week and the second student must guess the next friend's birth day correctly. The probability that this happens is $P(Y = 2) = (6/7)(1/7) = 0.1224$. Likewise, $P(Y = 3) = (6/7)(6/7)(1/7) = 0.1050$. In general, the probability that the first correct guess comes on the kth trial is $P(Y = k) = (6/7)^{k-1}(1/7)$. Let's summarize what we've learned about calculating a **geometric probability.**

Geometric Probability

If Y has the geometric distribution with probability p of success on each trial, the possible values of Y are 1, 2, 3, ... If k is any one of these values,

$$P(Y = k) = (1 - p)^{k-1}p$$

With our formula in hand, we can now compute any geometric probability.

EXAMPLE

The Birth Day Game

Calculating geometric probabilities

PROBLEM: **Let the random variable Y be defined as in the previous example.**

(a) Find the probability that the class receives 10 homework problems as a result of playing the Birth Day Game.

(b) Find $P(Y < 10)$ and interpret this value in context.

SOLUTION: Y = the number of attempts until a correct guess = the number of homework problems.

(a) $P(Y = 10) = (6/7)^9(1/7) = 0.0357$.

(b) $P(Y < 10) = P(Y = 1) + P(Y = 2) + P(Y = 3) + \cdots + P(Y = 9) = 1/7 + (6/7)(1/7) + (6/7)^2(1/7) + \cdots + (6/7)^8(1/7) = 0.7503$. There's about a 75% chance that the class will get less homework by playing the Birth Day Game.

For Practice *Try Exercise 97*

As you probably guessed, we used the calculator's `geometpdf` and `geometcdf` commands for the computations in the previous example. The following Technology Corner shows you how we did it.

TECHNOLOGY CORNER Geometric probability on the calculator

There are two handy commands on the TI-83/84 and TI-89 for finding geometric probabilities:

`geometpdf(p,k)` computes $P(Y = k)$
`geometcdf(p,k)` computes $P(Y \leq k)$

These two commands can be found in the distributions menu (2ND/VARS) on the TI-83/84 and in the CATALOG under Flash Apps on the TI-89.

For the Birth Day Game, with probability of success $p = 1/7$ on each trial:

$P(Y = 10) =$ `geometpdf(1/7,10)` $= 0.0357$

To find $P(Y < 10)$, use `geometcdf`:

$P(Y < 10) = P(Y \leq 9) =$ `geometcdf(1/7,9)` $= 0.7503$

```
geometpdf(1/7,10
)
            .0356763859
geometcdf(1/7,9)
            .7502652985
```

TI-Nspire instructions in Appendix B

The table below shows part of the probability distribution of Y. We can't show the entire distribution, because the number of trials it takes to get the first success could be an incredibly large number.

Value y_i:	1	2	3	4	5	6	7	8	9	...
Probability p_i:	0.143	0.122	0.105	0.090	0.077	0.066	0.057	0.049	0.042	

Figure 6.19 is a histogram of the probability distribution for values of Y from 1 to 26. Let's describe what we see.

Shape: The heavily right-skewed shape is characteristic of any geometric distribution. That's because the most likely value of a geometric random variable is 1. The probability of each successive value decreases by a factor of $(1 - p)$.

Center: The mean of Y is $\mu_Y = 7$. (Due to the infinite number of possible values of Y, the calculation of the mean is beyond the scope of this text.) If the class played

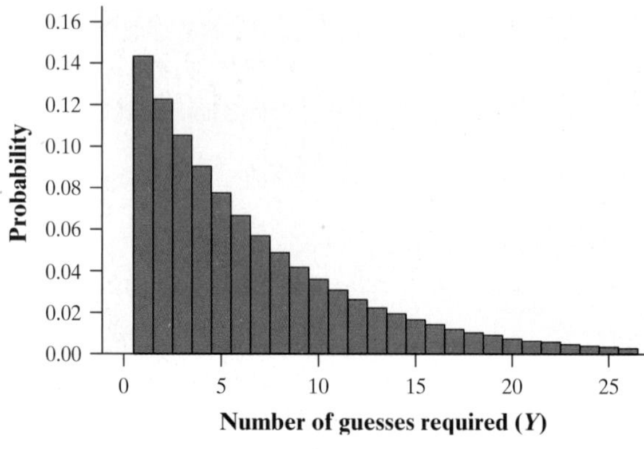

FIGURE 6.19 Histogram showing the probability distribution of the geometric random variable $Y =$ number of trials until students guess correctly in the Birth Day Game.

We didn't include the formula for the standard deviation of a geometric random variable in the summary box. That's because it is difficult to explain and doesn't appear on the AP Statistics course syllabus. In case you were wondering, the formula is

$$\sigma_Y = \sqrt{\frac{1-p}{p^2}}$$

If you substitute $p = 1/7$, you'll see how we got $\sigma_Y = 6.48$ for the Birth Day Game.

the Birth Day Game many times, they would receive an average of 7 homework problems. It's no coincidence that $p = 1/7$ and $\mu_Y = 7$. With probability of success 1/7 on each trial, we'd expect it to take an average of 7 trials to get the first success.

Spread: The standard deviation of Y is $\sigma_Y = 6.48$. If the class played the Birth Day game many times, the number of homework problems they receive would differ from 7 by an average of about 6.5. That could mean a lot of homework!

We can generalize the result for the mean of a geometric random variable.

Mean (Expected Value) of a Geometric Random Variable

If Y is a geometric random variable with probability of success p on each trial, then its **mean** (expected value) is $E(Y) = \mu_Y = \frac{1}{p}$. That is, the expected number of trials required to get the first success is $1/p$.

CHECK YOUR UNDERSTANDING

Suppose you roll a pair of fair, six-sided dice until you get doubles. Let T = the number of rolls it takes.

1. Show that T is a geometric random variable (check the BITS).
2. Find $P(T = 3)$. Interpret this result in context.
3. In the game of Monopoly, a player can get out of jail free by rolling doubles within 3 turns. Find the probability that this happens.

case closed

Does Therapeutic Touch Really Work?

Refer to the chapter-opening Case Study on page 339. The results from Emily's experiment were striking: the subjects correctly identified the location of Emily's hand in only 70 of 150 trials (47%). This is pretty much what we would expect if the subjects were just guessing. Figure 6.20 shows the binomial distribution with $n = 150$ and $p = 0.5$. The yellow area corresponds to the probability of guessing Emily's hand location correctly in 70 or more of the 150 trials: it's 0.815. With a little help from her mother (a nurse) and a statistician, Emily submitted an article summarizing her research to the *Journal of the American Medical Association*. It was published on April 1, 1998.

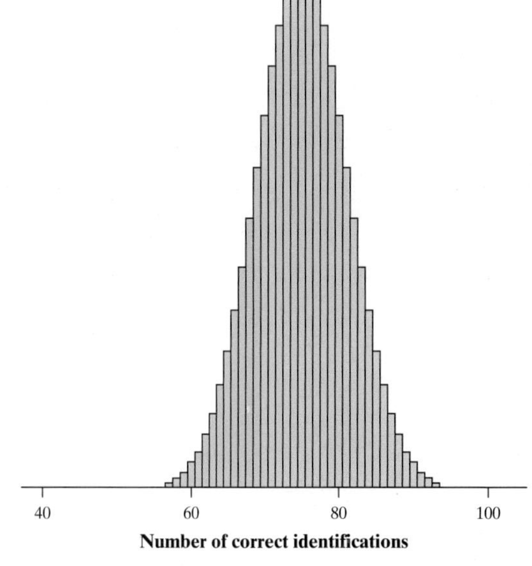

FIGURE 6.20 Binomial distribution of the number of correct guesses in 150 trials if $p = 0.5$.

SECTION 6.3 Summary

- A **binomial setting** consists of n independent trials of the same chance process, each resulting in a success or a failure, with probability of success p on each trial. The count X of successes is a **binomial random variable.** Its probability distribution is a **binomial distribution.**
- The **binomial coefficient**

$$\binom{n}{k} = {}_nC_k = \frac{n!}{k!(n-k)!}$$

counts the number of ways k successes can be arranged among n trials. The **factorial** of n is

$$n! = n(n-1)(n-2) \cdot \ldots \cdot (3)(2)(1)$$

for positive whole numbers n, and $0! = 1$.

- If X has the binomial distribution with parameters n and p, the possible values of X are the whole numbers $0, 1, 2, \ldots, n$. The **binomial probability** of observing k successes in n trials is

$$P(X = k) = \binom{n}{k} p^k (1-p)^{n-k}$$

Binomial probabilities in practice are best found using calculators or software.

- The **mean** and **standard deviation** of a binomial random variable X are

$$\mu_X = np$$
$$\sigma_X = \sqrt{np(1-p)}$$

- The binomial distribution with n trials and probability p of success gives a good approximation to the count of successes in an SRS of size n from a large population containing proportion p of successes. This is true as long as the sample size n is no more than 10% of the population size N.
- The **Normal approximation** to the binomial distribution* says that if X is a count having the binomial distribution with parameters n and p, then when n is large, X is approximately Normally distributed with mean np and standard deviation $\sqrt{np(1-p)}$. We will use this approximation when $np \geq 10$ and $n(1-p) \geq 10$.
- A **geometric setting** consists of repeated trials of the same chance process in which each trial results in a success or a failure; trials are independent;

*This topic is not required for the AP Statistics exam. Some teachers prefer to discuss this topic when presenting the sampling distribution of $\hat{p}$ (Chapter 7).

each trial has the same probability p of success; and the goal is to count the number of trials until the first success occurs. If $Y =$ the number of trials required to obtain the first success, then Y is a **geometric random variable.** Its probability distribution is called a **geometric distribution.**

- If X has the geometric distribution with probability of success p, the possible values of Y are the positive integers 1, 2, 3, ... The **geometric probability** that Y takes any value is

$$P(Y = k) = (1 - p)^{k-1}p$$

- The **mean** (expected value) of a geometric random variable X is $1/p$.

6.3 TECHNOLOGY CORNERS

TI-Nspire instructions in Appendix B

SECTION 6.3 Exercises

Exercises marked with an asterisk * involve the optional material on the Normal approximation to the binomial.

In Exercises 69 to 72, explain whether the given random variable has a binomial distribution.

69. **Sowing seeds** Seed Depot advertises that 85% of its flower seeds will germinate (grow). Suppose that the company's claim is true. Judy buys a packet with 20 flower seeds from Seed Depot and plants them in her garden. Let $X =$ the number of seeds that germinate.

70. **Long or short?** Put the names of all the students in your class in a hat. Mix them up, and draw four names without looking. Let $Y =$ the number whose last names have more than six letters.

pg 384

71. **Lefties** Exactly 10% of the students in a school are left-handed. Select students at random from the school, one at a time, until you find one who is left-handed. Let $V =$ the number of students chosen.

72. **Lefties** Exactly 10% of the students in a school are left-handed. Select 15 students at random from the school and define $W =$ the number who are left-handed.

pg 384

73. **Binomial setting?** A binomial distribution will be approximately correct as a model for one of these two sports settings and not for the other. Explain why by briefly discussing both settings.

(a) A National Football League kicker has made 80% of his field goal attempts in the past. This season he attempts 20 field goals. The attempts differ widely in distance, angle, wind, and so on.

(b) A National Basketball Association player has made 80% of his free-throw attempts in the past. This season he takes 150 free throws. Basketball free throws are always attempted from 15 feet away with no interference from other players.

74. **Binomial setting?** A binomial distribution will be approximately correct as a model for one of these two settings and not for the other. Explain why by briefly discussing both settings.

(a) When an opinion poll calls residential telephone numbers at random, only 20% of the calls reach a person. You watch the random digit dialing machine make 15 calls. X is the number that reach a person.

(b) When an opinion poll calls residential telephone numbers at random, only 20% of the calls reach a live person. You watch the random digit dialing machine make calls. Y is the number of calls until the first live person answers.

75. **Blood types** In the United States, 44% of adults have type O blood. Suppose we choose 7 U.S. adults at random. Let X = the number who have type O blood. Use the binomial probability formula to find $P(X = 4)$. Interpret this result in context.

76. **Rhubarb** Suppose you purchase a bundle of 10 bare-root rhubarb plants. The sales clerk tells you that on average you can expect 5% of the plants to die before producing any rhubarb. Assume that the bundle is a random sample of plants. Let Y = the number of plants that die before producing any rhubarb. Use the binomial probability formula to find $P(Y = 1)$. Interpret this result in context.

77. **Blood types** Refer to Exercise 75. How surprising would it be to get more than 4 adults with type O blood in the sample? Calculate an appropriate probability to support your answer.

78. **Rhubarb** Refer to Exercise 76. Would you be surprised if 3 or more of the plants in the bundle die before producing any rhubarb? Calculate an appropriate probability to support your answer.

79. **Sowing seeds** Refer to Exercise 69.

pg 388

(a) Find the probability that exactly 17 seeds germinate. Show your work.

(b) If only 12 seeds actually germinate, should Judy be suspicious that the company's claim is not true? Compute $P(X \leq 12)$ and use this result to support your answer.

80. **More lefties** Refer to Exercise 72.

(a) Find the probability that exactly 3 students in the sample are left-handed. Show your work.

(b) Would you be surprised if the random sample contained 4 or more left-handed students? Compute $P(W \geq 4)$ and use this result to support your answer.

81. **Random digit dialing** When an opinion poll calls residential telephone numbers at random, only 20% of the calls reach a live person. You watch the random digit dialing machine make 15 calls. Let X = the number of calls that reach a live person.

(a) Find and interpret μ_X.

(b) Find and interpret σ_X.

82. **Lie detectors** A federal report finds that lie detector tests given to truthful persons have probability about 0.2 of suggesting that the person is deceptive.[10] A company asks 12 job applicants about thefts from previous employers, using a lie detector to assess their truthfulness. Suppose that all 12 answer truthfully. Let X = the number of people who the lie detector says are being deceptive.

(a) Find and interpret μ_X.

(b) Find and interpret σ_X.

83. **Random digit dialing** Refer to Exercise 81. Let Y = the number of calls that *don't* reach a live person.

(a) Find the mean of Y. How is it related to the mean of X? Explain why this makes sense.

(b) Find the standard deviation of Y. How is it related to the standard deviation of X? Explain why this makes sense.

84. **Lie detectors** Refer to Exercise 82. Let Y = the number of people who the lie detector says are telling the truth.

(a) Find $P(Y \geq 10)$. How is this related to $P(X \leq 2)$? Explain.

(b) Calculate μ_Y and σ_Y. How do they compare with μ_X and σ_X? Explain why this makes sense.

85. **Aircraft engines** Engineers define reliability as the probability that an item will perform its function under specific conditions for a specific period of time. A certain model of aircraft engine is designed so that each engine has probability 0.999 of performing properly for an hour of flight. Company engineers test an SRS of 350 engines of this model. Let X = the number that operate for an hour without failure.

pg 392

(a) Explain why X is a binomial random variable.

(b) Find the mean and standard deviation of X. Interpret each value in context.

(c) Two engines failed the test. Are you convinced that this model of engine is less reliable than it's supposed to be? Compute $P(X \leq 348)$ and use the result to justify your answer.

86. **1 in 6 wins** As a special promotion for its 20-ounce bottles of soda, a soft drink company printed a message on the inside of each cap. Some of the caps said, "Please try again," while others said, "You're a winner!" The company advertised the promotion with the slogan "1 in 6 wins a prize." Suppose the company is telling the truth and that every 20-ounce bottle of soda it fills has a 1-in-6 chance of being a winner. Seven friends each buy one 20-ounce bottle of the soda at a local convenience store. Let X = the number who win a prize.

(a) Explain why X is a binomial random variable.

(b) Find the mean and standard deviation of X. Interpret each value in context.

(c) The store clerk is surprised when three of the friends win a prize. Is this group of friends just lucky, or is the company's 1-in-6 claim inaccurate? Compute $P(X \geq 3)$ and use the result to justify your answer.

pg 394

87. **Airport security** The Transportation Security Administration (TSA) is responsible for airport safety. On some flights, TSA officers randomly select passengers for an extra security check before boarding. One such flight had 76 passengers—12 in first class and 64 in coach class. Some passengers were surprised when none of the 10 passengers chosen for screening were seated in first class. Can we use a binomial distribution to approximate this probability? Justify your answer.

88. **Scrabble** In the game of Scrabble, each player begins by drawing 7 tiles from a bag containing 100 tiles. There are 42 vowels, 56 consonants, and 2 blank tiles in the bag. Cait chooses her 7 tiles and is surprised to discover that all of them are vowels. Can we use a binomial distribution to approximate this probability? Justify your answer.

89. **No replacement** To use a binomial distribution to approximate the count of successes in an SRS, why do we require that the sample size n be no more than 10% of the population size N?

90. * **Normal approximation** To use a Normal distribution to approximate binomial probabilities, why do we require that both np and $n(1 - p)$ be at least 10?

pg 396

91. * **On the Web** What kinds of Web sites do males aged 18 to 34 visit most often? Half of male Internet users in this age group visit an auction site such as eBay at least once a month.[11] A study of Internet use interviews a random sample of 500 men aged 18 to 34. Let X = the number in the sample who visit an auction site at least once a month.

(a) Show that X is approximately a binomial random variable.

(b) Check the conditions for using a Normal approximation in this setting.

(c) Use the Normal approximation to find the probability that at least 235 of the men in the sample visit an online auction site at least once a month.

92. * **Checking for survey errors** One way of checking the effect of undercoverage, nonresponse, and other sources of error in a sample survey is to compare the sample with known facts about the population. About 12% of American adults identify themselves as black. Suppose we take an SRS of 1500 American adults and let X be the number of blacks in the sample.

(a) Show that X is approximately a binomial random variable.

(b) Check the conditions for using a Normal approximation in this setting.

(c) Use the Normal approximation to find the probability that the sample will contain between 165 and 195 blacks. Show your work.

93. **Using Benford's law** According to Benford's law (Exercise 5, page 353), the probability that the first digit of the amount on a randomly chosen invoice is a 1 or a 2 is 0.477. Suppose you examine an SRS of 90 invoices from a vendor and find 29 that have first digits 1 or 2. Do you suspect that the invoice amounts are not genuine? Compute an appropriate probability to support your answer.

94. **A .300 hitter** In baseball, a .300 hitter gets a hit in 30% of times at bat. When a baseball player hits .300, fans tend to be impressed. Typical major leaguers bat about 500 times a season and hit about .260. A hitter's successive tries seem to be independent. Could a typical major leaguer hit .300 just by chance? Compute an appropriate probability to support your answer.

95. **Geometric or not?** Determine whether each of the following scenarios describes a geometric setting. If so, define an appropriate geometric random variable.

(a) A popular brand of cereal puts a card with one of five famous NASCAR drivers in each box. There is a 1/5 chance that any particular driver's card ends up in any box of cereal. Buy boxes of the cereal until you have all 5 drivers' cards.

(b) In a game of 4-Spot Keno, Lola picks 4 numbers from 1 to 80. The casino randomly selects 20 winning numbers from 1 to 80. Lola wins money if she picks 2 or more of the winning numbers. The probability that this happens is 0.259. Lola decides to keep playing games of 4-Spot Keno until she wins some money.

96. **Geometric or not?** Determine whether each of the following scenarios describes a geometric setting. If so, define an appropriate geometric random variable.

(a) Shuffle a standard deck of playing cards well. Then turn over one card at a time from the top of the deck until you get an ace.

(b) Lawrence is learning to shoot a bow and arrow. On any shot, he has about a 10% chance of hitting the bull's-eye. Lawrence's instructor makes him keep shooting until he gets a bull's-eye.

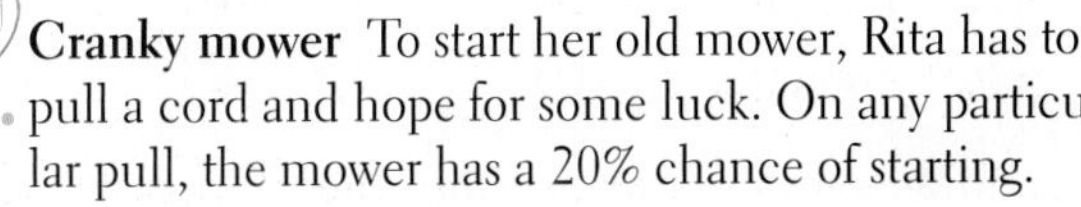

pg 399

97. **Cranky mower** To start her old mower, Rita has to pull a cord and hope for some luck. On any particular pull, the mower has a 20% chance of starting.

(a) Find the probability that it takes her exactly 3 pulls to start the mower. Show your work.

(b) Find the probability that it takes her more than 10 pulls to start the mower. Show your work.

98. **1-in-6 wins** Alan decides to use a different strategy for the 1-in-6 wins game of Exercise 86. He keeps buying one 20-ounce bottle of the soda at a time until he gets a winner.

(a) Find the probability that he buys exactly 5 bottles. Show your work.

(b) Find the probability that he buys no more than 8 bottles. Show your work.

99. **Roulette** Marti decides to keep placing a $1 bet on number 15 in consecutive spins of a roulette wheel until she wins. On any spin, there's a 1-in-38 chance that the ball will land in the 15 slot.

(a) How many spins do you expect it to take until Marti wins? Justify your answer.

(b) Would you be surprised if Marti won in 3 or fewer spins? Compute an appropriate probability to support your answer.

100. **Using Benford's law** According to Benford's law (Exercise 5, page 353), the probability that the first digit of the amount of a randomly chosen invoice is an 8 or a 9 is 0.097. Suppose you examine randomly selected invoices from a vendor until you find one whose amount begins with an 8 or a 9.

(a) How many invoices do you expect to examine until you get one that begins with an 8 or 9? Justify your answer.

(b) In fact, you don't get an amount starting with an 8 or 9 until the 40th invoice. Do you suspect that the invoice amounts are not genuine? Compute an appropriate probability to support your answer.

Multiple choice: Select the best answer for Exercises 101 to 105.

101. Joe reads that 1 out of 4 eggs contains salmonella bacteria. So he never uses more than 3 eggs in cooking. If eggs do or don't contain salmonella independently of each other, the number of contaminated eggs when Joe uses 3 chosen at random has the following distribution:

(a) binomial; $n = 4$ and $p = 1/4$

(b) binomial; $n = 3$ and $p = 1/4$

(c) binomial; $n = 3$ and $p = 1/3$

(d) geometric; $p = 1/4$

(e) geometric; $p = 1/3$

102. In the previous exercise, the probability that at least 1 of Joe's 3 eggs contains salmonella is about

(a) 0.84. (b) 0.68. (c) 0.58. (d) 0.42. (e) 0.30.

Exercises 103 and 104 refer to the following setting. Each entry in a table of random digits like Table D has probability 0.1 of being a 0, and digits are independent of each other.

103. The mean number of 0s in a line 40 digits long is

(a) 10. (b) 4. (c) 3.098. (d) 0.4. (e) 0.1.

104.* Ten lines in the table contain 400 digits. The count of 0s in these lines is approximately Normal with

(a) mean 40; standard deviation 36.

(b) mean 40; standard deviation 19.

(c) mean 40; standard deviation 6.

(d) mean 36; standard deviation 6.

(e) mean 10; standard deviation 19.

105.* In which of the following situations would it be appropriate to use a Normal distribution to approximate probabilities for a binomial distribution with the given values of n and p?

(a) $n = 10, p = 0.5$

(b) $n = 40, p = 0.88$

(c) $n = 100, p = 0.2$

(d) $n = 100, p = 0.99$

(e) $n = 1000, p = 0.003$

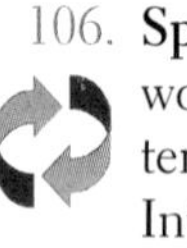

106. **Spoofing** (4.2) To collect information such as passwords, online criminals use "spoofing" to direct Internet users to fraudulent Web sites. In one study of Internet fraud, students were warned about spoofing and then asked to log in to their university account starting from the university's home page. In some cases, the login link led to the genuine dialog box. In others, the box looked genuine but in fact was linked to a different site that recorded the ID and password the student entered. An alert student could detect the fraud by looking at the true Internet address displayed in the browser status bar below the window, but most just entered their ID and password. Is this study an experiment? Why? What are the explanatory and response variables?

107. **Smoking and social class** (5.3) As the dangers of smoking have become more widely known, clear class differences in smoking have emerged. British government statistics classify adult men by occupation as "managerial and professional" (43% of the population), "intermediate" (34%), or "routine and manual" (23%). A survey finds that 20% of men in managerial and professional occupations smoke, 29% of the intermediate group smoke, and 38% in routine and manual occupations smoke.[12]

(a) Use a tree diagram to find the percent of all adult British men who smoke.

(b) Find the percent of male smokers who have routine and manual occupations.

Chapter 6 Chapter Review

A random variable summarizes the outcomes of a chance process with a single number. Its probability distribution specifies the values that the random variable can take and how often it takes those values. We can use the probability distribution to find the mean (expected value) and standard deviation of a discrete random variable, such as the number of heads in four tosses of a coin. For continuous random variables, which can take all values in an interval, we find probabilities using areas under a density curve (like the Normal curve).

In Section 6.2, we examined the effects of adding (or subtracting) a constant and multiplying (or dividing) by a constant on the probability distribution of a random variable. We also investigated the result of adding or subtracting random variables. Most of our calculations require that the random variables we are combining be independent.

Section 6.3 focused on two common discrete random variables: binomial and geometric. Both involve several trials of the same chance process, with a constant probability of getting a "success" on each trial. A binomial random variable counts the number of successes in n trials; a geometric random variable counts the number of trials required to obtain the first success.

Chapter Review Exercises

*These exercises are designed to help you review the important ideas and methods of the chapter. Relevant learning objectives are provided in bulleted form before each exercise. Exercises marked with an asterisk * involve the optional material on the Normal approximation to the binomial.*

- Use a probability distribution to answer questions about possible values of a random variable.
- Calculate the mean and standard deviation of a discrete random variable.

R6.1. Knees Patients receiving artificial knees often experience pain after surgery. The pain is measured on a subjective scale with possible values of 1 (low) to 5 (high). Let X be the pain score for a randomly selected patient. The following table gives part of the probability distribution for X.

Value:	1	2	3	4	5
Probability:	0.1	0.2	0.3	0.3	??

(a) Find $P(X = 5)$.

(b) If two patients who received artificial knees are chosen at random, what's the probability that both of them report pain scores of 1 or 2? Show your work.

(c) Compute the mean and standard deviation of X. Show your work.

- Describe the effects of transforming a random variable by adding or subtracting a constant and multiplying or dividing by a constant.
- Find the mean and standard deviation of the sum or difference of independent random variables.

R6.2. A glass act In a process for manufacturing glassware, glass stems are sealed by heating them in a flame. The temperature of the flame can be adjusted to one of five different settings. Here is the probability distribution of the flame temperature X (in degrees Celsius) for a randomly chosen glass:

Temperature:	540°	545°	550°	555°	560°
Probability:	0.1	0.25	0.3	0.25	0.1

$$\mu_X = 550°\text{C} \qquad \sigma_X = 5.7°\text{C}$$

(a) The target temperature is 550°C. What are the mean and standard deviation of the number of degrees off target, $D = X - 550$?

(b) A manager asks for results in degrees Fahrenheit. The conversion of X into degrees Fahrenheit is given by $y = \frac{9}{5}X + 32$. What are the mean μ_Y and the standard deviation σ_Y of the temperature of the flame in the Fahrenheit scale?

- Interpret the mean and standard deviation of a random variable.
- Find the mean and standard deviation of the sum or difference of independent random variables.

R6.3. Keno In a game of 4-Spot Keno, the player picks 4 numbers from 1 to 80. The casino randomly selects 20 winning numbers from 1 to 80. The table below shows the possible outcomes of the game and their probabilities, along with the amount of money (Payout) that the player wins for a $1 bet. If X = the payout for a single $1 bet, you can check that $\mu_X = \$0.70$ and $\sigma_X = \$6.58$.

Matches:	0	1	2	3	4
Probability:	0.308	0.433	0.213	0.043	0.003
Payout:	$0	$0	$1	$3	$120

(a) Interpret the values of μ_X and σ_X in context.

(b) Jerry places a single $5 bet on 4-Spot Keno. Find the expected value and the standard deviation of his winnings.

(c) Marla plays five games of 4-Spot Keno, betting $1 each time. Find the expected value and the standard deviation of her total winnings.

(d) Based on your answers to (b) and (c), which player would the casino prefer? Justify your answer.

- Determine whether two random variables are independent.
- Find probabilities involving the sum or difference of independent Normal random variables.

R6.4. Applying torque A machine fastens plastic screw-on caps onto containers of motor oil. If the machine applies more torque than the cap can withstand, the cap will break. Both the torque applied and the strength of the caps vary. The capping-machine torque T follows a Normal distribution with mean 7 inch-pounds and standard deviation 0.9 inch-pounds. The cap strength C (the torque that would break the cap) follows a Normal distribution with mean 10 inch-pounds and standard deviation 1.2 inch-pounds.

(a) Explain why it is reasonable to assume that the cap strength and the torque applied by the machine are independent.
(b) Let the random variable $D = C - T$. Find its mean and standard deviation.
(c) What is the probability that a cap will break while being fastened by the machine? Show your work.

Exercises R6.5 and R6.6 refer to the following setting. According to the Mars candy company, 20% of its plain M&M's candies are orange. Assume that the company's claim is true. Suppose that you reach into a large bag of plain M&M's (without looking) and pull out 8 candies. Let X = the number of orange candies you get.

- Determine whether the conditions for a binomial random variable are met.
- Calculate the mean and standard deviation of a binomial random variable. Interpret these values in context.

R6.5. Orange M&M's
(a) Explain why X is a binomial random variable.
(b) Find and interpret the expected value of X.
(c) Find and interpret the standard deviation of X.

- Compute and interpret probabilities involving binomial distributions.

R6.6. Orange M&M's
(a) Would you be surprised if none of the candies were orange? Compute an appropriate probability to support your answer.
(b) How surprising would it be to get 5 or more orange candies? Compute an appropriate probability to support your answer.

- Find probabilities involving geometric random variables.

R6.7. *Sushi Roulette* In the Japanese game show *Sushi Roulette,* the contestant spins a large wheel that's divided into 12 equal sections. Nine of the sections have a sushi roll, and three have a "wasabi bomb." When the wheel stops, the contestant must eat whatever food is on that section. To win the game, the contestant must eat one wasabi bomb. Find the probability that it takes 3 or more spins for the contestant to get a wasabi bomb. Show your method clearly.

- Determine whether the conditions for the Normal approximation to a binomial distribution are met.
- When appropriate, use the Normal approximation to estimate probabilities in a binomial setting.

R6.8.* Is this coin balanced? While he was a prisoner of war during World War II, John Kerrich tossed a coin 10,000 times. He got 5067 heads. If the coin is perfectly balanced, the probability of a head is 0.5.
(a) Find the mean and the standard deviation of the number of heads in 10,000 tosses, assuming the coin is perfectly balanced.
(b) Explain why the Normal approximation is appropriate for calculating probabilities involving the number of heads in 10,000 tosses.
(c) Is there reason to think that Kerrich's coin was not balanced? To answer this question, use a Normal distribution to estimate the probability that tossing a balanced coin 10,000 times would give a count of heads at least this far from 5000 (that is, at least 5067 heads or no more than 4933 heads).

Chapter 6 AP Statistics Practice Test

Section I: Multiple Choice *Select the best answer for each question.*

Questions T6.1 and T6.2 refer to the following setting. A psychologist studied the number of puzzles that subjects were able to solve in a five-minute period while listening to soothing music. Let X be the number of puzzles completed successfully by a subject. The psychologist found that X had the following probability distribution:

Value of X:	1	2	3	4
Probability:	0.2	0.4	0.3	0.1

T6.1. What is the probability that a randomly chosen subject completes at least 3 puzzles in the five-minute period while listening to soothing music?

(a) 0.3
(b) 0.4
(c) 0.6
(d) 0.9
(e) Cannot be determined

T6.2. Suppose that three randomly selected subjects solve puzzles for five minutes each. The expected value of the total number of puzzles solved by the three subjects is

(a) 1.8. (b) 2.3. (c) 2.5. (d) 6.9. (e) 7.5.

T6.3. Suppose a student is randomly selected from your school. Which of the following pairs of random variables are most likely independent?

(a) X = student's height; Y = student's weight

(b) X = student's IQ; Y = student's GPA

(c) X = student's PSAT Math score; Y = student's PSAT Verbal score

(d) X = average amount of homework the student does per night; Y = student's GPA

(e) X = average amount of homework the student does per night; Y = student's height

T6.4. A certain vending machine offers 20-ounce bottles of soda for $1.50. The number of bottles X bought from the machine on any day is a random variable with mean 50 and standard deviation 15. Let the random variable Y equal the total revenue from this machine on a given day. Assume that the machine works properly and that no sodas are stolen from the machine. What are the mean and standard deviation of Y?

(a) $\mu_Y = \$1.50, \sigma_Y = \$22.50$

(b) $\mu_Y = \$1.50, \sigma_Y = \$33.75$

(c) $\mu_Y = \$75, \sigma_Y = \$18.37$

(d) $\mu_Y = \$75, \sigma_Y = \$22.50$

(e) $\mu_Y = \$75, \sigma_Y = \$33.75$

Questions T6.5 and T6.6 refer to the following setting. The weight of tomatoes chosen at random from a bin at the farmer's market is a random variable with mean $\mu = 10$ ounces and standard deviation $\sigma = 1$ ounce. Suppose we pick four tomatoes at random from the bin and find their total weight T.

T6.5. The random variable T has a mean of

(a) 2.5 ounces.
(b) 4 ounces.
(c) 10 ounces.
(d) 40 ounces.
(e) 41 ounces.

T6.6. The random variable T has a standard deviation (in ounces) of

(a) 0.25. (b) 0.50. (c) 0.71. (d) 2. (e) 4.

T6.7. Which of the following random variables is geometric?

(a) The number of times I have to roll a die to get two 6s.

(b) The number of cards I deal from a well-shuffled deck of 52 cards until I get a heart.

(c) The number of digits I read in a randomly selected row of the random digits table until I find a 7.

(d) The number of 7s in a row of 40 random digits.

(e) The number of 6s I get if I roll a die 10 times.

T6.8. Seventeen people have been exposed to a particular disease. Each one independently has a 40% chance of contracting the disease. A hospital has the capacity to handle 10 cases of the disease. What is the probability that the hospital's capacity will be exceeded?

(a) 0.011 (b) 0.035 (c) 0.092 (d) 0.965 (e) 0.989

T6.9. The figure shows the probability distribution of a discrete random variable X. Which of the following best describes this random variable?

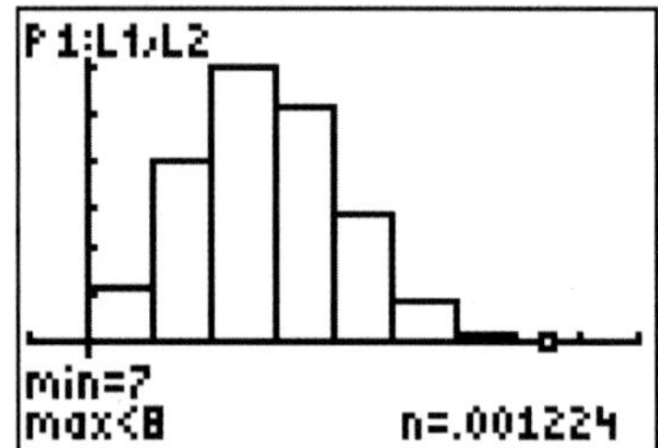

(a) Binomial with $n = 8, p = 0.1$

(b) Binomial with $n = 8, p = 0.3$

(c) Binomial with $n = 8, p = 0.8$

(d) Geometric with $p = 0.1$

(e) Geometric with $p = 0.2$

T6.10. A test for extrasensory perception (ESP) involves asking a person to tell which of 5 shapes—a circle, star, triangle, diamond, or heart—appears on a hidden computer screen. On each trial, the computer is equally likely to select any of the 5 shapes. Suppose researchers are testing a person who does not have ESP and so is just guessing on each trial. What is the probability that the person guesses the first 4 shapes incorrectly but gets the fifth correct?

(a) 1/5

(b) $\left(\frac{4}{5}\right)^4$

(c) $\left(\frac{4}{5}\right)^4 \cdot \left(\frac{1}{5}\right)$

(d) $\binom{5}{1} \cdot \left(\frac{4}{5}\right)^4 \cdot \left(\frac{1}{5}\right)$

(e) 4/5

Section II: Free Response *Show all your work. Indicate clearly the methods you use, because you will be graded on the correctness of your methods as well as on the accuracy and completeness of your results and explanations.*

T6.11. Let Y denote the number of broken eggs in a randomly selected carton of one dozen "store brand" eggs at a local supermarket. Suppose that the probability distribution of Y is as follows.

Value y_i:	0	1	2	3	4
Probability p_i:	0.78	0.11	0.07	0.03	0.01

(a) What is the probability that at least 10 eggs in a randomly selected carton are *unbroken*?

(b) Calculate and interpret μ_Y.

(c) Calculate σ_Y. Show your work.

(d) A quality control inspector at the store keeps looking at randomly selected cartons of eggs until he finds one with at least 2 broken eggs. Find the probability that this happens in one of the first three cartons he inspects.

T6.12. *Ladies Home Journal* magazine reported that 66% of all dog owners greet their dog before greeting their spouse or children when they return home at the end of the workday. Assume that this claim is true. Suppose 12 dog owners are selected at random. Let X = the number of owners who greet their dogs first.

(a) Explain why X is a binomial random variable.

(b) Only 4 of the owners in the random sample greeted their dogs first. Does this give convincing evidence against the *Ladies Home Journal* claim? Calculate an appropriate probability to support your answer.

T6.13. Ed and Adelaide attend the same high school, but are in different math classes. The time E that it takes Ed to do his math homework follows a Normal distribution with mean 25 minutes and standard deviation 5 minutes. Adelaide's math homework time A follows a Normal distribution with mean 50 minutes and standard deviation 10 minutes.

(a) Randomly select one math assignment of Ed's and one math assignment of Adelaide's. Let the random variable D be the difference in the amount of time each student spent on their assignments: $D = A - E$. Find the mean and the standard deviation of D. Show your work.

(b) Find the probability that Ed spent longer on his assignment than Adelaide did on hers. Show your work.

T6.14. According to the Census Bureau, 13% of American adults (aged 18 and over) are Hispanic. An opinion poll plans to contact an SRS of 1200 adults.

(a) What is the mean number of Hispanics in such samples? What is the standard deviation?

(b) Should we be suspicious if the sample selected for the opinion poll contains 15% Hispanic people? Compute an appropriate probability to support your answer.

Chapter 7

case study

Introduction

Sampling Distributions

Building Better Batteries

Everyone wants to have the latest technological gadget. That's why iPods, digital cameras, smartphones, Game Boys, and the Wii have sold millions of units. These devices require lots of power and can drain batteries quickly. Battery manufacturers are constantly searching for ways to build longer-lasting batteries.

A particular manufacturer produces AA batteries that are designed to last an average of 17 hours with a standard deviation of 0.8 hours. Quality control inspectors select a random sample of 30 batteries during each hour of production, and they then drain them under conditions that mimic normal use. Here are the lifetimes (in hours) of the batteries from one such sample:

16.91	18.83	17.58	15.84	17.42	17.65	16.63	16.84	15.63	16.37
15.80	15.93	15.81	17.45	16.85	16.33	16.22	16.59	17.13	17.10
16.96	16.40	17.35	16.37	15.98	16.52	17.04	17.07	15.73	16.74

Do these data suggest that the production process is working properly? Are plant managers safe sending out all the batteries produced in this hour for sale? In this chapter, you will develop the tools you need to help answer questions like this.

Introduction

The battery manufacturer in the Case Study could find the true mean lifetime μ of all the batteries produced in an hour. Quality control inspectors would simply measure the lifetime of each battery (by draining it) and then calculate the average. With this method, the company would know the truth about the population mean μ, but it would have no batteries left to sell!

Instead of taking a census, the manufacturer collects data from a random sample of 30 batteries produced that hour. The company's goal is to use the sample mean lifetime $\bar{x}$ to estimate the unknown population mean μ. This is an example of *statistical inference:* we use information from a sample to draw conclusions about a wider population.

To make such an inference, we need to know how close the sample mean $\bar{x}$ is likely to be to the population mean μ. After all, different random samples of 30 batteries from the same hour of production would yield different values of $\bar{x}$. How can we describe this *sampling distribution* of possible $\bar{x}$-values? We can think of $\bar{x}$ as a random variable because it takes numerical values that describe the outcomes of the random sampling process. As a result, we can examine its probability distribution using what we learned in Chapter 6.

This same reasoning applies to other types of inference settings. Here are a few examples.

- Each month, the Current Population Survey (CPS) interviews a random sample of individuals in about 55,000 U.S. households. The CPS uses the proportion of unemployed people in the sample $\hat{p}$ to estimate the national unemployment rate p.
- Tom is cooking a large turkey breast for a holiday meal. He wants to be sure that the turkey is safe to eat, which requires a minimum internal temperature of 165°F. Tom uses a thermometer to measure the temperature of the turkey meat at four randomly chosen points. If the minimum reading in the sample is 170°F, can Tom safely serve the turkey?
- How much do gasoline prices vary in a large city? To find out, a reporter records the price per gallon of regular unleaded gasoline at a random sample of 10 gas stations in the city on the same day. The range (maximum – minimum) of the prices in the sample is 25 cents. What can the reporter say about the range of gas prices at all the city's stations?

The following Activity gives you a chance to estimate an unknown population value based on data from a random sample.

ACTIVITY *The German tank problem*

MATERIALS: Tags or pieces of cardstock numbered 1 to *N*; small brown paper bag; index card and prelabeled graph grid for each team; prizes for the winners

During World War II, the Allies captured several German tanks. Each tank had a serial number on it. Allied commanders wanted to know how many tanks the Germans actually had so that they could allocate their forces appropriately. They sent the serial numbers of the captured tanks to a group of mathematicians in Washington, D.C., and asked for an estimate of the total number of German tanks N. In this Activity, you and your teammates will play the role of the mathematicians. To add some fun to the challenge, two prizes will be offered: one to the team that gets closest to the actual number of tanks and the other to the team whose method is judged to be the most likely to give a good estimate.

1. Your teacher will create tags numbered 1, 2, 3, . . . , N to represent the German tanks and place them in a bag. The class will be divided into teams of three or four students.

2. The teacher will mix the tags well, and ask four students to draw one tag each from the bag. Each selected tag represents the serial number of a captured German tank. All four numbers should be written on the board for everyone to see. Return the tags to the bag.

3. Each team will have 15 minutes to come up with a statistical formula for estimating the total number of tanks N in the bag. You should have time to try several ideas. When you are satisfied with your method, calculate your estimate of N. Write your team members' names, your formula, and your estimate on the index card provided.

4. When time is called, each team must give its index card to your teacher. The teacher will make a chart on the board showing the formulas and estimates. Each team will have one minute to explain why it chose the formula it did.

5. It's time for the first prize to be awarded! The teacher will reveal the actual number of tanks.

6. What if the Allies had captured four other German tanks? Which team's formula would consistently produce the best estimate? Students should help choose nine more simple random samples of four tanks from the bag. After each sample is taken, the four serial numbers chosen should be written on the board, and the tags should be returned to the bag and mixed thoroughly.

7. Each team should use its formula to estimate the total number of tanks N for each of the nine new samples. The team should then make a dotplot of its 10 estimates on the grid provided.

8. The teacher will now tape all the teams' dotplots one above the other on the board. As a class, decide which team used the best method for estimating the number of tanks.

More recently, people have used the mathematicians' method from the German tank problem to estimate the number of iPhones produced.

Comment: The mathematicians in Washington produced remarkably close estimates for the actual number of German tanks. Visit Wikipedia for more details.

Sampling distributions are the key to inference when data are produced by random sampling. Because the results of random samples include an element of chance, we can't guarantee that our inferences are correct. What we can guarantee is that our methods usually give correct answers. The reasoning of statistical inference rests on asking, "How often would this method give a correct answer if I used it very many times?" If our data come from random sampling, the laws of probability answer the question "What would happen if we did this many times?"

Section 7.1 presents the basic ideas of sampling distributions. The most common applications of statistical inference involve proportions and means. Section 7.2 focuses on sampling distributions of sample proportions. Section 7.3 investigates sampling distributions of sample means.

7.1

What Is a Sampling Distribution?

In Section 7.1, you'll learn about:

- Parameters and statistics
- Sampling variability
- Describing sampling distributions

What is the average income of American households? Each March, the government's Current Population Survey (CPS) asks detailed questions about income. The random sample of 54,100 households contacted in March 2009 had a mean "total money income" of \$68,424 in 2008.[1] (The median income was of course lower, at \$50,303.) That \$68,424 describes the sample, but we use it to estimate the mean income of all households.

Parameters and Statistics

As we begin to use sample data to draw conclusions about a wider population, we must be clear about whether a number describes a sample or a population. For the sample of 54,100 households contacted by the CPS, the mean income was $\bar{x}$ = \$68,424. The number \$68,424 is a **statistic** because it describes this one CPS sample. The population that the poll wants to draw conclusions about is all 117 million U.S. households. In this case, the **parameter** of interest is the mean income μ of all these households. We don't know the value of this parameter.

DEFINITION: Parameter, statistic

A **parameter** is a number that describes some characteristic of the population. In statistical practice, the value of a parameter is usually not known because we cannot examine the entire population.

A **statistic** is a number that describes some characteristic of a sample. The value of a statistic can be computed directly from the sample data. We often use a statistic to estimate an unknown parameter.

It is common practice to use Greek letters for parameters and Roman letters for statistics. In that case, the population proportion would be π (pi, the Greek letter for "p") and the sample proportion would be p. We'll stick with the notation that's used on the AP exam, however.

Remember **s** and **p**: statistics come from **s**amples, and **p**arameters come from **p**opulations. As long as we were just doing data analysis, the distinction between population and sample rarely came up. Now, however, it is essential. The notation we use should reflect this distinction. For instance, we write μ (the Greek letter mu) for the population mean and $\bar{x}$ for the sample mean. We use p to represent a *population proportion*. The *sample proportion* $\hat{p}$ is used to estimate the unknown parameter p.

EXAMPLE

From Ghosts to Cold Cabins

Parameters and statistics

PROBLEM: Identify the population, the parameter, the sample, and the statistic in each of the following settings.

(a) The Gallup Poll asked a random sample of 515 U.S. adults whether or not they believe in ghosts. Of the respondents, 160 said "Yes."[2]

(b) During the winter months, the temperatures outside the Starneses' cabin in Colorado can stay well below freezing (32°F, or 0°C) for weeks at a time. To prevent the pipes from freezing, Mrs. Starnes sets the thermostat at 50°F. She wants to know how low the temperature actually gets in the cabin. A digital thermometer records the indoor temperature at 20 randomly chosen times during a given day. The minimum reading is 38°F.

SOLUTION:

(a) The population is all U.S. adults, and the parameter of interest is p, the proportion of all U.S. adults who believe in ghosts. The sample is the 515 people who were interviewed in this Gallup Poll. The statistic is $\hat{p} = \frac{160}{515} = 0.31$, the proportion of the sample who say they believe in ghosts.

(b) The population is all times during the day in question; the parameter of interest is the minimum temperature in the cabin that day. The sample consists of the 20 temperature readings at randomly selected times. The statistic is the sample minimum, 38°F.

For Practice *Try Exercise 1*

CHECK YOUR UNDERSTANDING

Each boldface number in Questions 1 and 2 is the value of either a **parameter** *or a* **statistic**. *In each case, state which it is and use appropriate notation to describe the number.*

1. On Tuesday, the bottles of Arizona Iced Tea filled in a plant were supposed to contain an average of **20** ounces of iced tea. Quality control inspectors sampled 50 bottles at random from the day's production. These bottles contained an average of **19.6** ounces of iced tea.

2. On a New York–to–Denver flight, **8%** of the 125 passengers were selected for random security screening before boarding. According to the Transportation Security Administration, **10%** of passengers at this airport are chosen for random screening.

Sampling Variability

How can $\bar{x}$, based on a sample of only a few of the 117 million American households, be an accurate estimate of μ? After all, a second random sample taken at the same time would choose different households and likely produce a different value of $\bar{x}$. This basic fact is called **sampling variability:** the value of a statistic varies in repeated random sampling.

Sampling variability

To make sense of sampling variability, we ask, "What would happen if we took many samples?" Here's how to answer that question:

- Take a large number of samples from the same population.
- Calculate the statistic (like the sample mean $\bar{x}$ or sample proportion $\hat{p}$) for each sample.
- Make a graph of the values of the statistic.
- Examine the distribution displayed in the graph for shape, center, and spread, as well as outliers or other deviations.

The following Activity gives you a chance to see sampling variability in action.

ACTIVITY *Reaching for chips*

MATERIALS: 200 colored chips, including 100 of the same color; large bag or other container

Before class, your teacher will prepare a population of 200 colored chips, with 100 having the same color (say, red). The parameter is the actual proportion p of red chips in the population: $p = \frac{1}{2}$. In this Activity, you will investigate sampling variability by taking repeated random samples of size 20 from the population.

1. After your teacher has mixed the chips thoroughly, each student in the class should take a sample of 20 chips and note the sample proportion $\hat{p}$ of red chips. When finished, the student should return all the chips to the bag, stir them up, and pass the bag to the next student.

Note: If your class has fewer than 25 students, have some students take two samples.

2. Each student should record the $\hat{p}$-value in a chart on the board and plot this value on a class dotplot. Label the graph scale from 0.10 to 0.90 with tick marks spaced 0.05 units apart.

3. Describe what you see: shape, center, spread, and any unusual values.

When Mr. Caldwell's class did the "Reaching for Chips" Activity, his students produced the graph shown in Figure 7.1. Here's what the class said about its distribution of $\hat{p}$-values.

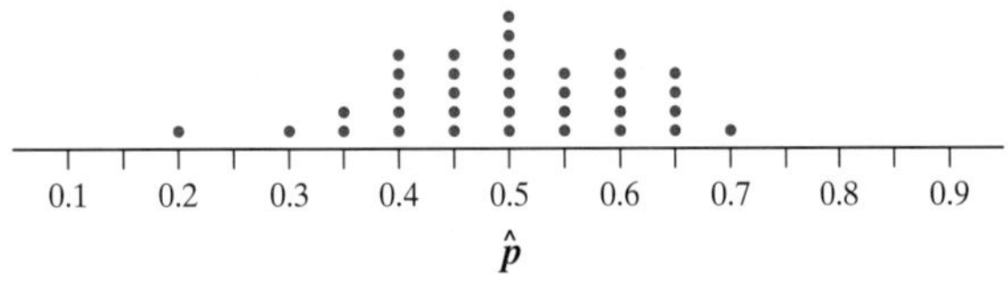

FIGURE 7.1 Dotplot of sample proportions obtained by Mr. Caldwell's class.

Shape: The graph is roughly symmetric with a single peak at 0.5.

Center: The mean of our sample proportions is 0.499. This is the balance point of the distribution.

Spread: The standard deviation of our sample proportions is 0.112. On average, the values of $\hat{p}$ are about 0.112 away from the mean.

Outliers: There are no obvious outliers or other unusual features.

Of course, the class only took 35 different simple random samples of 20 chips. There are many, many possible SRSs of size 20 from a population of size 200 (about $1.6 \cdot 10^{27}$, actually). If we took every one of those possible samples, calculated the value of $\hat{p}$ for each, and graphed all those $\hat{p}$-values, then we'd have a **sampling distribution.**

DEFINITION: Sampling distribution

The **sampling distribution** of a statistic is the distribution of values taken by the statistic in all possible samples of the same size from the same population.

In practice, it's usually too difficult to take all possible samples of size n to obtain the actual sampling distribution of a statistic. Instead, we can use simulation to imitate the process of taking many, many samples.

Reaching for Chips

Simulating a sampling distribution

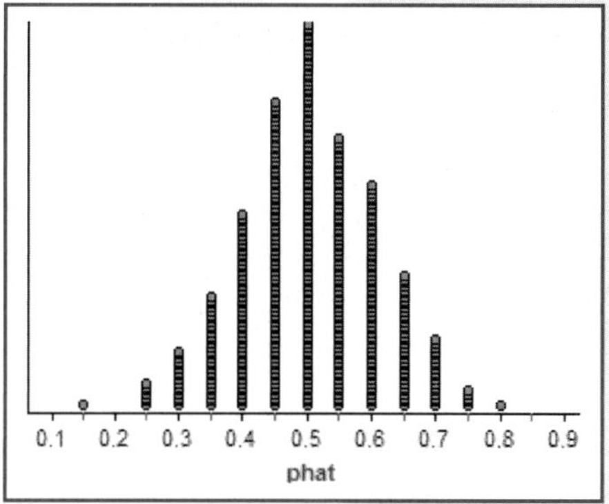

FIGURE 7.2 Dotplot of the sample proportion $\hat{p}$ of red chips in 500 simulated SRSs, created by Fathom software.

We used Fathom software to simulate choosing 500 SRSs of size $n = 20$ from a population of 200 chips, 100 red and 100 blue. Figure 7.2 is a dotplot of the values of $\hat{p}$, the sample proportion of red chips, from these 500 samples.

PROBLEM:

(a) Is this the sampling distribution of $\hat{p}$? Justify your answer.

(b) Describe the distribution. Are there any obvious outliers?

(c) Suppose your teacher prepares a bag with 200 chips and claims that half of them are red. A classmate takes an SRS of 20 chips; 17 of them are red. What would you conclude about your teacher's claim? Explain.

SOLUTION:

(a) This graph is *not* the sampling distribution of $\hat{p}$ because it doesn't show the values of the statistic in *all* possible samples.

(b) Shape: Symmetric, unimodal, and somewhat bell-shaped. Center: Around 0.5. Spread: The values of $\hat{p}$ fall mostly between 0.25 and 0.75. Outliers: One sample with $\hat{p} = 0.15$ stands out.

(c) It is very unlikely to obtain an SRS of 20 chips in which $\hat{p} = 17/20 = 0.85$ from a population of 200 chips, half of which are red. In fact, a value of $\hat{p}$ this large never occurred in 500 simulated samples. This student's result gives strong evidence against the teacher's claim. It seems doubtful that the bag contains half red chips.

For Practice *Try Exercise 9*

Strictly speaking, the sampling distribution is the ideal pattern that would emerge if we looked at *all* possible samples of size 20 from our population of chips. A distribution obtained from simulating a smaller number of random samples, like the 500 values of $\hat{p}$ in Figure 7.2, is only an approximation to the sampling distribution. One of the uses of probability theory in statistics is to obtain sampling distributions without simulation. We'll get to the theory later. The interpretation of a sampling distribution is the same, however, whether we obtain it by simulation or by the mathematics of probability.

Figure 7.3 on the next page illustrates the process of choosing many random samples of 20 chips and finding the sample proportion of red chips $\hat{p}$ for each one. Follow the flow of the figure from the population at the left, to choosing an SRS and finding the $\hat{p}$ for this sample, to collecting together the $\hat{p}$'s from many samples. The first sample has $\hat{p} = 0.40$. The second sample is a different group of chips, with $\hat{p} = 0.55$, and so on. The dotplot at the right of the figure shows the distribution of the values of $\hat{p}$ from 500 separate SRSs of size 20. This dotplot displays the approximate sampling distribution of the statistic $\hat{p}$.

Population distribution

As Figure 7.3 shows, there are actually three distinct distributions involved when we sample repeatedly and measure a variable of interest. The **population distribution** gives the values of the variable for all the individuals in the population. In this case, the individuals are the 200 chips and the variable we're recording is color. Our parameter of interest is the proportion of red chips in the population, $p = 0.50$. Each random sample that we take consists of 20 chips.

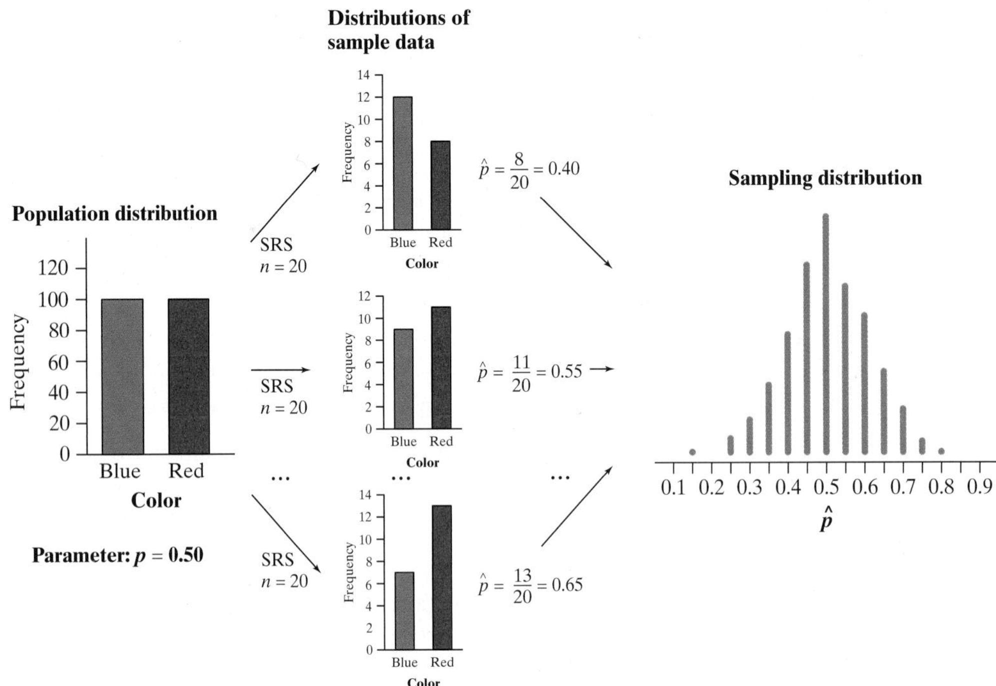

FIGURE 7.3 The idea of a sampling distribution: take many samples from the same population, collect the $\hat{p}$'s from all the samples, and display the distribution of the $\hat{p}$'s. The dotplot shows the results of 500 samples.

AP EXAM TIP Terminology matters. Don't say "sample distribution" when you mean sampling distribution. You will lose credit on free-response questions for misusing statistical terms.

The *distribution of sample data* shows the values of the variable "color" for the individuals in the sample. For each sample, we record a value for the statistic $\hat{p}$, the sample proportion of red chips. Finally, we collect the values of $\hat{p}$ from all possible samples of the same size and display them in the *sampling distribution.*

Be careful: The population distribution and the distribution of sample data describe individuals. *A sampling distribution describes how a* statistic *varies in many samples from the population.*

CHECK YOUR UNDERSTANDING

Mars, Incorporated, says that the mix of colors in its M&M'S® Milk Chocolate Candies is 24% blue, 20% orange, 16% green, 14% yellow, 13% red, and 13% brown. Assume that the company's claim is true. We want to examine the proportion of orange M&M'S in repeated random samples of 50 candies.

1. Graph the population distribution. Identify the individuals, the variable, and the parameter of interest.

2. Imagine taking an SRS of 50 M&M'S. Make a graph showing a possible distribution of the sample data. Give the value of the appropriate statistic for this sample.

3. Which of the graphs that follow could be the approximate sampling distribution of the statistic? Explain your choice.

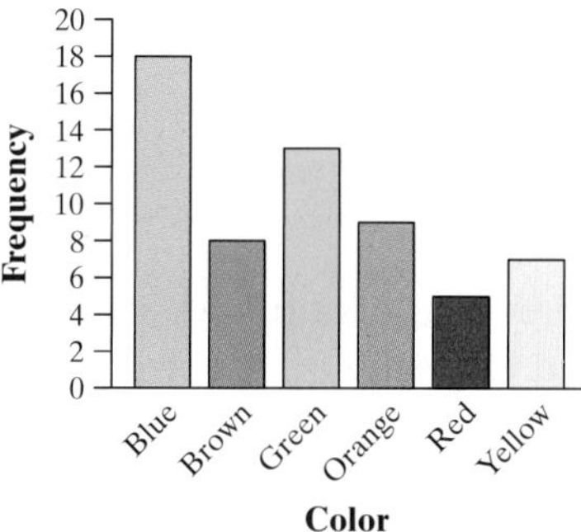

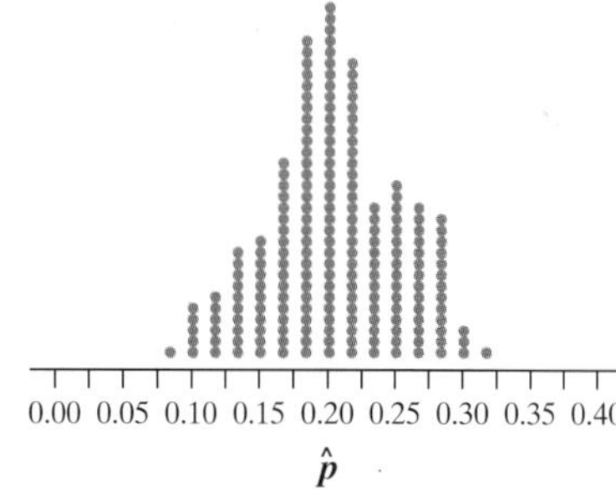

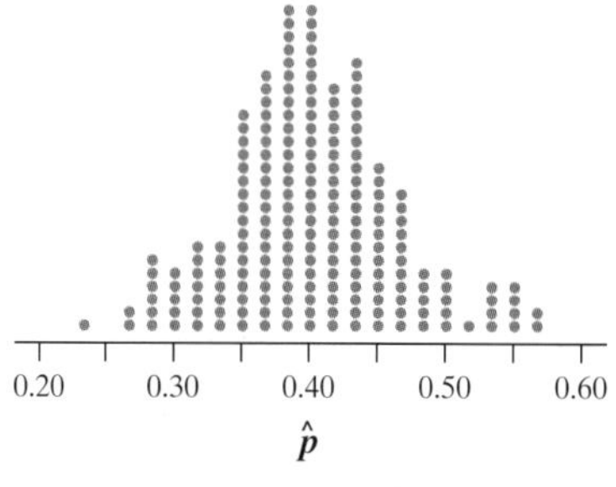

Describing Sampling Distributions

The fact that statistics from random samples have definite sampling distributions allows us to answer the question "How trustworthy is a statistic as an estimate of a parameter?" To get a complete answer, we consider the shape, center, and spread of the sampling distribution. For reasons that will be clear later, we'll save shape for last.

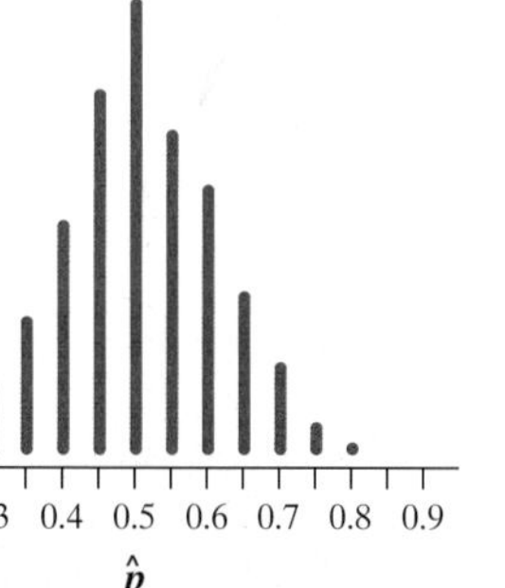

Center: Biased and unbiased estimators Let's return to the familiar chips example. How well does the sample proportion of red chips estimate the population proportion of red chips, $p = 0.5$? The dotplot in the margin shows the approximate sampling distribution of $\hat{p}$ once again. We noted earlier that the center of this distribution is very close to 0.5, the parameter value. In fact, if we took all possible samples of 20 chips from the population, calculated $\hat{p}$ for each sample, and then found the mean of all those $\hat{p}$-values, we'd get *exactly* 0.5. For this reason, we say that $\hat{p}$ is an **unbiased estimator** of p.

DEFINITION: Unbiased estimator

A statistic used to estimate a parameter is an **unbiased estimator** if the mean of its sampling distribution is equal to the true value of the parameter being estimated.

The value of an unbiased estimator will sometimes exceed the true value of the parameter and sometimes be less if we take many samples. Because its sampling distribution is centered at the true value, however, there is no systematic tendency to overestimate or underestimate the parameter. This makes the idea of lack of bias in the sense of "no favoritism" more precise.

As we will confirm in Section 7.2, the sample proportion $\hat{p}$ from an SRS is always an unbiased estimator of the population proportion p. That's a very helpful result if we're dealing with a categorical variable (like color). With quantitative variables, we might be interested in estimating the population mean, median, minimum, maximum, Q_1, Q_3, variance, standard deviation, IQR, or range. Which (if any) of these are unbiased estimators? The following Activity should shed some light on this question.

ACTIVITY *Sampling heights*

MATERIALS: Small piece of cardstock for each student; bag

In this Activity, you will use a population of quantitative data to investigate whether a given statistic is an unbiased estimator of its corresponding population parameter.

1. Each student should write his or her name and height (in inches) neatly on a small piece of cardstock and then place it in the bag.
2. After your teacher has mixed the cards thoroughly, each student in the class should take a sample of four cards and record the names and heights of the four chosen students. When finished, the student should return the cards to the bag, mix them up, and pass the bag to the next student.

Note: If your class has fewer than 25 students, have some students take two samples.

3. For your SRS of four students from the class, calculate the sample mean $\bar{x}$ and the sample range (maximum − minimum) of the heights. Then go to the board and record the initials of the four students in your sample, their heights, the sample mean $\bar{x}$, and the sample range in a chart like the one below.

Individuals	Height (in.)	Sample mean ($\bar{x}$)	Sample range (max − min)
ML, JM, FD, AS	62, 75, 68, 63	67	75 − 62 = 13

4. Plot the values of your sample mean and sample range on the two class dotplots drawn by your teacher.
5. Once everyone has finished, make a dotplot of the population distribution. Find the population mean μ and the population range.
6. Based on your approximate sampling distributions of $\bar{x}$ and the sample range, which statistic appears to be an unbiased estimator? Which appears to be a *biased estimator*?

When Mrs. Washington's class did the "Sampling Heights" Activity, they produced the three graphs shown in Figure 7.4. Her students concluded that the sample mean $\bar{x}$ is probably an unbiased estimator of the population mean μ. Their reason: the center of the approximate sampling distribution of $\bar{x}$, 65.67 inches, is close to the population mean of 66.07 inches. On the other hand,

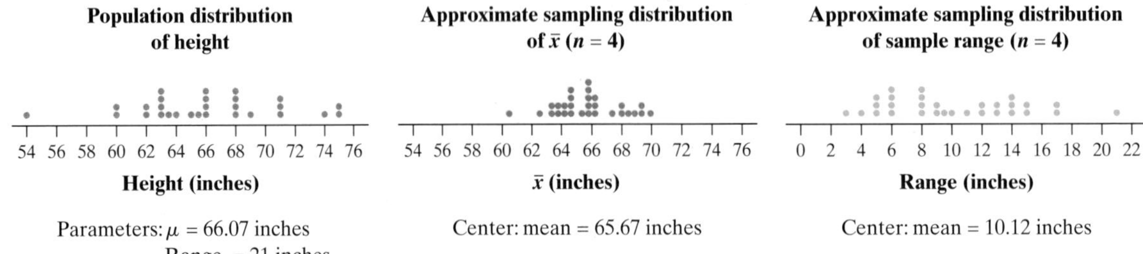

FIGURE 7.4 Results from Mrs. Washington's class. The sample mean appears to be an unbiased estimator. The sample range appears to be a biased estimator.

Biased estimator

Mrs. Washington's students decided that the sample range is a **biased estimator** of the population range. That's because the center of the sampling distribution for this statistic was 10.12 inches, much less than the corresponding parameter value of 21 inches.

To confirm the class's conclusions, we used Fathom software to simulate taking 250 SRSs of $n = 4$ students. For each sample, we plotted the sample mean height $\bar{x}$ and the range of the heights in the sample. Figure 7.5 shows the population distribution and the approximate sampling distributions for these two statistics. It looks like the class was right: $\bar{x}$ is an unbiased estimator of μ, but the sample range is clearly a biased estimator. The range of the sample heights tends to be much lower, on average, than the population range.

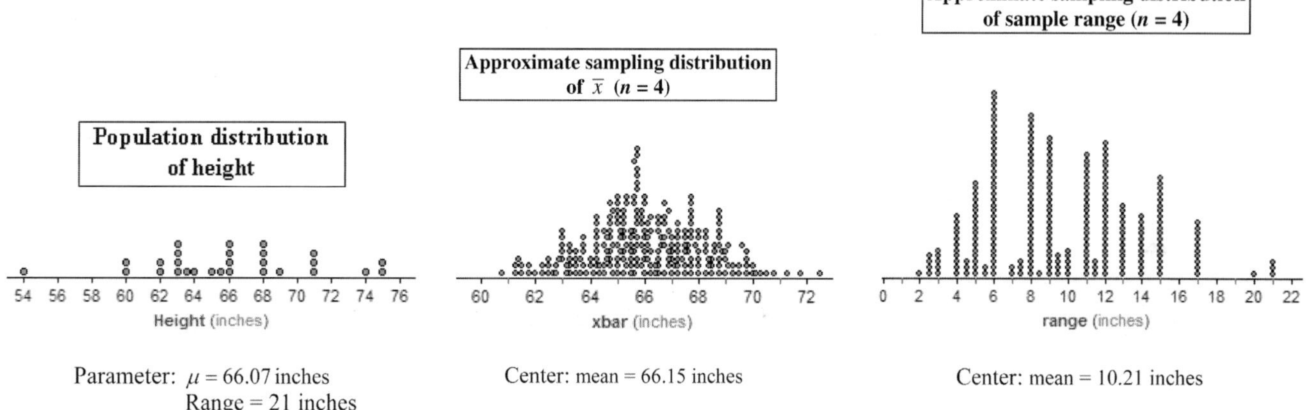

FIGURE 7.5 Results from a Fathom simulation of 250 SRSs of size $n = 4$ from the students in Mrs. Washington's class. The sample mean is an unbiased estimator. The sample range is a biased estimator.

Why do we divide by $n - 1$ when calculating the sample variance? In Chapter 1, we introduced the sample variance s_x^2 as a measure of spread for a set of quantitative data. The idea of s_x^2 is simple: it's a number that describes the "average" squared deviation of the values in the sample from their mean $\bar{x}$. It probably surprised you when we computed this average by dividing by $n - 1$ instead of n. Now we're ready to tell you why we defined $s_x^2 = \frac{1}{n-1}\sum(x_i - \bar{x})^2$.

In an inference setting involving a quantitative variable, we might be interested in estimating the variance σ^2 of the population distribution. The most logical choice for our estimator is the sample variance s_x^2. We used Fathom software to take 500 SRSs of size $n = 4$ from the population distribution of heights in Mrs. Washington's class. Note that the population variance is $\sigma^2 = 22.19$. For each sample, we recorded the value of two statistics:

$$\text{var} = s_x^2 = \frac{1}{n-1}\sum(x_i - \bar{x})^2 \quad \text{(the sample variance)}$$

$$\text{varn} = \frac{1}{n}\sum(x_i - \bar{x})^2$$

Figure 7.6 on the next page shows the approximate sampling distributions of these two statistics. (We used histograms to show the overall pattern more clearly.) The vertical lines mark the means of these two distributions.

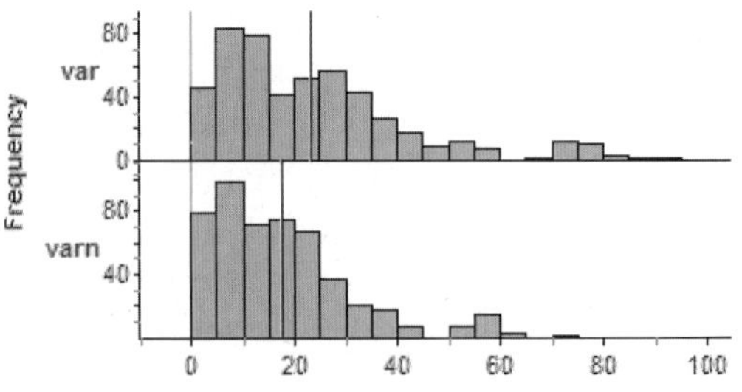

FIGURE 7.6 Results from a Fathom simulation of 500 SRSs of size $n = 4$ from the population distribution of heights in Mrs. Washington's class. The sample variance s_x^2 (labeled "var" in the figure) is an unbiased estimator. The "varn" statistic (dividing by n instead of $n - 1$) is a biased estimator.

We can see that "varn" is a *biased* estimator of the population variance. The center (mean) of its sampling distribution is clearly less than the value of the population parameter, 22.19. However, the statistic "var" (otherwise known as the sample variance s_x^2) is an unbiased estimator. Its values are centered at 22.19. That's why we divide by $n - 1$ and not n when calculating the sample variance: to get an unbiased estimator of the population variance.

Spread: Low variability is better! To get a trustworthy estimate of an unknown population parameter, start by using a statistic that's an unbiased estimator. This ensures that you won't tend to overestimate or underestimate. Unfortunately, using an unbiased estimator doesn't guarantee that the value of your statistic will be close to the actual parameter value. The following example illustrates what we mean.

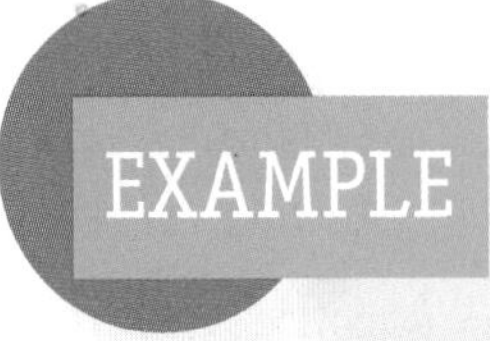

Who Watches Survivor?

Why sample size matters

Television executives and companies who advertise on TV are interested in how many viewers watch particular shows. According to Nielsen ratings, *Survivor* was one of the most-watched television shows in the United States during every week that it aired. Suppose that the true proportion of U.S. adults who have watched *Survivor* is $p = 0.37$.

Figure 7.7(a) shows the results of drawing 1000 SRSs of size $n = 100$ from a population with $p = 0.37$. We see that a sample of 100 people often gave a $\hat{p}$ quite

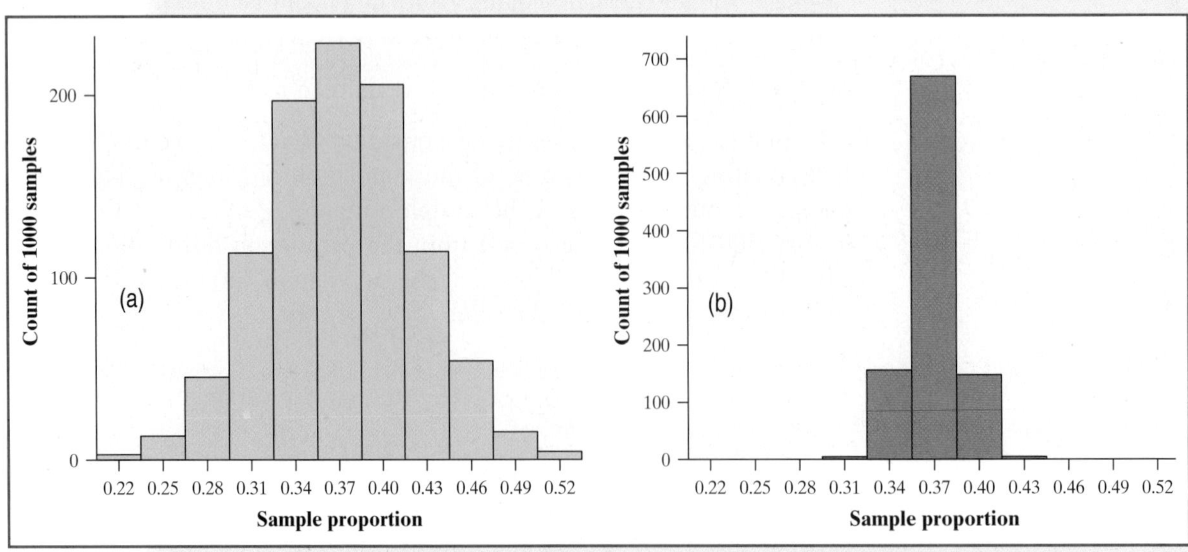

FIGURE 7.7 The approximate sampling distribution of the sample proportion $\hat{p}$ from SRSs of size (a) $n = 100$ and (b) $n = 1000$ drawn from a population with proportion $p = 0.37$ who have watched *Survivor*. Both histograms show the results of 1000 SRSs.

far from the population parameter. That is why a Gallup Poll asked not 100, but 1000 people whether they had watched *Survivor.* Let's repeat our simulation, this time taking 1000 SRSs of size $n = 1000$ from a population with proportion $p = 0.37$ who have watched *Survivor.* Figure 7.7(b) displays the distribution of the 1000 values of $\hat{p}$ from these new samples. Both graphs are drawn on the same horizontal scale to make comparison easy.

We can see that the *spread* of Figure 7.7(b) is much less than that of Figure 7.7(a). With samples of size 100, the values of $\hat{p}$ vary from 0.22 to 0.535. The standard deviation of the $\hat{p}$-values in Figure 7.7(a) is about 0.05. Using SRSs of size 1000, the values of $\hat{p}$ only vary from 0.321 to 0.421. In Figure 7.7(b), the standard deviation is about 0.016. Most random samples of 1000 people give a $\hat{p}$ that is within 0.03 of the actual population parameter, $p = 0.37$.

The sample proportion $\hat{p}$ from a random sample of any size is an unbiased estimator of the parameter p. As we can see from the previous example, though, larger samples have a clear advantage. They are much more likely to produce an estimate close to the true value of the parameter. Said another way, larger random samples give us more *precise* estimates than smaller random samples. That's because a large random sample gives us more information about the underlying population than a smaller sample does.

There are general rules for describing how the spread of the sampling distribution of a statistic decreases as the sample size increases. In Sections 7.2 and 7.3, we'll reveal these rules for the sampling distributions of $\hat{p}$ and $\bar{x}$. One important and surprising fact is that the variability of a statistic in repeated sampling does *not* depend very much on the size of the population.

Variability of a Statistic

The **variability of a statistic** is described by the spread of its sampling distribution. This spread is determined primarily by the size of the random sample. Larger samples give smaller spread. The spread of the sampling distribution does not depend on the size of the population, as long as the population is at least 10 times larger than the sample.

Why does the size of the population have little influence on the behavior of statistics from random samples? Imagine sampling harvested corn by thrusting a scoop into a large sack of corn kernels. The scoop doesn't know whether it is surrounded by a bag of corn or by an entire truckload. As long as the corn is well mixed (so that the scoop selects a random sample), the variability of the result depends only on the size of the scoop.

The fact that the variability of a statistic is controlled by the size of the sample has important consequences for designing samples. Suppose a researcher wants to estimate the proportion of all U.S. adults who blog regularly. A random sample of 1000 or 1500 people will give a fairly precise estimate of the parameter because the sample size is large. Now consider another researcher who wants to estimate the proportion of all Ohio State University students who blog regularly. It can take just as large an SRS to estimate the proportion of Ohio State University students who blog regularly as to estimate with equal precision the proportion of all U.S. adults who blog regularly. We can't expect to need a smaller SRS at Ohio State just because there are 50,000 Ohio State students and 235 million adults in the United States.

CAUTION

Taking a larger sample doesn't fix bias. Remember that even a very large voluntary response sample or convenience sample is worthless because of bias.

CHECK YOUR UNDERSTANDING

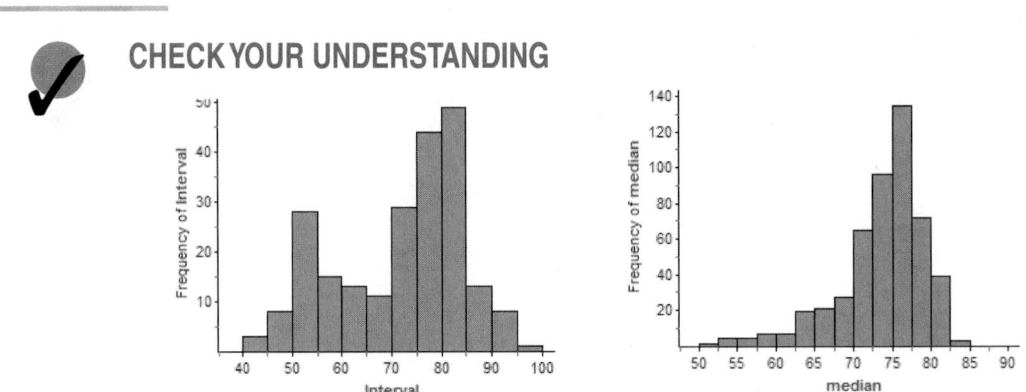

The histogram above left shows the intervals (in minutes) between eruptions of the Old Faithful geyser for all 222 recorded eruptions during a particular month. For this population, the median is 75 minutes. We used Fathom software to take 500 SRSs of size 10 from the population. The 500 values of the sample median are displayed in the histogram above right. The mean of the 500 sample median values is 73.5.

1. Is the sample median an unbiased estimator of the population median? Justify your answer.
2. Suppose we had taken samples of size 20 instead of size 10. Would the spread of the sampling distribution be larger, smaller, or about the same? Justify your answer.
3. Describe the shape of the sampling distribution. Explain what this means in terms of overestimating or underestimating the population median.

Bias, variability, and shape We can think of the true value of the population parameter as the bull's-eye on a target and of the sample statistic as an arrow fired at the target. Both bias and variability describe what happens when we take many shots at the target. *Bias* means that our aim is off and we consistently miss the bull's-eye in the same direction. Our sample values do not center on the population value. *High variability* means that repeated shots are widely scattered on the target. Repeated samples do not give very similar results. Figure 7.8 shows this target illustration of the two types of error.

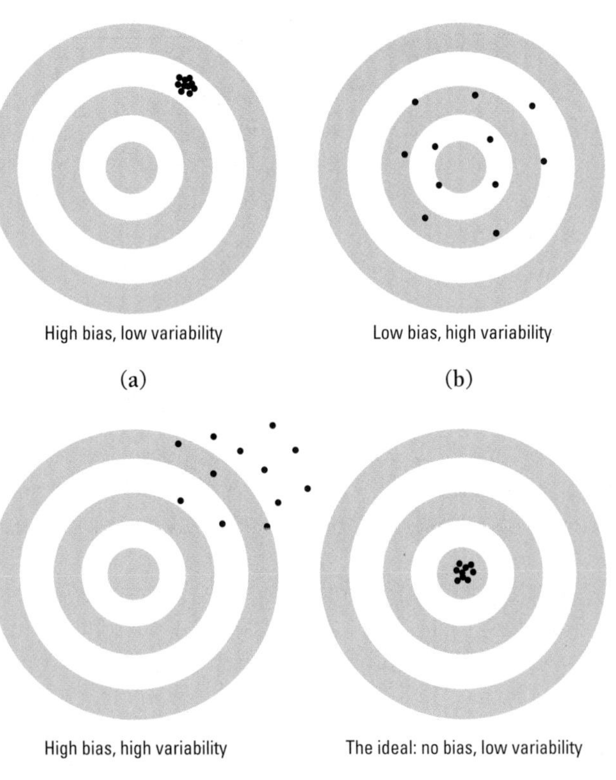

FIGURE 7.8 Bias and variability. (a) High bias, low variability. (b) Low bias, high variability. (c) High bias, high variability. (d) The ideal: no bias, low variability.

Notice that low variability (shots are close together) can accompany high bias (shots are consistently away from the bull's-eye in one direction). And low or no bias (shots center on the bull's-eye) can accompany high variability (shots are widely scattered). Ideally, we'd like our estimates to be *accurate* (unbiased) and *precise* (have low variability). See Figure 7.8(d).

The following example attempts to tie these ideas together in a familiar setting.

EXAMPLE

The German Tank Problem

Evaluating estimators: Shape, center, spread

Refer to the Activity on page 414. Mrs. Friedman's student teams came up with four different methods for estimating the number of tanks in the bag: (1) maxmin = maximum + minimum, (2) meanpl2sd = $\bar{x} + 2s_x$, (3) twicemean = $2\bar{x}$, and (4) twomedian = $2M$. She added one more method, called "partition." Figure 7.9 shows the results of taking 250 SRSs of 4 tanks and recording the value of the five statistics for each sample. The vertical line marks the actual value of the population parameter N: there were 342 tanks in the bag.

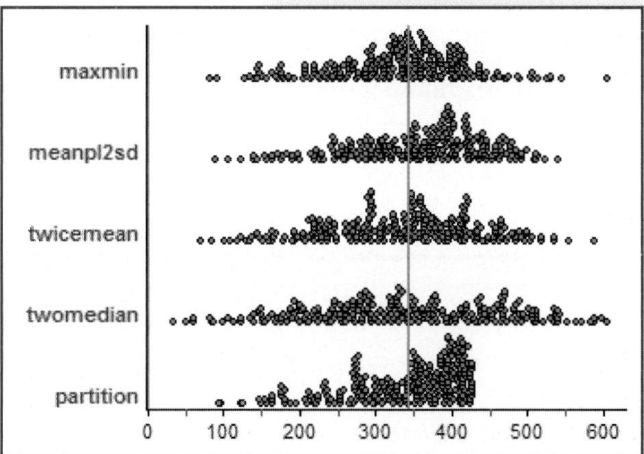

FIGURE 7.9 Results from a Fathom simulation of 250 SRSs of 4 tanks. The approximate sampling distributions of five different statistics are shown.

PROBLEM: Use the information in Figure 7.9 to help answer these questions.

(a) Which of the four statistics proposed by the student teams is the best estimator? Justify your answer.

(b) Why was the partition method, which uses the statistic (5/4) · maximum, recommended by the mathematicians in Washington, D.C.?

SOLUTION:

(a) Meanpl2sd is a biased estimator: the center of its sampling distribution is too high. This statistic produces consistent overestimates of the number of tanks. The other three statistics proposed by the students appear to be unbiased estimators. All three sampling distributions have roughly symmetric shapes, so these statistics are about equally likely to underestimate or overestimate the number of tanks. Since maxmin has the smallest variability among the three, it would generally produce estimates that are closer to the actual number of tanks. Among the students' proposed statistics, maxmin would be the best estimator.

(b) The partition method uses a statistic (5/4 · maximum) that is an unbiased estimator and that has much less variability than any of the student teams' statistics. Its sampling distribution is left-skewed, which means that the statistic is more likely to overestimate than underestimate the number of tanks. The mathematicians felt that it would be better to err on the side of caution and give the military commanders an estimate that's slightly too high.

For Practice *Try Exercise 19*

The lesson about center and spread is clear: given a choice of statistics to estimate an unknown parameter, choose one with no or low bias and minimum variability. Shape is a more complicated issue. We have seen sampling distributions that are left-skewed, right-skewed, roughly symmetric, and even approximately Normal. The same statistic can have sampling distributions with different shapes depending on the population distribution and the sample size. Our advice: be sure to consider the shape of the sampling distribution before doing inference.

SECTION 7.1 Summary

- A **parameter** is a number that describes a population. To estimate an unknown parameter, use a **statistic** calculated from a sample.
- The **population distribution** of a variable describes the values of the variable for all individuals in a population. The **sampling distribution** of a statistic describes the values of the statistic in all possible samples of the same size from the same population. Don't confuse the sampling distribution with a distribution of sample data, which gives the values of the variable for all individuals in a sample.
- A statistic can be an **unbiased estimator** or a **biased estimator** of a parameter. Bias means that the center (mean) of the sampling distribution is not equal to the true value of the parameter.
- The **variability** of a statistic is described by the spread of its sampling distribution. Larger samples give smaller spread.
- When trying to estimate a parameter, choose a statistic with low or no bias and minimum variability. Don't forget to consider the shape of the sampling distribution before doing inference.

SECTION 7.1 Exercises

For Exercises 1 to 4, identify the population, the parameter, the sample, and the statistic in each setting.

1. **Stop smoking!** A random sample of 1000 people who signed a card saying they intended to quit smoking were contacted nine months later. It turned out that 210 (21%) of the sampled individuals had not smoked over the past six months.

pg 416

2. **Unemployment** Each month, the Current Population Survey interviews a random sample of individuals in about 55,000 U.S. households. One of their goals is to estimate the national unemployment rate. In December 2009, 10.0% of those interviewed were unemployed.

3. **Hot turkey** Tom is cooking a large turkey breast for a holiday meal. He wants to be sure that the turkey is safe to eat, which requires a minimum internal temperature of 165°F. Tom uses a thermometer to measure the temperature of the turkey meat at four randomly chosen points. The minimum reading in the sample is 170°F.

4. **Gas prices** How much do gasoline prices vary in a large city? To find out, a reporter records the price per gallon of regular unleaded gasoline at a random sample of 10 gas stations in the city on the same day. The range (maximum – minimum) of the prices in the sample is 25 cents.

For each boldface number in Exercises 5 to 8, (1) state whether it is a parameter or a statistic and (2) use appropriate notation to describe each number; for example, $p = 0.65$.

5. **Get your bearings** A large container of ball bearings has mean diameter **2.5003** centimeters (cm). This is within the specifications for acceptance of the container by the purchaser. By chance, an inspector chooses 100 bearings from the container that have mean diameter **2.5009** cm. Because this is outside the specified limits, the container is mistakenly rejected.

6. **Florida voters** Florida has played a key role in recent presidential elections. Voter registration records show that **41%** of Florida voters are registered as Democrats. To test a random digit dialing device, you use it to call 250 randomly chosen residential telephones in Florida. Of the registered voters contacted, **33%** are registered Democrats.

7. **Unlisted numbers** A telemarketing firm in Los Angeles uses a device that dials residential telephone numbers in that city at random. Of the first 100 numbers dialed, **48%** are unlisted. This is not surprising because **52%** of all Los Angeles residential phones are unlisted.

8. **How tall?** A random sample of female college students has a mean height of **64.5** inches, which is greater than the **63**-inch mean height of all adult American women.

pg 419

9. **Doing homework** A school newspaper article claims that 60% of the students at a large high school did all their assigned homework last week. Some skeptical AP Statistics students want to investigate whether this claim is true, so they choose an SRS of 100 students from the school to interview. What values of the sample proportion $\hat{p}$ would be consistent with the claim that the population proportion of students who completed all their homework is $p = 0.60$? To find out, we used Fathom software to simulate choosing 250 SRSs of size $n = 100$ students from a population in which $p = 0.60$. The figure below is a dotplot of the sample proportion $\hat{p}$ of students who did all their homework.

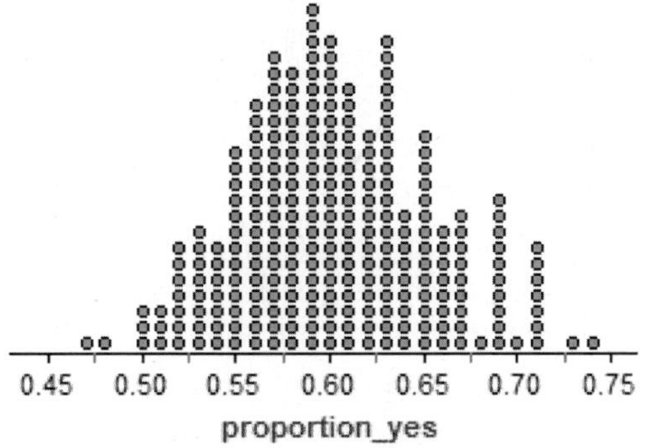

(a) Is this the sampling distribution of $\hat{p}$? Justify your answer.

(b) Describe the distribution. Are there any obvious outliers?

(c) Suppose that 45 of the 100 students in the actual sample say that they did all their homework last week. What would you conclude about the newspaper article's claim? Explain.

10. **Tall girls** According to the National Center for Health Statistics, the distribution of heights for 16-year-old females is modeled well by a Normal density curve with mean $\mu = 64$ inches and standard deviation $\sigma = 2.5$ inches. To see if this distribution applies at their high school, an AP Statistics class takes an SRS of 20 of the 300 16-year-old females at the school and measures their heights. What values of the sample mean $\bar{x}$ would be consistent with the population distribution being N(64, 2.5)? To find out, we used Fathom software to simulate choosing 250 SRSs of size $n = 20$ students from a population that is N(64, 2.5). The figure below is a dotplot of the sample mean height $\bar{x}$ of the students in the sample.

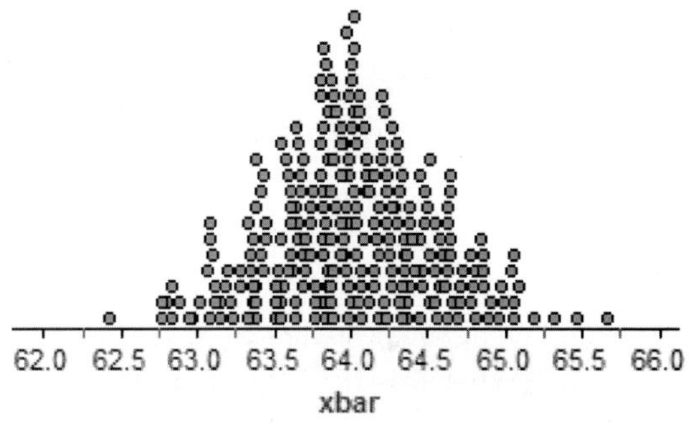

(a) Is this the sampling distribution of $\bar{x}$? Justify your answer.

(b) Describe the distribution. Are there any obvious outliers?

(c) Suppose that the average height of the 20 girls in the class's actual sample is $\bar{x} = 64.7$. What would you conclude about the population mean height μ for the 16-year-old females at the school? Explain.

11. **Doing homework** Refer to Exercise 9.

(a) Make a graph of the population distribution given that the newspaper's claim is correct and that there are 3000 students in the school.
(*Hint:* What type of variable is being measured?)

(b) Sketch a possible graph of the distribution of sample data for the SRS of size 100 taken by the AP Statistics students.

12. **Tall girls** Refer to Exercise 10.

(a) Make a graph of the population distribution.

(b) Sketch a possible graph of the distribution of sample data for the SRS of size 20 taken by the AP Statistics class.

Exercises 13 and 14 refer to the following setting. During the winter months, outside temperatures at the Starneses' cabin in Colorado can stay well below freezing (32°F, or 0°C) for weeks at a time. To prevent the pipes from freezing, Mrs. Starnes sets the thermostat at 50°F. The manufacturer claims that the thermostat allows variation in home temperature that follows a Normal distribution with $\sigma = 3$°F. To test this claim, Mrs. Starnes programs her digital thermometer to take an SRS of $n = 10$ readings during a 24-hour period. Suppose the thermostat is working properly and that the actual temperatures in the cabin vary according to a Normal distribution with mean $\mu = 50$°F and standard deviation $\sigma = 3$°F.

13. **Cold cabin?** The Fathom screen shot below shows the results of taking 500 SRSs of 10 temperature readings from a population distribution that's N(50, 3) and recording the sample variance s_x^2 each time.

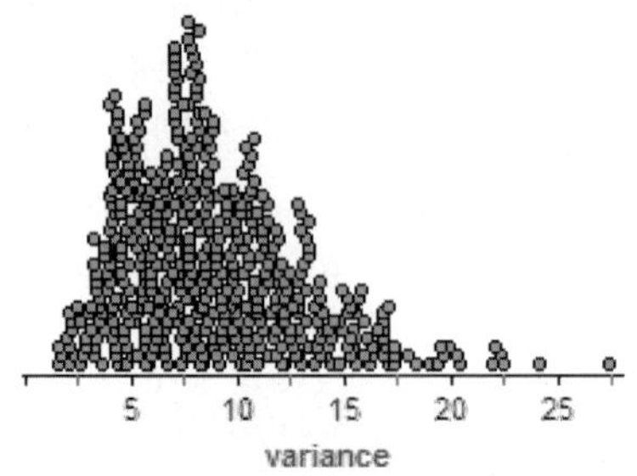

(a) Describe the approximate sampling distribution.

(b) Suppose that the variance from an actual sample is $s_x^2 = 25$. What would you conclude about the thermostat manufacturer's claim? Explain.

14. **Cold cabin?** The Fathom screen shot below shows the results of taking 500 SRSs of 10 temperature readings from a population distribution that's N(50, 3) and recording the sample minimum each time.

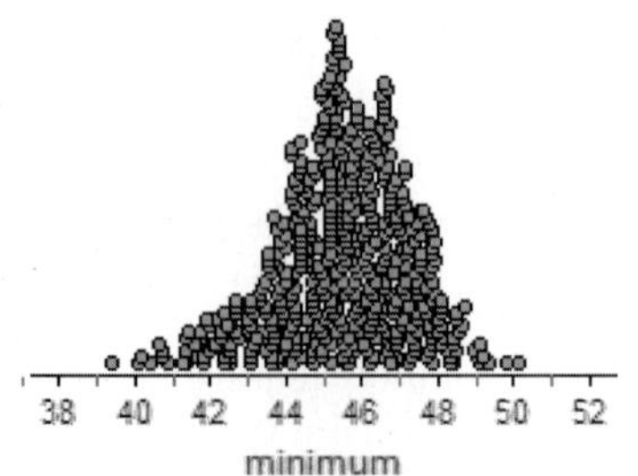

(a) Describe the approximate sampling distribution.

(b) Suppose that the minimum of an actual sample is 40°F. What would you conclude about the thermostat manufacturer's claim? Explain.

15. **Run a mile** During World War II, 12,000 able-bodied male undergraduates at the University of Illinois participated in required physical training. Each student ran a timed mile. Their times followed the Normal distribution with mean 7.11 minutes and standard deviation 0.74 minute. An SRS of 100 of these students has mean time $\bar{x} = 7.15$ minutes. A second SRS of size 100 has mean $\bar{x} = 6.97$ minutes. After many SRSs, the values of the sample mean $\bar{x}$ follow the Normal distribution with mean 7.11 minutes and standard deviation 0.074 minute.

(a) What is the population? Describe the population distribution.

(b) Describe the sampling distribution of $\bar{x}$. How is it different from the population distribution?

16. **Scooping beads** A statistics teacher fills a large container with 1000 white and 3000 red beads and then mixes the beads thoroughly. She then has her students take repeated SRSs of 50 beads from the container. After many SRSs, the values of the sample proportion $\hat{p}$ of red beads are approximated well by a Normal distribution with mean 0.75 and standard deviation 0.06.

(a) What is the population? Describe the population distribution.

(b) Describe the sampling distribution of $\hat{p}$. How is it different from the population distribution?

17. **IRS audits** The Internal Revenue Service plans to examine an SRS of individual federal income tax returns from each state. One variable of interest is the proportion of returns claiming itemized deductions. The total number of tax returns in each state varies from over 15 million in California to about 240,000 in Wyoming.

(a) Will the sampling variability of the sample proportion change from state to state if an SRS of 2000 tax returns is selected in each state? Explain your answer.

(b) Will the sampling variability of the sample proportion change from state to state if an SRS of 1% of all tax returns is selected in each state? Explain your answer.

18. **Predict the election** Just before a presidential election, a national opinion poll increases the size of its weekly random sample from the usual 1500 people to 4000 people.

(a) Does the larger random sample reduce the bias of the poll result? Explain.

(b) Does it reduce the variability of the result? Explain.

19. **Bias and variability** The figure below shows histograms of four sampling distributions of different statistics intended to estimate the same parameter.

pg 427

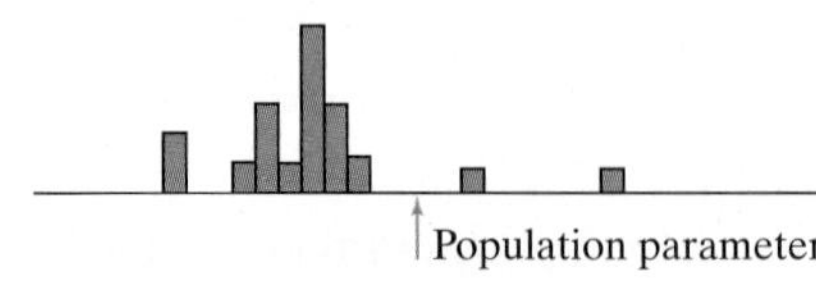

(a)

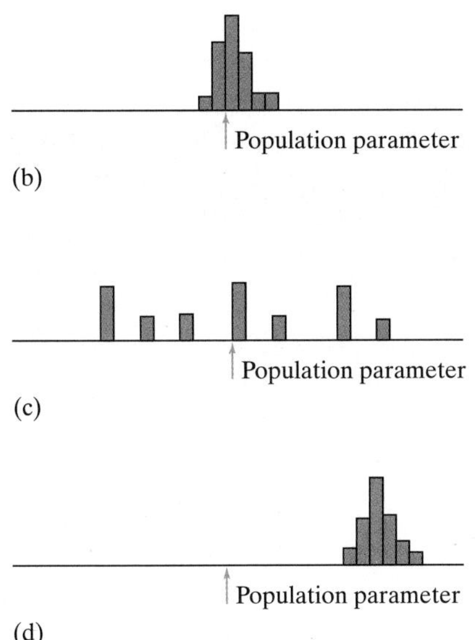

(a) Which statistics are unbiased estimators? Justify your answer.

(b) Which statistic does the best job of estimating the parameter? Explain.

20. **A sample of teens** A study of the health of teenagers plans to measure the blood cholesterol levels of an SRS of 13- to 16-year-olds. The researchers will report the mean $\bar{x}$ from their sample as an estimate of the mean cholesterol level μ in this population.

(a) Explain to someone who knows no statistics what it means to say that $\bar{x}$ is an unbiased estimator of μ.

(b) The sample result $\bar{x}$ is an unbiased estimator of the population mean μ no matter what size SRS the study chooses. Explain to someone who knows no statistics why a large random sample gives more trustworthy results than a small random sample.

Multiple choice: Select the best answer for Exercises 21 to 24.

21. A newspaper poll reported that 73% of respondents liked business tycoon Donald Trump. The number 73% is

(a) a population.

(b) a parameter.

(c) a sample.

(d) a statistic.

(e) an unbiased estimator.

22. The name for the pattern of values that a statistic takes when we sample repeatedly from the same population is

(a) the bias of the statistic.

(b) the variability of the statistic.

(c) the population distribution.

(d) the distribution of sample data.

(e) the sampling distribution of the statistic.

23. If we take a simple random sample of size $n = 500$ from a population of size 5,000,000, the variability of our estimate will be

(a) much less than the variability for a sample of size $n = 500$ from a population of size 50,000,000.

(b) slightly less than the variability for a sample of size $n = 500$ from a population of size 50,000,000.

(c) about the same as the variability for a sample of size $n = 500$ from a population of size 50,000,000.

(d) slightly greater than the variability for a sample of size $n = 500$ from a population of size 50,000,000.

(e) much greater than the variability for a sample of size $n = 500$ from a population of size 50,000,000.

24. Increasing the sample size of an opinion poll will

(a) reduce the bias of the poll result.

(b) reduce the variability of the poll result.

(c) reduce the effect of nonresponse on the poll.

(d) reduce the variability of opinions.

(e) all of the above.

25. **Dem bones** (2.2) Osteoporosis is a condition in which the bones become brittle due to loss of minerals. To diagnose osteoporosis, an elaborate apparatus measures bone mineral density (BMD). BMD is usually reported in standardized form. The standardization is based on a population of healthy young adults. The World Health Organization (WHO) criterion for osteoporosis is a BMD score that is 2.5 standard deviations below the mean for young adults. BMD measurements in a population of people similar in age and gender roughly follow a Normal distribution.

(a) What percent of healthy young adults have osteoporosis by the WHO criterion?

(b) Women aged 70 to 79 are, of course, not young adults. The mean BMD in this age group is about −2 on the standard scale for young adults. Suppose that the standard deviation is the same as for young adults. What percent of this older population has osteoporosis?

26. **Squirrels and their food supply** (3.2) Animal species produce more offspring when their supply of food goes up. Some animals appear able to anticipate unusual food abundance. Red squirrels eat seeds from pinecones, a food source that sometimes has very large crops. Researchers collected data on an index of the abundance of pinecones and the average number of offspring per female over 16 years.[3] Computer output from a least-squares regression on these data and a residual plot are shown on the next page.

Predictor	Coef	SE Coef	T	P
Constant	1.4146	0.2517	5.62	0.000
Cone index	0.4399	0.1016	4.33	0.001

S = 0.600309 R-Sq = 57.2% R-Sq(adj) = 54.2%

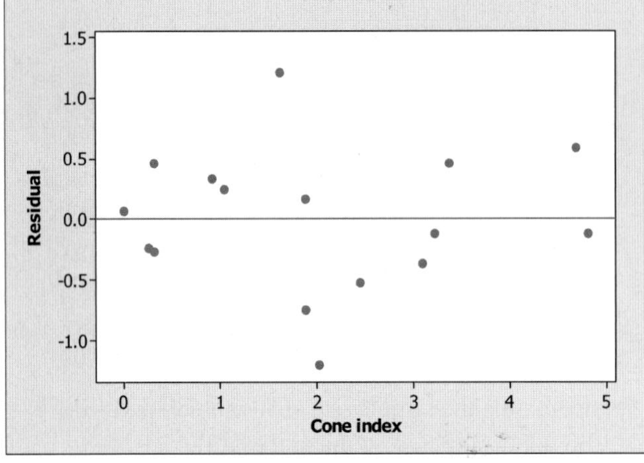

(a) Give the equation for the least-squares regression line. Define any variables you use.

(b) Explain what the residual plot tells you about how well the linear model fits the data.

(c) Interpret the values of r^2 and s in context.

Sample Proportions

In Section 7.2, you'll learn about:

- The sampling distribution of $\hat{p}$
- Using the Normal approximation for $\hat{p}$

What proportion of U.S. teens know that 1492 was the year in which Columbus "discovered" America? A Gallup Poll found that 210 out of a random sample of 501 American teens aged 13 to 17 knew this historically important date.[4] The sample proportion

$$\hat{p} = \frac{210}{501} = 0.42$$

is the statistic that we use to gain information about the unknown population parameter p. Since another random sample of 501 teens would likely result in a different estimate, we can only say that "about" 42% of U.S. teenagers know that Columbus discovered America in 1492. In this section, we'll use sampling distributions to clarify what "about" means.

The Sampling Distribution of $\hat{p}$

How good is the statistic $\hat{p}$ as an estimate of the parameter p? To find out, we ask, "What would happen if we took many samples?" The **sampling distribution of $\hat{p}$** answers this question. How do we determine the shape, center, and spread of the sampling distribution of $\hat{p}$? Let's start with a simulation.

ACTIVITY *The candy machine*

MATERIALS: Computer with Internet access—one for the class or one per pair of students

Imagine a very large candy machine filled with orange, brown, and yellow candies. When you insert money, the machine dispenses a sample of candies. In this Activity, you will use a Java applet to investigate the sample-to-sample variability in the proportion of orange candies dispensed by the machine.

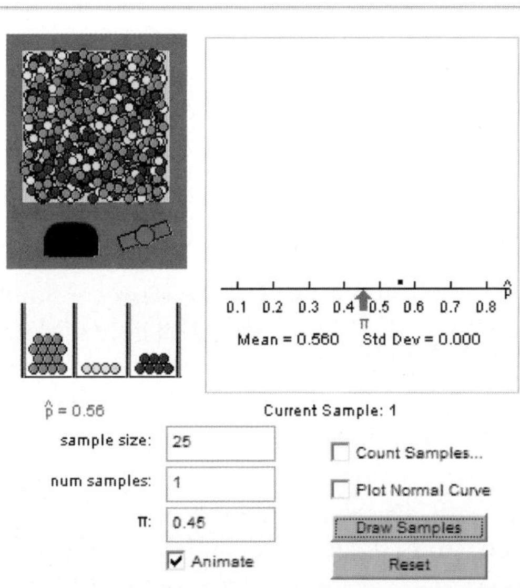

FIGURE 7.10 The result of taking one SRS of 25 candies from a large candy machine in which 45% of the candies are orange.

1. Launch the *Reese's Pieces®* applet at www.rossmanchance.com. Notice that the population proportion of orange candies is set to $p = 0.45$ (the applet calls this value π instead of p).

2. Click on the "Draw Samples" button. An animated simple random sample of $n = 25$ candies should be dispensed. Figure 7.10 shows the results of one such sample. Was your sample proportion of orange candies (look at the value of $\hat{p}$ in the applet window) close to the actual population proportion, $p = 0.45$?

3. Click "Draw Samples" 9 more times, so that you have a total of 10 sample results. Look at the dotplot of your $\hat{p}$-values. What is the mean of your 10 sample proportions? What is their standard deviation?

4. To take many more samples quickly, enter 390 in the "num samples" box. Click on the Animate box to turn the animation off. Then click "Draw Samples." You have now taken a total of 400 samples of 25 candies from the machine. Describe the shape, center, and spread of the approximate sampling distribution of $\hat{p}$ shown in the dotplot.

5. How would the sampling distribution of the sample proportion $\hat{p}$ change if the machine dispensed $n = 50$ candies each time instead of 25? "Reset" the applet. Take 400 samples of 50 candies. Describe the shape, center, and spread of the approximate sampling distribution.

6. How would the sampling distribution of $\hat{p}$ change if the proportion of orange candies in the machine was $p = 0.15$ instead of $p = 0.45$? Does your answer depend on whether $n = 25$ or $n = 50$? Use the applet to investigate these questions. Then write a brief summary of what you learned.

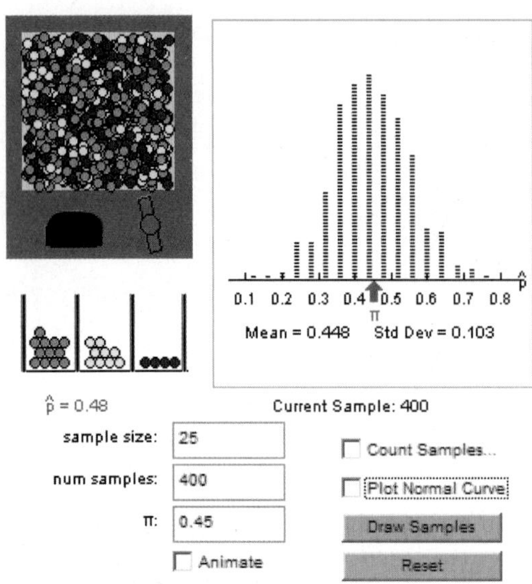

FIGURE 7.11 The result of taking 400 SRSs of 25 candies from a large candy machine in which 45% of the candies are orange. The dotplot shows the approximate sampling distribution of $\hat{p}$.

Figure 7.11 shows one set of possible results from Step 4 of "The Candy Machine" Activity. Let's describe what we see.

Shape: Roughly symmetric, unimodal, and bell-shaped. It looks like a Normal curve would approximate this distribution fairly well.
Center: The mean of the 400 sample proportions is 0.448. This is quite close to the actual population proportion, $p = 0.45$.
Spread: The standard deviation of the 400 values of $\hat{p}$ from these samples is 0.103.

The dotplot in Figure 7.11 is the approximate sampling distribution of $\hat{p}$. If we took all possible SRSs of $n = 25$ candies from the machine and graphed the value of $\hat{p}$ for each sample, then we'd have the actual sampling distribution of $\hat{p}$. We can get an idea of its shape, center, and spread from Figure 7.11.

Sampling Candies

Effect of n and p on shape, center, and spread

In a similar way, we can explore the sampling distribution of $\hat{p}$ when $n = 50$ (Step 5 of the Activity). As Figure 7.12 shows, the dotplot is once again roughly symmetric, unimodal, and bell-shaped. This graph is also centered at about 0.45. With samples of size 50, however, there is less spread in the values of $\hat{p}$. The standard deviation in Figure 7.12 is 0.068. For the samples of size 25 in Figure 7.11, it is 0.103. To repeat what we said earlier, larger samples give the sampling distribution a smaller spread.

What if the actual proportion of orange candies in the machine were $p = 0.15$? Figure 7.13(a) shows the approximate sampling distribution of $\hat{p}$ when $n = 25$. Notice that the dotplot is slightly right-skewed. The graph is centered close to the population parameter, $p = 0.15$. As for the spread, it's similar to the standard deviation in Figure 7.12, where $n = 50$ and $p = 0.45$. If we increase the sample size to $n = 50$, the sampling distribution of $\hat{p}$ should show less variability. The standard deviation in Figure 7.13(b) confirms this. (Note that we can't just visually compare the graphs because the horizontal scales are different.) The dotplot is more symmetrical than the graph in Figure 7.13(a) and is once again centered at a value that is close to $p = 0.15$.

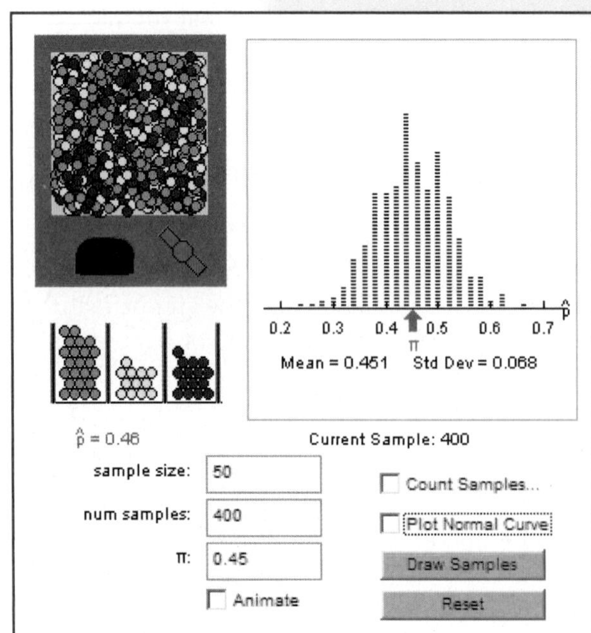

FIGURE 7.12 The approximate sampling distribution of $\hat{p}$ for SRSs of 50 candies from a population in which $p = 0.45$ of the candies are orange.

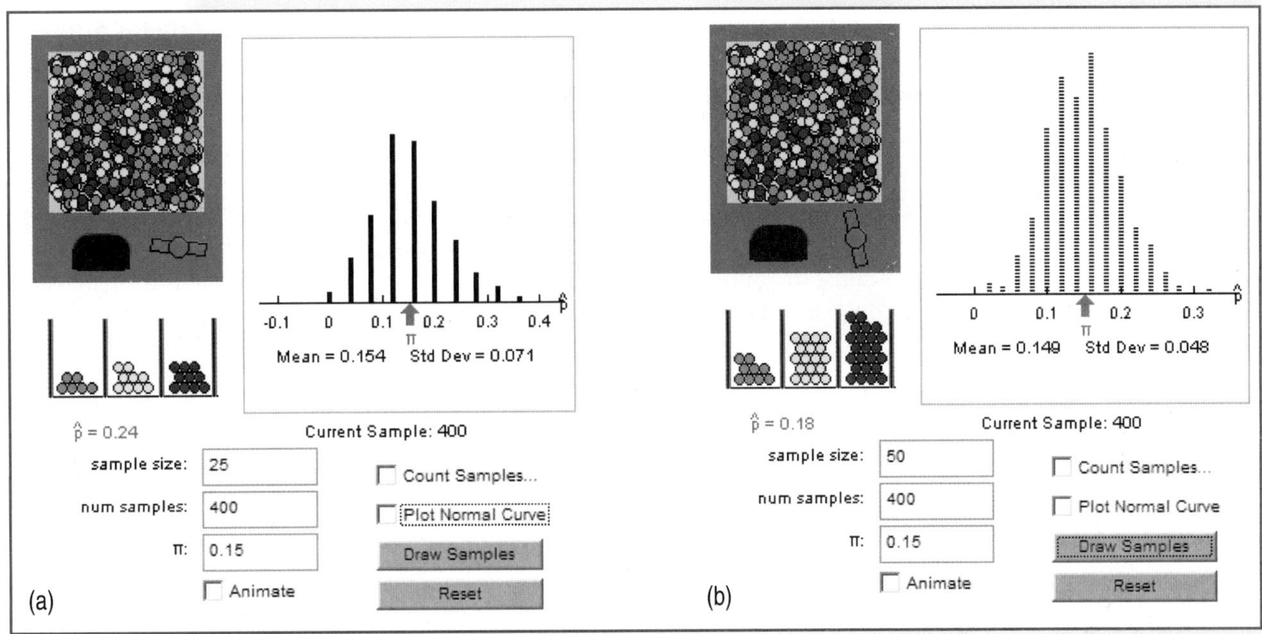

FIGURE 7.13 The result of taking 400 SRSs of (a) size $n = 25$ and (b) size $n = 50$ candies from a large candy machine in which 15% of the candies are orange. The dotplots show the approximate sampling distribution of $\hat{p}$ in each case.

What have we learned so far about the sampling distribution of $\hat{p}$?

Shape: In some cases, the sampling distribution of $\hat{p}$ can be approximated by a Normal curve. This seems to depend on both the sample size n and the population proportion p.
Center: The mean of the distribution is $\mu_{\hat{p}} = p$. This makes sense because the sample proportion $\hat{p}$ is an *unbiased estimator* of p.
Spread: For a specific value of p, the standard deviation $\sigma_{\hat{p}}$ gets smaller as n gets larger. The value of $\sigma_{\hat{p}}$ depends on both n and p.

To sort out the details of shape and spread, we need to make an important connection between the sample proportion $\hat{p}$ and the number of "successes" X in the sample.

In the candy machine example, we started by taking repeated SRSs of $n = 25$ candies from a population with proportion $p = 0.45$ of orange candies. For any such sample, we can think of each candy that comes out of the machine as a trial of this chance process. A "success" occurs when we get an orange candy. Let X = the number of orange candies obtained. As long as the number of candies in the machine is very large, X will have close to a binomial distribution with $n = 25$ and $p = 0.45$. (Refer to the sampling without replacement condition on page 394.) The sample proportion of successes is closely related to X:

$$\hat{p} = \frac{\text{count of successes in sample}}{\text{size of sample}} = \frac{X}{n}$$

How is the sampling distribution of $\hat{p}$ related to the binomial count X? From Chapter 6, we know that the mean and standard deviation of a binomial random variable X are

$$\mu_X = np \qquad \text{and} \qquad \sigma_X = \sqrt{np(1-p)}$$

Since $\hat{p} = X/n = (1/n)X$, we're just multiplying the random variable X by a constant $(1/n)$ to get the random variable $\hat{p}$. Recall from Chapter 6 that multiplying by a constant multiplies both the mean and the standard deviation of the new random variable by that constant. We have

$$\mu_{\hat{p}} = \frac{1}{n}(np) = p \quad \text{(confirming that } \hat{p} \text{ is an unbiased estimator of } p\text{)}$$

$$\sigma_{\hat{p}} = \frac{1}{n}\sqrt{np(1-p)} = \sqrt{\frac{np(1-p)}{n^2}} = \sqrt{\frac{p(1-p)}{n}}$$

(as sample size increases, spread decreases)

That takes care of center and spread. What about shape? Multiplying a random variable by a constant doesn't change the shape of the probability distribution. So the sampling distribution of $\hat{p}$ will have the same shape as the distribution of the binomial random variable X. If you studied the optional material in Chapter 6 about the Normal approximation to a binomial distribution, then you already know the punch line. Whenever np and $n(1-p)$ are at least 10, a Normal distribution can be used to approximate the sampling distribution of $\hat{p}$.

Here's a summary of the important facts about the sampling distribution of $\hat{p}$.

Sampling Distribution of a Sample Proportion

Choose an SRS of size n from a population of size N with proportion p of successes. Let $\hat{p}$ be the sample proportion of successes. Then:

- The **mean** of the sampling distribution of $\hat{p}$ is $\mu_{\hat{p}} = p$.
- The **standard deviation** of the sampling distribution of $\hat{p}$ is

$$\sigma_{\hat{p}} = \sqrt{\frac{p(1-p)}{n}}$$

as long as the *10% condition* is satisfied: $n \le \frac{1}{10}N$.

- As n increases, the sampling distribution of $\hat{p}$ becomes **approximately Normal.** Before you perform Normal calculations, check that the *Normal condition* is satisfied: $np \ge 10$ and $n(1-p) \ge 10$.

Figure 7.14 displays the facts in a form that helps you recall the big idea of a sampling distribution. The mean of the sampling distribution of $\hat{p}$ is the true value of the population proportion p. The standard deviation of $\hat{p}$ gets smaller as the sample size n increases. In fact, since the sample size n is under the square root sign, we'd have to take a sample four times as large to cut the standard deviation in half.

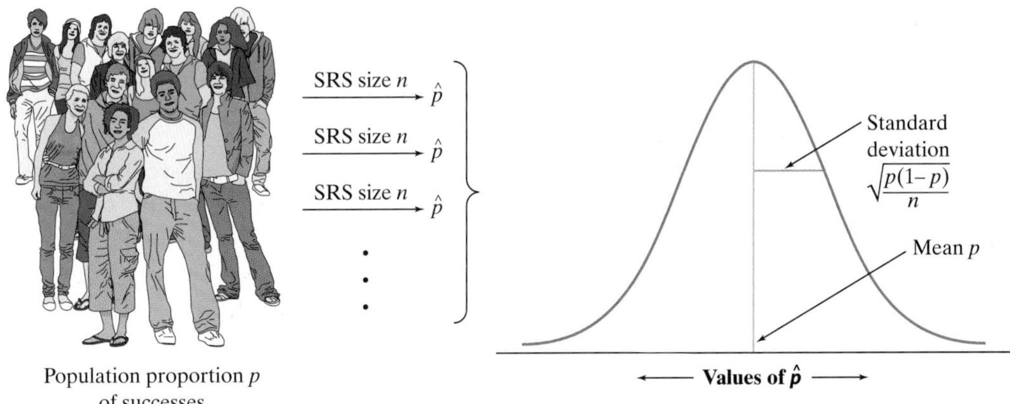

FIGURE 7.14 Select a large SRS from a population in which proportion p are successes. The sampling distribution of the proportion $\hat{p}$ of successes in the sample is approximately Normal. The mean is p and the standard deviation is $\sqrt{p(1-p)/n}$.

The two conditions in the preceding box are very important. (1) *10% condition:* When we're sampling without replacement from a (finite) population, the proportion p of successes is actually changing as we select each individual. But the standard deviation formula assumes that p remains constant. If we sample too large a fraction of the population, our calculated value of $\sigma_{\hat{p}}$ will be inaccurate. (2) *Normal condition:* If we assume that the sampling distribution of $\hat{p}$ is approximately Normal when it isn't, any calculations we make using a Normal distribution will be flawed.

Since larger random samples give better information, it sometimes makes sense to sample more than 10% of a population. In such a case, there's a more accurate formula for calculating the standard deviation $\sigma_{\hat{p}}$. It uses something called a *finite population correction* (FPC). We'll avoid situations that require the FPC in this text.

CHECK YOUR UNDERSTANDING

About 75% of young adult Internet users (ages 18 to 29) watch online video. Suppose that a sample survey contacts an SRS of 1000 young adult Internet users and calculates the proportion $\hat{p}$ in this sample who watch online video.

1. What is the mean of the sampling distribution of $\hat{p}$? Explain.
2. Find the standard deviation of the sampling distribution of $\hat{p}$. Check that the 10% condition is met.
3. Is the sampling distribution of $\hat{p}$ approximately Normal? Check that the Normal condition is met.
4. If the sample size were 9000 rather than 1000, how would this change the sampling distribution of $\hat{p}$?

Using the Normal Approximation for $\hat{p}$

Inference about a population proportion p is based on the sampling distribution of $\hat{p}$. When the sample size is large enough for np and $n(1-p)$ to both be at least 10 (the *Normal condition*), the sampling distribution of $\hat{p}$ is approximately Normal. In that case, we can use a Normal distribution to calculate the probability of obtaining an SRS in which $\hat{p}$ lies in a specified interval of values. Here is an example.

EXAMPLE Going to College

Normal calculations involving $\hat{p}$

STEP 4

A polling organization asks an SRS of 1500 first-year college students how far away their home is. Suppose that 35% of all first-year students actually attend college within 50 miles of home. What is the probability that the random sample of 1500 students will give a result within 2 percentage points of this true value? Follow the four-step process.

STATE: We want to find the probability that $\hat{p}$ falls between 0.33 and 0.37 (within 2 percentage points, or 0.02, of 0.35). In symbols, that's $P(0.33 \leq \hat{p} \leq 0.37)$.

PLAN: We have an SRS of size $n = 1500$ drawn from a population in which the proportion $p = 0.35$ attend college within 50 miles of home. What do we know about the sampling distribution of $\hat{p}$?

- Its mean is $\mu_{\hat{p}} = 0.35$.
- What about the standard deviation? We need to check the 10% condition. To use the standard deviation formula we derived, the population must contain at least $10(1500) = 15{,}000$ people. There are over 1.7 million first-year college students, so

$$\sigma_{\hat{p}} = \sqrt{\frac{p(1-p)}{n}} = \sqrt{\frac{(0.35)(0.65)}{1500}} = 0.0123$$

- Can we use a Normal distribution to approximate the sampling distribution of $\hat{p}$? Check the Normal condition: $np = 1500(0.35) = 525$ and $n(1-p) = 1500(0.65) = 975$. Both are much larger than 10, so the Normal approximation will be quite accurate.

Figure 7.15 shows the Normal distribution that we'll use. The area of the shaded region corresponds to the probability that $0.33 \leq \hat{p} \leq 0.37$.

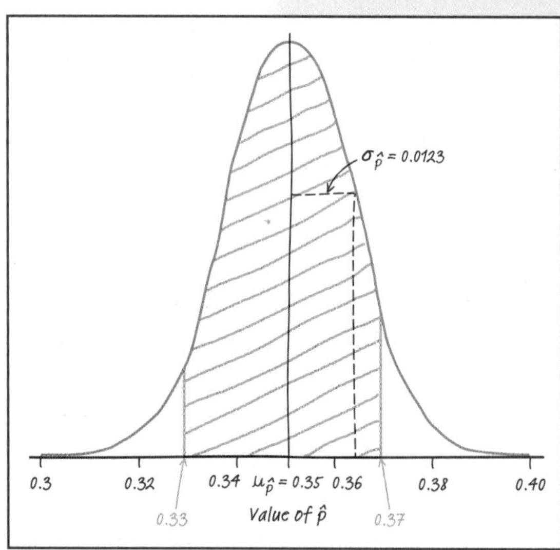

FIGURE 7.15 The Normal approximation to the sampling distribution of $\hat{p}$.

DO: We'll standardize and then use Table A to find the desired probability.

- *Standardize.* Find the standardized values (z-scores) of $\hat{p} = 0.33$ and $\hat{p} = 0.37$.

For $\hat{p} = 0.33$: $$z = \frac{0.33 - 0.35}{0.0123} = -1.63$$

For $\hat{p} = 0.37$: $$z = \frac{0.37 - 0.35}{0.0123} = 1.63$$

- *Use Table A.* Figure 7.16 shows the area under the standard Normal curve corresponding to these standardized values. The proportion of observations less than $z = -1.63$ is 0.0516, and the proportion of observations less than $z = 1.63$ is 0.9484. So the desired probability is

$$P(0.33 \leq \hat{p} \leq 0.37) = P(-1.63 \leq z \leq 1.63) = 0.9484 - 0.0516 = 0.8968$$

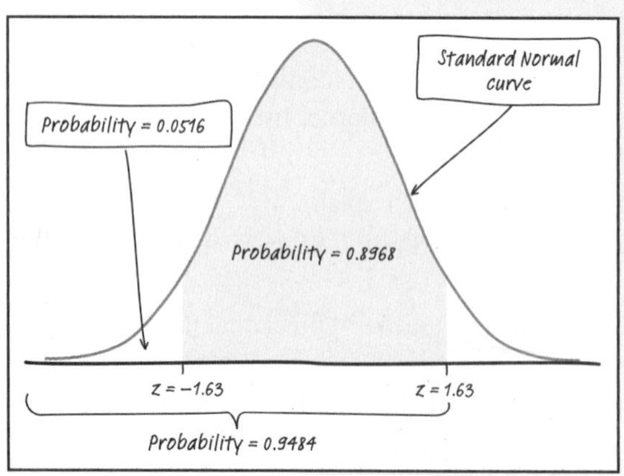

FIGURE 7.16 Probabilities as areas under the standard Normal curve.

CONCLUDE: About 90% of all SRSs of size 1500 will give a result within 2 percentage points of the truth about the population.

Using the *Normal Curve* applet or `normalcdf(0.33, 0.37, 0.35, 0.0123)` on the graphing calculator yields $P(0.33 < \hat{p} \leq 0.37) = 0.8961$.

For Practice *Try Exercise 39*

SECTION 7.2 Summary

- When we want information about the population proportion p of successes, we often take an SRS and use the sample proportion $\hat{p}$ to estimate the unknown parameter p. The **sampling distribution** of $\hat{p}$ describes how the statistic varies in all possible samples from the population.
- The **mean** of the sampling distribution of $\hat{p}$ is equal to the population proportion p. That is, $\hat{p}$ is an unbiased estimator of p.
- The **standard deviation** of the sampling distribution of $\hat{p}$ is $\sqrt{p(1-p)/n}$ for an SRS of size n. This formula can be used if the population is at least 10 times as large as the sample (the *10% condition*). The standard deviation of $\hat{p}$ gets smaller as the sample size n gets larger. Because of the square root, a sample four times larger is needed to cut the standard deviation in half.
- When the sample size n is large, the sampling distribution of $\hat{p}$ is close to a Normal distribution with mean p and standard deviation $\sqrt{p(1-p)/n}$. In practice, use this **Normal approximation** when both $np \geq 10$ and $n(1-p) \geq 10$ (the *Normal condition*).

SECTION 7.2 Exercises

27. **The candy machine** Suppose a large candy machine has 45% orange candies. Use Figures 7.11 and 7.12 (page 434) to help answer the following questions.

(a) Would you be surprised if a sample of 25 candies from the machine contained 8 orange candies (that's 32% orange)? How about 5 orange candies (20% orange)? Explain.

(b) Which is more surprising: getting a sample of 25 candies in which 32% are orange or getting a sample of 50 candies in which 32% are orange? Explain.

28. **The candy machine** Suppose a large candy machine has 15% orange candies. Use Figure 7.13 (page 435) to help answer the following questions.

(a) Would you be surprised if a sample of 25 candies from the machine contained 8 orange candies (that's 32% orange)? How about 5 orange candies (20% orange)? Explain.

(b) Which is more surprising: getting a sample of 25 candies in which 32% are orange or getting a sample of 50 candies in which 32% are orange? Explain.

29. **The candy machine** Suppose a large candy machine has 45% orange candies. Imagine taking an SRS of 25 candies from the machine and observing the sample proportion $\hat{p}$ of orange candies.

(a) What is the mean of the sampling distribution of $\hat{p}$? Why?

(b) Find the standard deviation of the sampling distribution of $\hat{p}$. Check to see if the 10% condition is met.

(c) Is the sampling distribution of $\hat{p}$ approximately Normal? Check to see if the Normal condition is met.

(d) If the sample size were 50 rather than 25, how would this change the sampling distribution of $\hat{p}$?

30. **The candy machine** Suppose a large candy machine has 15% orange candies. Imagine taking an SRS of 25 candies from the machine and observing the sample proportion $\hat{p}$ of orange candies.

(a) What is the mean of the sampling distribution of $\hat{p}$? Why?

(b) Find the standard deviation of the sampling distribution of $\hat{p}$. Check to see if the 10% condition is met.

(c) Is the sampling distribution of $\hat{p}$ approximately Normal? Check to see if the Normal condition is met.

(d) If the sample size were 75 rather than 25, how would this change the sampling distribution of $\hat{p}$?

31. **Airport security** The Transportation Security Administration (TSA) is responsible for airport safety. On some flights, TSA officers randomly select passengers for an extra security check before boarding. One such flight had 76 passengers—12 in first class and 64 in coach class. TSA officers selected an SRS of 10 passengers for screening. Let $\hat{p}$ be the proportion of first-class passengers in the sample.

(a) Is the 10% condition met in this case? Justify your answer.

(b) Is the Normal condition met in this case? Justify your answer.

32. **Scrabble** In the game of Scrabble, each player begins by drawing 7 tiles from a bag containing 100 tiles. There are 42 vowels, 56 consonants, and 2 blank tiles in the bag. Cait chooses an SRS of 7 tiles. Let $\hat{p}$ be the proportion of vowels in her sample.

(a) Is the 10% condition met in this case? Justify your answer.

(b) Is the Normal condition met in this case? Justify your answer.

In Exercises 33 and 34, explain why you cannot use the methods of this section to find the desired probability.

33. **Hispanic workers** A factory employs 3000 unionized workers, of whom 30% are Hispanic. The 15-member union executive committee contains 3 Hispanics. What would be the probability of 3 or fewer Hispanics if the executive committee were chosen at random from all the workers?

CHALLENGE: See if you can compute the probability using another method.

34. **Studious athletes** A university is concerned about the academic standing of its intercollegiate athletes. A study committee chooses an SRS of 50 of the 316 athletes to interview in detail. Suppose that 40% of the athletes have been told by coaches to neglect their studies on at least one occasion. What is the probability that at least 15 in the sample are among this group?

35. **Do you drink the cereal milk?** A *USA Today* Poll asked a random sample of 1012 U.S. adults what they do with the milk in the bowl after they have eaten the cereal. Of the respondents, 67% said that they drink it. Suppose that 70% of U.S. adults actually drink the cereal milk. Let $\hat{p}$ be the proportion of people in the sample who drink the cereal milk.

(a) What is the mean of the sampling distribution of $\hat{p}$? Why?

(b) Find the standard deviation of the sampling distribution of $\hat{p}$. Check to see if the 10% condition is met.

(c) Is the sampling distribution of $\hat{p}$ approximately Normal? Check to see if the Normal condition is met.

(d) Find the probability of obtaining a sample of 1012 adults in which 67% or fewer say they drink the cereal milk. Do you have any doubts about the result of this poll?

36. **Do you go to church?** The Gallup Poll asked a random sample of 1785 adults whether they attended church or synagogue during the past week. Of the respondents, 44% said they did attend. Suppose that 40% of the adult population actually went to church or synagogue last week. Let $\hat{p}$ be the proportion of people in the sample who attended church or synagogue.

(a) What is the mean of the sampling distribution of $\hat{p}$? Why?

(b) Find the standard deviation of the sampling distribution of $\hat{p}$. Check to see if the 10% condition is met.

(c) Is the sampling distribution of $\hat{p}$ approximately Normal? Check to see if the Normal condition is met.

(d) Find the probability of obtaining a sample of 1785 adults in which 44% or more say they attended church or synagogue last week. Do you have any doubts about the result of this poll?

37. **Do you drink the cereal milk?** What sample size would be required to reduce the standard deviation of the sampling distribution to one-half the value you found in Exercise 35(b)? Justify your answer.

38. **Do you go to church?** What sample size would be required to reduce the standard deviation of the sampling distribution to one-third the value you found in Exercise 36(b)? Justify your answer.

pg 438

39. **Students on diets** A sample survey interviews an SRS of 267 college women. Suppose (as is roughly true) that 70% of college women have been on a diet within the past 12 months. What is the probability that 75% or more of the women in the sample have been on a diet? Follow the four-step process.

40. **Who owns a Harley?** Harley-Davidson motorcycles make up 14% of all the motorcycles registered in the United States. You plan to interview an SRS of 500 motorcycle owners. How likely is your sample to contain 20% or more who own Harleys? Follow the four-step process.

41. **On-time shipping** Your mail-order company advertises that it ships 90% of its orders within three working days. You select an SRS of 100 of the 5000 orders received in the past week for an audit. The audit reveals that 86 of these orders were shipped on time.

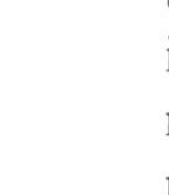

(a) If the company really ships 90% of its orders on time, what is the probability that the proportion in an SRS of 100 orders is as small as the proportion in your sample or smaller? Follow the four-step process.

(b) A critic says, "Aha! You claim 90%, but in your sample the on-time percentage is lower than that. So the 90% claim is wrong." Explain in simple language why your probability calculation in (a) shows that the result of the sample does not refute the 90% claim.

42. **Underage drinking** The Harvard College Alcohol Study finds that 67% of college students support efforts to "crack down on underage drinking." The study took a random sample of almost 15,000 students, so the population proportion who support a crackdown is close to $p = 0.67$.[5] The administration of a local college surveys an SRS of 100 students and finds that 62 support a crackdown on underage drinking.

(a) Suppose that the proportion of all students attending this college who support a crackdown is 67%, the same as the national proportion. What is the probability that the proportion in an SRS of 100 students is as small as or smaller than the result of the administration's sample? Follow the four-step process.

(b) A writer in the college's student paper says that "support for a crackdown is lower at our school than nationally." Write a short letter to the editor explaining why the survey does not support this conclusion.

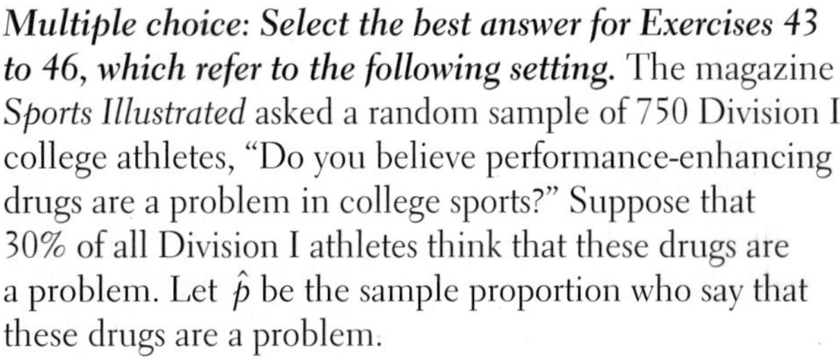

Multiple choice: Select the best answer for Exercises 43 to 46, which refer to the following setting. The magazine *Sports Illustrated* asked a random sample of 750 Division I college athletes, "Do you believe performance-enhancing drugs are a problem in college sports?" Suppose that 30% of all Division I athletes think that these drugs are a problem. Let $\hat{p}$ be the sample proportion who say that these drugs are a problem.

43. The sampling distribution of $\hat{p}$ has mean
(a) 225.
(b) 0.30.
(c) 0.017.
(d) 0.
(e) none of these.

44. The standard deviation of the sampling distribution is about
(a) 0.0006.
(b) 0.033.
(c) 0.017.
(d) 1.
(e) none of these.

45. Decreasing the sample size from 750 to 375 would multiply the standard deviation by
(a) 2.
(b) $\sqrt{2}$.
(c) 1/2.
(d) $1/\sqrt{2}$.
(e) none of these.

46. The sampling distribution of $\hat{p}$ is approximately Normal because
(a) there are at least 7570 Division I college athletes.
(b) $np = 225$ and $n(1 - p) = 525$.
(c) a random sample was chosen.
(d) a large sample size like $n = 750$ guarantees it.
(e) the sampling distribution of $\hat{p}$ always has this shape.

47. **Sharing music online** (5.2) A sample survey reports that 29% of Internet users download music files online, 21% share music files from their computers, and 12% both download and share music.[6] Make a Venn diagram that displays this information. What percent of Internet users neither download nor share music files?

48. **California's endangered animals** (4.1) The California Department of Fish and Game publishes a list of the state's endangered animals. The reptiles on the list are given across the bottom of the page.

(a) Describe how you would use Table D at line 111 to choose an SRS of 3 of these reptiles to study.

(b) Use your method from part (a) to select your sample. Identify the reptiles you chose.

Desert tortoise	Green sea turtle	Loggerhead sea turtle	Giant garter snake
Olive Ridley sea turtle	Leatherback sea turtle	Barefoot banded gecko	San Francisco garter snake
Island night lizard	Alameda whip snake	Coachella Valley fringe-toed lizard	
Flat-tailed horned lizard	Southern rubber boa	Blunt-nosed leopard lizard	

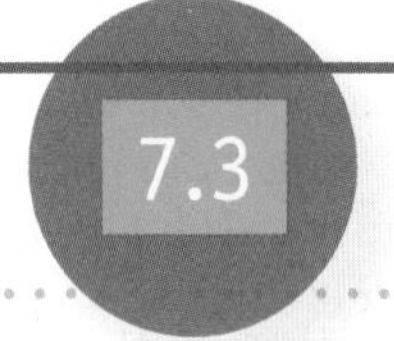

Sample Means

In Section 7.3, you'll learn about:

- The sampling distribution of $\bar{x}$: mean and standard deviation
- Sampling from a Normal population
- The central limit theorem

Sample proportions arise most often when we are interested in categorical variables. We then ask questions like "What proportion of U.S. adults have watched *Survivor: Guatemala*?" or "What percent of the adult population attended church last week?" But when we record quantitative variables—household income, lifetime of car brake pads, blood pressure—we are interested in other statistics, such as the median or mean or standard deviation of the variable. Because sample means are just averages of observations, they are among the most common statistics. This section describes the sampling distribution of the sample mean $\bar{x}$. The following Activity and the subsequent example give you a sense of what lies ahead.

ACTIVITY *Penny for your thoughts*

MATERIALS: Large container with several hundred pennies; chalkboard, whiteboard, or butcher paper for making comparative dotplots

Your teacher will assemble a large population of pennies of various ages.[7] In this Activity, your class will investigate the sampling distribution of the mean age $\bar{x}$ in a sample of pennies for SRSs of several different sizes. Then, you will compare these distributions of the mean age with the population distribution.

1. Make a dotplot of the population distribution of penny ages. To find the age, subtract the date on the penny from the current year.
2. Have each member of the class take an SRS of 5 pennies from the population and record the date on each penny. Be sure to replace these coins in the container before the next student takes a sample. If your class has fewer than 25 students, have each person take two samples.
3. Calculate the mean age $\bar{x}$ of the 5 pennies in your sample.
4. Make a class dotplot of the sample mean ages for SRSs of size 5 using the same scale as you did for the population distribution. Use $\bar{x}$'s instead of dots when making the graph.
5. Repeat the process in Steps 2 to 4 for samples of size 10. Use the same scale for your dotplot and place it directly beneath the graph for samples of size 5.
6. Repeat the process in Steps 2 to 4 for samples of size 25. Use the same scale for your dotplot and place it beside the graph for samples of size 10.
7. Compare the population distribution with the three approximate sampling distributions of $\bar{x}$. What do you notice about shape, center, and spread as the sample size increases?

Making Money

A first look at the sampling distribution of $\bar{x}$

Every March, the Current Population Survey contacts close to 100,000 U.S. households as part of its Annual Social and Economic Supplement. Figure 7.17(a) is a histogram of the earnings of the 61,742 households that had earned income greater than zero in a recent year.[8]

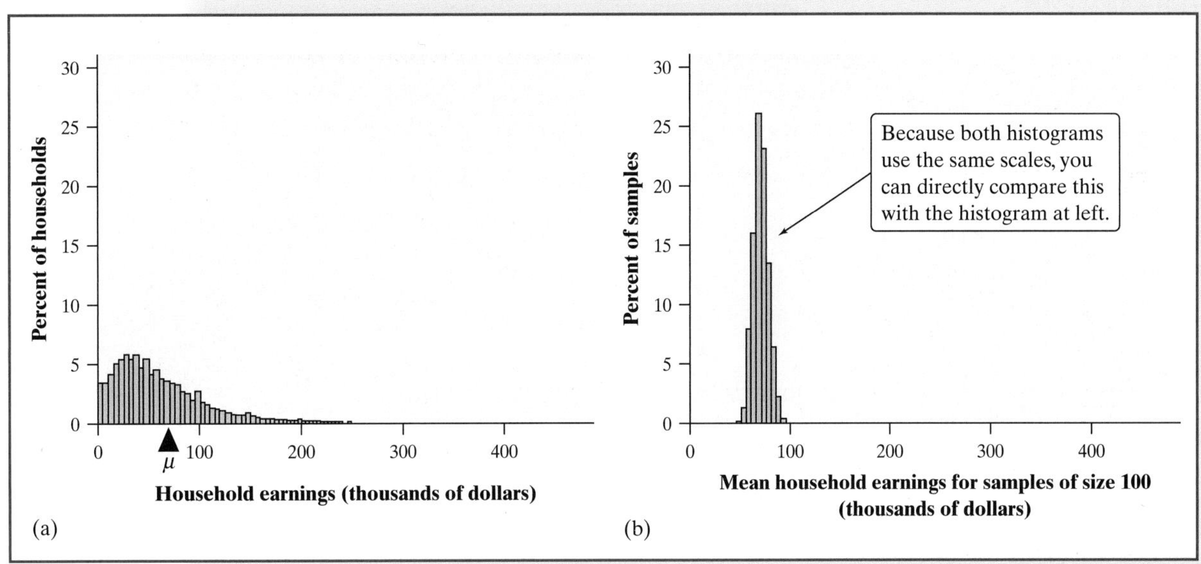

FIGURE 7.17 (a) The distribution of earned income in a population of 61,472 households. (b) The distribution of the mean earnings $\bar{x}$ for 500 SRSs of 100 households each from this population.

As we expect, the distribution of earned incomes is strongly skewed to the right and very spread out. The right tail of the distribution is even longer than the histogram shows because there are too few high incomes for their bars to be visible on this scale. In fact, we cut off the earnings scale at $400,000 to save space—a few households earned even more than $400,000. The mean earnings for these 61,742 households was $69,750.

Think of these 61,742 households as a population with mean $\mu = \$69,750$. Take an SRS of 100 households. The mean earnings in this sample is $\bar{x} = \$66,807$. That's less than the mean of the population. Take another SRS of size 100. The mean for this sample is $\bar{x} = \$70,820$. That's higher than the mean of the population. What would happen if we did this many times? Figure 7.17(b) is a histogram of the mean earnings for 500 samples, each of size $n = 100$. The scales in Figures 7.17(a) and 7.17(b) are the same, for easy comparison. Although the distribution of individual earnings is skewed and very spread out, the distribution of sample means is roughly symmetric and much less spread out.

This example illustrates an important fact that we will make precise in this section: averages are less variable than individual observations.

The Sampling Distribution of $\bar{x}$: Mean and Standard Deviation

Figure 7.17 suggests that when we choose many SRSs from a population, the sampling distribution of the sample mean is centered at the population mean μ and is less spread out than the population distribution. Here are the facts.

Mean and Standard Deviation of the Sampling Distribution of $\bar{x}$

Suppose that $\bar{x}$ is the mean of an SRS of size n drawn from a large population with mean μ and standard deviation σ. Then:

- The **mean** of the sampling distribution of $\bar{x}$ is $\mu_{\bar{x}} = \mu$.
- The **standard deviation** of the sampling distribution of $\bar{x}$ is

$$\sigma_{\bar{x}} = \frac{\sigma}{\sqrt{n}}$$

as long as the *10% condition* is satisfied: $n \leq \frac{1}{10}N$.

AP EXAM TIP Notation matters. The symbols $\hat{p}$, $\bar{x}$, p, μ, σ, $\mu_{\hat{p}}$, $\sigma_{\hat{p}}$, $\mu_{\bar{x}}$, and $\sigma_{\bar{x}}$ all have specific and different meanings. Either use notation correctly—or don't use it at all. You can expect to lose credit if you use incorrect notation.

The behavior of $\bar{x}$ in repeated samples is much like that of the sample proportion $\hat{p}$:

- The sample mean $\bar{x}$ is an *unbiased estimator* of the population mean μ.
- The values of $\bar{x}$ are less spread out for larger samples. Their standard deviation decreases at the rate $\sqrt{n}$, so you must take a sample four times as large to cut the standard deviation of $\bar{x}$ in half.
- You should use the formula $\sigma/\sqrt{n}$ for the standard deviation of $\bar{x}$ only when the population is at least 10 times as large as the sample (the 10% condition).

Notice that these facts about the mean and standard deviation of $\bar{x}$ are true *no matter what shape the population distribution has.*

EXAMPLE

This Wine Stinks

Mean and standard deviation of $\bar{x}$

Sulfur compounds such as dimethyl sulfide (DMS) are sometimes present in wine. DMS causes "off-odors" in wine, so winemakers want to know the odor threshold, the lowest concentration of DMS that the human nose can detect. Extensive studies have found that the DMS odor threshold of adults follows roughly a Normal distribution with mean $\mu = 25$ micrograms per liter and standard deviation $\sigma = 7$ micrograms per liter. Suppose we take an SRS of 10 adults and determine the mean odor threshold $\bar{x}$ for the individuals in the sample.

PROBLEM:

(a) What is the mean of the sampling distribution of $\bar{x}$? Explain.

(b) What is the standard deviation of the sampling distribution of $\bar{x}$? Check that the 10% condition is met.

SOLUTION:

(a) Since $\bar{x}$ is an unbiased estimator of μ, $\mu_{\bar{x}} = \mu = 25$ micrograms per liter.

(b) The standard deviation is $\sigma_{\bar{x}} = \frac{\sigma}{\sqrt{n}} = \frac{7}{\sqrt{10}} = 2.214$ since there are at least $10(10) = 100$ adults in the population.

For Practice Try Exercise 49

Can we confirm the formulas for the mean and standard deviation of $\bar{x}$? Choose an SRS of size n from a population, and measure a variable X on each individual in the sample. Call the individual measurements $X_1, X_2, \ldots, X_n$. If the population is large relative to the sample, we can think of these X_i's as independent random variables, each with mean μ and standard deviation σ. Since

$$\bar{x} = \frac{1}{n}(X_1 + X_2 + \cdots + X_n)$$

we can use the rules for random variables from Chapter 6 to find the mean and standard deviation of $\bar{x}$. If we let $T = X_1 + X_2 + \cdots + X_n$, then $\bar{x} = \frac{1}{n}T$.

Using the addition rules for means and variances

$$\mu_T = \mu_{X_1} + \mu_{X_2} + \cdots + \mu_{X_n} = \mu + \mu + \cdots + \mu = n\mu$$

$$\sigma_T^2 = \sigma_{X_1}^2 + \sigma_{X_2}^2 + \cdots + \sigma_{X_n}^2 = \sigma^2 + \sigma^2 + \cdots + \sigma^2 = n\sigma^2$$

$$\Rightarrow \sigma_T = \sqrt{n\sigma^2} = \sqrt{n}\sigma$$

Since $\bar{x}$ is just a constant multiple of the random variable T,

$$\mu_{\bar{x}} = \frac{1}{n}\mu_T = \frac{1}{n}(n\mu) = \mu$$

$$\sigma_{\bar{x}} = \frac{1}{n}\sigma_T = \frac{1}{n}\left(\sqrt{n}\sigma\right) = \sqrt{\frac{n}{n^2}}\sigma = \sqrt{\frac{1}{n}}\sigma = \frac{1}{\sqrt{n}}\sigma = \frac{\sigma}{\sqrt{n}}$$

Sampling from a Normal Population

We have described the mean and standard deviation of the sampling distribution of a sample mean $\bar{x}$ but not its shape. That's because the shape of the distribution of $\bar{x}$ depends on the shape of the population distribution. In one important case, there is a simple relationship between the two distributions. The following Activity shows what we mean.

ACTIVITY *Exploring the sampling distribution of $\bar{x}$ for a Normal population*

MATERIALS: Computer with Internet access—one for the class or one per pair of students

Professor David Lane of Rice University has developed a wonderful applet for investigating the sampling distribution of $\bar{x}$. It's dynamic, and it's pretty fun to play with. In this Activity, you'll use Professor Lane's applet to explore the shape of the sampling distribution when the population is Normally distributed.

1. Search for "Rice University sampling distributions applet" and go to the Web site. When the BEGIN button appears on the left side of the screen, click on it. You will then see a yellow page entitled "Sampling Distributions" like the one in the figure below.

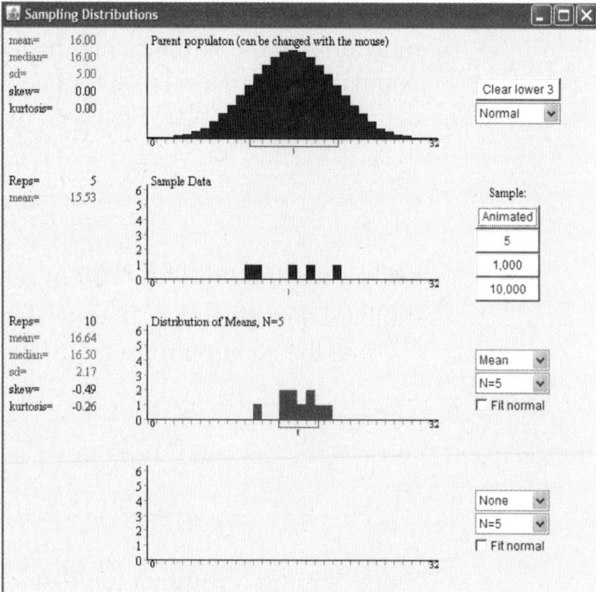

2. There are choices for the population distribution: Normal, uniform, skewed, and custom. The default is Normal. Click the "Animated" button. What happens? Click the button several more times. What do the black boxes represent? What is the blue square that drops down onto the plot below? What does the red horizontal band under the population histogram tell us?

Notice the left panel. Important numbers are displayed there. Did you notice that the colors of the numbers match up with the objects to the right? As you make things happen, the numbers change accordingly, like an automatic scorekeeper.

3. Click on "Clear lower 3" to start clean. Then click on the "1,000" button under "Sample:" repeatedly until you have simulated taking 10,000 SRSs of size $n = 5$ from the population (look for "Reps = 10000" on the left panel in black letters). Answer these questions:

- Does the approximate sampling distribution (blue bars) have a recognizable shape? Click the box next to "Fit normal."
- Compare the mean of the approximate sampling distribution with the mean of the population.

• How is the standard deviation of the approximate sampling distribution related to the standard deviation of the population?

4. Click "Clear lower 3." Use the drop-down menus to set up the bottom graph to display the mean for samples of size $n = 20$. Then sample 10,000 times. How do the two distributions of $\bar{x}$ compare: shape, center, and spread?

5. What have you learned about the shape of the sampling distribution of $\bar{x}$ when the population has a Normal shape?

As the previous Activity demonstrates, if the population distribution is Normal, then so is the sampling distribution of $\bar{x}$. *This is true no matter what the sample size is.*

Sampling Distribution of a Sample Mean from a Normal Population

Suppose that a population is Normally distributed with mean μ and standard deviation σ. Then the sampling distribution of $\bar{x}$ has the Normal distribution with mean μ and standard deviation $\sigma/\sqrt{n}$, provided that the 10% condition is met.

We already knew the mean and standard deviation of the sampling distribution. All we have added is the Normal shape. Now we have enough information to calculate probabilities involving $\bar{x}$ when the population distribution is Normal.

Young Women's Heights

Finding probabilities involving the sample mean

PROBLEM: The height of young women follows a Normal distribution with mean $\mu = 64.5$ inches and standard deviation $\sigma = 2.5$ inches.

(a) Find the probability that a randomly selected young woman is taller than 66.5 inches. Show your work.

(b) Find the probability that the mean height of an SRS of 10 young women exceeds 66.5 inches. Show your work.

SOLUTION:

(a) Let X be the height of a randomly selected young woman. The random variable X follows a Normal distribution with $\mu = 64.5$ inches and $\sigma = 2.5$ inches. To find $P(X > 66.5)$:

- *Standardize.* When $X = 66.5$, $z = \dfrac{66.5 - 64.5}{2.5} = 0.80$.
- *Use Table A.* $P(X > 66.5) = P(z > 0.80) = 1 - 0.7881 = 0.2119$.

The probability of choosing a young woman at random whose height exceeds 66.5 inches is about 0.21.

(b) For an SRS of 10 young women, the sampling distribution of their sample mean height $\bar{x}$ will have mean $\mu_{\bar{x}} = \mu = 64.5$ inches. The 10% condition is met since there are at least $10(10) = 100$ young women in the population. So the standard deviation is $\sigma_{\bar{x}} = \dfrac{\sigma}{\sqrt{n}} = \dfrac{2.5}{\sqrt{10}} = 0.79$. Since the population distribution is Normal, the values of $\bar{x}$ will follow an $N(64.5, 0.79)$ distribution. To find the probability that $\bar{x} > 66.5$ inches:

For part (a), using the *Normal Curve* applet or `normalcdf (66.5, 100, 64.5, 2.5)` on the graphing calculator yields $P(X > 66.5) = 0.2119$. For part (b), using the *Normal Curve* applet or `normalcdf (66.5, 100, 64.5, 0.79)` on the graphing calculator yields $P(\bar{x} > 66.5) = 0.0057$.

- *Standardize.* A sample mean of 66.5 inches yields

$$z = \frac{66.5 - 64.5}{0.79} = 2.53.$$

- *Use Table A.* $P(\bar{x} > 66.5) = P(z > 2.53) = 1 - 0.9943 = 0.0057.$

It is very unlikely (less than a 1% chance) that we would choose an SRS of 10 young women whose average height exceeds 66.5 inches.

Figure 7.18 compares the population distribution and the sampling distribution of $\bar{x}$. It also shows the areas corresponding to the probabilities that we just computed. You can see that it is much less likely for the average height of 10 randomly selected young women to exceed 66.5 inches than it is for the height of one randomly selected young woman to exceed 66.5 inches.

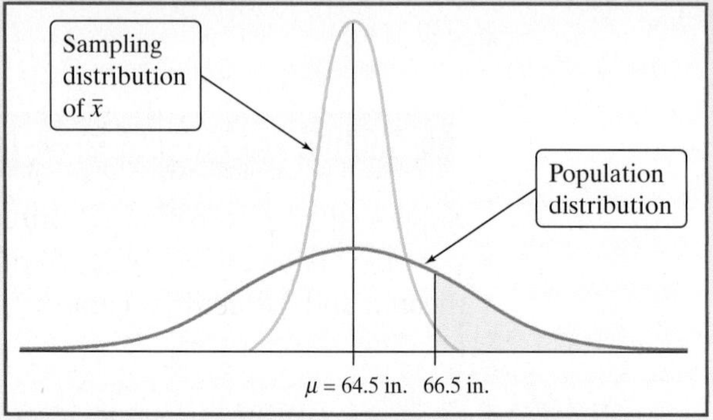

FIGURE 7.18 The sampling distribution of the mean height $\bar{x}$ for SRSs of 10 young women compared with the population distribution of young women's heights.

The fact that averages of several observations are less variable than individual observations is important in many settings. For example, it is common practice to repeat a measurement several times and report the average of the results. Think of the results of n repeated measurements as an SRS from the population of outcomes we would get if we repeated the measurement forever. The average of the n results (the sample mean $\bar{x}$) is less variable than a single measurement.

CHECK YOUR UNDERSTANDING

The length of human pregnancies from conception to birth varies according to a distribution that is approximately Normal with mean 266 days and standard deviation 16 days.

1. Find the probability that a randomly chosen pregnant woman has a pregnancy that lasts for more than 270 days. Show your work.

Suppose we choose an SRS of 6 pregnant women. Let $\bar{x}$ = the mean pregnancy length for the sample.

2. What is the mean of the sampling distribution of $\bar{x}$? Explain.

3. Compute the standard deviation of the sampling distribution of $\bar{x}$. Check that the 10% condition is met.

4. Find the probability that the mean pregnancy length for the women in the sample exceeds 270 days. Show your work.

The Central Limit Theorem

Most population distributions are not Normal. The household incomes in Figure 7.17(a) on page 443, for example, are strongly skewed. Yet Figure 7.17(b) suggests that the distribution of means for samples of size 100 is approximately Normal. What is the shape of the sampling distribution of $\bar{x}$ when the population distribution isn't Normal? The following Activity sheds some light on this question.

ACTIVITY *Exploring the sampling distribution of $\bar{x}$ for a non-Normal population*

MATERIALS: Computer with Internet access—one for the class or one per pair of students

Let's use the Rice University sampling distributions applet from the previous Activity to investigate what happens when we start with a non-Normal population distribution.

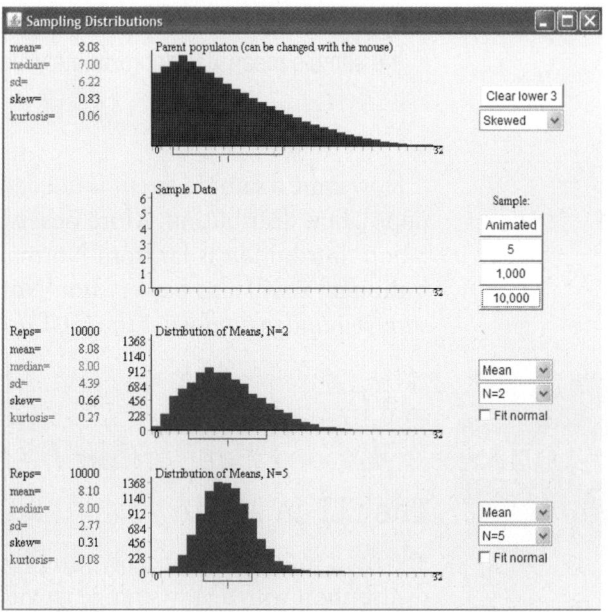

1. Go to the Web site and launch the applet. Select "Skewed" population. Set the bottom two graphs to display the mean—one for samples of size 2 and the other for samples of size 5. Click the Animated button a few times to be sure you see what's happening. Then "Clear lower 3" and take 10,000 SRSs. Describe what you see.

2. Change the sample sizes to $n = 10$ and $n = 16$ and repeat Step 1. What do you notice?

3. Now change the sample sizes to $n = 20$ and $n = 25$ and take 10,000 more samples. Did this confirm what you saw in Step 2?

4. Clear the page, and select "Custom" distribution. Click on a point on the population graph to insert a bar of that height. Or click on a point on the horizontal axis, and drag up to define a bar. Make a distribution that looks as strange as you can. (*Note:* You can shorten a bar or get rid of it completely by clicking on the top of the bar and dragging down to the axis.) Then repeat Steps 1 to 3 for your custom distribution. Cool, huh?

5. Summarize what you learned about the shape of the sampling distribution of $\bar{x}$.

It is a remarkable fact that as the sample size increases, the distribution of $\bar{x}$ changes shape: it looks less like that of the population and more like a Normal distribution. When the sample is large enough, the distribution of $\bar{x}$ is very close to Normal. This is true no matter what shape the population distribution has, as long as the population has a finite standard deviation σ. This famous fact of probability theory is called the **central limit theorem** (sometimes abbreviated as CLT).

DEFINITION: Central limit theorem (CLT)

Draw an SRS of size n from any population with mean μ and finite standard deviation σ. The **central limit theorem (CLT)** says that when n is large, the sampling distribution of the sample mean $\bar{x}$ is approximately Normal.

How large a sample size n is needed for $\bar{x}$ to be close to Normal depends on the population distribution. More observations are required if the shape of the population distribution is far from Normal. In that case, the sampling distribution of $\bar{x}$ will also be very non-Normal if the sample size is small. *Be sure you understand what the CLT does—and doesn't—say.*

CAUTION

A Strange Population Distribution

The CLT in action

We used the Rice University sampling distribution applet to create a population distribution with a very strange shape. Here is the population distribution.

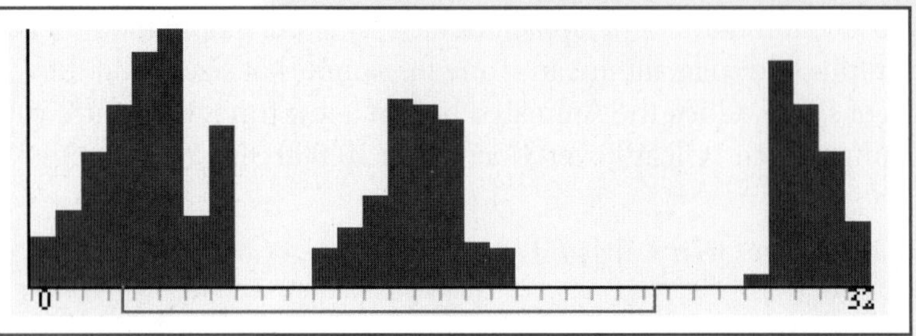

Figure 7.19 shows the approximate sampling distribution of the sample mean $\bar{x}$ for SRSs of size (a) $n = 2$, (b) $n = 5$, (c) $n = 10$, and (d) $n = 25$. As n increases, the shape becomes more Normal. The mean remains at $\mu = 13.5$, and the standard deviation decreases, taking the value $\sigma/\sqrt{n}$. For SRSs of size 2, the sampling distribution is very non-Normal. The distribution of $\bar{x}$ for 10 observations is slightly skewed to the right but already resembles a Normal curve. By $n = 25$, the sampling distribution is even more Normal. The contrast between the shapes of the population distribution and the distribution of the mean of 10 or 25 observations is striking.

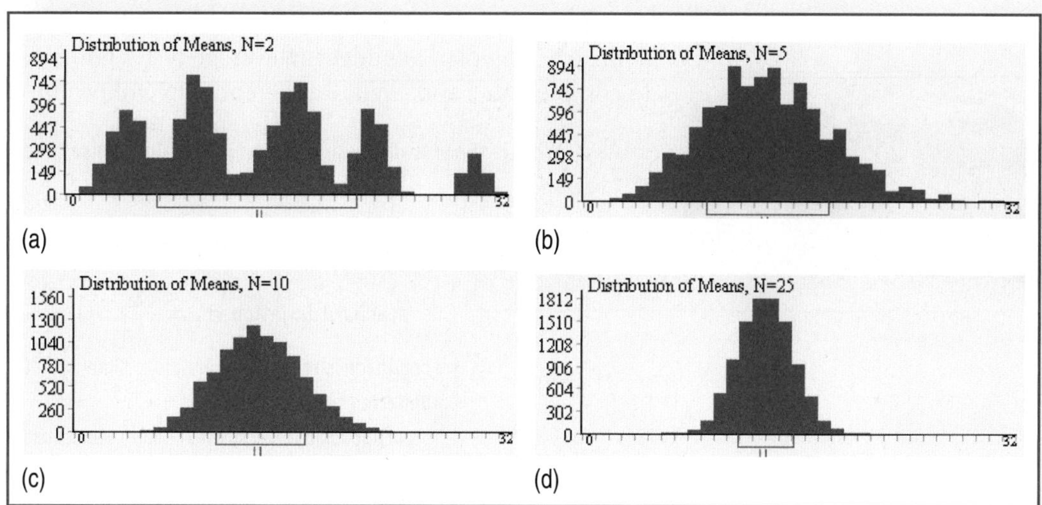

FIGURE 7.19 The central limit theorem in action: the distribution of sample means $\bar{x}$ from a strongly non-Normal population becomes more Normal as the sample size increases. (a) The distribution of $\bar{x}$ for 2 observations. (b) The distribution of $\bar{x}$ for 5 observations. (c) The distribution of $\bar{x}$ for 10 observations. (d) The distribution of $\bar{x}$ for 25 observations.

As the previous example illustrates, even when the population distribution is very non-Normal, the sampling distribution of $\bar{x}$ often looks approximately Normal with sample sizes as small as $n = 25$. To be safe, we'll require that n be at least 30 to invoke the CLT. With that issue settled, we can now state the *Normal condition* for sample means.

Normal Condition for Sample Means

- If the population distribution is Normal, then so is the sampling distribution of $\bar{x}$. This is true no matter what the sample size n is.
- If the population distribution is not Normal, the central limit theorem tells us that the sampling distribution of $\bar{x}$ will be approximately Normal in most cases if $n \geq 30$.

The central limit theorem allows us to use Normal probability calculations to answer questions about sample means from many observations even when the population distribution is not Normal.

Servicing Air Conditioners
Calculations using the CLT

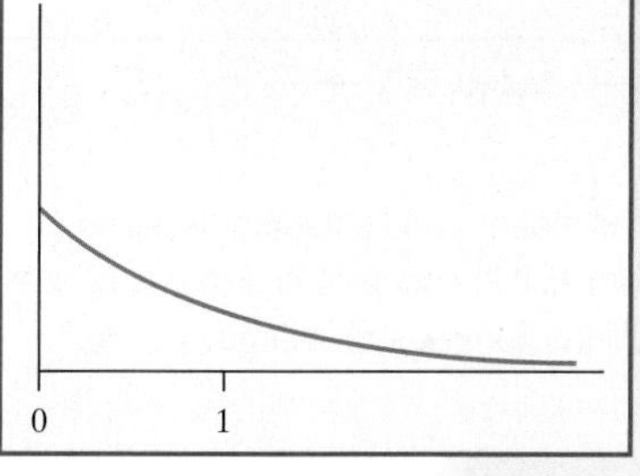

Your company has a contract to perform preventive maintenance on thousands of air-conditioning units in a large city. Based on service records from the past year, the time (in hours) that a technician requires to complete the work follows the distribution whose density curve is shown in the margin. This distribution is strongly right-skewed, and the most likely outcomes are close to 0. The mean time is $\mu = 1$ hour and the standard deviation is $\sigma = 1$ hour. In the

coming week, your company will service an SRS of 70 air-conditioning units in the city. You plan to budget an average of 1.1 hours per unit for a technician to complete the work. Will this be enough? Follow the four-step process.

STATE: What is the probability that the average maintenance time $\bar{x}$ for 70 units exceeds 1.1 hours?

PLAN: The sampling distribution of the sample mean time $\bar{x}$ spent working on 70 units has

- mean $\mu_{\bar{x}} = \mu = 1$ hour
- standard deviation $\sigma_{\bar{x}} = \frac{\sigma}{\sqrt{70}} = \frac{1}{\sqrt{70}} = 0.12$ since the 10% condition is met (there are more than $10(70) = 700$ air-conditioning units in the population)
- an approximately Normal shape since the Normal condition is met: $n = 70 \geq 30$

The distribution of $\bar{x}$ is therefore approximately $N(1, 0.12)$. We want to find $P(\bar{x} > 1.1)$. Figure 7.20 shows the Normal curve with the desired area shaded.

0.64 0.76 0.88 1 1.12 1.24 1.36

FIGURE 7.20 The Normal approximation from the central limit theorem for the average time needed to maintain an air conditioner.

DO: To determine whether it is safe to budget 1.1 hours, on average, the probability we want is $P(\bar{x} > 1.1)$.

- *Standardize.* $z = \frac{1.1 - 1}{0.12} = 0.83$.
- *Use Table A.* $P(\bar{x} > 1.1) = P(z > 0.83) = 1 - 0.7967 = 0.2033$.

CONCLUDE: If you budget 1.1 hours per unit, there is a 20% chance that the technicians will not complete the work within the budgeted time. You will have to decide if this risk is worth taking or if you should schedule more time for the work.

Using the *Normal curve* applet or `normalcdf(1.1, 100, 1, 0.12)` on the graphing calculator yields $P(\bar{x} > 1.1) = 0.2023$.

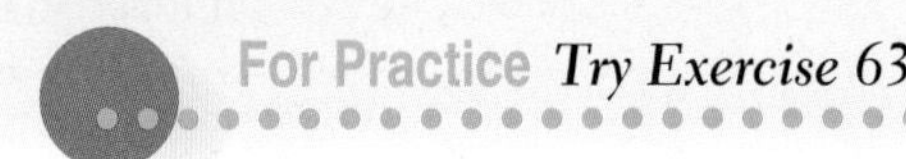

Figure 7.21 summarizes the facts about the sampling distribution of $\bar{x}$. It reminds us of the big idea of a sampling distribution. Keep taking random samples of size n from a population with mean μ. Find the sample mean $\bar{x}$ for each sample. Collect all the $\bar{x}$'s and display their distribution. That's the sampling distribution of $\bar{x}$. Sampling distributions are the key to understanding statistical inference. Keep this figure in mind for future reference.

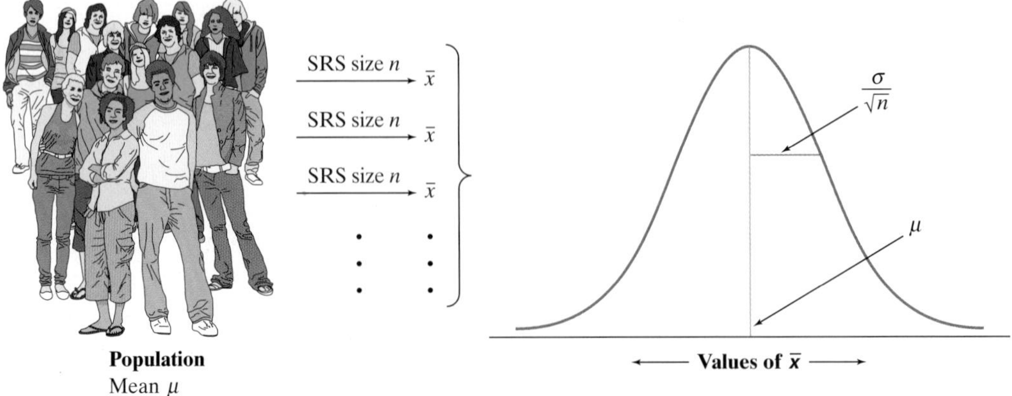

FIGURE 7.21 The sampling distribution of a sample mean $\bar{x}$ has mean μ and standard deviation $\sigma/\sqrt{n}$. It has a Normal shape if the population distribution is Normal. If the population distribution isn't Normal, the sampling distribution of $\bar{x}$ is approximately Normal if the sample size is large enough.

case closed

Building Better Batteries

Refer to the chapter-opening Case Study on page 413. Assuming the process is working properly, the population distribution of battery lifetimes has mean $\mu = 17$ hours and standard deviation $\sigma = 0.8$. We don't know the shape of the population distribution. What are the shape, center, and spread of the distribution of sample means for random samples of 30 batteries?

Shape: The sampling distribution of $\bar{x}$ will be approximately Normal since the Normal condition is met: the sample size is at least 30.

Center: The mean of the sampling distribution is $\mu_{\bar{x}} = \mu = 17$ hours.

Spread: As long as there are at least 300 batteries produced per hour (10% condition), the standard deviation of the sampling distribution is

$$\sigma_{\bar{x}} = \frac{\sigma}{\sqrt{n}} = \frac{0.8}{\sqrt{30}} = 0.1461 \text{ hours.}$$

The 30 batteries in this random sample had a mean lifetime of $\bar{x} = 16.7$ hours. If the process is working properly, what's the probability of getting a random sample of 30 batteries in which the sample mean lifetime is 16.7 hours or less?

- *Standardize.* $z = \dfrac{16.7 - 17}{0.1461} = -2.05.$
- *Use Table* A. $P(\bar{x} < 16.7) = P(z < -2.05) = 0.0202.$

There is only about a 2% chance of observing a sample mean as small or smaller than the one observed ($\bar{x} \le 16.7$) if the process is working properly. This small probability provides evidence that the process is not working properly.

SECTION 7.3 Summary

- When we want information about the population mean μ for some variable, we often take an SRS and use the sample mean $\bar{x}$ to estimate the unknown parameter μ. The **sampling distribution of** $\bar{x}$ describes how the statistic $\bar{x}$ varies in *all* possible samples of the same size from the population.
- The **mean** of the sampling distribution is μ, so that $\bar{x}$ is an unbiased estimator of μ.
- The **standard deviation** of the sampling distribution of $\bar{x}$ is $\sigma/\sqrt{n}$ for an SRS of size n if the population has standard deviation σ. That is, averages are less variable than individual observations. This formula can be used if the population is at least 10 times as large as the sample (*10% condition*).

- Choose an SRS of size n from a population with mean μ and standard deviation σ. If the population distribution is Normal, then so is the sampling distribution of the sample mean $\bar{x}$. If the population distribution is not Normal, the **central limit theorem (CLT)** states that when n is large, the sampling distribution of $\bar{x}$ is approximately Normal.
- We can use a Normal distribution to calculate approximate probabilities for events involving $\bar{x}$ whenever the *Normal condition* is met:
  - If the population distribution is Normal, so is the sampling distribution of $\bar{x}$.
  - If $n \geq 30$, the CLT tells us that the sampling distribution of $\bar{x}$ will be approximately Normal in most cases.

SECTION 7.3 Exercises

49. **Songs on an iPod** David's iPod has about 10,000 songs. The distribution of the play times for these songs is heavily skewed to the right with a mean of 225 seconds and a standard deviation of 60 seconds. Suppose we choose an SRS of 10 songs from this population and calculate the mean play time $\bar{x}$ of these songs. What are the mean and the standard deviation of the sampling distribution of $\bar{x}$? Explain.

pg 444

50. **Making auto parts** A grinding machine in an auto parts plant prepares axles with a target diameter $\mu = 40.125$ millimeters (mm). The machine has some variability, so the standard deviation of the diameters is $\sigma = 0.002$ mm. The machine operator inspects a random sample of 4 axles each hour for quality control purposes and records the sample mean diameter $\bar{x}$. Assuming that the process is working properly, what are the mean and standard deviation of the sampling distribution of $\bar{x}$? Explain.

51. **Songs on an iPod** Refer to Exercise 49. How many songs would you need to sample if you wanted the standard deviation of the sampling distribution of $\bar{x}$ to be 30 seconds? Justify your answer.

52. **Making auto parts** Refer to Exercise 50. How many axles would you need to sample if you wanted the standard deviation of the sampling distribution of $\bar{x}$ to be 0.0005 mm? Justify your answer.

53. **Larger sample** Suppose that the blood cholesterol level of all men aged 20 to 34 follows the Normal distribution with mean $\mu = 188$ milligrams per deciliter (mg/dl) and standard deviation $\sigma = 41$ mg/dl.

    (a) Choose an SRS of 100 men from this population. What is the sampling distribution of $\bar{x}$?

    (b) Find the probability that $\bar{x}$ estimates μ within ± 3 mg/dl. (This is the probability that $\bar{x}$ takes a value between 185 and 191 mg/dl.) Show your work.

    (c) Choose an SRS of 1000 men from this population. Now what is the probability that $\bar{x}$ falls within ± 3 mg/dl of μ? Show your work. In what sense is the larger sample "better"?

54. **Stop the car!** A car company has found that the lifetime of its disc brake pads varies from car to car according to a Normal distribution with mean $\mu = 55{,}000$ miles and standard deviation $\sigma = 4500$ miles. The company installs a new brand of brake pads on an SRS of 8 cars.

(a) If the new brand has the same lifetime distribution as the previous type of brake pad, what is the sampling distribution of the mean lifetime $\bar{x}$?

(b) The average life of the pads on these 8 cars turns out to be $\bar{x} = 51{,}800$ miles. Find the probability that the sample mean lifetime is 51,800 miles or less if the lifetime distribution is unchanged. What conclusion would you draw?

55. **Bottling cola** A bottling company uses a filling machine to fill plastic bottles with cola. The bottles are supposed to contain 300 milliliters (ml). In fact, the contents vary according to a Normal distribution with mean $\mu = 298$ ml and standard deviation $\sigma = 3$ ml.

pg 447

(a) What is the probability that an individual bottle contains less than 295 ml? Show your work.

(b) What is the probability that the mean contents of six randomly selected bottles is less than 295 ml? Show your work.

56. **ACT scores** The composite scores of individual students on the ACT college entrance examination in 2009 followed a Normal distribution with mean 21.1 and standard deviation 5.1.

(a) What is the probability that a single student randomly chosen from all those taking the test scores 23 or higher? Show your work.

(b) Now take an SRS of 50 students who took the test. What is the probability that the mean score $\bar{x}$ of these students is 23 or higher? Show your work.

57. **What does the CLT say?** Asked what the central limit theorem says, a student replies, "As you take larger and larger samples from a population, the histogram of the sample values looks more and more Normal." Is the student right? Explain your answer.

58. **The CLT applet** Go to the textbook Web site (www.whfreeman.com/tps4e) and click on "Statistical Applets." Launch the *Central Limit Theorem* applet. You should see a screen like the one shown here. Click and drag the slider to change the sample size, and watch how the density curve for the sampling distribution changes with it. Write a few sentences describing what is happening.

APPLET

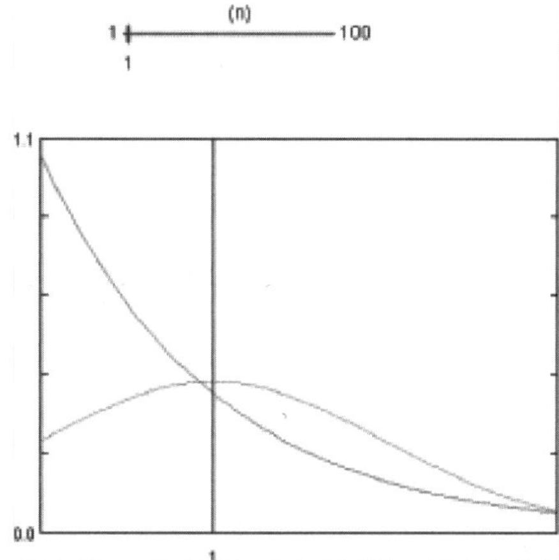

59. **Songs on an iPod** Refer to Exercise 49.

(a) Explain why you cannot safely calculate the probability that the mean play time $\bar{x}$ is more than 4 minutes (240 seconds) for an SRS of 10 songs.

(b) Suppose we take an SRS of 36 songs instead. Explain how the central limit theorem allows us to find the probability that the mean play time is more than 240 seconds. Then calculate this probability. Show your work.

60. **Lightning strikes** The number of lightning strikes on a square kilometer of open ground in a year has mean 6 and standard deviation 2.4. (These values are typical of much of the United States.) The National Lightning Detection Network (NLDN) uses automatic sensors to watch for lightning in a random sample of 10 one-square-kilometer plots of land.

(a) What are the mean and standard deviation of $\bar{x}$, the sample mean number of strikes per square kilometer?

(b) Explain why you cannot safely calculate the probability that $\bar{x} < 5$ based on a sample of size 10.

(c) Suppose the NLDN takes a random sample of $n = 50$ square kilometers instead. Explain how the central limit theorem allows us to find the probability that the mean number of lightning strikes per square kilometer is less than 5. Then calculate this probability. Show your work.

61. **Airline passengers get heavier** In response to the increasing weight of airline passengers, the Federal Aviation Administration (FAA) in 2003 told airlines to assume that passengers average 190 pounds in the summer, including clothes and carry-on baggage. But passengers vary, and the FAA did not specify a standard deviation. A reasonable standard deviation is 35 pounds. Weights are not Normally distributed, especially when the population includes both men and women, but they are not very non-Normal. A commuter plane carries 30 passengers.

(a) Explain why you cannot calculate the probability that a randomly selected passenger weighs more than 200 pounds.

(b) Find the probability that the total weight of the passengers on a full flight exceeds 6000 pounds. Show your work. (*Hint:* To apply the central limit theorem, restate the problem in terms of the mean weight.)

62. **How many people in a car?** A study of rush-hour traffic in San Francisco counts the number of people in each car entering a freeway at a suburban interchange. Suppose that this count has mean 1.5 and standard deviation 0.75 in the population of all cars that enter at this interchange during rush hours.

(a) Could the exact distribution of the count be Normal? Why or why not?

(b) Traffic engineers estimate that the capacity of the interchange is 700 cars per hour. Find the probability that 700 cars will carry more than 1075 people. Show your work. (*Hint:* Restate this event in terms of the mean number of people $\bar{x}$ per car.)

pg 451

63. **More on insurance** An insurance company knows that in the entire population of homeowners, the mean annual loss from fire is $\mu = \$250$ and the standard deviation of the loss is $\sigma = \$300$. The distribution of losses is strongly right-skewed: many policies have \$0 loss, but a few have large losses. If the company sells 10,000 policies, can it safely base its rates on the assumption that its average loss will be no greater than \$275? Follow the four-step process.

64. **Bad carpet** The number of flaws per square yard in a type of carpet material varies with mean 1.6 flaws per square yard and standard deviation 1.2 flaws per square yard. The population distribution cannot be Normal, because a count takes only whole-number values. An inspector studies 200 square yards of the material, records the number of flaws found in each square yard, and calculates $\bar{x}$, the mean number of flaws per square yard inspected. Find the probability that the mean number of flaws exceeds 2 per square yard. Follow the four-step process.

Multiple choice: Select the best answer for Exercises 65 to 68.

Exercises 65 and 66 refer to the following setting. Scores on the mathematics part of the SAT exam in a recent year were roughly Normal with mean 515 and standard deviation 114. You choose an SRS of 100 students and average their SAT Math scores. Suppose that you do this many, many times.

65. The mean of the average scores you get should be close to

(a) 515.
(b) $515/100 = 5.15$.
(c) $515/\sqrt{100} = 51.5$.
(d) 0.
(e) none of these.

66. The standard deviation of the average scores you get should be close to

(a) 114.
(b) $114/100 = 1.14$.
(c) $114/\sqrt{100} = 11.4$.
(d) 1.
(e) none of these.

67. A newborn baby has extremely low birth weight (ELBW) if it weighs less than 1000 grams. A study of the health of such children in later years examined a random sample of 219 children. Their mean weight at birth was $\bar{x} = 810$ grams. This sample mean is an *unbiased estimator* of the mean weight μ in the population of all ELBW babies, which means that

(a) in all possible samples of size 219 from this population, the mean of the values of $\bar{x}$ will equal 810.

(b) in all possible samples of size 219 from this population, the mean of the values of $\bar{x}$ will equal μ.

(c) as we take larger and larger samples from this population, $\bar{x}$ will get closer and closer to μ.

(d) in all possible samples of size 219 from this population, the values of $\bar{x}$ will have a distribution that is close to Normal.

(e) the person measuring the children's weights does so without any systematic error.

68. The number of hours a light bulb burns before failing varies from bulb to bulb. The distribution of burnout times is strongly skewed to the right. The central limit theorem says that

(a) as we look at more and more bulbs, their average burnout time gets ever closer to the mean μ for all bulbs of this type.

(b) the average burnout time of a large number of bulbs has a distribution of the same shape (strongly skewed) as the population distribution.

(c) the average burnout time of a large number of bulbs has a distribution with similar shape but not as extreme (skewed, but not as strongly) as the population distribution.

(d) the average burnout time of a large number of bulbs has a distribution that is close to Normal.

(e) the average burnout time of a large number of bulbs has a distribution that is exactly Normal.

Exercises 69 to 72 refer to the following setting. In the language of government statistics, you are "in the labor force" if you are available for work and either working or actively seeking work. The unemployment rate is the proportion of the labor force (not of the entire population) who are unemployed. Here are data from the Current Population Survey for the civilian population aged 25 years and over in a recent year. The table entries are counts in thousands of people.

Highest education	Total population	In labor force	Employed
Didn't finish high school	27,669	12,470	11,408
High school but no college	59,860	37,834	35,857
Less than bachelor's degree	47,556	34,439	32,977
College graduate	51,582	40,390	39,293

69. **Unemployment** (1.1) Find the unemployment rate for people with each level of education. How does the unemployment rate change with education?

70. **Unemployment** (5.1) What is the probability that a randomly chosen person 25 years of age or older is in the labor force? Show your work.

71. **Unemployment** (5.3) If you know that a randomly chosen person 25 years of age or older is a college graduate, what is the probability that he or she is in the labor force? Show your work.

72. **Unemployment** (5.3) Are the events "in the labor force" and "college graduate" independent? Justify your answer.

Chapter 7 Chapter Review

This chapter introduced sampling distributions of statistics. A sampling distribution describes the values a statistic would take in all possible samples of the same size from a given population. It's easiest to estimate an unknown population parameter with a statistic having no bias and low variability. Sections 7.2 and 7.3 gave details about the sampling distributions of two important statistics: the sample proportion $\hat{p}$ and the sample mean $\bar{x}$. These statistics behave in much the same way:

- They are unbiased estimators of their corresponding population parameters.
- The standard deviation of the statistic decreases by a factor of $\sqrt{n}$ as the sample size n increases.
- Their sampling distributions are approximately Normal if the sample is large enough.

This final similarity is a main reason why Normal distributions are so important in statistics.

Chapter Review Exercises

These exercises are designed to help you review the important ideas and methods of the chapter. Relevant learning objectives are provided in bulleted form before each exercise.

- Distinguish between a parameter and a statistic.

R7.1. Bad eggs Sale of eggs that are contaminated with salmonella can cause food poisoning in consumers. A large egg producer takes an SRS of 200 eggs from all the eggs shipped in one day. The laboratory reports that 9 of these eggs had salmonella contamination. Unknown to the producer, 0.1% (one-tenth of 1%) of all eggs shipped had salmonella. Identify the population, the parameter, the sample, and the statistic.

Exercises R7.2 and R7.3 refer to the following setting. Researchers in Norway analyzed data on the birth weights of 400,000 newborns over a six-year period. The distribution of birth weights is approximately Normal with a mean of 3668 grams and a standard deviation of 511 grams.[9] In this population, the range (maximum – minimum) of birth weights is 3417 grams. We used Fathom software to take 500 SRSs of size $n = 5$ and calculate the range (maximum – minimum) for each sample. The dotplot below shows the results.

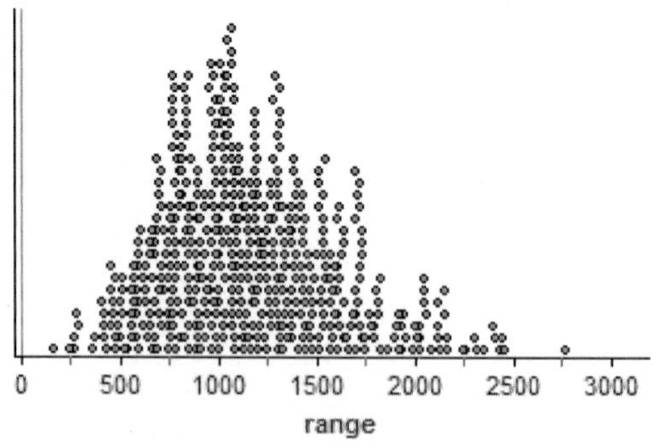

- Understand the definition of a sampling distribution.
- Distinguish between population distribution, sampling distribution, and the distribution of sample data.

R7.2. Birth weights

(a) Sketch a possible graph of the distribution of sample data for an SRS of size 5 with a range of 1000 grams.

(b) Explain why the dotplot of sample ranges above is *not* the actual sampling distribution of the sample range.

- Determine whether a statistic is an unbiased estimator of a population parameter.
- Understand the relationship between sample size and the variability of an estimator.

R7.3. Birth weights

(a) Is the sample range an unbiased estimator of the population range? Give evidence from the graph above to support your answer.

(b) Explain how we could decrease the variability of the sampling distribution of the sample range.

- Find the mean and standard deviation of the sampling distribution of a sample proportion $\hat{p}$ for an SRS of size n from a population having proportion p of successes.
- Check whether the 10% and Normal conditions are met in a given setting.
- Use Normal approximation to calculate probabilities involving $\hat{p}$.

R7.4. Do you jog? The Gallup Poll once asked a random sample of 1540 adults, "Do you happen to jog?" Suppose that in fact 15% of all adults jog.

(a) What is the mean of the sampling distribution of $\hat{p}$? Justify your answer.

(b) Find the standard deviation of the sampling distribution of $\hat{p}$. Check that the 10% condition is met.

(c) Is the sampling distribution of $\hat{p}$ approximately Normal? Justify your answer.

(d) Find the probability that between 13% and 17% of a random sample of 1540 adults are joggers. Show your work.

- Use the sampling distribution of $\hat{p}$ to evaluate a claim about a population proportion.

R7.5. Bag check Thousands of travelers pass through the airport in Guadalajara, Mexico, each day. Before leaving the airport, each passenger must pass through the Customs inspection area. Customs agents want to be sure that passengers do not bring illegal items into the country. But they do not have time to search every traveler's luggage. Instead, they require each person to press a button. Either a red or a green bulb lights up. If the red light shows, the passenger will be searched by Customs agents. A green light means "go ahead." Customs agents claim that the proportion of all travelers who will be stopped (red light) is 0.30, because the light has probability 0.30 of showing red on any push of the button. To test this claim, a concerned citizen watches a random sample of 100 travelers push the button. Only 20 get a red light.

(a) Assume that the Customs agents' claim is true. Find the probability that the proportion of travelers who get a red light is as small as or smaller than the result in this sample. Show your work.

(b) Based on your results in (a), do you believe the Customs agents' claim? Explain.

- Find the mean and standard deviation of the sampling distribution of a sample mean $\bar{x}$ from an SRS of size n.
- Calculate probabilities involving a sample mean $\bar{x}$ when the population distribution is Normal.
- Explain how the shape of the sampling distribution of $\bar{x}$ is related to the shape of the population distribution.

R7.6. IQ tests The Wechsler Adult Intelligence Scale (WAIS) is a common "IQ test" for adults. The distribution of WAIS scores for persons over 16 years of age is approximately Normal with mean 100 and standard deviation 15.

(a) What is the probability that a randomly chosen individual has a WAIS score of 105 or higher? Show your work.

(b) Find the mean and standard deviation of the sampling distribution of the average WAIS score $\bar{x}$ for an SRS of 60 people.

(c) What is the probability that the average WAIS score of an SRS of 60 people is 105 or higher? Show your work.

(d) Would your answers to any of parts (a), (b), or (c) be affected if the distribution of WAIS scores in the adult population were distinctly non-Normal? Explain.

- Use the central limit theorem to help find probabilities involving a sample mean $\bar{x}$.

R7.7. Detecting gypsy moths The gypsy moth is a serious threat to oak and aspen trees. A state agriculture department places traps throughout the state to detect the moths. When traps are checked periodically, the mean number of moths trapped is only 0.5, but some traps have several moths. The distribution of moth counts is discrete and strongly skewed, with standard deviation 0.7.

(a) What are the mean and standard deviation of the average number of moths $\bar{x}$ in 50 traps?

(b) Use the central limit theorem to help you find the probability that the average number of moths in 50 traps is greater than 0.6.

Chapter 7 AP Statistics Practice Test

Section I: Multiple Choice *Select the best answer for each question.*

T7.1. A study of voting chose 663 registered voters at random shortly after an election. Of these, **72%** said they had voted in the election. Election records show that only **56%** of registered voters voted in the election. Which of the following statements is true about the boldface numbers?

(a) 72% is a sample; 56% is a population.
(b) 72% and 56% are both statistics.
(c) 72% is a statistic and 56% is a parameter.
(d) 72% is a parameter and 56% is a statistic.
(e) 72% and 56% are both parameters.

T7.2. The Gallup Poll has decided to increase the size of its random sample of voters from about 1500 people to about 4000 people right before an election. The poll is designed to estimate the proportion of voters who favor a new law banning smoking in public buildings. The effect of this increase is to

(a) reduce the bias of the estimate.
(b) increase the bias of the estimate.
(c) reduce the variability of the estimate.
(d) increase the variability of the estimate.
(e) have no effect since the population size is the same.

T7.3. Suppose we select an SRS of size $n = 100$ from a large population having proportion p of successes. Let $\hat{p}$ be the proportion of successes in the sample. For which value of p would it be safe to use the Normal approximation to the sampling distribution of $\hat{p}$?

(a) 0.01 (b) 1/11 (c) 0.85 (d) 0.975 (e) 0.999

T7.4. The central limit theorem is important in statistics because it allows us to use the Normal distribution to make inferences concerning the population mean

(a) if the sample size is reasonably large (for any population).

(b) if the population is Normally distributed and the sample size is reasonably large.

(c) if the population is Normally distributed (for any sample size).

(d) if the population is Normally distributed and the population variance is known (for any sample size).

(e) if the population size is reasonably large (whether the population distribution is known or not).

T7.5. The number of undergraduates at Johns Hopkins University is approximately 2000, while the number at Ohio State University is approximately 40,000. At both schools, a simple random sample of about 3% of the undergraduates is taken. Each sample is used to estimate the proportion p of all students at that university who own an iPod. Suppose that, in fact, $p = 0.80$ at both schools. Which of the following is the best conclusion?

(a) The estimate from Johns Hopkins has less sampling variability than that from Ohio State.

(b) The estimate from Johns Hopkins has more sampling variability than that from Ohio State.

(c) The two estimates have about the same amount of sampling variability.

(d) It is impossible to make any statement about the sampling variability of the two estimates since the students surveyed were different.

(e) None of the above.

T7.6. A researcher initially plans to take an SRS of size n from a population that has mean 80 and standard deviation 20. If he were to double his sample size (to $2n$), the standard deviation of the sampling distribution of the sample mean would be multiplied by

(a) $\sqrt{2}$. (b) $1/\sqrt{2}$. (c) 2. (d) 1/2. (e) $1/\sqrt{2n}$.

T7.7. The student newspaper at a large university asks an SRS of 250 undergraduates, "Do you favor eliminating the carnival from the term-end celebration?" All in all, 150 of the 250 are in favor. Suppose that (unknown to you) 55% of all undergraduates favor eliminating the carnival. If you took a very large number of SRSs of size $n = 250$ from this population, the sampling distribution of the sample proportion $\hat{p}$ would be

(a) exactly Normal with mean 0.55 and standard deviation 0.03.

(b) approximately Normal with mean 0.55 and standard deviation 0.03.

(c) exactly Normal with mean 0.60 and standard deviation 0.03.

(d) approximately Normal with mean 0.60 and standard deviation 0.03.

(e) heavily skewed with mean 0.55 and standard deviation 0.03.

T7.8. Which of the following statements about the sampling distribution of the sample mean is *incorrect?*

(a) The standard deviation of the sampling distribution will decrease as the sample size increases.

(b) The standard deviation of the sampling distribution is a measure of the variability of the sample mean among repeated samples.

(c) The sample mean is an unbiased estimator of the true population mean.

(d) The sampling distribution shows how the sample mean will vary in repeated samples.

(e) The sampling distribution shows how the sample was distributed around the sample mean.

T7.9. A machine is designed to fill 16-ounce bottles of shampoo. When the machine is working properly, the mean amount poured into the bottles is 16.05 ounces with a standard deviation of 0.1 ounce. Assume that the machine is working properly. If four bottles are randomly selected each hour and the number of ounces in each bottle is measured, then 95% of the observations should occur in which interval?

(a) 16.05 to 16.15 ounces

(b) −0.30 to +0.30 ounces

(c) 15.95 to 16.15 ounces

(d) 15.90 to 16.20 ounces

(e) None of the above

T7.10. Suppose that you are a student aide in the library and agree to be paid according to the "random pay" system. Each week, the librarian flips a coin. If the coin comes up heads, your pay for the week is \$80. If it comes up tails, your pay for the week is \$40. You work for the library for 100 weeks. Suppose we choose an SRS of 2 weeks and calculate your average earnings $\bar{x}$. The shape of the sampling distribution of $\bar{x}$ will be

(a) Normal.

(b) approximately Normal.

(c) right-skewed.

(d) left-skewed.

(e) symmetric but not Normal.

Section II: Free Response *Show all your work. Indicate clearly the methods you use, because you will be graded on the correctness of your methods as well as on the accuracy and completeness of your results and explanations.*

T7.11. Below are histograms of the values taken by three sample statistics in several hundred samples from the same population. The true value of the population parameter is marked with an arrow on each histogram.

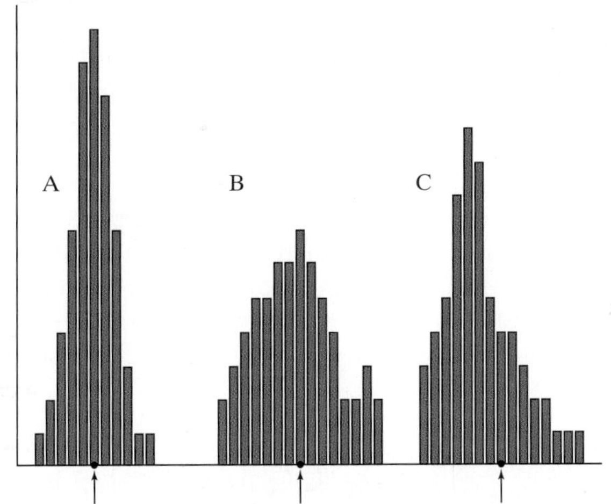

Which statistic would provide the best estimate of the parameter? Justify your answer.

T7.12. The amount that households pay service providers for access to the Internet varies quite a bit, but the mean monthly fee is \$28 and the standard deviation is \$10. The distribution is not Normal: many households pay about \$10 for limited dial-up access or about \$30 for unlimited dial-up access, but some pay much more for faster connections. A sample survey asks an SRS of 500 households with Internet access how much they pay. Let $\bar{x}$ be the mean amount paid.

(a) Explain why you can't determine the probability that the amount a randomly selected household pays for access to the Internet exceeds \$29.

(b) What are the mean and standard deviation of the sampling distribution of $\bar{x}$?

(c) What is the shape of the sampling distribution of $\bar{x}$? Justify your answer.

(d) Find the probability that the average fee paid by the sample of households exceeds \$29. Show your work.

T7.13. According to government data, 22% of American children under the age of six live in households with incomes less than the official poverty level. A study of learning in early childhood chooses an SRS of 300 children. Find the probability that more than 20% of the sample are from poverty households. Be sure to check that you can use the Normal approximation.

Cumulative AP Practice Test 2

Section I: Multiple Choice *Choose the best answer for each question.*

AP2.1. The five-number summary for a data set is given by min = 5, Q_1 = 18, M = 20, Q_3 = 40, max = 75. If you wanted to construct a modified boxplot for the data set (that is, one that would show outliers, if any existed), what would be the maximum possible length of the right-side "whisker"?

(a) 33 (b) 35 (c) 45 (d) 53 (e) 55

AP2.2. The probability distribution for the number of heads in four tosses of a coin is given by

Number of heads:	0	1	2	3	4
Probability:	0.0625	0.2500	0.3750	0.2500	0.0625

The probability of getting at least one *tail* in four tosses of a coin is

(a) 0.2500. (b) 0.3125. (c) 0.6875. (d) 0.9375. (e) none of these.

AP2.3. In a certain large population of adults, the distribution of IQ scores is strongly left-skewed with a mean of 122 and a standard deviation of 5. Suppose 200 adults are randomly selected from this population for a market research study. The distribution of the sample mean of IQ scores is

(a) left-skewed with mean 122 and standard deviation 0.35.

(b) exactly Normal with mean 122 and standard deviation 5.

(c) exactly Normal with mean 122 and standard deviation 0.35.

(d) approximately Normal with mean 122 and standard deviation 5.

(e) approximately Normal with mean 122 and standard deviation 0.35.

AP2.4. A 10-question multiple-choice exam offers 5 choices for each question. Jason just guesses the answers, so he has probability 1/5 of getting any one answer correct. You want to perform a simulation to determine the number of correct answers that Jason gets. One correct way to use a table of random digits to do this is the following:

(a) One digit from the random digit table simulates one answer, with 5 = right and all other digits = wrong. Ten digits from the table simulate 10 answers.

(b) One digit from the random digit table simulates one answer, with 0 or 1 = right and all other digits = wrong. Ten digits from the table simulate 10 answers.

(c) One digit from the random digit table simulates one answer, with odd = right and even = wrong. Ten digits from the table simulate 10 answers.

(d) Two digits from the random digit table simulate one answer, with 00 to 20 = right and 21 to 99 = wrong. Ten pairs of digits from the table simulate 10 answers.

(e) Two digits from the random digit table simulate one answer, with 00 to 05 = right and 06 to 99 = wrong. Ten pairs of digits from the table simulate 10 answers.

AP2.5. Suppose we roll a fair die four times. The probability that a 6 occurs on exactly one of the rolls is

(a) $4\left(\frac{1}{6}\right)^3\left(\frac{5}{6}\right)^1$ (c) $4\left(\frac{1}{6}\right)^1\left(\frac{5}{6}\right)^3$ (e) $6\left(\frac{1}{6}\right)^1\left(\frac{5}{6}\right)^3$

(b) $\left(\frac{1}{6}\right)^3\left(\frac{5}{6}\right)^1$ (d) $\left(\frac{1}{6}\right)^1\left(\frac{5}{6}\right)^3$

AP2.6. You want to take an SRS of 50 of the 816 students who live in a dormitory on a college campus. You label the students 001 to 816 in alphabetical order. In the table of random digits, you read the entries

95592 94007 69769 33547 72450 16632 81194 14873

The first three students in your sample have labels

(a) 955, 929, 400. (d) 929, 400, 769.
(b) 400, 769, 769. (e) 400, 769, 335.
(c) 559, 294, 007.

AP2.7. The number of unbroken charcoal briquets in a twenty-pound bag filled at the factory follows a Normal distribution with a mean of 450 briquets and a standard deviation of 20 briquets. The company expects that a certain number of the bags will be underfilled, so the company will replace for free the 5% of bags that have too few briquets. What is the minimum number of unbroken briquets the bag would have to contain for the company to avoid having to replace the bag for free?

(a) 404 (b) 411 (c) 418 (d) 425 (e) 448

AP2.8. You work for an advertising agency that is preparing a new television commercial to appeal to women. You have been asked to design an experiment to compare the effectiveness of three versions of the commercial. Each subject will be shown one of the three versions and then asked about her attitude toward the product. You think there may be large differences between women who are employed and those who are not. Because of these differences, you should use

(a) a block design, but not a matched pairs design.
(b) a completely randomized design.
(c) a matched pairs design.
(d) a simple random sample.
(e) a stratified random sample.

AP2.9. Suppose that you have torn a tendon and are facing surgery to repair it. The orthopedic surgeon explains the risks to you. Infection occurs in 3% of such operations, the repair fails in 14%, and both infection and failure occur together 1% of the time. What is the probability that the operation is successful for someone who has an operation that is free from infection?

(a) 0.0767 (c) 0.8400 (e) 0.9900
(b) 0.8342 (d) 0.8660

AP2.10. Social scientists are interested in the association between high school graduation rate (HSGR) and the percent of U.S. families living in poverty (POV). Data were collected from all 50 states and the District of Columbia, and a regression analysis was conducted. The resulting least-squares regression line is given by $\widehat{POV} = 59.2 - 0.620(HSGR)$ with $r^2 = 0.802$. Based on the information, which of the following is the best interpretation for the slope of the least-squares regression line?

(a) For each 1% increase in the graduation rate, the percent of families living in poverty is predicted to decrease by approximately 0.896.

(b) For each 1% increase in the graduation rate, the percent of families living in poverty is predicted to decrease by approximately 0.802.

(c) For each 1% increase in the graduation rate, the percent of families living in poverty is predicted to decrease by approximately 0.620.

(d) For each 1% increase in the percent of families living in poverty, the graduation rate is predicted to increase by approximately 0.802.

(e) For each 1% increase in the percent of families living in poverty, the graduation rate is predicted to decrease by approximately 0.620.

Here is a dotplot of the adult literacy rates in 177 countries in 2008, according to the United Nations. For example, the lowest literacy rate was 23.6%, in the African country of Burkina Faso. Use the dotplot below to answer Questions AP2.11 to AP2.13.

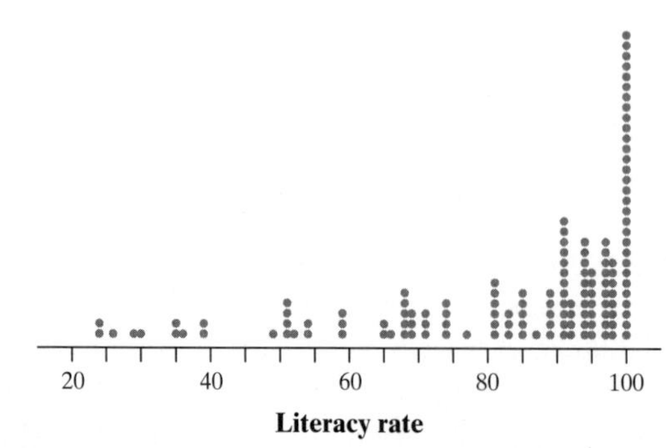

AP2.11. The overall shape of this distribution is
(a) clearly skewed to the right.
(b) clearly skewed to the left.
(c) roughly symmetric.
(d) uniform.
(e) There is no clear shape.

AP2.12. The mean of this distribution (*don't* try to find it) will be
(a) very close to the mode.
(b) greater than the median.
(c) less than the median.
(d) You can't say, because the median is random.
(e) You can't say, because the mean is random.

AP2.13. Based on the shape of this distribution, what numerical measures would best describe it?
(a) The five-number summary
(b) The mean and standard deviation
(c) The mean and the quartiles
(d) The median and the standard deviation
(e) It is not possible to determine which numerical values to use.

AP2.14. The correlation between the age and height of children under the age of 12 is found to be $r = 0.60$. Suppose we use the age x of a child to predict the height y of the child. What can we conclude?
(a) The height is generally 60% of a child's weight.
(b) About 60% of the time, age will accurately predict height.
(c) The fraction of the variation in heights explained by the least-squares regression line of y on x is 0.36.
(d) The least-squares regression line of y on x has a slope of 0.6.
(e) Thirty-six percent of the time, the least-squares regression line accurately predicts height.

AP2.15. An agronomist wants to test three different types of fertilizer (A, B, and C) on the yield of a new variety of wheat. The yield will be measured in bushels per acre. Six one-acre plots of land were randomly assigned to each of the three fertilizers. The treatment, experimental unit, and response variable are, respectively,
(a) a specific fertilizer, bushels per acre, a plot of land.
(b) a plot of land, bushels per acre, a specific fertilizer.
(c) random assignment, a plot of land, bushels per acre.
(d) a specific fertilizer, a plot of land, bushels per acre.
(e) a specific fertilizer, the agronomist, bushels per acre.

AP2.16. According to the U.S. census, the proportion of adults in a certain county who owned their own home was 0.71. An SRS of 100 adults in a certain section of the county found that 65 owned their home. Which one of the following represents the approximate probability of obtaining a sample of 100 adults in which fewer than 65 own their home, assuming that this section of the county has the same overall proportion of adults who own their home as does the entire county?

(a) $\binom{100}{65}(0.71)^{65}(0.29)^{35}$ (b) $\binom{100}{65}(0.29)^{65}(0.71)^{35}$

(c) $P\left(z < \dfrac{0.65 - 0.71}{\sqrt{\dfrac{(0.71)(0.29)}{100}}}\right)$ (d) $P\left(z < \dfrac{0.65 - 0.71}{\sqrt{\dfrac{(0.65)(0.35)}{100}}}\right)$

(e) $P\left(z < \dfrac{0.65 - 0.71}{\dfrac{(0.71)(0.29)}{\sqrt{100}}}\right)$

AP2.17. Which one of the following would be a correct interpretation if you have a z-score of $+2.0$ on an exam?
(a) It means that you missed two questions on the exam.
(b) It means that you got twice as many questions correct as the average student.
(c) It means that your grade was two points higher than the mean grade on this exam.
(d) It means that your grade was in the upper 2% of all grades on this exam.
(e) It means that your grade is two standard deviations above the mean for this exam.

AP2.18. Records from a random sample of dairy farms yielded the information below on the number of male and female calves born at various times of the day.

	Day	Evening	Night	Total
Males	129	15	117	261
Females	118	18	116	252
Total	247	33	233	513

What is the probability that a randomly selected calf was born in the night or was a female?

(a) $\frac{369}{513}$ (b) $\frac{485}{513}$ (c) $\frac{116}{513}$ (d) $\frac{116}{252}$ (e) $\frac{116}{233}$

AP2.19. When people order books from a popular online source, they are shipped in standard-sized boxes. Suppose that the mean weight of the boxes is 1.5 pounds with a standard deviation of 0.3 pounds, the mean weight of the packing material is 0.5 pounds with a standard deviation of 0.1 pounds, and the mean weight of the books shipped is 12 pounds with a standard deviation of 3 pounds. Assuming that the weights are independent, what is the standard deviation of the total weight of the boxes that are shipped from this source?
(a) 1.84 (c) 3.02 (e) 9.10
(b) 2.60 (d) 3.40

AP2.20. A grocery chain runs a prize game by giving each customer a ticket that may win a prize when the box is scratched off. Printed on the ticket is a dollar value ($500, $100, $25) or the statement, "This ticket is not a winner." Monetary prizes can be redeemed for groceries at the store. Here are the distribution of the prize values and the associated probabilities for each prize:

Amount won:	$500	$100	$25	$0
Probability:	0.01	0.05	0.20	0.74

Which of the following are the mean and standard deviation, respectively, of the winnings?

(a) $15.00, $2900.00
(b) $15.00, $53.85
(c) $15.00, $26.93
(d) $156.25, $53.85
(e) $156.25, $26.93

AP2.21. A large company is interested in improving the efficiency of its customer service and decides to examine the length of the business phone calls made to clients by its sales staff. A cumulative relative frequency graph is shown below from data collected over the past year. According to the graph, the shortest 80% of calls will take how long to complete?

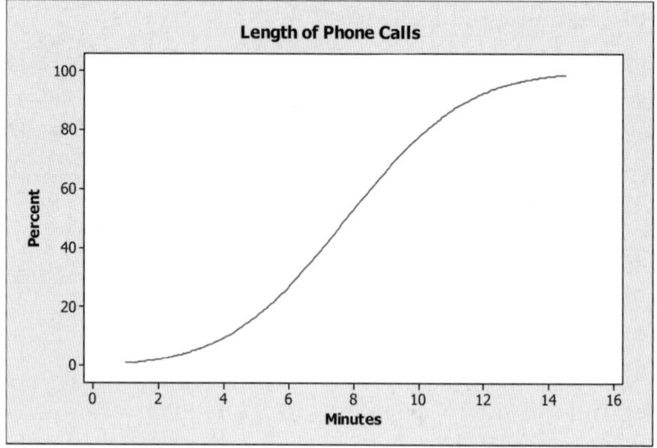

(a) Less than 10 minutes.
(b) At least 10 minutes.
(c) Exactly 10 minutes.
(d) At least 5.5 minutes.
(e) Less than 5.5 minutes.

Section II: Free Response *Show all your work. Indicate clearly the methods you use, because you will be graded on the correctness of your methods as well as on the accuracy and completeness of your results and explanations.*

AP2.22. A health worker is interested in determining if omega-3 fish oil can help reduce cholesterol in adults. She obtains permission to examine the health records of 200 people in a large medical clinic and classifies them according to whether or not they take omega-3 fish oil. She also obtains their latest cholesterol readings and finds that the mean cholesterol reading for those who are taking omega-3 fish oil is 18 points lower than the mean for the group not taking omega-3 fish oil.

(a) Is this an observational study or an experiment? Explain.

(b) Do these results provide convincing evidence that taking omega-3 fish oil lowers cholesterol?

(c) Explain the concept of confounding in the context of this study and give one example of a possible confounding variable.

AP2.23. There are four major blood types in humans: O, A, B, and AB. In a study conducted using blood specimens from the Blood Bank of Hawaii, individuals were classified according to blood type and ethnic group. The ethnic groups were Hawaiian, Hawaiian-White, Hawaiian-Chinese, and White. Suppose that a blood bank specimen is selected at random.

	Ethnic Group			
Blood type	Hawaiians	Hawaiian-White	Hawaiian-Chinese	White
O	1903	4469	2206	53,759
A	2490	4671	2368	50,008
B	178	606	568	16,252
AB	99	236	243	5001

(a) Find the probability that the specimen contains type O blood or comes from the Hawaiian-Chinese ethnic group. Show your work.

(b) What is the probability that the specimen contains type AB blood, given that it comes from the Hawaiian ethnic group? Show your work.

(c) Are the events "type B blood" and "Hawaiian ethnic group" independent? Give appropriate statistical evidence to support your answer.

Now suppose that two blood bank specimens are selected at random.

(d) Find the probability that at least one of the specimens contains type A blood from the White ethnic group.

AP2.24. Every 17 years, swarms of cicadas emerge from the ground in the eastern United States, live for about six weeks, and then die. (There are several different "broods," so we experience cicada eruptions more often than every 17 years.) There are so many cicadas that their dead bodies can serve as fertilizer and increase plant growth. In a study, a researcher added 10 dead cicadas under 39 randomly selected plants in a natural plot of American bellflowers on the forest floor, leaving other plants undisturbed. One of the response variables measured was the size of seeds produced by the plants. Here are the boxplots and summary statistics of seed mass (in milligrams) for 39 cicada plants and 33 undisturbed (control) plants:

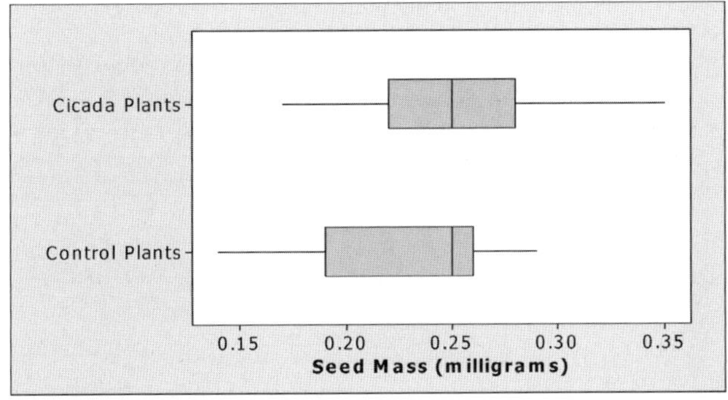

Variable:	n	Minimum	Q_1	Median	Q_3	Maximum
Cicada plants:	39	0.17	0.22	0.25	0.28	0.35
Control plants:	33	0.14	0.19	0.25	0.26	0.29

(a) Is this an observational study or an experiment? Explain.

(b) Based on the graphical displays, which distribution has the larger mean? Justify your answer.

(c) Do the data support the idea that dead cicadas can serve as fertilizer? Give graphical and numerical evidence to support your conclusion.

AP2.25. Five cards, each with a different symbol, are shuffled and you choose one. If it is the diamond, you win \$5.00. The cards are reshuffled after each draw. You must pay \$1.00 for each selection. You continue to play until you select the diamond. Is this a fair game (that is, on average will you win the same amount as you lose)?

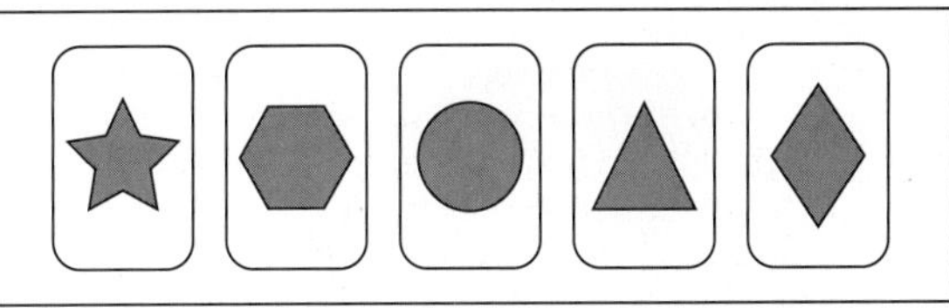

(a) Describe how you will carry out a simulation of this game using the random digit table below. Be sure to indicate what the digits will represent.

(b) Perform 10 repetitions of your simulation. Copy the random digit table onto your paper. Mark on or above the table so that someone can follow your work.

```
12975 13258 13048 45144 72321 81940 00360
02428 96767 35964 23822 96012 94591 65194
50842 53372 72829 50232 97892 63408 77919
44575 24870 04178 81565 42628 17797 49376
61762 16953 88604 12724 62964 88145 83083
69453 46109 59505 69680 00900 19687 12633
57857 95806 09931 02150 43163 58636
```

(c) Based on your simulation, what is the average number of cards you would need to draw in order to obtain a diamond? Justify your answer.

(d) Is this a fair game (that is, on average will you win the same amount as you lose)? Explain your reasoning.

Chapter 8

case study

Estimating with Confidence

Need Help? Give Us a Call!

If your cable television goes out, you phone the cable company to get it fixed. Does a real person answer your call? These days, probably not. It is far more likely that you will get an automated response. You will probably be offered several options, such as: to order cable service, press 1; for questions about your bill, press 2; to add new channels, press 3; (and finally) to speak with a customer service agent, press 4. Customers will get frustrated if they have to wait too long before speaking to a live person. So companies try hard to minimize the time required to connect to a customer service representative.

A large bank decided to study the call response times in its customer service department. The bank's goal was to have a representative answer an incoming call within 30 seconds. Figure 8.1 is a histogram of the response times in a random sample of 241 calls to the bank's customer service center in a given month. What does the graph suggest about how well the bank is meeting its goal?

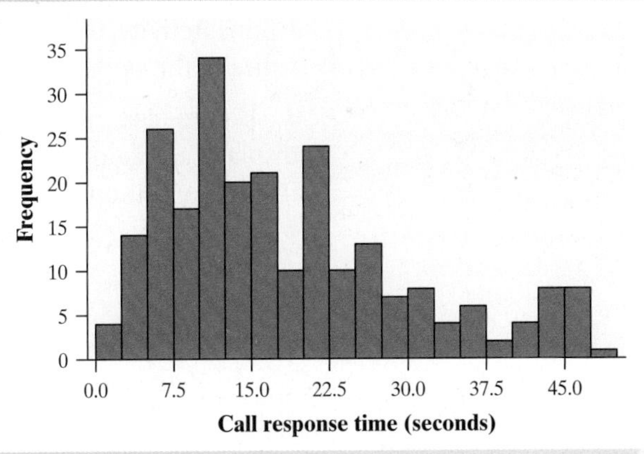

FIGURE 8.1 Histogram showing the response time (in minutes) at a bank's customer service center for a random sample of 241 calls.

By the end of this chapter, you will be able to use the sample data to make inferences about the proportion p of all calls to the customer service department that are answered within 30 seconds and the mean call response time μ.

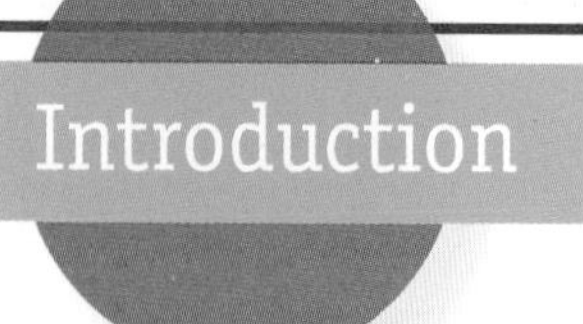

How long does a new model of laptop battery last? What proportion of college undergraduates have engaged in binge drinking? How much does the weight of a quarter-pound hamburger at a fast-food restaurant vary after cooking? These are the types of questions we would like to be able to answer.

It wouldn't be practical to determine the lifetime of *every* laptop battery, to ask *all* college undergraduates about their drinking habits, or to weigh *every* burger after cooking. Instead, we choose a sample of individuals (batteries, college students, burgers) to represent the population and collect data from those individuals. Our goal in each case is to use a sample statistic to estimate an unknown population parameter. From what we learned in Chapter 4, if we randomly select the sample, we should be able to generalize our results to the population of interest.

We cannot be certain that our conclusions are correct—a different sample would probably yield a different estimate. Statistical inference uses the language of probability to express the strength of our conclusions. Probability allows us to take chance variation due to random selection or random assignment into account. The following Activity gives you an idea of what lies ahead.

ACTIVITY *The mystery mean*

MATERIALS: TI-83/84 or TI-89 with display capability

In this Activity, each team of three to four students will try to estimate the mystery value of the population mean μ that your teacher entered before class.[1]

1. Before class, your teacher will store a value of μ (represented by M) in the display calculator as follows: on the home screen, type ### STO▶ M ENTER. The teacher will then clear the home screen so you can't see the value of M.

2. With the class watching, the teacher will execute the following command: `mean(randNorm(M,20,16))`.

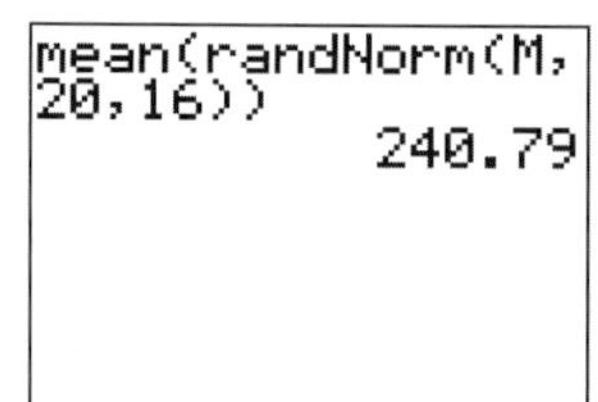

This tells the calculator to choose an SRS of 16 observations from a Normal population with mean M and standard deviation 20, and then compute the mean $\bar{x}$ of those 16 sample values.

3. Now for the challenge! Your group must determine an interval of *reasonable* values for the population mean μ. Use the result from Step 2 and what you learned about sampling distributions in the previous chapter.

4. Share your team's results with the class.

In this chapter and the next, we will meet the two most common types of formal statistical inference. Chapter 8 concerns *confidence intervals* for estimating the value of a parameter. Chapter 9 presents *significance tests*, which assess the evidence for a claim about a parameter. Both types of inference are based on the sampling distributions of statistics. That is, both report probabilities that state *what would happen if we used the inference method many times.*

Section 8.1 examines the idea of a confidence interval. We start by presenting the reasoning of confidence intervals in a general way that applies to estimating any unknown parameter. In Section 8.2, we show how to estimate a population proportion. Section 8.3 focuses on confidence intervals for a population mean.

8.1

Confidence Intervals: The Basics

In Section 8.1, you'll learn about:

- The idea of a confidence interval
- Interpreting confidence levels and confidence intervals
- Constructing a confidence interval
- Using confidence intervals wisely

Mr. Schiel's class did the mystery mean Activity from the Introduction. The TI screen shot displays the information that the students received about the unknown population mean μ. Here is a summary of what the class said about the calculator output:

- The population distribution is Normal and its standard deviation is $\sigma = 20$.
- A simple random sample of $n = 16$ observations was taken from this population.
- The sample mean is $\bar{x} = 240.79$.

```
mean(randNorm(M,
20,16))
              240.79
```

If we had to give a single number to estimate the value of M that Mr. Schiel chose, what would it be? (Such a value is known as a **point estimate.**) How about 240.79? That makes sense, because the sample mean $\bar{x}$ is an unbiased estimator of the population mean μ. We are using the statistic $\bar{x}$ as a **point estimator** of the parameter μ.

DEFINITION: Point estimator and point estimate

A **point estimator** is a statistic that provides an estimate of a population parameter. The value of that statistic from a sample is called a **point estimate.** Ideally, a point estimate is our "best guess" at the value of an unknown parameter.

As we saw in Chapter 7, the ideal point estimator will have no bias and low variability. Here's an example involving some of the more common point estimators.

From Batteries to Smoking

Point estimators

PROBLEM: In each of the following settings, determine the point estimator you would use and calculate the value of the point estimate.

(a) Quality control inspectors want to estimate the mean lifetime μ of the AA batteries produced in an hour at a factory. They select a random sample of 30 batteries during each hour of production and then drain them under conditions that mimic normal use. Here are the lifetimes (in hours) of the batteries from one such sample:

16.91 18.83 17.58 15.84 17.42 17.65 16.63 16.84 15.63 16.37 15.80 15.93
15.81 17.45 16.85 16.33 16.22 16.59 17.13 17.10 16.96 16.40 17.35 16.37
15.98 16.52 17.04 17.07 15.73 16.74

(b) What proportion p of U.S. high school students smoke? The 2007 Youth Risk Behavioral Survey questioned a random sample of 14,041 students in grades 9 to 12. Of these, 2808 said they had smoked cigarettes at least one day in the past month.

(c) The quality control inspectors in part (a) want to investigate the variability in battery lifetimes by estimating the population variance σ^2.

SOLUTION:

(a) Use the sample mean $\bar{x}$ as a point estimator for the population mean μ. For these data, our point estimate is $\bar{x} = 16.7$ hours.

(b) Use the sample proportion $\hat{p}$ as a point estimator for the population proportion p. For this survey, our point estimate is $\hat{p} = \frac{2808}{14{,}041} = 0.20$.

(c) Use the sample variance s_x^2 as a point estimator for the population variance σ^2. For the battery life data, our point estimate is $s_x^2 = 0.508$.

For Practice *Try Exercises 1 and 3*

The Idea of a Confidence Interval

Is the value of the population mean μ that Mr. Schiel entered in his calculator exactly 240.79? Probably not. Because $\bar{x} = 240.79$, we guess that μ is "somewhere around 240.79." How close to 240.79 is μ likely to be?

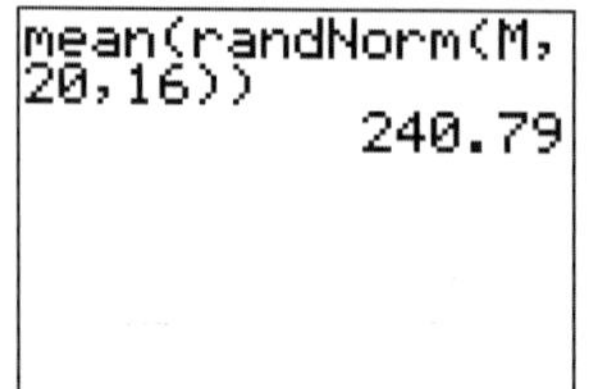

To answer this question, we ask another: *How would the sample mean $\bar{x}$ vary if we took many SRSs of size 16 from this same population?*

The sampling distribution of $\bar{x}$ describes how the values of $\bar{x}$ vary in repeated samples. Recall the facts about this sampling distribution from Chapter 7:

Shape: Because the population distribution is Normal, so is the sampling distribution of $\bar{x}$. Thus, the *Normal condition* is met.

Center: The mean of the sampling distribution of $\bar{x}$ is the same as the unknown mean μ of the entire population. That is, $\mu_{\bar{x}} = \mu$.
Spread: The standard deviation of $\bar{x}$ for an SRS of $n = 16$ observations is

$$\sigma_{\bar{x}} = \frac{\sigma}{\sqrt{n}} = \frac{20}{\sqrt{16}} = 5$$

because the *10% condition* is met—we are sampling from an infinite population in this case.

Figure 8.2 summarizes these facts.

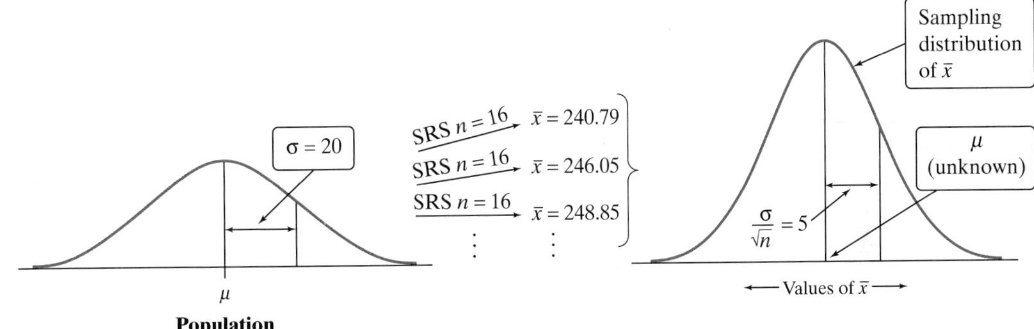

FIGURE 8.2 The sampling distribution of the mean score $\bar{x}$ for SRSs of 16 observations from a Normally distributed population with unknown mean μ and standard deviation $\sigma = 20$.

The next example gives the reasoning of statistical estimation in a nutshell.

The Mystery Mean

Moving beyond a point estimate

When Mr. Schiel's class discussed the results of the mystery mean Activity, students used the following logic to come up with an "interval estimate" for the unknown population mean μ.

```
mean(randNorm(M,
20,16))
                240.79
```

1. To estimate μ, use the mean $\bar{x} = 240.79$ of the random sample. (This is our *point estimate*.) We don't expect $\bar{x}$ to be exactly equal to μ, so we want to say how accurate this estimate is.
2. In repeated samples, the values of $\bar{x}$ follow a Normal distribution with mean μ and standard deviation 5, as in Figure 8.2.
3. The 95 part of the 68–95–99.7 rule for Normal distributions says that $\bar{x}$ is within 10 (that's two standard deviations) of the population mean μ in about 95% of all samples of size $n = 16$. See Figure 8.3 on the next page.

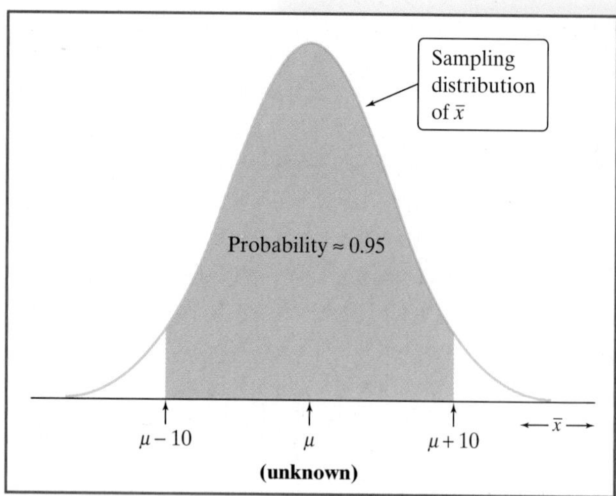

FIGURE 8.3 In 95% of all samples, $\bar{x}$ lies within ± 10 of the unknown population mean μ. So μ also lies within ± 10 of $\bar{x}$ in those samples.

4. Whenever $\bar{x}$ is within 10 points of μ, μ is within 10 points of $\bar{x}$. This happens in about 95% of all possible samples. So the interval from $\bar{x} - 10$ to $\bar{x} + 10$ "captures" the population mean μ in about 95% of all samples of size 16.

5. If we estimate that μ lies somewhere in the interval from

$$\bar{x} - 10 = 240.79 - 10 = 230.79$$

to

$$\bar{x} + 10 = 240.79 + 10 = 250.79$$

we'd be calculating this interval using a method that captures the true μ in about 95% of all possible samples of this size.

The big idea is that the sampling distribution of $\bar{x}$ tells us how close to μ the sample mean $\bar{x}$ is likely to be. Statistical estimation just turns that information around to say how close to $\bar{x}$ the unknown population mean μ is likely to be. In the mystery mean example, the value of μ is usually within 10 of $\bar{x}$ for SRSs of size 16. Since the class's sample mean was $\bar{x} = 240.79$, the interval 240.79 ± 10 gives an approximate **95% confidence interval** for μ.

There are several ways to write a confidence interval. We can give the interval for Mr. Schiel's mystery mean as 240.79 ± 10, as 230.79 to 250.79, as $230.79 \leq \mu \leq 250.79$, or as (230.79, 250.79).

All the confidence intervals we will meet have a form similar to this:

$$\text{estimate} \pm \text{margin of error}$$

The estimate ($\bar{x} = 240.79$ in our example) is our best guess for the value of the unknown parameter. The **margin of error,** 10, shows how close we believe our guess is, based on the variability of the estimate in repeated SRSs of size 16. We say that our **confidence level** is about 95% because the interval $\bar{x} \pm 10$ catches the unknown parameter in about 95% of all possible samples.

DEFINITION: Confidence interval, margin of error, confidence level

A **confidence interval** for a parameter has two parts:

- An interval calculated from the data, which has the form

$$\text{estimate} \pm \text{margin of error}$$

The **margin of error** tells how close the estimate tends to be to the unknown parameter in repeated random sampling.

- A **confidence level *C*,** which gives the overall success rate of the method for calculating the confidence interval. That is, in *C*% of all possible samples, the method would yield an interval that captures the true parameter value.

A confidence interval is sometimes referred to as an *interval estimate.* This is consistent with our earlier use of the term *point estimate.*

We usually choose a confidence level of 90% or higher because we want to be quite sure of our conclusions. The most common confidence level is 95%.

Interpreting Confidence Levels and Confidence Intervals

The following Activity gives you a chance to explore the meaning of the confidence level.

ACTIVITY *The* Confidence Interval *applet*

MATERIALS: Computer with Internet connection and display capability

The *Confidence Interval* applet at the book's Web site will quickly generate many confidence intervals. In this Activity, you will use the applet to investigate the idea of a confidence level.

1. Go to www.whfreeman.com/tps4e and launch the applet. The default setting for the confidence level is 95%. Change this to 90%.

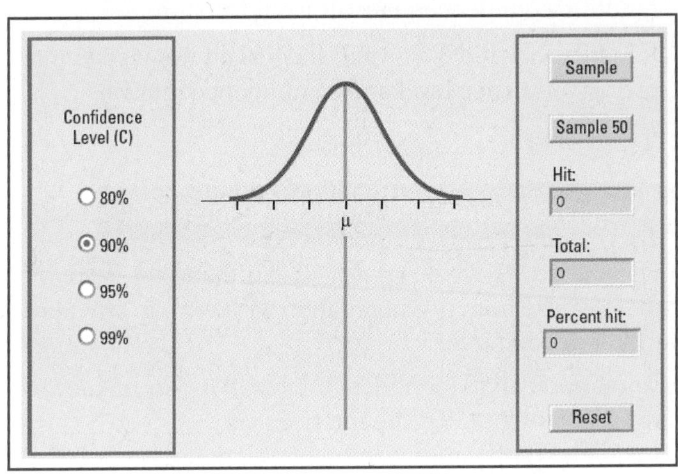

2. Click "Sample" to choose an SRS and display the resulting confidence interval. Did the interval capture the population mean μ (what the applet calls a "hit")? Do this a total of 10 times. How many of the intervals captured the population mean μ?

3. Reset the applet. Click "Sample 50" to choose 50 SRSs and display the confidence intervals based on those samples. How many captured the parameter μ? Keep clicking "Sample 50" and observe the value of "Percent hit." What do you notice?

As the Activity confirms, the confidence level is the overall capture rate if the method is used many times. Figure 8.4 illustrates the behavior of the confidence interval $\bar{x} \pm 10$ for Mr. Schiel's mystery mean. Starting with the population, imagine taking many SRSs of 16 observations. The first sample has $\bar{x} = 240.79$, the second has $\bar{x} = 246.05$, the third has $\bar{x} = 248.85$, and so on. The sample mean varies from sample to sample, but when we use the formula $\bar{x} \pm 10$ to get an interval based on each sample, 95% *of these intervals capture the unknown population mean* μ.

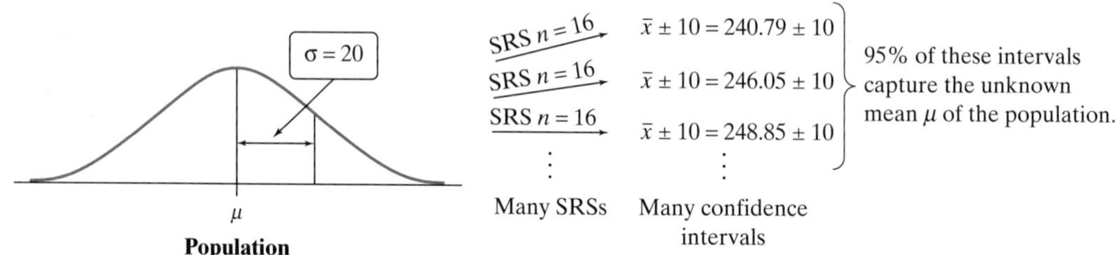

FIGURE 8.4 To say that $\bar{x} \pm 10$ is a 95% confidence interval for the population mean μ is to say that, in repeated samples, 95% of these intervals capture μ.

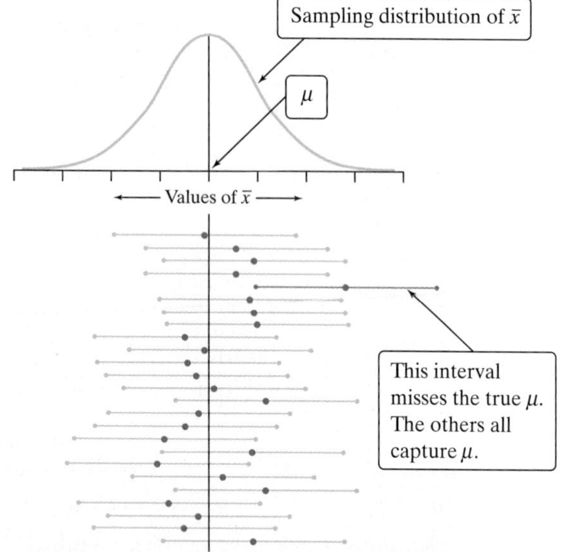

FIGURE 8.5 Twenty-five samples of the same size from the same population gave these 95% confidence intervals. In the long run, 95% of all samples give an interval that captures the population mean μ.

Figure 8.5 illustrates the idea of a confidence interval in a different form. It shows the result of drawing many SRSs from the same population and calculating a 95% confidence interval from each sample. The center of each interval is at $\bar{x}$ and therefore varies from sample to sample. The sampling distribution of $\bar{x}$ appears at the top of the figure to show the long-term pattern of this variation. The 95% confidence intervals from 25 SRSs appear below.

Here's what you should notice:

- The center $\bar{x}$ of each interval is marked by a dot.
- The distance from the dot to either endpoint of the interval is the margin of error.
- 24 of these 25 intervals (that's 96%) contain the true value of μ. If we took all possible samples, 95% of the resulting confidence intervals would capture μ.

Figures 8.4 and 8.5 give us the insight we need to interpret a confidence level and a confidence interval.

Interpreting Confidence Levels and Confidence Intervals

Confidence level: To say that we are 95% *confident* is shorthand for "95% of all possible samples of a given size from this population will result in an interval that captures the unknown parameter."

Confidence interval: To interpret a C% confidence interval for an unknown parameter, say, "We are C% confident that the interval from _____ to _____ captures the actual value of the [population parameter in context]."

Plausible does not mean the same thing as *possible*. Some would argue that just about any value of a parameter is possible. A plausible value of a parameter is a reasonable or believable value based on the data.

The confidence level tells us how likely it is that the method we are using will produce an interval that captures the population parameter *if we use it many times*. However, in practice we tend to calculate only a single confidence interval for a given situation. *The confidence level does not tell us the chance that a particular confidence interval captures the population parameter.* Instead, the confidence interval gives us a set of plausible values for the parameter.

CAUTION

EXAMPLE

The Mystery Mean

Interpreting a confidence level and a confidence interval

```
mean(randNorm(M,
20,16))
             240.79
```

The confidence level in the mystery mean example—roughly 95%—tells us that in about 95% of all SRSs of size 16 from Mr. Schiel's mystery population, the interval $\bar{x} \pm 10$ will contain the population mean μ. Mr. Schiel's SRS of 16 values gave $\bar{x} = 240.79$. The approximate 95% confidence interval is 240.79 ± 10. We say, "We are about 95% confident that the interval from 230.79 to 250.79 captures the mystery mean."

Be sure you understand the basis for our confidence. There are only two possibilities:

1. The interval from 230.79 to 250.79 contains the population mean μ.
2. The interval from 230.79 to 250.79 does *not* contain the population mean μ.

Our SRS was one of the few samples for which $\bar{x}$ is not within 10 points of the true μ. Only about 5% of all samples result in a confidence interval that fails to capture μ.

We cannot know whether our sample is one of the 95% for which the interval $\bar{x} \pm 10$ catches μ or whether it is one of the unlucky 5%. The statement that we are "95% confident" that the unknown μ lies between 230.79 and 250.79 is shorthand for saying, "We got these numbers by a method that gives correct results 95% of the time."

THINK ABOUT IT

What's the probability that our 95% confidence interval captures the parameter? It's *not* 95%! Before we execute the command `mean(randNorm(M,20,16))`, we have a 95% chance of getting a sample mean that's within $2\sigma_{\bar{x}}$ of the mystery μ, which would lead to a confidence interval that captures μ. Once we have chosen a random sample, the sample mean $\bar{x}$ either is or isn't within $2\sigma_{\bar{x}}$ of μ. And the resulting confidence interval either does or doesn't contain μ. *After* we construct a confidence interval, the probability that it captures the population parameter is either *1 (it does) or 0 (it doesn't).*

We interpret confidence levels and confidence intervals in much the same way whether we are estimating a population mean, proportion, or some other parameter.

EXAMPLE Do You Use Twitter?

Interpreting a confidence interval and a confidence level

In late 2009, the Pew Internet and American Life Project asked a random sample of 2253 U.S. adults, "Do you ever . . . use Twitter or another service to share updates about yourself or to see updates about others?" Of the sample, 19% said "Yes." According to Pew, the resulting 95% confidence interval is (0.167, 0.213).[2]

PROBLEM: **Interpret the confidence interval and the confidence level.**

SOLUTION:

Confidence interval: **We are 95% confident that the interval from 0.167 to 0.213 contains the actual proportion *p* of U.S. adults who use Twitter or another service for updates.**

Confidence level: **In 95% of all possible samples of 2253 U.S. adults, the resulting confidence interval would capture the actual population proportion of U.S. adults who use Twitter or another service for updates.**

For Practice *Try Exercise 15*

AP EXAM TIP On a given problem, you may be asked to interpret the confidence interval, the confidence level, or both. Be sure you understand the difference: the confidence level describes the long-run capture rate of the method, and the confidence interval gives a set of plausible values for the parameter.

Confidence intervals are statements about *parameters*. In the previous example, it would be wrong to say, "We are 95% confident that the interval from 0.167 to 0.213 contains the *sample* proportion of U.S. adults who use Twitter." Why? Because we *know* that the *sample* proportion, $\hat{p} = 0.19$, is in the interval. Likewise, in the mystery mean example, it would be wrong to say that "95% of the values are between 230.79 and 250.79," whether we are referring to the sample or the population. All we can say is, "Based on Mr. Schiel's sample, we believe that the *population mean* is somewhere between 230.79 and 250.79."

CHECK YOUR UNDERSTANDING

How much does the fat content of Brand X hot dogs vary? To find out, researchers measured the fat content (in grams) of a random sample of 10 Brand X hot dogs. A 95% confidence interval for the population standard deviation σ is 2.84 to 7.55.

1. Interpret the confidence interval.
2. Interpret the confidence level.
3. True or false: The interval from 2.84 to 7.55 has a 95% chance of containing the actual population standard deviation σ. Justify your answer.

Constructing a Confidence Interval

Why settle for 95% confidence when estimating an unknown parameter? The *Confidence Interval* applet might shed some light on this question.

ACTIVITY *The* Confidence Interval *applet*

MATERIALS: Computer with Internet connection and display capability

In this Activity, you will use the applet to explore the relationship between the confidence level and the confidence interval.

1. Go to www.whfreeman.com/tps4e and launch the applet. Set the confidence level at 95% and click "Sample 50."
2. Change the confidence level to 99%. What happens to the length of the confidence intervals? To the "Percent hit"?
3. Now change the confidence level to 90%. What happens to the length of the confidence intervals? To the "Percent hit"?
4. Finally, change the confidence level to 80%. What happens to the length of the confidence intervals? To the "Percent hit"?

As the Activity illustrates, the price we pay for greater confidence is a wider interval. If we're satisfied with 80% confidence, then our interval of plausible values for the parameter will be much narrower than if we insist on 90%, 95%, or 99% confidence. But we'll also be much less confident in our estimate. Taking this to an extreme, what if we want to estimate with 100% confidence the proportion p of all U.S. adults who use Twitter? That's easy: we're 100% confident that the interval from 0 to 1 captures the true population proportion!

```
mean(randNorm(M,
20,16))
                240.79
```

Let's look a bit more closely at the method we used earlier to calculate an approximate 95% confidence interval for Mr. Schiel's mystery mean. We started with

$$\text{estimate} \pm \text{margin of error}$$

Our (point) estimate came from the sample statistic $\bar{x} = 240.79$. What about the margin of error? Since the sampling distribution of $\bar{x}$ is Normal, about 95% of the values of $\bar{x}$ will lie within two standard deviations $(2\sigma_{\bar{x}})$ of the mystery mean μ. See the figure below. We could rewrite our interval as

$$240.79 \pm 2 \cdot 5 = \bar{x} \pm 2 \cdot \sigma_{\bar{x}}$$

This leads to the more general formula for a confidence interval:

$$\text{statistic} \pm (\text{critical value}) \cdot (\text{standard deviation of statistic})$$

Critical value

The **critical value** depends on both the confidence level C and the sampling distribution of the statistic.

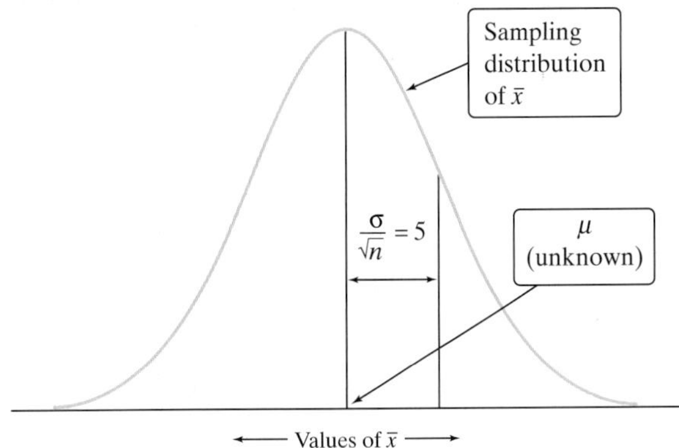

Calculating a Confidence Interval

The confidence interval for estimating a population parameter has the form

statistic ± (critical value) · (standard deviation of statistic)

where the statistic we use is the point estimator for the parameter.

The confidence interval for the mystery mean μ of Mr. Schiel's population illustrates several important properties that are shared by all confidence intervals in common use. The user chooses the confidence level, and the margin of error follows from this choice. We would like high confidence and also a small margin of error. High confidence says that our method almost always gives correct answers. A small margin of error says that we have pinned down the parameter quite precisely.

Our general formula for a confidence interval is

Statistic ± (critical value) · (standard deviation of statistic) ← Margin of error (the product of critical value and standard deviation of statistic)

Recall that $\sigma_{\bar{x}} = \frac{\sigma}{\sqrt{n}}$ and $\sigma_{\hat{p}} = \sqrt{\frac{p(1-p)}{n}}$. So as the sample size n increases, the standard deviation of the statistic decreases.

We can see that the margin of error depends on the critical value and the standard deviation of the statistic. The critical value is tied directly to the confidence level: greater confidence requires a larger critical value. The standard deviation of the statistic depends on the sample size n: larger samples give more precise estimates, which means less variability in the statistic.

So the margin of error gets smaller when:

- *The confidence level decreases.* There is a trade-off between the confidence level and the margin of error. To obtain a smaller margin of error from the same data, you must be willing to accept lower confidence. Earlier, we found that a 95% confidence interval for Mr. Schiel's mystery mean μ is 230.79 to 250.79. The 80% confidence interval for μ is 234.39 to 247.19. Figure 8.6 compares these two intervals.
- *The sample size n increases.* Increasing the sample size n reduces the margin of error for any fixed confidence level.

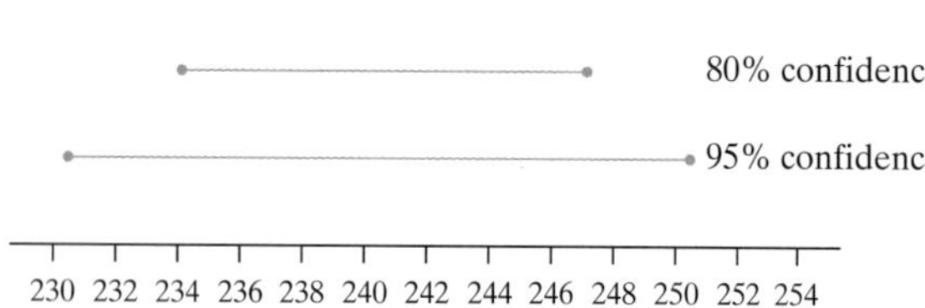

FIGURE 8.6 80% and 95% confidence intervals for Mr. Schiel's mystery mean. Higher confidence requires a longer interval.

Using Confidence Intervals Wisely

Our goal in this section has been to introduce you to the big ideas of confidence intervals without getting bogged down in technical details. You may have noticed that we only calculated intervals in a contrived setting: estimating an unknown population mean μ when we somehow knew the population standard deviation σ. In practice, when we don't know μ, we don't know σ either. We'll learn to construct confidence intervals for a population mean in this more realistic setting in Section 8.3. First, we will study confidence intervals for a population proportion p in Section 8.2. Although it is possible to estimate other parameters, confidence intervals for means and proportions are the most common tools in everyday use.

Before calculating a confidence interval for μ or p, there are three important **conditions** that you should check.

1. Random: The data should come from a well-designed random sample or randomized experiment. Otherwise, there's no scope for inference to a population (sampling) or inference about cause and effect (experiment). If we can't draw conclusions beyond the data at hand, then there's not much point constructing a confidence interval! Another important reason for random selection or random assignment is to introduce chance into the data production process. We can model chance behavior with a probability distribution, like the sampling distributions of Chapter 7. The probability distribution helps us calculate a confidence interval.

The experimental equivalent of a sampling distribution is called a randomization distribution.

2. Normal: The methods we use to construct confidence intervals for μ and p depend on the fact that the sampling distribution of the statistic ($\bar{x}$ or $\hat{p}$) is at least approximately Normal.

For means: The sampling distribution of $\bar{x}$ is exactly Normal if the population distribution is Normal. When the population distribution is not Normal, the central limit theorem tells us that the sampling distribution of $\bar{x}$ will be approximately Normal if n is sufficiently large (say, at least 30).

For proportions: We can use the Normal approximation to the sampling distribution of $\hat{p}$ as long as $np \geq 10$ and $n(1 - p) \geq 10$.

3. Independent: The procedures for calculating confidence intervals assume that individual observations are independent. Random sampling or random assignment can help ensure independent observations. However, our formulas for the standard deviation of the sampling distribution act as if we are sampling *with replacement* from a population. That is rarely the case. If we're sampling without replacement from a finite population, we should check the 10% condition: $n \leq \frac{1}{10}N$. Sampling more than 10% of a population can give us a more precise estimate of a population parameter but would require us to use a finite population correction. We'll avoid such situations in the text.

Conditions for Constructing a Confidence Interval

- **Random:** The data come from a well-designed random sample or randomized experiment.
- **Normal:** The sampling distribution of the statistic is approximately Normal.
- **Independent:** Individual observations are independent. When sampling without replacement, the sample size n should be no more than 10% of the population size N (the *10% condition*) to use our formula for the standard deviation of the statistic.

As you'll see in the next two sections, real-world studies rarely meet these conditions exactly. If any of the conditions is clearly violated, however, you should be reluctant to proceed with inference. For the rest of this chapter, we will limit ourselves to settings that involve random sampling. We will discuss inference for randomized experiments in Chapter 9.

Here are two other important reminders for constructing and interpreting confidence intervals.

- *Our method of calculation assumes that the data come from an SRS of size n from the population of interest.* Other types of random samples (stratified or cluster, say) might be preferable to an SRS in a given setting, but they require more complex calculations than the ones we'll use.

CAUTION

- *The margin of error in a confidence interval covers only chance variation due to random sampling or random assignment.* The margin of error is obtained from the sampling distribution (or randomization distribution in an experiment). It indicates how close our estimate is likely to be to the unknown parameter if we repeat the random sampling or random assignment process many times. Practical difficulties, such as undercoverage and nonresponse in a sample survey, can lead to additional errors that may be larger than this chance variation. Remember this unpleasant fact when reading the results of an opinion poll or other sample survey. The way in which a survey or experiment is conducted influences the trustworthiness of its results in ways that are not included in the announced margin of error.

CAUTION

SECTION 8.1 Summary

- To estimate an unknown population parameter, start with a statistic that provides a reasonable guess. The chosen statistic is a **point estimator** for the parameter. The specific value of the point estimator that we use gives a **point estimate** for the parameter.
- A **confidence interval** uses sample data to estimate an unknown population parameter with an indication of how precise the estimate is and of how confident we are that the result is correct.
- Any confidence interval has two parts: an interval computed from the data and a confidence level C. The interval has the form

$$\text{estimate} \pm \text{margin of error}$$

When calculating a confidence interval, it is common to use the form

$$\text{statistic} \pm (\text{critical value}) \cdot (\text{standard deviation of statistic})$$

- The **confidence level C** is the success rate of the method that produces the interval. If you use 95% confidence intervals often, in the long run 95% of your intervals will contain the true parameter value. You don't know whether a 95% confidence interval calculated from a particular set of data actually captures the true parameter value.
- Other things being equal, the **margin of error** of a confidence interval gets smaller as
  - the confidence level C decreases
  - the sample size *n* increases

- Before you calculate a confidence interval for a population mean or proportion, be sure to check conditions: **Random** sampling or random assignment, **Normal** sampling distribution, and **Independent** observations. Remember that the margin of error for a confidence interval includes only chance variation, not other sources of error like nonresponse and undercoverage.

SECTION 8.1 Exercises

In Exercises 1 to 4, determine the point estimator you would use and calculate the value of the point estimate.

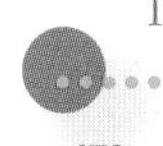

1. **Got shoes?** How many pairs of shoes, on average, do female teens have? To find out, an AP Statistics class conducted a survey. They selected an SRS of 20 female students from their school. Then they recorded the number of pairs of shoes that each student reported having. Here are the data:

pg 470

50	26	26	31	57	19	24	22	23	38
13	50	13	34	23	30	49	13	15	51

2. **Got shoes?** The class in Exercise 1 wants to estimate the variability in the number of pairs of shoes that female students have by estimating the population variance σ^2.

3. **Going to the prom** Tonya wants to estimate what proportion of the seniors in her school plan to attend the prom. She interviews an SRS of 50 of the 750 seniors in her school and finds that 36 plan to go to the prom.

pg 470

4. **Reporting cheating** What proportion of students are willing to report cheating by other students? A student project put this question to an SRS of 172 undergraduates at a large university: "You witness two students cheating on a quiz. Do you go to the professor?" Only 19 answered "Yes."[3]

5. **NAEP scores** Young people have a better chance of full-time employment and good wages if they are good with numbers. How strong are the quantitative skills of young Americans of working age? One source of data is the National Assessment of Educational Progress (NAEP) Young Adult Literacy Assessment Survey, which is based on a nationwide probability sample of households. The NAEP survey includes a short test of quantitative skills, covering mainly basic arithmetic and the ability to apply it to realistic problems. Scores on the test range from 0 to 500. For example, a person who scores 233 can add the amounts of two checks appearing on a bank deposit slip; someone scoring 325 can determine the price of a meal from a menu; a person scoring 375 can transform a price in cents per ounce into dollars per pound.[4]

   Suppose that you give the NAEP test to an SRS of 840 people from a large population in which the scores have mean 280 and standard deviation $\sigma = 60$. The mean $\bar{x}$ of the 840 scores will vary if you take repeated samples.

   (a) Describe the shape, center, and spread of the sampling distribution of $\bar{x}$.

   (b) Sketch the sampling distribution of $\bar{x}$. Mark its mean and the values one, two, and three standard deviations on either side of the mean.

   (c) According to the 68–95–99.7 rule, about 95% of all values of $\bar{x}$ lie within a distance m of the mean of the sampling distribution. What is m? Shade the region *on the axis* of your sketch that is within m of the mean.

   (d) Whenever $\bar{x}$ falls in the region you shaded, the population mean μ lies in the confidence interval $\bar{x} \pm m$. For what percent of all possible samples does the interval capture μ?

6. **Auto emissions** Oxides of nitrogen (called NOX for short) emitted by cars and trucks are important contributors to air pollution. The amount of NOX emitted by a particular model varies from vehicle to vehicle. For one light-truck model, NOX emissions vary with mean μ that is unknown and standard deviation $\sigma = 0.4$ gram per mile. You test an SRS of 50 of these trucks. The sample mean NOX level $\bar{x}$ estimates the unknown μ. You will get different values of $\bar{x}$ if you repeat your sampling.

(a) Describe the shape, center, and spread of the sampling distribution of $\bar{x}$.

(b) Sketch the sampling distribution of $\bar{x}$. Mark its mean and the values one, two, and three standard deviations on either side of the mean.

(c) According to the 68–95–99.7 rule, about 95% of all values of $\bar{x}$ lie within a distance m of the mean of the sampling distribution. What is m? Shade the region *on the axis* of your sketch that is within m of the mean.

(d) Whenever $\bar{x}$ falls in the region you shaded, the unknown population mean μ lies in the confidence interval $\bar{x} \pm m$. For what percent of all possible samples does the interval capture μ?

7. **NAEP scores** Refer to Exercise 5. Below your sketch, choose one value of $\bar{x}$ inside the shaded region and draw its corresponding confidence interval. Do the same for one value of $\bar{x}$ outside the shaded region. What is the most important difference between these intervals? (Use Figure 8.5, on page 474, as a model for your drawing.)

8. **Auto emissions** Refer to Exercise 6. Below your sketch, choose one value of $\bar{x}$ inside the shaded region and draw its corresponding confidence interval. Do the same for one value of $\bar{x}$ outside the shaded region. What is the most important difference between these intervals? (Use Figure 8.5, on page 474, as a model for your drawing.)

9. **How confident?** The figure below shows the result of taking 25 SRSs from a Normal population and constructing a confidence interval for each sample. Which confidence level—80%, 90%, 95%, or 99%—do you think was used? Explain.

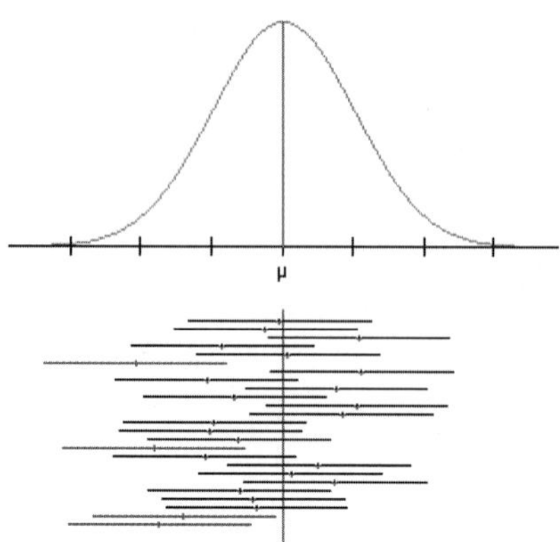

10. **How confident?** The figure at top right shows the result of taking 25 SRSs from a Normal population and constructing a confidence interval for each sample. Which confidence level—80%, 90%, 95%, or 99%—do you think was used? Explain.

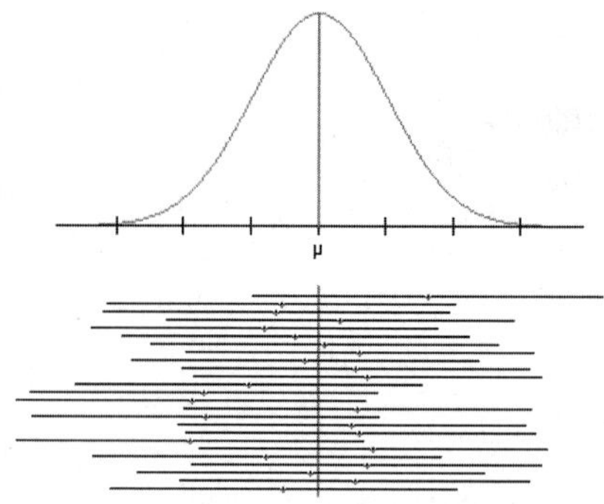

11. **Prayer in school** A *New York Times*/CBS News Poll asked the question, "Do you favor an amendment to the Constitution that would permit organized prayer in public schools?" Sixty-six percent of the sample answered "Yes." The article describing the poll says that it "is based on telephone interviews conducted from Sept. 13 to Sept. 18 with 1,664 adults around the United States, excluding Alaska and Hawaii. . . . The telephone numbers were formed by random digits, thus permitting access to both listed and unlisted residential numbers." The article gives the margin of error for a 95% confidence level as 3 percentage points.

(a) Explain what the margin of error means to someone who knows little statistics.

(b) State and interpret the 95% confidence interval.

(c) Interpret the confidence level.

12. **Losing weight** A Gallup Poll in November 2008 found that 59% of the people in its sample said "Yes" when asked, "Would you like to lose weight?" Gallup announced: "For results based on the total sample of national adults, one can say with 95% confidence that the margin of (sampling) error is ±3 percentage points."[5]

(a) Explain what the margin of error means in this setting.

(b) State and interpret the 95% confidence interval.

(c) Interpret the confidence level.

13. **Prayer in school** Refer to Exercise 11. The news article goes on to say: "The theoretical errors do not take into account a margin of additional error resulting from the various practical difficulties in taking any survey of public opinion." List some of the "practical difficulties" that may cause errors in addition to the ±3 percentage point margin of error. Pay particular attention to the news article's description of the sampling method.

14. **Losing weight** Refer to Exercise 12. As Gallup indicates, the 3 percentage point margin of error for this poll includes only sampling variability (what they

call "sampling error"). What other potential sources of error (Gallup calls these "nonsampling errors") could affect the accuracy of the 59% estimate?

15. **Shoes** The AP Statistics class in Exercise 1 also asked an SRS of 20 boys at their school how many shoes they have. A 95% confidence interval for the difference in the population means (girls – boys) is 10.9 to 26.5. Interpret the confidence interval and the confidence level.

pg 476

16. **Lying online** Many teens have posted profiles on sites such as Facebook and MySpace. A sample survey asked random samples of teens with online profiles if they included false information in their profiles. Of 170 younger teens (ages 12 to 14) polled, 117 said "Yes." Of 317 older teens (ages 15 to 17) polled, 152 said "Yes."[6] A 95% confidence interval for the difference in the population proportions (younger teens – older teens) is 0.120 to 0.297. Interpret the confidence interval and the confidence level.

17. **Explaining confidence** A 95% confidence interval for the mean body mass index (BMI) of young American women is 26.8 ± 0.6. Discuss whether each of the following explanations is correct.

(a) We are confident that 95% of all young women have BMI between 26.2 and 27.4.

(b) We are 95% confident that future samples of young women will have mean BMI between 26.2 and 27.4.

(c) Any value from 26.2 to 27.4 is believable as the true mean BMI of young American women.

(d) In 95% of all possible samples, the population mean BMI will be between 26.2 and 27.4.

(e) The mean BMI of young American women cannot be 28.

18. **Explaining confidence** The admissions director from Big City University found that (107.8, 116.2) is a 95% confidence interval for the mean IQ score of all freshmen. Comment on whether or not each of the following explanations is correct.

(a) There is a 95% probability that the interval from 107.8 to 116.2 contains μ.

(b) There is a 95% chance that the interval (107.8, 116.2) contains $\bar{x}$.

(c) This interval was constructed using a method that produces intervals that capture the true mean in 95% of all possible samples.

(d) 95% of all possible samples will contain the interval (107.8, 116.2).

(e) The probability that the interval (107.8, 116.2) captures μ is either 0 or 1, but we don't know which.

19. **Conditions** Explain briefly why each of the three conditions—Random, Normal, and Independent—is important when constructing a confidence interval.

20. **Plagiarizing** An online poll posed the following question:

It is now possible for school students to log on to Internet sites and download homework. Everything from book reports to doctoral dissertations can be downloaded free or for a fee. Do you believe that giving a student who is caught plagiarizing an F for their assignment is the right punishment?

Of the 20,125 people who responded, 14,793 clicked "Yes." That's 73.5% of the sample. Based on this sample, a 95% confidence interval for the percent of the population who would say "Yes" is 73.5% ± 0.61%. Which of the three inference conditions is violated? Why is this confidence interval worthless?

Multiple choice: Select the best answer for Exercises 21 to 24.

21. A researcher plans to use a random sample of n = 500 families to estimate the mean monthly family income for a large population. A 99% confidence interval based on the sample would be ______ than a 90% confidence interval.

(a) narrower and would involve a larger risk of being incorrect

(b) wider and would involve a smaller risk of being incorrect

(c) narrower and would involve a smaller risk of being incorrect

(d) wider and would involve a larger risk of being incorrect

(e) wider, but it cannot be determined whether the risk of being incorrect would be larger or smaller

22. In a poll,

I. Some people refused to answer questions.

II. People without telephones could not be in the sample.

III. Some people never answered the phone in several calls.

Which of these sources is included in the ±2% margin of error announced for the poll?

(a) I only (c) III only (e) None of these
(b) II only (d) I, II, and III

23. You have measured the systolic blood pressure of an SRS of 25 company employees. A 95% confidence

interval for the mean systolic blood pressure for the employees of this company is (122, 138). Which of the following statements gives a valid interpretation of this interval?

(a) 95% of the sample of employees have a systolic blood pressure between 122 and 138.

(b) 95% of the population of employees have a systolic blood pressure between 122 and 138.

(c) If the procedure were repeated many times, 95% of the resulting confidence intervals would contain the population mean systolic blood pressure.

(d) The probability that the population mean blood pressure is between 122 and 138 is 0.95.

(e) If the procedure were repeated many times, 95% of the sample means would be between 122 and 138.

24. A polling organization announces that the proportion of American voters who favor congressional term limits is 64%, with a 95% confidence margin of error of 3%. If the opinion poll had announced the margin of error for 80% confidence rather than 95% confidence, this margin of error would be

(a) 3%, because the same sample is used.

(b) less than 3%, because we require less confidence.

(c) less than 3%, because the sample size is smaller.

(d) greater than 3%, because we require less confidence.

(e) greater than 3%, because the sample size is smaller.

25. **Power lines and cancer** (4.2, 4.3) Does living near power lines cause leukemia in children? The National Cancer Institute spent 5 years and $5 million gathering data on this question. The researchers compared 638 children who had leukemia with 620 who did not. They went into the homes and actually measured the magnetic fields in children's bedrooms, in other rooms, and at the front door. They recorded facts about power lines near the family home and also near the mother's residence when she was pregnant. Result: no connection between leukemia and exposure to magnetic fields of the kind produced by power lines was found.[7]

(a) Was this an observational study or an experiment? Justify your answer.

(b) Does this study show that living near power lines doesn't cause cancer? Explain.

26. **Sisters and brothers** (3.1, 3.2) How strongly do physical characteristics of sisters and brothers correlate? Here are data on the heights (in inches) of 11 adult pairs:[8]

Brother:	71	68	66	67	70	71	70	73	72	65	66
Sister:	69	64	65	63	65	62	65	64	66	59	62

(a) Construct a scatterplot using brother's height as the explanatory variable. Describe what you see.

(b) Use your calculator to compute the least-squares regression line for predicting sister's height from brother's height. Interpret the slope in context.

(c) Damien is 70 inches tall. Predict the height of his sister Tonya.

(d) Do you expect your prediction in (c) to be very accurate? Give appropriate evidence to support your answer.

8.2 Estimating a Population Proportion

In Section 8.2, you'll learn about:

- Conditions for estimating p
- Constructing a confidence interval for p
- Putting it all together: the four-step process
- Choosing the sample size

In Section 8.1, we saw that a confidence interval can be used to estimate an unknown population parameter. We are often interested in estimating the proportion p of some outcome in the population. Here are some examples:

- What proportion of U.S. adults are unemployed right now?
- What proportion of high school students have cheated on a test?
- What proportion of pine trees in a national park are infested with beetles?
- What proportion of college students pray daily?
- What proportion of a company's laptop batteries last as long as the company claims?

This section shows you how to construct and interpret a confidence interval for a population proportion. The following Activity gives you a taste of what lies ahead.

ACTIVITY *The beads*

MATERIALS: Several thousand small plastic beads of at least two colors, thermos or other container, small cup for sampling, several small bowls

Before class, your teacher will prepare a large population of different-colored beads and put them into a container that you cannot see inside. Your goal is to estimate the actual proportion of beads in the population that have a particular color (say, red).

1. As a class, discuss how to use the cup provided to get a simple random sample of beads from the container. Think this through carefully, because you will get to take only one sample.
2. Have one student take an SRS of beads. Separate the beads into two groups: those that are red and those that aren't. Count the number of beads in each group.
3. Determine a point estimate for the unknown population proportion p of red beads in the container.
4. Now for the challenge: each team of three to four students will be given about 10 minutes to find a 90% confidence interval for the parameter p. Be sure to consider any conditions that are required for the methods you use.
5. Compare results with other teams in the class. Discuss any problems you encountered and how you dealt with them.

Conditions for Estimating p

When Mr. Vignolini's class did the beads Activity, they got 107 red beads and 144 white beads. Their point estimate for the unknown proportion p of red beads in the population is

$$\hat{p} = \frac{107}{251} = 0.426$$

How can the students in the class use this information to find a confidence interval for p?

As always, inference is based on the sampling distribution of a statistic. We described the sampling distribution of a sample proportion $\hat{p}$ in Section 7.2. Here is a brief review of its important properties:

Shape: If the sample size is large enough that both np and $n(1 - p)$ are at least 10 (*Normal condition*), the sampling distribution of $\hat{p}$ is approximately Normal.
Center: The mean is p. That is, the sample proportion $\hat{p}$ is an unbiased estimator of the population proportion p.
Spread: The standard deviation of the sampling distribution of $\hat{p}$ is

$$\sigma_{\hat{p}} = \sqrt{\frac{p(1-p)}{n}}$$

provided that the population is at least 10 times as large as the sample (10% condition).

Figure 8.7 displays this sampling distribution.

In practice, of course, we don't know the value of p. If we did, we wouldn't need to construct a confidence interval for it! So we cannot check whether np and $n(1 - p)$ are 10 or greater. In large samples, $\hat{p}$ will be close to p. Therefore, we replace p by $\hat{p}$ in checking the Normal condition.

Let's see how the conditions play out for Mr. Vignolini's class.

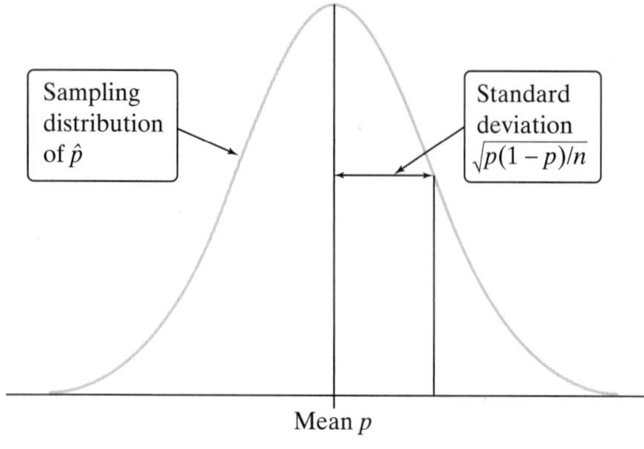

FIGURE 8.7 Select a large SRS from a population that contains proportion p of successes. The sampling distribution of the proportion $\hat{p}$ of successes in the sample is approximately Normal. The mean is p and the standard deviation is $\sqrt{\frac{p(1-p)}{n}}$.

The Beads

Checking conditions

Mr. Vignolini's class wants to construct a confidence interval for the proportion p of red beads in the container. Recall that the class's sample had 107 red beads and 144 white beads.

PROBLEM: **Check that the conditions for constructing a confidence interval for p are met.**

SOLUTION: **There are three conditions to check:**

- **Random: The class took an SRS of 251 beads from the container.**
- **Normal: To use a Normal approximation for the sampling distribution of $\hat{p}$, we need both np and $n(1 - p)$ to be at least 10. Since we don't know p, we check that**

$$n\hat{p} = 251\left(\frac{107}{251}\right) = 107 \quad \text{and} \quad n(1-\hat{p}) = 251\left(1-\frac{107}{251}\right) = 251\left(\frac{144}{251}\right) = 144$$

That is, the counts of successes (red beads) and failures (non-red beads) are both at least 10.

- **Independent: Since the class sampled without replacement, they need to check the 10% condition: at least 10(251) = 2510 beads need to be in the population. Mr. Vignolini reveals that there are 3000 beads in the container, so this condition is satisfied.**

Since all three conditions are met, it should be safe for the class to construct a confidence interval.

Notice that $n\hat{p}$ and $n(1 - \hat{p})$ should be whole numbers. You don't really need to calculate these values since they are just the number of successes and failures in the sample. In the previous example, we could address the Normal condition simply by saying, "The numbers of successes (107) and failures (144) in the sample are both at least 10."

CHECK YOUR UNDERSTANDING

In each of the following settings, check whether the conditions for calculating a confidence interval for the population proportion p are met.

1. An AP Statistics class at a large high school conducts a survey. They ask the first 100 students to arrive at school one morning whether or not they slept at least 8 hours the night before. Only 17 students say "Yes."

2. A quality control inspector takes a random sample of 25 bags of potato chips from the thousands of bags filled in an hour. Of the bags selected, 3 had too much salt.

Constructing a Confidence Interval for p

We can use the general formula from Section 8.1 to construct a confidence interval for an unknown population proportion p:

$$\text{statistic} \pm (\text{critical value}) \cdot (\text{standard deviation of statistic})$$

The sample proportion $\hat{p}$ is the statistic we use to estimate p. When the Independent condition is met, the standard deviation of the sampling distribution of $\hat{p}$ is

$$\sigma_{\hat{p}} = \sqrt{\frac{p(1-p)}{n}}$$

Since we don't know the value of p, we replace it with the sample proportion $\hat{p}$:

$$\sqrt{\frac{\hat{p}(1-\hat{p})}{n}}$$

Some books refer to $\sigma_{\hat{p}}$ as the "standard error" of $\hat{p}$ and to what we call the standard error as the "estimated standard error."

This quantity is called the **standard error (SE)** of the sample proportion $\hat{p}$. It describes how close the sample proportion $\hat{p}$ will be, on average, to the population proportion p in repeated SRSs of size n.

DEFINITION: Standard error

When the standard deviation of a statistic is estimated from data, the result is called the **standard error** of the statistic.

How do we get the critical value for our confidence interval? If the Normal condition is met, we can use a Normal curve. For the approximate 95% confidence intervals of Section 8.1, we used a critical value of 2 based on the 68–95–99.7 rule for Normal distributions. We can get a more accurate critical value from Table A or a calculator. As Figure 8.8 on the next page shows, the central 95% of the standard Normal distribution is marked off by two points, $z^* = 1.96$ and $-z^* = -1.96$. We use the * to remind you that this is a critical value, not a standardized score that has been calculated from data.

Critical value z^*

To find a level C confidence interval, we need to catch the central area C under the standard Normal curve. Here's an example that shows how to get the **critical value z^*** for a different confidence level.

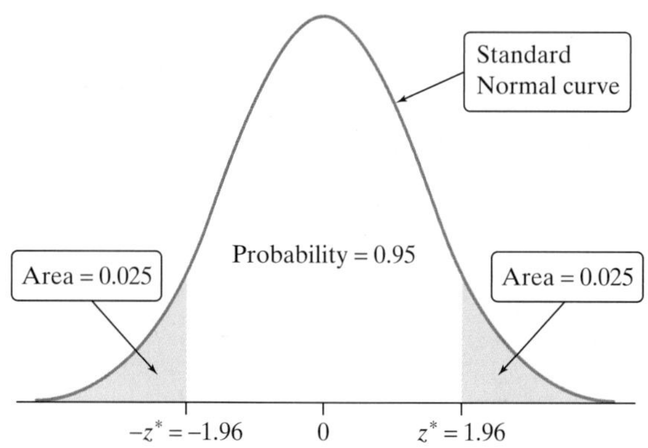

FIGURE 8.8 Finding the critical value for a 95% confidence interval: it's actually 1.96.

80% Confidence
Finding a critical value

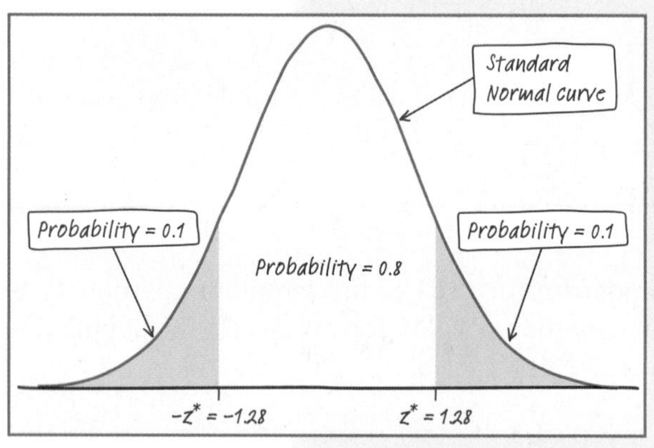

FIGURE 8.9 Finding the critical value for an 80% confidence interval.

PROBLEM: Use Table A to find the critical value z^* for an 80% confidence interval. Assume that the Normal condition is met.

SOLUTION: For an 80% confidence level, we need to catch the central 80% of the standard Normal distribution. In catching the central 80%, we leave out 20%, or 10% in each tail. So the desired critical value z^* is the point with area 0.1 to its right under the standard Normal curve. Figure 8.9 shows the details in picture form.

z	.07	.08	.09
−1.3	.0853	.0838	.0823
−1.2	.1020	.1003	.0985
−1.1	.1210	.1190	.1170

Search the body of Table A to find the point $-z^*$ with area 0.1 to its left. The closest entry is $z = -1.28$. (See the excerpt from Table A above.) So the critical value we want is $z^* = 1.28$.

For Practice *Try Exercise 31*

You can also find the critical value using the command `invNorm(0.9,0,1)`, which tells the calculator to find the z-value from the standard Normal curve that has area 0.9 to the *left* of it.

Once we find the critical value z^*, our confidence interval for the population proportion p is

$$\text{statistic} \pm (\text{critical value}) \cdot (\text{standard deviation of statistic})$$

$$= \hat{p} \pm z^* \sqrt{\frac{\hat{p}(1-\hat{p})}{n}}$$

Technically, the correct formula for a confidence interval is statistic ± (critical value) · (standard *error* of statistic). We are following the convention used on the AP Statistics exam formula sheet.

Notice that we replaced the standard deviation of $\hat{p}$ with the formula for its standard error. The resulting interval is sometimes called a **one-sample z interval for a population proportion.**

One-Sample z Interval for a Population Proportion

Choose an SRS of size n from a large population that contains an unknown proportion p of successes. An approximate level C **confidence interval for p** is

$$\hat{p} \pm z^* \sqrt{\frac{\hat{p}(1-\hat{p})}{n}}$$

where z^* is the critical value for the standard Normal curve with area C between $-z^*$ and z^*.

Use this interval only when the numbers of successes and failures in the sample are both at least 10 and the population is at least 10 times as large as the sample.

Now we can get the desired confidence interval for Mr. Vignolini's class.

The Beads

Calculating a confidence interval for p

PROBLEM: Mr. Vignolini's class took an SRS of beads from the container and got 107 red beads and 144 white beads.

(a) Calculate and interpret a 90% confidence interval for p.

(b) Mr. Vignolini claims that exactly half of the beads in the container are red. Use your result from (a) to comment on this claim.

SOLUTION: We checked conditions for calculating the interval earlier.

(a) Our confidence interval has the form

$$\text{statistic} \pm (\text{critical value}) \cdot (\text{standard deviation of statistic})$$

$$= \hat{p} \pm z^* \sqrt{\frac{\hat{p}(1-\hat{p})}{n}}$$

Earlier, we found that $\hat{p} = 107/251 = 0.426$. From Table A, we look for the point with area 0.05 to its left. As the excerpt from Table A shows, this point is between $z = -1.64$ and $z = -1.65$. The calculator's invNorm(0.05, 0, 1) gives $z = -1.645$. So we use $z^* = 1.645$ as our critical value.

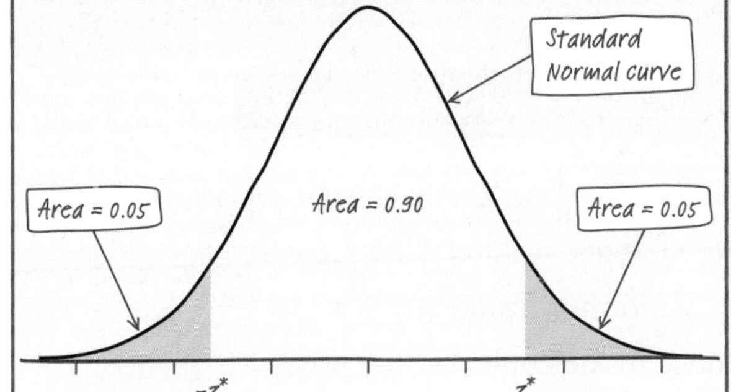

z	.03	.04	.05
−1.7	.0418	.0409	.0401
−1.6	.0516	.0505	.0495
−1.5	.0630	.0618	.0606

Computer studies have shown that a variation of our method for calculating a 95% confidence interval for p can result in closer to a 95% capture rate in the long run, especially for small sample sizes. This simple adjustment, first suggested by Edwin Bidwell Wilson in 1927, is sometimes called the "plus four" estimate. Just pretend we have four additional observations, two of which are successes and two of which are failures. Then calculate the "plus four interval" using the plus four estimate in place of $\hat{p}$ in our usual formula.

The resulting 90% confidence interval is

$$= 0.426 \pm 1.645\sqrt{\frac{(0.426)(1-0.426)}{251}}$$
$$= 0.426 \pm 0.051$$
$$= (0.375, 0.477)$$

We are 90% confident that the interval from 0.375 to 0.477 captures the actual proportion of red beads in Mr. Vignolini's container.

(b) The confidence interval in part (a) gives a range of plausible values for the population proportion of red beads. Since 0.5 is not contained in the interval, we have reason to doubt Mr. Vignolini's claim.

For Practice *Try Exercise 35*

CHECK YOUR UNDERSTANDING

Alcohol abuse has been described by college presidents as the number one problem on campus, and it is an important cause of death in young adults. How common is it? A survey of 10,904 randomly selected U.S. college students collected information on drinking behavior and alcohol-related problems.[9] The researchers defined "frequent binge drinking" as having five or more drinks in a row three or more times in the past two weeks. According to this definition, 2486 students were classified as frequent binge drinkers.

1. Identify the population and the parameter of interest.
2. Check conditions for constructing a confidence interval for the parameter.
3. Find the critical value for a 99% confidence interval. Show your method. Then calculate the interval.
4. Interpret the interval in context.

Putting It All Together: The Four-Step Process

Taken together, the examples about Mr. Vignolini's class and the beads Activity show you how to get a confidence interval for an unknown population proportion p. We can use the familiar four-step process whenever a problem asks us to construct and interpret a confidence interval.

Confidence Intervals: A Four-Step Process

State: What *parameter* do you want to estimate, and at what confidence level?

Plan: Identify the appropriate inference *method*. Check *conditions*.

Do: If the conditions are met, perform *calculations*.

Conclude: *Interpret* your interval in the context of the problem.

The next example illustrates the four-step process in action.

EXAMPLE Teens Say Sex Can Wait

Confidence interval for p

The Gallup Youth Survey asked a random sample of 439 U.S. teens aged 13 to 17 whether they thought young people should wait to have sex until marriage.[10] Of the sample, 246 said "Yes." Construct and interpret a 95% confidence interval for the proportion of all teens who would say "Yes" if asked this question.

STATE: We want to estimate the actual proportion p of all 13- to 17-year-olds in the United States who would say that young people should wait to have sex until they get married at a 95% confidence level.

PLAN: We should use a one-sample z interval for p if the *conditions* are satisfied.

- Random: Gallup surveyed a random sample of 439 U.S. teens.
- Normal: We check the counts of "successes" and "failures":

$$n\hat{p} = 246 \geq 10 \quad \text{and} \quad n(1-\hat{p}) = 193 \geq 10$$

- Independent: The responses of individual teens need to be independent. Since Gallup is sampling without replacement, we need to check the 10% condition: there are at least $10(439) = 4390$ U.S. teens aged 13 to 17.

DO: A 95% confidence interval for p is given by

$$\begin{aligned}\hat{p} \pm z^* \sqrt{\frac{\hat{p}(1-\hat{p})}{n}} &= 0.56 \pm 1.96\sqrt{\frac{(0.56)(0.44)}{439}} \\ &= 0.56 \pm 0.046 \\ &= (0.514, 0.606)\end{aligned}$$

CONCLUDE: We are 95% confident that the interval from 0.514 to 0.606 captures the true proportion of 13- to 17-year-olds in the United States who would say that teens should wait until marriage to have sex.

Another way to interpret this interval is "We are 95% confident that the true proportion of U.S. 13- to 17-year-olds who would say that teens should wait to have sex until marriage is between 0.514 and 0.606." We don't prefer this wording. It makes it sound like the interval is fixed and the population proportion is varying.

For Practice *Try Exercise 39*

CAUTION

Remember that the margin of error in a confidence interval includes only sampling variability! There are other sources of error that are not taken into account. As is the case with many surveys, we are forced to assume that the teens answered truthfully. If they didn't, then our estimate may be biased. Other problems like nonresponse and question wording can also affect the results of this or any other poll. *Lesson:* Sampling beads is much easier than sampling people!

Your calculator will handle the "Do" part of the four-step process, as the following Technology Corner illustrates. However, be sure to read the AP Exam Tip in the margin.

AP EXAM TIP If a free-response question asks you to construct and interpret a confidence interval, you are expected to do the entire four-step process. That includes clearly defining the parameter and checking conditions.

AP EXAM TIP You may use your calculator to compute a confidence interval on the AP exam. But there's a risk involved. If you just give the calculator answer with no work, you'll get either full credit for the "Do" step (if the interval is correct) or no credit (if it's wrong). We recommend showing the calculation with the appropriate formula and then checking with your calculator. If you opt for the calculator-only method, be sure to name the procedure (e.g., one-proportion z interval) and to give the interval (e.g., 0.514 to 0.606).

TECHNOLOGY CORNER Confidence interval for a population proportion

The TI-83/84 and TI-89 can be used to construct a confidence interval for an unknown population proportion. We'll demonstrate using the previous example. Of $n = 439$ teens surveyed, $X = 246$ said they thought that young people should wait to have sex until after marriage. To construct a confidence interval:

TI-83/84

- Press STAT, then choose TESTS and `A:1-PropZInt`.

TI-89

- In the Statistics/List Editor, press 2nd F2 ([F7]) and choose `5:1-PropZInt`.

- When the 1-PropZInt screen appears, enter $x = 246$, $n = 439$, and confidence level 0.95.

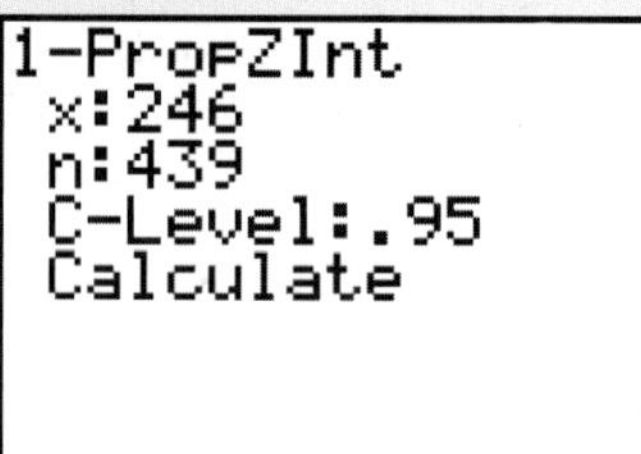

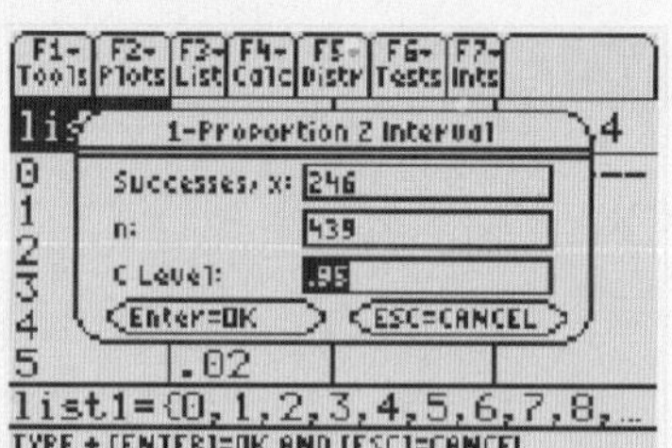

- Highlight "Calculate" and press ENTER. The 95% confidence interval for p is reported, along with the sample proportion $\hat{p}$ and the sample size, as shown here.

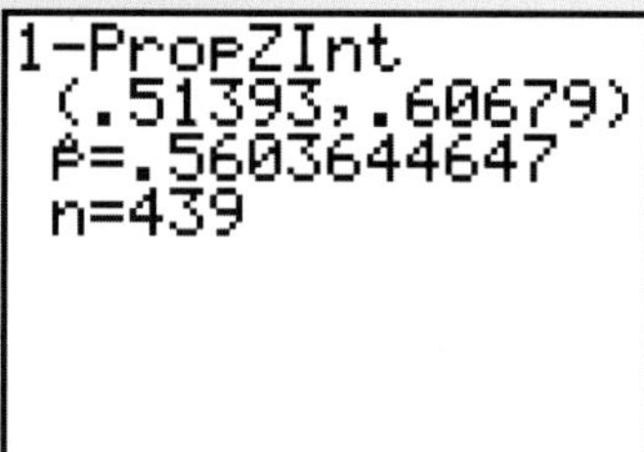

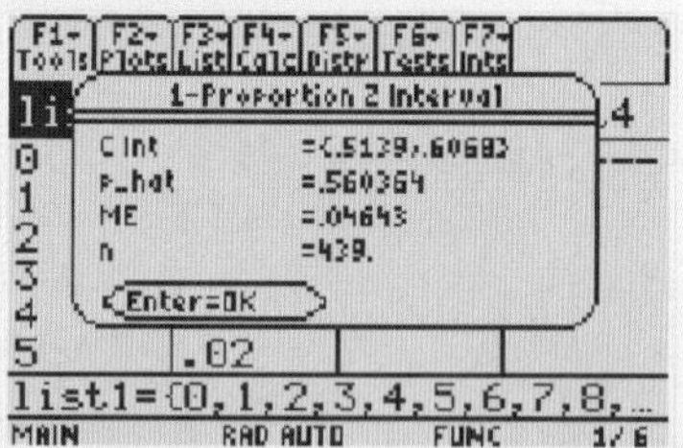

TI-Nspire instructions in Appendix B

Choosing the Sample Size

In planning a study, we may want to choose a sample size that allows us to estimate a population proportion within a given margin of error. National survey organizations like the Gallup Poll typically sample between 1000 and 1500 American adults, who are interviewed by telephone. Why do they choose such sample sizes?

The margin of error (ME) in the confidence interval for p is

$$ME = z^* \sqrt{\frac{\hat{p}(1-\hat{p})}{n}}$$

Here, z^* is the standard Normal critical value for the level of confidence we want. Because the margin of error involves the sample proportion of successes $\hat{p}$, we have to guess the latter value when choosing n. Here are two ways to do this:

1. Use a guess for $\hat{p}$ based on a pilot study or on past experience with similar studies. You should do several calculations that cover the range of $\hat{p}$-values you might get.

2. Use $\hat{p} = 0.5$ as the guess. The margin of error *ME* is largest when $\hat{p} = 0.5$, so this guess is conservative in the sense that if we get any other $\hat{p}$ when we do our study, we will get a margin of error smaller than planned.

Once you have a guess for $\hat{p}$, the formula for the margin of error can be solved to give the sample size n needed.

Sample Size for Desired Margin of Error

To determine the sample size n that will yield a level C confidence interval for a population proportion p with a maximum margin of error *ME*, solve the following inequality for n:

$$z^* \sqrt{\frac{\hat{p}(1-\hat{p})}{n}} \leq ME$$

where $\hat{p}$ is a guessed value for the sample proportion. The margin of error will always be less than or equal to *ME* if you take the guess $\hat{p}$ to be 0.5.

Here's an example that shows you how to determine the sample size.

EXAMPLE

Customer Satisfaction

Determining sample size

A company has received complaints about its customer service. The managers intend to hire a consultant to carry out a survey of customers. Before contacting the consultant, the company president wants some idea of the sample size that she will be required to pay for. One critical question is the degree of satisfaction with the company's customer service, measured on a five-point scale. The president wants to estimate the proportion p of customers who are satisfied (that is, who choose either "satisfied" or "very satisfied," the two highest levels on the five-point scale). She decides that she wants the estimate to be within 3% (0.03) at a 95% confidence level. How large a sample is needed?

PROBLEM: **Determine the sample size needed to estimate p within 0.03 with 95% confidence.**

SOLUTION: **The critical value for 95% confidence is $z^* = 1.96$. Since the company president wants a margin of error of no more than 0.03, we need to solve the equation**

$$1.96\sqrt{\frac{\hat{p}(1-\hat{p})}{n}} \leq 0.03$$

for n. Multiplying both sides by $\sqrt{n}$ and then dividing both sides by 0.03 yields

$$\frac{1.96}{0.03}\sqrt{\hat{p}(1-\hat{p})} \leq \sqrt{n}$$

Squaring both sides gives

$$\left(\frac{1.96}{0.03}\right)^2 \hat{p}(1-\hat{p}) \leq n$$

The largest value of n will be required if $\hat{p} = 0.5$, since that will make the product $\hat{p}(1-\hat{p})$ as large as possible. Because we have no idea of the true proportion p of satisfied customers, we decide to use $\hat{p} = 0.5$ as our guess. The sample size required is

$$\left(\frac{1.96}{0.03}\right)^2 (0.5)(1-0.5) \leq n$$

$$1067.111 \leq n$$

So we round up to 1068 respondents to ensure that the margin of error is no more than 3%.

For Practice *Try Exercise 43*

Why not round to the nearest whole number—in this case, 1067? Because a smaller sample size will result in a larger margin of error, possibly more than the desired 3% for the poll. If you want a 2.5% margin of error rather than 3%, then

$$n = \left(\frac{1.96}{0.025}\right)^2 (0.5)(1-0.5) = 1536.64 \Rightarrow n = 1537$$

For a 2% margin of error, the sample size you need is

$$n = \left(\frac{1.96}{0.02}\right)^2 (0.5)(1-0.5) = 2401$$

As usual, smaller margins of error call for larger samples.

News reports frequently describe the results of surveys with sample sizes between 1000 and 1500 and a margin of error of about 3%. These surveys generally use sampling procedures more complicated than a simple random sample, so the calculation of confidence intervals is more involved than what we have studied in this section. The calculations of the previous example still give you a rough idea of how such surveys are planned.

CHECK YOUR UNDERSTANDING

Refer to the previous example about the company's customer satisfaction survey.

1. In the company's prior-year survey, 80% of customers surveyed said they were "satisfied" or "very satisfied." Using this value as a guess for $\hat{p}$, find the sample size needed for a margin of error of 3% at a 95% confidence level.

2. What if the company president demands 99% confidence instead? Determine how this would affect your answer to Question 1.

SECTION 8.2 Summary

- Confidence intervals for a population proportion p are based on the sampling distribution of the sample proportion $\hat{p}$. When n is large enough that both np and $n(1-p)$ are at least 10, the sampling distribution of $\hat{p}$ is approximately Normal with mean p and standard deviation $\sqrt{p(1-p)/n}$.
- In practice, we use the sample proportion $\hat{p}$ to estimate the unknown parameter p. We therefore replace the standard deviation of $\hat{p}$ with its **standard error** when constructing a confidence interval. The level C **confidence interval for p** is

$$\hat{p} \pm z^* \sqrt{\frac{\hat{p}(1-\hat{p})}{n}}$$

where z^* is the standard Normal critical value.

- This inference procedure is approximately correct when these conditions are met:
  - **Random:** The data were produced by random sampling or random assignment.
  - **Normal:** The sample is large enough that both np and $n(1-p)$ are at least 10. Since we don't know the value of p, we check that $n\hat{p}$ and $n(1-\hat{p})$, the counts of successes and failures in the sample, are both at least 10.
  - **Independent:** Individual observations are independent. When sampling without replacement, we check the *10% condition:* the population is at least 10 times as large as the sample.
- When constructing a confidence interval, follow the familiar four-step process:

STATE: What *parameter* do you want to estimate, and at what *confidence level*?

PLAN: Identify the appropriate inference *method*. Check *conditions*.

DO: If the conditions are met, perform *calculations*.

CONCLUDE: *Interpret* your interval in the context of the problem.

- The sample size needed to obtain a confidence interval with approximate margin of error ME for a population proportion involves solving

$$z^* \sqrt{\frac{\hat{p}(1-\hat{p})}{n}} \leq ME$$

for n, where $\hat{p}$ is a guessed value for the sample proportion, and z^* is the critical value for the level of confidence you want. If you use $\hat{p} = 0.5$ in this formula, the margin of error of the interval will be less than or equal to ME.

8.2 TECHNOLOGY CORNER

TI-Nspire instructions in Appendix B

SECTION 8.2 Exercises

For Exercises 27 to 30, check whether each of the conditions is met for calculating a confidence interval for the population proportion p.

27. **Rating school food** Latoya wants to estimate what proportion of the seniors at her boarding high school like the cafeteria food. She interviews an SRS of 50 of the 175 seniors living in the dormitory. She finds that 14 think the cafeteria food is good.

pg 486

28. **High tuition costs** Glenn wonders what proportion of the students at his school think that tuition is too high. He interviews an SRS of 50 of the 2400 students at his college. Thirty-eight of those interviewed think tuition is too high.

29. **AIDS and risk factors** In the National AIDS Behavioral Surveys sample of 2673 adult heterosexuals, 0.2% had both received a blood transfusion and had a sexual partner from a group at high risk of AIDS. We want to estimate the proportion p in the population who share these two risk factors.

30. **Whelks and mussels** The small round holes you often see in sea shells were drilled by other sea creatures, who ate the former dwellers of the shells. Whelks often drill into mussels, but this behavior appears to be more or less common in different locations. Researchers collected whelk eggs from the coast of Oregon, raised the whelks in the laboratory, then put each whelk in a container with some delicious mussels. Only 9 of 98 whelks drilled into a mussel.[11] The researchers want to estimate the proportion p of Oregon whelks that will spontaneously drill into mussels.

31. **98% confidence** Find z^* for a 98% confidence interval using Table A or your calculator. Show your method.

pg 488

32. **93% confidence** Find z^* for a 93% confidence interval using Table A or your calculator. Show your method.

33. **Going to the prom** Tonya wants to estimate what proportion of her school's seniors plan to attend the prom. She interviews an SRS of 50 of the 750 seniors in her school and finds that 36 plan to go to the prom.

(a) Identify the population and parameter of interest.

(b) Check conditions for constructing a confidence interval for the parameter.

(c) Construct a 90% confidence interval for p. Show your method.

(d) Interpret the interval in context.

34. **Reporting cheating** What proportion of students are willing to report cheating by other students? A student project put this question to an SRS of 172 undergraduates at a large university: "You witness two students cheating on a quiz. Do you go to the professor?" Only 19 answered "Yes."[12]

(a) Identify the population and parameter of interest.

(b) Check conditions for constructing a confidence interval for the parameter.

(c) Construct a 99% confidence interval for p. Show your method.

(d) Interpret the interval in context.

35. **Abstain from drinking** In a Harvard School of Public Health survey, 2105 of 10,904 randomly selected U.S. college students were classified as abstainers (nondrinkers).

pg 489

(a) Construct and interpret a 99% confidence interval for p. Follow the four-step process.

(b) A newspaper article claims that 25% of U.S. college students are nondrinkers. Use your result from (a) to comment on this claim.

36. **Teens' online profiles** Over half of all American teens (ages 12 to 17 years) have an online profile, mainly on Facebook. A random sample of 487 teens with profiles found that 385 included photos of themselves.[13]

(a) Construct and interpret a 95% confidence interval for p. Follow the four-step process.

(b) Is it plausible that the true proportion of American teens with profiles who have posted photos of themselves is 0.75? Use your result from part (a) to support your answer.

37. **Abstain from drinking** Describe a possible source of error that is not included in the margin of error for the 99% confidence interval in Exercise 35.

38. **Teens' online profiles** Describe a possible source of error that is not included in the margin of error for the 95% confidence interval in Exercise 36.

39. **How common is SAT coaching?** A random sample of students who took the SAT college entrance examination twice found that 427 of the respondents had paid for coaching courses and that the remaining 2733 had not.[14] Construct and interpret a 99% confidence interval for the proportion of coaching among students who retake the SAT. Follow the four-step process.

pg 491

40. **2K10 begins** In January 2010, a Gallup Poll asked a random sample of adults, "In general, are you satisfied or dissatisfied with the way things are going in the United States at this time?" In all, 256 said that they were satisfied and the remaining 769 said they were not. Construct and interpret a 90% confidence interval for the proportion of adults who are satisfied with how things are going. Follow the four-step process.

41. **A TV poll** A television news program conducts a call-in poll about a proposed city ban on handgun ownership. Of the 2372 calls, 1921 oppose the ban. The station, following recommended practice, makes a confidence statement: "81% of the Channel 13 Pulse Poll sample opposed the ban. We can be 95% confident that the true proportion of citizens opposing a handgun ban is within 1.6% of the sample result."

(a) Is the station's quoted 1.6% margin of error correct? Explain.

(b) Is the station's conclusion justified? Explain.

42. **Equality for women?** Have efforts to promote equality for women gone far enough in the United States? A poll on this issue by the cable network MSNBC contacted 1019 adults. A newspaper article about the poll said, "Results have a margin of sampling error of plus or minus 3 percentage points."[15]

(a) The news article said that 65% of men, but only 43% of women, think that efforts to promote equality have gone far enough. Explain why we do not have enough information to give confidence intervals for men and women separately.

(b) Would a 95% confidence interval for women alone have a margin of error less than 0.03, about equal to 0.03, or greater than 0.03? Why? (You see that the news article's statement about the margin of error for poll results is a bit misleading.)

43. **Can you taste PTC?** PTC is a substance that has a strong bitter taste for some people and is tasteless for others. The ability to taste PTC is inherited. About 75% of Italians can taste PTC, for example. You want to estimate the proportion of Americans who have at least one Italian grandparent and who can taste PTC.

pg 493

(a) How large a sample must you test to estimate the proportion of PTC tasters within 0.04 with 90% confidence? Answer this question using the 75% estimate as the guessed value for $\hat{p}$.

(b) Answer the question in part (a) again, but this time use the conservative guess $\hat{p} = 0.5$. By how much do the two sample sizes differ?

44. **School vouchers** A national opinion poll found that 44% of all American adults agree that parents should be given vouchers that are good for education at any public or private school of their choice. The result was based on a small sample.

(a) How large an SRS is required to obtain a margin of error of 0.03 (that is, ±3%) in a 99% confidence interval? Answer this question using the previous poll's result as the guessed value for $\hat{p}$.

(b) Answer the question in part (a) again, but this time use the conservative guess $\hat{p} = 0.5$. By how much do the two sample sizes differ?

45. **Election polling** Gloria Chavez and Ronald Flynn are the candidates for mayor in a large city. We want to estimate the proportion p of all registered voters in the city who plan to vote for Chavez with 95% confidence and a margin of error no greater than 0.03. How large a random sample do we need? Show your work.

46. **Starting a nightclub** A college student organization wants to start a nightclub for students under the age of 21. To assess support for this proposal, they will select an SRS of students and ask each respondent if he or she would patronize this type of establishment. They expect that about 70% of the student body would respond favorably. What sample size is required to obtain a 90% confidence interval with an approximate margin of error of 0.04? Show your work.

47. **Teens and their TV sets** According to a Gallup Poll report, 64% of teens aged 13 to 17 have TVs in their rooms. Here is part of the footnote to this report:

These results are based on telephone interviews with a randomly selected national sample of 1028 teenagers in the Gallup Poll Panel of households, aged 13 to 17. For results based on this sample, one can say . . . that the maximum error attributable to sampling and other random effects is ±3 percentage points. In addition to sampling error, question wording and practical difficulties in conducting surveys can introduce error or bias into the findings of public opinion polls.[16]

(a) We omitted the confidence level from the footnote. Use what you have learned to determine the confidence level, assuming that Gallup took an SRS.

(b) Give an example of a "practical difficulty" that could lead to biased results for this survey.

48. **Gambling and the NCAA** Gambling is an issue of great concern to those involved in college athletics. Because of this concern, the National Collegiate Athletic Association (NCAA) surveyed randomly selected student-athletes concerning their gambling-related behaviors.[17] Of the 5594 Division I male athletes in the survey, 3547 reported participation in some gambling behavior. This includes playing cards, betting on games of skill, buying lottery tickets, betting on sports, and similar activities. A report of this study cited a 1% margin of error.

(a) The confidence level was not stated in the report. Use what you have learned to find the confidence level, assuming that the NCAA took an SRS.

(b) The study was designed to protect the anonymity of the student-athletes who responded. As a result, it was not possible to calculate the number of students who were asked to respond but did not. How does this fact affect the way that you interpret the results?

Multiple choice: Select the best answer for Exercises 49 to 52.

49. A Gallup Poll found that only 28% of American adults expect to inherit money or valuable possessions from a relative. The poll's margin of error was ±3 percentage points at a 95% confidence level. This means that

(a) the poll used a method that gets an answer within 3% of the truth about the population 95% of the time.

(b) the percent of all adults who expect an inheritance is between 25% and 31%.

(c) if Gallup takes another poll on this issue, the results of the second poll will lie between 25% and 31%.

(d) there's a 95% chance that the percent of all adults who expect an inheritance is between 25% and 31%.

(e) Gallup can be 95% confident that between 25% and 31% of the sample expect an inheritance.

50. Most people can roll their tongues, but many can't. The ability to roll the tongue is genetically determined. Suppose we are interested in determining what proportion of students can roll their tongues. We test a simple random sample of 400 students and find that 317 can roll their tongues. The margin of error for a 95% confidence interval for the true proportion of tongue rollers among students is closest to

(a) 0.008. (c) 0.03. (e) 0.208.
(b) 0.02. (d) 0.04.

51. You want to design a study to estimate the proportion of students at your school who agree with the statement, "The student government is an effective organization for expressing the needs of students to the administration." You will use a 95% confidence interval, and you would like the margin of error to be 0.05 or less. The minimum sample size required is

(a) 22. (b) 271. (c) 385. (d) 769. (e) 1795.

52. I collect an SRS of size n from a population and compute a 95% confidence interval for the population proportion. Which of the following would produce a new confidence interval with larger width (larger margin of error) based on these same data?

(a) Use a larger confidence level.

(b) Use a smaller confidence level.

(c) Increase the sample size.

(d) Use the same confidence level, but compute the interval n times. Approximately 5% of these intervals will be larger.

(e) Nothing can guarantee absolutely that you will get a larger interval. One can only say that the chance of obtaining a larger interval is 0.05.

Exercises 53 and 54 refer to the following setting. The table below displays the number of accidents at a factory during each hour of a 24-hour shift (1 = 1 A.M).

Hour	Number of accidents	Hour	Number of accidents
1	5	13	21
2	8	14	12
3	17	15	10
4	31	16	1
5	24	17	0
6	18	18	1
7	12	19	3
8	7	20	21
9	1	21	23
10	0	22	18
11	2	23	11
12	14	24	2

53. **Accidents happen** (1.2, 3.1)

(a) Construct a plot that displays the distribution of the number of accidents effectively.

(b) Construct a plot that shows the relationship between the number of accidents and the time when they occurred.

(c) Describe something that the plot in part (a) tells you about the data that the plot in part (b) does not.

(d) Describe something that the plot in part (b) tells you about the data that the plot in part (a) does not.

54. **Accidents happen** (1.3) Plant managers are concerned that the number of accidents may be significantly higher during the midnight to 8:00 A.M. shift than during the 4:00 P.M. to midnight shift. What would you tell them? Give appropriate statistical evidence to support your conclusion.

8.3

Estimating a Population Mean

In Section 8.3, you'll learn about:

- When σ is known: The one-sample z interval for a population mean
- Choosing the sample size
- When σ is unknown: The t distributions
- Constructing a confidence interval for μ
- Using t procedures wisely

Inference about a population proportion usually arises when we study *categorical* variables. We learned how to construct and interpret confidence intervals for an unknown parameter p in Section 8.2. To estimate a population mean, we have to record values of a *quantitative* variable for a sample of individuals. It makes sense to try to estimate the mean amount of sleep that students at a large high school got last night but not their mean eye color! In this section, we'll examine confidence intervals for a population mean μ.

When σ Is Known: The One-Sample z Interval for a Population Mean

Mr. Schiel's class did the mystery mean Activity (page 468) and got a value of $\bar{x} = 240.79$ from an SRS of size 16, as shown.

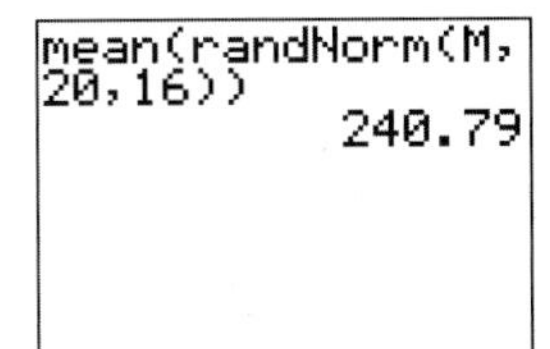

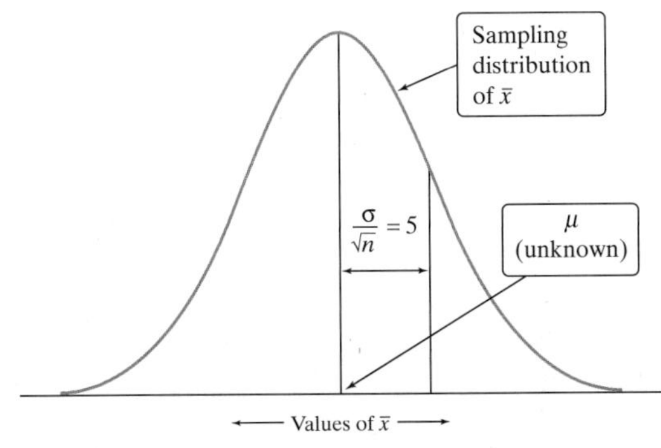

FIGURE 8.10 The Normal sampling distribution of $\bar{x}$ for the mystery mean Activity.

Their task was to estimate the unknown population mean μ. They knew that the population distribution was Normal and that its standard deviation was $\sigma = 20$. Their estimate was based on the sampling distribution of $\bar{x}$. Figure 8.10 shows this Normal sampling distribution once again.

To calculate a 95% confidence interval for μ, we use our familiar formula:

statistic $\pm$ (critical value) $\cdot$ (standard deviation of statistic)

The critical value, $z^* = 1.96$, tells us how many standardized units we need to go out to catch the middle 95% of the sampling distribution. Our interval is

$$\bar{x} \pm z^* \cdot \frac{\sigma}{\sqrt{n}} = 240.79 \pm 1.96 \cdot \frac{20}{\sqrt{16}} = 240.79 \pm 9.8 = (230.99, 250.59)$$

We call such an interval a **one-sample z interval for a population mean.** Whenever the conditions for inference (Random, Normal, Independent) are satisfied and the population standard deviation σ is known, we can use this method to construct a confidence interval for μ.

One-Sample z Interval for a Population Mean

Draw an SRS of size n from a population having unknown mean μ and known standard deviation σ. As long as the Normal and Independent conditions are met, a level C confidence interval for μ is

$$\bar{x} \pm z^* \frac{\sigma}{\sqrt{n}}$$

The critical value z^* is found from the standard Normal distribution.

This method isn't very useful in practice, however. In most real-world settings, if we don't know the population mean μ, then we don't know the population standard deviation σ either. But we can use the one-sample z interval for a population mean to estimate the sample size needed to achieve a specified margin of error. The process mimics what we did for a population proportion in Section 8.2.

Choosing the Sample Size

A wise user of statistics never plans data collection without planning the inference at the same time. You can arrange to have both high confidence and a small margin of error by taking enough observations. The margin of error ME of the confidence interval for the population mean μ is

$$z^* \frac{\sigma}{\sqrt{n}}$$

To determine the sample size for a desired margin of error ME, substitute the value of z^* for your desired confidence level. Use a reasonable estimate for the population standard deviation σ from a similar study that was done in the past or from a small-scale pilot study. Then set the expression for ME less than or equal to the specified margin of error and solve for n. Here is a summary of this strategy.

There are other methods of determining sample size that do not require us to use a known value of the population standard deviation σ. These methods are beyond the scope of this text. Our advice: consult with a statistician when planning your study!

Choosing Sample Size for a Desired Margin of Error When Estimating μ

To determine the sample size n that will yield a level C confidence interval for a population mean with a specified margin of error ME:

- Get a reasonable value for the population standard deviation σ from an earlier or pilot study.
- Find the critical value z^* from a standard Normal curve for confidence level C.
- Set the expression for the margin of error to be less than or equal to ME and solve for n:

$$z^* \frac{\sigma}{\sqrt{n}} \leq ME$$

The procedure is best illustrated with an example.

How Many Monkeys?

Determining sample size from margin of error

Researchers would like to estimate the mean cholesterol level μ of a particular variety of monkey that is often used in laboratory experiments. They would like their estimate to be within 1 milligram per deciliter (mg/dl) of the true value of μ at a 95% confidence level. A previous study involving this variety of monkey suggests that the standard deviation of cholesterol level is about 5 mg/dl.

PROBLEM: **Obtaining monkeys is time-consuming and expensive, so the researchers want to know the minimum number of monkeys they will need to generate a satisfactory estimate.**

SOLUTION: For 95% confidence, $z^* = 1.96$. We will use $\sigma = 5$ as our best guess for the standard deviation of the monkeys' cholesterol level. Set the expression for the margin of error to be at most 1 and solve for n:

$$1.96\frac{5}{\sqrt{n}} \leq 1$$

$$\frac{(1.96)(5)}{1} \leq \sqrt{n}$$

$$96.04 \leq n$$

Remember: always round up to the next whole number when finding n.

Because 96 monkeys would give a slightly larger margin of error than desired, the researchers would need 97 monkeys to estimate the cholesterol levels to their satisfaction. (On learning the cost of getting this many monkeys, the researchers might want to consider studying rats instead!)

For Practice *Try Exercise 55*

CAUTION

Taking observations costs time and money. The required sample size may be impossibly expensive. *Notice that it is the size of the sample that determines the margin of error. The size of the population does not influence the sample size we need.* This is true as long as the population is much larger than the sample.

CHECK YOUR UNDERSTANDING

To assess the accuracy of a laboratory scale, a standard weight known to weigh 10 grams is weighed repeatedly. The scale readings are Normally distributed with unknown mean (this mean is 10 grams if the scale has no bias). In previous studies, the standard deviation of the scale readings has been about 0.0002 gram. How many measurements must be averaged to get a margin of error of 0.0001 with 98% confidence? Show your work.

When σ Is Unknown: The t Distributions

When the sampling distribution of $\bar{x}$ is close to Normal, we can find probabilities involving $\bar{x}$ by standardizing:

$$z = \frac{\bar{x} - \mu}{\sigma/\sqrt{n}}$$

Recall that a statistic is a number computed from sample data. We know that the sample mean $\bar{x}$ is a statistic. So is the standardized value $z = \frac{\bar{x} - \mu}{\sigma/\sqrt{n}}$. The sampling distribution of z shows the values it takes in all possible SRSs of size n from the population.

Recall that the sampling distribution of $\bar{x}$ has mean μ and standard deviation $\sigma/\sqrt{n}$, as shown in Figure 8.11(a) on the next page. What are the shape, center, and spread of the sampling distribution of the new statistic z? From what we learned in Chapter 6, subtracting the constant μ from the values of the random variable $\bar{x}$ shifts the distribution left by μ units, making the mean 0. This transformation doesn't affect the shape or spread of the distribution. Dividing by the constant $\sigma/\sqrt{n}$ keeps the mean at 0, makes the standard deviation 1, and leaves the shape unchanged. As shown in Figure 8.11(b), z has the standard Normal distribution N(0, 1). Therefore, we can use Table A or a calculator to find the related probability involving z. That's how we have gotten the critical values for our confidence intervals so far.

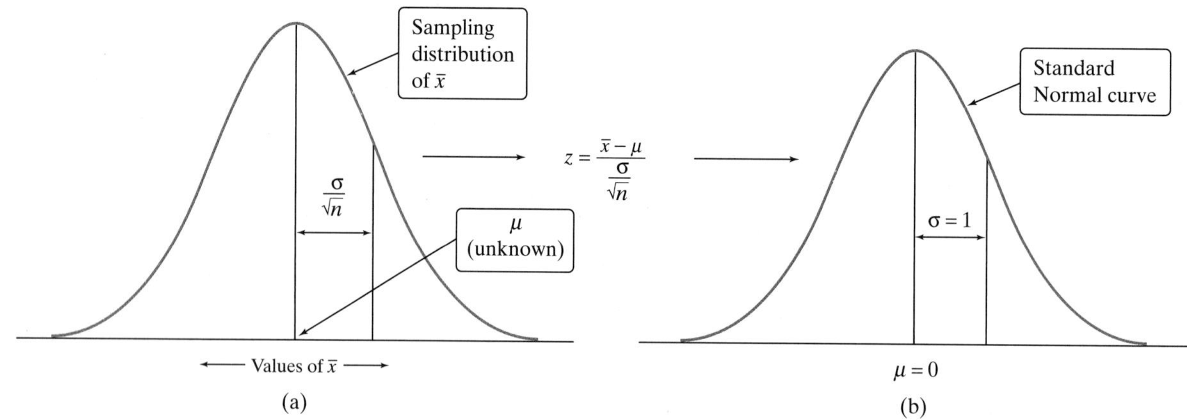

FIGURE 8.11 (a) Sampling distribution of $\bar{x}$ when the Normal condition is met. (b) Standardized values of $\bar{x}$ lead to the statistic z, which follows the standard Normal distribution.

When we don't know σ, we estimate it using the sample standard deviation s_x. What happens now when we standardize?

$$?? = \frac{\bar{x} - \mu}{s_x / \sqrt{n}}$$

As the following Activity shows, this new statistic does *not* have a standard Normal distribution.

ACTIVITY *Calculator bingo*

MATERIALS: TI-83/84 or TI-89 with display capability

When doing inference about a population mean μ, what happens when we use the sample standard deviation s_x to estimate the population standard deviation σ? In this Activity, you'll perform simulations on your calculator to help answer this question.[18] To make things easier, we'll start with a Normal population having mean $\mu = 100$ and standard deviation $\sigma = 5$.

1. Use the calculator to: (1) take an SRS of size 4 from the population; (2) compute the value of the sample mean $\bar{x}$; and (3) standardize the value of $\bar{x}$ using the "known" value $\sigma = 5$. You will use several commands joined together by colons (:).

TI-83/84: `randNorm(100,5,4)→L1:1-Var Stats L1:(x̄-100)/(5/√(4))`

- To get the `randNorm` command, press [MATH], arrow to PRB and choose `6:RandNorm(`.
- For one-variable statistics, press [STAT], arrow to CALC, and choose `1:1-Var Stats`.
- To get x̄, press [VARS], choose `5:Statistics` and `2:x̄`.

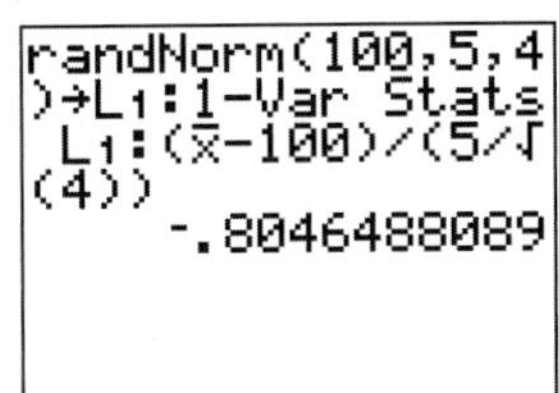

TI-89: `tistat.randnorm(100,5,4)→list1:tistat.onevar(list1):(statvars\x_bar-100)/(5/√(4))`

- To get the `tistat.randnorm` command, press CATALOG F3 (Flash Apps), press alpha 2 (R) to jump to the r's, and choose `randNorm(....`
- For one-variable statistics, press CATALOG F3 (Flash Apps), press alpha − (O) to jump to the o's, and choose `OneVar(....`
- To get $\bar{x}$, press 2nd − (VAR LINK), arrow down to STAT VARS, and choose `x_bar`.

2. Keep pressing ENTER to repeat the process in Step 1 until you have taken 100 SRSs. Say "Bingo!" any time you get a standardized value (z-score) that is less than -3 or greater than 3. Write down the value of z you get each time this happens.
3. According to the 68–95–99.7 rule, about how often should a "Bingo!" occur? How many times did you get a value of z that wasn't between -3 and 3 in your 100 repetitions of the simulation? Compare results with your classmates.
4. Now, let's see what happens when you standardize the value of $\bar{x}$ using the sample standard deviation s_x instead of the "known" σ. You will have to edit the calculator command as shown.

TI-83/84: `randNorm(100,5,4)→L1:1-Var Stats L1:(x̄-100)/(Sx/√(4))`

- Press 2nd ENTER (ENTRY) to recall the previous command.
- Use the arrow keys to position the cursor on the 1 in the last part of the command.

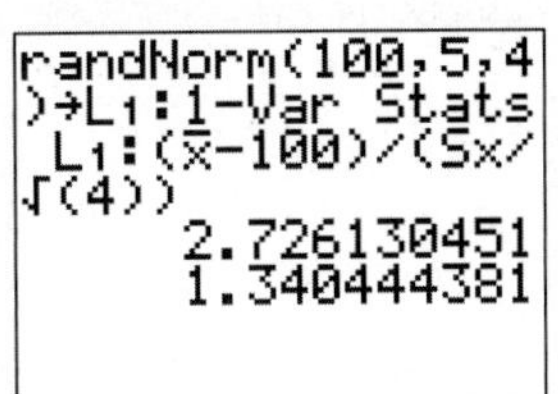

- To replace the population standard deviation (5) with the sample standard deviation s_x, press VARS, choose 5:Statistics and 3:Sx.

TI-89: `tistat.randnorm(100,5,4)→list1:tistat.onevar(list1):(statvars\x_bar-100)/(statvars\sx_/(√(4)))`

- Press 2nd ENTER (ENTRY) to recall the previous command.
- Use the arrow keys to position the cursor on the 1 in the last part of the command.

• To replace the population standard deviation (5) with the sample standard deviation s_x, press 2nd − (VAR LINK), arrow down to STAT VARS and choose sx_.

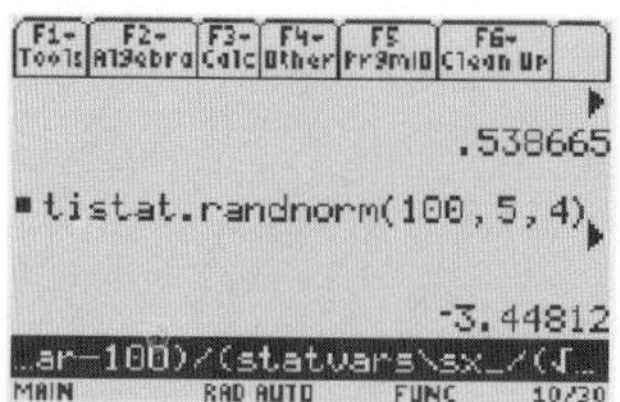

5. Keep pressing ENTER to repeat the process in Step 4 until you have taken 100 SRSs. Say "Bingo!" any time you get a standardized value that is less than −3 or greater than 3. Write down the value you get each time this happens.
6. Compare results with your classmates. How can you tell that the standardized values you are getting in Step 5 are *not* coming from a standard Normal curve?

Figure 8.12 shows the results of taking 500 SRSs of size $n = 4$ and standardizing the value of the sample mean $\bar{x}$ as described in the Activity. The values of z from Steps 1 and 2 follow a standard Normal distribution, as expected. The standardized values from Steps 4 and 5, using the sample standard deviation s_x in place of the population standard deviation σ, show much greater spread. In fact, in a few samples, the statistic

$$t = \frac{\bar{x} - \mu}{s_x/\sqrt{n}}$$

took values below −6 or above 6. This statistic has a distribution that is new to us, called a **t distribution.** It has a *different shape* than the standard Normal curve: still symmetric with a single peak at 0, but with much more area in the tails.

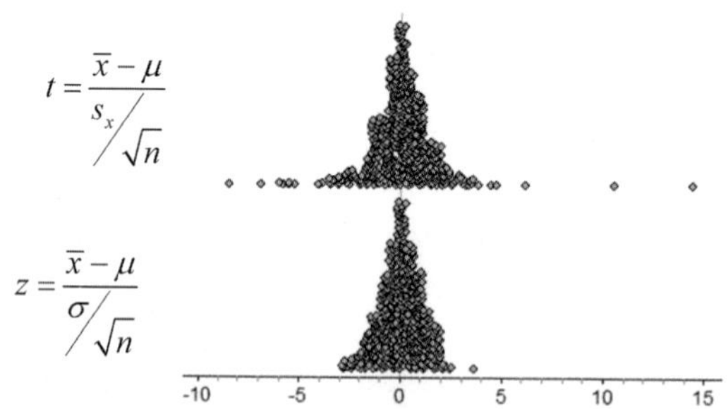

FIGURE 8.12 Fathom simulation showing standardized values of the sample mean $\bar{x}$ in 500 SRSs. The statistic z follows a standard Normal distribution. Replacing σ with s_x yields a statistic with much greater variability that doesn't follow the standard Normal curve.

The statistic t has the same interpretation as any standardized statistic: it says how far $\bar{x}$ is from its mean μ in standard deviation units. There is a different t distribution for each sample size. We specify a particular t distribution by giving its **degrees of freedom (df).** When we perform inference about a population mean μ using a t distribution, the appropriate degrees of freedom are found by subtracting

1 from the sample size n, making df $= n - 1$. We will write the t distribution with $n - 1$ degrees of freedom as t_{n-1} for short.

The *t* distribution and the *t* inference procedures were invented by William S. Gosset (1876–1937). Gosset worked for the Guinness brewery, and his goal in life was to make better beer. He used his new *t* procedures to find the best varieties of barley and hops. Gosset's statistical work helped him become head brewer. Because Gosset published under the pen name "Student," you will often see the *t* distribution called "Student's *t*" in his honor.

The *t* Distributions; Degrees of Freedom

Draw an SRS of size n from a large population that has a Normal distribution with mean μ and standard deviation σ. The statistic

$$t = \frac{\bar{x} - \mu}{s_x/\sqrt{n}}$$

has the ***t* distribution** with **degrees of freedom** df $= n - 1$. This statistic will have approximately a t_{n-1} distribution as long as the sampling distribution of $\bar{x}$ is close to Normal.

Figure 8.13 compares the density curves of the standard Normal distribution and the t distributions with 2 and 9 degrees of freedom. The figure illustrates these facts about the t distributions:

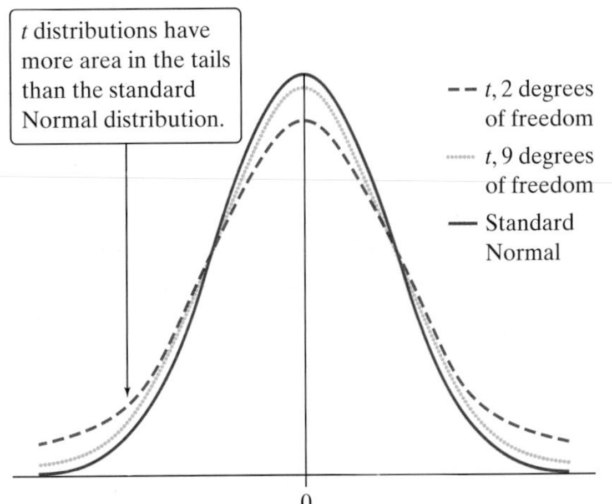

FIGURE 8.13 Density curves for the *t* distributions with 2 and 9 degrees of freedom and the standard Normal distribution. All are symmetric with center 0. The *t* distributions are somewhat more spread out.

- The density curves of the t distributions are similar in shape to the standard Normal curve. They are symmetric about 0, single-peaked, and bell-shaped.
- The spread of the t distributions is a bit greater than that of the standard Normal distribution. The t distributions in Figure 8.13 have more probability in the tails and less in the center than does the standard Normal. This is true because substituting the estimate s_x for the fixed parameter σ introduces more variation into the statistic.
- As the degrees of freedom increase, the t density curve approaches the standard Normal curve ever more closely. This happens because s_x estimates σ more accurately as the sample size increases. So using s_x in place of σ causes little extra variation when the sample is large.

Table B in the back of the book gives critical values t^* for the t distributions. Each row in the table contains critical values for the t distribution whose degrees of freedom appear at the left of the row. For convenience, several of the more common confidence levels C (in percents) are given at the bottom of the table. By looking down any column, you can check that the t critical values approach the Normal critical values z^* as the degrees of freedom increase.

Finding t^*

Using Table B

PROBLEM: Suppose you want to construct a 95% confidence interval for the mean μ of a Normal population based on an SRS of size $n = 12$. What critical value t^* should you use?

SOLUTION: In Table B, we consult the row corresponding to $df = n - 1 = 11$. We move across that row to the entry that is directly above 95% confidence level on the bottom of the chart. The desired critical value is $t^* = 2.201$.

	Upper-tail probability p			
df	.05	.025	.02	.01
10	1.812	2.228	2.359	2.764
11	1.796	2.201	2.328	2.718
12	1.782	2.179	2.303	2.681
z^*	1.645	1.960	2.054	2.326
	90%	95%	96%	98%
	Confidence level C			

For Practice *Try Exercise 57*

In the previous example, notice that the corresponding standard Normal critical value for 95% confidence is $z^* = 1.96$. We have to go out farther than 1.96 standard deviations to capture the central 95% of the t distribution with 11 degrees of freedom.

As with the standard Normal table, technology often makes Table B unnecessary.

TECHNOLOGY CORNER Inverse *t* on the calculator

Most newer TI-84 and TI-89 calculators allow you to find critical values t^* using the inverse t command. As with the calculator's inverse Normal command, you have to enter the *area to the left* of the desired critical value.

TI-84: Press 2nd VARS (DISTR) and choose `4:invT(`. Then complete the command `invT(.975,11)` and press ENTER.

TI-89: In the Statistics/List Editor, press F5, choose `2:Inverse` and `2:Inverse t...`. In the dialog box, enter Area: .975 and Deg of Freedom, df: 11, and then press ENTER.

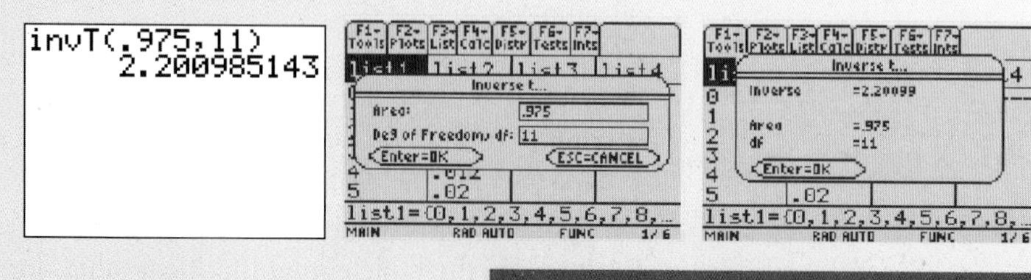

TI-Nspire instructions in Appendix B

CHECK YOUR UNDERSTANDING

Use Table B to find the critical value t^* that you would use for a confidence interval for a population mean μ in each of the following situations. If possible, check your answer with technology.

(a) A 98% confidence interval based on $n = 22$ observations.

(b) A 90% confidence interval from an SRS of 10 observations.

(c) A 95% confidence interval from a sample of size 7.

Constructing a Confidence Interval for μ

When the conditions for inference are satisfied, the sampling distribution of $\bar{x}$ has roughly a Normal distribution with mean μ and standard deviation $\sigma/\sqrt{n}$. Because we don't know σ, we estimate it by the sample standard deviation s_x. We then estimate the standard deviation of the sampling distribution by $\frac{s_x}{\sqrt{n}}$. This value is called the **standard error of the sample mean $\bar{x}$**, or just the standard error of the mean.

As with proportions, some books refer to the standard deviation of the sampling distribution of $\bar{x}$ as the "standard error" and what we call the standard error of the mean as the "estimated standard error." The standard error of the mean is often abbreviated SEM.

DEFINITION: Standard error of the sample mean

The **standard error of the sample mean** $\bar{x}$ is $\frac{s_x}{\sqrt{n}}$, where s_x is the sample standard deviation. It describes how far $\bar{x}$ will be from μ, on average, in repeated SRSs of size n.

To construct a confidence interval for μ, replace the standard deviation $\sigma/\sqrt{n}$ of $\bar{x}$ by its standard error $s_x/\sqrt{n}$ in the formula for the one-sample z interval for a population mean. Use critical values from the t distribution with $n - 1$ degrees of freedom in place of the z critical values. That is,

$$\text{statistic} \pm (\text{critical value}) \cdot (\text{standard deviation of statistic}) = \bar{x} \pm t^* \frac{s_x}{\sqrt{n}}$$

This **one-sample *t* interval for a population mean** is similar in both reasoning and computational detail to the one-sample z interval for a population proportion of Section 8.2. So we will now pay more attention to questions about using these methods in practice.

The One-Sample *t* Interval for a Population Mean

Choose an SRS of size n from a population having unknown mean μ. A level C confidence interval for μ is

$$\bar{x} \pm t^* \frac{s_x}{\sqrt{n}}$$

where t^* is the critical value for the t_{n-1} distribution. Use this interval only when (1) the population distribution is Normal *or* the sample size is large ($n \geq 30$), and (2) the population is at least 10 times as large as the sample.

As before, we have to verify three important conditions before we estimate a population mean. When we do inference in practice, verifying the conditions is often a bit more complicated.

Conditions for Inference about a Population Mean

- **Random:** The data come from a random sample of size n from the population of interest or a randomized experiment. This condition is very important.
- **Normal:** The population has a Normal distribution *or* the sample size is large ($n \geq 30$).
- **Independent:** The method for calculating a confidence interval assumes that individual observations are independent. To keep the calculations reasonably accurate when we sample without replacement from a finite population, we should check the *10% condition:* verify that the sample size is no more than 1/10 of the population size.

The following example shows you how to construct a confidence interval for a population mean when σ is unknown. By now, you should recognize the four-step process. *Since you are expected to include these four steps whenever you perform inference, we will stop saying "follow the four-step process" in examples and exercises. We will also limit our use of the icon to examples from this point forward.*

EXAMPLE

Video Screen Tension

Constructing a confidence interval for μ

A manufacturer of high-resolution video terminals must control the tension on the mesh of fine wires that lies behind the surface of the viewing screen. Too much tension will tear the mesh, and too little will allow wrinkles. The tension is measured by an electrical device with output readings in millivolts (mV). Some variation is inherent in the production process. Here are the tension readings from a random sample of 20 screens from a single day's production:

269.5	297.0	269.6	283.3	304.8	280.4	233.5	257.4	317.5	327.4
264.7	307.7	310.0	343.3	328.1	342.6	338.8	340.1	374.6	336.1

Construct and interpret a 90% confidence interval for the mean tension μ of all the screens produced on this day.

STATE: We want to estimate the true mean tension μ of all the video terminals produced this day at a 90% confidence level.

PLAN: If the *conditions* are met, we should use a one-sample t interval to estimate μ.

- Random: We are told that the data come from a random sample of 20 screens from the population of all screens produced that day.
- Normal: Since the sample size is small ($n = 20$), we must check whether it's reasonable to believe that the population distribution is Normal. So we examine the sample data. Figure 8.14 shows (a) a dotplot, (b) a boxplot, and (c) a Normal probability plot of the tension readings in the sample. Neither the dotplot nor the boxplot shows strong skewness or any outliers. The Normal probability plot looks roughly linear. These graphs give us no reason to doubt the Normality of the population.

When the sample size is small ($n < 30$), as in this example, the Normal condition is about the shape of the *population* distribution. We inspect the distribution to see if it's believable that these data came from a Normal population.

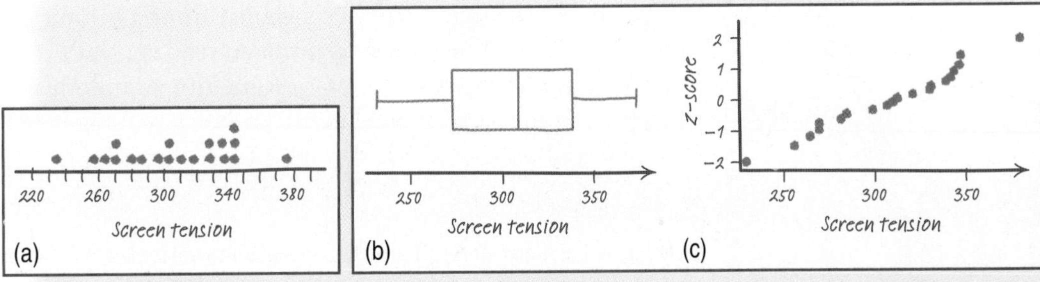

FIGURE 8.14 (a) A dotplot, (b) boxplot, and (c) Normal probability plot of the video screen tension readings.

In Chapter 2, we noted that a data set with an approximately Normal shape will have a Normal probability plot that's roughly linear. Now we're trying to get information about the shape of the population distribution from a Normal probability plot of the sample data. This is much harder, because even samples drawn from perfectly Normal populations don't always look Normal.

- **Independent:** Because we are sampling without replacement, we must check the 10% condition: we must assume that at least 10(20) = 200 video terminals were produced this day.

Upper-tail probability p			
df	.10	.05	.025
18	1.330	1.734	2.101
19	1.328	1.729	2.093
20	1.325	1.725	2.086
	80%	90%	95%
Confidence level C			

DO: We used our calculator to find the mean and standard deviation of the tension readings for the 20 screens in the sample: $\bar{x} = 306.32$ mV and $s_x = 36.21$ mV. We use the t distribution with df = 19 to find the critical value. For a 90% confidence level, the critical value is $t^* = 1.729$. So the 90% confidence interval for μ is

$$\bar{x} \pm t^* \frac{s_x}{\sqrt{n}} = 306.32 \pm 1.729 \frac{36.21}{\sqrt{20}} = 306.32 \pm 14.00 = (292.32, 320.32)$$

The calculator's `invT(.05,19)` gives $t = -1.729$ (to three decimal places).

CONCLUDE: We are 90% confident that the interval from 292.32 to 320.32 mV captures the true mean tension in the entire batch of video terminals produced that day.

For Practice *Try Exercise 63*

Now that we've calculated our first confidence interval for a population mean μ, it's time to make a simple observation. Inference for *proportions* uses z; inference for *means* uses t. That's one reason why distinguishing categorical from quantitative variables is so important.

When we use Table B to determine the correct value of t^* for a given confidence interval, all we need to know are the confidence level C and the degrees of freedom (df). Unfortunately, Table B does not include every possible sample size. *When the actual df does not appear in the table, use the greatest df available that is less than your desired df.* This guarantees a wider confidence interval than we need to justify a given confidence level. Better yet, use technology to find an accurate value of t^* for any df.

CAUTION

EXAMPLE

Auto Pollution

STEP 4

A one-sample t interval for μ

Environmentalists, government officials, and vehicle manufacturers are all interested in studying the auto exhaust emissions produced by motor vehicles.

The major pollutants in auto exhaust from gasoline engines are hydrocarbons, carbon monoxide, and nitrogen oxides (NOX). Researchers collected data on the NOX levels (in grams/mile) for a random sample of 40 light-duty engines of the same type. The mean NOX reading was 1.2675 and the standard deviation was 0.3332.[19]

PROBLEM:

(a) Construct and interpret a 95% confidence interval for the mean amount of NOX emitted by light-duty engines of this type.

(b) The Environmental Protection Agency (EPA) sets a limit of 1.0 gram/mile for NOX emissions. Are you convinced that this type of engine has a mean NOX level of 1.0 or less? Use your interval from (a) to support your answer.

SOLUTION:

(a) STATE: We want to estimate the true mean amount μ of NOX emitted by all light-duty engines of this type at a 95% confidence level.

PLAN: We should construct a one-sample t interval for μ if the conditions are met.

- Random: The data come from a "random sample" of 40 engines from the population of all light-duty engines of this type.
- Normal: We don't know whether the population distribution of NOX emissions is Normal. Because the sample size, $n = 40$, is large (at least 30), we should be safe using t procedures.
- Independent: We are sampling without replacement, so we need to check the 10% condition: we must assume that there are at least $10(40) = 400$ light-duty engines of this type.

DO: The formula for the one-sample t interval is

$$\bar{x} \pm t^* \frac{s_x}{\sqrt{n}}$$

From the information given, $\bar{x} = 1.2675$ g/mi and $s_x = 0.3332$ g/mi. To find the critical value t^*, we use the t distribution with df $= 40 - 1 = 39$. Unfortunately, there is no row corresponding to 39 degrees of freedom in Table B. We can't pretend we have a larger sample size than we actually do, so we use the more conservative df $= 30$.

Upper-tail probability *p*			
df	.05	.025	.02
29	1.699	2.045	2.150
30	1.697	2.042	2.147
40	1.684	2.021	2.123
	90%	95%	96%
	Confidence level *C*		

This is a good example of a situation in which using technology to find t^* will yield a more accurate result than using a printed table. At a 95% confidence level, the critical value is $t^* = 2.042$. So the 95% confidence interval for μ is

$$\bar{x} \pm t^* \frac{s_x}{\sqrt{n}} = 1.2675 \pm 2.042 \frac{0.3332}{\sqrt{40}} = 1.2675 \pm 0.1076 = (1.1599, 1.3751)$$

The command `invT(.025,39)` gives $t = -2.023$. Using the critical value $t^* = \pm 2.023$ for the 95% confidence interval gives

$$\bar{x} \pm t^* \frac{s_x}{\sqrt{n}}$$
$$= 1.2675 \pm 2.023 \frac{0.3332}{\sqrt{40}}$$
$$= 1.2675 \pm 0.1066$$
$$= (1.1609, 1{,}3741)$$

This interval is slightly narrower than the one found using Table B.

CONCLUDE: We are 95% confident that the interval from 1.1599 to 1.3751 grams/mile contains the true mean level of nitrogen oxides emitted by this type of light-duty engine.

(b) The confidence interval from (a) tells us that any value from 1.1599 to 1.3751 g/mi is a plausible value of the mean NOX level μ for this type of engine. Since the entire interval exceeds 1.0, it appears that this type of engine violates EPA limits.

For Practice *Try Exercise 67*

As we noted earlier, a confidence interval gives a range of plausible values for an unknown parameter. We can therefore use a confidence interval to determine whether a specified value of the parameter is reasonable, like the EPA's 1.0 limit in the example.

CHECK YOUR UNDERSTANDING

Biologists studying the healing of skin wounds measured the rate at which new cells closed a cut made in the skin of an anesthetized newt. Here are data from a random sample of 18 newts, measured in micrometers (millionths of a meter) per hour:[20]

29 27 34 40 22 28 14 35 26 35 12 30 23 18 11 22 23 33

We want to estimate the mean healing rate μ with a 95% confidence interval.

1. Define the parameter of interest.
2. What inference method will you use? Check that the conditions for using this procedure are met.
3. Construct a 95% confidence interval for μ. Show your method.
4. Interpret your interval in context.

Using *t* Procedures Wisely

The stated confidence level of a one-sample *t* interval for μ is exactly correct when the population distribution is exactly Normal. No population of real data is exactly Normal. The usefulness of the *t* procedures in practice therefore depends on how strongly they are affected by lack of Normality. Procedures that are not strongly affected when a condition for using them is violated are called **robust.**

DEFINITION: Robust procedures

An inference procedure is called **robust** if the probability calculations involved in that procedure remain fairly accurate when a condition for using the procedure is violated.

For confidence intervals, "robust" means that the stated confidence level is still pretty accurate. That is, if we use the procedure to calculate many 95% confidence intervals, about 95% of those intervals would capture the population mean μ. If the procedure isn't robust, then the actual capture rate might be very different from 95%.

If outliers are present in the sample, then the population may not be Normal. The *t* procedures are *not* robust against outliers, because $\bar{x}$ and s_x are not resistant to outliers.

More Auto Pollution

t procedures *not* robust against outliers

In an earlier example (page 509), we constructed a confidence interval for the mean level of NOX emitted by a specific type of light-duty car engine. The original random sample actually included 41 engines, but one of them recorded an unusually high amount (2.94 grams/mile) of NOX. Upon further inspection, this engine had a mechanical defect. So the researchers decided to remove this value from the data set. The Minitab computer output below gives some numerical summaries for NOX emissions in the original sample.

Descriptive Statistics: NOX

Variable	N	Mean	SE Mean	StDev	Minimum	Q1	Median	Q3	Maximum
NOX	41	1.3083	0.0656	0.4201	0.5700	1.1000	1.3100	1.4700	2.9400

Did you notice the SE Mean, 0.0656, in the computer output? As you probably guessed, this is the standard error of the mean, $\frac{s_x}{\sqrt{n}}$. You can check that $\frac{0.4201}{\sqrt{41}} = 0.0656$.

The confidence interval based on this sample of 41 engines would be (using df = 40 from Table B)

$$\bar{x} \pm t^* \frac{s_x}{\sqrt{n}} = 1.3083 \pm 2.021 \frac{0.4201}{\sqrt{41}} = 1.3083 \pm 0.1326 = (1.1757, 1.4409)$$

Our new confidence interval is wider and is centered at a higher value than our original interval of 1.1599 to 1.3751.

Fortunately, the t procedures are quite robust against non-Normality of the population except when outliers or strong skewness are present. Larger samples improve the accuracy of critical values from the t distributions when the population is not Normal. This is true for two reasons:

1. The sampling distribution of the sample mean $\bar{x}$ from a large sample is close to Normal (that's the central limit theorem). Normality of the individual observations is of little concern when the sample size is large.
2. As the sample size n grows, the sample standard deviation s_x will be an accurate estimate of σ whether or not the population has a Normal distribution.

Always make a plot to check for skewness and outliers before you use the t procedures for small samples. For most purposes, you can safely use the one-sample t procedures when $n \geq 15$ unless an outlier or strong skewness is present. Except in the case of small samples, *the condition that the data come from a random sample or randomized experiment is more important* than the condition that the population distribution is Normal. Here are practical guidelines for the Normal condition when performing inference about a population mean.[21]

CAUTION

Using One-Sample t Procedures: The Normal Condition

- *Sample size less than 15:* Use t procedures if the data appear close to Normal (roughly symmetric, single peak, no outliers). If the data are clearly skewed or if outliers are present, do not use t.
- *Sample size at least 15:* The t procedures can be used except in the presence of outliers or strong skewness.
- *Large samples:* The t procedures can be used even for clearly skewed distributions when the sample is large, roughly $n \geq 30$.

If your sample data would give a biased estimate for some reason, then you shouldn't bother computing a t interval. Or if the data you have *are* the entire population of interest, then there's no need to perform inference (because you would know the true parameter value).

People, Trees, and Flowers

Can we use t?

PROBLEM: **Determine whether we can safely use a one-sample t interval to estimate the population mean in each of the following settings.**

(a) Figure 8.15(a) is a histogram of the percent of each state's residents who are at least 65 years of age.

(b) Figure 8.15(b) is a stemplot of the force required to pull apart 20 pieces of Douglas fir.

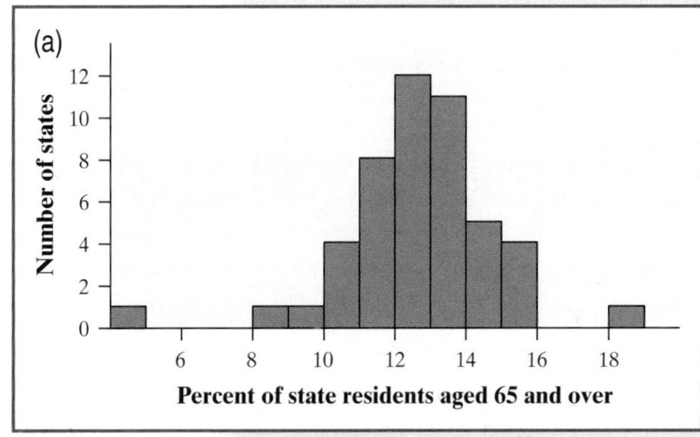

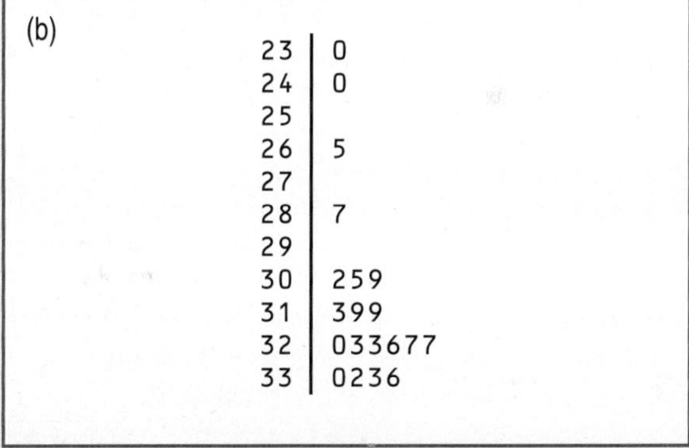

FIGURE 8.15 Can we use t procedures for these data? (a) Percent of residents aged 65 and over in the 50 states. (b) Force required to pull apart 20 pieces of Douglas fir.

(c) Figure 8.15(c) is a stemplot of the lengths of 23 specimens of the red variety of the tropical flower *Heliconia*.

(c)

```
37 | 489
38 | 00112289
39 | 268
40 | 67
41 | 5799
42 | 02
43 | 1
```

FIGURE 8.15 (c) Lengths of 23 tropical flowers of the same variety.

SOLUTION:

(a) No. We have data on the entire population of the 50 states, so formal inference makes no sense. We can calculate the exact mean for the population. There is no uncertainty due to having only a sample from the population, and no need for a confidence interval.

(b) No. The data are strongly skewed to the left with possible low outliers, so we cannot trust the t procedures for $n = 20$.

(c) Yes. The data are mildly skewed to the right and there are no outliers. We can use the t distributions for such data.

As you probably guessed, your calculator will compute a one-sample t interval for a population mean from sample data or summary statistics.

TECHNOLOGY CORNER One-sample *t* intervals for μ on the calculator

Confidence intervals for a population mean using *t* procedures can be constructed on the TI-83/84 and TI-89, thus avoiding the use of Table B. Here is a brief summary of the techniques when you have the actual data values and when you have only numerical summaries.

1. Using raw data Video screen tension example, page 508

Enter the 20 video screen tension readings data in L1(list1).

TI-83/84: From the home screen,

- Press STAT, arrow over to TESTS, and choose `8:TInterval....`
- On the TInterval screen, adjust your settings as shown and choose `Calculate.`

TI-89: From inside the Statistics/List Editor,

- Press 2nd F2 (`[F7]`) to go into the intervals (Ints) menu, then choose `2:TInterval....`
- Choose "Data" as the Data Input Method.
- On the T Interval screen, adjust your settings as shown.

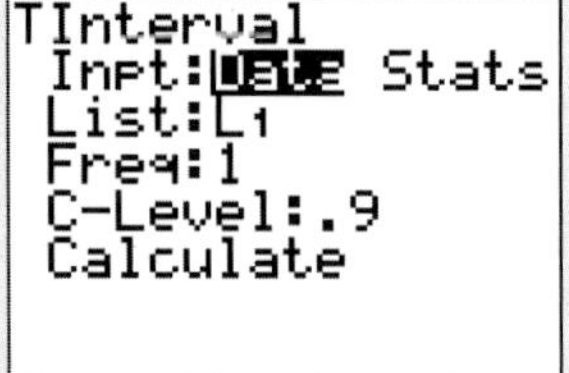

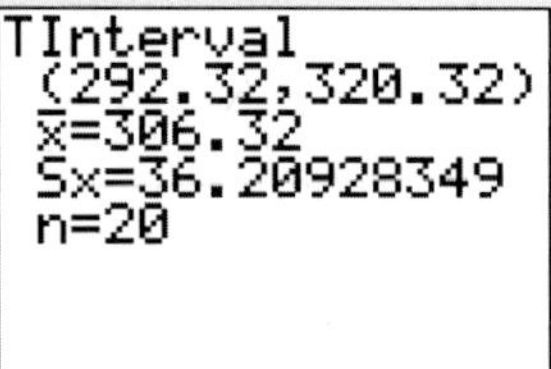

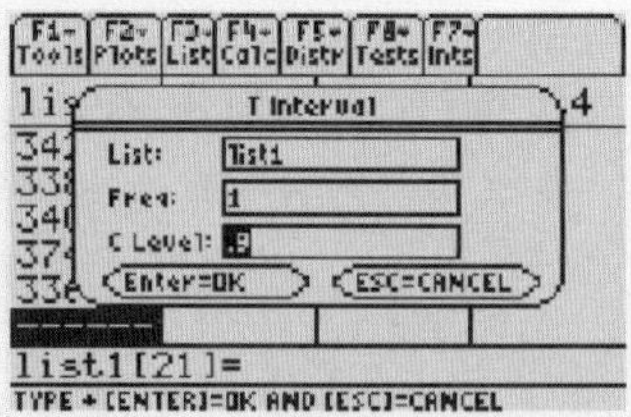

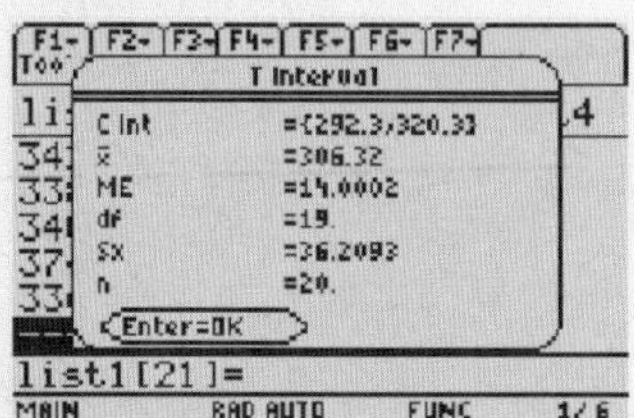

2. Using summary statistics Auto pollution example, page 509

This time, we have no data to enter into a list. Proceed to the TInterval screen as in Step 1, but choose Stats as the data input method. When you get to the TInterval screen, enter the inputs shown and calculate the interval.

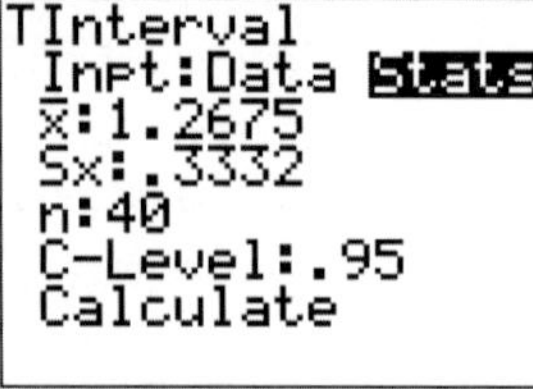

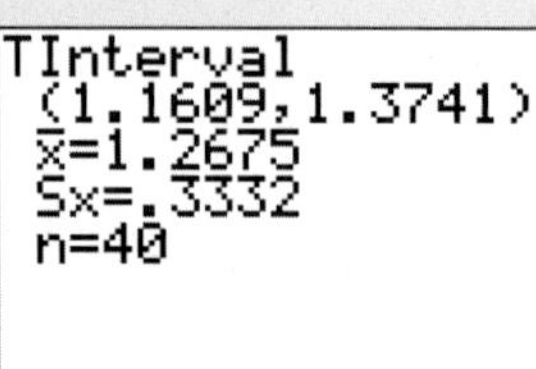

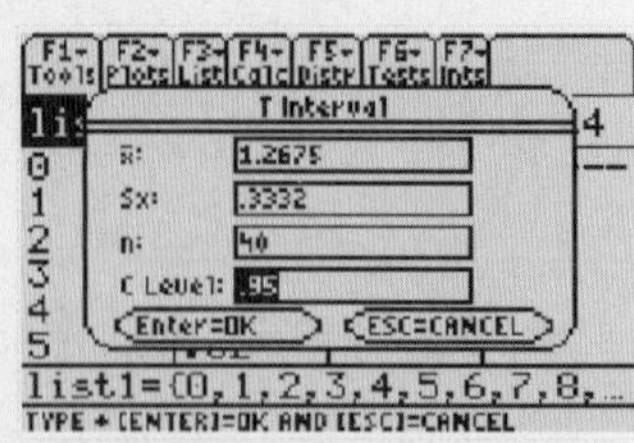

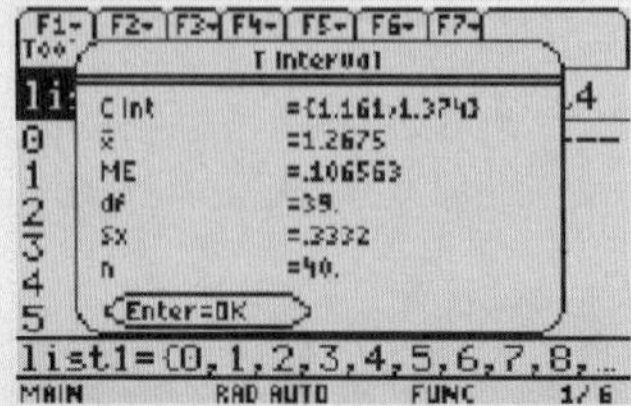

TI-Nspire instructions in Appendix B

The following Data Exploration asks you to use what you have learned about confidence intervals to analyze the safety of a company's product.

DATA EXPLORATION *I'm getting a headache!*

The makers of Aspro brand aspirin want to be sure that their tablets contain the right amount of active ingredient (acetylsalicylic acid). So they inspect a random sample of 36 tablets from a batch of production. When the production process is working properly, Aspro tablets have an average of $\mu = 320$ milligrams (mg) of active ingredient. Here are the amounts (in mg) of active ingredient in the 36 selected tablets:

319	328	321	324	322	320	324	321	320	324	322	317
321	320	322	318	326	316	316	326	325	320	316	319
319	321	322	319	326	320	324	320	318	321	322	318

What do these data tell us about the mean acetylsalicylic acid content of the tablets in this batch? Use what you have learned in this chapter to prepare a one-page response to this question. Be sure to include appropriate graphical and numerical evidence to support your answer.

case closed

Need Help? Give Us a Call!

Refer to the chapter-opening Case Study on page 467. The bank manager wants to know whether or not the bank's customer service agents generally met the goal of answering incoming calls within 30 seconds. We can approach this question in two ways: by estimating the proportion p of all calls that were answered within 30 seconds or by estimating the mean response time μ. An analysis of the data reveals that $\bar{x} = 18.353$ seconds, $s_x = 11.761$ seconds, and that 41 of the 241 call response times were 30 seconds or more.

Estimating p:

STATE: We want to estimate the true proportion p of all calls to the customer service center that were answered within 30 seconds at the 95% confidence level.

PLAN: If conditions are met, we should construct a one-sample z interval for p.

- **Random:** The data came from a random sample of 241 calls to the bank's customer service center.
- **Normal:** Both $n\hat{p} = 200$ and $n(1-\hat{p}) = 41$ are at least 10.
- **Independent:** Since we are sampling without replacement, there must be at least $10(241) = 2410$ calls to the customer service center in a given month.

DO: Our point estimate is $\hat{p} = \dfrac{200}{241} = 0.830$. The resulting 95% confidence interval for p is

$$0.830 \pm 1.96\sqrt{\frac{0.830(0.170)}{241}} = (0.783, 0.877)$$

CONCLUDE: We are 95% confident that the interval from 0.783 to 0.877 captures the actual proportion of calls to the bank's customer service center that were answered in less than 30 seconds.

Estimating μ:

STATE: This time, we want to estimate the actual mean call response time μ at the 95% confidence level.

PLAN: If conditions are met, we should construct a one-sample t interval for μ. We already checked the Random and Independent conditions above, so we need only to check the Normal condition.

Normal: Is the population distribution Normal? Graphical summaries of the call response times are shown below. The histogram (left) and boxplot (right) clearly show that the distribution of the call response times is skewed to the right. However, there are no outliers. We can rely on the robustness of the t procedures for the inference regarding the mean call response time because the sample size ($n = 241$) is large.

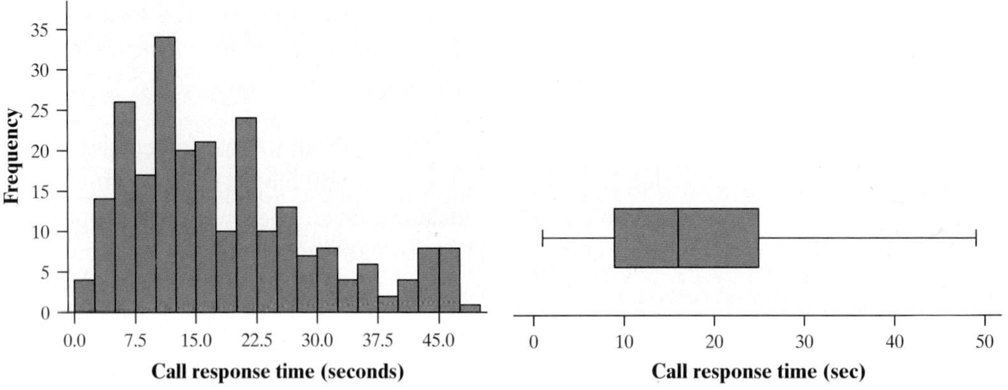

DO: Our point estimate is $\bar{x} = 18.353$ seconds. Using df $= 100$ and $t^* = 1.984$, our 95% confidence interval is

$$18.353 \pm 1.984 \frac{11.761}{\sqrt{241}} = (16.850, 19.856)$$

Software and the TI calculators give the interval from 16.861 to 19.845 seconds using df $= 240$.

CONCLUDE: We are 95% confident that the interval from 16.861 to 19.845 seconds contains the actual mean call response time.

Summary: It seems clear that most customers wait less than 30 seconds since the confidence interval for p is entirely above 0.5. The confidence interval for μ further suggests that the mean wait time for customers who call in is substantially less than 30 seconds.

SECTION 8.3 Summary

- **Confidence intervals for the mean μ of a Normal population** are based on the sample mean $\bar{x}$ of an SRS. Because of the central limit theorem, the resulting procedures are approximately correct for other population distributions when the sample is large.

- If we somehow know σ, we use the z critical value and the standard Normal distribution to help calculate confidence intervals. The sample size needed to obtain a confidence interval with approximate margin of error ME for a population mean involves solving

$$z^* \frac{\sigma}{\sqrt{n}} \leq ME$$

for n, where the standard deviation σ is a reasonable value from a previous or pilot study, and z^* is the critical value for the level of confidence we want.

- In practice, we usually don't know σ. Replace the standard deviation $\sigma/\sqrt{n}$ of the sampling distribution of $\bar{x}$ by the **standard error** $s_x/\sqrt{n}$ and use the ***t*** distribution with $n - 1$ **degrees of freedom (df).**
- There is a t distribution for every positive degrees of freedom. All are symmetric distributions similar in shape to the standard Normal distribution. The t distribution approaches the standard Normal distribution as the number of degrees of freedom increases.
- A level C confidence interval for the mean μ is given by the **one-sample *t* interval**

$$\bar{x} \pm t^* \frac{s_x}{\sqrt{n}}$$

The critical value t^* is chosen so that the t curve with $n - 1$ degrees of freedom has area C between $-t^*$ and t^*.

- This inference procedure is approximately correct when these conditions are met:
  - **Random:** The data were produced using random sampling or random assignment.
  - **Normal:** The population distribution is Normal or the sample size is large ($n \geq 30$).
  - **Independent:** Individual observations are independent. When sampling without replacement, we check the *10% condition:* the population is at least 10 times as large as the sample.

- Follow the four-step process—State, Plan, Do, Conclude—whenever you are asked to construct and interpret a confidence interval for a population mean.
- Remember: inference for proportions uses z; inference for means uses t.
- The t procedures are relatively **robust** when the population is non-Normal, especially for larger sample sizes. The t procedures are not robust against outliers, however.

8.3 TECHNOLOGY CORNERS

TI-Nspire instructions in Appendix B

SECTION 8.3 Exercises

Remember: We are no longer reminding you to use the four-step process in exercises that require you to perform inference.

55. **Estimating BMI** The body mass index (BMI) of all American young women is believed to follow a Normal distribution with a standard deviation of about 7.5. How large a sample would be needed to estimate the mean BMI μ in this population to within ± 1 with 99% confidence? Show your work.

pg 500

56. **The SAT again** High school students who take the SAT Math exam a second time generally score higher than on their first try. Past data suggest that the score increase has a standard deviation of about 50 points. How large a sample of high school students would be needed to estimate the mean change in SAT score to within 2 points with 95% confidence? Show your work.

57. **Critical values** What critical value t^* from Table B would you use for a confidence interval for the population mean in each of the following situations?

pg 506

(a) A 95% confidence interval based on $n = 10$ observations.

(b) A 99% confidence interval from an SRS of 20 observations.

58. **Critical values** What critical value t^* from Table B should be used for a confidence interval for the population mean in each of the following situations?

(a) A 90% confidence interval based on $n = 12$ observations.

(b) A 95% confidence interval from an SRS of 30 observations.

59. **Blood pressure** A medical study finds that $\bar{x} = 114.9$ and $s_x = 9.3$ for the seated systolic blood pressure of the 27 members of one treatment group. What is the standard error of the mean? Interpret this value in context.

60. **Travel time to work** A study of commuting times reports the travel times to work of a random sample of 20 employed adults in New York State. The mean is $\bar{x} = 31.25$ minutes, and the standard deviation is $s_x = 21.88$ minutes. What is the standard error of the mean? Interpret this value in context.

61. **Willows in Yellowstone** Writers in some fields summarize data by giving $\bar{x}$ and its standard error rather than $\bar{x}$ and s_x. Biologists studying willow trees in Yellowstone National Park reported their results in a table with columns labeled $\bar{x} \pm \text{SE}$. The table entry for the heights of willow trees (in centimeters) in one region of the park was 61.55 ± 19.03.[22] The researchers measured a total of 23 trees.

(a) Find the sample standard deviation s_x for these measurements. Show your work.

(b) Explain why the given interval is *not* a confidence interval for the mean height of willow trees in this region of the park.

62. **Blink** When two lights close together blink alternately, we "see" one light moving back and forth if the time between blinks is short. What is the longest interval of time between blinks that preserves the illusion of motion? Ask subjects to turn a knob that slows the blinking until they "see" two lights rather than one light moving. A report gives the results in the form "mean plus or minus the standard error of the mean."[23] Data for 12 subjects are summarized as 251 ± 45 (in milliseconds).

(a) Find the sample standard deviation s_x for these measurements. Show your work.

(b) Explain why the interval 251 ± 45 is *not* a confidence interval.

63. **Give it some gas!** Computers in some vehicles calculate various quantities related to performance. One of these is fuel efficiency, or gas mileage, usually expressed as miles per gallon (mpg). For one vehicle equipped in this way, the miles per gallon were recorded each time the gas tank was filled and the computer was then reset.[24] Here are the mpg values for a random sample of 20 of these records:

pg 508

15.8	13.6	15.6	19.1	22.4	15.6	22.5	17.2	19.4	22.6
19.4	18.0	14.6	18.7	21.0	14.8	22.6	21.5	14.3	20.9

Construct and interpret a 95% confidence interval for the mean fuel efficiency μ for this vehicle.

64. **Vitamin C content** Several years ago, the U.S. Agency for International Development provided 238,300 metric tons of corn-soy blend (CSB) for emergency relief in countries throughout the world. CSB is a highly nutritious, low-cost fortified food. As part of a study to evaluate appropriate vitamin C levels in this food, measurements were taken on samples of CSB produced in a factory.[25] The following data are the amounts of vitamin C,

measured in milligrams per 100 grams (mg/100 g) of blend, for a random sample of size 8 from one production run:

26 31 23 22 11 22 14 31

Construct and interpret a 95% confidence interval for the mean amount of vitamin C μ in the CSB from this production run.

65. **Critical value** What critical value t^* from Table B would you use for a 99% confidence interval for the population mean based on an SRS of size 58? If possible, use technology to find a more accurate value of t^*. What advantage does the more accurate df provide?

66. **Critical value** What critical value t^* from Table B would you use for a 90% confidence interval for the population mean based on an SRS of size 77? If possible, use technology to find a more accurate value of t^*. What advantage does the more accurate df provide?

67. **Bone loss by nursing mothers** Breast-feeding mothers secrete calcium into their milk. Some of the calcium may come from their bones, so mothers may lose bone mineral. Researchers measured the percent change in bone mineral content (BMC) of the spines of 47 randomly selected mothers during three months of breast-feeding.[26] The mean change in BMC was −3.587% and the standard deviation was 2.506%.

pg 509

(a) Construct and interpret a 99% confidence interval to estimate the mean percent change in BMC in the population.

(b) Based on your interval from (a), do these data give good evidence that on the average nursing mothers lose bone mineral? Explain.

68. **A big-toe problem** Hallux abducto valgus (call it HAV) is a deformation of the big toe that is fairly uncommon in youth and often requires surgery. Doctors used X-rays to measure the angle (in degrees) of deformity in a random sample of patients under the age of 21 who came to a medical center for surgery to correct HAV. The angle is a measure of the seriousness of the deformity. For these 21 patients, the mean HAV angle was 24.76 degrees and the standard deviation was 6.34 degrees. A dotplot of the data revealed no outliers or strong skewness.[27]

(a) Construct and interpret a 90% confidence interval for the mean HAV angle in the population of all such patients.

(b) Researchers omitted one patient with an HAV angle of 50 degrees from the analysis due to a measurement issue. What effect would including this outlier have on the confidence interval in (a)? Justify your answer.

69. **Paired tires** Researchers were interested in comparing two methods for estimating tire wear. The first method used the amount of weight lost by a tire. The second method used the amount of wear in the grooves of the tire. A random sample of 16 tires was obtained. Both methods were used to estimate the total distance traveled by each tire. The table below provides the two estimates (in thousands of miles) for each tire.[28]

(a) Construct and interpret a 95% confidence interval for the mean difference μ in the estimates from these two methods in the population of tires.

(b) Does your interval in part (a) give convincing evidence of a difference in the two methods of estimating tire wear? Justify your answer.

Tire	Weight	Groove	Tire	Weight	Groove
1	45.9	35.7	9	30.4	23.1
2	41.9	39.2	10	27.3	23.7
3	37.5	31.1	11	20.4	20.9
4	33.4	28.1	12	24.5	16.1
5	31.0	24.0	13	20.9	19.9
6	30.5	28.7	14	18.9	15.2
7	30.9	25.9	15	13.7	11.5
8	31.9	23.3	16	11.4	11.2

70. **Water** Trace metals found in wells affect the taste of drinking water, and high concentrations can pose a health risk. Researchers measured the concentration of zinc (in milligrams/liter) near the top and the bottom of 10 randomly selected wells in a large region. The data are provided in the table below.[29]

(a) Construct and interpret a 95% confidence interval for the mean difference μ in the zinc concentrations from these two locations in the wells.

(b) Does your interval in part (a) give convincing evidence of a difference in zinc concentrations at the top and bottom of wells in the region? Justify your answer.

Well:	1	2	3	4	5	6	7	8	9	10
Bottom:	0.430	0.266	0.567	0.531	0.707	0.716	0.651	0.589	0.469	0.723
Top:	0.415	0.238	0.390	0.410	0.605	0.609	0.632	0.523	0.411	0.612

71. **Pulling wood apart** How heavy a load (pounds) is needed to pull apart pieces of Douglas fir 4 inches long and 1.5 inches square? A random sample of 20 similar pieces of Douglas fir from a large batch was selected for a science class. The Fathom boxplot below shows the class's data. Explain why it would

not be wise to use a one-sample t interval to estimate the population mean μ.

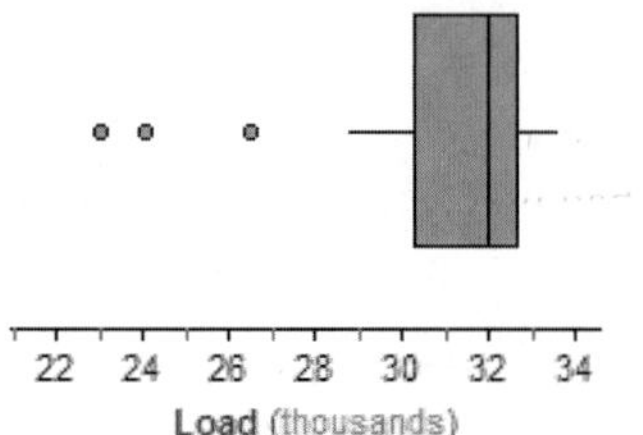

72. **Weeds among the corn** Velvetleaf is a particularly annoying weed in cornfields. It produces lots of seeds, and the seeds wait in the soil for years until conditions are right for sprouting. How many seeds do velvetleaf plants produce? The Fathom histogram below shows the counts from 28 plants that came up in a cornfield when no herbicide was used.[30] Explain why it would not be wise to use a one-sample t interval to estimate the mean number of seeds μ produced by velvetleaf plants.

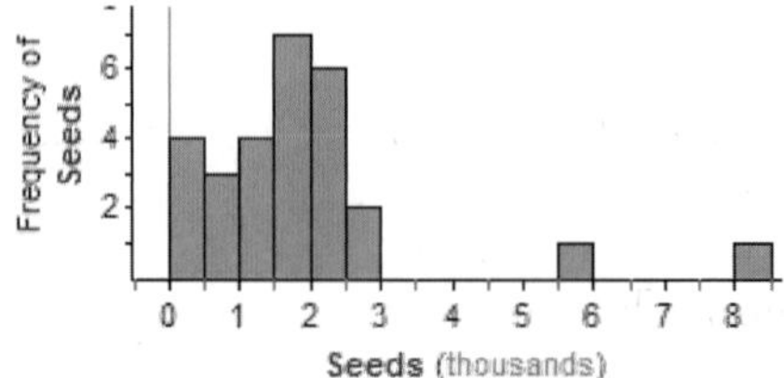

73. **Should we use t?** In each of the following situations, discuss whether it would be appropriate to construct a one-sample t interval to estimate the population mean.

pg 513

(a) We collect data from a random sample of adult residents in a state. Our goal is to estimate the overall percent of adults in the state who are college graduates.

(b) The coach of a college men's basketball team records the resting heart rates of the 15 team members. We use these data to construct a confidence interval for the mean resting heart rate of all male students at this college.

(c) Do teens text more than they call? To find out, an AP Statistics class at a large high school collected data on the number of text messages and calls sent or received by each of 25 randomly selected students. The Fathom boxplot below displays the difference (texts − calls) for each student.

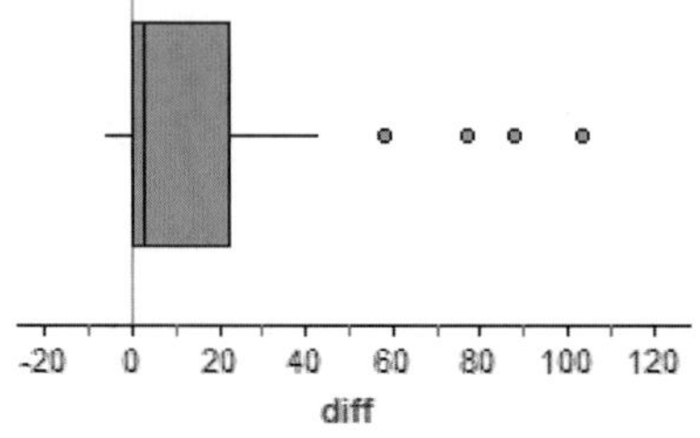

74. **Should we use t?** In each of the following situations, discuss whether it would be appropriate to construct a one-sample t interval to estimate the population mean.

(a) We want to estimate the average age at which U.S. presidents have died. So we obtain a list of all U.S. presidents who have died and their ages at death.

(b) How much time do students spend on the Internet? We collect data from the 32 members of our AP Statistics class and calculate the mean amount of time that each student spent on the Internet yesterday.

(c) Judy is interested in the reading level of a medical journal. She records the length of a random sample of 100 words from a multipage article. The Minitab histogram below displays the data.

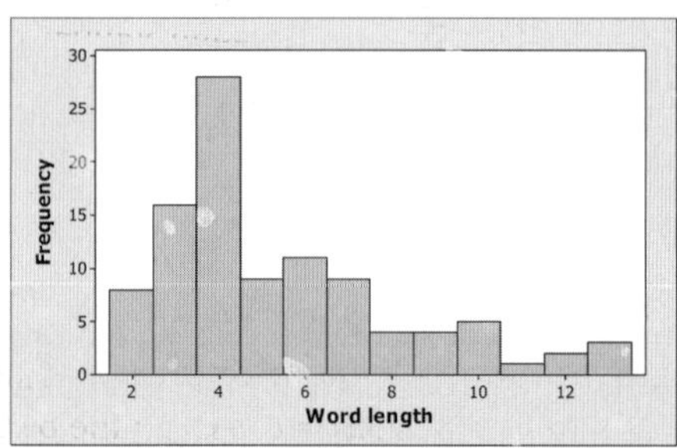

Multiple choice: Select the best answer for Exercises 75 to 78.

75. One reason for using a t distribution instead of the standard Normal curve to find critical values when calculating a level C confidence interval for a population mean is that

(a) z can be used only for large samples.

(b) z requires that you know the population standard deviation σ.

(c) z requires that you can regard your data as an SRS from the population.

(d) the standard Normal table doesn't include confidence levels at the bottom.

(e) a z critical value will lead to a wider interval than a t critical value.

76. You have an SRS of 23 observations from a Normally distributed population. What critical value would you use to obtain a 98% confidence interval for the mean μ of the population if σ is unknown?

(a) 2.508 (c) 2.326 (e) 2.177
(b) 2.500 (d) 2.183

77. A quality control inspector will measure the salt content (in milligrams) in a random sample of bags of potato chips from an hour of production. Which of

the following would result in the smallest margin of error in estimating the mean salt content μ?

(a) 90% confidence; $n = 25$

(b) 90% confidence; $n = 50$

(c) 95% confidence; $n = 25$

(d) 95% confidence; $n = 50$

(e) $n = 100$ at any confidence level

78. Scientists collect data on the blood cholesterol levels (milligrams per deciliter of blood) of a random sample of 24 laboratory rats. A 95% confidence interval for the mean blood cholesterol level μ is 80.2 to 89.8. Which of the following would cause the most worry about the validity of this interval?

(a) There is a clear outlier in the data.

(b) A stemplot of the data shows a mild right-skew.

(c) You do not know the population standard deviation σ.

(d) The population distribution is not exactly Normal.

(e) None of these would be a problem because the t procedures are robust.

79. **Watching TV** (6.1, 7.3) Choose a young person (aged 19 to 25) at random and ask, "In the past seven days, how many days did you watch television?" Call the response X for short. Here is the probability distribution for X:[31]

Days X:	0	1	2	3	4	5	6	7
Probability:	0.04	0.03	0.06	0.08	0.09	0.08	0.05	???

(a) What is the probability that X = 7? Justify your answer.

(b) Calculate the mean of the random variable X. Interpret this value in context.

(c) Suppose that you asked 100 randomly selected young people (aged 19 to 25) to respond to the question and found that the mean $\bar{x}$ of their responses was 4.96. Would this result surprise you? Justify your answer.

80. **Price cuts** (4.2) Stores advertise price reductions to attract customers. What type of price cut is most attractive? Experiments with more than one factor allow insight into interactions between the factors. A study of the attractiveness of advertised price discounts had two factors: percent of all foods on sale (25%, 50%, 75%, or 100%) and whether the discount was stated precisely (as in, for example, "60% off") or as a range (as in "40% to 70% off"). Subjects rated the attractiveness of the sale on a scale of 1 to 7.

(a) Describe a completely randomized design using 200 student subjects.

(b) Explain how you would use the partial table of random digits below to assign subjects to treatment groups. Then use your method to select the first 3 subjects for one of the treatment groups. Show your work clearly on your paper.

45740 41807 65561 33302 07051 93623 18132 09547
12975 13258 13048 45144 72321 81940 00360 02428

(c) The figure below shows the mean ratings for the eight treatments formed from the two factors.[32] Based on these results, write a careful description of how percent on sale and precise discount versus range of discounts influence the attractiveness of a sale.

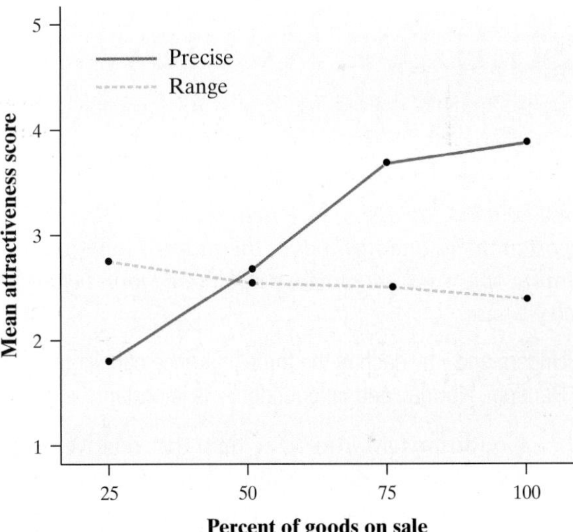

Chapter 8 Chapter Review

Statistical inference draws conclusions about a population on the basis of sample data and uses probability to indicate how reliable the conclusions are. A confidence interval estimates an unknown parameter.

The probabilities in confidence intervals tell us what would happen if we used the calculation method for the interval very many times. A confidence level describes the capture rate—the percent of all possible samples that will yield an interval containing the unknown parameter.

Section 8.1 focused on the underlying logic of confidence intervals. The one-sample z interval for a population proportion was presented in Section 8.2. Estimating an unknown population mean μ with a one-sample t interval was the subject of Section 8.3. A method for choosing a sample size to get an estimate within a specified margin of error was presented for both proportions and means.

Before you use any of the inference methods in this chapter, be sure to verify whether each of the three conditions—Random, Normal, and Independent—is satisfied.

Chapter Review Exercises

These exercises are designed to help you review the important ideas and methods of the chapter. Relevant learning objectives are provided in bulleted form before each exercise.

- Understand why each of the three inference conditions—Random, Normal, and Independent—is important.

R8.1. Conditions Martin says that the relative importance of the three conditions for performing inference is, in order from most to least important, Independent, Normal, and Random. Write a brief note to Martin explaining why he is incorrect.

- Determine critical values for calculating a confidence interval using a table or your calculator.

R8.2. It's critical Find the appropriate critical value for constructing a confidence interval in each of the following settings.

(a) Estimating a population proportion p at a 94% confidence level based on an SRS of size 125.

(b) Estimating a population mean μ at a 99% confidence level based on an SRS of size 58.

- Interpret a confidence level.
- Interpret a confidence interval in context.
- Determine sample statistics from a confidence interval.

R8.3. Batteries A company that produces AA batteries tests the lifetime of a random sample of 40 batteries using a special device designed to imitate real-world use. Based on the testing, the company makes the following statement: "Our AA batteries last an average of 430 to 470 minutes, and our confidence in that interval is 95%."[33]

(a) Determine the sample mean and standard deviation.

(b) A reporter translates the statistical announcement into "plain English" as follows: "If you buy one of this company's AA batteries, there is a 95% chance that it will last between 430 and 470 minutes." Comment on this interpretation.

(c) Your friend, who has just started studying statistics, claims that if you select 40 more AA batteries at random from those manufactured by this company, there is a 95% probability that the mean lifetime will fall between 430 and 470 minutes. Do you agree? Explain.

(d) Give a statistically correct interpretation of the confidence interval that could be published in a newspaper report.

- Carry out the steps in constructing a confidence interval for a population proportion: define the parameter; check conditions; perform calculations; interpret results in context.

R8.4. We love football! A recent Gallup Poll conducted telephone interviews with a random sample of adults aged 18 and older. Data were obtained for 1000 people. Of these, 37% said that football is their favorite sport to watch on television.

(a) Define the parameter p in this setting. Explain to someone who knows no statistics why we can't just say that 37% of all adults would say that football is their favorite sport to watch on television.
(b) Check conditions for constructing a confidence interval for p.
(c) Construct a 95% confidence interval for p.
(d) Interpret the interval in context.

- Carry out the steps in constructing a confidence interval for a population mean: define the parameter; check conditions; perform calculations; interpret results in context.

R8.5. Smart kids A school counselor wants to know how smart the students in her school are. She gets funding from the principal to give an IQ test to an SRS of 60 of the over 1000 students in the school. The mean IQ score was 114.98 and the standard deviation was 14.80.[34]

(a) Describe the parameter μ in this setting.
(b) Explain why the Normal condition is met in this case.
(c) Construct a 90% confidence interval for the mean IQ score of students at the school.
(d) Interpret your result from part (c) in context.

- Determine the sample size required to obtain a level C confidence interval for a population proportion with a specified margin of error.

R8.6. Do you go to church? The Gallup Poll plans to ask a random sample of adults whether they attended a religious service in the last 7 days. How large a sample would be required to obtain a margin of error of 0.01 in a 99% confidence interval for the population proportion who would say that they attended a religious service? Show your work.

- Construct and interpret a confidence interval for a population proportion.
- Explain how practical issues like nonresponse, undercoverage, and response bias can affect the interpretation of a confidence interval.

R8.7. Running red lights A random digit dialing telephone survey of 880 drivers asked, "Recalling the last ten traffic lights you drove through, how many of them were red when you entered the intersections?" Of the 880 respondents, 171 admitted that at least one light had been red.[35]

(a) Construct and interpret a 95% confidence interval for the population proportion.
(b) Nonresponse is a practical problem for this survey—only 21.6% of calls that reached a live person were completed. Another practical problem is that people may not give truthful answers. What is the likely direction of the bias: do you think more or fewer than 171 of the 880 respondents really ran a red light? Why? Are these sources of bias included in the margin of error?

- Construct and interpret a confidence interval for a population mean.
- Understand that a confidence interval gives a range of plausible values for the parameter.

R8.8. Engine parts Here are measurements (in millimeters) of a critical dimension on an SRS of 16 of the more than 200 auto engine crankshafts produced in one day:

224.120	224.001	224.017	223.982	223.989	223.961	223.960	224.089
223.987	223.976	223.902	223.980	224.098	224.057	223.913	223.999

(a) Construct and interpret a 95% confidence interval for the process mean at the time these crankshafts were produced.
(b) The process mean is supposed to be $\mu = 224$ mm but can drift away from this target during production. Does your interval from part (a) suggest that the process mean has drifted? Explain.

- Determine the sample size required to obtain a level C confidence interval for a population mean with a specified margin of error.

R8.9. Good wood? A lab supply company sells pieces of Douglas fir 4 inches long and 1.5 inches square for force experiments in science classes. From experience, the strength of these pieces of wood follows a Normal distribution with standard deviation 3000 pounds. You want to estimate the mean load needed to pull apart these pieces of wood to within 1000 pounds with 95% confidence. How large a sample is needed? Show your work.

- Understand how the margin of error of a confidence interval changes with the sample size and the level of confidence C.

R8.10. It's about *ME* Explain how each of the following would affect the margin of error of a confidence interval, if all other things remained the same.

(a) Increasing the confidence level
(b) Quadrupling the sample size

Chapter 8 AP Statistics Practice Test

Section I: Multiple Choice *Select the best answer for each question.*

T8.1. The Gallup Poll interviews 1600 people. Of these, 18% say that they jog regularly. The news report adds: "The poll had a margin of error of plus or minus three percentage points at a 95% confidence level." You can safely conclude that

(a) 95% of all Gallup Poll samples like this one give answers within ±3% of the true population value.

(b) the percent of the population who jog is certain to be between 15% and 21%.

(c) 95% of the population jog between 15% and 21% of the time.

(d) we can be 95% confident that the sample proportion is captured by the confidence interval.

(e) if Gallup took many samples, 95% of them would find that 18% of the people in the sample jog.

T8.2. The weights (in pounds) of three adult males are 160, 215, and 195. The standard error of the mean of these three weights is

(a) 190. (b) 27.84. (c) 22.73. (d) 16.07. (e) 13.13.

T8.3. In preparing to construct a one-sample t interval for a population mean, suppose we are not sure if the population distribution is Normal. In which of the following circumstances would we not be safe constructing the interval based on an SRS of size 24 from the population?

(a) A stemplot of the data is roughly bell-shaped.

(b) A histogram of the data shows slight skewness.

(c) A stemplot of the data has a large outlier.

(d) The sample standard deviation is large.

(e) The t procedures are robust, so it is always safe.

T8.4. Many television viewers express doubts about the validity of certain commercials. In an attempt to answer their critics, Timex Group USA wishes to estimate the proportion of consumers who believe what is shown in Timex television commercials. Let p represent the true proportion of consumers who believe what is shown in Timex television commercials. What is the smallest number of consumers that Timex can survey to guarantee a margin of error of 0.05 or less at a 99% confidence level?

(a) 550 (b) 600 (c) 650 (d) 700 (e) 750

T8.5. You want to compute a 90% confidence interval for the mean of a population with unknown population standard deviation. The sample size is 30. The value of t^* you would use for this interval is

(a) 1.645. (b) 1.699. (c) 1.697. (d) 1.96. (e) 2.045.

T8.6. A radio talk show host with a large audience is interested in the proportion p of adults in his listening area who think the drinking age should be lowered to eighteen. To find this out, he poses the following question to his listeners: "Do you think that the drinking age should be reduced to eighteen in light of the fact that eighteen-year-olds are eligible for military service?" He asks listeners to phone in and vote "Yes" if they agree the drinking age should be lowered and "No" if not. Of the 100 people who phoned in, 70 answered "Yes." Which of the following conditions for inference about a proportion using a confidence interval are violated?

I. The data are a random sample from the population of interest.

II. n is so large that both $n\hat{p}$ and $n(1-\hat{p})$ are at least 10.

III. The population is at least 10 times as large as the sample.

(a) I only (b) II only (c) III only (d) I and II only (e) I, II, and III

T8.7. A 90% confidence interval for the mean μ of a population is computed from a random sample and is found to be 9 ± 3. Which of the following *could* be the 95% confidence interval based on the same data?

(a) 9 ± 1.96 (b) 9 ± 2 (c) 9 ± 3 (d) 9 ± 4

(e) Without knowing the sample size, any of the above answers could be the 95% confidence interval.

T8.8. Suppose we want a 90% confidence interval for the average amount spent on books by freshmen in their first year at a major university. The interval is to have a margin of error of \$2. Based on last year's book sales, we estimate that the standard deviation of the amount spent will be close to \$30. The number of observations required is closest to

(a) 25. (b) 30. (c) 608. (d) 609. (e) 865.

T8.9. A telephone poll of an SRS of 1234 adults found that 62% are generally satisfied with their lives. The announced margin of error for the poll was 3%. Does the margin of error account for the fact that some adults do not have telephones?

(a) Yes. The margin of error includes all sources of error in the poll.

(b) Yes. Taking an SRS eliminates any possible bias in estimating the population proportion.

(c) Yes. The margin of error includes undercoverage but not nonresponse.

(d) No. The margin of error includes nonresponse but not undercoverage.

(e) No. The margin of error only includes sampling variability.

T8.10. A Census Bureau report on the income of Americans says that with 90% confidence the median income of all U.S. households in a recent year was \$57,005 with a margin of error of ±\$742. This means that

(a) 90% of all households had incomes in the range \$57,005 ± \$742.

(b) we can be sure that the median income for all households in the country lies in the range \$57,005 ± \$742.

(c) 90% of the households in the sample interviewed by the Census Bureau had incomes in the range \$57,005 ± \$742.

(d) the Census Bureau got the result \$57,005 ± \$742 using a method that will cover the true median income 90% of the time when used repeatedly.

(e) 90% of all possible samples of this same size would result in a sample median that falls within \$742 of \$57,005.

Section II: Free Response *Show all your work. Indicate clearly the methods you use, because you will be graded on the correctness of your methods as well as on the accuracy and completeness of your results and explanations.*

T8.11. The U.S. Forest Service is considering additional restrictions on the number of vehicles allowed to enter Yellowstone National Park. To assess public reaction, the service asks a random sample of 150 visitors if they favor the proposal. Of these, 89 say "Yes."

(a) Construct and interpret a 99% confidence interval for the proportion of all visitors to Yellowstone who favor the restrictions.

(b) Based on your work in part (a), can the U.S. Forest Service conclude that more than half of visitors to Yellowstone National Park favor the proposal? Justify your answer.

T8.12. How many people live in South African households? To find out, we collected data from an SRS of 48 out of the over 700,000 South African students who took part in the CensusAtSchool survey project. The mean number of people living in a household was 6.208; the standard deviation was 2.576.

(a) Is the Normal condition met in this case? Justify your answer.

(b) Maurice claims that a 95% confidence interval for the population mean is $6.208 \pm 1.96\frac{0.372}{\sqrt{47}}$. Explain why this interval is wrong. Then give the correct interval.

T8.13. A milk processor monitors the number of bacteria per milliliter in raw milk received at the factory. A random sample of 10 one-milliliter specimens of milk supplied by one producer gives the following data:

5370 4890 5100 4500 5260 5150 4900 4760 4700 4870

Construct and interpret a 90% confidence interval for the population mean μ.

Chapter 9

case study

Introduction

Testing a Claim

Do You Have a Fever?

Sometimes when you're sick, your forehead feels really warm. You might have a fever. How can you find out whether you do? By taking your temperature, of course. But what temperature should the thermometer show if you're healthy? Is this temperature the same for everyone?

Several years ago, researchers conducted a study to determine whether the "accepted" value for normal body temperature, 98.6°F, is accurate. They used an oral thermometer to measure the temperatures of a random sample of healthy men and women aged 18 to 40.

Exploratory data analysis revealed several interesting facts about this data set:[1]

- The distribution of temperature readings was approximately Normal.
- The mean temperature was $\overline{x} = 98.25$°F.
- The standard deviation of the temperature readings was $s_x = 0.73$°F.
- 62.3% of the temperature readings were less than 98.6°F.

Based on the results of this study, can we conclude that "normal" body temperature in the population of healthy 18- to 40-year-olds is *not* 98.6°F? By the end of this chapter, you will have developed the necessary tools for answering this question.

Introduction

Confidence intervals are one of the two most common types of statistical inference. Use a confidence interval when your goal is to estimate a population parameter. The second common type of inference, called *significance tests*, has a different goal: to assess the evidence provided by data about some claim concerning a population. Here is an Activity that illustrates the reasoning of statistical tests.

ACTIVITY *I'm a great free-throw shooter!*

MATERIALS: Computer with Internet access and projection capability

A basketball player claims to make 80% of the free throws that he attempts. We think he might be exaggerating. To test this claim, we'll ask him to shoot some free throws—virtually—using the *Test of Significance* applet at the book's Web site.

1. Go to www.whfreeman.com/tps4e and launch the applet.
2. Click "Show null hypothesis" so that the player's claim is visible in the graph.

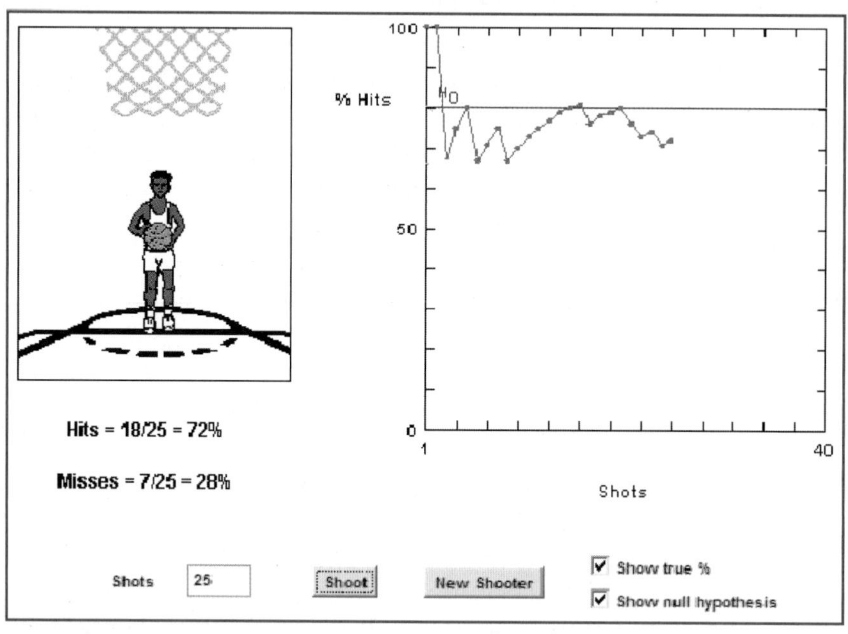

3. The applet is set to take 25 shots. Click "Shoot." How many of the 25 shots did the player make? Do you have enough data to decide whether the player's claim is valid?
4. Click "Shoot" again for 25 more shots. Keep doing this until *either* you are convinced that the player makes less than 80% of his shots *or* that the player's claim is true. How large a sample of shots did you need to make your decision?
5. Click "Show true %" to reveal the truth. Was your conclusion correct?
6. If time permits, choose a new shooter and repeat Steps 3 through 5. Is it easier to tell that the player is exaggerating when his actual proportion of free throws made is closer to 0.8 or farther from 0.8?

In the free-throw shooter Activity, the parameter of interest is the proportion p of free throws that the player will make if he shoots forever. Our player claims that $p = 0.80$. To test his claim, we let the applet simulate 25 shots. If the player makes only 10 of 25 free throws (a sample proportion of $\hat{p} = 0.40$), we have pretty strong evidence that he doesn't actually shoot 80%. But what if he makes 19 of 25 (a sample proportion of $\hat{p} = 0.76$)? This provides *some* evidence that his true long-term percent may be less than 80%, but it's not nearly as convincing as $\hat{p} = 0.40$. Statistical tests weigh the evidence *against* a claim like $p = 0.8$.

Section 9.1 focuses on the underlying logic of statistical tests. Once the foundation is laid, we consider the implications of using these tests to make decisions—about everything from free-throw shooting to the effectiveness of a new drug. In Section 9.2, we present the details of performing a test about a population proportion. Section 9.3 shows how to test a claim about a population mean. Along the way, we examine the connection between confidence intervals and tests.

9.1

Significance Tests: The Basics

In Section 9.1, you'll learn about:

- The reasoning of significance tests
- Stating hypotheses
- Interpreting P-values
- Statistical significance
- Type I and Type II errors
- Planning studies: The power of a statistical test

A **significance test** is a formal procedure for comparing observed data with a claim (also called a *hypothesis*) whose truth we want to assess. The claim is a statement about a parameter, like the population proportion p or the population mean μ. We express the results of a significance test in terms of a probability that measures how well the data and the claim agree. Now let's take a closer look at the underlying logic of significance tests.

The Reasoning of Significance Tests

Statistical tests deal with claims about a population. Tests ask if sample data give good evidence *against* a claim. A test might say, "If we took many random samples and the claim were true, we would rarely get a result like this." To get a numerical measure of how strong the sample evidence is, replace the vague term "rarely" by a probability. Here is an example of this reasoning at work.

A significance test is sometimes referred to as a *test of significance*, a *hypothesis test*, or a *test of hypotheses*.

I'm a Great Free-Throw Shooter!

The logic of statistical tests

Our virtual basketball player in the previous Activity claimed to be an 80% free-throw shooter. Suppose that he shoots 50 free throws and makes 32 of them. His sample proportion of made shots is $\hat{p} = \frac{32}{50} = 0.64$. What can we conclude about the player's claim based on the sample data? Let's perform a simulation to find out.

Our reasoning is based on asking what would happen if the player's claim (p = 0.80) were true and we observed many samples of 50 free throws. We used Fathom software to simulate 400 sets of 50 shots assuming that the player is really an 80% shooter. Figure 9.1 shows a dotplot of the results. Each dot on the graph represents the proportion of made shots in one set of 50 attempts. For example, if the player makes 43/50 shots in one trial, the dot would be placed at $\hat{p} = 0.86$.

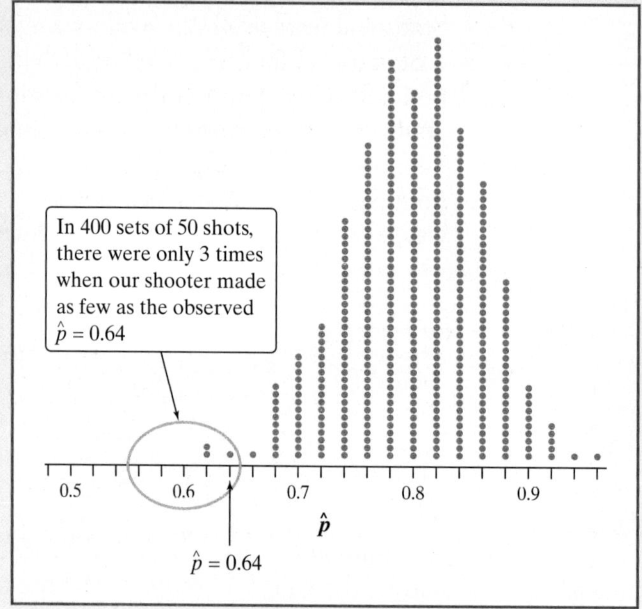

FIGURE 9.1 Fathom dotplot of the sample proportion $\hat{p}$ of free throws made by an 80% shooter in 400 sets of 50 shots.

You can say how strong the evidence against the player's claim is by giving the probability that he would make as few as 32 out of 50 free throws if he really makes 80% in the long run. Based on the simulation, our estimate of this probability is 3/400 = 0.0075. The observed statistic, $\hat{p} = 0.64$, is so unlikely if the actual parameter value is $p = 0.80$ that it gives convincing evidence that the player's claim is not true.

Be sure that you understand why this evidence is convincing. There are two possible explanations of the fact that our virtual player made only $\hat{p} = 32/50 = 0.64$ of his free throws:

1. The player's claim is correct ($p = 0.8$), and by bad luck, a very unlikely outcome occurred.

2. The population proportion is actually less than 0.8, so the sample result is not an unlikely outcome.

Explanation 1 might be correct—the result of our random sample of 50 shots *could* be due to chance alone. But the probability that such a result would occur by chance is so small (less than 1 in a 100) that we are quite confident that Explanation 2 is right.

Statistical tests use an elaborate vocabulary, but the basic idea is simple: *an outcome that would rarely happen if a claim were true is good evidence that the claim is not true.*

Stating Hypotheses

A significance test starts with a careful statement of the claims we want to compare. In our free-throw shooter example, the virtual player claims that his long-run proportion of made free throws is $p = 0.80$. This is the claim we seek evidence *against*. We call it the **null hypothesis,** abbreviated H_0. Usually, the null hypothesis is a statement of "no difference." For the free-throw shooter, no difference from what he claimed gives $H_0: p = 0.80$. The claim we hope or suspect to be true instead of the null hypothesis is called the **alternative hypothesis.** We abbreviate the alternative hypothesis as H_a. In this case, we believe the player might be exaggerating, so our alternative hypothesis is $H_a: p < 0.80$.

DEFINITION: Null hypothesis H_0, alternative hypothesis H_a

The claim tested by a statistical test is called the **null hypothesis (H_0).** The test is designed to assess the strength of the evidence *against* the null hypothesis. Often the null hypothesis is a statement of "no difference."

The claim about the population that we are trying to find evidence *for* is the **alternative hypothesis (H_a).**

Some people insist that all three possibilities—greater than, less than, and equal to—be accounted for in the hypotheses. For the free-throw shooter example, since the alternative hypothesis is H_a: $p < 0.80$, they would write the null hypothesis as $H_0: p \geq 0.80$. In spite of the mathematical appeal of covering all three cases, we use only the value $p = 0.80$ in our calculations. So we'll stick with $H_0: p = 0.80$.

In the free-throw shooter example, our hypotheses are

$$H_0: p = 0.80$$
$$H_a: p < 0.80$$

where p is the long-run proportion of made free throws. The alternative hypothesis is **one-sided** because we are interested only in whether the player is overstating his free-throw shooting ability. Because H_a expresses the effect that we hope to find evidence *for*, it is sometimes easier to begin by stating H_a and then set up H_0 as the statement that the hoped-for effect is not present. Here is an example in which the alternative hypothesis is **two-sided.**

Studying Job Satisfaction

Stating hypotheses

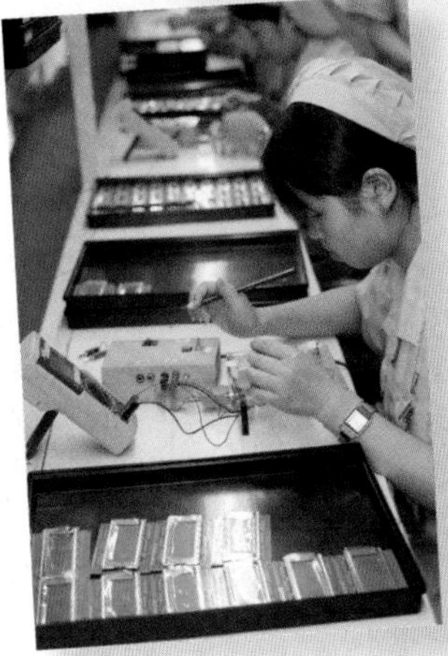

Does the job satisfaction of assembly-line workers differ when their work is machine-paced rather than self-paced? One study chose 18 subjects at random from a company with over 200 workers who assembled electronic devices. Half of the workers were assigned at random to each of two groups. Both groups did similar assembly work, but one group was allowed to pace themselves while the other group used an assembly line that moved at a fixed pace. After two weeks, all the workers took a test of job satisfaction. Then they switched work setups and took the test again after two more weeks. (This experiment uses a matched pairs design, which you learned about in Chapter 4.) The response variable is the difference in satisfaction scores, self-paced minus machine-paced.[2]

PROBLEM:

(a) Describe the parameter of interest in this setting.

(b) State appropriate hypotheses for performing a significance test.

SOLUTION:

(a) The parameter of interest is the mean μ of the differences (self-paced minus machine-paced) in job satisfaction scores in the population of all assembly-line workers at this company.

(b) Because the initial question asked whether job satisfaction differs, the alternative hypothesis is two-sided; that is, either $\mu < 0$ or $\mu > 0$. For simplicity, we write this as $\mu \neq 0$. The null hypothesis says that there is no difference in the scores. That is,

$$H_0: \mu = 0$$
$$H_a: \mu \neq 0$$

For Practice *Try Exercise 3*

CAUTION

The hypotheses should express the hopes or suspicions we have **before** *we see the data. It is cheating to look at the data first and then frame hypotheses to fit what the data show.* For example, the data for the job satisfaction study showed that the workers were more satisfied with self-paced work. But you should *not* change the alternative hypothesis to $H_a: \mu > 0$ after looking at the data. If you do not have a specific direction firmly in mind in advance, use a two-sided alternative.

DEFINITION: One-sided alternative hypothesis and two-sided alternative hypothesis

The alternative hypothesis is **one-sided** if it states that a parameter is *larger than* the null hypothesis value or if it states that the parameter is *smaller than* the null value. It is **two-sided** if it states that the parameter is *different from* the null hypothesis value (it could be either larger or smaller).

It is common to refer to a significance test with a one-sided alternative hypothesis as a *one-sided test* or *one-tailed test* and to a test with a two-sided alternative hypothesis as a *two-sided test* or *two-tailed test*.

In any significance test, the null hypothesis has the form H_0: parameter = value. The alternative hypothesis has one of the forms H_a: parameter < value, H_a: parameter > value, or H_a: parameter ≠ value. To determine the correct form of H_a, read the problem carefully.

CAUTION

Hypotheses always refer to a *population*, not to a sample. Be sure to state H_0 and H_a in terms of *population parameters*. It is *never* correct to write a hypothesis about a sample statistic, such as $\hat{p} = 0.64$ or $\bar{x} > 85$.

CHECK YOUR UNDERSTANDING

For each of the following settings, (a) describe the parameter of interest, and (b) state appropriate hypotheses for a significance test.

1. According to the Web site sleepdeprivation.com, 85% of teens are getting less than eight hours of sleep a night. Jannie wonders whether this result holds in her large high school. She asks an SRS of 100 students at the school how much sleep they get on a typical night. In all, 75 of the responders said less than 8 hours.

2. As part of its 2010 census marketing campaign, the U.S. Census Bureau advertised "10 questions, 10 minutes—that's all it takes." On the census form itself, we read, "The

U.S. Census Bureau estimates that, for the average household, this form will take about 10 minutes to complete, including the time for reviewing the instructions and answers." We suspect that the actual time it takes to complete the form may be longer than advertised.

Interpreting *P*-Values

The idea of stating a null hypothesis that we want to find evidence *against* seems odd at first. It may help to think of a criminal trial. The defendant is "innocent until proven guilty." That is, the null hypothesis is innocence and the prosecution must try to provide convincing evidence against this hypothesis. That's exactly how statistical tests work, although in statistics we deal with evidence provided by data and use a probability to say how strong the evidence is.

The null hypothesis H_0 states the claim that we are seeking evidence against. The probability that measures the strength of the evidence against a null hypothesis is called a **P-value**.

DEFINITION: *P*-value

The probability, computed assuming H_0 is true, that the statistic (such as $\hat{p}$ or $\bar{x}$) would take a value as extreme as or more extreme than the one actually observed is called the ***P*-value** of the test. The smaller the *P*-value, the stronger the evidence against H_0 provided by the data.

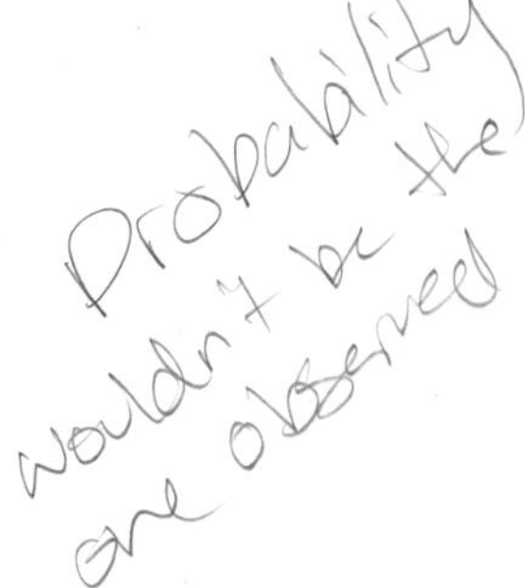

Small *P*-values are evidence against H_0 because they say that the observed result is unlikely to occur when H_0 is true. Large *P*-values fail to give convincing evidence against H_0 because they say that the observed result is likely to occur by chance when H_0 is true. We'll show you how to calculate *P*-values later. For now, let's focus on interpreting them.

EXAMPLE

I'm a Great Free-Throw Shooter!

Interpreting a *P*-value

The *P*-value is the probability of getting a sample result at least as extreme as the one we did if H_0 were true. Since the alternative hypothesis is $H_a: p < 0.80$, the sample results that count as "at least as extreme" are those with $\hat{p} \leq 0.64$. In other words, the *P*-value is the conditional probability $P(\hat{p} \leq 0.64 \mid p = 0.80)$. Earlier, we used a simulation to estimate this probability as $3/400 = 0.0075$. So if H_0 is true, and the player makes 80% of his free throws in the long run, there's less than a 1 in 100 chance that the player would make as few as 32 of 50 shots. The small probability gives strong evidence against H_0 and in favor of the alternative $H_a: p < 0.80$.

The alternative hypothesis sets the direction that counts as evidence against H_0. In the previous example, only values of $\hat{p}$ that are much less than 0.80 count as evidence against the null hypothesis because the alternative is one-sided on the low side. If the alternative is two-sided, both directions count.

Studying Job Satisfaction

Interpreting a *P*-value

For the job satisfaction study described on page 531, the hypotheses are

$$H_0: \mu = 0$$
$$H_a: \mu \neq 0$$

where μ is the mean difference in job satisfaction scores (self-paced − machine-paced) in the population of assembly-line workers at the company. Data from the 18 workers gave $\bar{x} = 17$ and $s_x = 60$. That is, these workers rated the self-paced environment, on average, 17 points higher.

PROBLEM: **Researchers performed a significance test using the sample data and obtained a *P*-value of 0.2302.**

(a) Explain what it means for the null hypothesis to be true in this setting.

(b) Interpret the *P*-value in context.

(c) Do the data provide convincing evidence against the null hypothesis? Explain.

SOLUTION:

(a) In this setting, $H_0: \mu = 0$ says that the mean difference in satisfaction scores (self-paced − machine-paced) for the entire population of assembly-line workers at the company is 0. If H_0 is true, then the workers don't favor one work environment over the other, on average.

(b) The *P*-value is the probability of observing a sample result as extreme or more extreme in the direction specified by H_a just by chance when H_0 is actually true. Because the alternative hypothesis is two-sided, the *P*-value is the probability of getting a value of $\bar{x}$ as far from 0 *in either direction* as the observed $\bar{x} = 17$ when H_0 is true. That is, an average difference of 17 or more points between the two work environments would happen 23% of the time just by chance in random samples of 18 assembly-line workers when the true population mean is $\mu = 0$.

(c) An outcome that would occur so often just by chance (almost 1 in every 4 random samples of 18 workers) when H_0 is true is *not* convincing evidence against H_0.

For Practice *Try Exercise 11*

The conclusion of the job satisfaction study is *not* that H_0 is true. The study looked for evidence against $H_0: \mu = 0$ and failed to find strong evidence. That is all we can say. *Failing to find evidence against H_0 means only that the data are* consistent *with H_0, not that we have clear evidence that H_0 is true.*

CAUTION

Statistical Significance

The final step in performing a significance test is to draw a conclusion about the competing claims you were testing. We will make one of two decisions based on the strength of the evidence against the null hypothesis (and in favor of the alternative hypothesis)—**reject H_0** or **fail to reject H_0**. If our sample result is too unlikely to have happened by chance assuming H_0 is true, then we'll reject H_0. Otherwise, we will fail to reject H_0.

This wording may seem unusual at first, but it's consistent with what happens in a criminal trial. Once the jury has weighed the evidence against the null hypothesis of innocence, they return one of two verdicts: "guilty" (reject H_0) or "not guilty" (fail to reject H_0). A not-guilty verdict doesn't guarantee that the defendant is innocent, just that there's not enough evidence of guilt. Likewise, a fail-to-reject H_0 decision in a significance test doesn't mean that H_0 is true. For that reason, *you should never "accept H_0" or use language implying that you believe H_0 is true.*

CAUTION

EXAMPLE

Free Throws and Job Satisfaction

Drawing conclusions

In the free-throw shooter example, the estimated *P*-value of 0.0075 is strong evidence against the null hypothesis $H_0: p = 0.80$. For that reason, we would reject H_0 in favor of the alternative $H_a: p < 0.80$. It appears that the virtual player makes fewer than 80% of his free throws.

For the job satisfaction experiment, however, the *P*-value of 0.2302 is not convincing evidence against $H_0: \mu = 0$. We therefore fail to reject H_0. Researchers cannot conclude that job satisfaction differs based on work conditions for the population of assembly-line workers at the company.

In a nutshell, our conclusion in a significance test comes down to

P-value small → reject H_0 → conclude H_a (in context)
P-value large → fail to reject H_0 → cannot conclude H_a (in context)

There is no rule for how small a *P*-value we should require in order to reject H_0—it's a matter of judgment and depends on the specific circumstances. But we can compare the *P*-value with a fixed value that we regard as decisive, called the **significance level.** We write it as α, the Greek letter alpha. If we choose $\alpha = 0.05$, we are requiring that the data give evidence against H_0 so strong that it would happen less than 5% of the time just by chance when H_0 is true. If we choose $\alpha = 0.01$, we are insisting on stronger evidence against the null hypothesis, a result that would occur less often than 1 in every 100 times in the long run if H_0 is true. When our *P*-value is less than the chosen α, we say that the result is **statistically significant.**

DEFINITION: Statistically significant

If the *P*-value is smaller than alpha, we say that the data are **statistically significant at level α.** In that case, we reject the null hypothesis H_0 and conclude that there is convincing evidence in favor of the alternative hypothesis H_a.

CAUTION

"Significant" in the statistical sense does not necessarily mean "important." It means simply "not likely to happen just by chance." The significance level α makes "not likely" more exact. Significance at level 0.01 is often expressed by the statement, "The results were significant ($P < 0.01$)."

Here P stands for the P-value. The actual P-value is more informative than a statement of significance because it allows us to assess significance at any level we choose. For example, a result with $P = 0.03$ is significant at the $\alpha = 0.05$ level but is not significant at the $\alpha = 0.01$ level. When we use a fixed significance level to draw a conclusion in a statistical test,

$$P\text{-value} < \alpha \rightarrow \text{reject } H_0 \rightarrow \text{conclude } H_a \text{ (in context)}$$
$$P\text{-value} \geq \alpha \rightarrow \text{fail to reject } H_0 \rightarrow \text{cannot conclude } H_a \text{ (in context)}$$

EXAMPLE

Better Batteries

Statistical significance

A company has developed a new deluxe AAA battery that is supposed to last longer than its regular AAA battery.[3] However, these new batteries are more expensive to produce, so the company would like to be convinced that they really do last longer. Based on years of experience, the company knows that its regular AAA batteries last for 30 hours of continuous use, on average. The company selects an SRS of 15 new batteries and uses them continuously until they are completely drained. A significance test is performed using the hypotheses

$$H_0: \mu = 30 \text{ hours}$$
$$H_a: \mu > 30 \text{ hours}$$

where μ is the true mean lifetime of the new deluxe AAA batteries. The resulting P-value is 0.0276.

PROBLEM: **What conclusion would you make for each of the following significance levels? Justify your answer.**

(a) $\alpha = 0.05$ (b) $\alpha = 0.01$

SOLUTION:

(a) **Since the *P*-value, 0.0276, is less than $\alpha = 0.05$, the sample result is statistically significant at the 5% level. We have sufficient evidence to reject H_0 and conclude that the company's deluxe AAA batteries last longer than 30 hours, on average.**

(B) **Since the *P*-value, 0.0276, is greater than $\alpha = 0.01$, the sample result is not statistically significant at the 1% level. We do not have enough evidence to reject H_0 in this case. Therefore, we cannot conclude that the deluxe AAA batteries last longer than 30 hours, on average.**

For Practice *Try Exercise 13*

AP EXAM TIP The conclusion to a significance test should always include three components: (1) an explicit comparison of the *P*-value to a stated significance level OR an interpretation of the *P*-value as a conditional probability, (2) a decision about the null hypothesis: reject or fail to reject H_0, and (3) an explanation of what the decision means in context.

In practice, the most commonly used significance level is $\alpha = 0.05$. This is mainly due to Sir Ronald A. Fisher, a famous statistician who worked on agricultural experiments in England during the early twentieth century. Fisher was the first to suggest deliberately using random assignment in an experiment. In a paper published in 1926, Fisher wrote that it is convenient to draw the line at about the level at which we can say: "Either there is something in the treatment, or a coincidence has occurred such as does not occur more than once in twenty trials."[4] The following Activity may shed some light on Fisher's choice of 0.05.

ACTIVITY *Pick a card!*

MATERIALS: Deck of playing cards (provided by your teacher) and cookies or some other prize for the winners

Your teacher will offer you the chance to compete for a reward.[5] All you have to do is get lucky! The teacher will shuffle a deck of cards and announce the winning color (red or black). Any student who wants to take the challenge may choose one card (without looking!) from the deck. If the card has the winning color, that student wins a prize. The chosen card will be replaced in the deck, and the cards will be shuffled again before the next student plays. Continue playing until all prizes have been awarded or all interested students have played.

Sometimes it may be preferable to choose $\alpha = 0.01$ or $\alpha = 0.10$, for reasons we will discuss shortly. *Warning:* if you are going to draw a conclusion based on statistical significance, then the significance level α should be stated *before* the data are produced. Otherwise, a deceptive user of statistics might set an α level after the data have been analyzed in an obvious attempt to manipulate the conclusion. This is just as inappropriate as choosing an alternative hypothesis to be one-sided in a particular direction *after* looking at the data.

CAUTION

THINK ABOUT IT

Why choose a significance level at all? The purpose of a significance test is to give a clear statement of the strength of evidence provided by the data against the null hypothesis. The *P*-value does this. But how small a *P*-value is convincing evidence against the null hypothesis? This depends mainly on two circumstances:

- *How plausible is H_0?* If H_0 represents an assumption that the people you must convince have believed for years, strong evidence (small *P*-value) will be needed to persuade them.
- *What are the consequences of rejecting H_0?* If rejecting H_0 in favor of H_a means making an expensive change of some kind, you need strong evidence that the change will be beneficial.

These criteria are a bit subjective. Different people will insist on different levels of significance. Giving the *P*-value allows each of us to decide individually if the evidence is sufficiently strong.

Users of statistics have often emphasized standard significance levels such as 10%, 5%, and 1%. For example, courts have tended to accept 5% as a standard in discrimination cases.[6] This emphasis reflects the time when tables of critical values rather than technology dominated statistical practice. The 5% level $\alpha = 0.05$ is particularly common (probably due to R. A. Fisher).

There is no practical distinction between the *P*-values 0.049 and 0.051. However, if we use an $\alpha = 0.05$ significance level, the former value will lead us to reject H_0 while the latter value will lead us to not reject H_0.

Beginning users of statistical tests generally find it easier to compare a *P*-value to a significance level than to interpret the *P*-value correctly in context. For that reason, we will include stating a significance level as a required part of every significance test. We'll also ask you to explain what a *P*-value means in a variety of settings.

Type I and Type II Errors

When we draw a conclusion from a significance test, we hope our conclusion will be correct. But sometimes it will be wrong. There are two types of mistakes we can make. We can reject the null hypothesis when it's actually true, known as a **Type I error,** or we can fail to reject a false null hypothesis, which is a **Type II error.**

DEFINITION: Type I error and Type II error

If we reject H_0 when H_0 is true, we have committed a **Type I error.**

If we fail to reject H_0 when H_0 is false, we have committed a **Type II error.**

		Truth about the population	
		H_0 true	H_0 false (H_a true)
Conclusion based on sample	Reject H_0	Type I error	Correct conclusion
	Fail to reject H_0	Correct conclusion	Type II error

FIGURE 9.2 The two types of errors in significance tests.

The possibilities are summed up in Figure 9.2. If H_0 is true, our conclusion is correct if we fail to reject H_0, but it is a Type I error if we reject H_0. If H_a is true, our conclusion is either correct or a Type II error. Only one error is possible at a time.

It is important to be able to describe Type I and Type II errors in the context of a problem. Considering the consequences of each of these types of error is also important, as the following example shows.

EXAMPLE *Perfect Potatoes*

Type I and Type II errors

A potato chip producer and its main supplier agree that each shipment of potatoes must meet certain quality standards. If the producer determines that more than 8% of the potatoes in the shipment have "blemishes," the truck will be sent away to get another load of potatoes from the supplier. Otherwise, the entire truckload will be used to make potato chips. To make the decision, a supervisor will inspect a random sample of potatoes from the shipment. The producer will then perform a significance test using the hypotheses

$$H_0: p = 0.08$$
$$H_a: p > 0.08$$

where p is the actual proportion of potatoes with blemishes in a given truckload.

PROBLEM: **Describe a Type I and a Type II error in this setting, and explain the consequences of each.**

SOLUTION: **A Type I error would occur if the producer concludes that the proportion of potatoes with blemishes is greater than 0.08 when the actual proportion is 0.08 (or less).** ***Consequence:*** **The potato-chip producer sends the truckload of acceptable potatoes away, which may result in lost revenue for the supplier. Furthermore, the producer will have to wait for another shipment of potatoes before producing the next batch of potato chips. A Type II error would occur if the producer does not send the truck away when more than 8% of the potatoes in the shipment have blemishes.** ***Consequence:*** **The producer uses the truckload of potatoes to make potato chips. More chips will be made with blemished potatoes, which may upset consumers.**

For Practice *Try Exercise 19*

Which is more serious: a Type I error or a Type II error? That depends on the situation. For the potato-chip producer, a Type II error could result in upset customers, leading to decreased sales. A Type I error, turning away a shipment even though less than 8% of the potatoes have blemishes, may not have much impact if additional shipments of potatoes can be obtained fairly easily.

CHECK YOUR UNDERSTANDING

Refer to the "Better Batteries" example on page 536.

1. Describe a Type I error in this setting.
2. Describe a Type II error in this setting.
3. Which type of error is more serious in this case? Justify your answer.

Error Probabilities We can assess the performance of a significance test by looking at the probabilities of the two types of error. That's because statistical inference is based on asking, "What would happen if I did this many times?" We cannot (without inspecting the whole truckload) guarantee that good shipments of potatoes will never be rejected and bad shipments will never be accepted. But we can think about our chances of making each of these mistakes.

EXAMPLE

Perfect Potatoes

Type I error probability

For the truckload of potatoes in the previous example, we were testing

$$H_0: p = 0.08$$
$$H_a: p > 0.08$$

where p is the actual proportion of potatoes with blemishes. Suppose that the potato-chip producer decides to carry out this test based on a random sample of 500 potatoes using a 5% significance level ($\alpha = 0.05$). A Type I error is to reject H_0 when H_0 is actually true. If our sample results in a value of $\hat{p}$ that is much larger than 0.08, we will reject H_0. How large would $\hat{p}$ need to be? The 5% significance level tells us to count results that could happen less than 5% of the time by chance if H_0 is true as evidence that H_0 is false.

Assuming $H_0: p = 0.08$ is true, the sampling distribution of $\hat{p}$ will have

Shape: Approximately Normal because $500(0.08) = 40$ and $500(0.92) = 460$ are both at least 10.

Center: $\mu_{\hat{p}} = p = 0.08$

Spread: $\sigma_{\hat{p}} = \sqrt{\dfrac{p(1-p)}{n}} = \sqrt{\dfrac{0.08(0.92)}{500}} = 0.0121$

Figure 9.3 on the next page shows the Normal curve that approximates this sampling distribution.

The shaded area in the right tail of Figure 9.3 is 5%. Values of $\hat{p}$ to the right of the green line at $\hat{p} = 0.10$ will cause us to reject H_0 even though H_0 is true. This will happen in 5% of all possible samples. That is, the probability of making a Type I error is 0.05.

Try to figure out where the value 0.0999 came from in Figure 9.3. You can use Table A and do some algebra or use `invNorm` on your calculator.

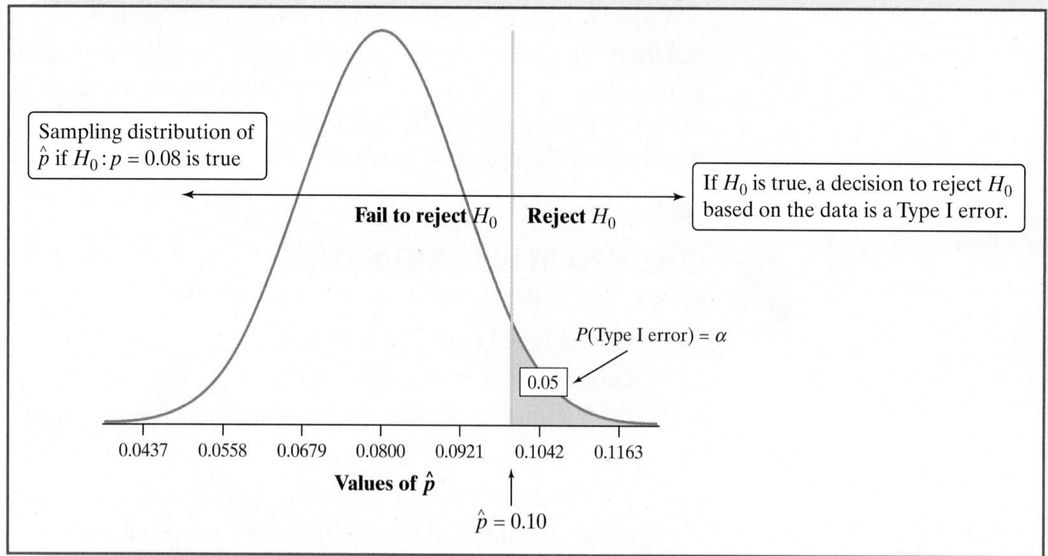

FIGURE 9.3 The probability of a Type I error (shaded area) is the probability of rejecting $H_0: p =$ 0.08 when H_0 is actually true.

The probability of a Type I error is the probability of rejecting H_0 when it is really true. As the previous example showed, this is exactly the significance level of the test.

Significance and Type I Error

The significance level α of any fixed level test is the probability of a Type I error. That is, α is the probability that the test will reject the null hypothesis H_0 when H_0 is in fact true. Consider the consequences of a Type I error before choosing a significance level.

What about Type II errors? A significance test makes a Type II error when it fails to reject a null hypothesis that really is false. There are many values of the parameter that satisfy the alternative hypothesis, so we concentrate on one value. A high probability of a Type II error for a particular alternative means that the test is not sensitive enough to usually detect that alternative. In the significance test setting, it is more common to report the probability that a test *does* reject H_0 when an alternative is true. This probability is called the **power** of the test against that specific alternative. The higher this probability is, the more sensitive the test is.

DEFINITION: Power

The **power** of a test against a specific alternative is the probability that the test will reject H_0 at a chosen significance level α when the specified alternative value of the parameter is true.

As the following example illustrates, Type II error and power are closely linked.

Perfect Potatoes

Type II error and power

The potato-chip producer wonders whether the significance test of $H_0: p = 0.08$ versus $H_a: p > 0.08$ based on a random sample of 500 potatoes has enough power to detect a shipment with, say, 11% blemished potatoes. In this case, a particular Type II error is to fail to reject $H_0: p = 0.08$ when $p = 0.11$. Figure 9.4 shows two sampling distributions of $\hat{p}$, one when $p = 0.08$ and the other when $p = 0.11$.

Earlier, we decided to reject H_0 if our sample yielded a value of $\hat{p}$ to the right of the green line. That decision was based on using a significance level (Type I error probability) of $\alpha = 0.05$. Now look at the sampling distribution for $p = 0.11$. The shaded area represents the probability of correctly rejecting H_0 when $p = 0.11$. That is, the *power* of this test to detect $p = 0.11$ is about 0.75. In other words, the potato-chip producer has roughly a 3-in-4 chance of rejecting a truckload with

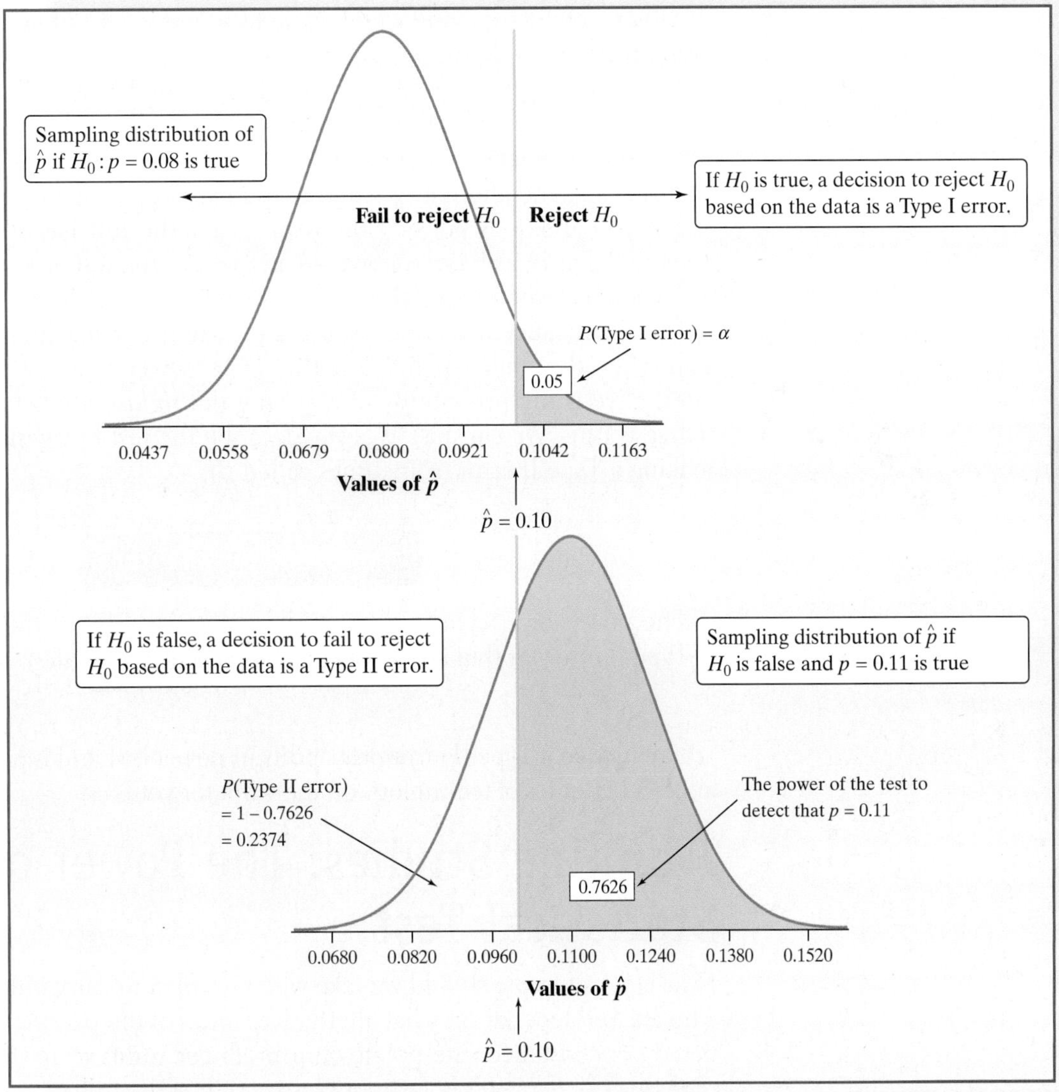

FIGURE 9.4 The power of the test (shaded area) is the probability that it correctly rejects $H_0: p = 0.08$ when the truth is $p = 0.11$. In this case, power $= 0.7626$. The probability of making a Type II error (white area) is $1 - 0.7626 = 0.2374$.

11% blemished potatoes based on a random sample of 500 potatoes from the shipment.

We would fail to reject H_0 if the sample proportion $\hat{p}$ falls to the left of the green line. The white area shows the probability of failing to reject H_0 when H_0 is false. This is the probability of a Type II error. The potato-chip producer has about a 1-in-4 chance of failing to send away a shipment with 11% blemished potatoes.

The potato-chip producer decided that it is important to detect when a shipment contains 11% blemished potatoes. Suppose the company had decided instead that detecting a shipment with 10% blemished potatoes is important. Obviously, our power calculations would have been different. (Can you tell whether the power of the test to detect $p = 0.10$ would be higher or lower than for $p = 0.11$?) Remember: the power of a test gives the probability of detecting a specific alternative value of the parameter. The choice of that alternative value is usually made by someone with a vested interest in the situation (like the potato-chip producer).

After reading the example, you might be wondering whether 0.76 is a high power or a low power. That depends on how certain the potato-chip producer wants to be to detect a shipment with 11% blemished potatoes. The power of a test against a specific alternative value of the parameter (like $p = 0.11$) is a number between 0 and 1. A power close to 0 means the test has almost no chance of detecting that H_0 is false. A power near 1 means the test is very likely to reject H_0 in favor of H_a when H_0 is false.

The significance level of a test is the probability of reaching the *wrong* conclusion when the null hypothesis is true. The power of a test to detect a specific alternative is the probability of reaching the *right* conclusion when that alternative is true. We can just as easily describe the test by giving the probability of making a Type II error (sometimes called β).

Power and Type II Error

The power of a test against any alternative is 1 minus the probability of a Type II error for that alternative; that is, power $= 1 - \beta$.

Calculating a Type II error probability or power by hand is possible but unpleasant. It's better to let technology do the work for you.

Planning Studies: The Power of a Statistical Test

How large a sample should we take when we plan to carry out a significance test? The answer depends on what alternative values of the parameter are important to detect. For instance, the potato-chip producer wants to have a good chance of rejecting $H_0: p = 0.08$ in favor of $H_a: p > 0.08$ if the true proportion of blemished potatoes in a shipment is $p = 0.11$. In the last example, we found that the power of the test to detect $p = 0.11$ using a random sample of size $n = 500$ and a significance level of $\alpha = 0.05$ is about 0.76.

Here are the questions we must answer to decide how many observations we need:

1. *Significance level.* How much protection do we want against a Type I error—getting a significant result from our sample when H_0 is actually true? By using $\alpha = 0.05$, the potato-chip producer has a 5% chance ($\alpha = 0.05$) of making a Type I error.
2. *Practical importance.* How large a difference between the hypothesized parameter value and the actual parameter value is important in practice? The chip producer feels that it's important to detect a shipment with 11% blemished potatoes—a difference of 3% from the hypothesized value of $p = 0.08$.
3. *Power.* How confident do we want to be that our study will detect a difference of the size we think is important?

Let's illustrate typical answers to these questions using an example.

Developing Stronger Bones

Planning a study

Can a six-month exercise program increase the total body bone mineral content (TBBMC) of young women? A team of researchers is planning a study to examine this question. The researchers would like to perform a test of

$$H_0: \mu = 0$$
$$H_a: \mu > 0$$

where μ is the true mean percent change in TBBMC due to the exercise program. To decide how many subjects they should include in their study, researchers begin by answering the three questions above.

1. *Significance level.* The researchers decide that $\alpha = 0.05$ gives enough protection against declaring that the exercise program increases bone mineral content when it really doesn't (a Type I error).
2. *Practical importance.* A mean increase in TBBMC of 1% would be considered important.
3. *Power.* The researchers want probability at least 0.9 that a test at the chosen significance level will reject the null hypothesis $H_0: \mu = 0$ when the truth is $\mu = 1$.

The following Activity gives you a chance to investigate the sample size needed to achieve a power of 0.9 in the bone mineral content study.

ACTIVITY *Investigating power*

MATERIALS: Computer with Internet connection and display capability

In this Activity, you will use the *Power* applet at the book's Web site to determine the sample size needed for the exercise study of the previous example. Based on the results of a previous study, researchers are willing to assume that $\sigma = 2$ for the percent change in TBBMC over the six-month period. We'll start by seeing whether or not 25 subjects are enough.

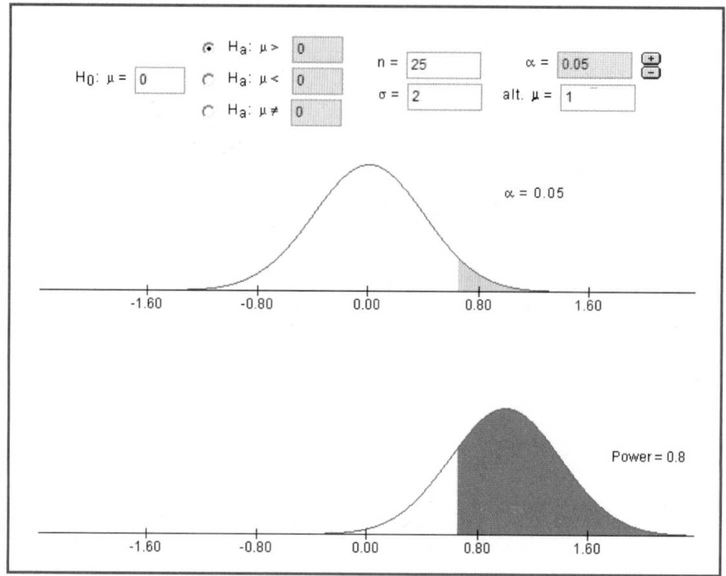

1. Enter the values: $H_0: \mu = 0$, $H_a: \mu > 0$, $n = 25$, $\sigma = 2$, $\alpha = 0.05$, and alternate $\mu = 1$. What is the power? As a class, discuss what this number means in simple terms.

2. Change the significance level to 0.01. What effect does this have on the power of the test to detect $\mu = 1$? Why?

3. The researchers decide that they are willing to risk only a 5% chance of making a Type I error. Change the significance level back to $\alpha = 0.05$. Now increase the sample size to 30. What happens to the power? Why?

4. Keep increasing the sample size until the power is at least 0.90. What minimum sample size should the researchers use for their study?

5. Would the researchers need a smaller or a larger sample size to detect a mean increase of 1.5% in TBBMC? A 0.85% increase? Use the applet to investigate.

6. Summarize what you have learned about ways to increase the power of a significance test.

Step 2 of the Activity reveals an important link between Type I and Type II error probabilities. Since $P(\text{Type I error}) = \alpha$, decreasing the significance level decreases the risk of making a Type I error. As the applet shows, this change also decreases the power of the test to detect $\mu = 1$. Since $P(\text{Type II error}) = 1 -$ power, lower power means a greater chance of making a Type II error. So decreasing the Type I error probability α increases the Type II error probability β. By the same logic, increasing the risk of a Type I error results in a lower chance of a Type II error.

Step 5 of the Activity makes it clear that smaller differences require larger samples to reach 90% power. Here is a summary of influences on "How large a sample do I need?"

- If you insist on a smaller significance level (such as 1% rather than 5%), you have to take a larger sample. A smaller significance level requires stronger evidence to reject the null hypothesis.

For a more in-depth look at the power of a test, search online for the WISE Statistical Power applet. Start with the demonstration guide offered at the site.

• If you insist on higher power (such as 99% rather than 90%), you will need a larger sample. Higher power gives a better chance of detecting a difference when it is really there.

• At any significance level and desired power, detecting a small difference requires a larger sample than detecting a large difference.

Our best advice for maximizing the power of a test is to choose as high an α level (Type I error probability) as you are willing to risk *and* as large a sample size as you can afford.

SECTION 9.1 Summary

- A **significance test** assesses the evidence provided by data against a **null hypothesis** H_0 in favor of an **alternative hypothesis** H_a.
- The hypotheses are stated in terms of population parameters. Often, H_0 is a statement of no change or no difference. H_a says that a parameter differs from its null hypothesis value in a specific direction (**one-sided alternative**) or in either direction (**two-sided alternative**).
- The reasoning of a significance test is as follows. Suppose that the null hypothesis is true. If we repeated our data production many times, would we often get data as inconsistent with H_0 as the data we actually have? If the data are unlikely when H_0 is true, they provide evidence against H_0.
- The ***P*-value** of a test is the probability, computed supposing H_0 to be true, that the statistic will take a value at least as extreme as that actually observed in the direction specified by H_a.
- Small *P*-values indicate strong evidence against H_0. To calculate a *P*-value, we must know the sampling distribution of the test statistic when H_0 is true. There is no universal rule for how small a *P*-value in a significance test provides convincing evidence against the null hypothesis.
- If the *P*-value is smaller than a specified value α (called the **significance level**), the data are **statistically significant** at level α. In that case, we can **reject** H_0. If the *P*-value is greater than or equal to α, we **fail to reject** H_0.
- A **Type I error** occurs if we reject H_0 when it is in fact true. A **Type II error** occurs if we fail to reject H_0 when it is actually false.
- In a fixed level α significance test, the probability of a Type I error is the significance level α.
- The **power** of a significance test against a specific alternative is the probability that the test will reject H_0 when the alternative is true. Power measures the ability of the test to detect an alternative value of the parameter. For a specific alternative, $P(\text{Type II error}) = 1 - \text{power}$.
- When you plan a statistical study, plan the inference as well. In particular, ask what sample size you need for successful inference. Increasing the size of the sample increases the power (reduces the probability of a Type II error) when the significance level remains fixed. We can also increase the power of a test by using a higher significance level (say, $\alpha = 0.10$ instead of $\alpha = 0.05$), which means increasing the risk of a Type I error.

SECTION 9.1 Exercises

In Exercises 1 to 6, each situation calls for a significance test. State the appropriate null hypothesis H_0 and alternative hypothesis H_a in each case. Be sure to define your parameter each time.

1. **Lefties** Simon reads a newspaper report claiming that 12% of all adults in the United States are left-handed. He wonders if 12% of the students at his large public high school are left-handed. Simon chooses an SRS of 100 students and records whether each student is right- or left-handed.

2. **Don't argue!** A Gallup Poll report on a national survey of 1028 teenagers revealed that 72% of teens said they seldom or never argue with their friends.[7] Yvonne wonders whether this national result would be true in her large high school. So she surveys a random sample of 150 students at her school.

pg 531

3. **Attitudes** The Survey of Study Habits and Attitudes (SSHA) is a psychological test that measures students' attitudes toward school and study habits. Scores range from 0 to 200. The mean score for U.S. college students is about 115. A teacher suspects that older students have better attitudes toward school. She gives the SSHA to an SRS of 45 of the over 1000 students at her college who are at least 30 years of age.

4. **Anemia** Hemoglobin is a protein in red blood cells that carries oxygen from the lungs to body tissues. People with less than 12 grams of hemoglobin per deciliter of blood (g/dl) are anemic. A public health official in Jordan suspects that Jordanian children are at risk of anemia. He measures a random sample of 50 children.

5. **Cold cabin?** During the winter months, the temperatures at the Colorado cabin owned by the Starnes family can stay well below freezing (32°F or 0°C) for weeks at a time. To prevent the pipes from freezing, Mrs. Starnes sets the thermostat at 50°F. The manufacturer claims that the thermostat allows variation in home temperature of $\sigma = 3$°F. Mrs. Starnes suspects that the manufacturer is overstating how well the thermostat works.

6. **Ski jump** When ski jumpers take off, the distance they fly varies considerably depending on their speed, skill, and wind conditions. Event organizers must position the landing area to allow for differences in the distances that the athletes fly. For a particular competition, the organizers estimate that the variation in distance flown by the athletes will be $\sigma = 10$ meters. An experienced jumper thinks that the organizers are underestimating the variation.

In Exercises 7 to 10, explain what's wrong with the stated hypotheses. Then give correct hypotheses.

7. **Better parking** A change is made that should improve student satisfaction with the parking situation at a local high school. Right now, 37% of students approve of the parking that's provided. The null hypothesis $H_0: p > 0.37$ is tested against the alternative $H_a: p = 0.37$.

8. **Better parking** A change is made that should improve student satisfaction with the parking situation at your school. Right now, 37% of students approve of the parking that's provided. The null hypothesis $H_0: \hat{p} = 0.37$ is tested against the alternative $H_a: \hat{p} \neq 0.37$.

9. **Birth weights** In planning a study of the birth weights of babies whose mothers did not see a doctor before delivery, a researcher states the hypotheses as

$$H_0: \bar{x} = 1000 \text{ grams}$$
$$H_a: \bar{x} < 1000 \text{ grams}$$

10. **Birth weights** In planning a study of the birth weights of babies whose mothers did not see a doctor before delivery, a researcher states the hypotheses as

$$H_0: \mu < 1000 \text{ grams}$$
$$H_a: \mu = 900 \text{ grams}$$

pg 534

11. **Lefties** Refer to Exercise 1. In Simon's SRS, 16 of the students were left-handed. A significance test yields a *P*-value of 0.2184.
    (a) Interpret this result in context.
    (b) Do the data provide convincing evidence against the null hypothesis? Explain.

12. **Don't argue!** Refer to Exercise 2. For Yvonne's survey, 96 students in the sample said they rarely or never argue with friends. A significance test yields a *P*-value of 0.0291.
    (a) Interpret this result in context.
    (b) Do the data provide convincing evidence against the null hypothesis? Explain.

13. **Attitudes** In the study of older students' attitudes from Exercise 3, the sample mean SSHA score was 125.7 and the sample standard deviation was 29.8. A significance test yields a *P*-value of 0.0101.

pg 536

(a) Interpret the *P*-value in context.

(b) What conclusion would you make if $\alpha = 0.05$? If $\alpha = 0.01$? Justify your answer.

14. **Anemia** For the study of Jordanian children in Exercise 4, the sample mean hemoglobin level was 11.3 g/dl and the sample standard deviation was 1.6 g/dl. A significance test yields a *P*-value of 0.0016.

(a) Interpret the *P*-value in context.

(b) What conclusion would you make if $\alpha = 0.05$? $\alpha = 0.01$? Justify your answer.

15. **Is this what *P* means?** When asked to explain the meaning of the *P*-value in Exercise 13, a student says, "This means there is only probability 0.01 that the null hypothesis is true." Explain clearly why the student's explanation is wrong.

16. **Statistical significance** Asked to explain the meaning of "statistically significant at the $\alpha = 0.05$ level," a student says, "This means that the probability that the null hypothesis is true is less than 0.05." Is this explanation correct? Why or why not?

17. **Statistical significance** Explain in plain language why a significance test that is significant at the 1% level must always be significant at the 5% level. If a test is significant at the 5% level, what can you say about its significance at the 1% level?

18. ***P*-values and statistical significance** Write a few sentences comparing what the *P*-values in Exercises 13 and 14 tell you about statistical significance in each of the two studies.

Exercises 19 and 20 refer to the following setting. Slow response times by paramedics, firefighters, and policemen can have serious consequences for accident victims. In the case of life-threatening injuries, victims generally need medical attention within 8 minutes of the accident. Several cities have begun to monitor emergency response times. In one such city, the mean response time to all accidents involving life-threatening injuries last year was $\mu = 6.7$ minutes. Emergency personnel arrived within 8 minutes after 78% of all calls involving life-threatening injuries last year. The city manager shares this information and encourages these first responders to "do better." At the end of the year, the city manager selects an SRS of 400 calls involving life-threatening injuries and examines the response times.

19. **Awful accidents**

pg 538

(a) State hypotheses for a significance test to determine whether the average response time has decreased. Be sure to define the parameter of interest.

(b) Describe a Type I error and a Type II error in this setting, and explain the consequences of each.

(c) Which is more serious in this setting: a Type I error or a Type II error? Justify your answer.

20. **Awful accidents**

(a) State hypotheses for a significance test to determine whether first responders are arriving within 8 minutes of the call more often. Be sure to define the parameter of interest.

(b) Describe a Type I error and a Type II error in this setting and explain the consequences of each.

(c) Which is more serious in this setting: a Type I error or a Type II error? Justify your answer.

(d) If you sustain a life-threatening injury due to a vehicle accident, you want to receive medical treatment as quickly as possible. Which of the two significance tests—$H_0: \mu = 6.7$ versus $H_a: \mu < 6.7$ or the one from part (a) of this exercise—would you be more interested in? Justify your answer.

21. **Opening a restaurant** You are thinking about opening a restaurant and are searching for a good location. From research you have done, you know that the mean income of those living near the restaurant must be over \$85,000 to support the type of upscale restaurant you wish to open. You decide to take a simple random sample of 50 people living near one potential location. Based on the mean income of this sample, you will decide whether to open a restaurant there.[8]

(a) State appropriate null and alternative hypotheses. Be sure to define your parameter.

(b) Describe a Type I and a Type II error, and explain the consequences of each.

(c) If you had to choose one of the "standard" significance levels for your significance test, would you choose $\alpha = 0.01, 0.05$, or 0.10? Justify your choice.

22. **Blood pressure screening** Your company markets a computerized device for detecting high blood pressure. The device measures an individual's blood pressure once per hour at a randomly selected time throughout a 12-hour period. Then it calculates the mean systolic (top number) pressure for the sample of measurements. Based on the sample results, the device determines whether there is significant evidence that the individual's actual mean systolic pressure is greater than 130. If so, it recommends that the person seek medical attention.

(a) State appropriate null and alternative hypotheses in this setting. Be sure to define your parameter.

(b) Describe a Type I and a Type II error, and explain the consequences of each.

(c) The blood pressure device can be adjusted to decrease one error probability at the cost of an increase in the other error probability. Which error probability would you choose to make smaller, and why?

23. **Error probabilities** You read that a statistical test at significance level $\alpha = 0.05$ has power 0.78. What are the probabilities of Type I and Type II errors for this test?

24. **Error probabilities** You read that a statistical test at the $\alpha = 0.01$ level has probability 0.14 of making a Type II error when a specific alternative is true. What is the power of the test against this alternative?

25. **Power** A drug manufacturer claims that fewer than 10% of patients who take its new drug for treating Alzheimer's disease will experience nausea. To test this claim, a significance test is carried out of

$$H_0: p = 0.10$$
$$H_a: p < 0.10$$

You learn that the power of this test at the 5% significance level against the alternative $p = 0.08$ is 0.64.

(a) Explain in simple language what "power = 0.64" means in this setting.

(b) You could get higher power against the same alternative with the same α by changing the number of measurements you make. Should you make more measurements or fewer to increase power? Explain.

(c) If you decide to use $\alpha = 0.01$ in place of $\alpha = 0.05$, with no other changes in the test, will the power increase or decrease? Justify your answer.

(d) If you shift your interest to the alternative $p = 0.07$ with no other changes, will the power increase or decrease? Justify your answer.

26. **What is power?** You manufacture and sell a liquid product whose electrical conductivity is supposed to be 5. You plan to make 6 measurements of the conductivity of each lot of product. If the product meets specifications, the mean of many measurements will be 5. You will therefore test

$$H_0: \mu = 5$$
$$H_a: \mu \neq 5$$

If the true conductivity is 5.1, the liquid is not suitable for its intended use. You learn that the power of your test at the 5% significance level against the alternative $\mu = 5.1$ is 0.23.

(a) Explain in simple language what "power = 0.23" means in this setting.

(b) You could get higher power against the same alternative with the same α by changing the number of measurements you make. Should you make more measurements or fewer to increase power?

(c) If you decide to use $\alpha = 0.10$ in place of $\alpha = 0.05$, with no other changes in the test, will the power increase or decrease? Justify your answer.

(d) If you shift your interest to the alternative $\mu = 5.2$, with no other changes, will the power increase or decrease? Justify your answer.

Multiple choice: Select the best answer for Exercises 27 to 30.

27. Experiments on learning in animals sometimes measure how long it takes mice to find their way through a maze. The mean time is 18 seconds for one particular maze. A researcher thinks that a loud noise will cause the mice to complete the maze faster. She measures how long each of 10 mice takes with a noise as stimulus. The sample mean is $\bar{x} = 16.5$ seconds. The appropriate hypotheses for the significance test are

(a) $H_0: \mu = 18$; $H_a: \mu \neq 18$.

(b) $H_0: \mu = 16.5$; $H_a: \mu < 18$.

(c) $H_0: \mu < 18$; $H_a: \mu = 18$.

(d) $H_0: \mu = 18$; $H_a: \mu < 18$.

(e) $H_0: \bar{x} = 18$; $H_a: \bar{x} < 18$.

28. You use technology to carry out a significance test and get a *P*-value of 0.031. The correct conclusion is

(a) accept H_a at the $\alpha = 0.05$ significance level.

(b) reject H_0 at the $\alpha = 0.05$ significance level.

(c) reject H_0 at the $\alpha = 0.01$ significance level.

(d) fail to reject H_0 at the $\alpha = 0.05$ significance level.

(e) fail to reject H_a at the $\alpha = 0.05$ significance level.

29. Vigorous exercise helps people live several years longer (on average). Whether mild activities like slow walking extend life is not clear. Suppose that the added life expectancy from regular slow walking is just 2 months. A statistical test is more likely to find a significant increase in mean life expectancy if

(a) it is based on a very large random sample and a 5% significance level is used.

(b) it is based on a very large random sample and a 1% significance level is used.

(c) it is based on a very small random sample and a 5% significance level is used.

(d) it is based on a very small random sample and a 1% significance level is used.

(e) the size of the sample doesn't have any effect on the significance of the test.

30. A researcher plans to conduct a significance test at the $\alpha = 0.01$ significance level. She designs her study to have a power of 0.90 at a particular alternative value of the parameter of interest. The probability

that the researcher will commit a Type II error for the particular alternative value of the parameter at which she computed the power is

(a) 0.01. (b) 0.10. (c) 0.89. (d) 0.90. (e) 0.99.

31. **Women in math** (5.3) Of the 16,701 degrees in mathematics given by U.S. colleges and universities in a recent year, 73% were bachelor's degrees, 21% were master's degrees, and the rest were doctorates. Moreover, women earned 48% of the bachelor's degrees, 42% of the master's degrees, and 29% of the doctorates.[9]

(a) How many of the mathematics degrees given in this year were earned by women? Justify your answer.

(b) Are the events "degree earned by a woman" and "degree was a master's degree" independent? Justify your answer using appropriate probabilities.

(c) If you choose 2 of the 16,701 mathematics degrees at random, what is the probability that at least 1 of the 2 degrees was earned by a woman? Show your work.

32. **Explaining confidence** (8.2) Here is an explanation from a newspaper concerning one of its opinion polls. Explain briefly but clearly in what way this explanation is incorrect.

For a poll of 1,600 adults, the variation due to sampling error is no more than three percentage points either way. The error margin is said to be valid at the 95 percent confidence level. This means that, if the same questions were repeated in 20 polls, the results of at least 19 surveys would be within three percentage points of the results of this survey.

Tests about a Population Proportion

In Section 9.2, you'll learn about:

- Carrying out a significance test
- The one-sample z test for a proportion
- Two-sided tests
- Why confidence intervals give more information

Confidence intervals and significance tests are based on the sampling distributions of statistics. That is, both use probability to say what would happen if we applied the inference method many times. Section 9.1 presented the reasoning of significance tests, including the idea of a *P*-value. In this section, we focus on the details of testing a claim about a population proportion.

Carrying Out a Significance Test

In Section 9.1, we met a virtual basketball player who claimed to make 80% of his free throws. We thought that he might be exaggerating. In an SRS of 50 shots, the player made only 32. His sample proportion of made free throws was therefore

$$\hat{p} = \frac{32}{50} = 0.64$$

This result is much lower than what he claimed. Does it provide *convincing* evidence against the player's claim? To find out, we need to perform a significance test of

$$H_0: p = 0.80$$
$$H_a: p < 0.80$$

where p = the actual proportion of free throws that the shooter makes in the long run.

Conditions In Chapter 8, we introduced three conditions that should be met before we construct a confidence interval for an unknown population proportion: Random, Normal, and Independent. These same three conditions must be verified before carrying out a significance test. The Normal condition for proportions requires that both np and $n(1-p)$ be at least 10. Since we assume H_0 is true when performing a significance test, we use the parameter value specified by the null hypothesis (sometimes called p_0) when checking the Normal condition. In this case, the Normal condition says that the expected number of successes np_0 and failures $n(1-p_0)$ are both at least 10.

EXAMPLE

I'm a Great Free-Throw Shooter!

Checking conditions

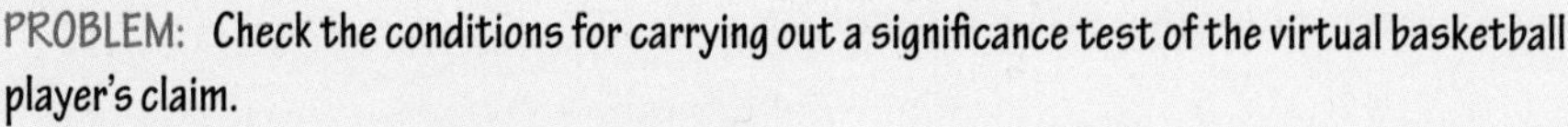

PROBLEM: Check the conditions for carrying out a significance test of the virtual basketball player's claim.

SOLUTION: The three required conditions are

- *Random* We can view this set of 50 shots as a simple random sample from the population of all possible shots that the virtual shooter takes.
- *Normal* Assuming H_0 is true, $p = 0.80$. Then $np = (50)(0.80) = 40$ and $n(1-p) = (50)(0.20) = 10$ are both at least 10, so the Normal condition is met.
- *Independent* The outcome of each shot is determined by the computer's random number generator, so individual observations are independent. We're not sampling without replacement from a finite population (since the applet can keep on shooting), so we don't need to check the 10% condition.

For Practice *Try Exercise 33*

If the null hypothesis $H_0: p = 0.80$ is true, then the player's sample proportion $\hat{p}$ of made free throws in an SRS of 50 shots would vary according to an approximately Normal sampling distribution with mean $\mu_{\hat{p}} = p = 0.80$ and standard deviation

$$\sigma_{\hat{p}} = \sqrt{\frac{p(1-p)}{n}} = \sqrt{\frac{(0.8)(0.2)}{50}} = 0.0566.$$

Figure 9.5 displays the sampling distribution. We have added the player's sample result, $\hat{p} = \frac{32}{50} = 0.64$.

Sampling distribution of $\hat{p}$ if $p = 0.8$

0.6302 0.6868 0.7434 0.8000 0.8566 0.9132 0.9698

$\hat{p} = 0.64$

FIGURE 9.5 The sampling distribution of the proportion $\hat{p}$ of made shots in random samples of 50 free throws by an 80% shooter.

Calculations: Test Statistic and *P*-Value A significance test uses sample data to measure the strength of evidence against H_0. Here are some principles that apply to most tests:

- The test compares a statistic calculated from sample data with the value of the parameter stated by the null hypothesis.
- Values of the statistic far from the null parameter value in the direction specified by the alternative hypothesis give evidence against H_0.

• To assess how far the statistic is from the parameter, standardize the statistic. This standardized value is called the **test statistic:**

$$\text{test statistic} = \frac{\text{statistic} - \text{parameter}}{\text{standard deviation of statistic}}$$

DEFINITION: Test statistic

A **test statistic** measures how far a sample statistic diverges from what we would expect if the null hypothesis H_0 were true, in standardized units. That is,

$$\text{test statistic} = \frac{\text{statistic} - \text{parameter}}{\text{standard deviation of statistic}}$$

The test statistic says how far the sample result is from the null parameter value, and in what direction, on a standardized scale. You can use the test statistic to find the *P*-value of the test, as the following example shows.

I'm a Great Free-Throw Shooter!

Computing the test statistic

PROBLEM: **In an SRS of 50 free throws, the virtual player made 32.**

(a) Calculate the test statistic.

(b) Find and interpret the *P*-value.

SOLUTION:

(a) His sample proportion of made shots, $\hat{p} = 0.64$, is pretty far below the hypothesized value $H_0: p = 0.80$. Standardizing, we get

$$\text{test statistic} = \frac{\text{statistic} - \text{parameter}}{\text{standard deviation of statistic}}$$

$$z = \frac{0.64 - 0.80}{0.0566} = -2.83$$

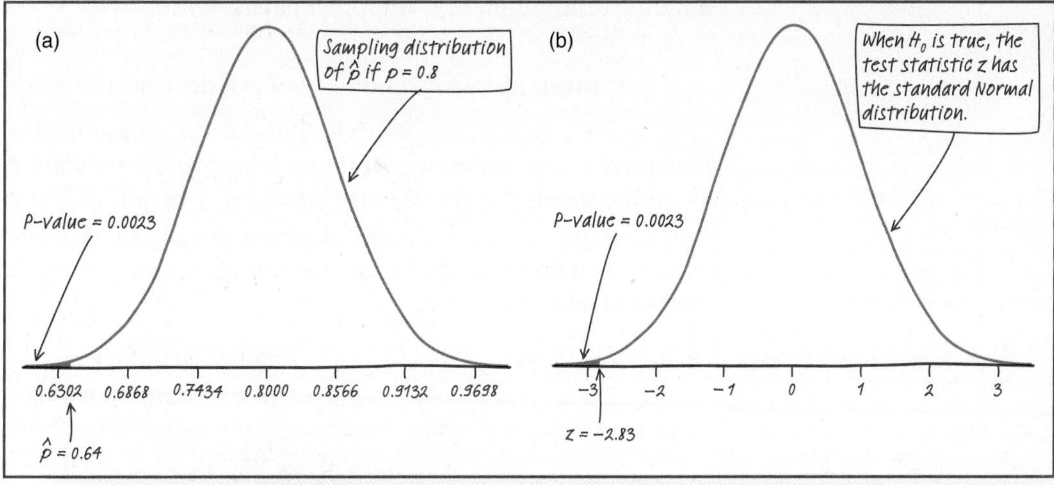

FIGURE 9.6 The shaded area shows the *P*-value for the player's sample proportion of made shots (a) on the sampling distribution of $\hat{p}$ from Figure 9.5 and (b) on the standard Normal curve.

You can also compute the P-value using the calculator's `normalcdf` command. To find $P(z \le -2.83)$ on the standard Normal distribution, use a value many standard deviations below the mean as the lower bound, like -100. The required command is `normalcdf(-100, -2.83,0,1)`, or just `normalcdf(-100, -2.83)`, since the calculator uses $\mu = 0$ and $\sigma = 1$ as the default values.

(b) The shaded area under the curve in Figure 9.6(a) shows the P-value. Figure 9.6(b) shows the corresponding area on the standard Normal curve, which displays the distribution of the z test statistic. Using Table A, we find that the P-value is $P(z \le -2.83) = 0.0023$.

So if H_0 is true, and the player makes 80% of his free throws in the long run, there's only about a 2-in-1000 chance that the player would make as few as 32 of 50 shots.

Earlier, we used simulation to estimate the P-value as 0.0075. As the example shows, the P-value is even smaller, 0.0023. Either way, it's very unlikely for an 80% free-throw shooter to make 32 or fewer shots in an SRS of 50 attempts. This result confirms our earlier decision to reject H_0 and to conclude that the player is exaggerating.

The One-Sample z Test for a Proportion

As you have probably guessed, we will follow the familiar four-step process whenever a problem asks us to perform a significance test.

Significance Tests: A Four-Step Process

State: What *hypotheses* do you want to test, and at what significance level? Define any *parameters* you use.

Plan: Choose the appropriate inference *method*. Check *conditions*.

Do: If the conditions are met, perform *calculations*.

- Compute the **test statistic.**
- Find the **P-value.**

Conclude: *Interpret* the results of your test in the context of the problem.

When the conditions are met—Random, Normal, and Independent—the sampling distribution of $\hat{p}$ is approximately Normal with

$$\text{mean } \mu_{\hat{p}} = p \quad \text{and} \quad \text{standard deviation } \sigma_{\hat{p}} = \sqrt{\frac{p(1-p)}{n}}$$

For confidence intervals, we substitute $\hat{p}$ for p in the standard deviation formula to obtain the standard error. When performing a significance test, however, the null hypothesis specifies a value for p, which we will call p_0. We assume that this value is correct when performing our calculations. If we standardize the statistic $\hat{p}$ by subtracting its mean and dividing by its standard deviation, we get the test statistic:

$$\text{test statistic} = \frac{\text{statistic} - \text{parameter}}{\text{standard deviation of statistic}}$$

$$z = \frac{\hat{p} - p_0}{\sqrt{\frac{p_0(1-p_0)}{n}}}$$

This z statistic has approximately the standard Normal distribution when H_0 is true. P-values therefore come from the standard Normal distribution. Here is a summary of the details for a **one-sample z test for a proportion.**

One-Sample z Test for a Proportion

Choose an SRS of size n from a large population that contains an unknown proportion p of successes. To test the hypothesis $H_0: p = p_0$, compute the z statistic

$$z = \frac{\hat{p} - p_0}{\sqrt{\frac{p_0(1-p_0)}{n}}}$$

Find the P-value by calculating the probability of getting a z statistic this large or larger in the direction specified by the alternative hypothesis H_a:

$H_a: p > p_0$ — z

$H_a: p < p_0$ — z

$H_a: p \neq p_0$ — $-|z|$, $|z|$

Use this test only when the expected numbers of successes and failures np_0 and $n(1 - p_0)$ are both at least 10 and the population is at least 10 times as large as the sample.

The AP Statistics course outline calls this test a *large-sample test for a proportion* because it is based on a Normal approximation to the sampling distribution of $\hat{p}$ that becomes more accurate as the sample size increases.

If the Normal condition is not satisfied, or if the population is less than 10 times as large as the sample, other procedures should be used. Here is an example of the test in action.

EXAMPLE

One Potato, Two Potato

Performing a significance test about p

STEP 4

The potato-chip producer of Section 9.1 has just received a truckload of potatoes from its main supplier. Recall that if the producer determines that more than 8% of the potatoes in the shipment have blemishes, the truck will be sent away to get another load from the supplier. A supervisor selects a random sample of 500 potatoes from the truck. An inspection reveals that 47 of the potatoes have blemishes. Carry out a significance test at the $\alpha = 0.10$ significance level. What should the producer conclude?

STATE: We want to perform a test at the $\alpha = 0.10$ significance level of

$$H_0: p = 0.08$$
$$H_a: p > 0.08$$

where p is the actual proportion of potatoes in this shipment with blemishes.

PLAN: If conditions are met, we should do a one-sample z test for the population proportion p.

- *Random* The supervisor took a random sample of 500 potatoes from the shipment.
- *Normal* Assuming $H_0: p = 0.08$ is true, the expected numbers of blemished and unblemished potatoes are $np_0 = 500(0.08) = 40$ and $n(1 - p_0) = 500(0.92) = 460$, respectively. Because both of these values are at least 10, we should be safe doing Normal calculations.
- *Independent* Because we are sampling without replacement, we need to check the *10% condition*. It seems reasonable to assume that there are at least $10(500) = 5000$ potatoes in the shipment.

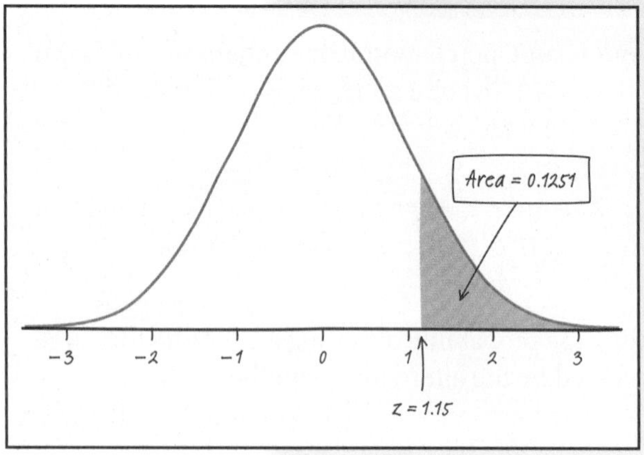

Area = 0.1251

−3 −2 −1 0 1 2 3

z = 1.15

FIGURE 9.7 The *P*-value for the one-sided test.

DO: The sample proportion of blemished potatoes is $\hat{p} = 47/500 = 0.094$

- Test statistic $z = \dfrac{\hat{p} - p_0}{\sqrt{\dfrac{p_0(1-p_0)}{n}}} = \dfrac{0.094 - 0.08}{\sqrt{\dfrac{0.08(0.92)}{500}}} = 1.15$
- *P*-value Figure 9.7 displays the *P*-value as an area under the standard Normal curve for this one-sided test. Using Table A or normalcdf(1.15, 100), the desired *P*-value is $P(z \geq 1.15) = 1 - 0.8749 = 0.1251$.

CONCLUDE: Since our *P*-value, 0.1251, is greater than the chosen significance level of $\alpha = 0.10$, we fail to reject H_0. There is not sufficient evidence to conclude that the shipment contains more than 8% blemished potatoes. The producer will use this truckload of potatoes to make potato chips.

For Practice *Try Exercise 41*

AP EXAM TIP When a significance test leads to a fail to reject H_0 decision, as in this example, be sure to interpret the results as "we don't have enough evidence to conclude H_a." Saying anything that sounds like you believe H_0 is (or might be) true will lead to a loss of credit. And don't write text-message-type responses, like "FTR the H_0."

The preceding example reminds us why significance tests are important. The sample proportion of blemished potatoes was $\hat{p} = 47/500 = 0.094$. This result gave evidence against H_0 in favor of H_a. To see whether such an outcome could occur just by chance when H_0 is true, we had to carry out a significance test. The *P*-value told us that a sample proportion this large or larger would occur in about 12.5% of all random samples of 500 potatoes when H_0 is true. That is, we can't rule out sampling variability as a plausible explanation for getting a sample proportion at least this much greater than 0.08.

THINK ABOUT IT

What happens when the data don't support H_a? Suppose the supervisor had inspected a random sample of 500 potatoes from the shipment and found 33 with blemishes. This yields a sample proportion of $\hat{p} = 33/500 = 0.066$. Such a sample doesn't give evidence that supports the alternative hypothesis $H_a: p > 0.08$! There's no need to continue with the significance test. The conclusion is clear: we should fail to reject $H_0: p = 0.08$. This truckload of potatoes will be used by the potato-chip producer.

If you weren't paying attention, you might end up carrying out the test. Let's see what would happen. The corresponding test statistic is

$$z = \frac{\hat{p} - p_0}{\sqrt{\dfrac{p_0(1-p_0)}{n}}} = \frac{0.066 - 0.08}{\sqrt{\dfrac{0.08(0.92)}{500}}} = -1.15$$

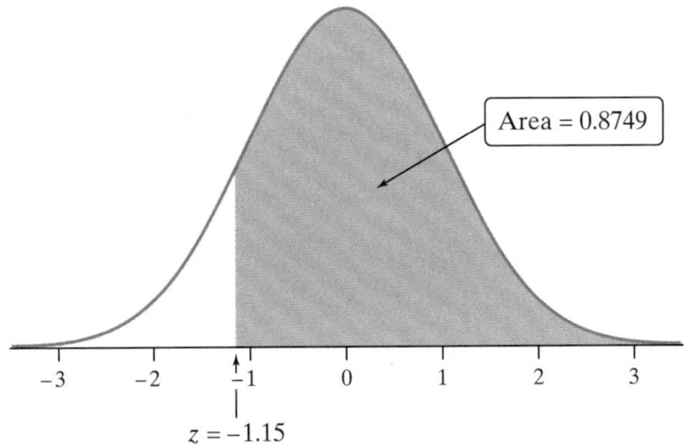

FIGURE 9.8 The *P*-value for the one-sided test.

What's the *P*-value? It's the probability of getting a *z* statistic this large or larger in the direction specified by H_a, $P(z \geq -1.15)$. Figure 9.8 shows this *P*-value as an area under the standard Normal curve. Using Table A or `normalcdf(−1.15,100)`, the *P*-value is $1 - 0.1251 = 0.8749$. There's about an 87.5% chance of getting a sample proportion as large as or larger than $\hat{p} = 0.066$ if $p = 0.08$. As a result, we would fail to reject H_0. Same conclusion, but with lots of unnecessary work!

Always check to see whether the data give evidence against H_0 in the direction specified by H_a before you do calculations.

CHECK YOUR UNDERSTANDING

According to the National Campaign to Prevent Teen and Unplanned Pregnancy, 20% of teens aged 13 to 19 say that they have electronically sent or posted sexually suggestive images of themselves.[10] The counselor at a large high school worries that the actual figure might be higher at her school. To find out, she gives an anonymous survey to a random sample of 250 of the school's 2800 students. All 250 respond, and 63 admit to sending or posting sexual images. Carry out a significance test at the $\alpha = 0.05$ significance level. What conclusion should the counselor draw?

Your calculator will handle the "Do" part of the four-step process, as the following Technology Corner illustrates. However, be sure to read the AP Exam Tip on the next page.

TECHNOLOGY CORNER One-proportion *z* test on the calculator

The TI-83/84 and TI-89 can be used to test a claim about a population proportion. We'll demonstrate using the previous example. In a random sample of size $n = 500$, the supervisor found $X = 47$ potatoes with blemishes. To perform a significance test:

TI-83/84	TI-89
Press STAT, then choose TESTS and `5:1-PropZTest`.	In the Statistics/List Editor, press 2nd F1 ([F6]) and choose `5:1-PropZTest`.

- On the 1-PropZTest screen, enter the values shown: $p_0 = 0.08$, $x = 47$, and $n = 500$. Specify the alternative hypothesis as "prop > p_0." *Note: x* is the number of successes and *n* is the number of trials. Both must be whole numbers!

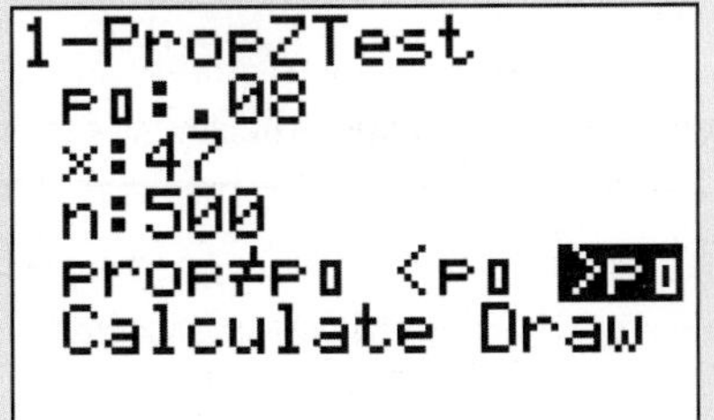

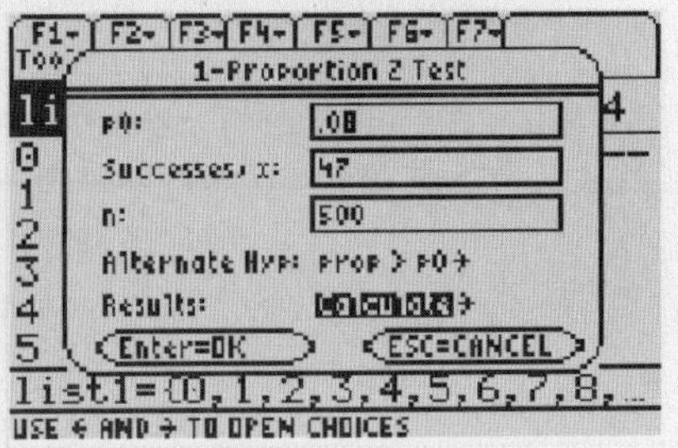

AP EXAM TIP You can use your calculator to carry out the mechanics of a significance test on the AP exam. But there's a risk involved. If you just give the calculator answer with no work, and one or more of your values is incorrect, you will probably get no credit for the "Do" step. We recommend doing the calculation with the appropriate formula and then checking with your calculator. If you opt for the calculator-only method, be sure to name the procedure (one-proportion z test) and to report the test statistic ($z = 1.15$) and P-value (0.1243).

• If you select the "Calculate" choice and press ENTER, you will see that the test statistic is $z = 1.15$ and the P-value is 0.1243.

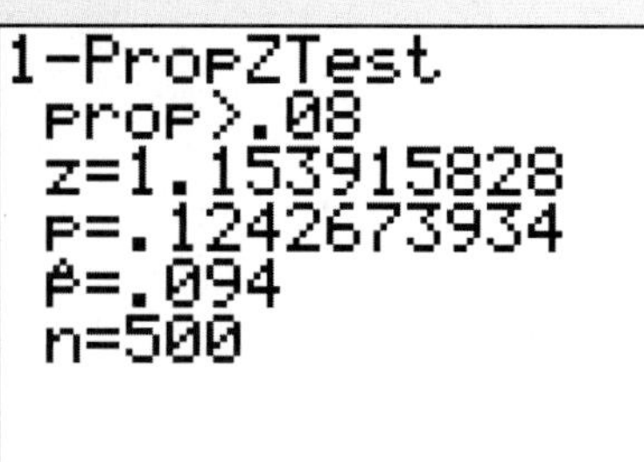

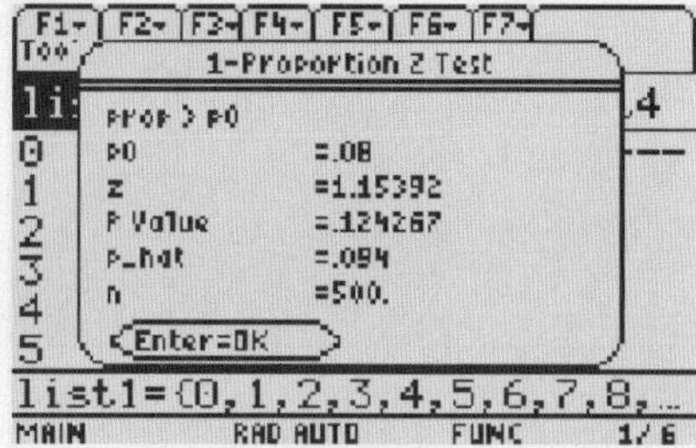

• If you select the "Draw" option, you will see the screen shown here. Compare these results with those in the example.

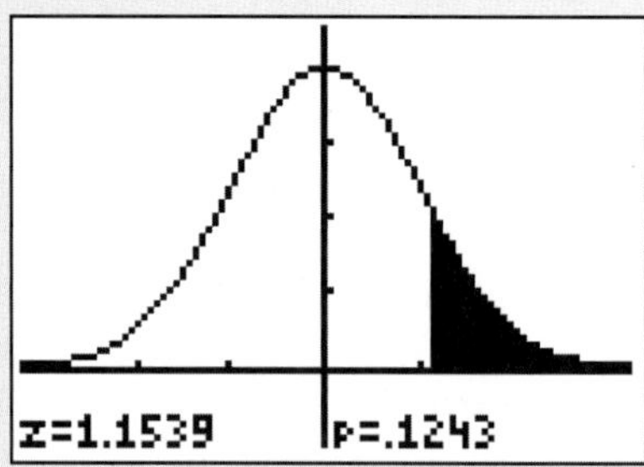

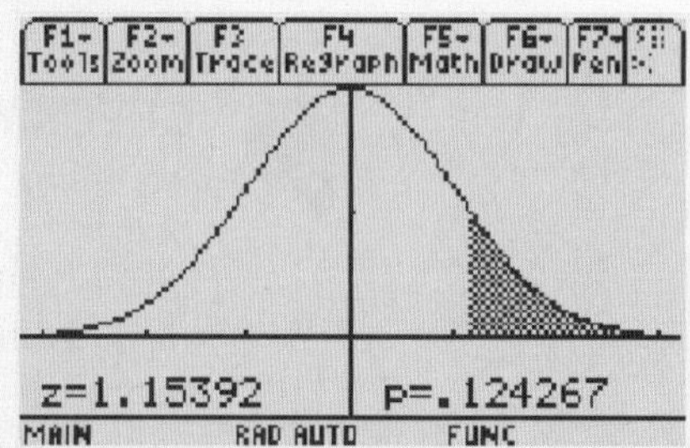

TI-Nspire instructions in Appendix B

Two-Sided Tests

Both the free-throw shooter and blemished-potato examples involved one-sided tests. Let's look at a situation that requires a two-sided test.

Nonsmokers

A two-sided test

According to the Centers for Disease Control and Prevention (CDC) Web site, 50% of high school students have never smoked a cigarette. Taeyeon wonders whether this national result holds true in his large, urban high school. For his AP Statistics class project, Taeyeon surveys an SRS of 150 students from his school. He gets responses from all 150 students, and 90 say that they have never smoked a cigarette. What should Taeyeon conclude? Give appropriate evidence to support your answer.

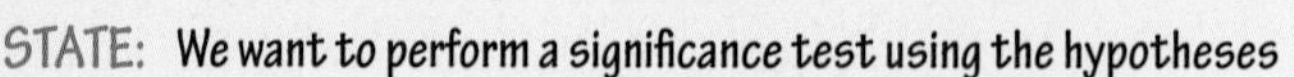

STATE: **We want to perform a significance test using the hypotheses**

$$H_0: p = 0.50$$
$$H_a: p \neq 0.50$$

where p = the actual proportion of students in Taeyeon's school who would say they have never smoked cigarettes. Because no significance level was stated, we will use $\alpha = 0.05$.

PLAN: **If conditions are met, we'll do a one-sample z test for the population proportion p.**

• *Random* **Taeyeon surveyed an SRS of 150 students from his school.**

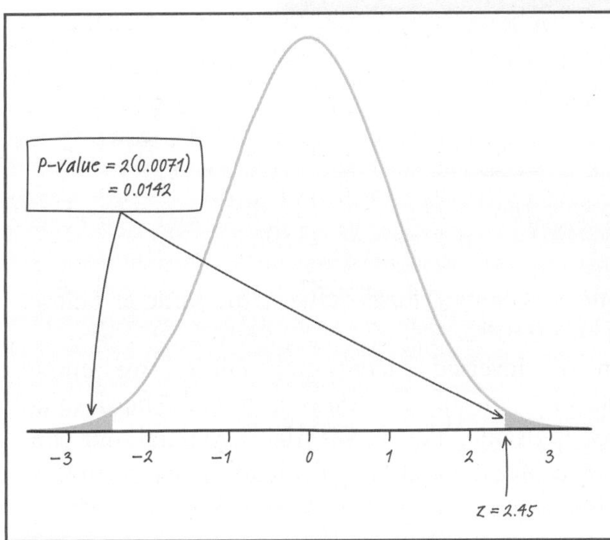

FIGURE 9.9 The *P*-value for the two-sided test.

- *Normal* Assuming $H_0: p = 0.50$ is true, the expected numbers of smokers and nonsmokers in the sample are $np_0 = 150(0.50) = 75$ and $n(1 - p_0) = 150(0.50) = 75$. Since both of these values are at least 10, we should be safe doing Normal calculations.
- *Independent* We are sampling without replacement, so we have to check the 10% condition. It seems reasonable to assume that there are at least 10(150) = 1500 students in a large high school.

DO: The sample proportion is $\hat{p} = 90/150 = 0.60$.

- *Test statistic*

$$z = \frac{\hat{p} - p_0}{\sqrt{\frac{p_0(1-p_0)}{n}}} = \frac{0.60 - 0.50}{\sqrt{\frac{0.50(0.50)}{150}}} = 2.45$$

- *P-value* Figure 9.9 displays the *P*-value as an area under the standard Normal curve for this two-sided test. To compute this *P*-value, we find the area in one tail and double it. Using Table A or normalcdf(2.45, 100) yields $P(z \geq 2.45) = 0.0071$ (the right-tail area). So the desired *P*-value is 2(0.0071) = 0.0142.

You can check the calculations in the "Do" step with a 1-PropZTest on the calculator. Here are the results: $z = 2.449$ and *P*-value = 0.0143.

CONCLUDE: Since our *P*-value, 0.0142, is less than $\alpha = 0.05$, we have sufficient evidence to reject H_0 and conclude that the proportion of students at Taeyeon's school who say they have never smoked differs from the national result of 0.50.

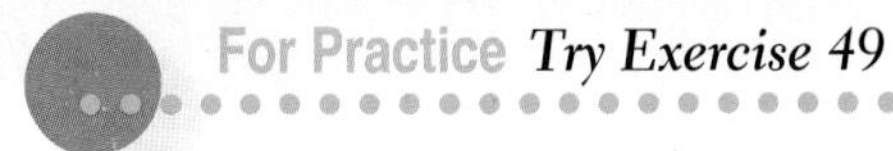
For Practice *Try Exercise 49*

AP EXAM TIP If you are asked to carry out a significance test on the AP exam but no significance level is given, we recommend using $\alpha = 0.05$.

You might have wondered why the potato-chip producer chose a significance level of $\alpha = 0.10$, but we opted for $\alpha = 0.05$ in the nonsmokers example. Recall from Section 9.1 that *P*(Type I error) = α. The potato-chip producer would rather make a Type I error—sending away a shipment that meets quality standards—than a Type II error—using a truckload with too many blemished potatoes to make chips. A greater risk of Type I error means a smaller risk of making a Type II error and higher power to detect a specific alternative value of the parameter. For Taeyeon's nonsmoker study, neither type of error seems very serious. We therefore used R. A. Fisher's default significance level, $\alpha = 0.05$.

THINK ABOUT IT

Does the way we define "success" matter? In the last example, we viewed a nonsmoker as a "success" and a smoker as a "failure." Suppose we reversed the two. Then the new parameter of interest would be p = the proportion of students at Taeyeon's school who have smoked cigarettes. If 90 students said they had never smoked a cigarette, then 60 students must have said they had smoked before. Let's repeat the significance test with this new definition of a success.

The national comparison value for the proportion of cigarette smokers is 0.50, because 50% who smoke means 50% who don't. Our hypotheses are therefore

$$H_0: p = 0.50$$
$$H_a: p \neq 0.50$$

where p = the actual proportion of students in Taeyeon's school who would say they have smoked a cigarette. We don't need to check the Normal condition

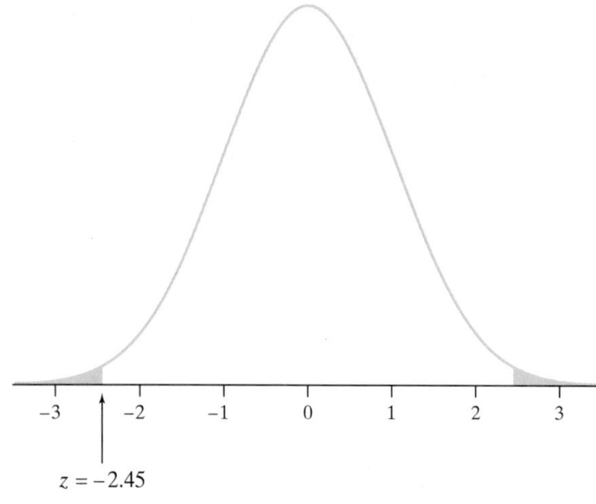

FIGURE 9.10 The *P*-value for the two-sided test.

again, since p_0 is the same as before, 0.50. The new sample proportion of successes is $\hat{p} = 60/150 = 0.40$, which makes the test statistic

$$z = \frac{\hat{p} - p_0}{\sqrt{\frac{p_0(1-p_0)}{n}}} = \frac{0.40 - 0.50}{\sqrt{\frac{0.50(0.50)}{150}}} = -2.45$$

As Figure 9.10 shows, the *P*-value is the same as before: $2(0.0071) = 0.0142$.

When we reversed success and failure, we simply changed the sign of the test statistic z. The *P*-value remained the same. This fact is true in general: our conclusion does not depend on our initial choice of success and failure.

CHECK YOUR UNDERSTANDING

According to the National Institute for Occupational Safety and Health, job stress poses a major threat to the health of workers. In a national survey of restaurant employees, 75% said that work stress had a negative impact on their personal lives.[11] Managers of a large restaurant chain wonder whether this national result holds for their employees. A random sample of 100 employees finds that 68 answer "Yes" when asked, "Does work stress have a negative impact on your personal life?" Is this good reason to think that the proportion of all employees in this chain who would say "Yes" differs from the national proportion of 0.75? Support your answer with a significance test.

Why Confidence Intervals Give More Information

The result of a significance test is basically a decision to reject H_0 or fail to reject H_0. In Taeyeon's smoking study, for instance, the data led us to reject $H_0: p = 0.50$ and conclude that the proportion of students at his school who would say they have never smoked cigarettes differs from the national value. We're left wondering what the actual proportion p might be. A confidence interval might shed some light on this issue.

Nonsmokers

A confidence interval gives more info

Taeyeon found that 90 of an SRS of 150 students said that they had never smoked a cigarette. We checked the conditions for performing the significance test earlier. Before we construct a confidence interval for the population proportion p, we should check that both $n\hat{p}$ and $n(1 - \hat{p})$ are at least 10. Since the number of successes and the number of failures in the sample are 90 and 60, respectively, we can proceed with calculations.

Our 95% confidence interval is

$$\hat{p} \pm z^* \sqrt{\frac{\hat{p}(1-\hat{p})}{n}} = 0.60 \pm 1.96\sqrt{\frac{0.60(0.40)}{150}} = 0.60 \pm 0.078 = (0.522, 0.678)$$

We are 95% confident that the interval from 0.522 to 0.678 captures the true proportion of students at Taeyeon's high school who would say that they have never smoked a cigarette.

The confidence interval in this example is much more informative than the significance test we performed earlier. The interval gives the values of p that are consistent with the sample data. We would not be surprised if the true proportion of students at Taeyeon's school who would say they have never smoked cigarettes was as low as 52.2% or as high as 67.8%. However, we would be surprised if the true proportion was 0.50 because this value is not contained in the confidence interval. Figure 9.11 gives computer output from Minitab software that includes both the results of the significance test and the confidence interval.

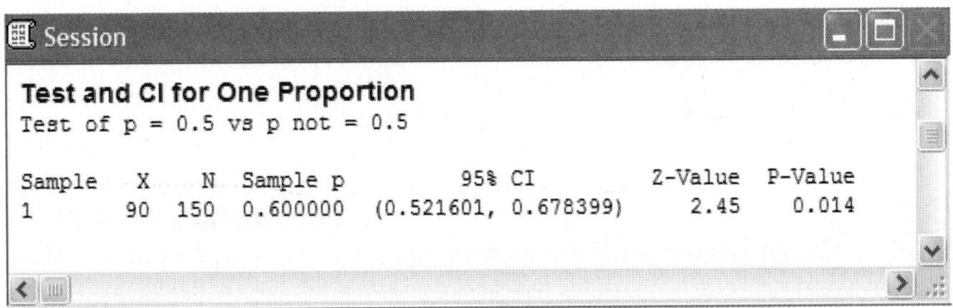
Session

Test and CI for One Proportion

Test of p = 0.5 vs p not = 0.5

Sample	X	N	Sample p	95% CI	Z-Value	P-Value
1	90	150	0.600000	(0.521601, 0.678399)	2.45	0.014

FIGURE 9.11 Minitab output for the two-sided significance test at $\alpha = 0.05$ and a 95% confidence interval for Taeyeon's smoking study.

There is a link between confidence intervals and two-sided tests. The 95% confidence interval (0.522, 0.678) gives an approximate range of p_0's that would not be rejected by a two-sided test at the $\alpha = 0.05$ significance level. With proportions, the link isn't perfect because the standard error used for the confidence interval is based on the sample proportion $\hat{p}$, while the denominator of the test statistic is based on the value p_0 from the null hypothesis.

Test statistic: $z = \dfrac{\hat{p} - p_0}{\sqrt{\dfrac{p_0(1-p_0)}{n}}}$ Confidence interval: $\hat{p} \pm z^* \sqrt{\dfrac{\hat{p}(1-\hat{p})}{n}}$

The big idea is still worth considering: a two-sided test at significance level α (say, 0.05) and a $100(1 - \alpha)\%$ confidence interval (a 95% confidence interval if $\alpha = 0.05$) give similar information about the population parameter.

THINK ABOUT IT

What's the connection between confidence intervals and *two-sided* tests? Earlier, we performed a test of

$$H_0: p = 0.50$$
$$H_a: p \neq 0.50$$

where p is the true proportion of students at Taeyeon's school who would say they have never smoked. Assuming H_0 is true, the sampling distribution of $\hat{p}$ will be approximately Normal with mean $p_0 = 0.5$ and standard deviation $\sqrt{\frac{0.5(0.5)}{150}} = 0.041$. Figure 9.12 shows this sampling distribution, and more.

If we use an $\alpha = 0.05$ significance level for our two-sided test, we will reject H_0 if the sample proportion falls far enough out in either tail of the sampling distribution. How far out would $\hat{p}$ have to be? Far enough from p_0 to give a tail probability less than 0.025, so the *P*-value for the two-sided test would be less than 0.05. To land in one of the reject H_0 regions in Figure 9.12, $\hat{p}$ would have to be more than 1.96 standard deviations from p_0. Because 1.96(0.041) is about 0.08, values of $\hat{p}$ that are less than 0.42 or greater than 0.58 would lead us to reject H_0.

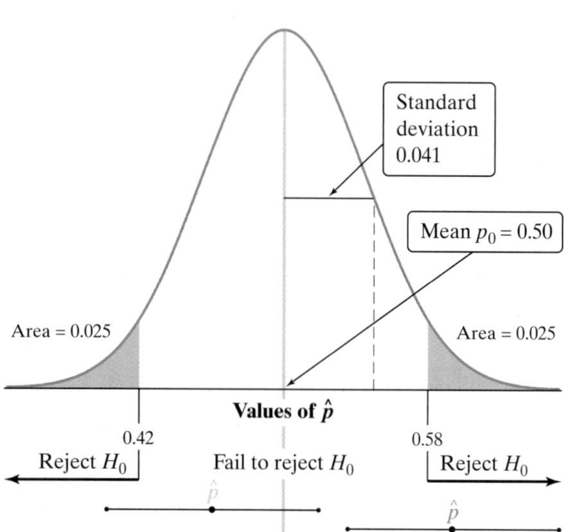

FIGURE 9.12 The connection between two-sided tests and confidence intervals. A value of $\hat{p}$ in one of the rejection regions will result in a confidence interval that doesn't contain p_0. A value of $\hat{p}$ in the fail to reject H_0 region will lead to a confidence interval that includes p_0 as a plausible value for the population proportion.

If the sample proportion falls in one of these rejection regions, like the blue value of $\hat{p}$ in the figure, a confidence interval centered at $\hat{p}$ would extend $z^* = 1.96$ standard deviations $\sqrt{\frac{\hat{p}(1-\hat{p})}{150}}$ in each direction. Note that $\hat{p}(1 - \hat{p})$ must be less than (0.5)(0.5), so this standard deviation must be less than our earlier value of 0.041. Such an interval would extend less than 0.08 in each direction from $\hat{p}$ and would therefore fail to include p_0 as a plausible value of the population proportion. So both the test and the confidence interval would lead to a reject H_0 conclusion.

If the sample proportion falls in the "fail to reject H_0" region, like the green value of $\hat{p}$ in Figure 9.12, the resulting 95% confidence interval would include p_0. In that case, both the significance test and the confidence interval would be unable to rule out p_0 as a plausible parameter value. However, if $\hat{p} = 0.58$, the 95% confidence interval does not contain 0.5 even though the significance test fails to reject H_0. The inconsistent results are due to the different standard deviations used to calculate the test statistic and the confidence interval.

The imperfect link we just described holds *only* for confidence intervals and two-sided tests. There is a connection between one-sided tests and confidence intervals, but it's more complicated and not nearly as useful.

CHECK YOUR UNDERSTANDING

The figure below shows Minitab output from a significance test and confidence interval for the restaurant worker data in the previous Check Your Understanding (page 558). Explain how the confidence interval is consistent with, but gives more information than, the test.

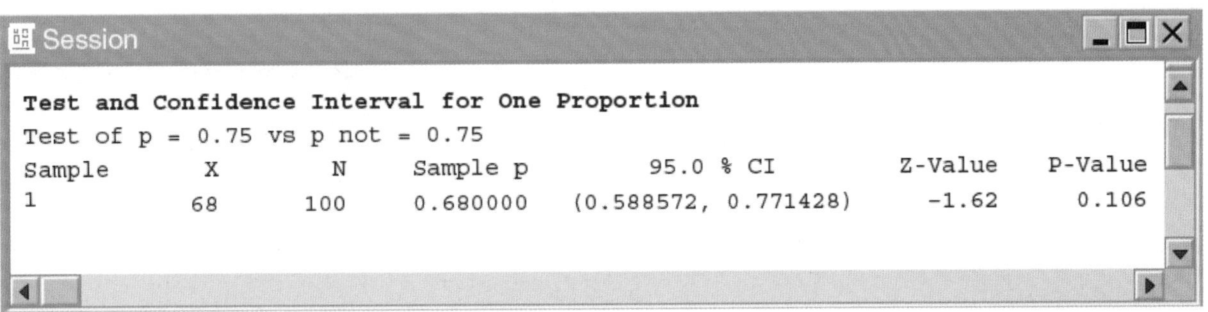

Session

```
Test and Confidence Interval for One Proportion
Test of p = 0.75 vs p not = 0.75
Sample      X      N     Sample p            95.0 % CI            Z-Value    P-Value
1          68    100     0.680000   (0.588572, 0.771428)          -1.62      0.106
```

SECTION 9.2 Summary

- As with confidence intervals, you should verify that the three conditions—Random, Normal, and Independent—are met before you carry out a significance test.
- Significance tests for $H_0: p = p_0$ are based on the **test statistic**

$$z = \frac{\hat{p} - p_0}{\sqrt{\frac{p_0(1 - p_0)}{n}}}$$

  with *P*-values calculated from the standard Normal distribution.
- The **one-sample *z* test for a proportion** is approximately correct when (1) the data were produced by random sampling or random assignment; (2) the population is at least 10 times as large as the sample; and (3) the sample is large enough to satisfy $np_0 \geq 10$ and $n(1 - p_0) \geq 10$ (that is, the expected numbers of successes and failures are both at least 10).
- Follow the four-step process when you are asked to carry out a significance test:

  STATE: What *hypotheses* do you want to test, and at what significance level? Define any *parameters* you use.

  PLAN: Choose the appropriate inference *method*. Check *conditions*.

  DO: If the conditions are met, perform *calculations*.

  - Compute the **test statistic.**
  - Find the ***P*-value.**

  CONCLUDE: *Interpret* the results of your test in the context of the problem.
- Confidence intervals provide additional information that significance tests do not—namely, a range of plausible values for the true population parameter *p*. A two-sided test of $H_0: p = p_0$ at significance level α gives roughly the same conclusion as a $100(1 - \alpha)\%$ confidence interval.

9.2 TECHNOLOGY CORNER

One-proportion z test on the calculator .... page 555

TI-Nspire instructions in Appendix B

SECTION 9.2 Exercises

In Exercises 33 and 34, check that the conditions for carrying out a one-sample z test for the population proportion p are met.

33. **Lefties** Simon reads a newspaper report claiming that 12% of all adults in the United States are left-handed. He wonders if 12% of the students at his large public high school are left-handed. Simon chooses an SRS of 100 students and records whether each student is right- or left-handed. pg 550

34. **Don't argue!** A Gallup Poll report on a national survey of 1028 teenagers revealed that 72% of teens said they rarely or never argue with their friends.[12] Yvonne wonders whether this national result would be true in her large high school, so she surveys a random sample of 150 students at her school.

In Exercises 35 and 36, explain why we aren't safe carrying out a one-sample z test for the population proportion p.

35. **No test** You toss a coin 10 times to test the hypothesis $H_0: p = 0.5$ that the coin is balanced.

36. **No test** A college president says, "99% of the alumni support my firing of Coach Boggs." You contact an SRS of 200 of the college's 15,000 living alumni to test the hypothesis $H_0: p = 0.99$.

37. **Lefties** Refer to Exercise 33. In Simon's SRS, 16 of the students were left-handed. pg 551

(a) Calculate the test statistic.

(b) Find the *P*-value using Table A. Show this result as an area under a standard Normal curve.

38. **Don't argue!** Refer to Exercise 34. For Yvonne's survey, 96 students in the sample said they seldom or never argue with friends.

(a) Calculate the test statistic.

(b) Find the *P*-value using Table A. Show this result as an area under a standard Normal curve.

39. **Significance tests** A test of $H_0: p = 0.5$ versus $H_a: p > 0.5$ has test statistic $z = 2.19$.

(a) What conclusion would you draw at the 5% significance level? At the 1% level?

(b) If the alternative hypothesis were $H_a: p \neq 0.5$, what conclusion would you draw at the 5% significance level? At the 1% level?

40. **Significance tests** A test of $H_0: p = 0.65$ against $H_a: p < 0.65$ has test statistic $z = -1.78$.

(a) What conclusion would you draw at the 5% significance level? At the 1% level?

(b) If the alternative hypothesis were $H_a: p \neq 0.65$, what conclusion would you draw at the 5% significance level? At the 1% level?

41. **Better parking** A local high school makes a change that should improve student satisfaction with the parking situation. Before the change, 37% of the school's students approved of the parking that was provided. After the change, the principal surveys an SRS of 200 of the over 2500 students at the school. In all, 83 students say that they approve of the new parking arrangement. The principal cites this as evidence that the change was effective. Perform a test of the principal's claim at the $\alpha = 0.05$ significance level. pg 553

42. **Side effects** A drug manufacturer claims that less than 10% of patients who take its new drug for treating Alzheimer's disease will experience nausea. To test this claim, researchers conduct an experiment. They give the new drug to a random sample of 300 out of 5000 Alzheimer's patients whose families have given informed consent for the patients to participate in the study. In all, 25 of the subjects experience nausea. Use these data to perform a test of the drug manufacturer's claim at the $\alpha = 0.05$ significance level.

43. **Better parking** Refer to Exercise 41.

(a) Describe a Type I error and a Type II error in this setting, and explain the consequences of each.

(b) The test has a power of 0.75 to detect that $p = 0.45$. Explain what this means.

(c) Identify two ways to increase the power in part (b).

44. **Side effects** Refer to Exercise 42.

(a) Describe a Type I error and a Type II error in this setting, and explain the consequences of each.

(b) The test has a power of 0.54 to detect that $p = 0.07$. Explain what this means.

(c) Identify two ways to increase the power in part (b).

45. **Are boys more likely?** We hear that newborn babies are more likely to be boys than girls. Is this true? A random sample of 25,468 firstborn children included 13,173 boys.[13] Boys do make up more than half of the sample, but of course we don't expect a perfect 50-50 split in a random sample.

(a) To what population can the results of this study be generalized: all children or all firstborn children? Justify your answer.

(b) Do these data give convincing evidence that boys are more common than girls in the population? Carry out a significance test to help answer this question.

46. **Fresh coffee** People of taste are supposed to prefer fresh-brewed coffee to the instant variety. On the other hand, perhaps many coffee drinkers just want their caffeine fix. A skeptic claims that only half of all coffee drinkers prefer fresh-brewed coffee. To test this claim, we ask a random sample of 50 coffee drinkers in a small city to take part in a study. Each person tastes two unmarked cups—one containing instant coffee and one containing fresh-brewed coffee—and says which he or she prefers. We find that 36 of the 50 choose the fresh coffee.

(a) We presented the two cups to each coffee drinker in a random order, so that some people tasted the fresh coffee first, while others drank the instant coffee first. Why do you think we did this?

(b) Do these results give convincing evidence that coffee drinkers favor fresh-brewed over instant coffee? Carry out a significance test to help answer this question.

47. **Bullies in middle school** A University of Illinois study on aggressive behavior surveyed a random sample of 558 middle school students. When asked to describe their behavior in the last 30 days, 445 students said their behavior included physical aggression, social ridicule, teasing, name-calling, and issuing threats. This behavior was not defined as bullying in the questionnaire.[14] Is this evidence that more than three-quarters of the students at that middle school engage in bullying behavior? To find out, Maurice decides to perform a significance test. Unfortunately, he made a few errors along the way. Your job is to spot the mistakes and correct them.

$$H_0: p = 0.75$$
$$H_a: \hat{p} > 0.797$$

where p = the true mean proportion of middle school students who engaged in bullying.

- A random sample of 558 middle school students was surveyed.
- $558(0.797) = 444.73$ is at least 10.

$$z = \frac{0.75 - 0.797}{\sqrt{\frac{0.797(0.203)}{445}}} = -2.46;\ P\text{-value} = 2(0.0069) = 0.0138$$

The probability that the null hypothesis is true is only 0.0138, so we reject H_0. This proves that more than three-quarters of the school engaged in bullying behavior.

48. **Is this coin fair?** The French naturalist Count Buffon (1707–1788) tossed a coin 4040 times. He got 2048 heads. That's a bit more than one-half. Is this evidence that Count Buffon's coin was not balanced? To find out, Luisa decides to perform a significance test. Unfortunately, she made a few errors along the way. Your job is to spot the mistakes and correct them.

$$H_0: \mu > 0.5$$
$$H_a: \bar{x} = 0.5$$

- **Independent** $4040(0.5) = 2020$ and $4040(1 - 0.5) = 2020$ are both at least 10.
- **Normal** There are at least 40,400 coins in the world.

$$t = \frac{0.5 - 0.507}{\sqrt{\frac{0.5(0.5)}{4040}}} = -0.89;\ P\text{-value} = 1 - 0.1867 = 0.8133$$

Reject H_0 because the P-value is so large and conclude that the coin is fair.

49. **Teen drivers** A state's Division of Motor Vehicles (DMV) claims that 60% of teens pass their driving test on the first attempt. An investigative reporter examines an SRS of the DMV records for 125 teens; 86 of them passed the test on their first try. Is this good evidence that the DMV's claim is incorrect? Carry out a test at the $\alpha = 0.05$ significance level to help answer this question.

pg 556

50. **We want to be rich** In a recent year, 73% of first-year college students responding to a national survey identified "being very well-off financially" as an important personal goal. A state university finds that 132 of an SRS of 200 of its first-year students say that this goal is important. Is there good evidence that the proportion of all first-year students at this university who think being very well-off is important differs from the national value, 73%? Carry out a test at the $\alpha =$ 0.05 significance level to help answer this question.

51. **Teen drivers** Refer to Exercise 49.

(a) Construct and interpret a 95% confidence interval for the proportion of all teens in the state who passed their driving test on the first attempt.

(b) Explain what the interval in part (a) tells you about the DMV's claim.

52. **We want to be rich** Refer to Exercise 50.

(a) Construct and interpret a 95% confidence interval for the true proportion p of all first-year students at the university who would identify being well-off as an important personal goal.

(b) Explain what the interval in part (a) tells you about whether the national value holds at this university.

53. **Do you Twitter?** In late 2009, the Pew Internet and American Life Project asked a random sample of U.S. adults, "Do you ever . . . use Twitter or another service to share updates about yourself or to see updates about others?" According to Pew, the resulting 95% confidence interval is $(0.167, 0.213)$.[15] Can we use this interval to conclude that the actual proportion of U.S. adults who would say they Twitter differs from 0.20? Justify your answer.

54. **Losing weight** A Gallup Poll found that 59% of the people in its sample said "Yes" when asked, "Would you like to lose weight?" Gallup announced: "For results based on the total sample of national adults, one can say with 95% confidence that the margin of (sampling) error is ± 3 percentage points."[16] Can we use this interval to conclude that the actual proportion of U.S. adults who would say they want to lose weight differs from 0.55? Justify your answer.

55. **Teens and sex** The Gallup Youth Survey asked a random sample of U.S. teens aged 13 to 17 whether they thought that young people should wait to have sex until marriage.[17] The Minitab output below shows the results of a significance test and a 95% confidence interval based on the survey data.

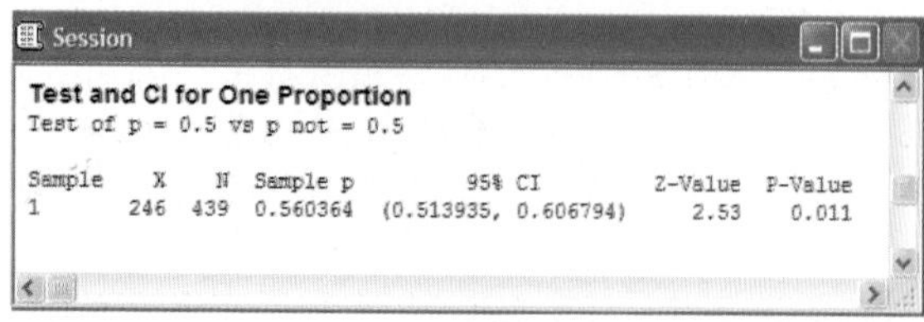

Session

Test and CI for One Proportion
Test of p = 0.5 vs p not = 0.5

Sample	X	N	Sample p	95% CI	Z-Value	P-Value
1	246	439	0.560364	(0.513935, 0.606794)	2.53	0.011

(a) Define the parameter of interest.

(b) Check that the conditions for performing the significance test are met in this case.

(c) Interpret the P-value in context.

(d) Do these data give convincing evidence that the actual population proportion differs from 0.5? Justify your answer with appropriate evidence.

56. **Reporting cheating** What proportion of students are willing to report cheating by other students? A student project put this question to an SRS of 172 undergraduates at a large university: "You witness two students cheating on a quiz. Do you go to the professor?" The Minitab output below shows the results of a significance test and a 95% confidence interval based on the survey data.[18]

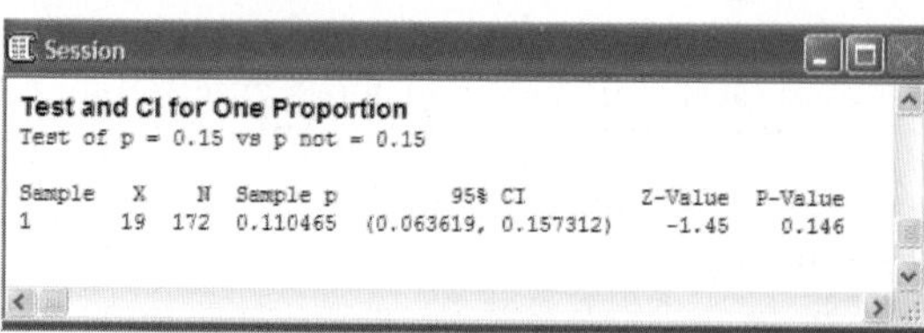

Session

Test and CI for One Proportion
Test of p = 0.15 vs p not = 0.15

Sample	X	N	Sample p	95% CI	Z-Value	P-Value
1	19	172	0.110465	(0.063619, 0.157312)	-1.45	0.146

(a) Define the parameter of interest.

(b) Check that the conditions for performing the significance test are met in this case.

(c) Interpret the P-value in context.

(d) Do these data give convincing evidence that the actual population proportion differs from 0.15? Justify your answer with appropriate evidence.

Multiple choice: Select the best answer for Exercises 57 to 60.

57. After once again losing a football game to the archrival, a college's alumni association conducted a survey to see if alumni were in favor of firing the coach. An SRS of 100 alumni from the population of all living alumni was taken, and 64 of the alumni in the sample were in favor of firing the coach. Suppose you wish to see if a majority of living alumni are in favor of firing the coach. The appropriate test statistic is

(a) $z = \dfrac{0.64 - 0.5}{\sqrt{\dfrac{0.64(0.36)}{100}}}$

(b) $t = \dfrac{0.64 - 0.5}{\sqrt{\dfrac{0.64(0.36)}{100}}}$

(c) $z = \dfrac{0.64 - 0.5}{\sqrt{\dfrac{0.5(0.5)}{100}}}$

(d) $z = \dfrac{0.64 - 0.5}{\sqrt{\dfrac{0.64(0.36)}{64}}}$

(e) $z = \dfrac{0.5 - 0.64}{\sqrt{\dfrac{0.5(0.5)}{100}}}$

58. Which of the following is *not* a condition for performing a significance test about an unknown population proportion p?

(a) The data should come from a random sample or randomized experiment.

(b) Individual measurements should be independent of one another.

(c) The population distribution should be approximately Normal, unless the sample size is large.

(d) Both np and $n(1-p)$ should be at least 10.

(e) If you are sampling without replacement from a finite population, then you should sample no more than 10% of the population.

59. The z statistic for a test of $H_0: p = 0.4$ versus H_a: $p > 0.4$ is $z = 2.43$. This test is

(a) not significant at either $\alpha = 0.05$ or $\alpha = 0.01$.

(b) significant at $\alpha = 0.05$ but not at $\alpha = 0.01$.

(c) significant at $\alpha = 0.01$ but not at $\alpha = 0.05$.

(d) significant at both $\alpha = 0.05$ and $\alpha = 0.01$.

(e) inconclusive because we don't know the value of $\hat{p}$.

60. Which of the following 95% confidence intervals would lead us to reject $H_0: p = 0.30$ in favor of $H_a: p \neq 0.30$ at the 5% significance level?

(a) (0.29, 0.38) (c) (0.27, 0.31) (e) None of these
(b) (0.19, 0.27) (d) (0.24, 0.30)

61. **Packaging CDs** (6.2, 5.3) A manufacturer of compact discs (CDs) wants to be sure that their CDs will fit inside the plastic cases they have bought for packaging. Both the CDs and the cases are circular. According to the supplier, the plastic cases vary Normally with mean diameter $\mu = 4.2$ inches and standard deviation $\sigma = 0.05$ inches. The CD manufacturer decides to produce CDs with mean diameter $\mu = 4$ inches. Their diameters follow a Normal distribution with $\sigma = 0.1$ inches.

(a) Let X = the diameter of a randomly selected CD and Y = the diameter of a randomly selected case. Describe the shape, center, and spread of the distribution of the random variable X − Y. What is the importance of this random variable to the CD manufacturer?

(b) Compute the probability that a randomly selected CD will fit inside a randomly selected case.

(c) The production process actually runs in batches of 100 CDs. If each of these CDs is paired with a randomly chosen plastic case, find the probability that all the CDs fit in their cases.

62. **Cash to find work?** (5.2) Will cash bonuses speed the return to work of unemployed people? The Illinois Department of Employment Security designed an experiment to find out. The subjects were 10,065 people aged 20 to 54 who were filing claims for unemployment insurance. Some were offered \$500 if they found a job within 11 weeks and held it for at least 4 months. Others could tell potential employers that the state would pay the employer \$500 for hiring them. A control group got neither kind of bonus.[19]

(a) Describe a completely randomized design for this experiment.

(b) How will you label the subjects for random assignment? Use Table D at line 127 to choose the first 3 subjects for the first treatment.

(c) Explain the purpose of a control group in this setting.

9.3 Tests about a Population Mean

In Section 9.3, you'll learn about:

- Carrying out a significance test for μ
- The one-sample t test
- Two-sided tests and confidence intervals
- Inference for means: Paired data
- Using tests wisely

Confidence intervals and significance tests for a population proportion p are based on z-values from the standard Normal distribution. Inference about a population mean μ uses a t distribution with $n-1$ degrees of freedom, except in the rare case when the population standard deviation σ is known. We learned how to construct confidence intervals for a population mean in Section 8.3. Now we'll examine the details of testing a claim about an unknown parameter μ.

Carrying Out a Significance Test for μ

In an earlier example, a company claimed to have developed a new AAA battery that lasts longer than its regular AAA batteries. Based on years of experience, the company knows that its regular AAA batteries last for 30 hours of continuous use, on average. An SRS of 15 new batteries lasted an average of 33.9 hours with a standard deviation of 9.8 hours. Do these data give convincing evidence that the new batteries last longer on average? To find out, we perform a test of

$$H_0: \mu = 30 \text{ hours}$$
$$H_a: \mu > 30 \text{ hours}$$

where μ is the true mean lifetime of the new deluxe AAA batteries.

Conditions By now, you are familiar with the three conditions that should be met before performing inference about a population mean: Random, Normal, and Independent. As in the previous chapter, the Normal condition for means is

population distribution is Normal or sample size is large ($n \geq 30$)

We often don't know whether the population distribution is Normal. But if the sample size is large ($n \geq 30$), we can safely carry out a significance test (due to the central limit theorem). If the sample size is small, we should examine the sample data for any obvious departures from Normality, such as skewness and outliers.

Recall from Chapter 8 that the t procedures are quite robust against non-Normality of the population except when outliers or strong skewness are present. For significance tests, "robust" means that the stated P-value is pretty accurate. Refer to the guidelines on page 512 when checking the Normal condition for a one-sample t test.

EXAMPLE

Better Batteries

Checking conditions

Figure 9.13 shows a dotplot, boxplot, and Normal probability plot of the battery lifetimes for an SRS of 15 batteries.

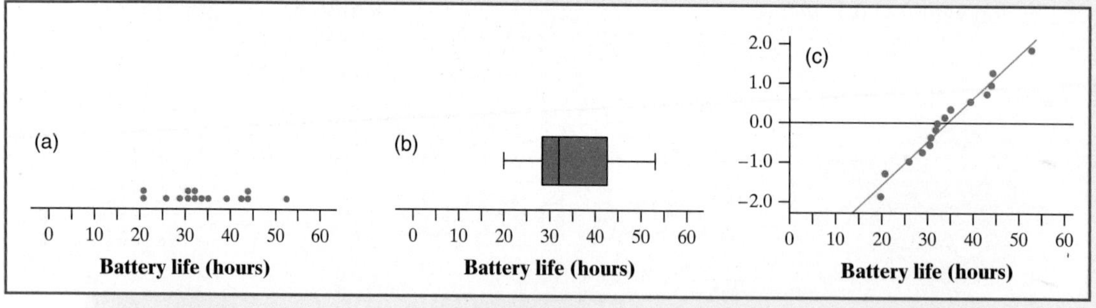

FIGURE 9.13 (a) A dotplot, (b) a boxplot, and (c) a Normal probability plot of the lifetime of a simple random sample of 15 AAA batteries.

PROBLEM: **Check the conditions for carrying out a significance test of the company's claim about its deluxe AAA batteries.**

SOLUTION:

- *Random* The company tested a simple random sample of 15 new AAA batteries.
- *Normal* We don't know if the population distribution of battery lifetimes for the company's new AAA batteries is Normal. With such a small sample size ($n = 15$), we need to inspect the data for any departures from Normality. The dotplot and boxplot show slight right-skewness but no outliers. The Normal probability plot is close to linear. We should be safe performing a test about the population mean lifetime μ.
- *Independent* Since the batteries are being sampled without replacement, we need to check the 10% condition: there must be at least $10(15) = 150$ new AAA batteries. This seems reasonable to believe.

For Practice *Try Exercise 63*

Calculations: Test Statistic and *P*-Value When performing a significance test, we do calculations assuming that the null hypothesis H_0 is true. The test statistic measures how far the sample result diverges from the parameter value specified by H_0, in standardized units. As before,

$$\text{test statistic} = \frac{\text{statistic} - \text{parameter}}{\text{standard deviation of statistic}}$$

For a test of $H_0: \mu = \mu_0$, our statistic is the sample mean $\bar{x}$. Its standard deviation is

$$\sigma_{\bar{x}} = \frac{\sigma}{\sqrt{n}}$$

In an ideal world, our test statistic would be

$$z = \frac{\bar{x} - \mu_0}{\sigma/\sqrt{n}}$$

Because the population standard deviation σ is usually unknown, we use the sample standard deviation s_x in its place. The resulting test statistic has the standard error of $\bar{x}$ in the denominator

$$t = \frac{\bar{x} - \mu_0}{s_x/\sqrt{n}}$$

As we saw earlier, when the Normal condition is met, this statistic has a t distribution with $n - 1$ degrees of freedom.

In Section 8.3, we used Table B to find critical values from the t distributions when constructing confidence intervals about an unknown population mean μ. Once we have calculated the test statistic, we can use Table B to find the *P*-value for a significance test about μ. The following example shows how this works.

EXAMPLE

Better Batteries

Computing the test statistic and *P*-value

The battery company wants to test $H_0: \mu = 30$ versus $H_a: \mu > 30$ based on an SRS of 15 new AAA batteries with mean lifetime $\bar{x} = 33.9$ hours and standard deviation $s_x = 9.8$ hours. The test statistic is

$$\text{test statistic} = \frac{\text{statistic} - \text{parameter}}{\text{standard deviation of statistic}}$$

$$t = \frac{\bar{x} - \mu_0}{s_x/\sqrt{n}} = \frac{33.9 - 30}{9.8/\sqrt{15}} = 1.54$$

The *P*-value is the probability of getting a result this large or larger in the direction indicated by H_a, that is, $P(t \geq 1.54)$. Figure 9.14 on the next page shows this probability as an area under the t distribution curve with df $= 15 - 1 = 14$. We can find this *P*-value using Table B.

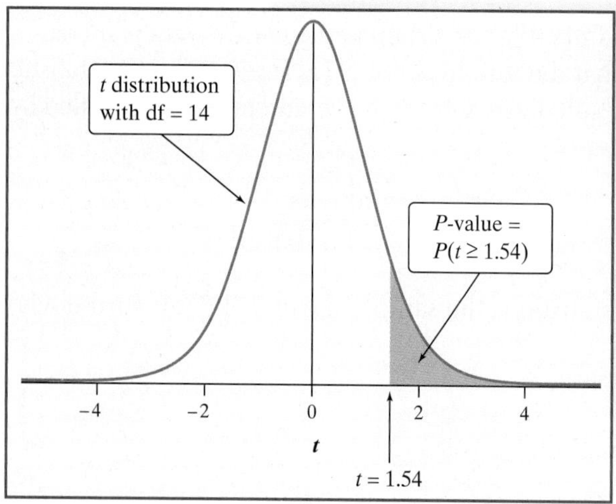

FIGURE 9.14 The *P*-value for a one-sided test with $t = 1.54$.

Go to the df = 14 row. The t statistic falls between the values 1.345 and 1.761. If you look at the top of the corresponding columns in Table B, you'll find that the "Upper-tail probability p" is between 0.10 and 0.05. (See the excerpt from Table B at right.) Since we are looking for $P(t \geq 1.54)$, this is the probability we seek. That is, the *P*-value for this test is between 0.05 and 0.10.

	Upper-tail probability *p*		
df	.10	.05	.025
13	1.350	1.771	2.160
14	1.345	1.761	2.145
15	1.341	1.753	2.131
	80%	90%	95%
	Confidence level *C*		

As you can see, Table B gives a range of possible *P*-values for a significance test. We can still draw a conclusion from the test in much the same way as if we had a single probability. In the case of the new AAA batteries, for instance, we don't have enough evidence to reject $H_0: \mu = 30$ because the *P*-value exceeds our default $\alpha = 0.05$ significance level. We can't conclude that the company's new AAA batteries last longer than 30 hours, on average.

Table B has other limitations for finding *P*-values. It includes probabilities only for t distributions with degrees of freedom from 1 to 30 and then skips to df = 40, 50, 60, 80, 100, and 1000. (The bottom row gives probabilities for df = ∞, which corresponds to the standard Normal curve.) In addition, Table B shows probabilities only for positive values of t. To find a *P*-value for a negative value of t, we use the symmetry of the t distributions. The next example shows how we deal with both of these issues.

EXAMPLE Two-Sided Tests, Negative t Values, and More

Using Table B wisely

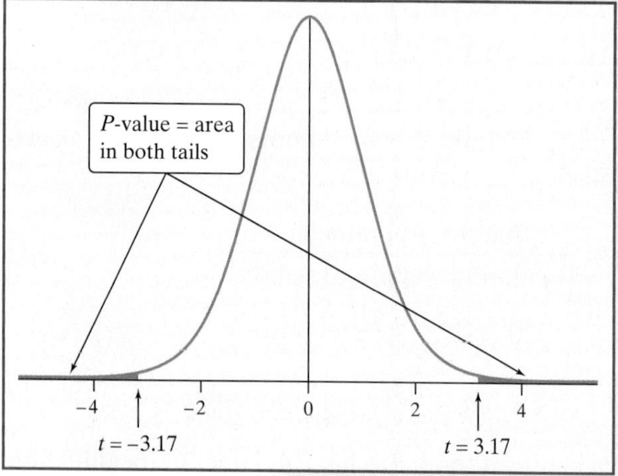

FIGURE 9.15 The *P*-value for a two-sided test with $t = -3.17$.

What if you were performing a test of $H_0: \mu = 5$ versus $H_a: \mu \neq 5$ based on a sample size of $n = 37$ and obtained $t = -3.17$? Since this is a two-sided test, you are interested in the probability of getting a value of t less than -3.17 or greater than 3.17. Figure 9.15 shows the desired *P*-value as an area under the t distribution curve with 36 degrees of freedom. Notice that $P(t \leq -3.17) = P(t \geq 3.17)$ due to the symmetric shape of the density curve. Since Table B shows only positive t-values, we will focus on $t = 3.17$.

	Upper-tail probability *p*		
df	.005	.0025	.001
29	2.756	3.038	3.396
30	2.750	3.030	3.385
40	2.704	2.971	3.307
	99%	99.5%	99.8%
	Confidence level *C*		

Since df = 37 − 1 = 36 is not available on the table,

use df = 30. You might be tempted to use df = 40, but doing so would result in a smaller *P*-value than you are entitled to with df = 36. (In other words, you'd be cheating!) Move across the df = 30 row, and notice that $t = 3.17$ falls between 3.030 and 3.385. The corresponding "Upper-tail probability *p*" is between 0.0025 and 0.001. (See the excerpt from Table B.) For this two-sided test, the corresponding *P*-value would be between 2(0.001) = 0.002 and 2(0.0025) = 0.005.

CAUTION

One point from the example deserves repeating: *if the df you need isn't provided in Table B, use the next lower df that is available*. It's no fair "rounding up" to a higher df. This is like pretending that your sample size is larger than it really is. Doing so would give you a smaller *P*-value than is actually true and would make you more likely to incorrectly reject H_0 when it's true (make a Type I error).

Given the limitations of Table B, our advice is to use technology to find *P*-values when carrying out a significance test about a population mean.

TECHNOLOGY CORNER Computing *P*-values from *t* distributions on the calculator

You can use the `tcdf` command on the TI-83/84 and TI-89 to calculate areas under a *t* distribution curve. The syntax is `tcdf(lower bound,upper bound,df)`. To access this command:

TI-83/84

Press 2nd VARS ([DISTR]) and choose `tcdf(`.

TI-89

From the home screen, press CATALOG F3 (Flash Apps) and choose `tCdf(`.

Let's use the `tcdf` command to compute the *P*-values from the last two examples.

- *Better batteries:* To find $P(t \geq 1.54)$, execute the command `tcdf(1.54,100,14)`.
- *Two-sided test:* To find the *P*-value for the two-sided test with df = 36 and $t = -3.17$, execute the command `2*tcdf(-100,-3.17,36)`.

```
tcdf(1.54,100,14
)
         .0729268628
2*tcdf(-100,-3.1
7,36)
         .0031080065
```

TI-Nspire instructions in Appendix B

There is a small number of real-world situations in which we might know the population standard deviation σ. When this is the case, the test statistic

$$z = \frac{\bar{x} - \mu_0}{\sigma / \sqrt{n}}$$

will follow a standard Normal distribution if the Normal condition is met. If so, then we can calculate *P*-values using Table A or technology. The TI-83/84 and TI-89's Z-Test option in the TESTS menu is designed for this special situation.

CHECK YOUR UNDERSTANDING

For the job satisfaction study described in Section 9.1, the hypotheses are

$$H_0: \mu = 0$$
$$H_a: \mu \neq 0$$

where μ is the mean difference in job satisfaction scores (self-paced − machine-paced) in the population of assembly-line workers at the company. Data from a random sample of 18 workers gave $\overline{x} = 17$ and $s_x = 60$.

1. Calculate the test statistic. Show your work.
2. Use Table B to find the *P*-value. What conclusion would you draw?
3. Now use your calculator to find the *P*-value as described in the Technology Corner. Is your result consistent with the value you obtained in Question 2?

The One-Sample *t* Test

When the conditions are met, we can test a claim about a population mean μ using a **one-sample *t* test**. Here are the details.

One-Sample *t* Test

Choose an SRS of size n from a large population with unknown mean μ. To test the hypothesis $H_0: \mu = \mu_0$, compute the one-sample t statistic

$$t = \frac{\overline{x} - \mu_0}{s_x / \sqrt{n}}$$

Find the *P*-value by calculating the probability of getting a t statistic this large or larger in the direction specified by the alternative hypothesis H_a in a t distribution with df $= n - 1$:

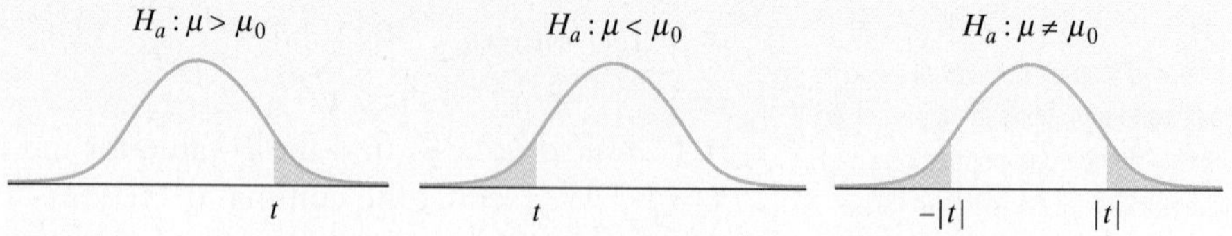

Use this test only when (1) the population distribution is Normal or the sample is large ($n \geq 30$), and (2) the population is at least 10 times as large as the sample.

Now we are ready to test a claim about an unknown population mean. Once again, we follow the four-step process.

Healthy Streams

Performing a significance test about μ

The level of dissolved oxygen (DO) in a stream or river is an important indicator of the water's ability to support aquatic life. A researcher measures the DO level at 15 randomly chosen locations along a stream. Here are the results in milligrams per liter (mg/l):

4.53	5.04	3.29	5.23	4.13	5.50	4.83	4.40
5.42	6.38	4.01	4.66	2.87	5.73	5.55	

A dissolved oxygen level below 5 mg/l puts aquatic life at risk.

PROBLEM:

(a) Can we conclude that aquatic life in this stream is at risk? Carry out a test at the $\alpha = 0.05$ significance level to help you answer this question.

(b) Given your conclusion in part (a), which kind of mistake—a Type I error or a Type II error—could you have made? Explain what this mistake would mean in context.

SOLUTION:

(a) We will follow the four-step process.

STATE: We want to test a claim about the actual mean dissolved oxygen level μ in this stream at the $\alpha = 0.05$ level. Our hypotheses are

$$H_0: \mu = 5$$
$$H_a: \mu < 5$$

PLAN: If the conditions are met, we should do a one-sample *t* test for μ.

- *Random* The researcher measured the DO level at 15 randomly chosen locations.
- *Normal* We don't know whether the population distribution of DO levels at all points along the stream is Normal. With such a small sample size ($n = 15$), we need to look at the data to see if it's safe to use *t* procedures. Figure 9.16 shows our hand sketches of a calculator histogram, boxplot, and Normal probability plot for these data. The histogram looks roughly symmetric; the boxplot shows no outliers; and the Normal probability plot is fairly linear. With no outliers or strong skewness, the *t* procedures should be pretty accurate even if the population distribution isn't Normal.

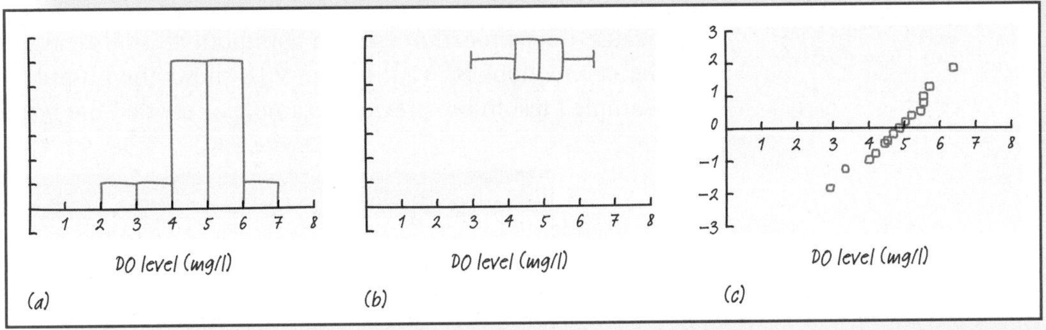

FIGURE 9.16 Sketches of (a) a histogram, (b) a boxplot, and (c) a Normal probability plot for the dissolved oxygen (DO) readings in the sample, in mg/l.

- *Independent* There is an infinite number of possible locations along the stream, so it isn't necessary to check the 10% condition. We do need to assume that individual DO measurements are independent.

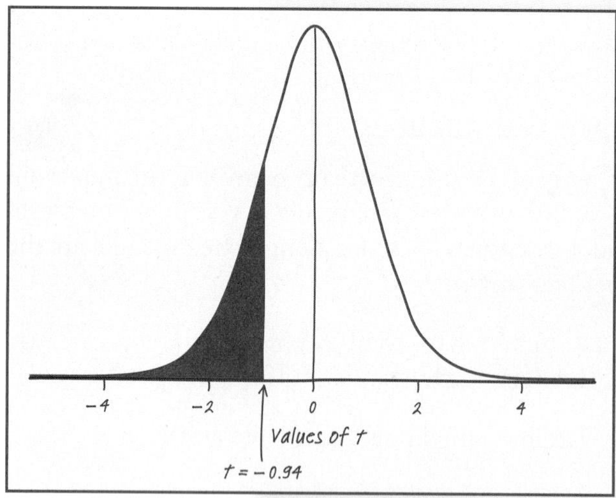

FIGURE 9.17 The *P*-value for a one-sided test when $t = -0.94$.

DO: We entered the data into our calculator and did 1-Var Stats (see screen shot). The sample mean is $\bar{x} = 4.771$ and the sample standard deviation is $s_x = 0.9396$.

```
1-Var Stats
x̄=4.771333333
Σx=71.57
Σx²=353.8441
Sx=.9395961645
σx=.907736134
↓n=15
```

- *Test statistic*

$$t = \frac{\bar{x} - \mu_0}{s_x/\sqrt{n}} = \frac{4.771 - 5}{0.9396/\sqrt{15}} = -0.94$$

- *P*-value The *P*-value is the area to the left of $t = -0.94$ under the *t* distribution curve with degrees of freedom df = 15 − 1 = 14. Figure 9.17 shows this probability.

Using technology: We can find the exact *P*-value using a calculator: tcdf(−100,−0.94,14) = 0.1816.

Using the table: Table B shows only areas in the upper tail of the distribution. Because the *t* distributions are symmetric, $P(t \le -0.94) = P(t \ge 0.94)$. Search the df = 14 row of Table B for entries that bracket $t = 0.94$ (see the excerpt at right). Because the observed *t* lies between 0.868 and 1.076, the *P*-value lies between 0.20 and 0.15. This is consistent with the calculator's 0.1816 result.

	Upper-tail probability *p*		
df	**.25**	**.20**	**.15**
13	.694	.870	1.079
14	.692	.868	1.076
15	.691	.866	1.074
	50%	60%	70%
	Confidence level *C*		

CONCLUDE: The *P*-value, 0.1816, is greater than our $\alpha = 0.05$ significance level. As a result, we fail to reject H_0. We don't have enough evidence to conclude that the mean DO level in the stream is less than 5 mg/l.

(b) Since we decided not to reject H_0 in part (a), we could have made a Type II error (failing to reject H_0 when H_0 is false). If we did, then the mean dissolved oxygen level μ in the stream is actually less than 5 mg/l, but we didn't detect that with our significance test.

Because the *t* procedures are so common, all statistical software packages will do the calculations for you. Figure 9.18 shows the output from Minitab for the one-sample *t* test in the previous example. Note that the results match!

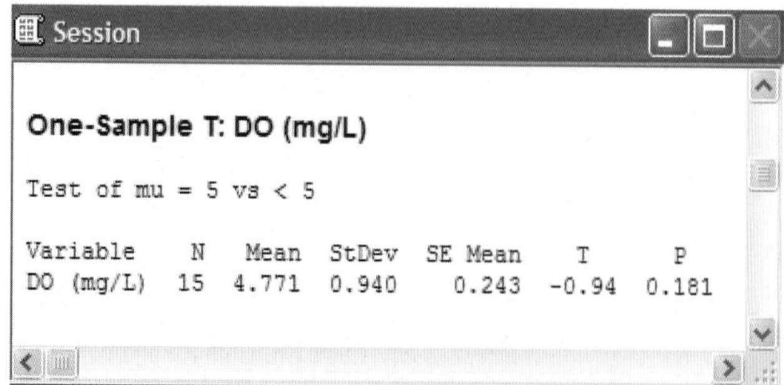

FIGURE 9.18 Minitab output for the one-sample *t* test from the dissolved oxygen example.

You can also use your calculator to carry out a one-sample t test. Once again, be sure to read the AP exam tip in the margin.

TECHNOLOGY CORNER One-sample *t* test on the calculator

You can perform a one-sample t test using either raw data or summary statistics on the TI-83/84 or TI-89. Let's use the calculator to carry out the test of H_0: $\mu = 5$ versus H_a: $\mu < 5$ from the dissolved oxygen example. Start by entering the sample data in L1/list1. Then, to do the test:

TI-83/84	TI-89
• Press STAT, choose TESTS and 2:T-Test.	• Press 2nd F1 ([F6]) and choose 2:T-Test.

• Adjust your settings as shown.

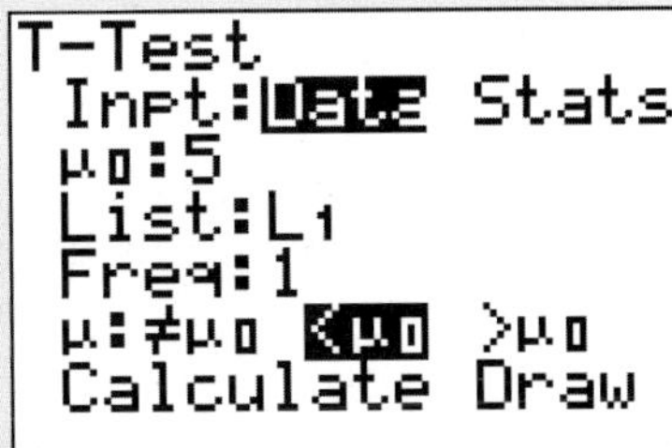

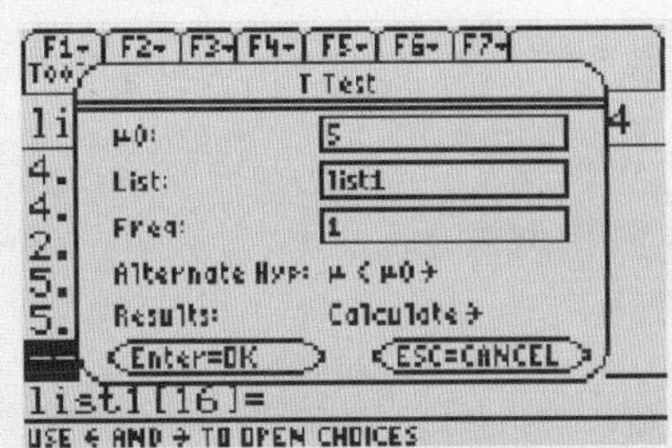

If you select "Calculate," the following screen appears:

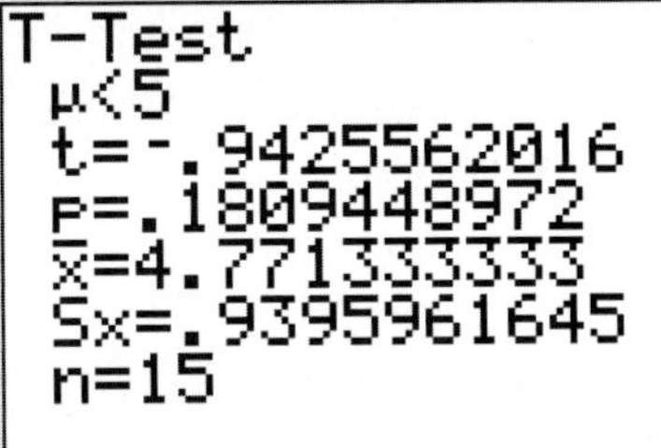

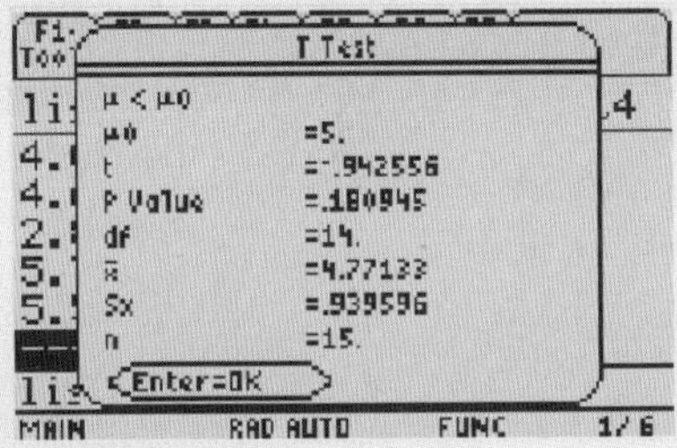

The test statistic is $t = -0.94$ and the P-value is 0.1809.

If you specify "Draw," you see a t distribution curve (df $= 14$) with the lower tail shaded.

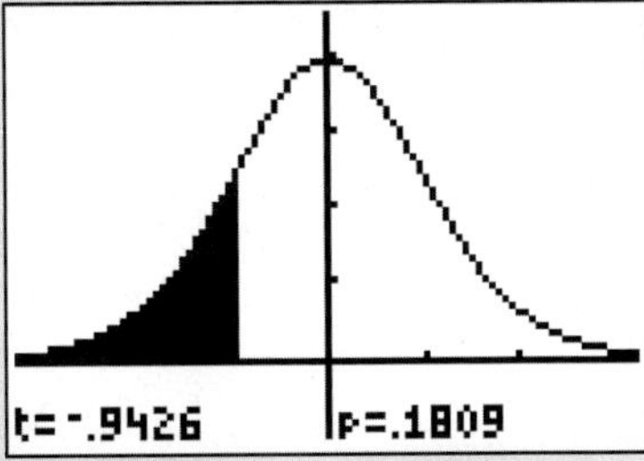

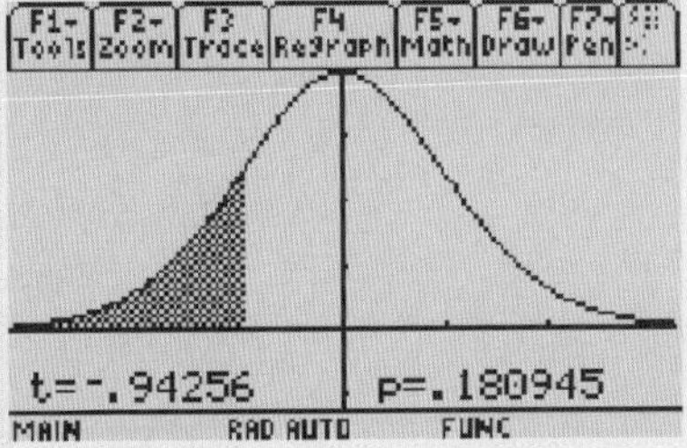

If you are given summary statistics instead of the original data, you would select the option "Stats" instead of "Data."

TI-Nspire instructions in Appendix B

AP EXAM TIP
Remember: if you just give calculator results with no work, and one or more values are wrong, you probably won't get any credit for the "Do" step. We recommend doing the calculation with the appropriate formula and then checking with your calculator. If you opt for the calculator-only method, name the procedure (t test) and report the test statistic ($t = -0.94$), degrees of freedom (df $= 14$), and P-value (0.1809).

CHECK YOUR UNDERSTANDING

A college professor suspects that students at his school are getting less than 8 hours of sleep a night, on average. To test his belief, the professor asks a random sample of 28 students, "How much sleep did you get last night?" Here are the data (in hours):

9 6 8 6 8 8 6 6.5 6 7 9 4 3 4 5 6 11 6 3 6 6 10 7 8 4.5 9 7 7

Do these data provide convincing evidence in support of the professor's suspicion? Carry out a significance test at the $\alpha = 0.05$ level to help answer this question.

Two-Sided Tests and Confidence Intervals

Now let's look at an example involving a two-sided test.

Juicy Pineapple

A two-sided test

At the Hawaii Pineapple Company, managers are interested in the sizes of the pineapples grown in the company's fields. Last year, the mean weight of the pineapples harvested from one large field was 31 ounces. A new irrigation system was installed in this field after the growing season. Managers wonder whether this change will affect the mean weight of future pineapples grown in the field. To find out, they select and weigh a random sample of 50 pineapples from this year's crop. The Minitab output below summarizes the data.

Descriptive Statistics: Weight (oz)

Variable	N	Mean	SE Mean	StDev	Minimum	Q1	Median	Q3	Maximum
Weight (oz)	50	31.935	0.339	2.394	26.491	29.990	31.739	34.115	35.547

PROBLEM:

(a) Determine whether there are any outliers. Show your work.

(b) Do these data suggest that the mean weight of pineapples produced in the field has changed this year? Give appropriate statistical evidence to support your answer.

(c) Can we conclude that the new irrigation system caused a change in the mean weight of pineapples produced? Explain your answer.

SOLUTION:

(a) From the Minitab output, the interquartile range is $IQR = Q_3 - Q_1 = 34.115 - 29.990 = 4.125$. Any data value greater than $Q_3 + 1.5(IQR)$ or less than $Q_1 - 1.5(IQR)$ is considered an outlier.

$$Q_3 + 1.5(IQR) = 34.115 + 1.5(4.125) = 40.3025 > 35.547$$
$$Q_1 - 1.5(IQR) = 29.990 - 1.5(4.125) = 23.8025 < 26.491$$

So there are no outliers. Here is another way to think about it: The maximum is only 1.432 from Q_3 and the minimum is only 3.499 from Q_1. Since both these distances are less than the IQR, there are no outliers.

(b) STATE: We want to test the hypotheses

$$H_0: \mu = 31$$
$$H_a: \mu \neq 31$$

where μ = the mean weight (in ounces) of all pineapples grown in the field this year. Since no significance level is given, we'll use $\alpha = 0.05$.

PLAN: If the conditions are met, we should conduct a one-sample t test for μ.

- *Random* The data came from a random sample of 50 pineapples from this year's crop.
- *Normal* We don't know whether the population distribution of pineapple weights this year is Normally distributed. But $n = 50 \geq 30$, so the large sample size (and the fact that there are no outliers) makes it OK to use t procedures.
- *Independent* There need to be at least 10(50) = 500 pineapples in the field because managers are sampling without replacement (10% condition). We would expect many more than 500 pineapples in a "large field."

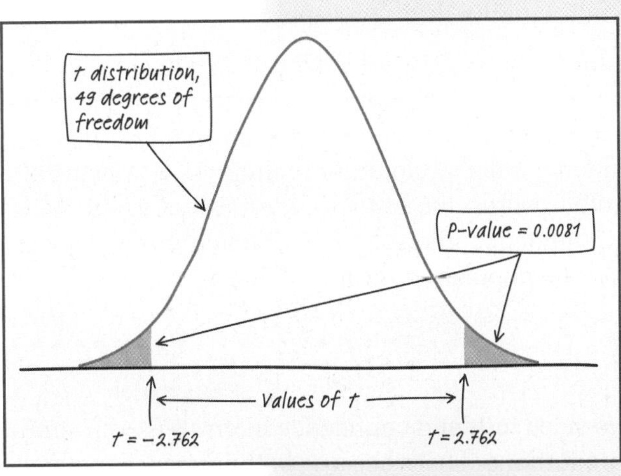

FIGURE 9.19 The P-value for the two-sided test with $t = 2.762$.

DO: From the Minitab output, $\bar{x} = 31.935$ ounces and $s_x = 2.394$ ounces.

- *Test statistic*

$$t = \frac{\bar{x} - \mu_0}{s_x / \sqrt{n}} = \frac{31.935 - 31}{2.394 / \sqrt{50}} = 2.762$$

- *P-value* Figure 9.19 displays the P-value for this two-sided test as an area under the t distribution curve with $50 - 1 = 49$ degrees of freedom.

Using technology: We can find the exact P-value using df = 49 with a calculator: `2*tcdf(2.762,100,49)` = 0.0081.

Using the table: Table B doesn't have an entry for df = 49, so we have to use the more conservative df = 40. Search the df = 40 row to find where $t = 2.762$ falls. As the excerpt at right shows, the observed t lies between 2.704 and 2.971, and the upper-tail probability is between 0.005 and 0.0025. So the desired P-value is between 2(0.005) and 2(0.0025), that is, between 0.01 and 0.005.

	Upper-tail probability p		
df	.005	.0025	.001
30	2.750	3.030	3.385
40	2.704	2.971	3.307
50	2.678	2.937	3.261
	99%	99.5%	99.8%
	Confidence level C		

CONCLUDE: The P-value, 0.0081, is less than our $\alpha = 0.05$ significance level, so we have enough evidence to reject H_0 and conclude that the mean weight of the pineapples in this year's crop is not 31 ounces.

(c) No. This was not a comparative experiment, so we cannot infer causation. It is possible that other things besides the irrigation system were different from last year's growing season. Maybe the weather was better this year, and that's why the pineapples were heavier.

The significance test in the previous example concludes that the mean weight μ of the pineapples grown in the field this year differs from last year's 31 ounces. Unfortunately, the test doesn't give us an idea of what the actual value of μ is. For that, we need a confidence interval.

EXAMPLE Juicy Pineapple

Confidence intervals give more information

Minitab output for a significance test and confidence interval based on the pineapple data is shown below. The test statistic and *P*-value match what we got earlier (up to rounding).

One-Sample T: Weight (oz)

```
Test of mu = 31 vs not = 31
Variable      N   Mean   StDev SE Mean        95% CI           T     P
Weight (oz)  50  31.935  2.394  0.339   (31.255, 32.616)  2.76  0.008
```

The 95% confidence interval for the mean weight of all the pineapples grown in the field this year is 31.255 to 32.616 ounces. We are 95% confident that this interval captures the true mean weight μ of this year's pineapple crop.

As with proportions, there is a link between a two-sided test at significance level α and a $100(1-\alpha)\%$ confidence interval for a population mean μ. For the pineapples, the two-sided test at $\alpha = 0.05$ rejects $H_0: \mu = 31$ in favor of $H_a: \mu \neq 31$. The corresponding 95% confidence interval does not include 31 as a plausible value of the parameter μ. In other words, the test and interval lead to the same conclusion about H_0. But the confidence interval provides much more information: a set of plausible values for the population mean.

The connection between two-sided tests and confidence intervals is even stronger for means than it was for proportions. That's because both inference methods for means use the standard error of $\bar{x}$ in the calculations.

$$\text{test statistic: } t = \frac{\bar{x} - \mu_0}{s_x / \sqrt{n}} \qquad \text{Confidence interval: } \bar{x} \pm t^* \frac{s_x}{\sqrt{n}}$$

When the two-sided significance test at level α rejects $H_0: \mu = \mu_0$, the $100(1-\alpha)\%$ confidence interval for μ will not contain the hypothesized value μ_0. And when the test fails to reject the null hypothesis, the confidence interval will contain μ_0.

THINK ABOUT IT

Is there a connection between one-sided tests and confidence intervals for a population mean? As you might expect, the answer is yes. But the link is more complicated. Consider a one-sided test of $H_0: \mu = 10$ versus $H_a: \mu > 10$ based on an SRS of 30 observations. With df $= 30 - 1 = 29$, Table B says that the test will reject H_0 at $\alpha = 0.05$ if the test statistic t is greater than 1.699. For this to happen, the sample mean $\bar{x}$ would have to exceed $\mu_0 = 10$ by more than 1.699 standardized units.

Table B also shows that $t^* = 1.699$ is the critical value for a 90% confidence interval. That is, a 90% confidence interval will extend 1.699 standardized units on either side of the sample mean $\bar{x}$. If $\bar{x}$ exceeds 10 by more than 1.699 standardized units, the resulting interval will not include 10. And the one-sided test will reject $H_0: \mu = 10$. There's the link: our one-sided test at $\alpha = 0.05$ gives the same conclusion about H_0 as a 90% confidence interval for μ.

CHECK YOUR UNDERSTANDING

The health director of a large company is concerned about the effects of stress on the company's middle-aged male employees. According to the National Center for Health Statistics, the mean systolic blood pressure for males 35 to 44 years of age is 128. The health director examines the medical records of a random sample of 72 male employees in this age group. The Minitab output below displays the results of a significance test and a confidence interval.

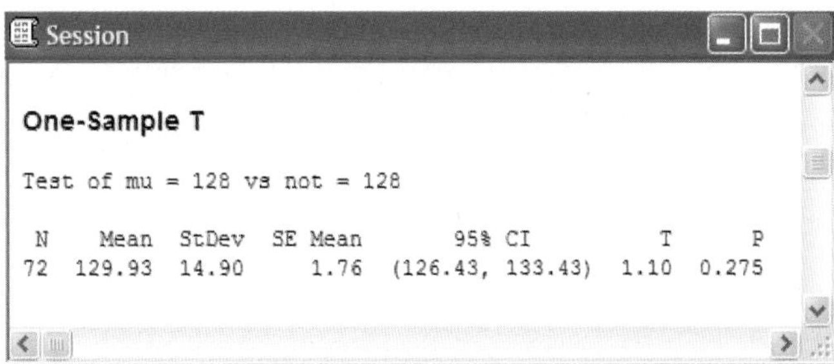

Session

One-Sample T

Test of mu = 128 vs not = 128

N	Mean	StDev	SE Mean	95% CI	T	P
72	129.93	14.90	1.76	(126.43, 133.43)	1.10	0.275

1. Do the results of the significance test allow us to conclude that the mean blood pressure for all the company's middle-aged male employees differs from the national average? Justify your answer.

2. Interpret the 95% confidence interval in context. Explain how the confidence interval leads to the same conclusion as in Question 1.

Inference for Means: Paired Data

Paired data

Comparative studies are more convincing than single-sample investigations. For that reason, one-sample inference is less common than comparative inference. Study designs that involve making two observations on the same individual, or one observation on each of two similar individuals, result in **paired data.** We saw an example of paired data in the job satisfaction study of Section 9.1. In this matched pairs experiment, each worker's satisfaction was recorded twice—after self-paced assembly work and after machine-paced assembly work.

Paired *t* procedures

When paired data result from measuring the same quantitative variable twice, as in the job satisfaction study, we can make comparisons by analyzing the differences in each pair. If the conditions for inference are met, we can use one-sample *t* procedures to perform inference about the mean difference μ_d. (These methods are sometimes called **paired *t* procedures.**) An example should help illustrate what we mean.

Is Caffeine Dependence Real?

Paired data and one-sample *t* procedures

Researchers designed an experiment to study the effects of caffeine withdrawal. They recruited 11 volunteers who were diagnosed as being caffeine dependent to serve as subjects. Each subject was barred from coffee, colas, and other substances with caffeine for the duration of the experiment. During one two-day period, subjects took capsules containing their normal caffeine intake. During another two-day period, they took placebo capsules. The order in which subjects took caffeine and the placebo was randomized. At the end of each two-day period, a test for depression was given to all 11 subjects. Researchers wanted to know whether being deprived of caffeine would lead to an increase in depression.[20]

The table below contains data on the subjects' scores on a depression test. Higher scores show more symptoms of depression.

Results of a caffeine-deprivation study		
Subject	**Depression (caffeine)**	**Depression (placebo)**
1	5	16
2	5	23
3	4	5
4	3	7
5	8	14
6	5	24
7	0	6
8	0	3
9	2	15
10	11	12
11	1	0

PROBLEM:

(a) *Why did researchers randomly assign the order in which subjects received placebo and caffeine?*

(b) *Carry out a test to investigate the researchers' question.*

SOLUTION:

(a) Researchers want to be able to conclude that any statistically significant change in depression score is due to the treatments themselves and not to some lurking variable. One such lurking variable is the order of the treatments. Maybe if caffeine were given first to all the subjects, it might lead to less depression during the placebo treatment than if the placebo were given first. Random assignment of the caffeine and placebo to the two time periods in the experiment should help ensure that no other variable is systematically affecting subjects' responses.

(b) We'll follow the four-step process.

STATE: If caffeine deprivation has no effect on depression, then we would expect the actual mean difference in depression scores to be 0. We therefore want to test the hypotheses

$$H_0: \mu_d = 0$$
$$H_a: \mu_d > 0$$

where μ_d is the true mean difference (placebo − caffeine) in depression score. (We chose this order of subtraction to get mostly positive values.) Since no significance level is given, we'll use $\alpha = 0.05$.

PLAN: If the conditions are met, we should conduct a paired *t* test for μ_d.

- *Random* Researchers randomly assigned the treatment order—placebo then caffeine, caffeine then placebo—to the subjects.
- *Normal* We don't know whether the actual distribution of difference in depression scores (placebo − caffeine) is Normal. With such a small sample size ($n = 11$), we need to examine the data to see if it's safe to use *t* procedures. Figure 9.20 shows hand sketches of a calculator histogram, boxplot, and Normal probability plot for these data. The histogram has an irregular shape with so few values; the boxplot shows some right-skewness but not outliers; and the Normal probability plot looks fairly linear. With no outliers or strong skewness, the *t* procedures should be pretty accurate.
- *Independent* We aren't sampling, so it isn't necessary to check the 10% condition. We will assume that the changes in depression scores for individual subjects are independent. This is reasonable if the experiment is conducted properly.

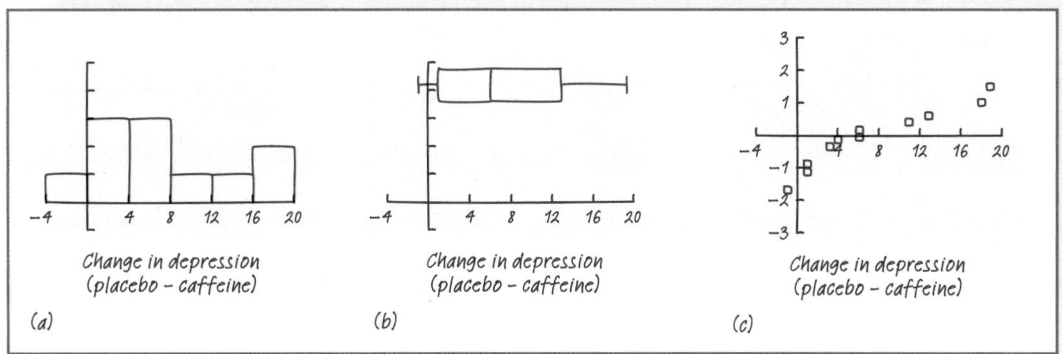

FIGURE 9.20 Sketches of (a) a histogram, (b) a boxplot, and (c) a Normal probability plot of the change in depression scores (placebo − caffeine) for the 11 subjects in the caffeine experiment.

DO: We entered the caffeine data in list1 and the placebo data in list2, and then we stored the difference in depression scores (list2 − list1) in list3. One-variable statistics on the differences gives $\bar{x}_d = 7.364$ and $s_d = 6.918$.

```
1-Var Stats
 x̄=7.363636364
 Σx=81
 Σx²=1075
 Sx=6.917697988
 σx=6.595766235
↓n=11
```

Using the calculator's *t* test command with the "Draw" option yields

- *Test statistic* $t = 3.53$
- *P-value* 0.0027, which is the area to the right of $t = 3.53$ on the *t* distribution curve with $df = 11 - 1 = 10$.

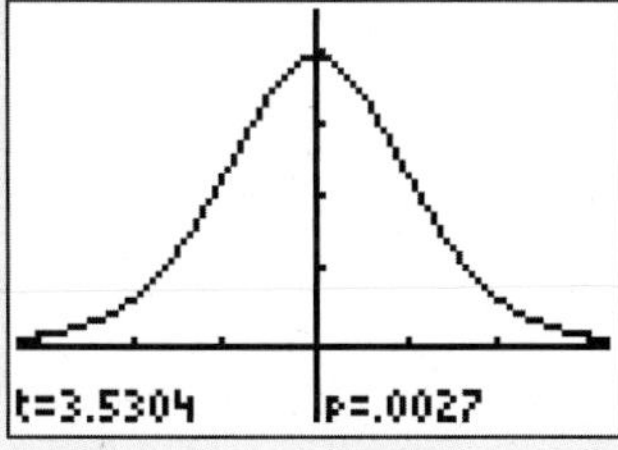

Just by looking at the data, it appears that the true mean change in depression score μ_d is greater than 0. However, it's possible that there has been no change and we got a result this much larger than $\mu_d = 0$ by the luck of the random assignment. The significance test tells us whether this explanation is plausible.

Note: The calculator doesn't report the degrees of freedom, but you should.

CONCLUDE: With a *P*-value of 0.0027, which is much less than our chosen $\alpha = 0.05$, we have convincing evidence to reject $H_0: \mu_d = 0$. We can therefore conclude that depriving these caffeine-dependent subjects of caffeine caused an average increase in depression scores.

For Practice *Try Exercise 89*

A few follow-up comments about this example are in order.

1. Although the differences in depression test scores for the 11 subjects are independent, the two depression scores for each subject are *not* independent. When you're working with paired data, the Independent condition refers to the differences.

2. We could have calculated the test statistic in the example using the formula

$$t = \frac{\bar{x}_d - \mu_0}{s_d / \sqrt{n}} = \frac{7.364 - 0}{6.918 / \sqrt{11}} = 3.53$$

and obtained the *P*-value using Table B or the calculator's `tcdf` command. Check with your teacher on whether the calculator-only method is acceptable. *Be sure to report the degrees of freedom with any t procedure, even if technology doesn't.*

CAUTION

3. The subjects in this experiment were *not* chosen at random from the population of caffeine-dependent individuals. As a result, we can't generalize our findings to *all* caffeine-dependent people—only to individuals like the ones who took part in this experiment. Because researchers randomly assigned the treatments, however, they can make an inference about cause and effect.

4. Remember: a confidence interval gives more information than a significance test. A 90% confidence interval for the true mean difference (placebo − caffeine) in depression score is

$$\bar{x}_d \pm t^* \frac{s_d}{\sqrt{n}} = 7.364 \pm 1.812 \frac{6.918}{\sqrt{11}} = 7.364 \pm 3.780 = (3.584, 11.144)$$

We are 90% confident that the interval from 3.584 to 11.144 captures the true mean change in depression score μ_d. That is, the data give convincing evidence that depriving individuals like the ones in this experiment of caffeine results in an average increase of between 3.5 and 11 points in depression test score.

Until now, we have only used one-sample t procedures in settings involving random sampling. The paired data in the caffeine example came from a matched pairs experiment, in which each subject received both treatments in a random order. A coin toss or some other chance process was used to carry out the random assignment. Why is it legitimate to use a t distribution to perform inference about the parameter μ in a randomized experiment? The answer to that question will have to wait until the next chapter.

DATA EXPLORATION *Nitrogen in tires—a lot of hot air?*

Most automobile tires are inflated with compressed air, which consists of about 78% nitrogen. Aircraft tires are filled entirely with nitrogen, which is safer than air in case of fire. Could filling automobile tires with nitrogen improve safety, performance, or both?

Consumers Union designed a study to test whether nitrogen-filled tires would maintain pressure better than air-filled tires. They obtained two tires from each of several brands and then filled one tire in each pair with air and one with nitrogen. (The article didn't say, but assume that these assignments were made at random.) All tires were inflated to a pressure of 30 pounds per square inch and then placed outside for a year. At the end of the year, Consumers Union measured the pressure in each tire.[21]

The amount of pressure lost (in pounds per square inch) during the year for the air-filled and nitrogen-filled tires of each brand is shown in the table below.

Brand	Air	Nitrogen	Brand	Air	Nitrogen
BF Goodrich Traction T/A HR	7.6	7.2	Pirelli P6 Four Seasons	4.4	4.2
Bridgestone HP50 (Sears)	3.8	2.5	Sumitomo HTR H4	1.4	2.1
Bridgestone Potenza EL400	2.1	1.0	Yokohama Avid H4S	4.3	3.0
Bridgestone Potenza G009	3.7	1.6	BF Goodrich Traction T/A V	5.5	3.4
Bridgestone Potenza RE950	4.7	1.5	Bridgestone Potenza RE950	4.1	2.8
Continental Premier Contact H	4.9	3.1	Continental ContiExtreme Contact	5.0	3.4
Cooper Lifeliner Touring SLE	5.2	3.5	Continental ContiProContact	4.8	3.3
Dayton Daytona HR	3.4	3.2	Cooper Lifeliner Touring SLE	3.2	2.5
Faken Ziex ZE-512	4.1	3.3	General Exclaim UHP	6.8	2.7
Fuzion Hrl	2.7	2.2	Hankook Ventus V4 H105	3.1	1.4
General Exclaim	3.1	3.4	Michelin Energy MXV4 Plus	2.5	1.5
Goodyear Assurance Tripletred	3.8	3.2	Michelin Pilot Exalto A/S	6.6	2.2
Hankook Optimo H418	3.0	0.9	Michelin Pilot HX MXM4	2.2	2.0
Kumho Solus KH16	6.2	3.4	Pirelli P6 Four Seasons	2.5	2.7
Michelin Energy MXV4 Plus	2.0	1.8	Sumitomo HTR+	4.4	3.7
Michelin Pilot XGT H4	1.1	0.7			

1. Make a scatterplot of the data using air pressure loss as the explanatory variable and nitrogen pressure loss as the response variable. How does the graph show that these are paired data?

2. Add the line $y = x$ to your scatterplot. What do you notice? Explain what this pattern means.

Now let's analyze the differences.

3. Check conditions for performing inference about the true mean difference μ in pressure loss (air − nitrogen).

Although the sample size is large ($n = 31$), there are two high outliers that will affect any results we get using t procedures. For that reason, you need to do Questions 4 and 5 *twice*—once with the outliers included and once with the outliers removed.

4. Carry out a significance test at the $\alpha = 0.05$ level. Do these data give convincing evidence that filling tires with nitrogen instead of air decreases pressure loss?

5. Construct and interpret a 95% confidence interval for μ.

6. Describe the effect that the two outliers had on the significance test and confidence interval.

Using Tests Wisely

Significance tests are widely used in reporting the results of research in many fields. New drugs require significant evidence of effectiveness and safety. Courts ask about statistical significance in hearing discrimination cases. Marketers want to know whether a new ad campaign significantly outperforms the old one, and medical researchers want to know whether a new therapy performs significantly better. In all these uses, statistical significance is valued because it points to an effect that is unlikely to occur simply by chance.

Carrying out a significance test is often quite simple, especially if you use a calculator or computer. Using tests wisely is not so simple. Here are some points to keep in mind when using or interpreting significance tests.

Statistical Significance and Practical Importance When a null hypothesis ("no effect" or "no difference") can be rejected at the usual levels ($\alpha = 0.05$ or $\alpha = 0.01$), there is good evidence of a difference. But that difference may be very small. When large samples are available, even tiny deviations from the null hypothesis will be significant.

EXAMPLE

Wound Healing Time

Significant doesn't mean important

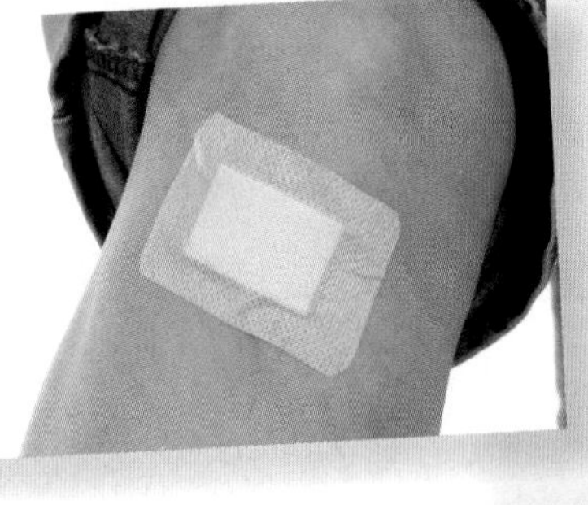

Suppose we're testing a new antibacterial cream, "Formulation NS," on a small cut made on the inner forearm. We know from previous research that with no medication, the mean healing time (defined as the time for the scab to fall off) is 7.6 days with a standard deviation of 1.4 days. The claim we want to test here is that Formulation NS speeds healing. We will use a 5% significance level.

Procedure: We cut a random sample of 25 college students and apply Formulation NS to the wounds. The mean healing time for these subjects is $\bar{x} = 7.1$ days and the standard deviation is $s_x = 1.4$ days.

Discussion: We want to test a claim about the mean healing time μ in the population of college students whose cuts are treated with Formulation NS. Our hypotheses are

$$H_0: \mu = 7.6 \text{ days}$$
$$H_a: \mu < 7.6 \text{ days}$$

An examination of the data reveals no outliers or strong skewness, so the conditions for performing a one-sample t test are met.We carry out the test and find that $t = -1.79$ and P-value $= 0.043$. Since 0.043 is less than $\alpha = 0.05$, we reject H_0 and conclude that Formulation NS's healing effect is statistically significant. However, this result is not practically important. Having your scab fall off half a day sooner is no big deal.

CAUTION

Remember the wise saying: *Statistical significance is not the same thing as practical importance*. The remedy for attaching too much importance to statistical significance is to pay attention to the actual data as well as to the P-value. Plot your data and examine them carefully. Are there outliers or other deviations from a consistent pattern? A few outlying observations can produce highly significant results if you blindly apply common significance tests. Outliers can also destroy the significance of otherwise-convincing data.

The foolish user of statistics who feeds the data to a calculator or computer without exploratory analysis will often be embarrassed. Is the difference you are seeking visible in your plots? If not, ask yourself whether the difference is large enough to be practically important. Give a confidence interval for the parameter in which you are interested. A confidence interval actually estimates the size of the difference rather than simply asking if it is too large to reasonably occur by chance alone. Confidence intervals are not used as often as they should be, whereas significance tests are perhaps overused.

Don't Ignore Lack of Significance There is a tendency to infer that there is no difference whenever a *P*-value fails to attain the usual 5% standard. A provocative editorial in the *British Medical Journal* entitled "Absence of Evidence Is Not Evidence of Absence" deals with this issue.[22] Here is one of the examples they cite.

Reducing HIV Transmission

Interpreting lack of significance

In an experiment to compare methods for reducing transmission of HIV, subjects were randomly assigned to a treatment group and a control group. Result: the treatment group and the control group had the same rate of HIV infection. Researchers described this as an "incident rate ratio" of 1.00. A ratio above 1.00 would mean that there was a greater rate of HIV infection in the treatment group, while a ratio below 1.00 would indicate a greater rate of HIV infection in the control group. The 95% confidence interval for the incident rate ratio was reported as 0.63 to 1.58.[23] Saying that the treatment has no effect on HIV infection is misleading. The confidence interval for the incident rate ratio indicates that the treatment may be able to achieve a 37% decrease in infection. It might also produce a 58% increase in infection. Clearly, more data are needed to distinguish between these possibilities.

The situation can be worse. Research in some fields has rarely been published unless significance at the 5% level is attained. For instance, a survey of four journals published by the American Psychological Association showed that of 294 articles using statistical tests, only 8 reported results that did not attain the 5% significance level.[24] That's too bad, because we can learn a great deal from studies that fail to find convincing evidence.

In some areas of research, small differences that are detectable only with large sample sizes can be of great practical significance. Data accumulated from a large number of patients taking a new drug may be needed before we can conclude that there are life-threatening consequences for a small number of people. *When planning a study, verify that the test you plan to use has a high probability (power) of detecting a difference of the size you hope to find.*

CAUTION

Statistical Inference Is Not Valid for All Sets of Data Badly designed surveys or experiments often produce invalid results. Formal statistical inference cannot correct basic flaws in the design. Each test is valid only in certain circumstances, with properly produced data being particularly important.

Does Music Increase Worker Productivity?

When inference isn't valid

You wonder whether background music would improve the productivity of the staff who process mail orders in your business. After discussing the idea with the workers, you add music and find a significant increase. You should not be impressed. In fact, almost any change in the work environment together with knowledge that a study is under way will produce a short-term productivity increase. This is the **Hawthorne effect,** named after the Western Electric manufacturing plant where it was first noted.

Hawthorne effect

The significance test correctly informs you that an increase has occurred that is larger than would often arise by chance alone. It does not tell you what other than chance caused the increase. The most plausible explanation is that workers change their behavior when they know they are being studied. Your experiment was uncontrolled, so the significant result cannot be interpreted. A randomized comparative experiment would isolate the actual effect of background music and so make significance meaningful.

Significance tests and confidence intervals are based on the laws of probability. Random sampling and random assignment ensure that these laws apply. Always ask how the data were produced. Don't be too impressed by *P*-values on a printout until you are confident that the data deserve a formal analysis.

By permission of xkcd.com

Beware of Multiple Analyses Statistical significance ought to mean that you have found a difference that you were looking for. The reasoning behind statistical significance works well if you decide what difference you are seeking, design a study to search for it, and use a significance test to weigh the evidence you get. In other settings, significance may have little meaning.

Cell Phones and Brain Cancer

Don't search for significance!

Might the radiation from cell phones be harmful to users? Many studies have found little or no connection between using cell phones and various illnesses. Here is part of a news account of one study:

> A hospital study that compared brain cancer patients and a similar group without brain cancer found no statistically significant association between cell phone use and

a group of brain cancers known as gliomas. But when 20 types of glioma were considered separately, an association was found between phone use and one rare form. Puzzlingly, however, this risk appeared to decrease rather than increase with greater mobile phone use.[25]

Think for a moment. Suppose that the 20 null hypotheses for these 20 significance tests are all true. Then each test has a 5% chance of being significant at the 5% level. That's what $\alpha = 0.05$ means: results this extreme occur only 5% of the time just by chance when the null hypothesis is true. We expect about 1 of 20 tests to give a significant result just by chance. Running one test and reaching the $\alpha = 0.05$ level is reasonably good evidence that you have found something; running 20 tests and reaching that level only once is not.

Searching data for patterns is certainly legitimate. Exploratory data analysis is an important part of statistics. But the reasoning of formal inference does not apply when your search for a striking effect in the data is successful. The remedy is clear. Once you have a hypothesis, design a study to search specifically for the effect you now think is there. If the result of this study is statistically significant, you have real evidence.

case closed

Do You Have a Fever?

At the beginning of the chapter, we described a study investigating whether "normal" human body temperature is really 98.6°F. As is often the case, the researchers did not provide their original data. Allen Shoemaker, from Calvin College, produced a data set with the same basic properties as the original temperature readings. His data set consists of one oral temperature reading for each of 130 randomly chosen, healthy 18- to 40-year-olds. Here is a summary of the details we provided in the chapter-opening Case Study (page 527).

- The mean temperature was $\bar{x} = 98.25$°F.
- The standard deviation of the temperature readings was $s_x = 0.73$°F.
- 62.3% of the temperature readings were less than 98.6°F.

Do these data provide convincing evidence that normal body temperature is *not* 98.6°F?

Let μ = the actual mean body temperature (in °F) in the population of healthy 18- to 40-year-olds. The Minitab output below shows the results of a one-sample t test and a 95% confidence interval for μ. (We'll ask you to check conditions in Exercise 107.)

One-Sample T

Test of mu = 98.6 vs not = 98.6

N	Mean	StDev	SE Mean	95% CI	T	P
130	98.2500	0.7300	0.0640	(98.1233, 98.3767)	−5.47	0.000

Since the *P*-value is almost 0, we have very strong evidence to reject H_0 and conclude that the mean body temperature μ for healthy adults of this age is different from 98.6. We are 95% confident that the interval from 98.12°F to 98.38°F contains the actual mean body temperature of healthy adults aged 18 to 40.

We could also do inference about the population proportion p of healthy 18- to 40-year-olds with body temperatures below 98.6°F. We'll ask you to do just that in Exercise 108.

SECTION 9.3 Summary

- Significance tests for the mean μ of a Normal population are based on the sampling distribution of the sample mean $\bar{x}$. Due to the central limit theorem, the resulting procedures are approximately correct for other population distributions when the sample is large.
- If we somehow know σ, we can use a z test statistic and the standard Normal distribution to perform calculations. In practice, we typically do not know σ. Then, we use the **one-sample *t* statistic**

$$t = \frac{\bar{x} - \mu_0}{s_x / \sqrt{n}}$$

  with *P*-values calculated from the t distribution with $n - 1$ degrees of freedom.
- The **one-sample *t* test** is approximately correct when

  Random The data were produced by random sampling or a randomized experiment.

  Normal The population distribution is Normal OR the sample size is large ($n \geq 30$).

  Independent Individual observations are independent. When sampling without replacement, check that the population is at least 10 times as large as the sample.
- Confidence intervals provide additional information that significance tests do not—namely, a range of plausible values for the parameter μ. A two-sided test of $H_0: \mu = \mu_0$ at significance level α gives the same conclusion as a $100(1 - \alpha)\%$ confidence interval for μ.
- Analyze **paired data** by first taking the difference within each pair to produce a single sample. Then use one-sample t procedures.
- Very small differences can be highly significant (small *P*-value) when a test is based on a large sample. A statistically significant difference need not be practically important. Plot the data to display the difference you are seeking, and use confidence intervals to estimate the actual values of parameters.

- On the other hand, lack of significance does not imply that H_0 is true. Even a large difference can fail to be significant when a test is based on a small sample.
- Significance tests are not always valid. Faulty data collection, outliers in the data, and other practical problems can invalidate a test. Many tests run at once will probably produce some significant results by chance alone, even if all the null hypotheses are true.

9.3 TECHNOLOGY CORNERS

Computing P-values from t distributions on the calculator .... page 569
One-sample t test on the calculator........ page 573

TI-Nspire instructions in Appendix B

SECTION 9.3 Exercises

63. **Attitudes** The Survey of Study Habits and Attitudes (SSHA) is a psychological test that measures students' attitudes toward school and study habits. Scores range from 0 to 200. The mean score for U.S. college students is about 115. A teacher suspects that older students have better attitudes toward school. She gives the SSHA to an SRS of 45 of the over 1000 students at her college who are at least 30 years of age. Check the conditions for carrying out a significance test of the teacher's suspicion.

pg 566

64. **Anemia** Hemoglobin is a protein in red blood cells that carries oxygen from the lungs to body tissues. People with fewer than 12 grams of hemoglobin per deciliter of blood (g/dl) are anemic. A public health official in Jordan suspects that Jordanian children are at risk of anemia. He measures a random sample of 50 children. Check the conditions for carrying out a significance test of the official's suspicion.

65. **Paying high prices?** A retailer entered into an exclusive agreement with a supplier who guaranteed to provide all products at competitive prices. The retailer eventually began to purchase supplies from other vendors who offered better prices. The original supplier filed a lawsuit claiming violation of the agreement. In defense, the retailer had an audit performed on a random sample of 25 invoices. For each audited invoice, all purchases made from other suppliers were examined and compared with those offered by the original supplier. The percent of purchases on each invoice for which an alternative supplier offered a lower price than the original supplier was recorded.[26] For example, a data value of 38 means that the price would be lower with a different supplier for 38% of the items on the invoice. A histogram and some computer output for these data are shown below. Explain why we should not carry out a one-sample t test in this setting.

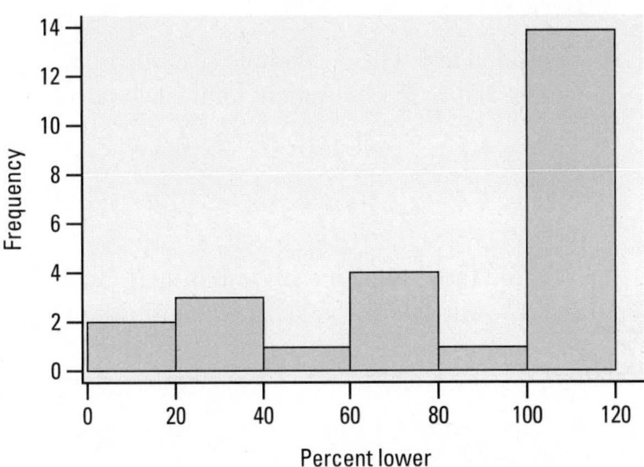

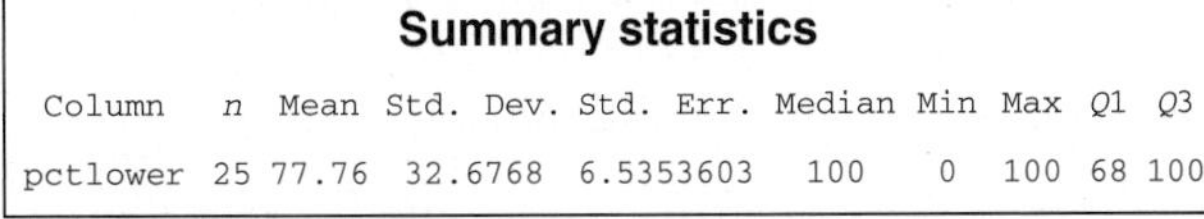

Summary statistics

Column	n	Mean	Std. Dev.	Std. Err.	Median	Min	Max	$Q1$	$Q3$
pctlower	25	77.76	32.6768	6.5353603	100	0	100	68	100

66. **Ancient air** The composition of the earth's atmosphere may have changed over time. To try to discover the nature of the atmosphere long ago, we can examine the gas in bubbles inside ancient amber. Amber is tree resin that has hardened and been trapped in rocks. The gas in bubbles within amber should be a sample of the atmosphere at the time the amber was formed. Measurements on 9 specimens of amber from the late Cretaceous era (75 to 95 million years ago) give these percents of nitrogen:[27]

63.4 65.0 64.4 63.3 54.8 64.5 60.8 49.1 51.0

Explain why we should not carry out a one-sample t test in this setting.

67. **Attitudes** In the study of older students' attitudes from Exercise 63, the sample mean SSHA score was 125.7 and the sample standard deviation was 29.8.

(a) Calculate the test statistic.

(b) Find the P-value using Table B. Then obtain a more precise P-value from your calculator.

68. **Anemia** For the study of Jordanian children in Exercise 64, the sample mean hemoglobin level was 11.3 mg/dl and the sample standard deviation was 1.6 mg/dl.

(a) Calculate the test statistic.

(b) Find the P-value using Table B. Then obtain a more precise P-value from your calculator.

69. **One-sided test** Suppose you carry out a significance test of $H_0: \mu = 5$ versus $H_a: \mu > 5$ based on a sample of size $n = 20$ and obtain $t = 1.81$.

(a) Find the P-value for this test using (i) Table B and (ii) your calculator. What conclusion would you draw at the 5% significance level? At the 1% significance level?

(b) Redo part (a) using an alternative hypothesis of $H_a: \mu \neq 5$.

70. **Two-sided test** The one-sample t statistic from a sample of $n = 25$ observations for the two-sided test of

$$H_0: \mu = 64$$
$$H_a: \mu \neq 64$$

has the value $t = -1.12$.

(a) Find the P-value for this test using (i) Table B and (ii) your calculator. What conclusion would you draw at the 5% significance level? At the 1% significance level?

(b) Redo part (a) using an alternative hypothesis of $H_a: \mu < 64$.

pg 571

71. **Sweetening colas** Cola makers test new recipes for loss of sweetness during storage. Trained tasters rate the sweetness before and after storage. From experience, the population distribution of sweetness losses will be close to Normal. Here are the sweetness losses (sweetness before storage minus sweetness after storage) found by tasters from a random sample of 10 batches of a new cola recipe:

2.0 0.4 0.7 2.0 −0.4 2.2 −1.3 1.2 1.1 2.3

Are these data good evidence that the cola lost sweetness? Carry out a test to help you answer this question.

72. **Heat through the glass** How well materials conduct heat matters when designing houses, for example. Conductivity is measured in terms of watts of heat power transmitted per square meter of surface per degree Celsius of temperature difference on the two sides of the material. In these units, glass has conductivity about 1. The National Institute of Standards and Technology provides exact data on properties of materials. Here are measurements of the heat conductivity of 11 randomly selected pieces of a particular type of glass:[28]

1.11 1.07 1.11 1.07 1.12 1.08 1.08 1.18 1.18 1.18 1.12

Is there convincing evidence that the conductivity of this type of glass is greater than 1? Carry out a test to help you answer this question.

73. **Healthy bones** The recommended daily allowance (RDA) of calcium for women between the ages of 18 and 24 years is 1200 milligrams (mg). Researchers who were involved in a large-scale study of women's bone health suspected that their participants had significantly lower calcium intakes than the RDA. To test this suspicion, the researchers measured the daily calcium intake of a random sample of 36 women from the study who fell in the desired age range. The Minitab output below displays descriptive statistics for these data, along with the results of a significance test.

Descriptive Statistics: Calcium intake (mg)

Variable	N	Mean	SE Mean	StDev	Min	Q1	Med	Q3	Maximum
Calcium	36	856.2	51.1	306.7	374.0	632.3	805.0	1090.5	1425.0

One-Sample T: Calcium intake (mg)

Test of mu = 1200 vs < 1200

Variable	N	Mean	StDev	SE Mean	T	P
Calcium	36	856.2	306.7	51.1	-6.73	0.000

(a) Determine whether there are any outliers. Show your work.

(b) Interpret the P-value in context.

(c) Do these data give convincing evidence to support the researchers' suspicion? Carry out a test to help you answer this question.

74. **Taking stock** An investor with a stock portfolio worth several hundred thousand dollars sued his broker due to the low returns he got from the portfolio at a time when the stock market did well overall. The investor's lawyer wants to compare the broker's performance against the market as a whole. He collects data on the broker's returns for a random sample of 36 weeks. Over the 10-year period that the broker has managed portfolios, stocks in the Standard & Poor's 500 index gained an average of 0.95% per month. The Minitab output below displays descriptive statistics for these data, along with the results of a significance test.

Descriptive Statistics: Return (percent)

Variable	N	Mean	SE Mean	StDev	Min	Q1	Median	Q3	Max
Return(%)	36	-1.441	0.802	4.810	-10.270	-3.418	-1.425	1.543	7.340

One-Sample T: Return (percent)

Test of mu = 0.95 vs < 0.95

Variable	N	Mean	StDev	SE Mean	T	P
Return (percent)	36	-1.441	4.810	0.802	-2.98	0.003

(a) Determine whether there are any outliers. Show your work.

(b) Interpret the *P*-value in context.

(c) Do these data give convincing evidence to support the lawyer's case? Carry out a test to help you answer this question.

pg 574

75. **Growing tomatoes** An agricultural field trial compares the yield of two varieties of tomatoes for commercial use. Researchers randomly select 10 Variety A and 10 Variety B tomato plants. Then the researchers divide in half each of 10 small plots of land in different locations. For each plot, a coin toss determines which half of the plot gets a Variety A plant; a Variety B plant goes in the other half. After harvest, they compare the yield in pounds for the plants at each location. The 10 differences (Variety A − Variety B) give $\bar{x} = 0.34$ and $s_x = 0.83$. A graph of the differences looks roughly symmetric and single-peaked with no outliers. Is there convincing evidence that Variety A has the higher mean yield? Perform a significance test using $\alpha = 0.05$ to answer the question.

76. **Study more!** A student group claims that first-year students at a university study 2.5 hours per night during the school week. A skeptic suspects that they study less than that on average. He takes a random sample of 30 first-year students and finds that $\bar{x} = 137$ minutes and $s_x = 45$ minutes. A graph of the data shows no outliers but some skewness. Carry out an appropriate significance test at the 5% significance level. What conclusion do you draw?

pg 571

77. **The power of tomatoes** The researchers who carried out the experiment in Exercise 75 suspect that the large *P*-value (0.114) is due to low power.

(a) Describe a Type I and a Type II error in this setting. Which type of error could you have made in Exercise 75? Why?

(b) Explain two ways that the researchers could have increased the power of the test to detect $\mu = 0.5$.

78. **Study more!** The significance test in Exercise 76 yields a *P*-value of 0.0622.

(a) Describe a Type I and a Type II error in this setting. Which type of error could you have made in Exercise 76? Why?

(b) Which of the following changes would give the test a higher power to detect $\mu = 120$ minutes: using $\alpha = 0.01$ or $\alpha = 0.10$? Explain.

79. **Pressing pills** A drug manufacturer forms tablets by compressing a granular material that contains the active ingredient and various fillers. The hardness of a sample from each batch of tablets produced is measured to control the compression process. The target value for the hardness is $\mu = 11.5$. The hardness data for a random sample of 20 tablets are

11.627	11.613	11.493	11.602	11.360
11.374	11.592	11.458	11.552	11.463
11.383	11.715	11.485	11.509	11.429
11.477	11.570	11.623	11.472	11.531

Is there significant evidence at the 5% level that the mean hardness of the tablets differs from the target value? Carry out an appropriate test to support your answer.

80. **Filling cola bottles** Bottles of a popular cola are supposed to contain 300 milliliters (ml) of cola. There is some variation from bottle to bottle because the filling machinery is not perfectly precise. From experience, the distribution of the contents is approximately Normal. An inspector measures the contents of six randomly selected bottles from a single day's production. The results are

299.4 297.7 301.0 298.9 300.2 297.0

Do these data provide convincing evidence that the mean amount of cola in all the bottles filled that day differs from the target value of 300 ml? Carry out an appropriate test to support your answer.

81. **Pressing pills** Refer to Exercise 79. Construct and interpret a 95% confidence interval for the population mean μ. What additional information does the confidence interval provide?

82. **Filling cola bottles** Refer to Exercise 80. Construct and interpret a 95% confidence interval for the population mean μ. What additional information does the confidence interval provide?

83. **Fast connection?** How long does it take for a chunk of information to travel from one server to another and back on the Internet? According to the site internettrafficreport.com, a typical response time is 200 milliseconds (about one-fifth of a second). Researchers collected data on response times of a random sample of 14 servers in Europe. A graph of the data reveals no strong skewness or outliers. The figure below displays Minitab output for a one-sample t interval for the population mean. Is there convincing evidence at the 5% significance level that the site's claim is incorrect? Use the confidence interval to justify your answer.

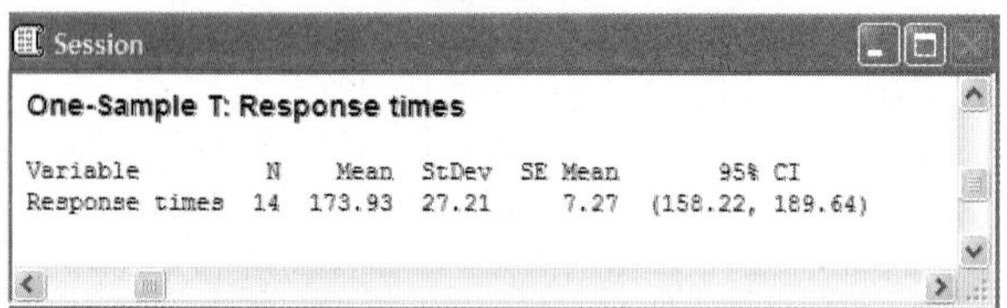

Session

One-Sample T: Response times

Variable	N	Mean	StDev	SE Mean	95% CI
Response times	14	173.93	27.21	7.27	(158.22, 189.64)

84. **Water!** A blogger claims that U.S. adults drink an average of five 8-ounce glasses of water per day. Skeptical researchers ask a random sample of 24 U.S. adults about their daily water intake. A graph of the data shows a roughly symmetric shape with no outliers. The figure below displays Minitab output for a one-sample t interval for the population mean. Is there convincing evidence at the 10% significance level that the blogger's claim is incorrect? Use the confidence interval to justify your answer.

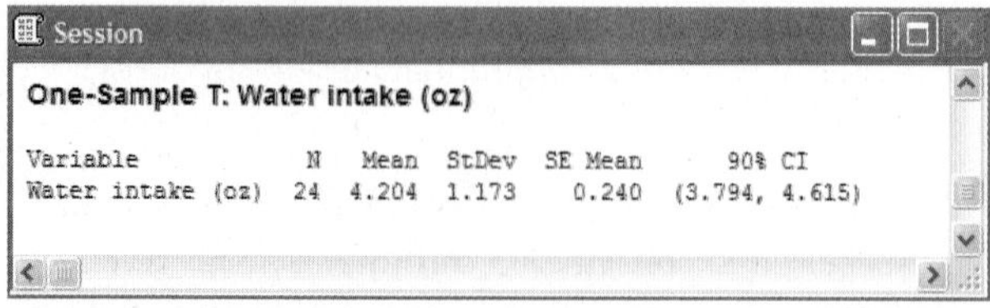

Session

One-Sample T: Water intake (oz)

Variable	N	Mean	StDev	SE Mean	90% CI
Water intake (oz)	24	4.204	1.173	0.240	(3.794, 4.615)

85. **Tests and CIs** The P-value for a two-sided test of the null hypothesis $H_0: \mu = 10$ is 0.06.

(a) Does the 95% confidence interval for μ include 10? Why or why not?

(b) Does the 90% confidence interval for μ include 10? Why or why not?

86. **Tests and CIs** The P-value for a one-sided test of the null hypothesis $H_0: \mu = 15$ is 0.03.

(a) Does the 99% confidence interval for μ include 15? Why or why not?

(b) Does the 95% confidence interval for μ include 15? Why or why not?

Exercises 87 and 88 refer to the following setting. An understanding of cockroach biology may lead to an effective control strategy for these annoying insects. Researchers studying the absorption of sugar by insects feed a random sample of cockroaches a diet containing measured amounts of sugar. After 10 hours, the cockroaches are killed and the concentration of the sugar in various body parts is determined by a chemical analysis. The paper that reports the research states that a 95% confidence interval for the mean amount (in milligrams) of sugar in the hindguts of cockroaches is 4.2 ± 2.3.[29]

87. **Cockroaches** Does this paper give convincing evidence that the mean amount of sugar in the hindguts under these conditions is not equal to 7 mg? Justify your answer.

88. **Cockroaches** Would the hypothesis that $\mu = 5$ mg be rejected at the 5% level in favor of a two-sided alternative? Justify your answer.

89. **Right versus left** The design of controls and instruments affects how easily people can use them. A student project investigated this effect by asking 25 right-handed students to turn a knob (with their right hands) that moved an indicator. There were two identical instruments, one with a right-hand thread (the knob turns clockwise) and the other with a left-hand thread (the knob must be turned counterclockwise). Each of the 25 students used both instruments in a random order. The following table gives the times in seconds each subject took to move the indicator a fixed distance:[30]

pg 577

Subject	Right thread	Left thread
1	113	137
2	105	105
3	130	133
4	101	108
5	138	115
6	118	170
7	87	103
8	116	145
9	75	78
10	96	107
11	122	84
12	103	148
13	116	147
14	107	87
15	118	166
16	103	146
17	111	123
18	104	135
19	111	112

Subject	Right thread	Left thread
20	89	93
21	78	76
22	100	116
23	89	78
24	85	101
25	88	123

(a) Explain why it was important to randomly assign the order in which each subject used the two knobs.

(b) The project designers hoped to show that right-handed people find right-hand threads easier to use. Carry out a significance test at the 5% significance level to investigate this claim.

90. **Floral scents and learning** We hear that listening to Mozart improves students' performance on tests. Maybe pleasant odors have a similar effect. To test this idea, 21 subjects worked two different but roughly equivalent paper-and-pencil mazes while wearing a mask. The mask was either unscented or carried a floral scent. Each subject used both masks, in a random order. The table below gives the subjects' times with both masks.[31]

Subject	Unscented	Scented
1	30.60	37.97
2	48.43	51.57
3	60.77	56.67
4	36.07	40.47
5	68.47	49.00
6	32.43	43.23
7	43.70	44.57
8	37.10	28.40
9	31.17	28.23
10	51.23	68.47
11	65.40	51.10
12	58.93	83.50
13	54.47	38.30
14	43.53	51.37
15	37.93	29.33
16	43.50	54.27
17	87.70	62.73
18	53.53	58.00
19	64.30	52.40
20	47.37	53.63
21	53.67	47.00

(a) Explain why it was important to randomly assign the order in which each subject used the two masks.

(b) Do these data provide convincing evidence that the floral scent improved performance? Carry out an appropriate test to support your answer.

91. **Music and memory** Does listening to music while studying help or hinder students' learning? Two AP Statistics students designed an experiment to find out. They selected a random sample of 30 students from their medium-sized high school to participate. Each subject was given 10 minutes to memorize two different lists of 20 words, once while listening to music and once in silence. The order of the two word lists was determined at random; so was the order of the treatments. A boxplot of the differences in the number of words recalled (music − silent) is shown below, along with some Minitab output from a one-sample t test. Perform a complete analysis of the students' data. Include a confidence interval.

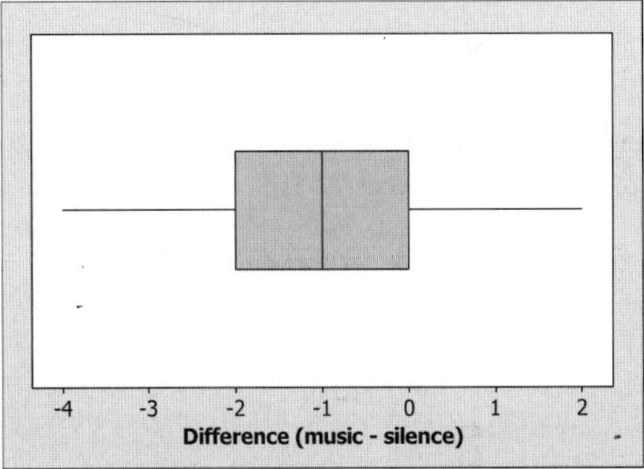

One-Sample T: Difference (music – silence)

Test of mu = 0 vs not = 0

Variable	N	Mean	StDev	SE Mean	T	P
Difference	30	-0.933	1.701	0.310	-3.01	0.005

92. **Darwin's plants** Charles Darwin, author of *On the Origin of Species* (1859), designed an experiment to compare the effects of cross-fertilization and self-fertilization on the size of plants. He planted pairs of very similar seedling plants, one self-fertilized and one cross-fertilized, in each of 15 pots at the same time. After a period of time, Darwin measured the heights (in inches) of all the plants. Here are the data:

Pair	Cross	Self	Pair	Cross	Self
1	23.5	17.4	9	18.3	16.5
2	12.0	20.4	10	21.6	18.0
3	21.0	20.0	11	23.3	16.3
4	22.0	20.0	12	21.0	18.0
5	19.1	18.4	13	22.1	12.8
6	21.5	18.6	14	23.0	15.5
7	22.1	18.6	15	12.0	18.0
8	20.4	15.3			

(a) Explain why it is not appropriate to perform a paired t test in this setting.

(b) A hasty student generates the Minitab output shown below. What conclusion should he draw at the $\alpha = 0.05$ significance level? Explain.

One-Sample T: Diff (cross – self)

```
Test of mu = 0 vs not = 0

Variable   N   Mean   StDev   SE Mean    T      P
Diff      15   2.61    4.71      1.22  2.14  0.050
```

93. **Is it significant?** For students without special preparation, SAT Math scores in recent years have varied Normally with mean $\mu = 518$. One hundred students go through a rigorous training program designed to raise their SAT Math scores by improving their mathematics skills. Use your calculator to carry out a test of

$$H_0: \mu = 518$$
$$H_a: \mu > 518$$

in each of the following situations.

(a) The students' scores have mean $\bar{x} = 536.7$ and standard deviation $s_x = 114$. Is this result significant at the 5% level?

(b) The students' scores have mean $\bar{x} = 537.0$ and standard deviation $s_x = 114$. Is this result significant at the 5% level?

(c) When looked at together, what is the intended lesson of (a) and (b)?

94. **Significance and sample size** A study with 5000 subjects reported a result that was statistically significant at the 5% level. Explain why this result might not be particularly large or important.

95. **Sampling shoppers** A marketing consultant observes 50 consecutive shoppers at a supermarket, recording how much each shopper spends in the store. Explain why it would not be wise to use these data to carry out a significance test about the mean amount spent by all shoppers at this supermarket.

96. **Ages of presidents** Joe is writing a report on the backgrounds of American presidents. He looks up the ages of all the presidents when they entered office. Because Joe took a statistics course, he uses these numbers to perform a significance test about the mean age of all U.S. presidents. Explain why this makes no sense.

97. **Do you have ESP?** A researcher looking for evidence of extrasensory perception (ESP) tests 500 subjects. Four of these subjects do significantly better ($P < 0.01$) than random guessing.

(a) Is it proper to conclude that these four people have ESP? Explain your answer.

(b) What should the researcher now do to test whether any of these four subjects have ESP?

98. **What is significance good for?** Which of the following questions does a significance test answer? Justify your answer.

(a) Is the sample or experiment properly designed?

(b) Is the observed effect due to chance?

(c) Is the observed effect important?

Multiple choice: Select the best answer for Exercises 99 to 104.

99. The reason we use t procedures instead of z procedures when carrying out a test about a population mean is that

(a) z can be used only for large samples.

(b) z requires that you know the population standard deviation σ.

(c) z requires you to regard your data as an SRS from the population.

(d) z applies only if the population distribution is perfectly Normal.

(e) z can be used only for confidence intervals.

100. You are testing $H_0: \mu = 10$ against $H_a: \mu < 10$ based on an SRS of 20 observations from a Normal population. The t statistic is $t = -2.25$. The P-value

(a) falls between 0.01 and 0.02.

(b) falls between 0.02 and 0.04.

(c) falls between 0.04 and 0.05.

(d) falls between 0.05 and 0.25.

(e) is greater than 0.25.

101. You are testing $H_0: \mu = 10$ against $H_a: \mu \neq 10$ based on an SRS of 15 observations from a Normal population. What values of the t statistic are statistically significant at the $\alpha = 0.005$ level?

(a) $t > 3.326$

(b) $t > 3.286$

(c) $t > 2.977$

(d) $t < -3.326$ or $t > 3.326$

(e) $t < -3.286$ or $t > 3.286$

102. After checking that conditions are met, you perform a significance test of $H_0: \mu = 1$ versus $H_a: \mu \neq 1$. You obtain a P-value of 0.022. Which of the following is true?

(a) A 95% confidence interval for μ will include the value 1.

(b) A 95% confidence interval for μ will include the value 0.

(c) A 99% confidence interval for μ will include the value 1.

(d) A 99% confidence interval for μ will include the value 0.

(e) None of these is necessarily true.

103. Does Friday the 13th have an effect on people's behavior? Researchers collected data on the numbers of shoppers at a sample of 45 different grocery stores on Friday the 6th and Friday the 13th in the same month. The dotplot and computer output below summarize the data on the difference in the number of shoppers at each store on these two days (subtracting in the order 6th minus 13th).[32]

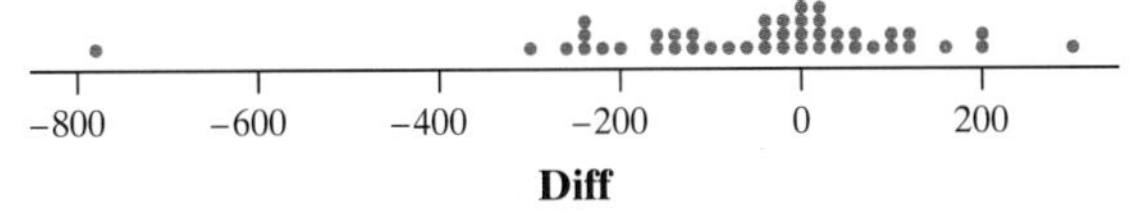

N	Mean	Median	TrMean	StDev	SEMean	Min	Max	Q1	Q3
45	-46.5	-11.0	-37.4	178.0	26.1	-774.0	302.0	-141.0	53.5

Researchers would like to carry out a test of H_0: $\mu_d = 0$ versus $H_a: \mu_d \neq 0$, where μ_d is the true mean difference in the number of grocery shoppers on these two days. Which of the following conditions for performing a paired t test is *not* met?

I. Random II. Normal III. Independent

(a) I only (b) II only (c) III only (d) I and II only (e) I, II, and III

104. The most important condition for sound conclusions from statistical inference is that

(a) the data come from a well-designed random sample or randomized experiment.

(b) the population distribution be exactly Normal.

(c) the data contain no outliers.

(d) the sample size be no more than 10% of the population size.

(e) the sample size be at least 30.

105. **Is your food safe?** (8.1) "Do you feel confident or not confident that the food available at most grocery stores is safe to eat?" When a Gallup Poll asked this question, 87% of the sample said they were confident.[33] Gallup announced the poll's margin of error for 95% confidence as ±3 percentage points. Which of the following sources of error are included in this margin of error? Explain.

(a) Gallup dialed landline telephone numbers at random and so missed all people without landline phones, including people whose only phone is a cell phone.

(b) Some people whose numbers were chosen never answered the phone in several calls or answered but refused to participate in the poll.

(c) There is chance variation in the random selection of telephone numbers.

106. **Spinning for apples** (6.3 or 7.3) In the "Ask Marilyn" column of *Parade* magazine, a reader posed this question: "Say that a slot machine has five wheels, and each wheel has five symbols: an apple, a grape, a peach, a pear, and a plum. I pull the lever five times. What are the chances that I'll get at least one apple?" Suppose that the wheels spin independently and that the five symbols are equally likely to appear on each wheel in a given spin.

(a) Find the probability that the slot player gets at least one apple in one pull of the lever. Show your method clearly.

(b) Now answer the reader's question. Show your method clearly.

Exercises 107 and 108 refer to the Case Closed about "normal" body temperature on page 585.

107. **Normal body temperature** (8.3) Check that the conditions are met for performing inference about the mean body temperature in the population of interest.

108. **Normal body temperature** (8.2) If "normal" body temperature really is 98.6°F, we would expect the proportion p of all healthy 18- to 40-year-olds who have body temperatures less than this value to be 0.5. Construct and interpret a 95% confidence interval for p. What conclusion would you draw?

Chapter 9 Chapter Review

A significance test shows how strong the evidence is for some claim about a parameter. In significance testing, the null hypothesis often says "no effect" or "no difference." The alternative hypothesis typically reflects what we hope to find evidence for.

The *P*-value is the probability that a random sample or randomized experiment would produce a result at least as extreme as the observed result if the null hypothesis really were true. That is, the *P*-value tells us how surprising the observed outcome is. Very surprising outcomes (small *P*-values) are good evidence that the null hypothesis is not true. If the *P*-value is less than a specified significance level α, we reject H_0 in favor of H_a. Otherwise, we fail to reject H_0 and do not have convincing evidence that H_a is true.

If we decide to reject H_0 when H_0 is true, we have committed a Type I error. The probability of making a Type I error is equal to the significance level α. If we fail to reject H_0 when H_0 is false, we have made a Type II error. The power of a significance test is the probability that the test will detect a specific alternative value of the parameter. Power = 1 − *P*(Type II error). We can increase the power of a test by increasing the sample size or by increasing our willingness to make a Type I error.

Section 9.1 presented the underlying logic and structure of significance tests. In Section 9.2, we looked at how to perform significance tests about a population proportion p. We also noted that confidence intervals give more information than a mere reject or fail to reject conclusion in a significance test. Section 9.3 discussed the one-sample *t* test for a population mean μ. We saw that paired data require a one-sample *t* test on the differences in each pair. Several important cautions for the wise use of significance tests were discussed at the end of Section 9.3.

All the methods of this chapter work well when the three important conditions—Random, Normal, and Independent—are satisfied. Be sure to verify these conditions before you proceed to calculations.

Chapter Review Exercises

These exercises are designed to help you review the important ideas and methods of the chapter. Relevant learning objectives are provided in bulleted form before each exercise.

- State correct hypotheses for a significance test about a population proportion or mean.

R9.1 Stating hypotheses State the appropriate null and alternative hypotheses in each of the following cases.

(a) The average height of 18-year-old American women is 64.2 inches. You wonder whether the mean height of this year's female graduates from a large local high school (over 3000 students) differs from the national average. You measure an SRS of 48 female graduates and find that $\bar{x} = 63.1$ inches.

(b) Mr. Starnes believes that less than 75% of the students at his school completed their math homework last night. The math teachers inspect the homework assignments from a random sample of students at the school to help Mr. Starnes test his claim.

- Check conditions for carrying out a test about a population proportion or mean.
- Interpret *P*-values in context.

R9.2 Eye black Athletes performing in bright sunlight often smear black grease under their eyes to reduce glare. Does eye black work? In one experiment, 16 randomly selected student subjects took a test of sensitivity to contrast after 3 hours facing into bright sun, both with and without eye black. Here are the differences in sensitivity, with eye black minus without eye black:[34]

0.07 0.64 −0.12 −0.05 −0.18 0.14 −0.16 0.03

0.05 0.02 0.43 0.24 −0.11 0.28 0.05 0.29

We want to know whether eye black increases sensitivity on the average.

(a) State hypotheses. Be sure to define the parameter.
(b) Check conditions for carrying out a significance test.
(c) The *P*-value of the test is 0.047. Interpret this value in context.

- **Interpret a Type I error and a Type II error in context, and give the consequences of each.**
- **Understand the relationship between the significance level of a test, *P*(Type II error), and power.**

R9.3 Strong chairs? A company that manufactures classroom chairs for high school students claims that the mean breaking strength of the chairs that they make is 300 pounds. One of the chairs collapsed beneath a 220-pound student last week. You wonder whether the manufacturer is exaggerating the breaking strength of the chairs.

(a) State null and alternative hypotheses in words and symbols.
(b) Describe a Type I error and a Type II error in this situation, and give the consequences of each.
(c) Would you recommend a significance level of 0.01, 0.05, or 0.10 for this test? Justify your choice.
(d) The power of this test to detect $\mu = 294$ is 0.71. Explain what this means to someone who knows little statistics.
(e) Explain two ways that you could increase the power of the test from (d).

- **If conditions are met, conduct a significance test about a population proportion.**

R9.4 Flu vaccine A drug company has developed a new vaccine for preventing the flu. The company claims that fewer than 5% of adults who use its vaccine will get the flu. To test the claim, researchers give the vaccine to a random sample of 1000 adults. Of these, 43 get the flu.

(a) Do these data provide convincing evidence to support the company's claim? Perform an appropriate test to support your answer.
(b) Which kind of mistake—a Type I error or a Type II error—could you have made in (a)? Explain.
(c) From the company's point of view, would a Type I error or Type II error be more serious? Why?

R9.5 Roulette An American roulette wheel has 18 red slots among its 38 slots. In a random sample of 50 spins, the ball lands in a red slot 31 times.

(a) Do the data give convincing evidence that the wheel is unfair? Carry out an appropriate test at the $\alpha = 0.05$ significance level to help answer this question.
(b) The casino manager uses your data to produce a 99% confidence interval for p and gets (0.44, 0.80). He says that this interval provides convincing evidence that the wheel is fair. How do you respond?

- **If conditions are met, conduct a one-sample *t* test about a population mean μ.**

R9.6 Improving health A large company's medical director launches a health promotion campaign to encourage employees to exercise more and eat better foods. One measure of the effectiveness of such a program is a drop in blood pressure. The director chooses a random sample of 50 employees and compares their blood pressures from physical exams given before the campaign and again a year later. The mean change (after − before) in systolic blood pressure for these 50 employees is −6 and the standard deviation is 19.8.

(a) Do these data provide convincing evidence of an average decrease in blood pressure among all of the company's employees during this year? Carry out a test at the $\alpha = 0.05$ significance level.
(b) Can we conclude that the health campaign caused a decrease in blood pressure? Why or why not?

R9.7 Fonts and reading ease Does the use of fancy type fonts slow down the reading of text on a computer screen? Adults can read four paragraphs of text in the common Times New Roman font in an average time of 22 seconds. Researchers asked a random sample of 24 adults to read this text in the ornate font named Gigi. Here are their times, in seconds:

23.2 21.2 28.9 27.7 29.1 27.3 16.1 22.6 25.6 34.2 23.9 26.8

20.5 34.3 21.4 32.6 26.2 34.1 31.5 24.6 23.0 28.6 24.4 28.1

Do these data provide good evidence that the mean reading time for Gigi is greater than 22 seconds? Carry out an appropriate test to help you answer this question.

- **Use a confidence interval to draw a conclusion for a two-sided test about a population mean.**

R9.8 Radon detectors Radon is a colorless, odorless gas that is naturally released by rocks and soils and may concentrate in tightly closed houses. Because radon is slightly radioactive, there is some concern that it may be a health hazard. Radon detectors are sold to homeowners worried about this risk, but the detectors may be inaccurate. University researchers placed a random sample of 11 detectors in a chamber where they were exposed to 105 picocuries per liter of radon over 3 days. A graph of the radon readings from the 11 detectors shows no strong skewness or outliers. The Minitab output below shows the results of a one-sample t interval.

Is there significant evidence at the 10% level that the mean reading μ differs from the true value 105? Give appropriate evidence to support your answer.

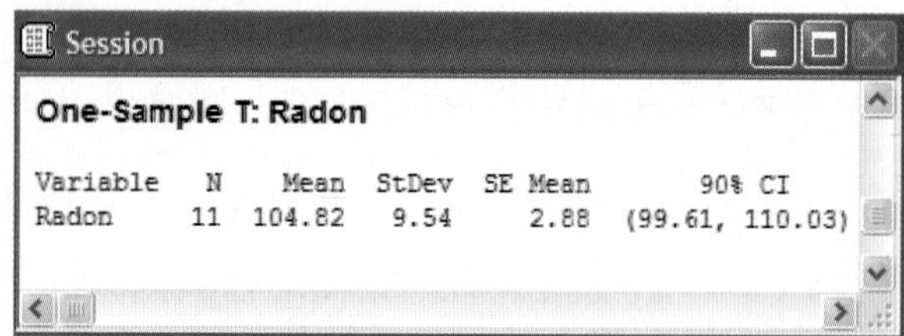
Session

One-Sample T: Radon

Variable	N	Mean	StDev	SE Mean	90% CI
Radon	11	104.82	9.54	2.88	(99.61, 110.03)

- Recognize paired data and use one-sample *t* procedures to perform significance tests for such data.

R9.9 Better barley Does drying barley seeds in a kiln increase the yield of barley? A famous experiment by William S. Gosset (who discovered the t distributions) investigated this question. Eleven pairs of adjacent plots were marked out in a large field. For each pair, regular barley seeds were planted in one plot and kiln-dried seeds were planted in the other. The following table displays the data on yield (lb/acre).[35]

Plot	Regular	Kiln
1	1903	2009
2	1935	1915
3	1910	2011
4	2496	2463
5	2108	2180
6	1961	1925
7	2060	2122
8	1444	1482
9	1612	1542
10	1316	1443
11	1511	1535

(a) How can the Random condition be satisfied in this study?

(b) Perform an appropriate test to help answer the research question. Assume that the Random condition is met. What conclusion would you draw?

Chapter 9 AP Statistics Practice Test

Section I: Multiple Choice *Select the best answer for each question.*

T9.1. An opinion poll asks a random sample of adults whether they favor banning ownership of handguns by private citizens. A commentator believes that more than half of all adults favor such a ban. The null and alternative hypotheses you would use to test this claim are

(a) $H_0: \hat{p} = 0.5; H_a: \hat{p} > 0.5$
(b) $H_0: p = 0.5; H_a: p > 0.5$
(c) $H_0: p = 0.5; H_a: p < 0.5$
(d) $H_0: p = 0.5; H_a: p \neq 0.5$
(e) $H_0: p > 0.5; H_a: p = 0.5$

T9.2. You are thinking of conducting a one-sample t test about a population mean μ using a 0.05 significance level. You suspect that the distribution of the population is not Normal and may be moderately skewed. Which of the following statements is correct?

(a) You should not carry out the test because the population does not have a Normal distribution.

(b) You can safely carry out the test if your sample size is large and there are no outliers.

(c) You can safely carry out the test if there are no outliers, regardless of the sample size.

(d) You can carry out the test only if the population standard deviation is known.

(e) The t procedures are robust—you can use them any time you want.

T9.3. To determine the reliability of experts who interpret lie detector tests in criminal investigations, a random sample of 280 such cases was studied. The results were

	Suspect's True Status	
Examiner's Decision	**Innocent**	**Guilty**
"Innocent"	131	15
"Guilty"	9	125

If the hypotheses are H_0: suspect is innocent versus H_a: suspect is guilty, then we could estimate the probability that experts who interpret lie detector tests will make a Type II error as

(a) 15/280. (c) 15/140. (e) 15/146.
(b) 9/280. (d) 9/140.

T9.4. A significance test allows you to reject a null hypothesis H_0 in favor of an alternative H_a at the 5% significance level. What can you say about significance at the 1% level?

(a) H_0 can be rejected at the 1% significance level.

(b) There is insufficient evidence to reject H_0 at the 1% significance level.

(c) There is sufficient evidence to accept H_0 at the 1% significance level.

(d) H_a can be rejected at the 1% significance level.

(e) The answer can't be determined from the information given.

T9.5. A random sample of 100 likely voters in a small city produced 59 voters in favor of Candidate A. The observed value of the test statistic for testing the null hypothesis $H_0: p = 0.5$ versus the alternative hypothesis $H_a: p > 0.5$ is

(a) $z = \dfrac{0.59 - 0.5}{\sqrt{\dfrac{0.59(0.41)}{100}}}$

(b) $z = \dfrac{0.59 - 0.5}{\sqrt{\dfrac{0.5(0.5)}{100}}}$

(c) $z = \dfrac{0.5 - 0.59}{\sqrt{\dfrac{0.59(0.41)}{100}}}$

(d) $z = \dfrac{0.5 - 0.59}{\sqrt{\dfrac{0.5(0.5)}{100}}}$

(e) $t = \dfrac{0.59 - 0.5}{\sqrt{\dfrac{0.5(0.5)}{100}}}$

T9.6. A researcher claims to have found a drug that causes people to grow taller. The coach of the basketball team at Brandon University has expressed interest but demands evidence. Over 1000 Brandon students volunteer to participate in an experiment to test this new drug. Fifty of the volunteers are randomly selected, their heights are measured, and they are given the drug. Two weeks later, their heights are measured again. The power of the test to detect an average increase in height of one inch could be increased by

(a) using only volunteers from the basketball team in the experiment.

(b) using $\alpha = 0.01$ instead of $\alpha = 0.05$.

(c) using $\alpha = 0.05$ instead of $\alpha = 0.01$.

(d) giving the drug to 25 randomly selected students instead of 50.

(e) using a two-sided test instead of a one-sided test.

T9.7. A 95% confidence interval for a population mean μ is calculated to be (1.7, 3.5). Assume that the conditions for performing inference are met. What conclusion can we draw for a test of $H_0: \mu = 2$ versus $H_a: \mu \neq 2$ at the $\alpha = 0.05$ level based on the confidence interval?

(a) None. We cannot carry out the test without the original data.

(b) None. We cannot draw a conclusion at the $\alpha = 0.05$ level since this test is connected to the 97.5% confidence interval.

(c) None. Confidence intervals and significance tests are unrelated procedures.

(d) We would reject H_0 at level $\alpha = 0.05$.

(e) We would fail to reject H_0 at level $\alpha = 0.05$.

T9.8. In a test of $H_0: p = 0.4$ against $H_a: p \neq 0.4$, a random sample of size 100 yields a test statistic of $z = 1.28$. The *P*-value of the test is approximately equal to

(a) 0.90.
(b) 0.40.
(c) 0.05.
(d) 0.20.
(e) 0.10.

T9.9. An SRS of 100 postal employees found that the average time these employees had worked at the postal service was 7 years with standard deviation 2 years. Do these data provide convincing evidence that the mean time of employment μ for the population of postal employees has changed from the value of 7.5 that was true 20 years ago? To determine this, we test the hypotheses $H_0: \mu = 7.5$ versus $H_a: \mu \neq 7.5$ using a one-sample *t* test. What conclusion should we draw at the 5% significance level?

(a) There is convincing evidence that the mean time working with the postal service has changed.

(b) There is not convincing evidence that the mean time working with the postal service has changed.

(c) There is convincing evidence that the mean time working with the postal service is still 7.5 years.

(d) There is convincing evidence that the mean time working with the postal service is now 7 years.

(e) We cannot draw a conclusion at the 5% significance level. The sample size is too small.

T9.10. Are TV commercials louder than their surrounding programs? To find out, researchers collected data on 50 randomly selected commercials in a given week. With the television's volume at a fixed setting, they measured the maximum loudness of each commercial and the maximum loudness in the first 30 seconds of regular programming that followed. Assuming conditions for inference are met, the most appropriate method for answering the question of interest is

(a) a one-proportion *z* test.
(b) a one-proportion *z* interval.
(c) a paired *t* test.
(d) a paired *t* interval.
(e) None of these.

Section II: Free Response *Show all your work. Indicate clearly the methods you use, because you will be graded on the correctness of your methods as well as on the accuracy and completeness of your results and explanations.*

T9.11. A software company is trying to decide whether to produce an upgrade of one of its programs. Customers would have to pay \$100 for the upgrade. For the upgrade to be profitable, the company needs to sell it to more than 20% of their customers. You contact a random sample of 60 customers and find that 16 would be willing to pay \$100 for the upgrade.

(a) Do the sample data give good evidence that more than 20% of the company's customers are willing to purchase the upgrade? Carry out an appropriate test at the $\alpha = 0.05$ significance level.

(b) Which would be a more serious mistake in this setting—a Type I error or a Type II error? Justify your answer.

(c) Other than increasing the sample size, describe one way to increase the power of the test in (a).

T9.12. "I can't get through my day without coffee" is a common statement from many students. Assumed benefits include keeping students awake during lectures and making them more alert for exams and tests. Students in a statistics class designed an experiment to measure memory retention with and without drinking a cup of coffee one hour before a test. This experiment took place on two different days in the same week (Monday and Wednesday). Ten students were used. Each student received no coffee or one cup of coffee, one hour before the test on a particular day. The test consisted of a series of words flashed on a screen, after which the student had to write down as many of the words as possible. On the other day, each student received a different amount of coffee (none or one cup).

(a) One of the researchers suggested that all the subjects in the experiment drink no coffee before Monday's test and one cup of coffee before Wednesday's test. Explain to the researcher why this is a bad idea *and* suggest a better method of deciding when each subject receives the two treatments.

(b) The data from the experiment are provided in the table below. Set up and carry out an appropriate test to determine whether there is convincing evidence that drinking coffee improves memory.

Student	No cup	One cup
1	24	25
2	30	31
3	22	23
4	24	24
5	26	27
6	23	25
7	26	28
8	20	20
9	27	27
10	28	30

T9.13. A government report says that the average amount of money spent per U.S. household per week on food is about \$158. A random sample of 50 households in a small city is selected, and their weekly spending on food is recorded. The Minitab output below shows the results of requesting a confidence interval for the population mean μ. An examination of the data reveals no outliers.

One-Sample T

N	Mean	StDev	SE Mean	95% CI
50	165.00	20.00	2.83	(159.32, 170.68)

(a) Explain why the Normal condition is met in this case.

(b) Can you conclude that the mean weekly spending on food in this city differs from the national figure of \$158? Give appropriate evidence to support your answer.

Chapter 10

case study

Comparing Two Populations or Groups

Fast-Food Frenzy!

More than $70 billion is spent each year in the drive-thru lanes of America's fast-food restaurants. Having quick, accurate, and friendly service at a drive-thru window translates directly into revenue for the restaurant. According to Jack Greenberg, former CEO of McDonald's, sales increase 1% for every six seconds saved at the drive-thru. So industry executives, stockholders, and analysts closely follow the ratings of fast-food drive-thru lanes that appear annually in *QSR*, a publication that reports on the quick-service restaurant industry.

The 2008 and 2009 *QSR* magazine drive-thru studies involved visits to a random sample of restaurants in the 20 largest fast-food chains in all 50 states. During each visit, the researcher ordered a modified main item (for example, a hamburger with no pickles), a side item, and a drink. If any item was not received as ordered, or if the restaurant failed to give the correct change or supply a straw and a napkin, then the order was considered "inaccurate." Service time, which is the time from when the car stopped at the speaker to when the entire order was received, was measured each visit.[1]

Here are some results from these two *QSR* studies:

- In 2008, 522 of the 558 Burger King visits resulted in accurate orders. In 2009, 477 of the 529 Burger King visits resulted in accurate orders.
- In 2009, McDonald's average service time for 596 drive-thru visits was 174.22 seconds with a standard deviation of 16.49 seconds. In 2008, McDonald's service time for 750 drive-thru visits had a mean of 158.77 seconds and a standard deviation of 17.06 seconds.
- Combining the different aspects of drive-thru quality, Chick-fil-A earned the highest overall ranking in 2009 for the sixth consecutive year.

Was there a change in accuracy at Burger King restaurants between 2008 and 2009? How much did average service time at McDonald's restaurants change in these two years? By the end of the chapter, you should have acquired the tools to help answer questions like these.

Introduction

Which of two popular drugs—Lipitor or Pravachol—helps lower "bad cholesterol" more? Researchers designed an experiment, called the PROVE-IT Study, to find out. They used about 4000 people with heart disease as subjects. These individuals were randomly assigned to one of two treatment groups: Lipitor or Pravachol. At the end of the study, researchers compared the proportion of subjects in each group who died, had a heart attack, or suffered other serious consequences within two years. For those using Pravachol, the proportion was 0.263; for those using Lipitor, it was 0.224.[2] Could such a difference have occurred purely by the chance involved in the random assignment? This is a question about *comparing two proportions.*

Who studies more in college—men or women? Researchers asked separate random samples of 30 males and 30 females at a large university how many minutes they studied on a typical weeknight. The females reported studying an average of 165.17 minutes; the male average was 117.17 minutes. How large a difference in the corresponding population means would we estimate? This is a question about *comparing two means.*

Comparing two proportions or means based on random sampling or a randomized experiment is one of the most common situations encountered in statistical practice. In the PROVE-IT experiment, the goal of inference is to determine whether the treatments (Lipitor and Pravachol) *caused* the observed difference in the proportion of subjects who experienced serious consequences in the two groups. For the college studying survey, the goal of inference is to draw a conclusion about the actual mean study times for *all* women and *all* men at the university.

The following Activity gives you a taste of what lies ahead in this chapter.

ACTIVITY *Is yawning contagious?*

MATERIALS: Deck of cards for each team of three to four students

In Chapter 4, we examined data from an experiment involving 50 subjects on the TV show *MythBusters* that investigated this question. Here's a brief recap. Each subject was placed in a booth for an extended period of time and monitored by hidden camera. Thirty-four subjects were given a "yawn seed" by one of the experimenters; that is, the experimenter yawned in the subject's presence before leaving the room. The remaining 16 subjects were given no yawn seed. What happened? The table below shows the results:[3]

	Yawn Seed?		
Subject Yawned?	**Yes**	**No**	**Total**
Yes	10	4	14
No	24	12	36
Total	34	16	50

Adam Savage and Jamie Hyneman, the cohosts of *MythBusters*, used these data to conclude that yawning is contagious. In this Activity, your class will investigate whether the results of the experiment were really statistically significant. Let's see what would happen just by chance if we randomly reassign the 50 people in this experiment to the two groups (yawn seed and no yawn seed) many times, *assuming the treatment received doesn't affect whether or not a person yawns.*

1. We need 50 cards from the deck to represent the original subjects. In the *MythBusters* experiment, 14 people yawned and 36 didn't. Because we're assuming that the treatment received won't change whether each subject yawns, we use 14 cards to represent people who yawn and 36 cards to represent those who don't. Remove the ace of spades and ace of clubs from the deck.

- Yawn: All the jacks, queens, kings, and aces (16 – 2 removed aces = 14)
- Don't yawn: All cards with denominations 2 through 10 (9 denominations × 4 suits = 36)

2. Shuffle and deal two piles of cards—one with 16 cards and one with 34 cards. The first pile represents the no yawn seed group and the second pile represents the yawn seed group. The shuffling reflects our assumption that the outcome for each subject is not affected by the treatment. Record the number of people who yawned in each group.
3. Calculate the difference in the proportions who yawned for the two groups (yawn seed – no yawn seed). For example, if you get 9 yawners in the yawn seed group and 5 yawners in the no yawn seed group, the resulting difference in proportions is

$$\frac{9}{34} - \frac{5}{16} = -0.048$$

A negative difference would mean that a smaller proportion of people in the yawn seed group yawned during the experiment than in the no yawn seed group.
4. Repeat Steps 2 and 3 four more times so that you have a total of 5 trials. Record your results in a table like this:

Trial	Number who yawned in yawn seed group	Number who yawned in no yawn seed group	Difference in proportions (yawn seed – no yawn seed)
1			

5. Make a class dotplot of the difference in proportions. In what percent of the class's trials did the difference in proportions equal or exceed $\frac{10}{34} - \frac{4}{16} = 0.0441$ (what the MythBusters got in their experiment)?
6. Based on the class's simulation results, how surprising would it be to get a result this large or larger simply due to the chance involved in the random assignment? Is the result statistically significant?
7. What conclusion would you draw about whether yawning is contagious? Explain.
8. Based on your conclusion in Step 7, could you have made a Type I error or a Type II error? Discuss this as a class.

10.1 Comparing Two Proportions

In Section 10.1, you'll learn about:

- The sampling distribution of a difference between two proportions
- Confidence intervals for $p_1 - p_2$
- Significance tests for $p_1 - p_2$
- Inference for experiments

Suppose we want to compare the proportions of individuals with a certain characteristic in Population 1 and Population 2. Let's call these parameters of interest p_1 and p_2. The ideal strategy is to take a separate random sample from each population and to compare the sample proportions $\hat{p}_1$ and $\hat{p}_2$ with that characteristic. What if we want to compare the effectiveness of Treatment 1 and Treatment 2 in a completely randomized experiment? This time, the parameters p_1 and p_2 that we want to compare are the true proportions of successful outcomes for each treatment. We use the proportions of successes in the two treatment groups, $\hat{p}_1$ and $\hat{p}_2$, to make the comparison. Here's a table that summarizes these two situations:

Population or treatment	Parameter	Statistic	Sample size
1	p_1	$\hat{p}_1$	n_1
2	p_2	$\hat{p}_2$	n_2

We compare the populations or treatments by doing inference about the difference $p_1 - p_2$ between the parameters. The statistic that estimates this difference is the difference between the two sample proportions, $\hat{p}_1 - \hat{p}_2$. To use $\hat{p}_1 - \hat{p}_2$ for inference, we must know its sampling distribution.

The Sampling Distribution of a Difference between Two Proportions

To explore the sampling distribution of $\hat{p}_1 - \hat{p}_2$, let's start with two populations having a known proportion of successes. Suppose that there are two large high schools, each with over 2000 students, in a certain town. At School 1, 70% of students did their homework last night. Only 50% of the students at School 2 did their homework last night. The counselor at School 1 takes an SRS of 100 students and records the proportion $\hat{p}_1$ that did the homework. School 2's counselor takes an SRS of 200 students and records the proportion $\hat{p}_2$ that did the homework. What can we say about the difference $\hat{p}_1 - \hat{p}_2$ in the sample proportions?

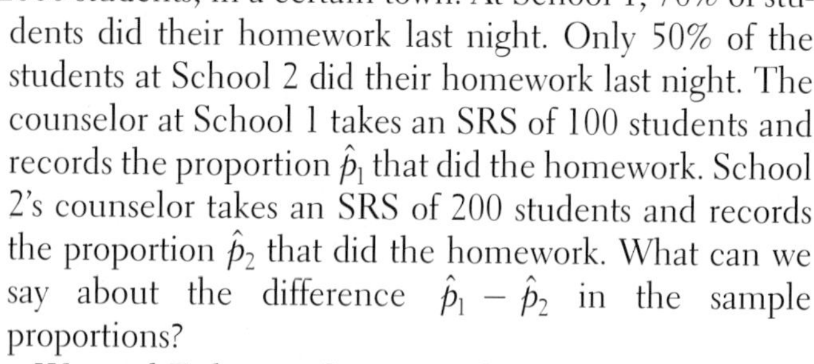

We used Fathom software to take an SRS of $n_1 = 100$ students from School 1 and a separate SRS of $n_2 = 200$ students from School 2 and to plot the values of $\hat{p}_1$, $\hat{p}_2$, and $\hat{p}_1 - \hat{p}_2$ from each sample. Our first set of simulated samples gave $\hat{p}_1 = 0.68$ and $\hat{p}_2 = 0.505$, so dots were placed above each of those values in Figure 10.1(a) and (b). The difference in the sample proportions is $\hat{p}_1 - \hat{p}_2 = 0.68 - 0.505 = 0.175$. A dot for this value appears in Figure 10.1(c). The three dotplots in Figure 10.1 show the results of repeating this process 1000 times. These are the approximate sampling distributions of $\hat{p}_1$, $\hat{p}_2$, and $\hat{p}_1 - \hat{p}_2$.

(a) Approximate sampling distribution of $\hat{p}_1$

0.55 0.60 0.65 0.70 0.75 0.80 0.85

phat1

Shape: Approximately Normal

Center: $\mu_{\hat{p}_1} = p_1 = 0.70$

Spread: $\sigma_{\hat{p}_1} = \sqrt{\frac{p_1(1-p_1)}{n_1}} = \sqrt{\frac{0.7(0.3)}{100}} = 0.0458$

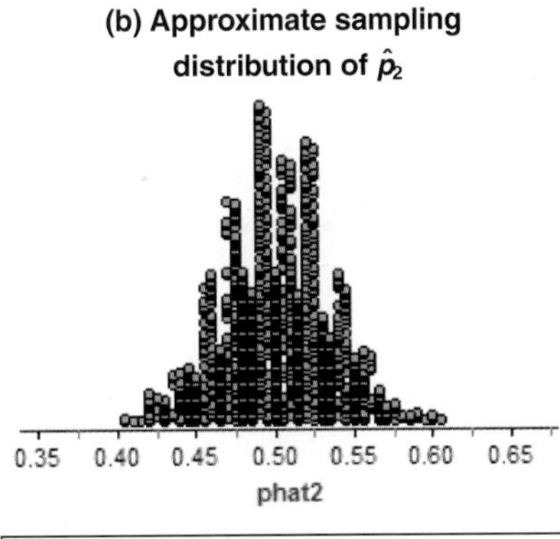

Shape: Approximately Normal

Center: $\mu_{\hat{p}_2} = p_2 = 0.50$

Spread: $\sigma_{\hat{p}_2} = \sqrt{\frac{p_2(1-p_2)}{n_2}} = \sqrt{\frac{0.5(0.5)}{200}} = 0.0354$

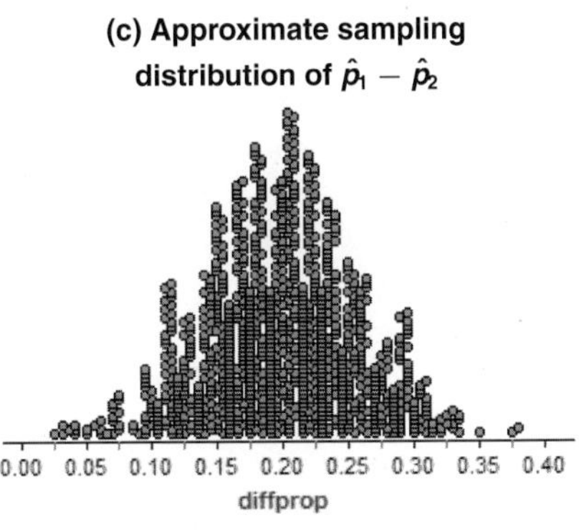

Shape: Approximately Normal

Center: $\mu_{\hat{p}_1 - \hat{p}_2} = 0.20$

Spread: $\sigma_{\hat{p}_1 - \hat{p}_2} = 0.058$

FIGURE 10.1 Simulated sampling distributions of (a) the sample proportion $\hat{p}_1$ of successes in 1000 SRSs of size $n_1 = 100$ from a population with $p_1 = 0.70$, (b) the sample proportion $\hat{p}_2$ of successes in 1000 SRSs of size $n_2 = 200$ from a population with $p_2 = 0.50$, and (c) the difference in sample proportions $\hat{p}_1 - \hat{p}_2$ for each of the 1000 repetitions.

In Chapter 7, we saw that the sampling distribution of a sample proportion $\hat{p}$ has the following properties:

- **Shape** Approximately Normal if $np \geq 10$ and $n(1-p) \geq 10$
- **Center** $\mu_{\hat{p}} = p$
- **Spread** $\sigma_{\hat{p}} = \sqrt{\frac{p(1-p)}{n}}$ if the sample is no more than 10% of the population (the 10% condition)

For the sampling distributions of $\hat{p}_1$ and $\hat{p}_2$ in this case:

	Sampling distribution of $\hat{p}_1$	**Sampling distribution of $\hat{p}_2$**
Shape	Approximately Normal; $n_1p_1 = 100(0.70) = 70 \geq 10$ and $n_1(1-p_1) = 100(0.30) = 30 \geq 10$	Approximately Normal; $n_2p_2 = 200(0.50) = 100 \geq 10$ and $n_2(1-p_2) = 200(0.50) = 100 \geq 10$
Center	$\mu_{\hat{p}_1} = p_1 = 0.70$	$\mu_{\hat{p}_2} = p_2 = 0.50$
Spread	$\sigma_{\hat{p}_1} = \sqrt{\frac{p_1(1-p_1)}{n_1}} = \sqrt{\frac{0.7(0.3)}{100}} = 0.0458$	$\sigma_{\hat{p}_2} = \sqrt{\frac{p_2(1-p_2)}{n_2}} = \sqrt{\frac{0.5(0.5)}{200}} = 0.0354$
	10% condition: School 1 has a population of over 10(100) = 1000 students.	*10% condition:* School 2 has a population of over 10(200) = 2000 students.

The dotplots in Figures 10.1(a) and (b) show these results.

What about the sampling distribution of $\hat{p}_1 - \hat{p}_2$? Figure 10.1(c) shows that it has an approximately Normal shape, is centered at 0.20, and has standard deviation 0.058. The shape makes sense because we are combining two independent Normal random variables, $\hat{p}_1$ and $\hat{p}_2$. How about the center? The true proportion of students who did last night's homework at School 1 is $p_1 = 0.70$ and at School 2 is $p_2 = 0.50$. We expect the difference $\hat{p}_1 - \hat{p}_2$ to center on the actual difference in the population proportions, $p_1 - p_2 = 0.70 - 0.50 = 0.20$. The spread, however, is a bit more complicated.

THINK ABOUT IT

How can we find formulas for the mean and standard deviation of the sampling distribution of $\hat{p}_1 - \hat{p}_2$? Both $\hat{p}_1$ and $\hat{p}_2$ are random variables. That is, their values would vary in repeated SRSs of size n_1 and n_2. The statistic $\hat{p}_1 - \hat{p}_2$ is the difference of these two random variables. In Chapter 6, we learned that for any two independent random variables X and Y,

$$\mu_{X-Y} = \mu_X - \mu_Y \quad \text{and} \quad \sigma^2_{X-Y} = \sigma^2_X + \sigma^2_Y$$

For the random variables $\hat{p}_1$ and $\hat{p}_2$, we have

$$\mu_{\hat{p}_1-\hat{p}_2} = \mu_{\hat{p}_1} - \mu_{\hat{p}_2} = p_1 - p_2$$

and

$$\sigma^2_{\hat{p}_1-\hat{p}_2} = \sigma^2_{\hat{p}_1} + \sigma^2_{\hat{p}_2} = \left(\sqrt{\frac{p_1(1-p_1)}{n_1}}\right)^2 + \left(\sqrt{\frac{p_2(1-p_2)}{n_2}}\right)^2 = \frac{p_1(1-p_1)}{n_1} + \frac{p_2(1-p_2)}{n_2}$$

$$\sigma_{\hat{p}_1-\hat{p}_2} = \sqrt{\frac{p_1(1-p_1)}{n_1} + \frac{p_2(1-p_2)}{n_2}}$$

For the school homework survey,

$$\sigma_{\hat{p}_1-\hat{p}_2} = \sqrt{\frac{p_1(1-p_1)}{n_1} + \frac{p_2(1-p_2)}{n_2}} = \sqrt{\frac{0.7(0.3)}{100} + \frac{0.5(0.5)}{200}} = 0.058$$

This confirms the result from the Fathom simulation.

Here are the facts we need.

The Sampling Distribution of $\hat{p}_1 - \hat{p}_2$

Choose an SRS of size n_1 from Population 1 with proportion of successes p_1 and an independent SRS of size n_2 from Population 2 with proportion of successes p_2.

- **Shape** When n_1p_1, $n_1(1 - p_1)$, n_2p_2, and $n_2(1 - p_2)$ are all at least 10, the sampling distribution of $\hat{p}_1 - \hat{p}_2$ is approximately Normal.
- **Center** The mean of the sampling distribution is $p_1 - p_2$. That is, the difference in sample proportions is an unbiased estimator of the difference in population proportions.
- **Spread** The standard deviation of the sampling distribution of $\hat{p}_1 - \hat{p}_2$ is

$$\sqrt{\frac{p_1(1-p_1)}{n_1} + \frac{p_2(1-p_2)}{n_2}}$$

as long as each sample is no more than 10% of its population (the 10% condition).

Figure 10.2 displays the sampling distribution of $\hat{p}_1 - \hat{p}_2$.

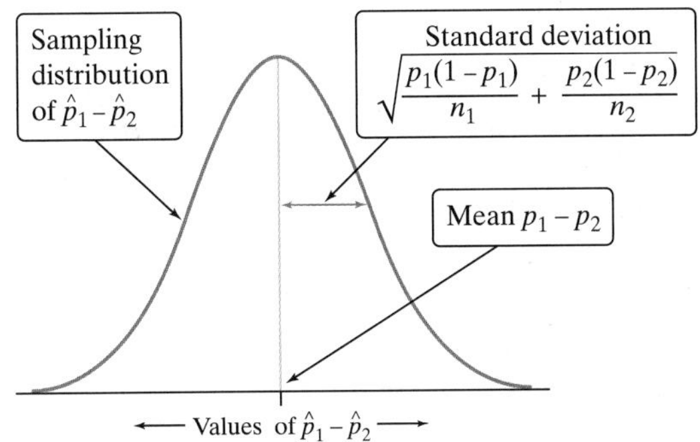

FIGURE 10.2 Select independent SRSs from two populations having proportions of successes p_1 and p_2. The proportions of successes in the two samples are $\hat{p}_1$ and $\hat{p}_2$. When the samples are large, the sampling distribution of the difference $\hat{p}_1 - \hat{p}_2$ is approximately Normal.

The formula for the standard deviation of the sampling distribution involves the unknown parameters p_1 and p_2. Just as in Chapters 8 and 9, we must replace these by estimates to do inference. And just as before, we do this a bit differently for confidence intervals and for tests. We'll get to inference shortly. For now, we can use what we know about the sampling distribution of $\hat{p}_1 - \hat{p}_2$ to compute probabilities, as the following example illustrates.

Who Does More Homework?

Finding probabilities with the sampling distribution

Suppose that there are two large high schools, each with more than 2000 students, in a certain town. At School 1, 70% of students did their homework last night. Only 50% of the students at School 2 did their homework last night. The counselor at School 1 takes an SRS of 100 students and records the proportion $\hat{p}_1$ that did homework. School 2's counselor takes an SRS of 200 students and records the proportion $\hat{p}_2$ that did homework. School 1's counselor and School 2's counselor meet to discuss the results of their homework surveys. After the meeting, they both report to their principals that $\hat{p}_1 - \hat{p}_2 = 0.10$.

PROBLEM:

(a) Describe the shape, center, and spread of the sampling distribution of $\hat{p}_1 - \hat{p}_2$.

(b) Find the probability of getting a difference in sample proportions $\hat{p}_1 - \hat{p}_2$ of 0.10 or less from the two surveys. Show your work.

(c) Does the result in part (b) give us reason to doubt the counselors' reported value? Explain.

SOLUTION:

(a) Because $n_1p_1 = 100(0.7) = 70$, $n_1(1 - p_1) = 100(0.3) = 30$, $n_2p_2 = 200(0.5) = 100$, and $n_2(1 - p_2) = 200(0.5) = 100$ are all at least 10, the sampling distribution of $\hat{p}_1 - \hat{p}_2$ is approximately Normal. Its mean is $p_1 - p_2 = 0.70 - 0.50 = 0.20$ and its standard deviation is $\sqrt{\frac{0.7(0.3)}{100} + \frac{0.5(0.5)}{200}} = 0.058$.

(b) Figure 10.3(a) shows the sampling distribution of $\hat{p}_1 - \hat{p}_2$ with the desired probability (area) shaded. To find this probability, we use the methods of Chapter 2.

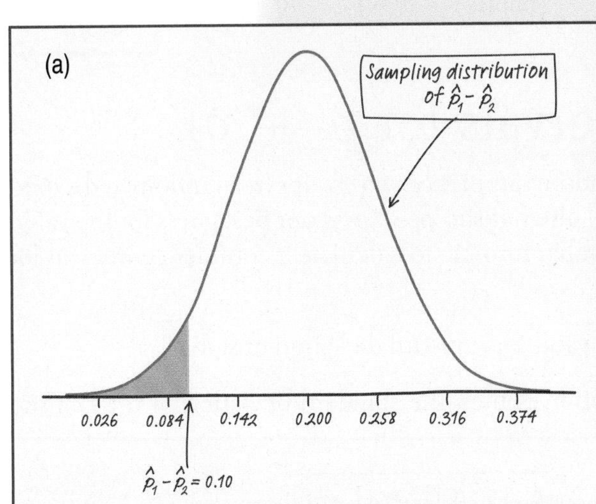

FIGURE 10.3 (a) Normal curve that approximates the sampling distribution of $\hat{p}_1 - \hat{p}_2$ for the homework surveys. We want to find $P(\hat{p}_1 - \hat{p}_2 \leq 0.10)$.

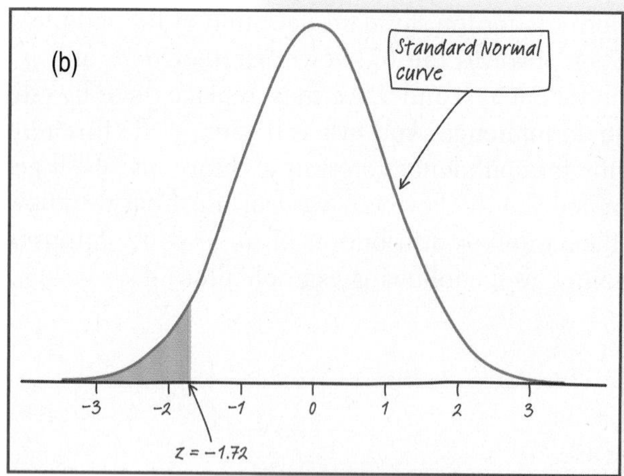

FIGURE 10.3 (b) The desired probability can also be expressed as $P(z < -1.72)$, which represents an area under the standard Normal curve.

- *Standardize.* When $\hat{p}_1 - \hat{p}_2 = 0.10$,

$$z = \frac{0.10 - 0.20}{0.058} = -1.72$$

- *Use Table A.* The area to the left of $z = -1.72$ under the standard Normal curve is 0.0427. This is the probability we seek. Figure 10.3(b) shows this result.

We could use technology to get the desired probability. Using `normalcdf(0,0.1,.2,.058)` or the *Normal Curve* applet gives 0.0421.

(c) There is only about a 4% chance of getting a difference in sample proportions as small as or smaller than the value of 0.10 reported by the counselors. This does seem suspicious!

For Practice *Try Exercise 5*

CHECK YOUR UNDERSTANDING

Your teacher brings two bags of colored goldfish crackers to class. She tells you that Bag 1 has 25% red crackers and Bag 2 has 35% red crackers. Each bag contains more than 500 crackers. Using a paper cup, your teacher takes an SRS of 50 crackers from Bag 1 and a separate SRS of 40 crackers from Bag 2. Let $\hat{p}_1 - \hat{p}_2$ be the difference in the sample proportions of red crackers.

1. What is the shape of the sampling distribution of $\hat{p}_1 - \hat{p}_2$? Why?
2. Find the mean and standard deviation of the sampling distribution. Show your work.
3. Find the probability that $\hat{p}_1 - \hat{p}_2$ is less than or equal to -0.02. Show your work.
4. Based on your answer to Question 3, would you be surprised if the difference in the proportion of red crackers in the two samples was $\hat{p}_1 - \hat{p}_2 = -0.02$? Explain.

Confidence Intervals for $p_1 - p_2$

When data come from two random samples or two groups in a randomized experiment (the Random condition), the statistic $\hat{p}_1 - \hat{p}_2$ is our best guess for the value of $p_1 - p_2$. We can use our familiar formula to calculate a confidence interval for $p_1 - p_2$:

$$\text{statistic} \pm (\text{critical value}) \cdot (\text{standard deviation of statistic})$$

When the Independent condition is met, the standard deviation of the statistic $\hat{p}_1 - \hat{p}_2$ is

$$\sigma_{\hat{p}_1 - \hat{p}_2} = \sqrt{\frac{p_1(1-p_1)}{n_1} + \frac{p_2(1-p_2)}{n_2}}$$

Because we don't know the values of the parameters p_1 and p_2, we replace them in the standard deviation formula with the sample proportions. The result is the

Standard error

standard error (also called the *estimated standard deviation*) of the statistic $\hat{p}_1 - \hat{p}_2$:

$$\sqrt{\frac{\hat{p}_1(1-\hat{p}_1)}{n_1} + \frac{\hat{p}_2(1-\hat{p}_2)}{n_2}}$$

If the Normal condition is met, we find the critical value z^* for the given confidence level from the standard Normal curve. Our confidence interval for $p_1 - p_2$ is therefore

$$\text{statistic} \pm (\text{critical value}) \cdot (\text{standard deviation of statistic})$$

$$(\hat{p}_1 - \hat{p}_2) \pm z^* \sqrt{\frac{\hat{p}_1(1-\hat{p}_1)}{n_1} + \frac{\hat{p}_2(1-\hat{p}_2)}{n_2}}$$

This is often called a **two-sample z interval for a difference between two proportions**.

Two-Sample *z* Interval for a Difference between Two Proportions

When the Random, Normal, and Independent conditions are met, an approximate level C confidence interval for $p_1 - p_2$ is

$$(\hat{p}_1 - \hat{p}_2) \pm z^* \sqrt{\frac{\hat{p}_1(1-\hat{p}_1)}{n_1} + \frac{\hat{p}_2(1-\hat{p}_2)}{n_2}}$$

where z^* is the critical value for the standard Normal curve with area C between $-z^*$ and z^*.

- **Random** The data are produced by a random sample of size n_1 from Population 1 and a random sample of size n_2 from Population 2 or by two groups of size n_1 and n_2 in a randomized experiment.
- **Normal** The counts of "successes" and "failures" in each sample or group—$n_1\hat{p}_1$, $n_1(1-\hat{p}_1)$, $n_2\hat{p}_2$, and $n_2(1-\hat{p}_2)$—are all at least 10.
- **Independent** Both the samples or groups themselves and the individual observations in each sample or group are independent. When sampling without replacement, check that the two populations are at least 10 times as large as the corresponding samples (the 10% condition).

The following example shows how to construct and interpret a confidence interval for a difference in proportions. As usual with inference problems, we follow the four-step process.

EXAMPLE

Teens and Adults on Social Networking Sites

Confidence interval for $p_1 - p_2$

As part of the Pew Internet and American Life Project, researchers conducted two surveys in late 2009. The first survey asked a random sample of 800 U.S. teens about their use of social media and the Internet. A second survey posed similar questions to a random sample of 2253 U.S. adults. In these two studies, 73% of teens and

47% of adults said that they use social-networking sites. Use these results to construct and interpret a 95% confidence interval for the difference between the proportion of all U.S. teens and adults who use social-networking sites.

STATE: Our *parameters* of interest are p_1 = the proportion of all U.S. teens who use social networking sites and p_2 = the proportion of all U.S. adults who use social-networking sites. We want to estimate the difference $p_1 - p_2$ at a 95% confidence level.

PLAN: We should use a two-sample z interval for $p_1 - p_2$ if the *conditions* are satisfied.

- *Random* The data come from a random sample of 800 U.S. teens and a separate random sample of 2253 U.S. adults.
- *Normal* We check the counts of "successes" and "failures":

$n_1\hat{p}_1 = 800(0.73) = 584$ $\qquad n_1(1-\hat{p}_1) = 800(1-0.73) = 216$

$n_2\hat{p}_2 = 2253(0.47) = 1058.91 \Rightarrow 1059$ $\qquad n_2(1-\hat{p}_2) = 2253(1-0.47) = 1194.09 \Rightarrow 1194$

Note that the counts have to be whole numbers! Because all four values are at least 10, the Normal condition is met.

- *Independent* We clearly have two independent samples—one of teens and one of adults. Individual responses in the two samples also have to be independent. The researchers are sampling without replacement, so we check the 10% condition: there are at least 10(800) = 8000 U.S. teens and at least 10(2253) = 22,530 U.S. adults.

DO: We know that $n_1 = 800$, $\hat{p}_1 = 0.73$, $n_2 = 2253$, and $\hat{p}_2 = 0.47$. For a 95% confidence level, the critical value is $z^* = 1.96$. So our 95% confidence interval for $p_1 - p_2$ is

$$(\hat{p}_1 - \hat{p}_2) \pm z^* \sqrt{\frac{\hat{p}_1(1-\hat{p}_1)}{n_1} + \frac{\hat{p}_2(1-\hat{p}_2)}{n_2}} = (0.73 - 0.47) \pm 1.96\sqrt{\frac{0.73(0.27)}{800} + \frac{0.47(0.53)}{2253}}$$
$$= 0.26 \pm 0.037$$
$$= (0.223, 0.297)$$

CONCLUDE: We are 95% confident that the interval from 0.223 to 0.297 captures the true difference in the proportion of all U.S. teens and adults who use social-networking sites. This interval suggests that more teens than adults in the United States engage in social networking by between 22.3 and 29.7 percentage points.

For Practice *Try Exercise 11*

The researchers in the previous example selected separate random samples from the two populations they wanted to compare. In practice, it's common to take one random sample that includes individuals from both populations of interest and then to separate the chosen individuals into two groups. The two-sample z procedures for comparing proportions are still valid in such situations, provided that the two groups can be viewed as independent samples from their respective populations of interest. This is an important fact about these methods.

You can use technology to perform the calculations in the "Do" step. Remember that this comes with potential benefits and risks on the AP exam.

AP EXAM TIP You may use your calculator to compute a confidence interval on the AP Exam. But there's a risk involved. If you give just the calculator answer with no work, you'll get either full credit for the "Do" step (if the interval is correct) or no credit (if it's wrong). If you opt for the calculator method, be sure to name the procedure (e.g., two-proportion z interval) and to give the interval (e.g., 0.223 to 0.297).

TECHNOLOGY CORNER Confidence interval for a difference in proportions

The TI-83/84 and TI-89 can be used to construct a confidence interval for $p_1 - p_2$. We'll demonstrate using the previous example. Of $n_1 = 800$ teens surveyed, $X = 584$ said they used social-networking sites. Of $n_2 = 2253$ adults surveyed, $X = 1059$ said they engaged in social networking. To construct a confidence interval:

TI-83/84

Press STAT, then choose TESTS and `B:2-PropZInt`.

TI-89

In the Statistics/List Editor, press 2nd F2 ([F7]) and choose `6:2-PropZInt`.

- When the 2-PropZInt screen appears, enter the values shown.

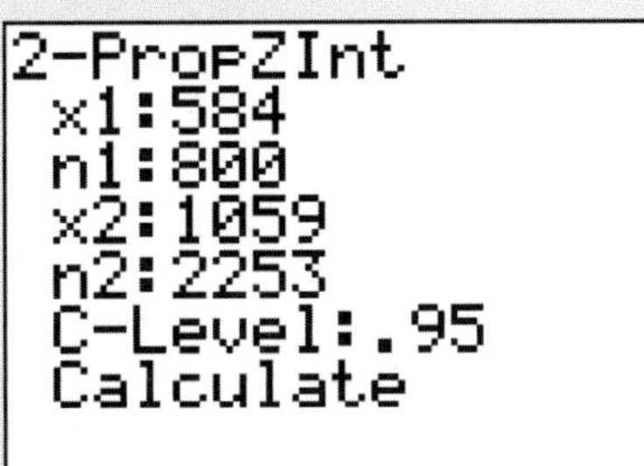

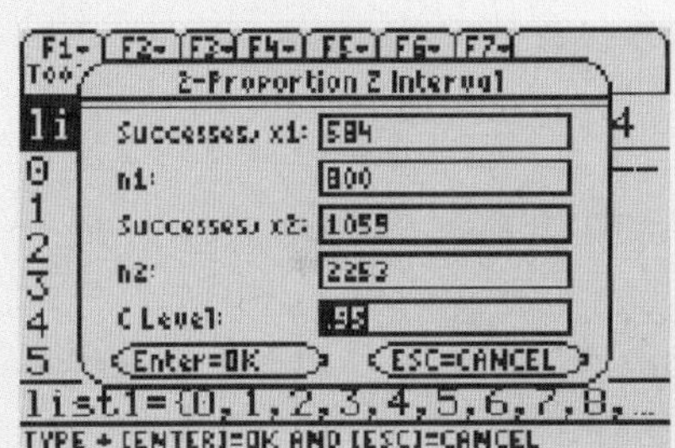

- Highlight "Calculate" and press ENTER.

2-PropZInt
(.22293,.29699)
$\hat{p}_1$=.73
$\hat{p}_2$=.4700399467
n_1=800
n_2=2253

TI-Nspire instructions in Appendix B

CHECK YOUR UNDERSTANDING

Are teens or adults more likely to go online daily? The Pew Internet and American Life Project asked a random sample of 800 teens and a separate random sample of 2253 adults how often they use the Internet. In these two surveys, 63% of teens and 68% of adults said that they go online every day. Construct and interpret a 90% confidence interval for $p_1 - p_2$.

Significance Tests for $p_1 - p_2$

An observed difference between two sample proportions can reflect an actual difference in the parameters, or it may just be due to chance variation in random sampling or random assignment. Significance tests help us decide which explanation makes more sense. The null hypothesis has the general form

$$H_0: p_1 - p_2 = \text{hypothesized value}$$

We'll restrict ourselves to situations in which the hypothesized difference is 0. Then the null hypothesis says that there is no difference between the two parameters:

$$H_0: p_1 - p_2 = 0 \text{ or, alternatively, } H_0: p_1 = p_2$$

The alternative hypothesis says what kind of difference we expect.

Hungry Children

Stating hypotheses

Researchers designed a survey to compare the proportions of children who come to school without eating breakfast in two low-income elementary schools. An SRS of 80 students from School 1 found that 19 had not eaten breakfast. At School 2, an SRS of 150 students included 26 who had not had breakfast. More than 1500 students attend each school. Do these data give convincing evidence of a difference in the population proportions?

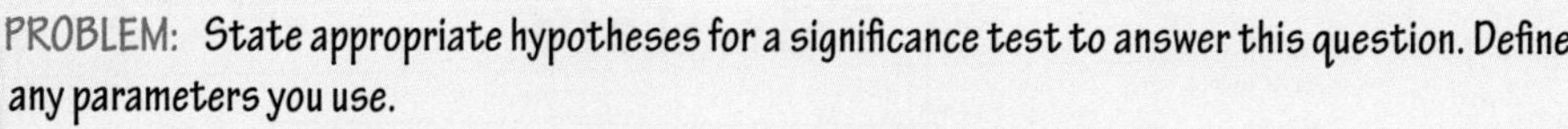

PROBLEM: **State appropriate hypotheses for a significance test to answer this question. Define any parameters you use.**

SOLUTION: **We should carry out a test of**

$$H_0: p_1 - p_2 = 0$$
$$H_a: p_1 - p_2 \neq 0$$

where p_1 = the true proportion of students at School 1 who did not eat breakfast, and p_2 = the true proportion of students at School 2 who did not eat breakfast.

If the Random, Normal, and Independent conditions are met, we can proceed with calculations. To do a test, standardize $\hat{p}_1 - \hat{p}_2$ to get a z statistic:

$$\text{test statistic} = \frac{\text{statistic} - \text{parameter}}{\text{standard deviation of statistic}}$$

$$z = \frac{(\hat{p}_1 - \hat{p}_2) - 0}{\text{standard deviation of statistic}}$$

If $H_0: p_1 = p_2$ is true, the two parameters are the same. We call their common value p. But now we need a way to estimate p, so it makes sense to combine the data from the two samples. This **pooled** (or **combined**) **sample proportion** is

Pooled (or combined) sample proportion

$$\hat{p}_C = \frac{\text{count of successes in both samples combined}}{\text{count of individuals in both samples combined}} = \frac{X_1 + X_2}{n_1 + n_2}$$

Recall that the standard deviation of $\hat{p}_1 - \hat{p}_2$ is

$$\sigma_{\hat{p}_1 - \hat{p}_2} = \sqrt{\frac{p_1(1-p_1)}{n_1} + \frac{p_2(1-p_2)}{n_2}}$$

We can use a little algebra to rewrite the denominator of the test statistic:

$$\sqrt{\frac{\hat{p}_C(1-\hat{p}_C)}{n_1}+\frac{\hat{p}_C(1-\hat{p}_C)}{n_2}} = \sqrt{\hat{p}_C(1-\hat{p}_C)\left(\frac{1}{n_1}+\frac{1}{n_2}\right)} = \sqrt{\hat{p}_C(1-\hat{p}_C)}\sqrt{\frac{1}{n_1}+\frac{1}{n_2}}$$

This latter formula looks like the one given on the AP exam formula sheet.

Use $\hat{p}_C$ in place of both p_1 and p_2 in this expression for the denominator of the test statistic:

$$z = \frac{(\hat{p}_1 - \hat{p}_2) - 0}{\sqrt{\frac{\hat{p}_C(1-\hat{p}_C)}{n_1}+\frac{\hat{p}_C(1-\hat{p}_C)}{n_2}}}$$

Since the Normal condition is met, this will yield a z statistic that has approximately the standard Normal distribution when H_0 is true. Here are the details for the **two-sample z test for the difference between two proportions**.

Two-Sample z Test for the Difference between Two Proportions

Suppose the Random, Normal, and Independent conditions are met. To test the hypothesis $H_0: p_1 - p_2 = 0$, first find the pooled proportion $\hat{p}_C$ of successes in both samples combined. Then compute the z statistic

$$z = \frac{(\hat{p}_1 - \hat{p}_2) - 0}{\sqrt{\frac{\hat{p}_C(1-\hat{p}_C)}{n_1}+\frac{\hat{p}_C(1-\hat{p}_C)}{n_2}}}$$

Find the P-value by calculating the probability of getting a z statistic this large or larger in the direction specified by the alternative hypothesis H_a:

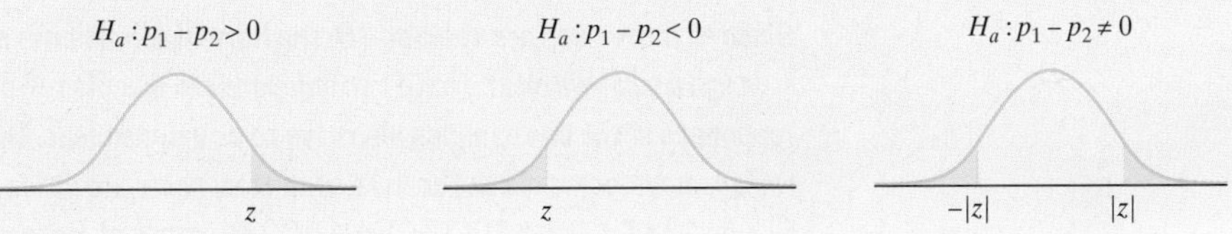

Conditions: Use this test when

- **Random** The data are produced by a random sample of size n_1 from Population 1 and a random sample of size n_2 from Population 2 or by two groups of size n_1 and n_2 in a randomized experiment.
- **Normal** The counts of "successes" and "failures" in each sample or group—$n_1\hat{p}_1$, $n_1(1-\hat{p}_1)$, $n_2\hat{p}_2$, and $n_2(1-\hat{p}_2)$—are all at least 10.
- **Independent** Both the samples or groups themselves and the individual observations in each sample or group are independent. When sampling without replacement, check that the two populations are at least 10 times as large as the corresponding samples (the 10% condition).

Note: It would also be correct to check the Normal condition using the pooled (combined) sample proportion $\hat{p}_C$. Just confirm that $n_1\hat{p}_C$, $n_1(1-\hat{p}_C)$, $n_2\hat{p}_C$, and $n_2(1-\hat{p}_C)$, are all at least 10. Some might argue that this approach is *more* correct because it's consistent with the assumption that $H_0: p_1 - p_2 = 0$ is true. But if the actual counts of successes and failures in the two samples are all at least 10, the expected counts of successes and failures computed using $\hat{p}_C$ will be, too. That's because $\hat{p}_1 \leq \hat{p}_C \leq \hat{p}_2$. We'll stick with the easier (and more conservative) method described in the summary box.

Now we can finish the test we started earlier.

Hungry Children

Significance test for $p_1 - p_2$

Researchers designed a survey to compare the proportions of children who come to school without eating breakfast in two low-income elementary schools. An SRS of 80 students from School 1 found that 19 had not eaten breakfast. At School 2, an SRS of 150 students included 26 who had not had breakfast. More than 1500 students attend each school. Do these data give convincing evidence of a difference in the population proportions? Carry out a significance test at the $\alpha = 0.05$ level to support your answer.

STATE: Our hypotheses are

$$H_0: p_1 - p_2 = 0$$
$$H_a: p_1 - p_2 \neq 0$$

where p_1 = the true proportion of students at School 1 who did not eat breakfast, and p_2 = the true proportion of students at School 2 who did not eat breakfast.

PLAN: If conditions are met, we should perform a two-sample z test for $p_1 - p_2$.

- *Random* The data were produced using two simple random samples—of 80 students from School 1 and 150 students from School 2.
- *Normal* We check the counts of "successes" and "failures":

$$n_1\hat{p}_1 = 19,\ n_1(1-\hat{p}_1) = 61,\ n_2\hat{p}_2 = 26,\ n_2(1-\hat{p}_2) = 124$$

Since all four values are at least 10, the Normal condition is met.

- *Independent* We clearly have two independent samples—one from each school. Individual responses in the two samples also have to be independent. The researchers are sampling without replacement, so we check the 10% condition: there are at least 10(80) = 800 students at School 1 and at least 10(150) = 1500 students at School 2.

DO: We know that $n_1 = 80$, $\hat{p}_1 = \frac{19}{80} = 0.2375$, $n_2 = 150$, and $\hat{p}_2 = \frac{26}{150} = 0.1733$. The pooled proportion of students who didn't eat breakfast in the two samples is

$$\hat{p}_C = \frac{19+26}{80+150} = \frac{45}{230} = 0.1957$$

- *Test statistic*

$$z = \frac{(\hat{p}_1 - \hat{p}_2) - 0}{\sqrt{\frac{\hat{p}_C(1-\hat{p}_C)}{n_1} + \frac{\hat{p}_C(1-\hat{p}_C)}{n_2}}} = \frac{(0.2375 - 0.1733) - 0}{\sqrt{\frac{0.1957(1-0.1957)}{80} + \frac{0.1957(1-0.1957)}{150}}} = 1.17$$

- *P-value* Figure 10.4 displays the *P*-value as an area under the standard Normal curve for this two-tailed test. Using Table A or `normalcdf`, the desired *P*-value is $2P(z \geq 1.17) = 2(1 - 0.8790) = 0.2420$.

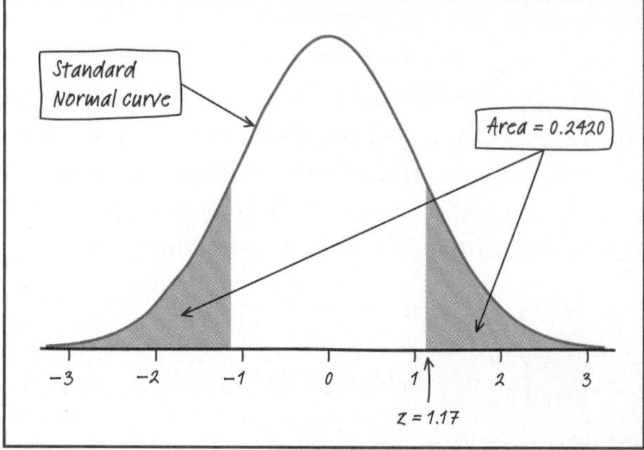

FIGURE 10.4 The *P*-value for the two-sided test.

CONCLUDE: *Since our P-value, 0.2420, is greater than the chosen significance level of $\alpha = 0.05$, we fail to reject H_0. There is not sufficient evidence to conclude that the proportions of students at the two schools who didn't eat breakfast are different.*

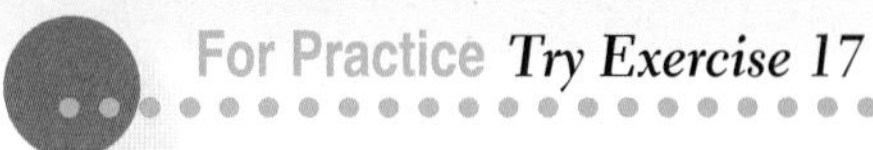

For Practice *Try Exercise 17*

The two-sample z test and two-sample z interval for the difference between two proportions don't always give consistent results. That's because the "standard deviation of the statistic" used in calculating the test statistic is

$$\sqrt{\frac{\hat{p}_C(1-\hat{p}_C)}{n_1}+\frac{\hat{p}_C(1-\hat{p}_C)}{n_2}}$$

but for the confidence interval, it's

$$\sqrt{\frac{\hat{p}_1(1-\hat{p}_1)}{n_1}+\frac{\hat{p}_2(1-\hat{p}_2)}{n_2}}.$$

Exactly what does the P-value in the previous example tell us? If we repeated the random sampling process many times, we'd get a difference in sample proportions as large as or larger than the one actually observed about 24% of the time when H_0: $p_1 - p_2 = 0$ is true. With such a high probability of getting a result like this just by chance when the null hypothesis is true, we don't have enough evidence to reject H_0.

We can get additional information about the difference between the population proportions at School 1 and School 2 with a confidence interval. The TI-84's `2-PropZInt` gives the 95% confidence interval for $p_1 - p_2$ as $(-0.047, 0.175)$. That is, we are 95% confident that the true difference in the proportions of students who ate breakfast at the two schools is between 4.7 percentage points lower at School 1 and 17.5 percentage points higher at School 1. This is consistent with our "fail to reject H_0" conclusion in the example because 0 is included in the interval of plausible values for $p_1 - p_2$.

Inference for Experiments

Most of the examples in this chapter have involved data that were produced by random sampling. However, many important statistical results come from randomized comparative experiments. Here is one such example that compares two proportions.

EXAMPLE

Cholesterol and Heart Attacks

Significance test in an experiment

High levels of cholesterol in the blood are associated with higher risk of heart attacks. Will using a drug to lower blood cholesterol reduce heart attacks? The Helsinki Heart Study recruited middle-aged men with high cholesterol but no history of other serious medical problems to investigate this question. The volunteer subjects were assigned at random to one of two treatments: 2051 men took the drug gemfibrozil to reduce their cholesterol levels, and a control group of 2030 men took a placebo. During the next five years, 56 men in the gemfibrozil group and 84 men in the placebo group had heart attacks. Is the apparent benefit of gemfibrozil statistically significant? Perform an appropriate test to find out.

STATE: *We hope to show that gemfibrozil reduces heart attacks, so we have a one-sided alternative:*

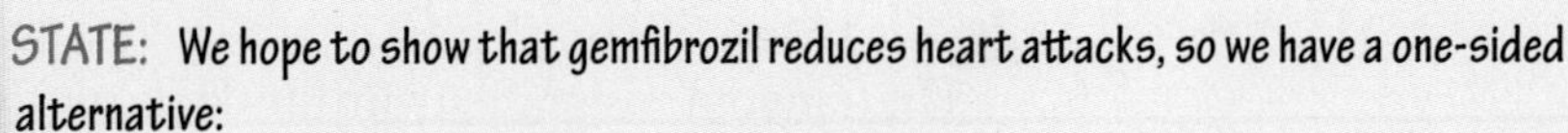

$$H_0: p_1 - p_2 = 0 \qquad \text{or, equivalently,} \qquad H_0: p_1 = p_2$$
$$H_a: p_1 - p_2 < 0 \qquad\qquad\qquad\qquad H_a: p_1 < p_2$$

where p_1 is the actual heart attack rate for middle-aged men like the ones in this study who take gemfibrozil, and p_2 is the actual heart attack rate for middle-aged men like the ones in this study who take only a placebo. No significance level was specified, so we'll use $\alpha = 0.01$

to reduce the risk of making a Type I error (concluding that gemfibrozil reduces heart attack risk when it actually doesn't).

PLAN: If conditions are met, we will do a two-sample z test for $p_1 - p_2$.

- *Random* The data come from two groups in a randomized experiment.
- *Normal* The number of successes (heart attacks!) and failures in the two groups are 56, 1995, 84, and 1946. These are all at least 10, so the Normal condition is met.
- *Independent* Due to the random assignment, these two groups of men can be viewed as independent. Individual observations in each group should also be independent: knowing whether one subject has a heart attack gives no information about whether another subject does.

Note that we did not need to check the 10% condition because the subjects in the experiment were not sampled without replacement from some larger population.

DO: The proportions of men who had heart attacks in each group are

$$\hat{p}_1 = \frac{56}{2051} = 0.0273 \text{ (gemfibrozil group) and } \hat{p}_2 = \frac{84}{2030} = 0.0414 \text{ (placebo group)}$$

The pooled proportion of heart attacks for the two groups is

$$\hat{p}_C = \frac{\text{count of heart attacks in both samples combined}}{\text{count of subjects in both samples combined}} = \frac{56+84}{2051+2030} = \frac{140}{4081} = 0.0343$$

- *Test statistic*

$$z = \frac{(\hat{p}_1 - \hat{p}_2) - 0}{\sqrt{\frac{\hat{p}_C(1-\hat{p}_C)}{n_1} + \frac{\hat{p}_C(1-\hat{p}_C)}{n_2}}} = \frac{(0.0273 - 0.0414) - 0}{\sqrt{\frac{0.0343(1-0.0343)}{2051} + \frac{0.0343(1-0.0343)}{2030}}} = -2.47$$

- *P-value* The one-sided *P*-value is the area under the standard Normal curve to the left of $z = -2.47$ (shown in Figure 10.5). Table A or `normalcdf(-100, -2.47)` gives 0.0068 as the *P*-value.

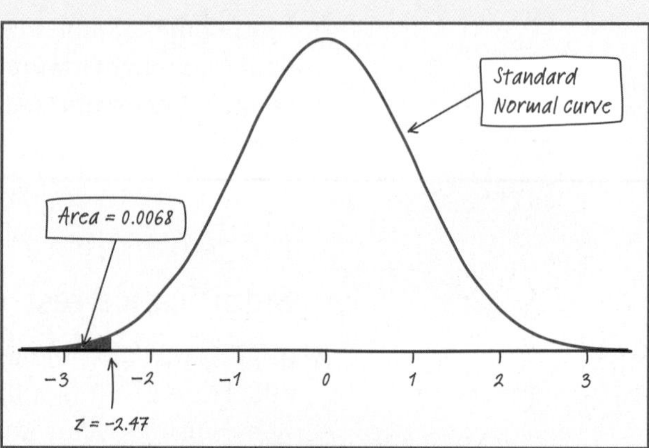

FIGURE 10.5 The *P*-value for the one-sided test.

CONCLUDE: Since the *P*-value, 0.0068, is less than 0.01, the results are statistically significant at the $\alpha = 0.01$ level. We can reject H_0 and conclude that there is convincing evidence of a lower heart attack rate for middle-aged men like these who take gemfibrozil than for those who take only a placebo.

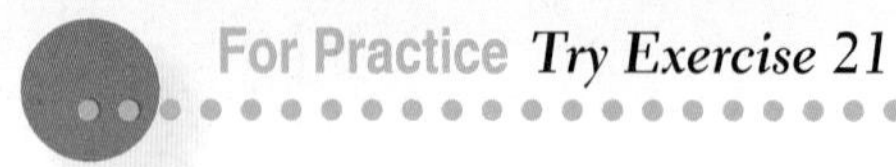

The random assignment in the Helsinki Heart Study allowed researchers to draw a cause-and-effect conclusion. They could say that gemfibrozil reduced the rate of heart attacks for middle-aged men like those who took part in the experiment. Because the subjects were not randomly selected from a larger population, researchers could not generalize the findings of this study any further. No conclusions could be drawn about the effectiveness of gemfibrozil at preventing heart attacks for all middle-aged men, for older men, or for women.

How much more effective is gemfibrozil than a placebo at reducing the risk of heart attack? In the experiment, 4.1% of the men in the placebo group but only 2.7% of the men who took the drug had heart attacks. A 95% confidence interval for the actual difference in the parameters is

$$(0.0414-0.0273)\pm 1.96\sqrt{\frac{0.0414(0.9586)}{2051}+\frac{0.0273(0.9727)}{2030}}=(0.0029,\ 0.0253)$$

Based on the results of this study, we are 95% confident that gemfibrozil reduces heart attack risk by 0.3 to 2.5 percentage points more than a placebo in middle-aged men like those in the Helsinki Heart Study.

THINK ABOUT IT

Why do the inference methods for random samples work for randomized experiments? Confidence intervals and tests for $p_1 - p_2$ are based on the sampling distribution of $\hat{p}_1 - \hat{p}_2$. But in experiments, we aren't sampling at random from any larger populations. We can think about what would happen if the random assignment were repeated many times under the assumption that $H_0\colon p_1 - p_2 = 0$ is true. That is, we assume that the specific treatment received doesn't affect an individual subject's response.

Let's see what would happen just by chance if we randomly reassign the 4081 subjects in the Helsinki Heart Study to the two groups many times, assuming the drug received *doesn't affect* whether or not each individual has a heart attack. We used Fathom software to redo the random assignment 1000 times. The approximate **randomization distribution** of $\hat{p}_1 - \hat{p}_2$ is shown in Figure 10.6. It has an approximately Normal shape with mean 0 and standard deviation 0.0058. These are roughly the same as the shape, center, and spread of the sampling distribution of $\hat{p}_1 - \hat{p}_2$ that we used to perform calculations in the previous example since

Randomization distribution

$$\sqrt{\frac{\hat{p}_C(1-\hat{p}_C)}{n_1}+\frac{\hat{p}_C(1-\hat{p}_C)}{n_2}}=\sqrt{\frac{0.0343(1-0.0343)}{2051}+\frac{0.0343(1-0.0343)}{2030}}=0.0057$$

In the Helsinki Heart Study, the difference in the proportions of subjects who had a heart attack in the gemfibrozil and placebo groups was $0.0273 - 0.0414 = -0.0141$. How likely is it that a difference this large or larger would happen just by chance when H_0 is true? Figure 10.6 provides a rough answer: 5 of the 1000 random reassignments yielded a difference in proportions less than or equal to -0.0141. That is, our estimate of the *P*-value is 0.005. This is quite close to the 0.0068 *P*-value that we calculated in the previous example.

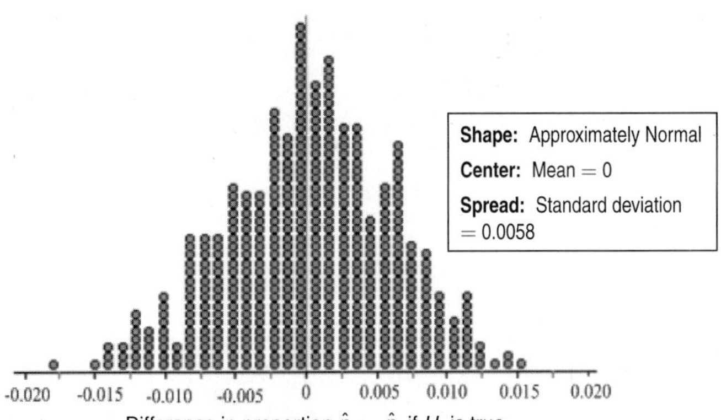

FIGURE 10.6 Fathom simulation showing the approximate randomization distribution of $\hat{p}_1 - \hat{p}_2$ from 1000 random reassignments of subjects to treatment groups in the Helsinki Heart Study.

Figure 10.7 shows the value of the z test statistic for each of the 1000 re-randomizations, calculated using our familiar formula

$$z=\frac{(\hat{p}_1-\hat{p}_2)-0}{\sqrt{\frac{\hat{p}_C(1-\hat{p}_C)}{n_1}+\frac{\hat{p}_C(1-\hat{p}_C)}{n_2}}}$$

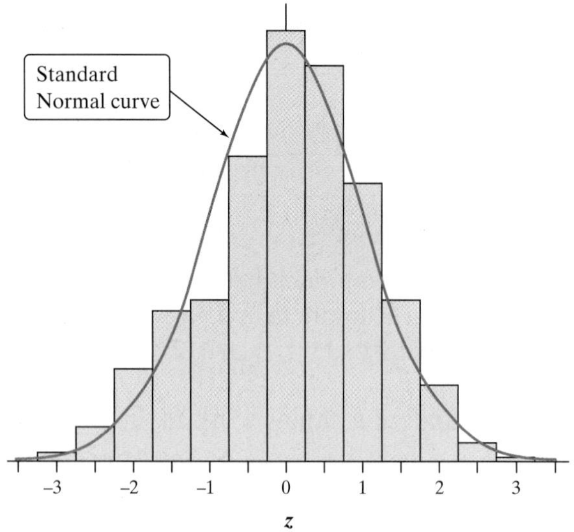

FIGURE 10.7 The distribution of the z test statistic for the 1000 random reassignments in Figure 10.6.

The standard Normal density curve is shown in blue. We can see that the z test statistic has approximately the standard Normal distribution in this case.

Whenever the Random, Normal, and Independent conditions are met, the randomization distribution of $\hat{p}_1 - \hat{p}_2$ looks much like its sampling distribution. We are therefore safe using two-sample z procedures for comparing two proportions.

The formulas for the two-sample z test and interval for $p_1 - p_2$ often lead to calculation errors by students. As a result, we recommend using the calculator's `2-PropZTest` and `2-PropZInt` features to do the dirty work.

TECHNOLOGY CORNER Significance test for a difference in proportions

The TI-83/84 and TI-89 can be used to perform significance tests for comparing two proportions. Here, we use the data from the Helsinki Heart Study. To perform a test of $H_0: p_1 - p_2 = 0$:

TI-83/84

Press [STAT], then choose TESTS and `6:2-PropZTest.`

TI-89

In the Statistics/List Editor, press [2nd] [F1] ([F6]) and choose `6:2-PropZTest.`

• When the 2-PropZTest screen appears, enter the values $x_1 = 56$, $n_1 = 2051$, $x_2 = 84$, $n_2 = 2030$. Specify the alternative hypothesis $p_1 < p_2$, as shown.

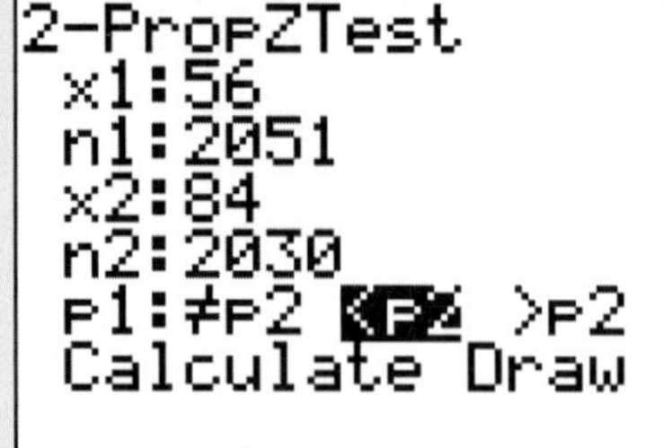

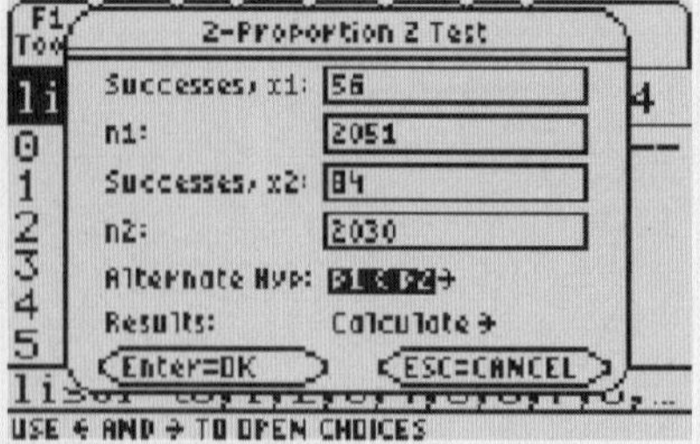

• If you select "Calculate" and press [ENTER], you are told that the z statistic is $z = -2.47$ and the P-value is 0.0068, at top right. These results agree with those from the previous example. Do you see the combined proportion of heart attacks?

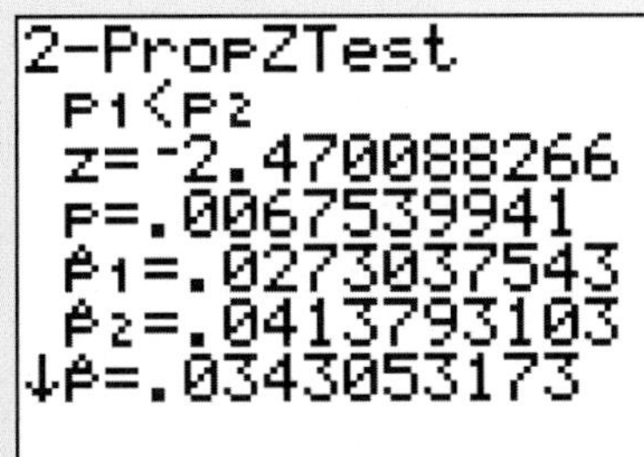

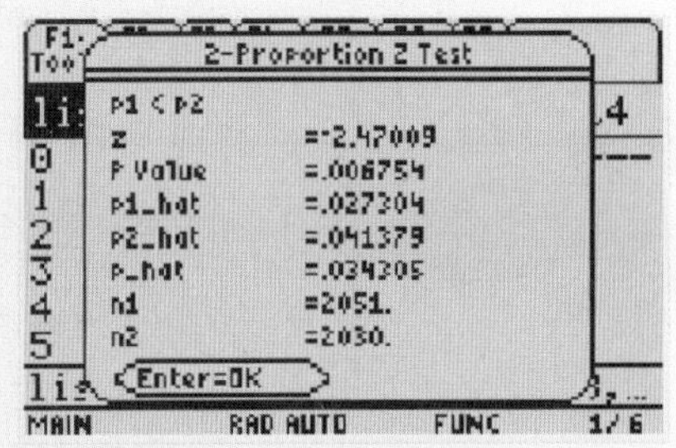

• If you select the "Draw" option, you will see the screen shown here.

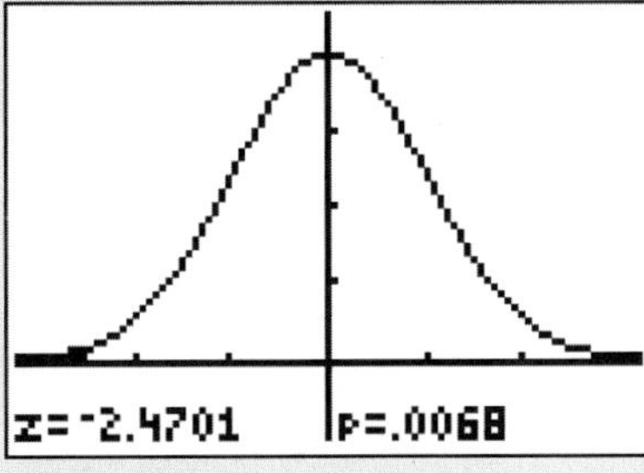

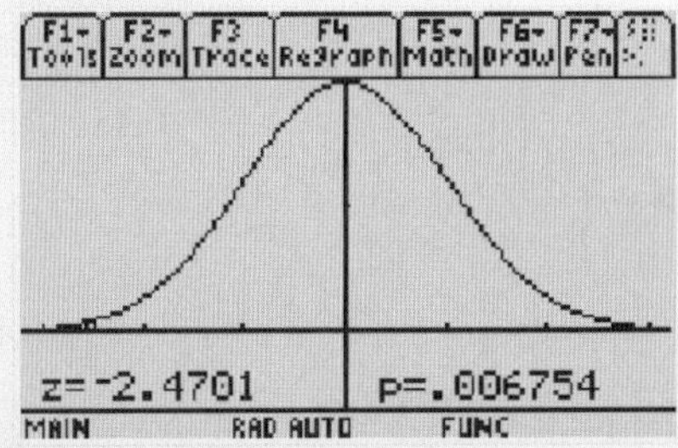

TI-Nspire instructions in Appendix B

CHECK YOUR UNDERSTANDING

To study the long-term effects of preschool programs for poor children, researchers designed an experiment. They recruited 123 children who had never attended preschool from low-income families in Michigan. Researchers randomly assigned 62 of the children to attend preschool (paid for by the study budget) and the other 61 to serve as a control group who would not go to preschool. One response variable of interest was the need for social services as adults. In the past 10 years, 38 children in the preschool group and 49 in the control group have needed social services.[4]

1. Does this study provide convincing evidence that preschool reduces the later need for social services? Carry out an appropriate test to help answer this question.

2. The Minitab output below gives a 95% confidence interval for $p_1 - p_2$. Interpret this interval in context. Then explain what additional information the confidence interval provides.

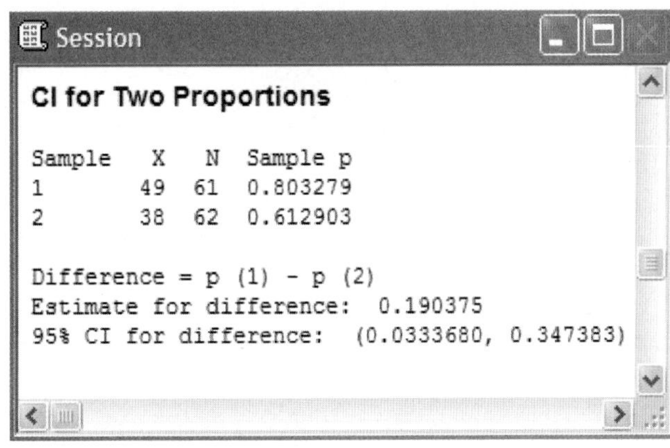

Session

CI for Two Proportions

Sample	X	N	Sample p
1	49	61	0.803279
2	38	62	0.612903

Difference = p (1) - p (2)
Estimate for difference: 0.190375
95% CI for difference: (0.0333680, 0.347383)

SECTION 10.1 Summary

- Choose independent SRSs of size n_1 from Population 1 with proportion of successes p_1 and of size n_2 from Population 2 with proportion of successes p_2. The sampling distribution of $\hat{p}_1 - \hat{p}_2$ has the following properties:
  - **Shape** Approximately Normal if the samples are large enough that n_1p_1, $n_1(1 - p_1)$, n_2p_2, and $n_2(1 - p_2)$ are all at least 10.
  - **Center** The mean is $p_1 - p_2$.
  - **Spread** As long as each sample is no more than 10% of its population (the 10% condition), the standard deviation is $\sqrt{\frac{p_1(1-p_1)}{n_1} + \frac{p_2(1-p_2)}{n_2}}$.
- Confidence intervals and tests to compare the proportions p_1 and p_2 of successes for two populations or treatments are based on the difference $\hat{p}_1 - \hat{p}_2$ between the sample proportions.
- When the Random, Normal, and Independent conditions are met, we can use two-sample z procedures to estimate and test claims about $p_1 - p_2$.
  - **Random** The data are produced by a random sample of size n_1 from Population 1 and a random sample of size n_2 from Population 2 or by two groups of size n_1 and n_2 in a randomized experiment.
  - **Normal** The counts of "successes" and "failures" in each sample or group—$n_1\hat{p}_1$, $n_1(1 - \hat{p}_1)$, $n_2\hat{p}_2$, and $n_2(1 - \hat{p}_2)$—are all at least 10.
  - **Independent** Both the samples or groups themselves and the individual observations in each sample or group are independent. When sampling without replacement, check that the two populations are at least 10 times as large as the corresponding samples (the 10% condition).
- An approximate level C confidence interval for $p_1 - p_2$ is

$$(\hat{p}_1 - \hat{p}_2) \pm z^* \sqrt{\frac{\hat{p}_1(1-\hat{p}_1)}{n_1} + \frac{\hat{p}_2(1-\hat{p}_2)}{n_2}}$$

  where z^* is the standard Normal critical value. This is called a **two-sample z interval** for $p_1 - p_2$.
- **Significance tests** of $H_0: p_1 - p_2 = 0$ use the **pooled (combined) sample proportion**

$$\hat{p}_C = \frac{\text{count of successes in both samples combined}}{\text{count of individuals in both samples combined}} = \frac{X_1 + X_2}{n_1 + n_2}$$

  The **two-sample z test for $p_1 - p_2$** uses the test statistic

$$z = \frac{(\hat{p}_1 - \hat{p}_2) - 0}{\sqrt{\frac{\hat{p}_C(1-\hat{p}_C)}{n_1} + \frac{\hat{p}_C(1-\hat{p}_C)}{n_2}}}$$

  with P-values calculated from the standard Normal distribution.

- Inference about the difference $p_1 - p_2$ in the effectiveness of two treatments in a completely randomized experiment is based on the **randomization distribution** of $\hat{p}_1 - \hat{p}_2$. When the Random, Normal, and Independent conditions are met, our usual inference procedures based on the sampling distribution of $\hat{p}_1 - \hat{p}_2$ will be approximately correct.

10.1 TECHNOLOGY CORNERS

TI-Nspire instructions in Appendix B

SECTION 10.1 Exercises

Note: We are no longer reminding you to use the four-step process in exercises that require you to perform inference.

1. **Toyota or Nissan?** Are Toyota or Nissan owners more satisfied with their vehicles? Let's design a study to find out. We'll select a random sample of 400 Toyota owners and a separate random sample of 400 Nissan owners. Then we'll ask each individual in the sample: "Would you say that you are generally satisfied with your (Toyota/Nissan) vehicle?"

   (a) Is this a problem about comparing means or comparing proportions? Explain.

   (b) What type of study design is being used to produce data?

2. **Binge drinking** Who is more likely to binge drink—male or female college students? The Harvard School of Public Health surveys random samples of male and female undergraduates at four-year colleges and universities about whether they have engaged in binge drinking.

   (a) Is this a problem about comparing means or comparing proportions? Explain.

   (b) What type of study design is being used to produce data?

3. **Computer gaming** Do experienced computer game players earn higher scores when they play with someone present to cheer them on or when they play alone? Fifty teenagers who are experienced at playing a particular computer game have volunteered for a study. We randomly assign 25 of them to play the game alone and the other 25 to play the game with a supporter present. Each player's score is recorded.

   (a) Is this a problem about comparing means or comparing proportions? Explain.

   (b) What type of study design is being used to produce data?

4. **Credit cards and incentives** A bank wants to know which of two incentive plans will most increase the use of its credit cards. It offers each incentive to a group of current credit card customers, determined at random, and compares the amount charged during the following six months.

   (a) Is this a problem about comparing means or comparing proportions? Explain.

   (b) What type of study design is being used to produce data?

pg 607

5. **I want red!** A candy maker offers Child and Adult bags of jelly beans with different color mixes. The company claims that the Child mix has 30% red jelly beans while the Adult mix contains 15% red jelly beans. Assume that the candy maker's claim is true. Suppose we take a random sample of 50 jelly beans from the Child mix and a separate random sample of 100 jelly beans from the Adult mix.

   (a) Find the probability that the proportion of red jelly beans in the Child sample is less than or equal to the proportion of red jelly beans in the Adult sample. Show your work.

(b) Suppose that the Child and Adult samples contain an equal proportion of red jelly beans. Based on your result in part (a), would this give you reason to doubt the company's claim? Explain.

6. **Literacy** A researcher reports that 80% of high school graduates but only 40% of high school dropouts would pass a basic literacy test.[5] Assume that the researcher's claim is true. Suppose we give a basic literacy test to a random sample of 60 high school graduates and a separate random sample of 75 high school dropouts.

(a) Find the probability that the proportion of graduates who pass the test is at least 0.20 higher than the proportion of dropouts who pass. Show your work.

(b) Suppose that the difference in the sample proportions (graduate – dropout) who pass the test is exactly 0.20. Based on your result in part (a), would this give you reason to doubt the researcher's claim? Explain.

Explain why the conditions for using two-sample z procedures to perform inference about $p_1 - p_2$ are not met in the settings of Exercises 7 through 10.

7. **Don't drink the water!** The movie *A Civil Action* (Touchstone Pictures, 1998) tells the story of a major legal battle that took place in the small town of Woburn, Massachusetts. A town well that supplied water to eastern Woburn residents was contaminated by industrial chemicals. During the period that residents drank water from this well, 16 of the 414 babies born had birth defects. On the west side of Woburn, 3 of the 228 babies born during the same time period had birth defects.

8. **In-line skaters** A study of injuries to in-line skaters used data from the National Electronic Injury Surveillance System, which collects data from a random sample of hospital emergency rooms. The researchers interviewed 161 people who came to emergency rooms with injuries from in-line skating. Wrist injuries (mostly fractures) were the most common.[6] The interviews found that 53 people were wearing wrist guards and 6 of these had wrist injuries. Of the 108 who did not wear wrist guards, 45 had wrist injuries.

9. **Shrubs and fire** Fire is a serious threat to shrubs in dry climates. Some shrubs can resprout from their roots after their tops are destroyed. One study of resprouting took place in a dry area of Mexico.[7] The investigators randomly assigned shrubs to treatment and control groups. They clipped the tops of all the shrubs. They then applied a propane torch to the stumps of the treatment group to simulate a fire. All 12 of the shrubs in the treatment group resprouted. Only 8 of the 12 shrubs in the control group resprouted.

10. **Broken crackers** We don't like to find broken crackers when we open the package. How can makers reduce breaking? One idea is to microwave the crackers for 30 seconds right after baking them. Breaks start as hairline cracks called "checking." Assign 65 newly baked crackers to the microwave and another 65 to a control group that is not microwaved. After one day, none of the microwave group and 16 of the control group show checking.[8]

11. **Who uses instant messaging?** Do younger people use online instant messaging (IM) more often than older people? A random sample of IM users found that 73 of the 158 people in the sample aged 18 to 27 said they used IM more often than email. In the 28 to 39 age group, 26 of 143 people used IM more often than email.[9] Construct and interpret a 90% confidence interval for the difference between the proportions of IM users in these age groups who use IM more often than email.

pg 609

12. **Listening to rap** Is rap music more popular among young blacks than among young whites? A sample survey compared 634 randomly chosen blacks aged 15 to 25 with 567 randomly selected whites in the same age group. It found that 368 of the blacks and 130 of the whites listened to rap music every day.[10] Construct and interpret a 95% confidence interval for the difference between the proportions of black and white young people who listen to rap every day.

13. **Young adults living at home** A surprising number of young adults (ages 19 to 25) still live in their parents' homes. A random sample by the National Institutes of Health included 2253 men and 2629 women in this age group.[11] The survey found that 986 of the men and 923 of the women lived with their parents.

(a) Construct and interpret a 99% confidence interval for the difference in population proportions (men minus women).

(b) Does your interval from part (a) give convincing evidence of a difference between the population proportions? Explain.

14. **Fear of crime** The elderly fear crime more than younger people, even though they are less likely to be victims of crime. One study recruited separate random samples of 56 black women and 63 black men over the age of 65 from Atlantic City, New Jersey. Of the women, 27 said they "felt vulnerable" to crime; 46 of the men said this.[12]

(a) Construct and interpret a 90% confidence interval for the difference in population proportions (men minus women).

(b) Does your interval from part (a) give convincing evidence of a difference between the population proportions? Explain.

15. **Who owns iPods?** As part of the Pew Internet and American Life Project, researchers surveyed a random sample of 800 teens and a separate random sample of 400 young adults. For the teens, 79% said that they own an iPod or MP3 player. For the young adults, this figure was 67%. Is there a significant difference between the population proportions? State appropriate hypotheses for a significance test to answer this question. Define any parameters you use.

pg 612

16. **Steroids in high school** A study by the National Athletic Trainers Association surveyed random samples of 1679 high school freshmen and 1366 high school seniors in Illinois. Results showed that 34 of the freshmen and 24 of the seniors had used anabolic steroids. Steroids, which are dangerous, are sometimes used to improve athletic performance.[13] Is there a significant difference between the population proportions? State appropriate hypotheses for a significance test to answer this question. Define any parameters you use.

17. **Who owns iPods?** Refer to Exercise 15.

(a) Carry out a significance test at the $\alpha = 0.05$ level.

pg 614

(b) Construct and interpret a 95% confidence interval for the difference between the population proportions. Explain how the confidence interval is consistent with the results of the test in part (a).

18. **Steroids in high school** Refer to Exercise 16.

(a) Carry out a significance test at the $\alpha = 0.05$ level.

(b) Construct and interpret a 95% confidence interval for the difference between the population proportions. Explain how the confidence interval is consistent with the results of the test in part (a).

19. **What's wrong?** "Would you marry a person from a lower social class than your own?" Researchers asked this question of a random sample of 385 black, never-married students at two historically black colleges in the South. Of the 149 men in the sample, 91 said "Yes." Among the 236 women, 117 said "Yes."[14] Is there reason to think that different proportions of men and women in this student population would be willing to marry beneath their class?

Holly carried out the significance test shown below to answer this question. Unfortunately, she made some mistakes along the way. Identify as many mistakes as you can, and tell how to correct each one.

State: I want to perform a test of

$$H_0: p_1 = p_2$$
$$H_a: p_1 \neq p_2$$

at the 95% confidence level.

Plan: If conditions are met, I'll do a one-sample z test for comparing two proportions.

- **Random** The data came from a random sample of 385 black, never-married students.
- **Normal** One student's answer to the question should have no relationship to another student's answer.
- **Independent** The counts of successes and failures in the two groups—91, 58, 117, and 119—are all at least 10.

Do: From the data, $\hat{p}_1 = \frac{91}{149} = 0.61$ and $\hat{p}_2 = \frac{117}{236} = 0.46$.

- **Test statistic**

$$z = \frac{(0.61 - 0.46) - 0}{\sqrt{\frac{0.61(0.39)}{149} + \frac{0.46(0.54)}{236}}} = 2.91$$

- **P-value** From Table A, $P(z \geq 2.91) = 1 - 0.9982 = 0.0018$.

Conclude: The *P*-value, 0.0018, is less than 0.05, so I'll reject the null hypothesis. This proves that a higher proportion of men than women are willing to marry someone from a social class lower than their own.

20. **What's wrong?** A driving school wants to find out which of its two instructors is more effective at preparing students to pass the state's driver's license exam. An incoming class of 100 students is randomly assigned to two groups, each of size 50. One group is taught by Instructor A; the other is taught by Instructor B. At the end of the course, 30 of Instructor A's students and 22 of Instructor B's students pass the

state exam. Do these results give convincing evidence that Instructor A is more effective?

Min Jae carried out the significance test shown below to answer this question. Unfortunately, he made some mistakes along the way. Identify as many mistakes as you can, and tell how to correct each one.

State: I want to perform a test of

$$H_0: p_1 - p_2 = 0$$
$$H_a: p_1 - p_2 > 0$$

where p_1 = the proportion of Instructor A's students that passed the state exam and p_2 = the proportion of Instructor B's students that passed the state exam. Since no significance level was stated, I'll use $\sigma = 0.05$.

Plan: If conditions are met, I'll do a two-sample z test for comparing two proportions.

- **Random** The data came from two random samples of 50 students.
- **Normal** The counts of successes and failures in the two groups—30, 20, 22, and 28—are all at least 10.
- **Independent** There are at least 1000 students who take this driving school's class.

Do: From the data, $\hat{p}_1 = \frac{20}{50} = 0.40$ and $\hat{p}_2 = \frac{30}{50} = 0.60$. So the pooled proportion of successes is

$$\hat{p}_C = \frac{22+30}{50+50} = 0.52$$

- **Test statistic**

$$z = \frac{(0.40 - 0.60) - 0}{\sqrt{\frac{0.52(0.48)}{100} + \frac{0.52(0.48)}{100}}} = -2.83$$

- **P-value** From Table A, $P(z \geq -2.83) = 1 - 0.0023 = 0.9977$.

Conclude: The P-value, 0.9977, is greater than $\alpha = 0.05$, so we fail to reject the null hypothesis. There is not convincing evidence that Instructor A's pass rate is higher than Instructor B's.

21. **Did the random assignment work?** A large clinical trial of the effect of diet on breast cancer assigned women at random to either a normal diet or a low-fat diet. To check that the random assignment did produce comparable groups, we can compare the two groups at the start of the study. Ask if there is a family history of breast cancer: 3396 of the 19,541 women in the low-fat group and 4929 of the 29,294 women in the control group said "Yes."[15] If the random assignment worked well, there should *not* be a significant difference in the proportions with a family history of breast cancer.

pg 615

(a) How significant is the observed difference? Carry out an appropriate test to help answer this question.

(b) Describe a Type I and a Type II error in this setting. Which is more serious? Explain.

22. **Preventing strokes** Aspirin prevents blood from clotting and so helps prevent strokes. The Second European Stroke Prevention Study asked whether adding another anticlotting drug, named dipyridamole, would be more effective for patients who had already had a stroke. Here are the data on strokes and deaths during the two years of the study:[16]

	Number of patients	Number of strokes
Aspirin alone	1649	206
Aspirin + dipyridamole	1650	157

The study was a randomized comparative experiment.

(a) Is there a significant difference in the proportion of strokes between these two treatments? Carry out an appropriate test to help answer this question.

(b) Describe a Type I and a Type II error in this setting. Which is more serious? Explain.

Exercises 23 through 26 involve the following setting. Some women would like to have children but cannot do so for medical reasons. One option for these women is a procedure called in vitro fertilization (IVF), which involves injecting a fertilized egg into the woman's uterus.

23. **Prayer and pregnancy** Two hundred women who were about to undergo IVF served as subjects in an experiment. Each subject was randomly assigned to either a treatment group or a control group. Women in the treatment group were intentionally prayed for by several people (called *intercessors*) who did not know them, a process known as intercessory prayer. The praying continued for three weeks following IVF. The intercessors did not pray for the women in the control group. Here are the results: 44 of the 88 women in the treatment group got pregnant, compared to 21 out of 81 in the control group.[17]

Is the pregnancy rate significantly higher for women who received intercessory prayer? To find out, researchers perform a test of $H_0: p_1 = p_2$ versus $H_a: p_1 > p_2$, where p_1 and p_2 are the actual pregnancy rates for women like those in the study who do and don't receive intercessory prayer, respectively.

(a) Name the appropriate test and check that the conditions for carrying out this test are met.

(b) The appropriate test from part (a) yields a P-value of 0.0007. Interpret this P-value in context.

(c) What conclusion should researchers draw at the $\alpha = 0.05$ significance level? Explain.

(d) The women in the study did not know if they were being prayed for. Explain why this is important.

24. **Acupuncture and pregnancy** A study reported in the medical journal *Fertility and Sterility* sought to determine whether the ancient Chinese art of acupuncture could help infertile women become pregnant.[18] One hundred sixty healthy women who planned to have IVF were recruited for the study. Half of the subjects (80) were randomly assigned to receive acupuncture 25 minutes before embryo transfer and again 25 minutes after the transfer. The remaining 80 women were assigned to a control group and instructed to lie still for 25 minutes after the embryo transfer. Results are shown in the table below.

	Acupuncture group	Control group
Pregnant	34	21
Not pregnant	46	59
Total	**80**	**80**

Is the pregnancy rate significantly higher for women who received acupuncture? To find out, researchers perform a test of $H_0: p_1 = p_2$ versus $H_a: p_1 > p_2$, where p_1 and p_2 are the actual pregnancy rates for women like those in the study who do and don't receive acupuncture, respectively.

(a) Name the appropriate test and check that the conditions for carrying out this test are met.

(b) The appropriate test from part (a) yields a P-value of 0.0152. Interpret this P-value in context.

(c) What conclusion should researchers draw at the $\alpha = 0.05$ significance level? Explain.

(d) What flaw in the design of the experiment prevents us from drawing a cause-and-effect conclusion? Explain.

25. **Prayer and pregnancy** Construct and interpret a 95% confidence interval for $p_1 - p_2$ in Exercise 23. Explain what additional information the confidence interval provides.

26. **Acupuncture and pregnancy** Construct and interpret a 95% confidence interval for $p_1 - p_2$ in Exercise 24. Explain what additional information the confidence interval provides.

27. **Children make choices** Many new products introduced into the market are targeted toward children. The choice behavior of children with regard to new products is of particular interest to companies that design marketing strategies for these products. As part of one study, randomly selected children in different age groups were compared on their ability to sort new products into the correct product category (milk or juice).[19] Here are some of the data:

Age group	N	Number who sorted correctly
4- to 5-year-olds	50	10
6- to 7-year-olds	53	28

Are these two age groups equally skilled at sorting? Use information from the Minitab output below to support your answer.

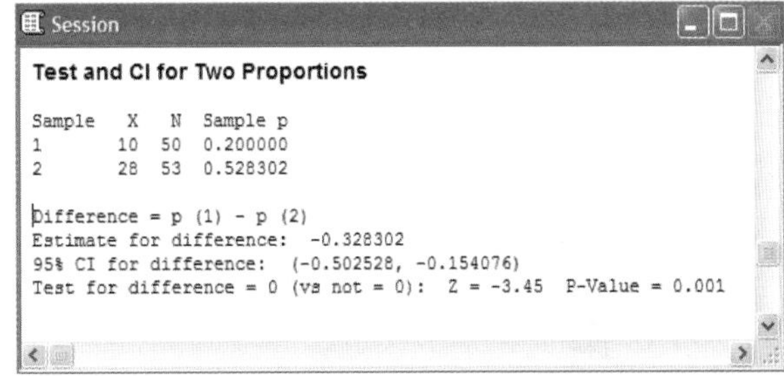

```
Session
Test and CI for Two Proportions

Sample   X   N  Sample p
1       10  50  0.200000
2       28  53  0.528302

Difference = p (1) - p (2)
Estimate for difference:  -0.328302
95% CI for difference:  (-0.502528, -0.154076)
Test for difference = 0 (vs not = 0):  Z = -3.45  P-Value = 0.001
```

28. **Police radar and speeding** Do drivers reduce excessive speed when they encounter police radar? Researchers studied the behavior of a sample of drivers on a rural interstate highway in Maryland where the speed limit was 55 miles per hour. They measured speed with an electronic device hidden in the pavement and, to eliminate large trucks, considered only vehicles less than 20 feet long. During some time periods (determined at random), police radar was set up at the measurement location. Here are some of the data:[20]

	Number of vehicles	Number over 65 mph
No radar	12,931	5,690
Radar	3,285	1,051

(a) The researchers chose a rural highway so that cars would be separated rather than in clusters, because some cars might slow when they see other cars slowing. Explain why this is important.

(b) Does the proportion of speeding drivers differ significantly when radar is being used and when it isn't? Use information from the Minitab computer output below to support your answer.

```
Session
Test and CI for Two Proportions

Sample     X      N  Sample p
1       5690  12931  0.440028
2       1051   3285  0.319939

Difference = p (1) - p (2)
Estimate for difference:  0.120089
95% CI for difference:  (0.101988, 0.138189)
Test for difference = 0 (vs not = 0):  Z = 12.47  P-Value = 0.000
```

Multiple choice: Select the best answer for Exercises 29 to 32.

A sample survey interviews SRSs of 500 female college students and 550 male college students. Each student is asked whether he or she worked for pay last summer. In all, 410 of the women and 484 of the men say "Yes." Exercises 29 to 31 are based on this survey.

29. Take p_M and p_F to be the proportions of all college males and females who worked last summer. We conjectured before seeing the data that men are more likely to work. The hypotheses to be tested are

(a) $H_0: p_M - p_F = 0$ versus $H_a: p_M - p_F \neq 0$.
(b) $H_0: p_M - p_F = 0$ versus $H_a: p_M - p_F > 0$.
(c) $H_0: p_M - p_F = 0$ versus $H_a: p_M - p_F < 0$.
(d) $H_0: p_M - p_F > 0$ versus $H_a: p_M - p_F = 0$.
(e) $H_0: p_M - p_F \neq 0$ versus $H_a: p_M - p_F = 0$.

30. The pooled sample proportion who worked last summer is about

(a) $\hat{p}_C = 1.70$.
(b) $\hat{p}_C = 0.89$.
(c) $\hat{p}_C = 0.88$.
(d) $\hat{p}_C = 0.85$.
(e) $\hat{p}_C = 0.82$.

31. The 95% confidence interval for the difference $p_M - p_F$ in the proportions of college men and women who worked last summer is about

(a) 0.06 ± 0.00095.
(b) 0.06 ± 0.043.
(c) 0.06 ± 0.036.
(d) -0.06 ± 0.043.
(e) -0.06 ± 0.036.

32. In an experiment to learn whether Substance M can help restore memory, the brains of 20 rats were treated to damage their memories. The rats were trained to run a maze. After a day, 10 rats (determined at random) were given M and 7 of them succeeded in the maze. Only 2 of the 10 control rats were successful. The two-sample z test for "no difference" against "a significantly higher proportion of the M group succeeds"

(a) gives $z = 2.25$, $P < 0.02$.
(b) gives $z = 2.60$, $P < 0.005$.
(c) gives $z = 2.25$, $P < 0.04$ but not < 0.02.
(d) should not be used because the Random condition is violated.
(e) should not be used because the Normal condition is violated.

Exercises 33 and 34 refer to the following setting. Thirty randomly selected seniors at Council High School were asked to report the age (in years) and mileage of their main vehicles. Here is a scatterplot of the data:

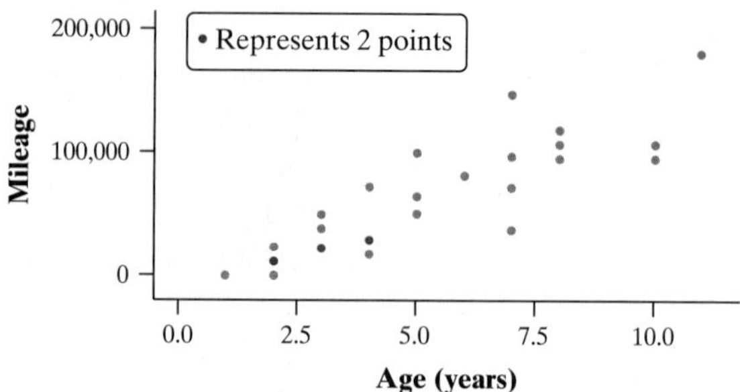

We used Minitab to perform a least-squares regression analysis for these data. Part of the computer output from this regression is shown below.

```
Predictor     Coef     Stdev    t-ratio       P
Constant    -13832     8773      -1.58     0.126
Age          14954     1546       9.67     0.000

s = 22723   R-sq = 77.0%    R-sq(adj) = 76.1%
```

33. **Drive my car** (3.2)

(a) What is the equation of the least-squares regression line? Be sure to define any symbols you use.

(b) Interpret the slope of the least-squares line in the context of this problem.

(c) One student reported that her 10-year-old car had 110,000 miles on it. Find the residual for this data value. Show your work.

34. **Drive my car** (3.2, 4.3)

(a) Explain what the value of r^2 tells you about how well the least-squares line fits the data.

(b) The mean age of the students' cars in the sample was $\bar{x} = 8$ years. Find the mean mileage of the cars in the sample. Show your work.

(c) Interpret the value of s in the context of this setting.

(d) Would it be reasonable to use the least-squares line to predict a car's mileage from its age for a Council High School teacher? Justify your answer.

Comparing Two Means

In Section 10.2, you'll learn about:

- The sampling distribution of a difference between two means
- The two-sample t statistic
- Confidence intervals for $\mu_1 - \mu_2$
- Significance tests for $\mu_1 - \mu_2$
- Using two-sample t procedures wisely

In the previous section, we developed methods for comparing two proportions. What if we want to compare the mean of some quantitative variable for the individuals in Population 1 and Population 2? Our parameters of interest are the population means μ_1 and μ_2. Once again, the best approach is to take separate random samples from each population and to compare the sample means $\bar{x}_1$ and $\bar{x}_2$. Suppose we want to compare the average effectiveness of two treatments in a completely randomized experiment. In this case, the parameters μ_1 and μ_2 are the true mean responses for Treatment 1 and Treatment 2, respectively. We use the mean response in the two groups, $\bar{x}_1$ and $\bar{x}_2$, to make the comparison. Here's a table that summarizes these two situations:

Population or treatment	Parameter	Statistic	Sample size
1	μ_1	$\bar{x}_1$	n_1
2	μ_2	$\bar{x}_2$	n_2

We compare the populations or treatments by doing inference about the difference $\mu_1 - \mu_2$ between the parameters. The statistic that estimates this difference is the difference between the two sample means, $\bar{x}_1 - \bar{x}_2$. To use $\bar{x}_1 - \bar{x}_2$ for inference, we must know its sampling distribution. Here is an Activity that gives you a preview of what lies ahead.

ACTIVITY *Does polyester decay?*

MATERIALS: 10 small pieces of card stock (or index cards) per pair of students

How quickly do synthetic fabrics such as polyester decay in landfills? A researcher buried polyester strips in the soil for different lengths of time, then dug up the strips and measured the force required to break them. Breaking strength is easy to measure and is a good indicator of decay. Lower strength means the fabric has decayed.

The researcher buried 10 strips of polyester fabric in well-drained soil in the summer. The strips were randomly assigned to two groups: 5 of them were buried for 2 weeks and the other 5 were buried for 16 weeks. Here are the breaking strengths in pounds:[21]

Group 1 (2 weeks):	118	126	126	120	129
Group 2 (16 weeks):	124	98	110	140	110

Do the data give good evidence that polyester decays more in 16 weeks than in 2 weeks?

1. The Fathom dotplot displays the data from the experiment. Discuss what this graph shows with your classmates.

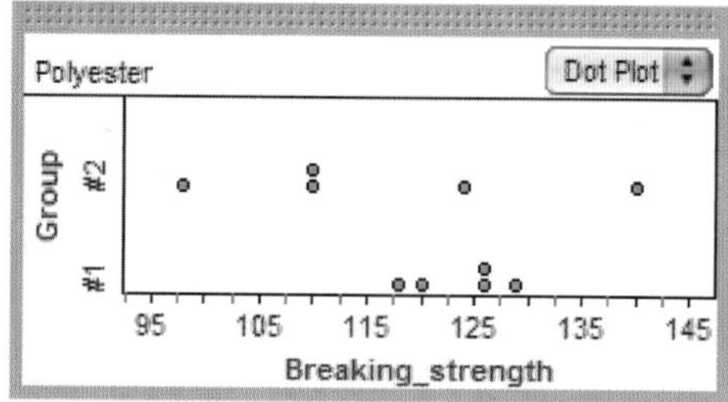

For Group 1, the mean breaking strength was $\bar{x}_1 = 123.8$ pounds. For Group 2, the mean breaking strength was $\bar{x}_2 = 116.4$ pounds. The observed difference in average breaking strength for the two groups is $\bar{x}_1 - \bar{x}_2 = 123.8 - 116.4 = 7.4$ pounds. Is it plausible that this difference is due to the chance involved in the random assignment and not to the treatments themselves? To find out, your class will perform a simulation.

Suppose that the length of time in the ground has no effect on the breaking strength of the polyester specimens. Then each specimen would have the same breaking strength regardless of whether it was assigned to Group 1 or Group 2. In that case, we could examine the results of repeated random assignments of the specimens to the two groups.

2. Write each of the 10 breaking-strength measurements on a separate card. Mix the cards well and deal them face down into two piles of 5 cards each. Be sure to decide which pile is Group 1 and which is Group 2 *before* you look at the cards. Calculate the difference in the mean breaking strength (Group 1 – Group 2). Record this value.

3. Repeat Step 2 four more times. When you have finished, plot your five values for difference in mean breaking strength on a class dotplot.

4. How often did the difference in group means from the class's random reassignments exceed the one found in this study (7.4 pounds)? What conclusion would you draw about whether there is convincing evidence that polyester decays when left in the ground for longer periods of time?

The Sampling Distribution of a Difference between Two Means

To explore the sampling distribution of $\bar{x}_1 - \bar{x}_2$, let's start with two Normally distributed populations having known means and standard deviations. Based on information from the U.S. National Health and Nutrition Examination Survey (NHANES), the heights of ten-year-old girls follow a Normal distribution with mean $\mu_f = 56.4$ inches and standard deviation $\sigma_f = 2.7$ inches. The heights of ten-year-old boys follow a Normal distribution with mean $\mu_m = 55.7$ inches and standard deviation $\sigma_m = 3.8$ inches.[22] Suppose we take independent SRSs of 12 girls and 8 boys of this age and measure their heights. What can we say about the difference $\bar{x}_f - \bar{x}_m$ in the average heights of the sample of girls and the sample of boys?

We used Fathom software to take an SRS of 12 ten-year-old girls and 8 ten-year-old boys and to plot the values of $\bar{x}_f$, $\bar{x}_m$, and $\bar{x}_f - \bar{x}_m$ for each sample. Our first set of simulated samples gave $\bar{x}_f = 56.09$ inches and $\bar{x}_m = 54.68$ inches, so dots were placed above each of those values in Figure 10.8(a) and (b). The difference in the sample means is $\bar{x}_f - \bar{x}_m = 56.09 - 54.68 = 1.41$ inches. A dot for this value appears in Figure 10.8(c). The three dotplots in Figure 10.8 show the results of repeating this process 1000 times. These are the approximate sampling distributions of $\bar{x}_f$, $\bar{x}_m$, and $\bar{x}_f - \bar{x}_m$.

In Chapter 7, we saw that the sampling distribution of a sample mean $\bar{x}$ has the following properties:

- **Shape** (1) If the population distribution is Normal, then so is the sampling distribution of $\bar{x}$; (2) if the population distribution isn't Normal, the sampling distribution of $\bar{x}$ will be approximately Normal if the sample size is large enough (say, $n \geq 30$) by the central limit theorem (CLT).
- **Center** $\mu_{\bar{x}} = \mu$
- **Spread** $\sigma_{\bar{x}} = \dfrac{\sigma}{\sqrt{n}}$ if the sample is no more than 10% of the population (the 10% condition)

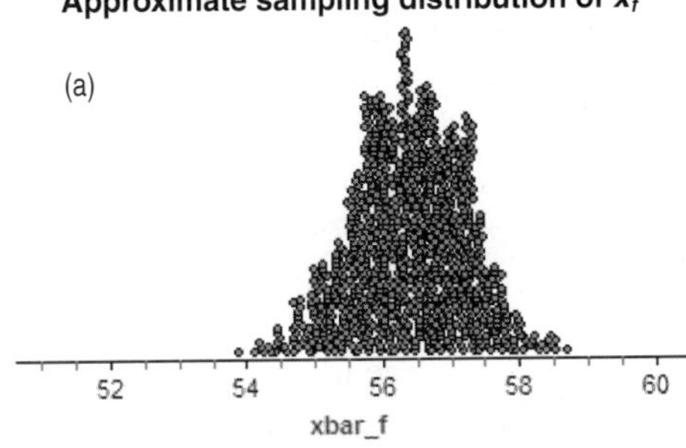

Shape: Normal

Center: $\mu_{\bar{x}_f} = \mu_f = 56.4$ inches

Spread: $\sigma_{\bar{x}_f} = \dfrac{\sigma_f}{\sqrt{n_f}} = \dfrac{2.7}{\sqrt{12}} = 0.78$ inches

Approximate sampling distribution of $\bar{x}_m$

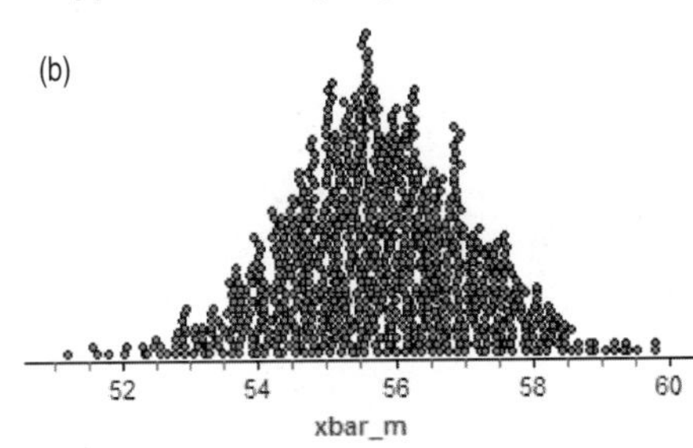

Shape: Normal

Center: $\mu_{\bar{x}_m} = \mu_m = 55.7$ inches

Spread: $\sigma_{\bar{x}_m} = \dfrac{\sigma_m}{\sqrt{n_m}} = \dfrac{3.8}{\sqrt{8}} = 1.34$ inches

Approximate sampling distribution of $\bar{x}_f - \bar{x}_m$

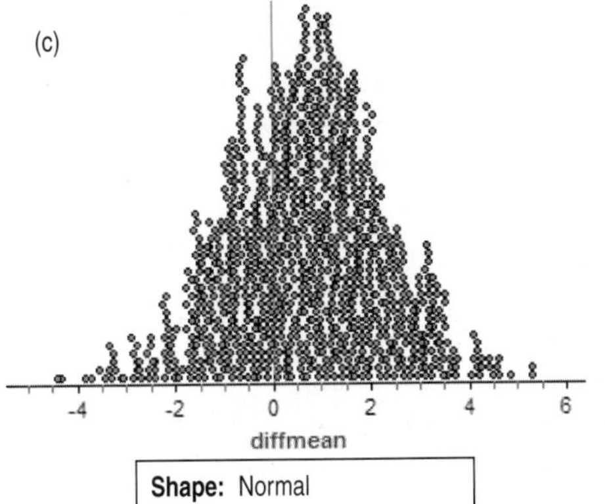

Shape: Normal

Center: $\mu_{\bar{x}_f - \bar{x}_m} = 0.7$ inches

Spread: $\sigma_{\bar{x}_f - \bar{x}_m} = 1.55$ inches

FIGURE 10.8 Simulated sampling distributions of (a) the sample mean height $\bar{x}_f$ in 1000 SRSs of size $n_1 = 12$ from the population of 10-year-old girls, (b) the sample mean height $\bar{x}_m$ in 1000 SRSs of size $n_2 = 8$ from the population of 10-year-old boys, and (c) the difference in sample means $\bar{x}_f - \bar{x}_m$ for each of the 1000 repetitions.

For the sampling distributions of $\bar{x}_f$ and $\bar{x}_m$ in this case:

	Sampling distribution of $\bar{x}_f$	**Sampling distribution of $\bar{x}_m$**
Shape	Normal, since the population distribution is Normal	Normal, since the population distribution is Normal
Center	$\mu_{\bar{x}_f} = \mu_f = 56.4$ inches	$\mu_{\bar{x}_m} = \mu_m = 55.7$ inches
Spread	$\sigma_{\bar{x}_f} = \frac{\sigma_f}{\sqrt{n_f}} = \frac{2.7}{\sqrt{12}} = 0.78$ inches *10% condition:* There are way more than $10(12) = 120$ ten-year-old girls in the United States.	$\sigma_{\bar{x}_m} = \frac{\sigma_m}{\sqrt{n_m}} = \frac{3.8}{\sqrt{8}} = 1.34$ inches *10% condition:* There are way more than $10(8) = 80$ ten-year-old boys in the United States.

The dotplots in Figures 10.8(a) and (b) confirm these results.

What about the sampling distribution of $\bar{x}_f - \bar{x}_m$? Figure 10.8(c) has a roughly Normal shape, is centered at 0.7 inches, and has standard deviation 1.55 inches. The shape makes sense because we are combining two independent Normal random variables, $\bar{x}_f$ and $\bar{x}_m$. How about the center? The actual mean height of ten-year-old girls is $\mu_f = 56.4$ inches. For ten-year-old boys, the actual mean height is $\mu_m = 55.7$ inches. We'd expect the difference $\bar{x}_f - \bar{x}_m$ to center on the actual difference in the population means, $\mu_f - \mu_m = 56.4 - 55.7 = 0.7$ inches. The spread, however, is a bit more complicated.

THINK ABOUT IT

How can we find formulas for the mean and standard deviation of the sampling distribution of $\bar{x}_1 - \bar{x}_2$? Both $\bar{x}_1$ and $\bar{x}_2$ are random variables. That is, their values would vary in repeated SRSs of size n_1 and n_2. The statistic $\bar{x}_1 - \bar{x}_2$ is the difference of these two random variables. In Chapter 6, we learned that for any two independent random variables X and Y,

$$\mu_{X-Y} = \mu_X - \mu_Y \quad \text{and} \quad \sigma^2_{X-Y} = \sigma^2_X + \sigma^2_Y$$

For the random variables $\bar{x}_1$ and $\bar{x}_2$, we have

$$\mu_{\bar{x}_1-\bar{x}_2} = \mu_{\bar{x}_1} - \mu_{\bar{x}_2} = \mu_1 - \mu_2$$

and

$$\sigma^2_{\bar{x}_1-\bar{x}_2} = \sigma^2_{\bar{x}_1} + \sigma^2_{\bar{x}_2} = \left(\frac{\sigma_1}{\sqrt{n_1}}\right)^2 + \left(\frac{\sigma_2}{\sqrt{n_2}}\right)^2 = \frac{\sigma_1^2}{n_1} + \frac{\sigma_2^2}{n_2}$$

$$\sigma_{\bar{x}_1-\bar{x}_2} = \sqrt{\frac{\sigma_1^2}{n_1} + \frac{\sigma_2^2}{n_2}}$$

For the observational study of the heights of ten-year-olds,

$$\sigma_{\bar{x}_f-\bar{x}_m} = \sqrt{\frac{\sigma_f^2}{n_f} + \frac{\sigma_m^2}{n_m}} = \sqrt{\frac{2.7^2}{12} + \frac{3.8^2}{8}} = 1.55$$

This confirms the result from the Fathom simulation.

Here are the facts we need.

The Sampling Distribution of $\bar{x}_1 - \bar{x}_2$

Choose an SRS of size n_1 from Population 1 with mean μ_1 and standard deviation σ_1 and an independent SRS of size n_2 from Population 2 with mean μ_2 and standard deviation σ_2.

- **Shape** When the population distributions are Normal, the sampling distribution of $\bar{x}_1 - \bar{x}_2$ is Normal. In other cases, the sampling distribution of $\bar{x}_1 - \bar{x}_2$ will be approximately Normal if the sample sizes are large enough ($n_1 \geq 30$ and $n_2 \geq 30$).
- **Center** The mean of the sampling distribution is $\mu_1 - \mu_2$. That is, the difference in sample means is an unbiased estimator of the difference in population means.
- **Spread** The standard deviation of the sampling distribution of $\bar{x}_1 - \bar{x}_2$ is

$$\sqrt{\frac{\sigma_1^2}{n_1} + \frac{\sigma_2^2}{n_2}}$$

as long as each sample is no more than 10% of its population (the 10% condition).

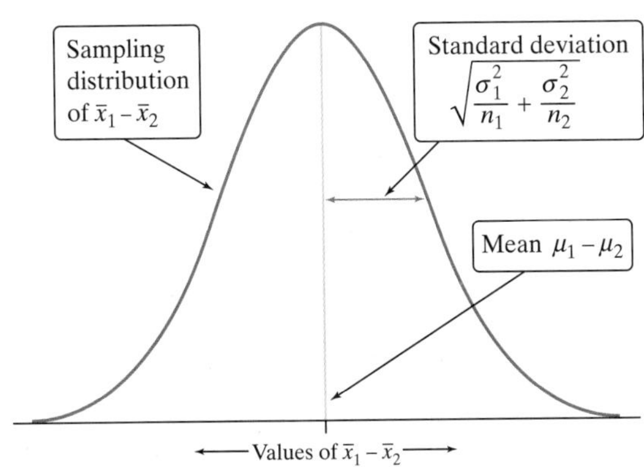

FIGURE 10.9 Select independent SRSs from two populations having means μ_1 and μ_2 and standard deviations σ_1 and σ_2. The two sample means are $\bar{x}_1$ and $\bar{x}_2$. If the population distributions are both Normal, the sampling distribution of the difference $\bar{x}_1 - \bar{x}_2$ is Normal. The sampling distribution will be approximately Normal in other cases if both samples are large enough ($n_1 \geq 30$ and $n_2 \geq 30$).

Figure 10.9 displays the sampling distribution of $\bar{x}_1 - \bar{x}_2$.

The formula for the standard deviation of the sampling distribution involves the parameters σ_1 and σ_2, which are usually unknown. Just as in Chapters 8 and 9, we must replace these by estimates to do inference. We'll get to confidence intervals and significance tests shortly. For now, we can use what we know about the sampling distribution of $\bar{x}_1 - \bar{x}_2$ to compute probabilities, as the following example illustrates.

Who's Taller at Ten: Boys or Girls?

Finding probabilities with the sampling distribution

Based on information from the U.S. National Health and Nutrition Examination Survey (NHANES), the heights of ten-year-old girls follow a Normal distribution with mean $\mu_f = 56.4$ inches and standard deviation $\sigma_f = 2.7$ inches. The heights of ten-year-old boys follow a Normal distribution with mean $\mu_m = 55.7$ inches and standard deviation $\sigma_m = 3.8$ inches. A researcher takes a random sample of 12 ten-year-old girls and a separate random sample of 8 ten-year-old boys in the United States. After analyzing the data, the researcher reports that the mean height $\bar{x}_m$ of the boys is larger than the mean height $\bar{x}_f$ of the girls.

PROBLEM:

(a) Describe the shape, center, and spread of the sampling distribution of $\bar{x}_f - \bar{x}_m$.

(b) Find the probability of getting a difference in sample means $\bar{x}_f - \bar{x}_m$ that's less than 0. Show your work.

(c) Does the result in part (a) give us reason to doubt the researcher's stated results? Explain.

SOLUTION:

(a) The sampling distribution of $\bar{x}_f - \bar{x}_m$ is Normal because both population distributions are Normal. Its mean is $\mu_1 - \mu_2 = 56.4 - 55.7 = 0.7$ inches, and its standard deviation is $\sqrt{\frac{2.7^2}{12} + \frac{3.8^2}{8}} = 1.55$ inches.

(b) Figure 10.10(a) shows the sampling distribution of $\bar{x}_f - \bar{x}_m$ with the desired probability (area) shaded. To find this probability, we use the methods of Chapter 2.

- *Standardize.* When $\bar{x}_f - \bar{x}_m = 0$,

$$z = \frac{0 - 0.70}{1.55} = -0.45$$

- *Use Table A.* The area to the left of $z = -0.45$ under the standard Normal curve is 0.3264. This is the probability we seek. Figure 10.10(b) shows this result.

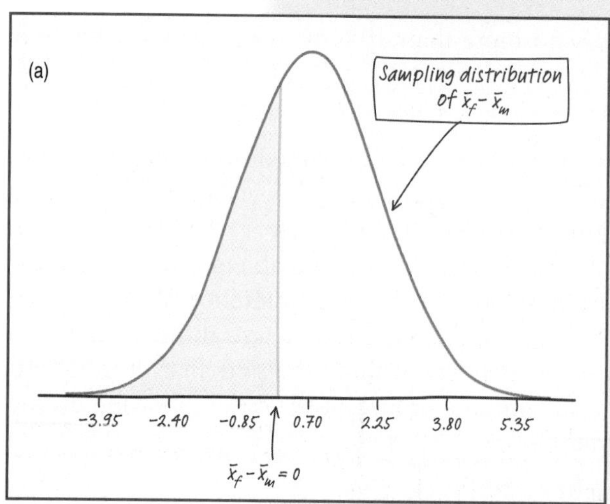

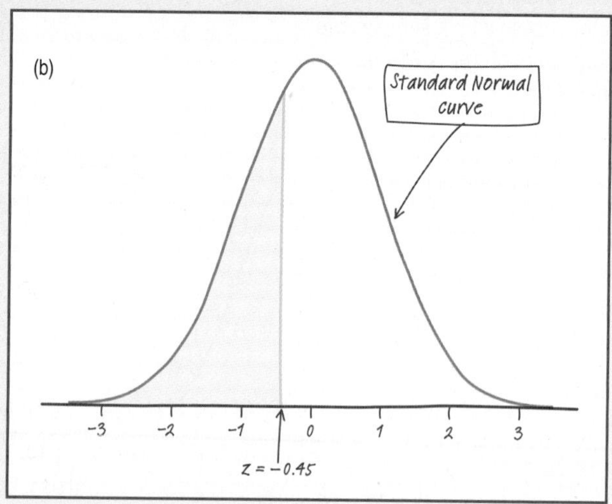

FIGURE 10.10 (a) The Normal sampling distribution of the difference $\bar{x}_f - \bar{x}_m$ in mean heights for a random sample of 12 girls and a random sample of 8 boys. We want to find $P(\bar{x}_f - \bar{x}_m < 0)$. (b) The desired probability can also be expressed as $P(z < -0.45)$, which represents an area under the standard Normal curve.

We could use technology to get the desired probability. Using `normalcdf(−100,0,.7,1.55)` or the *Normal curve* applet gives 0.3258.

(c) If the mean height of the boys is greater than the mean height of the girls, $\bar{x}_m > \bar{x}_f$. That is, $\bar{x}_f - \bar{x}_m < 0$. Part (b) shows that there's about a 33% chance of getting a difference in sample means that's negative just due to sampling variability. This gives us little reason to doubt the researcher's claim.

For Practice *Try Exercise 37*

CHECK YOUR UNDERSTANDING

A fast-food restaurant uses an automated filling machine to pour its soft drinks. The machine has different settings for small, medium, and large drink cups. According to the machine's manufacturer, when the large setting is chosen, the amount of liquid dispensed by the machine follows a Normal distribution with mean 27 ounces and standard deviation 0.8 ounces. When the medium setting is chosen, the amount of liquid dispensed follows a Normal distribution with mean 17 ounces and standard deviation 0.5 ounces. To test the manufacturer's claim, the restaurant manager measures the amount of liquid in a random sample of 25 cups filled with the medium setting and a separate random sample of 20 cups filled with the large setting. Let $\bar{x}_1 - \bar{x}_2$ be the difference in the sample mean amount of liquid under the two settings (large – medium).

1. What is the shape of the sampling distribution of $\bar{x}_1 - \bar{x}_2$? Why?
2. Find the mean and standard deviation of the sampling distribution.
3. Find the probability that $\bar{x}_1 - \bar{x}_2$ is more than 12 ounces. Show your work.
4. Based on your answer to Question 3, would you be surprised if the difference in the mean amount of liquid dispensed in the two samples was 12 ounces? Explain.

The Two-Sample t Statistic

When data come from two random samples or two groups in a randomized experiment (the Random condition), the statistic $\bar{x}_1 - \bar{x}_2$ is our best guess for the value of $\mu_1 - \mu_2$. If the Independent condition is met, the standard deviation of the sampling distribution of $\bar{x}_1 - \bar{x}_2$ is

$$\sigma_{\bar{x}_1 - \bar{x}_2} = \sqrt{\frac{\sigma_1^2}{n_1} + \frac{\sigma_2^2}{n_2}}$$

If the Normal condition is met, we can standardize the observed difference $\bar{x}_1 - \bar{x}_2$ to obtain a z statistic that follows a standard Normal distribution:

$$z = \frac{(\bar{x}_1 - \bar{x}_2) - (\mu_1 - \mu_2)}{\sqrt{\frac{\sigma_1^2}{n_1} + \frac{\sigma_2^2}{n_2}}}$$

Two-sample z statistic

We used this approach to calculate a probability involving heights of ten-year-olds in the previous example. In the unlikely event that both population standard deviations are known, this **two-sample z statistic** is the basis for inference about $\mu_1 - \mu_2$.

Standard error

Suppose now that the population standard deviations σ_1 and σ_2 are not known. We estimate them by the standard deviations s_1 and s_2 from our two samples. The result is the **standard error** (also called the *estimated standard deviation*) of $\bar{x}_1 - \bar{x}_2$:

$$\sqrt{\frac{s_1^2}{n_1} + \frac{s_2^2}{n_2}}$$

Two-sample t statistic

When we standardize the estimate, the result is the **two-sample t statistic:**

$$t = \frac{(\bar{x}_1 - \bar{x}_2) - (\mu_1 - \mu_2)}{\sqrt{\frac{s_1^2}{n_1} + \frac{s_2^2}{n_2}}}$$

The statistic t has the same interpretation as any z or t statistic: it says how far $\bar{x}_1 - \bar{x}_2$ is from its mean in standard deviation units.

The two-sample t statistic has approximately a t distribution. It does not have exactly a t distribution even if the populations are both exactly Normal. In practice, however, the approximation is very accurate. There are two practical options for using the two-sample t procedures. The two options are exactly the same except for the degrees of freedom used for t critical values and P-values.

Option 1 (Technology): Use the t distribution with degrees of freedom calculated from the data by a somewhat messy formula. With this option, the degrees of freedom may not be a whole number.

Option 2 (Conservative): Use the t distribution with degrees of freedom equal to the *smaller* of $n_1 - 1$ and $n_2 - 1$. With this option, the resulting confidence interval has a margin of error *as large as or larger than* is needed for the desired confidence level. The significance test using this option gives a P-value *equal to or greater than* the true P-value.

As the sample sizes increase, confidence levels and P-values from Option 2 become more accurate.[23]

Confidence Intervals for $\mu_1 - \mu_2$

If the Random, Normal, and Independent conditions are met, we can use our standard formula to construct a confidence interval for $\mu_1 - \mu_2$:

$$\text{statistic} \pm (\text{critical value}) \cdot (\text{standard deviation of statistic}) = (\bar{x}_1 - \bar{x}_2) \pm t^* \sqrt{\frac{s_1^2}{n_1} + \frac{s_2^2}{n_2}}$$

We can use either technology or the conservative approach with Table B to find the critical value t^* for the given confidence level. This method is called a **two-sample t interval for a difference between two means.**

Two-Sample t Interval for a Difference between Two Means

When the Random, Normal, and Independent conditions are met, an approximate level C confidence interval for $\bar{x}_1 - \bar{x}_2$ is

$$(\bar{x}_1 - \bar{x}_2) \pm t^* \sqrt{\frac{s_1^2}{n_1} + \frac{s_2^2}{n_2}}$$

Here t^* is the critical value for confidence level C for the t distribution with degrees of freedom from either Option 1 (technology) or Option 2 (the smaller of $n_1 - 1$ and $n_2 - 1$).

- **Random** The data are produced by a random sample of size n_1 from Population 1 and a random sample of size n_2 from Population 2 or by two groups of size n_1 and n_2 in a randomized experiment.
- **Normal** Both population distributions (or the true distributions of responses to the two treatments) are Normal OR both sample/group sizes are large ($n_1 \geq 30$ and $n_2 \geq 30$).
- **Independent** Both the samples/groups themselves and the individual observations in each sample/group are independent. When sampling without replacement, check that the two populations are at least 10 times as large as the corresponding samples (the 10% condition).

The following example shows how to construct and interpret a confidence interval for a difference in means. As usual with inference problems, we follow the four-step process.

Big Trees, Small Trees, Short Trees, Tall Trees

Confidence interval for $\mu_1 - \mu_2$

The Wade Tract Preserve in Georgia is an old-growth forest of longleaf pines that has survived in a relatively undisturbed state for hundreds of years. One question of interest to foresters who study the area is "How do the sizes of longleaf pine trees in the northern and southern halves of the forest compare?" To find out, researchers took random samples of 30 trees from each half and measured the diameter at breast height (DBH) in centimeters. Comparative boxplots of the data and summary statistics from Minitab are shown below.

Descriptive Statistics: North, South

Variable	N	Mean	StDev
North	30	23.70	17.50
South	30	34.53	14.26

PROBLEM:

(a) Based on the graph and numerical summaries, write a few sentences comparing the sizes of longleaf pine trees in the two halves of the forest.

(b) Construct and interpret a 90% confidence interval for the difference in the mean DBH of longleaf pines in the northern and southern halves of the Wade Tract Preserve.

SOLUTION:

(a) The distribution of DBH measurements in the northern sample is heavily skewed to the right, while the distribution of DBH measurements in the southern sample is heavily skewed to the left. It appears that trees in the southern half of the forest have larger diameters. The mean and median DBH for the southern sample are both much larger than the corresponding measures of center for the northern sample. In addition, the boxplots show that more than 75% of the southern trees have diameters that are above the northern sample's median. There is more variability in the diameters of the northern longleaf pines, as we can see from the larger range, *IQR*, and standard deviation for this sample. No outliers are present in either sample.

(b) STATE: Our *parameters* of interest are μ_1 = the true mean DBH of all trees in the southern half of the forest and μ_2 = the true mean DBH of all trees in the northern half of the forest. We want to estimate the difference $\mu_1 - \mu_2$ at a 90% confidence level.

PLAN: If conditions are met, we'll construct a two-sample *t* interval for $\mu_1 - \mu_2$.

- *Random* The data came from random samples of 30 trees each from the northern and southern halves of the forest.
- *Normal* The boxplots give us reason to believe that the population distributions of DBH measurements may not be Normal. However, since both sample sizes are at least 30, we are safe using *t* procedures.
- *Independent* Researchers took independent samples from the northern and southern halves of the forest. Because sampling without replacement was used, there have to be at least 10(30) = 300 trees in each half of the forest. This is pretty safe to assume.

	Upper-tail probability p		
df	.10	.05	.025
28	1.313	1.701	2.048
29	1.311	1.699	2.045
30	1.310	1.697	2.042
	80%	90%	95%
	Confidence level C		

DO: From the Minitab output, $\bar{x}_1 = 34.53$, $s_1 = 14.26$, $n_1 = 30$, $\bar{x}_2 = 23.70$, $s_2 = 17.50$, and $n_2 = 30$. We'll use the conservative df = the smaller of $n_1 - 1$ and $n_2 - 1$, which is 29. For a 90% confidence level the critical value from Table B is $t^* = 1.699$. So a 90% confidence interval for $\mu_1 - \mu_2$ is

$$(\bar{x}_1 - \bar{x}_2) \pm t^* \sqrt{\frac{s_1^2}{n_1} + \frac{s_2^2}{n_2}} = (34.53 - 23.70) \pm 1.699\sqrt{\frac{14.26^2}{30} + \frac{17.50^2}{30}}$$
$$= 10.83 \pm 7.00 = (3.83, 17.83)$$

CONCLUDE: We are 90% confident that the interval from 3.83 to 17.83 centimeters captures the difference in the actual mean DBH of the southern trees and the actual mean DBH of the northern trees. This interval suggests that the mean diameter of the southern trees is between 3.83 and 17.83 cm larger than the mean diameter of the northern trees.

The 90% confidence interval in the example does not include 0. This gives convincing evidence that the difference in the mean diameter of northern and southern trees in the Wade Tract Preserve isn't 0. However, the confidence interval provides more information than a simple reject or fail to reject H_0 conclusion. It gives a set of plausible values for $\mu_1 - \mu_2$.

We chose the parameters in the DBH example so that $\bar{x}_1 - \bar{x}_2$ would be positive. What would have happened if we had defined μ_1 as the mean DBH of the northern trees and μ_2 as the mean DBH of the southern trees? The 90% confidence interval for $\mu_1 - \mu_2$ would then be

$$(23.70 - 34.53) \pm 1.699\sqrt{\frac{17.50^2}{30} + \frac{14.26^2}{30}} = -10.83 \pm 7.00 = (-17.83, -3.83)$$

This interval suggests that the mean diameter of the northern trees is between 3.83 and 17.83 cm smaller than the mean diameter of the southern trees. Changing the order of subtraction doesn't change the result.

As with other inference procedures, you can use technology to perform the calculations in the "Do" step. Remember that technology comes with potential benefits and risks on the AP exam.

TECHNOLOGY CORNER Two-sample *t* intervals on the calculator

You can use the two-sample *t* interval command on the TI-83/84 or TI-89 to construct a confidence interval for the difference between two means. We'll show you the steps using the summary statistics from the pine trees example.

TI-83/84

Press STAT, then choose TESTS and `0:2-SampTInt....`

TI-89

Press 2nd F2 ([F7]) Ints and choose `4:2-SampTInt....`

- Choose Stats as the input method and enter the summary statistics as shown.

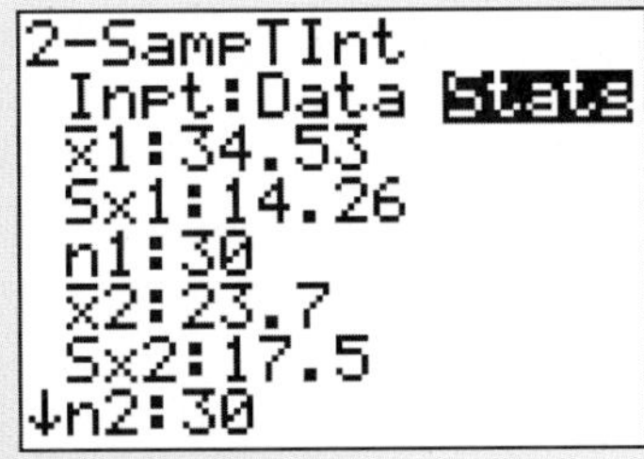

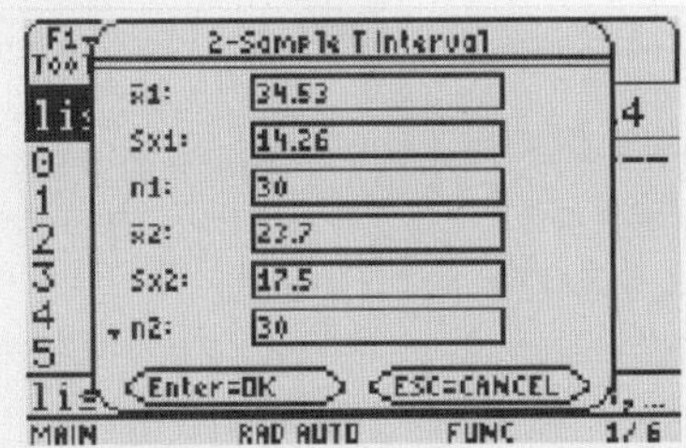

- Enter the confidence level: `C-level: .90`. For Pooled: choose "No." We'll discuss pooling later.
- Highlight Calculate and press ENTER.

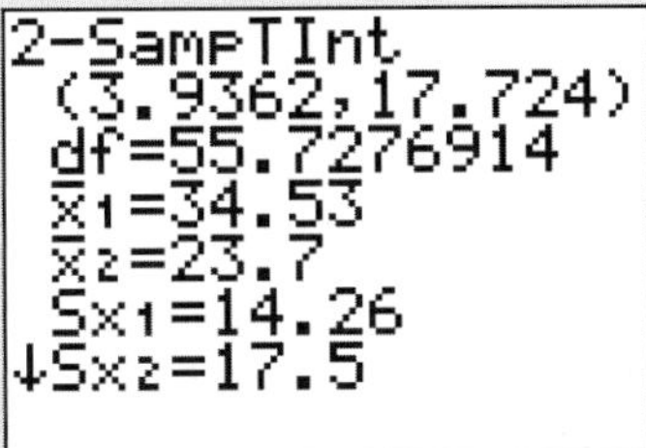

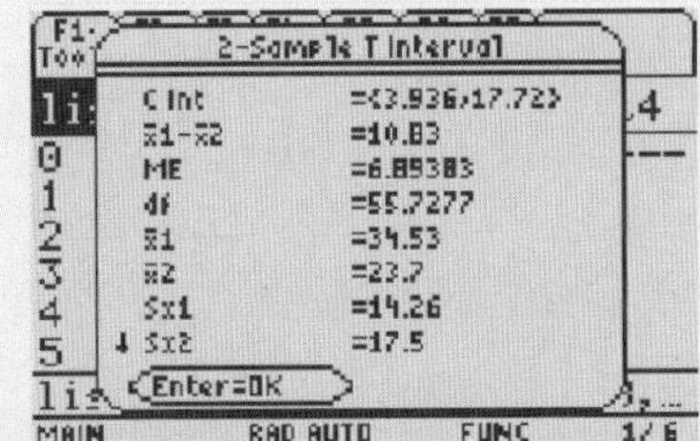

TI-Nspire instructions in Appendix B

The calculator's 90% confidence interval for $\mu_1 - \mu_2$ is (3.936, 17.724). This interval is narrower than the one we found earlier (3.83, 17.83). Why the difference? We used the conservative df = 29, but the calculator used df = 55.73. With more degrees of freedom, the calculator's critical value is smaller than our $t^* = 1.699$, which results in a smaller margin of error and a narrower interval.

THINK ABOUT IT

How does technology compute the degrees of freedom for two-sample t procedures? The exact distribution of the two-sample t statistic is not a t distribution. Moreover, the distribution changes as the unknown population standard deviations σ_1 and σ_2 change. However, the distribution of the two-sample t statistic can be approximated by a t distribution with degrees of freedom given by

$$\text{df} = \frac{\left(\dfrac{s_1^2}{n_1} + \dfrac{s_2^2}{n_2}\right)^2}{\dfrac{1}{n_1 - 1}\left(\dfrac{s_1^2}{n_1}\right)^2 + \dfrac{1}{n_2 - 1}\left(\dfrac{s_2^2}{n_2}\right)^2}$$

This is the approximation used by most statistical software. It is quite accurate when both sample sizes n_1 and n_2 are 5 or larger. Note that the number of df given by this formula is usually not a whole number. But it is always at least as large as the smaller of $n_1 - 1$ and $n_2 - 1$.

You can thank statisticians B. L. Welch and F. E. Satterthwaite for discovering this pretty remarkable formula.

CHECK YOUR UNDERSTANDING

The U.S. Department of Agriculture (USDA) conducted a survey to estimate the average price of wheat in July and in September of the same year. Independent random samples of wheat producers were selected for each of the two months. Here are summary statistics on the reported price of wheat from the selected producers, in dollars per bushel:[24]

Month	n	$\bar{x}$	s_x
July	90	\$2.95	\$0.22
September	45	\$3.61	\$0.19

Construct and interpret a 99% confidence interval for the difference in the mean wheat price in July and in September.

Significance Tests for $\mu_1 - \mu_2$

An observed difference between two sample means can reflect an actual difference in the parameters μ_1 and μ_2, or it may just be due to chance variation in random sampling or random assignment. Significance tests help us decide which explanation makes more sense. The null hypothesis has the general form

$$H_0: \mu_1 - \mu_2 = \text{hypothesized value}$$

We're often interested in situations in which the hypothesized difference is 0. Then the null hypothesis says that there is no difference between the two parameters:

$$H_0: \mu_1 - \mu_2 = 0 \quad \text{or, alternatively,} \quad H_0: \mu_1 = \mu_2$$

The alternative hypothesis says what kind of difference we expect.

If the Random, Normal, and Independent conditions are met, we can proceed with calculations. To do a test, standardize $\bar{x}_1 - \bar{x}_2$ to get a two-sample t statistic:

$$\text{test statistic} = \frac{\text{statistic} - \text{parameter}}{\text{standard deviation of statistic}}$$

$$t = \frac{(\bar{x}_1 - \bar{x}_2) - (\mu_1 - \mu_2)}{\sqrt{\dfrac{s_1^2}{n_1} + \dfrac{s_2^2}{n_2}}}$$

To find the P-value, use the t distribution with degrees of freedom given by Option 1 (technology) or Option 2 (df = smaller of $n_1 - 1$ and $n_2 - 1$). Here are the details for the **two-sample t test for the difference between two means.**

Two-Sample t Test for the Difference between Two Means

Suppose the Random, Normal, and Independent conditions are met. To test the hypothesis $H_0: \mu_1 - \mu_2 =$ hypothesized value, compute the two-sample t statistic

$$t = \frac{(\bar{x}_1 - \bar{x}_2) - (\mu_1 - \mu_2)}{\sqrt{\dfrac{s_1^2}{n_1} + \dfrac{s_2^2}{n_2}}}$$

Find the *P*-value by calculating the probability of getting a *t* statistic this large or larger in the direction specified by the alternative hypothesis H_a. Use the *t* distribution with degrees of freedom approximated by technology or the smaller of $n_1 - 1$ and $n_2 - 1$.

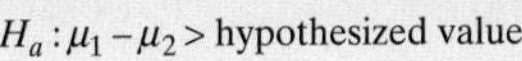

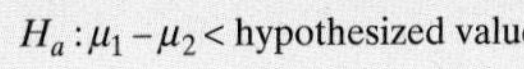

$H_a : \mu_1 - \mu_2 \neq$ hypothesized value

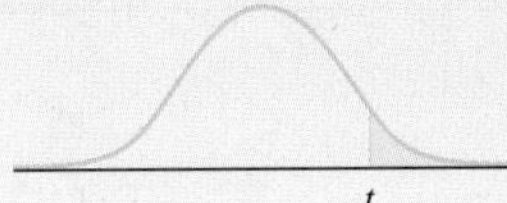

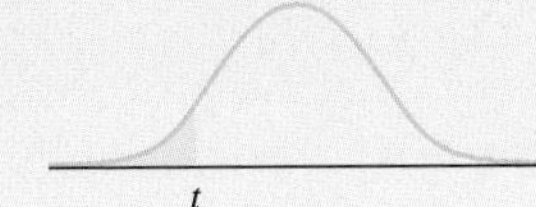

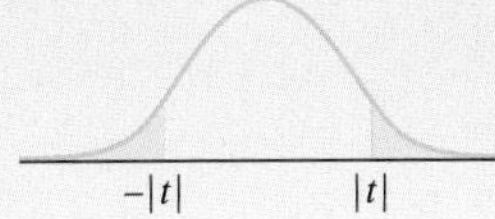

Conditions: Use this test when

- **Random** The data are produced by a random sample of size n_1 from Population 1 and a random sample of size n2 from Population 2 or by two groups of size n_1 and n_2 in a randomized experiment.
- **Normal** Both population distributions (or the true distributions of responses to the two treatments) are Normal OR both sample/group sizes are large ($n_1 \geq 30$ and $n_2 \geq 30$).
- **Independent** Both the samples/groups themselves and the individual observations in each sample/group are independent. When sampling without replacement, check that the two populations are at least 10 times as large as the corresponding samples (the 10% condition).

Did you notice? The conditions for performing a significance test and constructing a confidence interval to compare two means are exactly the same.

Here's an example that shows how to perform a two-sample *t* test.

Calcium and Blood Pressure

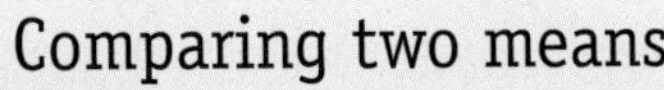

Comparing two means

STEP 4

Does increasing the amount of calcium in our diet reduce blood pressure? Examination of a large sample of people revealed a relationship between calcium intake and blood pressure. The relationship was strongest for black men. Such observational studies do not establish causation. Researchers therefore designed a randomized comparative experiment.

The subjects were 21 healthy black men who volunteered to take part in the experiment. They were randomly assigned to two groups: 10 of the men received a calcium supplement for 12 weeks, while the control group of 11 men received a placebo pill that looked identical. The experiment was double-blind. The response variable is the decrease in systolic (top number) blood pressure for a subject after 12 weeks, in millimeters of mercury. An increase appears as a negative response.[25] Here are the data:

Group 1 (calcium):	7	−4	18	17	−3	−5	1	10	11	−2	
Group 2 (placebo):	−1	12	−1	−3	3	−5	5	2	−11	−1	−3

PROBLEM:

(a) A Fathom dotplot of the data is shown below. Use the graph to answer these questions:

- Are the centers of the two groups similar or different? Explain.
- Are the spreads of the two groups similar or different? Explain.

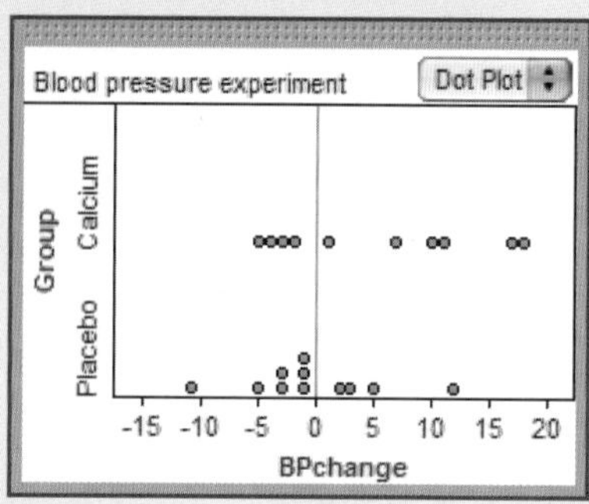

(b) Do the data provide sufficient evidence to conclude that a calcium supplement reduces blood pressure more than a placebo? Carry out an appropriate test to support your answer.

(c) Technology gives a *P*-value of 0.0644 for the test in part (b). Interpret this value in context.

SOLUTION:

(a) The centers are different. The calcium group is centered at a blood pressure decrease value greater than 0, while the placebo group is centered at a value less than 0. The spreads are similar. Both groups have a range of decreases in blood pressure of about 25 millimeters of mercury.

(b) STATE: We want to perform a test of

$$H_0: \mu_1 - \mu_2 = 0 \qquad H_a: \mu_1 - \mu_2 > 0 \qquad \text{or, equivalently,} \qquad H_0: \mu_1 = \mu_2 \qquad H_a: \mu_1 > \mu_2$$

where μ_1 is the true mean decrease in systolic blood pressure for healthy black men like the ones in this study who take a calcium supplement, and μ_2 is the true mean decrease in systolic blood pressure for healthy black men like the ones in this study who take a placebo. No significance level was specified, so we'll use $\alpha = 0.05$.

AP EXAM TIP When checking the Normal condition on an AP exam question involving inference about means, be sure to include a graph. Don't expect to receive credit for describing a graph that you made on your calculator but didn't put on paper.

PLAN: If conditions are met, we will carry out a two-sample *t* test for $\mu_1 - \mu_2$.

- *Random* The 21 subjects were randomly assigned to the two treatments.
- *Normal* With such small sample sizes, we need to examine the data to see if it's reasonable to believe that the actual distributions of differences in blood pressure when taking calcium or placebo are Normal. Figure 10.11 shows hand sketches of calculator boxplots and Normal probability plots for these data. The boxplots show no clear evidence of skewness and no outliers. The Normal probability plot of the placebo group's responses looks very linear, while the Normal probability plot of the calcium group's responses shows some slight curvature. With no outliers or clear skewness, the *t* procedures should be pretty accurate.
- *Independent* Due to the random assignment, these two groups of men can be viewed as independent. Individual observations in each group should also be independent: knowing one subject's change in blood pressure gives no information about another subject's response.

DO: From the data, we calculated summary statistics:

Group	Treatment	n	$\bar{x}$	s_x
1	Calcium	10	5.000	8.743
2	Placebo	11	−0.273	5.901

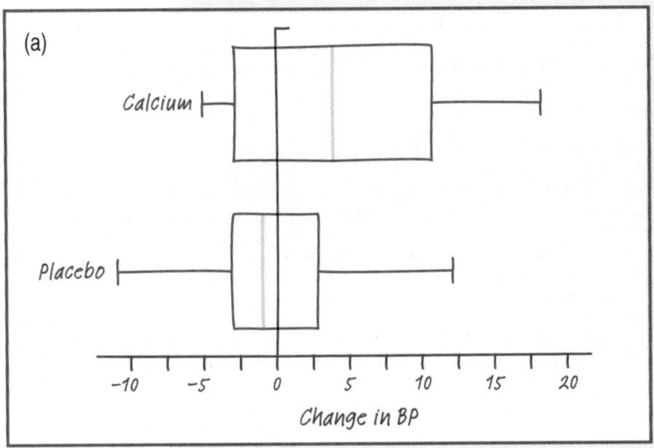

FIGURE 10.11 Sketches of (a) boxplots (b) and (c) Normal probability plots of the changes in blood pressure for the two groups of subjects in the calcium and blood pressure experiment.

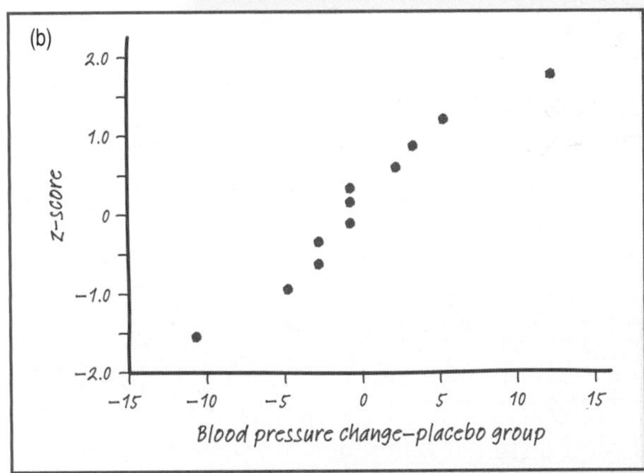

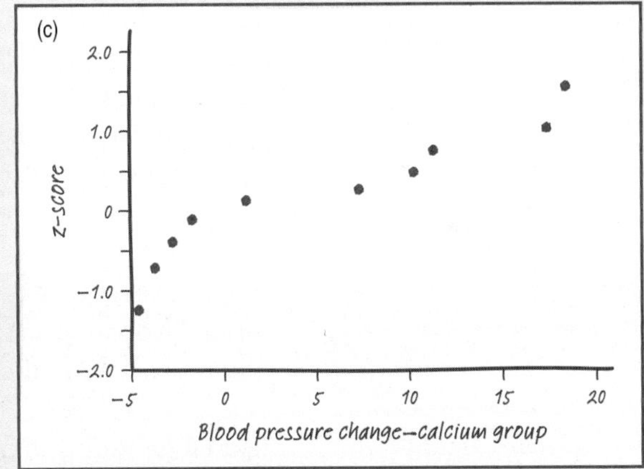

- *Test statistic*

$$t = \frac{(\bar{x}_1 - \bar{x}_2) - (\mu_1 - \mu_2)}{\sqrt{\frac{s_1^2}{n_1} + \frac{s_2^2}{n_2}}} = \frac{[5.000 - (-0.273)] - 0}{\sqrt{\frac{8.743^2}{10} + \frac{5.901^2}{11}}} = \frac{5.273}{3.2878} = 1.604$$

- *P-value* By the conservative method, the smaller of $n_1 - 1$ and $n_2 - 1$ gives df $= 9$. Because H_a counts only positive values of t as evidence against H_0, the P-value is the area to the right of $t = 1.604$ under the t distribution curve with df $= 9$. Figure 10.12 illustrates this P-value. Table B shows that the P-value lies between 0.05 and 0.10.

	Upper-tail probability *p*		
df	**.10**	**.05**	**.025**
8	1.397	1.860	2.306
9	1.383	1.833	2.262
10	1.372	1.812	2.228

FIGURE 10.12 The *P*-value for the one-sided test using the conservative method, which leads to the *t* distribution with 9 degrees of freedom.

CONCLUDE: Because the P-value is greater than $\alpha = 0.05$, we fail to reject H_0. The experiment provides some evidence that calcium reduces blood pressure, but the evidence is not convincing enough to conclude that calcium reduces blood pressure more than a placebo.

(c) Assuming H_0: $\mu_1 - \mu_2 = 0$ is true, the probability of getting a difference in mean blood pressure reduction for the two groups (calcium − placebo) of 5.273 or greater just by the chance involved in the random assignment is 0.0644.

AP EXAM TIP When a significance test leads to a fail to reject H_0 decision, as in the previous example, be sure to interpret the results as "We don't have enough evidence to conclude H_a." Saying anything that sounds like you believe H_0 is (or might be) true will lead to a loss of credit.

We can estimate the difference in the true mean decrease in blood pressure for the calcium and placebo treatments using a two-sample *t* interval for $\mu_1 - \mu_2$. To get results that are consistent with the one-tailed test at $\alpha = 0.05$ from the example, we'll use a 90% confidence level. The conditions for constructing a confidence interval are the same as the ones that we checked in the example before performing the two-sample *t* test. With df = 9, the critical value for a 90% confidence interval is $t^* = 1.833$. The interval is

$$(\bar{x}_1 - \bar{x}_2) \pm t^*\sqrt{\frac{s_1^2}{n_1} + \frac{s_2^2}{n_2}} = [5.000 - (-0.273)] \pm 1.833\sqrt{\frac{8.743^2}{10} + \frac{5.901^2}{11}}$$
$$= 5.273 \pm 6.027 = (-0.754, 11.300)$$

We are 90% confident that the interval from −0.754 to 11.300 captures the difference in true mean blood pressure reduction on calcium over a placebo. Because the 90% confidence interval includes 0 as a plausible value for the difference, we cannot reject $H_0: \mu_1 - \mu_2 = 0$ against the two-sided alternative at the $\alpha = 0.10$ significance level or against the one-sided alternative at the $\alpha = 0.05$ significance level.

THINK ABOUT IT

Why didn't researchers find a significant difference in the calcium and blood pressure experiment? The difference in mean systolic blood pressures for the two groups was 5.273 millimeters of mercury. This seems like a pretty large difference. With the small group sizes, however, this difference wasn't large enough to reject $H_0: \mu_1 - \mu_2 = 0$ in favor of the one-sided alternative. We suspect that larger groups might show a similar difference in mean blood pressure reduction, which would indicate that calcium has a significant effect. If so, then the researchers in this experiment made a Type II error—failing to reject a false H_0. (In fact, later analysis of data from an experiment with more subjects resulted in a *P*-value of 0.008.) *Sample size strongly affects the power of a test.* It is easier to detect an actual difference in the effectiveness of two treatments if both are applied to large numbers of subjects.

The formulas for the two-sample *t* test and interval for $\mu_1 - \mu_2$ often lead to calculation errors by students. As a result, we recommend using the calculator's `2-SampTInt` and `2-SampTTest` features to do the calculations.

TECHNOLOGY CORNER Two-sample *t* tests with computer software and calculators

Technology gives smaller *P*-values for two-sample *t* tests than the conservative method. That's because calculators and software use the more complicated formula on page 637 to obtain a larger number of degrees of freedom. Figure 10.13 gives computer output from Fathom and Minitab for the two-sample *t* test from the calcium experiment. The *P*-values differ slightly because Fathom uses 15.59 degrees of freedom while Minitab truncates to df = 15.

Let's look at what the calculator does.

- Enter the Group 1 (calcium) data in L1/list1 and the Group 2 (placebo) data in L2/list2.
- To perform the significance test, go to STAT/TESTS (Tests menu in the Statistics/List Editor on the TI-89) and choose `4:2-SampTTest`.

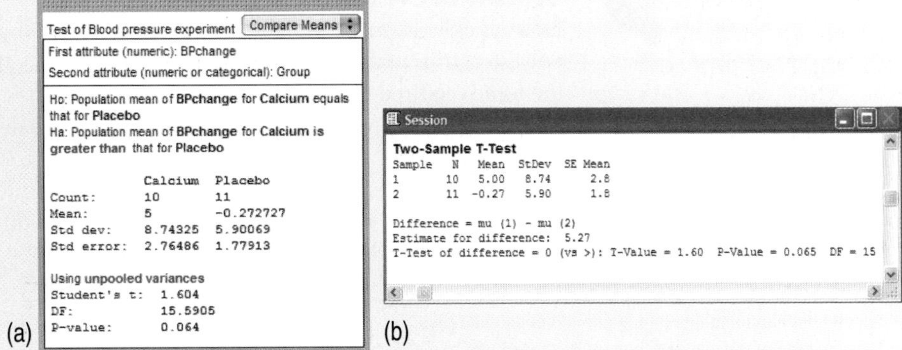

FIGURE 10.13 Computer output from (a) Fathom and (b) Minitab for a two-sample *t* test.

- In the 2-SampTTest screen, specify "Data" and adjust your other settings as shown.

TI-83/84 | TI-89

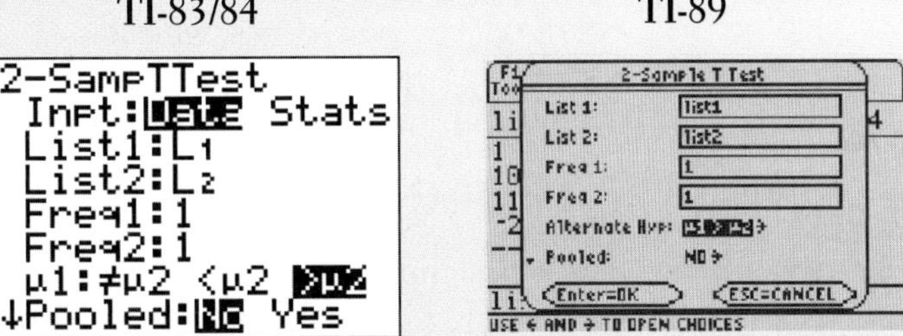

- Highlight "Calculate" and press ENTER. (The Pooled option will be discussed shortly.)

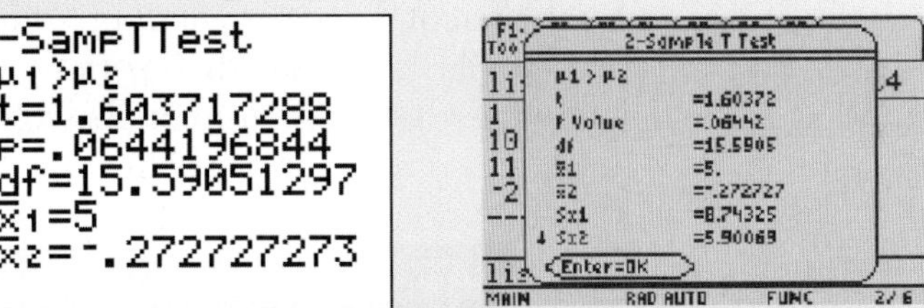

The results tell us that the two-sample *t* test statistic is $t = 1.604$, and the *P*-value is 0.0644. There is enough evidence against H_0 to reject it at the 10% significance level, but not at the 5% or 1% significance levels.

If you select "Draw" instead of "Calculate," the appropriate *t* distribution will be displayed, showing the test statistic and the shaded area corresponding to the *P*-value.

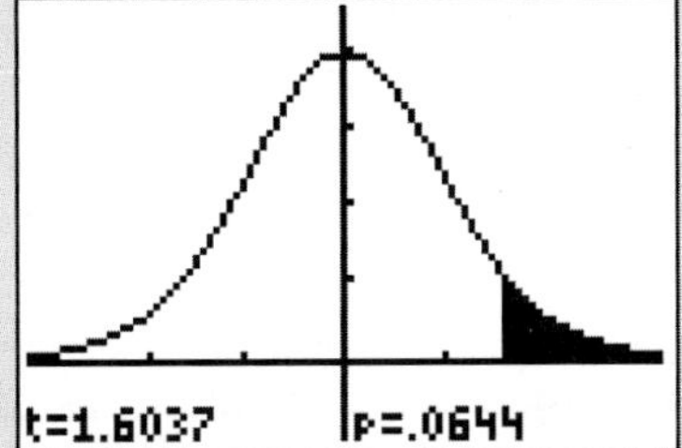

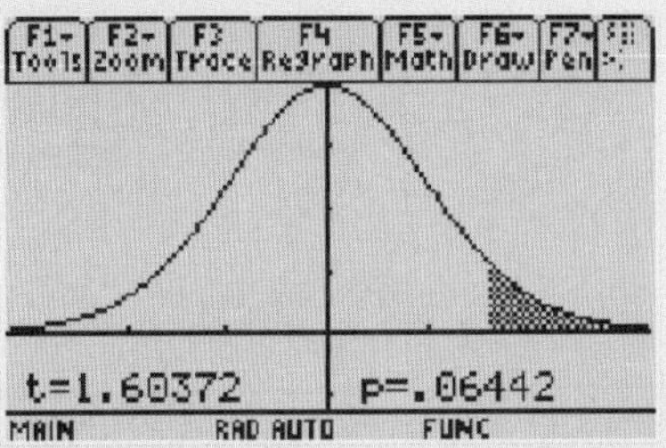

Note: The calculator's 90% confidence interval for the true difference $\mu_1 - \mu_2$ is $(-0.4767, 11.022)$. This is quite a bit narrower than the $(-0.754, 11.300)$ we obtained earlier.

TI-Nspire instructions in Appendix B

CHECK YOUR UNDERSTANDING

How quickly do synthetic fabrics such as polyester decay in landfills? A researcher buried polyester strips in the soil for different lengths of time, then dug up the strips and measured the force required to break them. Breaking strength is easy to measure and is a good indicator of decay. Lower strength means the fabric has decayed.

For one part of the study, the researcher buried 10 strips of polyester fabric in well-drained soil in the summer. The strips were randomly assigned to two groups: 5 of them were buried for 2 weeks and the other 5 were buried for 16 weeks. Here are the breaking strengths in pounds:[26]

Group 1 (2 weeks):	118	126	126	120	129
Group 2 (16 weeks):	124	98	110	140	110

Do the data give good evidence that polyester decays more in 16 weeks than in 2 weeks? Carry out an appropriate test to help answer this question.

Using Two-Sample *t* Procedures Wisely

The two-sample t procedures are more robust against non-Normality than the one-sample t methods. When the sizes of the two samples are equal and the two populations being compared have distributions with similar shapes, probability values from the t table are quite accurate for a broad range of distributions when the sample sizes are as small as $n_1 = n_2 = 5$.[27] If the two population distributions have different shapes, larger samples are needed. Here are practical guidelines for the Normal condition when performing inference about $\mu_1 - \mu_2$. These should look somewhat familiar from Chapter 8.

Using the Two-Sample *t* Procedures: The Normal Condition

- *Sample size less than 15:* Use two-sample t procedures if the data in both samples/groups appear close to Normal (roughly symmetric, single peak, no outliers). If the data are clearly skewed or if outliers are present, do not use t.
- *Sample size at least 15:* Two-sample t procedures can be used except in the presence of outliers or strong skewness.
- *Large samples:* The two-sample t procedures can be used even for clearly skewed distributions when both samples/groups are large, roughly $n \geq 30$.

In planning a two-sample study, choose equal sample sizes if you can. The two-sample t procedures are most robust against non-Normality in this case, and the conservative P-values (using Option 2) are most accurate.

The Pooled Two-Sample *t* Procedures (Don't Use Them!) Most software offers a choice of two-sample t statistics. One is often labeled "unequal" variances; the other, "equal" variances. The "unequal" variance procedure uses our two-sample t statistic. *This test is valid whether or not the population variances* *are equal*. The other choice is a special version of the two-sample t statistic that assumes that the two populations have the same variance. This procedure combines (the statistical term is *pools*) the two sample variances to estimate the common population variance. The resulting statistic is called the **pooled two-sample t statistic.**

Pooled two-sample t statistic

The pooled t statistic has exactly the t distribution with $n_1 + n_2 - 2$ degrees of freedom *if* the two population variances really are equal and the population distributions are exactly Normal. This method offers more degrees of freedom than Option 1 (technology), which leads to narrower confidence intervals and smaller P-values. The pooled t was in common use before software made it easy to use Option 1 for our two-sample t statistic. Of course, in the real world, distributions are not exactly Normal and population variances are not exactly equal. In practice, the Option 1 two-sample t procedures are almost always more accurate than the pooled procedures. Our advice: *Never use the pooled t procedures if you have software that will implement Option 1.*

CAUTION

Remember, we always use the pooled sample proportion $\hat{p}_C$ when performing a significance test for comparing two proportions. But we don't recommend pooling when comparing two means.

THINK ABOUT IT

How does pooling work? Suppose we have two Normally distributed populations with the same variance σ^2. If we take independent random samples from the two populations, both sample variances s_1^2 and s_2^2 estimate σ^2. The best way to combine (pool) these two estimates is to average them with weights equal to their degrees of freedom. This gives more weight to the information from the larger sample, which is reasonable. The resulting estimator of σ^2 is

$$s_p^2 = \frac{(n_1 - 1)s_1^2 + (n_2 - 1)s_2^2}{n_1 + n_2 - 2}$$

Pooled estimator of σ^2

This is called the **pooled estimator of σ^2** because it combines the information in both samples.

To perform inference about $\mu_1 - \mu_2$ in this setting, we start by simplifying our formula for the standard deviation of the statistic:

$$\sigma_{\bar{x}_1 - \bar{x}_2} = \sqrt{\frac{\sigma_1^2}{n_1} + \frac{\sigma_2^2}{n_2}} = \sqrt{\frac{\sigma^2}{n_1} + \frac{\sigma^2}{n_2}} = \sqrt{\sigma^2\left(\frac{1}{n_1} + \frac{1}{n_2}\right)} = \sigma\sqrt{\frac{1}{n_1} + \frac{1}{n_2}}$$

We then substitute the pooled estimate s_p for σ. The resulting pooled two-sample t statistic is

$$t = \frac{(\bar{x}_1 - \bar{x}_2) - (\mu_1 - \mu_2)}{s_p\sqrt{\frac{1}{n_1} + \frac{1}{n_2}}}$$

This statistic has a t distribution with $(n_1 - 1) + (n_2 - 1) = n_1 + n_2 - 2$ degrees of freedom.

Inference for Experiments Confidence intervals and tests for $\mu_1 - \mu_2$ are based on the sampling distribution of $\bar{x}_1 - \bar{x}_2$. But in experiments, we aren't sampling at random from any larger populations. We can think about what would happen if the random assignment were repeated many times under the assumption that $H_0: \mu_1 - \mu_2 = 0$ is true. That is, we assume that the specific treatment received doesn't affect an individual subject's response.

Let's see what would happen just by chance if we randomly reassign the 21 subjects in the calcium and blood pressure experiment to the two groups many times, assuming the drug received *doesn't affect* each individual's change in systolic blood pressure. We used Fathom software to redo the random assignment 1000 times. The approximate *randomization distribution* of $\bar{x}_1 - \bar{x}_2$ is shown in Figure 10.14. It has an approximately Normal shape with mean 0 (no difference) and standard deviation 3.42.

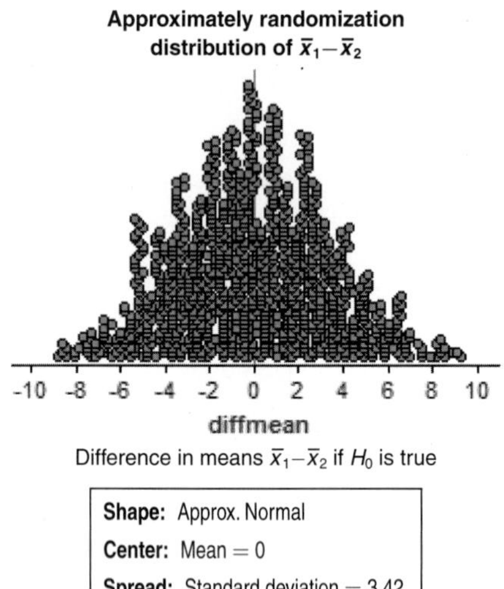

FIGURE 10.14 Fathom simulation showing the approximate randomization distribution of $\bar{x}_1 - \bar{x}_2$ from 1000 random reassignments of subjects to treatment groups in the calcium and blood pressure experiment.

In the actual experiment, the difference in the mean change in blood pressure in the calcium and placebo groups was $5.000 - (-0.273) = 5.273$. How likely is it that a difference this large or larger would happen just by chance when H_0 is true? Figure 10.14 provides a rough answer: 61 of the 1000 random reassignments yielded a difference in means greater than or equal to 5.273. That is, our estimate of the *P*-value is 0.061. This is quite close to the 0.0644 *P*-value that we obtained in the Technology Corner.

If Figure 10.14 displayed the results of *all* possible random reassignments of subjects to treatment groups, it would be the actual randomization distribution of $\bar{x}_1 - \bar{x}_2$. The *P*-value obtained from this distribution would be *exactly correct*. Using the two-sample *t* test to calculate the *P*-value gives only approximately correct results.

Figure 10.15 shows the value of the two-sample *t* test statistic for each of the 1000 re-randomizations, calculated using our familiar formula

$$t = \frac{(\bar{x}_1 - \bar{x}_2) - (\mu_1 - \mu_2)}{\sqrt{\frac{s_1^2}{n_1} + \frac{s_2^2}{n_2}}}$$

The density curve for the *t* distribution with df = 15.59 is shown in blue. We can see that the test statistic follows the *t* distribution quite closely in this case.

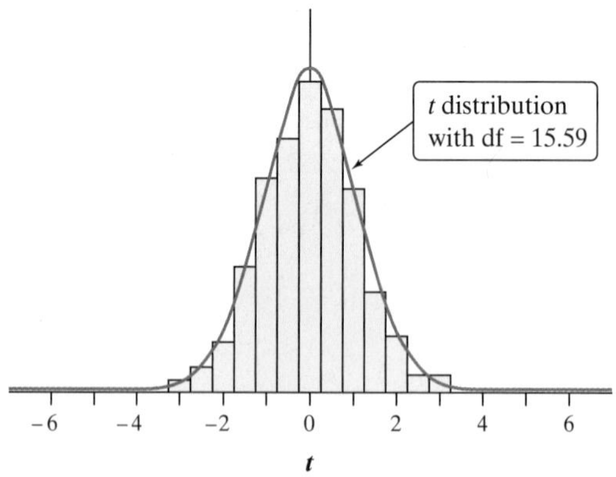

FIGURE 10.15 The distribution of the two-sample *t* test statistic for the 1000 random reassignments in Figure 10.14.

Whenever the Random, Normal, and Independent conditions are met, the randomization distribution of $\bar{x}_1 - \bar{x}_2$ looks much like its sampling distribution. We are therefore safe using two-sample *t* procedures for comparing two means in a randomized experiment.

DATA EXPLORATION *Can magnets help reduce pain?*

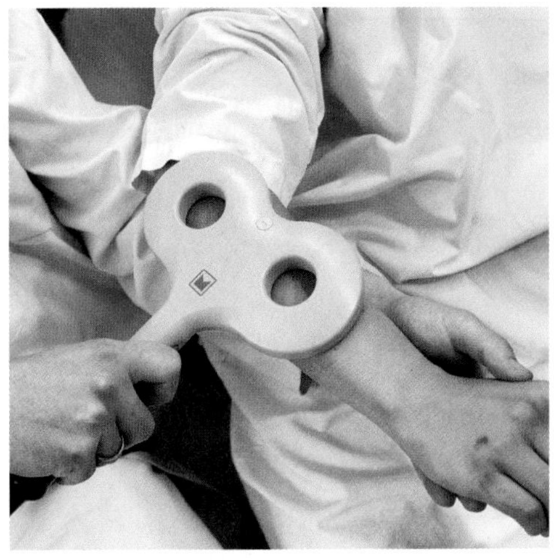

Some doctors have begun to use medical magnets to treat patients with chronic pain. Scientists wondered whether this type of therapy really worked. So they designed an experiment to find out. Fifty patients with chronic pain were recruited for the study. A doctor identified a painful site on each patient and asked him or her to rate the pain on a scale from 0 (mild pain) to 10 (severe pain). Then, the doctor selected a sealed envelope containing a magnet from a box that contained both active and inactive magnets. The chosen magnet was applied to the site of the pain for 45 minutes. After treatment, each patient was again asked to rate the level of pain from 0 to 10.

In all, 29 patients were given active magnets and 21 patients received inactive magnets. All but 1 of the patients rated their initial pain as an 8, 9, or 10. So scientists decided to focus on patients' final pain ratings. Here they are, grouped by the type of magnet used:[28]

Active:	0, 4, 7, 0, 4, 2, 5, 5, 3, 2, 2, 2, 3, 4, 6, 4, 3, 0, 2, 0, 4, 4, 5, 9, 10, 10, 10, 10, 7
Inactive:	4, 7, 5 8, 8, 6, 8, 10, 10, 6, 10, 8, 10, 10, 10, 10, 9, 9, 10, 10, 9

Did the active magnets help reduce pain? Give appropriate statistical evidence to support your answer.

Don't Use Two-Sample t Procedures on Paired Data In Chapter 9, we used paired t procedures to compare the mean change in depression scores for a group of caffeine-dependent individuals when taking caffeine and a placebo. The inference involved paired data because the same 11 subjects received both treatments. In this chapter, we used two-sample t procedures to compare the mean change in blood pressure for a group of healthy black men when taking calcium and a placebo. This time, the inference involved two independent groups of subjects. *The proper method of analysis depends on the design of the study.*

CAUTION

EXAMPLE *Comparing Tires and Comparing Workers*

Independent samples versus paired data

PROBLEM: **In each of the following settings, decide whether you should use paired t procedures or two-sample t procedures to perform inference.[29] Explain your choice.**

(a) To test the wear characteristics of two tire brands, A and B, one Brand A tire is mounted on one side of each car in the rear, while a Brand B tire is mounted on the other side. Which side gets which brand is determined by flipping a coin. The same procedure is used on the front.

(b) Can listening to music while working increase productivity? Twenty factory workers agree to take part in a study to investigate this question. Researchers randomly assign 10 workers to do a repetitive task while listening to music and the other 10 workers to do the task in silence.

SOLUTION:

(a) Paired *t* procedures. This is a matched pairs experiment, with the two treatments (Brand A and Brand B) being randomly assigned to the front pair and rear pair of wheels on each car.

(b) Two-sample *t* procedures. The data are being produced using two independent groups of workers in a randomized experiment.

Inference without Randomness Statistical tests amount to asking, "Is it plausible that the observed result happened just by chance when H_0 is true?" If data are produced by random sampling or random assignment, we can model the chance behavior with a sampling distribution or randomization distribution. But many interesting statistical questions involve data that were not produced by some chance process. In such cases, it still makes sense to ask, "How likely is it that such a result would occur by chance alone?" Here are some examples.

- Do college students who have volunteered for community service work differ from those who haven't? A study obtained data from 57 students who had done service and 17 who had not. One of the response variables was a measure of attachment to friends. An investigator compares the proportions who have scores above 100 on the Inventory of Parent and Peer Attachment in the two groups. If the two proportions are significantly different, what can the investigator conclude?
- Is there an association between class of travel and survival status on the *Titanic*? In this case, we have data on the entire population. These data suggest that the relationship between the two variables was too strong to be explained by chance. But what kind of inference can we perform?
- Are larger birds more likely to survive in bad weather? Herman Bumpus examined 24 adult male sparrows that lived and 35 adult male sparrows that died in a severe winter storm. He compared the means of various physical characteristics for the two groups. Suppose Bumpus finds that the birds that survived weighed significantly more on average than those that died. What conclusion can he draw?

In all three cases, we can say that the observed result is too unlikely to have happened by chance. However, because the data were not produced by random sampling, we can't generalize the results to any larger populations of interest. The lack of random assignment prevents us from establishing causation. *Due to the limited scope of inference, we will avoid such settings.*

case closed

Fast-Food Frenzy!

In the chapter-opening Case Study (page 601) about drive-thru service at fast-food restaurants, we presented some results from two *QSR* magazine studies of service time and order accuracy at randomly selected restaurants in the 50 states. Here is a brief recap.

1. In 2008, 522 of the 558 Burger King visits resulted in accurate orders. In 2009, 477 of the 529 Burger King visits resulted in accurate orders. Was there a significant change in accuracy at Burger King restaurants between 2008 and 2009?

STATE: We want to test $H_0: p_1 - p_2 = 0$ versus $H_a: p_1 - p_2 \neq 0$, where p_1 is the proportion of all Burger King orders in 2008 that were filled accurately, and p_2 is the proportion of all orders in 2009 that were filled accurately. Because no significance level was given, we'll use $\alpha = 0.05$.

PLAN: If the conditions are met, we should perform a two-sample z test for $p_1 - p_2$.

- *Random* The data came from visits to a random sample of Burger King restaurants in each of the two years.
- *Normal* The counts of successes and failures in the samples—522, 36, 477, and 52—are all at least 10.
- *Independent* We have independent random samples from 2008 and 2009. Because sampling without replacement was used, there have to be at least 10(558) = 5580 visits to Burger King drive-thrus in 2008 and at least 10(529) = 5290 visits in 2009. This is safe to assume.

DO: The calculator's `2-PropZTest` yields $z = 2.04$ and a P-value of 0.0413.

CONCLUDE: Because the P-value, 0.0413, is less than 0.05, we can reject H_0 and conclude that there is convincing evidence of a difference in order accuracy at Burger King restaurants in 2008 and 2009.

2. In 2009, McDonald's average service time for 596 drive-thru visits was 174.22 seconds with a standard deviation of 16.49 seconds. In 2008, McDonald's service time for 750 drive-thru visits had a mean of 158.77 seconds and a standard deviation of 17.06 seconds. How much did average service time at McDonald's restaurants change in these two years?

STATE: Our parameters of interest are μ_1 = the true mean service time at McDonald's drive-thrus in 2009 and μ_2 = the true mean service time at McDonald's drive-thrus in 2008. We want to estimate the difference $\mu_1 - \mu_2$ at a 95% confidence level.

PLAN: If conditions are met, we'll construct a two-sample t interval for $\mu_1 - \mu_2$.

- *Random* The data came from visits to random samples of McDonald's restaurants in 2008 and 2009.
- *Normal* We don't know whether the population distributions of drive-thru service times in these two years are Normally distributed. However, since both sample sizes (596 and 750) are much larger than 30, we are safe using t procedures.

• *Independent* *QSR* magazine took independent samples of restaurants in the two years. Because sampling without replacement was used, there have to be at least 10(750) = 7500 drive-thru visits in 2008 and 10(596) = 5960 drive-thru visits in 2009. This is pretty safe to assume.

DO: The calculator's `2SampTInt` gives a 95% confidence interval of (13.647, 17.253) using df = 1293.97.

CONCLUDE: We are 95% confident that the interval from 13.647 to 17.253 seconds captures the difference in the actual mean drive-thru service time in 2008 and the actual mean drive-thru service time in 2009. This interval suggests that the mean service time at all McDonald's drive-thrus was between 13.647 and 17.253 seconds longer in 2009 than in 2008.

SECTION 10.2 Summary

- Choose independent SRSs of size n_1 from Population 1 and size n_2 from Population 2. The sampling distribution of $\bar{x}_1 - \bar{x}_2$ has the following properties:
  - **Shape** Normal if both population distributions are Normal; approximately Normal otherwise if both samples are large enough ($n \geq 30$) by the central limit theorem.
  - **Center** Its mean is $\mu_1 - \mu_2$.
  - **Spread** As long as each sample is no more than 10% of its population (10% condition), its standard deviation is $\sqrt{\frac{\sigma_1^2}{n_1} + \frac{\sigma_2^2}{n_2}}$.
- Confidence intervals and tests for the difference between the means of two populations or the mean responses to two treatments μ_1 and μ_2 are based on the difference $\bar{x}_1 - \bar{x}_2$ between the sample means.
- If we somehow know the population standard deviations σ_1 and σ_2, we can use a z statistic and the standard Normal distribution to perform calculations.
- Since we almost never know the population standard deviations in practice, we use the **two-sample t statistic**

$$t = \frac{(\bar{x}_1 - \bar{x}_2) - (\mu_1 - \mu_2)}{\sqrt{\frac{s_1^2}{n_1} + \frac{s_2^2}{n_2}}}$$

  This statistic does not have exactly a t distribution. There are two options for using a t distribution to approximate the distribution of the two-sample t statistic:
  - **Option 1 (Technology)** Use the t distribution with degrees of freedom calculated from the data by a somewhat messy formula. The degrees of freedom may not be a whole number.

    - **Option 2 (Conservative)** Use the t distribution with degrees of freedom equal to the *smaller* of $n_1 - 1$ and $n_2 - 1$. This method gives wider confidence intervals and larger P-values than Option 1.

- When the Random, Normal, and Independent conditions are met, we can use two-sample t procedures to estimate and test claims about $\mu_1 - \mu_2$.
    - **Random** The data are produced by a random sample of size n_1 from Population 1 and a random sample of size n_2 from Population 2 or by two groups of size n_1 and n_2 in a randomized experiment.
    - **Normal** Both population distributions (or the true distributions of responses to the two treatments) are Normal OR both sample/group sizes are large ($n_1 \geq 30$ and $n_2 \geq 30$).
    - **Independent** Both the samples/groups themselves and the individual observations in each sample/group are independent. When sampling without replacement, check that the two populations are at least 10 times as large as the corresponding samples (the 10% condition).
- An approximate level C confidence interval for $\mu_1 - \mu_2$ is

$$(\bar{x}_1 - \bar{x}_2) \pm t^* \sqrt{\frac{s_1^2}{n_1} + \frac{s_2^2}{n_2}}$$

where t^* is the critical value for confidence level C for the t distribution with degrees of freedom from either Option 1 (technology) or Option 2 (the smaller of $n_1 - 1$ and $n_2 - 1$). This is called a **two-sample *t* interval for $\mu_1 - \mu_2$.**

- To test H_0: $\mu_1 - \mu_2 =$ hypothesized value, use a **two-sample *t* test for $\mu_1 - \mu_2$.** The test statistic is

$$t = \frac{(\bar{x}_1 - \bar{x}_2) - (\mu_1 - \mu_2)}{\sqrt{\frac{s_1^2}{n_1} + \frac{s_2^2}{n_2}}}$$

P-values are calculated using the t distribution with degrees of freedom from either Option 1 (technology) or Option 2 (the smaller of $n_1 - 1$ and $n_2 - 1$).

- The two-sample t procedures are quite robust against departures from Normality, especially when both sample/group sizes are large.
- Inference about the difference $\mu_1 - \mu_2$ in the effectiveness of two treatments in a completely randomized experiment is based on the **randomization distribution** of $\bar{x}_1 - \bar{x}_2$. When the Random, Normal, and Independent conditions are met, our usual inference procedures based on the sampling distribution of $\bar{x}_1 - \bar{x}_2$ will be approximately correct.
- Don't use two-sample t procedures to compare means for paired data.

10.2 TECHNOLOGY CORNERS

TI-Nspire instructions in Appendix B

SECTION 10.2 Exercises

In exercises that call for two-sample t procedures, you may use either of the two options for the degrees of freedom that we have discussed, unless you are told otherwise. Be sure to state what df you are using.

35. **Cholesterol** (6.2) The level of cholesterol in the blood for all men aged 20 to 34 follows a Normal distribution with mean 188 milligrams per deciliter (mg/dl) and standard deviation 41 mg/dl. For 14-year-old boys, blood cholesterol levels follow a Normal distribution with mean 170 mg/dl and standard deviation 30 mg/dl.

(a) Let M = the cholesterol level of a randomly selected 20- to 34-year-old man and B = the cholesterol level of a randomly selected 14-year-old boy. Describe the shape, center, and spread of the distribution of $M - B$.

(b) Find the probability that a randomly selected 14-year-old boy has higher cholesterol than a randomly selected man aged 20 to 34. Show your work.

36. **How tall?** (6.2) The heights of young men follow a Normal distribution with mean 69.3 inches and standard deviation 2.8 inches. The heights of young women follow a Normal distribution with mean 64.5 inches and standard deviation 2.5 inches.

(a) Let M = the height of a randomly selected young man and W = the height of a randomly selected young woman. Describe the shape, center, and spread of the distribution of $M - W$.

(b) Find the probability that a randomly selected young man is at least 2 inches taller than a randomly selected young woman. Show your work.

pg 631

37. **Cholesterol** Refer to Exercise 35. Suppose we select independent SRSs of 25 men aged 20 to 34 and 36 boys aged 14 and calculate the sample mean heights $\bar{x}_M$ and $\bar{x}_B$.

(a) Describe the shape, center, and spread of the sampling distribution of $\bar{x}_M - \bar{x}_B$.

(b) Find the probability of getting a difference in sample means $\bar{x}_M - \bar{x}_B$ that's less than 0 mg/dl. Show your work.

(c) Should we be surprised if the sample mean cholesterol level for the 14-year-old boys exceeds the sample mean cholesterol level for the men? Explain.

38. **How tall?** Refer to Exercise 36. Suppose we select independent SRSs of 16 young men and 9 young women and calculate the sample mean heights $\bar{x}_M$ and $\bar{x}_W$.

(a) Describe the shape, center, and spread of the sampling distribution of $\bar{x}_M - \bar{x}_W$.

(b) Find the probability of getting a difference in sample means $\bar{x}_M - \bar{x}_W$ that's greater than or equal to 2 inches. Show your work.

(c) Should we be surprised if the sample mean height for the young women is more than 2 inches less than the sample mean height for the young men? Explain.

In Exercises 39 to 42, determine whether or not the conditions for using two-sample t procedures are met.

39. **Shoes** How many pairs of shoes do teenagers have? To find out, a group of AP Statistics students conducted a survey. They selected a random sample of 20 female students and a separate random sample of 20 male students from their school. Then they recorded the number of pairs of shoes that each respondent reported having. The back-to-back stemplot below displays the data.

Females		Males
	0	4
	0	555677778
333	1	0000124
95	1	
4332	2	2
66	2	
410	3	
8	3	58
	4	
9	4	
100	5	
7	5	

Key: 2|2 represents a male student with 22 pairs of shoes.

40. **Household size** How do the numbers of people living in households in the United Kingdom (U.K.) and South Africa compare? To help answer this question, we used CensusAtSchool's random data selector to choose independent samples of 50 students from each country. Here is a Fathom dotplot of the household sizes reported by the students in the survey.

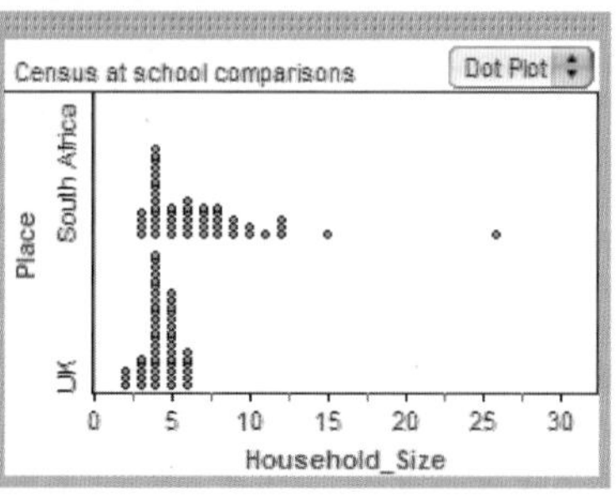

41. **Literacy rates** Do males have higher average literacy rates than females in Islamic countries? The table below shows the percent of men and women at least 15 years old who were literate in 2008 in the major Islamic nations. (We omitted countries with populations of less than 3 million.) Data for a few nations, such as Afghanistan and Iraq, were not available.[30]

Country	Female percent	Male percent
Algeria	66	94
Bangladesh	48	71
Egypt	58	88
Iran	77	97
Jordan	87	99
Kazakhstan	100	100
Lebanon	86	98
Libya	78	100
Malaysia	90	98
Morocco	43	84
Saudi Arabia	79	98
Syria	77	95
Tajikistan	100	100
Tunisia	69	97
Turkey	81	99
Uzbekistan	96	99
Yemen	41	93

42. **Long words** Mary was interested in comparing the mean word length in articles from a medical journal and an airline's in-flight magazine. She counted the number of letters in the first 200 words of an article in the medical journal and in the first 100 words of an article in the airline magazine. Mary then used Minitab statistical software to produce the histograms shown.

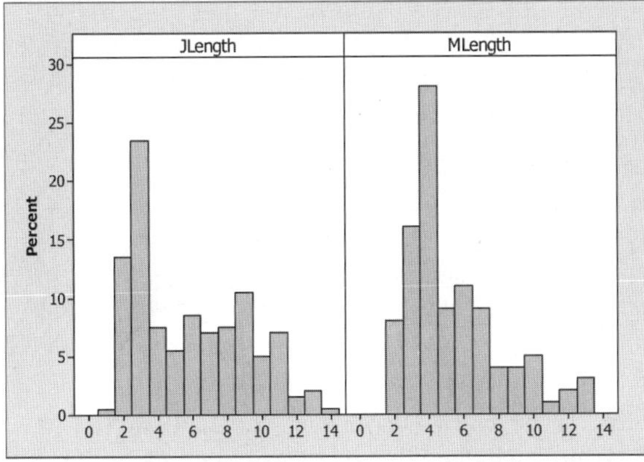

43. **Is red wine better than white wine?** Observational studies suggest that moderate use of alcohol by adults reduces heart attacks and that red wine may have special benefits. One reason may be that red wine contains polyphenols, substances that do good things to cholesterol in the blood and so may reduce the risk of heart attacks. In an experiment, healthy men were assigned at random to drink half a bottle of either red or white wine each day for two weeks. The level of polyphenols in their blood was measured before and after the two-week period. Here are the percent changes in level for the subjects in both groups:[31]

pg 635

Red wine:	3.5	8.1	7.4	4.0	0.7	4.9	8.4	7.0	5.5
White wine:	3.1	0.5	−3.8	4.1	−0.6	2.7	1.9	−5.9	0.1

(a) A Fathom dotplot of the data is shown below. Use the graph to answer these questions:

- Are the centers of the two groups similar or different? Explain.
- Are the spreads of the two groups similar or different? Explain.

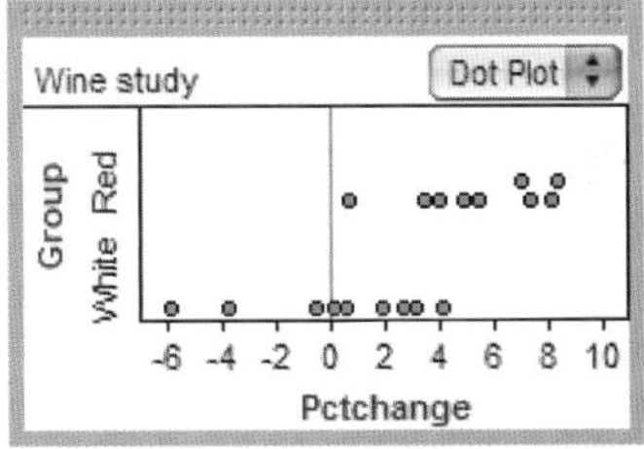

(b) Construct and interpret a 90% confidence interval for the difference in mean percent change in polyphenol levels for the red wine and white wine treatments.

(c) Does the interval in part (b) suggest that red wine is more effective than white wine? Explain.

44. **Tropical flowers** Different varieties of the tropical flower *Heliconia* are fertilized by different species of hummingbirds.

Researchers believe that over time, the lengths of the flowers and the forms of the hummingbirds' beaks have evolved to match each other. Here are data on the lengths in millimeters for random samples of two color varieties of the same species of flower on the island of Dominica:[32]

H. caribaea red							
41.90	42.01	41.93	43.09	41.17	41.69	39.78	40.57
39.63	42.18	40.66	37.87	39.16	37.40	38.20	38.07
38.10	37.97	38.79	38.23	38.87	37.78	38.01	

H. caribaea yellow							
36.78	37.02	36.52	36.11	36.03	35.45	38.13	37.1
35.17	36.82	36.66	35.68	36.03	34.57	34.63	

(a) A Fathom dotplot of the data is shown below. Use the graph to answer these questions:

- Are the centers of the two groups similar or different? Explain.
- Are the spreads of the two groups similar or different? Explain.

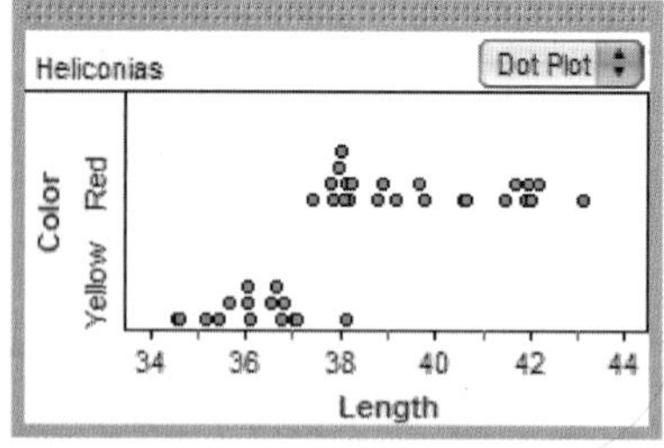

(b) Construct and interpret a 95% confidence interval for the difference in the mean lengths of these two varieties of flowers.

(c) Does the interval support the researchers' belief that the two flower varieties have different average lengths? Explain.

45. **Paying for college** College financial aid offices expect students to use summer earnings to help pay for college. But how large are these earnings? One large university studied this question by asking a random sample of 1296 students who had summer jobs how much they earned. The financial aid office separated the responses into two groups based on gender. Here are the data in summary form:[33]

Group	n	$\bar{x}$	s_x
Males	675	\$1884.52	\$1368.37
Females	621	\$1360.39	\$1037.46

(a) How can you tell from the summary statistics that the distribution of earnings in each group is strongly skewed to the right? A graph of the data reveals no outliers. The use of two-sample t procedures is still justified. Why?

(b) Construct and interpret a 90% confidence interval for the difference between the mean summer earnings of male and female students at this university.

(c) Interpret the 90% confidence level in the context of this study.

46. **Happy customers** As the Hispanic population in the United States has grown, businesses have tried to understand what Hispanics like. One study interviewed a random sample of customers leaving a bank. Customers were classified as Hispanic if they preferred to be interviewed in Spanish or as Anglo if they preferred English. Each customer rated the importance of several aspects of bank service on a 10-point scale.[34] Here are summary results for the importance of "reliability" (the accuracy of account records and so on):

Group	n	$\bar{x}$	s_x
Anglo	92	6.37	0.60
Hispanic	86	5.91	0.93

(a) The distribution of reliability ratings in each group is not Normal. A graph of the data reveals no outliers. The use of two-sample t procedures is justified. Why?

(b) Construct and interpret a 95% confidence interval for the difference between the mean ratings of the importance of reliability for Anglo and Hispanic bank customers.

(c) Interpret the 95% confidence level in the context of this study.

47. **Web business** You want to compare the daily sales for two different designs of Web pages for your Internet business. You assign the next 60 days to either Design A or Design B, 30 days to each.

(a) Describe how you would assign the days for Design A and Design B using the partial line of random digits provided below. Then use your plan to select the first three days for using Design A. Show your method clearly on your paper.

24005 52114 26224 39078

(b) Would you use a one-sided or a two-sided significance test for this problem? Explain your choice. Then set up appropriate hypotheses.

(c) If you plan to use Table B to calculate the P-value, what are the degrees of freedom?

(d) The t statistic for comparing the mean sales is 2.06. Using Table B, what P-value would you report? What would you conclude?

Exercises 48 to 50 refer to the following setting. Do birds learn to time their breeding? Blue titmice eat caterpillars. The birds would like lots of caterpillars around when they have young to feed, but they must breed much earlier. Do the birds learn from one year's experience when to time their breeding next year? Researchers randomly assigned 7 pairs of birds to have the natural caterpillar supply supplemented while feeding their young and another 6 pairs to serve as a control group relying on natural food supply. The next year, they measured how many days after the caterpillar peak the birds produced their nestlings.[35]

48. **Did the randomization produce similar groups?** First, compare the two groups of birds in the first year. The only difference should be the chance effect of the random assignment. The study report says: "In the experimental year, the degree of synchronization did not differ between food-supplemented and control females." For this comparison, the report gives $t = -1.05$.

(a) What type of t statistic (one-sample, paired, or two-sample) is this? Justify your answer.

(b) Explain how this value of t leads to the quoted conclusion.

49. **Did the treatment have an effect?** The investigators expected the control group to adjust their breeding date the next year, whereas the well-fed supplemented group had no reason to change. The report continues: "But in the following year, food-supplemented females were more out of synchrony with the caterpillar peak than the controls." Here are the data (days behind caterpillar peak):

Control:	4.6	2.3	7.7	6.0	4.6	−1.2	
Supplemented:	15.5	11.3	5.4	16.5	11.3	11.4	7.7

Carry out an appropriate test and show that it leads to the quoted conclusion.

50. **Year-to-year comparison** Rather than comparing the two groups in each year, we could compare the behavior of each group in the first and second years. The study report says: "Our main prediction was that females receiving additional food in the nestling period should not change laying date the next year, whereas controls, which (in our area) breed too late in their first year, were expected to advance their laying date in the second year."

Comparing days behind the caterpillar peak in Years 1 and 2 gave $t = 0.63$ for the control group and $t = -2.63$ for the supplemented group.

(a) What type of t statistic (one-sample, paired, or two-sample) are these? Justify your answer.

(b) What are the degrees of freedom for each t?

(c) Explain why these t-values do *not* agree with the prediction.

pg 639

51. **Teaching reading** An educator believes that new reading activities in the classroom will help elementary school pupils improve their reading ability. She recruits 44 third-grade students and randomly assigns them into two groups. One group of 21 students does these new activities for an 8-week period. A control group of 23 third-graders follows the same curriculum without the activities. At the end of the 8 weeks, all students are given the Degree of Reading Power (DRP) test, which measures the aspects of reading ability that the treatment is designed to improve. Comparative boxplots and summary statistics for the data from Fathom are shown below.[36]

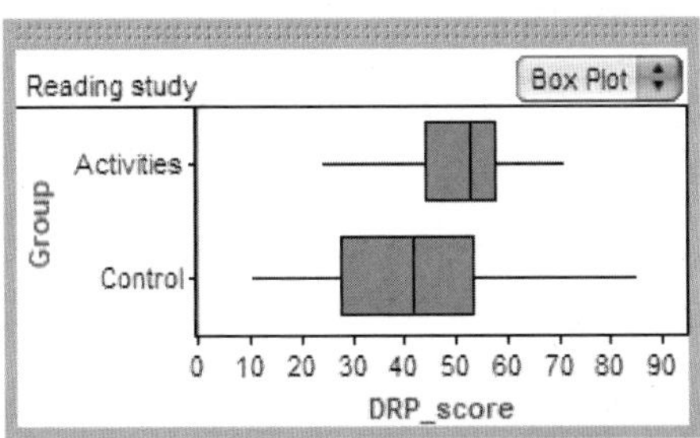

Reading study

	Group	
	Activities	Control
DRP_score	51.4762	41.5217
	21	23
	11.0074	17.1487

S1 = mean ( )
S2 = count ( )
S3 = stdDev ( )

(a) Based on the graph and numerical summaries, write a few sentences comparing the DRP scores for the two groups.

(b) Is the mean DRP score significantly higher for the students who did the reading activities? Carry out an appropriate test to support your answer.

(c) Can we conclude that the new reading activities caused an increase in the mean DRP score? Explain.

(d) Construct and interpret a 95% confidence interval for the difference in mean DRP scores. Explain how this interval provides more information than the significance test in part (b).

52. **Does breast-feeding weaken bones?** Breast-feeding mothers secrete calcium into their milk. Some of the calcium may come from their bones, so mothers may lose bone mineral. Researchers compared a random sample of 47 breast-feeding women with a random sample of 22 women of similar age who were neither pregnant nor lactating. They measured the percent change in the bone mineral content (BMC) of the women's spines over three months. Comparative

boxplots and summary statistics for the data from Fathom are shown below.[37]

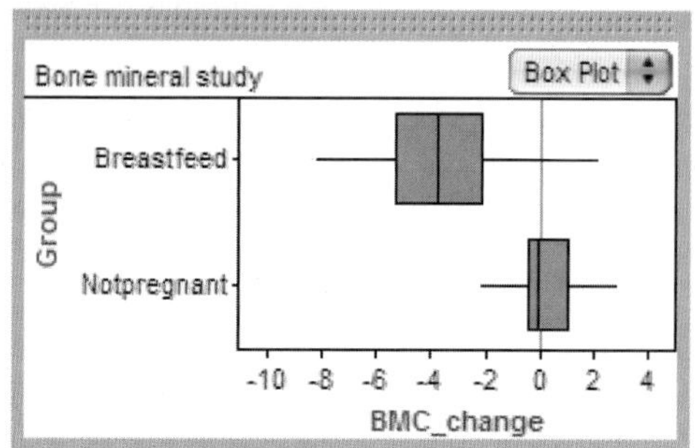

Bone mineral study

	Group	
	Breastfeed	Notpregnant
BMC_change	-3.58723	0.309091
	47	22
	2.50561	1.29832

S1 = mean ( )
S2 = count ( )
S3 = stdDev ( )

(a) Based on the graph and numerical summaries, write a few sentences comparing the percent changes in BMC for the two groups.

(b) Is the mean change in BMC significantly lower for the mothers who are breast-feeding? Carry out an appropriate test to support your answer.

(c) Can we conclude that breast-feeding causes a mother's bones to weaken? Why or why not?

(d) Construct and interpret a 95% confidence interval for the difference in mean bone mineral loss. Explain how this interval provides more information than the significance test in part (b).

53. **Who talks more—men or women?** Researchers equipped random samples of 56 male and 56 female students from a large university with a small device that secretly records sound for a random 30 seconds during each 12.5-minute period over two days. Then they counted the number of words spoken by each subject during each recording period and, from this, estimated how many words per day each subject speaks. The female estimates had a mean of 16,177 words per day with a standard deviation of 7520 words per day. For the male estimates, the mean was 16,569 and the standard deviation was 9108.

(a) Do these data provide convincing evidence of a difference in the average number of words spoken in a day by male and female students at this university? Carry out an appropriate test to support your answer.

(b) Interpret the *P*-value from part (a) in the context of this study.

54. **DDT in rats** Poisoning by the pesticide DDT causes convulsions in humans and other mammals. Researchers seek to understand how the convulsions are caused. In a randomized comparative experiment, they compared 6 white rats poisoned with DDT with a control group of 6 unpoisoned rats. Electrical measurements of nerve activity are the main clue to the nature of DDT poisoning. When a nerve is stimulated, its electrical response shows a sharp spike followed by a much smaller second spike. The researchers measured the height of the second spike as a percent of the first spike when a nerve in the rat's leg was stimulated.[38] For the poisoned rats the results were

12.207 16.869 25.050 22.429 8.456 20.589

The control group data were

11.074 9.686 12.064 9.351 8.182 6.642

Computer output for a two-sample *t* test on these data from SAS software is shown below. (Note that SAS provides two-sided *P*-values.)

```
                TTEST PROCEDURE
Variable: SPIKE
GROUP      N      MEAN          Std Dev      Std Error
DDT        6  17.60000000     6.34014839   2.58835474
CONTROL    6   9.49983333     1.95005932   0.79610839
Variances        T        DF     Prob>|T|
Unequal       2.9912     5.9      0.0247
Equal         2.9912    10.0      0.0135
```

(a) Do these data provide convincing evidence that DDT affects the mean height of the second spike's electrical response? Carry out a significance test to help answer this question.

(b) Interpret the *P*-value from part (a) in the context of this study.

55. **Competitive rowers** What aspects of rowing technique distinguish between novice and skilled competitive rowers? Researchers compared two randomly selected groups of female competitive rowers: a group of skilled rowers and a group of novices. The researchers measured many mechanical aspects of rowing style as the subjects rowed on a Stanford Rowing Ergometer. One important variable is the angular velocity of the knee, which describes the rate at which the knee joint opens as the legs push the body back on the sliding

seat. The data show no outliers or strong skewness. Here is the SAS computer output:[39]

TTEST PROCEDURE

Variable: KNEE

GROUP	N	Mean	Std Dev	Std Error
SKILLED	10	4.182	0.479	0.151
NOVICE	8	3.010	0.959	0.339

Variances	T	DF	Prob>|T|
Unequal	3.1583	9.8	0.0104
Equal	3.3918	16.0	0.0037

(a) The researchers believed that the knee velocity would be higher for skilled rowers. Use the computer output to carry out an appropriate test of this belief. (Note that SAS provides two-sided *P*-values.) What do you conclude?

(b) Use technology to construct and interpret a 90% confidence interval for the mean difference between the knee velocities of skilled and novice female rowers.

(c) If you had used Table B to construct the confidence interval described in part (b), how would the two results compare? Justify your answer *without* doing any calculations.

56. **Competitive rowers** The research in the previous exercise also wondered whether skilled and novice rowers differ in weight or other physical characteristics. Graphs of the data reveal no outliers or strong skewness. Here is the SAS computer output for weight in kilograms:

TTEST PROCEDURE

Variable: WEIGHT

GROUP	N	Mean	Std Dev	Std Error
SKILLED	10	70.37	6.10	1.93
NOVICE	8	68.45	9.04	3.20

Variances	T	DF	Prob>|T|
Unequal		11.8	0.6165
Equal	0.5376	16.0	0.5982

(a) Calculate the missing *t* statistic in the computer output. Show your work.

(b) Is there significant evidence of a difference in the mean weights of skilled and novice rowers? Justify your answer. (Note that SAS provides two-sided *P*-values.)

(c) If the *P*-value had been computed using the more conservative number of degrees of freedom, would it have been greater than, equal to, or less than 0.6165? Justify your answer *without* doing any calculations.

57. **Rewards and creativity** Dr. Teresa Amabile conducted a study involving 47 college students, who were randomly assigned to two treatment groups. The 23 students in one group were given a list of statements about external reasons (E) for writing, such as public recognition, making money, or pleasing their parents. The 24 students in the other group were given a list of statements about internal reasons (I) for writing, such as expressing yourself and enjoying playing with words. Both groups were then instructed to write a poem about laughter. Each student's poem was rated separately by 12 different poets using a creativity scale.[40] The 12 poets' ratings of each student's poem were averaged to obtain an overall creativity score.

We used Fathom software to randomly reassign the 47 subjects to the two groups 1000 times, assuming the treatment received doesn't affect each individual's average creativity rating. The dotplot shows the approximate randomization distribution of $\bar{x}_I - \bar{x}_E$.

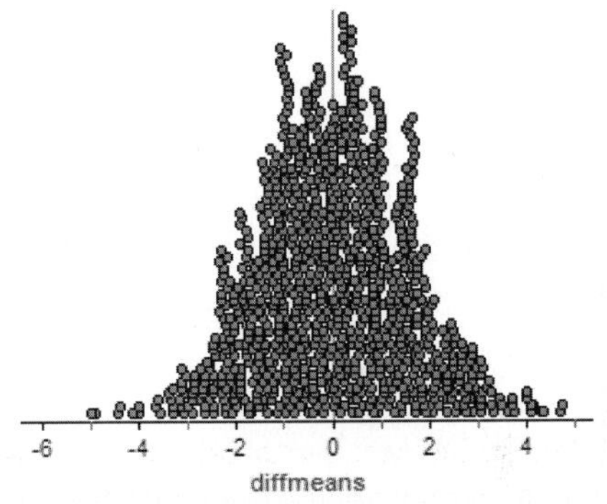

(a) Why did researchers randomly assign the subjects to the two treatment groups?

(b) In the actual experiment, $\bar{x}_I - \bar{x}_E = 4.15$. What conclusion would you draw? Justify your answer with appropriate evidence.

(c) Based on your conclusion in part (b), could you have made a Type I error or a Type II error? Justify your answer.

58. **Sleep deprivation** Does sleep deprivation linger for more than a day? Researchers designed a study using 21 volunteer subjects between the ages of 18 and 25. All 21 participants took a computer-based visual discrimination test at the start of the study. Then the subjects were randomly assigned into two groups. The 11 subjects in one group, D, were deprived of sleep for an entire night in a laboratory setting. The 10 subjects in the other group, A, were allowed unrestricted sleep for the night. Both groups were allowed as much sleep as they wanted for the next two nights. On Day 4, all the subjects took the same visual discrimination test on the computer. Researchers recorded the improvement in time (measured

in milliseconds) from Day 1 to Day 4 on the test for each subject.[41]

We used Fathom software to randomly reassign the 21 subjects to the two groups 1000 times, assuming the treatment received doesn't affect each individual's time improvement on the test. The dotplot shows the approximate randomization distribution of $\bar{x}_A - \bar{x}_D$.

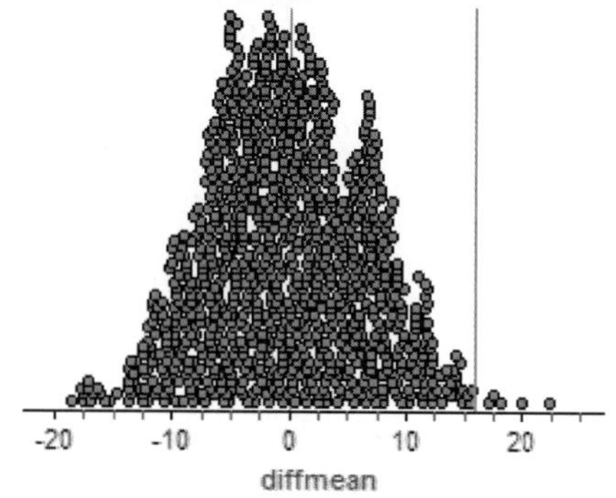

(a) Explain why the researchers didn't let the subjects choose whether to be in the sleep deprivation group or the unrestricted sleep group.

(b) In the actual experiment, $\bar{x}_A - \bar{x}_D = 15.92$. This value is marked with a blue line in the figure. What conclusion would you draw? Justify your answer with appropriate evidence.

(c) Based on your conclusion in part (b), could you have made a Type I error or a Type II error? Justify your answer.

59. **Paired or unpaired?** In each of the following settings, decide whether you should use paired *t* procedures or two-sample *t* procedures to perform inference. Explain your choice.[42]

pg 647

(a) To test the wear characteristics of two tire brands, A and B, each brand of tire is randomly assigned to 50 cars of the same make and model.

(b) To test the effect of background music on productivity, factory workers are observed. For one month, each subject works without music. For another month, the subject works while listening to music on an MP3 player. The month in which each subject listens to music is determined by a coin toss.

(c) A study was designed to compare the effectiveness of two weight-reducing diets. Fifty obese women who volunteered to participate were randomly assigned into two equal-sized groups. One group used Diet A and the other used Diet B. The weight of each woman was measured before the assigned diet and again after 10 weeks on the diet.

60. **Paired or unpaired?** In each of the following settings, decide whether you should use paired *t* procedures or two-sample *t* procedures to perform inference. Explain your choice.[43]

(a) To compare the average weight gain of pigs fed two different rations, nine pairs of pigs were used. The pigs in each pair were littermates. A coin toss was used to decide which pig in each pair got Ration A and which got Ration B.

(b) A random sample of college professors is taken. We wish to compare the average salaries of male and female teachers.

(c) To test the effects of a new fertilizer, 100 plots are treated with the new fertilizer, and 100 plots are treated with another fertilizer. A computer's random number generator is used to determine which plots get which fertilizer.

In each of Exercises 61 to 64, say which inference procedure from Chapter 8, 9, or 10 you would use. Be specific. For example, you might say, "Two-sample z test for the difference between two proportions." You do not need to carry out any procedures.

61. **Preventing drowning** Drowning in bathtubs is a major cause of death in children less than 5 years old. A random sample of parents was asked many questions related to bathtub safety. Overall, 85% of the sample said they used baby bathtubs for infants. Estimate the percent of all parents of young children who use baby bathtubs.

62. **Driving too fast** How seriously do people view speeding in comparison with other annoying behaviors? A large random sample of adults was asked to rate a number of behaviors on a scale of 1 (no problem at all) to 5 (very severe problem). Do speeding drivers get a higher average rating than noisy neighbors?

63. **Looking back on love** How do young adults look back on adolescent romance? Investigators interviewed 40 couples in their midtwenties. The female and male partners were interviewed separately. Each was asked about his or her current relationship and also about a romantic relationship that lasted at least two months when they were aged 15 or 16. One response variable was a measure on a numerical scale of how much the attractiveness of the adolescent partner mattered. You want to compare the men and women on this measure.

64. **Dropping out** You have data from interviews with a random sample of students who failed to graduate from a particular college in 7 years and also from a random sample of students who entered at the same time and did graduate. You will use these data to compare the percents of students from rural backgrounds among dropouts and graduates.

65. **A better drug?** In a pilot study, a company's new cholesterol-reducing drug outperforms the currently available drug. If the data provide convincing

evidence that the mean cholesterol reduction with the new drug is more than 10 milligrams per deciliter of blood (mg/dl) greater than with the current drug, the company will begin the expensive process of mass-producing the new drug. For the 14 subjects who were assigned at random to the current drug, the mean cholesterol reduction was 54.1 mg/dl with a standard deviation of 11.93 mg/dl. For the 15 subjects who were randomly assigned to the new drug, the mean cholesterol reduction was 68.7 mg/dl with a standard deviation of 13.3 mg/dl. Graphs of the data reveal no outliers or strong skewness.

(a) Carry out an appropriate significance test. What conclusion would you draw? (Note that the null hypothesis is *not* $H_0: \mu_1 - \mu_2 = 0$.)

(b) Based on your conclusion in part (a), could you have made a Type I error or a Type II error? Justify your answer.

66. **Down the toilet** A company that makes hotel toilets claims that its new pressure-assisted toilet reduces the average amount of water used by more than 0.5 gallon per flush when compared to its current model. To test this claim, the company randomly selects 30 toilets of each type and measures the amount of water that is used when each toilet is flushed once. For the current-model toilets, the mean amount of water used is 1.64 gal with a standard deviation of 0.29 gal. For the new toilets, the mean amount of water used is 1.09 gal with a standard deviation of 0.18 gal.

(a) Carry out an appropriate significance test. What conclusion would you draw? (Note that the null hypothesis is *not* $H_0: \mu_1 - \mu_2 = 0$.)

(b) Based on your conclusion in part (a), could you have made a Type I error or a Type II error? Justify your answer.

Multiple choice: Select the best answer for Exercises 67 to 70.

67. One major reason that the two-sample t procedures are widely used is that they are quite *robust.* This means that

(a) t procedures do not require that we know the standard deviations of the populations.

(b) t procedures work even when the Random, Normal, and Independent conditions are violated.

(c) t procedures compare population means, a comparison that answers many practical questions.

(d) confidence levels and P-values from the t procedures are quite accurate even if the population distribution is not exactly Normal.

(e) confidence levels and P-values from the t procedures are quite accurate even if outliers and strong skewness are present.

68. There are two common methods for measuring the concentration of a pollutant in fish tissue. Do the two methods differ on the average? You apply both methods to a random sample of 18 carp and use

(a) the paired t test for μ_d.

(b) the one-sample z test for p.

(c) the two-sample t test for $\mu_1 - \mu_2$.

(d) the two-sample z test for $p_1 - p_2$.

(e) none of these.

Exercises 69 and 70 refer to the following setting. A study of road rage asked samples of 596 men and 523 women about their behavior while driving. Based on their answers, each person was assigned a road rage score on a scale of 0 to 20. The participants were chosen by random digit dialing of telephone numbers.

69. We suspect that men are more prone to road rage than women. To see if this is true, test these hypotheses for the mean road rage scores of all male and female drivers:

(a) $H_0: \mu_M = \mu_F$ versus $H_a: \mu_M > \mu_F$.

(b) $H_0: \mu_M = \mu_F$ versus $H_a: \mu_M \neq \mu_F$.

(c) $H_0: \mu_M = \mu_F$ versus $H_a: \mu_M < \mu_F$.

(d) $H_0: \bar{x}_M = \bar{x}_F$ versus $H_a: \bar{x}_M > \bar{x}_F$.

(e) $H_0: \bar{x}_M = \bar{x}_F$ versus $H_a: \bar{x}_M < \bar{x}_F$.

70. The two-sample t statistic for the road rage study (male mean minus female mean) is $t = 3.18$. The P-value for testing the hypotheses from the previous exercise satisfies

(a) $0.001 < P < 0.005$.

(b) $0.0005 < P < 0.001$.

(c) $0.001 < P < 0.002$.

(d) $0.002 < P < 0.01$.

(e) $P > 0.01$.

Exercises 71 to 74 refer to the following setting. Coaching companies claim that their courses can raise the SAT scores of high school students. Of course, students who retake the SAT without paying for coaching generally raise their scores. A random sample of students who took the SAT twice found 427 who were coached and 2733 who were uncoached.[44] Starting with their Verbal scores on the first and second tries, we have these summary statistics:

		Try 1		Try 2		Gain	
	n	$\bar{x}$	s_x	$\bar{x}$	s_x	$\bar{x}$	s_x
Coached	427	500	92	529	97	29	59
Uncoached	2733	506	101	527	101	21	52

71. **Coaching and SAT scores** (10.1) What proportion of students who take the SAT twice are coached? To answer this question, Jannie decides to construct a 99% confidence interval. Her work is shown below. Explain what's wrong with Jannie's method.

$$\hat{p}_1 = \frac{427}{3160} = 0.135 = \text{proportion of students who were coached}$$

$$\hat{p}_2 = \frac{2733}{3160} = 0.865 = \text{proportion of students who weren't coached}$$

A 99% CI for $p_1 - p_2$ is

$$(0.135 - 0.865) \pm 2.575\sqrt{\frac{0.135(0.865)}{3160} + \frac{0.865(0.135)}{2733}} = -0.73 \pm 0.022 = (-0.752, -0.708)$$

We are 99% confident that the proportion of students taking the SAT twice who are coached is between 71 and 75 percentage points lower than students who aren't coached.

72. **Coaching and SAT scores** (8.3, 10.2) Let's first ask if students who are coached increased their scores significantly.

(a) You could use the information on the Coached line to carry out either a two-sample *t* test comparing Try 1 with Try 2 for coached students or a paired *t* test using Gain. Which is the correct test? Why?

(b) Carry out the proper test. What do you conclude?

(c) Construct and interpret a 99% confidence interval for the mean gain of all students who are coached.

73. **Coaching and SAT scores** (10.2) What we really want to know is whether coached students improve more than uncoached students, and whether any advantage is large enough to be worth paying for. Use the information above to answer these questions:

(a) Is there good evidence that coached students gained more on the average than uncoached students? Carry out a significance test to answer this question.

(b) How much more do coached students gain on the average? Construct and interpret a 99% confidence interval.

(c) Based on your work, what is your opinion: do you think coaching courses are worth paying for?

74. **Coaching and SAT scores: Critique** (4.1, 4.3) The data you used in the previous two exercises came from a random sample of students who took the SAT twice. The response rate was 63%, which is pretty good for nongovernment surveys.

(a) Explain how nonresponse could lead to bias in this study.

(b) We can't be sure that coaching actually *caused* the coached students to gain more than the uncoached students. Explain briefly but clearly why this is so.

75. **Quality control** (2.2, 5.3, 6.3) Many manufacturing companies use statistical techniques to ensure that the products they make meet standards. One common way to do this is to take a random sample of products at regular intervals throughout the production shift. Assuming that the process is working properly, the mean measurements from these random samples will vary Normally around the target mean μ, with a standard deviation of σ. For each question that follows, assume that the process is working properly.

(a) What's the probability that at least one of the next two sample means will fall more than 2σ from the target mean μ? Show your work.

(b) What's the probability that the first sample mean that is greater than $\mu + 2\sigma$ is the one from the fourth sample taken?

(c) Plant managers are trying to develop a criterion for determining when the process is not working properly. One idea they have is to look at the 5 most recent sample means. If at least 4 of the 5 fall outside the interval $(\mu - \sigma, \mu + \sigma)$, they will conclude that the process isn't working. Is this a reasonable criterion? Justify your answer with an appropriate probability.

76. **Information online** (8.2, 10.1) A random digit dialing sample of 2092 adults found that 1318 used the Internet.[45] Of the users, 1041 said that they expect businesses to have Web sites that give product information; 294 of the 774 nonusers said this.

(a) Construct and interpret a 95% confidence interval for the proportion of all adults who use the Internet.

(b) Construct and interpret a 95% confidence interval to compare the proportions of users and nonusers who expect businesses to have Web sites.

Chapter 10 Chapter Review

Problems that involve comparing two parameters are among the most common in statistical practice. Such problems arise from randomized comparative experiments and from observational studies that use random samples to compare two distinct populations. In either case, we get data from two independent samples or groups. This chapter presents two-sample inference methods for comparing proportions and means.

The conditions for inference are by now familiar: Random, Normal, and Independent. Checking the Random and Independent conditions is the same for both proportions and means. However, the Normal condition requires different thinking for the two settings. Before comparing two proportions, we examine the counts of successes and failures in both samples/groups. Before we compare two means, we check that both population distributions are Normal or both sample sizes are large ($n \geq 30$).

Confidence intervals and significance tests for comparing two proportions use two-sample z procedures for $p_1 - p_2$. There is a big computational difference between these two inference methods: confidence intervals use the two sample proportions $\hat{p}_1$ and $\hat{p}_2$ to estimate the standard deviation of the statistic, yet tests use the pooled proportion of successes $\hat{p}_C$ in the combined samples. Because of this difference, confidence intervals and tests for $p_1 - p_2$ don't always lead to the same conclusion.

We use two-sample t procedures for confidence intervals and significance tests about the difference $\mu_1 - \mu_2$. There are two options for the degrees of freedom: a more conservative value for use with Table B and a larger value used by technology. Like their one-sample counterparts, two-sample t procedures are fairly robust against non-Normality, especially when the sample sizes are large.

Chapter Review Exercises

These exercises are designed to help you review the important ideas and methods of the chapter. Relevant learning objectives are provided in bulleted form before each exercise.

- Determine the proper inference procedure to use in a given setting.

R10.1. Which procedure? For each of the following settings, say which inference procedure from Chapter 8, 9, or 10 you would use. Be specific. For example, you might say, "Two-sample z test for the difference between two proportions." You do not need to carry out any procedures.[46]

(a) Do people smoke less when cigarettes cost more? A random sample of 500 smokers was selected. The number of cigarettes each person smoked per day was recorded over a one-month period before a 30% cigarette tax was imposed and again for one month after the tax was imposed.

(b) How much greater is the percent of senior citizens who attend a play at least once per year than the percent of people in their twenties who do so? Random samples of 100 senior citizens and 100 people in their twenties were surveyed.

(c) You have data on rainwater collected at 16 locations in the Adirondack Mountains of New York State. One measurement is the acidity of the water, measured by pH on a scale of 0 to 14 (the pH of distilled water is 7.0). Estimate the average acidity of rainwater in the Adirondacks.

(d) Consumers Union wants to see which of two brands of calculator is easier to use. They recruit 100 volunteers and randomly assign them to two equal-sized groups. The people in one group use Calculator A and those in the other group use Calculator B. Researchers record the time required for each volunteer to carry out the same series of routine calculations (such as figuring discounts and sales tax, totaling a bill) on the assigned calculator.

- Construct and interpret a confidence interval to compare two proportions.

R10.2. Seat belt use The proportion of drivers who use seat belts depends on things like age (young people are more likely to go unbelted) and gender (women are more likely to use belts). It also depends on local law. In New York City, police can stop a driver who is not belted. In Boston at the time of the study, police could cite a driver for not wearing a seat belt only if the driver had been stopped for some other violation. Here are data from observing random samples of female Hispanic drivers in these two cities:[47]

City	Drivers	Belted
New York	220	183
Boston	117	68

(a) Is this an experiment or an observational study? Why?

(b) Construct and interpret a 95% confidence interval for the difference in the proportions of female Hispanic drivers in the two cities who wear seat belts.

(c) Based on the laws in the two cities, we would expect a smaller proportion of drivers to wear seat belts in Boston than in New York. Does the confidence interval in part (b) give good evidence that this is true for female Hispanic drivers? Justify your answer.

- Check conditions for using two-sample *t* procedures in a randomized experiment.
- Interpret the results of inference procedures in a randomized experiment.

R10.3. Expensive ads Consumers who think a product's advertising is expensive often also think the product must be of high quality. Can other information undermine this effect? To find out, marketing researchers did an experiment. The subjects were 90 women from the clerical and administrative staff of a large organization. All subjects read an ad that described a fictional line of food products called "Five Chefs." The ad also described the major TV commercials that would soon be shown, an unusual expense for this type of product. The 45 women who were randomly assigned to the control group read nothing else. The 45 in the "undermine group" also read a news story headlined "No Link between Advertising Spending and New Product Quality." All the subjects then rated the quality of Five Chefs products on a seven-point scale. The study report said, "The mean quality ratings were significantly lower in the undermine treatment ($\bar{x}_A = 4.56$) than in the control treatment ($\bar{x}_C = 5.05$; $t = 2.64$, $P < 0.01$)."[48]

(a) Explain why the Random and Independent conditions are met in this case.

(b) The distribution of individual responses is not Normal, because there is only a seven-point scale. Why is it still proper to use a two-sample *t* test?

(c) Interpret the *P*-value in context.

- Use two-sample *t* procedures to compare two means based on summary statistics.

R10.4. Men versus women The National Assessment of Educational Progress (NAEP) Young Adult Literacy Assessment Survey interviewed a random sample of 1917 people 21 to 25 years old. The sample contained 840 men and 1077 women.[49] The mean and standard deviation of scores on the NAEP's test of quantitative skills were $\bar{x}_1 = 272.40$ and $s_1 = 59.2$ for the men in the sample. For the women, the results were $\bar{x}_2 = 274.73$ and $s_2 = 57.5$. Is the difference between the mean scores for men and women significant at the 1% level? Give appropriate statistical evidence to justify your answer.

- Perform a significance test to compare two proportions.

R10.5. Treating AIDS The drug AZT was the first drug that seemed effective in delaying the onset of AIDS. Evidence for AZT's effectiveness came from a large randomized comparative experiment. The subjects were 870 volunteers who were infected with HIV, the virus that causes AIDS, but did not yet have AIDS. The study assigned 435 of the subjects at random to take 500 milligrams of AZT each day and another 435 to take a placebo. At the end of the study, 38 of the placebo subjects and 17 of the AZT subjects had developed AIDS.

(a) Researchers want to test the claim that taking AZT lowers the proportion of infected people who will develop AIDS in a given period of time. Carry out a test of this claim at the $\alpha = 0.05$ level.

(b) The experiment was double-blind. Explain why this is important.

(c) Describe a Type I and a Type II error in this setting, and give the consequences of each. Which is more serious?

- Determine whether the conditions for performing inference are met.

R10.6. Conditions Explain why it is not safe to use the methods of this chapter to perform inference in each of the following settings.

(a) Lyme disease is spread in the northeastern United States by infected ticks. The ticks are infected mainly by feeding on mice, so more mice result in more infected ticks. The mouse population in turn rises and falls with the abundance of acorns, their favored food. Experimenters studied two similar forest areas in a year when the acorn crop failed. They added hundreds of thousands of acorns to one area to imitate an abundant acorn crop, while leaving the other area untouched.

The next spring, 54 of the 72 mice trapped in the first area were in breeding condition, versus 10 of the 17 mice trapped in the second area.[50]

(b) Who texts more—males or females? For their final project, a group of AP Statistics students investigated their belief that females text more than males. They asked a random sample of 31 students—15 males and 16 females—from their school to record the number of text messages sent and received over a two-day period. Boxplots of their data are shown below.

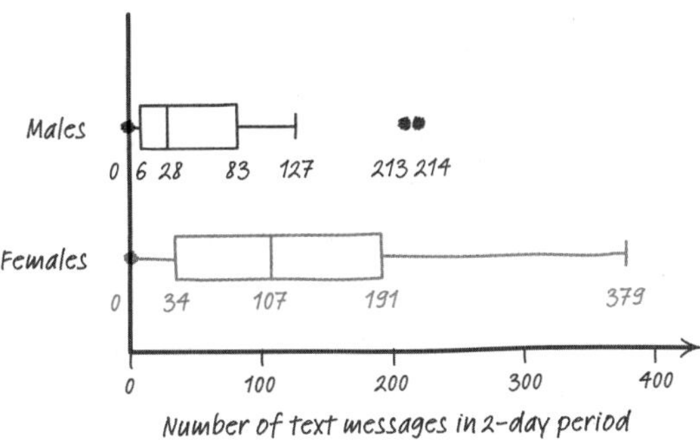

- Use two-sample t procedures to compare two means from raw data.

R10.7. Each day I am getting better in math A "subliminal" message is below our threshold of awareness but may nonetheless influence us. Can subliminal messages help students learn math? A group of 18 students who had failed the mathematics part of the City University of New York Skills Assessment Test agreed to participate in a study to find out. All received a daily subliminal message, flashed on a screen too rapidly to be consciously read. The treatment group of 10 students (assigned at random) was exposed to "Each day I am getting better in math." The control group of 8 students was exposed to a neutral message, "People are walking on the street." All 18 students participated in a summer program designed to raise their math skills, and all took the assessment test again at the end of the program. The table below gives data on the subjects' scores before and after the program.[51]

Treatment Group		Control Group	
Pretest	**Posttest**	**Pretest**	**Posttest**
18	24	18	29
18	25	24	29
21	33	20	24
18	29	18	26
18	33	24	38
20	36	22	27
23	34	15	22
23	36	19	31
21	34		
17	27		

(a) The Fathom boxplots below display the differences in pretest and posttest scores for the students in the control (Cdiff) and treatment (Tdiff) groups. Write a few sentences comparing the performance of these two groups.

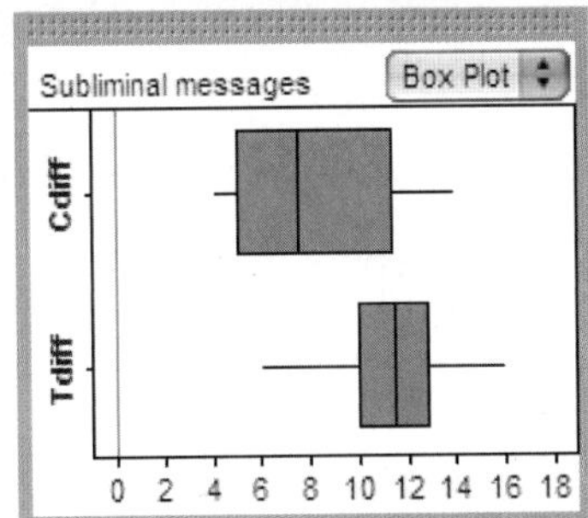

(b) Construct and interpret a 90% confidence interval for the mean difference in gains between treatment and control.

(c) Can we generalize these results to the population of all students who failed the mathematics part of the City University of New York Skills Assessment Test? Why or why not?

- Use confidence intervals to correctly estimate an unknown proportion.

R10.8. Golf club repairs The Ping Company makes custom-built golf clubs and competes in the $4 billion golf equipment industry. To improve its business process, Ping decided to study the time it took to repair golf clubs sent to the company by mail.[52] The company determined that 16% of a random sample of orders were sent back to the customers in 5 days or less. Ping examined the processing of repair orders and made changes. Following the changes, 90% of a random sample of orders were completed within 5 days. Assume that each of the estimated percents is based on a random sample of n orders.

(a) We used the sample data to construct two 90% confidence intervals for the proportion of orders completed within 5 days—one before and one after the changes at Ping. The two intervals are (0.858, 0.942) and (0.109, 0.211). Find the value of n.

(b) Explain why the interval (0.858 − 0.211, 0.942 − 0.109) = (0.647, 0.833) is *not* a 95% confidence interval for the improvement in the proportion of orders sent back to customers within 5 days following the change.

(c) Construct and interpret a correct 95% confidence interval to replace the one in part (b).

- Use an appropriate inference procedure to compare two proportions.

R10.9. Reducing nonresponse? Telephone surveys often have high rates of nonresponse. When the call is handled by voice mail, leaving a message might encourage people to respond when they are called again. Here are data from a study in which (at random) a message was or was not left when voice mail picked up the first call from a survey:[53]

	No contact	Contact
No message	42	58
Message	91	200

(a) Is this an experiment or an observational study? Justify your answer.

(b) Based on the data, can we conclude that leaving a message increases the proportion of households that are eventually interviewed? Give appropriate statistical evidence to support your answer.

- Interpret standard computer output for two-sample *t* procedures.

R10.10. Students' self-concept Here is SAS output for a study of the self-concept of a random sample of seventh-grade students. The variable SC is the score on the Piers-Harris Self-Concept Scale. The analysis was done to see if male and female students differ in mean self-concept score.[54]

```
                 TTEST PROCEDURE

Variable: SC

SEX          N      Mean     Std Dev     Std Error
F           31    55.516      12.696         2.280
M           47    57.915      12.265         1.789

Variances     T         DF      Prob>|T|
Unequal    -0.8276     62.8     0.4110
Equal      -0.8336     76.0     0.4071
```

Write a few sentences summarizing the comparison of females and males, as if you were preparing a report for publication.

Chapter 10 AP Statistics Practice Test

Section I: Multiple Choice *Select the best answer for each question.*

T10.1. A study of road rage asked separate random samples of 596 men and 523 women about their behavior while driving. Based on their answers, each respondent was assigned a road rage score on a scale of 0 to 20. Are the conditions for performing a two-sample *t* test satisfied?

(a) Maybe; we have independent random samples, but we need to look at the data to check Normality.

(b) No; road rage scores in a range between 0 and 20 can't be Normal.

(c) No; we don't know the population standard deviations.

(d) Yes; the large sample sizes guarantee that the corresponding population distributions will be Normal.

(e) Yes; we have two independent random samples and large sample sizes.

T10.2. Thirty-five people from a random sample of 125 workers from Company A admitted to using sick leave when they weren't really ill. Seventeen employees from a random sample of 68 workers from Company B admitted that they had used sick leave when they weren't ill. A 95% confidence interval for the difference in the proportions of workers at the two companies who would admit to using sick leave when they weren't ill is

(a) $0.03 \pm \sqrt{\frac{(0.28)(0.72)}{125} + \frac{(0.25)(0.75)}{68}}$

(b) $0.03 \pm 1.96\sqrt{\frac{(0.28)(0.72)}{125} + \frac{(0.25)(0.75)}{68}}$

(c) $0.03 \pm 1.645\sqrt{\frac{(0.28)(0.72)}{125} + \frac{(0.25)(0.75)}{68}}$

(d) $0.03 \pm 1.96\sqrt{\frac{(0.269)(0.731)}{125} + \frac{(0.269)(0.731)}{68}}$

(e) $0.03 \pm 1.645\sqrt{\frac{(0.269)(0.731)}{125} + \frac{(0.269)(0.731)}{68}}$

T10.3. The power takeoff driveline on tractors used in agriculture is a potentially serious hazard to operators of farm equipment. The driveline is covered by a shield in new tractors, but for a variety of reasons, the shield is often missing on older tractors. Two types of shields are the bolt-on and the flip-up. It was believed that the bolt-on shield was perceived as a nuisance by the operators and deliberately removed, but the flip-up shield is easily lifted for inspection and maintenance and may be left in place. In a study initiated by the U.S. National Safety Council, random samples of older tractors with both types of shields were taken to see what proportion of shields were removed. Of 183 tractors

Cumulative AP Practice Test 3

Section I: Multiple Choice *Choose the best answer.*

AP3.1. Suppose the probability that a softball player gets a hit in any single at-bat is .300. Assuming that her chance of getting a hit on a particular time at bat is independent of her other times at bat, what is the probability that she will not get a hit until her fourth time at bat in a game?

(a) $\binom{4}{3}(0.3)^1(0.7)^3$

(b) $\binom{4}{3}(0.3)^3(0.7)^1$

(c) $\binom{4}{1}(0.3)^3(0.7)^1$

(d) $(0.3)^3(0.7)^1$

(e) $(0.3)^1(0.7)^3$

AP3.2. The probability that Color Me Dandy wins a horse race at Batavia Downs given good track conditions is 0.60. The probability of good track conditions on any given day is 0.85. What is the probability that Color Me Dandy wins or the track conditions are good?

(a) 0.94

(b) 0.51

(c) 0.49

(d) 0.06

(e) The answer cannot be determined from the given information.

AP3.3. *Sports Illustrated* planned to ask a random sample of Division I college athletes, "Do you believe performance-enhancing drugs are a problem in college sports?" How many athletes must be interviewed to estimate the proportion concerned about use of drugs within ±2% with 90% confidence?

(a) 17

(b) 21

(c) 1680

(d) 1702

(e) 2401

AP3.4. The distribution of grade point averages for a certain college is approximately Normal with a mean of 2.5 and a standard deviation of 0.6. According to the 68–95–99.7 rule, within which interval would we expect to find approximately 81.5% of all GPAs for students at this college?

(a) (0.7, 3.1)

(b) (1.3, 3.7)

(c) (1.9, 3.7)

(d) (1.9, 4.2)

(e) (0.7, 4.2)

AP3.5. Which of the following will increase the power of a significance test?

(a) Increase the Type II error probability.

(b) Increase the sample size.

(c) Reject the null hypothesis only if the *P*-value is smaller than the level of significance.

(d) Decrease the significance level α.

(e) Select a value for the alternative hypothesis closer to the value of the null hypothesis.

AP3.6. You can find the Excite Poll online at poll.excite.com. You simply click on a response to become part of the sample. The poll question for June 19, 2005, was "Do you prefer watching first-run movies at a movie theater, or waiting until they are available on home video or pay-per-view?" In all, 8896 people responded, with only 12% (1118 people) saying they preferred theaters. You can conclude that

(a) American adults strongly prefer home videos or pay-per-view.

(b) the high nonresponse rate prevents us from drawing a conclusion.

(c) the sample is too small to draw any conclusion.

(d) the poll uses voluntary response, so the results tell us little about all American adults.

(e) American adults strongly prefer seeing movies at a movie theater.

AP3.7. A certain candy has different wrappers for various holidays. During Holiday 1 the candy wrappers are 30% silver, 30% red, and 40% pink. During Holiday 2 the wrappers are 50% silver and 50% blue. Forty pieces of candy are randomly selected from the Holiday 1 distribution, and 40 pieces are randomly selected from the Holiday 2 distribution. What are the expected value and standard deviation of the total number of silver wrappers?

(a) 32, 18.4

(b) 32, 6.06

(c) 32, 4.29

(d) 80, 18.4

(e) 80, 4.29

AP3.8. A beef rancher randomly sampled 42 cattle from her large herd to obtain a 95% confidence interval to estimate the mean weight of the cows in the herd. The interval obtained was (1010, 1321). If the rancher had used a 98% confidence interval instead, the interval would have been

(a) wider and would have less precision than the original estimate.

(b) wider and would have more precision than the original estimate.

(c) wider and would have the same precision as the original estimate.

(d) narrower and would have less precision than the original estimate.

(e) narrower and would have more precision than the original estimate.

AP3.9. School A has 400 students and School B has 2700 students. A local newspaper wants to compare the distributions of SAT scores for the two schools. Which of the following would be the most useful for making this comparison?

(a) Back-to-back stemplots for A and B

(b) A scatterplot of A versus B

(c) Dotplots for A and B drawn on the same scale

(d) Two relative frequency histograms of A and B drawn on the same scale

(e) Two frequency histograms for A and B drawn on the same scale

AP3.10. Let X represent the score when a fair six-sided die is rolled. For this random variable, $\mu_X = 3.5$ and $\sigma_X = 1.71$. If the die is rolled 100 times, what is the approximate probability that the total score is at least 375?

(a) 0.0000 (c) 0.0721 (e) 0.9279

(b) 0.0017 (d) 0.4420

AP3.11. Corn variety 1 yielded 140 bushels per acre last year at a research farm. This year, corn variety 2, planted in the same location, yielded only 110 bushels per acre. Unfortunately, we don't know whether the difference is due to the superiority of variety 1 or to the effect of this year's drought. This is an example of

(a) bias due to voluntary response.

(b) confounding.

(c) lack of blinding.

(d) sampling variability.

(e) the placebo effect.

AP3.12. The correlation between the heights of fathers and the heights of their (grownup) sons is $r = 0.52$, both measured in inches. If fathers' heights were measured in feet and sons' heights were measured in furlongs (one furlong equals 7920 inches), the correlation between heights of fathers and heights of sons would be

(a) much smaller than 0.52.

(b) slightly smaller than 0.52.

(c) unchanged; equal to 0.52.

(d) slightly larger than 0.52.

(e) much larger than 0.52.

AP3.13. A random sample of 200 New York State voters included 88 Republicans, while a random sample of 300 California voters produced 141 Republicans. Which of the following represents the 95% confidence interval that should be used to estimate the true difference in the proportions of Republicans in New York State and California?

(a) $(0.44 - 0.47) \pm 1.96 \dfrac{(0.44)(0.56) + (0.47)(0.53)}{\sqrt{200 + 300}}$

(b) $(0.44 - 0.47) \pm 1.96 \dfrac{(0.44)(0.56)}{\sqrt{200}} + \dfrac{(0.47)(0.53)}{\sqrt{300}}$

(c) $(0.44 - 0.47) \pm 1.96 \sqrt{\dfrac{(0.44)(0.56)}{200} + \dfrac{(0.47)(0.53)}{300}}$

(d) $(0.44 - 0.47) \pm 1.96 \sqrt{\dfrac{(0.44)(0.56) + (0.47)(0.53)}{200 + 300}}$

(e) $(0.44 - 0.47) \pm 1.96 \sqrt{(045)(0.55)\left(\dfrac{1}{200} + \dfrac{1}{300}\right)}$

AP3.14. Which of the following is *not* a property of a binomial setting?

(a) Outcomes of different trials are independent.

(b) The chance process consists of a fixed number of trials, n.

(c) The probability of success is the same for each trial.

(d) If we use a sample size of 30, the binomial distribution will be approximately Normal.

(e) Each trial can result in either a success or a failure.

AP3.15. Mrs. Woods and Mrs. Bryan are avid vegetable gardeners. They use different fertilizers, and each claims that hers is the best fertilizer to use when growing tomatoes. Both agree to do a study using the weight of their tomatoes as the response variable. They had each planted the same varieties of tomatoes on the same day and fertilized the plants on the same schedule throughout the growing season. At harvest time, they each randomly select 15 tomatoes from their respective gardens and weigh them. After performing a two-sample t test on the difference in mean weights of tomatoes, they get $t = 5.24$ and $P = 0.0008$. Can the gardener with the larger mean claim that her fertilizer caused her tomatoes to be heavier?

(a) No, because the soil conditions in the two gardens is a potential confounding variable.

(b) No, because there was no replication.

(c) Yes, because a different fertilizer was used on each garden.

(d) Yes, because random samples were taken from each garden.

(e) Yes, because the P-value is so low.

AP3.16. The Environmental Protection Agency is charged with monitoring industrial emissions that pollute the atmosphere and water. So long as emission levels stay within specified guidelines, the EPA does not take action against the polluter. If the polluter is in violation of the regulations, the offender can be fined, forced to clean up the problem, or possibly closed. Suppose that for a particular industry the acceptable emission level has been set at no more than 5 parts per million (5 ppm). The null and alternative hypotheses are $H_0: \mu = 5$ versus $H_a: \mu > 5$. Which of the following describes a Type II error?

(a) The EPA fails to find evidence that emissions exceed acceptable limits when, in fact, they are within acceptable limits.

(b) The EPA concludes that emissions exceed acceptable limits when, in fact, they are within acceptable limits.

(c) The EPA concludes that emissions exceed acceptable limits when, in fact, they do exceed acceptable limits.

(d) The EPA takes more samples to ensure that they make the correct decision.

(e) The EPA fails to find evidence that emissions exceed acceptable limits when, in fact, they do exceed acceptable limits.

AP3.17. Which of the following is *false*?

(a) A measure of center alone does not completely describe the characteristics of a set of data. Some measure of spread is also needed.

(b) If the original measurements are expressed in inches, the standard deviation would be expressed in square inches.

(c) One of the disadvantages of a histogram is that it doesn't show each data value.

(d) Between the range and the interquartile range, the *IQR* is a better measure of spread if there are outliers.

(e) If a distribution is skewed, the median and interquartile range should be reported rather than the mean and standard deviation.

AP3.18. A 96% confidence interval for the proportion of the labor force that is unemployed in a certain city is (0.07, 0.10). Which of the following statements about this interval is true?

(a) The probability is 0.96 that between 7% and 10% of the labor force is unemployed.

(b) About 96% of the intervals constructed by this method will contain the true proportion of unemployed in the city.

(c) The probability is 0.04 that less than 7% or more than 10% of the labor force is unemployed.

(d) The true rate of unemployment lies within this interval 96% of the time.

(e) Between 7% and 10% of the labor force is unemployed 96% of the time.

AP3.19. A large toy company introduces a lot of new toys to its product line each year. The company wants to predict the demand as measured by y, first-year sales (in millions of dollars) using x, awareness of the product (as measured by the proportion of customers who had heard of the product by the end of the second month after its introduction). A random sample of 65 new products was taken, and a correlation of 0.96 was computed. Which of the following is a correct interpretation of this value?

(a) Ninety-six percent of the time, the least-squares regression line accurately predicts first-year sales.

(b) About 92% of the time, the proportion of people who have heard of the product by the end of the second month will correctly predict first-year sales.

(c) About 92% of first-year sales can be explained by the proportion of people who have heard of the product by the end of the second month.

(d) The least-squares regression line relating the proportion of people who have heard of the product by the end of the second month and first-year sales will have a slope of 0.96.

(e) Ninety-two percent of the variation in first-year sales can be explained by the least-squares regression line with proportion of people who have heard of the product by the end of the second month as the explanatory variable.

AP3.20. Final grades for a class are approximately Normally distributed with a mean of 76 and a standard deviation of 8. A professor says that the top 10% of the class will receive an A, the next 20% a B, the next 40% a C, the next 20% a D, and the bottom 10% an F. What is the approximate maximum average a student could attain and still receive an F for the course?

(a) 70
(b) 69.27
(c) 65.75
(d) 62.84
(e) 57

AP3.21. National Park rangers keep data on the bears that inhabit their park. Below is a histogram of the weights of 143 bears measured in a recent year.

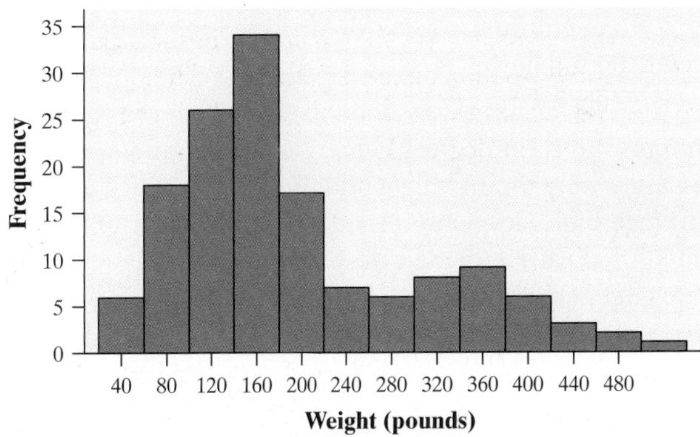

Which statement below is correct?

(a) The median will lie in the interval (140, 180), and the mean will lie in the interval (180, 220).

(b) The median will lie in the interval (140, 180), and the mean will lie in the interval (260, 300).

(c) The median will lie in the interval (100, 140), and the mean will lie in the interval (180, 220).

(d) The mean will lie in the interval (140, 180), and the median will lie in the interval (260, 300).

(e) The mean will lie in the interval (100, 140), and the median will lie in the interval (180, 220).

AP3.22. A random sample of size n will be selected from a population, and the proportion of those in the sample who have a Facebook page will be calculated. How would the margin of error for a 95% confidence interval be affected if the sample size were increased from 50 to 200?

(a) It remains the same.

(b) It is multiplied by 2.

(c) It is multiplied by 4.

(d) It is divided by 2.

(e) It is divided by 4.

AP3.23. A scatterplot and a least-squares regression line are shown in the figure below. If the labeled point **P** (20, 24) is removed from the data set, which of the following statements is true?

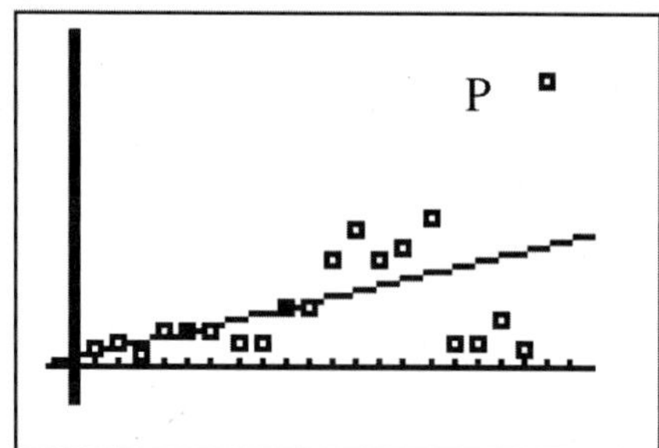

(a) The slope will decrease, and the correlation will decrease.

(b) The slope will decrease, and the correlation will increase.

(c) The slope will increase, and the correlation will increase.

(d) The slope will increase, and the correlation will decrease.

(e) No conclusion can be drawn since the other coordinates are unknown.

AP3.24. The following dotplots show the average high temperatures (in degrees Fahrenheit) for a sample of tourist cities from around the world. Both the January and July average high temperatures are shown. What is one statement that can be made with certainty from an analysis of the graphical display?

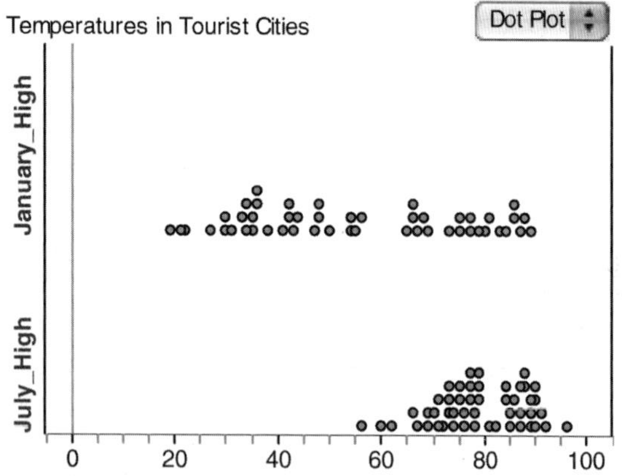

(a) Every city has a larger average high temperature in July than in January.

(b) The distribution of temperatures in July is skewed right, while the distribution of temperatures in January is skewed left.

(c) The median average high temperature for January is higher than the median average high temperature for July.

(d) There appear to be outliers in the average high temperatures for January and July.

(e) There is more variability in average high temperatures in January than in July.

AP3.25. Suppose the null and alternative hypotheses for a significance test are defined as

$$H_0: \mu = 40$$
$$H_a: \mu < 40$$

Which of the following specific values for H_a will give the highest power?

(a) $\mu = 38$ (c) $\mu = 40$ (e) $\mu = 42$

(b) $\mu = 39$ (d) $\mu = 41$

AP3.26. A large university is considering the establishment of a schoolwide recycling program. To gauge interest in the program by means of a questionnaire, the university takes separate random samples of undergraduate students, graduate students, faculty, and staff. This is an example of what type of sampling design?

(a) Simple random sample

(b) Stratified random sample

(c) Convenience sample

(d) Cluster sample

(e) Systematic sample

AP3.27. Suppose the true proportion of people who use public transportation to get to work in the Washington, D.C., area is 0.45. In a simple random sample of 250 people who work in Washington, what is the standard deviation of the sampling distribution of $\hat{p}$?

(a) 0.4975 (c) 0.0315 (e) 0.617

(b) 0.2475 (d) 0.0009

Questions 28 and 29 refer to the following setting. According to sleep researchers, if you are between the ages of 12 and 18 years old, you need 9 hours of sleep to be fully functional. A simple random sample of 28 students was chosen from a large high school, and these students were asked how much sleep they got the previous night. The mean of the responses was 7.9 hours, with a standard deviation of 2.1 hours.

AP3.28. If we are interested in whether students at this high school are getting too little sleep, which of the following represents the appropriate null and alternative hypotheses?

(a) $H_0: \mu = 7.9$ and $H_a: \mu < 7.9$

(b) $H_0: \mu = 7.9$ and $H_a: \mu \neq 7.9$

(c) $H_0: \mu = 9$ and $H_a: \mu \neq 9$

(d) $H_0: \mu = 9$ and $H_a: \mu < 9$

(e) $H_0: \mu \leq 9$ and $H_a: \mu \geq 9$

AP3.29. Which of the following is the test statistic for the hypothesis test?

(a) $t = \dfrac{7.9 - 9}{\frac{2.1}{\sqrt{28}}}$ (c) $t = \dfrac{7.9 - 9}{\sqrt{\frac{2.1}{28}}}$ (e) $t = \dfrac{9 - 7.9}{\frac{2.1}{\sqrt{27}}}$

(b) $t = \dfrac{9 - 7.9}{\frac{2.1}{\sqrt{28}}}$ (d) $t = \dfrac{7.9 - 9}{\frac{2.1}{\sqrt{27}}}$

AP3.30. Shortly before the 2008 presidential election, a survey was taken by the school newspaper at a very large state university. Randomly selected students were asked,

"Whom do you plan to vote for in the upcoming presidential election?" Here is a two-way table of the responses by political persuasion for 1850 students:

Candidate of choice	Political Persuasion: Democrat	Republican	Independent	Total
Obama	925	78	26	1029
McCain	78	598	19	695
Other	2	8	11	21
Undecided	32	28	45	105
Total	1037	712	101	1850

Which of the following statements about these data is true?

(a) The percent of Republicans among the respondents is 41%.

(b) The marginal distribution of the variable choice of candidate is given by Obama: 55.6%; McCain: 37.6%; Other: 1.1%; Undecided: 5.7%.

(c) About 11.2% of Democrats reported that they planned to vote for McCain.

(d) About 44.6% of those who are undecided are Independents.

(e) The conditional distribution of political persuasion among those for whom McCain is the candidate of choice is Democrat: 7.5%; Republican: 84.0%; Independent: 18.8%

Section II: Free Response *Show all your work. Indicate clearly the methods you use, because you will be graded on the correctness of your methods as well as on the accuracy and completeness of your results and explanations.*

AP3.31. A researcher wants to determine whether or not a five-week crash diet is effective over a long period of time. A random sample of 15 dieters is selected. Each person's weight is recorded before starting the diet and one year after it is concluded. Based on the data shown below (weight in pounds), can we conclude that the diet has a long-term effect, that is, that dieters managed to not regain the weight they lost? Include appropriate statistical evidence to justify your answer.

	1	2	3	4	5	6	7	8
Before	158	185	176	172	164	234	258	200
After	163	182	188	150	161	220	235	191

	9	10	11	12	13	14	15
Before	228	246	198	221	236	255	231
After	228	237	209	220	222	268	234

AP3.32 Starting in the 1970s, medical technology allowed babies with very low birth weight (VLBW, less than 1500 grams, or about 3.3 pounds) to survive without major handicaps. It was noticed that these children nonetheless had difficulties in school and as adults. A long study has followed 242 randomly selected VLBW babies to age 20 years, along with a control group of 233 randomly selected babies from the same population who had normal birth weight.[55]

(a) Is this an experiment or an observational study? Why?

(b) At age 20, 179 of the VLBW group and 193 of the control group had graduated from high school. Is the graduation rate among the VLBW group significantly lower than for the normal-birth-weight controls? Give appropriate statistical evidence to justify your answer.

AP3.33. A nuclear power plant releases water into a nearby lake every afternoon at 4:51 P.M. Environmental researchers are concerned that fish are being driven away from the area around the plant. They believe that the temperature of the water discharged may be a factor. The scatterplot below shows the temperature of the water (°C) released by the plant and the measured distance (in meters) from the outflow pipe of the plant to the nearest fish found in the water on eight randomly chosen afternoons.

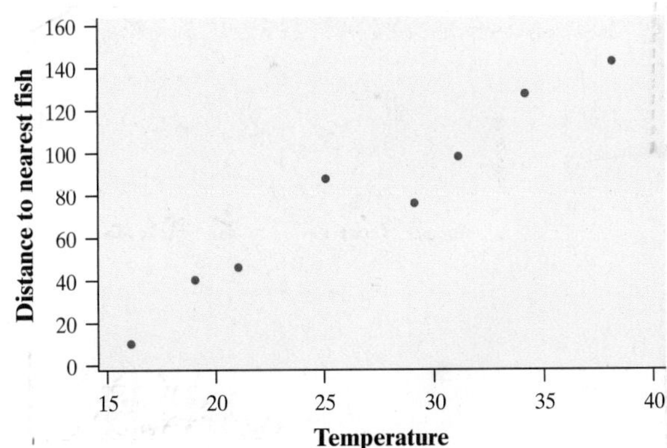

Computer output from a least-squares regression analysis on these data is shown below.

Predictor	Coef	SE Coef	T	P
Constant	−73.64	15.48	−4.76	0.003
Temperature	5.7188	0.5612	−10.19	0.000

S = 11.4175 R-Sq = 94.5% R-Sq(adj) = 93.6%

(a) Write the equation of the least-squares regression line. Define any variables you use.

(b) Interpret the slope of the regression line in context.

(c) A residual plot for the regression is shown on the next page. Is a linear model appropriate for describing the relationship between temperature and distance to the nearest fish? Justify your answer.

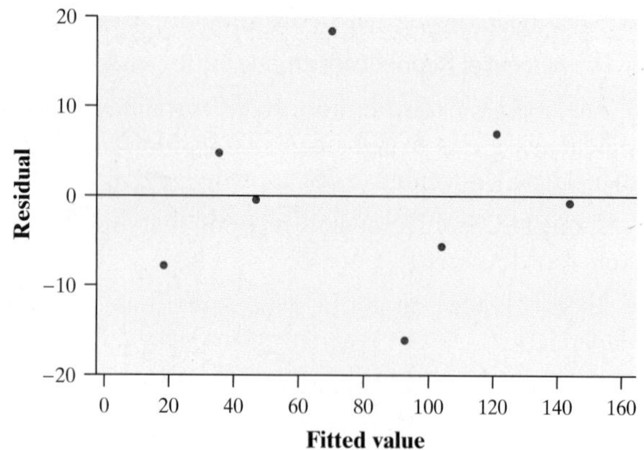

(d) Does the linear model in part (a) overpredict or underpredict the measured distance from the outflow pipe to the nearest fish found in the water for a temperature of 29°? Explain your reasoning.

AP3.34. The Candy Shoppe assembles gift boxes that contain 8 chocolate truffles and two handmade caramel nougats. The truffles have a mean weight of 2 ounces with a standard deviation of 0.5 ounce, and the nougats have a mean weight of 4 ounces with a standard deviation of 1 ounce. The empty boxes weigh 3 ounces with a standard deviation of 0.2 ounce.

(a) Assuming that the weights of the truffles, nougats, and boxes are independent, what are the mean and standard deviation of the weight of a box of candy?

(b) Assuming that the weights of the truffles, nougats, and boxes are approximately Normally distributed, what is the probability that a randomly selected box of candy will weigh more than 30 ounces?

(c) If five gift boxes are randomly selected, what is the probability that at least one of them will weigh more than 30 ounces?

(d) If five gift boxes are randomly selected, what is the probability that the mean weight of the five boxes will be more than 30 ounces?

AP3.35. An investor is comparing two stocks, A and B. She wants to know if over the long run, there is a significant difference in the return on investment as measured by the percent increase or decrease in the price of the stock from its date of purchase. The investor takes a random sample of 50 daily returns over the past five years for each stock. The data are summarized below.

Stock	Mean return	Standard deviation
A	11.8%	12.9%
B	7.1%	9.6%

(a) Is there a significant difference in the mean return on investment for the two stocks? Support your answer with appropriate statistical evidence. Use a 5% significance level.

(b) The investor believes that although the return on investment for Stock A usually exceeds that of Stock B, Stock A represents a riskier investment, where the risk is measured by the price volatility of the stock. The standard deviation is a statistical measure of the price volatility and indicates how much an investment's actual performance during a specified period varies from its average performance over a longer period. Do the price fluctuations in Stock A significantly exceed those of Stock B, as measured by their standard deviations? Identify an appropriate set of hypotheses that the investor is interested in testing.

(c) To measure this, we will construct a test statistic defined as

$$F = \frac{\text{larger sample variance}}{\text{smaller sample variance}}$$

What value(s) of the statistic would indicate that the price fluctuations in Stock A significantly exceed those of Stock B? Explain.

(d) Calculate the value of the F statistic using the information given in the table.

(e) Two hundred simulated values of this test statistic, F, were calculated assuming no difference in the standard deviations of the returns for the two stocks. The results of the simulation are displayed in the following dotplot.

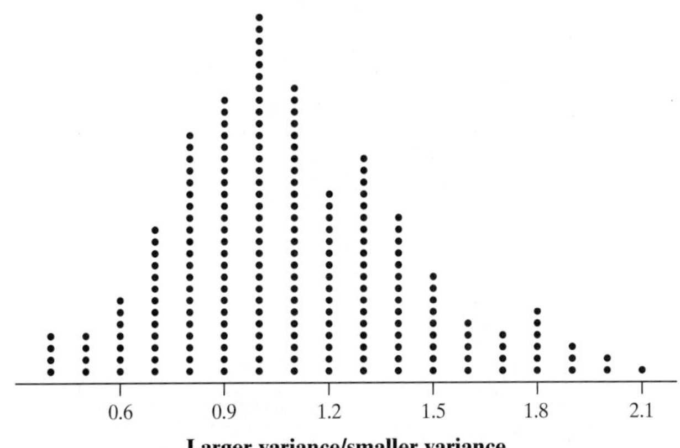

Use these simulated values and the test statistic that you calculated in part (d) to determine whether the observed data provide evidence that Stock A is a riskier investment than Stock B. Explain your reasoning.

Chapter 11

case study

Inference for Distributions of Categorical Data

Do Dogs Resemble Their Owners?

Some people look a lot like their pets. Maybe they deliberately choose animals that match their appearance. Or maybe we're perceiving similarities that aren't really there. Researchers at the University of California, San Diego, decided to investigate. They designed an experiment to test whether dogs resemble their owners. The researchers believed that resemblance between dog and owner might differ for purebred and mixed-breed dogs.

A random sample of 45 dogs and their owners was photographed separately at three dog parks. Then, researchers "constructed triads of pictures, each consisting of one owner, that owner's dog, and one other dog photographed at the same park." The subjects in the experiment were 28 undergraduate psychology students. Each subject was presented with the individual sets of photographs and asked to identify which dog belonged to the pictured owner. A dog was classified as resembling its owner if more than half of the 28 undergraduate students matched dog to owner.[1]

The table below summarizes the results. There is some support for the researchers' belief that resemblance between dog and owner might differ for purebred and mixed-breed dogs.

	Resemble owner	**Don't resemble**
Purebred dogs	16	9
Mixed-breed dogs	7	13

Do these data provide convincing evidence of an association between dogs' breed status and whether or not they resemble their owners? By the end of this chapter, you will have developed the tools you need to answer this question.

Introduction

In the previous chapter, we discussed inference procedures for comparing the proportion of successes for two populations or treatments. Sometimes we want to examine the distribution of a single categorical variable in a population. The *chi-square goodness-of-fit test* allows us to determine whether a hypothesized distribution seems valid. This method is useful in a field like genetics, where the laws of probability give the expected proportion of outcomes in each category.

We can decide whether the distribution of a categorical variable differs for two or more populations or treatments using a *chi-square test for homogeneity*. In doing so, we will often organize our data in a two-way table. It is also possible to use the information in a two-way table to study the relationship between two categorical variables. The *chi-square test for association/independence* allows us to determine if there is convincing evidence of an association between the variables in the population at large.

The methods of this chapter help us answer questions such as these:

- Are births evenly distributed across the days of the week?
- Does background music influence customer purchases?
- Is there an association between anger level and heart disease?

Of course, we have to do a careful job of describing the data before we proceed to statistical inference. In Chapter 1, we discussed graphical and numerical methods of data analysis for categorical variables. You may want to quickly review Section 1.1 now.

Here's an Activity that gives you a taste (pardon the pun) of what lies ahead.

ACTIVITY *The candy man can*

MATERIALS: Small bag of M&M'S® Milk Chocolate Candies for each student or pair of students (1.69-ounce size works well)

Mars, Incorporated, which is headquartered in McLean, Virginia, makes milk chocolate candies. Here's what the company's Consumer Affairs Department says about the color distribution of its M&M'S Milk Chocolate Candies:

> *On average, the new mix of colors of M&M'S Milk Chocolate Candies will contain 13 percent of each of browns and reds, 14 percent yellows, 16 percent greens, 20 percent oranges and 24 percent blues.*

The purpose of this Activity is to compare the color distribution of M&M'S in your individual bag with the advertised distribution. We want to see if there is enough evidence to dispute the company's claim. For the time being, assume that your bag is a random sample of M&M'S Milk Chocolate Candies from the population of candies produced in a particular batch.

1. Open your bag and carefully count the number of whole M&M'S of each color as well as the total number of M&M'S in the bag. *Don't consume the data until your teacher tells you to.*

Make a table on your paper like the one shown on the next page. In the "Observed" column, fill in the counts, by color, and the total number of M&M'S in your bag. Leave the other columns blank for now.

Color	Observed	Expected	Observed − Expected	(Observed − Expected)^2	$\frac{(\text{Observed} - \text{Expected})^2}{\text{Expected}}$
Blue					
Orange					
Green					
Yellow					
Red					
Brown					
Sum					

2. Assuming that the company's claim is true, how many of each color would you expect in *your* bag? For example, suppose that your bag has 60 M&M'S. How many blue M&M'S should you expect? If 24% of all M&M'S Milk Chocolate Candies are blue, then you'd expect $(0.24)(60) = 14.40$ blue M&M'S, on average, in a bag with 60 M&M'S. In a similar way, you can calculate the *expected counts* for the other five colors.

Compute the expected counts (to two decimal places) for your bag. Record your results in the "Expected" column of your table. Check that the sum of the expected counts is equal to the number of M&M'S in your bag.

3. How close are your observed counts to the expected counts? To answer this question, calculate the difference for each color: Observed − Expected. Record these differences in the appropriate column. Find the sum of the values in this column. What do you notice about $\sum$(Observed − Expected)?

In Step 3, the sum of the difference values should be 0 (up to roundoff error). Does that mean your individual bag matches the color distribution claimed by Mars, Inc., perfectly? Of course not. You get a total difference of 0 because the positive and negative values cancel each other. How can we fix this problem? We could use absolute values, but instead, we choose to square the difference values (does this remind you of anything earlier in the text?).

4. Compute the values of (Observed − Expected)^2 and fill in the next column of your table. Find the sum of the values in this column. Compare your results with those of your classmates.

5. In the last column of your table, divide the values in the (Observed − Expected)^2 column by the corresponding expected counts. Then find the sum. This final result is called the *chi-square statistic* and is denoted by χ^2 (the lowercase Greek letter chi, pronounced "kye"). Keep this number handy—you will use it later in the chapter.

6. If your sample reflects the distribution advertised by Mars, Inc., there should be very little difference between the observed counts and the expected counts. Therefore, the calculated values making up the sum χ^2 should be very small. Are the entries in the last row all about the same, or do any values stand out because they are much larger than the others? Did you get way more or way fewer of a particular color than you expected?

7. Compare your value of χ^2 with the results obtained by your classmates. How do we know if a particular χ^2 statistic is large enough to provide convincing evidence against the company's claim? We need to learn more about the distribution of the chi-square statistic before we can answer this question.

Chi-Square Goodness-of-Fit Tests

In Section 11.1, you'll learn about:

- Comparing observed and expected counts: The chi-square statistic
- The chi-square distributions and *P*-values
- Carrying out a test
- Follow-up analysis

Jerome did the M&M'S Activity with his class. The **one-way table** below summarizes the data from Jerome's bag of M&M'S Milk Chocolate Candies. In general, one-way tables display the distribution of a categorical variable for the individuals in a sample.

One-way table

Color:	Blue	Orange	Green	Yellow	Red	Brown	**Total**
Count:	9	8	12	15	10	6	**60**

The sample proportion of blue M&M'S is $\hat{p} = \frac{9}{60} = 0.15$. Since the company claims that 24% of all M&M'S Milk Chocolate Candies are blue, Jerome might believe that something fishy is going on. We could use the one-sample z test for a proportion from Chapter 9 to test the hypotheses

$$H_0: p = 0.24$$
$$H_a: p \neq 0.24$$

where p is the true population proportion of blue M&M'S. We could then perform additional significance tests for each of the remaining colors. Not only would this method be pretty inefficient, but it would also lead to the problem of multiple comparisons, which we'll discuss in Section 11.2. More important, this approach wouldn't tell us how likely it is to get a random sample of 60 candies with a color distribution that differs as much from the one claimed by the company as Jerome's bag does (taking all the colors into consideration at one time). For that, we need a new kind of significance test, called a **chi-square goodness-of-fit test.**

Chi-square goodness-of-fit test

Comparing Observed and Expected Counts: The Chi-Square Statistic

As with any test, we begin by stating hypotheses. The null hypothesis in a chi-square goodness-of-fit test should state a claim about the distribution of a single categorical variable in the population of interest. In the case of the M&M'S Milk Chocolate Candies, the categorical variable we're measuring is color. The appropriate null hypothesis is

H_0: The company's stated color distribution for M&M'S Milk Chocolate Candies is correct.

The alternative hypothesis in a chi-square goodness-of-fit test is that the categorical variable does not have the specified distribution. For the M&M'S, our alternative hypothesis is

H_a: The company's stated color distribution for M&M'S Milk Chocolate Candies is not correct.

Actually, we could write the alternative hypothesis as H_a: At least two of the p_i's are incorrect. If the stated proportion in one category is wrong, then the stated proportion in at least one other category must be wrong because the sum of the p_i's must be 1.

We can also write the hypotheses in symbols as

$$H_0: p_{\text{blue}} = 0.24,\ p_{\text{orange}} = 0.20,\ p_{\text{green}} = 0.16,$$
$$p_{\text{yellow}} = 0.14,\ p_{\text{red}} = 0.13,\ p_{\text{brown}} = 0.13,$$

$$H_a\text{: At least one of the } p_i\text{'s is incorrect}$$

where p_{color} = the true population proportion of M&M'S Milk Chocolate Candies of that color.

CAUTION

Don't state the alternative hypothesis in a way that suggests that *all* the proportions in the hypothesized distribution are wrong. For instance, it would be *incorrect* to write

$$H_a: p_{\text{blue}} \neq 0.24,\ p_{\text{orange}} \neq 0.20,\ p_{\text{green}} \neq 0.16,$$
$$p_{\text{yellow}} \neq 0.14,\ p_{\text{red}} \neq 0.13,\ p_{\text{brown}} \neq 0.13$$

Observed counts

Expected counts

The idea of the chi-square goodness-of-fit test is this: we compare the **observed counts** from our sample with the counts that would be expected if H_0 is true. (*Remember*: we always assume that H_0 is true when performing a significance test.) The more the observed counts differ from the **expected counts,** the more evidence we have against the null hypothesis. In general, the expected counts can be obtained by multiplying the proportion of the population distribution in each category by the sample size. Here's an example that illustrates the process.

Return of the M&M'S

Computing expected counts

PROBLEM: **Jerome's bag of M&M'S Milk Chocolate Candies contained 60 candies. Calculate the expected counts for each color. Show your work.**

SOLUTION: **Assuming that the color distribution stated by Mars, Inc., is true, 24% of all M&M'S Milk Chocolate Candies produced are blue. For random samples of 60 candies, the average number of blue M&M'S should be (0.24)(60) = 14.40. This is our expected count of blue M&M'S. Using this same method, we find the expected counts for the other color categories:**

Orange: (0.20)(60) = 12.00

Green: (0.16)(60) = 9.60

Yellow: (0.14)(60) = 8.40

Red: (0.13)(60) = 7.80

Brown: (0.13)(60) = 7.80

Color	Observed	Expected
Blue	9	14.40
Orange	8	12.00
Green	12	9.60
Yellow	15	8.40
Red	10	7.80
Brown	6	7.80

Did you notice that the expected count sounds a lot like the expected value of a random variable from Chapter 6? That's no coincidence. The number of M&M'S of a specific color in a random sample of 60 candies is a random variable. Its expected value is the average number of candies of this color in many bags of 60 M&M'S Milk Chocolate Candies. The expected value is not likely to be a whole number.

To see if the data give convincing evidence against the null hypothesis, we compare the observed counts from our sample with the expected counts assuming H_0 is true. If the observed counts are far from the expected counts, that's the evidence we were seeking. The table in the example gives the observed and expected counts for Jerome's sample of 60 M&M'S. Figure 11.1 on the next page shows a side-by-side bar graph comparing the observed and expected counts.

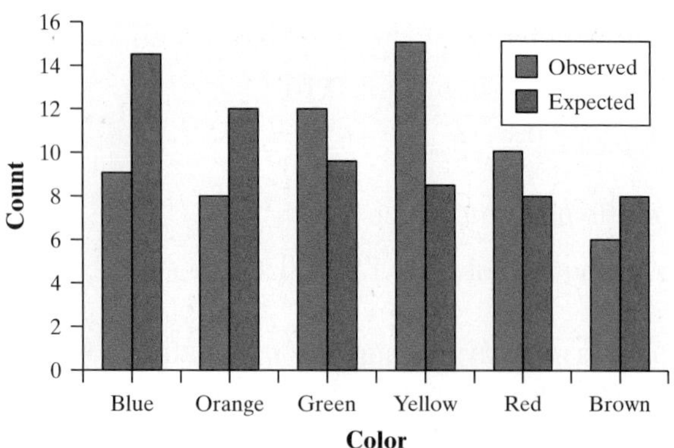

FIGURE 11.1 Bar graph comparing observed and expected counts for Jerome's bag of 60 M&M'S Milk Chocolate Candies.

We see some fairly large differences between the observed and expected counts in several color categories. How likely is it that differences this large or larger would occur just by chance in random samples of size 60 from the population distribution claimed by Mars, Inc.? To answer this question, we calculate a statistic that measures how far apart the observed and expected counts are. The statistic we use to make the comparison is the **chi-square statistic**

$$\chi^2 = \sum \frac{(\text{Observed} - \text{Expected})^2}{\text{Expected}}$$

(The symbol χ is the lowercase Greek letter chi, pronounced "kye.")

DEFINITION: Chi-square statistic

The **chi-square statistic** is a measure of how far the observed counts are from the expected counts. The formula for the statistic is

$$\chi^2 = \sum \frac{(\text{Observed} - \text{Expected})^2}{\text{Expected}}$$

where the sum is over all possible values of the categorical variable.

Let's use this formula to compare the observed and expected counts for Jerome's sample.

Return of the M&M'S

Calculating the chi-square statistic

The table shows the observed and expected counts for Jerome's random sample of 60 M&M'S Milk Chocolate Candies.

Color	Observed	Expected
Blue	9	14.40
Orange	8	12.00
Green	12	9.60
Yellow	15	8.40
Red	10	7.80
Brown	6	7.80

PROBLEM: Calculate the chi-square statistic.

SOLUTION: The formula for the chi-square statistic is

$$\chi^2 = \sum \frac{(\text{Observed} - \text{Expected})^2}{\text{Expected}}$$

For Jerome's data, we add six terms—one for each color category:

$$\begin{aligned}\chi^2 &= \frac{(9-14.40)^2}{14.40} + \frac{(8-12.00)^2}{12.00} + \frac{(12-9.60)^2}{9.60} \\ &\quad + \frac{(15-8.40)^2}{8.40} + \frac{(10-7.80)^2}{7.80} + \frac{(6-7.80)^2}{7.80} \\ &= 2.025 + 1.333 + 0.600 + 5.186 + 0.621 + 0.415 = 10.180\end{aligned}$$

For Practice *Try Exercise 3*

Think of χ^2 as a measure of the distance of the observed counts from the expected counts. Like any distance, it is always zero or positive, and it is zero only when the observed counts are exactly equal to the expected counts. Large values of χ^2 are stronger evidence against H_0 because they say that the observed counts are far from what we would expect if H_0 were true. Small values of χ^2 suggest that the data are consistent with the null hypothesis. Is $\chi^2 = 10.180$ a large value? You know the drill: compare the observed value 10.180 against the sampling distribution that shows how χ^2 would vary in repeated random sampling if the null hypothesis were true.

THINK ABOUT IT

Why do we divide by the expected count when calculating the chi-square statistic? Suppose you obtain a random sample of 60 M&M'S Milk Chocolate Candies. Which would be more surprising: getting 18 blue candies or 12 yellow candies in the sample? Earlier, we computed the expected counts for these two categories as 14.4 and 8.4, respectively. The difference in the observed and expected counts for the two colors would be

$$\text{Blue: } 18 - 14.4 = 3.6 \qquad \text{Yellow: } 12 - 8.4 = 3.6$$

In both cases, the actual number of M&M'S in the sample exceeds the expected count by the same amount. But it's actually much more surprising to be off by 3.6 out of an expected 8.4 yellow candies (almost a 50% discrepancy) than to be off by 3.6 out of an expected 14.4 blue candies (a 25% discrepancy). For that reason, we want the category with a larger *relative* difference to contribute more heavily to the evidence against H_0 measured by the χ^2 statistic.

If we computed (Observed − Expected)^2 for each category instead, the contributions of these two color categories would be the same:

$$\text{Blue: } (18 - 14.40)^2 = 12.96 \qquad \text{Yellow: } (12 - 8.40)^2 = 12.96$$

By using $\dfrac{(\text{Observed} - \text{Expected})^2}{\text{Expected}}$, we guarantee that the color category with the larger relative difference will contribute more heavily to the total:

$$\text{Blue: } \frac{(18-14.40)^2}{14.40} = 0.90 \qquad \text{Yellow: } \frac{(12-8.40)^2}{8.40} = 1.54$$

CHECK YOUR UNDERSTANDING

Mars, Inc., reports that their M&M'S Peanut Chocolate Candies are produced according to the following color distribution: 23% each of blue and orange, 15% each of green and yellow, and 12% each of red and brown. Joey bought a bag of Peanut Chocolate Candies and counted the colors of the candies in his sample: 12 blue, 7 orange, 13 green, 4 yellow, 8 red, and 2 brown.

1. State appropriate hypotheses for testing the company's claim about the color distribution of peanut M&M'S.

2. Calculate the expected count for each color, assuming that the company's claim is true. Show your work.

3. Calculate the chi-square statistic for Joey's sample. Show your work.

The Chi-Square Distributions and *P*-Values

We used Fathom software to simulate taking 500 random samples of size 60 from the population distribution of M&M'S Milk Chocolate Candies given by Mars, Inc. Figure 11.2 shows a dotplot of the values of the chi-square statistic for these 500 samples. The blue vertical line is plotted at the observed result of $\chi^2 = 10.180$. Recall that larger values of χ^2 give more convincing evidence against H_0. According to Fathom, 37 of the 500 simulated samples resulted in a chi-square statistic of 10.180 or higher. Our estimated *P*-value is $37/500 = 0.074$. Since the *P*-value exceeds the default $\alpha = 0.05$ significance level, we do not have enough evidence to refute the company's claim about the color distribution.

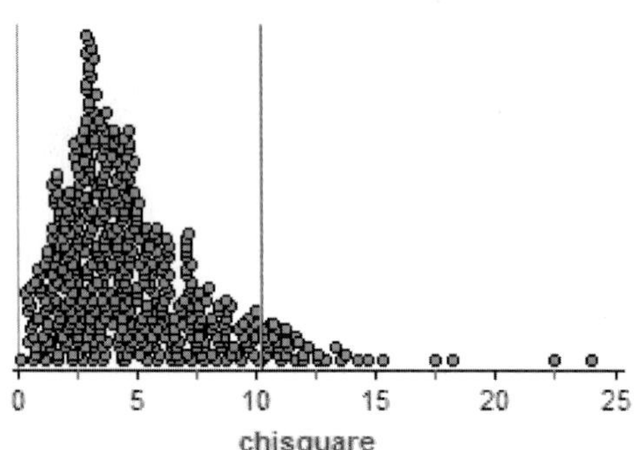

FIGURE 11.2 Fathom dotplot showing values of the chi-square statistic in 500 simulated samples of size $n = 60$ from the population distribution of M&M'S Milk Chocolate Candies stated by the company.

As Figure 11.2 suggests, the sampling distribution of the chi-square statistic is *not* a Normal distribution. It is a right-skewed distribution that allows only positive values because χ^2 can never be negative. The sampling distribution of χ^2 differs depending on the number of possible values for the categorical variable (that is, on the number of categories).

When the expected counts are all at least 5, the sampling distribution of the χ^2 statistic is close to a **chi-square distribution** with degrees of freedom (df) equal to the number of categories minus 1. As with the *t* distributions, there is a different chi-square distribution for each possible value of df. Here are the facts.

The Chi-Square Distributions

The chi-square distributions are a family of distributions that take only positive values and are skewed to the right. A particular chi-square distribution is specified by giving its degrees of freedom. The chi-square goodness-of-fit test uses the chi-square distribution with degrees of freedom = the number of categories − 1.

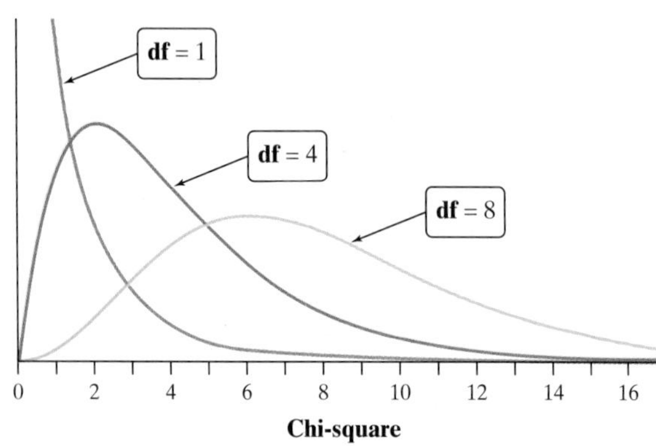

FIGURE 11.3 The density curves for three members of the chi-square family of distributions. The sampling distributions of chi-square statistics belong to this family.

Figure 11.3 shows the density curves for three members of the chi-square family of distributions. As the degrees of freedom (df) increase, the density curves become less skewed, and larger values become more probable. Here are two other interesting facts about the chi-square distributions:

- The mean of a particular chi-square distribution is equal to its degrees of freedom.
- For df > 2, the mode (peak) of the chi-square density curve is at df − 2.

When df = 8, for example, the chi-square distribution has a mean of 8 and a mode of 6.

To get *P*-values from a chi-square distribution, we can use Table C in the back of the book or technology. The following example shows how to use the table.

Return of the M&M'S

Finding the *P*-value

In the last example, we computed the chi-square statistic for Jerome's random sample of 60 M&M'S Milk Chocolate Candies: $\chi^2 = 10.180$. Now let's find the *P*-value. Because all the expected counts are at least 5, the χ^2 statistic will follow a chi-square distribution reasonably well when H_0 is true. There are 6 color categories for M&M'S Milk Chocolate Candies, so df = 6 − 1 = 5.

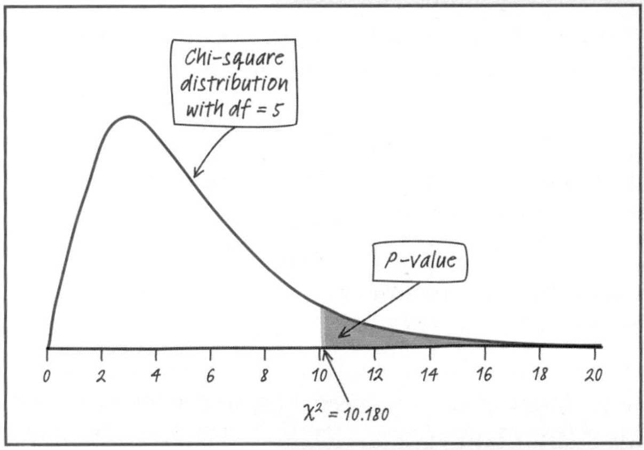

FIGURE 11.4 The *P*-value for a chi-square goodness-of-fit test using Jerome's M&M'S data.

The *P*-value is the probability of getting a value of χ^2 as large as or larger than 10.180 when H_0 is true. Figure 11.4 shows this probability as an area under the chi-square density curve with 5 degrees of freedom.

To find the *P*-value using Table C, look in the df = 5 row. The value $\chi^2 = 10.180$ falls between the critical values 9.24 and 11.07. The corresponding areas in the right tail of the chi-square distribution with 5 degrees of freedom are 0.10 and 0.05. (See the excerpt from Table C above right.) So the *P*-value for a test based on Jerome's data is between 0.05 and 0.10.

	P		
df	**.15**	**.10**	**.05**
4	6.74	7.78	9.49
5	8.12	9.24	11.07
6	9.45	10.64	12.59

Now let's look at how to find the *P*-value with your calculator.

TECHNOLOGY CORNER Finding *P*-values for chi-square tests on the calculator

To find the *P*-value in the M&M'S example with your calculator, use the `χ²cdf(` (`chi2cdf` on the TI-89) command. You'll find this command in the distributions (DISTR) menu on the TI-83/84 and in the CATALOG under FlashApps on the TI-89. We ask for the area between $\chi^2 = 10.180$ and a very large number (we'll use 1000) under the chi-square density curve with 5 degrees of freedom. The command that does this is `χ²cdf(10.180,1000,5)`. As the calculator screen shots show, this method gives a more precise *P*-value than Table C.

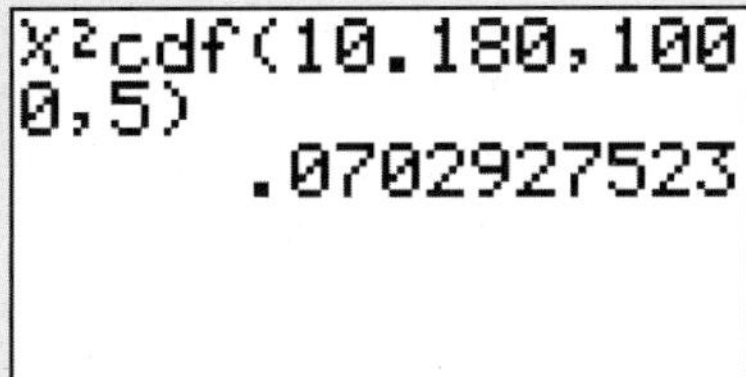

TI-Nspire instructions in Appendix B

Table C gives us an interval in which the *P*-value falls. The calculator's χ^2`cdf` (`chi2cdf` on the TI-89) command gives a result that is consistent with Table C but more precise. For that reason, we recommend using your calculator to compute *P*-values from a chi-square distribution.

Based on Jerome's sample, what conclusion can we draw about H_0: the company's stated color distribution for M&M'S Milk Chocolate Candies is correct? Since our *P*-value of 0.07 is greater than $\alpha = 0.05$, we fail to reject H_0. We don't have sufficient evidence to conclude that the company's claimed color distribution is incorrect.

THINK ABOUT IT

Why might we reject H_0 too often when working with M&M'S? Suppose we give a 1.69-ounce bag of M&M'S Milk Chocolate Candies to every member of a 30-student class and ask each student to perform a chi-square goodness-of-fit test at the $\alpha = 0.05$ level. If the color distribution claimed by Mars, Inc., is true, we'd expect about 5% of the students to reject H_0 even though H_0 is true. That's because the $\alpha = 0.05$ significance level gives the probability of making a Type I error.

When one teacher did the "Candy Man Can" Activity (page 676) with her class of 30 students, 6 of them rejected the null hypothesis. How could this happen? Maybe the color distribution claimed by the company is wrong. Let's give Mars, Inc., the benefit of the doubt for the time being. The students could have made a calculation error of some kind. But that wasn't the case. Maybe the class got very unlucky. We can use a binomial distribution with $n = 30$ and $p = 0.05$ to find the probability of making 6 or more Type I errors just by chance. It's 0.0033. That doesn't seem like a plausible explanation either.

It turns out that the problem is with one of the inference conditions. We treated each bag as a random sample of M&M'S Milk Chocolate Candies from the larger population. According to the customer service department at Mars, Inc., that's not such a good idea. Here's how they responded to the many teachers and students who wrote, called, and emailed concerning the claimed color distribution: "While we mix the colors as thoroughly as possible, the (stated) ratios may vary somewhat, especially in the smaller bags. This is because we combine the various colors in large quantities for the last production stage (printing). The bags are then filled on high-speed packaging machines by weight, not by count."

The moral of the story is simple: conditions matter!

CHECK YOUR UNDERSTANDING

Let's continue our analysis of Joey's sample of M&M'S Peanut Chocolate Candies from the previous Check Your Understanding (page 681).

1. Confirm that the expected counts are large enough to use a chi-square distribution. Which distribution (specify the degrees of freedom) should we use?
2. Sketch a graph like Figure 11.4 that shows the *P*-value.
3. Use Table C to find the *P*-value. Then use your calculator's χ^2`cdf` command.
4. What conclusion would you draw about the company's claimed color distribution for M&M'S Peanut Chocolate Candies? Justify your answer.

Carrying Out a Test

Like our test for a population proportion, the chi-square goodness-of-fit test uses some approximations that become more accurate as we take more observations. Our rule of thumb is that all expected counts must be at least 5. This **Large Sample Size condition** takes the place of the Normal condition for z and t procedures. To use the chi-square goodness-of-fit test, we must also check that the Random and Independent conditions are met.

Large Sample Size condition

The Chi-Square Goodness-of-Fit Test

Suppose the Random, Large Sample Size, and Independent conditions are met. To determine whether a categorical variable has a specified distribution, expressed as the proportion of individuals falling into each possible category, perform a test of

H_0: The specified distribution of the categorical variable is correct.

H_a: The specified distribution of the categorical variable is not correct.

We can also write these hypotheses symbolically using p_i to represent the proportion of individuals that fall in category i:

$$H_0: p_1 = \_\_\_, p_2 = \_\_\_, \ldots, p_k = \_\_\_.$$

H_a: At least one of the p_i's is incorrect.

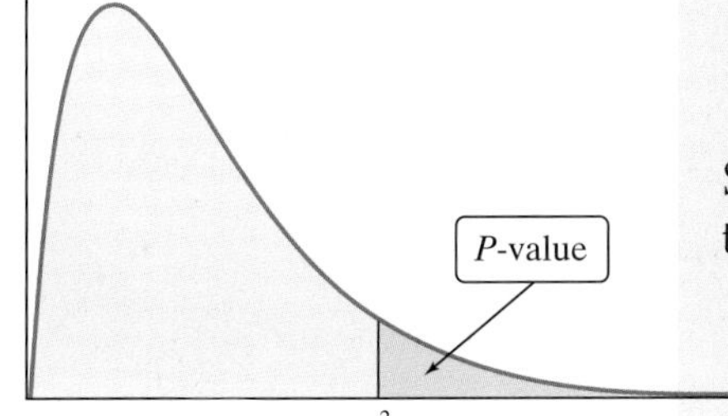

Start by finding the expected count for each category assuming that H_0 is true. Then calculate the chi-square statistic

$$\chi^2 = \sum \frac{(\text{Observed} - \text{Expected})^2}{\text{Expected}}$$

where the sum is over the k different categories. The P-value is the area to the right of χ^2 under the density curve of the chi-square distribution with $k - 1$ degrees of freedom.

Conditions: Use this test when

- **Random** The data come from a random sample or a randomized experiment.
- **Large Sample Size** All expected counts are at least 5.
- **Independent** Individual observations are independent. When sampling without replacement, check that the population is at least 10 times as large as the sample (the 10% condition).

Before we start using the chi-square goodness-of-fit test, we have two important cautions to offer.

1. The chi-square test statistic compares observed and expected *counts*. Don't try to perform calculations with the observed and expected *proportions* in each category.
2. When checking the Large Sample Size condition, be sure to examine the *expected* counts, not the observed counts.

The next two examples show the chi-square goodness-of-fit test in action. As always, we follow the four-step process when performing inference.

When Were You Born?

A test for equal proportions

STEP 4

Are births evenly distributed across the days of the week? The one-way table below shows the distribution of births across the days of the week in a random sample of 140 births from local records in a large city:

Day:	Sun.	Mon.	Tue.	Wed.	Thu.	Fri.	Sat.
Births:	13	23	24	20	27	18	15

Do these data give significant evidence that local births are not equally likely on all days of the week?

STATE: We want to perform a test of

H_0: Birth days in this local area are evenly distributed across the days of the week.

H_a: Birth days in this local area are not evenly distributed across the days of the week.

The null hypothesis says that the proportions of births are the same on all days. In that case, all 7 proportions must be 1/7. So we could also write the hypotheses as

$$H_0: p_{Sun} = p_{Mon} = p_{Tues} = \cdots = p_{Sat} = 1/7.$$

H_a: At least one of the proportions is not 1/7.

No significance level was specified, so we'll use $\alpha = 0.05$.

PLAN: If the conditions are met, we should conduct a chi-square goodness-of-fit test.

- *Random* The data came from a random sample of local births.
- *Large Sample Size* We check that all expected counts are at least 5. Assuming H_0 is true, we would expect one-seventh of the births to occur on each day of the week. For the sample of 140 births, the expected count for all 7 days would be $1/7(140) = 20$ births.
- *Independent* Individual births in the random sample should occur independently (assuming no twins). Because we are sampling without replacement, there need to be at least $10(140) = 1400$ births in the local area. This should be the case in a large city.

DO: We begin by calculating the chi-square statistic.

- *Test statistic*

$$\chi^2 = \sum \frac{(\text{Observed} - \text{Expected})^2}{\text{Expected}}$$

$$= \frac{(13-20)^2}{20} + \frac{(23-20)^2}{20} + \frac{(24-20)^2}{20} + \frac{(20-20)^2}{20} + \frac{(27-20)^2}{20} + \frac{(18-20)^2}{20} + \frac{(15-20)^2}{20}$$

$$= 2.45 + 0.45 + 0.80 + 0.00 + 2.45 + 0.20 + 1.25 = 7.60$$

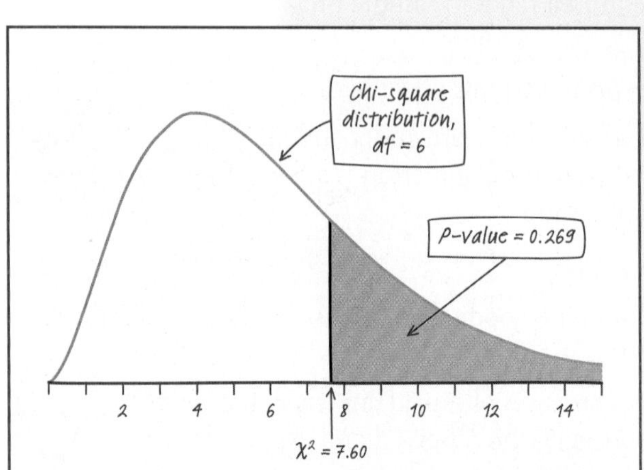

FIGURE 11.5 The P-value for the chi-square goodness-of-fit test with $\chi^2 = 7.60$ and df $= 6$.

- *P-value* Figure 11.5 displays the P-value for this test as an area under the chi-square distribution with $7 - 1 = 6$ degrees of freedom.

	p	
df	.25	.20
5	6.63	7.29
6	7.84	8.56
7	9.04	9.80

- *Using Table C:* As the excerpt at right shows, $\chi^2 = 7.60$ is less than the smallest entry in the df = 6 row, which corresponds to tail area 0.25. The *P*-value is therefore greater than 0.25.
- *Using technology:* We can find the exact *P*-value with a calculator: χ^2`cdf(7.60,1000,6)` = 0.269.

CONCLUDE: Because the *P*-value, 0.269, is greater than $\alpha = 0.05$, we fail to reject H_0. These 140 births don't provide enough evidence to say that all local births in this area are not evenly distributed across the days of the week.

Failing to reject H_0 does not mean that the null hypothesis is correct! That is, we can't conclude that birth days in this local area *are* evenly distributed across the days of the week. In fact, a careful inspection of the entire population of local records reveals that this isn't the case. Fewer babies are born on Saturday and Sunday than on other days. But the data in the example did not provide enough evidence to reject the null hypothesis. As a result, we made a Type II error—failing to reject a false H_0. With such a small number of births, the test did not have enough power to detect the difference between the hypothesized and the actual distribution of birth days.

You can use your calculator to carry out the "Do" step for a chi-square goodness-of-fit test. Remember that this comes with potential benefits and risks on the AP exam.

TECHNOLOGY CORNER Chi-square goodness-of-fit test on the calculator

You can use the TI-83/84 or TI-89 to perform the calculations for a chi-square goodness-of-fit test. We'll use Jerome's M&M'S Milk Chocolate Candies data from earlier in the chapter to illustrate the steps.

1. Enter the observed counts and expected counts in two separate lists.

   - Clear L1/list1 and L2/list2.
   - Enter the observed counts in L1/list1. Calculate the expected counts separately and enter them in L2/list2.

Color	Observed	Expected
Blue	9	14.40
Orange	8	12.00
Green	12	9.60
Yellow	15	8.40
Red	10	7.80
Brown	6	7.80

2. Perform a chi-square goodness-of-fit test.

Note: Some older TI-83s and TI-84s don't have this test. TI-84 users can get this functionality by upgrading their operating systems.

TI-83/84: Press [STAT], arrow over to `TESTS` and choose `D:`χ^2`GOF-Test...`.

TI-89: In the Stats/List Editor APP, press [2nd] [F1] ([F6]) and choose `7:Chi2GOF...`.

Enter the inputs shown on the next page. If you choose `Calculate`, you'll get a screen with the test statistic, *P*-value, and df. If you choose the `Draw` option, you'll get a picture of the appropriate chi-square distribution with the test statistic marked and shaded area corresponding to the *P*-value.

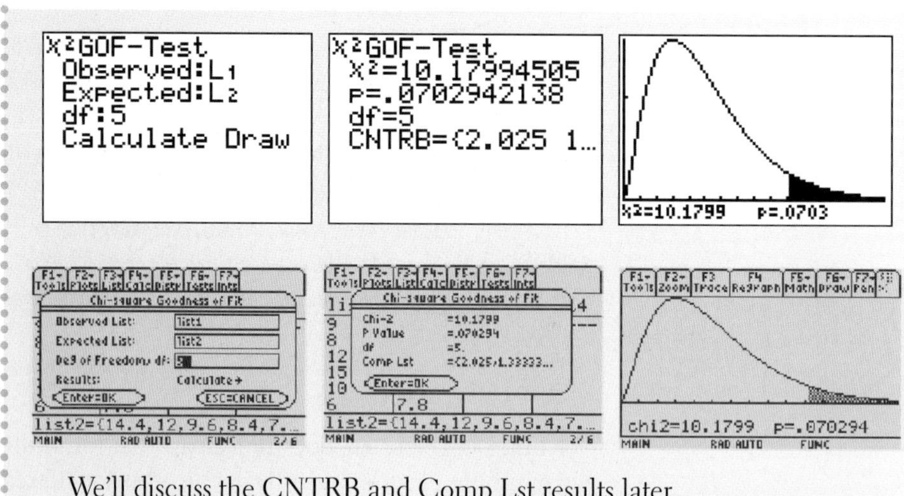

We'll discuss the CNTRB and Comp Lst results later.

TI-Nspire instructions in Appendix B

One of the most common applications of the chi-square goodness-of-fit test is in the field of genetics. Researchers want to investigate the traits inherited by offspring that result from mating (also called *crossing*) parents with known genetic makeups. Probability rules for the inheritance of dominant and recessive genes help researchers predict the proportion of offspring that will fall in each possible genetic category. Then, the researchers mate the parents and classify the resulting offspring. The chi-square goodness-of-fit test helps the scientists assess the validity of their hypothesized distribution of proportions.

Inherited Traits

Testing a genetic model

STEP 4

Biologists wish to cross pairs of tobacco plants having genetic makeup Gg, indicating that each plant has one dominant gene (G) and one recessive gene (g) for color. Each offspring plant will receive one gene for color from each parent. The following table, often called a *Punnett square*, shows the possible combinations of genes received by the offspring:

		Parent 2 passes on:	
		G	**g**
Parent 1 passes on:	**G**	GG	Gg
	g	Gg	gg

The Punnett square suggests that the expected ratio of green (GG) to yellow-green (Gg) to albino (gg) tobacco plants should be 1:2:1. In other words, the biologists predict that 25% of the offspring will be green, 50% will be yellow-green, and 25% will be albino.

To test their hypothesis about the distribution of offspring, the biologists mate 84 randomly selected pairs of yellow-green parent plants. Of 84 offspring, 23 plants were green, 50 were yellow-green, and 11 were albino. Do these data differ significantly from what the biologists have predicted? Carry out an appropriate test at the $\alpha = 0.05$ level to help answer this question.

STATE: We want to perform a test of

H_0: The biologists' predicted color distribution for tobbaco plant offspring is correct. That is, $p_{green} = 0.25$, $p_{yellow\text{-}green} = 0.5$, $p_{albino} = 0.25$

H_a: The biologists' predicted color distribution isn't correct. That is, at least one of the stated proportions is incorrect.

at the $\alpha = 0.05$ significance level.

PLAN: If the conditions are met, we should conduct a chi-square goodness-of-fit test.

- *Random* The data were produced using a random sample of tobacco plants.
- *Large Sample Size* We check that all expected counts are at least 5. Assuming H_0 is true, the expected counts for the different colors of offspring are

green: $(0.25)(84) = 21$; yellow-green: $(0.50)(84) = 42$; albino: $(0.25)(84) = 21$

The complete table of observed and expected counts is shown below.

- *Independent* Individual offspring inherit their traits independently from one another. Since we are sampling without replacement, there would need to be at least $10(84) = 840$ tobacco plants in the population. This seems reasonable to believe.

Offspring color	Observed	Expected
Green	23	21
Yellow-green	50	42
Albino	11	21

DO: We'll use the calculator's chi-square goodness-of-fit test to complete this step.

- *Test statistic*

$$\chi^2 = \sum \frac{(\text{Observed} - \text{Expected})^2}{\text{Expected}} = \frac{(23-21)^2}{21} + \frac{(50-42)^2}{42} + \frac{(11-21)^2}{21} = 6.476$$

- *P-value* Note that df = number of categories $- 1 = 3 - 1 = 2$. Using df $= 2$, the *P*-value from the calculator is 0.0392.

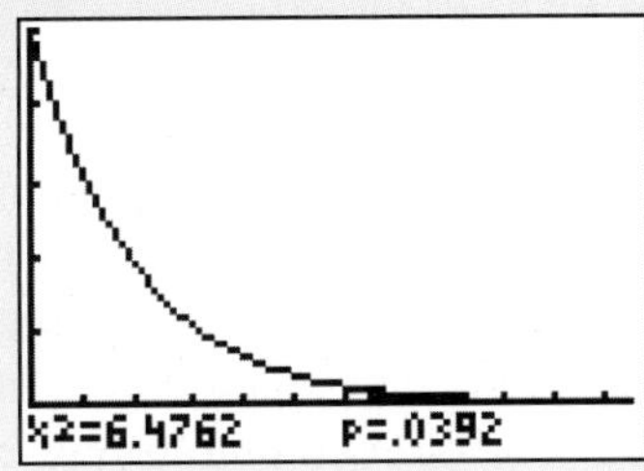

CONCLUDE: Our *P*-value, 0.0392, is less than $\alpha = 0.05$, so we will reject H_0. We have convincing evidence that the biologists' hypothesized distribution for the color of tobacco plant offspring is incorrect.

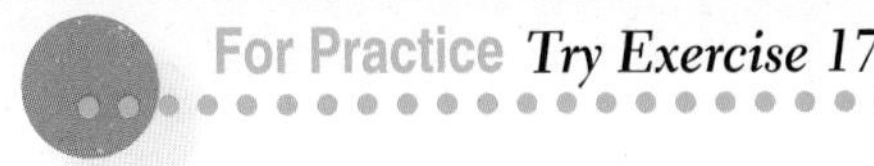
For Practice *Try Exercise 17*

Remember, there are only two possibilities for our significance test. Either our conclusion is correct and the genetic theory doesn't hold in this case, or we made a Type I error by rejecting H_0 when it is actually true.

CHECK YOUR UNDERSTANDING

Biologists wish to mate pairs of fruit flies having genetic makeup RrCc, indicating that each has one dominant gene (R) and one recessive gene (r) for eye color, along with one dominant (C) and one recessive (c) gene for wing type. Each offspring will receive one gene for each of the two traits from each parent. The following Punnett square shows the possible combinations of genes received by the offspring:

	Parent 2 passes on:			
Parent 1 passes on:	**RC**	**Rc**	**rC**	**rc**
RC	RRCC (x)	RRCc (x)	RrCC (x)	RrCc (x)
Rc	RRCc (x)	RRcc (y)	RrCc (x)	Rrcc (y)
rC	RrCC (x)	RrCc (x)	rrCC (z)	rrCc (z)
rc	RrCc (x)	Rrcc (y)	rrCc (z)	rrcc (w)

Any offspring receiving an R gene will have red eyes, and any offspring receiving a C gene will have straight wings. So based on this Punnett square, the biologists predict a ratio of 9 red-eyed, straight-winged (x):3 red-eyed, curly-winged (y):3 white-eyed, straight-winged (z):1 white-eyed, curly-winged (w) offspring.

To test their hypothesis about the distribution of offspring, the biologists mate a random sample of pairs of fruit flies. Of 200 offspring, 99 had red eyes and straight wings, 42 had red eyes and curly wings, 49 had white eyes and straight wings, and 10 had white eyes and curly wings. Do these data differ significantly from what the biologists have predicted? Carry out a test at the $\alpha = 0.01$ significance level.

Follow-up Analysis

In the chi-square goodness-of-fit test, we test the null hypothesis that a categorical variable has a specified distribution. If the sample data lead to a statistically significant result, we can conclude that our variable has a distribution different from the specified one. When this happens, start by examining which categories of the variable show large deviations between the observed and expected counts. Then look at the individual terms $\frac{(\text{Observed} - \text{Expected})^2}{\text{Expected}}$ that are added together to produce the test statistic χ^2. These **components** show which terms contribute most to the chi-square statistic.

Components

Let's return to the tobacco plant example. The table of observed and expected counts for the 84 offspring is repeated below. Looking at these counts, we see that there were many more yellow-green tobacco offspring than expected and many fewer albino plants than expected.

Offspring color	Observed	Expected
Green	23	21
Yellow-green	50	42
Albino	11	21

The chi-square statistic is

$$\chi^2 = \frac{(23-21)^2}{21} + \frac{(50-42)^2}{42} + \frac{(11-21)^2}{21}$$
$$= 0.190 + 1.524 + 4.762 = 6.476$$

We see that the component for the albino offspring, 4.762, made the largest contribution to the chi-square statistic.

Note: When we ran the chi-square goodness-of-fit test on the calculator, a list of these individual components was stored. On the TI-83/84, the list is called CNTRB (for contribution). On the TI-89, it's called Comp Lst (component list).

SECTION 11.1 Summary

- A **one-way table** is often used to display the distribution of a categorical variable for a sample of individuals.
- The **chi-square goodness-of-fit test** tests the null hypothesis that a categorical variable has a specified distribution.
- This test compares the **observed count** in each category with the counts that would be expected if H_0 were true. The **expected count** for any category is found by multiplying the specified proportion of the population distribution in that category by the sample size.
- The **chi-square statistic** is

$$\chi^2 = \sum \frac{(\text{Observed} - \text{Expected})^2}{\text{Expected}}$$

  where the sum is over all possible categories.
- The test compares the value of the statistic χ^2 with critical values from the **chi-square distribution** with degrees of freedom df = number of categories − 1. Large values of χ^2 are evidence against H_0, so the *P*-value is the area under the chi-square density curve to the right of χ^2.
- The chi-square distribution is an approximation to the sampling distribution of the statistic χ^2. You can safely use this approximation when all expected cell counts are at least 5 (**Large Sample Size condition**).
- Be sure to check that the Random, Large Sample Size, and Independent conditions are met before performing a chi-square goodness-of-fit test.
- If the test finds a statistically significant result, do a follow-up analysis that compares the observed and expected counts and that looks for the largest **components** of the chi-square statistic.

11.1 TECHNOLOGY CORNERS

TI-Nspire instructions in Appendix B

SECTION 11.1 Exercises

pg 679

1. **Aw, nuts!** A company claims that each batch of its deluxe mixed nuts contains 52% cashews, 27% almonds, 13% macadamia nuts, and 8% brazil nuts. To test this claim, a quality control inspector takes a random sample of 150 nuts from the latest batch. The one-way table below displays the sample data.

Nut:	Cashew	Almond	Macadamia	Brazil
Count:	83	29	20	18

(a) State appropriate hypotheses for performing a test of the company's claim.

(b) Calculate the expected counts for each type of nut. Show your work.

2. **Roulette** Casinos are required to verify that their games operate as advertised. American roulette wheels have 38 slots—18 red, 18 black, and 2 green. In one casino, managers record data from a random sample of 200 spins of one of their American roulette wheels. The one-way table below displays the results.

Color:	Red	Black	Green
Count:	85	99	16

(a) State appropriate hypotheses for testing whether these data give convincing evidence that the distribution of outcomes on this wheel is not what it should be.

(b) Calculate the expected counts for each color. Show your work.

3. **Aw, nuts!** Calculate the chi-square statistic for the data in Exercise 1. Show your work.

pg 680 4. **Roulette** Calculate the chi-square statistic for the data in Exercise 2. Show your work.

5. **Aw, nuts!** Refer to Exercises 1 and 3.

(a) Confirm that the expected counts are large enough to use a chi-square distribution. Which distribution (specify the degrees of freedom) should you use?

(b) Sketch a graph like Figure 11.4 (page 683) that shows the *P*-value.

(c) Use Table C to find the *P*-value. Then use your calculator's χ^2`cdf` command.

(d) What conclusion would you draw about the company's claimed distribution for its deluxe mixed nuts? Justify your answer.

6. **Roulette** Refer to Exercises 2 and 4.

(a) Confirm that the expected counts are large enough to use a chi-square distribution. Which distribution (specify the degrees of freedom) should you use?

(b) Sketch a graph like Figure 11.4 (page 683) that shows the *P*-value.

(c) Use Table C to find the *P*-value. Then use your calculator's χ^2`cdf` command.

(d) What conclusion would you draw about whether the roulette wheel is operating correctly? Justify your answer.

7. **Birds in the trees** Researchers studied the behavior of birds that were searching for seeds and insects in an Oregon forest. In this forest, 54% of the trees were Douglas firs, 40% were ponderosa pines, and 6% were other types of trees. At a randomly selected time during the day, the researchers observed 156 red-breasted nuthatches: 70 were seen in Douglas firs, 79 in ponderosa pines, and 7 in other types of trees.[2] Do these data suggest that nuthatches prefer particular types of trees when they're searching for seeds and insects? Carry out a chi-square goodness-of-fit test to help answer this question.

8. **Seagulls by the seashore** Do seagulls show a preference for where they land? To answer this question, biologists conducted a study in an enclosed outdoor space with a piece of shore whose area was made up of 56% sand, 29% mud, and 15% rocks. The biologists chose 200 seagulls at random. Each seagull was released into the outdoor space on its own and observed until it landed somewhere on the piece of shore. In all, 128 seagulls landed on the sand, 61 landed in the mud, and 11 landed on the rocks. Carry out a chi-square goodness-of-fit test. What do you conclude?

9. **No chi-square** A school's principal wants to know if students spend about the same amount of time on homework each night of the week. She asks a random sample of 50 students to keep track of their homework time for a week. The following table

displays the average amount of time (in minutes) students reported per night:

Night:	Sunday	Monday	Tuesday	Wednesday	Thursday	Friday	Saturday
Average time:	130	108	115	104	99	37	62

Explain carefully why it would not be appropriate to perform a chi-square goodness-of-fit test using these data.

10. **No chi-square** The principal in Exercise 9 also asked the random sample of students to record whether they did all of the homework that was assigned on each of the five school days that week. Here are the data:

School day:	Monday	Tuesday	Wednesday	Thursday	Friday
No. who did homework:	34	29	32	28	19

Explain carefully why it would not be appropriate to perform a chi-square goodness-of-fit test using these data.

11. **Benford's law** Faked numbers in tax returns, invoices, or expense account claims often display patterns that aren't present in legitimate records. Some patterns are obvious and easily avoided by a clever crook. Others are more subtle. It is a striking fact that the first digits of numbers in legitimate records often follow a model known as Benford's law.[3] Call the first digit of a randomly chosen record X for short. Benford's law gives this probability model for X (note that a first digit can't be 0):

First digit (X):	1	2	3	4	5	6	7	8	9
Probability:	0.301	0.176	0.125	0.097	0.079	0.067	0.058	0.051	0.046

A forensic accountant who is familiar with Benford's law inspects a random sample of 250 invoices from a company that is accused of committing fraud. The table below displays the sample data.

First digit:	1	2	3	4	5	6	7	8	9
Count:	61	50	43	34	25	16	7	8	6

(a) Are these data inconsistent with Benford's law? Carry out an appropriate test at the $\alpha = 0.05$ level to support your answer. If you find a significant result, perform a follow-up analysis.

(b) Describe a Type I error and a Type II error in this setting, and give a possible consequence of each. Which do you think is more serious?

12. **Housing** According to the Census Bureau, the distribution by ethnic background of the New York City population in a recent year was

Hispanic: 28% Black: 24% White: 35%

Asian: 12% Others: 1%

The manager of a large housing complex in the city wonders whether the distribution by race of the complex's residents is consistent with the population distribution. To find out, she records data from a random sample of 800 residents. The table below displays the sample data.[4]

Race:	Hispanic	Black	White	Asian	Other
Count:	212	202	270	94	22

Are these data significantly different from the city's distribution by race? Carry out an appropriate test at the $\alpha = 0.05$ level to support your answer. If you find a significant result, perform a follow-up analysis.

13. **Skittles** Statistics teacher Jason Molesky contacted Mars, Inc., to ask about the color distribution for Skittles candies. Here is an excerpt from the response he received: "The original flavor blend for the SKITTLES BITE SIZE CANDIES is lemon, lime, orange, strawberry and grape. They were chosen as a result of consumer preference tests we conducted. The flavor blend is 20 percent of each flavor."

(a) State appropriate hypotheses for a significance test of the company's claim.

(b) Find the expected counts for a bag of Skittles with 60 candies.

(c) How large a χ^2 statistic would you need to get in order to have significant evidence against the company's claim at the $\alpha = 0.05$ level? At the $\alpha = 0.01$ level?

(d) Create a set of observed counts for a bag with 60 candies that gives a *P*-value between 0.01 and 0.05. Show the calculation of your chi-square statistic.

14. **Is your random number generator working?** Use your calculator's RandInt function to generate 200 digits from 0 to 9 and store them in a list.

(a) State appropriate hypotheses for a chi-square goodness-of-fit test to determine whether your calculator's random number generator gives each digit an equal chance to be generated.

(b) Carry out the test. Report your observed counts, expected counts, chi-square statistic, *P*-value, and your conclusion.

15. **What's your sign?** The University of Chicago's General Social Survey (GSS) is the nation's most important social science sample survey. For reasons known only to social scientists, the GSS regularly asks a random sample of people their astrological sign. Here are the counts of responses from a recent GSS:

pg 686

Sign:	Aries	Taurus	Gemini	Cancer	Leo	Virgo
Count:	321	360	367	374	383	402

Sign:	Libra	Scorpio	Sagittarius	Capricorn	Aquarius	Pisces
Count:	392	329	331	354	376	355

If births are spread uniformly across the year, we expect all 12 signs to be equally likely. Are these data inconsistent with that belief? Carry out an appropriate test to support your answer. If you find a significant result, perform a follow-up analysis.

16. **Munching Froot Loops** Kellogg's Froot Loops cereal comes in six fruit flavors: orange, lemon, cherry, raspberry, blueberry, and lime. Charise poured out her morning bowl of cereal and methodically counted the number of cereal pieces of each flavor. Here are her data:

Flavor:	Orange	Lemon	Cherry	Raspberry	Blueberry	Lime
Count:	28	21	16	25	14	16

Test the null hypothesis that the population of Froot Loops produced by Kellogg's contains an equal proportion of each flavor. If you find a significant result, perform a follow-up analysis.

17. **Mendel and the peas** Gregor Mendel (1822–1884), an Austrian monk, is considered the father of genetics. Mendel studied the inheritance of various traits in pea plants. One such trait is whether the pea is smooth or wrinkled. Mendel predicted a ratio of 3 smooth peas for every 1 wrinkled pea. In one experiment, he observed 423 smooth and 133 wrinkled peas. The data were produced in such a way that the Random and Independent conditions are met. Carry out a chi-square goodness-of-fit test based on Mendel's prediction. What do you conclude?

pg 688

18. **You say tomato** The paper "Linkage Studies of the Tomato" (*Transactions of the Canadian Institute*, 1931) reported the following data on phenotypes resulting from crossing tall cut-leaf tomatoes with dwarf potato-leaf tomatoes. We wish to investigate whether the following frequencies are consistent with genetic laws, which state that the phenotypes should occur in the ratio 9:3:3:1.

Phenotype:	Tall cut	Tall potato	Dwarf cut	Dwarf potato
Frequency:	926	288	293	104

The data were produced in such a way that the Random and Independent conditions are met. Carry out a chi-square goodness-of-fit test using these data. What do you conclude?

Multiple choice: Select the best answer for Exercises 19 to 22.

Exercises 19 to 21 refer to the following setting. Researchers wondered whether the trees in a longleaf pine forest in Georgia are randomly distributed. To find out, they divided the forest into four equal quadrants. Then the researchers took a random sample of 100 trees and counted the number in each quadrant. Here are their data:

Quadrant:	1	2	3	4
Count:	18	22	39	21

19. An appropriate null hypothesis to test whether the trees in the forest are randomly distributed is

(a) $H_0: \mu = 25$, where $\mu =$ the mean number of trees in each quadrant.

(b) $H_0: p = 0.25$, where $p =$ the proportion of all trees in the forest that are in Quadrant 1.

(c) $H_0: n_1 = n_2 = n_3 = n_4 = 25$, where n_i is the number of trees from the sample in Quadrant i.

(d) $H_0: p_1 = p_2 = p_3 = p_4 = 0.25$, where p_i is the actual proportion of trees in the forest that are in Quadrant i.

(e) $H_0: \hat{p}_1 = \hat{p}_2 = \hat{p}_3 = \hat{p}_4 = 0.25$, where $\hat{p}_i$ is the proportion of trees in the sample that are in Quadrant i.

20. The chi-square statistic is

(a) $\frac{(18-25)^2}{25} + \frac{(22-25)^2}{25} + \frac{(39-25)^2}{25} + \frac{(21-25)^2}{25}$

(b) $\frac{(25-18)^2}{18} + \frac{(25-22)^2}{22} + \frac{(25-39)^2}{39} + \frac{(25-21)^2}{21}$

(c) $\frac{(18-25)}{25} + \frac{(22-25)}{25} + \frac{(39-25)}{25} + \frac{(21-25)}{25}$

(d) $\frac{(18-25)^2}{100} + \frac{(22-25)^2}{100} + \frac{(39-25)^2}{100} + \frac{(21-25)^2}{100}$

(e) $\frac{(0.18-0.25)^2}{0.25} + \frac{(0.22-0.25)^2}{0.25} + \frac{(0.39-0.25)^2}{0.25} + \frac{(0.21-0.25)^2}{0.25}$

21. The *P*-value for a chi-square goodness-of-fit test is 0.0129. The correct conclusion is

(a) reject H_0 at $\alpha = 0.05$; there is strong evidence that the trees are randomly distributed.

(b) reject H_0 at $\alpha = 0.05$; there is not strong evidence that the trees are randomly distributed.

(c) reject H_0 at $\alpha = 0.05$; there is strong evidence that the trees are not randomly distributed.

(d) fail to reject H_0 at $\alpha = 0.05$; there is not strong evidence that the trees are randomly distributed.

(e) fail to reject H_0 at $\alpha = 0.05$; there is strong evidence that the trees are randomly distributed.

22. Your teacher prepares a large container full of colored beads. She claims that 1/8 of the beads are red, 1/4 are blue, and the remainder are yellow. Your class will take a simple random sample of beads from the container to test the teacher's claim. The smallest number of beads you can take so that the conditions for performing inference are met is

(a) 15. (d) 40.
(b) 16. (e) 80.
(c) 30.

Exercises 23 through 25 refer to the following setting. Do students who read more books for pleasure tend to earn higher grades in English? The boxplots below show data from a simple random sample of 79 students at a large high school. Students were classified as light readers if they read fewer than 3 books for pleasure per year. Otherwise, they were classified as heavy readers. Each student's average English grade for the previous two marking periods was converted to a GPA scale where 4.0 = A, 3.0 = B, and so on.

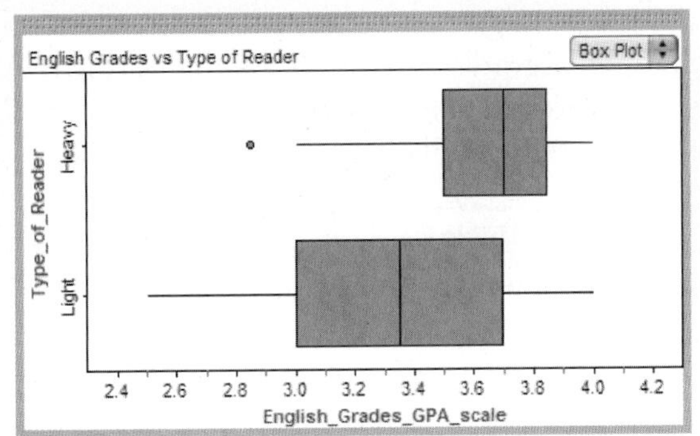

23. **Reading and grades** (1.3) Write a few sentences comparing the distributions of English grades for light and heavy readers.

24. **Reading and grades** (10.2) Summary statistics for the two groups from Minitab are provided below.

Type of reader	N	Mean	StDev	SE Mean
Heavy	47	3.640	0.324	0.047
Light	32	3.356	0.380	0.067

(a) Explain why it is acceptable to use two-sample *t* procedures in this setting.

(b) Construct and interpret a 95% confidence interval for the difference in the mean English grade for light and heavy readers.

(c) Does the interval in part (b) provide convincing evidence that reading more causes an increase in students' English grades? Justify your answer.

25. **Reading and grades** (3.2) The Fathom scatterplot below shows the number of books read and the English grade for all 79 students in the study. A least-squares regression line has been added to the graph.

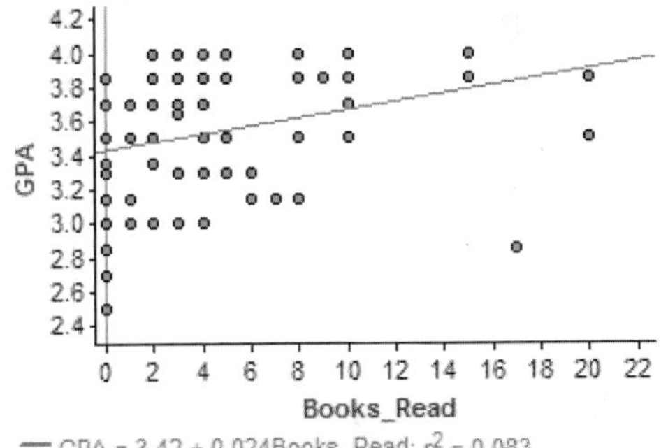

(a) Interpret the meaning of the *y* intercept in context.

(b) The student who reported reading 17 books for pleasure had an English GPA of 2.85. Find this student's residual. Show your work.

(c) How strong is the relationship between English grades and number of books read? Give appropriate evidence to support your answer.

26. **Yahtzee** (5.3, 6.3) In the game of Yahtzee, 5 six-sided dice are rolled simultaneously. To get a Yahtzee, the player must get the same number on all 5 dice.

(a) Luis says that the probability of getting a Yahtzee in one roll of the dice is $\left(\frac{1}{6}\right)^5$. Explain why Luis is wrong.

(b) Nassir decides to keep rolling all 5 dice until he gets a Yahtzee. He is surprised when he still hasn't gotten a Yahtzee after 25 rolls. Should he be? Calculate an appropriate probability to support your answer.

Inference for Relationships

In Section 11.2, you'll learn about:

- Comparing distributions of a categorical variable
- Expected counts and the chi-square statistic
- The chi-square test for homogeneity
- Follow-up analysis
- Comparing several proportions
- Relationships between two categorical variables
- The chi-square test for association/independence
- Using chi-square tests wisely

The two-sample z procedures of Chapter 10 allow us to compare the proportions of successes in two populations or for two treatments. What if we want to compare more than two samples or groups? More generally, what if we want to compare the distributions of a single categorical variable across several populations or treatments? We need a new statistical test. The new test starts by presenting the data in a two-way table.

Two-way tables have more general uses than comparing distributions of a single categorical variable. As we saw in Section 1.1, they can be used to describe relationships between any two categorical variables. In this section, we will start by developing a test to determine whether the distribution of a categorical variable is the same for each of several populations or treatments. Then we'll examine a related test to see whether there is an *association* between the row and column variables in a two-way table.

Comparing Distributions of a Categorical Variable

EXAMPLE

Does Background Music Influence What Customers Buy?

Comparing conditional distributions

Market researchers suspect that background music may affect the mood and buying behavior of customers. One study in a supermarket compared three randomly assigned treatments: no music, French accordion music, and Italian string music. Under each condition, the researchers recorded the numbers of bottles of French, Italian, and other wine purchased.[5] Here is a table that summarizes the data:

	Music			
Wine	**None**	**French**	**Italian**	**Total**
French	30	39	30	**99**
Italian	11	1	19	**31**
Other	43	35	35	**113**
Total	**84**	**75**	**84**	**243**

PROBLEM:

(a) Calculate the conditional distribution (in proportions) of the type of wine sold for each treatment.

(b) Make an appropriate graph for comparing the conditional distributions in part (a).

(c) Are the distributions of wine purchases under the three music treatments similar or different? Give appropriate evidence from parts (a) and (b) to support your answer.

SOLUTION:

(a) When no music was playing, the distribution of wine purchases was

$$\text{French: } \frac{30}{84} = 0.357 \quad \text{Italian: } \frac{11}{84} = 0.131 \quad \text{Other: } \frac{43}{84} = 0.512$$

When French accordion music was playing, the distribution of wine purchases was

$$\text{French: } \frac{39}{75} = 0.520 \quad \text{Italian: } \frac{1}{75} = 0.013 \quad \text{Other: } \frac{35}{75} = 0.467$$

When Italian string music was playing, the distribution of wine purchases was

$$\text{French: } \frac{30}{84} = 0.357 \quad \text{Italian: } \frac{19}{84} = 0.226 \quad \text{Other: } \frac{35}{84} = 0.417$$

(b) The bar graphs in Figure 11.6 compare the distributions of types of wine sold for each of the three music treatments.

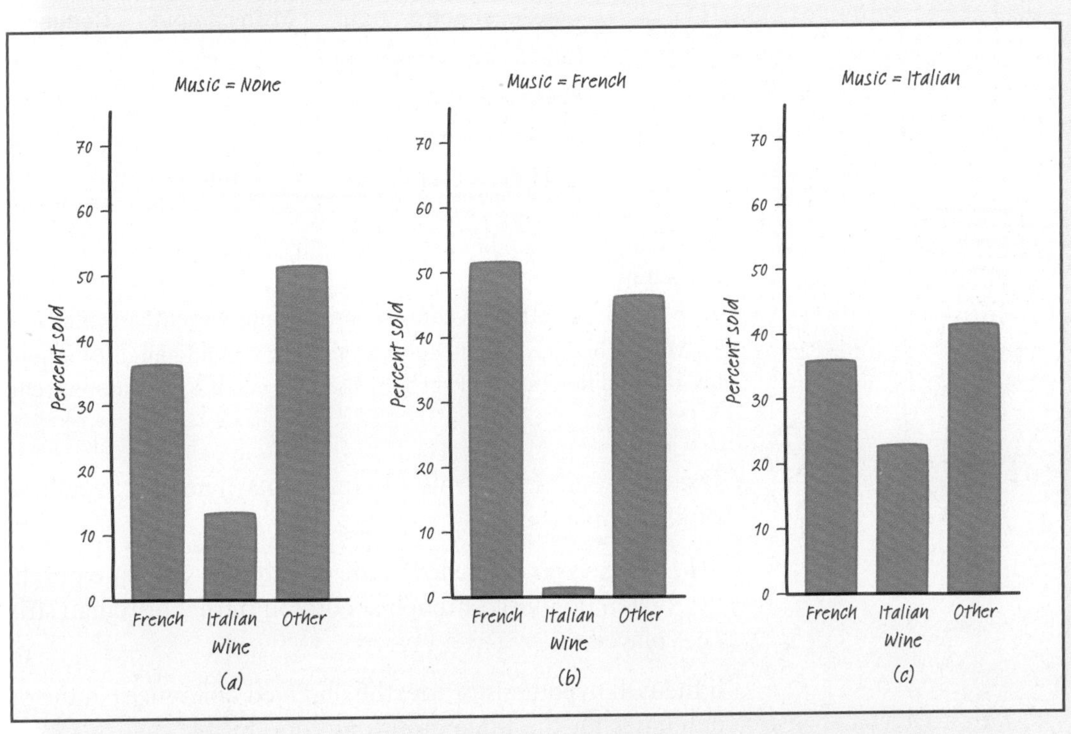

FIGURE 11.6 Bar graphs comparing the distributions of type of wine sold for different music conditions.

(c) The type of wine that customers buy seems to differ considerably across the three music treatments. Sales of Italian wine are very low (1.3%) when French music is playing but are higher when Italian music (22.6%) or no music (13.1%) is playing. French wine appears popular in this market, selling well under all music conditions but notably better when French music is playing. For all three music treatments, the percent of Other wine purchases was similar.

For Practice *Try Exercise 27*

In the example, we summarized the observed relationship between the type of music being played and the type of wine bought. The researchers expected that music would influence sales, so music type is the explanatory variable and the type of wine bought is the response variable. In general, the clearest way to describe this relationship is to compare the conditional distributions of the response variable for each value of the explanatory variable. That's why we compared the conditional distributions of wine purchases for each type of music played.

CHECK YOUR UNDERSTANDING

The Pennsylvania State University has its main campus in the town of State College and more than 20 smaller "commonwealth campuses" around the state. The Penn State Division of Student Affairs polled separate random samples of undergraduates from the main campus and commonwealth campuses about their use of online social networking. Facebook was the most popular site, with more than 80% of students having an account. Here is a comparison of Facebook use by undergraduates at the main campus and commonwealth campuses who have a Facebook account:[6]

Use Facebook	Main campus	Commonwealth
Several times a month or less	55	76
At least once a week	215	157
At least once a day	640	394
Total Facebook users	910	627

1. Calculate the conditional distribution (in proportions) of Facebook use for each campus setting.
2. Why is it important to compare proportions rather than counts in Question 1?
3. Make a bar graph that compares the two conditional distributions. What are the most important differences in Facebook use between the two campus settings?

The Problem of Multiple Comparisons The null hypothesis in the music and wine example is

H_0: There is *no difference* in the distributions of wine purchases at this store when no music, French accordion music, or Italian string music is played.

If the null hypothesis is true, the observed differences in the distributions of wine purchases for the three groups are due to the chance involved in the random assignment of treatments. The alternative hypothesis says that there *is* a difference but does not specify the nature of that difference:

H_a: There is a difference in the distributions of wine purchases at this store when no music, French accordion music, or Italian string music is played.

Any difference among the three actual distributions of wine purchases when no music, French accordion music, or Italian string music is played means that the null hypothesis is false and the alternative hypothesis is true. The alternative hypothesis is not one-sided or two-sided. We might call it "many-sided" because it allows any kind of difference.

With only the methods we already know, we might start by comparing the proportions of French wine purchases when no music and French accordion music are played. We could similarly compare other pairs of proportions, ending up with many tests and many *P*-values. This is a bad idea. The *P*-values belong to each test separately, not to the collection of all the tests together.

	Music			
Wine	**None**	**French**	**Italian**	**Total**
French	30	39	30	**99**
Italian	11	1	19	**31**
Other	43	35	35	**113**
Total	**84**	**75**	**84**	**243**

Think of the distinction between the probability that a virtual basketball player makes a free throw and the probability that she makes all her free throws in a game. *When we do many individual tests or construct many confidence intervals, the individual P-values and confidence levels don't tell us how confident we can be in all the inferences taken together.*

CAUTION

Because of this, it's cheating to pick out one large difference from the two-way table and then perform a significance test as if it were the only comparison we had in mind. For example, the proportions of French wine bought under the no music and French accordion music treatments are $30/84 = 0.357$ and $39/75 = 0.520$, respectively. A two-sample z test shows that the difference between the proportions is statistically significant ($z = 2.06, P = 0.039$) if we make just this one comparison. But we could also pick a comparison that is not significant. For example, the proportions of Italian wine purchased for the no music and Italian string music treatment are $11/84 = 0.131$ and $19/84 = 0.226$, respectively. These two proportions do not differ significantly ($z = 1.61$, $P = 0.107$). Individual comparisons can't tell us whether the three *distributions* of the categorical variable (in this case, wine bought) are significantly different.

The problem of how to do many comparisons at once with an overall measure of confidence in all our conclusions is common in statistics. This is the problem of **multiple comparisons.** Statistical methods for dealing with multiple comparisons usually have two parts:

Multiple comparisons

1. An *overall test* to see if there is good evidence of *any* differences among the parameters that we want to compare.
2. A detailed *follow-up analysis* to decide which of the parameters differ and to estimate how large the differences are.

The overall test uses the familiar chi-square statistic and distributions. But in this new setting the test will be used to compare the distribution of a categorical variable for several populations or treatments. The follow-up analysis can be quite elaborate. We will concentrate on the overall test and use data analysis to describe the nature of the differences.

Expected Counts and the Chi-Square Statistic

It would also be correct to state the null hypothesis as H_0: The distribution of a categorical variable is the same for each of several populations or treatments. We prefer the "no difference" wording because it's more consistent with the language we used in the significance tests of Chapter 10.

To perform a test of

H_0: There is no difference in the distribution of a categorical variable for several populations or treatments.

H_a: There is a difference in the distribution of a categorical variable for several populations or treatments.

we compare the observed counts in a two-way table with the counts we would expect if H_0 were true. Finding the expected counts is not that difficult, as the following example illustrates.

Does Background Music Influence What Customers Buy?

Computing expected counts

The null hypothesis in the wine and music experiment is that there's no difference in the distribution of wine purchases in the store when no music, French accordion music, or Italian string music is played. To find the expected counts, we start by assuming that H_0 is true. We can see from the two-way table that 99 of the 243 bottles of wine bought during the study were French wines.

Observed Counts

	Music			
Wine	**None**	**French**	**Italian**	**Total**
French	30	39	30	**99**
Italian	11	1	19	**31**
Other	43	35	35	**113**
Total	**84**	**75**	**84**	**243**

If the specific type of music that's playing has no effect on wine purchases, the proportion of French wine sold under each music condition should be 99/243 = 0.407. For instance, there were 84 bottles of wine bought when no music was playing. We would expect

$$\frac{99}{243} \cdot 84 = 34.22$$

of those bottles to be French wine, on average. The expected counts of French wine bought under the other two music conditions can be found in a similar way:

$$\text{French music: } \frac{99}{243} \cdot 75 = 30.56 \qquad \text{Italian music: } \frac{99}{243} \cdot 84 = 34.22$$

We repeat the process to find the expected counts for the other two types of wine. The overall proportion of Italian wine bought during the study was 31/243 = 0.128. So the expected counts of Italian wine bought under each treatment are

$$\text{No music: } \frac{31}{243} \cdot 84 = 10.72 \qquad \text{French music: } \frac{31}{243} \cdot 75 = 9.57$$

$$\text{Italian music: } \frac{31}{243} \cdot 84 = 10.72$$

The overall proportion of Other wine bought during the experiment was 113/243 = 0.465. So the expected counts of Other wine bought for each treatment are

$$\text{No music: } \frac{113}{243} \cdot 84 = 39.06 \qquad \text{French music: } \frac{113}{243} \cdot 75 = 34.88$$

$$\text{Italian music: } \frac{113}{243} \cdot 84 = 39.06$$

The following table summarizes the expected counts for all three treatments. Note that the values for no music and Italian music are the same because 84 bottles of wine were bought under each condition. We can check our work by adding the expected

Although any count of bottles of wine sold must be a whole number, an expected count need not be. The expected count gives the average number of bottles sold if H_0 is true and the random assignment process is repeated many times.

counts to obtain the row and column totals, as in the table. These should be the same as those in the table of observed counts except for small roundoff errors, such as 75.01 rather than 75 for the total number of bottles of French wine sold.

Expected Counts

	Music			
Wine	**None**	**French**	**Italian**	**Total**
French	34.22	30.56	34.22	**99**
Italian	10.72	9.57	10.72	**31**
Other	39.06	34.88	39.06	**113**
Total	**84**	**75**	**84**	**243**

Let's take a look at the two-way table from the wine and music study one more time. In the example, we found the expected count of French wine bought when no music was playing as follows:

$$\frac{99}{243} \cdot 84 = 34.22$$

Observed Counts

	Music			
Wine	**None**	**French**	**Italian**	**Total**
French	30	39	30	**99**
Italian	11	1	19	**31**
Other	43	35	35	**113**
Total	**84**	**75**	**84**	**243**

We marked the three numbers used in this calculation in the table. These values are the row total for French wine bought, the column total for wine purchased when no music was playing, and the table total of wine bought during the experiment. We can rewrite the original calculation as

$$\frac{99 \cdot 84}{243} = 34.22$$

This suggests a more general formula for the expected count in any cell of a two-way table:

$$\frac{\text{row total} \cdot \text{column total}}{\text{table total}}$$

Convince yourself that this alternate version works for the other cells in the table.

Finding Expected Counts

The expected count in any cell of a two-way table when H_0 is true is

$$\text{expected count} = \frac{\text{row total} \cdot \text{column total}}{\text{table total}}$$

All the expected counts in the music and wine study are at least 5. This satisfies the Large Sample Size condition. The Random condition is met because the treatments were assigned at random. What about the Independent condition?

We're comparing three independent groups in a randomized experiment. But are individual observations (each wine bottle sold) independent? If a customer buys several bottles of wine at the same time, knowing that one bottle is French wine might give us additional information about the other bottles bought by this customer. In that case, the Independent condition would be violated. But if each customer buys only one bottle of wine, this condition is probably met. We can't be sure, so we'll proceed to inference with caution.

Just as we did with the chi-square goodness-of-fit test, we compare the observed counts with the expected counts using the statistic

$$\chi^2 = \sum \frac{(\text{Observed} - \text{Expected})^2}{\text{Expected}}$$

This time, the sum is over all cells (not including the totals!) in the two-way table.

EXAMPLE Does Background Music Influence What Customers Buy?

The chi-square statistic

PROBLEM: The tables below show the observed and expected counts for the wine and music experiment. Calculate the chi-square statistic. Show your work.

Observed Counts

	Music			
Wine	**None**	**French**	**Italian**	**Total**
French	30	39	30	**99**
Italian	11	1	19	**31**
Other	43	35	35	**113**
Total	**84**	**75**	**84**	**243**

Expected Counts

	Music			
Wine	**None**	**French**	**Italian**	**Total**
French	34.22	30.56	34.22	**99**
Italian	10.72	9.57	10.72	**31**
Other	39.06	34.88	39.06	**113**
Total	**84**	**75**	**84**	**243**

SOLUTION: For French wine with no music, the observed count is 30 bottles and the expected count is 34.22. The contribution to the χ^2 statistic for this cell is

$$\frac{(\text{Observed} - \text{Expected})^2}{\text{Expected}} = \frac{(30-34.22)^2}{34.22} = 0.52$$

The χ^2 statistic is the sum of nine such terms:

$$\chi^2 = \sum \frac{(\text{Observed} - \text{Expected})^2}{\text{Expected}} = \frac{(30-34.22)^2}{34.22} + \frac{(39-30.56)^2}{30.56} + \cdots + \frac{(35-39.06)^2}{39.06}$$
$$= 0.52 + 2.33 + \cdots + 0.42 = 18.28$$

For Practice Try Exercise 29

AP EXAM TIP In the "Do" step, you aren't required to show every term in the chi-square statistic. Writing the first few terms of the sum and the last term, separated by ellipsis, is considered as "showing work." We suggest that you do this and then let your calculator tackle the computations.

As in the goodness-of-fit test, you should think of the chi-square statistic χ^2 as a measure of how much the observed counts deviate from the expected counts. Once again, large values of χ^2 are evidence against H_0. The *P*-value measures the strength of this evidence. To find the *P*-value corresponding to $\chi^2 = 18.28$ in the example, we need to know which chi-square distribution to use. We'll get to that shortly.

CHECK YOUR UNDERSTANDING

In the previous Check Your Understanding (page 698), we presented data on the use of Facebook by two randomly selected groups of Penn State students. Here are the data once again.

Use Facebook	Main campus	Commonwealth
Several times a month or less	55	76
At least once a week	215	157
At least once a day	640	394
Total Facebook users	910	627

Do these data provide convincing evidence of a difference in the distributions of Facebook use among students in the two campus settings?

1. State appropriate null and alternative hypotheses for a significance test to help answer this question.
2. Calculate the expected counts. Show your work.
3. Calculate the chi-square statistic. Show your work.

The Chi-Square Test for Homogeneity

In Section 11.1, we used a chi-square goodness-of-fit test to test a hypothesized model for the distribution of a categorical variable. Our *P*-values came from a chi-square distribution with df = the number of categories − 1. When the Random, Large Sample Size, and Independent conditions are met, the χ^2 statistic calculated from a two-way table can be used to perform a test of H_0: There is no difference in the distribution of a categorical variable for several populations or treatments. *P*-values for this test come from a chi-square distribution with df = (number of rows − 1)(number of columns − 1). This new procedure is known as a **chi-square test for homogeneity.**

This test is also known as a chi-square test for homogeneity of proportions. We prefer the simpler name.

Chi-Square Test for Homogeneity

Suppose the Random, Large Sample Size, and Independent conditions are met. You can use the **chi-square test for homogeneity** to test

H_0: There is no difference in the distribution of a categorical variable for several populations or treatments.

H_a: There is a difference in the distribution of a categorical variable for several populations or treatments.

Start by finding the expected counts. Then calculate the chi-square statistic

$$\chi^2 = \sum \frac{(\text{Observed} - \text{Expected})^2}{\text{Expected}}$$

where the sum is over all cells (not including totals) in the two-way table. If H_0 is true, the χ^2 statistic has approximately a chi-square distribution with degrees of freedom = (number of rows − 1)(number of columns − 1). The *P*-value is the area to the right of χ^2 under the corresponding chi-square density curve.

P-value

χ^2

Conditions: Use this test when

- **Random** The data come from separate random samples from each population of interest or from the groups in a randomized experiment.
- **Large Sample Size** All expected counts are at least 5.
- **Independent** Both the samples or groups themselves and the individual observations in each sample or group are independent. When sampling without replacement, check that the individual populations are at least 10 times as large as the corresponding samples (the 10% condition).

Now we're ready to complete the significance test for the music and wine study.

Does Background Music Influence What Customers Buy?

P-value and conclusion

Earlier, we started a significance test of

H_0: There is no difference in the distributions of wine purchases at this store when no music, French accordion music, or Italian string music is played.

H_a: There is a difference in the distributions of wine purchases at this store when no music, French accordion music, or Italian string music is played.

Observed Counts

	Music			
Wine	**None**	**French**	**Italian**	**Total**
French	30	39	30	**99**
Italian	11	1	19	**31**
Other	43	35	35	**113**
Total	**84**	**75**	**84**	**243**

We decided to proceed with caution because, although the Random and Large Sample Size conditions are met, we aren't sure that individual observations (type of wine bought) are independent. Our calculated test statistic is $\chi^2 = 18.28$.

PROBLEM:

(a) Use Table C to find the *P*-value. Then use your calculator's χ^2`cdf` command.

(b) Interpret the *P*-value from the calculator in context.

(c) What conclusion would you draw? Justify your answer.

SOLUTION:

(a) Because the two-way table has three rows and three columns that contain the data from the study, we use a chi-square distribution with df $= (3-1)(3-1) = 4$ to find the *P*-value.

- *Table:* Look at the df = 4 row in Table C. The calculated value $\chi^2 = 18.28$ lies between the critical values 16.42 and 18.47. The corresponding *P*-value is between 0.001 and 0.0025.

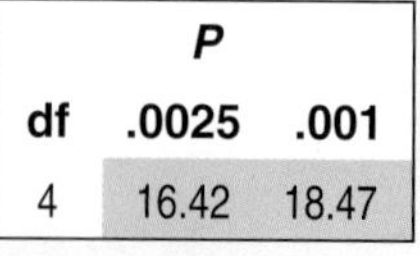

	P	
df	**.0025**	**.001**
4	16.42	18.47

- *Calculator:* The command χ^2`cdf`(18.28,1000,4) gives 0.0011.

(b) Assuming that there is no difference in the actual distributions of wine purchases in this store when no music, French accordion music, or Italian string music is played, the probability of observing a difference in the distributions of wine purchases among the three treatment groups as large as or larger than the one in this study is about 1 in 1000.

(c) The small P-value gives us convincing evidence to reject H_0 and conclude that there is a difference in the distributions of wine purchases at this store when no music, French accordion music, or Italian string music is played. Furthermore, the random assignment allows us to say that the difference is caused by the music that's played.

CHECK YOUR UNDERSTANDING

Refer to the previous Check Your Understanding (page 703).

1. Use Table C to find the P-value. Then use your calculator's χ^2cdf command.
2. Interpret the P-value from the calculator in context.
3. What conclusion would you draw? Justify your answer.

Calculating the expected counts and then the chi-square statistic by hand is a bit time-consuming. As usual, technology saves time and gets the arithmetic right.

TECHNOLOGY CORNER Chi-square tests for two-way tables on the calculator

You can use the TI-83/84 or TI-89 to perform calculations for a chi-square test for homogeneity. We'll use the data from the wine and music study to illustrate the process.

1. Enter the observed counts in matrix [A].

TI-83/84

- Press 2nd x⁻¹ (MATRIX), arrow to EDIT, and choose `1:A`.
- Enter the dimensions of the matrix: 3 × 3.

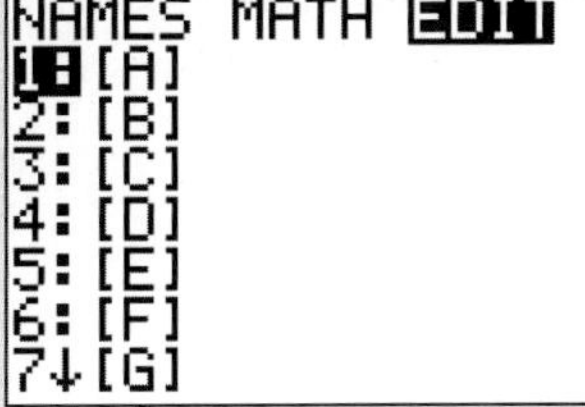

TI-89

- Press APPS, select `6:Data/Matrix Editor` and then `3:New. . . .`
- Adjust your settings to match those shown.

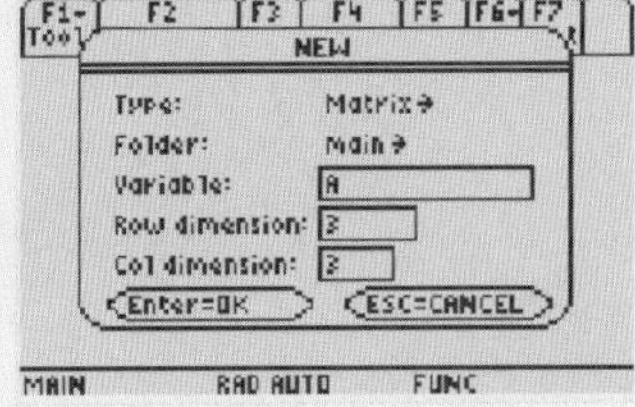

- Enter the observed counts from the two-way table in the same locations in the matrix.

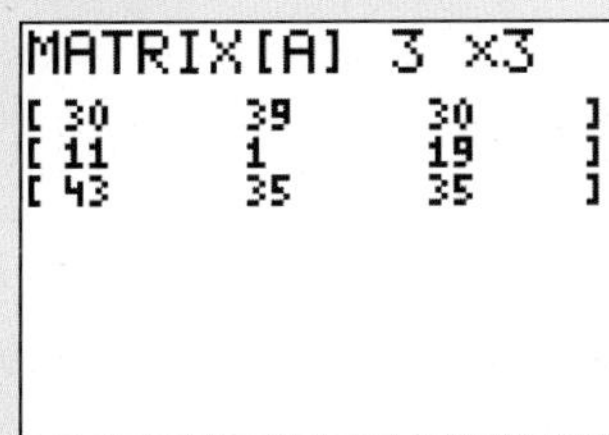

MAT 3x3	c1	c2	c3
1	30	39	30
2	11	1	19
3	43	35	35
4			

r3c3=35

2. Specify the chi-square test, the matrix where the observed counts are found, and the matrix where the expected counts will be stored.

- Press STAT, arrow to TESTS, and choose `C:χ²-Test`.
- Adjust your settings as shown.

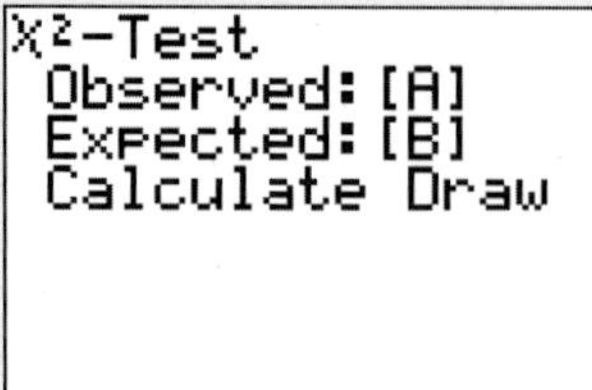

- In the Statistics/List Editor, press 2nd F1 ([F6]), and choose `8:Chi2 2-way...`.
- Adjust your settings as shown.

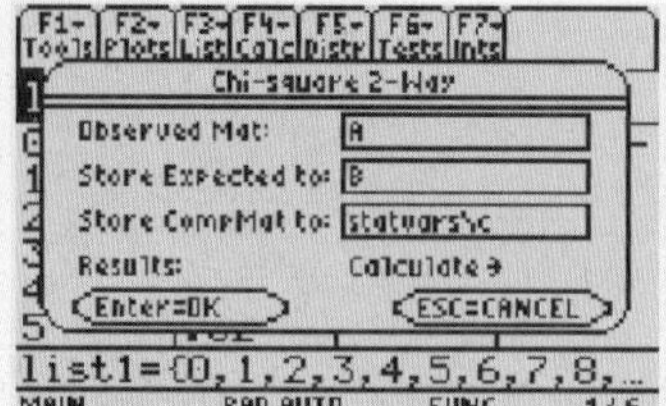

3. Choose "Calculate" or "Draw" to carry out the test. If you choose "Calculate," you should get the test statistic, *P*-value, and df shown below. If you specify "Draw," the chi-square distribution with 4 degrees of freedom will be drawn, the area in the tail will be shaded, and the *P*-value will be displayed.

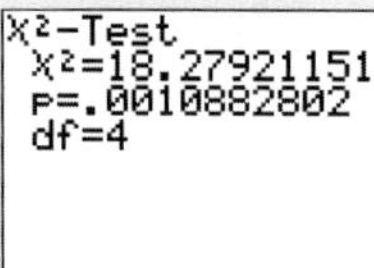

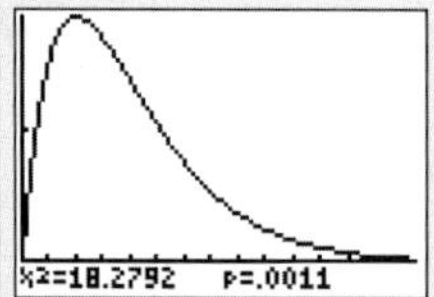

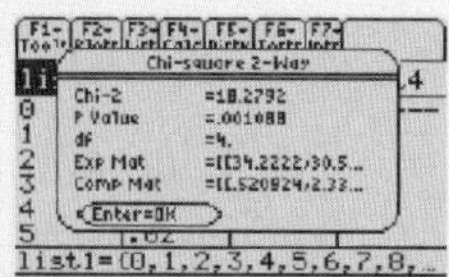

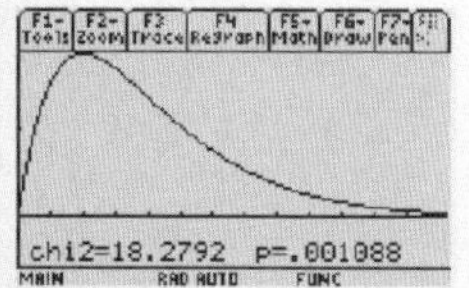

4. To see the expected counts, go to the home screen and ask for a display of the matrix [B].

- Press 2nd x^{-1} (MATRIX), and choose 2:[B].

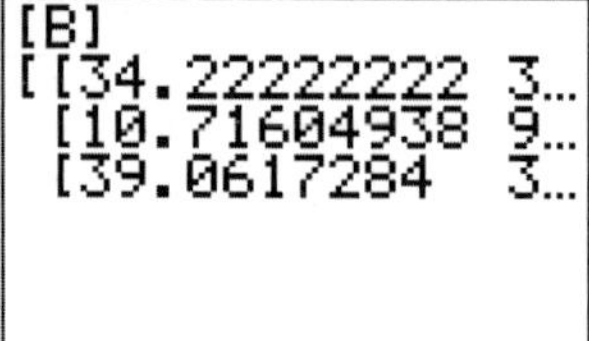

- Press 2nd − (Var-LINK) and choose B.

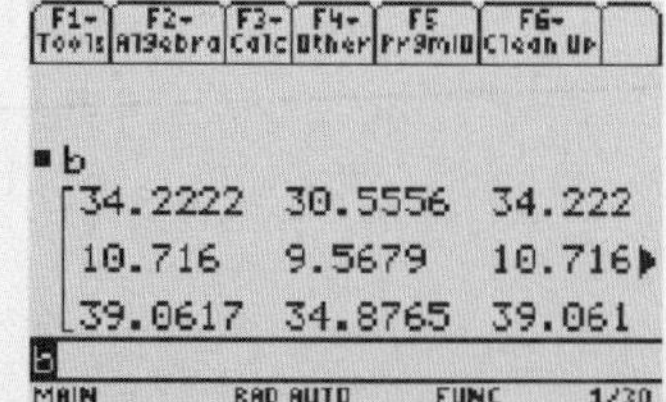

TI-Nspire instructions in Appendix B

Let's look at an example of a chi-square test for homogeneity from start to finish. As usual, we follow the four-step process when performing a significance test.

Are Cell-Only Telephone Users Different?

The chi-square test for homogeneity

Random digit dialing telephone surveys used to exclude cell phone numbers. If the opinions of people who have only cell phones differ from those of people who have landline service, the poll results may not represent the entire adult population. The Pew Research Center interviewed separate random samples of cell-only and landline telephone users who were less than 30 years old. Here's

what the Pew survey found about how these people describe their political party affiliation:[7]

	Cell-only sample	Landline sample
Democrat or lean Democratic	49	47
Refuse to lean either way	15	27
Republican or lean Republican	32	30
Total	96	104

PROBLEM:

(a) Construct an appropriate graph to compare the distributions of political party affiliation for cell-only and landline phone users.

(b) Do these data provide convincing evidence that the distribution of party affiliation differs in the cell-only and landline user populations? Carry out a significance test at the $\alpha = 0.05$ level.

SOLUTION:

(a) Because the sample sizes are different, we should compare the proportions of individuals in each political affiliation category in the two samples. The table shows the conditional distributions of political party affiliation for cell-only and landline phone users. We made a segmented bar graph to compare these two distributions.

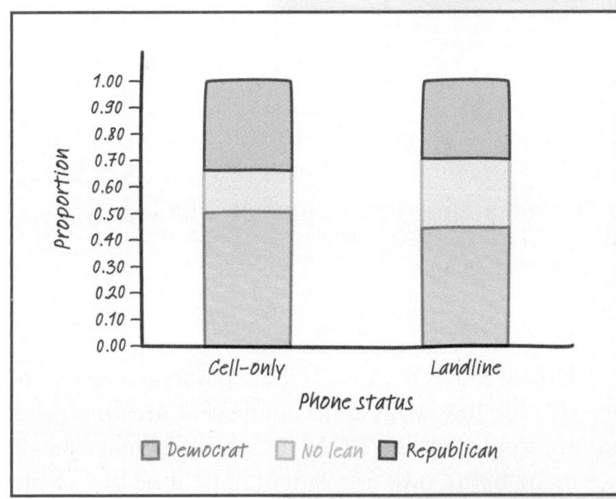

	Phone Status	
Political affiliation	Cell only	Landline
Democrat	0.51	0.45
No lean	0.16	0.26
Republican	0.33	0.29

(b) STATE: We want to perform a test of

H_0: There is no difference in the distribution of party affiliation in the cell-only and landline populations.

H_a: There is a difference in the distribution of party affiliation in the cell-only and landline populations.

at the $\alpha = 0.05$ level.

PLAN: If conditions are met, we should use a chi-square test for homogeneity.

- *Random* The data came from separate random samples of 96 cell-only and 104 landline users.
- *Large Sample Size* We followed the steps in the Technology Corner to get the expected counts. The calculator screen shot confirms that all expected counts are at least 5.

```
[B]
 [[46.08 49.92]
  [20.16 21.84]
  [29.76 32.24]]
```

- *Independent* Researchers took independent samples of cell-only and landline phone users. Sampling without replacement was used, so there need to be at least $10(96) = 960$ cell-only users under age 30 and at least $10(104) = 1040$ landline users under age 30. This is safe to assume.

DO: A chi-square test on the calculator gave

- *Test statistic*

$$\chi^2 = \sum \frac{(\text{Observed} - \text{Expected})^2}{\text{Expected}}$$

$$= \frac{(49 - 46.08)^2}{46.08} + \frac{(47 - 49.92)^2}{49.92} + \cdots + \frac{(30 - 32.24)^2}{32.24} = 3.22$$

- *P-value* Using df = (number of rows − 1)(number of columns − 1) = (3 − 1)(2 − 1) = 2, the *P*-value is 0.20.

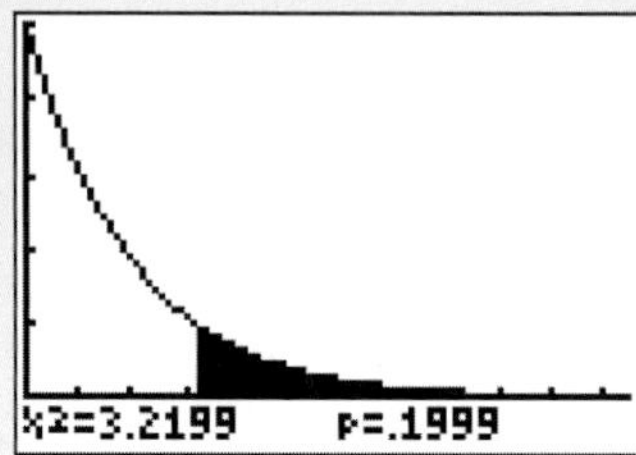

CONCLUDE: Because our *P*-value, 0.20, is greater than $\alpha = 0.05$, we fail to reject H_0. There is not enough evidence to conclude that the distribution of party affiliation differs in the cell-only and landline user populations.

For Practice *Try Exercise 33*

It is common practice to describe a two-way table by its number of rows and columns. For instance, the data in the previous example were given in a 3 × 2 table. The following Check Your Understanding involves a 5 × 2 table.

CHECK YOUR UNDERSTANDING

Canada has universal health care. The United States does not but often offers more elaborate treatment to patients with access. How do the two systems compare in treating heart attacks? Researchers compared random samples of 2600 U.S. and 400 Canadian heart attack patients. One key outcome was the patients' own assessment of their quality of life relative to what it had been before the heart attack. Here are the data for the patients who survived a year:

Quality of life	Canada	United States
Much better	75	541
Somewhat better	71	498
About the same	96	779
Somewhat worse	50	282
Much worse	19	65
Total	311	2165

1. Construct an appropriate graph to compare the distributions of opinion about quality of life among heart attack patients in Canada and the United States.
2. Is there a significant difference between the two distributions of quality-of-life ratings? Carry out an appropriate test at the $\alpha = 0.01$ level.

Follow-up Analysis

The chi-square test for homogeneity allows us to compare the distribution of a categorical variable for any number of populations or treatments. If the test allows us to reject the null hypothesis of no difference, we then want to do a follow-up analysis that examines the differences in detail. Start by examining which cells in the two-way table show large deviations between the observed and expected counts. Then look at the individual components $\frac{(\text{Observed} - \text{Expected})^2}{\text{Expected}}$ to see which terms contribute most to the chi-square statistic.

Our earlier music and wine study found significant differences among the distributions of wines purchased under each of the three music conditions. We enter the two-way table for the study into Minitab software package and request a chi-square test. The output appears in Figure 11.7. Minitab repeats the two-way table of observed counts and puts the expected count for each cell below the observed count. Finally, the software prints the 9 individual components that contribute to the χ^2 statistic.

```
Chi-Square Test: None, French, Italian

Expected counts are printed below observed counts
Chi-Square contributions are printed below expected counts

           None      French     Italian     Total
1            30          39          30        99
          34.22       30.56       34.22
          0.521       2.334       0.521

2            11           1          19        31
          10.72        9.57       10.72
          0.008       7.672       6.404

3            43          35          35       113
          39.06       34.88       39.06
          0.397       0.000       0.422

Total        84          75          84       243

Chi-Sq = 18.279, DF = 4, P-Value = 0.001
```

FIGURE 11.7 Minitab output for the two-way table in the wine and music study. The output gives the observed counts, the expected counts, and the individual components of the chi-square statistic.

Looking at the output, we see that just two of the nine components that make up the chi-square statistic contribute about 14 (almost 77%) of the total $\chi^2 = 18.28$. Comparing the observed and expected counts in these two cells, we see that sales of Italian wine are much below expectation when French music is playing and well above expectation when Italian music is playing. We are led to a specific conclusion: sales of Italian wine are strongly affected by Italian and French music. More advanced methods provide tests and confidence intervals that make this follow-up analysis more complete.

Comparing Several Proportions

Many studies involve comparing the proportion of successes for each of several populations or treatments. The two-sample z test from Chapter 10 allows us to test the null hypothesis $H_0: p_1 = p_2$, where p_1 and p_2 are the actual proportions of successes for the two populations or treatments. The chi-square test for homogeneity allows us to test $H_0: p_1 = p_2 = \cdots = p_k$. This null hypothesis says that there is no difference in the proportions of successes for the k populations or treatments. The alternative hypothesis is H_a: at least two of the p_i's are different. Many students *incorrectly state* H_a as "all the proportions are different." Think about it this way: the opposite of "all the proportions are equal" is "some of the proportions are not equal."

CAUTION

Cocaine Addiction Is Hard to Break

Comparing proportions of successes

Cocaine addicts need cocaine to feel any pleasure, so perhaps giving them an antidepressant drug will help. A three-year study with 72 chronic cocaine users compared an antidepressant drug called desipramine with lithium (a standard drug to treat cocaine addiction) and a placebo. One-third of the subjects were randomly assigned to receive each treatment. Here are the results:[8]

		Cocaine Relapse?	
Group	Treatment	Yes	No
1	Desipramine	10	14
2	Lithium	18	6
3	Placebo	20	4

PROBLEM:

(a) Make a graph to compare the rates of cocaine relapse for the three treatments. Describe what you see.

(b) Are the differences between the three groups statistically significant at the 1% level? Give appropriate evidence to support your answer.

SOLUTION:

(a) The distributions of cocaine relapse outcomes for the three treatment groups are

Desipramine:	Yes: 10/24 = 0.417	No: 14/24 = 0.583
Lithium:	Yes: 18/24 = 0.75	No: 6/24 = 0.25
Placebo:	Yes: 20/24 = 0.833	No: 4/24 = 0.167

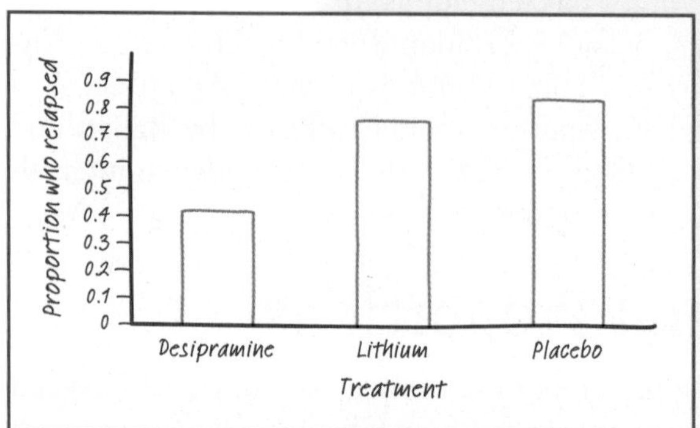

FIGURE 11.8 Bar graph comparing the proportions of cocaine addicts who relapsed under each of three treatments.

The bar graph in Figure 11.8 compares the proportion of subjects who relapsed in the three groups. A much smaller proportion of subjects relapsed in the desipramine group than in either the lithium or placebo groups.

(b) As usual, we follow the four-step process.

STATE: We want to perform a test of

H_0: $p_1 = p_2 = p_3$ There is no difference in the relapse rate for the three treatments.

H_a: at least two of the p_i's are different There is a difference in the relapse rate for the three treatments.

where p_i = the actual proportion of chronic cocaine users like the ones in this experiment who would relapse under treatment i. We'll use $\alpha = 0.01$ as instructed.

PLAN: If conditions are met, we should carry out a chi-square test for homogeneity.

- *Random* The subjects were randomly assigned to the three treatment groups.
- *Large Sample Size* We can use our familiar formula to find the expected counts from the two-way table assuming that H_0 is true:

$$\text{expected count} = \frac{\text{row total} \cdot \text{column total}}{\text{table total}}$$

The row total is 24 for all three groups, and the column totals are Yes = 48 and No = 24. So the expected count of subjects who relapse under each treatment is $\frac{24 \cdot 48}{72} = 16$. The expected count of subjects who don't relapse under each treatment is $\frac{24 \cdot 24}{72} = 8$. All the expected counts are at least 5, so this condition is met.

- *Independent* The random assignment helps create three independent groups. If the experiment is conducted properly, then knowing one subject's relapse status should give us no information about another subject's outcome. So individual observations are independent.

DO: We used the calculator's chi-square test to do the calculations.

- *Test statistic*

$$\chi^2 = \sum \frac{(\text{Observed} - \text{Expected})^2}{\text{Expected}} = 10.5$$

- *P-value* With df = (number of rows − 1)(number of columns − 1) = (3 − 1)(2 − 1) = 2, the calculator gives $P = 0.0052$.

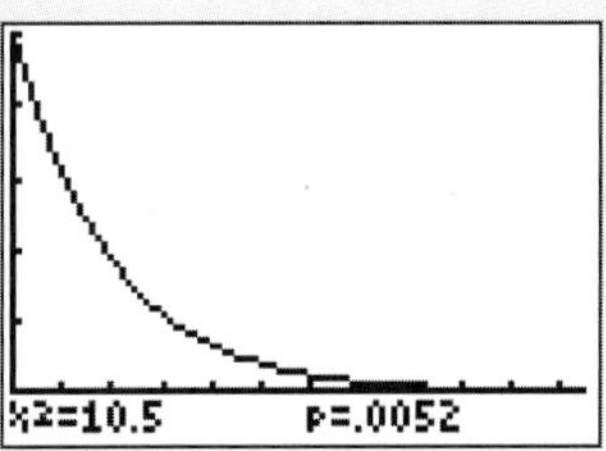

Because the variable "relapse status" has only two categories, comparing the distribution of this variable in the three treatment groups amounts to comparing the proportions of subjects in each group who relapse.

CONCLUDE: Because the *P*-value, 0.0052, is less than $\alpha = 0.01$, we have sufficient evidence to reject H_0 and conclude that the true relapse rates for the three treatments are not all the same.

THINK ABOUT IT

What if we only want to compare two proportions? We have two ways to do this: the two-sample z test from Section 10.1 and the chi-square test for homogeneity based on a 2×2 table with 1 degree of freedom. In both cases, our null hypothesis is $H_0: p_1 = p_2$. This hypothesis says that there is no difference in the proportion of successes for the two populations or treatments. The alternative hypothesis for the chi-square test is that these proportions are different, which we write as $H_a: p_1 \neq p_2$. That is, the chi-square test is equivalent to the two-sample z test with a two-sided alternative. When the conditions are met, these two tests give the same *P*-value. In fact, the chi-square statistic χ^2 is just the square of the two-sample z statistic.

We cannot use a chi-square test for a one-sided alternative hypothesis. The two-sample z procedures allow us to perform one-sided tests and to construct confidence intervals for the difference between proportions. For that reason, we recommend the Chapter 10 methods for comparing two proportions whenever you are given a choice.

The following Data Exploration gives you a chance to examine the connections between these two different methods of comparing two proportions.

DATA EXPLORATION *Secondhand stores*

Shopping at secondhand stores is becoming more popular and has even attracted the attention of business schools. A study of customers' attitudes toward secondhand stores interviewed separate random samples of shoppers at two secondhand stores of the same chain in two cities. The table shows the breakdown of respondents by gender.[9]

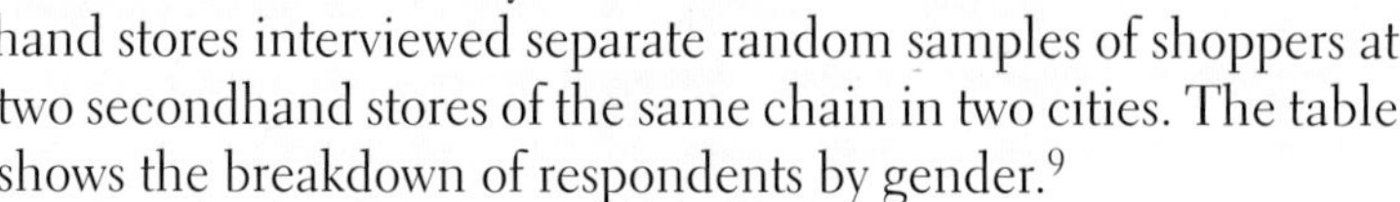

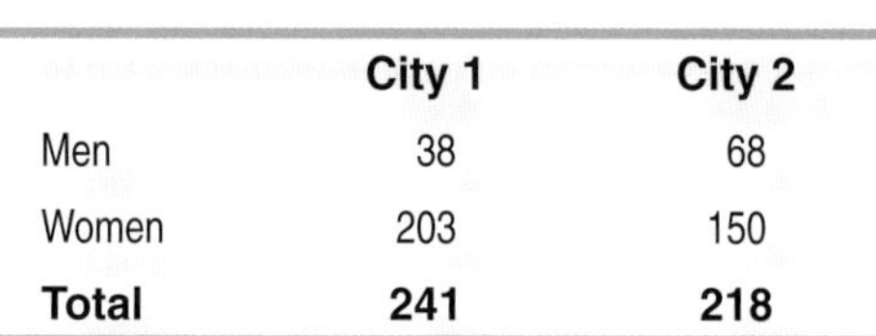

	City 1	City 2
Men	38	68
Women	203	150
Total	**241**	**218**

Minitab output from a chi-square test and from two-sample *z* procedures for comparing two proportions is shown below.

```
Chi-Square Test: City 1, City 2
Expected counts are printed below observed counts
Chi-Square contributions are printed below
expected counts
            City 1      City 2      Total
   1            38          68        106
             55.66       50.34
             5.601       6.192

   2           203         150        353
            185.34      167.66
             1.682       1.859
 Total         241         218        459

Chi-Sq = 15.334, DF = 1, P-Value = 0.000
```

```
Test and CI for Two Proportions
Sample     X     N    Sample p
1        203   241    0.842324
2        150   218    0.688073

Difference = p (1) - p (2)
Estimate for difference: 0.154250
95% CI for difference: (0.0774449,
0.231056)
Test for difference = 0 (vs not = 0):
z = 3.916 P-value = 0.000
```

1. Confirm that the chi-square statistic is actually the square of the two-sample *z* statistic.
2. Use technology to find the *P*-values accurate to 5 decimal places for both significance tests. Confirm that these values are the same.
3. Use all the information provided to prepare a brief report for the regional manager of this chain comparing the actual proportions of women who shop at these two stores.

CHECK YOUR UNDERSTANDING

Sample surveys on sensitive issues can give different results depending on how the question is asked. A University of Wisconsin study randomly divided 2400 respondents into three groups. All participants were asked if they had ever used cocaine. One group of 800 was interviewed by phone; 21% said they had used cocaine. Another 800 people were asked the question in a one-on-one personal interview; 25% said "Yes." The remaining 800 were allowed to make an anonymous written response; 28% said "Yes."[10]

1. Was this an experiment or an observational study? Justify your answer.
2. Make a two-way table of responses about cocaine use by how the survey was administered.
3. Are the differences between the three groups statistically significant? Give appropriate evidence to support your answer.

Relationships between Two Categorical Variables

Two-way tables can arise in several ways. The music and wine experiment compared three music treatments using independent groups of observations. The phone use and political party affiliation observational study compared independent random samples from the cell-only and landline user populations. In both cases, we are comparing the distributions of a categorical variable in several populations or treatments. We use the chi-square test for homogeneity to perform inference in such settings.

Another common situation that leads to a two-way table is when a *single* random sample of individuals is chosen from a *single* population and then classified according to two categorical variables. In that case, our goal is to analyze the relationship between the variables. The next example describes a study of this type.

Do Angry People Have More Heart Disease?

Relationships between two categorical variables

A study followed a random sample of 8474 people with normal blood pressure for about four years.[11] All the individuals were free of heart disease at the beginning of the study. Each person took the Spielberger Trait Anger Scale test, which measures how prone a person is to sudden anger. Researchers also recorded whether each individual developed coronary heart disease (CHD). This includes people who had heart attacks and those who needed medical treatment for heart disease. Here is a two-way table that summarizes the data:

	Low anger	Moderate anger	High anger	Total
CHD	53	110	27	190
No CHD	3057	4621	606	8284
Total	3110	4731	633	8474

PROBLEM:

(a) Calculate appropriate conditional distributions (in proportions) to describe the relationship between anger level and CHD status.

(b) Make a well-labeled graph that compares the conditional distributions in part (a).

(c) Write a few sentences describing the relationship between the two variables.

SOLUTION:

(a) Since we're interested in whether angrier people tend to get heart disease more often, we can compare the percents of people who did and did not get heart disease in each of the three anger categories:

Low anger: CHD $\frac{53}{3110} = 0.0170 = 1.70\%$ no CHD $\frac{3057}{3110} = 0.9830 = 98.30\%$

Moderate anger: CHD $\frac{110}{4731} = 0.0233 = 2.33\%$ no CHD $\frac{4621}{4731} = 0.9767 = 97.67\%$

High anger: CHD $\frac{27}{633} = 0.0427 = 4.27\%$ no CHD $\frac{606}{633} = 0.9573 = 95.73\%$

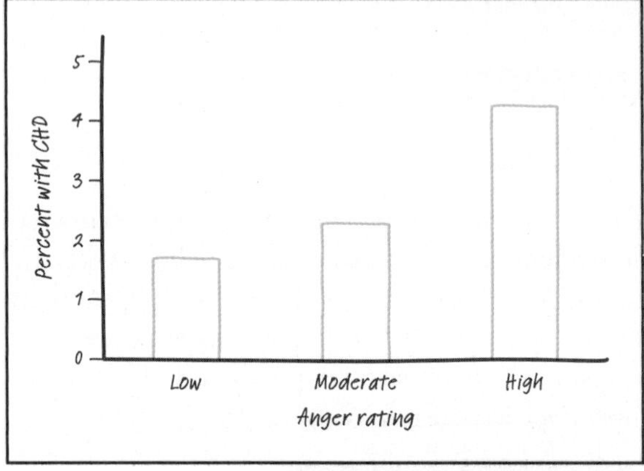

FIGURE 11.9 Bar graph comparing the percents of people in each anger category who got coronary heart disease (CHD).

(b) The bar graph in Figure 11.9 shows the percent of people in each of the three anger categories who developed CHD.

(c) There is a clear trend: as the anger score increases, so does the percent who suffer heart disease. A much higher percent of people in the high anger category developed CHD (4.27%) than in the moderate (2.33%) and low (1.70%) anger categories.

For Practice *Try Exercise 45*

Anger rating on the Spielberger scale is a categorical variable that takes three possible values: low, medium, and high. Whether or not someone gets heart disease is also a categorical variable. The two-way table in the example shows the relationship between anger rating and heart disease for a random sample of 8474 people. Do these data provide convincing evidence of an *association* between the variables in the larger population? To answer that question, we work with a new significance test.

The Chi-Square Test for Association/Independence

We often gather data from a random sample and arrange them in a two-way table to see if two categorical variables are associated. The sample data are easy to investigate: turn them into percents and look for a relationship between the variables. Is the association in the sample evidence of an association between these variables in the entire population? Or could the sample association easily arise just from the luck of random sampling? This is a question for a significance test.

Our null hypothesis is that there is *no association* between the two categorical variables. The alternative hypothesis is that there *is* an association between the variables. For the observational study of anger level and coronary heart disease, we want to test the hypotheses

H_0: There is no association between anger level and heart disease in the population of people with normal blood pressure.

We could substitute the word "dependent" in place of "not independent" in the alternative hypothesis. We'll avoid this practice, however, because saying that two variables are dependent sounds too much like saying that changes in one variable *cause* changes in the other. "Dependent" implies only that knowing the value of one variable gives us additional information about the value of the other.

H_a: There is an association between anger level and heart disease in the population of people with normal blood pressure.

No association between two variables means that the values of one variable do not tend to occur in common with values of the other. That is, the variables are independent. An equivalent way to state the hypotheses is therefore

H_0: Anger and heart disease are independent in the population of people with normal blood pressure.
H_a: Anger and heart disease are not independent in the population of people with normal blood pressure.

As with the two previous types of chi-square tests, we begin by comparing the observed counts in a two-way table with the expected counts if H_0 is true.

Do Angry People Have More Heart Disease?

Finding expected counts

The null hypothesis is that there is no association between anger level and heart disease in the population of interest. If we assume that H_0 is true, then anger level and heart disease are independent. We can find the expected cell counts in the two-way table using the definition of independent events from Chapter 5: $P(A \mid B) = P(A)$. Let's start by considering the events "CHD" and "low anger."

	Low anger	Moderate anger	High anger	Total
CHD	53	110	27	190
No CHD	3057	4621	606	8284
Total	3110	4731	633	8474

We see from the two-way table that 190 of the 8474 people in the study had CHD. If we imagine choosing one of these people at random, $P(\text{CHD}) = 190/8474$. Because anger level and heart disease are independent, knowing that the selected individual is low anger gives us no additional information about the chance that this person has heart disease. That is, $P(\text{CHD} \mid \text{low anger}) = P(\text{CHD}) = 190/8474$. Of the 3110 low-anger people in the study, we'd expect

$$\frac{190}{8474} \cdot 3110 = 69.73$$

to get CHD. We could also write this as

$$\frac{190 \cdot 3110}{8474}$$

You can see that the general formula we developed earlier for a test for homogeneity applies in this situation also:

$$\text{expected count} = \frac{\text{row total} \cdot \text{column total}}{\text{table total}}$$

We find the expected counts for all the cells in the two-way table in a similar way.

CHD, Low	**CHD, Moderate**	**CHD, High**
$\frac{190 \cdot 3110}{8474} = 69.73$	$\frac{190 \cdot 4731}{8474} = 106.08$	$\frac{190 \cdot 633}{8474} = 14.19$
no CHD, Low	**no CHD, Moderate**	**no CHD, High**
$\frac{8284 \cdot 3110}{8474} = 3040.27$	$\frac{8284 \cdot 4731}{8474} = 4624.92$	$\frac{8284 \cdot 633}{8474} = 618.81$

If the Random, Large Sample Size, and Independent conditions are met, the χ^2 statistic calculated from a two-way table can be used to perform a test of H_0: There is no association between two categorical variables in the population of interest. *P*-values for this test come from a chi-square distribution with df = (number of rows − 1)(number of columns − 1). This new procedure is known as a **chi-square test for association/independence.**

Chi-Square Test for Association/Independence

Suppose the Random, Large Sample Size, and Independent conditions are met. You can use the **chi-square test for association/independence** to test

H_0: There is no association between two categorical variables in the population of interest.

H_a: There is an association between two categorical variables in the population of interest.

Or, alternatively

H_0: Two categorical variables are independent in the population of interest.

H_a: Two categorical variables are not independent in the population of interest.

Start by finding the expected counts. Then calculate the chi-square statistic

$$\chi^2 = \sum \frac{(\text{Observed} - \text{Expected})^2}{\text{Expected}}$$

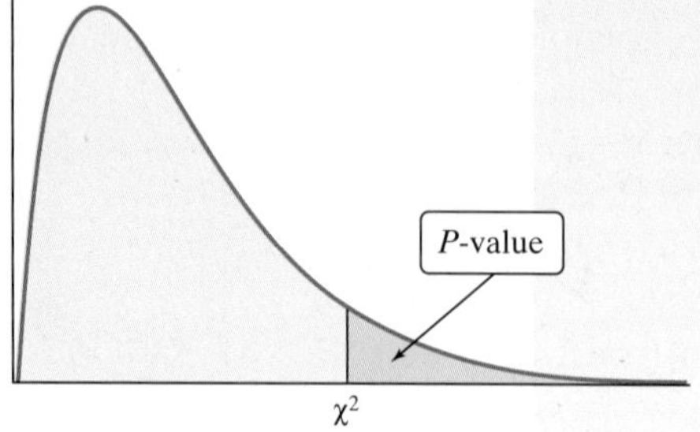

where the sum is over all cells in the two-way table. If H_0 is true, the χ^2 statistic has approximately a chi-square distribution with degrees of freedom = (number of rows − 1)(number of columns − 1). The *P*-value is the area to the right of χ^2 under the corresponding chi-square density curve.

Conditions: Use this test when

- **Random** The data come from a random sample or randomized experiment.
- **Large Sample Size** All expected counts are at least 5.
- **Independent** Individual observations are independent. When sampling without replacement, check that the population is at least 10 times as large as the sample (the 10% condition).

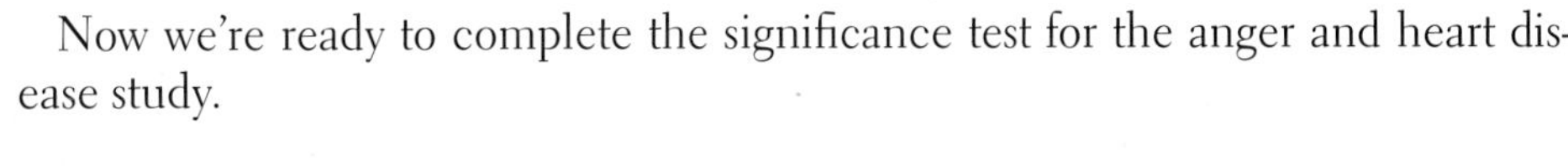

Now we're ready to complete the significance test for the anger and heart disease study.

EXAMPLE

Do Angry People Have More Heart Disease?

Chi-square test for association/independence

Here is the complete table of observed and expected counts for the CHD and anger study side by side:

	Observed			Expected		
	Low	Moderate	High	Low	Moderate	High
CHD	53	110	27	69.73	106.08	14.19
No CHD	3057	4621	606	3040.27	4624.92	618.81

Do the data provide convincing evidence of an association between anger level and heart disease in the population of interest? Carry out an appropriate test to help answer this question.

STATE: We want to perform a test of

H_0: There is no association between anger level and heart disease in the population of people with normal blood pressure.

H_a: There is an association between anger level and heart disease in the population of people with normal blood pressure.

Since no significance level was stated, we'll use $\alpha = 0.05$.

PLAN: If conditions are met, we should carry out a chi-square test of association/independence.

- *Random* The data came from a random sample of 8474 people with normal blood pressure.
- *Large Sample Size* All the expected counts are at least 5, so this condition is met.
- *Independent* Because we are sampling without replacement, we need to check that the total number of people in the population with normal blood pressure is at least 10(8474) = 84,740. This seems reasonable to assume. Thus, knowing the values of both variables for one person in the study gives us no meaningful information about the values of the variables for another person. So individual observations are independent.

DO: We perform calculations assuming H_0 is true.

- *Test statistic*

$$\chi^2 = \sum \frac{(\text{Observed} - \text{Expected})^2}{\text{Expected}}$$

$$= \frac{(53 - 69.73)^2}{69.73} + \frac{(110 - 106.08)^2}{106.08} + \cdots + \frac{(606 - 618.81)^2}{618.81}$$

$$= 4.014 + 0.145 + \cdots + 0.265 = 16.077$$

- *P-value* The two-way table of anger level versus heart disease has 2 rows and 3 columns. We will use the chi-square distribution with df = (2 − 1)(3 − 1) = 2 to find the *P*-value.

Table: Look at the df = 2 line in Table C. The observed statistic $\chi^2 = 16.077$ is larger than the critical value 15.20 for $\alpha = 0.0005$. So the *P*-value is less than 0.0005.

Technology: The command χ^2`cdf(16.077,1000,2)` gives 0.00032.

CONCLUDE: Because the *P*-value is clearly less than $\alpha = 0.05$, we have sufficient evidence to reject H_0 and conclude that anger level and heart disease are associated in the population of people with normal blood pressure.

For Practice *Try Exercise 47*

CAUTION

Can we conclude that proneness to anger *causes* heart disease? No. The anger and heart disease study is an observational study, not an experiment. It isn't surprising that some lurking variables are confounded with anger. For example, people prone to anger are more likely than others to be men who drink and smoke. We don't know whether the increased rate of heart disease among those with higher anger levels in the study is due to their anger or perhaps to their drinking and smoking or maybe even to gender.

CHECK YOUR UNDERSTANDING

Many popular businesses are franchises—think of McDonald's. The owner of a local franchise benefits from the brand recognition, national advertising, and detailed guidelines provided by the franchise chain. In return, he or she pays fees to the franchise firm and agrees to follow its policies. The relationship between the local owner and the franchise firm is spelled out in a detailed contract.

One clause that the contract may or may not contain is the entrepreneur's right to an exclusive territory. This means that the new outlet will be the only representative of the franchise in a specified territory and will not have to compete with other outlets of the same chain. How does the presence of an exclusive-territory clause in the contract relate to the survival of the business?

A study designed to address this question collected data from a random sample of 170 new franchise firms. Two categorical variables were measured for each franchisor. First, the franchisor was classified as successful or not based on whether or not it was still offering franchises as of a certain date. Second, the contract each franchisor offered to franchisees was classified according to whether or not there was an exclusive-territory clause. Here are the count data, arranged in a two-way table:[12]

	Exclusive Territory		
Success	**Yes**	**No**	**Total**
Yes	108	15	123
No	34	13	47
Total	**142**	**28**	**170**

Do these data provide convincing evidence of an association between an exclusive-territory clause and business survival? Carry out an appropriate test at the $\alpha = 0.01$ level.

Using Chi-Square Tests Wisely

CAUTION

Both the chi-square test for homogeneity and the chi-square test for association/independence start with a two-way table of observed counts. They even calculate the test statistic, degrees of freedom, and *P*-value in the same way. *The questions that these two tests answer are different, however.* A chi-square test for homogeneity tests whether the distribution of a categorical variable is the same for each of several populations or treatments. The chi-square test for association/independence tests whether two categorical variables are associated in some population of interest. Unfortunately, it is quite common to see questions asking about association when a test for homogeneity applies and questions asking about differences between proportions or the distribution of a variable when a test of association applies. Sometimes, people avoid the distinction altogether and pose questions about the "relationship" between two variables.

AP EXAM TIP If you have trouble distinguishing the two types of chi-square tests for two-way tables, you're better off just saying "chi-square test" than choosing the wrong type. Better yet, learn to tell the difference!

Instead of focusing on the question asked, it's much easier to look at how the data were produced. If the data come from two or more independent random samples or treatment groups in a randomized experiment, then do a chi-square test for homogeneity. If the data come from a single random sample, with the individuals classified according to two categorical variables, use a chi-square test for association/independence.

EXAMPLE

Scary Movies and Fear

Choosing the right type of chi-square test

Are men and women equally likely to suffer lingering fear from watching scary movies as children? Researchers asked a sample of 117 college students to write narrative accounts of their exposure to scary movies before the age of 13. More than one-fourth of the students said that some of the fright symptoms are still present when they are awake.[13] The following table breaks down these results by gender.

	Gender		
Fright symptoms?	**Male**	**Female**	**Total**
Yes	7	29	36
No	31	50	81
Total	**38**	**79**	**117**

Minitab output for a chi-square test using these data is shown below.

```
Chi-Square Test: Male, Female
Expected counts are printed below observed counts
Chi-Square contributions are printed below expected counts
          Male      Female     Total
1            7          29        36
         11.69       24.31
         1.883       0.906
2           31          50        81
         26.31       54.69
         0.837       0.403
Total       38          79       117
Chi-Sq = 4.028, DF = 1, P-Value = 0.045
```

PROBLEM: **Assume that the conditions for performing inference are met.**

(a) Explain why a chi-square test for association/independence and not a chi-square test for homogeneity should be used in this setting.

(b) Researchers decided to use the null hypothesis H_0: Gender and ongoing fright symptoms are independent in the population of interest. State the correct alternative hypothesis.

(c) Which cell contributes most to the chi-square statistic? In what way does this cell differ from what the null hypothesis suggests?

(d) Interpret the *P*-value in context. What conclusion would you draw at $\alpha = 0.01$?

SOLUTION:

(a) The data were produced using a single random sample of college students, who were then classified by gender and whether or not they had lingering fright symptoms. The chi-square test for homogeneity requires independent random samples from each population.

(b) The alternative hypothesis is H_a: Gender and ongoing fright symptoms are not independent in the population of interest.

(c) Men who admit to having lingering fright symptoms account for the largest component of the chi-square statistic: 1.883 of the total 4.028. Far fewer men in the sample admitted to fright symptoms (7) than we would expect if H_0 were true (11.69).

(d) If gender and ongoing fright symptoms really are independent in the population of interest, the probability of obtaining a random sample of 117 students that gives a chi-square statistic of 4.028 or higher is 0.045. At the $\alpha = 0.01$ significance level, we would fail to reject H_0 because the P-value is greater than 0.01. We lack evidence to conclude that gender and fright symptoms are not independent in the population of interest.

For Practice *Try Exercise 51*

As we mentioned in Chapter 1, it is possible to convert a quantitative variable to a categorical variable by grouping intervals of values together. For instance, the study of shoppers in secondhand stores described in the Data Exploration (page 712) also compared the income distributions of shoppers in the two stores. Here is a two-way table of counts:

Income	City 1	City 2
Under \$10,000	70	62
\$10,000 to \$19,999	52	63
\$20,000 to \$24,999	69	50
\$25,000 to \$34,999	22	19
\$35,000 or more	28	24

Personal income is a quantitative variable. But by grouping the values of this variable, we create a categorical variable. We could use these data to carry out a chi-square test for homogeneity because the data came from independent random samples of shoppers at the two stores. Comparing the distributions of income for shoppers at the two stores would give more information than simply comparing their mean incomes.

THINK ABOUT IT

What can we do if the expected cell counts aren't all at least 5? Let's look at a situation where this is the case. A sample survey asked a random sample of young adults, "Where do you live now? That is, where do you stay most often?" A two-way table of all 2984 people in the sample (both men and women) classified by their age and by where they lived is provided below.[14] Living arrangement is a categorical variable. Even though age is quantitative, the two-way table treats age as dividing the young adults into four categories. The table gives the observed counts for all 20 combinations of age and living arrangement.

	Age (years)				
Living arrangement	**19**	**20**	**21**	**22**	**Total**
Parents' home	324	378	337	318	1357
Another person's home	37	47	40	38	162
Your own place	116	279	372	487	1254
Group quarters	58	60	49	25	192
Other	5	2	3	9	19
Total	540	766	801	877	2984

Our null hypothesis is H_0: There is no association between age and where young people live. The table below shows the expected counts assuming H_0 is true. We can see that two of the expected counts (circled in red) are less than 5. This violates the Large Sample Size condition.

	Age (years)				
Living arrangement	**19**	**20**	**21**	**22**	**Total**
Parents' home	245.57	348.35	364.26	398.82	1357
Another person's home	29.32	41.59	43.49	47.61	162
Your own place	226.93	321.90	336.61	368.55	1254
Group quarters	34.75	49.29	51.54	56.43	192
Other	3.44	4.88	5.10	5.58	19
Total	540	766	801	877	2984

A clever strategy is to "collapse" the table by combining two or more rows or columns. In this case, it might make sense to combine the 19- and 20-year-olds because they are still of typical college age. Doing so and then running a chi-square test in Minitab gives the following output. Notice that the Large Sample Size condition is now met.

```
Chi-Square Test: 19 or 20, 21, 22

Expected counts are printed below observed counts
Chi-Square contributions are printed below expected counts

      19 or 20        21        22     Total
1          702       337       318      1357
        593.91    364.26    398.82
        19.670     2.040    16.379

2           84        40        38       162
         70.90     43.49     47.61
         2.420     0.279     1.940

3          395       372       487      1254
        548.84    336.61    368.55
        43.119     3.720    38.068

4          118        49        25       192
         84.03     51.54     56.43
        13.731     0.125    17.505

5            7         3         9        19
          8.32      5.10      5.58
         0.208     0.865     2.090

Total     1306       801       877      2984

Chi-Sq = 162.159, DF = 8, P-Value = 0.000
```

case closed

Do Dogs Resemble Their Owners?

In the chapter-opening Case Study (page 675), we described a study that investigated whether dogs really resemble their owners. Here is a two-way table summarizing the results:

	Resembles owner	Doesn't resemble
Purebred dogs	16	9
Mixed-breed dogs	7	13

Do these data provide convincing evidence of an association between dog breed and resemblance to owner at the $\alpha = 0.05$ significance level?

STATE: We want to perform a test of

H_0: There is no association between dog breed and resemblance to owner in the population.

H_a: There is an association between dog breed and resemblance to owner in the population.

at the $\alpha = 0.05$ significance level.

	Resembles owner	Doesn't resemble
Purebred dogs	12.78	12.22
Mixed-breed dogs	10.22	9.78

PLAN: If conditions are met, we should perform a chi-square test for association/independence.

- **Random** The data came from a random sample of 45 dog–owner pairs.
- **Large Sample Size** Assuming H_0 is true, the expected counts are all at least 5, as shown in the table.
- **Independent** Individual observations should be independent since knowing the status of one dog–owner pair provides no additional information about the status of another dog–owner pair. Due to the sampling without replacement, we should feel confident that there are at least $10(45) = 450$ dog–owner pairs. This is reasonable to assume.

DO: Using the calculator's chi-square test, we obtain

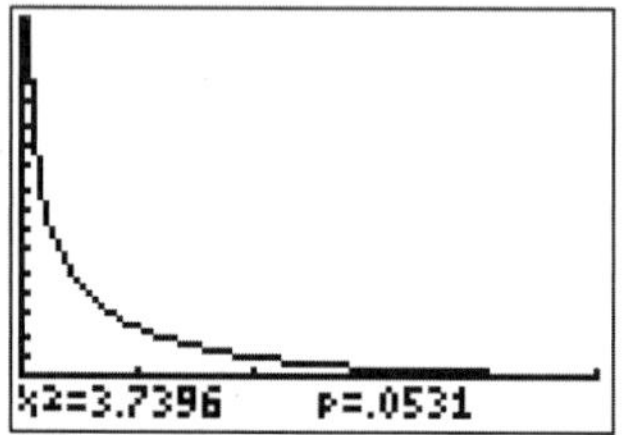

- ***Test statistic***

$$\chi^2 = \sum \frac{(\text{Observed} - \text{Expected})^2}{\text{Expected}}$$

$$= \frac{(16-12.78)^2}{12.78} + \frac{(9-12.22)^2}{12.22} + \frac{(7-10.22)^2}{10.22} + \frac{(13-9.78)^2}{9.78} = 3.73$$

- ***P-value*** Using df $= (2-1)(2-1) = 1$, the calculator gives $P = 0.053$.

CONCLUDE: Because the *P*-value, 0.053, is greater than $\alpha = 0.05$, we fail to reject H_0. We do not have enough evidence to conclude that there is an association between breed and resemblance to owner in the population of dog–owner pairs.

Remark: Recall that some people prefer not to use fixed significance levels when performing tests because they represent arbitrary cutoffs for statistical significance. This Case Study shows why. The *P*-value of 0.053 indicates moderately strong evidence against H_0. It would be reasonable to argue that this result gives some indication of an association between breed status and resemblance to owner.

SECTION 11.2 Summary

- We can use a two-way table to summarize data on the relationship between two categorical variables. To analyze the data, we first compute percents or proportions that describe the relationship of interest. Then we turn to formal inference. Two different ways of producing data for two-way tables lead to two different types of chi-square tests.
- If data are produced using independent random samples from each of several populations of interest or the treatment groups in a randomized comparative experiment, then each observation is classified according to a categorical variable of interest. The null hypothesis is that the distribution of this categorical variable is the same for all the populations or treatments. We use the **chi-square test for homogeneity** to test this hypothesis. One common use of the chi-square test for homogeneity is to compare the proportions of successes for each of several populations or treatments.
- If data are produced using a single random sample from a population of interest, then each observation is classified according to two categorical variables. The **chi-square test of association/independence** tests the null hypothesis that there is no association between the two categorical variables in the population of interest. Another way to state the null hypothesis is H_0: The two categorical variables are independent in the population of interest.
- The **expected count** in any cell of a two-way table when H_0 is true is

$$\text{expected count} = \frac{\text{row total} \cdot \text{column total}}{\text{table total}}$$

- The **chi-square statistic** is

$$\chi^2 = \sum \frac{(\text{Observed} - \text{Expected})^2}{\text{Expected}}$$

  where the sum is over all cells in the two-way table.
- The chi-square test compares the value of the statistic χ^2 with critical values from the chi-square distribution with df = (number of rows − 1)(number of columns − 1). Large values of χ^2 are evidence against H_0, so the P-value is the area under the chi-square density curve to the right of χ^2.

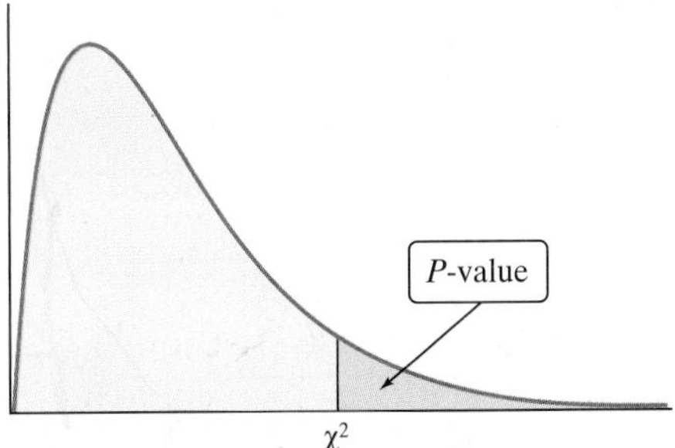

- The chi-square distribution is an approximation to the distribution of the statistic χ^2. You can safely use this approximation when all expected cell counts are at least 5 (the Large Sample Size condition).
- Be sure to check that the Random, Large Sample Size, and Independent conditions are met before performing a chi-square test for a two-way table.

- If the test finds a statistically significant result, do a follow-up analysis that compares the observed and expected counts and that looks for the largest components of the chi-square statistic.

11.2 TECHNOLOGY CORNERS

SECTION 11.2 Exercises

27. **Why men and women play sports** Do men and women participate in sports for the same reasons? One goal for sports participants is social comparison—the desire to win or to do better than other people. Another is mastery—the desire to improve one's skills or to try one's best. A study on why students participate in sports collected data from independent random samples of 67 male and 67 female undergraduates at a large university.[15] Each student was classified into one of four categories based on his or her responses to a questionnaire about sports goals. The four categories were high social comparison–high mastery (HSC-HM), high social comparison–low mastery (HSC-LM), low social comparison–high mastery (LSC-HM), and low social comparison–low mastery (LSC-LM). One purpose of the study was to compare the goals of male and female students. Here are the data displayed in a two-way table:

pg 696

Observed Counts for Sports Goals

	Sex	
Goal	**Female**	**Male**
HSC-HM	14	31
HSC-LM	7	18
LSC-HM	21	5
LSC-LM	25	13

(a) Calculate the conditional distribution (in proportions) of the reported sports goals for each gender.

(b) Make an appropriate graph for comparing the conditional distributions in part (a).

(c) Write a few sentences comparing the distributions of sports goals for male and female undergraduates.

28. **How are schools doing?** The nonprofit group Public Agenda conducted telephone interviews with three randomly selected groups of parents of high school children. There were 202 black parents, 202 Hispanic parents, and 201 white parents. One question asked was "Are the high schools in your state doing an excellent, good, fair, or poor job, or don't you know enough to say?" Here are the survey results:[16]

	Black parents	Hispanic parents	White parents
Excellent	12	34	22
Good	69	55	81
Fair	75	61	60
Poor	24	24	24
Don't know	22	28	14
Total	202	202	201

(a) Calculate the conditional distribution (in proportions) of responses for each group of parents.

(b) Make an appropriate graph for comparing the conditional distributions in part (a).

(c) Write a few sentences comparing the distributions of responses for the three groups of parents.

29. **Why women and men play sports** Refer to Exercise 27. Do the data provide convincing evidence of a difference in the distributions of sports goals for male and female undergraduates at the university?

pg 702

(a) State appropriate null and alternative hypotheses for a significance test to help answer this question.

(b) Calculate the expected counts. Show your work.

(c) Calculate the chi-square statistic. Show your work.

30. **How are schools doing?** Refer to Exercise 28. Do the data provide convincing evidence of a difference in the distributions of opinions about how high schools are doing among black, Hispanic, and white parents?

(a) State appropriate null and alternative hypotheses for a significance test to help answer this question.

(b) Calculate the expected counts. Show your work.

(c) Calculate the chi-square statistic. Show your work.

31. **Why women and men play sports** Refer to Exercises 27 and 29.

pg 704

(a) Check that the conditions for performing the chi-square test are met.

(b) Use Table C to find the *P*-value. Then use your calculator's χ^2cdf command.

(c) Interpret the *P*-value from the calculator in context.

(d) What conclusion would you draw? Justify your answer.

32. **How are schools doing?** Refer to Exercises 28 and 30.

(a) Check that the conditions for performing the chi-square test are met.

(b) Use Table C to find the *P*-value. Then use your calculator's χ^2cdf command.

(c) Interpret the *P*-value from the calculator in context.

(d) What conclusion would you draw? Justify your answer.

33. **Python eggs** How is the hatching of water python eggs influenced by the temperature of the snake's nest? Researchers randomly assigned newly laid eggs to one of three water temperatures: hot, neutral, or cold. Hot duplicates the extra warmth provided by the mother python, and cold duplicates the absence of the mother. Here are the data on the number of eggs and the number that hatched:[17]

pg 706

	Eggs	Hatched
Cold	27	16
Neutral	56	38
Hot	104	75

(a) Make a two-way table of temperature by outcome (hatched or not). Calculate the proportion of eggs in each group that hatched. The researchers believed that eggs would not hatch in cold water. Do the data support that belief?

(b) Are the differences between the three groups statistically significant? Give appropriate evidence to support your answer.

34. **Equivalent groups?** After randomly assigning subjects to treatments in a randomized comparative experiment, we can compare the treatment groups to see how well the random assignment worked. We hope to find no significant differences among the groups. A study of how to provide premature infants with a substance essential to their development assigned infants at random to receive one of four types of supplement, called PBM, NLCP, PL-LCP, and TG-LCP. The subjects were 77 premature infants. In the experiment, 20 were assigned to the PBM group and 19 to each of the other treatments.[18]

(a) The random assignment resulted in 9 females in the TG-LCP group and 11 females in each of the other groups. Make a two-way table of group by gender. Calculate the proportion of females in each treatment group. Does it appear that the random assignment roughly balanced the groups by gender? Explain.

(b) Are the differences between the groups statistically significant? Give appropriate evidence to support your answer.

35. **Sorry, no chi-square** How do U.S. residents who travel overseas for leisure differ from those who travel for business? The following is the breakdown by occupation:[19]

Occupation	Leisure travelers (%)	Business travelers (%)
Professional/technical	36	39
Manager/executive	23	48
Retired	14	3
Student	7	3
Other	20	7
Total	100	100

Explain why we can't use a chi-square test to learn whether these two distributions differ significantly.

36. **Sorry, no chi-square** Does eating chocolate trigger headaches? To find out, women with chronic headaches followed the same diet except for eating chocolate bars and carob bars that looked and tasted the same. Each subject ate both chocolate and carob bars in random order with at least three days between. Each woman then reported whether or not she had a headache within 12 hours of eating the bar. Here is a two-way table of the results for the 64 subjects:[20]

	No headache	Headache
Chocolate	53	11
Carob (placebo)	38	26

The researchers carried out a chi-square test on this table to see if the two types of bar differ in triggering headaches. Explain why this test is incorrect.

pg 710

37. **How to quit smoking** It's hard for smokers to quit. Perhaps prescribing a drug to fight depression will work as well as the usual nicotine patch. Perhaps combining the patch and the drug will work better than either treatment alone. Here are data from a randomized, double-blind trial that compared four treatments.[21] A "success" means that the subject did not smoke for a year following the beginning of the study.

Group	Treatment	Subjects	Successes
1	Nicotine patch	244	40
2	Drug	244	74
3	Patch plus drug	245	87
4	Placebo	160	25

(a) Summarize these data in a two-way table.

(b) Make a graph to compare the success rates for the four treatments. Describe what you see.

(c) Explain in words what the null hypothesis H_0: $p_1 = p_2 = p_3 = p_4$ says about subjects' smoking habits.

(d) Find the expected counts if H_0 is true, and display them in a two-way table similar to the table of observed counts.

38. **Preventing strokes** Aspirin prevents blood from clotting and so helps prevent strokes. The Second European Stroke Prevention Study asked whether adding another anticlotting drug named dipyridamole would be more effective for patients who had already had a stroke. Here are the data on strokes during the two years of the study:[22]

Group	Treatment	Number of patients	Number of strokes
1	Placebo	1649	250
2	Aspirin	1649	206
3	Dipyridamole	1654	211
4	Both	1650	157

(a) Summarize these data in a two-way table.

(b) Make a graph to compare the rates of strokes for the four treatments. Describe what you see.

(c) Explain in words what the null hypothesis H_0: $p_1 = p_2 = p_3 = p_4$ says about the incidence of strokes.

(d) Find the expected counts if H_0 is true, and display them in a two-way table similar to the table of observed counts.

39. **How to quit smoking** Refer to Exercise 37. Do the data provide convincing evidence of a difference in the effectiveness of the four treatments? Carry out an appropriate test at the $\alpha = 0.05$ significance level.

40. **Preventing strokes** Refer to Exercise 38. Do the data provide convincing evidence of a difference in the effectiveness of the four treatments? Carry out an appropriate test at the $\alpha = 0.05$ significance level.

41. **How to quit smoking** Perform a follow-up analysis of the test in Exercise 39 by finding the individual components of the chi-square statistic. Which cell(s) contributed most to the final result?

42. **Preventing strokes** Perform a follow-up analysis of the test in Exercise 40 by finding the individual components of the chi-square statistic. Which cell(s) contributed most to the final result?

43. **Treating ulcers** Gastric freezing was once a recommended treatment for ulcers in the upper intestine. Use of gastric freezing stopped after experiments showed it had no effect. One randomized comparative experiment found that 28 of the 82 gastric-freezing patients improved, while 30 of the 78 patients in the placebo group improved.[23] We can test the hypothesis of "no difference" in the effectiveness of the treatments in two ways: with a two-sample z test or with a chi-square test.

(a) Minitab output for a chi-square test is shown below. State appropriate hypotheses and interpret the P-value in context. What conclusion would you draw?

```
Chi-Square Test: Gastric freezing, Placebo
Expected counts are printed below observed
counts
Chi-Square contributions are printed below
expected counts
        Gastric freezing    Placebo    Total
1                     28         30       58
                   29.73      28.27
                   0.100      0.105

2                     54         48      102
                   52.27      49.73
                   0.057      0.060
Total                 82         78      160
Chi-Sq = 0.322, DF = 1, P-Value = 0.570
```

(b) Minitab output for a two-sample z test is shown below. Explain how these results are consistent with the test in part (a).

```
Test for Two Proportions
Sample     X      N     Sample p
1         28     82     0.341463
2         30     78     0.384615
Difference = p (1) - p (2)
Estimate for difference: -0.0431520
Test for difference = 0 (vs not = 0):
Z = -0.57 P-Value = 0.570
```

44. **Opinions about the death penalty** The General Social Survey asked a random sample of adults, "Do you favor or oppose the death penalty for persons convicted of murder?" The following table gives the responses of people whose highest educa-

tion was a high school degree and of people with a bachelor's degree:

	Favor	Oppose
High school	1010	369
Bachelor's	319	185

We can test the hypothesis of "no difference" in support for the death penalty among people in these educational categories in two ways: with a two-sample z test or with a chi-square test.

(a) Minitab output for a chi-square test is shown below. State appropriate hypotheses and interpret the P-value in context. What conclusion would you draw?

```
Chi-Square Test: C1, C2
Expected counts are printed below
observed counts
Chi-Square contributions are printed
below expected counts
            C1        C2     Total
1         1010       369      1379
        973.28    405.72
         1.385     3.323

2          319       185       504
        355.72    148.28
         3.790     9.092
Total     1329       554      1883
Chi-Sq = 17.590, DF = 1, P-Value = 0.000
```

(b) Minitab output for a two-sample z test is shown below. Explain how these results are consistent with the test in part (a).

```
Test for Two Proportions
Sample       X       N    Sample p
1         1010    1379    0.732415
2          319     504    0.632937

Difference = p (1) - p (2)
Estimate for difference: 0.0994783
Test for difference = 0 (vs not = 0):
Z = 4.19 P-Value = 0.000
```

45. **Attitudes toward recycled products** Some people think recycled products are lower in quality than other products, a fact that makes recycling less practical. Here are data on attitudes toward coffee filters made of recycled paper from a random sample of adults:[24]

pg 713

	Think the Quality of the Recycled Product Is . . .		
	Higher	Same	Lower
Buyers	20	7	9
Nonbuyers	29	25	43

(a) Make a bar graph that compares buyers' and nonbuyers' opinions about recycled filters. Describe what you see.

(b) Minitab output for a chi-square test using these data is shown below. Carry out the test. What conclusion do you draw?

```
Session

             Higher    Same    Lower      All

Buyer            20       7        9       36
              55.56   19.44    25.00   100.00
              13.26    8.66    14.08    36.00

Nonbuyer         29      25       43       97
              29.90   25.77    44.33   100.00
              35.74   23.34    37.92    97.00

All              49      32       52      133
              36.84   24.06    39.10   100.00
              49.00   32.00    52.00   133.00

Cell Contents:         Count
                       % of Row
                       Expected count

Chi-Square = 7.638,  DF = 2, P-Value = 0.022
```

46. **Is astrology scientific?** The General Social Survey asked a random sample of adults their opinion about whether astrology is very scientific, sort of scientific, or not at all scientific. Here is a two-way table of counts for people in the sample who had three levels of higher education:[25]

	Degree Held		
	Associate's	Bachelor's	Master's
Not at all scientific	169	256	114
Very or sort of scientific	65	65	18

(a) Make a bar graph that compares opinions about astrology for the three education categories. Describe what you see.

(b) Minitab output for a chi-square test using these data is shown below. Carry out the test. What conclusion do you draw?

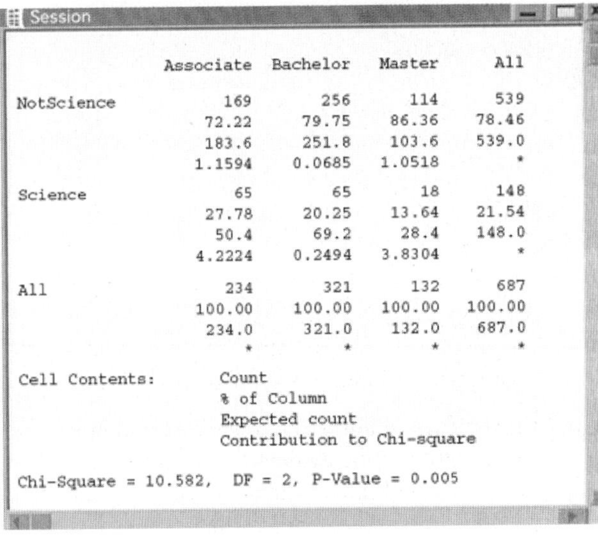

```
Session

             Associate  Bachelor  Master      All

NotScience         169       256     114      539
                 72.22     79.75   86.36    78.46
                 183.6     251.8   103.6    539.0
                1.1594    0.0685  1.0518        *

Science             65        65      18      148
                 27.78     20.25   13.64    21.54
                  50.4      69.2    28.4    148.0
                4.2224    0.2494  3.8304        *

All                234       321     132      687
                100.00    100.00  100.00   100.00
                 234.0     321.0   132.0    687.0
                     *         *       *        *

Cell Contents:          Count
                        % of Column
                        Expected count
                        Contribution to Chi-square

Chi-Square = 10.582,  DF = 2, P-Value = 0.005
```

47. **Extracurricular activities and grades** North Carolina State University studied student performance in a course required by its chemical engineering major. One question of interest was the relationship between time spent in extracurricular activities and whether a student earned a C or better in the course. Here are the data for the 119 students who answered a question about extracurricular activities:[26]

pg 717

	Extracurricular Activities (hours per week)		
	<2	2 to 12	>12
C or better	11	68	3
D or F	9	23	5

(a) Calculate percents and draw a bar graph that describes the nature of the relationship between time spent on extracurricular activities and performance in the course. Give a brief summary in words.

(b) Explain why you should not perform a chi-square test in this setting.

48. **Removing warts** A recently reported study looked at the use of oral zinc supplements to get rid of warts. The treatment group took an oral zinc supplement. Of these 32 people, 19 were rid of their warts after one month, and 6 additional patients were rid of them after two months. The 23 subjects in a control group were given a placebo. Of these, 1 was rid of his warts after one month and 2 additional patients were rid of their warts after two months. The reported results of the study included this table:[27]

	Treatment	Control
After 1 month	19	1
After 2 months	6	2
Total	25	3

(a) The researchers say they did a chi-square test on this table. Explain why that makes no sense. Then make a correct two-way table displaying the results of the experiment.

(b) Explain why it is not appropriate to use a chi-square test in this setting even with the correct two-way table.

49. **Regulating guns** The National Gun Policy Survey asked a random sample of adults, "Do you think there should be a law that would ban possession of handguns except for the police and other authorized persons?" Here are the responses, broken down by the respondent's level of education:[28]

Education	Yes	No
Less than high school	58	58
High school graduate	84	129
Some college	169	294
College graduate	98	135
Postgraduate degree	77	99

(a) How do opinions about banning handgun ownership seem to be related to level of education? Make an appropriate graph to display this relationship. Describe what you see.

(b) Determine whether or not the sample provides convincing evidence that education level and opinion about a handgun ban are dependent in the adult population.

50. **Market research** Before bringing a new product to market, firms carry out extensive studies to learn how consumers react to the product and how best to advertise its advantages. Here are data from a study of a new laundry detergent.[29] The participants are a random sample of people who don't currently use the established brand that the new product will compete with. Give subjects free samples of both detergents. After they have tried both for a while, ask which they prefer. The answers may depend on other facts about how people do laundry.

	Laundry Practices			
	Soft water, warm wash	Soft water, hot wash	Hard water, warm wash	Hard water, hot wash
Prefer standard product	53	27	42	30
Prefer new product	63	29	68	42

(a) How are laundry practices (water hardness and wash temperature) related to the choice of detergent? Make an appropriate graph to display this relationship. Describe what you see.

(b) Determine whether or not the sample provides convincing evidence that laundry practices and product preference are independent in the population of interest.

51. **Where do young adults live?** A survey by the National Institutes of Health asked a random sample of young adults (aged 19 to 25 years), "Where do you live now? That is, where do you stay most often?" Here is the full two-way table (omitting a few who refused to answer and one who claimed to be homeless):[30]

pg 719

	Female	Male
Parents' home	923	986
Another person's home	144	132
Own place	1294	1129
Group quarters	127	119

(a) Should we use a chi-square test for homogeneity or a chi-square test of association/independence in this setting? Justify your answer.

(b) State appropriate hypotheses for performing the type of test you chose in part (a). Minitab output from a chi-square test is shown below.

```
Chi-Square Test: Female, Male
Expected counts are printed below observed
counts
Chi-Square contributions are printed below
expected counts
         Female      Male   Total
  1         923       986    1909
         978.49    930.51
          3.147     3.309

  2         144       132     276
         141.47    134.53
          0.045     0.048

  3        1294      1129    2423
        1241.95   1181.05
          2.181     2.294

  4         127       119     246
         126.09    119.91
          0.007     0.007

Total      2488      2366    4854
Chi-Sq = 11.038, DF = 3, P-Value = 0.012
```

(c) Check that the conditions for carrying out the test are met.

(d) Interpret the P-value in context. What conclusion would you draw?

52. **Students and catalog shopping** What is the most important reason that students buy from catalogs? The answer may differ for different groups of students. Here are results for separate random samples of American and Asian students at a large midwestern university:[31]

	American	Asian
Save time	29	10
Easy	28	11
Low price	17	34
Live far from stores	11	4
No pressure to buy	10	3

(a) Should we use a chi-square test for homogeneity or a chi-square test of association/independence in this setting? Justify your answer.

(b) State appropriate hypotheses for performing the type of test you chose in part (a). Minitab output from a chi-square test is shown below.

```
Chi-Square Test: American, Asian
Expected counts are printed below observed
counts
Chi-Square contributions are printed below
expected counts
          American     Asian    Total
  1             29        10       39
             23.60     15.40
             1.236     1.894

  2             28        11       39
             23.60     15.40
             0.821     1.258

  3             17        34       51
             30.86     20.14
             6.225     9.538

  4             11         4       15
              9.08      5.92
             0.408     0.625

  5             10         3       13
              7.87      5.13
             0.579     0.887

Total           95        62      157
Chi-Sq = 23.470, DF = 4, P-Value = 0.0001
```

(c) Check that the conditions for carrying out the test are met.

(d) Interpret the P-value in context. What conclusion would you draw?

Multiple choice: Select the best answer for Exercises 53 to 58.

The National Longitudinal Study of Adolescent Health interviewed a random sample of 4877 teens (grades 7 to 12). One question asked was "What do you think are the chances you will be married in the next ten years?" Here is a two-way table of the responses by gender:[32]

	Female	Male
Almost no chance	119	103
Some chance, but probably not	150	171
A 50–50 chance	447	512
A good chance	735	710
Almost certain	1174	756

53. The appropriate null hypothesis for performing a chi-square test is that

(a) equal proportions of female and male teenagers are almost certain they will be married in 10 years.

(b) there is no difference between female and male teenagers in this sample in their distributions of opinions about marriage.

(c) there is no difference between female and male teenagers in the population in their distributions of opinions about marriage.

(d) there is no association between gender and opinion about marriage in the sample.

(e) there is no association between gender and opinion about marriage in the population.

54. The expected count of females who respond "almost certain" is

(a) 464.6. (b) 891.2. (c) 1038.8. (d) 1174. (e) None of these.

55. The degrees of freedom for the chi-square test for this two-way table are

(a) 4. (b) 8. (c) 10. (d) 20. (e) None of these.

56. The cell in the table that contributes the most to the chi-square statistic is

(a) Female, 50–50 chance.

(b) Male, 50–50 chance.

(c) Female, almost certain.

(d) Male, almost certain.

(e) All the cells contribute equally to the test statistic.

57. Software gives test statistic $\chi^2 = 69.8$ and *P*-value close to 0. The correct interpretation of this result is

(a) the probability of getting a random sample of 4877 teens that yields a value of χ^2 of 69.8 or larger is basically 0.

(b) the probability of getting a random sample of 4877 teens that yields a value of χ^2 of 69.8 or larger if H_0 is true is basically 0.

(c) the probability of making a Type I error is basically 0.

(d) the probability of making a Type II error is basically 0.

(e) it's very unlikely that these data are true.

58. Which of the following explains why one of the conditions for performing the chi-square test is met in this case?

(a) The sample is large, 4877 teenagers in all.

(b) The sample is random.

(c) All the observed counts are greater than 100.

(d) We used software to do the calculations.

(e) Both variables are categorical.

For Exercises 59 and 60, you may find the inference summary chart inside the back cover helpful.

59. **Inference recap** (8.1 to 11.2) In each of the following settings, say which inference procedure from Chapter 8, 9, 10, or 11 you would use. Be specific. For example, you might say "two-sample *z* test for the difference between two proportions." You do not need to carry out any procedures.[33]

(a) What is the average voter turnout during an election? A random sample of 38 cities was asked to report the percent of registered voters who actually voted in the most recent election.

(b) Are blondes more likely to have a boyfriend than the rest of the single world? Independent random samples of 300 blondes and 300 nonblondes were asked whether they have a boyfriend.

60. **Inference recap** (8.1 to 11.2) In each of the following settings, say which inference procedure from Chapter 8, 9, 10, or 11 you would use. Be specific. For example, you might say "two-sample *z* test for the difference between two proportions." You do not need to carry out any procedures.[34]

(a) Is there a relationship between attendance at religious services and alcohol consumption? A random sample of 1000 adults was asked whether they regularly attend religious services and whether they drink alcohol daily.

(b) Separate random samples of 75 college students and 75 high school students were asked how much time, on average, they spend watching television each week. We want to estimate the difference in the average amount of TV watched by high school and college students.

Exercises 61 to 64 refer to the following setting. For their final project, a group of AP Statistics students investigated the following question: "Will changing the rating scale on a survey affect how respondents answer the question?" To find out, the group took an SRS of 90 students from an alphabetical roster of the school's just over 1000 students. The first 45 students chosen were asked to rate the cafeteria food on a scale of 1 (terrible) to 5 (excellent). The remaining 45 students were asked to rate the cafeteria food on a scale of 0 (terrible) to 4 (excellent). Here are the data:

1 to 5 scale

Rating	1	2	3	4	5
Frequency	2	3	1	13	3

0 to 4 scale

Rating	0	1	2	3	4
Frequency	0	0	2	18	8

61. **Design** (4.2) Was this an observational study or an experiment? Justify your answer.

62. **Sorry, no chi-square** (11.1) Explain why it would not be appropriate to perform a chi-square goodness-of-fit test in this setting.

63. **Average ratings** (1.3, 10.2) The students decided to compare the average ratings of the cafeteria food on the two scales.

(a) Find the mean and standard deviation of the ratings for the students who were given the 1-to-5 scale.

(b) For the students who were given the 0-to-4 scale, the ratings have a mean of 3.21 and a standard deviation of 0.568. Since the scales differ by one point, the group decided to add 1 to each of these ratings. What are the mean and standard deviation of the adjusted ratings?

(c) Would it be appropriate to compare the means from parts (a) and (b) using a two-sample *t* test? Justify your answer.

64. **No answer** (4.1) Explain carefully how nonresponse could lead to bias in this project.

Chapter 11 Chapter Review

This chapter examines inference methods for distributions of categorical data. A *chi-square goodness-of-fit test* can be used to test whether the distribution of a categorical variable differs from some specified distribution. The test begins by finding the expected counts for each category, assuming that the specified distribution is correct. The chi-square statistic measures how much the observed counts deviate from the expected counts:

$$\chi^2 = \sum \frac{(\text{Observed} - \text{Expected})^2}{\text{Expected}}$$

Large values of the chi-square statistic provide evidence to reject the null hypothesis of no difference. When the Random, Large Sample Size, and Independent conditions are met, *P*-values can be found using the chi-square distribution with degrees of freedom (df) = the number of categories − 1.

The *chi-square test for homogeneity* tests the null hypothesis that the distribution of a categorical variable is the same for each of several populations or treatments. This test can be used when data come from independent random samples from the respective populations of interest or from treatment groups in a randomized comparative experiment. The same chi-square statistic is used to compare observed and expected counts for each cell in the two-way table of possible outcomes. However, the degrees of freedom become

$$\text{df} = (\text{number of rows} - 1)(\text{number of columns} - 1)$$

Two-way tables also arise when a random sample is taken from a single population, and each individual is classified according to two categorical variables. In this setting, use a *chi-square test of association/independence*. This procedure tests the null hypothesis that there is no association between the two variables in the population.

Whenever a chi-square test yields a significant result, use bar graphs along with appropriate percents or proportions to summarize the data. Then examine the individual components of the chi-square statistic as part of a follow-up analysis.

Chapter Review Exercises

These exercises are designed to help you review the important ideas and methods of the chapter. Relevant learning objectives are provided in bulleted form before each exercise.

- Use a chi-square goodness-of-fit test to determine whether sample data are consistent with a specified distribution of a categorical variable.

R11.1. Representative sample? For a class project, a group of statistics students is required to take an SRS of students from their large high school to take part in a survey. The students' sample consists of 54 freshmen, 66 sophomores, 56 juniors, and 30 seniors. The school roster shows that 29% of the students enrolled at the school are freshmen, 27% are sophomores, 25% are juniors, and 19% are seniors.

(a) Construct a well-labeled bar graph that shows the distribution of grade levels (in percents) for the sample data. Do these data give you any reason to suspect that the statistics students' sample is unusual? Explain.

(b) Use an appropriate test to determine whether the sample data differ significantly from the actual distribution of students by grade level at the school.

- **Check the Random, Large Sample Size, and Independent conditions before performing a chi-square test.**

R11.2. Sorry, no chi-square We would prefer to learn from teachers who know their subject. Perhaps even preschool children are affected by how knowledgeable they think teachers are. Assign 48 three- and four-year-olds at random to be taught the name of a new toy by either an adult who claims to know about the toy or an adult who claims not to know about it. Then ask the children to pick out a picture of the new toy in a set of pictures of other toys and say its name. The response variable is the count of right answers in four tries. Here are the data:[35]

	Correct Answers				
	0	1	2	3	4
Knowledgeable teacher	5	1	6	3	9
Ignorant teacher	20	0	3	0	1

The researchers report that children taught by the teacher who claimed to be knowledgeable did significantly better ($\chi^2 = 20.24$, $P < 0.05$). Explain why this result isn't valid.

- **Use a chi-square test for homogeneity to determine whether the distribution of a categorical variable differs for several populations or treatments.**

R11.3. Stress and heart attacks You read a newspaper article that describes a study of whether stress management can help reduce heart attacks. The 107 subjects all had reduced blood flow to the heart and so were at risk of a heart attack. They were assigned at random to three groups. The article goes on to say:

> One group took a four-month stress management program, another underwent a four-month exercise program, and the third received usual heart care from their personal physicians. In the next three years, only three of the 33 people in the stress management group suffered "cardiac events," defined as a fatal or non-fatal heart attack or a surgical procedure such as a bypass or angioplasty. In the same period, seven of the 34 people in the exercise group and 12 out of the 40 patients in usual care suffered such events.[36]

(a) Use the information in the news article to make a two-way table that describes the study results.

(b) What are the success rates of the three treatments in preventing cardiac events?

(c) Is there a significant difference among the success rates for the three treatments? Give appropriate statistical evidence to support your answer.

- **Distinguish between the three types of chi-square tests.**
- **Interpret computer output for a chi-square test based on a two-way table.**
- **Know how to compute expected counts, conditional distributions, and contributions to the chi-square statistic.**

R11.4. Sexy magazine ads? Researchers looked at a random sample of 1509 full-page ads that show a model in magazines aimed at young men, at young women, or at young adults in general. They classified the ads as "not sexual" or "sexual," depending on how the model was dressed (or not dressed). Here are the data:[37]

	Readers		
	Men	Women	General
Sexual	105	225	66
Not sexual	514	351	248

The figure below displays Minitab output for a chi-square test using these data.

```
Session

            Men    Women  General     All

Sex         105      225       66     396
          16.96    39.06    21.02   26.24
          162.4    151.2     82.4   396.0
         20.312   36.074    3.265       *

notsexy     514      351      248    1113
          83.04    60.94    78.98   73.76
          456.6    424.8    231.6  1113.0
          7.227   12.835    1.162       *

All         619      576      314    1509
         100.00   100.00   100.00  100.00
          619.0    576.0    314.0  1509.0
              *        *        *       *

Cell Contents:       Count
                     % of Column
                     Expected count
                     Contribution to Chi-square

Chi-Square = 80.874,  DF = 2, P-Value = 0.00
```

(a) Which type of chi-square test should be used in this case? Justify your answer.

(b) State an appropriate pair of hypotheses for the test you chose in part (a).

(c) Show how each of the numbers 60.94, 424.8, and 12.835 was obtained for the "notsexy, Women" cell.

(d) Assuming that the conditions for performing inference are met, what conclusion would you draw? Explain.

- Use a chi-square test for association/independence to determine whether there is convincing evidence of an association between two categorical variables.
- Examine individual components of the chi-square statistic as part of a follow-up analysis.

R11.5. Popular kids Who were the popular kids at your elementary school? Did they get good grades or have good looks? Were they good at sports? A study was performed to examine the factors that determine social status for children in grades 4, 5, and 6. Researchers administered a questionnaire to a random sample of 478 students in these grades. One of the questions they asked was "What would you most like to do at school: make good grades, be good at sports, or be popular?" The two-way table below summarizes the students' responses.[38]

	Goal		
Gender	Grades	Popular	Sports
Female	130	91	30
Male	117	50	60

(a) Construct an appropriate graph to compare male and female responses. Write a few sentences describing the relationship between gender and goals.

(b) Is there convincing evidence of an association between gender and goals for elementary school students? Carry out a test at the $\alpha = 0.05$ level and report your conclusion.

(c) Which cell contributes most to the chi-square statistic in part (b)? Explain.

- Show that the two-sample z test for comparing two proportions and the chi-square test for a 2-by-2 two-way table give equivalent results.

R11.6. Identity theft A study of identity theft looked at how well consumers protect themselves from this increasingly prevalent crime. The behaviors of 61 randomly selected college students were compared with the behaviors of 59 randomly selected nonstudents.[39] One of the questions was "When asked to create a password, I have used either my mother's maiden name, or my pet's name, or my birth date, or the last four digits of my social security number, or a series of consecutive numbers." For the students, 22 agreed with this statement while 30 of the nonstudents agreed.

(a) Display the data in a two-way table and perform the appropriate chi-square test. Summarize the results.

(b) Reanalyze the data using the methods for comparing two proportions that we studied in Chapter 10. Compare the results and verify that the chi-square statistic is the square of the z statistic.

Chapter 11 AP Statistics Practice Test

Section I: Multiple Choice *Select the best answer for each question.*

T11.1. A chi-square goodness-of-fit test is used to test whether a 0 to 9 spinner is "fair" (that is, the outcomes are all equally likely). The spinner is spun 100 times, and the results are recorded. The degrees of freedom for the test will be

(a) 8.
(b) 9.
(c) 10.
(d) 99.
(e) None of these.

Exercises T11.2 and T11.3 refer to the following setting. Recent revenue shortfalls in a midwestern state led to a reduction in the state budget for higher education. To offset the reduction, the largest state university proposed a 25% tuition increase. It was determined that such an increase was needed simply to compensate for the lost support from the state. Separate random samples of 50 freshmen, 50 sophomores, 50 juniors, and 50 seniors from the university were asked whether or not they were strongly opposed to the increase, given that it was the minimum increase necessary to maintain the university's budget at current levels. The results are given in the following table.

	Year			
Strongly Opposed?	Freshman	Sophomore	Junior	Senior
Yes	39	36	29	18
No	11	14	21	32

T11.2. Which hypotheses would be appropriate for performing a chi-square test?

(a) The null hypothesis is that the closer students get to graduation, the less likely they are to be opposed to tuition increases. The alternative is that how close students are to graduation makes no difference in their opinion.

(b) The null hypothesis is that the mean number of students who are strongly opposed is the same for each of the four years. The alternative is that the mean is different for at least two of the four years.

(c) The null hypothesis is that the distribution of student opinion about the proposed tuition increase is the same for each of the four years at this university. The alternative is that the distribution is different for at least two of the four years.
(d) The null hypothesis is that year in school and student opinion about the tuition increase in the sample are independent. The alternative is that these variables are dependent.
(e) The null hypothesis is that there is an association between year in school and opinion about the tuition increase at this university. The alternative hypothesis is that these variables are not associated.

T11.3. The conditions for carrying out the chi-square test in exercise T11.2 are

I. Separate random samples from the populations of interest.
II. Expected counts large enough.
III. The samples themselves and the individual observations in each sample are independent.

Which of the conditions is (are) satisfied in this case?

(a) I only
(b) II only
(c) I and II only
(d) II and III only
(e) I, II, and III

Exercises T11.4 to T11.6 refer to the following setting. A random sample of traffic tickets given to motorists in a large city is examined. The tickets are classified according to the race of the driver. The results are summarized in the following table:

Race:	White	Black	Hispanic	Other
Number of tickets:	69	52	18	9

The proportion of this city's population in each of the racial categories listed above is as follows:

Race:	White	Black	Hispanic	Other
Proportion:	0.55	0.30	0.08	0.07

We wish to test H_0: The racial distribution of traffic tickets in the city is the same as the racial distribution of the city's population.

T11.4. Assuming H_0 is true, the expected number of Hispanic drivers who would receive a ticket is

(a) 8.
(b) 10.36.
(c) 11.
(d) 11.84.
(e) 12.

T11.5. We compute the value of the χ^2 statistic to be 6.58. Assuming that the conditions for inference are met, the *P*-value of our test is

(a) greater than 0.20.
(b) between 0.10 and 0.20.
(c) between 0.05 and 0.10.
(d) between 0.01 and 0.05.
(e) less than 0.01.

T11.6. The category that contributes the largest component to the χ^2 statistic is

(a) White.
(b) Black.
(c) Hispanic.
(d) Other.
(e) The answer cannot be determined since this is only a sample.

Exercises T11.7 to T11.10 refer to the following setting. All current-carrying wires produce electromagnetic (EM) radiation, including the electrical wiring running into, through, and out of our homes. High-frequency EM radiation is thought to be a cause of cancer. The lower frequencies associated with household current are generally assumed to be harmless. To investigate this, researchers visited the addresses of a random sample of children who had died of some form of cancer (leukemia, lymphoma, or some other type) and classified the wiring configuration outside the dwelling as either a high-current configuration (HCC) or a low-current configuration (LCC). Here are the data:

	Leukemia	Lymphoma	Other cancers
HCC	52	10	17
LCC	84	21	31

Computer software was used to analyze the data. The output is given below. It includes the cell counts, some of the expected cell counts, and the value of the χ^2 statistic. In the table, expected counts are printed below observed counts and enclosed within parentheses.

	Leukemia	Lymphoma	Other cancers	Total
HCC	52 (49.97)	10	17 (17.64)	79
LCC	84 (86.03)	21	31 (30.36)	136
Total	136	31	48	215

$$\chi^2 = 0.082 + 0.170 + 0.023 + 0.048 + 0.099 + 0.013 = 0.435$$

T11.7. The appropriate degrees of freedom for the χ^2 statistic is

(a) 1. (b) 2. (c) 3. (d) 4. (e) 5.

T11.8. The expected count of cases with lymphoma in homes with an HCC is

(a) $\frac{79 \cdot 31}{215}$. (b) $\frac{10 \cdot 21}{215}$. (c) $\frac{79 \cdot 31}{10}$. (d) $\frac{136 \cdot 31}{215}$.

(e) None of these.

T11.9. Which of the following may we conclude, based on the test results?

(a) There is strong evidence of an association between wiring configuration and the chance that a child will develop some form of cancer.

(b) HCC either causes cancer directly or is a major contributing factor to the development of cancer in children.

(c) Leukemia is the most common type of cancer among children.

(d) There is not much evidence of an association between wiring configuration and the type of cancer that caused the deaths of children in the study.

(e) There is weak evidence that HCC causes cancer in children.

T11.10. A Type I error would occur if we conclude that

(a) HCC wiring caused cancer when it actually didn't.

(b) HCC wiring didn't cause cancer when it actually did.

(c) there is no association between the type of wiring and the form of cancer when there actually is an association.

(d) there is an association between the type of wiring and the form of cancer when there actually is no association.

(e) the type of wiring and the form of cancer have a positive correlation when they actually don't.

Section II: Free Response *Show all your work. Indicate clearly the methods you use, because you will be graded on the correctness of your methods as well as on the accuracy and completeness of your results and explanations.*

T11.11. A large distributor of gasoline claims that 60% of all cars stopping at their service stations choose regular unleaded gas and that premium and supreme are each selected 20% of the time. To investigate this claim, researchers collected data from a random sample of drivers who put gas in their vehicles at the distributor's service stations in a large city. The results were as follows:

Gasoline Selected

Regular	Premium	Supreme
261	51	88

Carry out a significance test of the distributor's claim. Use a 5% significance level.

T11.12. A study conducted in Charlotte, North Carolina, tested the effectiveness of three police responses to spouse abuse: (1) advise and possibly separate the couple, (2) issue a citation to the offender, and (3) arrest the offender. Police officers were trained to recognize eligible cases. When presented with an eligible case, a police officer called the dispatcher, who would randomly assign one of the three available treatments to be administered. There were a total of 650 cases in the study. Each case was classified according to whether the abuser was subsequently arrested within six months of the original incident.[40]

Treatment	No subsequent arrest	Subsequent arrest
1. Advise and separate	187	25
2. Citation	181	43
3. Arrest	175	39

(a) Explain the purpose of the random assignment in the design of this study.

(b) Construct a well-labeled graph that is suitable for comparing the effectiveness of the three treatments.

(c) We want to use these data to perform a test of H_0: $p_1 = p_2 = p_3$, where p_i = the true proportion of spouse abusers like the ones in this study who would be arrested again within six months after receiving treatment i. State an appropriate alternative hypothesis.

(d) Assume that all the conditions for performing the test in part (b) are met. The test yields $\chi^2 = 5.063$ and a P-value of 0.0796. Interpret this P-value in context. What conclusion should we draw from the study?

T11.13. In the United States, there is a strong relationship between education and smoking: well-educated people are less likely to smoke. Does a similar relationship hold in France? To find out, researchers recorded the level of education and smoking status of a random sample of 459 French men aged 20 to 60 years.[41] The two-way table below displays the data.

	Smoking Status			
Education	**Nonsmoker**	**Former**	**Moderate**	**Heavy**
Primary school	56	54	41	36
Secondary school	37	43	27	32
University	53	28	36	16

(a) Is the relationship between smoking status and educational level statistically significant? Give appropriate evidence to support your answer.

(b) Which cell in the table contributes most to the relationship in part (a)? Justify your answer.

Chapter 12

case study

More about Regression

Do Longer Drives Mean Lower Scores on the PGA Tour?

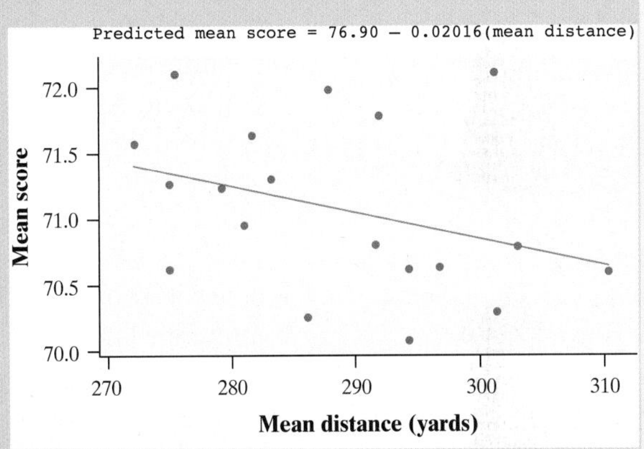

FIGURE 12.1 Scatterplot and least-squares regression analysis of mean score versus mean drive distance for a random sample of 19 players on the 2008 PGA Tour.

Recent advances in technology have led to golf balls that fly farther, clubs that generate more speed at impact, and swings that have been perfected through computer video analysis. In addition, today's professional golfers are fitter than ever. The net result is many more players who routinely hit drives traveling 300 yards or more. Does greater distance off the tee translate to better (lower) scores?

Data on mean drive distance (in yards) and mean score per round from an SRS of 19 of the 197 players on the 2008 Professional Golfers Association (PGA) Tour are shown in the table below. Figure 12.1 is a scatterplot of the data with results from a least-squares regression analysis added. The graph shows that there is a moderately weak negative linear relationship between mean drive distance and mean score for the 19 players in the sample.[1]

Player	Mean distance (yards)	Mean score per round	Player	Mean distance (yards)	Mean score per round	Player	Mean distance (yards)	Mean score per round
Paul Claxton	275.4	72.1	Robert Gamez	281.6	71.64	Pat Perez	294.2	70.08
K. J. Choi	286.1	70.26	Nick Watney	302.9	70.79	Ryan Palmer	294.2	70.62
John Huston	291.5	70.81	Tom Scherrer	287.7	71.98	Olin Browne	272.1	71.58
Cliff Kresge	280.9	70.95	J. B. Holmes	310.3	70.6	Charles Warren	301.1	72.11
Davis Love III	301.3	70.3	Bill Haas	296.7	70.64	Omar Uresti	274.9	71.27
Chris Riley	279.1	71.24	Glen Day	274.9	70.62			
Nick Flanagan	291.8	71.79	Scott McCarron	283.1	71.31			

The slope of the least-squares regression line for the sample data is about −0.02. This line predicts a 0.02 decrease in average score per round for each 1 yard increase in average driving distance. But a slope of −0.02 is very close to 0. Do these data really give convincing evidence that the slope of the true regression line for *all* 197 PGA Tour golfers is negative? By the end of this chapter, you'll have developed the tools you need to answer this question.

When a scatterplot shows a linear relationship between a quantitative explanatory variable x and a quantitative response variable y, we can use the least-squares line fitted to the data to predict y for a given value of x. If the data are a random sample from a larger population, we need statistical inference to answer questions like these:

- Is there really a linear relationship between x and y in the population, or could the pattern we see in the scatterplot plausibly happen just by chance?
- In the population, how much will the predicted value of y change for each increase of 1 unit in x? What's the margin of error for this estimate?

If the data come from a randomized experiment, the values of the explanatory variable correspond to the levels of some factor that is being manipulated by the researchers. For instance, researchers might want to investigate how temperature affects the life span of mosquitoes. They could set up several tanks at each of several different temperatures and then randomly assign hundreds of mosquitoes to each of the tanks. The response variable of interest is the average time (in days) from hatching to death. Suppose that a scatterplot of average life span versus temperature has a linear form. We need statistical inference to decide if it's plausible that there is no linear relationship between the variables, and that the pattern observed in the graph is due simply to the chance involved in the random assignment.

It is conventional to refer to a scatterplot of the points (x, y) as a graph of y versus x. So a scatterplot of life span versus temperature uses life span as the response variable and temperature as the explanatory variable.

In Section 12.1, we will show you how to estimate and test claims about the slope of the population (true) regression line that describes the relationship between two quantitative variables. The following Activity gives you a preview of inference for linear regression.

ACTIVITY *The helicopter experiment*

MATERIALS: 50 copies of the George Box helicopter template from the *Teacher's Resource Binder* (or the book's Web site), scissors, tape measures, stopwatches

Is there a linear relationship between the height from which a paper helicopter is released and the time it takes to hit the ground? In this Activity, your class will perform an experiment to investigate this question.[2]

1. Follow the directions provided with the template to construct 50 long-rotor helicopters.

2. Randomly assign 10 helicopters to each of five different drop heights. (The experiment works best for drop heights of 5 feet or more.)

3. Work in teams to release the helicopters from their assigned drop heights and record the descent times.

4. Make a scatterplot of the data in Step 3. Find the least-squares regression line for predicting descent time from drop height.

5. Interpret the slope of the regression line from Step 4 in context. What is your best guess for the increase in descent time for each additional foot of drop height?

6. Does it seem plausible that there is no linear relationship between descent time and drop height and that the observed slope happened just by chance due to the random assignment? Discuss this as a class.

Sometimes a scatterplot reveals that the relationship between two quantitative variables has a strong curved form. One strategy is to transform one or both variables so that the graph shows a linear pattern. Then we can use least-squares regression to fit a linear model to the data. Section 12.2 examines methods of transforming data to achieve linearity.

12.1 Inference for Linear Regression

In Section 12.1, you'll learn about:

- Sampling distribution of b
- Conditions for regression inference
- Estimating the parameters
- Constructing a confidence interval for the slope
- Performing a significance test for the slope

In Chapter 3, we examined data on eruptions of the Old Faithful geyser. Figure 12.2 is a scatterplot of the duration and interval of time until the next eruption for all 222 recorded eruptions in a single month. The least-squares regression line for this population of data has been added to the graph. It has slope 10.36 and y intercept 33.97. We call this the **population regression line** (or *true regression line*) because it uses all the observations that month.

Population (true) regression line

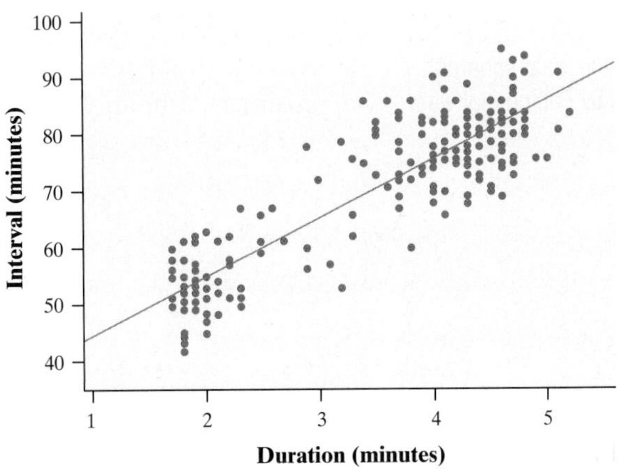

FIGURE 12.2 Scatterplot of the duration and interval between eruptions of Old Faithful for all 222 eruptions in a single month. The population least-squares line is shown in blue.

Sample regression line

Suppose we take an SRS of 20 eruptions from the population and calculate the least-squares regression line $\hat{y} = a + bx$ for the sample data. How does the slope b of the **sample regression line** (also called the *estimated regression line*) relate to the slope of the population regression line? Figure 12.3 shows the results of taking three different SRSs of 20 Old Faithful eruptions in this month. Each graph displays the selected points and the least-squares regression line for that sample.

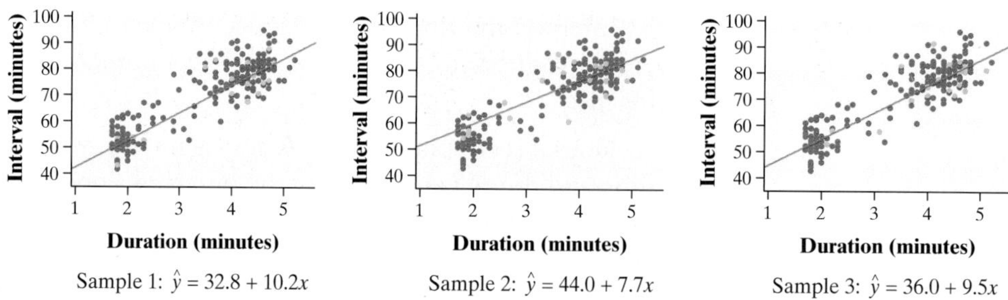

FIGURE 12.3 Scatterplots and least-squares regression lines for three different SRSs of 20 Old Faithful eruptions.

Notice that the slopes of the sample regression lines—10.2, 7.7, and 9.5—vary quite a bit from the slope of the population regression line, 10.36. The pattern of variation in the slope b is described by its sampling distribution.

Sampling Distribution of b

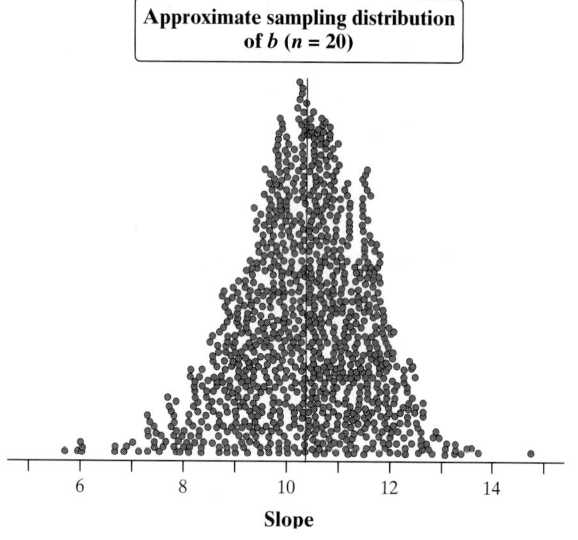

FIGURE 12.4 Dotplot of the slope b of the least-squares regression line in 1000 simulated SRSs by Fathom software.

Confidence intervals and significance tests about the slope of the population regression line are based on the sampling distribution of b, the slope of the sample regression line. We used Fathom software to simulate choosing 1000 SRSs of $n = 20$ from the Old Faithful data, each time calculating the equation $\hat{y} = a + bx$ of the least-squares regression line for the sample. Figure 12.4 displays the values of the slope b for the 1000 sample regression lines. We have added a vertical line at 10.36 corresponding to the actual slope of the population regression line. Let's describe this approximate sampling distribution of b.

Shape: We can see that the distribution of b-values is roughly symmetric and unimodal. Figure 12.5(a) is a Normal probability plot of these sample regression line slopes. The strong linear pattern in the graph tells us that the approximate sampling distribution of b is close to Normal.

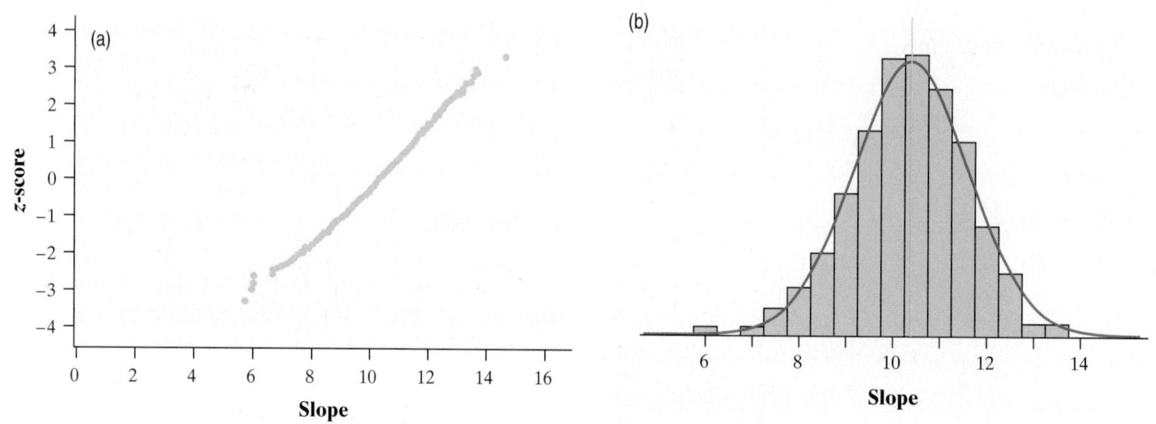

FIGURE 12.5 (a) Normal probability plot and (b) histogram of the 1000 sample regression line slopes from Figure 12.4. The red density curve in Figure 12.5(b) is for a Normal distribution with mean 10.36 (marked by the yellow line) and standard deviation 1.30.

Center: The mean of the 1000 b-values is 10.32. This value is quite close to the slope of the population (true) regression line, 10.36.

Spread: The standard deviation of the 1000 b-values is 1.31. Later, we will see that the standard deviation of the sampling distribution of b is actually 1.30.

Figure 12.5(b) is a histogram of the b-values from the 1000 simulated SRSs. We have superimposed the density curve for a Normal distribution with mean 10.36 and standard deviation 1.30. This curve models the approximate sampling distribution of the slope quite well.

Conditions for Regression Inference

We can fit a least-squares line to *any* data relating two quantitative variables, but the results are useful only if the scatterplot shows a linear pattern. Inference about regression requires more detailed conditions. Because the conclusions of inference often concern some population, the conditions describe the population and how the data are produced from it. The slope b and intercept a of the least-squares line are *statistics*. That is, we calculate them from the sample data. These statistics would take somewhat different values if we repeated the data production process. To do inference, think of a and b as estimates of unknown *parameters* α and β that describe the population of interest.

Note that the symbols α and β here refer to the intercept and slope, respectively, of the population regression line. They are in no way related to Type I and Type II error probabilities, which are sometimes designated by these same symbols.

Here are the conditions for performing inference about the linear regression model.

Conditions for Regression Inference

Suppose we have n observations on an explanatory variable x and a response variable y. Our goal is to study or predict the behavior of y for given values of x.

- **Linear** The actual relationship between x and y is linear. For any fixed value of x, the mean response μ_y falls on the population (true) regression line $\mu_y = \alpha + \beta x$. The slope β and intercept α are usually unknown parameters.
- **Independent** Individual observations are independent of each other.
- **Normal** For any fixed value of x, the response y varies according to a Normal distribution.
- **Equal variance** The standard deviation of y (call it σ) is the same for all values of x. The common standard deviation σ is usually an unknown parameter.
- **Random** The data come from a well-designed random sample or randomized experiment.

The acronym LINER should help you remember the conditions for inference about regression.

Figure 12.6 shows the regression model when the conditions are met in picture form. The line in the figure is the population regression line $\mu_y = \alpha + \beta x$. For each possible value of the explanatory variable x, the mean of the responses $\mu(y \mid x)$ moves along this line. The Normal curves show how y will vary when x is held fixed at different values. All the curves have the same standard deviation σ, so the variability of y is the same for all values of x. The value of σ determines whether the points fall close to the population regression line (small σ) or are widely scattered (large σ).

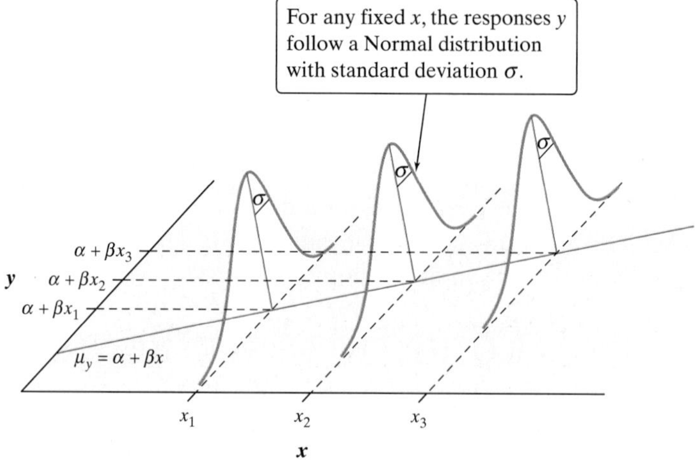

FIGURE 12.6 The regression model when the conditions for inference are met. The line is the population (true) regression line, which shows how the mean response μ_y changes as the explanatory variable x changes. For any fixed value of x, the observed response y varies according to a Normal distribution having mean μ_y and standard deviation σ.

THINK ABOUT IT

What does the regression model in Figure 12.6 tell us? Consider the population of all eruptions of the Old Faithful geyser in a given year. For each eruption, let x be the duration (in minutes) and y be the interval of time (in minutes) until the next eruption. Suppose that the conditions for regression inference are met for this data set, that the population regression line is $\mu_y = 34 + 10.4x$, and that the spread around the line is given by $\sigma = 6$. Let's focus on the eruptions that lasted $x = 2$ minutes. For this "subpopulation":

- The average interval of time until the next eruption is $\mu_y = 34 + 10.4(2) = 54.8$ minutes.
- The intervals of time until the next eruption follow a Normal distribution with mean 54.8 minutes and standard deviation 6 minutes.
- For about 95% of these eruptions, the interval of time y until the next eruption is between $54.8 - 2(6) = 42.8$ minutes and $54.8 + 2(6) = 66.8$ minutes.

You should always check the conditions before doing inference about the regression model.

Checking the Conditions Although the conditions for regression inference are a bit complicated, it is not hard to check for major violations. Most of the conditions involve the population regression line and the deviations of responses from this line. We usually can't observe the population line, but the least-squares line estimates it, and the residuals estimate the deviations from the population line. We can check several of the conditions for regression inference by looking at graphs of the residuals. Start by making a histogram or Normal probability plot of the residuals and also a residual plot. Here's a summary of how to check the conditions one by one.

How to Check the Conditions for Regression Inference

- **Linear** Examine the scatterplot to check that the overall pattern is roughly linear. Look for curved patterns in the residual plot. Check to see that the residuals center on the "residual = 0" line at each x-value in the residual plot.
- **Independent** Look at how the data were produced. Random sampling and random assignment help ensure the independence of individual observations. If sampling is done without replacement, remember to check that the population is at least 10 times as large as the sample (10% condition).
- **Normal** Make a stemplot, histogram, or Normal probability plot of the residuals and check for clear skewness or other major departures from Normality.
- **Equal variance** Look at the scatter of the residuals above and below the "residual = 0" line in the residual plot. The amount of scatter should be roughly the same from the smallest to the largest x-value.
- **Random** See if the data were produced by random sampling or a randomized experiment.

Let's look at an example that illustrates the process of checking conditions.

The Helicopter Experiment

Checking conditions

Mrs. Barrett's class did a variation of the helicopter experiment on page 738. Students randomly assigned 14 helicopters to each of five drop heights: 152 centimeters (cm), 203 cm, 254 cm, 307 cm, and 442 cm. Teams of students released the 70 helicopters in a predetermined random order and measured the flight times in seconds. The class used Minitab to carry out a least-squares regression analysis for these data. A scatterplot, residual plot, histogram, and Normal probability plot of the residuals are shown below.

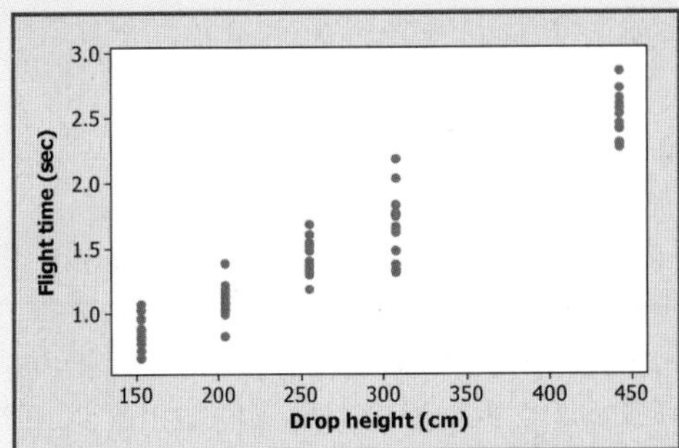

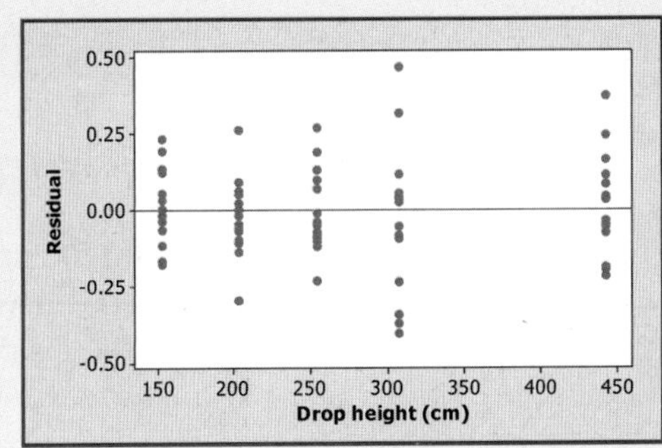

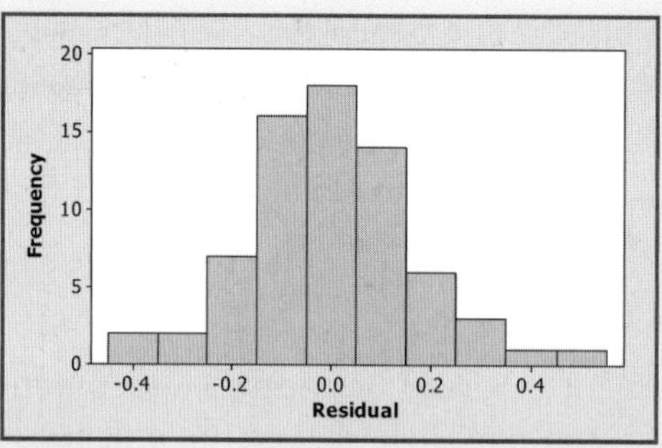

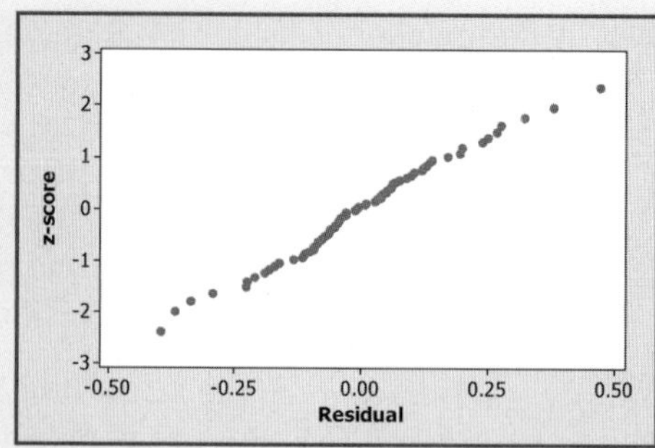

PROBLEM: Check whether the conditions for performing inference about the regression model are met.

SOLUTION: We'll use our LINER acronym!

- *Linear* The scatterplot shows a clear linear form. For each drop height used in the experiment, the residuals are centered on the horizontal line at 0. The residual plot shows a random scatter about the horizontal line.
- *Independent* Because the helicopters were released in a random order and no helicopter was used twice, knowing the result of one observation should give no additional information about another observation.
- *Normal* The histogram of the residuals is single-peaked, unimodal, and somewhat bell-shaped. In addition, the Normal probability plot is very close to linear.
- *Equal variance* The residual plot shows a similar amount of scatter about the residual = 0 line for the 152, 203, 254, and 442 cm drop heights. Flight times (and the corresponding residuals) seem to vary more for the helicopters that were dropped from a height of 307 cm.
- *Random* The helicopters were randomly assigned to the five possible drop heights.

Except for a slight concern about the equal-variance condition, we should be safe performing inference about the regression model in this setting.

You will always see some irregularity when you look for Normality and fixed standard deviation in the residuals, especially when you have few observations. Don't overreact to minor violations of the conditions. Inference for regression is not very sensitive to lack of Normality, especially when we have many observations. *Beware of influential observations* (see Section 3.2), which can greatly affect the results of inference. You will sometimes find that, as the response y gets larger, so does the scatter of the residuals. You cannot trust the results of inference when this happens.

CAUTION

AP EXAM TIP The AP exam formula sheet gives $\hat{y} = b_0 + b_1x$ for the equation of the sample (estimated) regression line. We will stick with our simpler notation, $\hat{y} = a + bx$, which is also used by TI calculators. Just remember: the coefficient of x is always the slope, no matter what symbol is used.

Estimating the Parameters

When the conditions are met, we can do inference about the regression model $\mu_y = \alpha + \beta x$. The first step is to estimate the unknown parameters. If we calculate the least-squares regression line $\hat{y} = a + bx$, the slope b is an unbiased estimator of the true slope β, and the y intercept a is an unbiased estimator of the true y intercept α. The remaining parameter is the standard deviation σ, which describes the variability of the response y about the population (true) regression line.

The least-squares regression line computed from the sample data estimates the population (true) regression line. So the residuals estimate how much y varies about the population line. Because σ is the standard deviation of responses about the population (true) regression line, we estimate it by the standard deviation of the residuals

Since s is estimated from data, it is sometimes called the *regression standard error* or the *root mean squared error*.

$$s = \sqrt{\frac{\sum \text{residuals}^2}{n-2}} = \sqrt{\frac{\sum (y_i - \hat{y}_i)^2}{n-2}}$$

Recall from Chapter 3 that s describes the size of a "typical" prediction error.

The Helicopter Experiment

Estimating the parameters

Computer output from the least-squares regression analysis on the helicopter data for Mrs. Barrett's class is shown below.

Regression Analysis: Flight time (sec) versus Drop height (cm)

Predictor	Coef	SE Coef	T	P
Constant	−0.03761	0.05838	−0.64	0.522
Drop height (cm)	0.0057244	0.0002018	28.37	0.000

S = 0.168181 R-Sq = 92.2% R-Sq(adj) = 92.1%

When we compute the least-squares regression line based on a random sample of data, we can think about doing inference for the population regression line. When our least-squares regression line is based on data from a randomized experiment, as in this example, the resulting inference is about the *true* regression line relating the explanatory and response variables. We'll follow this convention from now on.

The least-squares regression line for these data is

$$\widehat{\text{flight time}} = -0.03761 + 0.0057244(\text{drop height})$$

The slope β of the true regression line says how much the average flight time of the paper helicopters increases when the drop height increases by 1 centimeter. Because $b = 0.0057244$ estimates the unknown β, we estimate that, on average, flight time increases by about 0.0057244 seconds for each additional centimeter of drop height.

We need the intercept $a = -0.03761$ to draw the line and make predictions, but it has no statistical meaning in this example. No helicopter was dropped from less than 150 cm, so we have no data near $x = 0$. We might expect the actual y intercept α of the true regression line to be 0 because it should take no time for a helicopter to fall no distance. The y intercept of the sample regression line is −0.03761, which is pretty close to 0.

Our estimate for the standard deviation σ of flight times about the true regression line at each x-value is $s = 0.168$ seconds. This is also the size of a typical prediction error if we use the least-squares regression line to predict the flight time of a helicopter from its drop height.

It is possible to do inference about any of the three parameters in the regression model: α, β, or σ. However, the slope β of the population (true) regression line is usually the most important parameter in a regression problem. So we'll restrict our attention to inference about the slope.

The Sampling Distribution of b Again Let's return to our earlier exploration of Old Faithful eruptions. For all 222 eruptions in a single month, the population regression line for predicting the interval of time until the next eruption y from the duration of the previous eruption x is $\mu_y = 33.97 + 10.36x$. The standard deviation of responses about this line is given by $\sigma = 6.159$. Figure 12.7 shows the approximate sampling distribution of b that we generated before.

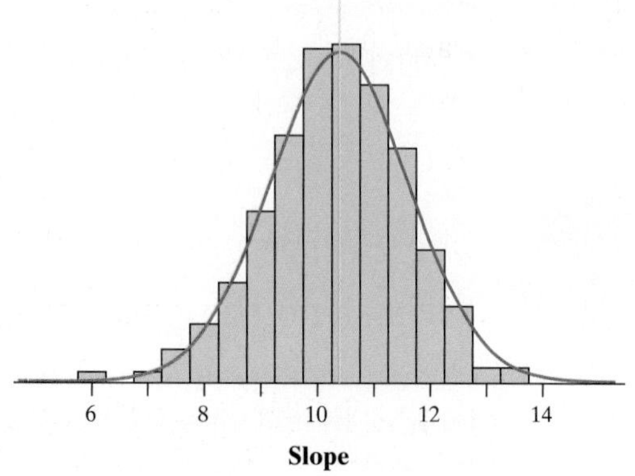

FIGURE 12.7 The approximate sampling distribution of the slope b for samples of size $n = 20$ eruptions. This distribution has a roughly Normal shape with mean about 10.36 and standard deviation 1.30.

If we take *all* possible SRSs of 20 eruptions from the population, we get the actual sampling distribution of b. Can you guess its shape, center, and spread?

Shape: Normal

Center: $\mu_b = \beta = 10.36$ (b is an unbiased estimator of β)

Spread: $\sigma_b = \dfrac{\sigma}{s_x\sqrt{n-1}} = \dfrac{6.159}{1.083\sqrt{20-1}} = 1.30$ (very close to the 1.31 from our simulation—we'd be a little surprised if you guessed the formula for the spread!)

Standard error of the slope

In practice, we don't know σ for the population regression line. So we estimate it with the standard deviation of the residuals, s. Then we estimate the spread of the sampling distribution of b with the **standard error of the slope**

$$\text{SE}_b = \frac{s}{s_x\sqrt{n-1}}$$

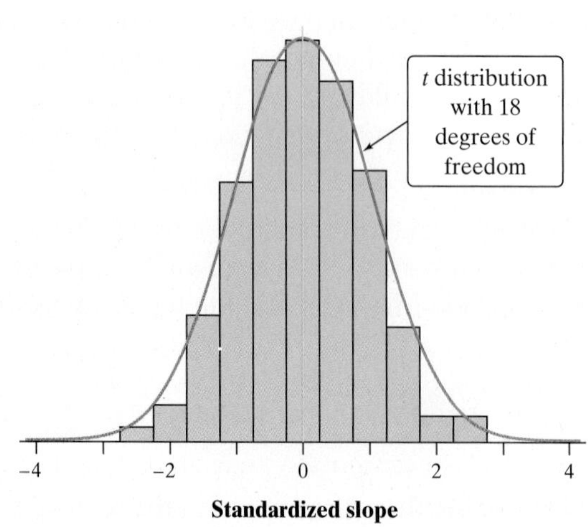

FIGURE 12.8 The sampling distribution of the standardized slope values has a t distribution with df $= n - 2$.

What happens if we transform the values of b by standardizing? Since the sampling distribution of b is Normal, the statistic

$$z = \frac{b - \beta}{\sigma_b}$$

has the standard Normal distribution. Replacing the standard deviation σ_b of the sampling distribution with its standard error gives the statistic

$$t = \frac{b - \beta}{\text{SE}_b}$$

which has a t distribution with $n - 2$ degrees of freedom. (The explanation of why df $= n - 2$ is beyond the scope of this book.) Figure 12.8 shows the result of standardizing the values in the sampling distribution of b. The superimposed curve is a t distribution with df $= 20 - 2 = 18$.

Constructing a Confidence Interval for the Slope

The slope β of the population (true) regression line $\mu_y = \alpha + \beta x$ is the rate of change of the mean response as the explanatory variable increases. We often want to estimate β. The slope b of the sample regression line is our point estimate for β. A confidence interval is more useful than the point estimate because it shows how precise the estimate b is likely to be. The confidence interval for β has the familiar form

$$\text{statistic} \pm (\text{critical value}) \cdot (\text{standard deviation of statistic})$$

Because we use the statistic b as our estimate, the confidence interval is

$$b \pm t^* \text{SE}_b$$

We call this a ***t* interval for the slope.** Here are the details.

AP EXAM TIP The AP exam formula sheet gives the formula for the standard error of the slope as

$$s_{b_1} = \frac{\sqrt{\dfrac{\sum (y_i - \hat{y}_i)^2}{n-2}}}{\sqrt{\sum (x_i - \bar{x})^2}}$$

The numerator is just a fancy way of writing the standard deviation of the residuals s. Can you show that the denominator of this formula is the same as ours?

t Interval for the Slope of a Least-Squares Regression Line

When the conditions for regression inference are met, a level C confidence interval for the slope β of the population (true) regression line is

$$b \pm t^* \text{SE}_b$$

In this formula, the standard error of the slope is

$$\text{SE}_b = \frac{s}{s_x \sqrt{n-1}}$$

and t^* is the critical value for the t distribution with df $= n - 2$ having area C between $-t^*$ and t^*.

Although we give the formula for the standard error of b, you should rarely have to calculate it by hand. Computer output gives the standard error SE_b along with b itself. However we get it, SE_b estimates how much the slope of the sample regression line typically varies from the slope of the population (true) regression line if we repeat the data production process many times.

The Helicopter Experiment

A confidence interval for β

Earlier, we used Minitab to perform a least-squares regression analysis on the helicopter data for Mrs. Barrett's class. Recall that the data came from dropping 70 paper helicopters from various heights and measuring the flight times. Some computer output from this regression is shown on the next page. We checked conditions for performing inference earlier.

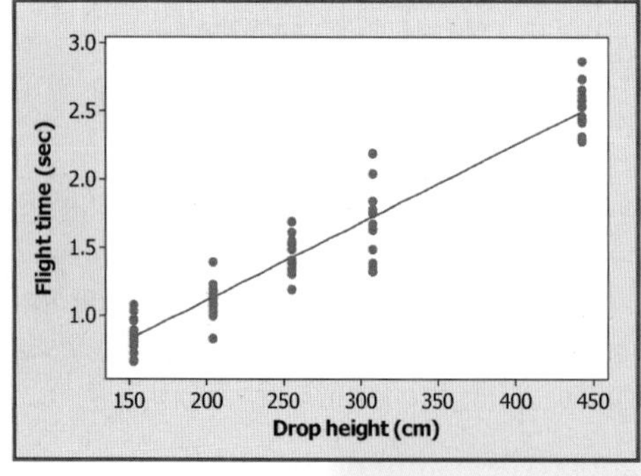

Regression Analysis: Flight time versus Drop height

Predictor	Coef	SE Coef	T	P
Constant	−0.03761	0.05838	−0.64	0.522
Drop height (cm)	0.0057244	0.0002018	28.37	0.000

S = 0.168181 R-Sq = 92.2% R-Sq(adj) = 92.1%

PROBLEM:

(a) Identify the standard error of the slope SE_b from the computer output. Interpret this value in context.

(b) Find the critical value for a 95% confidence interval for the slope of the true regression line. Then calculate the confidence interval. Show your work.

(c) Interpret the interval from part (b) in context.

(d) Explain the meaning of "95% confident" in context.

SOLUTION:

(a) We got the value of the standard error of the slope, 0.0002018, from the "SE Coef" column in the computer output. If we repeated the random assignment many times, the slope of the sample regression line would typically vary by about 0.0002 from the slope of the true regression line for predicting flight time from drop height.

(b) Because the conditions are met, we can calculate a t interval for the slope β based on a t distribution with df $= n - 2 = 70 - 2 = 68$. Using the more conservative df $= 60$ from Table B gives $t^* = 2.000$. The 95% confidence interval is

$$b \pm t^* SE_b = 0.0057244 \pm 2.000(0.0002018) = 0.0057244 \pm 0.0004036$$
$$= (0.0053208, 0.0061280)$$

Note: From `invT(.025,68)`, we get $t^* = 1.995$. The resulting 95% confidence interval is

$$0.0057244 \pm 1.995(0.0002018) = 0.0057244 \pm 0.0004026$$
$$= (0.0053218, 0.0061270)$$

This interval is slightly narrower due to the more precise t^* critical value.

(c) We are 95% confident that the interval from 0.0053208 to 0.0061280 seconds per cm captures the slope of the true regression line relating the flight time y and drop height x of paper helicopters.

(d) If we repeat the experiment many, many times, and use the method in part (a) to construct a confidence interval each time, about 95% of the resulting intervals will capture the slope of the true regression line relating flight time y and drop height x of paper helicopters.

The values of t given in the computer regression output are *not* the critical values for a confidence interval. They come from carrying out a significance test about the y intercept or slope of the population (true) regression line. We'll discuss tests in more detail shortly.

You can find a confidence interval for the y intercept α of the population (true) regression line in the same way, using a and SE_a from the "Constant" row of the Minitab output. However, we are usually interested only in the point estimate for α that's provided in the computer output.

Does Fidgeting Keep You Slim?

Confidence interval for a slope

In Chapter 3, we examined data from a study that investigated why some people don't gain weight even when they overeat. Perhaps fidgeting and other "nonexercise activity" (NEA) explains why—some people may spontaneously increase nonexercise activity when fed more. Researchers deliberately overfed a random sample of 16 healthy young adults for 8 weeks. They measured fat gain (in kilograms) and change in energy use (in calories) from activity other than deliberate exercise—fidgeting, daily living, and the like—for each subject. Here are the data:

NEA change (cal):	−94	−57	−29	135	143	151	245	355
Fat gain (kg):	4.2	3.0	3.7	2.7	3.2	3.6	2.4	1.3

NEA change (cal):	392	473	486	535	571	580	620	690
Fat gain (kg):	3.8	1.7	1.6	2.2	1.0	0.4	2.3	1.1

Minitab output from a least-squares regression analysis for these data is shown below.

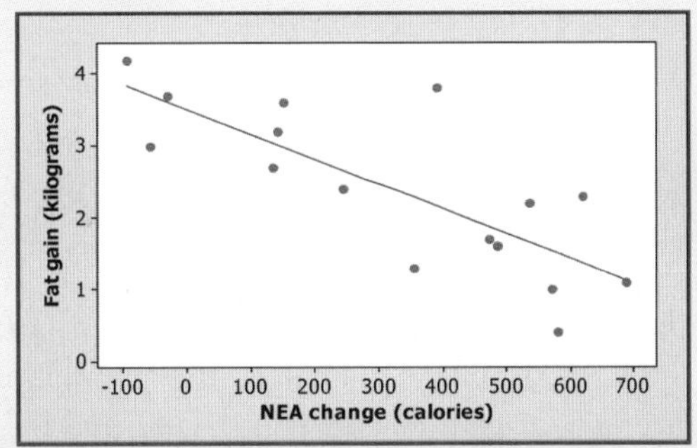

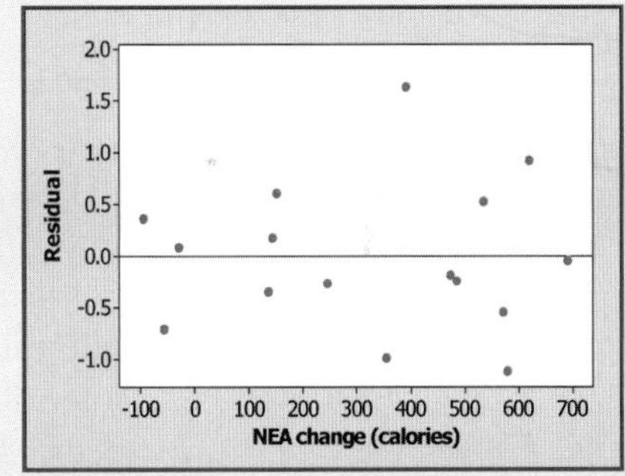

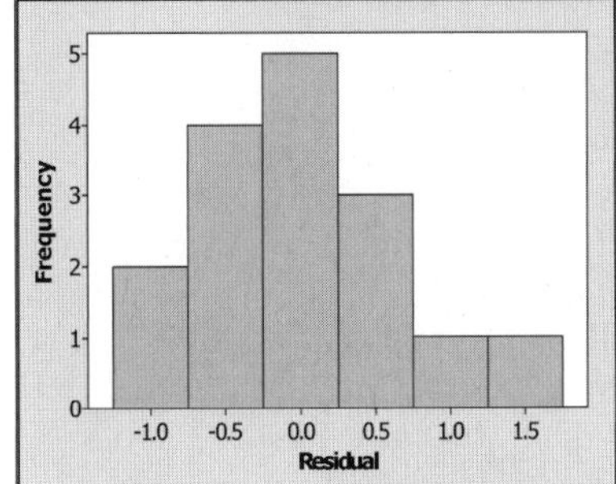

Regression Analysis: Fat gain versus NEA change

```
Predictor     Coef         SE Coef      T       P
Constant      3.5051       0.3036       11.54   0.000
NEA change    −0.0034415   0.0007414    −4.64   0.000
S = 0.739853 R-Sq = 60.6% R-Sq(adj) = 57.8%
```

PROBLEM: Construct and interpret a 90% confidence interval for the slope of the population regression line.

SOLUTION: We will follow the familiar four-step process.

STATE: We want to estimate the true slope β of the population regression line relating NEA change to fat gain at the 90% confidence level.

PLAN: If the conditions are met, we will use a t interval for the slope to estimate β.

- *Linear* The scatterplot shows a clear linear pattern. Also, the residual plot shows a random scatter of points about the "residual = 0" line.
- *Independent* Individual observations of fat gain should be independent if the study is carried out properly. Because researchers sampled without replacement, there have to be at least $10(16) = 160$ healthy young adults in the population of interest.
- *Normal* The histogram of the residuals is roughly symmetric and single-peaked, so there are no obvious departures from Normality.
- *Equal variance* It is hard to tell from so few points whether the scatter of points around the residual = 0 line is about the same at all x-values.
- *Random* The subjects in this study were randomly selected to participate.

With no obvious violation of any condition, we should be safe doing inference for regression.

DO: We use the t distribution with $16 - 2 = 14$ degrees of freedom to find the critical value. For a 90% confidence level, the critical value is $t^* = 1.761$. So the 90% confidence interval for β is

$$b \pm t^* SE_b = -0.0034415 \pm 1.761(0.0007414) = -0.0034415 \pm 0.0013056$$
$$= (-0.004747, -0.002136)$$

Since the data in this example came from a random sample, it makes sense to talk about the population regression line relating NEA change to fat gain for young adults.

CONCLUDE: We are 90% confident that the interval from -0.004747 to -0.002136 kg captures the actual slope of the population regression line relating NEA change to fat gain for healthy young adults.

For Practice *Try Exercise 9*

The predicted change in average fat gain is quite small for a 1-calorie increase in NEA. What if NEA increased by 100 calories? We can just multiply both endpoints of the confidence interval in the example by 100 to get a 90% confidence interval for the corresponding predicted change in average fat gain. The resulting interval is (−0.4747, −0.2136). That is, the population regression line predicts a decrease in average fat gain of between 0.4747 and 0.2136 kg for each 100-calorie increase in NEA.

CHECK YOUR UNDERSTANDING

In Chapter 3, we examined data on the body weights and backpack weights of a group of eight randomly selected ninth-grade students at the Webb Schools. Some Minitab output from a least-squares regression analysis for these data is shown.

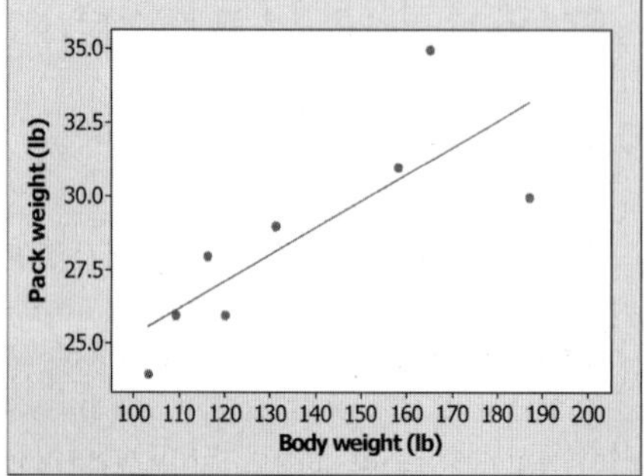

Regression Analysis: Pack weight versus Body weight

Predictor	Coef	SE Coef	T	P
Constant	16.265	3.937	4.13	0.006
Body weight	0.09080	0.02831	3.21	0.018

1. What conditions must be met for regression inference to be appropriate?

2. With such a small sample size, it is difficult to check several of the conditions for regression inference. Assume that the conditions are met. Construct and interpret a 95% confidence interval for the slope of the population regression line.

Performing a Significance Test for the Slope

When the conditions for inference are met, we can use the slope b of the sample regression line to construct a confidence interval for the slope β of the population (true) regression line. We can also perform a significance test to determine whether a specified value of β is plausible. The null hypothesis has the general form $H_0: \beta =$ hypothesized value. To do a test, standardize b to get the test statistic:

$$\text{test statistic} = \frac{\text{statistic} - \text{parameter}}{\text{standard deviation of statistic}}$$

$$t = \frac{b - \beta_0}{\text{SE}_b}$$

To find the *P*-value, use a t distribution with $n - 2$ degrees of freedom. Here are the details for the ***t* test for the slope.**

t Test for the Slope of the Population Regression Line

Suppose the conditions for inference are met. To test the hypothesis $H_0: \beta =$ hypothesized value, compute the test statistic

$$t = \frac{b - \beta_0}{\text{SE}_b}$$

Find the *P*-value by calculating the probability of getting a t statistic this large or larger in the direction specified by the alternative hypothesis H_a. Use the t distribution with df $= n - 2$.

$H_a: \beta >$ hypothesized value

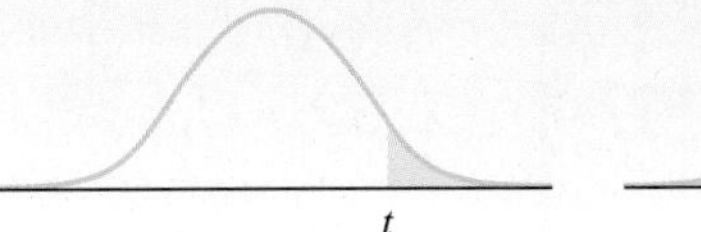

$H_a: \beta <$ hypothesized value

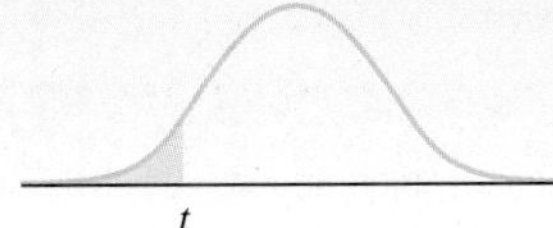

$H_a: \beta \neq$ hypothesized value

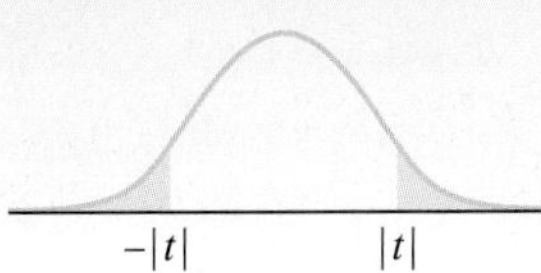

If sample data suggest a linear relationship between two variables, how can we determine whether this happened just by chance or whether there is actually a linear relationship between x and y in the population? By performing a test of $H_0: \beta = 0$. A regression line with slope 0 is horizontal. That is, the mean of y does not change at all when x changes. So H_0 says that there is *no linear relationship* between x and y in the population. Put another way, H_0 says that *linear regression of y on x is of no value for predicting y.*

Regression output from statistical software usually gives t and its two-sided *P*-value for a test of $H_0: \beta = 0$. For a one-sided test in the proper direction, just divide the *P*-value in the output by 2. The following example shows what we mean.

Crying and IQ

Significance test for β

Infants who cry easily may be more easily stimulated than others. This may be a sign of higher IQ. Child development researchers explored the relationship between the crying of infants 4 to 10 days old and their later IQ test scores. A snap of a rubber band on the sole of the foot caused the infants to cry. The researchers recorded the crying and measured its intensity by the number of peaks in the most active 20 seconds. They later measured the children's IQ at age three years using the Stanford-Binet IQ test. The table below contains data from a random sample of 38 infants.[3]

Crying	IQ	Crying	IQ	Crying	IQ	Crying	IQ
10	87	20	90	17	94	12	94
12	97	16	100	19	103	12	103
9	103	23	103	13	104	14	106
16	106	27	108	18	109	10	109
18	109	15	112	18	112	23	113
15	114	21	114	16	118	9	119
12	119	12	120	19	120	16	124
20	132	15	133	22	135	31	135
16	136	17	141	30	155	22	157
33	159	13	162				

AP EXAM TIP When you see a list of data values on an exam question, don't just start typing the data into your calculator. Read the question first. Often, additional information is provided that makes it unnecessary for you to enter the data at all. This can save you valuable time on the AP exam.

PROBLEM:

(a) Here is a scatterplot of the data with the least-squares line added. Describe what this graph tells you about the relationship between these two variables.

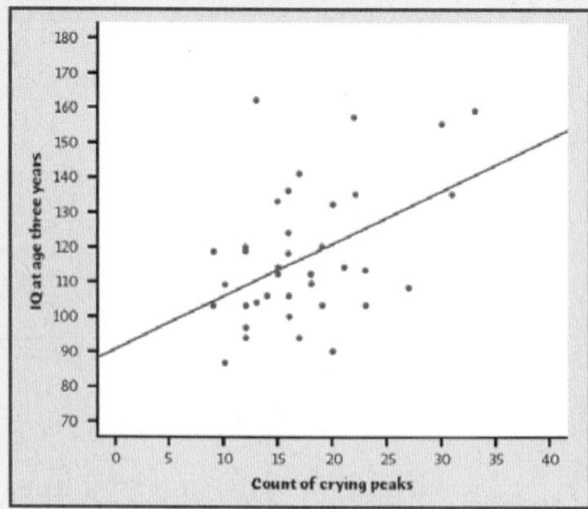

Minitab output from a linear regression on these data is shown below.

Regression Analysis: IQ versus Crycount

Predictor	Coef	SE Coef	T	P
Constant	91.268	8.934	10.22	0.000
Crycount	1.4929	0.4870	3.07	0.004

S = 17.50 R-Sq = 20.7% R-Sq(adj) = 18.5%

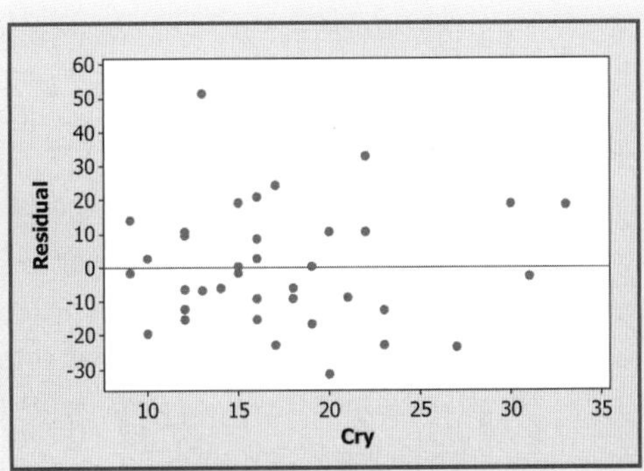

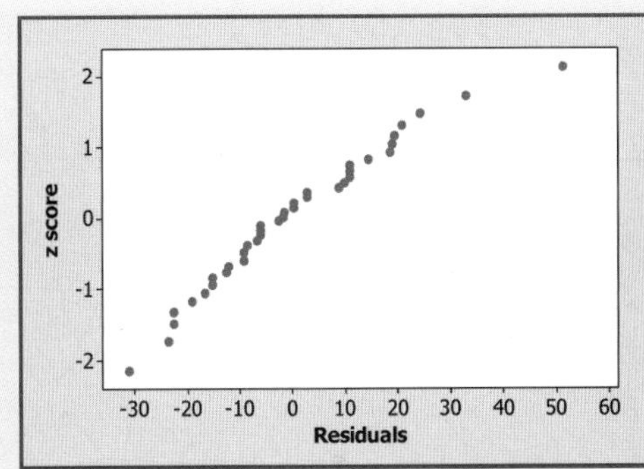

(b) What is the equation of the least-squares regression line for predicting IQ at age 3 from the number of crying peaks (crying count)? Define any variables you use.

(c) Interpret the slope and *y* intercept of the regression line in context.

(d) Do these data provide convincing evidence that there is a positive linear relationship between crying counts and IQ in the population of infants? Carry out an appropriate test to help answer this question.

SOLUTION:

(a) There is a moderately weak positive linear relationship between crying count and later IQ score, with no extreme outliers or potentially influential observations. It appears that children who cry more vigorously tend to have higher IQs.

(b) The equation of the least-squares line is

$$\text{predicted IQ score} = 91.268 + 1.4929\,(\text{cry count})$$

(c) *Slope:* For each additional crying peak in the most active 20 seconds, the regression line predicts an increase of about 1.5 IQ points. *y intercept:* The model predicts that an infant who doesn't cry when flicked with a rubber band will have a later IQ score of about 91. Because there were no infants who recorded fewer than 9 crying peaks in their most active 20 seconds, it is a risky extrapolation to use this line to predict the value of *y* when $x = 0$.

(d) We'll follow the four-step process.

STATE: We want to perform a test of

$$H_0: \beta = 0$$
$$H_a: \beta > 0$$

where β is the true slope of the population regression line relating crying count to IQ score. No significance level was given, so we'll use $\alpha = 0.05$.

PLAN: If the conditions are met, we will do a *t* test for the slope β.

- *Linear* The scatterplot suggests a moderately weak positive linear relationship between crying peaks and IQ. The residual plot shows a random scatter of points about the residual = 0 line.
- *Independent* Later IQ scores of individual infants should be independent. Due to sampling without replacement, there have to be at least $10(38) = 380$ infants in the population from which these children were selected.
- *Normal* The Normal probability plot of the residuals shows a slight curvature, which suggests that the responses may not be Normally distributed about the line at each *x*-value. With such a large sample size ($n = 38$), however, the *t* procedures are robust against departures from Normality.
- *Equal variance* The residual plot shows a fairly equal amount of scatter around the horizontal line at 0 for all *x*-values.
- *Random* We are told that these 38 infants were randomly selected.

With no obvious violations of the conditions, we proceed to inference.

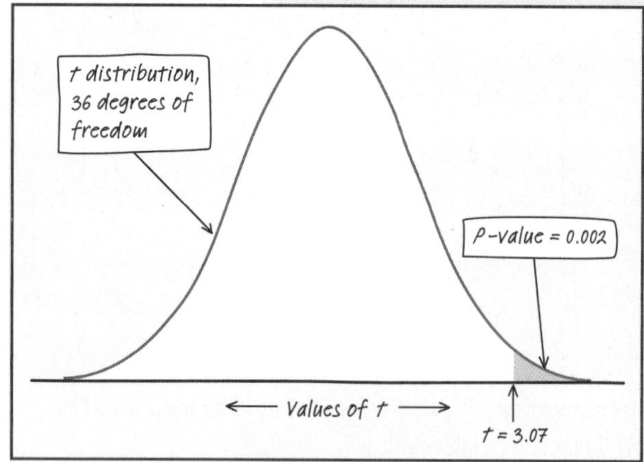

FIGURE 12.9 The P-value for the one-sided test.

DO: We can get the test statistic and P-value from the Minitab output.

- *Test statistic* $t = \frac{b - \beta_0}{SE_b} = \frac{1.4929 - 0}{0.4870} = 3.07$ (look in the "T" column of the computer output across from "Crycount")
- *P-value* Figure 12.9 displays the P-value for this one-sided test as an area under the t distribution curve with $38 - 2 = 36$ degrees of freedom. The Minitab output gives $P = 0.004$ as the P-value for a two-sided test. The P-value for the one-sided test is half of this, $P = 0.002$.

CONCLUDE: The P-value, 0.002, is less than our $\alpha = 0.05$ significance level, so we have enough evidence to reject H_0 and conclude that there is a positive linear relationship between intensity of crying and IQ score in the population of infants.

For Practice Try Exercise 13

Based on the results of the crying and IQ study, should we ask doctors and parents to make infants cry more so that they'll be smarter later in life? Hardly. This study gives statistically significant evidence of a positive linear relationship between the two variables. However, we can't conclude that more intense crying as an infant *causes* an increase in IQ. Maybe infants who cry more are more alert to begin with and tend to score higher on intelligence tests.

Confidence Intervals Give More Information We know from previous chapters that confidence intervals give more information than tests do. Let's construct a 90% confidence interval for the slope β in the crying and IQ example. The Minitab output from the linear regression analysis for these data gives $b = 1.4929$ and $SE_b = 0.4870$. We need to find the critical value t^* for 90% confidence from the t distribution with 36 degrees of freedom. Because Table B does not have a row for df = 36, we must use either technology or the next smaller degrees of freedom in the table, df = 30.

Reminder: The value of t in the "Crycount" row of the Minitab output is a test statistic, not a critical value for a confidence interval.

The TI-84 command `invT(0.05,36)` yields $t^* = 1.688$. The resulting 90% confidence interval for the slope β is

$$b \pm t^* SE_b = 1.4929 \pm 1.688(0.4870) = 1.4929 \pm 0.822 = (0.6709, 2.3149)$$

We are 90% confident that mean IQ increases roughly by between 0.67 and 2.31 points for each additional peak in crying. Note that 0 is not included in this interval of plausible values for the parameter β, which is consistent with the reject H_0 conclusion from the example.

As with inference about means, a *two-sided test* at significance level α and a $100(1 - \alpha)\%$ confidence interval for the slope β lead to the same conclusion about H_0. That's because both inference methods use the standard error of the slope SE_b in the calculations. The connection between *one-sided tests* and confidence intervals for β is familiar, too. Our one-sided test at $\alpha = 0.05$ in the crying and IQ example gives the same conclusion about H_0 as the 90% confidence interval for β that we calculated above.

Magnetic Fields and Healing

A confidence interval gives more information

Our bodies have a natural electrical field that is known to help wounds heal. How does changing the field strength affect healing? A series of experiments with newts investigated this question. In one experiment, the two hind limbs of 12 newts were assigned at random to either experimental or control groups. The electrical field in the experimental limbs was reduced to zero by applying a voltage. The control limbs were left alone. Researchers measured the rates at which new cells closed a razor cut in each limb, in micrometers per hour.[4]

PROBLEM: An examination of the data suggests that the conditions for performing regression inference are met. A 95% confidence interval for the slope β of the true regression line for predicting healing rates in newts' limbs when exposed to zero voltage from healing rates in their limbs with normal electrical fields is $(-0.42, 0.72)$. Based on this interval, what conclusion should we draw from a test of $H_0: \beta = 0$ versus $H_a: \beta \neq 0$ at the $\alpha = 0.05$ significance level? Justify your answer.

SOLUTION: The 95% confidence interval includes 0 as a plausible value for the slope β of the true regression line. At the $\alpha = 0.05$ level, we would fail to reject $H_0: \beta = 0$ and conclude that this experiment provides insufficient evidence of a linear relationship between healing rates of newts' limbs when exposed to zero voltage and healing rates of their limbs with normal electrical fields.

For Practice *Try Exercise 17*

CHECK YOUR UNDERSTANDING

The previous Check Your Understanding (page 750) described some results from a study of body weights and backpack weights. Here, again, is the Minitab output from a least-squares regression analysis for these data.

Regression Analysis: Pack weight versus Body weight

Predictor	Coef	SE Coef	T	P
Constant	16.265	3.937	4.13	0.006
Body weight	0.09080	0.02831	3.21	0.018

1. Do these data provide convincing evidence of a linear relationship between pack weight and body weight in the population of ninth-grade students at the school? Carry out a test at the $\alpha = 0.05$ significance level. Assume that the conditions for regression inference are met.

2. Would a 99% confidence interval for the slope β of the population regression line include 0? Justify your answer. (You don't need to do any calculations to answer this question.)

So far, we have used computer regression output when performing inference about the slope of a population regression line. The TI-83/84, TI-89, and TI-Nspire can do the calculations for inference when you are provided with the sample data.

TECHNOLOGY CORNER Regression inference on the calculator

Let's use the data from the nonexercise activity (NEA) and fat gain study on page 749 to illustrate significance tests and confidence intervals on the TI-83/84 and TI-89.

NEA change (cal):	−94	−57	−29	135	143	151	245	355
Fat gain (kg):	4.2	3.0	3.7	2.7	3.2	3.6	2.4	1.3

NEA change (cal):	392	473	486	535	571	580	620	690
Fat gain (kg):	3.8	1.7	1.6	2.2	1.0	0.4	2.3	1.1

Enter the x-values (NEA change) into L1/list1 and the y-values (Fat gain) into L2/list2.

To do a significance test:

TI-83/84

- Press STAT, then choose TESTS and `F:LinRegTTest....`
- In the LinRegTTest screen, adjust the inputs as shown. Then highlight "Calculate" and press ENTER.

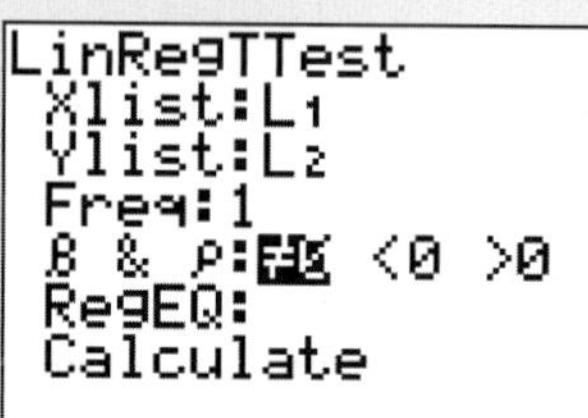

TI-89

- Press 2nd F1 ([F6]) and choose `A:LinRegTTest....`
- In the Linear Regression T Test screen, adjust the inputs as shown and press ENTER.

- The linear regression t test results take two screens to present.

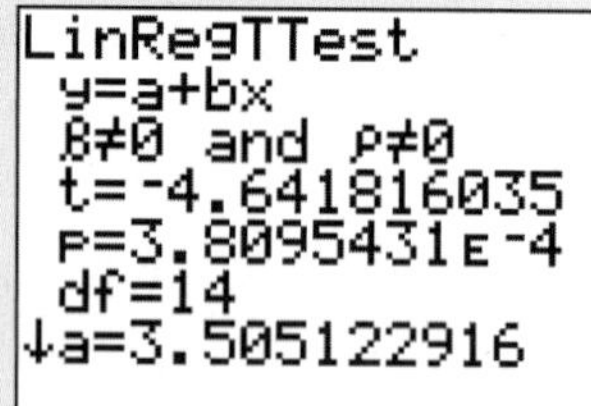

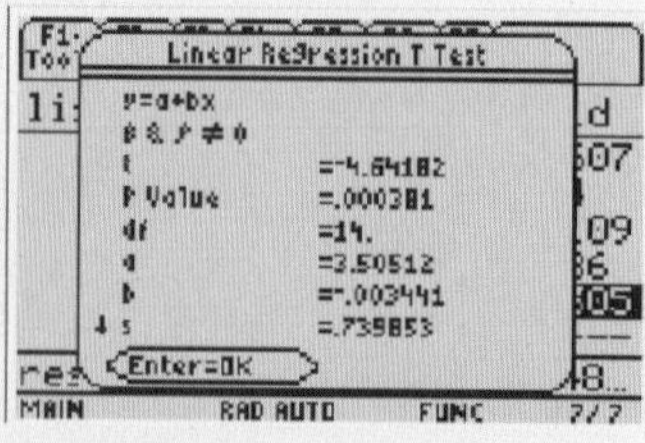

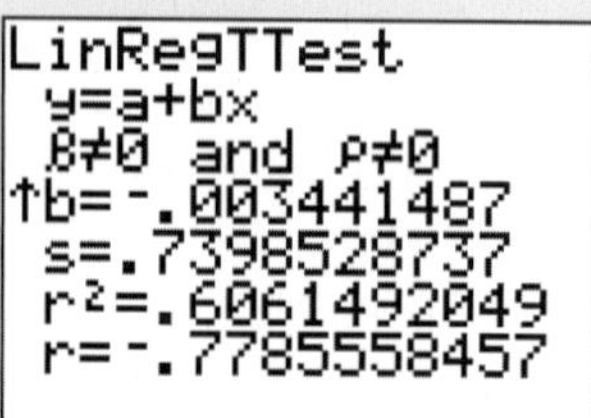

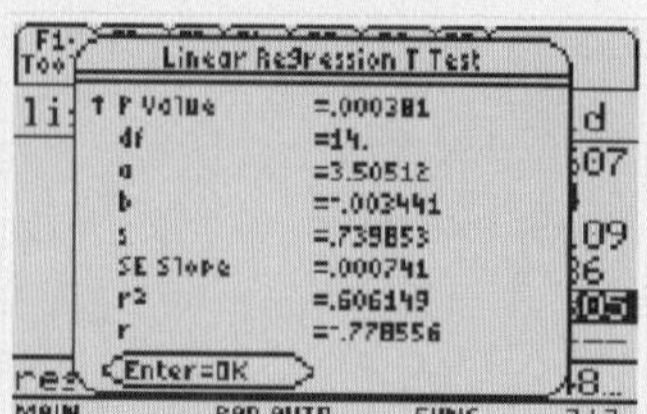

Compare these results with the Minitab regression output below.

Regression Analysis: Fat gain versus NEA change

Predictor	Coef	SE Coef	T	P
Constant	3.5051	0.3036	11.54	0.000
NEA change	−0.0034415	0.0007414	−4.64	0.000

S = 0.739853 R-Sq = 60.6% R-Sq(adj) = 57.8%

To construct a confidence interval:

TI-84 with recent OS

- Press [STAT], then choose TESTS and `G:LinRegTInt....`
- In the LinRegTInt screen, adjust the inputs as shown. Then highlight "Calculate" and press [ENTER].

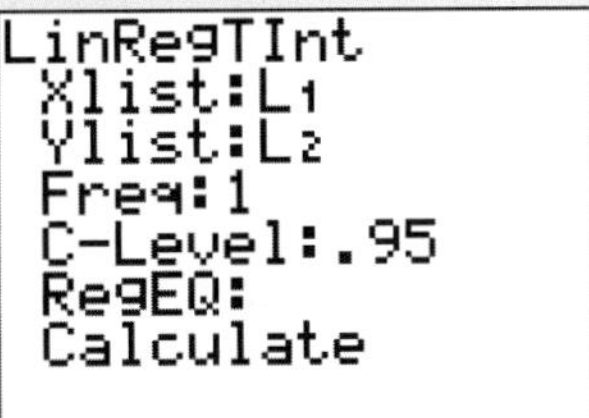

TI-89

- Press [2nd] [F2] ([F7]) and choose `7:LinRegTInt....`
- In the Linear Regression T Interval screen, adjust the inputs as shown and press [ENTER].

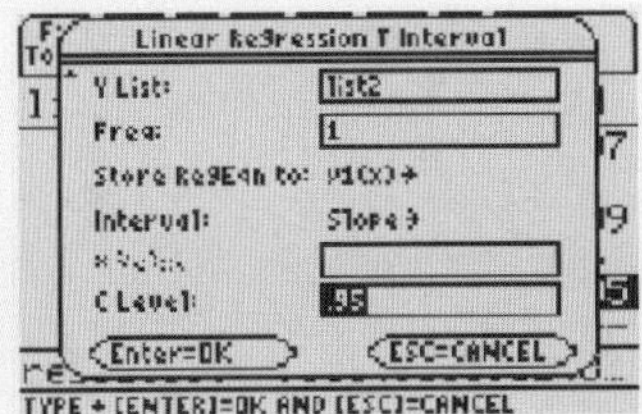

- The linear regression t interval results take two screens to present. We show only the first screen.

```
LinRegTInt
 y=a+bx
 (-.005,-.0019)
 b=-.003441487
 df=14
 s=.7398528737
↓a=3.505122916
```

TI-Nspire instructions in Appendix B

THINK ABOUT IT

What's with that $\rho \neq 0$ in the LinRegTTest screen? The slope b of the least-squares regression line is closely related to the correlation r between the explanatory and response variables x and y. (Recall that $b = r\frac{s_y}{s_x}$.) In the same way, the slope β of the population regression line is closely related to the correlation ρ (the lowercase Greek letter rho) between x and y in the population. In particular, the slope is 0 exactly when the correlation is 0.

Testing the null hypothesis $H_0: \beta = 0$ is, therefore, exactly the same as testing that there is *no correlation* between x and y in the population from which we drew our data. You can use the test for zero slope to test the hypothesis $H_0: \rho = 0$ of zero correlation between any two quantitative variables. That's a useful trick. Because correlation also makes sense when there is no explanatory-response distinction, it is handy to be able to test correlation without doing regression.

SECTION 12.1 Summary

- Least-squares regression fits a straight line of the form $\hat{y} = a + bx$ to data to predict a response variable y from an explanatory variable x. Inference in this setting uses the **sample regression line** to estimate or test a claim about the **population (true) regression line.**
- The conditions for regression inference are
  - **Linear** The actual relationship between x and y is linear. For any fixed value of x, the mean response μ_y falls on the population (true) regression line $\mu_y = \alpha + \beta x$.
  - **Independent** Individual observations are independent.
  - **Normal** For any fixed value of x, the response y varies according to a Normal distribution.
  - **Equal variance** The standard deviation of y (call it σ) is the same for all values of x.
  - **Random** The data are produced from a well-designed random sample or randomized experiment.
- The slope b and intercept a of the least-squares line estimate the slope β and intercept α of the population (true) regression line. To estimate σ, use the standard deviation s of the residuals.
- Confidence intervals and significance tests for the slope β of the population regression line are based on a t distribution with $n - 2$ degrees of freedom.
- The ***t* interval for the slope** β has the form $b \pm t^* \mathrm{SE}_b$, where the **standard error of the slope** is $\mathrm{SE}_b = \dfrac{s}{s_x \sqrt{n-1}}$.
- To test the null hypothesis $H_0: \beta =$ hypothesized value, carry out a ***t* test for the slope.** This test uses the statistic $t = \dfrac{b - \beta_0}{\mathrm{SE}_b}$. The most common null hypothesis is $H_0: \beta = 0$, which says that there is no linear relationship between x and y in the population.

12.1 TECHNOLOGY CORNER

TI-Nspire instructions in Appendix B

SECTION 12.1 Exercises

1. **Oil and residuals** Exercise 59 on page 193 (Chapter 3) examined data on the depth of small defects in the Trans-Alaska Oil Pipeline. Researchers compared the results of measurements on 100 defects made in the field with measurements of the same defects made in the laboratory.[5] The figure below shows a residual plot for the least-squares regression line based on these data. Are the conditions for performing inference about the slope β of the population regression line met? Justify your answer.

pg 743

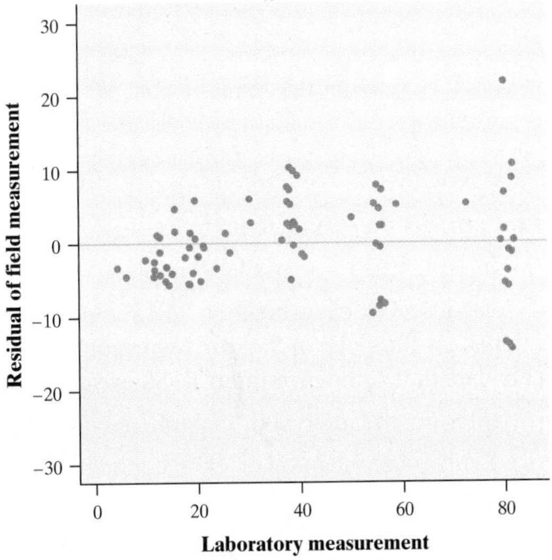

2. **SAT Math scores** In Chapter 3, we examined data on the percent of high school graduates in each state who took the SAT and the state's mean SAT Math score in a recent year. The figure below shows a residual plot for the least-squares regression line based on these data. Are the conditions for performing inference about the slope β of the population regression line met? Justify your answer.

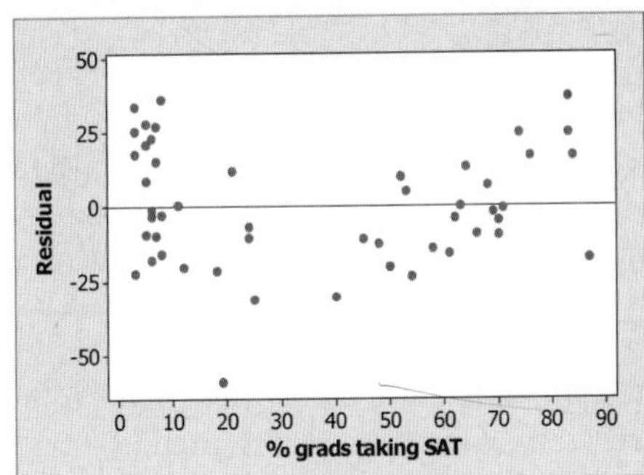

3. **Prey attracts predators** Here is one way in which nature regulates the size of animal populations: high population density attracts predators, which remove a higher proportion of the population than when the density of the prey is low. One study looked at kelp perch and their common predator, the kelp bass. The researcher set up four large circular pens on sandy ocean bottoms off the coast of southern California. He chose young perch at random from a large group and placed 10, 20, 40, and 60 perch in the four pens. Then he dropped the nets protecting the pens, allowing bass to swarm in, and counted the perch left after two hours. Here are data on the proportions of perch eaten in four repetitions of this setup:[6]

Number of Perch	Proportion Killed			
10	0.0	0.1	0.3	0.3
20	0.2	0.3	0.3	0.6
40	0.075	0.3	0.6	0.725
60	0.517	0.55	0.7	0.817

The explanatory variable is the number of perch (the prey) in a confined area. The response variable is the proportion of perch killed by bass (the predator) in two hours when the bass are allowed access to the perch. A scatterplot of the data shows a linear relationship.

We used Minitab software to carry out a least-squares regression analysis for these data. A residual plot and a histogram of the residuals are shown below. Check whether the conditions for performing inference about the regression model are met.

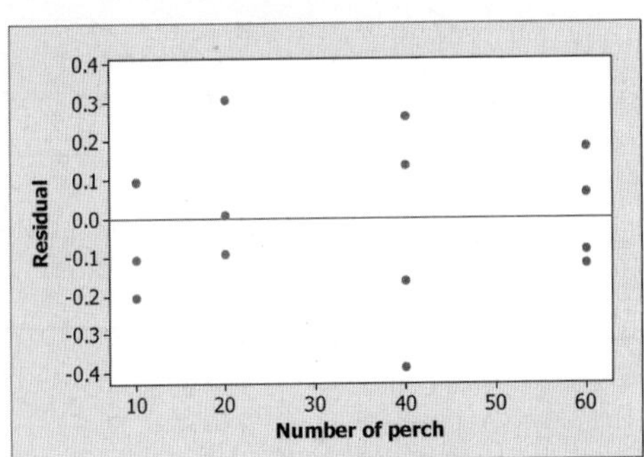

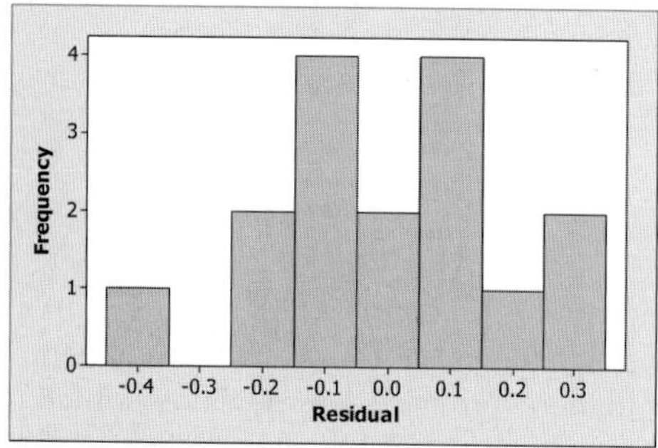

The model for regression inference has three parameters: α, β, and σ. Explain what each parameter represents in context. Then provide an estimate for each.

4. **Beer and BAC** How well does the number of beers a person drinks predict his or her blood alcohol content (BAC)? Sixteen volunteers with an initial BAC of 0 drank a randomly assigned number of cans of beer. Thirty minutes later, a police officer measured their BAC. Least-squares regression was performed on the data. A residual plot and a histogram of the residuals are shown below. Check whether the conditions for performing inference about the regression model are met.

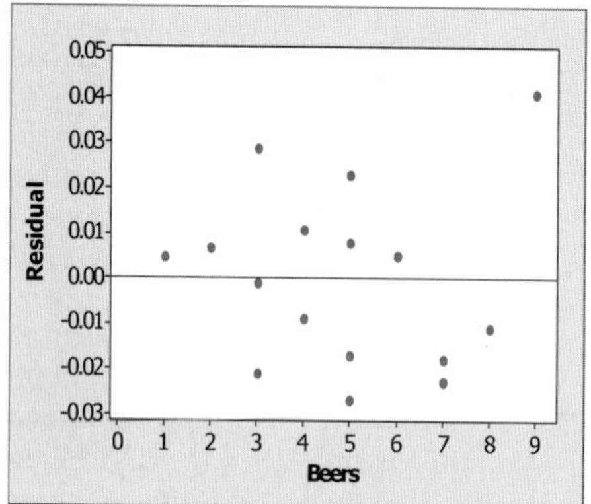

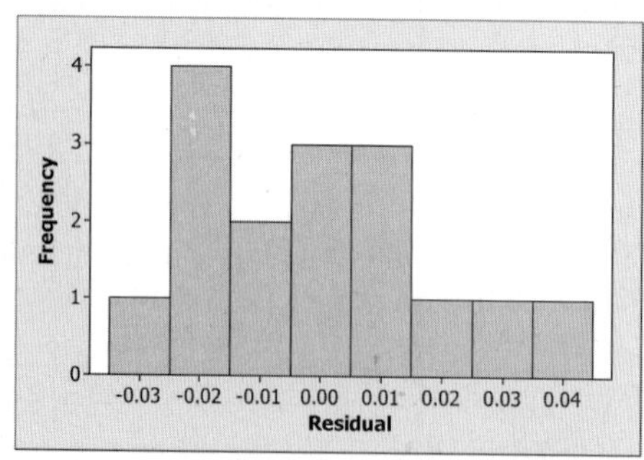

5. **Prey attracts predators** Refer to Exercise 3. Computer output from the least-squares regression analysis on the perch data is shown below.

Predictor	Coef	Stdev.	t-ratio	p
Constant	0.12049	0.09269	1.30	0.215
Perch	0.008569	0.002456	3.49	0.004

S = 0.1886 R-Sq = 46.5% R-Sq(adj) = 42.7%

6. **Beer and BAC** Refer to Exercise 4. Computer output from the least-squares regression analysis on the beer and blood alcohol data is shown below.

Dependent variable is: **BAC**

No Selector

R squared = 80.0% R squared (adjusted) = 78.6%

s = 0.0204 with 16 − 2 = 14 degrees of freedom

Variable	Coefficient	s.e. of Coeff	t-ratio	prob
Constant	−0.012701	0.0126	−1.00	0.3320
Beers	0.017964	0.0024	7.84	≤0.0001

The model for regression inference has three parameters: α, β, and σ. Explain what each parameter represents in context. Then provide an estimate for each.

7. **Prey attracts predators** Refer to Exercise 5.

pg 747

(a) Interpret the value of SE_b in context.

(b) Find the critical value for a 90% confidence interval for the slope of the true regression line. Then calculate the confidence interval. Show your work.

(c) Interpret the interval from part (b) in context.

(d) Explain the meaning of "90% confident" in context.

8. **Beer and BAC** Refer to Exercise 6.

(a) Interpret the value of SE_b in context.

(b) Find the critical value for a 99% confidence interval for the slope of the true regression line. Then calculate the confidence interval. Show your work.

(c) Interpret the interval from part (b) in context.

(d) Explain the meaning of "99% confident" in context.

9. **Beavers and beetles** Do beavers benefit beetles? Researchers laid out 23 circular plots, each four meters in diameter, at random in an area where beavers were cutting down cottonwood trees. In each plot, they counted the number of stumps from trees cut by beavers and the number of clusters of beetle larvae. Ecologists think that the new sprouts from stumps are more tender than other cottonwood growth, so that beetles prefer them. If so, more stumps should produce more beetle larvae.[7]

pg 749

Minitab output for a regression analysis on these data is shown below. Construct and interpret a 99% confidence interval for the slope of the population regression line. Assume that the conditions for performing inference are met.

Regression Analysis: Beetle larvae versus Stumps

```
Predictor    Coef    SE Coef      T        P
Constant   -1.286    2.853    -0.45    0.657
Stumps     11.894    1.136    10.47    0.000
S = 6.41939 R-Sq = 83.9% R-Sq(adj) = 83.1%
```

10. **Ideal proportions** The students in Mr. Shenk's class measured the arm spans and heights (in inches) of a random sample of 18 students from their large high school. Some computer output from a least-squares regression analysis on these data is shown below. Construct and interpret a 90% confidence interval for the slope of the population regression line. Assume that the conditions for performing inference are met.

```
Predictor     Coef      Stdev    t-ratio     p
Constant    11.547      5.600       2.06   0.056
armspan      0.84042  0.08091      10.39   0.000
S = 1.613 R-Sq = 87.1% R-Sq(adj) = 86.3%
```

11. **Beavers and beetles** Refer to Exercise 9.

(a) How many clusters of beetle larvae would you predict in a circular plot with 5 tree stumps cut by beavers? Show your work.

(b) About how far off do you expect the prediction in part (a) to be from the actual number of clusters of beetle larvae? Justify your answer.

12. **Ideal proportions** Refer to Exercise 10.

(a) What height would you predict for a student with an arm span of 76 inches? Show your work.

(b) About how far off do you expect the prediction in part (a) to be from the student's actual height? Justify your answer.

13. **Weeds among the corn** Lamb's-quarter is a common weed that interferes with the growth of corn. An agriculture researcher planted corn at the same rate in 16 small plots of ground and then weeded the plots by hand to allow a fixed number of lamb's-quarter plants to grow in each meter of corn row. The decision of how many of these plants to leave in each plot was made at random. No other weeds were allowed to grow. Here are the yields of corn (bushels per acre) in each of the plots:[8]

pg 752

Weeds per meter	Corn yield	Weeds per meter	Corn yield
0	166.7	3	158.6
0	172.2	3	176.4
0	165.0	3	153.1
0	176.9	3	156.0
1	166.2	9	162.8
1	157.3	9	142.4
1	166.7	9	162.8
1	161.1	9	162.4

(a) A scatterplot of the data with the least-squares line added is shown below. Describe what this graph tells you about the relationship between these two variables.

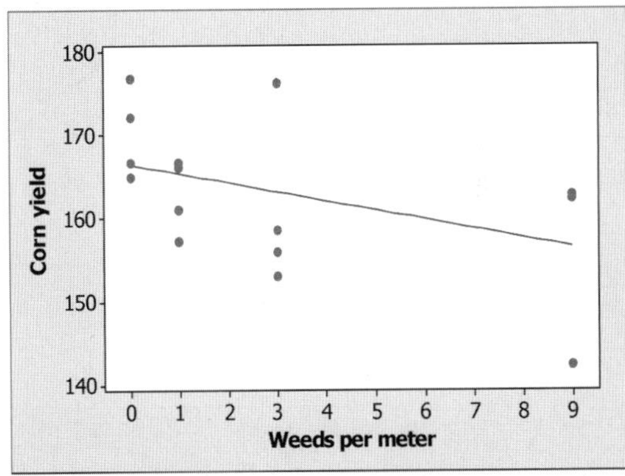

Minitab output from a linear regression on these data is shown below.

```
Predictor      Coef      SE Coef      T        P
Constant    166.483      2.725    61.11   0.000
Weeds per   -1.0987     0.5712    -1.92   0.075
 meter
S = 7.97665 R-Sq = 20.9% R-Sq(adj) = 15.3%
```

(b) What is the equation of the least-squares regression line for predicting corn yield from the number of lamb's quarter plants per meter? Define any variables you use.

(c) Interpret the slope and y intercept of the regression line in context.

(d) Do these data provide convincing evidence that more weeds reduce corn yield? Carry out an appropriate test at the $\alpha = 0.05$ level to help answer this question.

14. **Time at the table** Does how long young children remain at the lunch table help predict how much they eat? Here are data on a random sample of 20 toddlers observed over several months.[9] "Time" is the average number of minutes a child spent at the table when lunch was served. "Calories" is the

average number of calories the child consumed during lunch, calculated from careful observation of what the child ate each day.

Time:	21.4	30.8	37.7	33.5	32.8	39.5	22.8
Calories:	472	498	465	456	423	437	508
Time:	34.1	33.9	43.8	42.4	43.1	29.2	31.3
Calories:	431	479	454	450	410	504	437
Time:	28.6	32.9	30.6	35.1	33.0	43.7	
Calories:	489	436	480	439	444	408	

(a) A scatterplot of the data with the least-squares line added is shown below. Describe what this graph tells you about the relationship between these two variables.

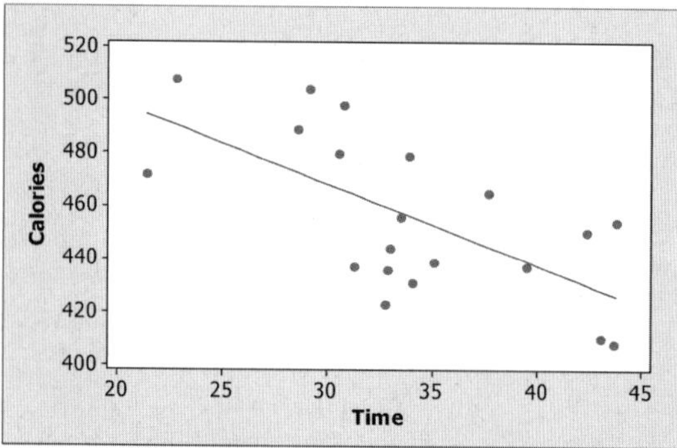

Minitab output from a linear regression on these data is shown below.

```
Predictor     Coef     SE Coef      T       P
Constant    560.65     29.37     19.09   0.000
Time        -3.0771    0.8498    -3.62   0.002
S = 23.3980 R-Sq = 42.1% R-Sq(adj) = 38.9%
```

(b) What is the equation of the least-squares regression line for predicting calories consumed from time at the table? Define any variables you use.

(c) Interpret the slope of the regression line in context. Does it make sense to interpret the y intercept in this case? Why or why not?

(d) Do these data provide convincing evidence of a negative linear relationship between time at the table and calories consumed in the population of toddlers? Carry out an appropriate test at the $\alpha = 0.01$ level to help answer this question.

15. **Weeds among the corn** Refer to Exercise 13.

(a) Construct and interpret a 90% confidence interval for the slope of the true regression line. Explain how your results are consistent with the significance test in Exercise 13.

(b) Interpret each of the following in context:

(i) s

(ii) r^2

(iii) The standard error of the slope

16. **Time at the table** Refer to Exercise 14.

(a) Construct and interpret a 98% confidence interval for the slope of the population regression line. Explain how your results are consistent with the significance test in Exercise 14.

(b) Interpret each of the following in context:

(i) s

(ii) r^2

(iii) The standard error of the slope

17. **Paired tires** Exercise 69 in Chapter 8 (page 519) compared two methods for estimating tire wear. The first method used the amount of weight lost by a tire. The second method used the amount of wear in the grooves of the tire. A random sample of 16 tires was obtained. Both methods were used to estimate the total distance traveled by each tire. The scatterplot below displays the two estimates (in thousands of miles) for each tire.[10]

pg 755

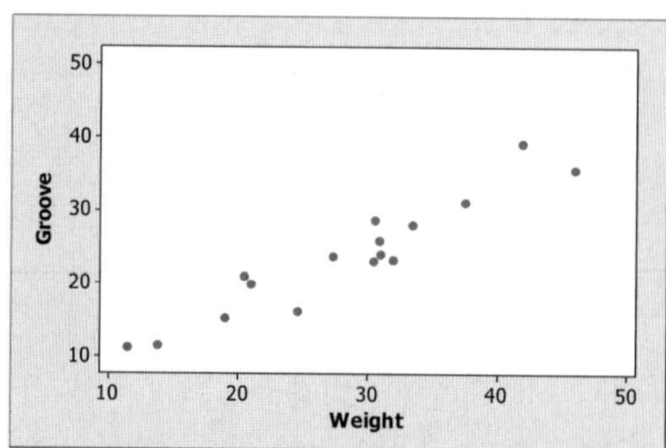

Computer output from a least-squares regression analysis of these data is shown below. Assume that the conditions for regression inference are met.

```
Predictor   Coef      SE Coef     T       P
Constant    1.351     2.105      0.64   0.531
Weight      0.79021   0.07104   11.12   0.000
S = 2.62078 R-Sq = 89.8% R-Sq(adj) = 89.1%
```

(a) Verify that the 99% confidence interval for the slope of the population regression line is (0.5785, 1.001).

(b) Researchers want to test whether there is a difference in the two methods of estimating tire wear. Explain why an appropriate pair of hypotheses for this test is $H_0: \beta = 1$ versus $H_a: \beta \neq 1$.

(c) What conclusion would you draw for this significance test based on your interval in part (a)? Justify your answer.

18. **Stats teachers' cars** A random sample of AP Statistics teachers was asked to report the age (in years) and mileage of their primary vehicles. A scatterplot of the data is shown below.

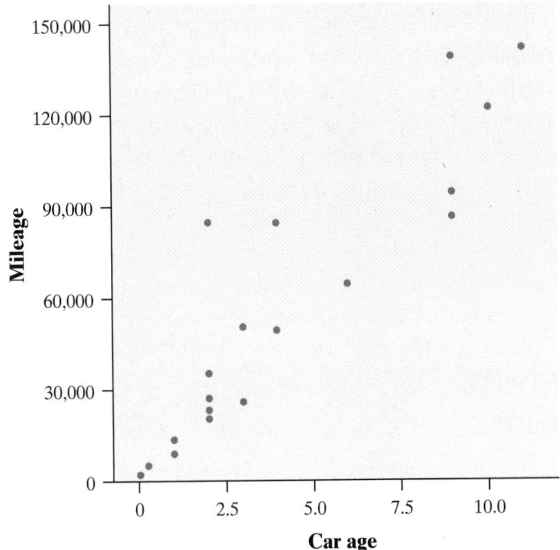

Computer output from a least-squares regression analysis of these data is shown below. Assume that the conditions for regression inference are met.

```
R squared = 82.0% R squared (adjusted) = 81.1%
s = 19280 with 21 - 2 = 19 deg of freedom
Variable    Coef     SE Coef  t-ratio    prob
Constant   7288.54    6591     1.11    <0.2826
Car age   11630.6     1249     9.31    <0.0001
```

(a) Verify that the 95% confidence interval for the slope of the population regression line is (9016.4, 14,244.8).

(b) A national automotive group claims that the typical driver puts 15,000 miles per year on his or her main vehicle. We want to test whether AP Statistics teachers are typical drivers. Explain why an appropriate pair of hypotheses for this test is $H_0: \beta = 15{,}000$ versus $H_a: \beta \neq 15{,}000$.

(c) What conclusion would you draw for this significance test based on your interval in part (a)? Justify your answer.

19. **Is wine good for your heart?** A researcher from the University of California, San Diego, collected data on average per capita wine consumption and heart disease death rate in a random sample of 19 countries for which data were available. The following table displays the data.[11]

Alcohol from wine (liters/year)	Heart disease death rate (per 100,000)	Alcohol from wine (liters/year)	Heart disease death rate (per 100,000)
2.5	211	7.9	107
3.9	167	1.8	167
2.9	131	1.9	266
2.4	191	0.8	227
2.9	220	6.5	86
0.8	297	1.6	207
9.1	71	5.8	115
2.7	172	1.3	285
0.8	211	1.2	199
0.7	300		

(a) Is there statistically significant evidence of a negative linear relationship between wine consumption and heart disease deaths in the population of countries? Carry out an appropriate significance test at the $\alpha = 0.05$ level.

(b) Calculate and interpret a 95% confidence interval for the slope β of the population regression line.

20. **The professor swims** Here are data on the time (in minutes) Professor Moore takes to swim 2000 yards and his pulse rate (beats per minute) after swimming on a random sample of 23 days:

Time:	34.12	35.72	34.72	34.05	34.13	35.72
Pulse:	152	124	140	152	146	128
Time:	36.17	35.57	35.37	35.57	35.43	36.05
Pulse:	136	144	148	144	136	124
Time:	34.85	34.70	34.75	33.93	34.60	34.00
Pulse:	148	144	140	156	136	148
Time:	34.35	35.62	35.68	35.28	35.97	
Pulse:	148	132	124	132	139	

(a) Is there statistically significant evidence of a negative linear relationship between Professor Moore's swim time and his pulse rate in the population of days on which he swims 2000 yards? Carry out an appropriate significance test at the $\alpha = 0.05$ level.

(b) Calculate and interpret a 95% confidence interval for the slope β of the population regression line.

Multiple choice: Select the best answer for Exercises 21 to 26.

Exercises 21 to 26 are based on the following information. To determine property taxes, Florida reappraises real estate every year, and the county appraiser's Web site lists the current "fair market value" of each piece of property. Property usually sells for somewhat more than the appraised market value.

We collected data on the appraised market values x and actual selling prices y (in thousands of dollars) of a random sample of 16 condominium units. A scatterplot of the data shows a fairly strong positive linear relationship between the variables. Here is part of the Minitab output from a least-squares regression analysis using these data.[12]

```
Predictor     Coef     SE Coef      T       P
Constant     127.27     79.49     1.60   0.132
appraisal    1.0466    0.1126     9.29   0.000
S = 69.7299 R-Sq = 86.1% R-Sq(adj) = 85.1%
```

21. The equation of the least-squares regression line for predicting selling price from appraised value is
(a) $\widehat{\text{price}} = 79.49 + 0.1126$ (appraised value).
(b) $\widehat{\text{price}} = 0.1126 + 1.0466$ (appraised value).
(c) $\widehat{\text{price}} = 127.27 + 1.0466$ (appraised value).
(d) $\widehat{\text{price}} = 1.0466 + 127.27$ (appraised value).
(e) $\widehat{\text{price}} = 1.0466 + 69.7299$ (appraised value).

22. What is the correlation between selling price and appraised value?
(a) 0.1126 (b) 0.861 (c) −0.861 (d) 0.928 (e) −0.928

23. The slope β of the population regression line describes
(a) the exact increase in the selling price of an individual unit when its appraised value increases by $1000.
(b) the average increase in the appraised value in a population of units when selling price increases by $1000.
(c) the average increase in selling price in a population of units when appraised value increases by $1000.
(d) the average selling price in a population of units when a unit's appraised value is 0.
(e) the average increase in appraised value in a sample of 16 units when selling price increases by $1000.

24. Is there significant evidence that selling price increases as appraised value increases? To answer this question, test the hypotheses
(a) $H_0: \beta = 0$ versus $H_a: \beta > 0$.
(b) $H_0: \beta = 0$ versus $H_a: \beta < 0$.
(c) $H_0: \beta = 0$ versus $H_a: \beta \neq 0$.
(d) $H_0: \beta > 0$ versus $H_a: \beta = 0$.
(e) $H_0: \beta = 1$ versus $H_a: \beta > 1$.

25. Confidence intervals and tests for these data use the t distribution with degrees of freedom
(a) 9.29. (b) 14. (c) 15. (d) 16. (e) 30.

26. A 95% confidence interval for the population slope β is
(a) 1.0466 ± 149.5706.
(b) 1.0466 ± 0.2415.
(c) 1.0466 ± 0.2387.
(d) 1.0466 ± 0.1983.
(e) 1.0466 ± 0.1126.

Exercises 27 to 30 refer to the following setting. Does the color in which words are printed affect your ability to read them? Do the words themselves affect your ability to name the color in which they are printed? Mr. Starnes designed a study to investigate these questions using the 16 students in his AP Statistics class as subjects. Each student performed two tasks in a random order while a partner timed: (1) read 32 words aloud as quickly as possible, and (2) say the color in which each of 32 words is printed as quickly as possible. Try both tasks for yourself using the word list below.

YELLOW	RED	BLUE	GREEN
RED	GREEN	YELLOW	YELLOW
GREEN	RED	BLUE	BLUE
YELLOW	BLUE	GREEN	RED
BLUE	YELLOW	RED	RED
RED	BLUE	YELLOW	GREEN
BLUE	GREEN	GREEN	BLUE
GREEN	YELLOW	RED	YELLOW

27. **Color words** (4.2) Let's review the design of the study.
(a) Explain why this was an experiment and not an observational study.
(b) Did Mr. Starnes use a completely randomized design or a randomized block design? Why do you think he chose this experimental design?
(c) Explain the purpose of the random assignment in the context of the study.

The data from Mr. Starnes's experiment are shown below. For each subject, the time to perform the two tasks is given to the nearest second.

Subject	Words	Colors	Subject	Words	Colors
1	13	20	9	10	16
2	10	21	10	9	13
3	15	22	11	11	11
4	12	25	12	17	26
5	13	17	13	15	20
6	11	13	14	15	15
7	14	32	15	12	18
8	16	21	16	10	18

28. **Color words** (1.3) Do the data provide evidence of a difference in the average time required to perform the two tasks? Include an appropriate graph and numerical summaries in your answer.

29. **Color words** (9.3) Explain why it is not safe to use paired *t* procedures to do inference about the difference in the mean time to complete the two tasks.

30. **Color words** (3.2, 12.1) Can we use a student's word task time to predict his or her color task time?

(a) Make an appropriate scatterplot to help answer this question. Describe what you see.

(b) Use your calculator to find the equation of the least-squares regression line. Define any symbols you use.

(c) What is the residual for the student who completed the word task in 9 seconds? Show your work.

(d) Assume that the conditions for performing inference about the slope of the true regression line are met. The *P*-value for a test of $H_0: \beta = 0$ versus $H_a: \beta > 0$ is 0.0215. Explain what this value means in context.

Note: John Ridley Stroop is often credited with the discovery in 1935 of the fact that the color in which "color words" are printed interferes with people's ability to identify the color. The so-called Stroop Effect, though, was originally published by German researchers in 1929.

Exercises 31 and 32 refer to the following setting. Yellowstone National Park surveyed a random sample of 1526 winter visitors to the park. They asked each person whether they owned, rented, or had never used a snowmobile. Respondents were also asked whether they belonged to an environmental organization (like the Sierra Club). The two-way table summarizes the survey responses.

	Environmental Clubs		
	No	Yes	Total
Never used	445	212	657
Snowmobile renter	497	77	574
Snowmobile owner	279	16	295
Total	1221	305	1526

31. **Snowmobiles** (5.2, 5.3)

(a) If we choose a survey respondent at random, what's the probability that this individual

(i) is a snowmobile owner?

(ii) belongs to an environmental organization or owns a snowmobile?

(iii) has never used a snowmobile given that the person belongs to an environmental organization?

(b) Are the events "is a snowmobile owner" and "belongs to an environmental organization" independent for the members of the sample? Justify your answer.

(c) If we choose two survey respondents at random, what's the probability that

(i) both are snowmobile owners?

(ii) at least one of the two belongs to an environmental organization?

32. **Snowmobiles** (11.2) Do these data provide convincing evidence of an association between environmental club membership and snowmobile use for the population of visitors to Yellowstone National Park? Carry out an appropriate test at the 5% significance level.

Transforming to Achieve Linearity

In Section 12.2, you'll learn about:

- Transforming with powers and roots
- Transforming with logarithms

In Chapter 3, we learned how to analyze relationships between two quantitative variables that showed a linear pattern. When two-variable data show a curved relationship, we must develop new techniques for finding an appropriate model. This section describes several simple *transformations* of data that can straighten a nonlinear pattern. Once the data have been transformed to achieve linearity, we can use least-squares regression to generate a useful model for making predictions. And if the conditions for regression inference are met, we can estimate or test a claim about the slope of the population (true) regression line using the transformed data.

EXAMPLE *Health and Wealth*

Straightening out a curved pattern

The Gapminder Web site, www.gapminder.org, provides loads of data on the health and well-being of the world's inhabitants. Figure 12.10 is a scatterplot of data from Gapminder.[13] The individuals are all the world's nations for which data are available. The explanatory variable is a measure of how rich a country is: income per person. The response variable is life expectancy at birth.

We expect people in richer countries to live longer because they have better access to medical care and typically lead healthier lives. The overall pattern of the scatterplot does show this, but the relationship is not linear. Life expectancy rises very quickly as income per person increases and then levels off. People in very rich countries such as the United States live no longer than people in poorer but not extremely poor nations. In some less wealthy countries, people live longer than in the United States.

Four African nations are outliers. Their life expectancies are similar to those of their neighbors, but their income per person is higher. Gabon and Equatorial Guinea produce oil, and South Africa and Botswana produce diamonds. It may be that income from mineral exports goes mainly to a few people and so pulls up income per person without much effect on either the income or the life expectancy of ordinary citizens. That is, income per person is a mean, and we know that mean income can be much higher than median income.

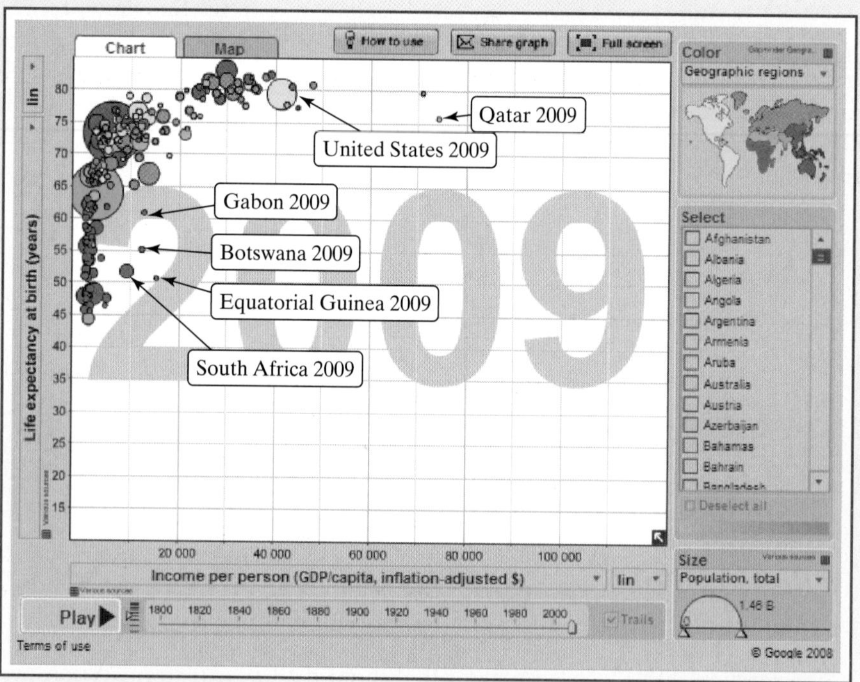

FIGURE 12.10 Scatterplot of the life expectancy of people in many nations against each nation's income per person. The color of each circle indicates the geographic region in which that country is located. The size of each circle is based on the population of the country—bigger circles indicate larger populations.

The scatterplot in Figure 12.10 shows a curved pattern. We can straighten things out using logarithms. Figure 12.11 plots the logarithm of income per person against life expectancy for these same countries. The effect is almost magical. This graph has a clear, linear pattern.

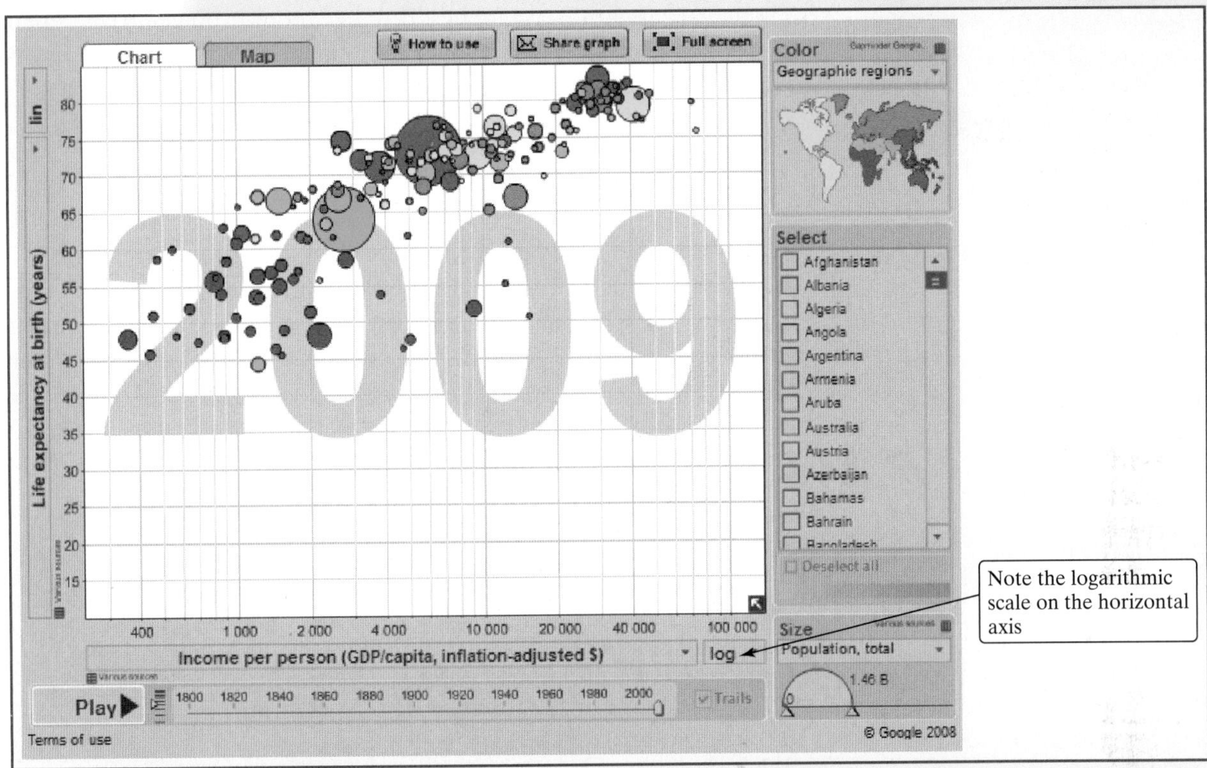

FIGURE 12.11 Scatterplot of life expectancy against the logarithm of income per person for many nations.

Transforming

Applying a function such as the logarithm or square root to a quantitative variable is called **transforming** the data. We will see in this section that understanding how simple functions work helps us choose and use transformations to straighten nonlinear patterns.

Transforming data amounts to changing the scale of measurement that was used when the data were collected. We can choose to measure temperature in degrees Fahrenheit or in degrees Celsius, distance in miles or in kilometers. These changes of units are *linear transformations*, discussed in Chapter 2. Linear transformations cannot straighten a curved relationship between two variables. To do that, we resort to functions that are not linear. The logarithm function, applied in the "Health and Wealth" example, is a nonlinear function. We'll return to transformations involving logarithms later.

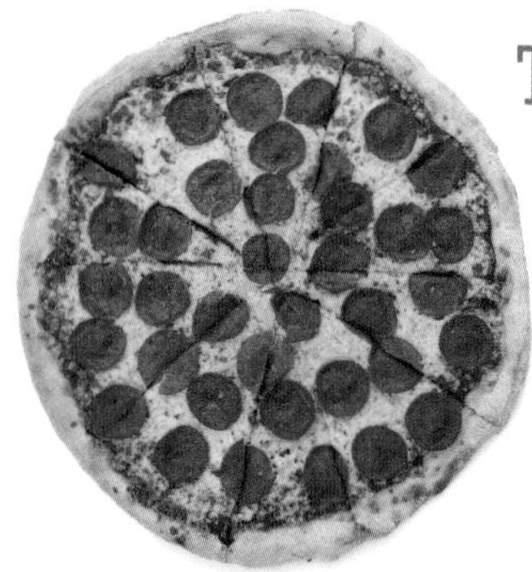

Transforming with Powers and Roots

When you visit a pizza parlor, you order a pizza by its diameter—say, 10 inches, 12 inches, or 14 inches. But the amount you get to eat depends on the area of the pizza. The area of a circle is π times the square of its radius r. So the area of a round pizza with diameter x is

$$\text{area} = \pi r^2 = \pi\left(\frac{x}{2}\right)^2 = \pi\left(\frac{x^2}{4}\right) = \frac{\pi}{4}x^2$$

Power model

This is a **power model** of the form $y = ax^p$ with $a = \pi/4$ and $p = 2$.

When we are dealing with things of the same general form, whether circles or fish or people, we expect area to go up with the square of a dimension such as diameter or height. Volume should go up with the cube of a linear dimension. That is, geometry tells us to expect power models in some settings. There are other physical relationships between two variables that are described by power models. Here are some examples from science.

- The distance that an object dropped from a given height falls is related to time since release by the model

$$\text{distance} = a(\text{time})^2$$

- The time it takes a pendulum to complete one back-and-forth swing (its period) is related to its length by the model

$$\text{period} = a\sqrt{\text{length}} = a(\text{length})^{1/2}$$

- The intensity of a light bulb is related to distance from the bulb by the model

$$\text{intensity} = \frac{a}{\text{distance}^2} = a(\text{distance})^{-2}$$

Although a power model of the form $y = ax^p$ describes the relationship between x and y in each of these settings, there is a *linear* relationship between x^p and y. If we transform the values of the explanatory variable x by raising them to the p power, and graph the points (x^p, y), the scatterplot should have a linear form. The following example shows what we mean.

EXAMPLE

Go Fish!

Transforming with powers

Imagine that you have been put in charge of organizing a fishing tournament in which prizes will be given for the heaviest Atlantic Ocean rockfish caught. You know that many of the fish caught during the tournament will be measured and released. You are also aware that using delicate scales to try to weigh a fish that is flopping around in a moving boat will probably not yield very accurate results. It would be much easier to measure the length of the fish while on the boat. What you need is a way to convert the length of the fish to its weight.

You contact the nearby marine research laboratory, and they provide reference data on the length (in centimeters) and weight (in grams) for Atlantic Ocean rockfish of several sizes.[14]

Length:	5.2	8.5	11.5	14.3	16.8	19.2	21.3	23.3	25.0	26.7
Weight:	2	8	21	38	69	117	148	190	264	293

Length:	28.2	29.6	30.8	32.0	33.0	34.0	34.9	36.4	37.1	37.7
Weight:	318	371	455	504	518	537	651	719	726	810

Figure 12.12 is a scatterplot of the data. Note the clear curved shape.

Because length is one-dimensional and weight (like volume) is three-dimensional, a power model of the form weight $= a$ (length)^3 should describe the relationship. What happens if we cube the lengths in the data table and then graph weight versus length^3? Figure 12.13 gives us the answer. This transformation of the explanatory variable helps us produce a graph that is quite linear.

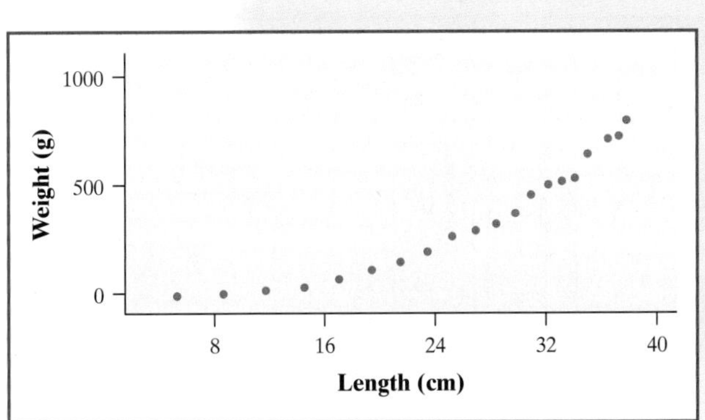

FIGURE 12.12 Scatterplot of Atlantic Ocean rockfish weight versus length.

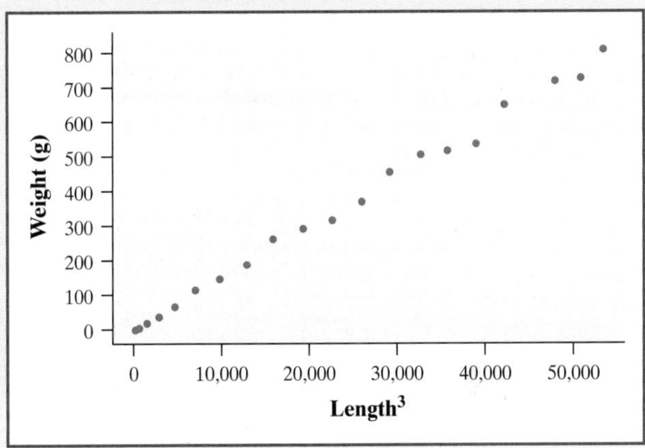

FIGURE 12.13 The scatterplot of weight versus length^3 is linear.

There's another way to transform the data in the example to achieve linearity. We can take the cube root of the weight values and graph $\sqrt[3]{\text{weight}}$ versus length. Figure 12.14 shows that the resulting scatterplot has a linear form. Why does this transformation work? Start with weight $= a(\text{length})^3$ and take the cube root of both sides of the equation:

$$\sqrt[3]{\text{weight}} = \sqrt[3]{a(\text{length})^3}$$
$$\sqrt[3]{\text{weight}} = \sqrt[3]{a}(\text{length})$$

That is, there is a linear relationship between length and $\sqrt[3]{\text{weight}}$.

Once we straighten out the curved pattern in the original scatterplot, we fit a least-squares line to the transformed data. This linear model can be used to predict values of the response variable y. As in Chapter 3, the residuals and the r^2-value tell us how well the regression line fits the data.

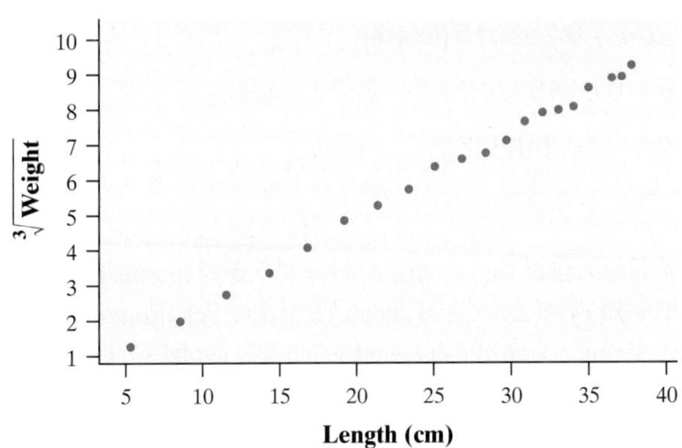

FIGURE 12.14 The scatterplot of $\sqrt[3]{\text{weight}}$ versus length is linear.

EXAMPLE Go Fish!

Here is Minitab output from separate regression analyses of the two sets of transformed Atlantic Ocean rockfish data.

Transformation 1: (length^3, weight)

Predictor	Coef	SE Coef	T	P
Constant	4.066	6.902	0.59	0.563
Length^3	0.0146774	0.0002404	61.07	0.000

S = 18.8412 R-Sq = 99.5% R-Sq(adj) = 99.5%

Transformation 2: (length, $\sqrt[3]{\text{weight}}$)

Predictor	Coef	SE Coef	T	P
Constant	−0.02204	0.07762	−0.28	0.780
Length	0.246616	0.002868	86.00	0.000

S = 0.124161 R-Sq = 99.8% R-Sq(adj) = 99.7%

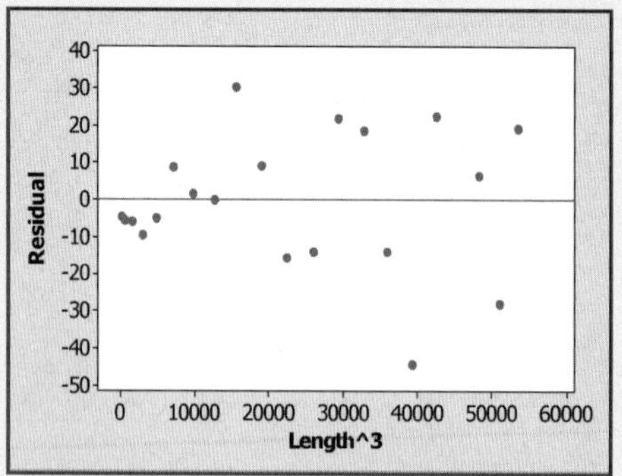

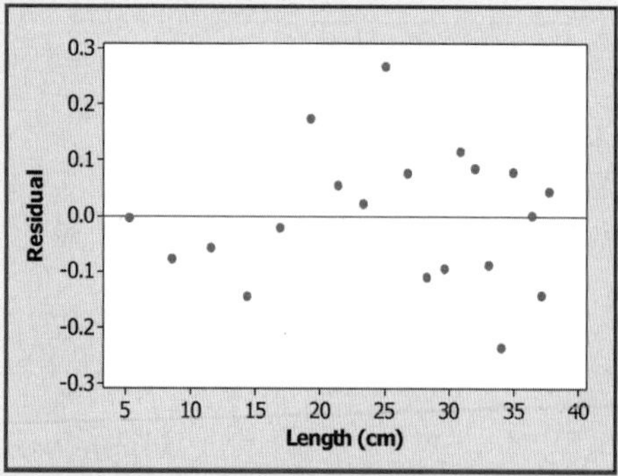

PROBLEM: *Do each of the following for both transformations.*

(a) *Give the equation of the least-squares regression line. Define any variables you use.*

(b) *Suppose a contestant in the fishing tournament catches an Atlantic Ocean rockfish that's 36 centimeters long. Use the model from part (a) to predict the fish's weight. Show your work.*

(c) *Interpret the value of s in context.*

SOLUTION:

(a) Transformation 1: $\widehat{\text{weight}} = 4.066 + 0.0146774\,(\text{length}^3)$

Transformation 2: $\widehat{\sqrt[3]{\text{weight}}} = -0.02204 + 0.246616\,(\text{length})$

(b) Transformation 1: $\widehat{\text{weight}} = 4.066 + 0.0146774(36^3) = 688.9$ grams

Transformation 2: $\widehat{\sqrt[3]{\text{weight}}} = -0.02204 + 0.246616(36) = 8.856$

$\widehat{\text{weight}} = 8.856^3 = 694.6$ grams

(c) For Transformation 1, the standard deviation of the residuals is $s = 18.841$ grams. Predictions of fish weight using this model will be off by an average of about 19 grams. For Transformation 2, $s = 0.12$. That is, predictions of the cube root of fish weight using this model will be off by an average of about 0.12.

For Practice Try Exercise 35

When experience or theory suggests that the relationship between two variables is described by a power model of the form $y = ax^p$, you now have two strategies for transforming the data to achieve linearity.

1. Raise the values of the explanatory variable x to the p power and plot the points (x^p, y).
2. Take the pth root of the values of the response variable y and plot the points $(x, \sqrt[p]{y})$.

What if you have no idea what power to choose? You could guess and test until you find a transformation that works. Some technology comes with built-in sliders that allow you to dynamically adjust the power and watch the scatterplot change shape as you do.

It turns out that there is a much more efficient method for linearizing a curved pattern in a scatterplot. Instead of transforming with powers and roots, we use logarithms. This more general method works when the data follow an unknown power model or any of several other common mathematical models.

Transforming with Logarithms

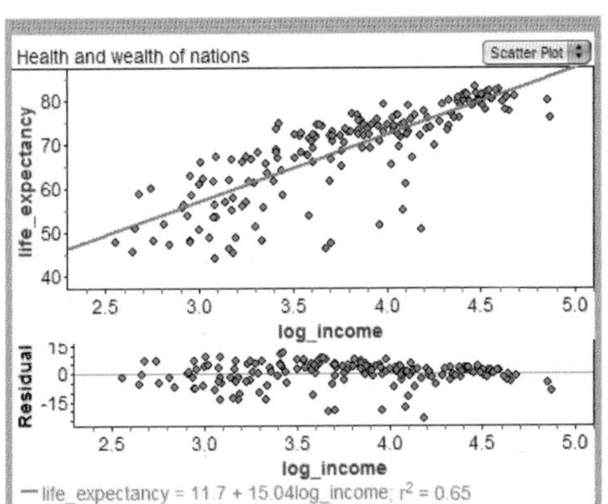

FIGURE 12.15 Scatterplot with least-squares line added and residual plot from Fathom for the transformed data about the health and wealth of nations.

Not all curved relationships are described by power models. For instance, in the "Health and Wealth" example (page 766), a graph of life expectancy versus the logarithm (base 10) of income per person showed a linear pattern. We used Fathom software to fit a least-squares regression line to the transformed data and to make a residual plot. Figure 12.15 shows the results.

The regression line is

$$\text{predicted life expectancy} = 11.7 + 15.04 \log(\text{income})$$

How well does this model fit the data? The residual plot shows a random scatter of prediction errors about the residual = 0 line. Also, because $r^2 = 0.65$, about 65% of the variation in life expectancy is accounted for by the linear model using log(income) as the explanatory variable.

Logarithmic model

The relationship between life expectancy and income per person is described by a **logarithmic model** of the form $y = a + b \log x$. We can use this model to predict how long a country's citizens will live from how much money they make. For the United States, which has income per person of \$41,256.10,

$$\text{predicted life expectancy} = 11.7 + 15.04 \log(41{,}256.1) = 81.117 \text{ years}$$

The actual U.S. life expectancy in 2009 was 79.43 years.

Taking the logarithm of the income per person values straightened out the curved pattern in the original scatterplot. The logarithm transformation can also help achieve linearity when the relationship between two variables is described by a power model or an *exponential model.*

Exponential model

Exponential Models A linear model has the form $y = a + bx$. The value of y increases (or decreases) at a constant rate as x increases. The slope b describes the constant rate of change of a linear model. That is, for each 1 unit increase in x, the model predicts an increase of b units in y. You can think of a linear model as describing the repeated addition of a constant amount. Sometimes the relationship between y and x is based on repeated *multiplication* by a constant factor. That is, each time x increases by 1 unit, the value of y is multiplied by b. An **exponential model** of the form $y = ab^x$ describes such multiplicative growth.

Populations of living things tend to grow exponentially if not restrained by outside limits such as lack of food or space. More pleasantly (unless we're talking about credit card debt!), money also displays exponential growth when interest is compounded each time period. Compounding means that last period's income earns income in the next period.

Money, Money, Money

Understanding exponential growth

Suppose that you invest \$100 in a savings account that pays 6% interest compounded annually. After a year, you will have earned \$100(0.06) = \$6.00 in interest. Your new account balance is the initial deposit plus the interest earned: \$100 + (\$100)(0.06), or \$106. We can rewrite this as \$100(1 + 0.06), or more simply as \$100(1.06). That is, 6% annual interest means that any amount on deposit for the entire year is multiplied by 1.06. If you leave the money invested for a second year, your new balance will be [\$100(1.06)](1.06) = $\$100(1.06)^2$ = \$112.36. Notice that you earn \$6.36 in interest during the second year. That's another \$6 in interest from your initial \$100 deposit plus the interest on your \$6 interest earned for Year 1. After x years, your account balance y is given by the exponential model $y = 100(1.06)^x$.

The accompanying table shows the balance in your savings account at the end of each of the first six years. Figure 12.16 shows the growth in your investment over 100 years. It is characteristic of exponential growth that the increase appears slow for a long period and then seems to explode.

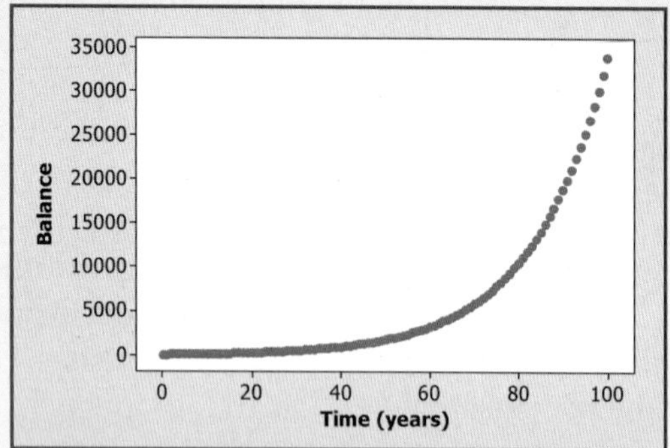

FIGURE 12.16 Scatterplot of the growth of a \$100 investment in a savings account paying 6% interest, compounded annually.

Time x (years)	Account balance y
0	\$100.00
1	\$106.00
2	\$112.36
3	\$119.10
4	\$126.25
5	\$133.82
6	\$141.85
...	...

If an exponential model of the form $y = ab^x$ describes the relationship between x and y, we can use logarithms to transform the data to produce a linear relationship. Start by taking the logarithm (we'll use base 10, but the natural logarithm ln using base e would work just as well). Then use algebraic properties of logarithms to simplify the resulting expressions. Here are the details:

$\log y = \log (ab^x)$	taking the logarithm of both sides
$\log y = \log a + \log (b^x)$	using the property $\log(mn) = \log m + \log n$
$\log y = \log a + x \log b$	using the property $\log m^p = p \log m$

We can then rearrange the final equation as $\log y = \log a + (\log b)x$. Notice that $\log a$ and $\log b$ are constants because a and b are constants. So the equation gives a linear model relating the explanatory variable x to the transformed variable $\log y$. Thus, if the relationship between two variables follows an exponential model, and we plot the logarithm (base 10 or base e) of y against x, we should observe a *straight-line pattern* in the transformed data.

If we fit a least-squares regression line to the transformed data, we can find the predicted value of the logarithm of y for any value of the explanatory variable x by substituting our x-value into the equation of the line. To obtain the corresponding prediction for the response variable y, we have to "undo" the logarithm transformation to return to the original units of measurement. One way of doing this is to use the definition of a logarithm as an exponent:

On the TI-83/84, you can "undo" the logarithm using the [2nd] function keys. To solve $\log y = 2$, press [2nd] [LOG] [2] [ENTER]. To solve $\ln y = 2$, press [2nd] [LN] [2] [ENTER].

$$\log_b a = x \Rightarrow b^x = a$$

For instance, if we have $\log y = 2$, then

$$\log y = 2 \Rightarrow \log_{10} y = 2 \Rightarrow 10^2 = y \Rightarrow 100 = y$$

If instead we have $\ln y = 2$, then

$$\ln y = 2 \Rightarrow \log_e y = 2 \Rightarrow e^2 = y \Rightarrow 7.389 = y$$

EXAMPLE *Moore's Law and Computer Chips*

Logarithm transformations and exponential models

Gordon Moore, one of the founders of Intel Corporation, predicted in 1965 that the number of transistors on an integrated circuit chip would double every 18 months. This is Moore's law, one way to measure the revolution in computing. Here are data on the dates and number of transistors for Intel microprocessors:[15]

Processor	Date	Transistors
4004	1971	2,250
8008	1972	2,500
8080	1974	5,000
8086	1978	29,000
286	1982	120,000
386	1985	275,000
486 DX	1989	1,180,000
Pentium	1993	3,100,000
Pentium II	1997	7,500,000
Pentium III	1999	24,000,000
Pentium 4	2000	42,000,000
Itanium 2	2003	220,000,000
Itanium 2 w/9MB cache	2004	592,000,000
Dual-core Itanium 2	2006	1,700,000,000
Six-core Xeon 7400	2008	1,900,000,000
8-core Xeon Nehalem-EX	2010	2,300,000,000

FIGURE 12.17 Scatterplot of the number of transistors on a computer chip from 1971 to 2010.

Figure 12.17 shows the growth in the number of transistors on a computer chip from 1971 to 2010. Notice that we used "years since 1970" as the explanatory variable. We'll explain this later. If Moore's law is correct, then an exponential model should describe the relationship between the variables.

PROBLEM:

(a) A scatterplot of the natural logarithm (log base *e* or ln) of the number of transistors on a computer chip versus years since 1970 is shown. Based on this graph, explain why it would be reasonable to use an exponential model to describe the relationship between number of transistors and years since 1970.

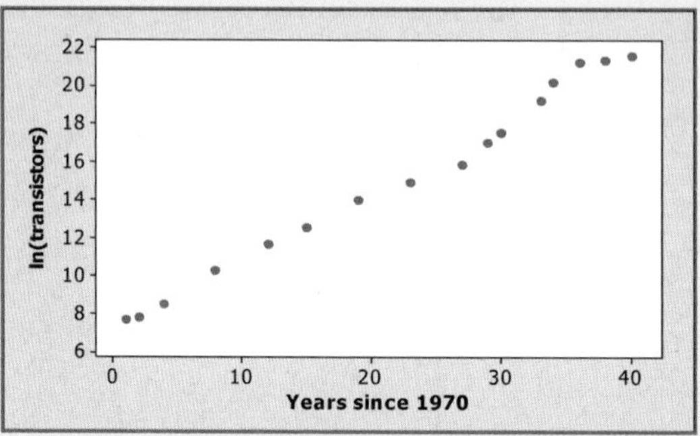

(b) Minitab output from a linear regression analysis on the *transformed* data is shown below. Give the equation of the least-squares regression line. Be sure to define any variables you use.

Predictor	Coef	SE Coef	T	P
Constant	7.0647	0.2672	26.44	0.000
Years since 1970	0.36583	0.01048	34.91	0.000
S = 0.544467 R-Sq = 98.9% R-Sq(adj) = 98.8%				

(c) Use your model from part (b) to predict the number of transistors on an Intel computer chip in 2020. Show your work.

(d) A residual plot for the linear regression in part (b) is shown below. Discuss what this graph tells you about the appropriateness of the model.

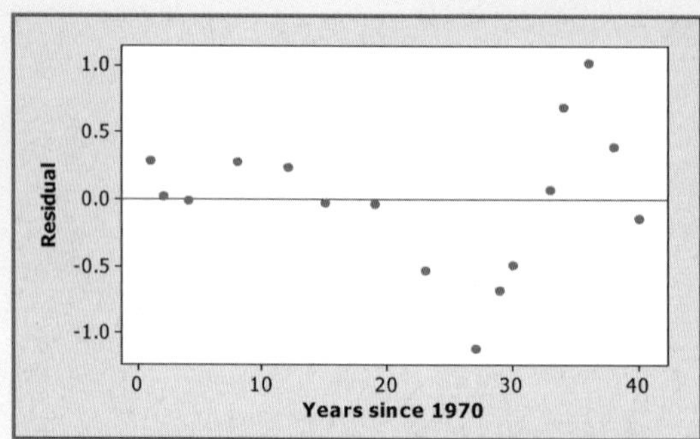

SOLUTION:

(a) If an exponential model describes the relationship between two variables x and y, then we expect a scatterplot of $(x, \ln y)$ to be roughly linear. The scatterplot of ln(transistors) versus years since 1970 has a fairly linear pattern, especially through the year 2000. So an exponential model seems reasonable here.

(b) $\widehat{\ln(\text{transistors})} = 7.0647 + 0.36583(\text{years since } 1970)$

(c) Because 2020 is 50 years since 1970, we have

$$\widehat{\ln(\text{transistors})} = 7.0647 + 0.36583(50) = 25.3562$$

To find the predicted number of transistors, we use the definition of a logarithm as an exponent: $\log_b a = x \Rightarrow b^x = a$. (Or use 2nd LN on the TI-83/84.)

$$\widehat{\ln(\text{transistors})} = 25.3562 \Rightarrow \widehat{\log_e(\text{transistors})} = 25.362$$

$$\widehat{\text{transistors}} = e^{25.362} \approx 1.028 \cdot 10^{11}$$

This model predicts that an Intel chip made in 2020 will have about 100 billion transistors.

(d) The residual plot shows a distinct pattern, with the residuals going from positive to negative to positive as we move from left to right. But the residuals are small in size relative to the transformed y-values. Also, the scatterplot of the transformed data is much more linear than the original scatterplot. We feel reasonably comfortable using this model to make predictions about the number of transistors on a computer chip.

For Practice *Try Exercise 39*

Make sure that you understand the big idea here. The necessary transformation is carried out by taking the logarithm of the response variable. Your calculator and most statistical software will calculate the logarithms of all the values of a variable with a single command. The crucial property of the logarithm for our purposes is that *if a variable grows exponentially, its logarithm grows linearly.*

THINK ABOUT IT

How do we find the exponential model for predicting y from x? The least-squares line for the transformed data in the computer chip example is

$$\widehat{\ln(\text{transistors})} = 7.0647 + 0.36583\,(\text{years since } 1970)$$

If we use the definition of the logarithm as an exponent, we can rewrite this equation as

$$\widehat{\text{transistors}} = e^{7.0647 + 0.36583(\text{years since } 1970)}$$

Using properties of exponents, we can simplify this as follows:

$\widehat{\text{transistors}} = e^{7.0647} \cdot e^{0.36583(\text{years since } 1970)}$	using the fact that $b^m b^n = b^{m+n}$
$\widehat{\text{transistors}} = e^{7.0647} \cdot (e^{0.36583})^{(\text{years since } 1970)}$	using the fact that $(b^m)^n = b^{mn}$
$\widehat{\text{transistors}} = 1169.93 \cdot (1.44171)^{(\text{years since } 1970)}$	simplifying

This equation is now in the familiar form of an exponential model $y = ab^x$ with $a = 1169.93$ and $b = 1.44171$.

We could use the exponential model to predict the number of transistors on an Intel chip in 2020: $\widehat{\text{transistors}} = 1169.93(1.44171)^{50} \approx 1.0281 \cdot 10^{11}$. This is the same prediction we got earlier. How does this compare with the prediction from Moore's law? Suppose the number of transistors on an Intel computer chip doubles every 18 months (1.5 years). Then in the 49 years from 1971 to 2020, the number of transistors would double 49/1.5 = 32.67 times. So the predicted number of transistors on an Intel chip in 2020 would be

$$\widehat{\text{transistors}} = 2250(2)^{32.67} = 1.54 \cdot 10^{13}$$

Moore's law predicts more rapid exponential growth than our model does.

The calculation at the end of the Think about It feature might give you some idea of why we used years since 1970 as the explanatory variable in the example. To make a prediction, we substituted the value $x = 50$ into the equation for the exponential model. This value is the exponent in our calculation. If we had used years as the explanatory variable, our exponent would have been 2020. Such a large exponent can lead to overflow errors on a calculator.

CHECK YOUR UNDERSTANDING

The following table gives the resident population of the United States from 1790 to 1880, in millions of people:

Year	Population	Year	Population
1790	3.9	1840	17.1
1800	5.3	1850	23.2
1810	7.2	1860	31.4
1820	9.6	1870	39.8
1830	12.9	1880	50.2

The graph at the left below is a scatterplot of these data using years since 1700 as the explanatory variable. A plot of the natural logarithms of the population values against years since 1700 is shown at the right below.

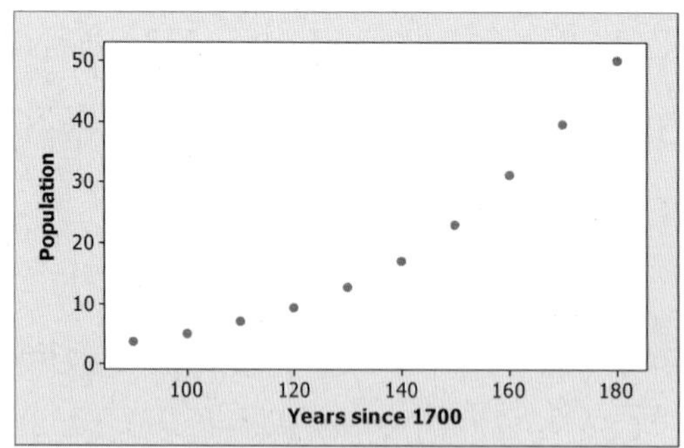

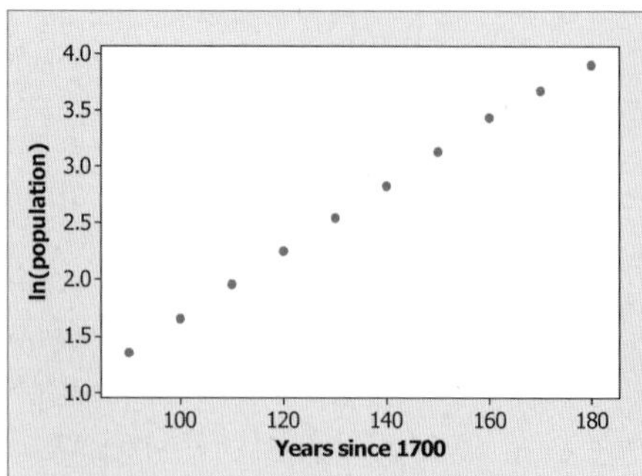

1. Explain why it would be reasonable to use an exponential model to describe the relationship between the U.S. population in the years 1790 to 1880 and the number of years since 1700.

Here is some Minitab output from a linear regression analysis on the transformed data.

```
Predictor    Coef        SE Coef       T       P
Constant    −1.19311    0.04848     −24.61   0.000
Year         0.0287280  0.0003513    81.78   0.000
S = 0.0319068 R-Sq = 99.9% R-Sq(adj) = 99.9%
```

2. Give the equation of the least-squares regression line. Define any variables you use.
3. Use this linear model to predict the U.S. population in 1890. Show your work.
4. Based on the residual plot, do you expect your prediction to be too high or too low? Justify your answer.

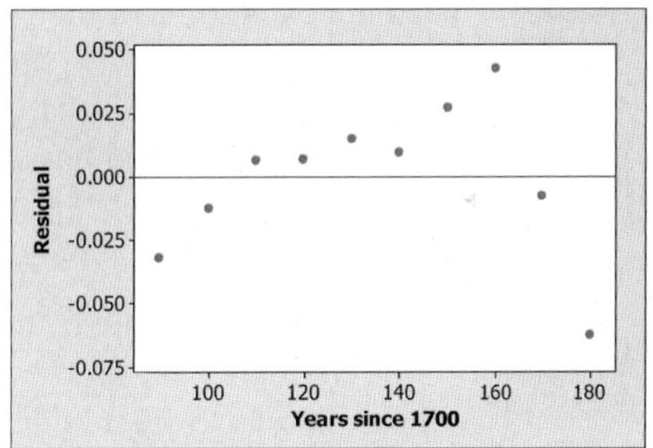

Power Models Again Biologists have found that many characteristics of living things are described quite closely by power models. There are more mice than elephants, and more flies than mice—the abundance of species follows a power model with body weight as the explanatory variable. So do pulse rate, length of life, the number of eggs a bird lays, and so on.

Sometimes the powers can be predicted from geometry, but sometimes they are mysterious. Why, for example, does the rate at which animals use energy go up as the 3/4 power of their body weight? Biologists call this relationship Kleiber's law. It has been found to work all the way from bacteria to whales. The search goes on for some physical or geometrical explanation for why life follows power laws.

When we apply the logarithm transformation to the response variable y in an exponential model, we produce a linear relationship. To achieve linearity from a power model, we apply the logarithm transformation to *both* variables. Here are the details:

1. A power model has the form $y = ax^p$, where a and p are constants.

2. Take the logarithm of both sides of this equation. Using properties of logarithms,

$$\log y = \log(ax^p) = \log a + \log(x^p) = \log a + p \log x$$

The equation $\log y = \log a + p \log x$ shows that taking the logarithm of both variables results in a linear relationship between $\log x$ and $\log y$.

3. Look carefully: the *power* p in the power model becomes the *slope* of the straight line that links $\log y$ to $\log x$.

If a power model describes the relationship between two variables, a scatterplot of the logarithms of both variables should produce a linear pattern. Then we can fit a least-squares regression line to the transformed data and use the linear model to make predictions. Here's an example.

What's a Planet, Anyway?

Power models and logarithm transformations

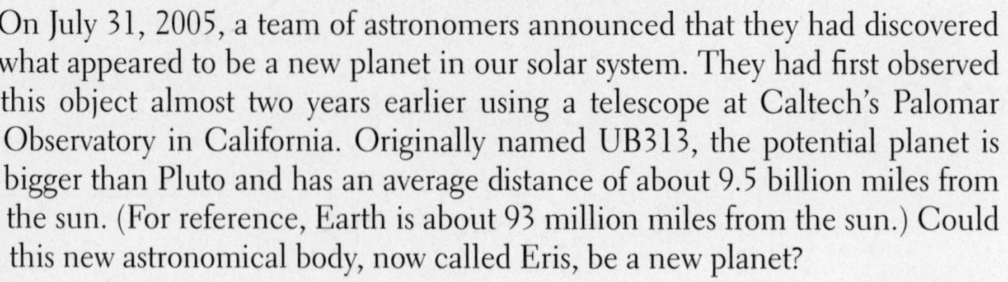

On July 31, 2005, a team of astronomers announced that they had discovered what appeared to be a new planet in our solar system. They had first observed this object almost two years earlier using a telescope at Caltech's Palomar Observatory in California. Originally named UB313, the potential planet is bigger than Pluto and has an average distance of about 9.5 billion miles from the sun. (For reference, Earth is about 93 million miles from the sun.) Could this new astronomical body, now called Eris, be a new planet?

At the time of the discovery, there were nine known planets in our solar system. Here are data on the distance from the sun and period of revolution of those planets. Note that distance is measured in astronomical units (AU), the number of earth distances the object is from the sun.[16]

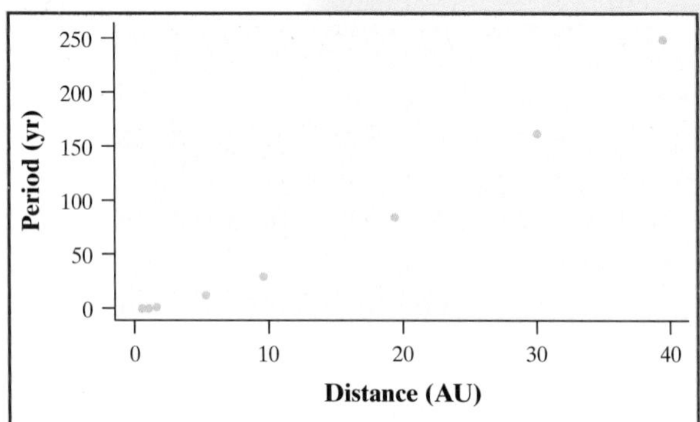

FIGURE 12.18 Scatterplot of planetary distance from the sun and period of revolution.

Planet	Distance from sun (astronomical units)	Period of revolution (Earth years)
Mercury	0.387	0.241
Venus	0.723	0.615
Earth	1.000	1.000
Mars	1.524	1.881
Jupiter	5.203	11.862
Saturn	9.539	29.456
Uranus	19.191	84.070
Neptune	30.061	164.810
Pluto	39.529	248.530

Figure 12.18 is a scatterplot of the planetary data. There appears to be a strong curved relationship between distance from the sun and period of revolution.

PROBLEM: The graphs below show the results of two different transformations of the data. Figure 12.19(a) plots the natural logarithm of period against distance from the sun for all 9 planets. Figure 12.19(b) plots the natural logarithm of period against the natural logarithm of distance from the sun for the 9 planets.

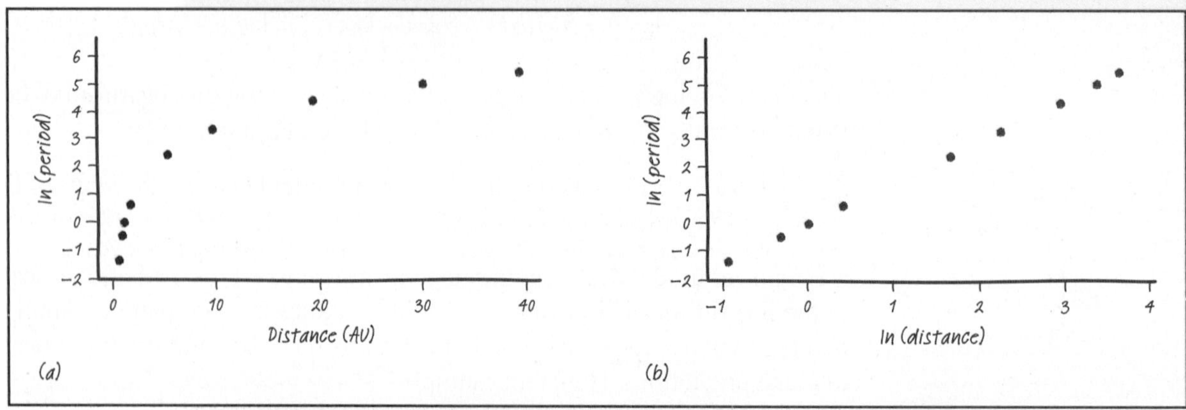

FIGURE 12.19 (a) A scatterplot of ln(period) versus distance. (b) A scatterplot of ln(period) versus ln(distance).

(a) Explain why a power model would provide a more appropriate description of the relationship between period of revolution and distance from the sun than an exponential model.

(b) Minitab output from a linear regression analysis on the transformed data in Figure 12.19(b) is shown below. Give the equation of the least-squares regression line. Be sure to define any variables you use.

Predictor	Coef	SE Coef	T	P
Constant	0.0002544	0.0001759	1.45	0.191
ln(distance)	1.49986	0.00008	18598.27	0.000

S = 0.000393364 R-Sq = 100.0% R-Sq(adj) = 100.0%

(c) Use your model from part (b) to predict the period of revolution for Eris, which is 9,500,000,000/93,000,000 = 102.15 AU from the sun. Show your work.

(d) A residual plot for the linear regression in part (b) is shown below. Do you expect your prediction in part (c) to be too high, too low, or just right? Justify your answer.

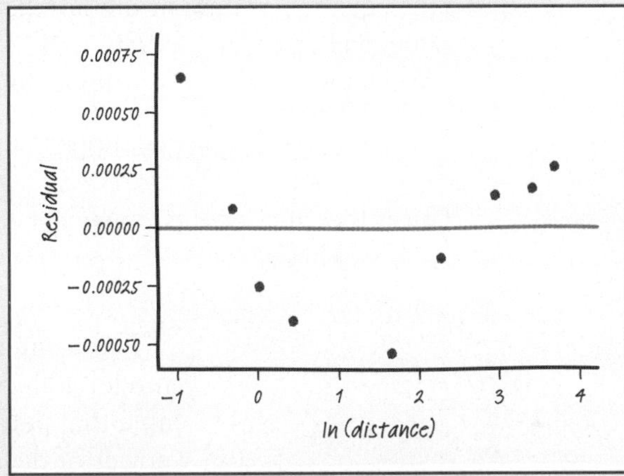

SOLUTION:

(a) The scatterplot of ln(period) versus distance is clearly curved, so an exponential model would not be appropriate. However, the graph of ln(period) versus ln(distance) has a strong linear pattern, indicating that a power model would be more appropriate.

(b) $\widehat{\ln(\text{period})} = 0.0002544 + 1.49986\ \ln(\text{distance})$

(c) Eris's average distance from the sun is 102.15 AU. Using this value for distance in our model from part (b) gives

$$\widehat{\ln(\text{period})} = 0.0002544 + 1.49986\ \ln(102.15) = 6.939$$

To predict the period, we have to undo the logarithm transformation:

$$\widehat{\text{period}} = e^{6.939} \approx 1032 \text{ years}$$

We wouldn't want to wait for Eris to make a full revolution to see if our prediction is accurate!

(d) Eris's value for ln(distance) is 6.939, which would fall at the far right of the residual plot, where all the residuals are positive. Because residual = actual y − predicted y seems likely to be positive, we would expect our prediction to be too low.

In August 2006, the International Astronomical Union agreed on a new definition of "planet." Both Pluto and Eris were classified as "dwarf planets."

For Practice *Try Exercise 43*

How do we find the power model for predicting y from x? The least-squares line for the transformed planet data is

$$\widehat{\ln(\text{period})} = 0.0002544 + 1.49986 \ln(\text{distance})$$

If we use the definition of the logarithm as an exponent, we can rewrite this equation as

$$\widehat{\text{period}} = e^{0.0002544+1.49986\ln(\text{distance})}$$

Using properties of exponents, we can simplify this as follows:

$$\widehat{\text{period}} = e^{0.0002544} \cdot e^{1.49986\ln(\text{distance})} \quad \text{using the fact that } b^m b^n = b^{m+n}$$

$$\widehat{\text{period}} = e^{0.0002544} \cdot e^{\ln(\text{distance})^{1.49986}} \quad \text{using the fact that } p \ln x = \ln x^p$$

$$\widehat{\text{period}} = 1.0002544(\text{distance})^{1.49986} \quad \text{using the fact that } e^{\ln x} = x$$

This equation is now in the familiar form of a power model $y = ax^p$, with $a = 1.0002544$ and $b = 1.49986$.

We could use the power model to predict Eris's period of revolution:

$$\widehat{\text{period}} = 1.0002544(102.15)^{1.49986} \approx 1032 \text{ years}$$

This is the same prediction we got earlier without all the messy algebra.

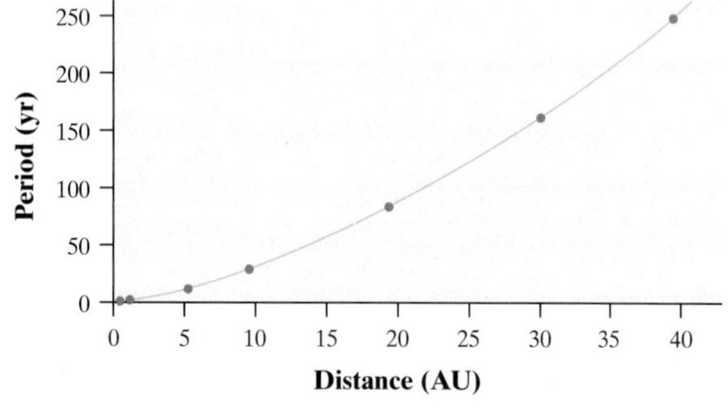

FIGURE 12.20 Planetary data with power model.

The scatterplot of the original data with the power model added appears in Figure 12.20. It seems remarkable that period of revolution is closely related to the 1.5 power of distance from the sun. Johannes Kepler made this fascinating discovery many years ago without the aid of modern technology—a result known as Kepler's third law.

Now it's time to put what you have learned to use in tackling the following Data Exploration.

DATA EXPLORATION *It's a matter of life and death*

One sad fact about life is that we'll all die someday. Many adults plan ahead for their eventual passing by purchasing life insurance. Many different types of life insurance policies are available. Some provide coverage throughout an individual's life (whole life), while others last only for a specified number of years (term life). The policyholder makes regular payments (premiums) to the insurance company in return for the coverage. When the insured person dies, a payment is made to designated family members or other beneficiaries.

How do insurance companies decide how much to charge for life insurance? They rely on a staff of highly trained actuaries—people with expertise in probability, statistics, and advanced mathematics—to establish premiums. For an individual who wants to buy life insurance, the premium will depend on the type and amount of the policy as well as on personal characteristics like age, sex, and health status.

The table shows monthly premiums for a 10-year term-life insurance policy worth $1,000,000.[17]

Age (years)	Monthly premium
40	$29
45	$46
50	$68
55	$106
60	$157
65	$257

Your goal in this Data Exploration is to choose among three possible models for predicting monthly premium from age. Option 1 is based on the original data, while Options 2 and 3 involve transformations of the original data. Each of the Fathom screen shots below displays a scatterplot with a least-squares regression line added and a residual plot for one of the three options.

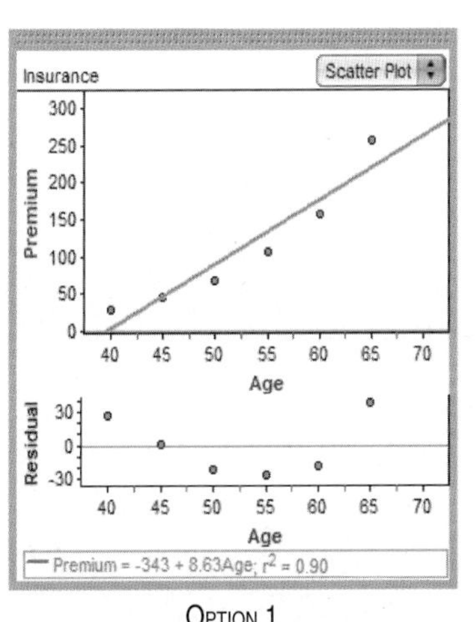

OPTION 1

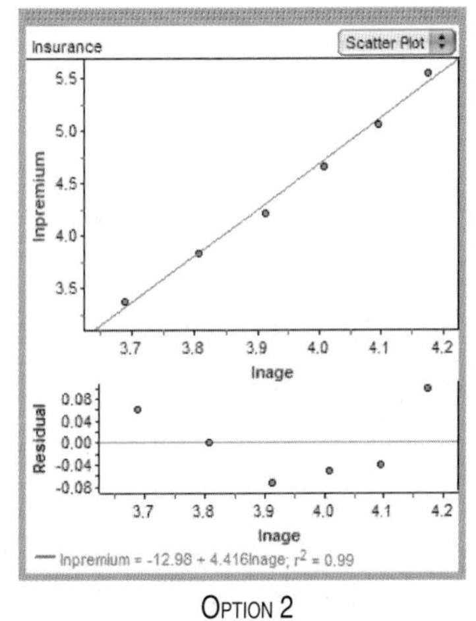

OPTION 2

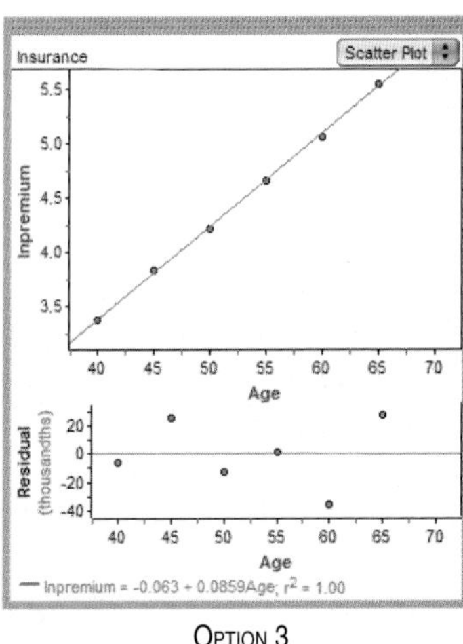

OPTION 3

1. Use each model to predict how much a 58-year-old would pay for such a policy. Show your work.

2. Which of the three predictions do you expect to be closest to the true monthly premium? Justify your answer.

3. What type of function—linear, power, or exponential—best describes the relationship between age and monthly premium? Explain.

We have used statistical software to do all the transformations and linear regression analysis in this section so far. Now let's look at how the process works on a graphing calculator.

TECHNOLOGY CORNER Transforming to achieve linearity on the calculator

We'll use the planet data to illustrate a general strategy for performing transformations with logarithms on the TI-83/84 and TI-89. A similar approach could be used for transforming data with powers and roots.

• Enter the values of the explanatory variable in L1/list1 and the values of the response variable in L2/list2. Make a scatterplot of y versus x and confirm that there is a curved pattern.

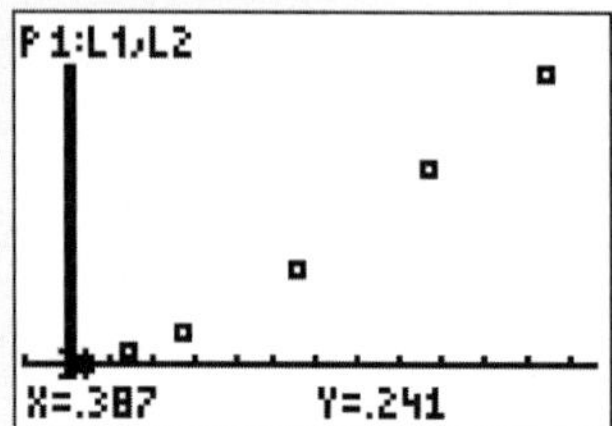

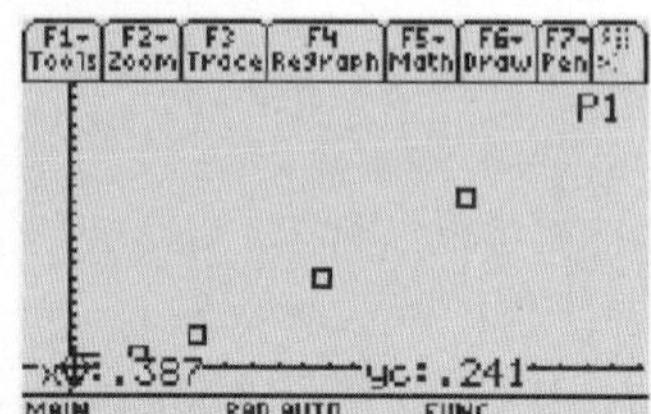

• Define L3/list 3 to be the logarithm (base 10 or e) of L1/list1 and L4/list4 to be the logarithm (same base) of L2/list2. To see whether an exponential model fits the original data, make a plot of L4/list4 versus L1/list1 and look for linearity. To see whether a power model fits the original data, make a plot of L4/list4 versus L3/list3 and look for linearity. (We used ln to match the example.)

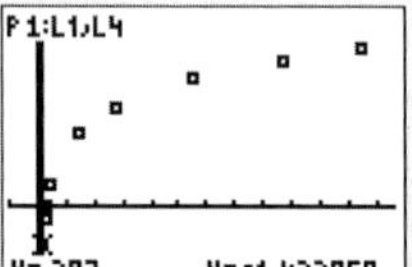

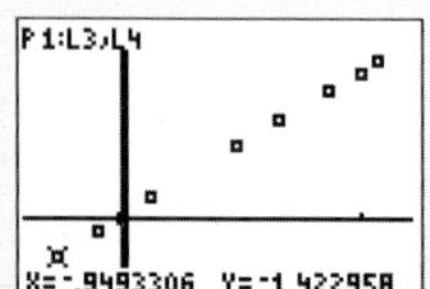

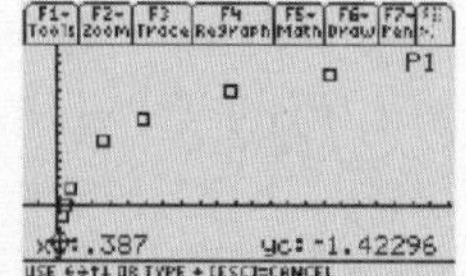

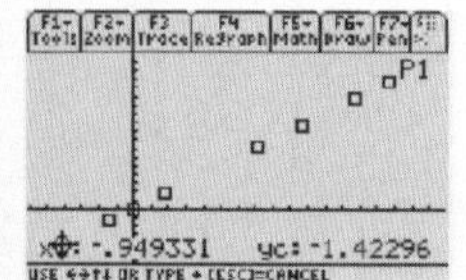

• If a linear pattern is present, calculate the equation of the least-squares regression line and store it in Y1. For the planet data, we executed the command `LinReg(a+bx)L3,L4,Y1`.

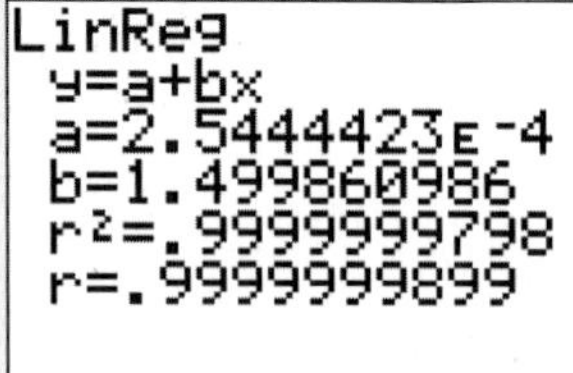

Construct a residual plot to look for any departures from the linear pattern. For Xlist, enter the list you used as the explanatory variable in the linear regression calculation. For Ylist, use the RESID list stored in the calculator. For the planet data, we used L3/list 3 as the Xlist.

When you run a linear regression on the TI–83/84 or TI–89, the calculator stores the residuals automatically in a list called RESID. You can find the RESID list on the List menu (2nd STAT) on the TI–83/84 and in VAR-LINK (2nd −) under STATVARS on the TI–89. Too bad we didn't tell you about this in Chapter 3!

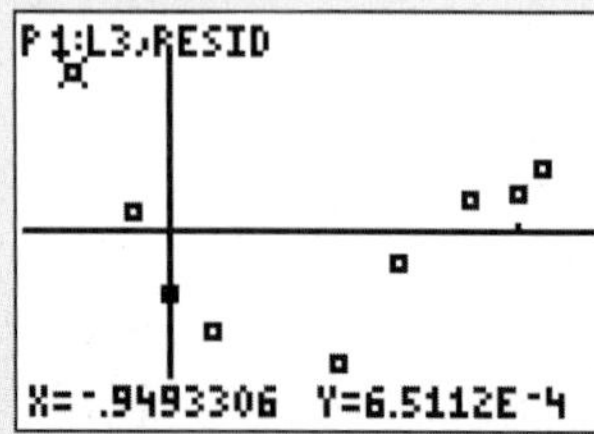

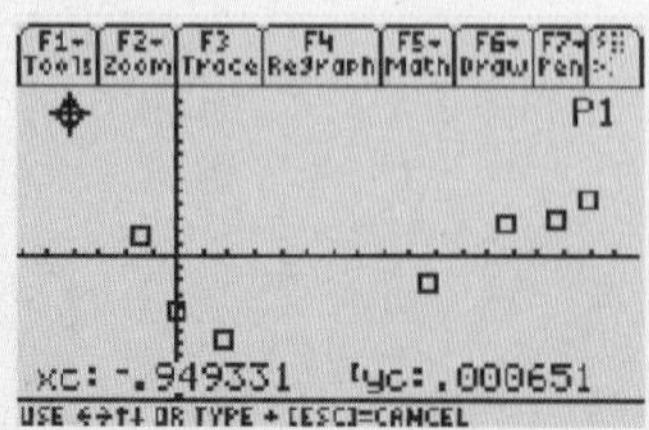

- To make a prediction for a specific value of the explanatory variable $x = k$, modify the regression equation in Y1 by changing x to $\log x$ or $\ln x$, if appropriate. Then use Y1(k) to obtain the predicted value of $\log y$ or $\ln y$. To get the predicted value of y, use 10^Ans or e^Ans to undo the logarithm transformation. Here's our prediction of the period of revolution for Eris, which is at a distance of 102.15 AU from the sun:

```
Y1(102.15)
        6.939274784
e^(Ans)
        1032.021505
```

```
F1-   F2-     F3-  F4-   F5     F6-
Tools Algebra Calc Other PrgmIO Clean Up

■ y1(102.15)                 6.93927
■ e^6.9392747840111
                             1032.02
e^(ans(1))
MAIN        RAD AUTO     FUNC    2/30
```

TI-Nspire instructions in Appendix B

Note: The calculator's ExpReg or PwrReg command will give you the exponential model or power model that results from transforming the data with logarithms, fitting a least-squares regression line to the transformed data, and then undoing the logarithm transformation to get back to the original variables. For instance, here is the result of executing the command `PwrReg L1,L2` for the planet data.

```
PwrReg
 y=a*x^b
 a=1.000254477
 b=1.499860986
 r²=.9999999798
 r=.9999999899
```

The desired power model is $\widehat{\text{period}} = 1.0002544(\text{distance})^{1.49986}$, which matches what we got in the Think about It feature on page 780. Unfortunately, *the calculator may report residuals from the power model itself instead of from the least-squares line*. Because the primary goal of this section is transforming data to achieve linearity, you should use only linear regression on your calculator.

CAUTION

Do Longer Drives Mean Lower Scores on the PGA Tour?

In the chapter-opening Case Study (page 737), we examined data on the mean drive distance (in yards) and mean score per round for an SRS of 19 of the 197 players on the 2008 PGA Tour. Here is some Minitab output from a least-squares regression analysis on these data:

Predcited mean score = 76.90 - 0.02016(mean distance)

Predictor	Coef	SE Coef	T	P
Constant	76.904	3.808	20.20	0.000
Avg. distance	−0.02016	0.01319	−1.53	0.145
S = 0.618396 R-Sq = 12.1% R-Sq(adj) = 6.9%				

Do these data give convincing evidence that the slope of the population regression line is negative? For one final time, we'll follow the four-step process.

STATE: We want to perform a test of

$$H_0: \beta = 0$$
$$H_a: \beta < 0$$

where β is the actual slope of the regression line for predicting mean score from mean distance for all 197 players on the 2008 PGA Tour. Because no significance level was stated, we'll use $\alpha = 0.05$.

PLAN: If conditions are met, we should carry out a *t* test for the slope β.

- **Linear** The scatterplot suggests that there is a weak negative linear relationship between mean score and mean distance in the population. At each *x*-value, the values of *y* seem to be centered on the regression line.
- **Independent** The individual observations of mean score for 19 randomly selected players should be independent. Because we're sampling without replacement, we need to check the 10% condition: there are at least 10(19) = 190 players on the 2008 PGA Tour.
- **Normal** A histogram of the residuals is single-peaked and roughly symmetric.
- **Equal variance** The residual plot shows that the scatter of points about the regression line is roughly the same at all *x*-values.

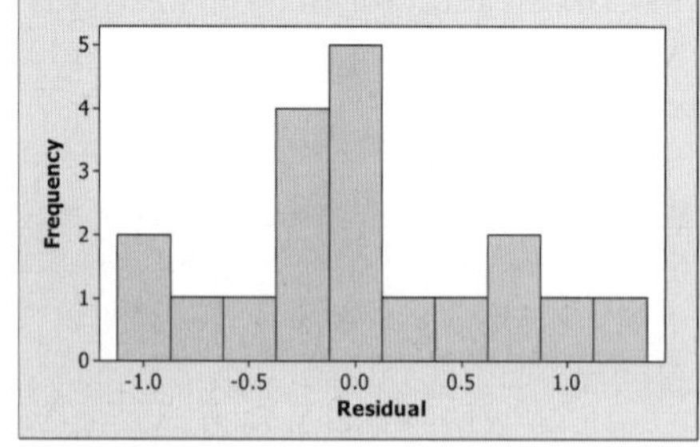

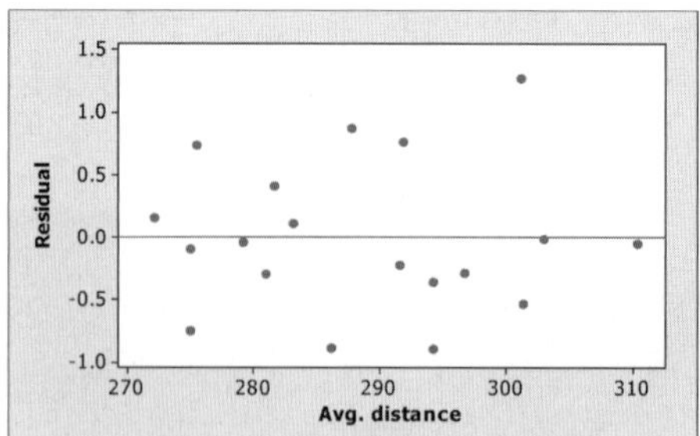

- **Random** The data came from a simple random sample of 19 players.

DO: The computer regression output gives the results from a two-sided test.

- **Test statistic** $t = -1.53$
- ***P*-value** Using df $= 19 - 2 = 17$, the *P*-value for the one-sided test is $0.145/2 = 0.0725$.

CONCLUDE: Because the *P*-value, 0.0725, is greater than $\alpha = 0.05$, we fail to reject H_0. We do not have enough evidence to conclude that there is a negative linear relationship between average drive distance and average score in the population of 197 players.

Remark: In fact, the population regression line for all the players on the 2008 PGA Tour is $\mu_y = 71.97 - 0.00367x$. So there *is* a negative linear relationship between average drive distance and average score. We made a Type II error in this case by failing to reject $H_0: \beta = 0$ when $\beta = -0.00367$. It seems likely that our test did not have enough power to detect such a small difference with a sample size of only $n = 19$.

SECTION 12.2 Summary

- Nonlinear relationships between two quantitative variables can sometimes be changed into linear relationships by **transforming** one or both of the variables. Transformation is particularly effective when there is reason to think that the data are governed by some nonlinear mathematical model.
- When theory or experience suggests that the relationship between two variables follows a **power model** of the form $y = ax^p$, there are two transformations involving powers and roots that can linearize a curved pattern in a scatterplot. Option 1: Raise the values of the explanatory variable x to the power p, then look at a graph of (x^p, y). Option 2: Take the pth root of the values of the response variable y, then look at a graph of $(x, \sqrt[p]{y})$.
- In a linear model of the form $y = a + bx$, the values of the response variable are predicted to increase by a constant amount b for each increase of 1 unit in the explanatory variable. For an **exponential model** of the form $y = ab^x$, the predicted values of the response variable are multiplied by an additional factor of b for each increase of one unit in the explanatory variable.
- A useful strategy for straightening a curved pattern in a scatterplot is to take the **logarithm** of one or both variables. To achieve linearity when the relationship between two variables follows an exponential model, plot the logarithm (base 10 or base e) of y against x. When a power model describes the relationship between two variables, a plot of $\log y$ ($\ln y$) versus $\log x$ ($\ln x$) should be linear.
- Once we transform the data to achieve linearity, we can fit a least-squares regression line to the transformed data and use this linear model to make predictions.

12.2 TECHNOLOGY CORNER

TI-Nspire instructions in Appendix B

SECTION 12.2 Exercises

33. **The swinging pendulum** Mrs. Hanrahan's precalculus class collected data on the length (in centimeters) of a pendulum and the time (in seconds) the pendulum took to complete one back-and-forth swing (called its period). Here are their data:

Length (cm)	Period (s)
16.5	0.777
17.5	0.839
19.5	0.912
22.5	0.878
28.5	1.004
31.5	1.087
34.5	1.129
37.5	1.111
43.5	1.290
46.5	1.371
106.5	2.115

(a) Make a reasonably accurate scatterplot of the data by hand, using length as the explanatory variable. Describe what you see.

(b) The theoretical relationship between a pendulum's length and its period is

$$\text{period} = \frac{2\pi}{\sqrt{g}}\sqrt{\text{length}}$$

where g is a constant representing the acceleration due to gravity (in this case, $g = 980$ cm/s^2). Use the graph below to identify the transformation that was used to linearize the curved pattern in part (a).

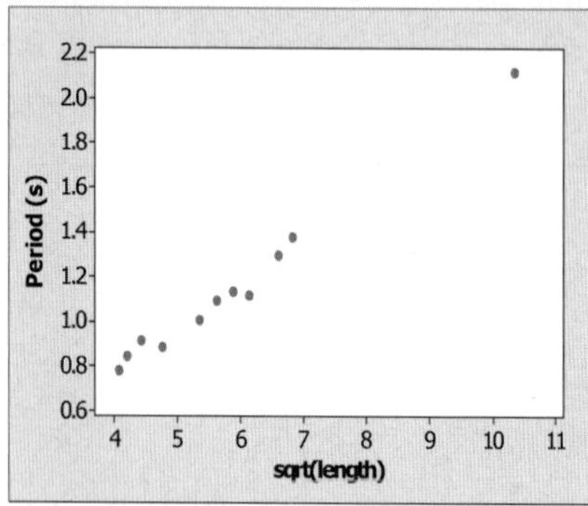

(c) Use the following graph to identify the transformation that was used to linearize the curved pattern in part (a).

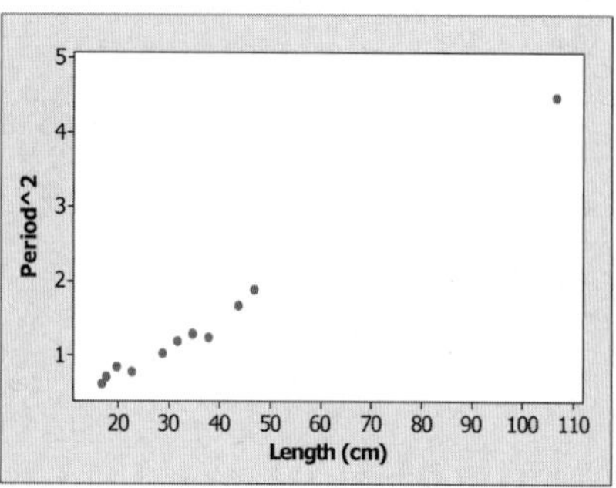

34. **Boyle's law** If you have taken a chemistry class, then you are probably familiar with Boyle's law: for gas in a confined space kept at a constant temperature, pressure times volume is a constant (in symbols, $PV = k$). Students collected the following data on pressure and volume using a syringe and a pressure probe.

Volume (cubic centimeters)	Pressure (atmospheres)
6	2.9589
8	2.4073
10	1.9905
12	1.7249
14	1.5288
16	1.3490
18	1.2223
20	1.1201

(a) Make a reasonably accurate scatterplot of the data by hand using volume as the explanatory variable. Describe what you see.

(b) If the true relationship between the pressure and volume of the gas is $PV = k$, we can divide both sides of this equation by V to obtain the theoretical model $P = k/V$, or $P = k(1/V)$. Use the graph below to identify the transformation that was used to linearize the curved pattern in part (a).

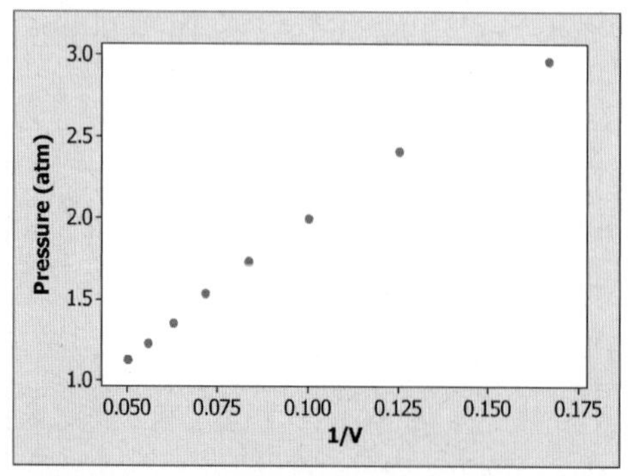

(c) Use the graph below to identify the transformation that was used to linearize the curved pattern in part (a).

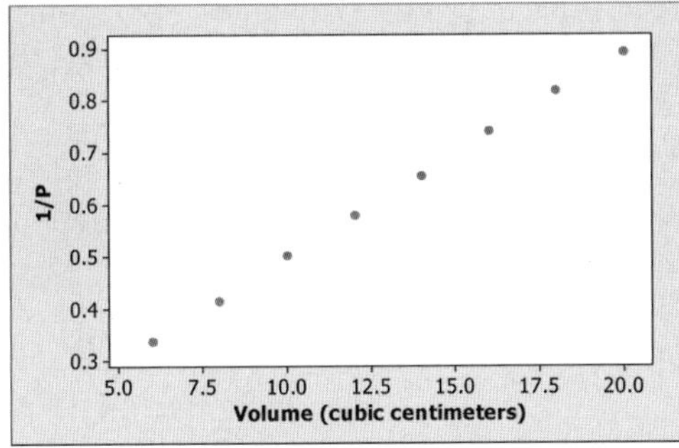

35. **The swinging pendulum** Refer to Exercise 33. Here is Minitab output from separate regression analyses of the two sets of transformed pendulum data:

pg 770

Transformation 1: ($\sqrt{\text{length}}$, period)

Predictor	Coef	SE Coef	T	P
Constant	−0.08594	0.05046	−1.70	0.123
sqrt (length)	0.209999	0.008322	25.23	0.000

S = 0.0464223 R-Sq = 98.6% R-Sq(adj) = 98.5%

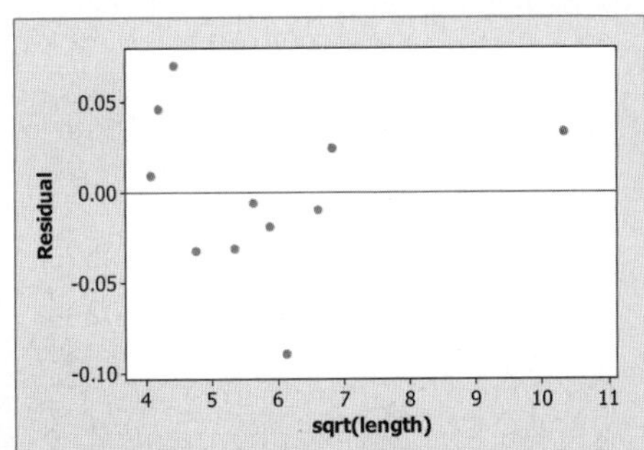

Transformation 2: (length, period^2)

Predictor	Coef	SE Coef	T	P
Constant	−0.15465	0.05802	−2.67	0.026
Length (cm)	0.042836	0.001320	32.46	0.000

S = 0.105469 R-Sq = 99.2% R-Sq(adj) = 99.1%

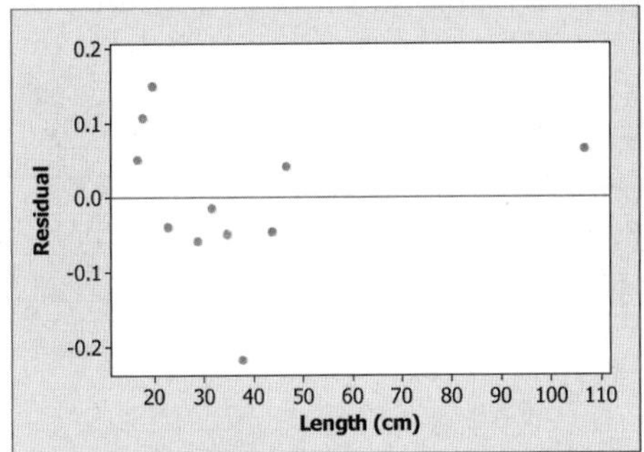

Do each of the following for *both* transformations.

(a) Give the equation of the least-squares regression line. Define any variables you use.

(b) Use the model from part (a) to predict the period of a pendulum with length 80 centimeters. Show your work.

(c) Interpret the value of *s* in context.

36. **Boyle's law** Refer to Exercise 34. Here is Minitab output from separate regression analyses of the two sets of transformed pressure data:

Transformation 1: $\left(\frac{1}{\text{volume}}, \text{pressure}\right)$

Predictor	Coef	SE Coef	T	P
Constant	0.36774	0.04055	9.07	0.000
1/V	15.8994	0.4190	37.95	0.000

S = 0.044205 R-Sq = 99.6% R-Sq(adj) = 99.5%

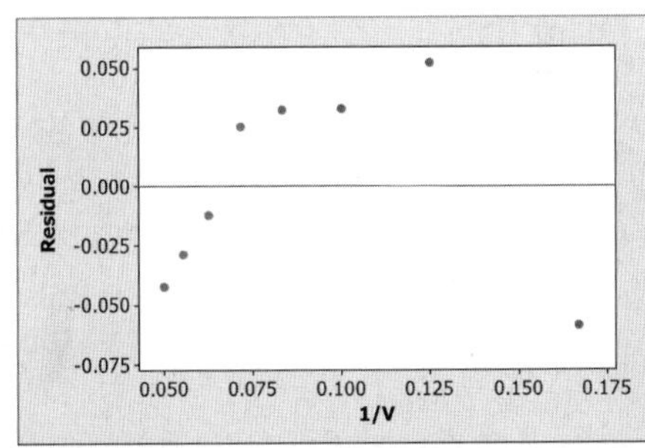

Transformation 2: $\left(\text{volume}, \frac{1}{\text{pressure}}\right)$

Predictor	Coef	SE Coef	T	P
Constant	0.100170	0.003779	26.51	0.000
Volume	0.0398119	0.0002741	145.23	0.000

S = 0.003553 R-Sq = 100.0% R-Sq(adj) = 100.0%

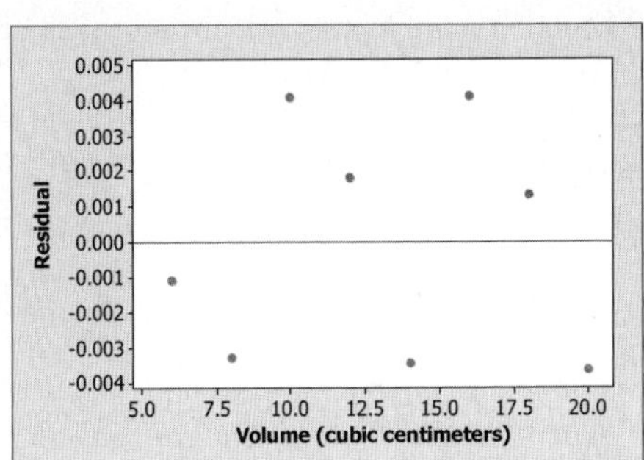

Do each of the following for *both* transformations.

(a) Give the equation of the least-squares regression line. Define any variables you use.

(b) Use the model from part (a) to predict the pressure in the syringe when the volume is 17 cubic centimeters. Show your work.

(c) Interpret the value of *s* in context.

37. **Killing bacteria** Expose marine bacteria to X-rays for time periods from 1 to 15 minutes. Here are the number of surviving bacteria (in hundreds) on a culture plate after each exposure time:[18]

Time t	Count y	Time t	Count y
1	355	9	56
2	211	10	38
3	197	11	36
4	166	12	32
5	142	13	21
6	106	14	19
7	104	15	15
8	60		

(a) Make a reasonably accurate scatterplot of the data by hand, using time as the explanatory variable. Describe what you see.

(b) A scatterplot of the natural logarithm of the number of surviving bacteria versus time is shown below. Based on this graph, explain why it would be reasonable to use an exponential model to describe the relationship between count of bacteria and time.

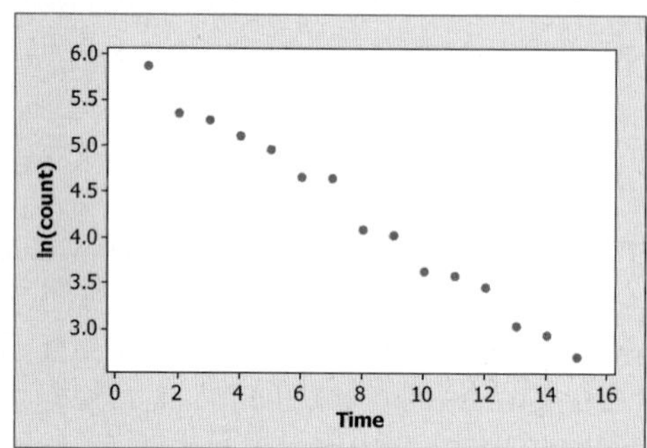

(c) Minitab output from a linear regression analysis on the transformed data is shown below.

```
Predictor     Coef       SE Coef      T       P
Constant    5.97316    0.05978     99.92  0.000
Time       -0.218425   0.006575   -33.22  0.000
S = 0.110016 R-Sq = 98.8% R-Sq(adj) = 98.7%
```

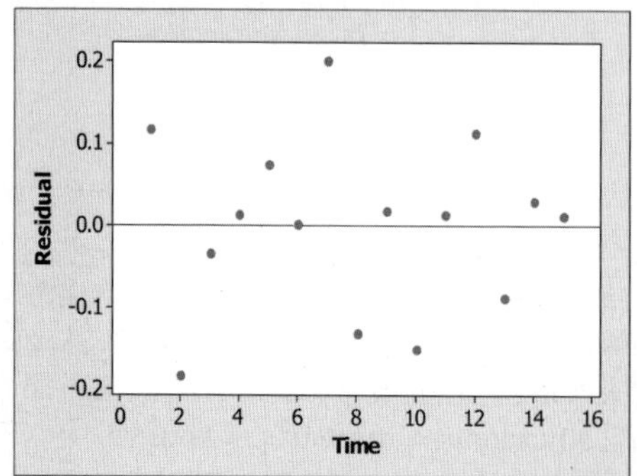

Give the equation of the least-squares regression line. Be sure to define any variables you use.

(d) Use your model to predict the number of surviving bacteria after 17 minutes. Show your work. Do you expect this prediction to be too high, too low, or about right? Explain.

38. **Light through the water** Some college students collected data on the intensity of light at various depths in a lake. Here are their data:

Depth (meters)	Light intensity (lumens)
5	168.00
6	120.42
7	86.31
8	61.87
9	44.34
10	31.78
11	22.78

(a) Make a reasonably accurate scatterplot of the data by hand, using depth as the explanatory variable. Describe what you see.

(b) A scatterplot of the natural logarithm of light intensity versus depth is shown below. Based on this graph, explain why it would be reasonable to use an exponential model to describe the relationship between light intensity and depth.

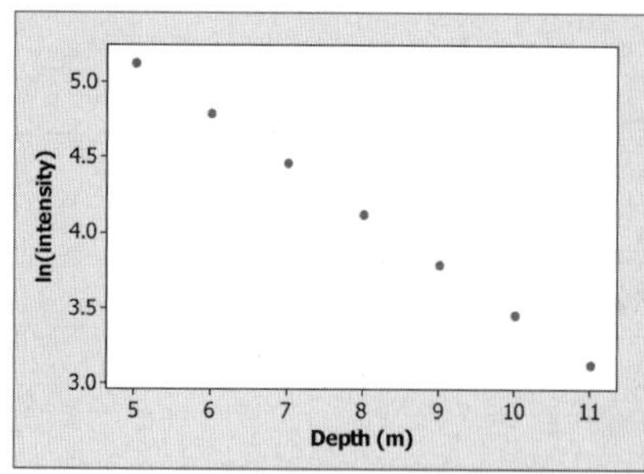

(c) Minitab output from a linear regression analysis on the transformed data is shown below.

```
Predictor     Coef       SE Coef       T          P
Constant     6.78910    0.00009     78575.46  0.000
Depth (m)   -0.333021   0.000010   -31783.44  0.000
S = 0.000055 R-Sq = 100.0% R-Sq(adj) = 100.0%
```

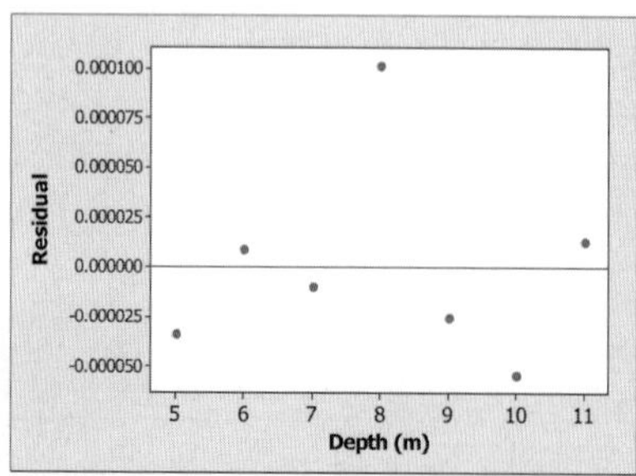

Give the equation of the least-squares regression line. Be sure to define any variables you use.

(d) Use your model to predict the light intensity at a depth of 12 meters. The actual light intensity reading at that depth was 16.2 lumens. Does this surprise you? Explain.

39. **Brawn versus brain** How is the weight of an animal's brain related to the weight of its body? Researchers collected data on the brain weight (in grams) and body weight (in kilograms) for 96 species of mammals.[19] The figure below is a scatterplot of the logarithm of brain weight against the logarithm of body weight for all 96 species. The least-squares regression line for the transformed data is

pg 773

$$\widehat{\log y} = 1.01 + 0.72 \log x$$

Based on footprints and some other sketchy evidence, some people think that a large apelike animal, called Sasquatch or Bigfoot, lives in the Pacific Northwest. His weight is estimated to be about 280 pounds, or 127 kilograms. How big is Bigfoot's brain? Show your method clearly.

40. **Determining tree biomass** It is easy to measure the "diameter at breast height" of a tree. It's hard to measure the total "aboveground biomass" of a tree, because to do this you must cut and weigh the tree. The biomass is important for studies of ecology, so ecologists commonly estimate it using a power model. Combining data on 378 trees in tropical rain forests gives this relationship between biomass y measured in kilograms and diameter x measured in centimeters:[20]

$$\widehat{\ln y} = -2.00 + 2.42 \ln x$$

Use this model to estimate the biomass of a tropical tree 30 centimeters in diameter. Show your work.

41. **Counting carnivores** Ecologists look at data to learn about nature's patterns. One pattern they have found relates the size of a carnivore (body mass in kilograms) to how many of those carnivores there are in an area. The right measure of "how many" is to count carnivores per 10,000 kilograms (kg) of their prey in the area. The table below gives data for 25 carnivore species.[21]

Carnivore species	Body mass (kg)	Abundance (per 10,000 kg of prey)
Least weasel	0.14	1656.49
Ermine	0.16	406.66
Small Indian mongoose	0.55	514.84
Pine marten	1.3	31.84
Kit fox	2.02	15.96
Channel Island fox	2.16	145.94
Arctic fox	3.19	21.63
Red fox	4.6	32.21
Bobcat	10.0	9.75
Canadian lynx	11.2	4.79
European badger	13.0	7.35
Coyote	13.0	11.65
Ethiopian wolf	14.5	2.70
Eurasian lynx	20.0	0.46
Wild dog	25.0	1.61
Dhole	25.0	0.81
Snow leopard	40.0	1.89
Wolf	46.0	0.62
Leopard	46.5	6.17
Cheetah	50.0	2.29
Puma	51.9	0.94
Spotted hyena	58.6	0.68
Lion	142.0	3.40
Tiger	181.0	0.33
Polar bear	310.0	0.60

Here is a scatterplot of the data.

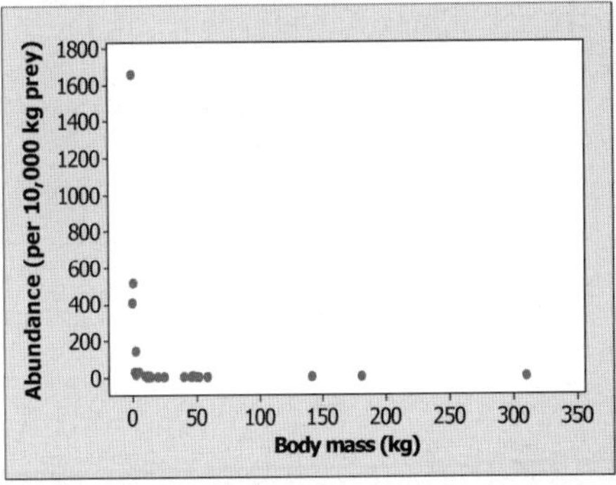

(a) The following graphs show the results of two different transformations of the data. Would an exponential model or a power model provide a better description of the relationship between body mass and abundance? Justify your answer.

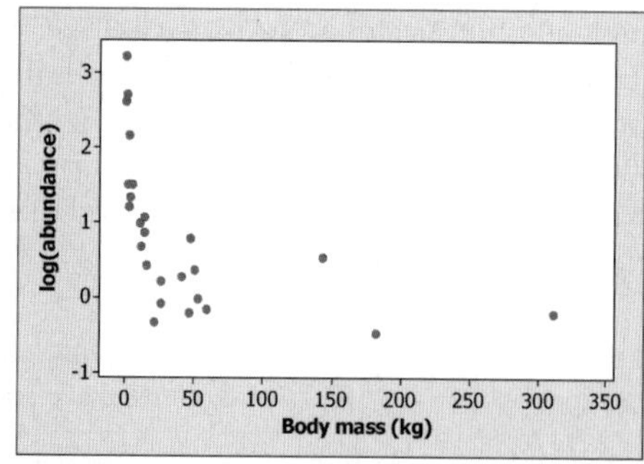

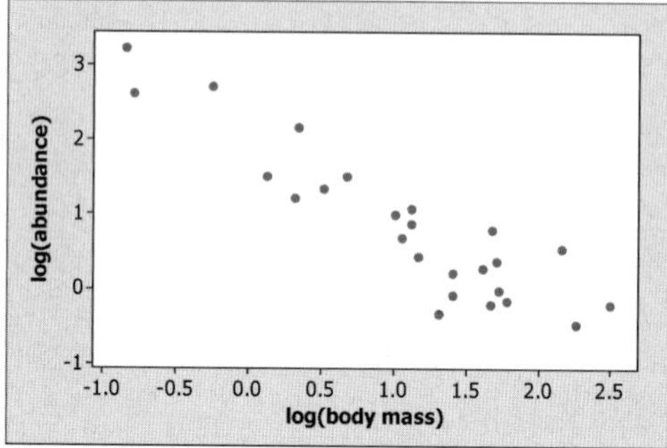

(b) Minitab output from a linear regression analysis on the transformed data of log(abundance) versus log(body mass) is shown below. Give the equation of the least-squares regression line. Be sure to define any variables you use.

```
Predictor       Coef     SE Coef       T      P
Constant      1.9503     0.1342     14.53  0.000
log(body    -1.04811     0.09802   -10.69  0.000
 mass)

S = 0.423352 R-Sq = 83.3% R-Sq(adj) = 82.5%
```

(c) Use your model from part (b) to predict the abundance of black bears, which have a body mass of 92.5 kilograms. Show your work.

(d) A residual plot for the linear regression in part (b) is shown below. Explain what this graph tells you about how well the model fits the data.

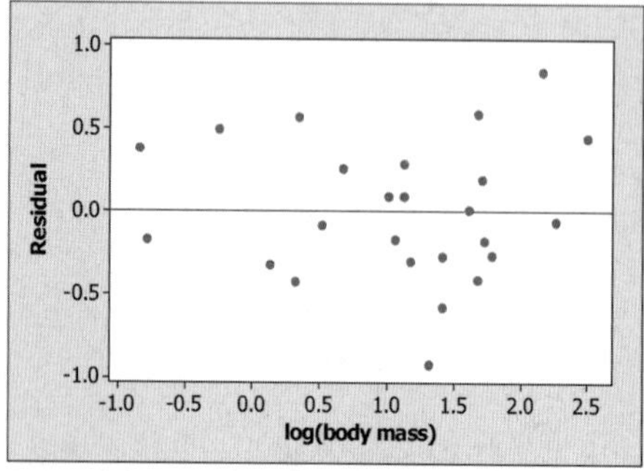

42. **Follow the bouncing ball** Students in Mr. Handford's class dropped a kickball beneath a motion detector. The detector recorded the height of the ball as it bounced up and down several times. Here are the heights of the ball at the highest point on the first five bounces:

Bounce number	Height (feet)
1	2.240
2	1.620
3	1.235
4	0.958
5	0.756

Here is a scatterplot of the data:

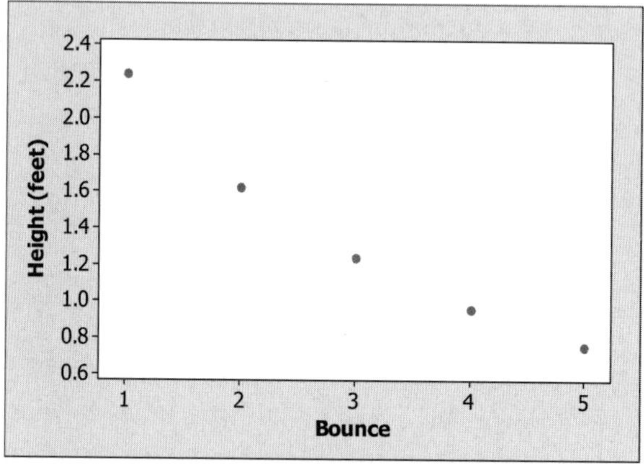

(a) The following graphs show the results of two different transformations of the data. Would an exponential model or a power model provide a better description of the relationship between bounce number and height? Justify your answer.

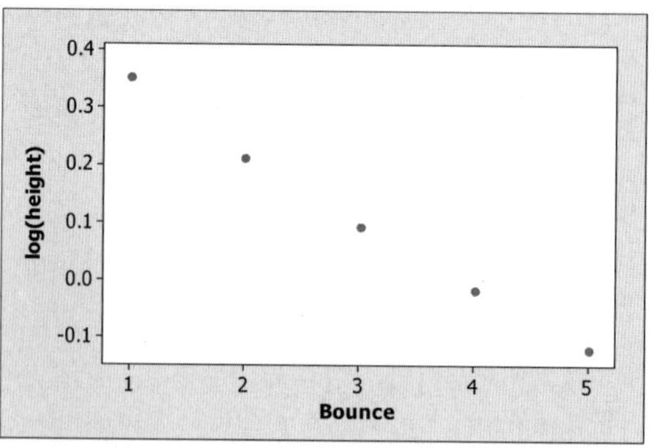

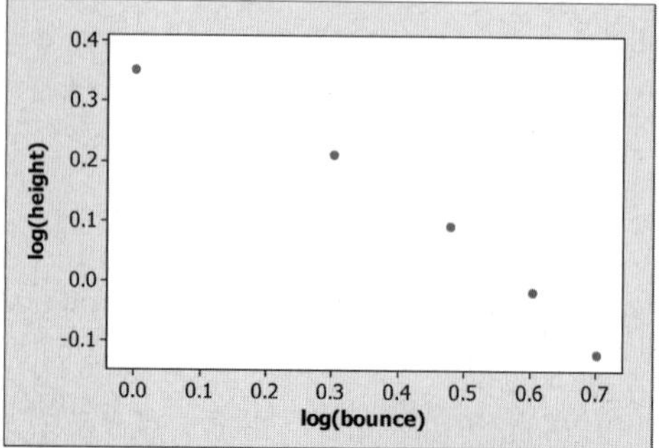

(b) Minitab output from a linear regression analysis on the transformed data of log(height) versus bounce number is shown below. Give the equation of the least-squares regression line. Be sure to define any variables you use.

```
Predictor     Coef       SE Coef        T       P
Constant    0.45374    0.01385     32.76  0.000
Bounce     -0.117160   0.004176   -28.06  0.000
S = 0.0132043  R-Sq = 99.6%  R-Sq(adj) = 99.5%
```

(c) Use your model from part (b) to predict the highest point the ball reaches on its seventh bounce. Show your work.

(d) A residual plot for the linear regression in part (b) is shown below. Do you expect your prediction part (c) to be too high, too low, or just right? Justify your answer.

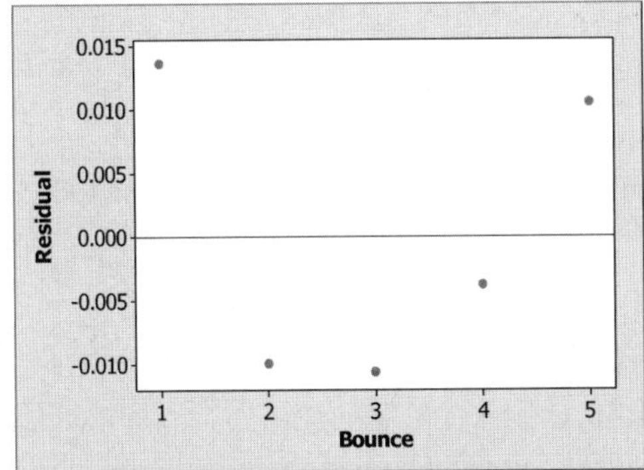

43. **Galileo's experiment** Galileo studied motion by rolling balls down ramps. He rolled a ball down a ramp with a horizontal shelf at the end of it so that the ball was moving horizontally when it started to fall off the shelf. The top of the ramp was placed at different heights above the floor (that is, the length of the ramp varied), and Galileo measured the horizontal distance the ball traveled before it hit the floor. Here are Galileo's data. (We won't try to describe the obsolete seventeenth-century units Galileo used to measure distance and height.)[22]

pg 778

Distance	Height
1500	1000
1340	828
1328	800
1172	600
800	300

(a) Make an appropriate scatterplot for predicting horizontal distance traveled from ramp height. Describe what you see.

(b) Use transformations to linearize the relationship. Does the relationship between distance and height seem to follow an exponential model or a power model? Justify your answer.

(c) Perform least-squares regression on the transformed data. Give the equation of your regression line. Define any variables you use.

(d) Use your model from part (c) to predict the horizontal distance a ball would travel if the ramp height was 700. Show your work.

44. **Heart weights of mammals** Here are some data on the hearts of various mammals.[23]

Mammal	Heart weight (grams)	Length of cavity of left ventricle (centimeters)
Mouse	0.13	0.55
Rat	0.64	1.0
Rabbit	5.8	2.2
Dog	102	4.0
Sheep	210	6.5
Ox	2030	12.0
Horse	3900	16.0

(a) Make an appropriate scatterplot for predicting heart weight from length. Describe what you see.

(b) Use transformations to linearize the relationship. Does the relationship between heart weight and length seem to follow an exponential model or a power model? Justify your answer.

(c) Perform least-squares regression on the transformed data. Give the equation of your regression line. Define any variables you use.

(d) Use your model from part (c) to predict the heart weight of a human who has a left ventricle 6.8 centimeters long. Show your work.

Multiple Choice: Select the best answer for Exercises 45 to 48.

45. Suppose that the relationship between a response variable y and an explanatory variable x is modeled by $y = 2.7(0.316)^x$. Which of the following scatterplots would approximately follow a straight line?

(a) A plot of y against x

(b) A plot of y against $\log x$

(c) A plot of $\log y$ against x

(d) A plot of $\log y$ against $\log x$

(e) None of (a) through (d)

Exercises 46 to 48 refer to the following setting. Some high school physics students dropped a ball and measured the distance fallen (in centimeters) at various times (in seconds) after its release. If you have studied physics, then you probably know that the theoretical relationship between the variables is distance = 490(time)^2. A scatterplot of the students' data showed a clear curved pattern.

46. At 0.68 seconds after release, the ball had fallen 220.4 centimeters. How much more or less did the ball fall than the theoretical model predicts?
    (a) More by 226.576 centimeters
    (b) More by 6.176 centimeters
    (c) No more and no less
    (d) Less by 226.576 centimeters
    (e) Less by 6.176 centimeters

47. Which of the following single transformations should linearize the relationship?
    I. time^2 II. distance^2 III. $\sqrt{\text{distance}}$
    (a) I only (c) III only (e) I and III only
    (b) II only (d) I and II only

48. A scatterplot of which of the following should have a roughly linear pattern?
    (a) [time, ln(distance)]
    (b) [ln(time), distance]
    (c) [ln(time), ln(distance)]
    (d) [ln(distance), time]
    (e) [distance, ln(time)]

49. **Shower time** (1.3, 2.2, 6.3, 7.3) Marcella takes a shower every morning when she gets up. Her time in the shower varies according to a Normal distribution with mean 4.5 minutes and standard deviation 0.9 minutes.
    (a) If Marcella took a 7-minute shower, would it be classified as an outlier? Justify your answer.
    (b) Suppose we choose 10 days at random and record the length of Marcella's shower each day. What's the probability that her shower time is 7 minutes or higher on at least 2 of the days? Show your work.
    (c) Find the probability that the *mean* length of her shower times on these 10 days exceeds 5 minutes. Show your work.

50. **Tattoos** (8.2) What percent of U.S. adults have one or more tattoos? The Harris Poll conducted an online survey of 2302 adults during January 2008. According to the published report, "Respondents for this survey were selected from among those who have agreed to participate in Harris Interactive surveys."[24] The pie chart at top right summarizes the responses from those who were surveyed. Explain why it would *not* be appropriate to use these data to construct a 95% confidence interval for the proportion of all U.S. adults who have tattoos.

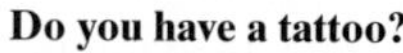

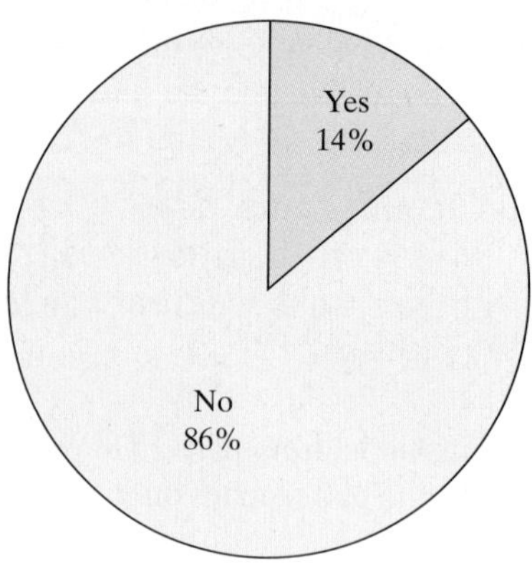

Exercises 51 and 52 refer to the following setting. About 1100 high school teachers attended a weeklong summer institute for teaching AP classes. After hearing about the survey in Exercise 50, the teachers in the AP Statistics class wondered whether the results of the tattoo survey would be similar for teachers. They designed a survey to find out. The class opted for a sample size of 100 teachers. One of the questions on the survey was

Do you have any tattoos on your body?

(Circle one) YES NO

51. **Tattoos** (4.1) One of the first decisions the class had to make was what kind of sampling method to use.
    (a) They knew that a simple random sample was the "preferred" method. With 1100 teachers in 40 different sessions, the class decided not to use an SRS. Give at least two reasons why you think they made this decision.
    (b) The AP Statistics class believed that there might be systematic differences in the proportions of teachers who had tattoos based on the subject areas that they taught. What sampling method would you recommend to account for this possibility? Explain a statistical advantage of this method over an SRS.

52. **Tattoos** (8.2, 9.2) Of the 98 teachers who responded, 23.5% said that they had one or more tattoos.
    (a) Construct and interpret a 95% confidence interval for the actual proportion of teachers at the AP institute who would say they had tattoos.
    (b) Does the interval in part (a) provide convincing evidence that the proportion of teachers at the institute with tattoos is not 0.14 (the value cited in the Harris Poll report)? Justify your answer.
    (c) Two of the selected teachers refused to respond to the survey. If both of these teachers had responded, could your answer to part (b) have changed? Justify your answer.

Chapter 12 Chapter Review

When a scatterplot shows a straight-line relationship between an explanatory variable x and a response variable y, we often fit a least-squares regression line to describe the relationship and use this line to predict y from x. Statistical inference in the regression setting, however, requires more than just an overall linear pattern on a scatterplot. Section 12.1 examined inference for regression.

The regression model says that there is a true straight-line relationship between x and the mean response μ_y. We usually can't observe this population (true) regression line $\mu_y = \alpha + \beta x$. The responses y that we do observe vary if we take several observations at the same x. The regression model requires that for any fixed x, the responses have a Normal distribution. Moreover, the standard deviation σ of this distribution must be the same for all values of x. The standard deviation σ describes how much variation there is in responses y when x is held fixed. Since we usually don't know σ, we use the standard deviation of the residuals, s, to estimate it.

We can do these types of inference for the linear regression model:

- Give confidence intervals for the slope β of the population (true) regression line.
- Perform a significance test about the slope β of the population (true) regression line. Most often, we test the null hypothesis that the slope is zero. This hypothesis says that a straight-line relationship between x and y is of no value for predicting y.

In Section 12.2, we learned how to use transformations to straighten out curved relationships between two variables. If theory or experience tells us that data should follow a power model of the form $y = ax^p$, we can achieve linearity using powers or roots. In most cases involving nonlinear relationships, including power models and exponential models of the form $y = ab^x$, our best bet is to take the logarithm of one or both variables. Once we obtain a linear relationship, we fit a least-squares line to the transformed data and use the resulting linear model to make predictions.

Chapter Review Exercises

These exercises are designed to help you review the important ideas and methods of the chapter. Relevant learning objectives are provided in bulleted form before each exercise.

Exercises R12.1 to R12.3 refer to the following setting. In the casting of metal parts, molten metal flows through a "gate" into a die that shapes the part. The gate velocity (the speed at which metal is forced through the gate) plays a critical role in die casting. A firm that casts cylindrical aluminum pistons examined a random sample of 12 pistons formed from the same alloy of metal. What is the relationship between the cylinder wall thickness (inches) and the gate velocity (feet per second) chosen by the skilled workers who do the casting? If there is a clear pattern, it can be used to direct new workers or to automate the process. A scatterplot of the data is shown below.[25]

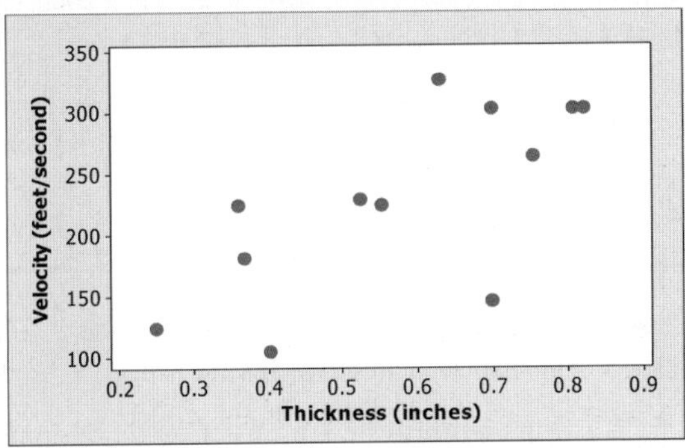

A least-squares regression analysis was performed on the data. Some computer output and a residual plot are shown below. A Normal probability plot of the residuals (not shown) is roughly linear.

```
Predictor     Coef    SE Coef     T       P
Constant      70.44   52.90      1.33   0.213
Thickness    274.78   88.18      ***     ***
S = 56.3641 R-Sq = 49.3% R-Sq(adj) = 44.2%
```

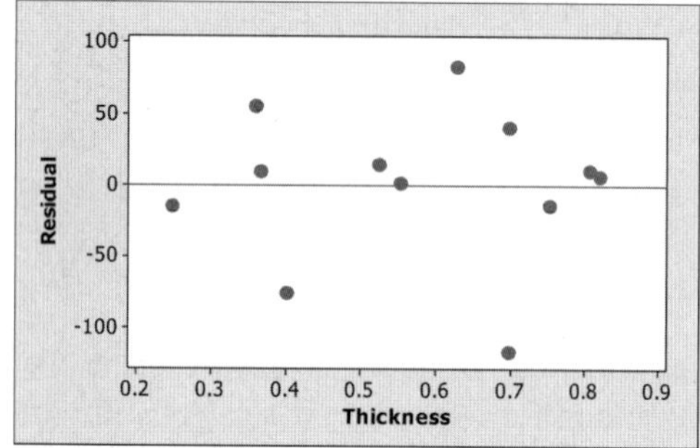

- Interpret computer output from a least-squares regression analysis.

R12.1. Casting aluminum

(a) Describe what the scatterplot tells you about the relationship between cylinder wall thickness and gate velocity.

(b) What is the equation of the least-squares regression line? Define any variables you use.

(c) One of the cylinders in the sample had a wall thickness of 0.4 inches. The gate velocity chosen for this cylinder was 104.8 feet per second. Does the regression line in part (b) overpredict or underpredict the gate velocity for this cylinder? By how much? Show your work.

(d) Is a linear model appropriate in this setting? Justify your answer with appropriate evidence.

(e) Interpret each of the following in context:
(i) The slope
(ii) s
(iii) r^2
(iv) The standard error of the slope

- Perform a significance test about the slope β of a population (true) regression line.

R12.2. Casting aluminum Do these data provide convincing evidence of a straight-line relationship between thickness and gate velocity in the population of pistons formed from this alloy of metal? Carry out an appropriate significance test at the $\alpha = 0.05$ level.

- Construct and interpret a confidence interval for the slope β of the population (true) regression line.

R12.3. Casting aluminum Construct and interpret a 95% confidence interval for the slope of the population regression line. Explain how this interval is consistent with the results of Exercise R12.2.

- Check conditions for performing inference about the slope β of the population (true) regression line.

R12.4. SAT essay—is longer better? Following the debut of the new SAT Writing test in March 2005, Dr. Les Perelman from the Massachusetts Institute of Technology collected data from a set of sample essays provided by the College Board. A least-squares regression analysis was performed on these data. The two graphs below display the results of that analysis. Explain why the conditions for performing inference are not met in this setting.

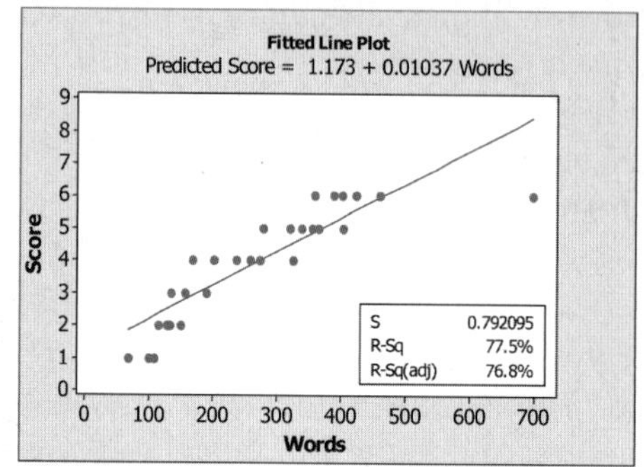

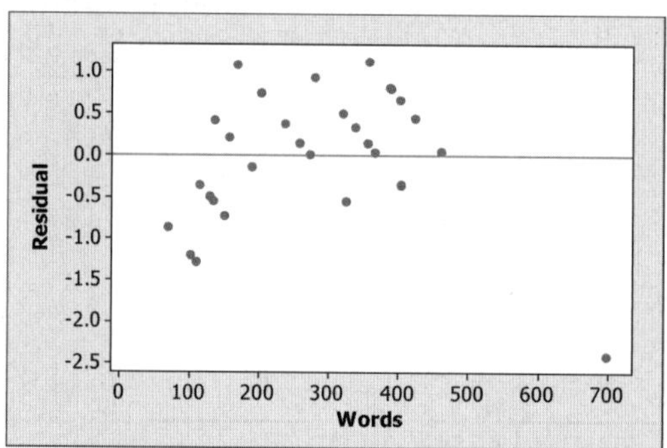

- Use transformations involving powers and roots to achieve linearity for a relationship between two variables.
- Make predictions from a least-squares regression line involving transformed data.

R12.5. Light intensity In physics class, the intensity of a 100-watt light bulb was measured by a sensor at various distances from the light source. A scatterplot of the data is shown below. Note that a candela (cd) is a unit of luminous intensity in the International System of Units.

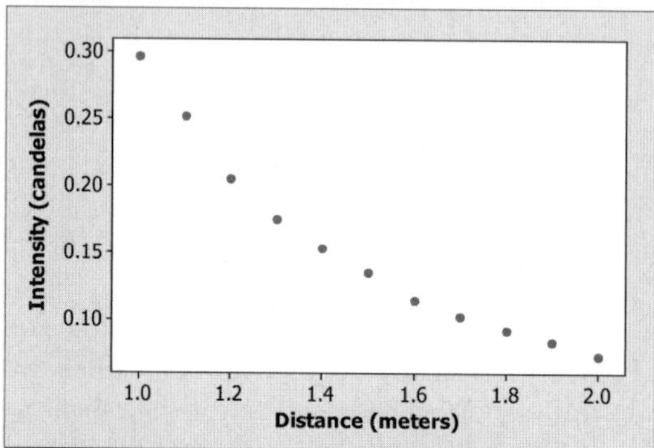

Physics textbooks suggest that the relationship between light intensity y and distance x should follow an "inverse square law," that is, a power law model of the form $y = ax^{-2}$. We transformed the distance measurements by squaring them and then taking their reciprocals. Some computer output and a residual plot from a least-squares regression analysis on the transformed data are shown below. Note that the horizontal axis on the residual plot displays predicted light intensity.

```
Predictor          Coef       SE Coef    T      P
Constant       -0.000595  0.001821  -0.33  0.751
Distance^(-2)   0.299624  0.003237  92.56  0.000
S = 0.00248369 R-Sq = 99.9% R-Sq(adj) = 99.9%
```

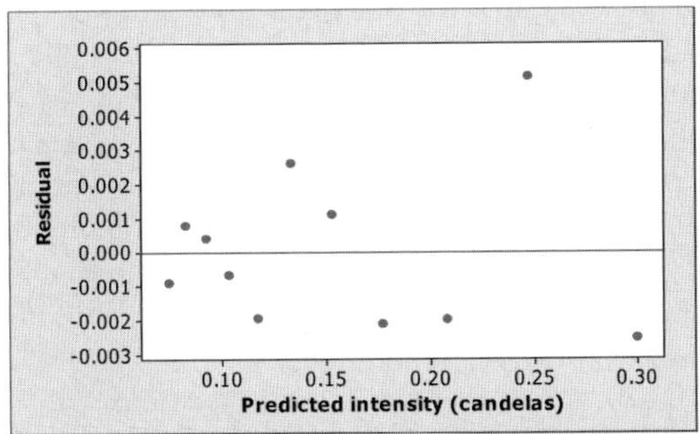

(a) Did this transformation achieve linearity? Give appropriate evidence to justify your answer.

(b) What would you predict for the intensity of a 100-watt bulb at a distance of 2.1 meters? Show your work.

- **Use transformations involving logarithms to achieve linearity for a relationship between two variables.**
- **Determine which of several transformations does a better job of producing a linear relationship.**

R12.6. An experiment was conducted to determine the effect of practice time (in seconds) on the percent of unfamiliar words recalled. Here is a Fathom scatterplot of the results with a least-squares regression line superimposed.

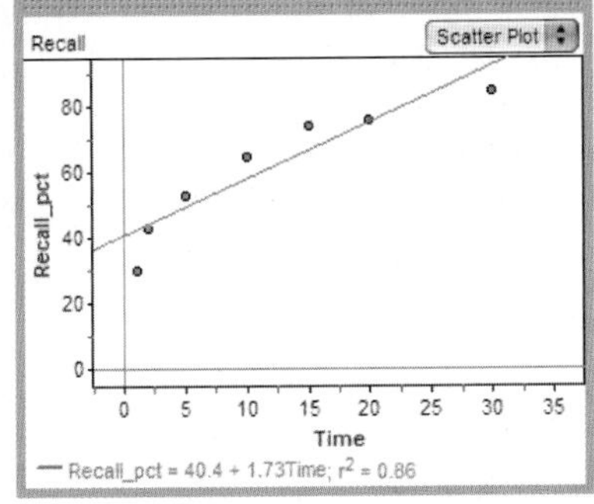

(a) Sketch a residual plot. Be sure to label your axes.

(b) Explain why a linear model is not appropriate for describing the relationship between practice time and percent of words recalled.

(c) We used Fathom to transform the data in hopes of achieving linearity. The screen shots below show the results of two different transformations. Would an exponential model or a power model describe the relationship better? Justify your answer.

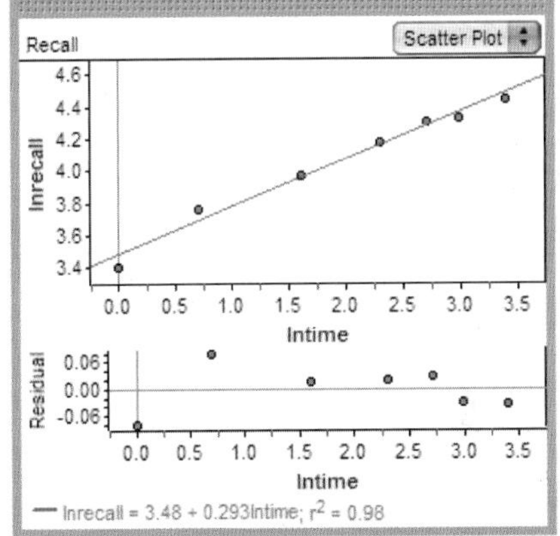

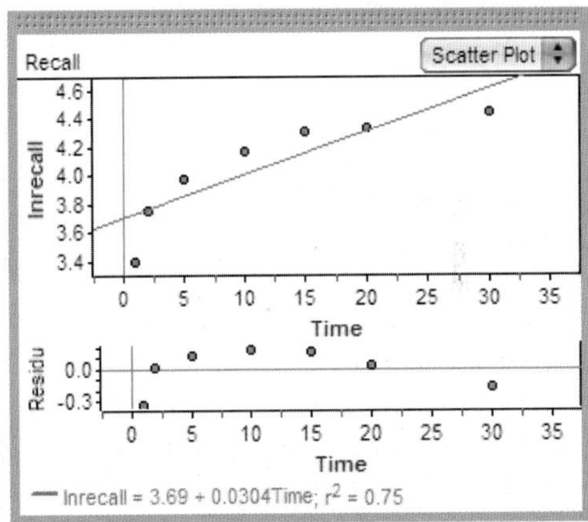

(d) Use the model you chose in part (c) to predict word recall for 25 seconds of practice. Show your method.

Chapter 12 AP Statistics Practice Test

Section I: Multiple Choice *Select the best answer for each question.*

T12.1. Which of the following is *not* one of the conditions that must be satisfied in order to perform inference about the slope of a least-squares regression line?

(a) For each value of x, the population of y-values is Normally distributed.

(b) The standard deviation σ of the population of y-values corresponding to a particular value of x is always the same, regardless of the specific value of x.

(c) The sample size—that is, the number of paired observations (x, y)—exceeds 30.

(d) There exists a straight line $y = \alpha + \beta x$ such that, for each value of x, the mean μ_y of the corresponding population of y-values lies on that straight line.

(e) The data come from a random sample or a randomized experiment.

T12.2. Students in a statistics class drew circles of varying diameters and counted how many Cheerios could be placed in the circle. The scatterplot shows the results.

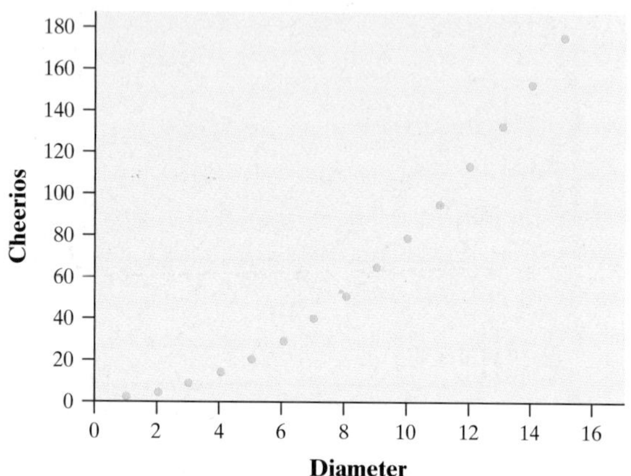

The students want to determine an appropriate equation for the relationship between diameter and the number of Cheerios. The students decide to transform the data to make it appear more linear before computing a least-squares regression line. Which of the following single transformations would be reasonable for them to try?

I. Take the square root of the number of Cheerios.

II. Cube the number of Cheerios.

III. Take the log of the number of Cheerios.

IV. Take the log of the diameter.

(a) I and II (b) I and III (c) II and III (d) II and IV (e) I and IV

T12.3. Inference about the slope β of a least-squares regression line is based on which of the following distributions?

(a) The t distribution with $n - 1$ degrees of freedom

(b) The standard Normal distribution

(c) The chi-square distribution with $n - 1$ degrees of freedom

(d) The t distribution with $n - 2$ degrees of freedom

(e) The Normal distribution with mean μ and standard deviation σ

Exercises T12.4 through T12.8 refer to the following setting. An old saying in golf is "You drive for show and you putt for dough." The point is that good putting is more important than long driving for shooting low scores and hence winning money. To see if this is the case, data from a random sample of 69 of the nearly 1000 players on the PGA Tour's world money list are examined. The average number of putts per hole and the player's total winnings for the previous season are recorded. A least-squares regression line was fitted to the data. The following results were obtained from statistical software.

```
Predictor       Coef    SE Coef     T       P
Constant     7897179    3023782  6.86   0.000
Avg. Putts  -4139198    1698371  ****    ****
S = 281777 R-Sq = 8.1% R-Sq(adj) = 7.8%
```

T12.4. The correlation between total winnings and average number of putts per hole for these players is

(a) −0.285. (b) −0.081. (c) −0.007. (d) 0.081. (e) 0.285.

T12.5. Suppose that the researchers test the hypotheses $H_0: \beta = 0$, $H_a: \beta < 0$. The value of the t statistic for this test is

(a) 2.61. (b) 2.44. (c) 0.081. (d) −2.44. (e) −20.24.

T12.6. The P-value for the test in Question T12.5 is 0.0087. A correct interpretation of this result is that

(a) the probability that there is no linear relationship between average number of putts per hole and total winnings for these 69 players is 0.0087.

(b) the probability that there is no linear relationship between average number of putts per hole and total winnings for all players on the PGA Tour's world money list is 0.0087.

(c) if there is no linear relationship between average number of putts per hole and total winnings for the players in the sample, the probability of getting a random sample of 69 players that yields a least-squares regression line with a slope of −4139198 or less is 0.0087.

(d) if there is no linear relationship between average number of putts per hole and total winnings for the players on the PGA Tour's world money list, the probability of getting a random sample of 69 players that yields a least-squares regression line with a slope of −4139198 or less is 0.0087.

(e) the probability of making a Type II error is 0.0087.

T12.7. A 95% confidence interval for the slope β of the population regression line is

(a) 7,897,179 ± 3,023,782.
(b) 7,897,179 ± 6,047,564.
(c) −4,139,198 ± 1,698,371.
(d) −4,139,198 ± 3,328,807.
(e) −4,139,198 ± 3,396,742.

T12.8. A residual plot from the least-squares regression is shown below. Which of the following statements is supported by the graph?

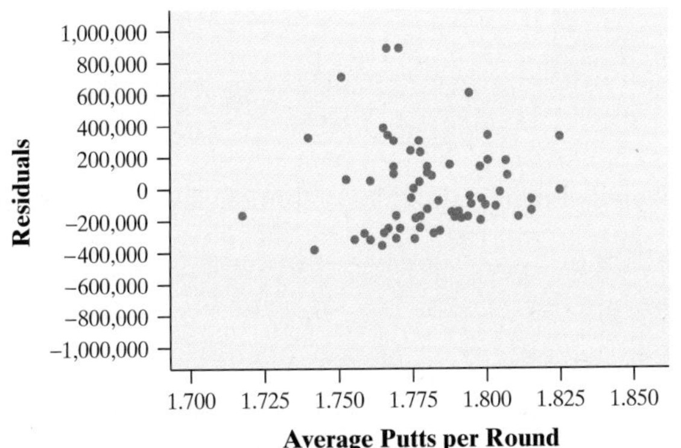

(a) The residual plot contains dramatic evidence that the standard deviation of the response about the population regression line increases as the average number of putts per round increases.

(b) The sum of the residuals is not 0. Obviously, there is a major error present.

(c) Using the regression line to predict a player's total winnings from his average number of putts almost always results in errors of less than $200,000.

(d) For two players, the regression line underpredicts their total winnings by more than $800,000.

(e) The residual plot reveals a strong positive correlation between average putts per round and prediction errors from the least-squares line for these players.

T12.9. Which of the following would provide evidence that a power law model of the form $y = ax^b$, where $b \neq 0$ and $b \neq 1$, describes the relationship between a response variable y and an explanatory variable x?

(a) A scatterplot of y versus x looks approximately linear.

(b) A scatterplot of ln y versus x looks approximately linear.

(c) A scatterplot of y versus ln x looks approximately linear.

(d) A scatterplot of ln y versus ln x looks approximately linear.

(e) None of these

T12.10. We record data on the population of a particular country from 1960 to 2010. A scatterplot reveals a clear curved relationship between population and year. However, a different scatterplot reveals a strong linear relationship between the logarithm (base 10) of the population and the year. The least-squares regression line for the transformed data is

$$\widehat{\log(\text{population})} = -13.5 + 0.01(\text{year})$$

Based on this equation, the population of the country in the year 2020 should be about

(a) 6.7.
(b) 812.
(c) 5,000,000.
(d) 6,700,000.
(e) 8,120,000.

Section II: Free Response *Show all your work. Indicate clearly the methods you use, because you will be graded on the correctness of your methods as well as on the accuracy and completeness of your results and explanations.*

T12.11. Growth hormones are often used to increase the weight gain of chickens. In an experiment using 15 chickens, five different doses of growth hormone (0, 0.2, 0.4, 0.8, and 1.0 milligrams) were injected into chickens (3 chickens were randomly assigned to each dose), and the subsequent weight gain (in ounces) was recorded. A researcher plots the data and finds that a linear relationship appears to hold. Computer output from a least-squares regression analysis for these data is shown below.

```
Predictor    Coef     SE Coef    T       P
Constant     4.5459   0.6166     7.37   <0.0001
Dose         4.8323   1.0164     4.75    0.0004
S = 3.135 R-Sq = 38.4% R-Sq(adj) = 37.7%
```

(a) What is the equation of the least-squares regression line for these data? Define any variables you use.

(b) Interpret each of the following in context:

(i) The slope
(ii) The y intercept
(iii) s
(iv) The standard error of the slope
(v) r^2

(c) Assume that the conditions for performing inference about the slope β of the true regression line are met. Do the data provide convincing evidence of a linear relationship between dose and weight gain? Carry out a significance test at the $\alpha = 0.05$ level.

(d) Construct and interpret a 95% confidence interval for the slope parameter.

T12.12. Foresters are interested in predicting the amount of usable lumber they can harvest from various tree species. They collect data on the diameter at breast height (DBH) in inches and the yield in board feet of a random sample of 20 Ponderosa pine trees that have been harvested. (Note that a board foot is defined as a piece of lumber 12 inches by 12 inches by 1 inch.) A scatterplot of the data is shown below.

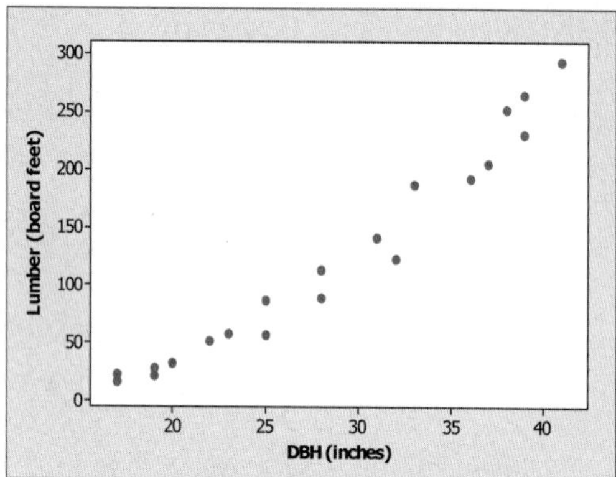

(a) Some computer output and a residual plot from a least-squares regression on these data appear below. Explain why a linear model may not be appropriate in this case.

```
Predictor        Coef    SE Coef      T      P
Constant      -191.12     16.98  -11.25  0.000
DBH (inches)  11.0413    0.5752   19.19  0.000
S = 20.3290 R-Sq = 95.3% R-Sq(adj) = 95.1%
```

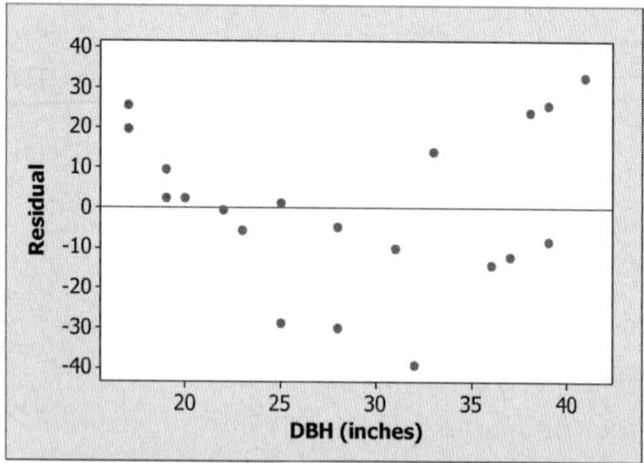

The foresters are considering two possible transformations of the original data: (1) cubing the diameter values or (2) taking the natural logarithm of the yield measurements. After transforming the data, a least-squares regression analysis is performed. Some computer output and a residual plot for each of the two possible regression models follow.

Option 1: Cubing the diameter values

```
Predictor   Coef       SE Coef       T      P
Constant    2.078      5.444        0.38  0.707
DBH^3       0.0042597  0.0001549   27.50  0.000
S = 14.3601 R-Sq = 97.7% R-Sq(adj) = 97.5%
```

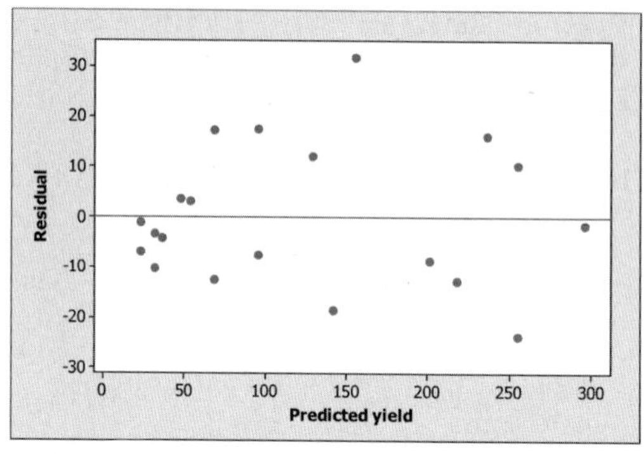

Option 2: Taking natural logarithm of yield measurements

```
Predictor      Coef      SE Coef     T      P
Constant       1.2319    0.1795     6.86  0.000
DBH (inches)   0.113417  0.006081  18.65  0.000
S = 0.214894 R-Sq = 95.1% R-Sq(adj) = 94.8%
```

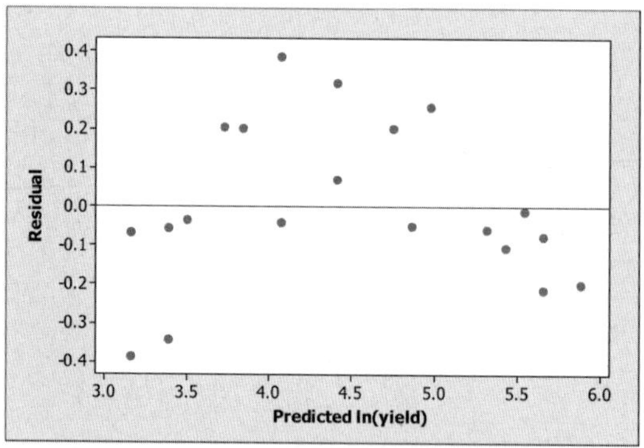

(b) Use both models to predict the amount of usable lumber from a Ponderosa pine with diameter 30 inches. Show your work.

(c) Which of the predictions in part (b) seems more reliable? Give appropriate evidence to support your choice.

Cumulative AP Practice Test 4

Section I: Multiple Choice *Choose the best answer.*

AP4.1. A major agricultural company is testing a new variety of wheat to determine whether it is more resistant to certain insects than is the current wheat variety. The proportion of a current wheat crop lost to insects is 4%. Thus, the company wishes to test the following hypotheses:

$$H_0: p = 0.04$$
$$H_a: p < 0.04$$

Which of the following significance levels and sample sizes would lead to the highest power for this test?

(a) $n = 200$ and $\alpha = 0.01$

(b) $n = 400$ and $\alpha = 0.05$

(c) $n = 400$ and $\alpha = 0.01$

(d) $n = 500$ and $\alpha = 0.01$

(e) $n = 500$ and $\alpha = 0.05$

AP4.2. If $P(A) = 0.24$ and $P(B) = 0.52$ and events A and B are independent, what is $P(A \text{ or } B)$?

(a) 0.1248 (c) 0.6352

(b) 0.28 (d) 0.76

(e) The answer cannot be determined from the given information.

AP4.3. As part of a bear population study, data were gathered on a sample of black bears in the western United States to examine the relationship between the bear's neck girth (distance around the neck) and the weight of the bear. A scatterplot of the data reveals a straight-line pattern. The r^2-value from a least-squares regression analysis was determined to be $r^2 = 0.963$. Which one of the following is the correct value and corresponding interpretation for the correlation?

(a) The correlation is −0.963, and 96.3% of the variation in a bear's weight can be explained by the least-squares regression line using neck girth as the explanatory variable.

(b) The correlation is 0.963. There is a strong positive linear relationship between a bear's neck girth and its weight.

(c) The correlation is 0.981, and 98.1% of the variation in a bear's weight can be explained by the least-squares regression line using neck girth as the explanatory variable.

(d) The correlation is −0.981. There is a strong negative linear relationship between a bear's neck girth and its weight.

(e) The correlation is 0.981. There is a strong positive linear relationship between a bear's neck girth and its weight.

AP4.4. The school board in a certain school district obtained a random sample of 200 residents and asked if they were in favor of raising property taxes to fund the hiring of more statistics teachers. The resulting confidence interval for the true proportion of residents in favor of raising taxes was (0.183, 0.257). The margin of error for this confidence interval is

(a) 0.037. (c) 0.220. (e) 0.740.

(b) 0.183. (d) 0.257.

AP4.5. Suppose a name-brand drug has been deemed effective for reducing hypertension (high blood pressure). The developing company gets to keep a patent on the drug for a specific period of time before other companies can develop a generic form of the drug. Suppose the patent period is about to expire, and another company produces a generic version of this drug. The Food and Drug Administration (FDA) wants to know whether the generic drug is at least as effective as the name-brand drug in reducing blood pressure. The following hypotheses will be used:

$$H_0: \mu_G = \mu_N \text{ versus } H_a: \mu_G < \mu_N$$

where

μ_G = true mean reduction in blood pressure using the generic drug

μ_N = true mean reduction in blood pressure using the name-brand drug.

In the context of this situation, which of the following describes a Type I error?

(a) The FDA finds sufficient evidence that the generic drug does not reduce blood pressure as much as the name-brand drug when, in fact, it does not.

(b) The FDA finds sufficient evidence that the generic drug does not reduce blood pressure as much as the name-brand drug when, in fact, it does.

(c) The FDA finds sufficient evidence that the generic drug does reduce blood pressure as much as the name-brand drug when, in fact, it does not.

(d) The FDA finds sufficient evidence that the generic drug does reduce blood pressure as much as the name-brand drug when, in fact, it does.

(e) The FDA does not find sufficient evidence that the generic drug is as effective in reducing blood pressure as the name-brand drug when, in fact, it is.

AP4.6. Which of the following sampling plans for estimating the proportion of all adults in a medium-sized town who favor a tax increase to support the local school system does *not* suffer from undercoverage bias?

(a) A random sample of 250 names from the local phone book

(b) A random sample of 200 parents whose children attend one of the local schools

(c) A sample consisting of 500 people from the city who take an online survey about the issue

(d) A random sample of 300 homeowners in the town

(e) A random sample of 100 people from an alphabetical list of all adults who live in the town

AP4.7. Which of the following is a categorical variable?

(a) The weight of automobiles

(b) The time required to complete the Olympic marathon

(c) The average gas mileage of a hybrid car

(d) The brand of shampoo purchased by shoppers in a grocery store

(e) The average closing price of a particular stock on the New York Stock Exchange

AP4.8. A large machine is filled with thousands of small pieces of candy, 40% of which are orange. When money is deposited, the machine dispenses 60 randomly selected pieces of candy. The machine will be recalibrated if a group of 60 candies contains fewer than 18 that are orange. What is the approximate probability that this will happen?

(a) $P\left(z < \dfrac{0.3-0.4}{\sqrt{\dfrac{(0.4)(0.6)}{60}}}\right)$

(b) $P\left(z < \dfrac{0.4-0.3}{\sqrt{\dfrac{(0.3)(0.7)}{60}}}\right)$

(c) $P\left(z < \dfrac{0.3-0.4}{\dfrac{\sqrt{(0.4)(0.6)}}{60}}\right)$

(d) $P\left(z < \dfrac{0.3-0.4}{\dfrac{(0.4)(0.6)}{\sqrt{60}}}\right)$

(e) $P\left(z < \dfrac{0.4-0.3}{\dfrac{\sqrt{(0.3)(0.7)}}{60}}\right)$

AP4.9. A random sample of 900 students at a very large university was asked which social-networking site they used most often during a typical week. Their responses are shown in the table below.

Networking site	Male	Female	Total
Facebook	221	283	504
Twitter	42	38	80
LinkedIn	108	87	195
Flixster	23	26	49
BlackPlanet	29	43	72
Total	423	477	900

Assuming that gender and preferred networking site are independent, which of the following is the expected count for female and LinkedIn?

(a) 18.85 (b) 46.11 (c) 87.00 (d) 91.65 (e) 103.35

AP4.10. Insurance adjusters are always vigilant about being overcharged for accident repairs. The adjusters suspect that Repair Shop 1 quotes higher estimates than Repair Shop 2. To check their suspicion, the adjusters randomly select 12 cars that were recently involved in an accident and then take each of the cars to both repair shops to obtain separate estimates of the cost to fix the vehicle. The estimates are given below in hundreds of dollars.

Car:	1	2	3	4	5	6
Shop 1:	21.2	25.2	39.0	11.3	15.0	18.1
Shop 2:	21.3	24.1	36.8	11.5	13.7	17.6

Car:	7	8	9	10	11	12
Shop 1:	25.3	23.2	12.4	42.6	27.6	12.9
Shop 2:	24.8	21.3	12.1	42.0	26.7	12.5

Assuming that the conditions for inference are reasonably met, which of the following significance tests could legitimately be used to determine whether the adjusters' suspicion is correct?

I. A paired *t* test

II. A two-sample *t* test

III. A *t* test to see if the slope of the population regression line is 0.

(a) I only

(b) II only

(c) I and III

(d) II and III

(e) I, II, and III

AP4.11. A survey firm wants to ask a random sample of adults in Ohio if they support an increase in the state sales tax from 5% to 6%, with the additional revenue going to education. Let $\hat{p}$ denote the proportion in the sample who say that they support the increase. Suppose that 40% of all adults in Ohio support the increase. How large a sample would be needed to guarantee that the standard deviation of $\hat{p}$ is no more than 0.01?

(a) 1500 (b) 2400 (c) 2401 (d) 2500 (e) 9220

AP4.12. A set of 10 cards consists of 5 red cards and 5 black cards. The cards are shuffled thoroughly, and you choose one at random, observe its color, and replace it in the set. The cards are thoroughly reshuffled, and you again choose a card at random, observe its color, and replace it in the set. This is done a total of four times. Let X be the number of red cards observed in these four trials. The random variable X has which of the following probability distributions?

(a) The Normal distribution with mean 2 and standard deviation 1

(b) The binomial distribution with $n = 10$ and $p = 0.5$

(c) The binomial distribution with $n = 5$ and $p = 0.5$

(d) The binomial distribution with $n = 4$ and $p = 0.5$

(e) The geometric distribution with $p = 0.5$

AP4.13. A study of road rage asked random samples of 596 men and 523 women about their behavior while driving. Based on their answers, each respondent was assigned a road rage score on a scale of 0 to 20. The respondents were chosen by random digit dialing of telephone numbers. Are the conditions for two-sample *t* inference satisfied?

(a) Maybe. The data came from independent random samples, but we need to examine the data to check for Normality.

(b) No. Road rage scores in a range between 0 and 20 can't be Normal.

(c) No. A paired *t* test should be used in this case.

(d) Yes. The large sample sizes guarantee that the corresponding population distributions will be Normal.

(e) Yes. We have two independent random samples and large sample sizes, and the 10% condition is met.

AP4.14. Do hummingbirds prefer store-bought food made from concentrate or a simple mixture of sugar and water? To find out, a researcher obtains 10 identical hummingbird feeders and fills 5, chosen at random, with store-bought food from concentrate and the other 5 with a mixture of sugar and water. The feeders are then randomly assigned to 10 possible hanging locations in the researcher's yard. Which inference procedure should you use to test whether hummingbirds show a preference for store-bought food based on amount consumed?

(a) A one-sample z test for a proportion

(b) A two-sample z test for a proportion

(c) A chi-square test for association/independence

(d) A two-sample *t* test

(e) A paired *t* test

AP4.15. A Harris Poll found that 54% of American adults don't think that human beings developed from earlier species. The poll's margin of error for 95% confidence was 3%. This means that

(a) there is a 95% chance that the interval (51%, 57%) contains the true percent of American adults who do not think that human beings developed from earlier species.

(b) the poll used a method that provides an estimate within 3% of the truth about the population 95% of the time.

(c) if Harris takes another poll using the same method, the results of the second poll will lie between 51% and 57%.

(d) Harris should have surveyed more people so that the margin of error would be 0% instead of 3%.

(e) the poll used a method that would result in an interval that contains 54% in 95% of all possible samples of the same size from this population.

AP4.16. Two six-sided dice are rolled and the sum of the faces showing is recorded after each roll. Let X = the number of rolls required until a sum greater than 7 is obtained. If 100 trials are conducted, which of the following is most likely to be part of the probability distribution of X?

(a)

Number of rolls	Number of trials
1	34
2	20
3	16
4	10
5	6
6	6
7	3
8	2
9	1
10	0
11	1
12	0
13	1

(b)

Number of rolls	Number of trials
0	34
1	20
2	16
3	10
4	6
5	6
6	3
7	2
8	1
9	0
10	1
11	0
12	1

(c)

Number of rolls	Number of trials
1	18
2	23
3	26
4	15
5	9
6	6
7	1
8	0
9	1
10	0
11	0
12	0
13	1

(d)

Number of rolls	Number of trials
1	10
2	9
3	10
4	12
5	7
6	13
7	10
8	7
9	9
10	10
11	2
12	1

(e)

Number of rolls	Number of trials
1	2
2	2
3	5
4	10
5	11
6	15
7	22
8	17
9	9
10	4
11	2
12	0
13	1

AP4.17. Women who are severely overweight suffer enormous social and economic consequences, a new study has shown. They are 20% less likely to marry, they have household incomes that are an average of $6710 lower, and they are 10% more likely to live in poverty. The findings are from an eight-year observational study of 10,039 randomly selected women who were 16 to 24 years old when the research began. Does this study give strong evidence that being severely overweight causes a woman to have a lower income?

(a) Yes. The study included both women who were severely overweight and women who were not.

(b) Yes. The subjects in the study were selected at random.

(c) No. The study showed that there is no connection between income and being severely overweight.

(d) No. The study suggests an association between income and being severely overweight, but we can't draw a cause-and-effect conclusion.

(e) There is not enough information to answer this question.

Questions AP 4.18 and 4.19 refer to the following situation. Could mud wrestling be the cause of a rash contracted by University of Washington students? Two physicians at the University of Washington student health center wondered about this when one male and six female students complained of rashes after participating in a mud-wrestling event. Questionnaires were sent to a random sample of students who participated in the event. The results, by gender, are summarized in the following table.

	Men	Women
Developed rash	12	12
No rash	38	12

Some Minitab output for the previous table is given below. The output includes the observed counts, the expected counts, and the chi-square statistic.

```
Expected counts are printed below observed
counts

                    MEN      WOMEN    Total
Developed rash      12       12       24
                    16.22    7.78

No rash             38       12       50
                    33.78    16.22

Total               50       24       74
ChiSq = 5.002
```

AP4.18. The cell that contributes most to the chi-square statistic is

(a) men who developed a rash.

(b) men who did not develop a rash.

(c) women who developed a rash.

(d) women who did not develop a rash.

(e) both (a) and (d).

AP4.19. From the chi-square test performed in this study, we may conclude that

(a) there is convincing evidence of an association between the gender of an individual participating in the event and development of a rash.

(b) mud wrestling causes a rash, especially for women.

(c) there is absolutely no evidence of any relation between the gender of an individual participating in the event and the subsequent development of a rash.

(d) development of a rash is a real possibility if you participate in mud wrestling, especially if you do so on a regular basis.

(e) the gender of the individual participating in the event and the development of a rash are independent.

AP4.20. Random assignment is part of a well-designed comparative experiment because

(a) it is more fair to the subjects.

(b) it helps create roughly equivalent groups before treatments are imposed on the subjects.

(c) it allows researchers to generalize the results of their experiment to a larger population.

(d) it helps eliminate any possibility of bias in the experiment.

(e) it prevents the placebo effect from occurring.

AP4.21. The following back-to-back stemplots compare the ages of players from two minor-league hockey teams (1|7 = 17 years).

Team A		Team B
98777	1	788889
44333221	2	00123444
7766555	2	556679
521	3	023
86	3	55

Which of the following *cannot* be justified from the plots?

(a) Team A has the same number of players in their 30s as does Team B.

(b) The median age of both teams is the same.

(c) Both age distributions are skewed to the right.

(d) The age ranges of both teams are similar.

(e) There are no outliers by the 1.5*IQR* rule in either distribution.

AP4.22. A distribution that represents the number of cars parked in a randomly selected residential driveway on any night is given by

x_i:	0	1	2	3	4
p_i:	0.1	0.2	0.35	0.25	0.15

Which of the following statements is correct?

(a) This is a legitimate probability distribution because each of the p_i-values is between 0 and 1.

(b) This is a legitimate probability distribution because $\sum x_i$ is exactly 10.

(c) This is a legitimate probability distribution because each of the p_i-values is between 0 and 1 and the $\sum x_i$ is exactly 10.

(d) This is not a legitimate probability distribution because $\sum x_i$ is not exactly 10.

(e) This is not a legitimate probability distribution because $\sum p_i$ is not exactly 1.

AP4.23. Which sampling method was used in each of the following settings, in order from I to IV?

I. A student chooses for a survey the first 20 students to arrive at school.

II. The name of each student in a school is written on a card, the cards are well mixed, and 10 names are drawn.

III. A state agency randomly selects 50 people from each of the state's senatorial districts.

IV. A city council randomly selects eight city blocks and then surveys all the voting-age residents of those blocks.

(a) Voluntary response, SRS, cluster, stratified

(b) Convenience, SRS, stratified, cluster

(c) Convenience, cluster, SRS, stratified

(d) Convenience, SRS, cluster, stratified

(e) Cluster, SRS, stratified, convenience

AP4.24. Western lowland gorillas, whose main habitat is the central African continent, have a mean weight of 275 pounds with a standard deviation of 40 pounds. Capuchin monkeys, whose main habitat is Brazil and a few other parts of Latin America, have a mean weight of 6 pounds with a standard deviation of 1.1 pounds. Both weight distributions are approximately Normally distributed. If a particular western lowland gorilla is known to weigh 345 pounds, approximately how much would a capuchin monkey have to weigh, in pounds, to have the same standardized weight as the lowland gorilla?

(a) 4.08 (c) 7.93

(b) 7.27 (d) 8.20

(e) There is not enough information to determine the weight of a capuchin monkey.

AP4.25. Suppose that the mean weight of a certain type of pig is 280 pounds with a standard deviation of 80 pounds. The weight distribution of pigs tends to be somewhat skewed to the right. A random sample of 100 pigs is taken. Which of the following statements about the sampling distribution of the sample mean weight $\bar{x}$ is true?

(a) It will be Normally distributed with a mean of 280 pounds and a standard deviation of 80 pounds.

(b) It will be Normally distributed with a mean of 280 pounds and a standard deviation of 8 pounds.

(c) It will be approximately Normally distributed with a mean of 280 pounds and a standard deviation of 80 pounds.

(d) It will be approximately Normally distributed with a mean of 280 pounds and a standard deviation of 8 pounds.

(e) There is not enough information to determine the mean and standard deviation of the sampling distribution.

AP4.26. Which of the following statements about the *t* distribution with degrees of freedom df is (are) true?

I. It is symmetric.

II. It has more variability than the *t* distribution with df + 1 degrees of freedom.

III. As df increases, the *t* distribution approaches the standard Normal distribution.

(a) I only (c) III only (e) I, II, and III

(b) II only (d) I and II

AP4.27. A company has been running television commercials for a new children's product on five different family programs during the evening hours in a large city over a one-month period. A random sample of families is taken, and they are asked to indicate the one program they viewed most often and their rating of the advertised product. The results are summarized in the following table.

	Family program				
Product rating	A	B	C	D	E
Excellent	23	29	42	48	51
Good	25	33	44	53	49
Fair	31	29	25	16	10
Poor	38	32	25	18	12

The advertiser decided to use a chi-square test to see if there is a relationship between the family program viewed and the product's rating. What would be the degrees of freedom for this test?

(a) 3 (c) 12 (e) 19
(b) 4 (d) 18

Questions AP4.28 and AP4.29 refer to the following situation. Park rangers are interested in estimating the weight of the bears that inhabit their state. The rangers have data on weight (in pounds) and neck girth (distance around the neck in inches) for 10 randomly selected bears. Some regression output for these data is shown below.

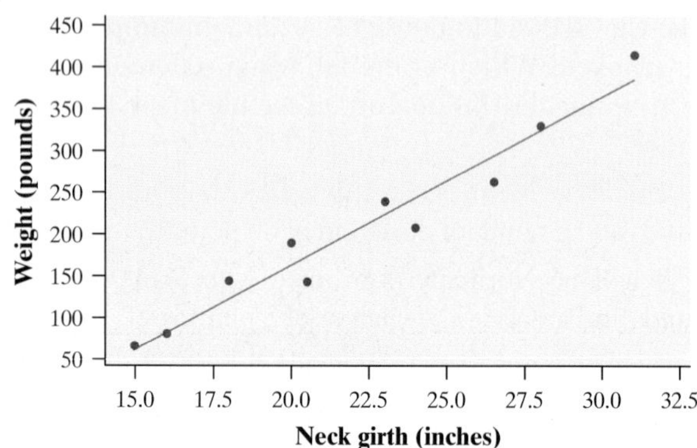

Predictor	Coef	SE Coef	T	P
Constant	−241.70	38.57	−6.27	0.000
Neck Girth	20.230	1.695	11.93	0.000
S = 26.7565	R-Sq = 94.7%			

AP4.28. Which of the following represents a 95% confidence interval for the true slope of the least-squares regression line relating the weight of a bear and its neck girth?

(a) -241.710 ± 38.57
(b) -241.70 ± 88.94
(c) 20.230 ± 1.695
(d) 20.230 ± 3.83
(e) 20.230 ± 3.91

AP4.29. A bear was recently captured whose neck girth was 35 inches and whose weight was 466.35 pounds. If this bear were added to the data set given above, what would be the effect on the value of r^2?

(a) It would decrease the value of r^2 because the added point is an outlier.

(b) It would increase the value of r^2 because any point added to the data would increase the percent of variation in bear weight that can be explained by the least-squares regression line.

(c) It would increase the value of r^2 because the added point lies on the least-squares regression line and is far from the point $(\bar{x}, \bar{y})$.

(d) It would have no effect on the value of r^2 because the added point lies far from the point $(\bar{x}, \bar{y})$.

(e) It would have no effect on the value of r^2 because it lies on the least-squares regression line.

AP4.30. An experimenter wishes to test whether or not two types of fish food (a standard fish food and a new product) work equally well at producing fish of equal weight after a two-month feeding program. The experimenter has two identical fish tanks (1 and 2) to put fish in and is considering how to assign 40 fish, each of which has a numbered tag, to the tanks. The best way to do this would be to

(a) put all the odd-numbered fish in one tank, the even in the other, and give the standard food type to the odd-numbered ones.

(b) obtain pairs of fish whose weights are virtually equal at the start of the experiment and randomly assign one to Tank 1 and the other to Tank 2, with the feed assigned at random to the tanks.

(c) proceed as in part (b), but put the heavier of the pair into Tank 2.

(d) assign the fish at random to the two tanks and give the standard feed to Tank 1.

(e) not proceed as in part (b) because using the initial weight in part (b) is a nonrandom process. Use the initial length of the fish instead.

AP4.31. A city wants to conduct a poll of taxpayers to determine the level of support for constructing a new city-owned baseball stadium. Which of the following is the primary reason for using a large sample size in constructing a confidence interval to estimate the proportion of city taxpayers who would support such a project?

(a) To increase the confidence level
(b) To eliminate any confounding variables
(c) To reduce nonresponse bias
(d) To increase the precision of the estimate
(e) To reduce undercoverage

AP4.32. A standard deck of playing cards contains 52 cards, of which 4 are aces and 13 are hearts. You are offered a choice of the following two wagers:

I. Draw one card at random from the deck. You win \$10 if the card drawn is an ace. Otherwise, you lose \$1.

II. Draw one card at random from the deck. If the card drawn is a heart, you win \$2. Otherwise, you lose \$1.

Which of the two wagers should you prefer?

(a) Wager 1, because it has a higher expected value

(b) Wager 2, because it has a higher expected value

(c) Wager 1, because it has a higher probability of winning

(d) Wager 2, because it has a higher probability of winning

(e) Both wagers are equally favorable.

AP4.33. Below are boxplots of SAT Critical Reading and Math scores for a randomly selected group of female juniors at a highly competitive suburban school.

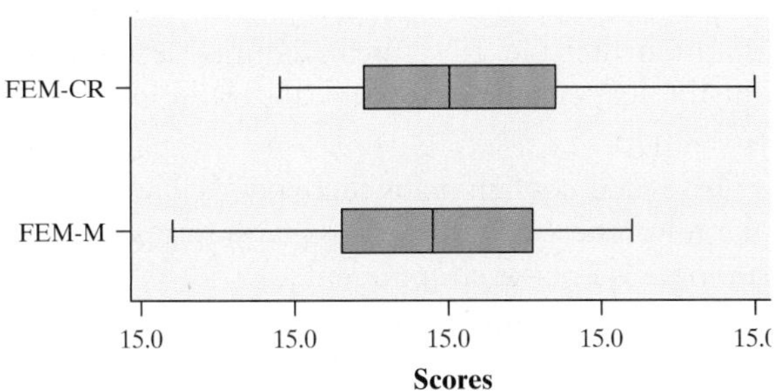

Which of the following *cannot* be justified by the plots shown above?

(a) The maximum Critical Reading score is higher than the maximum Math score.

(b) Critical Reading scores are skewed to the right, whereas Math scores are somewhat skewed to the left.

(c) The median Critical Reading score for females is slightly higher than the median Math score.

(d) There appear to be no outliers in the SAT score distributions.

(e) The mean Critical Reading score and the mean Math score for females are about the same.

AP4.34. A distribution of exam scores has mean 60 and standard deviation 18. If each score is doubled, and then 5 is subtracted from that result, what will be the mean and standard deviation, respectively, of the new scores?

(a) mean = 115 and standard deviation = 31

(b) mean = 115 and standard deviation = 36

(c) mean = 120 and standard deviation = 6

(d) mean = 120 and standard deviation = 31

(e) mean = 120 and standard deviation = 36

AP4.35. In a clinical trial, 30 patients with a certain blood disease are randomly assigned to two groups. One group is then randomly assigned the currently marketed medicine, and the other group receives the experimental medicine. Each week, patients report to the clinic where blood tests are conducted. The lab technician is unaware of the kind of medicine the patient is taking, and the patient is also unaware of which medicine he or she has been given. This design can be described as

(a) a double-blind, completely randomized experiment, with the currently marketed medicine and the experimental medicine as the two treatments.

(b) a single-blind, completely randomized experiment, with the currently marketed medicine and the experimental medicine as the two treatments.

(c) a double-blind, matched pairs design, with the currently marketed medicine and the experimental medicine forming a pair.

(d) a double-blind, block design that is not a matched pairs design, with the currently marketed medicine and the experimental medicine as the two blocks.

(e) a double-blind, randomized observational study.

AP4.36. A local investment club that meets monthly has 200 members ranging in age from 27 to 81. A cumulative relative frequency graph is shown below. Approximately how many members of the club are more than 60 years of age?

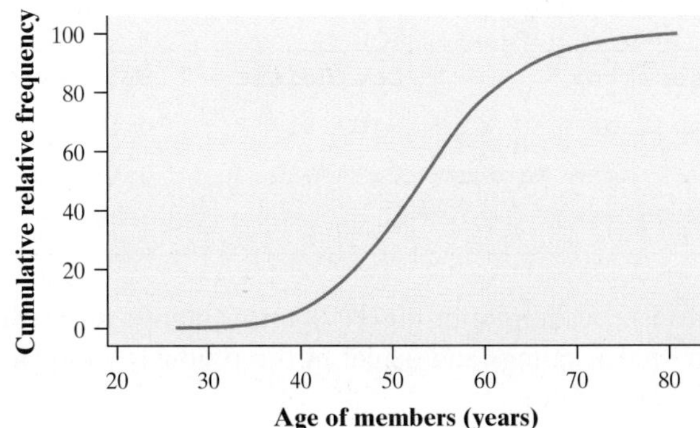

(a) 20 (b) 44 (c) 78 (d) 90 (e) 110

AP4.37. A manufacturer of electronic components is testing the durability of a newly designed integrated circuit to determine whether its life span is longer than that of the earlier model, which has a mean life span of 58 months. The company takes a simple random sample of 120 integrated circuits and simulates normal use until they stop working. The null and alternative hypotheses used for the significance test are given by $H_0: \mu = 58$ and $H_a: \mu > 58$. The *P*-value for the resulting one-sample *t* test is 0.035. Which of the following best describes what the *P*-value measures?

(a) The probability that the new integrated circuit has the same life span as the current model is 0.035.

(b) The probability that the test correctly rejects the null hypothesis in favor of the alternative hypothesis is 0.035.

(c) The probability that a single new integrated circuit will not last as long as one of the earlier circuits is 0.035.

(d) The probability of getting a sample statistic as far or farther from the null value if there really is no difference between the new and the old circuits is 0.035.

(e) The probability of getting a sample mean for the new integrated circuit that is lower than the mean for the earlier model is 0.035.

Questions AP4.38 and AP4.39 refer to the following situation. Do children's fear levels change over time and, if so, in what ways? Little research has been done on the prevalence and persistence of fears in children. In 1989, two researchers surveyed a group of 94 third- and fourth-grade children, asking them to rate their level of fearfulness about a variety of situations. Two years later, the children again completed the same survey. The researchers computed the mean fear rating for each child in both years and were interested in the relationship between these ratings. They then assumed that the true regression line was

$$\mu_{1991 \text{ mean rating}} = \alpha + \beta(1989 \text{ mean rating})$$

and that the assumptions for regression inference were satisfied. This model was fitted to the data using least-squares regression. The following results were obtained from statistical software.

Predictor	Coefficient	St. Dev.
Constant	0.877917	0.1184
1989 Mean Rating	0.397911	0.0676
S = 0.2374	R-Sq = 0.274	

Here is a scatterplot of the 1991 mean ratings versus the 1989 mean ratings and a plot of the residuals versus the 1989 mean ratings.

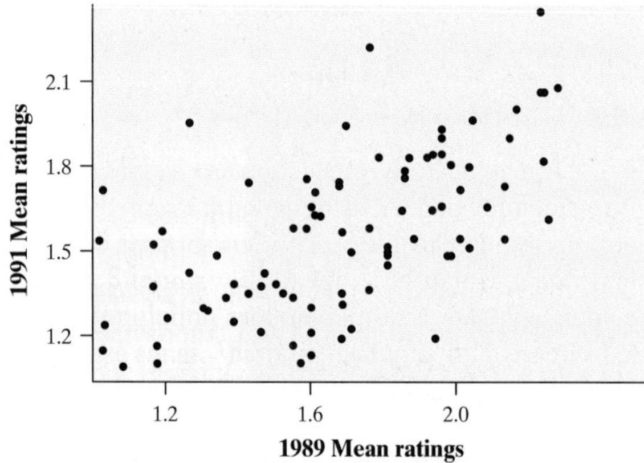

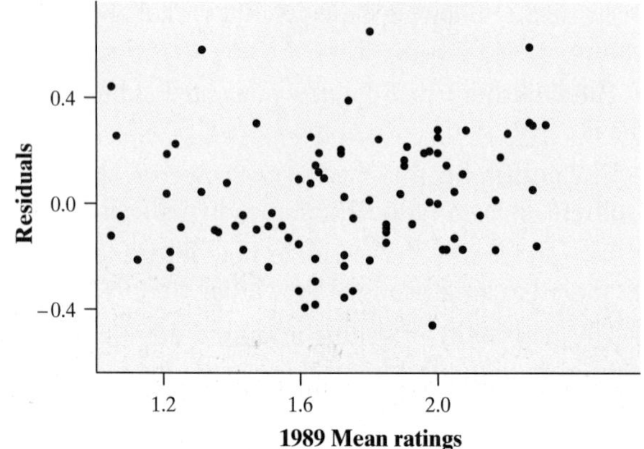

AP4.38. Which of the following statements is supported by these plots?

(a) There is no striking evidence that the assumptions for regression inference are violated.

(b) The abundance of outliers and influential observations in the plots means that the assumptions for regression are clearly violated.

(c) These plots contain dramatic evidence that the standard deviation of the response about the true regression line is not approximately the same everywhere.

(d) These plots call into question the validity of the assumption that the 1991 mean ratings vary Normally about the least-squares line for each value of the 1989 mean ratings.

(e) These plots contain many more points than were used by the researchers to fit the least-squares regression. Obviously, there is a major error present.

AP4.39. Which of the following conclusions seems most justified?

(a) The mean fear ratings increased from 1989 to 1991, because the slope of the least-squares line is positive.

(b) It appears that children with high fear ratings in 1989 had lower ratings in 1991.

(c) It appears that children with high fear ratings in 1989 had higher ratings in 1991.

(d) There is no evidence of a significant relationship between the mean fear ratings in 1989 and in 1991.

(e) There are sufficient inconsistencies in the data to make it impossible to draw any conclusions.

AP4.40. The table below provides data on the political affiliation and opinion about the death penalty of 850 randomly selected voters from a congressional district.

	Favor	Oppose	Total
Republican	299	98	397
Democrat	77	171	248
Other	118	87	205
Total	494	356	850

Which of the following does *not* support the conclusion that being a Republican and favoring the death penalty are not independent?

(a) $\frac{299}{494} \neq \frac{98}{356}$

(b) $\frac{299}{494} \neq \frac{397}{850}$

(c) $\frac{494}{850} \neq \frac{299}{397}$

(d) $\frac{299}{397} \neq \frac{77}{248}$

(e) $\frac{(397)(494)}{850} \neq 299$

Section II: Free Response *Show all your work. Indicate clearly the methods you use, because you will be graded on the correctness of your methods as well as on the accuracy and completeness of your results and explanations.*

AP4.41. The body's natural electrical field helps wounds heal. If diabetes changes this field, it might explain why people with diabetes heal more slowly. A study of this idea compared randomly selected normal mice and randomly selected mice bred to spontaneously develop diabetes. The investigators attached sensors to the right hip and front feet of the mice and measured the difference in electrical potential (in millivolts) between these locations. Graphs of the data for each group reveal no outliers or strong skewness. The computer output below provides numerical summaries of the data.[26]

Variable	N	Mean	StDev	Minimum
Diabetic Mice	24	13.090	4.839	1.050
Normal Mice	18	10.022	2.915	4.950

Q1	Median	Q3	Maximum
10.038	12.650	17.038	22.600
8.238	9.250	12.375	16.100

The researchers want to know if there is evidence of a significant difference in mean electrical potentials between normal mice and mice with diabetes. Carry out a test using a 5% level of significance and report your conclusion.

AP4.42. Can physical activity in youth lead to mental sharpness in old age? A 2010 study investigating this question involved 9344 randomly selected, mostly white women over age 65 from four U.S. states. These women were asked about their levels of physical activity during their teenage years, 30s, 50s, and later years. Those who reported being physically active as teens enjoyed the lowest level of cognitive decline—only 8.5% had cognitive impairment—compared with 16.7% of women who reported not being physically active at that time.

(a) State an appropriate pair of hypotheses that the researchers could use to test whether the proportion of women who suffered a cognitive decline was significantly lower for women who were physically active in their youth than for women who were not physically active at that time. Be sure to define any parameters you use.

(b) Assuming the conditions for performing inference are met, what inference method would you use to test the hypotheses you identified in part (b)? Do *not* carry out the test.

(c) Suppose the test in part (b) shows that the proportion of women who suffered a cognitive decline was significantly lower for women who were physically active in their youth than for women who were not physically active at that time. Can we generalize the results of this study to all women aged 65 and older? Justify your answer.

(d) We cannot conclude that being physically active as a teen *causes* a lower level of cognitive decline for women over 65, due to possible confounding with other variables. Explain the concept of confounding and give an example of a potential confounding variable in this study.

AP4.43. In a recent poll, randomly selected New York State residents at various fast-food restaurants were asked if they supported or opposed a "fat tax" on nondiet sugared soda. Thirty-one percent said that they were in favor of such a tax and 66% were opposed. But when asked if they would support such a tax if the money raised were used to fund health care given the high incidence of obesity in the United States, 48% said that they were in favor and 49% were opposed.

(a) In this situation, explain how bias may have been introduced based on the way the questions were worded *and* suggest a way that they could have been worded differently in order to avoid this bias.

(b) In this situation, explain how bias may have been introduced based on the way the sample was taken *and* suggest a way that the sample could have been obtained in order to avoid this bias.

(c) This poll was conducted only in New York State. Suppose the pollsters wanted to ensure that estimates for the proportion of people who would support a tax on nondiet sugared soda were available for each state as well as an overall estimate for the nation as a whole. Identify a sampling method that would achieve this goal *and* briefly describe how the sample would be taken.

AP4.44. Each morning, coffee is brewed in the school workroom by one of three faculty members, depending on who arrives first at work. Mr. Worcester arrives first 10% of the time, Dr. Currier arrives first 50% of the time, and Mr. Legacy arrives first on the remaining mornings. The probability that the coffee is strong when brewed by Dr. Currier is 0.1, while the corresponding probabilities when it is brewed by Mr. Legacy and Mr. Worcester are 0.2 and 0.3, respectively. Mr. Worcester likes strong coffee!

(a) What is the probability that on a randomly selected morning the coffee will be strong? Show your work.

(b) If the coffee is strong on a randomly selected morning, what is the probability that it was brewed by Dr. Currier? Show your work.

AP4.45. The following table gives data on the mean number of seeds produced in a year by several common tree species and the mean weight (in milligrams) of the seeds produced. (Some species appear twice because their seeds were counted in two locations.) We might expect that trees with heavy seeds produce fewer of them, but what mathematical model best describes the relationship?[27]

Tree species	Seed count	Seed weight (mg)
Paper birch	27,239	0.6
Yellow birch	12,158	1.6
White spruce	7,202	2.0
Engelmann spruce	3,671	3.3
Red spruce	5,051	3.4
Tulip tree	13,509	9.1
Ponderosa pine	2,667	37.7
White fir	5,196	40.0
Sugar maple	1,751	48.0
Sugar pine	1,159	216.0
American beech	463	247
American beech	1,892	247
Black oak	93	1,851
Scarlet oak	525	1,930
Red oak	411	2,475
Red oak	253	2,475
Pignut hickory	40	3,423
White oak	184	3,669
Chestnut oak	107	4,535

(a) Based on the scatterplot below, is a linear model appropriate to describe the relationship between seed count and seed weight? Explain.

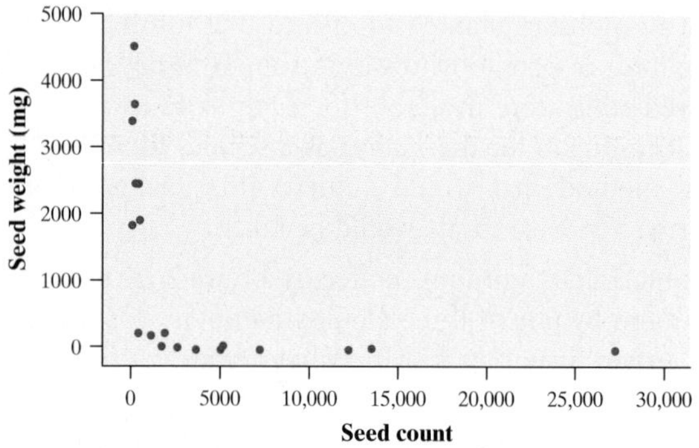

(b) Two alternative models based on transforming the original data are proposed to predict the seed weight from the seed count. Graphs and computer output from a least-squares regression analysis on the transformed data are shown below.

Model A:

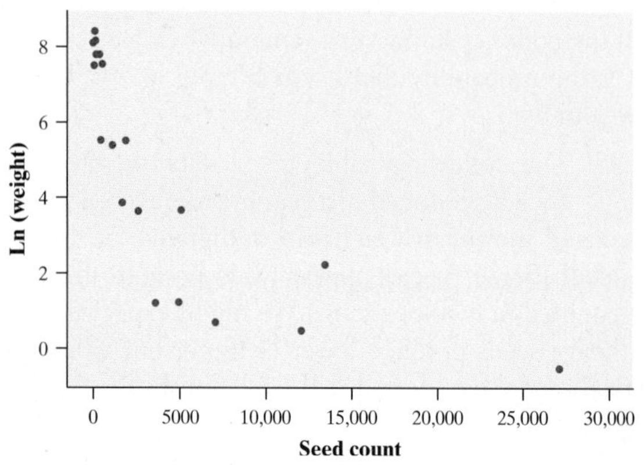

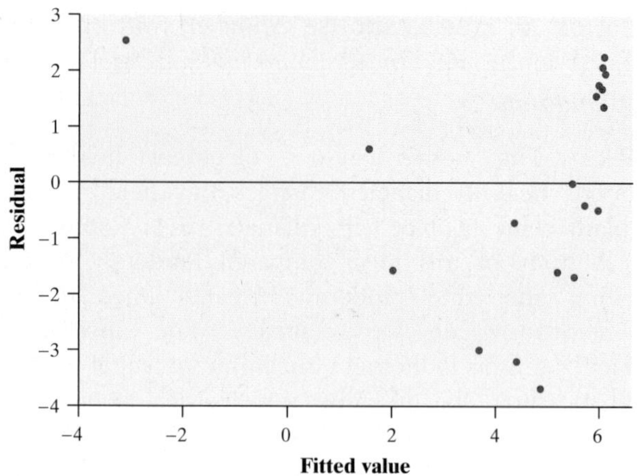

Predictor	Coef	SE Coef	T	P
Constant	6.1394	0.5726	10.72	0.000
Seed Count	−0.00033869	0.00007187	−4.71	0.000

S = 2.08100 R-Sq = 56.6% R-Sq(adj) = 54.1%

Model B:

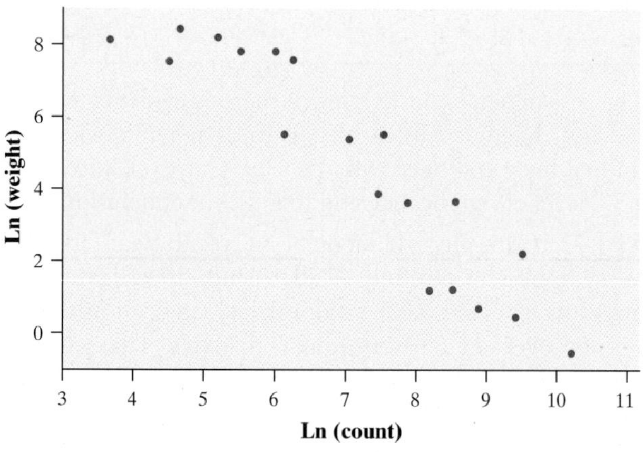

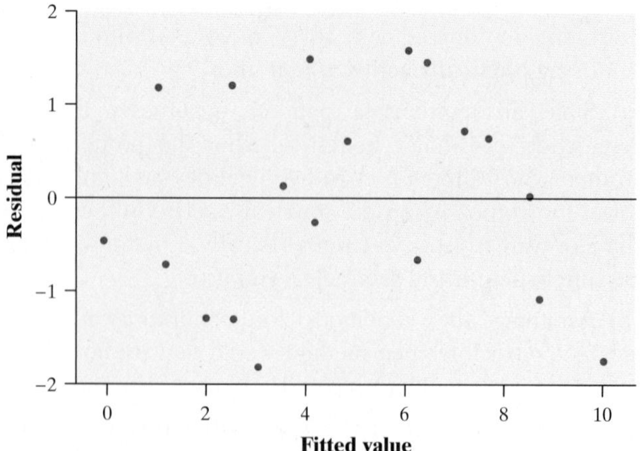

Predictor	Coef	SE Coef	T	P
Constant	15.491	1.081	14.33	0.000
Ln (count)	−1.5222	0.1470	−10.35	0.000

S = 1.16932 R-Sq = 86.3% R-Sq(adj) = 85.5%

Which model, A or B, is more appropriate for predicting seed weight from seed count? Justify your answer.

(c) Using the model you chose in part (b), predict the seed weight if the seed count is 3700.

(d) Interpret the value of r^2 for your model.

AP4.46. Suppose a company manufactures plastic lids for disposable coffee cups. When the manufacturing process is working correctly, the diameters of the lids are approximately Normally distributed with a mean diameter of 4 inches and a standard deviation of 0.02 inches. To make sure the machine is not producing lids that are too big or too small, each hour a random sample of 25 lids is selected and the sample mean is calculated.

(a) Describe the shape, center, and spread of the sampling distribution of the sample mean diameter, assuming the machine is working properly.

The company decides that it will shut down the machine if the sample mean diameter is less than 3.99 inches or greater than 4.01 inches, since this indicates that some lids will be too small or too large for the cups. If the sample mean is less than 3.99 or greater than 4.01, all the lids from that hour are thrown away since the company does not want to sell bad products.

(b) Assuming that the machine is working properly, what is the probability that a random sample of 25 lids will have a mean diameter less than 3.99 inches or greater than 4.01 inches? Show your work.

Additionally, to look for any trends, each hour the company records the value of the sample mean on a chart, like the one below.

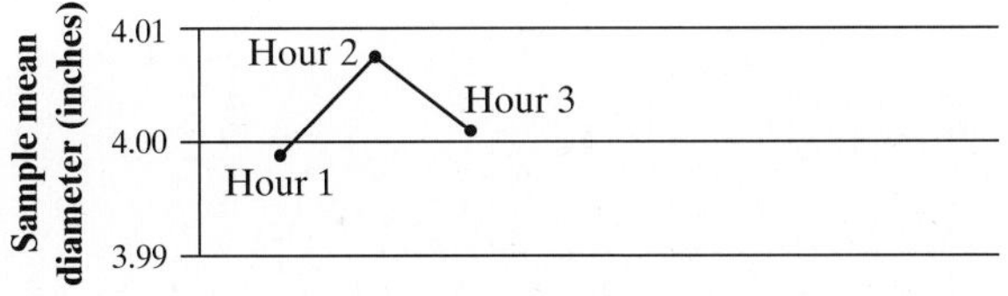

One benefit of using this type of chart is that out-of-control production trends can be noticed before it is too late and lids have to be thrown away. For example, if the sample mean increased in 3 consecutive samples, this would suggest that something might be wrong with the machine. If this trend can be noticed before the sample mean gets larger than 4.01, then the machine can be fixed without having to throw away any lids.

(c) Assuming that the manufacturing process is working correctly, what is the probability that the sample mean diameter will be above the desired mean of 4.00 but below the upper boundary of 4.01? Show your work.

(d) Assuming that the manufacturing process is working correctly, what is the probability that in 5 consecutive samples, 4 or 5 of the sample means will be above the desired mean of 4.00 but below the upper boundary of 4.01? Show your work.

(e) Which of the following results gives more convincing evidence that the machine needs to be shut down? Explain.

1. Getting a single sample mean below 3.99 or above 4.01

OR

2. Taking 5 consecutive samples and having at least 4 of the sample means be between 4.00 and 4.01.

(f) Suggest a different rule (other than 1 and 2 stated in part (e)) for stopping the machine before it starts producing lids that have to be thrown away. Assuming that the machine is working properly, calculate the probability that the machine will be shut down when using your rule.

Photo Credits

p. viii (top left) Anne Tabor;
(top right) Envisage studios
p. viii (middle left) Frank Lopez Photography;
(upper middle right) Fred T. Miller
p. viii (bottom left) Barbara Elkins;
(lower middle right) Tom Houser
p. viii (bottom right) Robert Sanborn
p. xv istockphoto
p. xvi istockphoto
p. xx (top) istockphoto
p. xx (bottom) istockphoto
p. xxi istockphoto
p. xxii (top) istockphoto
p. xxii (bottom) istockphoto

Chapter 1
p. xxiv istockphoto
p. 2 istockphoto
p. 3 Courtesy RSSCSE
p. 4 © The New Yorker Collection, 1976, Dana Fradon. From cartoonbank.com
p. 5 istockphoto
p. 10 (top) © 2009 Scott Adams/Distributed by UFS Inc.
p. 10 (bottom) istockphoto
p. 12 istockphoto
p. 15 istockphoto
p. 19 The Kobal Collection/20th Century Fox/Paramount
p. 27 AP
p. 28 Alamy
p. 32 istockphoto
p. 33 istockphoto
p. 35 istockphoto
p. 51 istockphoto
p. 52 W. H. Freeman and Worth Publishers
p. 53 istockphoto
p. 59 Photo by Sara Wolfram/Getty Images
p. 60 istockphoto
p. 62 © 2008 Zits Partnership. Distributed by King Features Syndicate
p. 66 istockphoto
p. 68 istockphoto

Chapter 2
p. 82 Photo by Chris Hondros/Getty Images
p. 84 istockphoto
p. 85 istockphoto
p. 88 AP Photo/Jae C. Hong
p. 91 istockphoto
p. 96 istockphoto
p. 98 NASA and The Hubble Heritage Team
p. 101 FOXTROT © 2005 Bill Amend. Universal Press Syndicate.
p. 110 © Dana Fradon from cartoonbank.com. All rights reserved.
p. 113 istockphoto
p. 117 (top) By permission of John L. Hart FLP and Creators Syndicate, Inc.
p. 117 (bottom) istockphoto
p. 127 istockphoto
p. 129 (top) istockphoto
p. 129 (bottom) Photo by Chris Hondros/Getty Images

Chapter 3
p. 140 Alamy
p. 142 istockphoto
p. 143 istockphoto
p. 145 Alamy
p. 146 © 2005 Zits Partnership, distributed by King Features Syndicate
p. 148 istockphoto
p. 149 Mother Goose & Grimm-Grimmy, Inc. King Features Syndicate
p. 153 Photo by Ben Martin
p. 156 AP Photo/Eric Gay
p. 164 istockphoto
p. 167 istockphoto
p. 168 istockphoto
p. 173 istockphoto
p. 182 Alamy
p. 183 istockphoto
p. 185 istockphoto
p. 188 istockphoto
p. 189 Alamy

Chapter 4
p. 204 Markus Dlouhy/Das Fotoarchiv/Peter Arnold
p. 206 © MBI/Alamy
p. 207 istockphoto
p. 208 (top) © Charles O. Cecil/Alamy
p. 208 (middle) istockphoto
p. 209 © Image Source/Alamy
p. 211 Cartoon by Landers, cartoons@landers.co.uk
p. 213 istockphoto
p. 214 DILBERT reprinted by permission of United Feature Syndicate, Inc.
p. 215 epa.gov
p. 216 © Li Ding/Alamy
p. 218 © Richard Levine/Alamy
p. 219 © David Hodges/Alamy
p. 223 © 1995 Watterson. Universal Press Syndicate. Reprinted with permission. All rights reserved.
p. 231 istockphoto
p. 233 © Golden Pixels LLC/Alamy
p. 234 istockphoto
p. 235 istockphoto
p. 236 istockphoto
p. 237 © moodboard/Alamy
p. 241 istockphoto
p. 243 By permission of John L. Hart FLP and Creators Syndicate, Inc.
p. 245 istockphoto
p. 246 © Martin Wierink/Alamy
p. 255 Courtesy Blake Suttle
p. 261 istockphoto
p. 264 istockphoto
p. 268 Markus Dlouhy/Das Fotoarchiv/Peter Arnold

Chapter 5
p. 280 istockphoto
p. 282 Reproduced by permission of Bob Schochet
p. 283 Creatas/Thinkstock
p. 287 © Eyebyte/Alamy
p. 288 Mitchell Layton/Getty Images
p. 291 AP Photo/Tom Strattman
p. 299 © Judith Collins/Alamy
p. 302 © Murray Lee/Photolibrary
p. 303 © Big Cheese Photo/SuperStock
p. 306 istockphoto
p. 307 © Flint/Corbis
p. 313 istockphoto
p. 316 Graeme Harris/Getty Images
p. 320 istockphoto
p. 323 © 2008 Chris Browne 2010 King Features Syndicate, Inc.
p. 324 istockphoto
p. 326 AP Photo/Mark J. Terrill
p. 327 istockphoto

Chapter 6
p. 338 Alamy
p. 340 istockphoto
p. 341 istockphoto
p. 343 Picture Partners/Photo Researchers, Inc.
p. 345 istockphoto
p. 346 Scott Adams/Distributed by United Feature Syndicate, Inc.
p. 351 Alamy
p. 359 Alamy
p. 363 istockphoto
p. 366 istockphoto
p. 368 istockphoto
p. 369 Alamy
p. 374 istockphoto
p. 375 istockphoto
p. 378 istockphoto
p. 384 istockphoto
p. 385 istockphoto
p. 392 Dreamstime
p. 394 istockphoto

p. 395 istockphoto
p. 396 istockphoto
p. 398 istockphoto
p. 401 Alamy

Chapter 7

p. 412 © 2010 Photolibrary.com
p. 414 istockphoto
p. 417 istockphoto
p. 418 istockphoto
p. 422 Alamy
p. 425 David Young-Wolff/Photo Edit
p. 427 istockphoto
p. 432 FOXTROT © 2005 Bill Amend. Universal Press Syndicate.
p. 437 Dreamstime
p. 442 istockphoto
p. 447 Alamy
p. 452 istockphoto
p. 453 © 2010 Photolibrary.com

Chapter 8

p. 466 Dreamstime
p. 468 istockphoto
p. 470 istockphoto
p. 473 Garfield © 1999 Paws, Inc. Dist. By Universal Uclick. Reprinted by permission. All rights reserved
p. 476 (top) Alamy
p. 476 (bottom) istockphoto
p. 485 istockphoto
p. 489 istockphoto
p. 491 Alamy
p. 493 istockphoto
p. 500 Alamy
p. 502 Alamy
p. 508 istockphoto
p. 510 Alamy
p. 511 istockphoto
p. 515 Dreamstime

Chapter 9

p. 526 istockphoto
p. 529 Otnaydur/Dreamstime.com
p. 531 Alamy
p. 533 istockphoto
p. 535 istockphoto
p. 536 istockphoto
p. 537 istockphoto
p. 539 fotosearch
p. 541 istockphoto
p. 543 istockphoto
p. 550 fotosearch
p. 553 istockphoto
p. 556 istockphoto
p. 567 istockphoto
p. 571 istockphoto
p. 574 Alamy
p. 577 Alamy
p. 580 Courtesy Judy Starnes
p. 582 istockphoto
p. 585 istockphoto

Chapter 10

p. 600 Spencer Grant/Photo Edit
p. 602 istockphoto
p. 604 Alamy
p. 609 Photo by Peter Macdiarmid/Getty Images
p. 612 Alamy
p. 614 istockphoto
p. 615 Photolibrary
p. 623 By permission of Dave Coverly and Creators Syndicate, Inc.
p. 627 istockphoto
p. 629 Jeff Greenberg/Photo Edit
p. 631 Alamy
p. 635 istockphoto
p. 639 Alamy
p. 647 (top) Markus Dlouhy/Das Fotoarchiv/ Peter Arnold
p. 647 (bottom) Alamy
p. 648 istockphoto
p. 649 Spencer Grant/Photo Edit
p. 653 istockphoto

Chapter 11

p. 674 istockphoto
p. 676 R Rivera Moret
p. 679 (top) R Rivera Moret
p. 679 (bottom) R Rivera Moret
p. 681 R Rivera Moret
p. 684 R Rivera Moret
p. 688 Alamy
p. 696 Alamy
p. 700 Alamy
p. 707 Alamy
p. 712 Alamy
p. 713 Alamy
p. 717 Alamy
p. 719 Alamy
p. 722 istockphoto

Chapter 12

p. 736 © Theodore A. Wagner/Icon SMI/ Corbis
p. 738 fotosearch
p. 747 istockphoto
p. 749 istockphoto
p. 752 fotosearch
p. 755 © David A. Northcott/Corbis
p. 768 (top) istockphoto
p. 768 (bottom) Alamy
p. 772 istockphoto
p. 773 Copyright © Intel Corporation. All rights reserved.
p. 778 NASA
p. 780 istockphoto
p. 783 © Theodore A. Wagner/Icon SMI/ Corbis

Notes and Data Sources

OVERVIEW: WHAT IS STATISTICS?

1. Joachim Schüz et al., "Cellular telephone use and cancer risk: update of a nationwide Danish cohort," *Journal of the National Cancer Institute*, 98 (2006), pp. 1707–1713.
2. Nikhil Swaminathan, "Gender Jabber: Do Women Talk More than Men?" *Scientific American*, July 6, 2007.
3. Data from the World Bank's *1999 World Development Indicators*. Life expectancy is estimated for 1997, and GDP per capita (purchasing power parity basis) for 1998.
4. Data from the Josephson Institute's "2008 Report Card of American Youth," accessed online at charactercounts.org, April 10, 2010.

CHAPTER 1

1. T. M. Amabile, "Children's artistic creativity: detrimental effects of competition in a field setting," *Personality and Social Psychology Bulletin*, 8 (1982), pp. 573–578.
2. Roller coaster data for 2009 from www.redb.com.
3. Arbitron, *Radio Today*, www.arbitron.com.
4. Arbitron, *Internet and Multimedia*, www.arbitron.com.
5. We got the idea for this example from David Lane's case study "Who is buying iMacs?" which we found online at www.ruf.rice.edu/~lane/case_studies/imac/index.html.
6. The National Longitudinal Study of Adolescent Health interviewed a stratified random sample of 27,000 adolescents, then reinterviewed many of the subjects six years later, when most were aged 19 to 25. These data are from the Wave III reinterviews in 2000 and 2001, found at the Web site of the Carolina Population Center, www.cpc.unc.edu.
7. Siem Oppe and Frank De Charro, "The effect of medical care by a helicopter trauma team on the probability of survival and the quality of life of hospitalized victims," *Accident Analysis and Prevention*, 33 (2001), pp. 129–138. The authors give the data in this example as a "theoretical example" to illustrate the need for more elaborate analysis of actual data using severity scores for each victim.
8. 2008 DuPont Automotive Color Survey, at www.automotive.dupont.com.
9. Found at spam-filter-review.toptenreviews.com, which claims to have compiled data "from a number of different reputable sources."
10. Centers for Disease Control and Prevention, *Births: Final Data for 2005*, National Vital Statistics Reports, 56, No. 6, 2007, at www.cdc.gov. These are the most recent data available as of early 2008, but the numbers change only slightly from year to year.
11. National Center for Health Statistics, *Deaths: Preliminary Data for 2005*, at www.cdc.gov/nchs.
12. *U.S. Hispanic Population 2006*, at www.census.gov, based on the March 2006 Current Population Survey, Annual Social and Economic Supplement.
13. Data for 2005 from the 2007 *Statistical Abstract of the United States* at the Census Bureau Web site, www.census.gov.
14. Arbitron, *Internet and Media*, at www.arbitron.com.
15. *Statistical Abstract of the United States*, 2004–2005, Table 1233.
16. Lien-Ti Bei, "Consumers' purchase behavior toward recycled products: an acquisition-transaction utility theory perspective," MS thesis, Purdue University, 1993.
17. S. V. Zagona (ed.), *Studies and Issues in Smoking Behavior*, University of Arizona Press, 1967, pp. 157–180.
18. From the Web site of the Carolina Population Center, www.cpc.unc.edu.
19. M. Radelet, "Racial characteristics and imposition of the death penalty," *American Sociological Review*, 46 (1981), pp. 918–927.
20. Data on the U.S. Women's National Soccer Team are from the United States Soccer Federation Web site, www.ussoccer.com.
21. From the 2006 American Community Survey, at factfinder.census.gov.
22. James T. Fleming, "The measurement of children's perception of difficulty in reading materials," *Research in the Teaching of English*, 1 (1967), pp. 136–156.
23. See Note 13.
24. The original paper is T. M. Amabile, "Motivation and creativity: effects of motivational orientation on creative writers," *Journal of Personality and Social Psychology*, 48, No. 2 (February 1985), pp. 393–399. The data for Exercise 1.43 came from Fred Ramsey and Dan Schafer, *The Statistical Sleuth*, 2nd ed., Duxbury, 2002.
25. The cereal data came from the Data and Story Library, http://lib.stat.cmu.edu/DASL/.
26. From a plot in K. Krishna Kumar et al., "Unraveling the mystery of Indian monsoon failure during El Niño," *Science*, 314 (2006), pp. 115–119.
27. Basketball scores from the *Los Angeles Times*, February 16, 2001.
28. Monthly stock returns from the Web site of Professor Kenneth French of Dartmouth, mba.tuck.dartmouth.edu/pages/faculty/ken.french. A fine point: the data are actually the "excess returns" on stocks, the actual returns less the small monthly returns on Treasury bills.
29. C. B. Williams, *Style and Vocabulary: Numerological Studies*, Griffin, 1970.
30. From the 2006 American Community Survey, at factfinder.census.gov.
31. International Energy Agency, *Key World Energy Statistics 2007*, at www.iea.org.
32. Maribeth Cassidy Schmitt, "The effects of an elaborated directed reading activity on the metacomprehension skills of third graders," PhD thesis, Purdue University, 1987.
33. These data were collected by students as a class project.
34. Advanced Placement exam results from AP Central Web site, apcentral.collegeboard.com.
35. Arbitron, *Internet and Media 2006*, www.arbitron.com.
36. H. Lindberg, H. Roos, and P. Gardsell, "Prevalence of coxarthritis in former soccer players," *Acta Orthopedica Scandinavica*, 64 (1993), pp. 165–167.

37. From the 2003 American Community Survey, at the Census Bureau Web site, www.census.gov. The data are a subsample of the 13,194 individuals in the ACS North Carolina sample who had travel times greater than zero.

38. Pets data set from *Guidelines for Assessment and Instruction in Statistics Education: Pre-K–12 Report,* American Statistical Association, 2007.

39. This isn't a mathematical theorem. The mean can be less than the median in right-skewed distributions that take only a few values, many of which lie exactly at the median. The rule almost never fails for distributions taking many values, and most counterexamples don't appear clearly skewed in graphs even though they may be slightly skewed according to technical measures of skewness. See Paul T. von Hippel, "Mean, median, and skew: correcting a textbook rule," *Journal of Statistics Education,* 13, No. 2 (2005), online journal, www.amstat.org/PUBLICATIONS/JSE/.

40. Thanks to Sanderson Smith of Santa Barbara City College for suggesting this Activity.

41. From the National Association of Realtors, www.realtor.org.

42. Tom Lloyd et al., "Fruit consumption, fitness, and cardiovascular health in female adolescents: the Penn State Young Women's Health Study," *American Journal of Clinical Nutrition,* 67 (1998), pp. 624–630.

43. The daily returns for the year ending November 8, 2007, for the CREF Equity Index Fund and the TIAA Real Estate Fund. Daily price data for these funds are at www.tiaa-cref.org. Returns can be easily calculated because dividends are incorporated in the daily prices rather than given separately.

44. Data from the most recent Annual Demographic Supplement can be found at www.bls.census.gov/cps/datamain.htm.

45. Ethan J. Temeles and W. John Kress, "Adaptation in a plant-hummingbird association," *Science,* 300 (2003), pp. 630–633. I thank Ethan J. Temeles for providing the data.

46. 2007 CIRP Freshman Survey; data from the Higher Education Research Institute's report "The American freshman: national norms for fall 2007," published January 2008.

47. From the Internet Movie data base, www.imdb.

48. Movie ratings data for 2007 found at informationplease.com.

49. The report "Cell phones key to teens' social lives, 47% can text with eyes closed" is published online by Harris Interactive, July 2008, www.marketingcharts.com.

50. A. Karpinski and A. Duberstein, "A description of Facebook use and academic performance among undergraduate and graduate students," paper presented at the American Educational Research Association annual meeting, April 2009. Thanks to Aryn Karpinski for providing us with some original data from the study.

51. S. M. Stigler, "Do robust estimators work with real data?" *Annals of Statistics,* 5 (1977), pp. 1055–1078.

52. T. Bjerkedal, "Acquisition of resistance in guinea pigs infected with different doses of virulent tubercle bacilli," *American Journal of Hygiene,* 72 (1960), pp. 130–148.

53. Data from the Bureau of Labor Statistics, Annual Demographic Supplement for 2002, www.bls.census.gov.

54. Data from the report "Is our tuna family-safe?" prepared by Defenders of Wildlife, 2006.

CHAPTER 2

1. Will Shortz, "A few words about Sudoku, which has none," *New York Times,* August 28, 2005.

2. Data found online at www.exploringdata.cqu.edu.au.

3. S. M. Stigler, "Do robust estimators work with real data?" *Annals of Statistics,* 5 (1977), pp. 1055–1078.

4. Data from Gary Community School Corporation, courtesy of Celeste Foster, Department of Education, Purdue University.

5. Sheldon Ross, *Introduction to Probability and Statistics for Engineers and Scientists,* 3rd ed., Academic Press, 2004.

6. Stephen Jay Gould, "Entropic homogeneity isn't why no one hits .400 anymore," *Discover,* August 1986, pp. 60–66. Gould does not standardize but gives a speculative discussion instead.

7. Information on bone density in the reference populations was found at www.courses.washington.edu/bonephys/opbmd.html.

8. Data from *USA Today* sports salary data base at www.usatoday.com.

9. Summary information on PSAT scores of juniors was obtained from the College Board Web site at www.collegeboard.com.

10. Detailed data appear in P. S. Levy et al., *Total Serum Cholesterol Values for Youths 12–17 Years,* Vital and Health Statistics, Series 11, No. 155, National Center for Health Statistics, 1976.

11. Data found at Bureau of Labor Statistics, www.bls.gov.

12. Data provided by Chris Olsen, who found the information in *Scuba News and Skin Diver* magazines.

13. See Note 3.

14. The data set was constructed based on information provided in P. D. Wood et al., "Plasma lipoprotein distributions in male and female runners," in P. Milvey (ed.), *The Marathon: Physiological, Medical, Epidemiological, and Psychological Studies,* New York Academy of Sciences, 1977.

15. Data found online at www.earthtrends.wri.org.

16. R. Shine, T. R. L. Madsen, M. J. Elphick, and P. S. Harlow, "The influence of nest temperature and maternal brooding on hatchling phenotypes in water python," *Ecology,* 78 (1997), pp. 1713–1721.

17. We found the information on birth weights of Norwegian children on the National Institute of Environmental Health Sciences Web site: http://eb.niehs.nih.gov/bwt/subcfreq.htm.

18. From a graph in L. Partridge and M. Farquhar, "Sexual activity reduces lifespan of male fruit flies," *Nature,* 294 (1981), pp. 580–582. Provided by Brigitte Baldi.

CHAPTER 3

1. The Florida Boating Accident Statistical Report for each year (at research.myfwc.com/law/boating) gives the number of registered vessels. The Florida Wildlife Commission maintains a manatee death data base at research.myfwc.com/manatees.

2. Data provided by Darlene Gordon, Purdue University.

3. Government data for 2005, presented in "An accident waiting to happen?" *Consumer Reports,* March 2007, pp. 16–19.

4. From a graph in Bernt-Erik Saether, Steiner Engen, and Erik Mattysen, "Demographic characteristics and population dynamical patterns of solitary birds," *Science,* 295 (2002), pp. 2070–2073.

5. Based on T. N. Lam, "Estimating fuel consumption from engine size," *Journal of Transportation Engineering,* 111 (1985), pp. 339–357. The data for 10 to 50 km/h are measured; those for 60 and higher are calculated from a model given in the paper and are therefore smoothed.

6. From a graph in Christer G. Wiklund, "Food as a mechanism of density-dependent regulation of breeding numbers in the merlin *Falco columbarius,*" *Ecology,* 82 (2001), pp. 860–867.

7. From a graph in Naomi I. Eisenberger, Matthew D. Lieberman, and Kipling D. Williams, "Does rejection hurt? An fMRI study of social exclusion," *Science*, 302 (2003), pp. 290–292.

8. M. A. Houck et al., "Allometric scaling in the earliest fossil bird, *Archaeopteryx lithographica*," *Science*, 247 (1990), pp. 195–198. The authors conclude from a variety of evidence that all specimens represent the same species.

9. *Consumer Reports*, June 1986, pp. 366–367.

10. G. A. Sacher and E. F. Staffelt, "Relation of gestation time to brain weight for placental mammals: implications for the theory of vertebrate growth," *American Naturalist*, 108 (1974), pp. 593–613. We found the data in Fred L. Ramsey and Daniel W. Schafer, *The Statistical Sleuth: A Course in Methods of Data Analysis*, Duxbury, 1997, p. 228.

11. Table 1 of E. Thomassot et al., "Methane-related diamond crystallization in the earth's mantle: stable isotopes evidence from a single diamond-bearing xenolith," *Earth and Planetary Science Letters*, 257 (2007), pp. 362–371.

12. National Center for Education Statistics, *Debt Burden: A Comparison of 1992–93 and 1999–2000 Bachelor's Degree Recipients a Year after Graduating*, 2005, nces.ed.gov.

13. Data from a plot in James A. Levine, Norman L. Eberhart, and Michael D. Jensen, "Role of non-exercise activity thermogenesis in resistance to fat gains in humans," *Science*, 283 (1999), pp. 212–214.

14. P. Goldblatt (ed.), *Longitudinal Study: Mortality and Social Organization*, Her Majesty's Stationery Office, 1990. At least, so claims Richard Conniff, in *The Natural History of the Rich*, Norton, 2002, p. 45. We have not been able to access the Goldblatt report.

15. Michael Winerip, "On education: SAT essay test rewards length and ignores errors," *New York Times*, May 4, 2005.

16. W. M. Lewis and M. C. Grant, "Acid precipitation in the western United States," *Science*, 207 (1980), pp. 176–177.

17. Gary Smith, "Do statistics test scores regress toward the mean?" *Chance*, 10, No. 4 (1997), pp. 42–45.

18. Data from National Institute of Standards and Technology, *Engineering Statistics Handbook*, www.itl.nist.gov/div898/handbook. The analysis there does not comment on the bias of field measurements.

19. Debora L. Arsenau, "Comparison of diet management instruction for patients with non–insulin dependent diabetes mellitus: learning activity package vs. group instruction," MS thesis, Purdue University, 1993.

20. From the Web site of the Tropical Meteorology Project, hurricane.atmos.colostate.edu.

21. From a graph in G. D. Martinsen, E. M. Driebe, and T. G. Whitham, "Indirect interactions mediated by changing plant chemistry: beaver browsing benefits beetles," *Ecology*, 79 (1998), pp. 192–200.

22. A. K. Yousafzai et al., "Comparison of armspan, arm length and tibia length as predictors of actual height of disabled and nondisabled children in Dharavi, Mumbai, India," *European Journal of Clinical Nutrition*, 57 (2003), pp. 1230–1234. In fact, $r^2 = 0.93$.

23. David M. Fergusson and L. John Horwood, "Cannabis use and traffic accidents in a birth cohort of young adults," *Accident Analysis and Prevention*, 33 (2001), pp. 703–711.

24. *The World Almanac and Book of Facts* (2009).

25. G. L. Kooyman et al., "Diving behavior and energetics during foraging cycles in king penguins," *Ecological Monographs*, 62 (1992), pp. 143–163.

26. We found the data on cherry blossoms in the paper "Linear equations and data analysis," which was posted on the North Carolina School of Science and Mathematics Web site, www.ncssm.edu.

27. From a graph in Craig Packer et al., "Ecological change, group territoriality, and population dynamics in Serengeti lions," *Science*, 307 (2005), pp. 390–393.

CHAPTER 4

1. Carlos Vallbona et al., "Response of pain to static magnetic fields in postpolio patients, a double blind pilot study," *Archives of Physical Medicine and Rehabilitation*, 78 (1997), pp. 1200–1203.

2. "Traffic Safety Facts 2009," published by the National Highway Traffic Safety Administration on their Web site, www.nhtsa.dot.gov.

3. Results from the Cox Communications Teen Online and Wireless Safety Survey, in Partnership with the National Center for Missing and Exploited Children® (NCMEC) and John Walsh, published May 2009, www.cox.com/takecharge/safe_teens_2009.

4. Sheldon Cohen, William J. Doyle, Cuneyt M. Alper, Denise Janicki-Deverts, and Ronald B. Turner, "Sleep habits and susceptibility to the common cold," *Archives of Internal Medicine*, 169, No. 1 (2009), pp. 62–67.

5. Rebecca Smith and Alison Wessner, The Lawrenceville School, "Does listening to music while studying affect performance on learning assessments?" May 2009.

6. Based on a suggested Science Fair project entitled "Do the Eyes Have It?" at the Science Buddies Web site, www.sciencebuddies.org.

7. *The ASCAP Survey and Your Royalties*, ASCAP, undated.

8. For information on the 2006 American Community Survey of households (there is a separate sample of group quarters), go to www.census.gov/acs.

9. The Pew press release and the full report "Polls face growing resistance, but still representative" (April 20, 2004) are at people-press.org/reports.

10. For more detail on the limits of memory in surveys, see N. M. Bradburn, L. J. Rips, and S. K. Shevell, "Answering autobiographical questions: the impact of memory and inference on surveys," *Science*, 236 (1987), pp. 157–161.

11. Cynthia Crossen, "Margin of error: studies galore support products and positions, but are they reliable?" *Wall Street Journal*, November 14, 1991.

12. Robert C. Parker and Patrick A. Glass, "Preliminary results of double-sample forest inventory of pine and mixed stands with high- and low-density LiDAR," in Kristina F. Connoe (ed.), *Proceedings of the 12th Biennial Southern Silvicultural Research Conference*, U.S. Department of Agriculture, Forest Service, Southern Research Station, 2004. The researchers actually sampled every tenth plot. This is a systematic sample.

13. Gary S. Foster and Craig M. Eckert, "Up from the grave: a socio-historical reconstruction of an African American community from cemetery data in the rural Midwest," *Journal of Black Studies*, 33 (2003), pp. 468–489.

14. Bryan E. Porter and Thomas D. Berry, "A nationwide survey of self-reported red light running: measuring prevalence, predictors,

and perceived consequences," *Accident Analysis and Prevention*, 33 (2001), pp. 735–741.

15. Giuliana Coccia, "An overview of non-response in Italian telephone surveys," *Proceedings of the 99th Session of the International Statistics Institute*, 1993, Book 3, pp. 271–272.

16. Mario A. Parada et al., "The validity of self-reported seatbelt use: Hispanic and non-Hispanic drivers in El Paso," *Accident Analysis and Prevention*, 33 (2001), pp. 139–143.

17. These figures are cited by Dr. William Dement at "The Sleep Well" Web site, www.stanford.edu/~dement/.

18. See, for example, Martin Enserink, "The vanishing promises of hormone replacement," *Science*, 297 (2002), pp. 325–326; and Brian Vastag, "Hormone replacement therapy falls out of favor with expert committee," *Journal of the American Medical Association*, 287 (2002), pp. 1923–1924. A National Institutes of Health panel's comprehensive report is *International Position Paper on Women's Health and Menopause*, NIH Publication 02-3284.

19. "Family dinner linked to better grades for teens: survey finds regular meal time yields additional benefits," written by John Mackenzie for ABC News's *World News Tonight*, September 13, 2005.

20. Information found online at http//ssw.unc.edu/abou/news/careerstart_1-13-09.

21. Simplified from Arno J. Rethans, John L. Swasy, and Lawrence J. Marks, "Effects of television commercial repetition, receiver knowledge, and commercial length: a test of the two-factor model," *Journal of Marketing Research*, 23 (February 1986), pp. 50–61.

22. Martin J. Bergee and Lecia Cecconi-Roberts, "Effects of small-group peer interaction on self-evaluation of music performance," *Journal of Research in Music Education*, 50 (2002), pp. 256–268.

23. Steering Committee of the Physicians' Health Study Research Group, "Final report on the aspirin component of the ongoing Physicians' Health Study," *New England Journal of Medicine*, 321 (1989), pp. 129–135.

24. The placebo effect examples are from Sandra Blakeslee, "Placebos prove so powerful even experts are surprised," *New York Times*, October 13, 1998.

25. The "three-quarters" estimate is cited by Martin Enserink, "Can the placebo be the cure?" *Science*, 284 (1999), pp. 238–240. An extended treatment is Anne Harrington (ed.), *The Placebo Effect: An Interdisciplinary Exploration*, Harvard University Press, 1997.

26. The flu trial quotation is from Kristin L. Nichol et al., "Effectiveness of live, attenuated intranasal influenza virus vaccine in healthy, working adults," *Journal of the American Medical Association*, 282 (1999), pp. 137–144.

27. From the Electronic Encyclopedia of Statistical Examples and Exercises (EESEE), "Anecdotes of Placebos."

28. David L. Strayer, Frank A. Drews, and William A. Johnston, "Cell phone–induced failures of visual attention during simulated driving," *Journal of Experimental Psychology: Applied*, 9 (2003), pp. 23–32.

29. We obtained the tire pressure loss data from the Consumer Reports Web site: http://blogs.consumerreports.org/cars/2007/10/tires-nitrogen-.html.

30. *Early Human Development*, 76, No. 2 (February 2004), pp. 139–145.

31. National Institute of Child Health and Human Development, Study of Early Child Care and Youth Development. The article appears in the July 2003 issue of *Child Development*. The quotation is from the summary on the NICHD Web site, www.nichd.nih.gov.

32. J. E. Muscat et al., "Handheld cellular telephone use and risk of brain cancer," *Journal of the American Medical Association*, 284 (2000), pp. 3001–3007.

33. Naomi D. L. Fisher, Meghan Hughes, Marie Gerhard-Herman, and Norman K. Hollenberg, "Flavonol-rich cocoa induces nitric-oxide-dependent vasodilation in healthy humans," *Journal of Hypertension*, 21, No. 12 (2003), pp. 2281–2286.

34. Joel Brockner et al., "Layoffs, equity theory, and work performance: further evidence of the impact of survivor guilt," *Academy of Management Journal*, 29 (1986), pp. 373–384.

35. K. B. Suttle, Meredith A. Thomsen, and Mary E. Power, "Species interactions reverse grassland responses to changing climate," *Science*, 315 (2007), pp. 640–642.

36. Christopher Anderson, "Measuring what works in health care," *Science*, 263 (1994), pp. 1080–1082.

37. Esther Duflo, Rema Hanna, and Stephan Ryan, "Monitoring works: getting teachers to come to school," November 21, 2007, at econ-mit.edu/files/2066.

38. Details of the Carolina Abecedarian Project, including references to published work, can be found online at www.fpg.unc.edu/~abc/.

39. See Note 27.

40. Marielle H. Emmelot-Vonk et al., "Effect of testosterone supplementation on functional mobility, cognition, and other parameters in older men," *Journal of the American Medical Association*, 299 (2008), pp. 39–52.

41. W. E. Paulus et al., "Influence of acupuncture on the pregnancy rate in patients who undergo assisted reproductive therapy," *Fertility and Sterility*, 77, No. 4 (2002), pp. 721–724.

42. Linda Stern et al., "The effects of low-carbohydrate versus conventional weight loss diets in severely obese adults: one-year follow up of a randomized trial," *Annals of Internal Medicine*, 140, No. 10 (May 2004), pp. 778–785.

43. Mary O. Mundinger et al., "Primary care outcomes in patients treated by nurse practitioners or physicians," *Journal of the American Medical Association*, 238 (2000), pp. 59–68.

44. Sterling C. Hilton et al., "A randomized controlled experiment to assess technological innovations in the classroom on student outcomes: an overview of a clinical trial in education," manuscript, no date. A brief report is Sterling C. Hilton and Howard B. Christensen, "Evaluating the impact of multimedia lectures on student learning and attitudes," *Proceedings of the 6th International Conference on the Teaching of Statistics*, at www.stat.aukland.ac.nz.

45. Study conducted by cardiologists at Athens Medical School, Greece, and announced at a European cardiology conference in February 2004.

46. Evan H. DeLucia et al., "Net primary production of a forest ecosystem with experimental CO_2 enhancement," *Science*, 284 (1999), pp. 1177–1179. The investigators used the block design.

47. Niels Juel-Nielsen, *Individual and Environment: Monozygotic Twins Reared Apart*, International Universities Press, 1980.

48. The sleep deprivation study is described in R. Stickgold, L. James, and J. Hobson, "Visual discrimination learning requires post-training sleep," *Nature Neuroscience*, 2000, pp. 1237–1238. We obtained the data from Allan Rossman, who got it courtesy of the authors.

49. The idea for this chart came from Fred Ramsey and Daniel Schafer, *The Statistical Sleuth: A Course in Methods of Data Analysis*, 2nd edition, Duxbury Press, 2002.

50. Thanks to Dan Teague for providing this example.

51. *The Health Consequences of Smoking: 1983*, U.S. Health Service, 1983.

52. See the details on the Web site of the Office for Human Research Protections of the Department of Health and Human Services, hhs.gov/ohrp.
53. Thanks to Josh Tabor for contributing this Activity.
54. Declaration of Helsinki, www.cirp.org/library/ethics/helsinki/.
55. Charles A. Nelson III et al., "Cognitive recovery in socially deprived young children: the Bucharest Early Intervention Project," *Science*, 318 (2007), pp. 1937–1940.
56. The article describing this study is Nikhil Swaminathan, "Gender jabber: do women talk more than men?" *Scientific American*, July 6, 2007.
57. Marilyn Ellis, "Attending church found factor in longer life," *USA Today*, August 9, 1999.
58. Scott DeCarlo with Michael Schubach and Vladimir Naumovski, "A decade of new issues," *Forbes*, March 5, 2001, www.forbes.com.
59. Warren McIsaac and Vivek Goel, "Is access to physician services in Ontario equitable?" Institute for Clinical Evaluative Sciences in Ontario, October 18, 1993.
60. L. E. Moses and F. Mosteller, "Safety of anesthetics," in J. M. Tanur et al. (eds.), *Statistics: A Guide to the Unknown*, 3rd ed., Wadsworth, 1989, pp. 15–24.
61. Jo Phelan et al., "The stigma of homelessness: the impact of the label 'homeless' on attitudes towards poor persons," *Social Psychology Quarterly*, 60 (1997), pp. 323–337.
62. R. C. Shelton et al., "Effectiveness of St. John's wort in major depression," *Journal of the American Medical Association*, 285 (2001), pp. 1978–1986.
63. E. M. Peters et al., "Vitamin C supplementation reduces the incidence of post-race symptoms of upper respiratory tract infection in ultra-marathon runners," *American Journal of Clinical Nutrition*, 57 (1993), pp. 170–174.
64. Based on a news item "Bee off with you," *Economist*, November 2, 2002, p. 78.
65. Jay Schreiber and Megan Thee, "Fans concerned about steroid use and believe it's widespread, poll shows," *New York Times*, March 31, 2008.
66. From the Electronic Encyclopedia of Statistical Examples and Exercises (EESEE) case study "Is Caffeine Dependence Real?"

Addendum for Activity "How large is a typical U.S. state?" (page 215)

How large is the typical state?

	State	Total area (sq. mi.)		State	Total area (sq. mi.)
1	Alabama	52,419	26	Montana	147,042
2	Alaska	663,267	27	Nebraska	77,354
3	Arizona	113,998	28	Nevada	110,561
4	Arkansas	53,179	29	New Hampshire	9,350
5	California	163,696	30	New Jersey	8,721
6	Colorado	104,094	31	New Mexico	121,589
7	Connecticut	5,543	32	New York	54,556
8	Delaware	2,489	33	North Carolina	53,819
9	Florida	65,755	34	North Dakota	70,700
10	Georgia	59,425	35	Ohio	44,825
11	Hawaii	10,931	36	Oklahoma	69,898
12	Idaho	83,570	37	Oregon	98,381
13	Illinois	57,914	38	Pennsylvania	46,055
14	Indiana	36,418	39	Rhode Island	1,545
15	Iowa	56,272	40	South Carolina	32,020
16	Kansas	82,277	41	South Dakota	77,116
17	Kentucky	40,409	42	Tennessee	42,143
18	Louisiana	51,840	43	Texas	268,581
19	Maine	35,385	44	Utah	84,899
20	Maryland	12,407	45	Vermont	9,614
21	Massachusetts	10,555	46	Virginia	42,774
22	Michigan	96,716	47	Washington	71,300
23	Minnesota	86,939	48	West Virginia	24,230
24	Mississippi	48,430	49	Wisconsin	65,498
25	Missouri	69,704	50	Wyoming	97,814
				TOTAL	3,794,017

Addendum for Activity "Sampling sunflowers" (page 216)

Sampling sunflowers										
	A	B	C	D	E	F	G	H	I	J
1	106	104	106	105	105	105	108	108	107	106
2	103	102	103	104	103	103	103	104	103	105
3	101	99	102	101	100	100	102	101	102	101
4	98	99	99	99	100	99	98	101	99	101
5	98	100	100	100	99	99	99	98	100	98
6	97	98	98	98	99	99	99	99	97	98
7	103	102	102	104	101	102	102	102	104	102
8	104	103	103	103	102	102	103	104	102	102
9	106	106	104	106	102	107	104	103	106	106
10	107	108	109	110	106	107	109	107	106	107

CHAPTER 5

1. From the EESEE story "Baby Hearing Screening."
2. Note that pennies have rims that make spinning more stable. The probability of a head in spinning a coin depends on the type of coin and also on the surface. *Chance News 11.02*, at www.dartmouth.edu/~chance, reports about 45% heads in more than 20,000 spins of American pennies by Robin Lock's students at Saint Lawrence University.
3. R. Vallone and A. Tversky, "The hot hand in basketball: on the misperception of random sequences," *Cognitive Psychology*, 17 (1985), pp. 295–314.
4. From the Web site of the Gallup Organization, www.gallup.com. Individual poll reports remain on this site for only a limited time.
5. Information for the 1999–2000 academic year is from the 2003 *Statistical Abstract of the United States*, Table 286.
6. Data for 2006 from the Web site of Statistics Canada, www.statcan.ca.
7. Gail Burrill, "Two-way tables: introducing probability using real data," paper presented at the Mathematics Education into the Twenty-first Century Project, Czech Republic, September 2003. Burrill cites as her source H. Kranendonk, P. Hopfensperger, and R. Scheaffer, *Exploring Probability*, Dale Seymour Publications, 1999.
8. From the EESEE story "Is It Tough to Crawl in March?"
9. Pierre J. Meunier et al., "The effects of strontium ranelate on the risk of vertebral fracture in women with postmenopausal osteoporosis," *New England Journal of Medicine*, 350 (2004), pp. 459–468.
10. The table closely follows the grade distributions for these three schools at the University of New Hampshire in the fall of 2000, found in a self-study document at www.unh.edu/academic-affairs/neasc/. The counts of grades mirror the proportions of UNH students in these schools. The table is simplified to refer to a university with only these three schools.
11. Information about Internet users comes from sample surveys carried out by the Pew Internet and American Life Project, at www.pewinternet.org.
12. Thanks to Michael Legacy for providing these data.
13. This is one of several tests discussed in Bernard M. Branson, "Rapid HIV testing: 2005 update," a presentation by the Centers for Disease Control and Prevention, at www.cdc.gov. The Malawi clinic result is reported by Bernard M. Branson, "Point-of-care rapid tests for HIV antibody," *Journal of Laboratory Medicine*, 27 (2003), pp. 288–295.
14. The National Longitudinal Study of Adolescent Health interviewed a stratified random sample of 27,000 adolescents, then reinterviewed many of the subjects six years later, when most were aged 19 to 25. These data are from the Wave III reinterviews in 2000 and 2001, found at the Web site of the Carolina Population Center, www.cpc.unc.edu.
15. Information about Internet users comes from sample surveys carried out by the Pew Internet and American Life Project, found online at www.pewinternet.org. The music-downloading data were collected in 2003.
16. These probabilities came from studies by the sociologist Harry Edwards, reported in the *New York Times*, February 25, 1986.
17. We got these data from the Energy Information Administration on their Web site at http://tonto.eia.doe.gov/dnav/pet/pet_sum_mkt_dcu_sct_m.htm.
18. F. J. G. M. Klaassen and J. R. Magnus, "How to reduce the service dominance in tennis? Empirical results from four years at Wimbledon," in S. J. Haake and A. O. Coe (eds.), *Tennis Science and Technology*, Blackwell, 2000, pp. 277–284.
19. Robert P. Dellavalle et al., "Going, going, gone: lost Internet references," *Science*, 302 (2003), pp. 787–788.
20. From the National Institutes of Health's National Digestive Diseases Information Clearinghouse, found at wrongdiagnosis.com.
21. The probabilities given are realistic, according to the fund-raising firm SCM Associates, at scmassoc.com.
22. Probabilities from trials with 2897 people known to be free of HIV antibodies and 673 people known to be infected are reported in J. Richard George, "Alternative specimen sources: methods for confirming positives," 1998 Conference on the Laboratory Science of HIV, found online at the Centers for Disease Control and Prevention, www.cdc.gov.
23. Margaret A. McDowell et al., "Anthropometric reference data for children and adults: U.S. population, 1999–2002," *National Center for Health Statistics, Advance Data from Vital and Health Statistics*, No. 361 (2005), at www.cdc.gov/nchs.
24. The General Social Survey exercises in this chapter present tables constructed using the search function at the GSS archive, sda.berkeley.edu/archive.htm.
25. National population estimates for July 1, 2006, at the Census Bureau Web site www.census.gov. The table omits people who consider themselves to belong to more than one race.
26. Data provided by Patricia Heithaus and the Department of Biology at Kenyon College.
27. Thanks to Tim Brown, The Lawrenceville School, for providing the idea for this exercise.

CHAPTER 6

1. In most applications, X takes a finite number of possible values. The same ideas, implemented with more advanced mathematics, apply to random variables with an infinite but still countable collection of values.
2. The Apgar score data came from National Center for Health Statistics, *Monthly Vital Statistics Reports*, Vol. 30, No. 1, Supplement, May 6, 1981.
3. Information from www.records.ncsu.edu/cgi-bin/grddist.pl.

4. The mean of a continuous random variable X with density function $f(x)$ can be found by integration:

$$\mu_X = \int x f(x)\,dx$$

This integral is a kind of weighted average, analogous to the discrete-case mean

$$\mu_X = \sum x_i p_i$$

The variance of a continuous random variable X is the average squared deviation of the values of X from their mean, found by the integral

$$\sigma_X^2 = \int (x - \mu)^2 f(x)\,dx$$

5. You can find a mathematical explanation of Benford's law in Ted Hill, "The first-digit phenomenon," *American Scientist*, 86 (1996), pp. 358–363; and Ted Hill, "The difficulty of faking data," *Chance*, 12, No. 3 (1999), pp. 27–31. Applications in fraud detection are discussed in the second paper by Hill and in Mark A. Nigrini, "I've got your number," *Journal of Accountancy*, May 1999, available online at www.aicpa.org/pubs/jofa/joaiss.htm.

6. The National Longitudinal Study of Adolescent Health interviewed a stratified random sample of 27,000 adolescents, then reinterviewed many of the subjects six years later, when most were aged 19 to 25. These data are from the Wave III reinterviews in 2000 and 2001, found at the Web site of the Carolina Population Center, www.cpc.unc.edu.

7. Data from the Census Bureau's 1998 American Housing Survey.

8. Thomas K. Cureton et al., *Endurance of Young Men*, Monographs of the Society for Research in Child Development, Vol. 10, No. 1, 1945.

9. The survey question is reported in Trish Hall, "Shop? Many say 'Only if I must,'" *New York Times*, November 28, 1990. In fact, 66% (1650 of 2500) in the sample said "Agree."

10. Office of Technology Assessment, *Scientific Validity of Polygraph Testing: A Research Review and Evaluation*, Government Printing Office, 1983.

11. John Schwartz, "Leisure pursuits of today's young men," *New York Times*, March 29, 2004. The source cited is comScore Media Matrix.

12. Population base from the 2006 Labor Force Survey, at www.statistics.gov.uk. Smoking data from Action on Smoking and Health, *Smoking and Health Inequality*, at www.ash.org.uk.

CHAPTER 7

1. Data on income and education from the March 2008 CPS supplement, downloaded from the Census Bureau Web site using the government's FERRET system. These are raw survey data, including many people with no income or negative income. To simplify comparisons, we report only four education classes.

2. This and similar results of Gallup polls are from the Gallup Organization Web site, www.gallup.com.

3. From a graph in Stan Boutin et al., "Anticipatory reproduction and population growth in seed predators," *Science*, 314 (2006), pp. 1928–1930.

4. See Note 2.

5. Henry Wechsler et al., *Binge Drinking on America's College Campuses*, Harvard School of Public Health, 2001.

6. Amanda Lenhart and Mary Madden, "Music downloading, file-sharing and copyright," Pew Internet and American Life Project, 2003, at www.pewinternet.org.

7. This Activity is suggested in Richard L. Schaeffer, Ann Watkins, Mrudulla Gnanadeskian, and Jeffrey A. Witmer, *Activity-Based Statistics*, Springer, 1996.

8. See Note 1.

9. We found the information on birth weights of Norwegian children on the National Institute of Environmental Health Sciences Web site: http://eb.niehs.nih.gov/bwt/subcfreq.htm.

CHAPTER 8

1. Thanks to Floyd Bullard for sharing the idea for this Activity.

2. This and similar results from the Pew Internet and American Life Project, 2009, can be found at www.pewinternet.org.

3. Michele L. Head, "Examining college students' ethical values," Consumer Science and Retailing honors project, Purdue University, 2003.

4. Francisco L. Rivera-Batiz, "Quantitative literacy and the likelihood of employment among young adults," *Journal of Human Resources*, 27 (1992), pp. 313–328.

5. This and similar results of Gallup polls are from the Gallup Organization Web site, www.gallup.com.

6. Amanda Lenhart and Mary Madden, "Teens, privacy and online social networks," Pew Internet and American Life Project, 2007, at www.pewinternet.org.

7. E. W. Campion, "Editorial: power lines, cancer, and fear," *New England Journal of Medicine*, 337, No. 1 (1997), pp. 44–46. The study report is M. S. Linet et al., "Residential exposure to magnetic fields and acute lymphoblastic leukemia in children," pp. 1–8 in the same issue. See also G. Taubes, "Magnetic field–cancer link: will it rest in peace?" *Science*, 277 (1997), p. 29.

8. Karl Pearson and A. Lee, "On the laws of inheritance in man," *Biometrika*, 2 (1902), p. 357. These data also appear in D. J. Hand et al., *A Handbook of Small Data Sets*, Chapman & Hall, 1994. This book offers more than 500 data sets that can be used in statistical exercises.

9. Data from the College Alcohol Study Web site, www.hsph.harvard.edu/cas/.

10. Linda Lyons, "Teens: sex can wait," December 14, 2004, from the Gallup Poll Web site, www.gallup.com.

11. Eric Sanford et al., "Local selection and latitudinal variation in a marine predator–prey interaction," *Science*, 300 (2003), pp. 1135–1137.

12. See Note 3.

13. See Note 6.

14. Wayne J. Camera and Donald Powers, "Coaching and the SAT I," *TIP*, July 1999 (online journal at www.siop.org/tip).

15. "Poll: men, women at odds on sexual equality," Associated Press dispatch appearing in the *Lafayette (Ind.) Journal and Courier*, October 20, 1997.

16. Linda Lyons, "Most Teens have in-room entertainment," February 22, 2005, www.gallup.com/poll/14989/Most-Teens-InRoom-Entertainment.

17. Based on information in "NCAA 2003 national study of collegiate sports wagering and associated health risks," which can be found on the NCAA Web site, www.ncaa.org.

18. See Note 1.

19. T. J. Lorenzen, "Determining statistical characteristics of a vehicle emissions audit procedure," *Technometrics*, 22 (1980), pp. 483–493.

20. Data provided by Drina Iglesia, Purdue University. The data are part of a larger study reported in D. D. S. Iglesia, E. J. Cragoe, Jr., and J. W. Vanable, "Electric field strength and epithelization in the newt (*Notophthalmus viridescens*)," *Journal of Experimental Zoology*, 274 (1996), pp. 56–62.

21. For a qualitative discussion explaining why skewness is the most serious violation of the Normal shape condition, see Dennis D. Boos and Jacqueline M. Hughes-Oliver, "How large does n have to be for the Z and t intervals?" *American Statistician*, 54 (2000), pp. 121–128. Our recommendations are based on extensive computer work. See, for example, Harry O. Posten, "The robustness of the one-sample t-test over the Pearson system," *Journal of Statistical Computation and Simulation*, 9 (1979), pp. 133–149; and E. S. Pearson and N. W. Please, "Relation between the shape of population distribution and the robustness of four simple test statistics," *Biometrika*, 62 (1975), pp. 223–241.

22. Lisa M. Baril et al., "Willow-bird relationships on Yellowstone's northern range," *Yellowstone Science*, 17, No. 3 (2009), pp. 19–26.

23. Helen E. Staal and D. C. Donderi, "The effect of sound on visual apparent movement," *American Journal of Psychology*, 96 (1983), pp. 95–105.

24. The vehicle is a 1997 Pontiac Transport Van owned by George P. McCabe.

25. These data are from "Results report on the vitamin C pilot program," prepared by SUSTAIN (Sharing United States Technology to Aid in the Improvement of Nutrition) for the U.S. Agency for International Development. The report was used by the Committee on International Nutrition of the National Academy of Sciences/Institute of Medicine (NAS/IOM) to make recommendations on whether or not the vitamin C content of food commodities used in U.S. food aid programs should be increased. The program was directed by Peter Ranum and Françoise Chomé.

26. M. Ann Laskey et al., "Bone changes after 3 mo of lactation: influence of calcium intake, breast-milk output, and vitamin D–receptor genotype," *American Journal of Clinical Nutrition*, 67 (1998), pp. 685–692.

27. Alan S. Banks et al., "Juvenile hallux abducto valgus association with metatarsus adductus," *Journal of the American Podiatric Medical Association*, 84 (1994), pp. 219–224.

28. R. D. Stichler, G. G. Richey, and J. Mandel, "Measurement of treadware of commercial tires," *Rubber Age*, 73, No. 2 (May 1953).

29. Data from Pennsylvania State University Stat 500 Applied Statistics online course, www.stat.psu.edu/online.

30. Harry B. Meyers, "Investigations of the life history of the velvetleaf seed beetle, *Althaeus folkertsi* Kingsolver," MS thesis, Purdue University, 1996. The 95% t interval is 1227.9 to 2507.6. A 95% bootstrap BCa interval is 1444 to 2718, confirming that t inference is inaccurate for these data.

31. Based on interviews in 2000 and 2001 by the National Longitudinal Study of Adolescent Health. Found at the Web site of the Carolina Population Center, www.cpc.unc.edu.

32. Simplified from Sanjay K. Dhar, Claudia Gonzalez-Vallejo, and Dilip Soman, "Modeling the effects of advertised price claims: tensile versus precise pricing," *Marketing Science*, 18 (1999), pp. 154–177.

33. From program 19, "Confidence Intervals," in the *Against All Odds* video series.

34. James T. Fleming, "The measurement of children's perception of difficulty in reading materials," *Research in the Teaching of English*, 1 (1967), pp. 136–156.

35. Bryan E. Porter and Thomas D. Berry, "A nationwide survey of self-reported red light running: measuring prevalence, predictors, and perceived consequences," *Accident Analysis and Prevention*, 33 (2001), pp. 735–741.

CHAPTER 9

1. P. A. Mackowiak, S. S. Wasserman, and M. M. Levine, "A critical appraisal of 98.6 degrees F, the upper limit of the normal body temperature, and other legacies of Carl Reinhold August Wunderlich," *Journal of the American Medical Association*, 268 (1992), pp. 1578–1580.

2. Based on G. Salvendy et al., "Impact of personality and intelligence on job satisfaction of assembly line and bench work—an industrial study," *Applied Ergonomics*, 13 (1982), pp. 293–299.

3. Thanks to Josh Tabor for suggesting the idea for this example.

4. R. A. Fisher, "The arrangement of field experiments," *Journal of the Ministry of Agriculture of Great Britain*, 33 (1926), p. 504, quoted in Leonard J. Savage, "On rereading R. A. Fisher," *Annals of Statistics*, 4 (1976), p. 471. Fisher's work is described in a biography by his daughter: Joan Fisher Box, *R. A. Fisher: The Life of a Scientist*, Wiley, 1978.

5. Thanks to Roxy Peck for sharing the idea for this Activity.

6. For a discussion of statistical significance in the legal setting, see D. H. Kaye, "Is proof of statistical significance relevant?" *Washington Law Review*, 61 (1986), pp. 1333–1365. Kaye argues: "Presenting the P-value without characterizing the evidence by a significance test is a step in the right direction. Interval estimation, in turn, is an improvement over P-values."

7. Julie Ray, "Few teens clash with friends," May 3, 2005, on the Gallup Organization Web site, www.galluppoll.com.

8. The idea for this exercise was provided by Michael Legacy and Susan McGann.

9. Projections from the 2002 *Digest of Education Statistics*, found online at nces.ed.gov.

10. From the report "Sex and tech: results from a study of teens and young adults," published by the National Campaign to Prevent Teen and Unplanned Pregnancy, www.thenationalcampaign.org/sextech.

11. National Institute for Occupational Safety and Health, *Stress at Work*, 2000, available online at www.cdc.gov/niosh/stresswk.html. Results of this survey were reported in *Restaurant Business*, September 15, 1999, pp. 45–49.

12. See Note 7.

13. Matthew A. Carlton and William D. Stansfield, "Making babies by the flip of a coin?" *The American Statistician*, 59 (2005), pp. 180–182.

14. Dorothy Espelage et al., "Factors associated with bullying behavior in middle school students," *Journal of Early Adolescence*, 19, No. 3 (August 1999), pp. 341–362.

15. Amanda Lenhart et al., "Social media and mobile Internet use among teens and young adults," published by the Pew Internet and American Life Project, February 3, 2010.

16. This and similar results of Gallup polls are from the Gallup Organization Web site, www.gallup.com.

17. Linda Lyons, "Teens: sex can wait," December 14, 2004, from the Gallup Organization Web site, www.galluppoll.com.
18. Michele L. Head, "Examining college students' ethical values," Consumer Science and Retailing honors project, Purdue University, 2003.
19. Based on Stephen A. Woodbury and Robert G. Spiegelman, "Bonuses to workers and employers to reduce unemployment: randomized trials in Illinois," *American Economic Review*, 77 (1987), pp. 513–530.
20. E. C. Strain et al., "Caffeine dependence syndrome: evidence from case histories and experimental evaluation," *Journal of the American Medical Association*, 272 (1994), pp. 1604–1607.
21. We obtained the tire pressure loss data from the Consumer Reports Web site: http://blogs.consumerreports.org/cars/2007/10/tires-nitrogen-.html.
22. The editorial was written by Phil Anderson. See *British Medical Journal*, 328 (2004), pp. 476–477. A letter to the editor on this topic by Doug Altman and J. Martin Bland appeared shortly after. See "Confidence intervals illuminate absence of evidence," *British Medical Journal*, 328 (2004), pp. 1016–1017.
23. A. Kamali et al., "Syndromic management of sexually transmitted infections and behavior change interventions on transmission of HIV-1 in rural Uganda: a community randomised trial," *Lancet*, 361 (2003), pp. 645–652.
24. T. D. Sterling, "Publication decisions and their possible effects on inferences drawn from tests of significance—or vice versa," *Journal of the American Statistical Association*, 54 (1959), pp. 30–34. Related comments appear in J. K. Skipper, A. L. Guenther, and G. Nass, "The sacredness of 0.05: a note concerning the uses of statistical levels of significance in social science," *American Sociologist*, 1 (1967), pp. 16–18.
25. Warren E. Leary, "Cell phones: questions but no answers," *New York Times*, October 26, 1999.
26. This exercise is based on events that are real. The data and details have been altered to protect the privacy of the individuals involved.
27. R. A. Berner and G. P. Landis, "Gas bubbles in fossil amber as possible indicators of the major gas composition of ancient air," *Science*, 239 (1988), pp. 1406–1409. The 95% t confidence interval is 54.78 to 64.40. A bootstrap BCa interval is 55.03 to 62.63. So t is reasonably accurate despite the skew and the small sample.
28. From the NIST Web site, www.nist.gov/srd/online.htm.
29. D. L. Shankland et al., "The effect of 5-thio-D-glucose on insect development and its absorption by insects," *Journal of Insect Physiology*, 14 (1968), pp. 63–72.
30. Data provided by Timothy Sturm.
31. A. R. Hirsch and L. H. Johnston, "Odors and learning," *Journal of Neurological and Orthopedic Medicine and Surgery*, 17 (1996), pp. 119–126.
32. From the story "Friday the 13th," at the Data and Story Library, libstat.cmu.edu/DASL.
33. From the Gallup Organization Web site, www.gallup.com. The poll was taken in December 2006.
34. Data simulated from a Normal distribution based on information in Brian M. DeBroff and Patricia J. Pahk, "The ability of periorbitally applied antiglare products to improve contrast sensitivity in conditions of sunlight exposure," *Archives of Ophthalmology*, 121 (2003), pp. 997–1001.
35. W. S. Gosset, "The probable error of a mean," *Biometrika*, 6 (1908), 1–25.

CHAPTER 10

1. This Case Study is based on the story "Drive-Thru Competition" in the Electronic Encyclopedia of Statistical Examples and Exercises (EESEE). Updated data were obtained from the *QSR* magazine Web site: www.qsrmagazine.com/drive-thru.
2. C. P. Cannon et al., "Intensive versus moderate lipid lowering with statins after acute coronary syndromes," *New England Journal of Medicine*, 350 (2004), pp. 1495–1504.
3. This Activity is based on an experiment conducted on the March 9, 2005, episode of the TV show *MythBusters*. You may be able to find a video clip from this program at the Discovery Channel Web site, www.yourdiscovery.com.
4. The study is reported in William Celis III, "Study suggests Head Start helps beyond school," *New York Times*, April 20, 1993. See www.highscope.org.
5. The idea for this exercise was inspired by an example in David M. Lane's *Hyperstat Online* textbook at http://davidmlane.com/hyperstat.
6. Modified from Richard A. Schieber et al., "Risk factors for injuries from in-line skating and the effectiveness of safety gear," *New England Journal of Medicine*, 335 (1996), Internet summary at content.nejm.org.
7. Francisco Lloret et al., "Fire and resprouting in Mediterranean ecosystems: insights from an external biogeographical region, the Mexican shrubland," *American Journal of Botany*, 88 (1999), pp. 1655–1661.
8. Saiyad S. Ahmed, "Effects of microwave drying on checking and mechanical strength of low-moisture baked products," MS thesis, Purdue University, 1994.
9. Eulynn Shiu and Amanda Lenhart, "How Americans use instant messaging," Pew Internet and American Life Project, 2004, at www.pewinternet.org.
10. From the Web site of the Black Youth Project, blackyouthproject.uchicago.edu.
11. The National Longitudinal Study of Adolescent Health interviewed a stratified random sample of 27,000 adolescents, then reinterviewed many of the subjects six years later, when most were aged 19 to 25. These data are from the Wave III reinterviews in 2000 and 2001, found at the Web site of the Carolina Population Center, www.cpc.unc.edu.
12. Janice Joseph, "Fear of crime among black elderly," *Journal of Black Studies*, 27 (1997), pp. 698–717.
13. National Athletic Trainers Association, press release dated, September 30, 1994.
14. Louie E. Ross, "Mate selection preferences among African American college students," *Journal of Black Studies*, 27 (1997), pp. 554–569.
15. Ross L. Prentice et al., "Low-fat dietary pattern and risk of invasive breast cancer," *Journal of the American Medical Association*, 295 (2006), pp. 629–642.
16. Martin Enserink, "Fraud and ethics charges hit stroke drug trial," *Science*, 274 (1996), pp. 2004–2005.
17. Kwang Y. Cha, Daniel P. Wirth, and Rogerio A. Lobo, "Does prayer influence the success of *in vitro* fertilization–embryo transfer?" *Journal of Reproductive Medicine*, 46 (2001), pp. 781–787.
18. W. E. Paulus et al., "Influence of acupuncture on the pregnancy rate in patients who undergo assisted reproductive therapy," *Fertility and Sterility*, 77, No. 4 (2002), pp. 721–724.

19. Based on Deborah Roedder John and Ramnath Lakshmi-Ratan, "Age differences in children's choice behavior: the impact of available alternatives," *Journal of Marketing Research,* 29 (1992), pp. 216–226.
20. These are some of the data from the EESEE story "Radar Detectors and Speeding." The study is reported in N. Teed, K. L. Adrian, and R. Knoblouch, "The duration of speed reductions attributable to radar detectors," *Accident Analysis and Prevention,* 25 (1991), pp. 131–137.
21. Sapna Aneja, "Biodeterioration of textile fibers in soil," MS thesis, Purdue University, 1994.
22. We obtained the National Health and Nutrition Examination Survey data from the Centers for Disease Control and Prevention Web site at www.cdc.gov/nchs/nhanes.htm.
23. Detailed information about the conservative *t* procedures can be found in Paul Leaverton and John J. Birch, "Small sample power curves for the two sample location problem," *Technometrics,* 11 (1969), pp. 299–307; Henry Scheffe, "Practical solutions of the Behrens-Fisher problem," *Journal of the American Statistical Association,* 65 (1970), pp. 1501–1508; and D. J. Best and J. C. W. Rayner, "Welch's approximate solution for the Behrens-Fisher problem," *Technometrics,* 29 (1987), pp. 205–210.
24. Based on R. G. Hood, "Results of the prices received by farmers for grain quality assurance project," U.S. Department of Agriculture Report SRB-95-07, 1995.
25. This study is reported in Roseann M. Lyle et al., "Blood pressure and metabolic effects of calcium supplementation in normotensive white and black men," *Journal of the American Medical Association,* 257 (1987), pp. 1772–1776. The data were provided by Dr. Lyle.
26. See Note 21.
27. See the extensive simulation studies in Harry O. Posten, "The robustness of the two-sample *t*-test over the Pearson system," *Journal of Statistical Computation and Simulation,* 6 (1978), pp. 295–311; and in Harry O. Posten, H. Yeh, and Donald B. Owen, "Robustness of the two-sample *t*-test under violations of the homogeneity assumption," *Communications in Statistics,* 11 (1982), pp. 109–126.
28. See Note 1 in Chapter 4.
29. The idea for this example was provided by Robert Hayden.
30. United Nations data on literacy were found at www.earthtrends.wri.org.
31. Shailija V. Nigdikar et al., "Consumption of red wine polyphenols reduces the susceptibility of low-density lipoproteins to oxidation in vivo," *American Journal of Clinical Nutrition,* 68 (1998), pp. 258–265.
32. Ethan J. Temeles and W. John Kress, "Adaptation in a plant–hummingbird association," *Science,* 300 (2003), pp. 630–633. We thank Ethan J. Temeles for providing the data.
33. Data for 1982, provided by Marvin Schlatter, Division of Financial Aid, Purdue University.
34. Gabriela S. Castellani, "The effect of cultural values on Hispanics' expectations about service quality," MS thesis, Purdue University, 2000.
35. From a graph in Fabrizio Grieco, Arie J. van Noordwijk, and Marcel E. Visser, "Evidence for the effect of learning on timing of reproduction in blue tits," *Science,* 296 (2002), pp. 136–138.
36. Adapted from Maribeth Cassidy Schmitt, "The effects of an elaborated directed reading activity on the metacomprehension skills of third graders," PhD dissertation, Purdue University, 1987.
37. M. Ann Laskey et al., "Bone changes after 3 mo of lactation: influence of calcium intake, breast-milk output, and vitamin D–receptor genotype," *American Journal of Clinical Nutrition,* 67 (1998), pp. 685–692.
38. This example is loosely based on D. L. Shankland, "Involvement of spinal cord and peripheral nerves in DDT poisoning syndrome in albino rats," *Toxicology and Applied Pharmacology,* 6 (1964), pp. 197–213.
39. Based on W. N. Nelson and C. J. Widule, "Kinematic analysis and efficiency estimate of intercollegiate female rowers," unpublished manuscript, 1983.
40. See Note 24 in Chapter 1.
41. The data for this exercise came from Rossman, Cobb, Chance, and Holcomb's National Science Foundation project shared at JMM 2008 in San Diego. Their original source was Robert Stickgold, LaTanya James, and J. Allan Hobson, "Visual discrimination learning requires sleep after training," *Nature Neuroscience,* 3 (2000), pp. 1237–1238.
42. The idea for this exercise was provided by Robert Hayden.
43. See Note 42.
44. Wayne J. Camera and Donald Powers, "Coaching and the SAT I," *TIP* (online journal at www.siop.org/tip), July 1999.
45. Pew Internet Project, "Counting on the Internet," December 29, 2002, www.pewinternet.org.
46. Thanks to Larry Green, Lake Tahoe Community College, for allowing us to use some of the contexts from his "Classifying Statistics Problems" Web site, www.ltcconline.net/greenl/java/Statistics/StatsMatch/StatsMatch.htm.
47. JoAnn K. Wells, Allan F. Williams, and Charles M. Farmer, "Seat belt use among African Americans, Hispanics, and whites," *Accident Analysis and Prevention,* 34 (2002), pp. 523–529.
48. Based on Amna Kirmani and Peter Wright, "Money talks: perceived advertising expense and expected product quality," *Journal of Consumer Research,* 16 (1989), pp. 344–353.
49. Francisco L. Rivera-Batiz, "Quantitative literacy and the likelihood of employment among young adults," *Journal of Human Resources,* 27 (1992), pp. 313–328.
50. Clive G. Jones et al., "Chain reactions linking acorns to gypsy moth outbreaks and Lyme disease risk," *Science,* 279 (1998), pp. 1023–1026.
51. Data provided by Warren Page, New York City Technical College, from a study done by John Hudesman.
52. Based on Robert T. Driescher, "A quality swing with Ping," *Quality Progress,* August 2001, pp. 37–41.
53. These are some of the data from the EESEE story "Leave Survey after the Beep." The study is reported in M. Xu, B. J. Bates, and J. C. Schweitzer, "The impact of messages on survey participation in answering machine households," *Public Opinion Quarterly,* 57 (1993), pp. 232–237.
54. Data provided by Darlene Gordon, School of Education, Purdue University.
55. Maureen Hack et al., "Outcomes in young adulthood for very-low-birth-weight infants," *New England Journal of Medicine,* 346 (2002), pp. 149–157. Exercise AP3.32 is simplified, in that the measures reported in this paper have been statistically adjusted for "sociodemographic status."

CHAPTER 11

1. M. M. Roy and N. J. S. Christenfeld, "Do dogs resemble their owners?" *Psychological Science*, 15, No. 5 (May 2004), pp. 361–363.
2. R. W. Mannan and E. C. Meslow, "Bird populations and vegetation characteristics in managed and old-growth forests, northwestern Oregon," *Journal of Wildlife Management*, 48 (1984), pp. 1219–1238.
3. You can find a mathematical explanation of Benford's law in Ted Hill, "The first-digit phenomenon," *American Scientist*, 86 (1996), pp. 358–363; and Ted Hill, "The difficulty of faking data," *Chance*, 12, No. 3 (1999), pp. 27–31. Applications in fraud detection are discussed in the second paper by Hill and in Mark A. Nigrini, "I've got your number," *Journal of Accountancy*, May 1999, available online at www.aicpa.org/pubs/jofa/joaiss.htm.
4. The idea for this exercise came from a post to the AP Statistics electronic discussion group by Joshua Zucker.
5. C. M. Ryan et al., "The effect of in-store music on consumer choice of wine," *Proceedings of the Nutrition Society*, 57 (1998), p. 1069A.
6. Pennsylvania State University Division of Student Affairs, "Net behaviors, November 2006,"*Penn State Pulse*, at www.sa.psu.edu.
7. Pew Research Center for the People and the Press, "The cell phone challenge to survey research," news release for May 15, 2006, at www.people-press.org.
8. D. M. Barnes, "Breaking the cycle of addiction," *Science*, 241 (1988), pp. 1029–1030.
9. William D. Darley, "Store-choice behavior for preowned merchandise," *Journal of Business Research*, 27 (1993), pp. 17–31.
10. See Note 8.
11. Janice E. Williams et al., "Anger proneness predicts coronary heart disease risk," *Circulation*, 101 (2000), pp. 63–95.
12. P. Azoulay and S. Shane, "Entrepreneurs, contracts, and the failure of young firms," *Management Science*, 47 (2001), pp. 337–358.
13. J. Cantor, "Long-term memories of frightening media often include lingering trauma symptoms," poster paper presented at the Association for Psychological Science Convention, New York, May 26, 2006.
14. The National Longitudinal Study of Adolescent Health interviewed a stratified random sample of 27,000 adolescents, then reinterviewed many of the subjects six years later, when most were aged 19 to 25. These data are from the Wave III reinterviews in 2000 and 2001, found at www.cpc.unc.edu.
15. Joan L. Duda, "The relationship between goal perspectives, persistence, and behavioral intensity among male and female recreational sports participants," *Leisure Sciences*, 10 (1988), pp. 95–106.
16. Data compiled from a table of percents in "Americans view higher education as key to the American dream," press release by the National Center for Public Policy and Higher Education, www.highereducation.org, May 3, 2000.
17. R. Shine et al., "The influence of nest temperatures and maternal brooding on hatchling phenotypes in water pythons," *Ecology*, 78 (1997), pp. 1713–1721.
18. Virgilio P. Carnielli et al., "Intestinal absorption of long chain polyunsaturated fatty acids in preterm infants fed breast milk or formula," *American Journal of Clinical Nutrition*, 67 (1998), pp. 97–103.
19. U.S. Department of Commerce, Office of Travel and Tourism Industries, in-flight survey, 2007, at http://tinet.ita.doc.gov.
20. Simplified from D. A. Marcus et al., "A double-blind provocative study of chocolate as a trigger of headache," *Cephalalgia*, 17 (1997), pp. 855–862.
21. Douglas E. Jorenby et al., "A controlled trial of sustained-release bupropion, a nicotine patch, or both for smoking cessation," *New England Journal of Medicine*, 340 (1990), pp. 685–691.
22. Martin Enserink, "Fraud and ethics charges hit stroke drug trial," *Science*, 274 (1996), pp. 2004–2005.
23. Lillian Lin Miao, "Gastric freezing: an example of the evaluation of medical therapy by randomized clinical trials," in John P. Bunker, Benjamin A. Barnes, and Frederick Mosteller (eds.), *Costs, Risks, and Benefits of Surgery*, Oxford University Press, 1977, pp. 198–211.
24. Lien-Ti Bei, "Consumers' purchase behavior toward recycled products: an acquisition-transaction utility theory perspective," MS thesis, Purdue University, 1993.
25. All General Social Survey exercises in this chapter present tables constructed using the search function at the GSS archive, http://sda.berkeley.edu/archive.htm.
26. Richard M. Felder et al., "Who gets it and who doesn't: a study of student performance in an introductory chemical engineering course," in American Society for Engineering Education, *1992 ASEE Annual Conference Proceedings*, 1992, pp. 1516–1519.
27. The idea for this exercise was provided by Robert Hayden.
28. Based closely on Susan B. Sorenson, "Regulating firearms as a consumer product," *Science*, 286 (1999), pp. 1481–1482. Because the results in the paper were "weighted to the U.S. population," we have changed some counts slightly for consistency.
29. Data produced by Ries and Smith, found in William D. Johnson and Gary G. Koch, "A note on the weighted least squares analysis of the Ries-Smith contingency table data," *Technometrics*, 13 (1971), pp. 438–447.
30. See Note 14.
31. Mei-Hui Chen, "An exploratory comparison of American and Asian consumers' catalog patronage behavior," MS thesis, Purdue University, 1994.
32. See Note 14.
33. Thanks to Larry Green, Lake Tahoe Community College, for giving us permission to use several of the contexts from his Web site at www.ltcconline.net/greenl/java/Statistics/StatsMatch/StatsMatch.htm.
34. See Note 33.
35. Mark A. Sabbagh and Dare A. Baldwin, "Learning words from knowledgeable versus ignorant speakers: links between preschoolers' theory of mind and semantic development," *Child Development*, 72 (2001), pp. 1054–1070. Many statistical software packages offer "exact tests" that are valid even when there are small expected counts.
36. Brenda C. Coleman, "Study: heart attack risk cut 74% by stress management," Associated Press dispatch appearing in the *Lafayette (Ind.) Journal and Courier*, October 20, 1997.
37. Tom Reichert, "The prevalence of sexual imagery in ads targeted to young adults," *Journal of Consumer Affairs*, 37 (2003), pp. 403–412.
38. Based on the EESEE story "What Makes Pre-teens Popular?
39. George R. Milne, "How well do consumers protect themselves from identity theft?" *Journal of Consumer Affairs*, 37 (2003), pp. 388–402.

40. Based on the EESEE story "Domestic Violence."

41. Karine Marangon et al., "Diet, antioxidant status, and smoking habits in French men," *American Journal of Clinical Nutrition*, 67 (1998), pp. 231–239.

CHAPTER 12

1. Data obtained from the PGA Tour Web site, www.pgatour.com, by Kevin Stevick, The Lawrenceville School.

2. The idea for this Activity came from Gloria Barrett, Floyd Bullard, and Dan Teague at the North Carolina School of Science and Math.

3. Samuel Karelitz et al., "Relation of crying activity in early infancy to speech and intellectual development at age three years," *Child Development*, 35 (1964), pp. 769–777.

4. Data provided by Drina Iglesia, Purdue University. The data are part of a larger study reported in D. D. S. Iglesia, E. J. Cragoe, Jr., and J. W. Vanable, "Electrical field strength and epithelization in the newt (*Notophthalmus viridescens*)," *Journal of Experimental Zoology*, 274 (1996), pp. 56–62.

5. Data from National Institute of Standards and Technology, *Engineering Statistics Handbook*, www.itl.nist.gov/div898/handbook. The analysis there does not comment on the bias of field measurements.

6. Todd W. Anderson, "Predator responses, prey refuges, and density-dependent mortality of a marine fish," *Ecology*, 81 (2001), pp. 245–257.

7. Based on a plot in G. D. Martinsen, E. M. Driebe, and T. G. Whitham, "Indirect interactions mediated by changing plant chemistry: beaver browsing benefits beetles," *Ecology*, 79 (1998), pp. 192–200.

8. Data provided by Samuel Phillips, Purdue University.

9. Based on Marion E. Dunshee, "A study of factors affecting the amount and kind of food eaten by nursery school children," *Child Development*, 2 (1931), pp. 163–183. This article gives the means, standard deviations, and correlation for 37 children but does not give the actual data.

10. See Note 28 in Chapter 8.

11. M. H. Criqui, University of California, San Diego, reported in the *New York Times*, December 28, 1994.

12. Data for the building at 1800 Ben Franklin Drive, Sarasota, Florida, starting in March 2003. From the Web site of the Sarasota County Property Appraiser, www.sarasotaproperty.net.

13. Information about the sources used to obtain the data can be found under "Documentation" at the Gapminder Web site, www.gapminder.org.

14. Gordon L. Swartzman and Stephen P. Kaluzny, *Ecological Simulation Primer*, Macmillan, 1987, p. 98.

15. From the Intel Web site, www.intel.com/technology/silicon/index.htm.

16. Planetary data from http://hyperphysics.phy-astr.gsu.edu/hbase/solar/soldata2.html.

17. Sample Pennsylvania female rates provided by Life Quotes, Inc., in *USA Today*, December 20, 2004.

18. S. Chatterjee and B. Price, *Regression Analysis by Example*, Wiley, 1977.

19. G. A. Sacher and E. F. Staffelt, "Relation of gestation time to brain weight for placental mammals: implications for the theory of vertebrate growth," *American Naturalist*, 108 (1974), pp. 593–613. We found these data in F. L. Ramsey and D. W. Schafer, *The Statistical Sleuth: A Course in Methods of Data Analysis*, Duxbury, 1997.

20. Jérôme Chave, Bernard Riéra, and Marc-A. Dubois, "Estimation of biomass in a neotropical forest of French Guiana: spatial and temporal variability," *Journal of Tropical Ecology*, 17 (2001), pp. 79–96.

21. Chris Carbone and John L. Gittleman, "A common rule for the scaling of carnivore density," *Science*, 295 (2002), pp. 2273–2276.

22. Stillman Drake, *Galileo at Work*, University of Chicago Press, 1978. We found these data in D. A. Dickey and J. T. Arnold, "Teaching statistics with data of historic significance," *Journal of Statistics Education*, 3 (1995), www.amstat.org/publications/jse/.

23. Data originally from A. J. Clark, *Comparative Physiology of the Heart*, Macmillan, 1927, p. 84. Obtained from Frank R. Giordano and Maurice D. Weir, *A First Course in Mathematical Modeling*, Brooks/Cole, 1985, p. 56.

24. Tattoo survey data from the Harris Poll®, No. 15, www.harrisinteractive.com, February 12, 2008.

25. Peter H. Chen, Neftali Herrera, and Darren Christiansen, "Relationships between gate velocity and casting features among aluminum round castings," no date. Provided by Darren Christiansen.

26. Data provided by Corinne Lim, Purdue University, from a student project supervised by Professor Joseph Vanable.

27. Data from many studies compiled in D. F. Greene and E. A. Johnson, "Estimating the mean annual seed production of trees," *Ecology*, 75 (1994), pp. 642–647.

Solutions

CHAPTER 1
Introduction

Answers to Check Your Understanding

Page 5 **1.** The cars in the student parking lot. **2.** He measured the car's model (categorical), year (quantitative), color (categorical), number of cylinders (quantitative), gas mileage (quantitative), whether it has a navigation system (categorical), and weight (quantitative).

Answers to Odd-Numbered Introduction Exercises

1. Categorical: Type of wood, type of water repellent, paint color. Quantitative: Paint thickness, weathering time.
3. **(a)** AP Statistics students who completed a questionnaire on the first day of class. **(b)** Categorical: gender, handedness, favorite type of music. Quantitative: height (in inches), amount of time the student is expecting to spend on homework (in minutes), the total value of coins in a student's pocket (in cents). **(c)** Female, right-handed, 58 inches tall, spends 60 minutes on homework, prefers Alternative music, had 76 cents in her pocket.
5. Student answers will vary. Quantitative examples: retention rate (percent of students), graduation rate (percent of students). Categorical examples: region of the country and type of institution (two-year college, four-year college, university).
7. b

Section 1.1

Answers to Check Your Understanding

Page 14 **1.** There are 2367 females, 2459 males, and 4826 total in the data set. This means that $\left(\frac{2367}{4826}\right)100\% = 49\%$ are female and 51% are male. **2.** The percents of males and females in the bar graph are very similar.

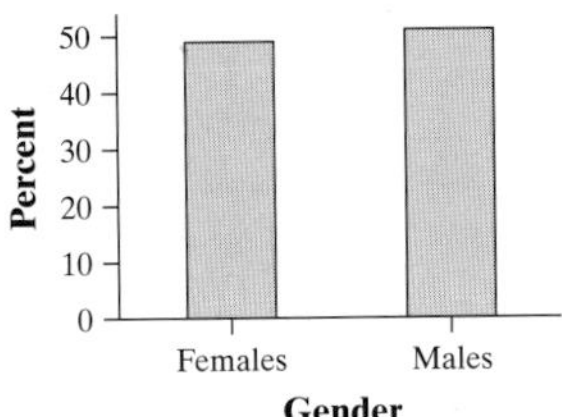

Page 17 **1.** A 50-50 chance: Female $= \left(\frac{696}{1416}\right)100\% = 49.15\%$ and Male $= \left(\frac{720}{1416}\right)100\% = 50.85\%$. The rest are shown in the table at top right.

2.

	Female	Male	All
A 50-50 chance	696 49.15	720 50.85	**1416** **100.00**
A good chance	663 46.66	758 53.34	**1421** **100.00**
Almost certain	486 44.88	597 55.12	**1083** **100.00**
Almost no chance	96 49.48	98 50.52	**194** **100.00**
Some chance but probably not	426 59.83	286 40.17	**712** **100.00**
All	**2367** **49.05**	**2459** **50.95**	4826 100.00

Cell contents: Count
% of row

Answers to Odd-Numbered Section 1.1 Exercises

9. **(a)** 0% **(b)** Here is a bar graph.

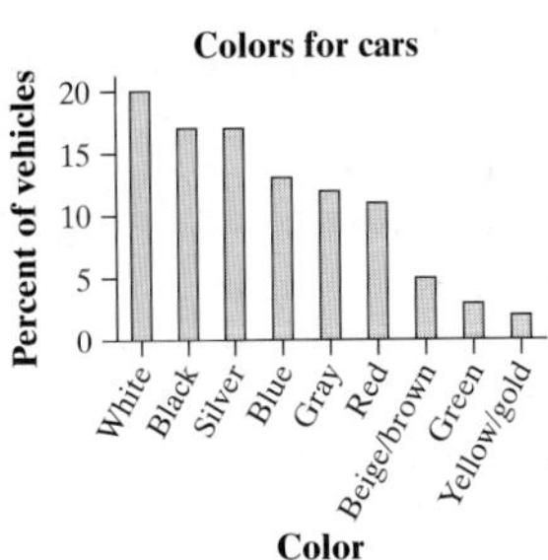

(c) Yes; we have all possible colors listed here.
11. **(a)** A pie chart would also be appropriate since all days are accounted for in the data set.

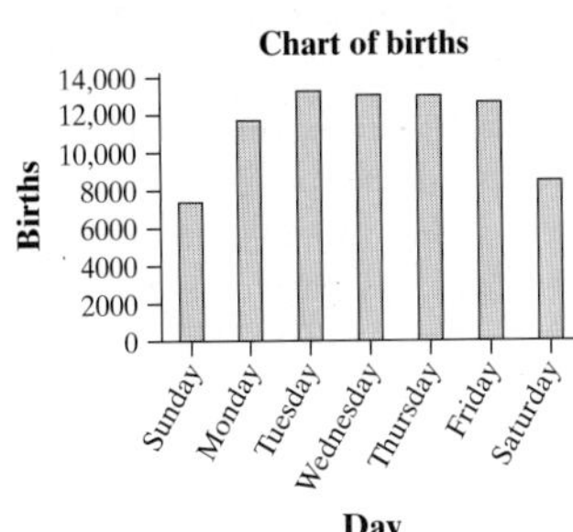

(b) Perhaps induced or C-section births are often scheduled for weekdays.
13. Estimates will vary but should be close: 64% Mexican, 9% Puerto Rican.

15. **(a)** The given percents represent fractions of different age groups rather than parts of a single whole. **(b)** Here is a bar graph.

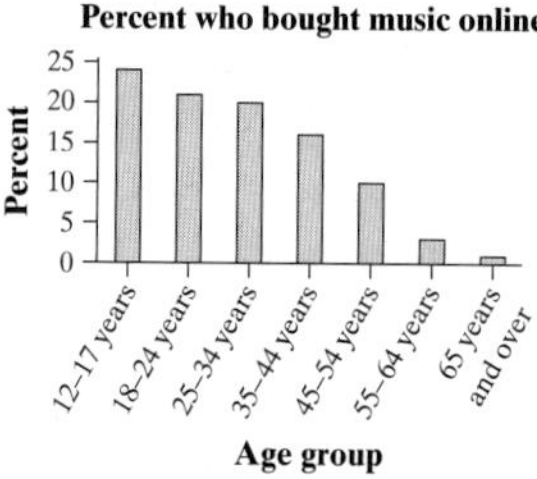

17. **(a)** The pictures should be proportional to the number of students they represent. **(b)** Here is a bar graph.

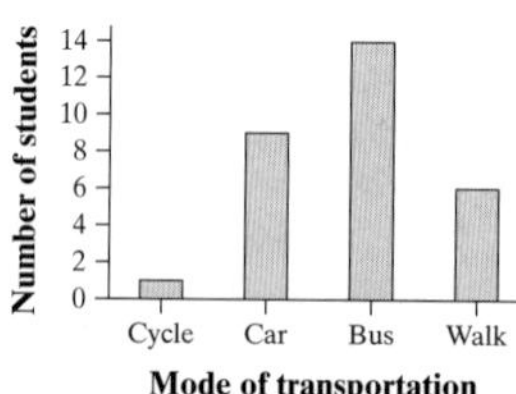

19. **(a)** 133 people; 36 buyers of coffee filters made of recycled paper. **(b)** 36.84% said "higher," 24.06% said "the same," and 39.10% said "lower." 60.9% responded "the same" or "higher."

21. The two conditional distributions of opinion are shown in the table below. Buyers are much more likely to consider recycled filters higher in quality, though 25% still think they are lower in quality.

	Higher	The same	Lower
Buyers	55.56%	19.44%	25.00%
Nonbuyers	29.90%	25.77%	44.33%

23. Americans are much more likely to choose white/pearl, while Europeans are much more likely to choose silver. Also, Europeans are more likely to choose black or gray than Americans, while Americans are more likely to choose red than Europeans.

25. **State:** What is the relationship between whether someone belongs to an environmental organization and their use of a snowmobile when visiting Yellowstone National Park? **Plan:** We suspect that belonging to an environmental group will reduce the chances that someone will use a snowmobile, so we'll compare the conditional distributions of snowmobile use for those who belong to an environmental organization and for those who don't. **Do:** We'll make side-by-side bar graphs to compare the snowmobile use of the two groups of people. **Conclude:** Those who are members of an environmental group are much more likely to have never used a snowmobile (nearly 70% of this group have never used a snowmobile), whereas only about 35% of the other group have never used a snowmobile. Those in an environmental group are less likely to have rented or owned a snowmobile than those not in an environmental group.

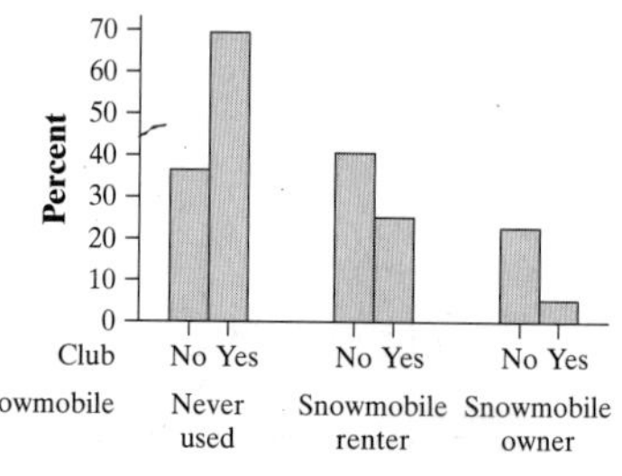

27. d

29. d

31. e

33. Answers will vary. Two possibilities are given below.

10	40
50	0

30	20
30	20

35. **(a)** The two-way table is shown below. **(b)** Overall, 11.88% of white defendants and 10.24% of black defendants receive the death penalty. For white victims, 12.58% of white defendants and 17.46% of black defendants receive the death penalty. For black victims, 0% of white defendants and 5.83% of black defendants receive the death penalty. **(c)** The death penalty is more likely when the victim was white (14.02%) rather than black (5.36%). Because most convicted killers are of the same race as their victims, whites are more often sentenced to death.

	Death penalty	No death penalty
White defendant	19	141
Black defendant	17	149

Section 1.2

Answers to Check Your Understanding

Page 31 **1.** Skewed to the right. **2.** The median is 1.5 and the mean is 1.75. **3.** The number of siblings varies from 0 to 6. **4.** There are two potential outliers: those students reporting 5 and 6 siblings. *Page 34* **1.** In general, it appears that females have more pairs of shoes than males. The median report for the males was 9 pairs while the female median was 26. The females also have a larger range of $57-13=44$ in comparison to the range of males, which is $38-4=34$. Finally, both males and females have distributions that are skewed to the right, though the distribution for the males is more heavily skewed, as evidenced by the three likely outliers at 22, 35, and 38. The females do not have any likely outliers. **2.** b **3.** b **4.** b *Page 39* **1.** Here is a histogram.

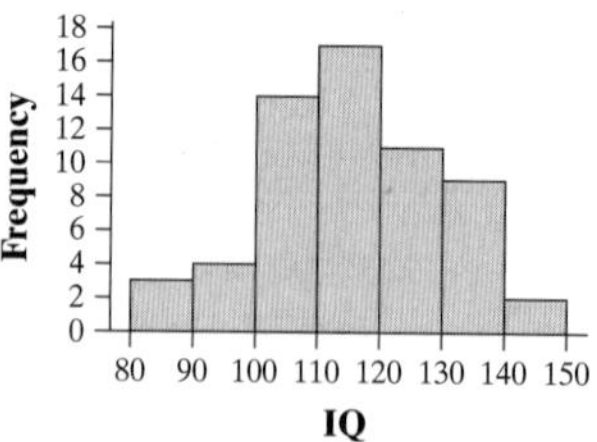

2. The distribution is roughly symmetric and bell-shaped. The median IQ appears to be between 110 and 120, and the IQs vary from 80 to 150. There do not appear to be any outliers. *Page 41* **1.** Here is a histogram.

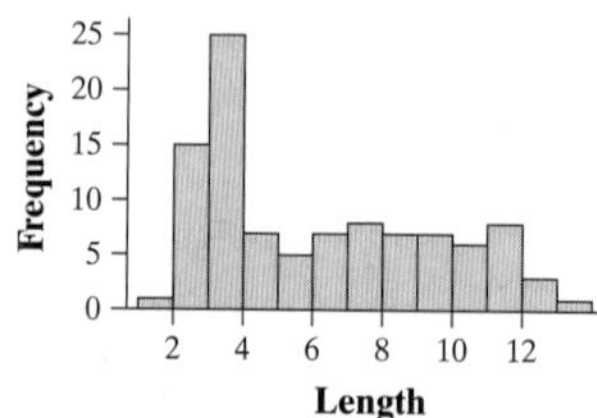

2. Here is a histogram.

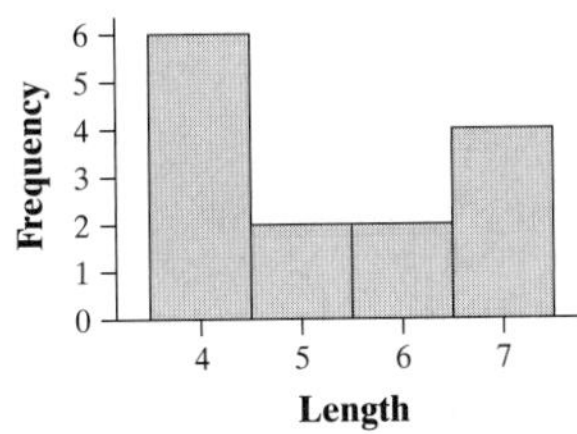

3. This is a bar graph because the horizontal axis divides the observations into categories. 4. No; the x variable is categorical, not quantitative.

Answers to Odd-Numbered Section 1.2 Exercises

37. (a) Here is a dotplot.

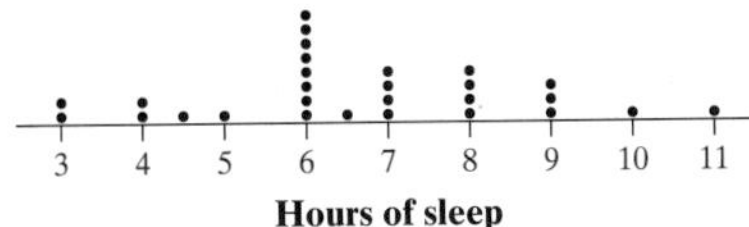

(b) Roughly symmetric with a center of 6 hours and range of $11-3=8$ hours. There do not appear to be any outliers.
39. (a) Games where the opposing team won by two goals. (b) Only 2 of the 34 differentials are negative, which indicates that the U.S. women's soccer team had a very good season. The team scored at least as many goals as their opponents in 32 of 34 games. In one game, they beat the other team by 8 goals, a very unusual event in soccer.
41. (a) Answers will vary.

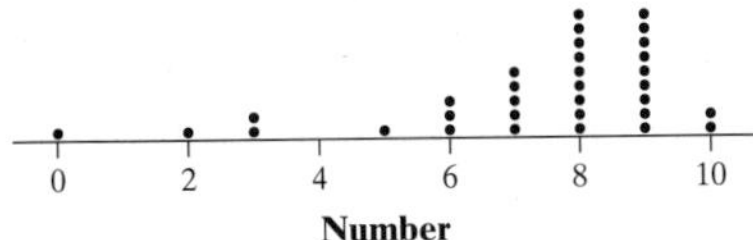

(b) As coins get older, they get taken out of circulation and new coins are introduced. So most coins in someone's pocket will be from recent years, but there may be a few from previous years.
43. The average ratings were higher for the students in the internal reasons group. Both distributions are roughly symmetric, with a range of about 20. But the internal reasons distribution has a center of 21, whereas the external reasons distribution has a center near 17.5.
45. (a) If we had not split the stems, most of the data would appear on just a few stems. (b) 16.0%. The high percent for Utah may be due to the Mormon Church. (c) Roughly symmetric with a center of 13% and a spread of roughly 3.5%.
47. (a) The stemplots are given below:

Without splitting stems

6	03557
7	0124488999
8	113667
9	06

With splitting stems

6	03
6	557
7	01244
7	88999
8	113
8	667
9	0
9	6

Student preferences may vary, but the split stems on the right show more detail. (b) Relatively symmetric, with center near 780 mm (the median is 784 mm) and range of $957-604=353$ mm. (c) El Niño appears to reduce monsoon rainfall. Monsoon rainfall was below average in 18 of the 23 El Niño years and exceeded 900 mm in only 1 of those years.
49. (a) Not only are most responses multiples of 10, but many are multiples of 30 and 60. Most people will round their answers when asked to give an estimate like this. The students who claimed 360 minutes (about 6 hours) and 300 minutes (about 5 hours) may have been exaggerating. (b) Yes; the approximate centers are 175 minutes for women and 120 minutes for men.

Women		Men
	0	0 3 3 3 3 4
9 6	0	6 6 6 7 9 9 9 9
2 2 2 2 2 2 2	1	2 2 2 2 2 2 2
8 8 8 8 8 8 8 8 8 8 7 5 5 5 5	1	5 5 8
4 4 4 0	2	0 0 3 4 4
	2	
	3	0
6	3	

51. (a) The distribution is slightly skewed to the left. (b) Answers will vary. The center is between 0% and 2.5%. (c) The highest return was between 10% and 12.5%. Ignoring the low outliers, the lowest return was between −12.5% and −10%. (d) About 37% of these months (102 out of 273) had negative returns.
53. (a) Here is a histogram.

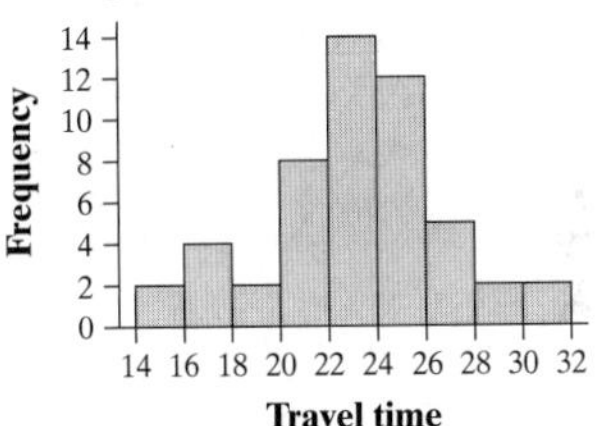

(b) Roughly symmetric. The center is near 23 minutes, and the range is $30.9-15.5=15.4$ minutes. There do not appear to be any outliers.
55. The distribution is somewhat skewed to the left with center at 35. The smallest DRP score is 14 and the largest DRP score is 54, so the scores have a range of 40. There are no gaps or outliers.

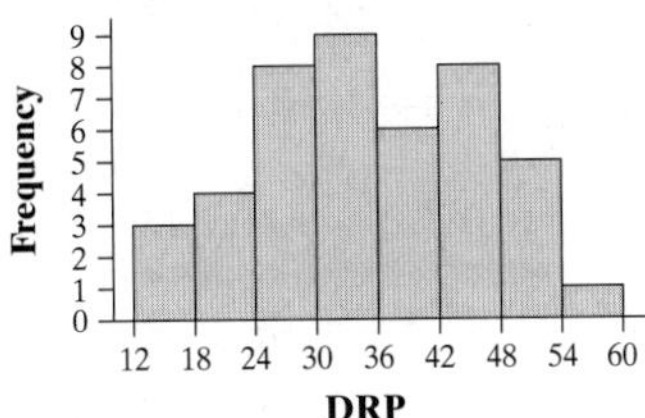

57. (a) The distribution is skewed to the right with a single peak. The center is at 4 letters, with a range of $15-1=14$ letters. There are no gaps or outliers.

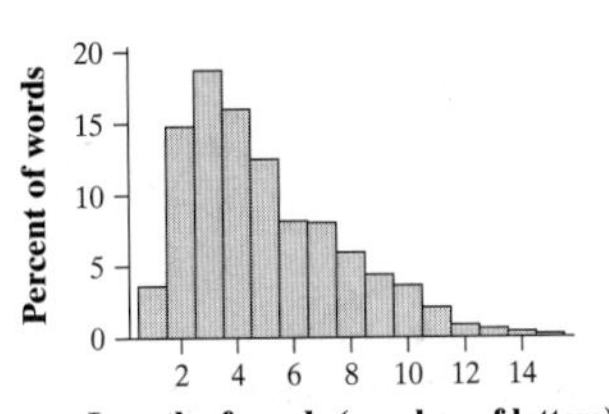

(b) There are more 2-, 3-, and 4-letter words in Shakespeare's plays and more very long words in *Popular Science* articles.

59. The scale on the x axis is very different from one graph to the other.

61. Answers will vary.

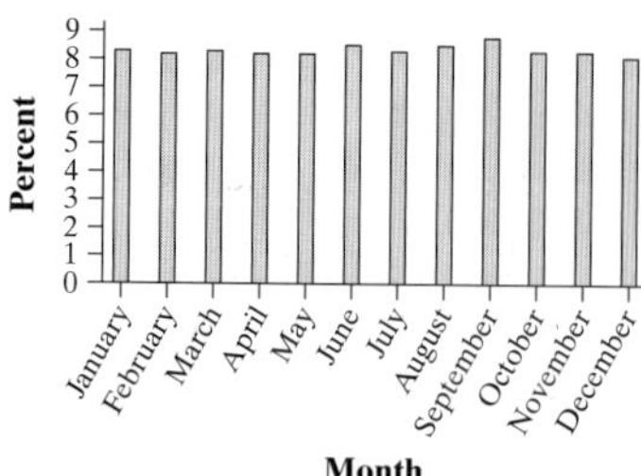

63. (a) Roundoff error. (b) Relative frequency histograms are appropriate because there are considerably more men than women.

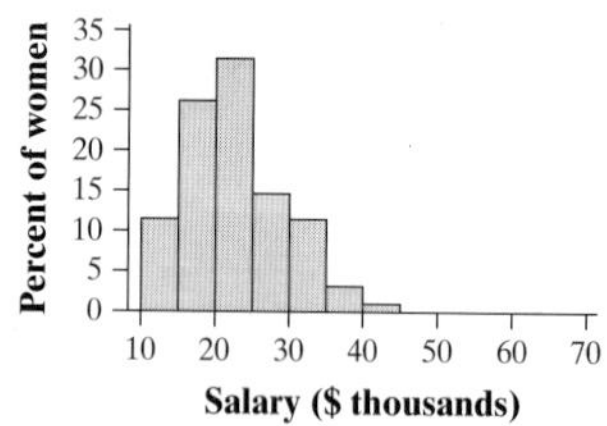

(c) Both histograms are skewed to the right, with the women's salaries generally lower than the men's. The peak for women is the interval from \$20,000 to \$25,000, and the peak for men is the interval from \$25,000 to \$30,000. The range of salaries is the same, with salaries in the smallest and largest intervals for both genders.

65. (a) Two bar graphs can be used.

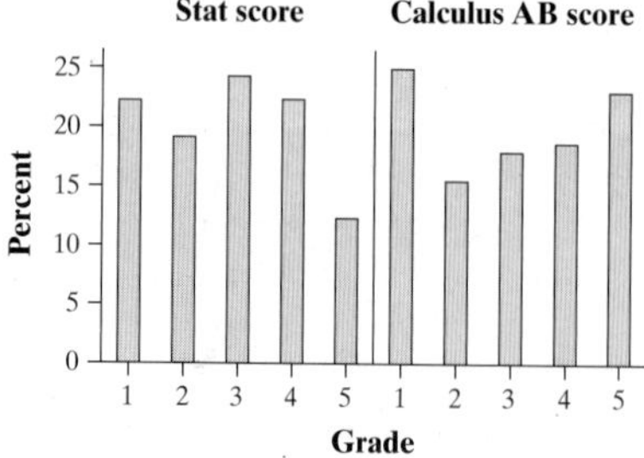

(b) The two distributions are very different. The distribution of scores on the Statistics exam is roughly symmetric with a peak at 3. The distribution of scores on the AB Calculus exam shows a very different pattern, with a peak at 1 and another, slightly lower peak at 5. The College Board considers "3 or above" to be a passing score. The percents of students "passing" the exams are very close (59.5% for Calculus AB and 58.8% for Statistics). From these two distributions it is impossible to tell which exam is "easier."

67. (a) Amount of studying. We expect most students to do some studying, but not a huge amount. Any outliers would likely be high outliers, leading to a right-skewed distribution. (b) Right-versus left-handed variable. About 90% of the population is right-handed, and since 0 represents right-handed people, we expect a much higher bar at 0 than at 1. (c) Gender. We expect a more even distribution among the males and females than we would for the right-handed and left-handed students. (d) Heights. The distribution of heights is usually symmetric and bell-shaped.

69. a

71. c

73. b

75. (a) Major League Baseball players who were on the roster on opening day of the 2009 season. (b) Six variables besides name. Categorical: team, position. Quantitative: age, height, weight, salary. (c) Age is measured in years, height in feet and inches, weight in pounds, and salary in dollars.

77. (a) 8.28% were elite soccer players. 5.01% of the people had arthritis. (b) 14.1% of the elite soccer players had arthritis. 23.3% of those with arthritis were elite soccer players.

Section 1.3

Answers to Check Your Understanding

Page 55 1. Since the distribution is skewed to the right, we expect the mean to be larger than the median. 2. The mean is $\frac{5+10+10+15+\cdots+85}{20}=31.25$ minutes, which is bigger than the median, which is 22.5 minutes. 3. If we divided the travel time evenly among all 20 people, each would have a 31.25-minute travel time. 4. In this case, since the distribution is skewed, the median would be a better measure of the center of the distribution. *Page 61* 1. The data, when ordered, are

307 311 311 313 317 318 318 326 338 353

Therefore, the minimum is 307 and the maximum is 353. The median is the average of the 5th and 6th observations and is $\frac{317+318}{2}=317.5$ The first quartile is the median of the bottom 5 observations, or the 3rd observation: 311. The third quartile is the median of the top 5 observations, or the 8th observation: 326. So the five-number summary is

307 311 317.5 326 353

2. The IQR is $326 - 311 = 15$. This is the range of the middle half of the weights of Dallas Cowboys offensive linemen in 2009. 3. $1.5IQR = 1.5(15) = 22.5$. Any outliers occur below $311 - 22.5 = 288.5$ or above $326 + 22.5 = 348.5$. There are no observations below 288.5. However, there is one observation above 348.5: the value 353. It is an outlier. 4. Here is a boxplot.

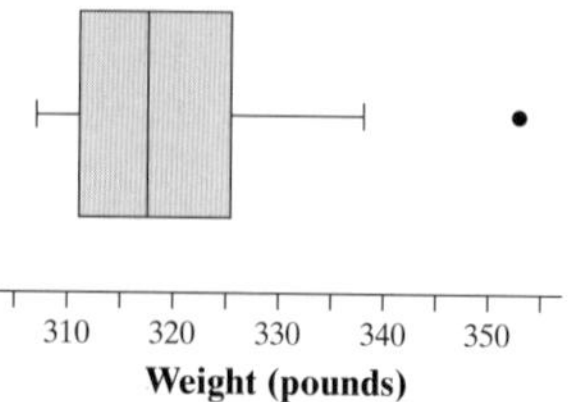

Page 64 1. The mean is $\frac{67+72+76+76+84}{5}=75$. If the total of all the heights was the same, but all the players were the same height, they would each be 75 inches tall.

2.

Observation	Deviation	Squared deviation
67	$67 - 75 = -8$	$(-8)^2 = 64$
72	$72 - 75 = -3$	$(-3)^2 = 9$
76	$76 - 75 = 1$	$1^2 = 1$
76	$76 - 75 = 1$	$1^2 = 1$
84	$84 - 75 = 9$	$9^2 = 81$
Total	**0**	**156**

3. The variance is the sum of the squared deviations (taken from the "Total" row) divided by 4 (the number of observations minus 1). In this case, that means that the variance is $\frac{156}{4} = 39$ inches squared. The standard deviation is the square root of the variance. In this case, that is 6.24 inches. 4. The players' heights vary about 6.24 inches from the mean height of 75 inches on average.

Answers to Odd-Numbered Section 1.3 Exercises

79. The mean of Joey's first 14 quiz scores is $\frac{86 + 84 + \cdots + 93}{14} = \frac{1190}{14} = 85$. If Joey had scored the same number of total points on the first 14 quizzes, but the scores had all been the same, then he would have scored an 85 on each quiz.

81. **(a)** Putting the scores in order:

74 75 76 78 80 82 84 86 87 90 91 93 96 98

Since there are 14 scores, the median is the mean of the 7th and 8th scores. Therefore, the median is $\frac{84 + 86}{2} = 85$ About half of the scores are smaller than 85 and about half are larger than 85. **(b)** If Joey had a 0 for the 15th quiz, then the sum of his quiz scores would still be 1190, leading to a mean of $\frac{1190}{15} = 79.33$ To find the median, we add the 0 to the beginning of the list in part (a). Since there are now 15 measurements, the median is the 8th measurement, which is 84. Notice that the median did not change much but the mean did. This shows that the mean is not resistant to outliers, but the median is.

83. Mean: \$60,954. Median: \$48,097. The distribution is likely to be quite right-skewed because of a few people who have very large incomes. When a distribution is skewed to the right, the mean is bigger since the tail values pull the mean toward them.

85. \$30 million. No; you would not be able to calculate the team's annual payroll from the median, because you cannot determine the sum of all 25 salaries from the median.

87. **(a)** Estimate the frequencies of the bars (from left to right): 10, 40, 42, 58, 105, 60, 58, 38, 27, 18, 20, 10, 5, 5, 1, and 3. We estimate the mean by first adding 2 ten times, 3 forty times, . . . , and 17 three times. This gives us a sum of 3504. The mean is then estimated by dividing by the number of responses: $\bar{x} = \frac{3504}{500} = 7.01$. We estimate the median by finding the average of the 250th and 251st values. The median is 6. **(b)** Since the median is less than the mean, we would use the median to argue that shorter domain names are more popular.

89. **(a)** Putting the data in order we get:

74 75 76 78 80 82 84 86 87 90 91 93 96 98

There are 14 observations here so the first quartile is the median of the bottom 7 observations. This means that it is the value of the 4th observation. It is 78. The third quartile is the median of the top 7 observations, so it is the value of the 11th observation. It is 91. So $IQR = 13$. The middle 50% of the data have a spread of 13 points. (b) Any outliers are below $Q_1 - 1.5IQR$ or above $Q_3 + 1.5IQR$. These limits are computed to be $78 - 1.5(13) = 58.5$ and $91 + 1.5(13) = 110.5$. There are no points outside these bounds, so there are no outliers.

91. **(a)** Here is a boxplot.

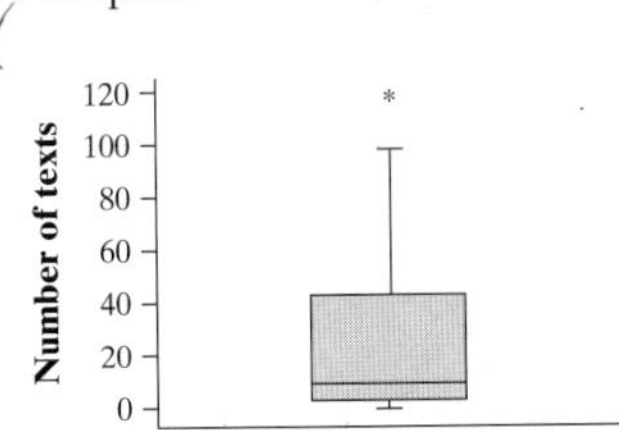

(b) The article claims that teens send 1742 texts a month. This works out to be about 58 texts a day (assuming a 30-day month). That seems pretty high given this data set. Twenty-one of the 25 students sent fewer than that; in fact, half of the students sent fewer than 10 messages (about 1/6 of the number claimed in the article). The mean number of texts per day was 27.48, less than half of 58.

93. **(a)** Since the data are recorded as *texts* – *calls*, positive numbers indicate students who had more text messages than calls. It appears from the boxplot as though the first quartile is 0, which means that approximately 25% made more calls than they texted. But the remaining 75% had more texts than calls. So this does support the article's conclusion. **(b)** No; we cannot make any more general conclusions. The sample was not a random sample, and there may be some commonality among his students that affected their responses to this question.

95. **(a)** The stock fund varied between about −3.5% and 3%. **(b)** The median return for the stock fund was slightly positive, about 0.1%, while the median real estate fund return appears to be close to 0%. **(c)** The stock fund is much more variable. It has higher positive returns, but also higher negative returns.

97. **(a)** $s_x = 0.6419$ mg/dl. **(b)** The typical phosphate level is an average of 0.6419 mg/dl different from the mean level.

99. **(a)** It looks like the distribution is skewed to the right because the mean is much larger than the median. **(b)** The standard deviation is \$21.70. The average distance between the individual amounts spent and the mean amount spent is about \$21.70. **(c)** The first quartile is 19.27 and the third quartile is 45.4, so the IQR is $45.4 - 19.27 = 26.13$. Any points below $19.27 - 1.5(26.13) = -19.925$ or above $45.4 + 1.5(26.13) = 84.595$ are outliers. Since the maximum point is 93.34, there are outliers.

101. Yes, the IQR is resistant. Answers will vary. Consider the simple data set 1, 2, 3, 4, 5, 6, 7, 8. Median = 4.5, $Q_1 = 2.5$, $Q_3 = 6.5$, and $IQR = 4$. Changing any value outside the interval

between Q_1 and Q_3 will have no effect on the *IQR*. For example, if 8 is changed to 88, the *IQR* will still be 4.

103. **(a)** One possible answer is 1, 1, 1, 1. **(b)** 0, 0, 10, 10. **(c)** For (a), any set of four identical numbers will have $s_x = 0$. For (b), the answer is unique; here is a rough description of why. We want to maximize the "spread-out-ness" of the numbers (which is what the standard deviation measures), so 0 and 10 seem to be reasonable choices based on that idea. We also want to make each individual squared deviation—$(x_1 - \bar{x})^2$, $(x_2 - \bar{x})^2$, $(x_3 - \bar{x})^2$, and $(x_4 - \bar{x})^2$—as large as possible. If we choose 0, 10, 10, 10—or 10, 0, 0, 0—we make the first squared deviation 7.5^2, but the other three are only 2.5^2. Our best choice is two values at each extreme, which makes all four squared deviations equal to 5^2.

105. **State:** Do the data indicate that women have better study habits and attitudes toward learning than men? **Plan:** We will draw side-by-side boxplots for each group. We will compute the five-number summary, the mean, and the standard deviation for the scores of each group. Then we will compare the groups using both graphical and numerical summaries. **Do:** Create boxplots and numerical summaries.

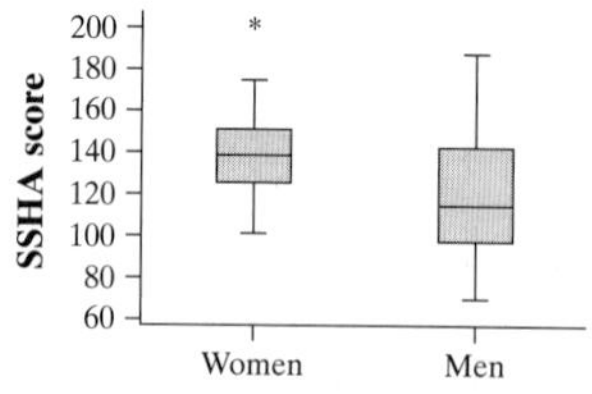

Variable	*N*	Mean	St. dev.	Min.	Q_1	*M*	Q_3	Max.
Women	18	141.06	26.44	101.00	126.00	138.50	154.00	200.00
Men	20	121.25	32.85	70.00	98.00	114.50	143.00	187.00

Conclude: It appears from the boxplots and the numerical summaries that the women have higher values for all the components of the five-number summary. They also have a higher mean and a smaller standard deviation. This means that their scores are not only higher but have less variability.

107. d

109. a

111. **(a)** Yes; the categories (method of communication) are mutually exclusive (each student chose one method by which they most often communicated with friends) and are parts of a whole. **(b)** This is a bar graph of categorical data. Skewness describes only quantitative data.

113. Women appear to be more likely to engage in behaviors that are indicative of "habits of mind." They are especially more likely to revise papers to improve their writing (about 55% of females report this, as opposed to about 37% of males). The difference is a little less for seeking feedback on their work. In that case, about 49% of the females did this, as opposed to about 38% of males.

Answers to Chapter 1 Review Exercises

R1.1 **(a)** Movies. **(b)** Categorical: name, rating, genre. Quantitative: year (in years), time (in minutes), box office sales (in dollars). **(c)** *Avatar,* released in 2009, rated PG-13, runs 162 minutes, is an action film, and had box office sales of $2,714,767,458.

R1.2 Here is a bar graph.

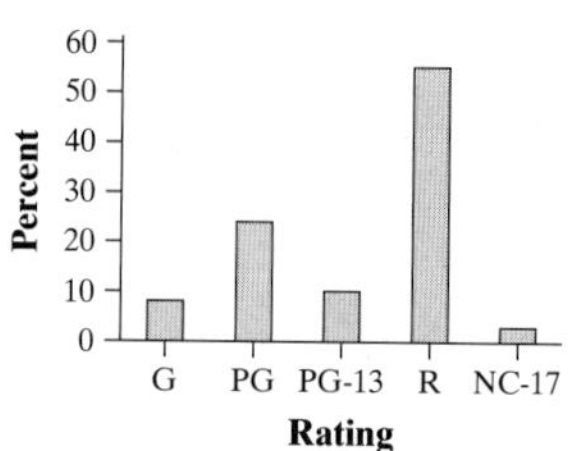

R1.3 **(a)** It is the areas of the phones that should be in proportion, not just the heights. For example, the picture for "send/receive text messages" should be roughly twice the size of the picture for "camera," but it is actually much more than twice as large. **(b)** No; they do not describe parts of a whole. Students were free to answer in more than one category. **(c)** Here is a bar graph.

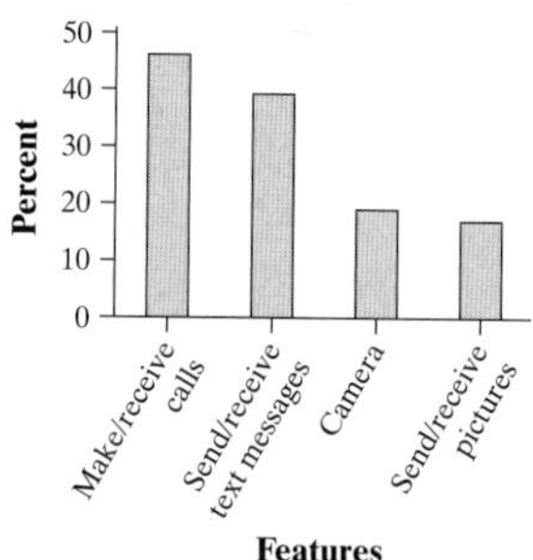

R1.4 **(a)** 67.6% were Facebook users. This is part of a marginal distribution because it compares the total in one column to the overall total in the table. **(b)** The percent of younger students who are Facebook users is 95.1%. 52.7% of Facebook users are younger.

R1.5 There does appear to be an association between age and Facebook status. Looking at the conditional distributions of Facebook status for each of the age groups, we get the following table:

	Facebook User?	
Age (years)	**Yes**	**No**
Younger (18–22)	95.1%	4.9%
Middle (23–27)	70.0%	30.0%
Older (28 and up)	31.3%	68.7%

This can also be seen in a bar graph. From both the table and the graph, we can see that the older the student is, the less likely the student is to be a member of Facebook. For younger students, about 95% are members. That drops to 70% for middle students and drops even further to 31.3% for older students.

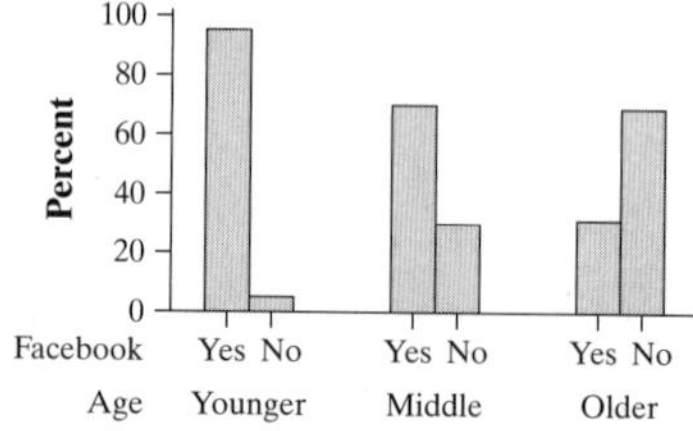

R1.6 (a)

```
Stem-and-leaf of density N = 29
Leaf unit = 0.010

48 | 8
49 |
50 | 7
51 | 0
52 | 6799
53 | 04469
54 | 2467
55 | 03578
56 | 12358
57 | 59
58 | 5
```

(b) The distribution is roughly symmetric, with one value (4.88) that is somewhat low. The center of the distribution is between 5.4 and 5.5. The densities have a range of $5.85 - 4.88 = 0.97$ and there are no outliers. (c) Since the distribution is roughly symmetric, we can use the mean to estimate Earth's density to be about 5.45 in these units.

R1.7 (a) The survival times are right-skewed, as expected. The median survival time is 102.5 days, and the range of survival times is $598 - 43 = 555$ days. There are several high outliers, with survival times above 500.

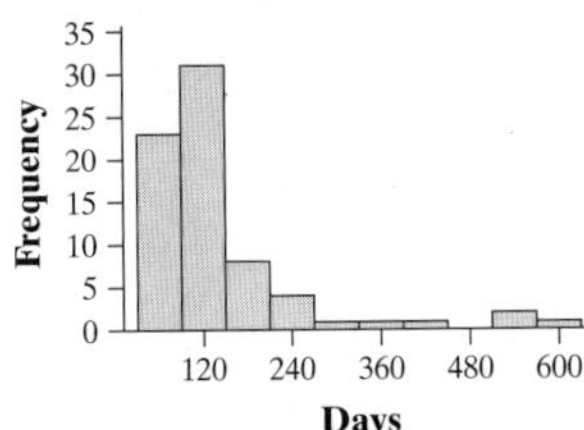

(b) Here is a boxplot.

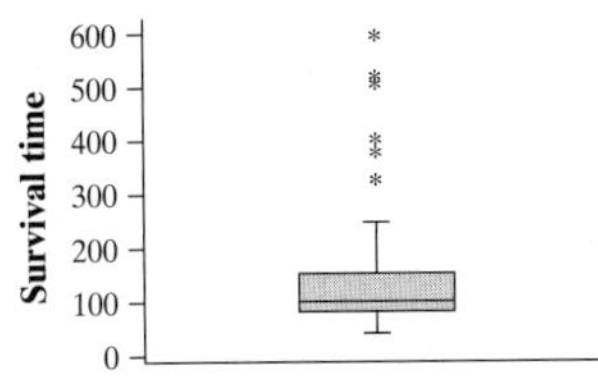

R1.8 (a) About 20% of low-income and 33% of high-income households consisted of two people. (b) The majority of low-income households, but only about 7% of high-income households, consist of one person. One-person households often have less income because they include many young people who have no job or have only recently started working. (Income generally increases with age.)

R1.9 (a) A typical measurement of mercury per can of tuna will be about 0.300 from the mean. (b) Any point below -0.3925 or above 0.8435 would be considered an outlier. Since the smallest value is 0.012, there are no low outliers. But the largest value is 1.50, which is larger than 0.8435. From the histogram, we can see that there are high outliers. (c) The distribution of the amount of mercury in cans of tuna is highly skewed to the right. The median is 0.18 ppm and the *IQR* is 0.309 ppm.

R1.10 Albacore tuna generally has more mercury. Its minimum, first quartile, median, and third quartile are all larger than the respective values for light tuna. But that doesn't mean that light tuna is always better. It has a much bigger spread of values, and some cans have as much as twice the amount of mercury as the largest amount in the albacore tuna. Both distributions are skewed to the right with at least a couple outliers on the right. The albacore tuna has a median of 0.4 ppm and the light tuna has a median of 0.16 ppm. But the albacore has less variation, with an *IQR* of 0.167, whereas the light tuna has an *IQR* of 0.288.

Answers to Chapter 1 AP Statistics Practice Test

T1.1 d
T1.2 e
T1.3 b
T1.4 b
T1.5 c
T1.6 c
T1.7 b
T1.8 c
T1.9 e
T1.10 b
T1.11 d
T1.12 (a) Here is a histogram.

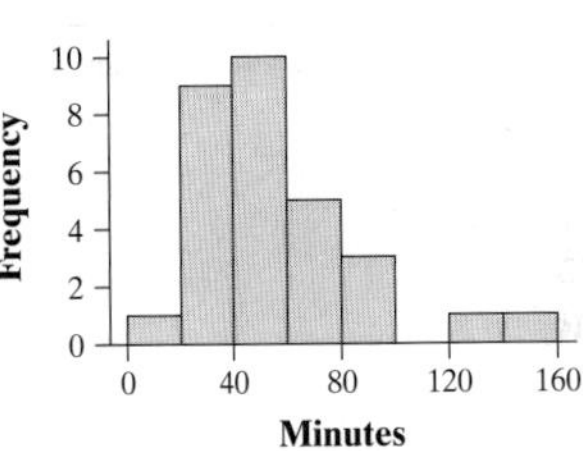

(b) The first quartile is the median of the bottom 15 data points, or the 8th data value. Therefore, it is 30 minutes. The third quartile is the 23rd data point (the median of the top 15 data points), which is 77. So $IQR = 77 - 30 = 47$. Any point below $30 - 1.5(47) = -40.5$ or above $77 + 1.5(47) = 147.5$ is an outlier. So the observation of 151 minutes is an outlier. (c) It would be better to use the median and *IQR* to describe the center and spread of this distribution because it is skewed to the right.

T1.13 (a)

Birth Defects	Diabetic Status			
	Nondiabetic	Prediabetic	Diabetic	Total
None	754	362	38	**1154**
One or more	31	13	9	**53**
Total	**785**	**375**	**47**	**1207**

(b)

	Diabetic Status		
Birth Defects	**Nondiabetic**	**Prediabetic**	**Diabetic**
None	96.1%	96.5%	80.9%
One or more	3.9%	3.5%	19.1%

(c) Here is a bar graph.

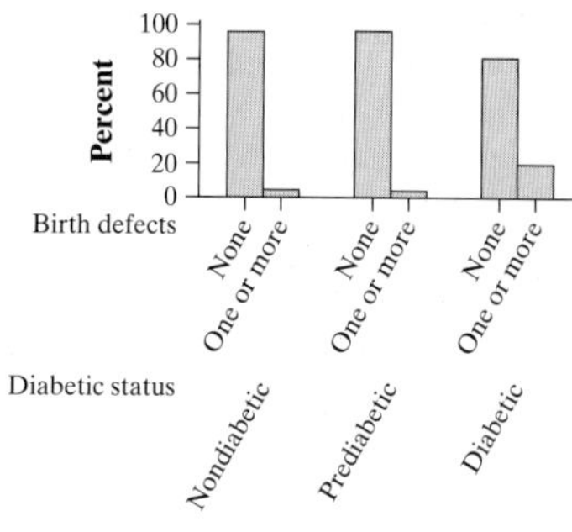

(d) There does appear to be an association. Nondiabetics and prediabetics appear to have babies with birth defects at about the same rate. However, women with diabetes have a much higher rate of babies with birth defects.

T1.14 (a) The longest that any battery lasted was between 550 and 559 hours. (b) Someone might prefer to use Brand X because it has a higher minimum lifetime. (c) On the other hand, some might prefer Brand Y because it has a higher median lifetime.

T1.15 Side-by-side boxplots. Given below are descriptive statistics for both the American League and the National League.

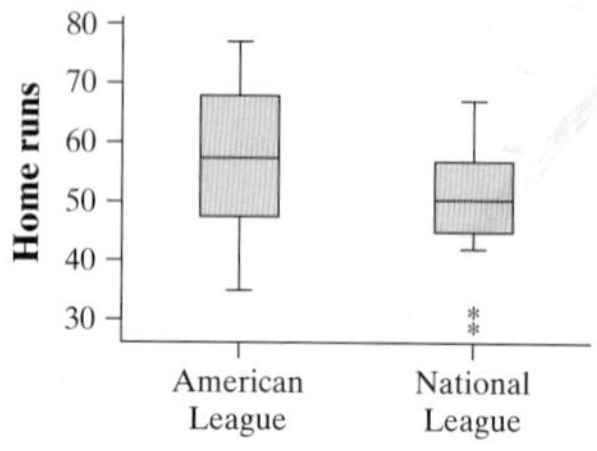

Variable	N	Mean	St. dev.	Min.	Q_1	M	Q_3	Max.
American	14	56.93	12.69	35.00	49.00	57.50	68.00	77.00
National	14	50.14	11.13	29.00	46.00	50.50	55.00	67.00

The data suggest that the number of home runs is somewhat less in the National League. All five numbers in the five-number summary and the mean are less for the National League teams than for the American League teams. However, there is more variability among the American League teams, which have a standard deviation of 12.69, compared with 11.13 for the National League. Both distributions are reasonably symmetric. The American League has no outliers, but the teams that hit 29 and 31 home runs are outliers in the National League.

CHAPTER 2

Section 2.1

Answers to Check Your Understanding

Page 89 **1.** c **2.** Her daughter weighs more than 87% of girls her age and she is taller than 67% of girls her age. **3.** About 65% of calls lasted less than 30 minutes. About 35% of calls lasted 30 minutes or longer. **4.** The first quartile (25th percentile) is at about 14 minutes. The third quartile (75th percentile) is at about 32 minutes. IQR = 32 − 14 = 18 minutes. *Page 91* **1.** $z = \dfrac{65-67}{4.29} = -0.466$. Lynette's height is about 0.47 standard deviation below the mean height of the class. **2.** Brent's z-score is $z = \dfrac{74-67}{4.29} = 1.63$. Brent's height is about 1.63 standard deviations above the mean height of the class. **3.** Since Brent's z-score is −0.85, we know that $-0.85 = \dfrac{74-76}{\sigma}$. Solving this for σ we find that $\sigma = 2.35$ inches. *Page 97* **1.** The shape will not change. The center and spread will be multiplied by 2.54. **2.** The shape and spread will not change. The center will have 6 inches added to it. **3.** The shape will not change. The mean will change to 0 and the standard deviation will change to 1. *Page 103* **1.** The curve is positive everywhere and it has total area of 1.0. **2.** About 12% of the observations lie between 7 and 8. **3.** On the graph, A = median and B = mean.

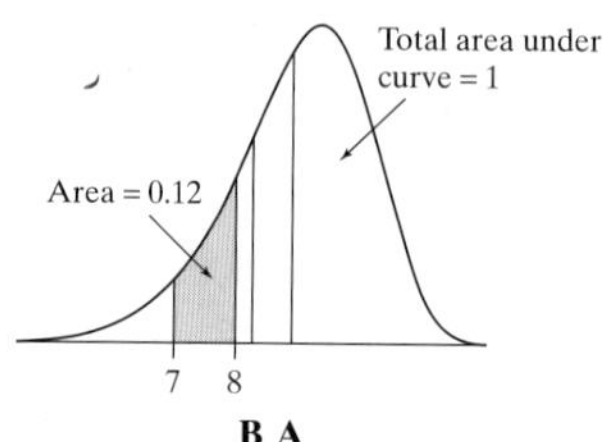

4. The mean is less than the median in this case because the distribution is skewed to the left.

Answers to Odd-Numbered Section 2.1 Exercises

1. (a) The girl with 22 pairs of shoes is the 6th smallest. Her percentile is 0.25. 25% of girls have fewer pairs of shoes. (b) The boy with 22 pairs has more shoes than 17 people. His percentile is 0.85. 85% of boys have fewer pairs of shoes. (c) The boy is more unusual because only 15% of the boys have as many or more than he has, while the girl has a value that is more centered in the distribution. 25% have fewer and 75% have as many or more.

3. According to the *Los Angeles Times*, the speed limits on California highways are such that 85% of the vehicle speeds on those stretches of road are less than the speed limit.

5. The girl in question weighs more than 48% of girls her age, but is taller than 78% of the girls her age. Since she is taller than 78% of girls, but only weighs more than 48% of girls, she is probably fairly skinny.

7. (a) The highlighted student sent about 212 text messages in the two-day period which placed her at about the 80th percentile. (b) The median number of texts is the same as the 50th percentile. Locate 50% on the y axis, read over to the points and then find the relevant place on the x axis. The median is approximately 115 text messages.

9. (a) First find the quartiles. The first quartile is the 25th percentile. Find 25 on the y axis, read over to the line and then down to the x axis to get about \$19. The 3rd quartile is the 75th percentile. Find 75 on the y axis, read over to the line and then down to the x axis to get about \$50. So the interquartile range is \$50 − \$19 = \$31. (b) Approximately the 26th percentile. (c) Here is a histogram.

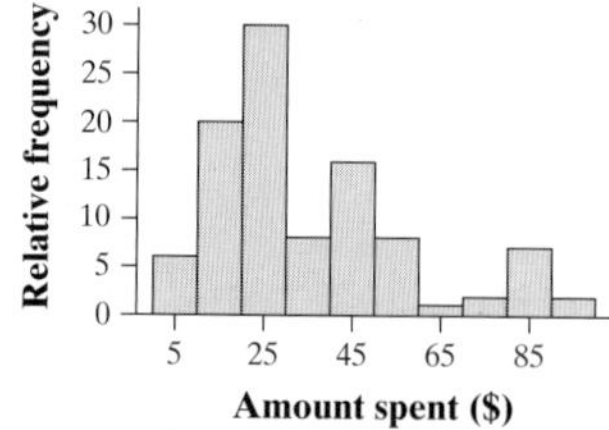

11. Eleanor's standardized score, $z = 1.8$, is higher than Gerald's standardized score, $z = 1.5$.
13. **(a)** Judy's bone density score is about one and a half standard deviations below the average score for all women her age. The fact that your standardized score is negative indicates that your bone density is below the average for your peer group. The magnitude of the standardized score tells us how many standard deviations you are below the average (about 1.5). **(b)** $\sigma = 5.52$ grams/cm^2.
15. **(a)** Since 22 salaries were less than Lidge's salary, his salary is at the 75.86 percentile. **(b)** $z = 0.79$. Lidge's salary was 0.79 standard deviations above the mean salary of $3,388,617.
17. **(a)** In the national group, about 94.8% of the test takers scored below 65. Scott's percentiles, 94.8th among the national group and 68th within the school, indicate that he did better among all test takers than he did among the 50 boys at his school. **(b)** Scott's z-scores are $z = 1.57$ among the national group and $z = 0.62$ among the 50 boys at his school.
19. **(a)** The mean and the median both increase by 18 so the mean is 87.188 and the median is 87.5. The distribution of heights just shifts by 18 inches. **(b)** The standard deviation and IQR do not change. For the standard deviation, note that although the mean increased by 18, the observations each increased by 18 as well so that the deviations did not change. For the IQR, Q_1 and Q_3 both increase by 18 so that their difference remains the same as in the original data set.
21. **(a)** To give the heights in feet, not inches, we would divide each observation by 12 (12 inches = 1 foot). Thus the mean and median are divided by 12. The new mean is 5.77 feet and the new median is 5.79 feet. **(b)** To find the standard deviation in feet, note that each deviation in terms of feet is found by dividing the original deviation by 12.

$$\text{standard deviation}_{\text{New}} = \sqrt{\frac{(\text{first deviation (ft)})^2 + \cdots + ((\text{last deviation (ft)})^2)}{n-1}}$$
$$= \sqrt{\frac{\left(\frac{\text{first deviation (in)}}{12}\right)^2 + \cdots + \left(\frac{\text{last deviation (in)}}{12}\right)^2}{n-1}}$$
$$= \frac{1}{12} * \text{standard deviation}_{\text{Old}} = \frac{3.2}{12} = 0.27 \text{ feet}$$

The first and third quartiles are still the medians of the first and second halves of the data: these values must simply be converted to feet. To do this, divide the first and third quartiles of the original data set by 12: $Q_1 = \frac{67.75}{12} = 5.65$ feet and $Q_3 = \frac{71}{12} = 5.92$ feet. So the interquartile range is $IQR = 5.92 - 5.65 = 0.27$ feet.
23. Mean in degrees Fahrenheit is 77. Standard deviation in degrees Fahrenheit is 3.6.
25. Sketches will vary.
27. **(a)** It is on or above the horizontal axis everywhere and the area beneath the curve is 1. **(b)** $\frac{1}{3}$ **(c)** Since $(1.1 - 0.8) \cdot \frac{1}{3} = 0.1$, one-tenth of accidents occur next to Sue's property.
29. Both are 1.5.
31. **(a)** Mean is C, median is B. **(b)** Mean is B, median is B.
33. c
35. c
37. d
39. The distribution is skewed to the right since most of the values are 25 minutes or less, but the values stretch out up to about 90 minutes. The data are centered roughly around 20 minutes and the range of the distribution is close to 90 minutes. The two largest values appear to be outliers.

Section 2.2

Answers to Check Your Understanding

Page 114 **1.** Here is the graph.

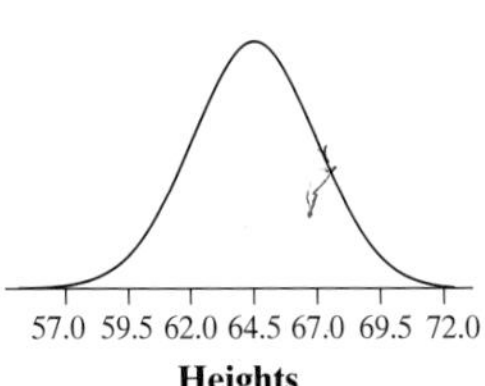

2. Since 67 inches is one standard deviation above the mean, approximately $\frac{1-0.68}{2} = 16\%$ of young women have heights greater than 67 inches. **3.** Since 62 is one standard deviation below the mean and 72 is three standard deviations above the mean, approximately $\frac{0.68}{2} + \frac{0.997}{2} = 84\%$ of young women have heights between 62 and 72 inches. *Page 119* **1.** The proportion is 0.9177. **2.** The proportion is 0.9842. **3.** The proportion is $0.9649 - 0.2877 = 0.6772$. **4.** The z-score for the 20th percentile is -0.84. **5.** The 55th percentile is the z-value where 45% are greater than that value. $z = 0.13$.

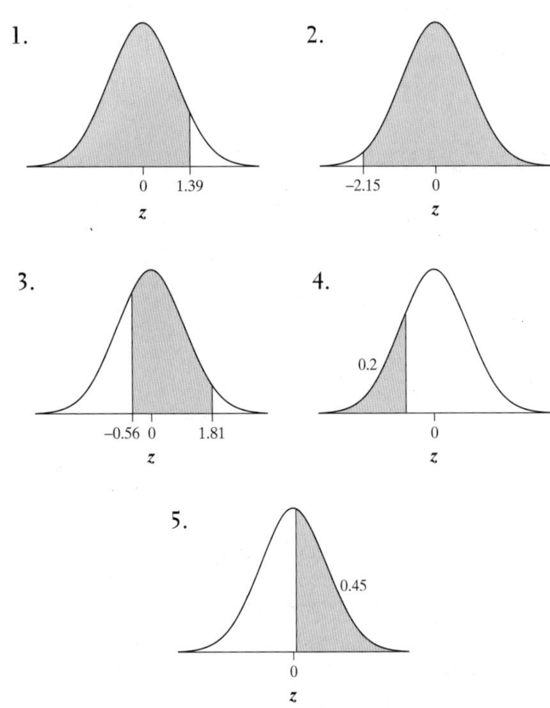

Page 124 **1.** For 14-year old boys with cholesterol 240, the z-score is $z = \frac{240-170}{30} = 2.33$. The proportion of z-scores above 2.33 is

$1 - 0.9901 = 0.0099$. **2.** The z-score for a cholesterol level of 200 is $z = \frac{200 - 170}{30} = 1$. The proportion of z-scores between 1 and 2.33 is $0.9901 - 0.8413 = 0.1488$. **3.** The 80th percentile of the standard Normal distribution is 0.84. This means that the distance, x, of Tiger Woods's drive lengths that satisfies the 80th percentile is the solution to $0.84 = \frac{x - 304}{8}$. Solving for x, we get 310.72 yards.

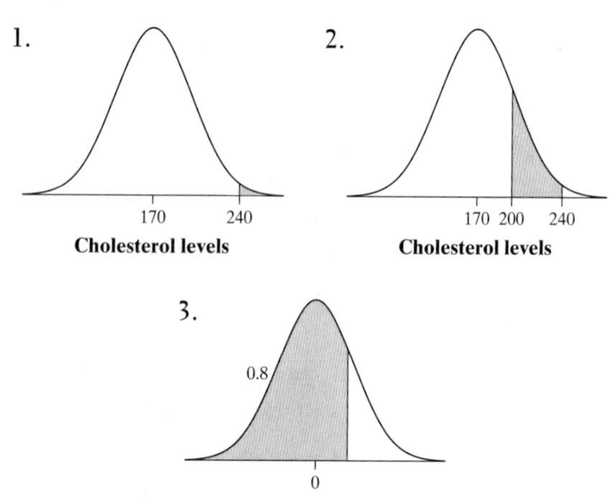

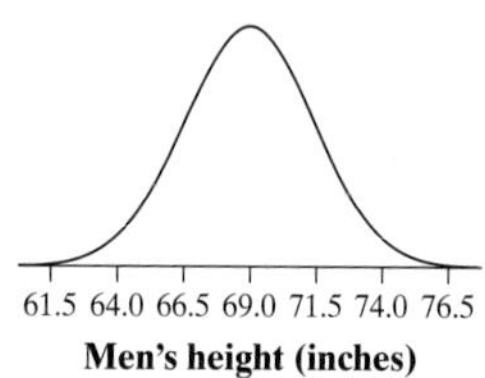

Answers to Odd-Numbered Section 2.2 Exercises

41. Here is the graph.

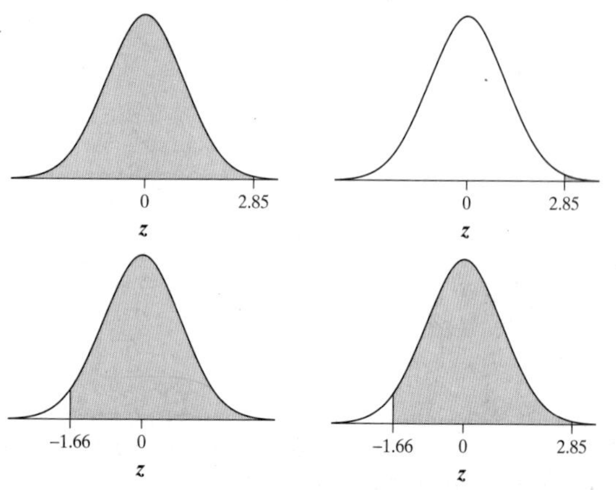

43. **(a)** Approximately 2.5% **(b)** 64 and 74 inches **(c)** Approximately 13.5% **(d)** 84th percentile.

45. The standard deviation is approximately 0.2 for the tall, more concentrated one and 0.5 for the short, less concentrated one.

47. **(a)** 0.9978 **(b)** 0.0022 **(c)** 0.9515 **(d)** 0.9493

49. **(a)** 0.8587 **(b)** 0.2718

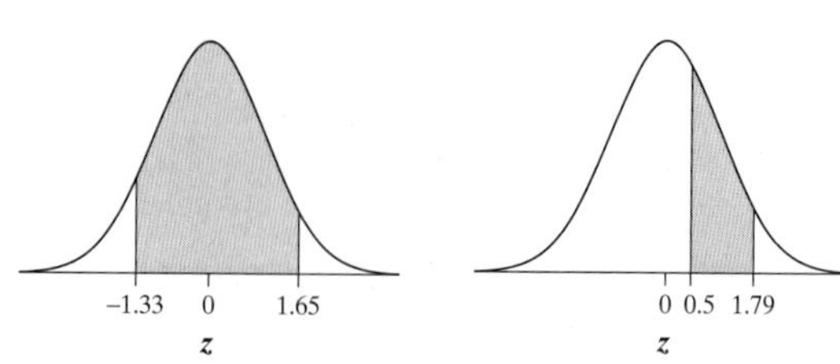

51. **(a)** $z = -1.28$ **(b)** $z = 0.41$.

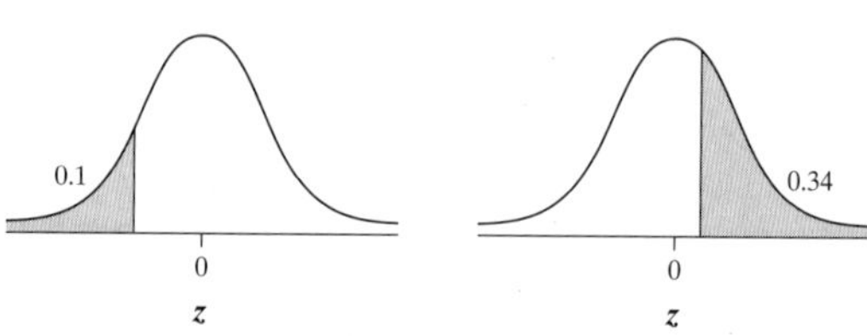

53. **(a) State:** Let x = the length of pregnancies. The variable x has a Normal distribution with $\mu = 266$ days and $\sigma = 16$ days. We want the proportion of pregnancies that last less than 240 days. **Plan:** The proportion of pregnancies lasting less than 240 days is shown in the graph. **Do:** For $x = 240$ we have $z = -1.63$, so $x < 240$ corresponds to $z < -1.63$. Using Table A, we see that the proportion of observations less than -1.63 is 0.0516 or about 5.2%. **Conclude:** About 5.2% of pregnancies last less than 240 days which means that 240 is approximately the 5th percentile. **(b) Do:** From part (a) we have that for $x = 240$, $z = -1.63$. For $x = 270$, we have $z = 0.25$. Using Table A we see that the proportion of observations less than 0.25 is 0.5987. So the proportion of observations between -1.63 and 0.25 is about 55%. **Conclude:** Approximately 55% of pregnancies last between 240 and 270 days.

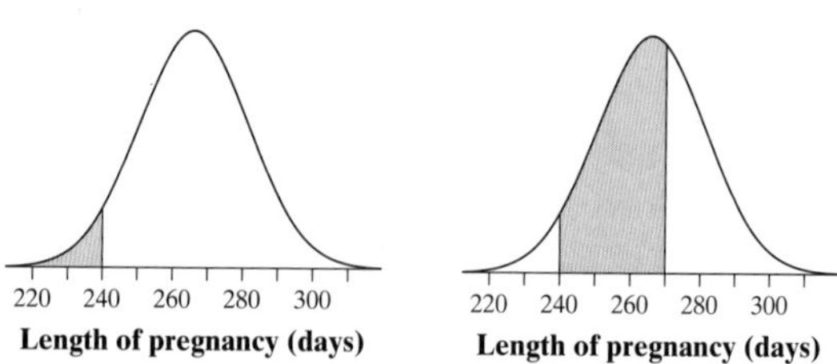

(c) Do: Using Table A, the 80th percentile for the standard Normal distribution is 0.84. Therefore, the 80th percentile for the length of human pregnancies can be found by solving the equation $0.84 = \frac{x - 266}{16}$ for x. Thus, $x = (0.84)16 + 266 = 279.44$. **Conclude:** The longest 20% of pregnancies last approximately 279 or more days.

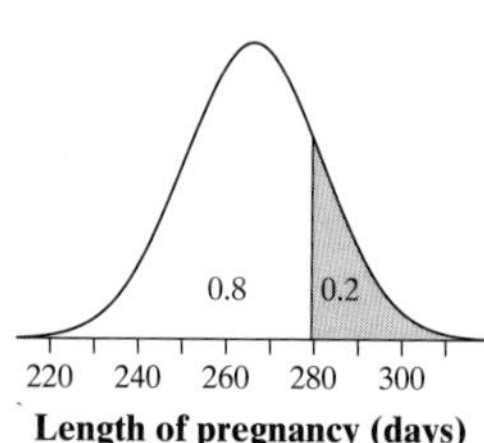

55. (a)

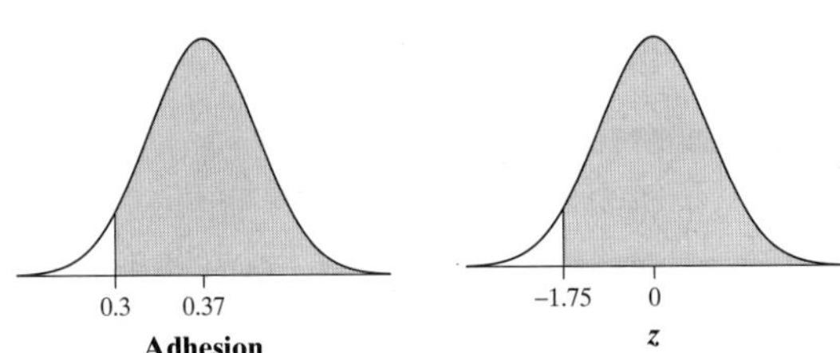

We would expect trains to arrive on time about 96% of the time. (b) $z = 3.25$. We would expect trains to arrive early 0.06% of the time.

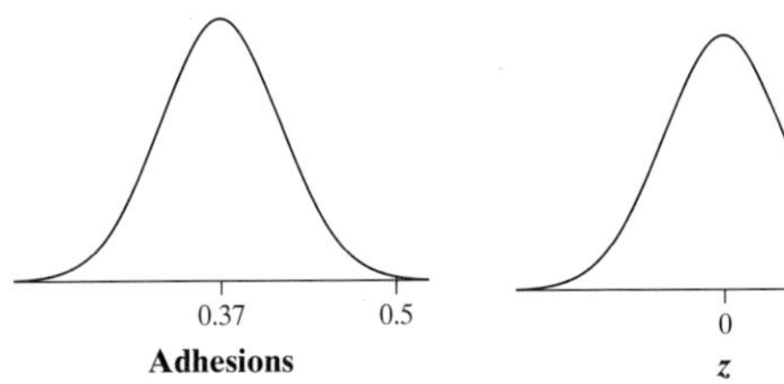

(c) It makes sense to try to have the value found in part (a) larger. We want the train to arrive at its destination on time, but not to arrive at the switch point early.

57. (a) Solve $z = \frac{0.30 - \mu}{0.04} = -2.05$ for μ. The mean adhesion should be 0.382. (b) Solve $z = \frac{0.30 - 0.37}{\sigma} = -2.05$ for σ. The standard deviation of the adhesion values should be 0.034. (c) To compare the options, we want to find the area under the Normal distribution to the right of 0.50. Under option (a), $z = 2.95$ and the area is $1 - 0.9984 = 0.0016$. Under option (b), $z = 3.82$ and the area is $1 - 0.9999 = 0.0001$. Therefore, we prefer option (b).

59. (a) ± 1.28 (b) $-1.28(2.5) + 64.5 = 61.3$ inches and $1.28(2.5) + 64.5 = 67.7$ inches

61. $\mu = 41.43$ minutes; $\sigma = 17.86$ minutes

63. (a) Below are descriptive statistics and a histogram.

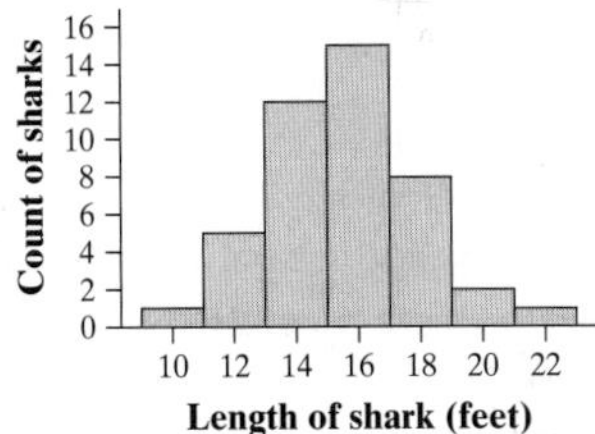

Variable	N	Mean	StDev	Minimum	Q_1	Median	Q_3	Maximum
shlength	44	15.586	2.550	9.400	13.55	15.750	17.200	22.800

The distribution of shark lengths is roughly symmetric with a peak at 16 and varies from 9.4 feet to 22.8 feet. (b) 68.2% of the lengths fall within one standard deviation of the mean, 95.5% of the lengths fall within two standard deviations of the mean, and 100% of the lengths fall within 3 standard deviations of the mean. These are very close to the 68–95–99.7 rule. (c) A Normal probability plot is shown.

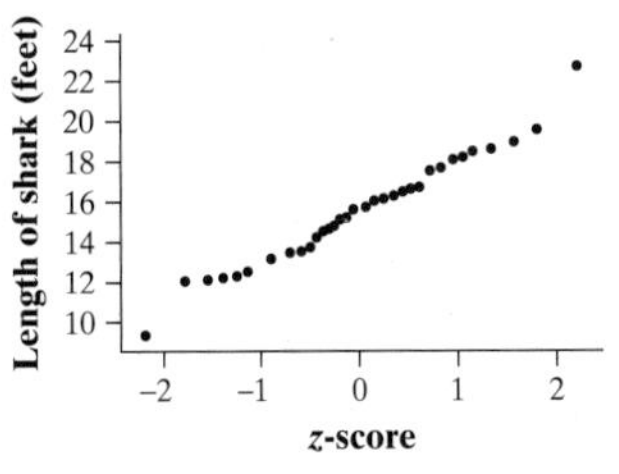

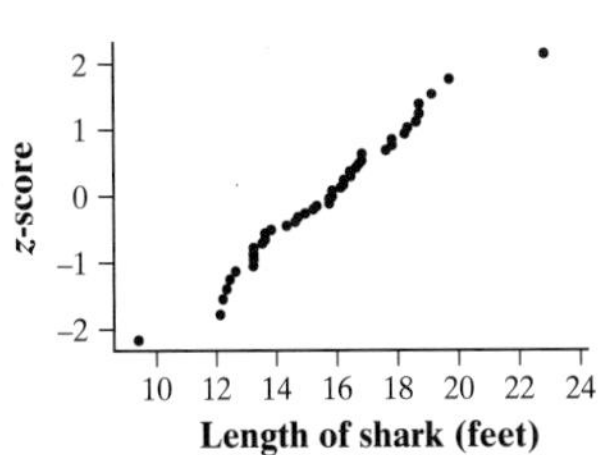

Except for one small shark and one large shark, the plot is fairly linear, indicating that the Normal distribution is appropriate. (d) The graphical display in (a), check of the 68–95–99.7 rule in (b), and Normal probability plot in (c) indicate that shark lengths are approximately Normal.

65. The plot is nearly linear. The smallest value is smaller than we would expect and the largest two values are larger than we would expect. There is also a cluster of points around 125 bpm that are a little larger than expected.

67. (a) 11.47% (b) 15.34% (c) 12.30%

69. d

71. b

73. c

75. For both kinds of cars, the highway miles per gallon is higher than the city miles per gallon. The two-seater cars have a wider spread of miles per gallon values than the minicompact cars do, both on the highway and in the city. Also, the miles per gallon values are slightly lower for the two-seater cars than for the minicompact cars. This difference is greater on the highway than it is in the city. All four distributions are roughly symmetric.

Answers to Chapter 2 Review Exercises

R2.1 (a) The only way to obtain a z-score of 0 is if the x-value equals the mean. Thus, the mean is 170 cm. To find the standard deviation, use the fact that a z-score of 1 corresponds to a height of 177.5. Then $1 = \frac{177.5 - 170}{\sigma}$ so $\sigma = 7.5$. Thus, the standard deviation of the height distribution for 15-year-old males is 7.5 cm. (b) Since $2.5 = \frac{x - 170}{7.5}$, $x = 2.5(7.5) + 170 = 188.75$ cm. A height of 188.75 cm has a z-score of 2.5.

R2.2 (a) $z = 1.20$. Paul is somewhat taller than average for his age. His height is 1.20 standard deviations above the average male height for his age. (b) 85% of boys Paul's age are shorter than Paul's height.

R2.3 (a) Approximately the 70th percentile. (b) The median (50th percentile) is about 5, Q_1 (25th percentile) is about 2.5, and Q_3 (75th percentile) is about 11. There are outliers, according to the 1.5(IQR) rule, because values exceeding $Q_3 + 1.5(IQR) = 23.75$ clearly exist.

R2.4 (a) The shape would not change. The new mean would be 13.32 meters, the median would be 12.80 meters, the standard deviation would be 3.81 meters, and the IQR would be 3.81 meters.

(b) The mean error would be 43.7 − 42.6 = 1.1 feet. The standard deviation of the errors would be the same as the standard deviation of the guesses, 12.5 feet, because we have just shifted the distribution, but not changed its width by subtracting 42.6 from each guess.

R2.5 (a) Answers will vary. Line A on the graph. (b) Answers will vary. Line B on the graph.

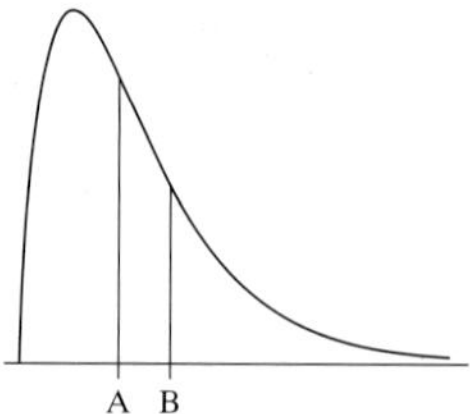

R2.6 (a) (327,345) (b) 339 is one standard deviation above the mean, so 16% of the horse pregnancies last longer than 339 days. This is because 68% are within one standard deviation of the mean, so 32% are more than one standard deviation from the mean and half of those are greater than 339.

R2.7 (a) 0.0122 (b) 0.9878 (c) 0.0384 (d) 0.9494

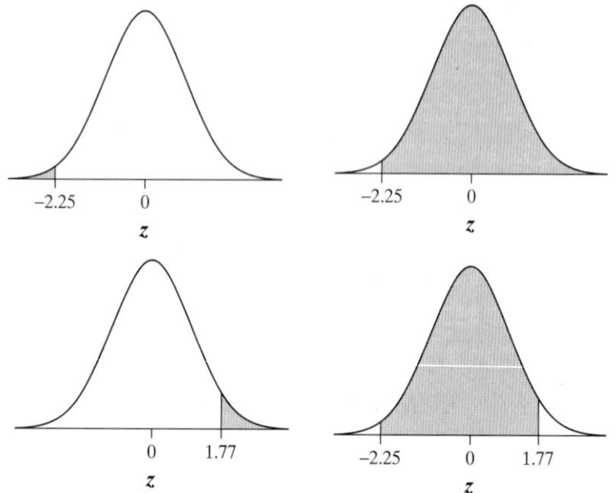

R2.8 (a) $z = 0.84$ (b) $z = 0.39$

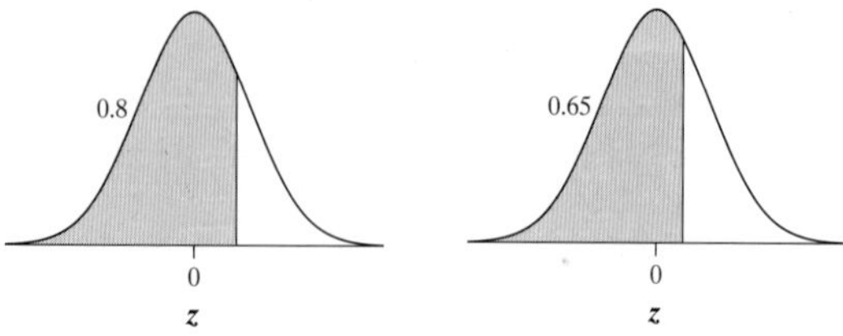

R2.9 (a) $z = -2.29$. The z-value that corresponds to a baby weight less than 2500 grams at birth is −2.29. The percent of babies weighing less than this is the area to the left. According to Table A, this is 0.0110. So, approximately 1% of babies will be identified as having low birth weight. (b) The z-values corresponding to the quartiles are −0.67 and 0.67. To find the x-values corresponding to the quartiles, we solve the following equations for x.

$$-0.67 = \frac{x - 3668}{511} \Rightarrow x = -0.67(511) + 3668 = 3325.63$$

$$0.67 = \frac{x - 3668}{511} \Rightarrow x = 0.67(511) + 3668 = 4010.37$$

The quartiles of the birth weight distribution are 3325.6 and 4010.4.

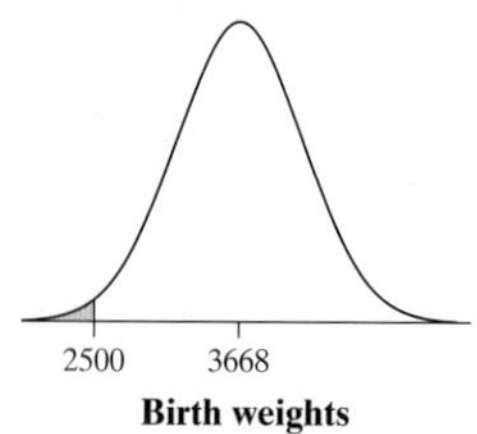

R2.10 If the distribution is Normal, it must be symmetric about its mean—and in particular, the 10th and 90th percentiles must be equal distances above and below the mean—so the mean is 250 points. If 225 points below (above) the mean is the 10th (90th) percentile, this is 1.28 standard deviations below (above) the mean, so the distribution's standard deviation is $\frac{225}{1.28} = 175.8$ points.

R2.11 A histogram and Normal probability plot both indicate that the data are not exactly Normally distributed. The histogram is roughly symmetric. The descriptive statistics given below indicate that the mean and median are very similar, which is consistent with rough symmetry. For most purposes, these data can be considered approximately Normal.

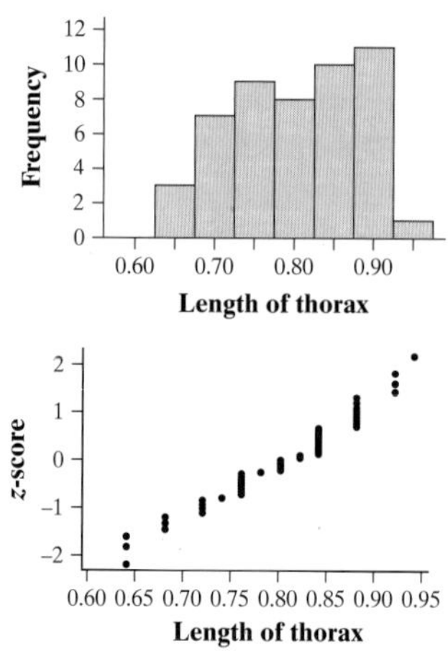

Variable	N	Mean	StDev	Minimum	Q_1	Median	Q_3	Maximum
length of thorax	49	0.8004	0.0782	0.6400	0.7600	0.8000	0.8600	0.9400

R2.12 The steep, nearly vertical portion at the bottom and the bow upward indicate that the distribution of the data is right-skewed with several outliers. In other words, these data are not Normally distributed.

Answers to Chapter 2 AP Statistics Practice Test

T2.1 e
T2.2 d
T2.3 b
T2.4 b
T2.5 a
T2.6 d
T2.7 c
T2.8 e
T2.9 e
T2.10 c

T2.11 (a) Jane's performance was better. She did more curl-ups than 85% of girls her age. This means that she qualified for both the Presidential award and for the National award. Matt did more curl-ups than 50% of boys his age. This means that less than 50% of the boys his age did better than he did, whereas less than 15% of the girls her age did better than Jane. Matt qualified for the National award, but did not qualify for the Presidential award. (b) Since Jane's position in her distribution is so much higher than Matt's position in his distribution, Jane's z-score is likely to be bigger than Matt's z-score.

T2.12 (a) The z-value that corresponds to this soldier's head circumference is $z = \frac{23.9 - 22.8}{1.1} = 1$. So the proportion of observations lower than this is 0.8413 (using Table A). This means that this soldier's head circumference is at approximately the 84th percentile.

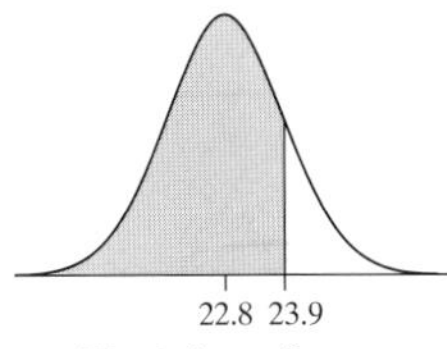

Head circumferences

(b) Standardizing the left endpoint we get $z = \frac{20 - 22.8}{1.1} = -2.55$. Using Table A, the area below −2.55 is 0.0054. Standardizing the right endpoint we get $z = \frac{26 - 22.8}{1.1} = 2.91$. The area below 2.91 (using Table A) is 0.9982, so the area above 2.91 is 1 − 0.9982 = 0.0018. This means that the area in both tails is 0.0054 + 0.0018 = 0.0072. So approximately 0.7% of soldiers require custom helmets.

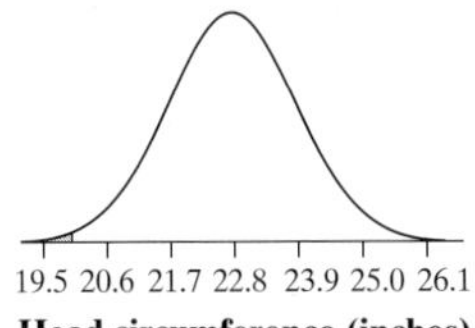

Head circumference (inches)

(c) The quartiles of a standard Normal distribution are −0.67 and 0.67. To find the quartiles of the head circumference distribution, we solve the following equations for x.

$$-0.67 = \frac{x - 22.8}{1.1} \Rightarrow x = -0.67(1.1) + 22.8 = 22.063$$

$$0.67 = \frac{x - 22.8}{1.1} \Rightarrow x = 0.67(1.1) + 22.8 = 23.537$$

This means that $Q_1 = 22.063$ and $Q_3 = 23.537$. So $IQR = Q_3 - Q_1 = 23.537 - 22.063 = 1.474$ inches.

T2.13 No, these data do not seem to follow a Normal distribution. First, there is a large difference between the mean and the median. The mean is 48.25 and the median is 37.80. The Normal distribution is symmetric so the mean and median should be quite close in a Normally distributed data set. This data set appears to be highly skewed to the right. This can be seen by the fact that the mean is so much larger than the median. It can also be seen by the fact that the distance between the minimum and the median is 37.80 − 2 = 35.80, but the distance between the median and the maximum is 204.90 − 37.80 = 167.10.

CHAPTER 3

Section 3.1

Answers to Check Your Understanding

Page 144 **1.** Explanatory variable: number of cans of beer. Response variable: blood alcohol level. **2.** There are two explanatory variables: amount of debt and income. Response variable: stress caused by college debt. *Page 149* **1.** The relationship is positive. The longer the duration of the eruption, the longer the wait between eruptions. One reason for this may be that if the geyser erupted for longer, it expended more energy and it will take longer to build up the energy needed to erupt again. **2.** The form is roughly linear with two clusters. The clusters indicate that in general there are two types of eruptions: one shorter, the other somewhat longer. **3.** The relationship is fairly strong. Two points define a line, and in this case we could think of each cluster as a point, so the two clusters seem to define a line. **4.** There are a few outliers around the clusters, but not many and not very distant from the main grouping of points. **5.** The Starnes family needs to know how long the last eruption lasted in order to predict how long until the next one. *Page 154* **1.** Estimates of r will vary. **(a)** The correlation is about 0.9. There is a strong, positive linear relationship between the number of boats registered in Florida and the number of manatees killed. **(b)** The correlation is about 0.5. There is a moderate, positive linear relationship between the number of named storms predicted and the actual number of named storms. **(c)** The correlation is about 0.3. There is a weak, positive linear relationship between the healing rate of the two front limbs of the newts. **(d)** The correlation is about −0.1. There is a weak, negative linear relationship between last year's percent return and this year's percent return in the stock market. **2.** The correlation would decrease. This point has the effect of strengthening the observed linear relationship that we see.

Answers to Odd-Numbered Section 3.1 Exercises

1. Explanatory variable: water temperature. Response variable: weight change (growth). Both are quantitative.

3. **(a)** Students with higher IQs tend to have higher GPAs, and those with lower IQs generally have lower GPAs. The plot does show a positive association. **(b)** Roughly linear, because a line through the scatterplot of points would provide a good summary. The positive association is moderately strong because most of the points would be close to the line. **(c)** An IQ of about 103, a GPA of about 0.4.

5. Here is a scatterplot.

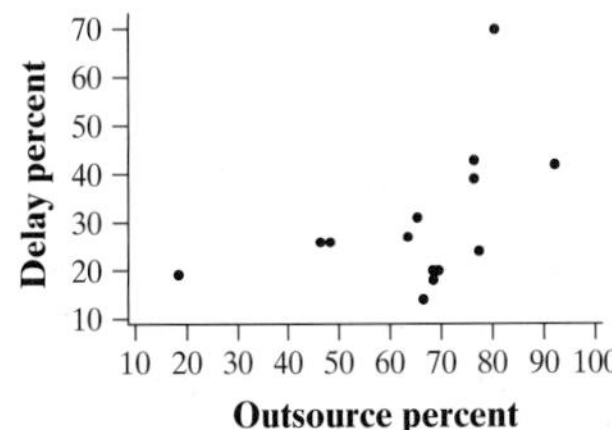

7. **(a)** Positive, somewhat curved, moderately weak association. **(b)** The outlier is Hawaiian Airlines. Without this outlier, the relationship is more linear but still not very strong.

9. (a) Here is a scatterplot.

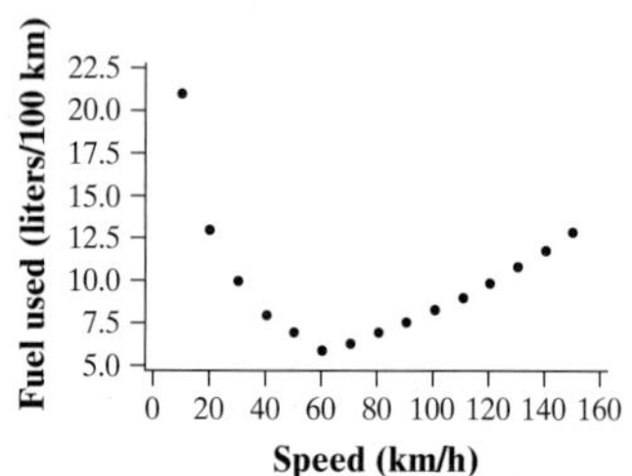

(b) The relationship is curved. High amounts of fuel were used for low and high values of speed, and low amounts of fuel were used for moderate speeds. That's because the best fuel efficiency is obtained by driving at moderate speeds. (c) Both are present. At low speeds the graph shows a negative association. At higher speeds the association is positive. (d) Very strong, with little deviation from a curve that can be drawn through the points.
11. (a) Several southern states lie at the lower edges of their clusters. Students in the southern states do not do as well as their counterparts in other portions of the country. (b) It has a much lower mean SAT Math score than the other states with a similar percent of students taking the exam.
13. **State:** Is the relationship between the number of breeding pairs of merlins and the percent of males who return the next season negative? **Plan:** Begin with a scatterplot, and compute the correlation if appropriate. **Do:** A scatterplot of the percent returning against the number of breeding pairs shows the expected negative association. Though slightly curved, it is reasonable to compute $r = -0.7943$ as a measure of the strength of the linear association. **Conclude:** This supports the theory: a smaller percent of birds survive following a successful breeding season.

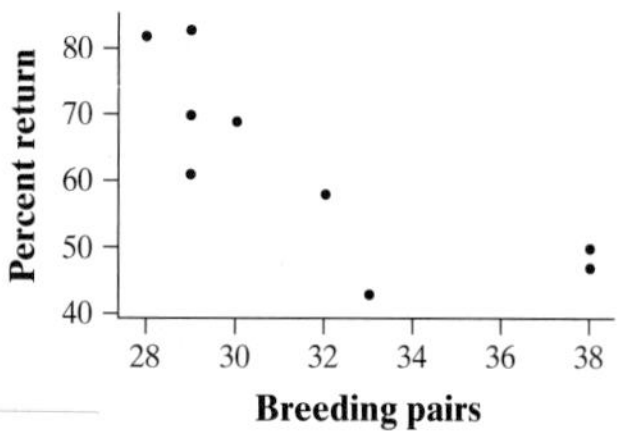

15. (a) $r = 0.9$. (b) $r = 0$. (c) $r = 0.7$. (d) $r = -0.3$. (e) $r = -0.9$.
17. (a) Gender is a categorical variable. (b) r is at most 1. (c) r has no units.
19. (a) The scatterplot shows a strong positive linear relationship between the two measurements. Thus, all five specimens appear to be from the same species.

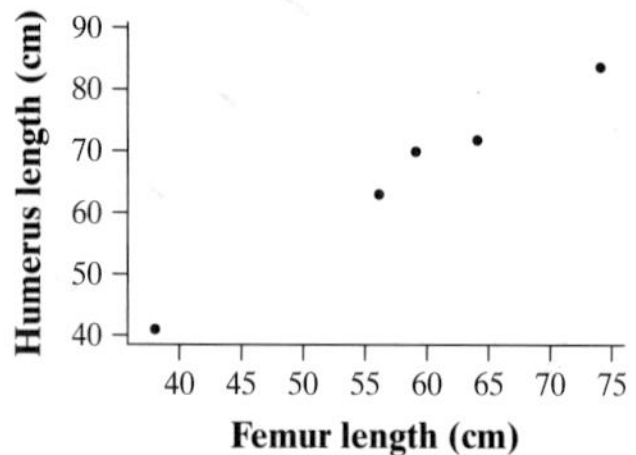

(b) The femur measurements have mean 58.2 and standard deviation 13.2. The humerus measurements have mean 66 and standard deviation 15.89. The sum of the products of the standardized values is 3.97659, so the correlation coefficient is $r = 0.9941$.

21. (a) There is a strong, positive linear association between the salt content and calories of hot dogs. (b) It would tend to decrease the strength of the linear relationship, and thus, the correlation.
23. (a) The correlation would not change. It does not have units associated with it, so a change in units for either variable (or both) will not change the correlation. (b) The correlation would not change. The correlation does not distinguish between the explanatory and response variables.
25. Here is a scatterplot. The one unusual point (10, 1) is responsible for reducing the correlation.

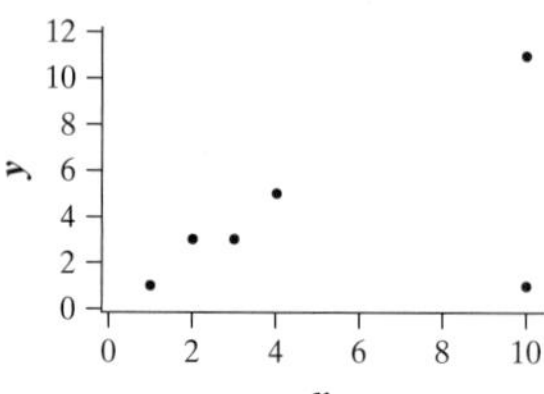

27. a
29. d
31. c
33. A histogram is shown below. The distribution is sharply right-skewed, with several possible high outliers. The five-number summary is 0.1, 3.5, 5.4, 9, 33.8.

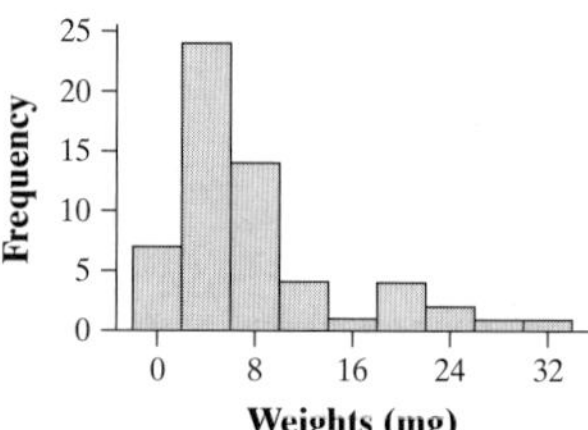

Section 3.2

Answers to Check Your Understanding

Page 167 **1.** The slope is 40. We predict that a rat will gain 40 grams of weight per week. **2.** The y intercept is 100. This suggests that we expect a rat at birth to be 100 grams. **3.** We predict the rat's weight to be 740 grams. **4.** The time is measured in weeks for this equation, so 2 years becomes 104 weeks. We then predict the rat's weight to be 4260 grams, which is equivalent to 9.4 pounds (about the weight of a large newborn human). This is unreasonable and is the result of extrapolation. ***Page 176*** 1. We predict the fat gain for this person to be 1.3722 kg. So the residual is 2.3 – 1.3722 = 0.9278. **2.** The residual says that this person gained 0.9278 kg more than we would have predicted using the least-squares line as a model. **3.** The line overpredicted the fat gain the most for the person who had an NEA change of 580 and a fat gain of 0.4. This person's predicted fat gain is 1.51, and so the residual is –1.11. Based on the list of residuals given in the previous example, this is the largest negative residual and therefore the point for which the fat gain was most overpredicted. ***Page 179*** **1.** There is a moderate positive linear relationship with one outlier in the bottom-right corner of the plot. **2.** The average error (residual) in predicting the backpack weight is 2.27 lb using the least-squares regression line. ***Page 181*** **1.** c **2.** d

Answers to Odd-Numbered Section 3.2 Exercises

35. The equation is $\hat{y} = 80 - 6x$ where $\hat{y} =$ the estimated weight of the soap and $x =$ the number of days since the bar was new.

37. **(a)** The slope is 1.109. We predict highway mileage will increase by 1.109 mpg for each 1 mpg increase in city mileage. **(b)** The intercept is 4.62 mpg. This is not statistically meaningful, because this would represent the highway mileage for a car that gets 0 mpg in the city. **(c)** With city mpg of 16, the predicted highway mpg is $4.62 + 1.109(16) = 22.36$ mpg. With city mpg of 28, the predicted highway mpg is $4.62 + 1.109(28) = 35.67$ mpg.

39. **(a)** The slope is -0.0053; the pH decreased by 0.0053 units per week on average. **(b)** The y intercept is 5.43, and it provides an estimate for the pH level at the beginning of the study. **(c)** The pH is predicted to be 4.635 at the end of the study.

41. No. The data was collected weekly for 150 weeks. 1000 months corresponds to roughly 4000 weeks, which is well outside the observed time period. This constitutes extrapolation.

43. The dotted line in the scatterplot is the line $\hat{y} = 1 - x$ and the solid line is the line $\hat{y} = 3 - 2x$. The dotted line comes closer to all the data points. Thus, the line $\hat{y} = 1 - x$ fits the data better.

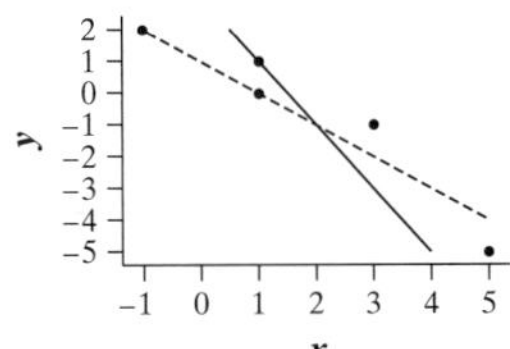

45. The residual is -0.085. The line predicted a pH value for that week that was 0.085 too large.

47. **(a)** The equation for predicting $y =$ husband's height from $x =$ wife's height is $\hat{y} = 33.67 + 0.54x$. **(b)** The predicted height is 69.85 inches. 67 inches is one standard deviation above the mean for women. So the predicted value for husband's height would be $\bar{y} + rs_y = 69.85$.

49. **(a)** $r^2 = 0.25$. Thus, the straight-line relationship explains 25% of the variation in husbands' heights. **(b)** The average error (residual) when using the line for prediction is 1.2 inches.

51. **(a)** The regression line is $\hat{y} = -3.5519 + 0.101x$. **(b)** $r^2 = 0.4016$. Thus, 40.16% of the variation in GPA is accounted for by the linear relationship with IQ. **(c)** The predicted GPA for this student is $\hat{y} = 6.8511$ and the residual is -6.3211. The student had a GPA that was 6.3211 points worse than expected for someone with an IQ of 103.

53. **(a)** Here is a scatterplot.

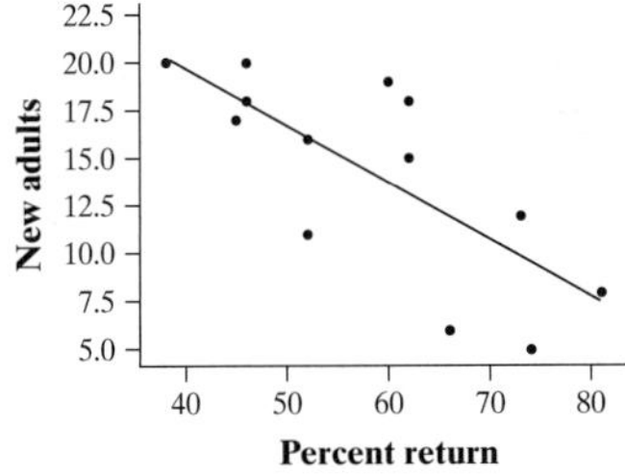

(b) The least-squares regression line is $\hat{y} = 31.9 - 0.304x$. Minitab output is shown at top right.

```
The regression equation is
newadults = 31.9 - 0.304 %returning

Predictor   Coef      SE Coef      T      P
Constant    31.934    4.838      6.60  0.000
%returning  -0.30402  0.08122   -3.74  0.003

S = 3.66689  R-Sq = 56.0% R-Sq(adj) = 52.0%
```

(c) The slope tells us that as the percent of returning birds increases by 1, we predict the number of new birds will decrease by 0.304. The y intercept provides a prediction that we will see 31.9 new adults in a colony when the percent of returning birds is 0. This is extrapolation. **(d)** The predicted value for the number of new adults is 13.66, or about 14.

55. **(a)** A residual plot suggests that the line is a decent fit. The points are all scattered around a residual value of 0.

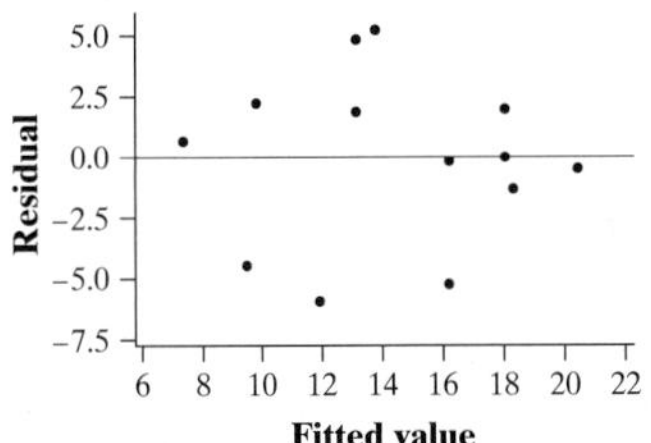

(b) The point with the largest residual has a residual of about -6. This means that the line overpredicted the number of new adults by 6.

57. 56% of the variation in the number of new adult birds is explained by the straight-line relationship. The typical error when using the line for prediction is 3.67 new adults.

59. **(a)** There is a positive linear association between the two variables. There is more variation in the field measurements for larger laboratory measurements. **(b)** The points for the larger depths fall systematically below the line $y = x$, which means that the field measurements are too small compared with the laboratory measurements. **(c)** The slope would decrease and the intercept would increase.

61. No; the data show a clearly curved pattern in the residual plot.

63. **(a)** The regression line is $\hat{y} = 157.68 - 2.99x$. Following a season with 30 breeding pairs, we predict that about 68% of males will return. **(b)** The linear relationship explains 63.1% of the variation in the percent of returning males. **(c)** $r = -0.79$; the sign is negative because it has the same sign as the slope coefficient. **(d)** Since $s = 9.46$, the typical error when using the line to predict the return rate of males is about 9.46%.

65. **(a)** Here is a scatterplot (with regression lines).

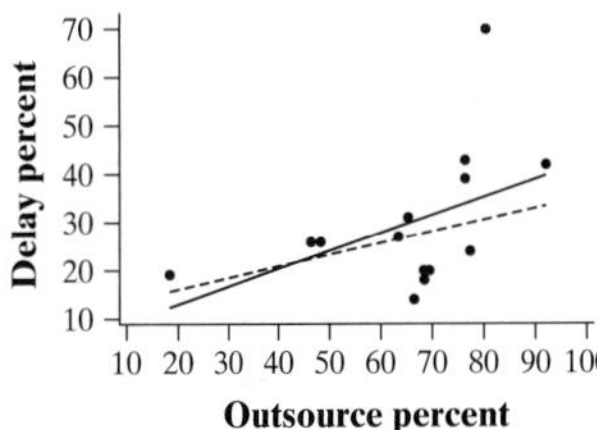

(b) The correlation is $r = 0.4765$ with all points. It rises slightly to 0.4838 when the outlier is removed; this is too small a change to consider the outlier influential for correlation. **(c)** With all points, $\hat{y} = 4.73 + 0.3868x$ (the solid line), and the prediction for $x = 76$ is

34.13%. With Hawaiian Airlines removed, $\hat{y} = 10.878 + 0.2495x$ (the dotted line), and the prediction is 29.84%. This difference in prediction indicates that the outlier is influential for regression.

67. **(a)** Here is the scatterplot with two new points. Point A is a horizontal outlier. Point B is a vertical outlier.

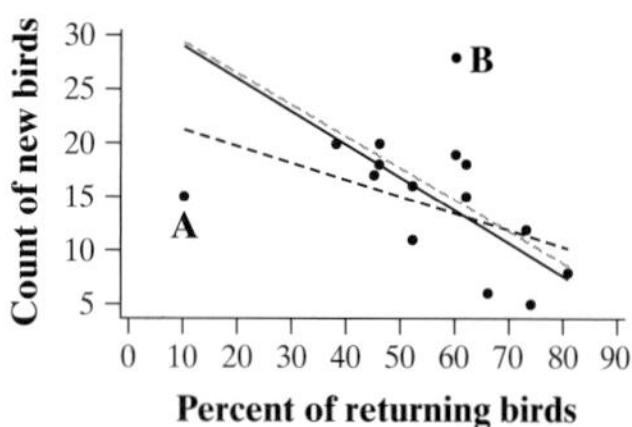

(b) The three regression formulas are $\hat{y} = 31.9 - 0.304x$ (the original data, solid line); $\hat{y} = 22.8 - 0.156x$ (with Point A, dashed line); $\hat{y} = 32.3 - 0.293x$ (with Point B, gray dashed line). Adding Point B has little impact. Point A is influential; it pulls the line down.

69. **State:** How accurate are Dr. Gray's forecasts? **Plan:** Construct a scatterplot with Gray's forecast as the explanatory variable and, if appropriate, find the regression equation. Then make a residual plot and calculate r^2 and s. **Do:** The scatterplot shows a moderate positive association; the regression line is $\hat{y} = 1.688 + 0.9154x$ with $r^2 = 0.30$ and $s = 4.0$. The relationship is strengthened by the large number of storms in the 2005 season, but it is weakened by 2006 and 2007, when Gray's forecasts were the highest, but the actual numbers of storms were unremarkable. As an indication of the influence of the 2005 season, we might find the regression line without that point; it is $\hat{y} = 3.977 + 0.6699x$, with $r^2 = 0.265$ and $s = 3.14$.

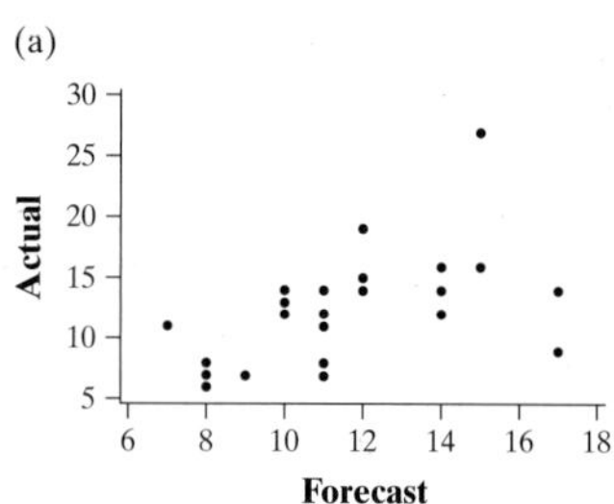

Finally, the residual plot does not indicate any problems with fitting the linear equation.

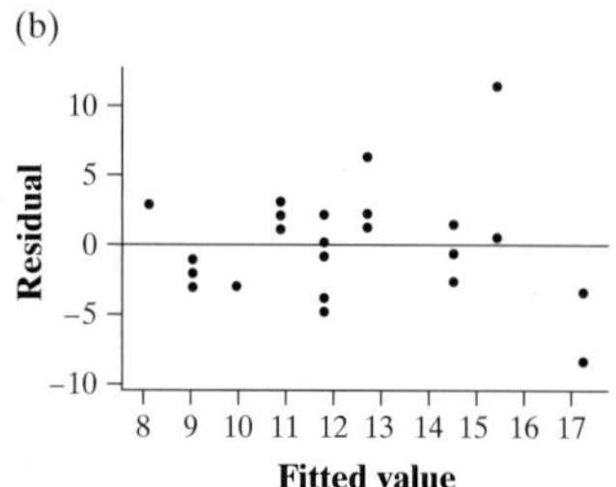

Conclude: If Gray forecasts $x = 16$ tropical storms, we expect 16.33 storms in that year. However, we do not have very much confidence in this estimate, because the regression line explains only 30% of the variation in tropical storms and the typical error we should expect when using this line for prediction is 4 storms. (If we exclude 2005, the prediction is 14.7 storms, but this estimate is less reliable than the first.)

71. b

73. b

75. b

77. d

79. About 92.92%

81. **(a)** There is evidence of an association between accident rate and marijuana use. Those people who use marijuana more are more likely to have caused accidents.

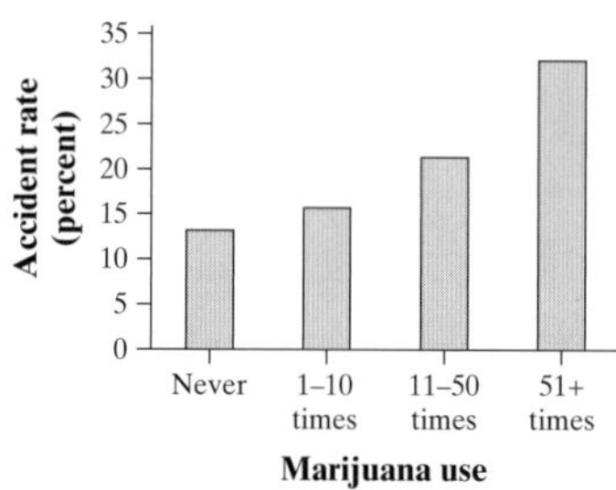

(b) This was an observational study. If we wanted to see whether using marijuana *caused* more accidents, then we would have to set up an experiment where we randomly assigned people to use more or less marijuana.

Answers to Chapter 3 Review Exercises

R3.1 **(a)** Explanatory variable: weight of a person. Response variable: mortality rate. **(b)** No. Since obese people tend to be poor, their higher mortality rate may be due to low-quality medical care and not to their weight.

R3.2 **(a)** The direction of the scatterplot is positive but appears curved, not linear. The strength of the association is moderate. **(b)** The hippopotamus is unusual because its lifespan is longer than would be expected given its gestation period. The Asian elephant is unusual because it has the second-longest gestation time and has a longer lifetime than expected. The observation for the giraffe tends to follow the curvilinear shape, with possibly a little shorter lifespan than expected based on the pattern in the remaining data.

R3.3 **(a)** The slope is 0.0138 minutes per meter. We predict that if the depth of the dive is increased by one meter, it will add 0.0138 minutes (about 0.83 seconds) to the time spent underwater. **(b)** When Depth = 200, the regression line estimates DiveDuration to be $\hat{y} = 5.45$ minutes. **(c)** The intercept suggests that a dive of no depth would last an average of 2.69 minutes; this does not make any sense.

R3.4 **(a)** The least-squares regression line is $\hat{y} = 7288.54 + 11630.6x$ where y represents the mileage of the cars and x represents the age. **(b)** The residual for this car is -12072.14. **(c)** The slope of the line is 11,630.6. We expect that cars will be driven 11,630.6 miles per year on average. **(d)** Since the slope is positive, the correlation $r = 0.906$. This shows that there is a strong, linear relationship between the age of cars and their mileage. **(e)** The line fits reasonably well. The residual plot shows no large pattern. The typical error (residual) when using this line to predict mileage is 19,280. This suggests that although the line fits reasonably well, it would not be very useful in practice since our predictions would be off by an average of approximately 20,000 miles.

R3.5 (a) The association is negative, linear, and fairly strong.

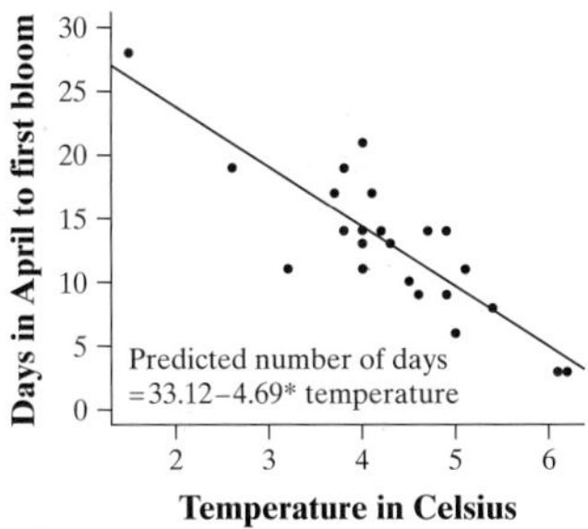

(b) The least-squares regression equation is $\hat{y} = 33.12 - 4.69x$ where y represents the number of days and x represents the temperature. For every 1 degree increase in average March temperature, we predict the number of days in April until first bloom to decrease by 4.69. The y intercept is outside the range of data and therefore has no meaningful interpretation. (c) Predicted number of days until first bloom is 16.7. We predict the first cherry blossom to appear on April 17. (d) Predicted number of days until first bloom is $\hat{y} = 12.015$. The observed value was 10. The residual is then -2.015. (e) The plot is the residuals versus the explanatory variable. There is no discernible pattern in the residuals. They are clustered about 0 in a random fashion.

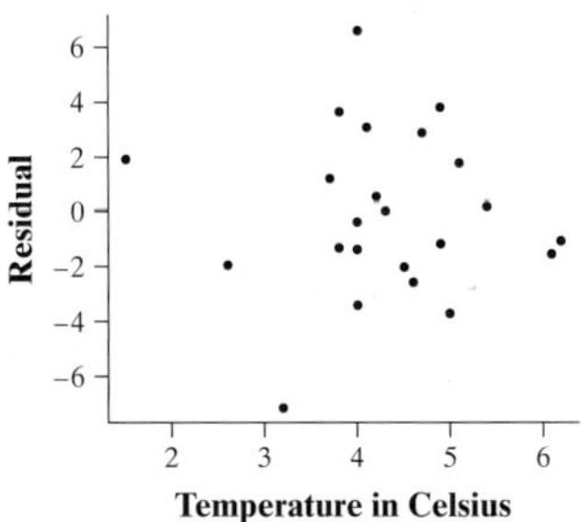

(f) $r^2 = 0.72$ and $s = 3.02$. 72% of the variation in the number of days in April until the first cherry blossom appears is explained by the linear relationship with the average temperature (in Celsius) in March. We expect a typical prediction error of 3.02 days.

R3.6 (a) The regression equation is $\hat{y} = 0.16x + 30.2$, where $x =$ pre-exam total and $y =$ final exam score. For every extra point earned on the pre-exam total, we predict that the score on the final exam will increase by about 0.16. (b) Julie's predicted final-exam score is $\hat{y} = 78.2$. (c) $r^2 = 0.36$, so only 36% of the variability in the final-exam scores is accounted for by the linear relationship with pre-exam totals. Julie has a good reason to think that this is not a good estimate.

R3.7 (a) The correlation would decrease because Hawaii is far above average for both the maximum 24-hour precipitation and the maximum annual precipitation. (b) The blue line is the line calculated with all 50 states. Hawaii's point is influential and pulls the line up toward it. The other line is the one with all states except Hawaii. (c) The correlation will not change since it does not have units; s will decrease since it would now be measured in feet as well; the slope of the regression line would not change since both x and y are measured in the same units, leading to a slope without units, but the y intercept would decrease since it changes units from inches to feet. If we switch the explanatory and response variables, the correlation will not change, but the standard error and the least-squares line will.

Answers to Chapter 3 AP Statistics Practice Test

T3.1 d
T3.2 e
T3.3 c
T3.4 a
T3.5 a
T3.6 c
T3.7 b
T3.8 e
T3.9 b
T3.10 c
T3.11 (a) Here is a scatterplot, with regression line added.

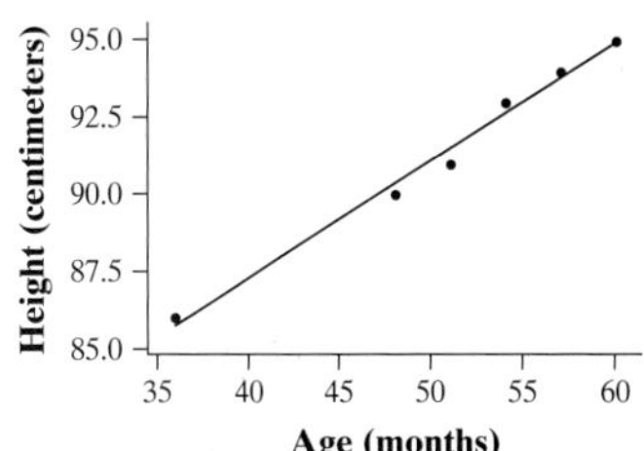

(b) The regression line for predicting $y =$ height from $x =$ age is $\hat{y} = 71.95 + 0.3833x$. (c) At age 480 months, we would predict Sarah's height to be $\hat{y} = 71.95 + 0.3833(480) = 255.934$ cm. Her height in inches would be $\frac{255.934}{2.54} = 100.76$ in. (d) This height is impossibly large (about 8 feet, 4 inches) because we used extrapolation. Obviously, the linear trend does not continue all the way out to 40 years. Our data were based only on the first 5 years of life.

T3.12 (a) The unusual point is the one in the upper-right corner with isotope value about -19.3 and silicon value about 345. This point is unusual in that it has such a high silicon value for the given isotope value. (b) (i) If the point were removed, the correlation would increase because this point does not follow the linear pattern of the other points. (ii) And since this point has a higher silicon value, if it were removed, the slope of the regression line would decrease and the y intercept would increase.

T3.13 (a) The regression equation is $\hat{y} = 92.29 - 0.05762x$. The variable y represents the percent of the grass burned, and x represents the number of wildebeest. (b) The slope of the regression line suggests that for every increase of 1000 wildebeest (this is a 1 unit increase in x since x is measured in terms of 1000s of wildebeest), we predict that the percent of grass area burned will decrease by about 0.058. (c) $r = -0.804$. There is a moderately strong, negative linear relationship between wildebeest abundance and percent of grass area burned. (d) The linear model is appropriate for describing the relationship between wildebeest abundance and percent of grass area burned. The residual plot shows a fairly "random" scatter of points around the "residual = 0" line. There is one large positive residual at 1249 thousand wildebeest. Since $r^2 = 0.646$, 64.6% of the variation in percent of grass area burned is explained by the least-squares regression of percent of grass area burned on wildebeest abundance. That leaves 35.4% of the variation in percent of grass area burned unexplained by the linear relationship.

CHAPTER 4

Section 4.1

Answers to Check Your Understanding

Page 211 **1.** Convenience sampling. This could lead him to think that the oranges are of better quality than they really are, if the farmer puts the best oranges on top. **2.** Voluntary response sampling. Only those who feel particularly strongly about the issue are likely to respond. In this case, those who are happy that the United Nations has its headquarters in the United States already have what they want and so are less likely to worry about responding to the question. *Page 219* **1.** You would have to identify 200 different seats, go to those seats in the arena, and find the people who are sitting there. This means going to 200 different locations throughout the arena, which would take time. Other problems are that people are not always in their seats throughout the game, and not all seats will necessarily be occupied in any given game. **2.** Use lettered rows as the strata because each lettered row is the same distance from the court and so would contain only seats with the same (or nearly the same) ticket price. This means that all people in any given stratum would have paid roughly the same amount for their tickets. **3.** Use numbered sections as the clusters because they include all different seat prices. Each section contains seats with many different ticket prices, so the people in a section would mirror the characteristics of the population as a whole. *Page 224* **1.** **(a)** Sampling error. This will result in undercoverage because those who are not listed in the phone book (those who do not have a phone or have only a cell phone) do not have the opportunity to be chosen. **(b)** Nonsampling error. This did not occur because of the way the sample was chosen but rather was an effect of the way the survey was administered. **(c)** Sampling error. Who you find will depend on where (in what neighborhood, etc.) you are standing. This has to do with how you choose your sample. **2.** By making it sound like disposable diapers are not a problem in the landfill, this question will result in fewer people suggesting that we should ban them. The author of the question highlights several other items that take up more space in the landfill, which makes it look like disposable diapers are really not a problem.

Answers to Odd-Numbered Section 4.1 Exercises

1. Population: all local businesses. Sample: the 73 businesses that return the questionnaire.

3. Population: 1000 envelopes stuffed during a given hour. Sample: 40 envelopes selected.

5. Only persons with a strong opinion on the subject, strong enough that they are willing to spend the time and money, will respond to this advertisement.

7. This is a voluntary response sample. It is biased in favor of those who feel most strongly about the issue being surveyed.

9. **(a)** Convenience sample. **(b)** It is unlikely that the first 100 students to arrive at school are representative of the student population in general. 7.2 hours is probably higher since you might expect that the students who arrive first are those who got a good night's sleep the night before. Students who received less sleep the night before are probably more likely to run late the next morning.

11. **(a)** Number the 40 students from 01 to 40 alphabetically. Go to the random number table, pick a starting point, and record two-digit numbers until you have 5 unique numbers between 01 and 40. **(b)** Starting at line 107, the sample is Johnson (20), Drasin (11), Washburn (38), Rider (31), and Calloway (07).

13. **(a)** Number the plots from 0001 to 1410. Go to the random number table, pick a starting point, and record four-digit numbers until you have 141 unique numbers between 0001 and 1410. **(b)** Starting at line 131, the sample is plots 0500, 0419, and 0418.

15. If you always begin at the same place, the results will not be random. You will end up using the same sample in every case.

17. **(a)** To obtain a random sample, the phones must be numbered in some way. Keeping track of the ordering of 1000 phones may be difficult. **(b)** It is possible that the quality of the phones changes over the course of the day so that the last phones manufactured are not representative of the day's production. **(c)** Each sample of 20 phones does not have the same probability of being selected. In fact, the 20 phones that are sampled will be the 50th, 100th, . . . , 1000th—the others have no chance of being sampled.

19. Assign 01 to 30 to the students (in alphabetical order). Starting on line 123 gives 08-Ghosh, 15-Jones, 07-Fisher, and 27-Shaw. Assigning 0–9 to the faculty members gives 1-Besicovitch and 0-Andrews.

21. **(a)** Use the three types of seats as the three strata since ticket prices will be similar within each stratum but different among the three strata. **(b)** It might be easier to obtain a cluster sample because a stratified random sample will still likely choose seats all over the stadium, which would make it very time-consuming to get to everyone. A cluster sample would be easier to obtain, because there would be many people sitting all together who would be part of the sample. In this case, one would use the section numbers for the clusters.

23. No. Some samples of size 250 have no chance of being selected (e.g., a sample containing 250 women).

25. **(a)** Cluster sampling. **(b)** Answers will vary. Label each block from 01 through 65; beginning at line 142, the 5 blocks are 02, 32, 26, 34, and 08.

27. Households without telephones or with unlisted numbers are omitted from this frame. Such households would likely be made up of poor individuals (who cannot afford a phone), those who choose not to have phones, and those who do not wish to have their phone number published. If the variable being measured tends to have different values for those excluded by this sampling method, then our sample result would be off in a particular direction from the truth about the population of households.

29. **(a)** You are sampling only from the lower-priced ticket holders. **(b)** Sampling error. The sampling frame differs from the population of interest (undercoverage).

31. **(a)** 89.1%. **(b)** It is likely that the high amount of nonresponse gave the researchers a lower mean number of miles driven because those who drive more are at home less to answer the phone.

33. More than 171 respondents have run red lights. We would not expect very many people to claim they *have* run red lights when they have not, but some people will deny running red lights when they have.

35. **(a)** The wording is clear. The question is slanted in favor of warning labels. **(b)** The question is clear, but it is clearly slanted in favor of national health insurance by asserting it would reduce administrative costs. **(c)** The wording is too technical for many people to understand, and for those who do understand the question, it is slanted because it suggests reasons why one should support recycling. It could be rewritten to something like "Do you support economic incentives to promote recycling?"

37. c
39. d
41. e
43. The predicted sleep debt for a 5-day school week, based on the least-squares regression equation, is $2.23 + 3.17(5) = 18.08$ hours, a little more than 3 hours greater than what was found in the research study. Based on their collected data, the students have reason to be skeptical of the research study's reported results.

Section 4.2

Answers to Check Your Understanding

Page 233 **1.** An experiment because a treatment (brightness of screen) was imposed on the laptops. **2.** An observational study. Students were not assigned to a particular number of meals to eat with their family per week. **3.** Explanatory variable: the number of meals per week eaten with their family. Response variable: probably their GPA. **4.** This is an observational study and there may well be lurking variables that are actually influencing the response variable. For instance, families that eat more meals together may also be families where the parents show more interest in their children's education and therefore help them to do better in school. *Page 240* **1.** Here is an outline of the design.

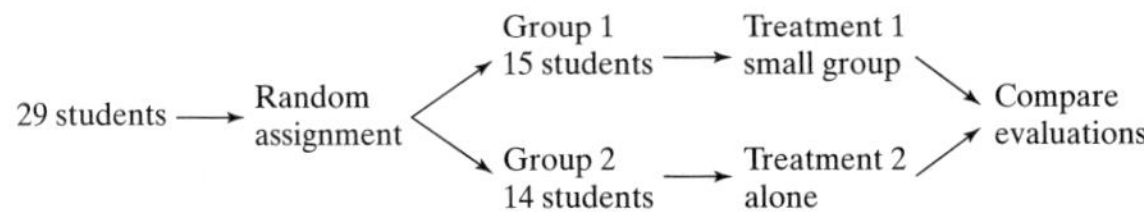

2. Using an alphabetical list of the students, assign each student a number between 01 and 29. Pick a line of the random number table and read off two-digit numbers until you have 15 numbers between 01 and 29. These students belong in the treatment group, where students will meet in small groups. The other students will view the videos alone. **3.** The purpose of the control group is to have a group with which to compare the treatment group. Presumably, the students have been evaluating their own performances by themselves. If you incorporate such a group into your experiment, you can evaluate whether the group work is actually better. *Page 244* **1.** No. It is possible that women who thought they were getting an ultrasound would have different reactions to pregnancy than those who knew that they hadn't received an ultrasound. **2.** No. While the people weighing the babies at birth may not have known whether that particular mother had an ultrasound or not, the mothers did know whether they had had an ultrasound or not. This means that the mothers may have affected the outcome since they knew whether they had received the treatment or not. **3.** An improved design would be one in which all mothers are treated as if they had an ultrasound, but for some mothers the ultrasound machine just wasn't turned on (a fact that would not be obvious to the women). This means that the ultrasound would have to be done in such a way that the women could not see the screen.

Answers to Odd-Numbered Section 4.2 Exercises

45. **(a)** Observational study. No treatment was imposed on the mothers. The researchers simply asked them to report both their chocolate consumption and their babies' temperament. **(b)** Explanatory variable: mother's chocolate consumption. Response variable: baby's temperament. **(c)** No. This study is an observational study, so we cannot make a conclusion about cause and effect. There could be a lurking variable that is actually causing the difference in temperament.

47. **(a)** Experiment. Students were randomly assigned to the different teaching methods. **(b)** Since this was an experiment with proper randomization, the teacher can conclude that using the computer software appears to result in higher increases in test scores.

49. Answers may vary. Private versus public schools. Private schools tend to have smaller classes, and private school students might tend to earn higher scores. There might be something else about the private schools, however, that leads to that success other than the small class sizes. So final success could be dependent on either of these two variables.

51. Experimental units: pine seedlings. Explanatory variable: light intensity. Treatments: full light, 25% light, and 5% light. Response variable: dry weight at the end of the study.

53. Subjects: the individuals who were called. Explanatory variables: (1) information provided by interviewer (2) whether caller offered survey results. Treatments: (1) giving name/no survey results; (2) identifying university/no survey results; (3) giving name and university/no survey results; (4) giving name/offering to send survey results; (5) identifying university/offering to send survey results; (6) giving name and university/offering to send survey results. Response variable: whether or not the interview was completed.

55. Experimental units: fabric specimens. Explanatory variables: (1) roller type; (2) dyeing-cycle time; (3) temperature. Treatments: (1) metal, 30 minutes, 150 degrees; (2) natural, 30 minutes, 150 degrees; (3) metal, 40 minutes, 150 degrees; (4) natural, 40 minutes, 150 degrees; (5) metal, 30 minutes, 175 degrees; (6) natural, 30 minutes, 175 degrees; (7) metal, 40 minutes, 175 degrees; (8) natural, 40 minutes, 175 degrees. Response variable: a quality measurement.

57. There was no control group for comparison purposes. We don't know if this was a placebo effect or if the flavonols actually affected the blood flow.

59. **(a)** Write all names on slips of paper, place in a container, and mix thoroughly. Select one slip and assign that name to Treatment 1. Select another name and assign to Treatment 2. Assign third name to Treatment 3. Rotate through the treatments until all have been assigned. **(b)** Assign the students numbers between 001 and 120. Pick a spot on Table D, read the first 40 numbers between 001 and 120, skipping repeats or any that are not between 001 and 120, and assign to Treatment 1. Assign next 40 numbers to Treatment 2. Assign remainder to Treatment 3. (c) Assign the students numbers between 001 and 120. Using the RandInt function on the calculator and ignoring all repeats, assign the first 40 numbers to Treatment 1, the next 40 to Treatment 2, and the remainder to Treatment 3.

61. **(a)** Completely randomized design.

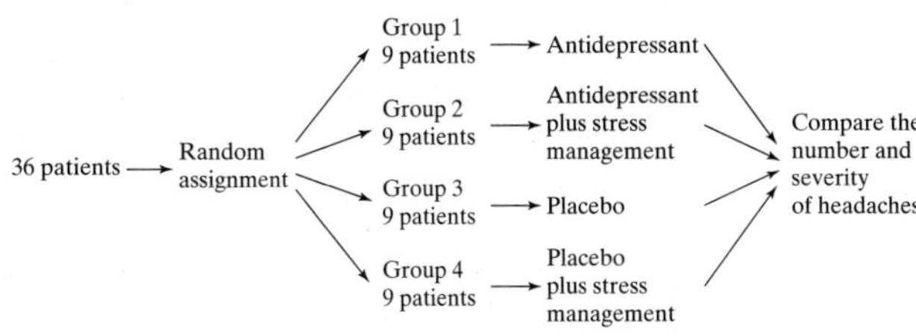

(b) Write patient names on 36 identical slips of paper, put them in a hat, and mix them well. Draw out 9 slips. The corresponding patients will be in Group 1. Draw out 9 more slips. Those patients will be in Group 2. The next 9 slips drawn will be the patients in Group 3, and the remaining 9 patients will be assigned to Group 4. Random assignments will vary.

63. (a) Expense, condition of the patient, etc. In a serious case, when the patient has little chance of surviving, a doctor might choose not to recommend surgery; it might be seen as an unnecessary measure, bringing expense and a hospital stay with little benefit to the patient. (b) Randomly assign the patients to two groups of 150 each. One group will receive the traditional surgery, and the other group will receive the new method of treatment. At the end of the study, measure how many patients survived.

65. (a) Random assignment. If players are allowed to choose which treatment they get, those who choose one particular treatment over the other may be different in a fundamental way. For example, maybe the weaker players will be more likely to choose the new method, and the stronger players would stick with weight lifting. (b) The response variable of the number of push-ups that a player can do could be part of what the coach should measure, but this measures only one kind of upper-body strength, and he should probably combine this with other measures.

67. (a) First, we need to control for the effects of lurking variables and to use at least two groups for comparison purposes. Next we need random assignment to help create roughly equivalent groups before the treatments are administered. Finally, we need replication to ensure that a difference in response between the two groups is due to the treatments and not to chance variation. (b) There were two groups in the study: one in which the children were assigned to an intensive preschool program and one in which they weren't. All of the children were given nutritional supplements and help from social workers. The children were assigned at random to the two groups, and there were a total of 111 children in the experiment.

69. (a) Harmless leaf. (b) The subjects developed rashes on the arm exposed to the placebo (a harmless leaf) simply because they thought they were being exposed to the active treatment (a poison ivy leaf).

71. Because the experimenter knew which subjects had learned the meditation techniques, he or she may have had some expectations about the outcome of the experiment: if the experimenter believed that meditation was beneficial, he or she may subconsciously rate that group as being less anxious.

73. (a) Control: The effects of lurking variables on the response, whether the woman became pregnant, were controlled by controlling the manner in which the treatments were applied: half of the women received acupuncture treatment 25 minutes before embryo transfer and again 25 minutes after the transfer, the other half lay still for 25 minutes after the transfer. Random assignment: Randomly assigning the women to the two treatments should eliminate any systematic bias in assigning the subjects and should also balance out the effects of any lurking variables across the two treatment groups. Replication: 80 women were assigned to each treatment. These groups are large enough to ensure that the differences in the pregnancy rates of the two groups are due to the treatments themselves and not to chance variation in the random assignment. (b) The difference in the percent of women who received acupuncture and became pregnant and those who lay still and became pregnant was large enough to conclude that the difference was most likely due to the treatments rather than to chance. (c) We should be cautious about drawing conclusions based on the results of one study. The way this study was designed, it's possible that the observed difference is due in part to the placebo effect since the women were aware of which treatment they received. If possible, another study should be done in which the control group receives a fake acupuncture treatment.

75. (a) Write "yawn" on 14 slips of paper and "no yawn" on 36 slips of paper. Mix the slips and draw 16 of them. These will be the people subjected to the treatment with no yawn seed. The remainder will be in the yawn-seed treatment group. (b) We would not conclude that yawning is contagious. In our 50 random reassignments, getting 10 yawns out of 34 people was not at all unusual.

77. (a) The blocks are the different diagnoses (e.g., asthma). Within any given diagnosis, we are looking for differences in patients' health and satisfaction with medical care between doctors and nurse-practitioners. (b) A completely randomized design would assign patients to two treatment groups (doctors and nurses) without regard to their diagnosis. This ignores differences between patients with asthma, diabetes, and high blood pressure, which would probably result in a great deal of variability in measures of health and satisfaction in both groups. That would make it harder to compare the effectiveness of nurses and doctors. Blocking will control for the variability in subjects' responses due to their diagnosis. This will allow researchers to look separately at health and satisfaction for patients with each of the three diagnoses, as well as to better assess the relative effectiveness of nurses and doctors.

79. (a) The difference in soil fertility among the plots is a potential lurking variable. A completely randomized design could assign one of the varieties of corn to more fertile plots just by chance. If those plots produced extremely high yields, we wouldn't know if it was due to the corn variety or to soil fertility. Blocking will allow researchers to control for the variability in yield due to soil fertility. (b) Use rows as the blocks. All plots in the same row have the same amount of fertility and so are as similar as possible. (c) Let the digits 1–5 correspond to the five corn varieties A through E. Begin with, say, line 110, and assign the letters to the rows from west to east (left to right). Use a new line (111, 112, 113, and 114) for each row. For example, for Block 1, we obtain 3, 4, 1, 2, corresponding to varieties C, D, A, B, and lastly E being planted in the first row from west to east (ignoring repeats and digits 0, 6–9). The remaining rows are assigned using this same process. The results of this assignment are shown in the following table.

Block 1:	C	D	A	B	E
Block 2:	A	D	E	C	B
Block 3:	E	C	D	A	B
Block 4:	B	E	D	C	A
Block 5:	D	E	A	C	B

81. (a) If all rats from litter 1 were fed Diet A and we found Diet A to be better, we would not know if this was because of the diet itself or because litter 1 was different from litter 2. This is what it means for the genetics and the diet to be confounded. We cannot separate out the effects of one variable from the effects of the other. (b) A better design would be a randomized block design with the litters as the blocks. In this case, each diet would be given to some rats of each litter.
83. (a) Completely randomized design. (b) Have the students be the blocks to control the variability in driving ability among individual students. Have each student drive the simulator twice–once with a hands-free cell phone and once without, in random order. (c) To control for any possible effects of treatment order.
85. (a) Here is an outline of the design.

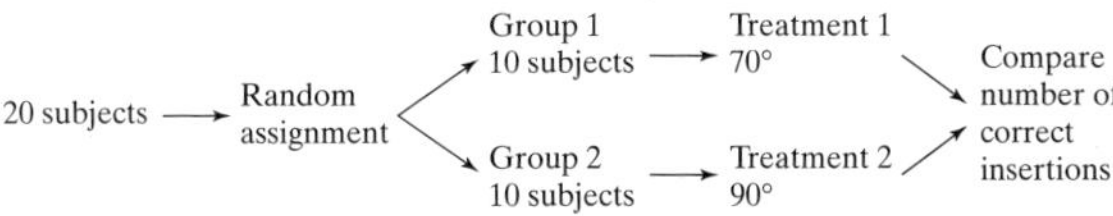

(b) All subjects will perform the task twice; once in each temperature condition. Randomly choose which temperature each subject works in first by flipping a coin.
87. (a) This experiment confounds gender with deodorant. If the students find a difference between the two groups, they will not know if it is a gender difference or due to the deodorant. (b) A better design would be a matched pairs design in which each student would have one armpit randomly assigned to receive deodorant A and the other deodorant B. Have each student rate the difference between their own armpits at the end of the day. This design prevents confounding by gender since each subject is compared to him- or herself.
89. Take the 50 volunteers and randomly assign them to one of two equal groups (25 volunteers each). Now randomly select one of the two groups. In this group, give men Razor A to use. Give Razor B to the men in the other group. Measure how close the razor shaves. On the next morning, give the men the other razor and measure how close the razor shaves. Analyze the difference in closeness.
91. c
93. b
95. c
97. c
99. (a) Since we know the weights of seeds of a variety of winged bean are approximately Normal, we can use the Normal model to find the percent of seeds that weigh more than 500 mg. First, we standardize 500 mg:

$$z = \frac{x-\mu}{\sigma} = \frac{500-525}{110} = \frac{-25}{110} = -0.23$$

Using Table A, we find the proportion of the standard Normal curve that lies to the left of $z = -0.23$ to be 0.4090, which means that $1 - 0.4090 = 0.5910$ lies to the right of $z = -0.23$. Thus, 59.1% of seeds weigh more than 500 mg. (b) We need to find the z-score with 10% (or 0.10) to its left. The value $z = -1.28$ has proportion 0.1003 to its left, which is the closest proportion to 0.10. Now, we need to find the value of x for the seed weights that gives us $z = -1.28$:

$$-1.28 = \frac{x-525}{110}$$
$$-1.28(110) = x - 525$$
$$525 - 1.28(110) = x$$
$$384.2 = x$$

If we discard the lightest 10% of these seeds, the smallest weight among the remaining seeds is 384.2 mg.
101. If we subtract Twin B – Twin A, and look at a dotplot of the differences, we see that all but one of the differences are positive. This suggests that in most cases, Twin B (the one living in the higher-income homes) tends to have a higher IQ. This is confirmed by computing the mean and standard deviation. The mean difference is 5.83 and the standard deviation is 3.93. This says that the IQ of the twin living in a home with a higher income has, typically, an IQ that is 5.83 points higher than his or her corresponding twin and that the average difference for sets of twins is about 3.93 points away from 5.83.

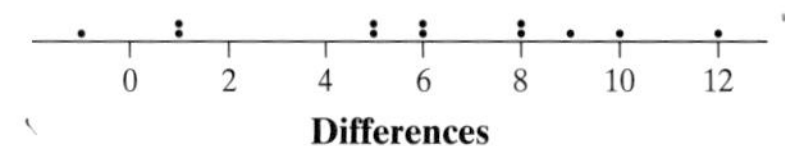

Section 4.3

Answers to Odd-Numbered Section 4.3 Exercises

103. Since this study involved random assignment to the treatments, we can infer cause and effect. Therefore, we can conclude that living in foster care in Romania is better than living in an institution.
105. It is difficult to draw a useful conclusion from this study. Since this study did not involve random assignment to a treatment, we cannot infer cause and effect. Also, since the individuals were not randomly chosen, we cannot generalize to a larger population.
107. Daytime running lights may be effective because they catch the attention of other drivers. As they become more common, they may be less effective at catching the attention of other drivers because people may simply get used to them. We also need to pay attention to how much reduction there is from using daytime running lights. If it's only a very small amount, the cost of installing them may not be justified.
109. Answers will vary. Possible answers include the following. (a) Many people would consider pricking a finger to be of minimal risk. (b) Fewer people would consider drawing blood from an arm to be of minimal risk. (c) It is unlikely that very many people would consider inserting a tube into the arm that remains there to be of minimal risk.
111. Answers will vary. Possible answers include the following. (a) Many would consider this to be an appropriate use of collecting data without participants' knowledge because the data are, in effect, anonymous. (b) Many would consider this to be appropriate because the meetings are public and the psychologist is not misleading the participants. (c) Most would consider this to be inappropriate because the psychologist is misleading the other participants and attending private meetings.
113. Confidential. The person taking the survey knows who is answering the questions but will not share the results of individuals with anyone else.
115. In this case, the subjects were not able to give informed consent. They did not know what was happening to them, and they were not old enough to understand the ramifications in any event.
117. From the given two-way table of response by gender, find and compare the conditional distributions of response for men alone and women alone. These values are in the table below. To find the conditional distributions, divide each entry in the table by its column total.

Response	Male	Female
Strongly agree	14.7%	9.3%
Agree	52.3%	38.8%
Neither	16.9%	21.9%
Disagree	11.8%	19.3%
Strongly disagree	4.3%	10.7%

From the table and a bar graph, we see that men are more likely to view animal testing as justified if it might save human lives: over two-thirds of men agree or strongly agree with this statement, compared with slightly less than half of the women. The percents who disagree or strongly disagree tell a similar story: 16% of men versus 30% of women.

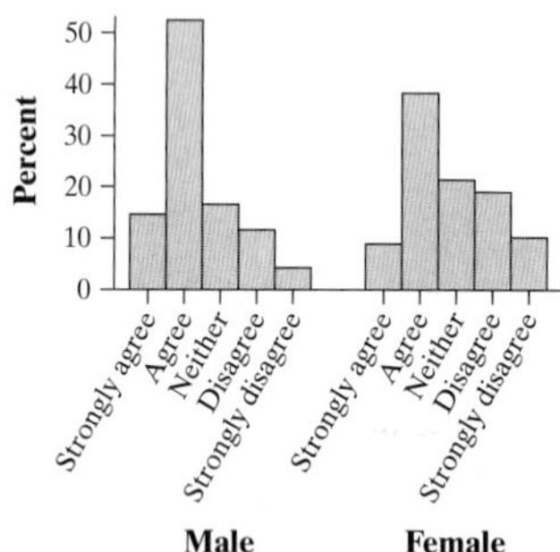

Percent within gender.

Answers to Chapter 4 Review Exercises

R4.1 **(a)** Population: Ontario residents. Sample: 61,239 people interviewed. **(b)** The sample size is very large, so if there were large numbers of both sexes in the sample—this is a safe assumption since we are told this is a "random sample"—these two numbers should be fairly accurate reflections of the values for the whole population.

R4.2 Answers will vary. One possible answer is: **(a)** Announce in daily bulletin that there is a survey concerning student parking available in the main office for students who want to respond. Since voluntary surveys are generally responded to only by those who feel strongly about the issue, these results will likely be biased. **(b)** Personally interview a group of students having lunch in the center quad. Convenience samples are not generally representative of the population and lead to biased results.

R4.3 **(a)** Alphabetically associate each name with a two-digit number: Agarwal = 01, Andrews = 02, . . . , Wilson = 25. Move from left to right reading pairs of digits until you find three different pairs between 01 and 25. **(b)** Using the numbers given, choose 17 09 23. These numbers correspond to Musselman, Fuhrmann, and Smith.

R4.4 A stratified random sample would probably be best here; one could select 50 faculty members from each type of institution. If a large proportion of faculty in your state work at a particular class of institution, it may be useful to stratify unevenly. If, for example, about 50% teach at Class I institutions, you may want half your sample to come from Class I institutions. A simple random sample might miss faculty from one particular type of institution, especially if there are not many faculty at that type of institution. A cluster sample might introduce bias. If the clusters are taken to be different schools, faculty at one school may have a different opinion than faculty at other schools, depending on their particular student body.

R4.5 **(a)** People may not remember how many movies they watched in a movie theater in the past year. It might help the polling organization to shorten the amount of time that they ask about, perhaps to 3 or 6 months. **(b)** The poll contacted people through "residential phone numbers." Since more and more people (especially younger adults) are using only a cell phone (and do not have a residential phone), the poll omitted these people from the sampling frame. These same people might be more likely to watch movies in a movie theater. The polling organization should include cell phone numbers in their list of possible numbers to call.

R4.6 **(a)** The data were collected after the anesthesia was administered. Hospital records were used to "observe" the death rates, rather than imposing different anesthetics. **(b)** One possible confounding variable is type of surgery. If one anesthesia is used more often with a type of surgery that has a higher death rate anyway, we wouldn't know if the death rate was higher because of the anesthesia type or the surgery type.

R4.7 The experimental units are the potatoes used in the experiment. The explanatory variables are storage time and time from slicing until cooking. There are six treatments: (1) fresh picked and cooked immediately, (2) fresh picked and cooked after an hour, (3) stored at room temperature and cooked immediately, (4) stored at room temperature and cooked after an hour, (5) stored in refrigerator and cooked immediately, (6) stored in refrigerator and cooked after an hour. The response variables are ratings of color and flavor.

R4.8 Assign each person a number and using a random number table or technology, choose half to be in one group and the other half to be in the second group. When a person from the first group is called, use the first description. When a person in the second group is called, use the second description. Record their responses.

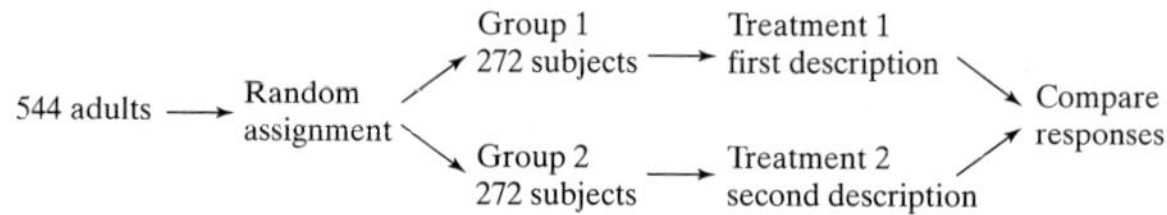

R4.9 **(a)** It gave some patients a treatment that should have no effect at all but looks, tastes, and feels like the Saint-John's-wort. These people think that they are being treated, but in fact, they are not. So if they get better, this would be a case of the placebo effect. **(b)** The study should be double-blind. The subjects should not know which treatment they are receiving so that the researchers can measure how much placebo effect there is. But the researchers should also be blinded so that they cannot influence how they measure the results.

R4.10 **(a)** This is a randomized block design. Blocking helps control for the variability in response due to people's running habits. **(b)** To make sure that the two groups of people in each block are as similar as possible before the treatments are administered. **(c)** A difference in rate of infection may have been due to the effects of the treatments, or it may simply have been due to random chance. Saying that the placebo rate of 68% is "significantly more" than the vitamin C rate of 33% means that the observed difference is too large to have occurred by chance alone. In other words, vitamin C appears to have played a role in lowering the infection rate of runners.

R4.11 **(a)** Randomly assign 15 students to Group 1 (easy mazes) and the other 15 to Group 2 (hard mazes). Compare the time estimates of Group 1 with those of Group 2. **(b)** Each student does the activity twice, once with the easy mazes and once with the hard mazes. Randomly decide (for each student) which set of mazes is used first. Compare each student's "easy" and "hard" time estimate (e.g., by looking at each "hard" minus "easy" difference). **(c)** The matched pairs design would be more likely to detect a difference because it controls for the variability between subjects.

R4.12 **(a)** No. The subjects did not know the nature of the experiment before they agreed to participate. **(b)** All individual data should be kept confidential, and the experiment should go before an institutional review board before being implemented. **(c)** This would allow for inference about cause and effect if the students are randomly assigned to the two treatments.

Answers to Chapter 4 AP Statistics Practice Test

T4.1 c
T4.2 e
T4.3 d
T4.4 c
T4.5 b
T4.6 b
T4.7 d
T4.8 d
T4.9 d
T4.10 b
T4.11 d

T4.12 **(a)** The experimental units are the acacia trees. The treatments are placing either active beehives, empty beehives, or nothing in the trees. The response variable is the damage caused by elephants to the trees. **(b)** Randomly assign 24 of the acacia trees to have active beehives placed in them, 24 randomly to have empty beehives placed in them, and the remaining 24 to remain empty. To do this, assign the trees numbers from 01 to 72 and use a random number table to pick 24 two-digit numbers in this range. Those trees will get the active beehives. The trees associated with the next 24 two-digit numbers will get the empty beehives, and the remaining 24 trees will remain empty. Compare the damage caused by elephants to the trees with active beehives, those with empty beehives, and those with no beehives.

T4.13 **(a)** It is not a simple random sample because not all samples were possible. For instance, given their method, they could not have had all respondents from the East Coast. **(b)** One adult was chosen at random to control for lurking variables. Perhaps the household members who generally answer the phone have a different opinion from those who don't generally answer the phone. **(c)** There was undercoverage in this survey. Those who do not have telephones and those who have only cell phones were not part of the sampling frame. So their opinions would not have been measured. Since cell-phone-only users tend to be younger, the results of the survey may not accurately reflect the entire population's opinion.

T4.14 **(a)** Each of the 11 individuals was a block in a matched pairs design. Each participant took the caffeine tablets on one of the two-day sessions and the placebo on the other. The order in which they took the caffeine or the placebo was decided randomly. The tapping test was administered at the end of each two-day trial. The results to be compared are the differences between the caffeine and placebo scores on the tapping test. The blocking was done to control for individual differences in dexterity. **(b)** The order was randomized to control for any possible influence of the order in which the treatments were administered on the subject's tapping speed. **(c)** It is possible to carry out this experiment in a double-blind manner. This means that neither the subjects nor the people who come in contact with them during the experiment (including those who record the number of taps) have knowledge of the order in which the caffeine or placebo was administered.

Solutions to Cumulative AP Practice Test 1

AP1.1 d
AP1.2 e
AP1.3 b
AP1.4 c
AP1.5 a
AP1.6 c
AP1.7 e
AP1.8 e
AP1.9 d
AP1.10 d
AP1.11 d
AP1.12 b
AP1.13 b
AP1.14 a

AP1.15 **(a)** When comparing data sets it is expected that center, spread, and shape will be addressed. The median of the distribution of cardiovascular fitness measures for Machine A (28) is lower than the median of the distribution of cardiovascular fitness measures for Machine B (38). The range of the distribution of cardiovascular fitness measures for Machine A is substantially smaller than the range of the distribution of cardiovascular fitness measures for Machine B ($32 < 57$). (Alternatively, the *IQR* for Machine A ($37 - 22 = 15$) is smaller than the *IQR* for Machine B ($47 - 25 = 22$).) The distribution of cardiovascular fitness measures for Machine A is reasonably symmetric in shape, while the distribution of cardiovascular fitness measures for Machine B is skewed to the left (toward the lower values). In general, the cardiovascular fitness measures for Machine B tend to be higher than those for Machine A. **(b)** The company should choose Machine B if they want to advertise it as achieving the highest overall gain in cardiovascular fitness. The median for Machine B is higher than it is for Machine A, as is the mean ($\bar{x}_B = 35.4$ versus $\bar{x}_A = 28.9$). **(c)** The company should choose Machine A if they want to advertise it as achieving the most consistent gains in cardiovascular fitness. Machine A exhibits less variation in gains than does Machine B. The *IQR* for Machine A is 15, while the *IQR* for Machine B is 22. Additionally, the standard deviation for Machine A is 9.38, while the standard deviation for Machine B is 16.19. **(d)** Volunteers were used for the experiment and these volunteers may be different in some way from the general population of those who are interested in cardiovascular fitness. Another reason is that the experiment was conducted at only *one* fitness center. Results may vary at other fitness centers in this city and in other cities.

If the company had taken a random sample of *all* people who were interested in cardiovascular fitness, it could make inferences about this much larger population. If the company had randomly selected a number of fitness centers from across the country and then randomly selected members of each fitness center to participate in the experiment, the company would be using a stratified random sampling method with the fitness centers serving as the strata.

AP1.16 (a) Assign each retail sales district a number from 1 to 60 using a random number generator. Order the sales districts numerically. The first 30 are assigned to the monetary incentives group and the remaining 30 to the intangible incentives group. After a specified period of time, compare the mean change in sales for each of the treatment groups to see if there is a difference. (b) Assign each retail sales district a number between 01 and 60. Go through the random number table taking two digits at a time. The first 30 two-digit numbers between 01 and 60 to come up are assigned to the monetary incentives group. The remainder will be assigned to the intangible incentives group.

07511 88915 41267 16853 84569 79367 32337 03316

The districts labeled 07, 51, and 18 are the first three to be assigned to the monetary incentives group. (c) It would be better to use a matched pairs design. There could be a large variation among the sales figures for the various districts due to the various sizes of those districts across the different regions of the United States. Matching the districts based on their size reduces the effect of variation among the experimental units due to their size on the response variable—sales volume. Pair the two largest districts in size, the next two largest, down to the two smallest districts. For each pair, pick one of the districts and flip a coin. If the flip is "heads," this district is assigned to the monetary incentives group. If it is "tails," this district is assigned to the intangible incentives group. The other district in the pair is assigned to the other group. After a specified period of time, compare the mean change in sales for each of the treatment groups to see if there is a difference.
AP1.17 (a) Yes. A linear model is reasonable since the scatterplot shows a strong positive linear association between shelf length and weekly sales (in dollars). (b) $\hat{y} = 317.94 + 152.68x$, where $y =$ weekly sales (in dollars) and $x =$ shelf length (in feet). (c) $\hat{y} = 317.94 + 152.68(5) = 1081.34$. A shelf length of 5 feet would, on average, yield weekly sales of \$1081.34. (d) The value $s = 22.9212$ represents the standard deviation of the residuals. On average, there is a \$22.92 difference between the predicted weekly sales and the actual weekly sales. (e) $r^2 = 0.982$. 98.2% of the variation in weekly sales revenue can be explained by the linear regression using shelf length allocated to the house brand as the predictor. (f) It would be inappropriate to interpret the intercept, since the data represent sales based on shelf lengths of 3 to 6 feet, and 0 feet falls substantially outside that domain. We would be trying to extrapolate beyond the set of data.

CHAPTER 5

Section 5.1

Answers to Check Your Understanding

Page 286 1. (a) This probably means that if you asked a large sample of U.S. adults whether they usually eat breakfast, about 61% of them will answer "Yes." (b) In a random sample of 100 adults, we would expect that around 61 of them will usually eat breakfast. However, the exact number will vary from sample to sample. 2. (a) 0. If an outcome can never occur, then it will occur in 0% of cases. (b) 1. If an outcome will occur on every trial, then it will occur in 100% of cases. (c) 0.01. This means that the event occurs in about 1% of trials. (d) 0.6. An outcome that will occur more often than not will have a probability that is greater than 0.50. This leaves us with two choices: 0.60 and 0.99. The wording suggests that the event occurs often but not nearly every time. This suggests something that occurs 60% of the time. *Page 292* 1. Assign the members of the AP Statistics class the numbers 01–28 and the rest of the students numbers 29–95. Ignore the numbers 96–99 and 00. In Table D, read off four two-digit numbers, ignoring duplicates. Record whether all four numbers are between 01 and 28 or not. Do this many times and compute the percent of the time that all four students are part of the AP Statistics class. 2. Assign the numbers 1–10 to Jeff Gordon, 11–40 to Dale Earnhardt Jr., 41–60 to Tony Stewart, 61–85 to Danica Patrick, and 86–99 and 100 to Jimmie Johnson. Do RandInt (1,100) until you have at least one number (box) for each of the five drivers. Count how many numbers (boxes) you had to sample in order to get at least one of each.

Answers to Odd-Numbered Section 5.1 Exercises

1. (a) If we use a polygraph machine on many, many people who are all telling the truth, about 8% of the time the machine will say that the people are lying. (b) Answers will vary. The U.S. judicial system is set up to think that a false positive would be worse—that is, saying that someone is guilty (lying) who is not is worse than finding someone to be truthful when they are, in fact, guilty (lying).
3. (a) If we look at many families where the husband and wife both carry this gene, in approximately 25% of them the firstborn child will develop cystic fibrosis. (b) If the family has 4 children, this constitutes a sample of size 4, which is very small. For the probability to be closely reflected in the sample, the sample size must be very large.
5. (a) Answers will vary. We obtained 16 tails and 9 heads so $P(\text{heads}) = 0.36$. (b) Spin the coin many more times.
7. In the short run, there was quite a bit of variability in the percent of free throws made. In fact, the basketball player did not do as well early on, but his percent both increased somewhat and became less variable as time went on.
9. No. The TV commentator is incorrectly applying the law of large numbers to a small number of at-bats for the player.
11. (a) There are 10,000 four-digit numbers (0000, 0001, 0002, . . . , 9999), and each is equally likely to be chosen. (b) Most people would say that 2873 is more likely than 9999 to be randomly chosen. To many, it somehow "looks" more random—we don't "expect" to get the same number four times in a row. Choose a number most would avoid.
13. (a) Let 1, 2, and 3 represent the player making the free throw and 4 represent a miss. If 5 or 6 comes up, ignore it and roll again. (b) Let the two-digit numbers 01–75 represent making the free throw and 76–99 and 00 represent missing. Read a two-digit number from Table D. (c) Let diamonds, spades, and clubs represent making a free throw and hearts represent missing. Deal one card from the deck.
15. (a) There are 19 numbers between 00 and 18, 19 numbers between 19 and 37, and 3 numbers between 38 and 40. This changes the proportions between the three different outcomes. (b) Do not skip numbers that have already been encountered in the table. These numbers just represent the handedness, not a particular individual to select for the sample.
17. (a) Legitimate. The chance of rolling a 1, 2, or 3 is 75% on a four-sided die, and the rolls are independent of each other. (b) Not valid. The chance of heads (assuming the coin is fair) is 50% rather than the 60% that she hits the center of the target.

19. **(a)** What is the probability that in a random selection of 10 passengers, none from first class are chosen? **(b)** Number the first-class passengers 01–12 and the other passengers 13–76. Ignore all other numbers. Look up two-digit numbers in Table D until you have 10 unique numbers. Count the number of two-digit numbers between 01 and 12. **(c)** Sample is 71 48 70 29 07 63 61 68 34 52. There is one person among the 10 selected who is in first class in this sample. **(d)** Since in 15% of the samples no first-class passenger was chosen, it seems plausible that the actual selection was random.

21. **(a)** Read off 30 three-digit numbers from the table, ignoring numbers greater than 365 and 000. Record whether there were any repeats in the sample or not. **(b)** Answers will vary. We used Minitab to select five samples. There were repeats in all five samples. **(c)** Answers will vary. After the simulation we would not be surprised that the probability is 0.71 since we found repeats in 100% of our samples.

23. **(a)** 43 of the 200 samples yielded at least 55% who say they recycle. This means that, according to the simulation, if 50% recycle, we would see at least 55% of the sample saying they recycle in about 21.5% of samples. This is not particularly unusual. **(b)** However, only 1 of the 200 samples yielded at least 63% who said that they recycle. This means that it would happen in about 0.5% of samples. This seems rather unusual. It would be much more likely that the actual percent who recycle is larger than 50%.

25. **State:** How many men would we expect to choose in order to find one who is red-green color-blind? **Plan:** We'll use technology to simulate choosing men. We'll label the numbers 01–07 as color-blind men and all other two-digit numbers as non-color-blind men. Use technology to produce two-digit numbers until a number between 01 and 07 appears. Count how many two-digit numbers there are in the sample. **Do:** We did 50 repetitions of the simulation using technology. Here is the first repetition: 17 33 49 41 **02**. The number in bold is the stopping point. For this repetition we chose 5 men in order to get the one color-blind man. In 50 repetitions, the average number of men needed was 16.88. **Conclude:** Based on our simulation, we would suggest that we would need to sample about 17 men, on average.

0 10 20 30 40 50 60 70
Men

27. **State:** What is the probability that, in a sample of 4 randomly selected U.S. adult males, at least 1 of them is red-green color-blind? **Plan:** We'll use Table D to simulate selecting samples of 4 men. Since 7% of men are red-green color-blind, let 01–07 denote a color-blind man and 08–99 and 00 denote a non-color-blind man. Read 4 two-digit numbers from the table for each sample and record whether the sample had at least 1 red-green color-blind man in it. **Do:** We did 50 repetitions. We started on row 109 of Table D. The first sample (Sample 1) is 36 00 91 93 (no color-blind). In our 50 samples, 15 had at least 1 color-blind man in them. **Conclude:** Based on our 50 samples, we would suggest that about 30% of samples would have at least 1 color-blind man in them.

29. **State:** What is the probability that a random assignment of 8 men and 12 women to two groups of 10 each will result in at least 6 men in the same group? **Plan:** Number the men 01–08 and the women 09–20. Use technology to pick 10 numbers between 01 and 20 for one group; the remaining 10 numbers are in the other group. Record whether at least 6 of the numbers between 01 and 08 are in one group. **Do:** We did 50 repetitions. The first repetition resulted in Group 1: 17 **05 04** 14 11 10 **03** 15 **08 01**; Group 2: **02** 19 12 **06** 18 09 16 13 **07** 20. The men are shown in bold. There were 5 men in the first group and 3 men in the second group, so neither group had 6 or more men. In our 50 repetitions, 9 had one group with 6 or more men in it. **Conclude:** Based on our 50 random assignments, the researchers should not have been too surprised. 18% of our repetitions had one group with at least 6 men in it.

31. c

33. d

35. c

37. **(a)** The population is adult U.S. residents, and the sample is the 178,545 adults who were interviewed. **(b)** Since the interviews were conducted using the telephone, those people who do not have a telephone were excluded. In general, people who do not have phones tend to be poorer and may experience more stress in their lives than the population as a whole.

Section 5.2

Answers to Check Your Understanding

Page 303 **1.** A person cannot have a cholesterol level of both 240 or above and between 200 and 239 at the same time. **2.** A person has either a cholesterol level of 240 or above or a cholesterol level between 200 and 239. $P(A \text{ or } B) = P(A) + P(B) = 0.45$. **3.** $P(C) = 1 - P(A \text{ or } B) = 0.55$ *Page 305* **1.**

	Face card	Not face card
Heart	3	10
Not heart	9	30

2. $P(\text{face card and heart}) = \frac{3}{52}$ **3.** Since A and B are not disjoint, this is not equal to the sum of the probability of A and the probability of B. $P(A \text{ or } B) = P(A) + P(B) - P(A \text{ and } B) = 0.423$.

Answers to Odd-Numbered Section 5.2 Exercises

39. **(a)** {(1,1), (1,2), (1,3), (1,4), (2,1), (2,2), (2,3), (2,4), (3,1), (3,2), (3,3), (3,4), (4,1), (4,2), (4,3), (4,4)}. **(b)** Each outcome has probability $\frac{1}{16}$.

41. $\frac{1}{4}$

43. **(a)** Legitimate. **(b)** Not legitimate: the total is more than 1. **(c)** Legitimate (even if the deck of cards is not!).

45. **(a)** The given probabilities have sum 0.96, so $P(\text{type } AB) = 1 - 0.96 = 0.04$. **(b)** 0.96 **(c)** 0.69.

47. **(a)** The given probabilities have sum 0.72, so this probability must be 0.28. **(b)** $P(\text{at least a high school education}) = 1 - P(\text{has not finished HS}) = 0.87$.

49. **(a)** Individuals: the students in the urban school. Variables: gender, whether or not they eat breakfast regularly. **(b)** $P(\text{female}) = \frac{275}{595}$. $P(\text{Eats breakfast regularly}) = \frac{300}{595}$.

$P(\text{Female and eats breakfast regularly}) = \frac{110}{595}$.

$P(\text{Female or eats breakfast regularly}) = \frac{465}{595}$.

51. (a)

	Black	Not black	Total
Even	10	10	**20**
Not even	8	10	**18**
Total	**18**	**20**	**38**

(b) $P(B)=\frac{18}{38}$; $P(E)=\frac{20}{38}$. (c) The ball lands in a black, even spot. $\frac{10}{38}$ (d) If we added the separate probabilities for B and E, we would double-count the probability that it lands in a black, even spot. $P(B \text{ or } E)=\frac{28}{38}$.

53. (a) Here is a Venn diagram.

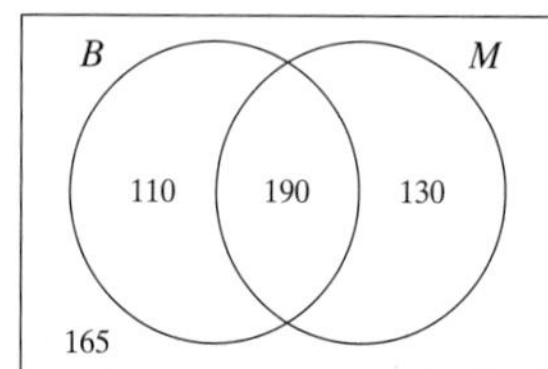

(b) $\frac{430}{595}$. The probability of being either a male or someone who eats breakfast, or both. (c) $\frac{165}{595}$. The probability of being neither male nor someone who eats breakfast.

55. (a)

	Facebook	Not Facebook	Total
MySpace	8,400,000	2,400,000	**10,800,000**
Not MySpace	8,600,000	600,000	**9,200,000**
Total	**17,000,000**	**3,000,000**	**20,000,000**

(b) Here is a Venn diagram.

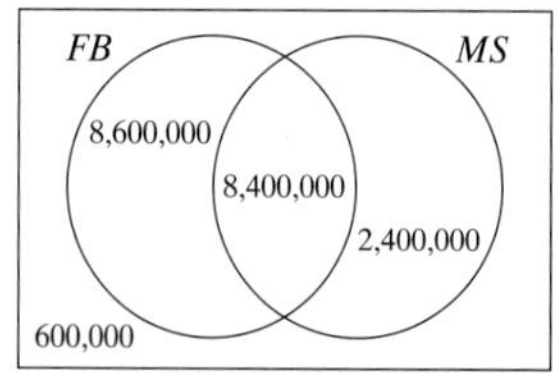

(c) $P(\text{FB} \cup \text{MS})$ (d) $\frac{19{,}400{,}000}{20{,}000{,}000}$

57. c

59. c

61. Here is a scatterplot.

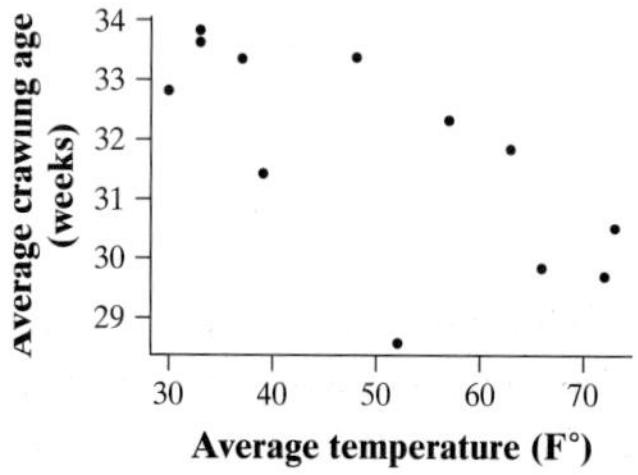

In the scatterplot, we see that there appears to be a moderately strong negative linear relationship. In fact, the correlation bears this out: −0.70. The equation for the least-squares regression line is $\widehat{\text{age}} = 35.7-0.077\text{temp}$. We predict that babies will walk 0.077 week earlier for every degree warmer it gets.

Section 5.3

Answers to Check Your Understanding

Page 314 **1.** 0.3656. 36.56% of grades are lower than a B. **2.** $P(E \mid L)=\frac{800}{3656}=0.219$. $P(L \mid E)=\frac{800}{1600}=0.50$. $P(L \mid E)$. This gives the probability of getting a lower grade if the student is an engineer. We would want to compare that to the probability of getting a lower grade if the student is in liberal arts and social sciences. *Page 317* 1. Independent. Since we are putting the first card back and then reshuffling the cards before drawing the second card, knowing what the first card was will not tell us anything about what the second card will be. **2.** Not independent. Once we know the suit of the first card, then the probability of getting a heart on the second card will change, depending on what the first card was. **3.** Independent. Once we know that the chosen person is female, this does not tell us anything more about whether she is right-handed or not. Overall, $\frac{24}{28}=\frac{6}{7}$ of the students are right-handed. And, among the women, $\frac{18}{21}=\frac{6}{7}$ are right-handed. So $P(\text{right-handed})=P(\text{right-handed} \mid \text{female})$. *Page 321* 1. Here is a tree diagram.

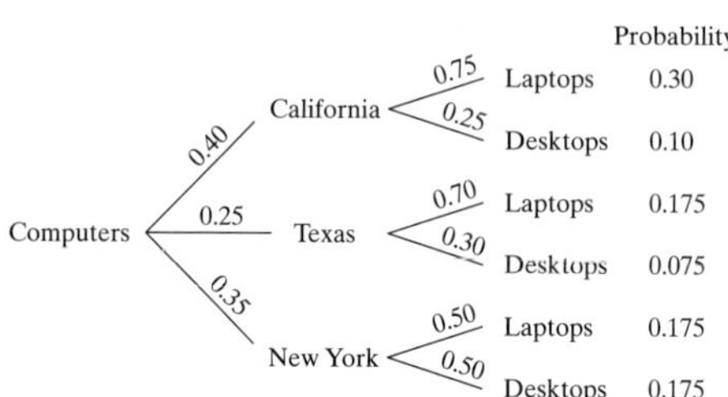

2. $P(\text{laptop}) = P(\text{laptop} \cap \text{CA}) + P(\text{laptop} \cap \text{TX}) + P(\text{laptop} \cap \text{NY}) = 0.65$ *Page 323* **1.** $P(\text{one returned safely}) = 0.95$. Assuming independence, $P(\text{safe return on all 20 missions}) = 0.95^{20} = 0.3585$. **2.** No. Whether one is a college student and one's age are not independent events. Far more younger people than older people are college students.

Answers to Odd-Numbered Section 5.3 Exercises

63. **(a)** 0.2428 **(b)** 0.5983

65. **(a)** 0.2801 **(b)** 0.2944 **(c)** Not independent since the two probabilities in parts (a) and (b) are not the same.

67. **(a)** 0.7647. 76.47% of the females are Democrats. **(b)** 0.2167. 21.67% of the Democrats are female.

69. $P(D) = \frac{60}{100} = 0.60$. From Exercise 67, we saw that $P(D \mid F) = 0.7647$. Since these two probabilities are not the same, D and F are not independent.

71. **(a)** 0.41 **(b)** 0.6341

73. $P(B) < P(B \mid T) < P(T) < P(T \mid B)$. There are very few professional basketball players, so $P(B)$ should be the smallest probability. It's much more likely to be over 6 feet tall than it is to be a professional basketball player if you're over 6 feet tall. Lastly, if you are a professional basketball player, it is quite

likely that you are tall (greater than the probability that a randomly selected individual is over 6 feet tall).

75. P(sum of 7 | green is 4) = 0.1667. P(sum of 7) = 0.1667. Since these two probabilities are the same, the events are independent.

77. (a) Here is a tree diagram.

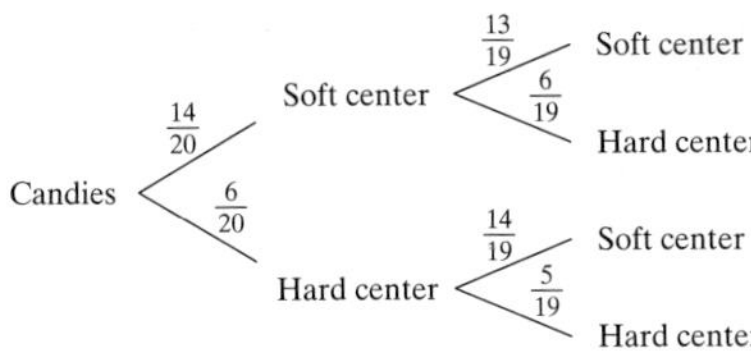

(b) 0.4421

79. 0.29(0.67) = 0.1943

81. 0.05(0.017)(0.4) = 0.0003

83. Here is a tree diagram.

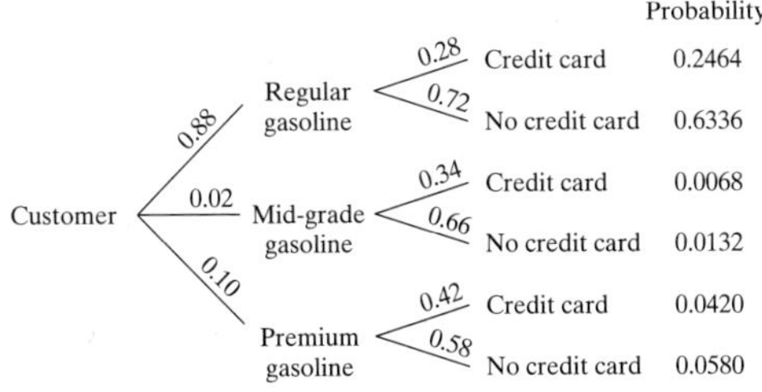

State: What is the probability that a randomly chosen customer paid with a credit card? **Plan:** We need to find the probability P(credit card). **Do:** Look at the tree diagram. P(credit card) is the sum of the three branches that involve people using their credit card. So P(credit card) = 0.2952. **Conclude:** About 29.5% of customers use their credit card to pay for gasoline purchases.

85. 0.1423

87. (a) Here is a tree diagram.

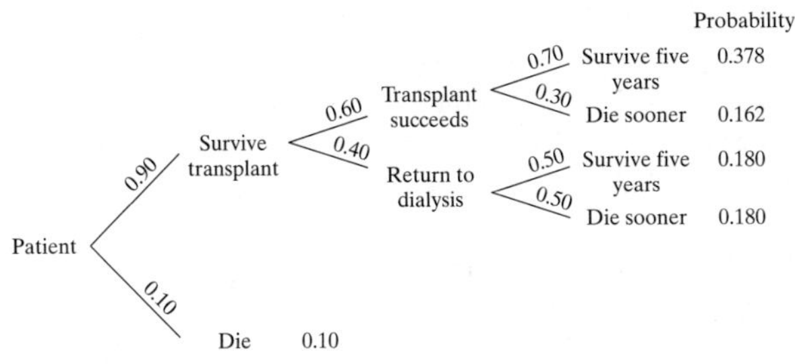

(b) P(survive) = 0.378 + 0.180 = 0.558

89. 0.6676

91. **State:** If 10 people appear at random to give blood, what is the probability that at least 1 of them is a universal donor? **Plan:** Since the people are appearing at random, it is reasonable to assume that their blood types are independent of each other. Use the fact that P(at least one universal donor) = 1 − P(none are universal donors). **Do:** The probability that any 1 person is not a universal donor is 1 − 0.072 = 0.928. So P(none are universal donors) = $(0.928)^{10}$ = 0.4737. This means that P(at least one universal donor) = 0.5263. **Conclude:** There is about a 53% chance that at least 1 of 10 random blood donors is a universal donor.

93. We cannot simply multiply the probabilities together. If the first of the three consecutive shows starts late, it is much less likely that the next show will start on time. These events are not independent of each other.

95. 0.42/0.85 = 0.4941

97. (a) Here is a tree diagram. 0.257.

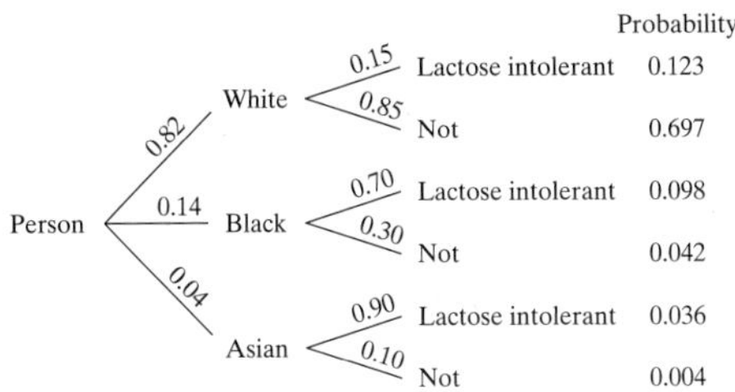

(b) 0.1401

99. (a) Here is a tree diagram.

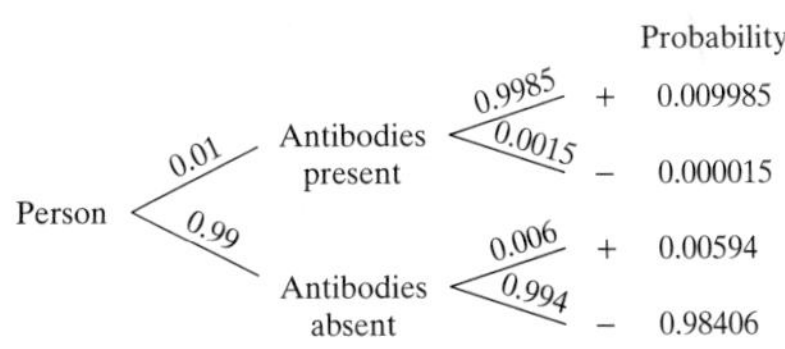

(b) 0.015925 (c) 0.6270

101. (a), (b) These probabilities are

$$P(\text{1st card } \spadesuit) = 0.25$$
$$P(\text{2nd card } \spadesuit \mid \spadesuit \text{ picked}) = 0.2353$$
$$P(\text{3rd card } \spadesuit \mid 2\spadesuit\text{s picked}) = 0.22$$
$$P(\text{4th card } \spadesuit \mid 3\spadesuit\text{s picked}) = 0.2041$$
$$P(\text{5th card } \spadesuit \mid 4\spadesuit\text{s picked}) = 0.1875$$

(c) The product of these conditional probabilities gives the probability of a flush in spades by the extended multiplication rule: we must draw a spade and then another and then a third, a fourth, and a fifth. The product of these probabilities is about 0.0004952. (d) 0.001981

103. (a) 0.333 (b) 0.5

105. e

107. The z-score corresponding to a BMI of 18.5 is $\dfrac{18.5 - 26.8}{7.4} = -1.12$. Using Table A, the percent with z-score lower than −1.12 is 0.131.

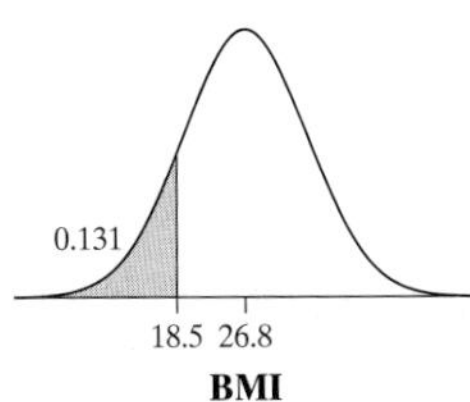

109. Generally, more education means more freedom. For example, 45% of those with bachelor's degrees felt free to organize their work, compared to 29.7% of those who finished high school and

24% of those who did not finish high school. This can also be seen in the bar graph.

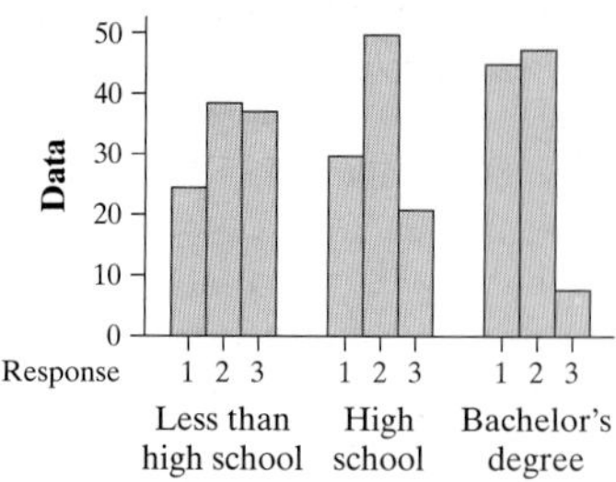

Answers to Chapter 5 Review Exercises

R5.1 When the weather conditions are like those seen today, it has rained on the following day about 30% of the time.

R5.2 (a) Let the numbers 01–15 represent drivers who do not have their seat belts on and 16–99 and 00 represent drivers who do have their seat belts on. Using the table of random digits, read 10 sets of two-digit numbers. For each set of 10 two-digit numbers, record whether there are two consecutive numbers between 01 and 15 or not. (b) The first sample is 29 **07** 71 48 63 61 68 34 70 52. Those not wearing seat belts are in bold. In this sample, there was only one not wearing a seat belt, so there were not two consecutive people not wearing their seat belts. The second sample is 62 22 45 **10** 25 95 **05** 29 **09 08**, wherein there were two consecutive drivers not wearing their seatbelts (represented by 09 and 08). The third sample is 73 59 27 51 86 87 **13** 69 57 61, wherein there were not two consecutive drivers not wearing their seat belts. Based on these three simulations, we'd suggest that the probability of finding two consecutive people out of a sample of 10 not wearing their seat belts to be about 0.33.

R5.3 Assign each of the eight teams a single-digit number. In particular, let the Sandblasters be the number 1. Now use technology to pick five different random numbers between 1 and 8. These are the teams that had their sculpture blown up last year. Pick five more numbers between 1 and 8 for those teams to have their sculpture blown up this year. Record whether the number 1 was missing in both years or not. Repeat this many times. The first sample in our simulation was 1 2 4 3 8 for the first year and 6 8 3 2 5 for the second year. In this sample, Team 1 was chosen in the first year. In our 50 simulations, Team 1 was chosen in at least one year 47 times. Only 3 times out of 50, or 6% of the time, was Team 1 not chosen in either year. The other teams should probably be suspicious.

R5.4 (a)

Difference:	1	5	−3
Probability:	$\frac{1}{2}$	$\frac{1}{6}$	$\frac{1}{3}$

(b) Die A is more likely to be higher. Half of the time, Die B rolls a 1 and Die A is automatically higher. If Die B rolls a 5, there are still some times that Die A is higher—when it rolls a 6. So more than half of the time, Die A is higher.

R5.5 (a) It is legitimate because every person must fall into exactly one category, the probabilities are all between 0 and 1, and they add up to 1. (b) The probability that a randomly chosen person is Hispanic is 0.149. (c) The probability that a randomly chosen person is not a non-Hispanic white is 0.326. (d) Being white and being Hispanic are not mutually exclusive events. So $P(\text{white or Hispanic}) = 0.823$.

R5.6 (a) Here is a Venn diagram.

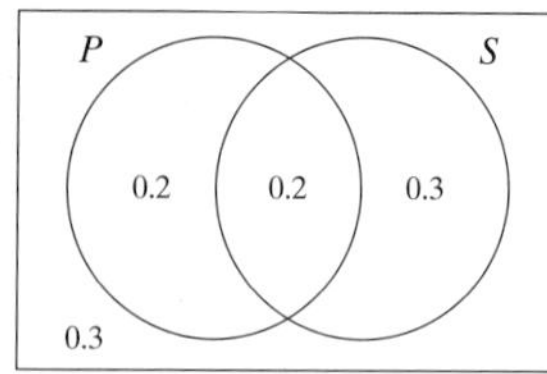

(b) 0.3 (c) 0.7: $P(P \cup S) = P(P) + P(S) - P(P \cap S) = 0.4 + 0.5 - 0.2 = 0.7$.

R5.7 (a) Here is a tree diagram.

(b) $0.10(0.95) + 0.90(0.03) = 0.122$ (c) $0.10(0.95)/0.122 = 0.7787$

R5.8 (a) These events are not independent, because $P(\text{pizza with mushrooms}) = \frac{4}{7}$, but $P(\text{pizza with mushrooms} \mid \text{thick crust}) = \frac{2}{3}$. If the events had been independent, these probabilities would have been equal. (b) With the eighth pizza $P(\text{mushrooms}) = \frac{4}{8} = \frac{1}{2}$, and $P(\text{mushrooms} \mid \text{thick crust}) = \frac{2}{4} = \frac{1}{2}$, so these events are independent.

R5.9 (a) 0.24 (b) $P(\text{damage} \mid \text{no cover}) = 0.2844$, $P\left(\text{damage} \,\middle|\, < \frac{1}{3}\right) = 0.3248$, $P\left(\text{damage} \,\middle|\, \frac{1}{3} \text{ to } \frac{2}{3}\right) = 0.1991$, and $P\left(\text{damage} \,\middle|\, > \frac{2}{3}\right) = 0.1415$. (c) Yes. Knowing the amount of cover does change the probability of deer damage. It appears that deer do much more damage when there is no cover or less than 1/3 cover and much less damage when there is more cover than that.

R5.10 (a) 0.274625 (b) 0.3175

R5.11 (a) Here is a tree diagram.

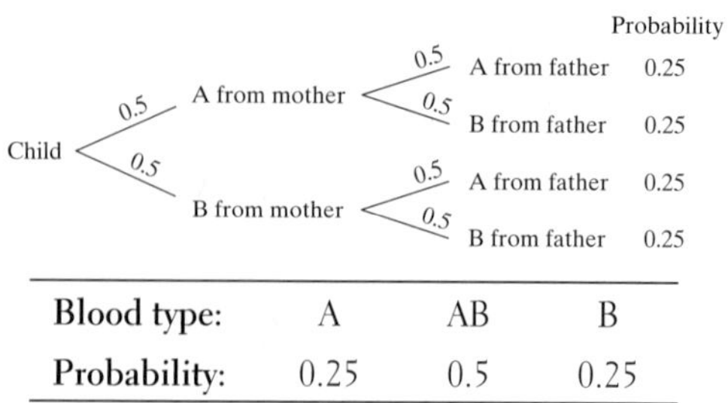

Blood type:	A	AB	B
Probability:	0.25	0.5	0.25

(b) 0.4375.

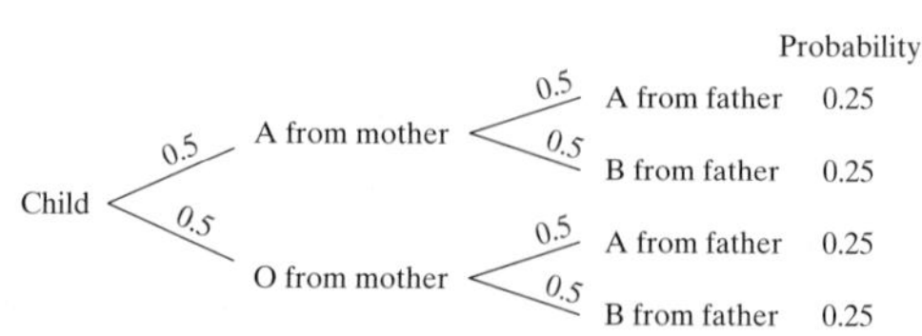

Answers to Chapter 5 AP Statistics Practice Test

T5.1 c

T5.2 d

T5.3 c

T5.4 b

T5.5 b

T5.6 c

T5.7 e

T5.8 e

T5.9 b

T5.10 c

T5.11 **(a)** Since each outcome is equally likely and there are 48 outcomes, each outcome has probability $\frac{1}{48}$. There are 27 ways in which the teacher wins (all boxes above and to the right of the diagonal line indicating ties). So the probability that the teacher wins is $\frac{27}{48}$. **(b)** We use the fact that $P(A \cup B) = P(A) + P(B) - P(A \cap B)$. From part (a), $P(A) = \frac{27}{48}$. There are 8 outcomes in which "you get a 3," so $P(B) = \frac{8}{48}$ and there are 5 outcomes in which "you roll a 3" and the teacher still wins, so $P(A \cap B) = \frac{5}{48}$. Putting all this together gives $P(A \cup B) = \frac{27}{48} + \frac{8}{48} - \frac{5}{48} = \frac{5}{8}$. **(c)** From part (b), we have $P(A) = \frac{27}{48}$, $P(B) = \frac{8}{48}$, and $P(A \cap B) = \frac{5}{48}$. Since $P(A \cap B) = \frac{5}{48} \neq \left(\frac{27}{48}\right)\left(\frac{8}{48}\right) = P(A)P(B)$, A and B are not independent.

T5.12 **(a)** Here is a tree diagram.

Probability
Part
0.60 A: 0.10 Defective 0.06; 0.90 OK 0.54
0.30 B: 0.30 Defective 0.09; 0.70 OK 0.21
0.10 C: 0.40 Defective 0.04; 0.60 OK 0.06

(b) To get the probability that a part randomly chosen from all parts produced by this machine is defective, add the probabilities from all branches in the tree that end in a defective part. $P(\text{defective}) = 0.06 + 0.09 + 0.04 = 0.19$. **(c)** First compute the conditional probabilities that the part was produced on a particular machine given that it is defective. $P(A \mid \text{defective}) = 0.3158$. $P(B \mid \text{defective}) = 0.4737$. $P(C \mid \text{defective}) = 0.2105$. Since the largest of these three conditional probabilities is for Machine B, given that a part is defective, it is most likely to have come from Machine B.

T5.13 **(a)** $P(\text{gets cancer} \mid \text{smoker}) = \frac{P(\text{gets cancer} \cap \text{smoker})}{P(\text{smoker})} = 0.32$. **(b)** The event that an individual either smokes or gets cancer is the complement of the event that the individual neither smokes nor gets cancer. So $P(\text{smokes} \cup \text{gets cancer}) = 1 - P(\text{does not smoke} \cap \text{does not get cancer}) = 0.29$. (c) First find the probability that one person gets cancer. From part (b) we know that $P(\text{smokes} \cup \text{gets cancer}) = 0.29$. But we also know that $P(\text{smokes}) = 0.25$ and $P(\text{smokes} \cap \text{gets cancer}) = 0.08$. Now, using $P(S \cup C) = P(S) + P(C) - P(S \cap C)$ we get $0.29 = 0.25 + P(\text{gets cancer}) - 0.08$. Solving for $P(\text{gets cancer})$ we get 0.12. Finally, $P(\text{at least one gets cancer}) = 1 - P(\text{neither gets cancer}) = 0.2256$.

T5.14 **(a)** Assign the numbers 01–17 to represent cars with out-of-state plates and the remaining two-digit numbers between 00 and 99 to represent other cars. Start reading two-digit numbers from a random number table until you have found two numbers between 01 and 17. Record how many two-digit numbers you had to read in order to get two numbers between 01 and 17. Repeat many times for the simulation. **(b)** The first sample is 41 **05 09**. The numbers in bold represent cars with out-of-state plates. In this sample, it took 3 cars to find 2 with out-of-state plates. The second sample is 20 31 **06** 44 90 50 59 59 88 43 18 80 53 **11**, wherein it took 14 cars to find 2 with out-of-state plates. The third sample is 58 44 69 94 86 85 79 67 **05** 81 18 45 **14**, wherein it took 13 cars to find 2 with out-of-state plates.

CHAPTER 6

Section 6.1

Answers to Check Your Understanding

Page 344 **1.** The probability that the student gets either an A or a B. 0.68. **2.** $P(X < 2) = 0.12$ **3.** The histogram is left-skewed. Higher grades are more likely, but there are a few lower grades.

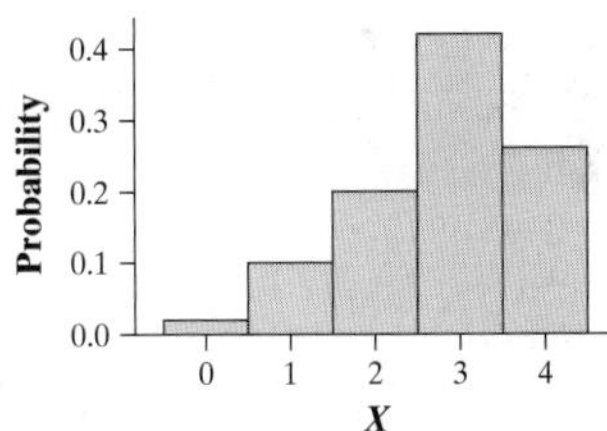

Page 349 **1.** $\mu_X = 1.1$. The long-run average, over many Friday mornings, will be about 1.1 cars sold. **2.** $\sigma_X = 0.943$. On average, the number of cars sold on a randomly selected Friday will differ from the mean (1.1) by 0.943 cars sold.

Answers to Odd-Numbered Section 6.1 Exercises

1. **(a)**

Value:	0	1	2	3	4
Probability:	0.0625	0.25	0.375	0.25	0.0625

(b) The histogram shows that this distribution is symmetric with a center at 2.

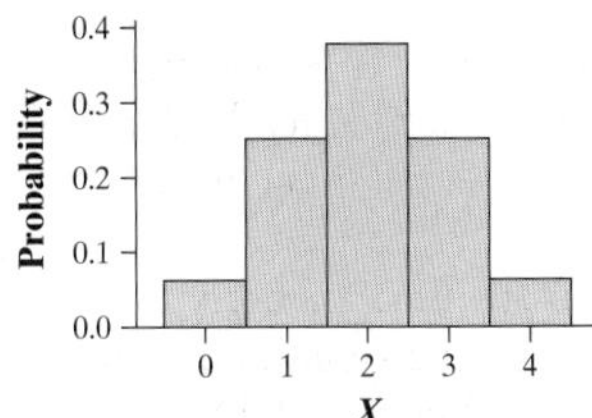

(c) 0.9375. There is a 93.75% chance that you will get three or fewer heads on 4 tosses of a fair coin.
3. (a) The event $\{X \geq 1\}$ or $\{X > 0\}$. 0.9. (b) No more than two nonword errors. $P(X \leq 2) = 0.6$; $P(X < 2) = 0.3$
5. (a) All the probabilities are between 0 and 1, and they sum to 1. (b) This is a right-skewed distribution with the largest amount of probability on the digit 1.

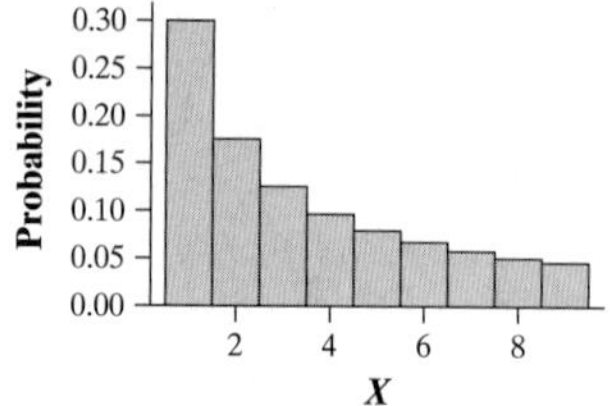

(c) The first digit in a randomly chosen record is a 6 or higher. 0.222. (d) The event $\{X \leq 5\}$. 0.778.
7. (a) {7, 8, 9}; $P(A) = 0.155$ (b) {1, 3, 5, 7, 9}; $P(B) = 0.609$ (c) {1, 3, 5, 7, 8, 9}; $P(A \text{ or } B) = P(A \cup B) = 0.66$. This is not the same as $P(A) + P(B)$ because A and B are not mutually exclusive.
9. (a)

Value:	\$0	\$3
Probability:	0.75	0.25

(b) $\mu_X = \$0.75$. In the long run, for every \$1 the player bets, he only gets \$0.75 back.
11. 2.1. On average, undergraduates make 2.1 nonword errors per 250-word essay.
13. (a) This distribution is symmetric and 5 is located at the center. (b) Following Benford's law, $\mu_X = 3.441$. The average of first digits following Benford's law is 3.441. To detect a fake expense report, compute the sample mean of the first digits and see if it is near 5 or near 3.441. (c) Under the equally likely assumption, $P(Y > 6) = 0.333$. Under Benford's law, $P(X > 6) = 0.155$. When looking at a suspect report, find the percent of figures that start with numbers higher than 6. If that percent is closer to 33% than to 15%, it is probably fake.
15.

$$\sigma_X = \sqrt{(0-2.1)^2(0.1) + (1-2.1)^2(0.2) + (2-2.1)^2(0.3) + (3-2.1)^2(0.3) + (4-2.1)^2(0.1)}$$

$= 1.1358$. On average, the number of nonword errors in a randomly selected essay will differ from the mean (2.1) by 1.14 words.
17. (a) $\sigma_Y = 2.58$ (b) $\sigma_X = 2.4618$. This would not be the best way to tell the difference between a fake and a real expense report because the standard deviations are not too different from one another.
19. (a) Probability histograms. The distribution of the number of rooms is roughly symmetric for owners and skewed to the right for renters. The center is slightly over 6 units for owners and slightly over 4 for renters. Overall, renter-occupied units tend to have fewer rooms than owner-occupied units.

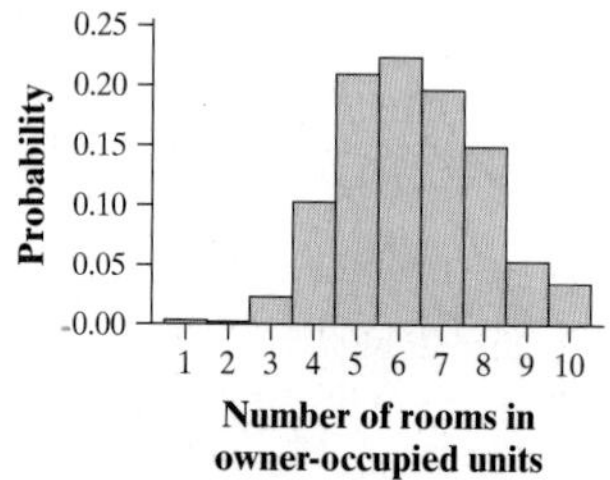

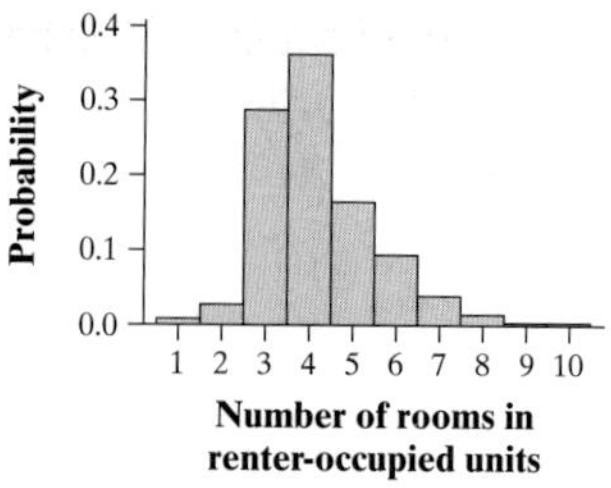

(b) The mean for owner-occupied units is $\mu_x = 6.284$ rooms. The mean for renter-occupied units is $\mu_x = 4.187$ rooms. A comparison of the centers ($6.284 > 4.187$) matches the observation in part (a) that the number of rooms for owner-occupied units tends to be higher than the number of rooms for renter-occupied units. (c) We would expect the owner distribution to have a slightly wider spread than the renter distribution. Even though the distribution of renter-occupied units is skewed to the right, it is more concentrated (contains less variability) about the "peak" than the symmetric distribution for owner-occupied units. $\sigma_X = 1.6399$ rooms and $\sigma_Y = 1.3077$ rooms.
21. (a) 0.51 (b) 0.51 (c) 0.34
23. **State:** What is the probability that a randomly chosen student scores a 9 or better on the ITBS? **Plan:** The score X of the randomly chosen student has the N(6.8, 1.6) distribution. We want to find $P(X \geq 9)$. We'll standardize the scores and find the area shaded in the Normal curve.

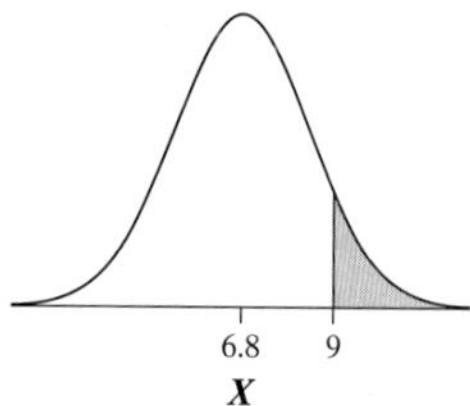

Do: The standardized score for the test is $z = \dfrac{9 - 6.8}{1.6} = 1.38$. $P(Z \geq 1.38) = 0.0838$. **Conclude:** There is about an 8% chance that the chosen student's score is 9 or higher.
25. (a) 0.9652 (using Table A) (b) $P(V \geq 0.72) \approx 0$. If people answered truthfully, it would be virtually impossible to get a sample in which 72% said they had voted.
27. b
29. c
31. Yes. Students did have higher scores, in general, after participating in the chess program. If we look at the differences in the scores, the mean difference was 5.38 and the median was 3. This means that at least half of the students (though less than three quarters since Q_1 was negative) improved their reading scores.
33. predicted posttest = 17.897 + 0.78301(pretest)

Section 6.2

Answers to Check Your Understanding

Page 362 1. $\mu_Y = \$550$; $\sigma_Y = \$471.5$ 2. $\mu_T = \$475$; $\sigma_T = \$471.5$ *Page 370* 1. $\mu_T = 1.8$. On average, this dealership sells or leases 1.8 cars in the first hour of business on Fridays. 2. $\sigma_T = \sqrt{0.943^2 + 0.64^2} = 1.14$ 3. $\mu_B = 500(1.1) + 300(0.7) = \$760$; $\sigma_B = \sqrt{500^2(0.943)^2 + 300^2(0.64)^2} = \$509.09$ *Page 372* 1. $\mu_D = 0.4$. On average, this dealership sells 0.4 cars more than it leases during the first hour of business on Fridays. 2. $\sigma_D = \sqrt{0.943^2 + 0.64^2} = 1.14$ 3. $\mu_B = 500(1.1) - 300(0.7) = \$340$; $\sigma_B = \sqrt{500^2(0.943)^2 + 300^2(0.64)^2} = \$509.09$.

Answers to Odd-Numbered Section 6.2 Exercises

35. $\mu_Y = 3.048$ cm and $\sigma_Y = 0.635$ cm
37. (a) The graph is skewed to the left. Most of the time, the ferry makes \$20 or \$25.

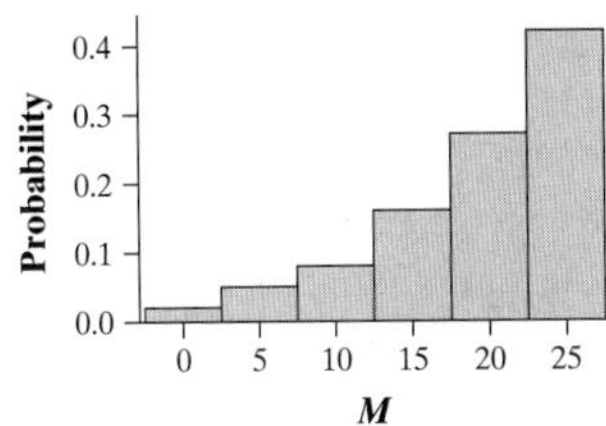

(b) μ_M = \$19.35. The ferry makes \$19.35 per trip on average. (c) σ_M = \$6.45. The individual amounts made on the ferry trips will vary by about \$6.45 from the mean (\$19.35) on average.
39. (a) $\mu_G = 7.6(10) = 76$ (b) $\sigma_G = 1.32(10) = 13.2$ (c) Since the variance of G is the square of the standard deviation, $\sigma_G^2 = (10\sigma_X)^2 = 100\sigma_X^2$. In other words, we need to multiply the variance of X by the square of the constant 10 to get the variance of G.
41. (a) The mean of Y is 20 less than the mean of M. Thus, the ferry company loses an average of \$0.65 per trip. (b) The standard deviation of Y is the same as the standard deviation of M. The individual profits made on the ferry trips will vary by about \$6.45 from the mean (–\$0.65), on average.
43. $\mu_Y = 6(3.87) - 20 = \$3.22$; $\sigma_Y = 6(1.29) = \$7.74$
45. (a) μ_Y = 47.3 degrees Fahrenheit; σ_Y = 4.05 degrees Fahrenheit (b) $P(T < 40) = P(Z < -1.80) = 0.0359$
47. (a)

Value:	3	4	5	6	7	9
Probability:	0.14	0.35	0.06	0.15	0.21	0.09

(b) From the distribution of T, we get $\mu_T = 3(0.14) + 4(0.35) + 5(0.06) + 6(0.15) + 7(0.21) + 9(0.09) = 5.3$. Now using the relationship between T, X, and Y, we get $\mu_T = \mu_X + \mu_Y = 2.7 + 2.6 = 5.3$, which is the same. (c) From the distribution of T, we get $\sigma_T^2 = (3 - 5.3)^2(0.14) + (4 - 5.3)^2(0.35) + \cdots + (9 - 5.3)^2(0.09) = 3.25$. Now using the relationship between T, X, and Y, we get $\sigma_T^2 = \sigma_X^2 + \sigma_Y^2 = (1.55)^2 + (0.917)^2 = 3.24$, which is only different due to rounding error. Note that $\sigma_T = 1.80$ but $\sigma_X + \sigma_Y = 1.55 + 0.917 = 2.467$. These last two values are not the same.
49. (a) Dependent: since the cards are being drawn from the deck without replacement, the nature of the third card will depend upon the nature of the first two cards that were drawn. (b) Independent: X relates to the outcome of the first roll, Y to the outcome of the second roll, and individual dice rolls are independent (the dice have no memory).
51. (a) Yes. The mean of a sum is *always* equal to the sum of the means. (b) No. The variance of the sum is not equal to the sum of the variances because it is not reasonable to assume that X and Y are independent.
53. (a) Let T represent the total payoff: $\mu_T = \$0.50 + \$0.50 = \$1.00$; $\sigma_T = \sqrt{15.8^2 + 15.8^2} = \$22.34$. (b) Let W represent total winnings: $\mu_W = -\$1.00$; $\sigma_W = \$22.34$.
55. $\mu_{Y-X} = 0.001$ g; $\sigma_{Y-X} = 0.002236$ g. On average, the difference between the readings of the two scales will be 0.001 g. Individual difference will vary from this by 0.002236 g on average.
57. $\mu_W = \$303.35$; $\sigma_W = \$6{,}864.29$.
59. (a) $\mu = 31$ seconds; $\sigma = 4.4721$ seconds (b) P(total time < 30 sec) $= P(Z < -0.22) = 0.4129$
61. **State:** What is the probability that the difference between the NOX levels for two randomly selected cars is at least 0.80 g/mi? **Plan:** Since both cars have NOX levels that follow a N(1.4, 0.3) distribution, the distribution of the difference X – Y will be $N(0, \sqrt{0.3^2 + 0.3^2}) \approx N(0, 0.4243)$. Use this to find $P(X - Y > 0.8$ or $X - Y < -0.8)$. **Do:** $P(X - Y > 0.8$ or $X - Y < -0.8) = 0.0588$. **Conclude:** There is only about a 6% chance of randomly finding two cars with as big a difference in NOX levels as the attendant found.
63. **State:** What is the probability that the total team swim time is less than 220 seconds? **Plan:** Let T stand for the total team swim time and X_1 be Wendy's time, X_2 be Jill's time, X_3 be Carmen's time, and X_4 be Latrice's time. Then $T = X_1 + X_2 + X_3 + X_4$. The mean and variance are $\mu_T = 224.2$ seconds and $\sigma_T^2 = 30.89$ seconds. This means that $\sigma_T = \sqrt{30.89} = 5.56$ seconds. Since each individual's time is approximately Normally distributed, that means that T is approximately Normally distributed. Putting this all together, we have that T has a N(224.2, 5.56) distribution. Use this to find $P(T < 220)$. **Do:** $P(T < 220) = 0.2236$. **Conclude:** There is approximately a 22% chance that the team's swim time will be less than 220 seconds.
65. c
67. (a) Fidelity Technology Fund is more closely tied to the stock market as a whole because its correlation is larger. (b) No. Correlations tell nothing about the absolute size of the variables, only how they are related to each other.

Section 6.3

Answers to Check Your Understanding

Page 385 **1.** Check the BINS. *Binary?* "Success" = get an ace; "Failure" = don't get an ace. *Independent?* Because you are replacing the card in the deck and shuffling each time, the result of one trial does not affect the outcome of any other trial. *Number?* The number of trials is set at 10 in advance. *Success?* The probability of success is $\frac{4}{52}$ for each trial. This is a binomial setting. Since X counts the number of successes, it is a binomial random variable with $n = 10$ and $p = \frac{4}{52}$. **2.** Check the BINS. *Binary?* "Success" = over 6 feet; "Failure" = not over 6 feet. *Independent?* Because we are selecting without replacement from a small number of students, the observations are not independent. *Number?* The number of trials is set at 3 in advance. *Success?* The probability of success will not change from person to person. This is not a binomial setting. **3.** Check the BINS. *Binary?* "Success" = roll a 5; "Failure" = don't roll a 5. *Independent?* Because you are rolling a die, the outcome of any one trial does not affect the outcome of any other trial. *Number?* The number of trials is set at 5 in advance. *Success?* The probability of success will change depending on whether a six-sided or an eight-sided die is used. This is not a binomial setting. *Page 390* **1.** Check the BINS. *Binary?* "Success" = question answered "correctly"; "Failure" = question not answered "correctly." *Independent?* The computer randomly assigned correct answers to the questions, so one trial should not affect any other trial. *Number?* There were 10 questions. *Success?* The probability of success is 0.20 for each trial. This is a binomial setting. Since X counts the number of successes, it is a binomial random variable with $n = 10$ and $p = 0.20$. **2.** 0.2013. There is a 20% chance that Patti will answer exactly 3 questions correctly. **3.** 0.0064. There is only a 0.64% chance that a student would pass, so we would be quite surprised if Patti was able to pass. *Page 393* **1.** $\mu_X = 2$. We would expect the average student to get 2 answers correct on such a quiz. **2.** $\sigma_X = 1.265$. We would expect individual students' scores to vary from a mean of 2 correct answers by an average of 1.265 correct answers. **3.** $P(X > 4.53) = 1 - P(X \leq 4) = 0.0328$ *Page 401* **1.** Check the BITS: *Binary?* "Success" = roll doubles; "Failure" = don't roll doubles. *Independent?* Rolling dice is an independent process. *Trials?* We are counting trials up to and including the first success. *Success?* The probability of success is $\frac{1}{6}$ for each roll (there are 6 ways to get doubles out of a total of 36 possible rolls). This is a geometric setting. Since X is measuring the number of trials to get the first success, it is a geometric random variable with $p = \frac{1}{6}$. **2.** 0.1157. There is about a 12% chance that you will get the first set of doubles on the third roll of the dice. **3.** 0.4213

Answers to Odd-Numbered Section 6.3 Exercises

69. *Binary?* "Success" = seed germinates; "Failure" = seed does not germinate. *Independent?* Possibly, although it is possible that if one seed does not germinate, it's more likely that others in the packet will not grow either. *Number?* 20 seeds. *Success?* The probability that each seed germinates is 85%, assuming the advertised percent is true. Assuming independence does hold, this is a binomial setting, and X would have a binomial distribution.

71. *Binary?* "Success" = person is left-handed; "Failure" = person is right-handed. *Independent?* Since students are selected randomly, their handedness is independent. *Number?* There is not a fixed number of trials for this chance process since you continue until you find a left-handed student. Since the number of trials is not fixed, this is not a binomial setting and V is not a binomial random variable.

73. **(a)** A binomial distribution is *not* an appropriate choice for field goals made by the NFL player, because given the different situations the kicker faces, his probability of success is likely to change from one attempt to another. **(b)** It would be reasonable to use a binomial distribution for free throws made by the NBA player because we have $n = 150$ attempts, presumably independent (or at least approximately so), with chance of success $p = 0.8$ each time.

75. 0.2304. There is about a 23% chance that exactly 4 of the 7 people chosen have blood type O.

77. 0.1402. There is about a 14% chance that more than 4 people of the 7 chosen will have blood type O. This is not particularly surprising.

79. **(a)** $P(X = 17) = \binom{20}{17}(0.85)^{17}(0.15)^{3} = 0.2428$ **(b)** 0.0059. This would indeed be surprising and should make Judy suspicious.

81. **(a)** $\mu_X = 3$. You would expect to reach a live person in an average of 3 phone calls when making 15 calls. **(b)** $\sigma_X = 1.55$. In actual practice, you would expect the number of live persons you reach to vary from 3 per 15 calls by 1.55 on average.

83. **(a)** $\mu_Y = 12$. Notice that $\mu_X = 3$ and that $12 + 3 = 15$. In other words, if we reach an average of 3 live people in our 15 calls, we must not reach a live person in an average of 12 calls. **(b)** $\sigma_Y = 1.55$. This is the same thing as σ_X because we have just switched the definitions of p and $1 - p$.

85. **(a)** *Binary?* "Success" = operates for an hour without failure; "Failure" = does not operate for an hour without failure. *Independent?* Whether one engine operates for an hour or not should not affect whether any other engine does or does not. *Number?* We are looking at 350 engines. *Success?* Each engine has the same probability of success: 0.999. This is a binomial setting so since X is counting the number of successes, it has a binomial distribution. **(b)** $\mu_X = 349.65$. If we were to test 350 engines over and over again, we would expect that, on average, 349.65 of them would operate for an hour without failure. $\sigma_X = 0.591$. In individual tests, we would expect to find the number of engines that operate for an hour without failure to vary from 349.65 by an average of 0.591. **(c)** 0.0486. There is less than a 5% chance that 2 or more engines will fail (or 348 engines or fewer will operate correctly). Since there is such a small chance of 2 or more engines failing, it seems likely that they are less reliable than they are supposed to be.

87. We cannot use the binomial distribution here because the sample size (10) is more than 10% of the population (76). This means that the probability of success changes too much as we take each observation out of the population and do not replace it.

89. If the sample size is more than 10% of the population, the amount of change to the makeup of the population is too much. The probability of success changes too much to be considered constant.

91. **(a)** *Binary?* "Success" = visit an auction site at least once a month; "Failure" = don't visit an auction site at least once a month. *Independent?* The sample was random, so one person's actions shouldn't affect anyone else's, and even though we are sampling without replacement, the sample size (500) is far less than 10% of all males aged 18 to 34. *Number?* Sampled 500 men. *Success?* The probability of success for any one individual is 0.50. This is a binomial setting so since X measures the number of successes, X is approximately a binomial random variable. **(b)** $np = 250$ and $n(1 - p) = 250$ are both greater than 10. **(c)** 0.9099

93. Let X be the number of 1s and 2s. Using technology, we find $P(X \leq 29) = 0.0021$. Using the Normal approximation, 0.0016. Either way, the probability is quite small, so we have reason to be suspicious.

95. **(a)** *Binary?* "Success" = get a card you don't already have; "Failure" = get a card you already have. *Independent?* Assuming the cards are put into the boxes randomly at the factory, what card you get in a particular box should be independent of what card you get in any other box. *Trials?* We are not counting trials until the first success. This is not a geometric setting. **(b)** *Binary?* "Success" = Lola wins some money; "Failure" = Lola does not win money. *Independent?* The outcomes of the games are independent of each other. *Trials?* We are counting the number of games until she wins one. *Success?* The probability of success on any given game is 0.259. This is a geometric setting. Let Y be the number of games Lola plays up to and including her first win. This is a geometric random variable.

97. **(a)** $(0.8)^2(0.2) = 0.128$ **(b)** $(0.8)^{10} = 0.1074$

99. **(a)** $\mu_X = 38$. We would expect it to take 38 games to win if one wins in 1 out of 38 games. **(b)** $P(X \leq 3) = 0.0769$. While this is not a usual occurrence, it would happen about 8% of the time, so it is not completely surprising.

101. b

103. b

105. c

107. **(a)** 27.2% of British men smoke.

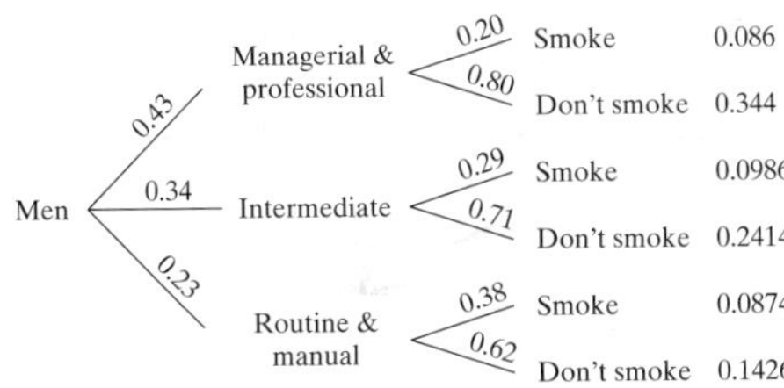

(b) About 32% of the smokers are routine and manual laborers.

Answers to Chapter 6 Review Exercises

R6.1 (a) 0.1 (b) 0.09 (c) $\mu_X = 3.1$; $\sigma_X = 1.136$
R6.2 (a) $\mu_D = 0°C$, $\sigma_D = 5.7°C$; (b) $\mu_Y = 1022°F$; $\sigma_Y = 10.26°F$
R6.3 (a) The average payout on a \$1 bet is \$0.70, and the amount that individual payouts vary from this is \$6.58, on average. (b) Let Y be the amount of Jerry's payout: μ_Y = \$3.50 and σ_Y = \$32.90. (c) Let W be the amount of Marla's payout: μ_W = \$3.50 and σ_W = \$14.71. (d) The casino would probably prefer Marla since there is less variability in her strategy. They are less likely to get great amounts from her but also less likely to have to pay great amounts to her.
R6.4 (a) The machine that makes the caps and the machine that applies the torque are not the same. (b) $\mu_D = 3$ inch-pounds and $\sigma_D = 1.5$ inch-pounds (c) $P(T > C) = 0.0228$
R6.5 (a) *Binary?* "Success" = candy is orange; "Failure" = candy is not orange. *Independent?* Assuming the candies are well mixed, the color of one candy chosen should not tell us anything about the color of any other candy chosen. *Number?* We are choosing a fixed sample of 8 candies. *Success?* Since the sample is far smaller than 10% of the population, the probability of success remains approximately constant at 0.20 even though we are sampling without replacement. This is a binomial setting, so X has a binomial distribution. (b) $\mu_X = 1.6$. We expect to find 1.6 orange candies in samples of size 8, on average. (c) $\sigma_X = 1.13$. Individual samples of size 8 will have the number of orange candies varying from 1.6 by 1.13 on average.
R6.6 (a) $P(X = 0) = 0.1678$. Since the probability is about 17%, it would not be surprising to get no orange M&M's. (b) $P(X \geq 5) = 0.0104$. Since the probability is about 1%, it would be somewhat surprising to find 5 or more orange M&M's.
R6.7 Let Y be the number of spins to get a "wasabi bomb." $P(Y \geq 3) = 0.5625$.
R6.8 (a) Let X be the number of heads in 10,000 tosses: $\mu_X = 5{,}000$ and $\sigma_X = 50$. (b) The conditions are met to approximate the binomial distribution by the Normal distribution: both $np = 5000$ and $n(1-p) = 5000$ are far larger than 10. (c) $P(X \leq 4933$ or $X \geq 5067) = 0.1802$. If the coin were balanced, finding 5067 heads or more, or 4933 heads or fewer, would happen 18% of the time in 10,000 tosses. This is not particularly surprising, so we don't have strong evidence that the coin was not balanced.

Answers to Chapter 6 AP Statistics Practice Test

T6.1 b
T6.2 d
T6.3 e
T6.4 d
T6.5 d
T6.6 d
T6.7 c
T6.8 b
T6.9 b
T6.10 c
T6.11 (a) 0.96. There is a 96% chance that at least 10 eggs are unbroken in a randomly selected carton of "store brand" eggs. (b) $\mu_Y = 0.38$. We expect to find 0.38 broken eggs on average in a carton of a dozen eggs. (c) $\sigma_Y = 0.8219$. The number of broken eggs in individual cartons will vary from 0.38 broken eggs by 0.8219 broken eggs, on average. (d) 0.2950
T6.12 (a) *Binary?* "Success" = dog owner greets dog first; "Failure" = dog owner does not greet dog first. *Independent?* Dog owners are selected at random, so their behaviors should be independent. *Number?* We have a fixed number of trials (12). *Success?* The probability of success is constant for all trials (0.66). This is a binomial setting, and since X counts the number of successes in the 12 trials, it is a binomial random variable. (b) 0.0213. If the *Ladies Home Journal* is correct, we would have only about a 2% chance of getting a sample where 4 or fewer owners greeted their dogs first when coming home. This is reasonably unlikely, so we would suggest that the *Ladies Home Journal* may be wrong.
T6.13 (a) $\mu_D = 25$ minutes. Assuming that the amount of time they spend on homework is independent of each other, σ_D = 11.18 minutes. (b) 0.0125
T6.14 (a) Let X stand for the number of Hispanics in the sample. $\mu_X = 156$ and $\sigma_X = 11.6499$. (b) If the sample contains 15% Hispanics, that means that there were $1200(0.15) = 180$ Hispanics in the sample. $P(X \geq 180) = 0.0235$ using the binomial distribution. If we use the Normal approximation to the binomial distribution, $P(X \geq 180) = 0.0197$. These probabilities are low, so we should be suspicious of the opinion poll.

CHAPTER 7

Section 7.1

Answers to Check Your Understanding

Page 417 1. The parameter is $\mu = 20$ ounces of iced tea. The statistic is $\bar{x} = 19.6$ ounces of iced tea. 2. The parameter is $p = 0.10$, or 10% of passengers. The statistic is $\hat{p} = 0.08$, or 8% of the sample of passengers.
Page 420 1. The individuals are the M&M'S® Milk Chocolate Candies, the variable is color, and the parameter of interest is the proportion of orange M&M'S.

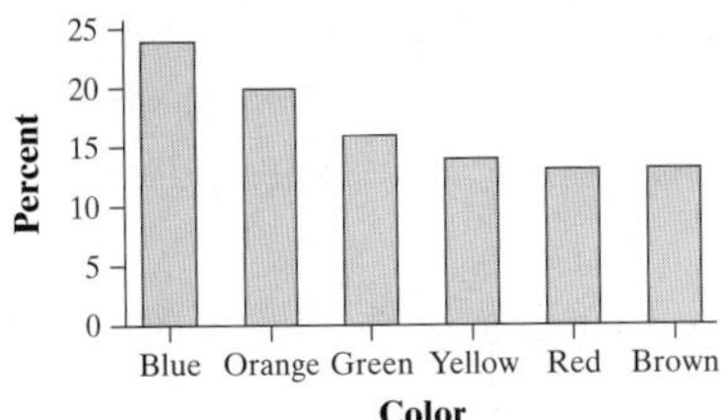

2. For this sample, there are 11 orange M&M'S, so $\hat{p} = \frac{11}{50} = 0.22$.

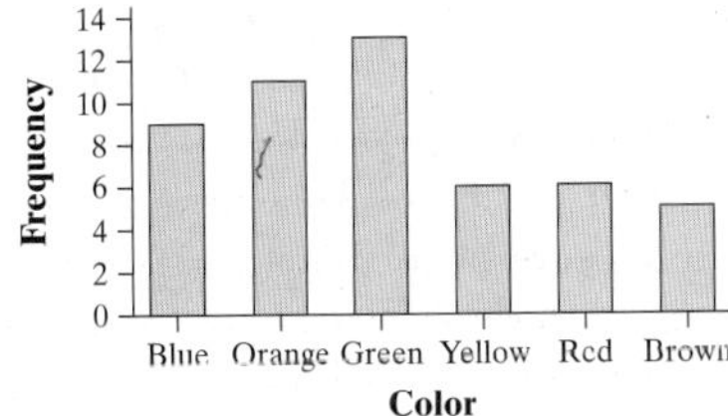

3. The middle graph is the approximate sampling distribution of $\hat{p}$. The statistic measures the proportion of oranges in samples of

M&M'S Milk Chocolate Candies. Assuming that the company is correct, 20% of the M&M'S are orange, so the center of the distribution of $\hat{p}$ should be at approximately 0.20. The first graph shows the distribution of the colors for one sample, rather than the distribution of $\hat{p}$ from many samples, and the third graph is centered at the wrong spot. *Page 426* **1.** The median does not appear to be an unbiased estimator of the population median. The mean of the 500 sample medians is 73.5, whereas the median of the population is 75. **2.** With larger samples, the spread of the sampling distribution is smaller, so increasing the sample size from 10 to 20 will decrease the spread of the sampling distribution. **3.** The sampling distribution is skewed to the left. This means that, in general, underestimates of the population median will be greater than overestimates.

Answers to Odd-Numbered Section 7.1 Exercises

1. Population: people who signed a card saying that they intend to quit smoking. Parameter of interest: proportion of the population who signed the card saying they would not smoke who actually quit smoking. Sample: a random sample of 1000 people who signed the cards. Sample statistic: $\hat{p} = 0.21$.

3. Population: all the turkey meat. Parameter of interest: minimum temperature. Sample: 4 randomly chosen points in the turkey. Sample statistic: sample minimum = 170°F.

5. $\mu = 2.5003$ is a parameter and $\bar{x} = 2.5009$ is a statistic.

7. $\hat{p} = 0.48$ is a statistic and $p = 0.52$ is a parameter.

9. **(a)** This is not the exact sampling distribution, because that would require a value of $\hat{p}$ for all possible samples of size 100. However, it is an approximation of the sampling distribution that we created through simulation. **(b)** The distribution is centered at 0.60 and is reasonably symmetric and bell-shaped. Values vary from about 0.47 to 0.74. The values at 0.47, 0.73, and 0.74 are outliers. **(c)** If we found that only 45 students said they did all their homework last week, we would be skeptical of the newspaper's claim that 60% of students did their homework last week. None of the simulated samples had a proportion this low.

11. **(a)**

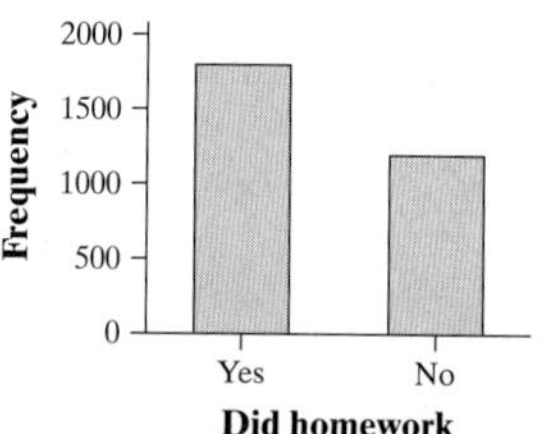

(b) Answers will vary. An example bar graph is given.

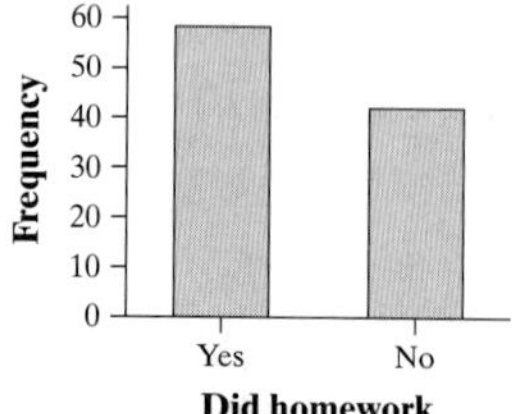

13. **(a)** The approximate sampling distribution is skewed to the right with a center at $9(°F)^2$. The values vary from about 2 to $27.5(°F)^2$. **(b)** A sample variance of 25 is quite large compared with what we would expect, since only one out of 500 SRSs had a variance that high. It suggests that the manufacturer's claim is false and that the thermostat actually has more variability than claimed.

15. **(a)** The population is the 12,000 students; the population distribution (Normal with mean 7.11 minutes and standard deviation 0.74 minutes) describes the time it takes randomly selected individuals to run a mile. **(b)** The sampling distribution (Normal with mean of 7.11 minutes and standard deviation of 0.074 minutes) describes the average mile-time for 100 randomly selected students. This is different from the population distribution in that it has a smaller standard deviation and it describes the mean of 100 mile-times rather than individual mile-times.

17. **(a)** Since the smallest number of total tax returns (i.e., the smallest population) is still more than 10 times the sample size, the variability of the sample proportion will be (approximately) the same for all states. **(b)** Yes. It will change—the sample taken from Wyoming will be about the same size, but the sample from, for example, California will be considerably larger, and therefore, the variability of the sample proportion will be smaller.

19. **(a)** Graph (c) shows an unbiased estimator because the mean of the distribution is very close to the population parameter. **(b)** The graph in part (b) shows the statistic that does the best job at estimating the parameter. Although it is biased, the bias is small and the statistic has very little variability.

21. d

23. c

25. **(a)** This is the same thing as asking what percent of Normal scores are more than 2.5 standard deviations below the mean. In other words, what is $P(z < -2.5)$? Using Table A, this value is 0.0062. **(b)** The distribution for the older women, based on the standard scale for younger women, is Normal with mean −2 and standard deviation 1. So the question is asking for the probability of getting a standard score of less than −2.5. This is $P(X < -2.5) = P\left(z < \frac{-2.5-(-2)}{1}\right) = P(z < -0.5) = 0.3085$. So, based on this criterion, about 31% of women aged 70–79 have osteoporosis.

Section 7.2

Answers to Check Your Understanding

Page 437 **1.** The mean of the sampling distribution is the same thing as the population proportion. In this case, $\mu_{\hat{p}} = 0.75$. **2.** The standard deviation of the sampling distribution is $\sigma_{\hat{p}} = \sqrt{\frac{p(1-p)}{n}} = \sqrt{\frac{0.75(0.25)}{1000}} = 0.0137$. There are more than 10,000 young adult Internet users, so the 10% condition has been met. **3.** The sampling distribution of $\hat{p}$ is approximately Normal. Both $np = 1000(0.75) = 750$ and $n(1-p) = 1000(0.25) = 250$ are greater than 10. **4.** If the sample size were 9000 instead of 1000, the sampling distribution would still be approximately Normal with mean 0.75. But the standard deviation of the sampling distribution would be smaller by a factor of 3. In this case, $\sigma_{\hat{p}} = \sqrt{\frac{p(1-p)}{n}} = \sqrt{\frac{0.75(0.25)}{9000}} = 0.0046$.

Answers to Odd-Numbered Section 7.2 Exercises

27. **(a)** We would not be surprised to find 8 (32%) orange candies. As shown in Figure 7.11, there were a fair number of simulations in which there were 8 or fewer orange candies. On the other hand, there were only a couple of simulations in which there were 5 (20%) or fewer, so if this occurred, we should be surprised. **(b)** It is more surprising to get 32% orange candies in a sample of 50 than it is in a sample of 25. Comparing the graphs in Figures 7.11 and 7.12, there were a fair number of simulations in

Figure 7.11 (sample size 25) with 32% or less, but very few in Figure 7.12 (sample size 50) with 32% or less.
29. **(a)** The mean of the sampling distribution is the same as the population proportion, so $\mu_{\hat{p}} = p = 0.45$. **(b)** The standard deviation of the sampling distribution is $\sigma_{\hat{p}} = \sqrt{\frac{p(1-p)}{n}} = \sqrt{\frac{0.45(0.55)}{25}} = 0.0995$. In this case, the 10% condition is met because it is very likely true that there are more than 250 candies. **(c)** The sampling distribution is approximately Normal because $np = 25(0.45) = 11.25$ and $n(1-p) = 25(0.55) = 13.75$ are both at least 10. **(d)** If the sample size were 50 rather than 25, the sampling distribution would still be approximately Normal with mean 0.45, but the standard deviation would be $\sigma_{\hat{p}} = \sqrt{\frac{p(1-p)}{n}} = \sqrt{\frac{0.45(0.55)}{50}} = 0.0704$.
31. **(a)** The 10% condition is not met here. Out of the population of 76 passengers, 10 people were screened (13%). This means that they sampled more than 10% of the population. **(b)** No. Since the total sample size was 10, both np and $n(1-p)$ will be less than 10.
33. The Normal condition is not met here: $np = 15(0.3) = 4.5 < 10$. *Challenge:* Let X be the number of Hispanic workers in the sample. X has an approximate binomial distribution with $n = 15$ and $p = 0.3$. $P(X \le 3) = \binom{15}{0}(0.3)^0(1-0.3)^{15} + \cdots + \binom{15}{3}(0.3)^3(1-0.3)^{12} = 0.2969$.
35. **(a)** The mean of the sampling distribution is the same as the population proportion, so it is $\mu_{\hat{p}} = p = 0.70$. **(b)** $\sigma_{\hat{p}} = \sqrt{\frac{p(1-p)}{n}} = \sqrt{\frac{0.7(0.3)}{1012}} = 0.0144$. The population (all U.S. adults) is clearly at least 10 times as large as the sample (the 1012 surveyed adults), so the 10% condition is met. **(c)** The sampling distribution is approximately Normal since $np = 1012(0.70) = 708.4$ and $n(1-p) = 1012(0.30) = 303.6$ are both at least 10. **(d)** $P(\hat{p} \le 0.67) = P(z \le -2.08) = 0.0188$. This is a fairly unusual result if 70% of the population actually drink the cereal milk.
37. 4048 adults. Since the standard deviation is found by dividing by $\sqrt{n}$, using $4n$ for the sample size halves the standard deviation ($\sqrt{4n} = 2\sqrt{n}$).
39. **Plan:** $\mu_{\hat{p}} = 0.70$. Since the population clearly contains more than $267(10) = 2670$ college women, $\sigma_{\hat{p}} = \sqrt{\frac{0.7(0.3)}{267}} = 0.0280$. Since $np = 267(0.7) = 186.9$ and $n(1-p) = 267(0.3) = 80.1$ are both at least 10, the distribution of $\hat{p}$ can be approximated by a Normal distribution. **Do:** $P(\hat{p} \ge 0.75) = P\left(z \ge \frac{0.75 - 0.7}{0.0280}\right) = P(z \ge 1.79) = 0.0367$. **Conclude:** About 3.67% of all SRSs of size 267 will give a sample proportion that is 0.75 or greater.
41. **(a) Plan:** $\mu_{\hat{p}} = 0.90$. Since the population contains more than $100(10) = 1000$ orders in the last week, $\sigma_{\hat{p}} = \sqrt{\frac{0.9(0.1)}{100}} = 0.03$. Since $np = 100(0.9) = 90$ and $n(1-p) = 100(0.1) = 10$ are both at least 10, the distribution of $\hat{p}$ can be approximated by a Normal distribution. **Do:** $P(\hat{p} \le 0.86) = P\left(z \le \frac{0.86 - 0.9}{0.03}\right) = P(z \le -1.33) = 0.0918$. **Conclude:** There is a 9.18% chance that we would get a sample in which 86% or fewer of the orders were shipped within three working days. **(b)** If the claim is correct, then we can expect to observe 86% or fewer orders shipped on time in about 9.18% of the samples of this size. Getting a sample proportion at or below 0.86 is not an unlikely event.
43. b
45. b
47.

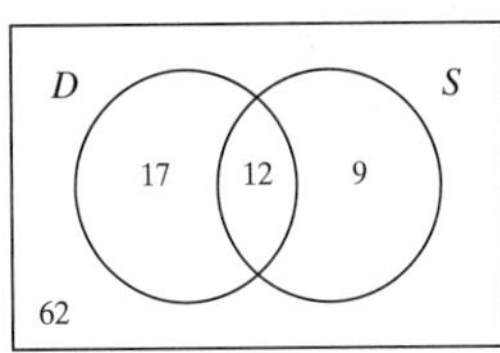

62% neither download nor share music files.

Section 7.3

Answers to Check Your Understanding

Page 448 1. $P(X > 270) = P\left(z > \frac{270 - 266}{16}\right) = P(z > 0.25) = 0.4013$
2. The mean of the sampling distribution of $\bar{x}$ is the same as the mean of the distribution of X, so $\mu_{\bar{x}} = \mu_X = 266$ days.
3. First, we check the 10% condition. We are taking a sample of 6 pregnant women. There are clearly more than $10(6) = 60$ pregnant women, so this condition is met. Therefore, the standard deviation of the sampling distribution is $\sigma_{\bar{x}} = \frac{\sigma_X}{\sqrt{n}} = \frac{16}{\sqrt{6}} = 6.532$ days.
4. $P(\bar{x} > 270) = P\left(z > \frac{270 - 266}{6.532}\right) = P(z > 0.61) = 0.2709$

Answers to Odd-Numbered Section 7.3 Exercises

49. The mean is $\mu_{\bar{x}} = \mu_X = 225$ seconds, and the standard deviation is $\sigma_{\bar{x}} = \frac{\sigma_X}{\sqrt{n}} = \frac{60}{\sqrt{10}} = 18.974$ seconds. These results do not depend on the shape of the distribution of individual play times.
51. If we want $\sigma_{\bar{x}} = 30$, then we need to solve the following equation for n: $30 = \frac{60}{\sqrt{n}} \to \sqrt{n} = \frac{60}{30} = 2 \to n = 4$. So we need a sample of size 4.
53. **(a)** The sampling distribution of $\bar{x}$ is Normal with $\mu_{\bar{x}} = \mu_X = 188$ mg/dl and $\sigma_{\bar{x}} = \frac{\sigma_X}{\sqrt{n}} = \frac{41}{\sqrt{100}} = 4.1$ mg/dl **(b)** $P(185 \le \bar{x} \le 191) = P\left(\frac{185 - 188}{4.1} \le z \le \frac{191 - 188}{4.1}\right) = P(-0.73 \le z \le 0.73) = 0.7673 - 0.2327 = 0.5346$. **(c)** In this case, $\sigma_{\bar{x}} = \frac{\sigma_X}{\sqrt{n}} = \frac{41}{\sqrt{1000}} = 1.30$ mg/dl. So now the probability becomes $P(185 \le \bar{x} \le 191) = P\left(\frac{185 - 188}{1.30} \le z \le \frac{191 - 188}{1.30}\right) = P(-2.31 \le z \le 2.31) = 0.9896 - 0.0104 = 0.9792$. The larger sample is better since it is more likely to produce a sample mean within 3 mg/dl of the population mean.
55. **(a)** Let X denote the amount of cola in a bottle. $P(X < 295) = P\left(z < \frac{295 - 298}{3}\right) = P(z < -1) = 0.1587$. **(b)** If $\bar{x}$ is the mean contents of six bottles, then $\bar{x}$ has a Normal distribution with $\mu_{\bar{x}} = \mu_X = 298$ ml and $\sigma_{\bar{x}} = \frac{\sigma_X}{\sqrt{n}} = \frac{3}{\sqrt{6}} = 1.2247$ ml (10% condition OK since there are more than 60 bottles in the population). $P(\bar{x} < 295) = P\left(z < \frac{295 - 298}{1.2247}\right) = P(z < -2.45) = 0.0071$.
57. No. The histogram of the sample values will look like the population distribution, whatever it might happen to be. The central limit theorem says that the histogram of the distribution of *sample means* (from many large samples) will look more and more Normal.
59. **(a)** Since the distribution of the play times of the population of songs is heavily skewed to the right, a sample size of 10 will not

be enough for the Normal approximation to be appropriate. **(b)** With a sample size of 36, we now have enough observations in our sample for the central limit theorem to apply.

$$P(\bar{x} > 240) = P\left(z > \frac{240 - 225}{60/\sqrt{36}}\right) = P(z > 1.5) = 0.0668$$

61. **(a)** This probability cannot be calculated, because we do not know the shape of the distribution of the weights. **(b)** If W is total weight and $\bar{x} = W/30$, the central limit theorem says that $\bar{x}$ is approximately Normal with mean 190 lb and standard deviation $\sigma_{\bar{x}} = \frac{35}{\sqrt{30}} = 6.3901$ lb, assuming the weights are independent. Thus, $P(W > 6000) = P(\bar{x} > 200) = P\left(z > \frac{200 - 190}{6.3901}\right) = P(z > 1.56) = 0.0594$. There is about a 6% chance that the total weight exceeds the limit of 6000 lb.

63. **State:** What is the probability that the average loss will be no greater than \$275? **Plan:** The sampling distribution of the sample mean loss $\bar{x}$ has mean $\mu_{\bar{x}} = \mu_X = \$250$ and standard deviation $\sigma_{\bar{x}} = \frac{300}{\sqrt{10{,}000}} = \$3$ (10% condition is met, assuming there are at least 100,000 policies). Since the sample size is so large (10,000 > 30), we can safely use the Normal distribution as an approximation for the sampling distribution of $\bar{x}$. **Do:** $P(\bar{x} > 275) = P\left(z > \frac{275 - 250}{3}\right) = P(z > 8.33) \approx 0$. **Conclude:** It is very, very unlikely that the company would have an average loss of more than \$275.

65. a

67. b

69. The unemployment rates for each level of education are

$$P(\text{unemployed} \mid \text{didn't finish HS}) = \frac{(12{,}470 - 11{,}408)}{12{,}470} = \frac{1062}{12{,}470} = 0.0852$$

$$P(\text{unemployed} \mid \text{HS but no college}) = \frac{1977}{37{,}834} = 0.0523$$

$$P(\text{unemployed} \mid \text{less than bachelor's degree}) = \frac{1462}{34{,}439} = 0.0425$$

$$P(\text{unemployed} \mid \text{college graduate}) = \frac{1097}{40{,}390} = 0.0272$$

The unemployment rate decreases with additional education.

71. $P(\text{in labor force} \mid \text{college graduate}) = \frac{40{,}390}{51{,}582} = 0.7830$

Answers to Chapter 7 Review Exercises

R7.1 The population is the set of all eggs shipped on the day in question. The sample consists of the 200 eggs examined. The parameter is the proportion $p = 0.001$ of eggs shipped that day that had salmonella. The statistic is the sample proportion $\hat{p} = \frac{9}{200} = 0.045$ of eggs in the sample that had salmonella.

R7.2 **(a)** Answers will vary. An example dotplot is given.

2800 2940 3080 3220 3360 3500 3640 3780

Birth weight (grams)

(b) The dotplot does not show the range of every possible sample of size 5 from the population. Instead, it shows the ranges from 500 SRSs from the population. This is a very small subset of the values that make up the sampling distribution.

R7.3 **(a)** The sample range is not an unbiased estimator of the population range. If it were unbiased, then the sampling distribution would have 3417 (the actual range) as its mean. But according to the graph, none of the 500 observations were greater than 3000 (and certainly none were greater than 3417). The mean in the graph is closer to 1200, which means that the sample range underestimates the value of the population range. **(b)** If we want to reduce the variability of the sampling distribution of the sample range, we should take larger samples.

R7.4 **(a)** The mean is $\mu_{\hat{p}} = p = 0.15$. **(b)** Since the population (all adults) is considerably larger than 10 times the sample size ($n = 1540$), the standard deviation is $\sigma_{\hat{p}} = \sqrt{\frac{0.15(0.85)}{1540}} = 0.0091$. **(c)** Since $np = 1540(0.15) = 231$ and $n(1 - p) = 1540(0.85) = 1309$ are both at least 10, the sampling distribution is approximately Normal. **(d)** $P(0.13 \le \hat{p} \le 0.17) = P\left(\frac{0.13 - 0.15}{0.0091} \le z \le \frac{0.17 - 0.15}{0.0091}\right) = P(-2.20 \le z \le 2.20) = 0.9722$

R7.5 **(a)** We are looking for the probability that, in a random sample of 100 travelers, 20 or fewer get a red light. This is equivalent to finding $P(\hat{p} \le 0.20)$. First, we need the sampling distribution of $\hat{p}$. The mean is $\mu_{\hat{p}} = p = 0.3$. Since the population (all travelers passing through Guadalajara, Mexico) is considerably larger than 10 times the sample size ($n = 100$), the standard deviation is $\sigma_{\hat{p}} = \sqrt{\frac{0.3(0.7)}{100}} = 0.0458$. Since $np = 100(0.3) = 30$ and $n(1 - p) = 100(0.7) = 70$ are both at least 10, the sampling distribution is approximately Normal. This means that the probability can be computed as $P(\hat{p} \le 0.20) = P\left(z \le \frac{0.20 - 0.30}{0.0458}\right) = P(z \le -2.18) = 0.0146$. **(b)** The claim is unlikely to be true. There is only a 1.5% chance that we would find a sample with as few red lights as we saw in our sample if their claim was true.

R7.6 (a) Let X denote the WAIS score for a randomly selected individual. $P(X \ge 105) = P\left(z \ge \frac{105 - 100}{15}\right) = P(z \ge 0.33) = 0.3707$. **(b)** The mean is $\mu_{\bar{x}} = 100$ and the standard deviation is $\sigma_{\bar{x}} = \frac{15}{\sqrt{60}} = 1.9365$. **(c)** $P(\bar{x} \ge 105) = P\left(z \ge \frac{105 - 100}{1.9365}\right) = P(z \ge 2.582) = 0.0049$ **(d)** The answer to part (a) could be quite different. The answer to part (b) would be the same because the mean and standard deviation do not depend on the shape of the population. Because of the large sample size, the answer we gave for part (c) would still be fairly reliable because of the central limit theorem.

R7.7 **(a)** The mean is $\mu_{\bar{x}} = 0.5$ and the standard deviation is $\sigma_{\bar{x}} = \frac{0.7}{\sqrt{50}} = 0.0990$. **(b)** Because we have a large sample size (larger than 30), the central limit theorem applies and $P(\bar{x} \ge 0.6) = P\left(z \ge \frac{0.6 - 0.5}{0.0990}\right) = P(z \ge 1.01) = 0.1562$.

Answers to Chapter 7 AP Statistics Practice Test

T7.1 c. The statistic is a measure of the sample, and the parameter is a measure of the population.

T7.2 c. Sample size has no effect on the bias of an estimate, but larger samples will reduce the variability of an estimate.

T7.3 c. To use the Normal approximation, both np and $n(1 - p)$ must be at least 10. In all options other than (c), this condition is not met.

T7.4 a. The central limit theorem is not needed when the original distribution is Normal; the distribution of the sample mean is always Normal in that case.
T7.5 b. Since a sample of 3% of the undergraduates from Ohio State University consists of approximately 1200 students, whereas a sample of 3% of the undergraduates from Johns Hopkins consists of just 60 students, the estimate from Ohio State University will have less sampling variability.
T7.6 b. The variation in the sample mean is related to the square root of the sample size, so if you double the sample size, the variation is reduced by the square root of 2.
T7.7 b. The sampling distribution would be only approximately Normal with mean equal to the population proportion (0.55 in this case) and standard deviation equal to $\sqrt{\frac{(0.55)(0.45)}{250}} = 0.03$.
T7.8 e. The sampling distribution has information about how the sample mean varies from sample to sample, not what any sample itself looks like.
T7.9 e. We do not have a sample size large enough to use the central limit theorem, and we do not know that the distribution of fill amounts is Normally distributed.
T7.10 e. The distribution of the average amount of pay will not be Normal, because there are only three possible outcomes, $\bar{x} =$ 40, 60, or 80.
T7.11 Sample statistic A provides the best estimate of the parameter. Both statistics A and B appear unbiased, while statistic C appears to be biased (low). In addition, statistic A has lower variability than statistic B. In this situation, we want low bias and low variability, so statistic A is the best choice.
T7.12 **(a)** We cannot calculate this probability because we do not know the shape of the distribution of the amount that individual households pay for Internet service. For instance, we know that "many households pay about \$10," but we don't know what percent of households are in this category. **(b)** The mean of the sampling distribution of the sample mean is the same as the mean of the original distribution. Therefore, $\mu_{\bar{x}} = \mu_X = \$28$. Also, we know that $\sigma_{\bar{x}} = \frac{100}{\sqrt{500}} = 0.4472$ because there are at least $10(500) =$ 5000 households with Internet access. **(c)** Since the sample size is large (much more than 30), the central limit theorem tells us that the sampling distribution of the sample mean is approximately Normal. **(d)** $P(\bar{x} > 29) = P\left(z > \frac{29-28}{0.4472}\right) = P(z > 2.24) = 0.0125$. There is about a 1.25% chance of getting a sample of 500 in which the average amount paid for the Internet is more than \$29.
T7.13 We know that $\mu_{\hat{p}} = p = 0.22$. Since the population size (all children under 6 years old) is much more than 10 times the sample size (300), we know that $\sigma_{\hat{p}} = \sqrt{\frac{0.22(0.78)}{300}} = 0.0239$. Finally, since $np = 300(0.22) = 66$ and $n(1-p) = 300(0.78) =$ 234 are both at least 10, the sampling distribution of the sample proportion is approximately Normal. This leads to the following probability calculation: $P(\hat{p} > 0.2) = P\left(z > \frac{0.2 - 0.22}{0.0239}\right) = P(z > -0.84) = 0.7995$. There is about an 80% chance that a sample of 300 children will yield more than 20% who live in households with incomes less than the official poverty level.

Answers to Cumulative AP Practice Test 2

AP2.1 a
AP2.2 d
AP2.3 e
AP2.4 b
AP2.5 c
AP2.6 e
AP2.7 c
AP2.8 a
AP2.9 d
AP2.10 c
AP2.11 b
AP2.12 c
AP2.13 a
AP2.14 c
AP2.15 d
AP2.16 c
AP2.17 e
AP2.18 a
AP2.19 c
AP2.20 b
AP2.21 a
AP2.22 **(a)** This is an observational study. No treatments were imposed on the subjects. **(b)** No. Since this was an observational study and not a randomized controlled experiment, no cause-and-effect conclusions are possible. **(c)** Two variables are confounded when their effects on the cholesterol level cannot be distinguished from one another. For example, people who take omega-3 fish oil might also be more health conscious in general and do other things such as eat more healthfully or exercise more. Researchers would not know whether it was the omega-3 fish oil or the more healthy food consumption or exercise that was the real explanation of lower cholesterol. In other words, the effects of omega-3 fish oil consumption are mixed up with the effects of the other activities that more health conscious people do.
AP2.23 **(a)** $P(\text{type O or Hawaiian-Chinese}) = 65{,}516/145{,}057 = 0.452$ **(b)** $P(\text{type AB} \mid \text{Hawaiian}) = 99/4670 = 0.021$ **(c)** $P(\text{Hawaiian}) = 0.032$; $P(\text{Hawaiian} \mid \text{type B}) = 0.010$. Since these probabilities are not equal, the two events are not independent. **(d)** The probability of randomly selecting a specimen that contains type A blood from the white ethnic group is $50{,}008/145{,}057 = 0.345$. The probability that at least one of the two samples matches this description $= 1 - P(\text{neither are type A from white ethnic group}) = 1 - (1 - 0.345)^2 = 0.571$.
AP2.24 **(a)** This was an experiment since a treatment was imposed. The researchers added dead cicadas to some plants, while the others served as a control group. **(b)** The distribution of seed mass for the cicada plants has the higher mean. The distribution of seed mass for the cicada plants is slightly skewed to the right, which will pull the mean above its median and toward the higher values. The distribution of seed mass for the control plants is skewed to the left, which will pull the mean of this distribution below its median toward the lower values. Since the medians of both distributions are equal, the mean for the cicada plants lies above the mean for the control plants. **(c)** As stated in part (b), the mean seed mass for the cicada plants is higher than the mean seed mass for the control plants. However, the median seed mass for the two groups is the same. The boxplots show a great deal of overlap between the seed masses of the plants in the two groups. There is some evidence that cicadas can be used effectively as fertilizer, but the difference in the mean seed weights for the two groups isn't large enough to rule out the chance involved in the random assignment as a plausible explanation.

AP2.25 **(a)** Since $P(\text{diamond}) = 1/5$, let 8 and 9 represent drawing the card with a diamond and the digits 0 through 7 represent drawing a card without the diamond. Moving left to right across a row, examine one digit at a time. Stop when you get a diamond. Count the number of cards drawn. **(b)** The results: 3, 7, 5, 11, 2, 12, 1, 7, 5, 3 **(c)** Expected number of cards to get the diamond: $\frac{56}{10} = 5.6$. On average, you will draw 5.6 cards in order to win. **(d)** At $1.00 per card, you will expect to pay $5.60, on average, in order to win $5.00. The game doesn't appear to be fair.

CHAPTER 8

Section 8.1

Answers to Check Your Understanding

Page 476 **1.** We are 95% confident that the interval from 2.84 to 7.55 g captures the true standard deviation of the fat content of Brand X hot dogs. **2.** In 95% of all possible samples of 10 Brand X hot dogs, the resulting confidence interval would capture the true standard deviation. **3.** False. The probability is either 1 (if the interval contains the true standard deviation) or 0 (if it doesn't).

Answers to Odd-Numbered Section 8.1 Exercises

1. Mean number of pairs of shoes. $\bar{x} = 30.35$.

3. Sample proportion of those planning to attend the prom. $\hat{p} = 0.72$.

5. **(a)** Approximately Normal; $\mu_{\bar{x}} = 280$, $\sigma_{\bar{x}} = 2.0702$.
(b)

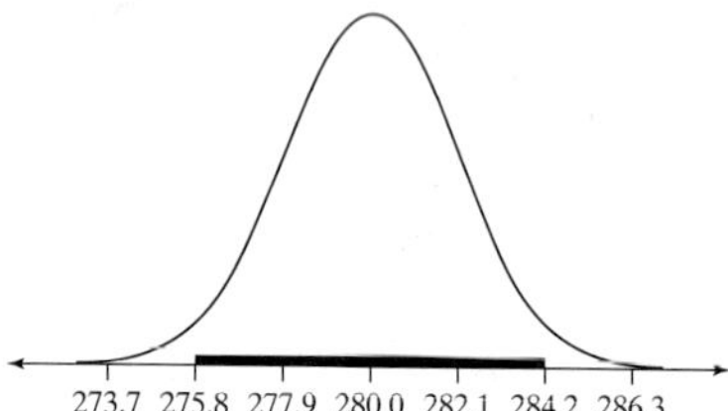

(c) $m = 4.2$ (see graph above) **(d)** About 95%

7. Both will have the same length, but the interval with the value of $\bar{x}$ in the shaded region will contain the true population mean, whereas the other will not.

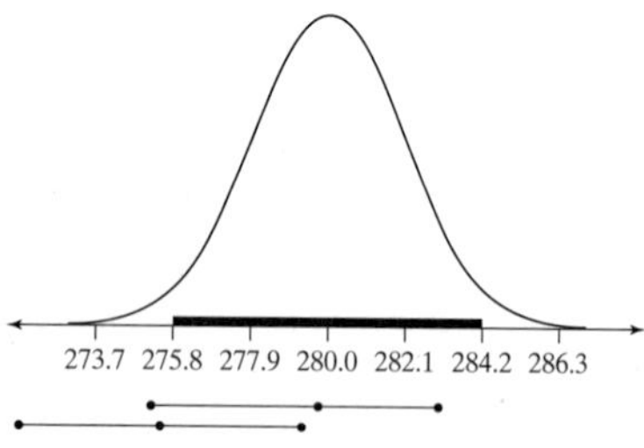

9. 80%, because only 21/25 (84%) of the intervals captured the true mean of the population.

11. **(a)** If we were to repeat the sampling procedure many times, the sample proportion would be within 3 percentage points of the true proportion in 95% of samples. **(b)** (0.63, 0.69). We are 95% confident that the interval from 0.63 to 0.69 captures the true proportion of those who favor an amendment to the Constitution that would permit organized prayer in public schools. **(c)** If we were to repeat the sampling procedure many times, about 95% of the confidence intervals computed would contain the true proportion of those who favor an amendment to the Constitution that would permit organized prayer in public schools.

13. Nonresponse and undercoverage of those who do not have telephones or have only cell phones.

15. We are 95% confident that the interval from 10.9 to 26.5 captures the true difference in the average number of pairs of shoes owned by girls and boys (girls − boys). If this sampling method were employed many times, approximately 95% of the resulting confidence intervals would capture the true difference between the average number of pairs of shoes owned by girls and boys.

17. **(a)** Incorrect. The interval is for the mean, not individual values. **(b)** Incorrect. This would be true only if we knew that $\mu = 26.8$. **(c)** Correct. A confidence interval provides an interval of plausible values for a parameter. **(d)** Incorrect. The population mean is a constant. It is either in this interval or it isn't. **(e)** Incorrect. Even though this value is outside the interval, it is always possible that a confidence interval doesn't capture the true value.

19. *Random:* We can generalize our results to a larger population or make inferences about cause and effect. *Normal:* Know the sampling distribution of the statistic, which, in turn, leads to the computation of the confidence interval. *Independent:* For calculating the appropriate standard deviations.

21. b

23. c

25. **(a)** Observational study because no treatments were imposed on the children. **(b)** No. Because this was an observational study, the effects of lurking variables may be obscuring a possible cause-and-effect relationship.

Section 8.2

Answers to Check Your Understanding

Page 487 **1.** *Random:* not met. This was a convenience sample. *Normal:* met. $n\hat{p} = 17$ and $n(1 - \hat{p}) = 83$ are both at least 10. *Independent:* met. A large high school has more then $10(100) = 1000$ students. **2.** *Random:* met. *Normal:* not met. $n\hat{p} = 3$ is not at least 10. *Independent:* met. There are thousands of bags produced per hour, so the sample is less than 10% of the population.
Page 490 **1.** Population: U.S. college students. Parameter: true proportion who are classified as binge drinkers. **2.** *Random:* students were chosen randomly. *Normal:* 2486 successes and 8418 failures, both are at least 10. *Independent:* 10,904 is clearly less than 10% of all U.S. college students. All conditions are met.
3. $z^* = 2.576$. (0.218, 0.238). **4.** We are 99% confident that the interval from 21.8% to 23.8% captures the true proportion of U.S. college students who would be classified as binge drinkers.

Page 494 **1.** $1.96\sqrt{\frac{0.8(1-0.8)}{n}} = 0.03 \rightarrow n = 683$ **2.** This will require a larger sample size. $2.576\sqrt{\frac{0.8(1-0.8)}{n}} = 0.03 \rightarrow n = 1179$

Answers to Odd-Numbered Section 8.2 Exercises

27. *Random:* met, an SRS is interviewed. *Normal:* met, $n\hat{p} = 14$ and $n(1 - \hat{p}) = 36$ are both at least 10. *Independent:* not met because 50 is more than 10% of 175.

29. *Random:* may not be met because we don't know how the sample was obtained. *Normal:* not met because $n\hat{p} = 5$ is not at least 10. *Independent:* met, there are more than $10(2673) = 26{,}730$ adult heterosexuals.

31. 2.33

33. **(a)** Population: the seniors at Tonya's high school. Parameter: true proportion who plan to attend the prom. **(b)** *Random:* The sample is an SRS. *Normal:* There are 36 successes and 14 failures. Both of these numbers are at least 10. *Independent:* The sample

consists of less than 10% of the population. **(c)** $0.72 \pm 1.645(0.0635) = (0.616, 0.824)$ **(d)** We are 90% confident that the interval from 0.616 to 0.824 captures the true proportion of seniors who plan to attend the prom.

35. **(a) State:** We want to construct a 99% confidence interval for the actual proportion p of all U.S. college students who are abstainers. **Plan:** One-sample z interval for p. *Random:* The students were selected randomly. *Normal:* $2105 \geq 10, 8799 \geq 10$. *Independent:* The sample is less than 10% of the population of all college students. **Do:** $0.193 \pm 2.576\sqrt{\frac{0.193(0.807)}{10904}} = (0.183, 0.203)$. **Conclude:** We are 99% confident that the interval from 0.183 to 0.203 captures the true proportion of U.S. college students who are abstainers. **(b)** The value 25% does not appear in our 99% confidence interval, so it would be very surprising if the claim was correct.

37. Whether the students told the truth in the survey or not.

39. **State:** We want to estimate the actual proportion of all students taking the SAT twice who use coaching at a 99% confidence level. **Plan:** We should use a one-sample z interval for p if the conditions are satisfied. *Random:* The students were selected randomly. *Normal:* There were 427 successes and 2733 failures. Both are at least 10. *Independent:* The sample is less than 10% of the population of all students taking the SAT twice. The conditions are met. **Do:** A 99% confidence interval is given by $0.135 \pm 2.576 \sqrt{\frac{0.135(0.865)}{3160}} = (0.119, 0.151)$. **Conclude:** We are 99% confident that the interval from 0.119 to 0.151 captures the true proportion of students taking the SAT twice who receive coaching.

41. **(a)** No. They should not calculate a margin of error, because the Random condition wasn't met. **(b)** No. This was a voluntary response sample, so no inference should be made about any population.

43. **(a)** 318 **(b)** 423. The sample sizes differ by 105.

45. 1068

47. **(a)** 95% **(b)** Nonresponse bias

49. a

51. c

53. **(a)** A histogram of the number of accidents per hour.

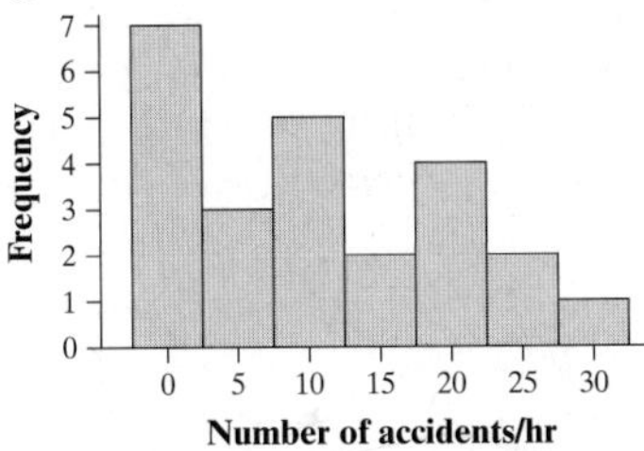

(b) A time plot of the number of accidents.

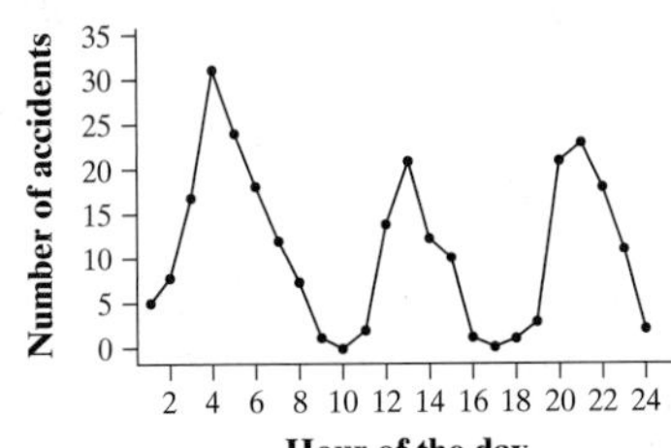

(c) The histogram in part (a) shows that the number of accidents has a distribution that is skewed to the right. **(d)** The time plot in part (b) shows that there is a cyclical nature to the number of accidents.

Section 8.3

Answers to Check Your Understanding

Page 501 $2.33\frac{0.0002}{\sqrt{n}} = 0.0001 \rightarrow n = 22$ *Page 507* **(a)** $t^* = 2.518$ **(b)** $t^* = 1.833$ **(c)** $t^* = 2.447$ *Page 511* **1.** Population mean healing rate. **2.** One-sample t interval for μ. *Random:* The description says that the newts were randomly chosen. *Normal:* We do not know if the data are Normal and there are fewer than 30 observations, so we graph the data. The histogram shows that the data are reasonably symmetric with no outliers, so this condition is met. *Independent:* We have data on 18 newts. There are clearly more than 180 newts, so this condition is met.

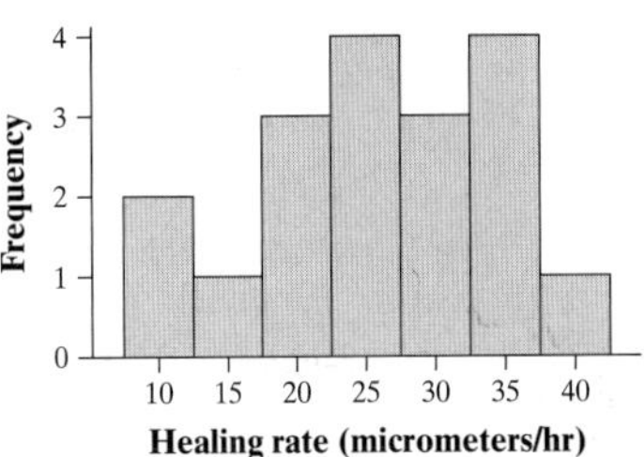

3. $25.67 \pm 2.110\frac{8.32}{\sqrt{18}} = (21.53, 29.81)$ **4.** We are 95% confident that the interval from 21.53 to 29.81 micrometers per hour captures the true mean healing time for newts.

Answers to Odd-Numbered Section 8.3 Exercises

55. $2.576\frac{7.5}{\sqrt{n}} = 1 \rightarrow 374$

57. **(a)** df = 9, $t^* = 2.262$ **(b)** df = 19, $t^* = 2.861$

59. 1.7898. In repeated sampling, the average distance between the sample means and the population mean will be about 1.7898 units.

61. **(a)** $19.03(\sqrt{23}) = 91.26$ **(b)** This interval is not a confidence interval, because it has no t^* as part of the calculation.

63. **State:** Given in the stem. **Plan:** One-sample t interval for μ. *Random:* The data come from a random sample. *Normal:* The histogram does not show any strong skewness or outliers, so this condition is met. *Independent:* We have less than 10% of the possible records for mpg.

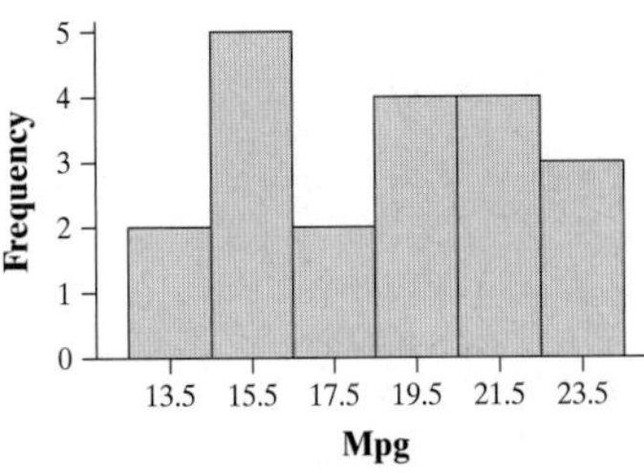

Do: $18.48 \pm 2.093\left(\frac{3.116}{\sqrt{20}}\right) = (17.022, 19.938)$. **Conclude:** We are 95% confident that the interval from 17.022 to 19.938 captures the true mean miles per gallon for this car.

65. $t^* = 2.678$. Using technology: $t^* = 2.665$. If we are able to use the exact degrees of freedom, we will have a slightly shorter interval with the same level of confidence.

67. **(a) State:** Given in the stem. **Plan:** One-sample t interval for μ. *Random:* The data come from a random sample. *Normal:* We have a sample size that is larger than 30. *Independent:* We have data from less than 10% of nursing mothers. **Do:** $-3.587 \pm 2.704\left(\frac{2.506}{\sqrt{47}}\right) = (-4.575, -2.599)$. **Conclude:** We are

99% confident that the interval from −4.575 to −2.599 captures the true mean percent change in BMC. **(b)** Yes. The interval includes only negative numbers, which represent bone mineral loss, so we are quite confident that nursing mothers lose bone mineral.

69. **(a) State:** Given in the stem. **Plan:** One-sample *t* interval for μ. *Random:* The data come from a random sample. *Normal:* The graph of the differences indicates that there is no strong skewness or outliers. *Independent:* We have data from less than 10% of tires.

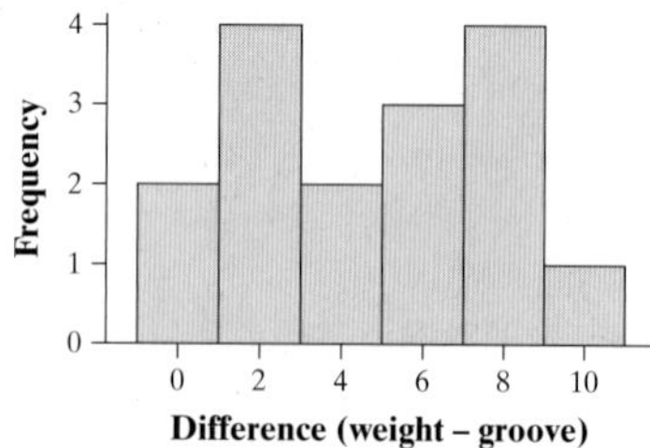

Do: $4.556 \pm 2.131\left(\frac{3.226}{\sqrt{16}}\right) = (2.837, 6.275)$. **Conclude:** We are 95% confident that the interval from 2.837 to 6.275 thousands of miles captures the true mean difference in measured tire wear. **(b)** Since the entire interval is positive, and we measured weight – groove, we are reasonably sure that the measurement by the weight method is giving larger numbers than the measurement by the groove method, on average.

71. The sample size is small and there are several outliers.

73. **(a)** No. The goal is to estimate a population proportion, not a population mean. **(b)** No. The sample was not selected randomly from all male students at this college. **(c)** No. The sample size is small and there are several outliers.

75. b

77. b

79. **(a)** $1 - 0.43 = 0.57$ **(b)** 5.44. If we were to randomly select many young people and ask them this question, we would expect the average number of days to be about 5.44. **(c)** $P(\bar{x} < 4.96) = P\left(z < \frac{4.96 - 5.44}{2.14/\sqrt{100}}\right) = P(z < -2.24) = 0.0125$. We would expect to see results as extreme as or more extreme than ours about 1.25% of the time. The average number of days spent watching TV seems unusually low for this group.

Answers to Chapter 8 Review Exercises

R8.1 Random is the most important condition because it assures us that the sample is chosen in such a way that it will typically be representative of the population. This means that we can use the data to make inferences about the population (sampling) or about cause and effect (experiment). Otherwise, we are stuck making conclusions about only the data at hand, which means the confidence interval does us no good. Normality is the second most important condition because our stated confidence level will not be correct if this condition is violated. Finally, Independent is the third most important condition. Our methods of computing the standard error require independence of observations. If this condition is violated because we are sampling without replacement, the standard error we are using is actually larger than the true standard error. Note that there are other methods that can be used if either the Normal condition or the Independent condition is not met. They are not covered in this text.

R8.2 **(a)** $z^* = 1.88$ **(b)** From Table B: $t^* = 2.678$. Using technology, $t^* = 2.665$.

R8.3 **(a)** $\bar{x} = 450$ minutes. $s_x = 62.527$ minutes. **(b)** Incorrect. The confidence interval provided gives an interval estimate for the mean lifetime of batteries produced by this company, not individual lifetimes. **(c)** No. A confidence interval provides a statement about an unknown population mean, not another sample mean. **(d)** We are 95% confident that the interval from 430 to 470 minutes captures the true mean lifetime of all AA batteries produced by this company.

R8.4 **(a)** The parameter p refers to the proportion of all adults aged 18 and older who say that football is their favorite sport to watch on television. If we take a different sample, then we will probably get a different estimate. There is variability from sample to sample. **(b)** *Random:* The sample was random. *Normal:* There were 370 successes and 630 failures, which are both at least 10. *Independent:* We have less than 10% of all adults in our sample. **(c)** $0.37 \pm 1.96\sqrt{\frac{0.37(0.63)}{1000}} = (0.340, 0.400)$ **(d)** We are 95% confident that the interval from 0.340 to 0.400 captures the true proportion of all adults who say football is their favorite sport to watch on television.

R8.5 **(a)** The parameter μ refers to the mean IQ score for the 1000 students in the school. **(b)** We have a sample size well over 30, so the Normal condition is met. **(c)** Using df = 59, $114.98 \pm 1.671\left(\frac{14.8}{\sqrt{60}}\right) = (111.79, 118.17)$. **(d)** We are 95% confident that the interval from 111.79 to 118.17 captures the true mean IQ score for the students in the school.

R8.6 $2.576\sqrt{\frac{0.5(1-0.5)}{n}} = 0.01 \rightarrow n = 16{,}590$.

R8.7 **(a) State:** We want to construct a 95% confidence interval for the actual proportion p of drivers who would admit having run a red light. **Plan:** One-sample *z* interval for p. *Random:* The drivers were selected randomly. *Normal:* There were 171 successes and 709 failures. Both are at least 10. *Independent:* The sample is less than 10% of the population of all drivers. The conditions are met. **Do:** $0.194 \pm 1.96\sqrt{\frac{0.194(0.806)}{880}} = (0.168, 0.220)$. **Conclude:** We are 95% confident that the interval from 0.168 to 0.220 captures the true proportion of all drivers who would say that they had run at least one red light. **(b)** More than 171 respondents have run red lights. We would not expect very many people to claim they *have* run red lights when they have not, but some people will deny running red lights when they have. The margin of error does not account for these sources of bias, only sampling variability.

R8.8 **(a) State:** Given in the stem. **Plan:** One-sample *t* interval for μ. *Random:* The data come from a random sample. *Normal:* The graph indicates that there is no strong skewness or outliers. *Independent:* We have data from less than 10% of the crankshafts produced that day.

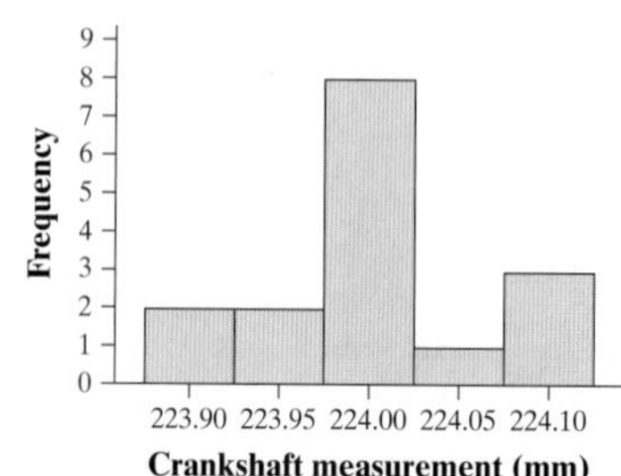

Do: $224.002 \pm 2.131\left(\frac{0.0618}{\sqrt{16}}\right) = (223.969, 224.035)$. **Conclude:** We are 95% confident that the interval from 223.969 to 224.035 mm captures the true mean measurement of these crankshafts. **(b)** Since 224 is in this interval, it is certainly a plausible value, so we don't have convincing evidence that the process mean has drifted.

R8.9 $1.96\frac{3000}{\sqrt{n}} = 1000 \rightarrow n = 35$

R8.10 (a) Increase (b) Decrease by a factor of 2

Answers to Chapter 8 AP Statistics Practice Test

T8.1 a
T8.2 d
T8.3 c
T8.4 d
T8.5 b
T8.6 a
T8.7 d
T8.8 d
T8.9 e
T8.10 d

T8.11 **(a) State:** We want to estimate the actual proportion of all visitors to Yellowstone who favor the restrictions at a 99% confidence level. **Plan:** We should use a one-sample z interval for p if the conditions are satisfied. *Random:* The visitors were selected randomly. *Normal:* There were 89 successes and 61 failures. Both are at least 10. *Independent:* The sample is less than 10% of the population of all visitors to Yellowstone. The conditions are met. **Do:** A 99% confidence interval is given by $0.593 \pm 2.576\sqrt{\frac{0.593(0.407)}{150}} = (0.490, 0.696)$. **Conclude:** We are 99% confident that the interval from 0.490 to 0.696 captures the true proportion of all visitors who would say that they favor the restrictions. **(b)** The U.S. Forest Service cannot conclude that more than half of visitors to Yellowstone National Park favor the proposal. It is plausible that only 49% favor the proposal.

T8.12 **(a)** Since the sample size is relatively large (larger than 30), the Normal condition is met. **(b)** Maurice's interval uses a z^* instead of a t^*. This would be appropriate only if we knew the population standard deviation. Also, Maurice used a wrong value for the standard deviation and used $n - 1$ in the denominator instead of n. The correct interval is (5.460, 6.956).

T8.13 **State:** We want to estimate the true mean number of bacteria per milliliter of raw milk μ for milk received at the factory at the 90% confidence level. **Plan:** We should construct a one-sample t interval for μ if the conditions are met. *Random:* The data come from a random sample. *Normal:* The graph indicates that there is no strong skewness or outliers. *Independent:* We have data from less than 10% of possible samples of milk received at the factory. The conditions are met.

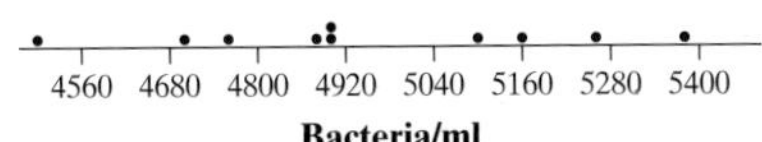

Do: We compute from the data that $\bar{x} = 4950.0$ and $s = 268.5$, and we have a sample of $n = 10$ observations. This means that we have 9 df and $t^* = 1.833$. The confidence interval, then, is $4950.0 \pm 1.833\left(\frac{268.5}{\sqrt{10}}\right) = (4794.4, 5105.6)$. **Conclude:** We are 90% confident that the interval from 4794.4 to 5105.6 bacteria/ml captures the true mean number of bacteria in the milk received at this factory.

CHAPTER 9

Section 9.1

Answers to Check Your Understanding

Page 532 **1. (a)** p = proportion of students at Jannie's high school who get less than 8 hours of sleep at night. **(b)** $H_0: p = 0.85$ and $H_a: p \neq 0.85$. **2. (a)** μ = mean amount of time that it takes to complete the census form. **(b)** $H_0: \mu = 10$ and $H_a: \mu > 10$. *Page 539* **1.** Conclude that the new batteries last longer than 30 hours when, in fact, their average lifetime is 30 hours. **2.** Conclude that the new batteries do not last longer than 30 hours when, in fact, they do last longer than 30 hours. **3.** Answers will vary. One possible answer is: Type I error. The company would be wasting its money producing a new battery that wasn't any better.

Answers to Odd-Numbered Section 9.1 Exercises

1. $H_0: p = 0.12; H_a: p \neq 0.12$. p = proportion of lefties in his school.
3. $H_0: \mu = 115; H_a: \mu > 115$. μ = mean score on the SSHA for students at least 30 years of age at the teacher's college.
5. $H_0: \sigma = 3; H_a: \sigma > 3$. σ = standard deviation of the temperature in the cabin.
7. The alternative hypothesis gives the current situation rather than what we are looking for evidence for. $H_0: p = 0.37; H_a: p > 0.37$.
9. The hypotheses are about the sample statistic. $H_0: \mu = 1000$ grams; $H_a: \mu < 1000$ grams.
11. **(a)** If the proportion of lefties at Simon's school is really 0.12, there is a 21.84% chance of finding a sample of 100 people with a value of $\hat{p}$ that is as far from 0.12 as the sample value in either direction. **(b)** No. Something that happens over 20% of the time just by chance when H_0 is true is not strong evidence against H_0.
13. **(a)** If the mean score on the SSHA for older students at this school is really 115, there is a 1.01% chance of finding a sample of 45 older students with a mean score of at least 125.7. **(b)** Reject H_0 if $\alpha = 0.05$; fail to reject H_0 if $\alpha = 0.01$ since $0.01 < 0.0101 < 0.05$.
15. Either H_0 is true (probability that H_0 is true is 1) or H_0 is false (probability that H_0 is true is 0). A *P*-value of 0.01 means that if H_0 is true, then the chance of observing a test statistic with the value we obtained or with a value more extreme is 1%.
17. Significance at the 1% level means that the *P*-value for the test is less than 0.01 and thus less than 0.05. If a test is significant at the 5% level, then we know that the *P*-value is less than 0.05, but it may or may not be less than 0.01.
19. **(a)** μ = mean response time for all accidents involving life-threatening injuries in the city. $H_0: \mu = 6.7; H_a: \mu < 6.7$. **(b)** Type I error: The city concludes that the response time has improved when it really hasn't. Type II error: The city concludes that the response time has not improved when it really has. **(c)** A Type I error would be worse. The city may stop trying to improve its response times because they think they have met the goal when, in fact, they haven't. More people could die.
21. **(a)** $H_0: \mu = \$85,000; H_a: \mu > \$85,000$. μ = the mean income of residents near the restaurant. **(b)** Type I error: Open the restaurant in a location where the residents will not be able to support it. Type II error: Do not open your restaurant in a location where the residents would have been able to support it. **(c)** $\alpha = 0.01$ to minimize the risk of a Type I error.

23. Type I: 0.05; Type II: 0.22
25. (a) If $p = 0.08$, the probability of (correctly) rejecting the null hypothesis is 0.64. (b) More (c) Decrease (d) Increase
27. d
29. a
31. (a) 16,701(0.73)(0.48) + 16,701(0.21)(0.42) + 16,701(0.06)(0.29) = 7616 (b) No, since P(woman) = 0.456 but P(woman | master's) = 0.42. (c) $1 - (0.456)^2 = 0.7041$

Section 9.2

Answers to Check Your Understanding

Page 555 **State:** We want to perform a test at the $\alpha = 0.05$ significance level of $H_0: p = 0.20$ versus $H_a: p > 0.20$ where p is the actual proportion of teens at the school who would say they have electronically sent or posted sexually suggestive images of themselves. **Plan:** If conditions are met, we should do a one-sample z test for the population proportion p. *Random:* The sample was randomly selected. *Normal:* The expected number of successes $np_0 = 50$ and failures $n(1 - p_0) = 200$ are both at least 10. *Independent:* There were 250 in the sample and 2800 in the school, so the sample is less than 10% of the population. All conditions have been met. **Do:** The sample proportion is $\hat{p} = 0.252$. The corresponding test statistic is $z = 2.06$. Since this is a one-sided test, the *P*-value is 0.0197. **Conclude:** Since our *P*-value is less than 0.05, we reject the null hypothesis. It appears that the counselor's concerns are valid and that more than 20% of the teens in her school would say they have electronically sent or posted sexually suggestive images of themselves. *Page 558* **State:** We want to perform a test at the $\alpha = 0.05$ significance level of $H_0: p = 0.75$ versus $H_a: p \neq 0.75$, where p is the actual proportion of restaurant employees who say that work stress has a negative impact on their personal lives. **Plan:** If conditions are met, we should do a one-sample z test for the population proportion p. *Random:* The sample was randomly selected. *Normal:* The expected number of successes $np_0 = 75$ and failures $n(1 - p_0) = 25$ are both at least 10. *Independent:* There were 100 in the sample, and since this is a large restaurant chain, it is very likely that there are more than 1000 workers in the population, so the sample is less than 10% of the population. All conditions have been met. **Do:** The sample proportion is $\hat{p} = 0.68$. The corresponding test statistic is $z = -1.62$. Since this is a two-sided test, the *P*-value is 0.1052. **Conclude:** Since our *P*-value is greater than 0.05, we fail to reject the null hypothesis. We do not have enough evidence to conclude that the proportion of restaurant employees at this large restaurant chain who say that work stress has a negative impact on their personal lives is different from the national proportion of 0.75. *Page 561* In the previous Check Your Understanding, we failed to reject the null hypothesis that the proportion of restaurant employees at this chain who say that work stress has a negative impact on their personal lives is the same as the national proportion of 0.75. The confidence interval given in the output includes 0.75, which means that 0.75 is a plausible value for the population proportion that we are seeking. So both the hypothesis test (which didn't rule out 0.75 as the proportion) and the confidence interval (which gave 0.75 as a plausible value) give the same conclusion. The confidence interval, however, gives more information in that it gives a whole range of plausible values, whereas the hypothesis test concentrates only on the one value as a possibility for the population proportion (0.75 here).

Answers to Odd-Numbered Section 9.2 Exercises

33. *Random:* The sample was randomly selected. *Normal:* The expected numbers of successes (12) and failures (88) are at least 10. *Independent:* It is very likely that there are more than 10(100) = 1000 students in the population.
35. Normal condition not met.
37. (a) $z = 1.23$ (b) 0.2186
39. (a) Reject at the 5% significance level. Fail to reject the null hypothesis at the 1% significance level. (b) Same as part (a).
41. **State:** $H_0: p = 0.37$ versus $H_a: p > 0.37$, where p is the actual proportion of students who are satisfied with the parking situation. **Plan:** One-sample z test for p. *Random:* The sample was randomly selected. *Normal:* The expected number of successes $np_0 = 74$ and failures $n(1 - p_0) = 126$ are both at least 10. *Independent:* There were 200 in the sample, and since there are 2500 students in the population, the sample is less than 10% of the population. **Do:** $z = 1.32$, *P*-value = 0.0934. **Conclude:** Since our *P*-value is greater than 0.05, we fail to reject the null hypothesis. We do not have enough evidence to conclude that the new parking arrangement increased student satisfaction with parking at this school.
43. (a) Type I error: Conclude that more than 37% of students were satisfied with the new parking arrangement when, in reality, only 37% were satisfied. *Consequence:* the principal believes that students are satisfied and takes no further action. Type II error: Say that we do not have enough evidence to conclude that more than 37% are satisfied with the parking arrangements when, in fact, more than 37% are satisfied. *Consequence:* the principal takes further action on parking when none is needed. (b) If $p = 0.45$, the probability of (correctly) rejecting the null hypothesis is 0.75. (c) Increase the sample size or the significance level.
45. (a) Firstborn children, due to the sampling method. (b) **State:** $H_0: p = 0.50$, $H_a: p > 0.50$ **Plan:** One-sample z test for p. *Random:* The sample was randomly selected. *Normal:* $np_0 = 12{,}734$ and $n(1 - p_0) = 12{,}734$ are at least 10. *Independent:* Population more than 254,680. **Do:** $z = 5.50$, *P*-value ≈ 0. **Conclude:** Since our *P*-value is smaller than 0.05, we reject H_0. It appears that boys are more prevalent among newborn firstborn children.
47. Corrections: Let p = the true proportion of middle school students who engage in bullying behavior. $H_0: p = 0.75$ and $H_a: p > 0.75$. *Random:* The sample was randomly selected. *Normal:* $np_0 = 418.5$ and $n(1 - p_0) = 139.5$ are at least 10. *Independent:* Population more than 5580. $z = 2.59$, *P*-value = 0.0048. Since the *P*-value is small, we reject H_0 and conclude that more than 75% of middle school students engage in bullying behavior.
49. **State:** $H_0: p = 0.60$, $H_a: p \neq 0.60$ **Plan:** One-sample z test for p. *Random:* The sample was randomly selected. *Normal:* $np_0 = 75$ and $n(1 - p_0) = 50$ are at least 10. *Independent:* Population more than 1250. **Do:** $z = 2.01$, *P*-value = 0.0444. **Conclude:** Since our *P*-value is less than 0.05, we reject H_0. It appears that a proportion other than 0.60 of teens pass the driving test on their first attempt.
51. (a) **State:** We want to estimate the actual proportion p of all teens who pass the driving test on the first try at a 95% confidence level. **Plan:** One-sample z interval for p. *Random:* The teens were selected randomly. *Normal:* 86 successes and 39 failures are both at least 10. *Independent:* Population more than 1250. **Do:** (0.607, 0.769). **Conclude:** We are 95% confident that the interval from 0.607 to 0.769 captures the true proportion of teens who pass the driving test on the first try. (b) The interval doesn't contain 0.60 as a plausible value of p, which gives convincing evidence against the DMV's claim.
53. No, because 0.20 is contained in the interval.
55. (a) p = the true proportion of teens who think that young people should wait to have sex until marriage. (b) *Random:* The sample was randomly selected. *Normal:* $np_0 = 219.5$ and $n(1 - p_0) = 219.5$ are

at least 10. *Independent:* There are more than 4390 U.S. teens. **(c)** If the true proportion of teens who think that young people should wait to have sex until marriage is 0.50, there is a 1.1% chance of getting a sample of 439 teens that is as different from that proportion as the sample we found. **(d)** Yes. Since the *P*-value is less than 0.05, we reject the null hypothesis and conclude that the actual proportion of teens who think that young people should wait is not 0.50.
57. c
59. d
61. **(a)** The random variable X − Y has a Normal distribution with mean $\mu_{X-Y} = -0.2$ and standard deviation $\sigma_{X-Y} = 0.112$. This random variable is important because for a CD to fit in the case, the variable must take on a negative number. **(b)** $P(X - Y < 0) = P(z < 1.79) = 0.9633$ **(c)** $(0.9633)^{100} = 0.0238$

Section 9.3

Answers to Check Your Understanding

Page 570 **1.** $t = (17-0)/(60/\sqrt{18}) = 1.20$ **2.** 0.20 < *P*-value < 0.30. Since the *P*-value is greater than 0.05, we fail to reject H_0. There is not enough evidence to conclude that the job satisfaction of assembly-line workers differs when their work is machine-paced rather than self-paced. **3.** 0.2466. Yes. *Page 574* **State:** We want to perform a test of $H_0: \mu = 8$ versus $H_a: \mu < 8$, where μ is the actual mean amount of sleep that students at the professor's school get each night. We will perform the test at the $\alpha = 0.05$ significance level. **Plan:** If conditions are met, we should do a one-sample *t* test for the population mean μ. *Random:* The sample was randomly selected. *Normal:* There were only 28 students, so we need to examine the sample data. The histogram below indicates that there is not much skewness and no outliers. *Independent:* There were 28 students from this university in the sample. There are likely many more than 280 students in the university overall, so the sample is less than 10% of the population. All conditions have been met.

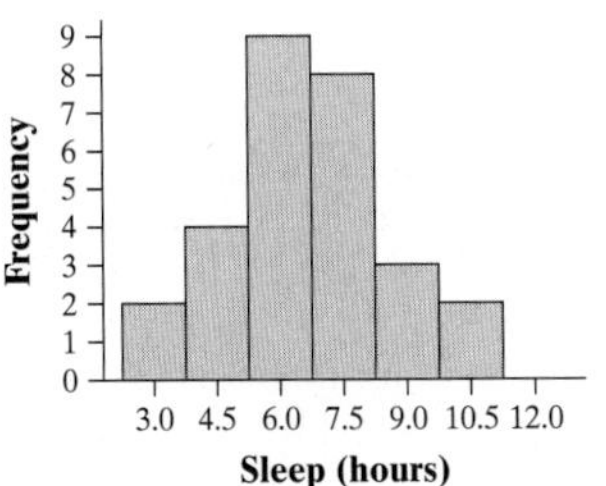

Do: The sample mean and standard deviation are $\bar{x} = 6.643$ and $s_x = 1.981$. The corresponding test statistic is $t = -3.625$. Since this is a one-sided test with df = 27, the *P*-value is 0.0006. **Conclude:** Since our *P*-value is less than 0.05, we reject the null hypothesis. It appears that students at this university get less than 8 hours of sleep on average. *Page 577* **1.** No, since 128 is contained in the confidence interval. **2.** We are 95% confident that the interval from 126.43 to 133.43 captures the true mean systolic blood pressure for the company's middle-aged male employees. The value of 128 is in this interval and therefore is a plausible mean systolic blood pressure for the males 35 to 44 years of age. This agrees with the hypothesis test, which did not rule out a mean value of 128.

Answers to Odd-Numbered Section 9.3 Exercises

63. *Random:* The sample was randomly selected. *Normal:* The sample size is at least 30. *Independent:* The sample size is 45, which is less than 10% of the 1000 students at the college who are at least 30 years of age.
65. The Normal condition is not met.
67. **(a)** $t = 2.409$ **(b)** Using Table B and 40 df, 0.01 < *P*-value < 0.02. Using technology, the *P*-value is 0.0101.
69. **(a)** Table B: 0.025 < *P*-value < 0.05. Technology: the *P*-value is 0.043. Reject H_0 at the 5% significance level. Fail to reject H_0 at the 1% significance level. **(b)** Table B: 0.05 < *P*-value < 0.10. Technology: the *P*-value is 0.086. Fail to reject H_0 at both levels.
71. **State:** $H_0: \mu = 0$ versus $H_a: \mu > 0$, where μ is the actual mean amount of sweetness loss. **Plan:** One-sample *t* test for μ. *Random:* The sample was randomly selected. *Normal:* Previous experience tells us that sweetness losses will be close to Normal. *Independent:* There are at least 100 batches of the new soda available. **Do:** $t = 2.70$, *P*-value = 0.0122. **Conclude:** Since our *P*-value is less than 0.05, we reject H_0. It appears that there is an average loss of sweetness for this cola.
73. **(a)** No. *IQR* = 458.2, which is greater than max − Q_3 and Q_1 − min. **(b)** If the mean daily calcium intake for women 18 to 24 years of age is really 1200 mg, then the likelihood of getting a sample of 36 women with a mean intake of 856.2 mg or smaller is roughly 0. **(c) State:** $H_0: \mu = 1200$ versus $H_a: \mu < 1200$, where μ is the actual mean daily calcium intake of women 18 to 24 years of age. **Plan:** One-sample *t* test for μ. *Random:* The sample was randomly selected. *Normal:* The sample size was 36, which is at least 30. *Independent:* There are clearly many more than 360 women in the United States. **Do:** $t = -6.73$, *P*-value is approximately 0. **Conclude:** Since our *P*-value is less than 0.05, we reject H_0. It appears that women in this age range are getting less than 1200 mg of calcium daily, on average.
75. **State:** $H_0: \mu = 0$, $H_a: \mu > 0$ **Plan:** *Random:* Random assignment. *Normal:* Graph of the data is roughly symmetric with no outliers. *Independent:* There are more than 100 plants of each variety. **Do:** $t = 1.295$, *P*-value = 0.1138. **Conclude:** Since the *P*-value is greater than 0.05, we fail to reject H_0. We do not have enough evidence to conclude that Variety A has a higher mean yield than Variety B.
77. **(a)** Type I error: experts conclude that Variety A has a higher mean yield when it actually doesn't. Type II error: experts conclude that there is no mean difference in yields when, in fact, Variety A has a higher mean yield. Type II error since we failed to reject H_0. **(b)** Increasing the significance level, decreasing the standard deviation σ or increasing the sample size.
79. **State:** $H_0: \mu = 11.5$, $H_a: \mu \neq 11.5$ **Plan:** *Random:* Random sample. *Normal:* The histogram indicates that the distribution is roughly symmetric with no outliers. *Independent:* There are more than 200 tablets.

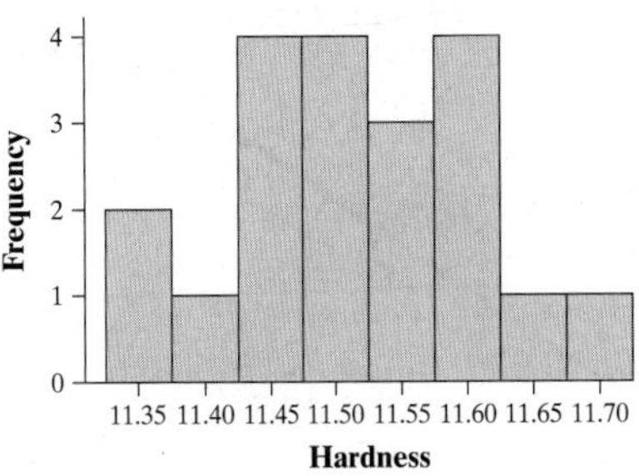

Do: $t = 0.772$, *P*-value = 0.4494. **Conclude:** Since the *P*-value is greater than 0.05, we fail to reject H_0. We do not have enough evidence to conclude that the hardness of these tablets is something other than 11.5.

81. **(a) State:** We want to estimate the true mean hardness μ for this type of pill at a 95% confidence level. **Plan:** One-sample t interval for μ. We checked the conditions in Exercise 79 and all were met. **Do:** (11.472, 11.561) **Conclude:** We are 95% confident that the interval from 11.472 to 11.561 captures the true hardness measurement for this type of pill. **(b)** The confidence interval agrees with the test done in Exercise 79. Both give 11.5 as a plausible value. The confidence interval, however, gives other plausible values as well.

83. **State:** $H_0: \mu = 200$ milliseconds versus $H_a: \mu \neq 200$ milliseconds, where μ is the actual mean amount of response time on the Internet. **Plan:** One-sample t test for μ. *Random:* The servers were selected randomly. *Normal:* The sample size was only 14, but a graph of the data reveals no strong skewness or outliers. *Independent:* There were 14 response times measured. This is less than 10% of all possible response times. **Do:** (158.22, 189.64) **Conclude:** Since our 95% confidence interval does not contain 200 milliseconds, we reject the null hypothesis at the $\alpha = 0.05$ significance level. We have enough evidence to conclude that the mean response time on the Internet is different from 200 milliseconds.

85. **(a)** Yes, since P-value > 0.05. **(b)** No, since P-value is < 0.10.

87. Yes, since 7 isn't included in the interval.

89. **(a)** So that we average out any effect due to doing the activity better the second time no matter which knob is used second. **(b) State:** $H_0: \mu_d = 0$ seconds versus $H_a: \mu_d > 0$ seconds, where μ_d is the actual mean difference (left − right) in the time it takes to turn the knob with the left-hand thread and with the right-hand thread. **Plan:** Paired t test for μ_d. *Random:* This was a randomized experiment. *Normal:* The sample size was only 25, but the histogram below indicates no strong skewness or outliers. *Independent:* We aren't sampling, so it isn't necessary to check the 10% condition. The difference in times for individual subjects should be independent if the experiment is conducted properly.

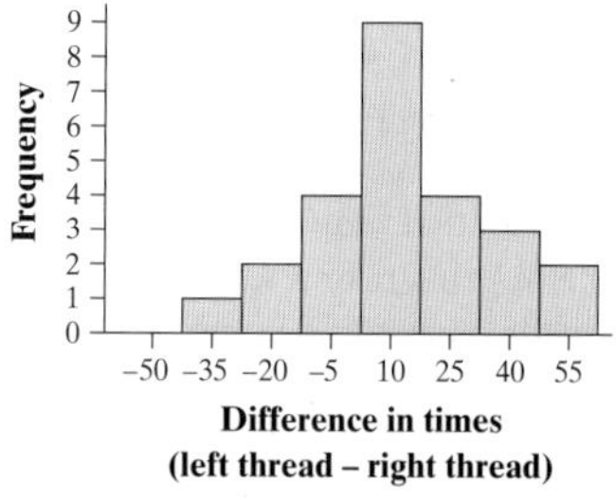

Do: $t = 2.904$, P-value $= 0.0039$. **Conclude:** Since our P-value < 0.05, we reject H_0. We have enough evidence to conclude that it takes longer for right-handed students to complete the task when the knob has a left-hand thread, on average.

91. **State:** We want to estimate the true mean difference μ_d in the word recall (music − silence) between memorizing with music and in silence for students at this high school at a 95% confidence level. **Plan:** Paired t interval for μ_d. *Random:* This was a randomized experiment. *Normal:* The boxplot of the differences does not show strong skewness or outliers, and there are more than 30 observations. *Independent:* There were 30 students tested. A medium-sized high school would have more than 300 students. **Do:** (−1.568, −0.298) **Conclude:** We are 95% confident that the interval from −1.568 to −0.298 captures the true mean difference (music − silence) in number of words recalled by students. Since the entire interval is negative, we have convincing evidence that more words are recalled when memorized in silence.

93. **(a)** No **(b)** Yes **(c)** A very small difference in measurement could lead to a statistically significant result that isn't practically important.

95. Any number of things could go wrong with this convenience sample. Depending on the time of day or the day of the week, certain types of shoppers would or would not be present.

97. **(a)** No. We'd expect about 5 of 500 subjects who don't have ESP to do better than random guessing just by chance. **(b)** The researcher should repeat the procedure on these four to see if they again perform well.

99. b

101. d

103. a

105. **(a)** Not included **(b)** Not included **(c)** Included

107. *Random:* The adults measured were chosen at random. *Normal:* There were many more than 30 measurements in the sample. *Independent:* The sample size was 130, which is clearly less than 10% of the healthy 18- to 40-year-olds in the United States. The conditions are met.

Answers to Chapter 9 Review Exercises

R9.1 **(a)** $H_0: \mu = 64.2$; $H_a: \mu \neq 64.2$ **(b)** $H_0: p = 0.75$; $H_a: p < 0.75$

R9.2 **(a)** $H_0: \mu_d = 0$; $H_a: \mu_d > 0$. μ_d = true mean difference (with eye black − without eye black) in sensitivity among athletes. **(b)** *Random:* The students were chosen randomly. *Normal:* The histogram shows that the distribution is somewhat right-skewed with no outliers. We should be safe using t procedures. *Independent:* The sample size is 16, which is much less than 10% of all possible students. The conditions are met.

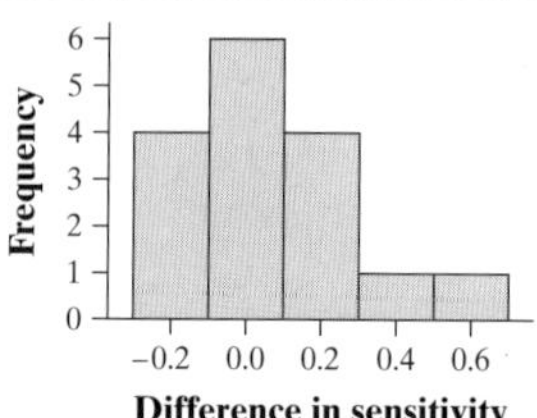

(c) If eye black does not increase sensitivity, on average, the chance of finding a sample where the difference is as big as ours is 4.7%.

R9.3 **(a)** $H_0: \mu = 300$ (the company's claim is true) versus $H_a: \mu < 300$ (the mean breaking strength is less than the company's claim). **(b)** Type I error: conclude that the company's claim is incorrect ($\mu < 300$) when, in fact, it is legitimate ($\mu = 300$). We might falsely accuse the company of exaggerating its claim. Type II error: conclude that the company's claim is legitimate when, in fact, it is exaggerated. This could result in more injuries from broken chairs. **(c)** Since a Type II error is more serious, we would like to reduce the likelihood of that error. This means increasing the likelihood of a Type I error. This suggests that we should use a significance level of 0.10. **(d)** If, in fact, the mean breaking strength is 294 lb, there is a 71% chance that we will reject the null hypothesis in our test. That is, there is a 71% chance that we will decide that the mean breaking strength is less than 300 lb. **(e)** Increase the sample size or the significance level.

R9.4 **(a) State:** $H_0: p = 0.05$ versus $H_a: p < 0.05$, where p is the actual proportion of adults who use the vaccine and who will get the flu if they use the vaccine. **Plan:** One-sample z test for p. *Random:* The sample was randomly selected. *Normal:* The expected number of successes $np_0 = 50$ and failures $n(1 - p_0) = 950$ are both at least 10. *Independent:* There were 1000 adults, which is clearly less than 10% of the population. **Do:** The sample proportion is

$\hat{p} = 0.043$. The corresponding test statistic is $z = -1.02$. Since this is a one-sided test, the *P*-value is $P(z < -1.02) = 0.1539$. **Conclude:** Since the *P*-value is greater than 0.05, we fail to reject H_0. We do not have enough evidence to conclude that fewer than 5% of adults who receive this vaccine will get the flu. **(b)** Type II error, since we failed to reject H_0. **(c)** Type I error. This would mean concluding that the vaccine is more effective than it really is.

R9.5 **(a) State:** $H_0: p = 0.4737$ versus $H_a: p \neq 0.4737$, where p is the actual proportion of times the ball lands in a red slot. **Plan:** One-sample z test for p. *Random:* The sample was randomly selected. *Normal:* The expected number of successes $np_0 = 23.685$ and failures $n(1 - p_0) = 26.315$ are both at least 10. *Independent:* There were 50 spins, which is clearly less than 10% of all possible spins. **Do:** The sample proportion is $\hat{p} = 0.62$. The corresponding test statistic is $z = 2.07$. Since this is a two-sided test, the *P*-value is 0.0384. **Conclude:** Since *P*-value < 0.05, we reject H_0. We have enough evidence to conclude that the wheel is unfair. **(b)** By using a 99% confidence level, the manager has obtained a very wide interval of plausible values that includes 0.4737. You might ask why he didn't choose a 95% confidence level!

R9.6 **(a) State:** $H_0: \mu_d = 0$ versus $H_a: \mu_d < 0$, where μ_d is the actual mean difference (after − before) in blood pressure among the company's employees between after the campaign and before it. **Plan:** One-sample t test for μ. *Random:* The employees were chosen at random. *Normal:* The sample size was 50, which is at least 30. *Independent:* There were 50 people in the sample. Since the company is described as being a large company, there are very likely more than 500 employees, so this is less than 10% of all employees. **Do:** The sample mean and standard deviation are $\bar{x} = -6$ and $s_x = 19.8$. The corresponding test statistic is $t = -2.143$. The *P*-value from technology is 0.0186. **Conclude:** Since *P*-value < 0.05, we reject H_0. We have enough evidence to conclude that the actual mean change in blood pressure among the company's employees is negative. People's blood pressure was reduced over the year, on average. **(b)** No. This was an observational study with no comparison group.

R9.7 **State:** $H_0: \mu = 22$ versus $H_a: \mu > 22$, where μ is the actual mean amount of time it takes adults to read the paragraph in Gigi font. **Plan:** One-sample t test for μ. *Random:* The adults were chosen at random. *Normal:* The sample size was only 24, so we look at the histogram below. It shows that the distribution is roughly symmetric with no outliers. *Independent:* There were 24 people in the sample. This is clearly less than 10% of all adults.

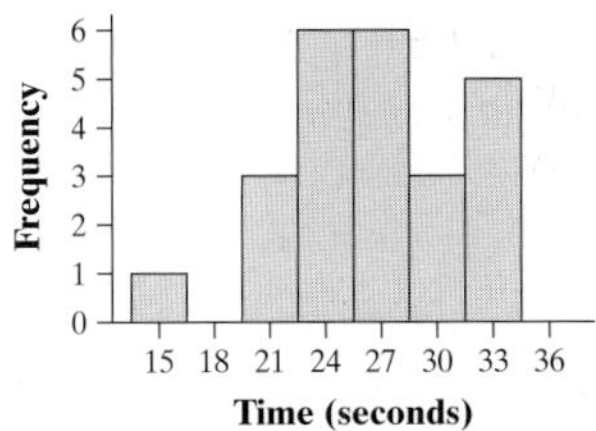

Do: The sample mean and standard deviation are $\bar{x} = 26.496$ and $s_x = 4.728$. The corresponding test statistic is $t = 4.659$. The *P*-value from technology using df = 23 is approximately 0. **Conclude:** Since *P*-value < 0.05, we reject H_0. We have enough evidence to conclude that the mean time for adults to read the paragraph in Gigi font is more than 22 seconds.

R9.8 **State:** We want to perform a test of $H_0: \mu = 105$ versus $H_a: \mu \neq 105$, where μ is the actual mean reading from radon detectors. We will perform the test at the $\alpha = 0.10$ significance level. **Plan:** If conditions are met, we should do a one-sample t test for the population mean μ. *Random:* The radon detectors were chosen at random. *Normal:* The sample size was only 11, but a graph of the data shows no strong skewness or outliers. *Independent:* There were 11 radon detectors in the sample. This is clearly less than 10% of all possible radon detectors. All conditions have been met. **Do:** The 90% confidence interval is (99.61, 110.03). **Conclude:** Since the 90% confidence interval contains 105, we do not reject the null hypothesis. We do not have enough evidence to conclude that the radon detectors are inaccurate.

R9.9 **(a)** By randomly allocating which plot got the regular barley seeds and which one got the kiln-dried seeds. **(b) State:** We want to perform a test of $H_0: \mu_d = 0$ versus $H_a: \mu_d < 0$, where μ_d is the actual mean difference (regular − kiln) in yield between regular barley seeds and kiln-dried barley seeds. We will perform the test at the $\alpha = 0.05$ significance level. **Plan:** If conditions are met, we should do a paired t test for the population mean μ. *Random:* We are told that this was a randomized experiment. *Normal:* The sample size was 11, so we check a graph. The histogram below shows no strong skewness or outliers in the data. *Independent:* There were 11 pairs of plots used in the sample. This is clearly less than 10% of all possible plots. All conditions have been met.

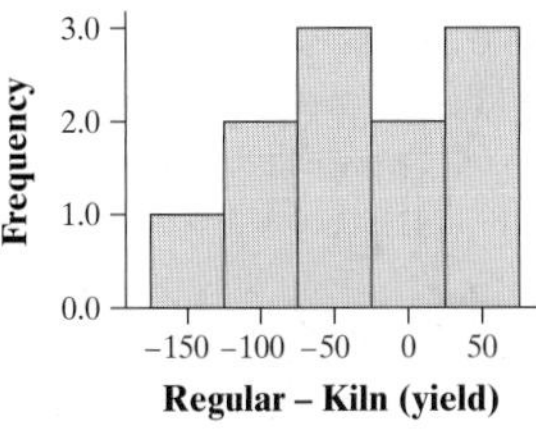

Do: The sample mean and standard deviation are $\bar{x} = -33.7$ and $s_x = 66.2$. The corresponding test statistic is $t = -1.688$. The *P*-value from technology using df = 10 is 0.0612. **Conclude:** Since our *P*-value is greater than 0.05, we fail to reject the null hypothesis. We do not have enough evidence to conclude that kiln-dried barley seeds produce a larger yield.

Answers to Chapter 9 AP Statistics Practice Test

T9.1 b
T9.2 b
T9.3 c
T9.4 e
T9.5 b
T9.6 c
T9.7 e
T9.8 d
T9.9 a
T9.10 c

T9.11 **(a) State:** We want to perform a test at the $\alpha = 0.05$ significance level of $H_0: p = 0.20$ versus $H_a: p > 0.20$, where p is the actual proportion of customers who would pay \$100 for the upgrade. **Plan:** If conditions are met, we should do a one-sample z test for the population proportion p. *Random:* The sample was randomly selected. *Normal:* The expected number of successes $np_0 =$ 12 and failures $n(1 - p_0) = 48$ are both at least 10. *Independent:* There were 60 customers, which is presumably less than 10% of all

possible customers (there are very likely more than 600 customers). All conditions have been met. **Do:** The sample proportion is $\hat{p} = 0.267$. The corresponding test statistic is $z = 1.30$. Since this is a one-sided test, the *P*-value is 0.0968. **Conclude:** Since our *P*-value is greater than 0.05, we fail to reject the null hypothesis. We do not have enough evidence to conclude that more than 20% of customers would pay $100 for the upgrade. **(b)** A Type I error would be concluding that more than 20% of customers would pay when, in fact, they would not. A Type II error would be to conclude that not more than 20% of customers would pay when, in fact, they would. For the company a Type I error is worse because they would go ahead with the upgrade when it would not be profitable for them. **(c)** Increase the significance level.
T9.12 **(a)** Students may improve from Monday to Wednesday just because they have already done the task once. Also, something external may affect scores on one of the days (perhaps students didn't get as much sleep the night before because of some scheduled activity). A better way to run the experiment would be to randomly assign half the students to get one cup of coffee on Monday and the other half to get the cup of coffee on Wednesday. **(b) State:** We want to perform a test of $H_0: \mu_d = 0$ versus $H_a: \mu_d < 0$, where μ_d is the actual mean difference (no coffee − coffee) in the number of words recalled without coffee and with coffee. We will perform the test at the $\alpha = 0.05$ significance level. **Plan:** If conditions are met, we should do a paired *t* test for the population mean μ. *Random:* This was a randomized experiment. *Normal:* The sample size was only 10, but the graph below shows a symmetric distribution with no outliers. *Independent:* The difference in scores for individual subjects should be independent if the experiment is conducted properly. All conditions have been met.

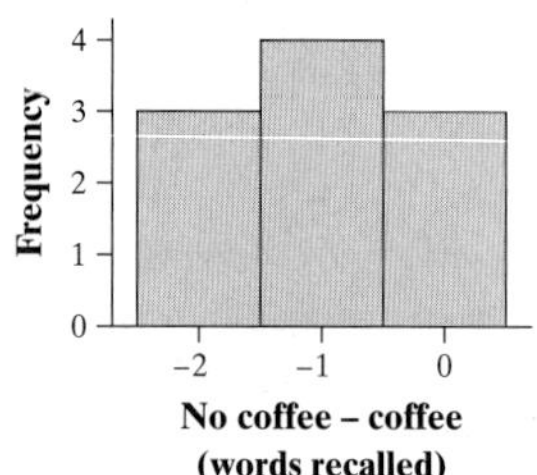

Do: The sample mean and standard deviation are $\bar{x} = -1$ and $s_x = 0.816$. The corresponding test statistic is $t = -3.875$. The *P*-value from technology using df = 9 is 0.0019. **Conclude:** Since our *P*-value is smaller than 0.05, we reject the null hypothesis. We have enough evidence to conclude that the mean difference (no coffee − coffee) in word recall is negative. Students performed better on the test with coffee than without, on average.
T9.13 **(a)** We have a sample of size 50 with no outliers. **(b) State:** We want to perform a test of $H_0: \mu = \$158$ versus $H_a: \mu \neq \$158$, where μ is the actual mean amount spent on food by households in this city. We will perform the test at the $\alpha = 0.05$ significance level. **Plan:** If conditions are met, we should do a one-sample *t* test for the population mean μ. *Random:* The households were chosen at random. *Normal:* The sample size was 50 and there were no outliers. *Independent:* There were 50 households in the sample. A small city would contain more than 500 households, so this is less than 10% of all possible households. All conditions have been met. **Do:** The 95% confidence interval is (159.32, 170.68). **Conclude:** Since the 95% confidence interval does not contain 158, we reject the null hypothesis. We conclude that the mean amount spent on food per household in this city is different from the U.S. value.

CHAPTER 10

Section 10.1

Answers to Check Your Understanding

Page 608 **1.** Normal. Bag 1: $np = 12.5$ and $n(1 - p) = 37.5$. Bag 2: $np = 14$ and $n(1 - p) = 26$. All at least 10. **2.** $\mu_{\hat{p}_1 - \hat{p}_2} = 0.25 - 0.35 = -0.10$. $\sigma_{\hat{p}_1 - \hat{p}_2} = \sqrt{\frac{0.25(0.75)}{50} + \frac{0.35(0.65)}{40}} = 0.0971$. **3.** $P(\hat{p}_1 - \hat{p}_2 \leq -0.02) = P(z \leq 0.82) = 0.7939$ **4.** No. In about 79% of samples, the value of $\hat{p}_1 - \hat{p}_2$ will be less than or equal to −0.02. *Page 611* **State:** Our parameters are p_1 = proportion of teens who go online every day and p_2 = proportion of adults who go online every day. We want to estimate $p_1 - p_2$ at a 90% confidence level. **Plan:** Use a two-sample *z* interval for $p_1 - p_2$ if the conditions are satisfied. *Random:* Both samples were selected randomly. *Normal:* The number of successes and the number of failures in both groups are at least 10 (teens: 504 successes, 296 failures; adults: 1532 successes, 721 failures). *Independent:* Both samples are less than 10% of their respective populations (there are more than 8000 teens and 22,530 adults in the United States). **Do:** From the data, $n_1 = 800$, $\hat{p}_1 = 0.63$, $n_2 = 2253$, and $\hat{p}_2 = 0.68$. The 90% confidence interval is $(0.63 - 0.68) \pm 1.645\sqrt{\frac{0.63(0.37)}{800} + \frac{0.68(0.32)}{2253}} = (-0.0824, -0.0176)$. **Conclude:** We are 90% confident that the interval from −0.0824 to −0.0176 captures the true difference in the proportion of all U.S. teens and adults who go online every day. This interval suggests that between 1.76% and 8.24% more adults are online everyday than teens. *Page 619* **1. State:** We want to perform a test at the $\alpha = 0.05$ significance level of $H_0: p_1 - p_2 = 0$ versus $H_a: p_1 - p_2 > 0$, where p_1 is the actual proportion of children who did not attend preschool that used social services later, and p_2 is the actual proportion of children who did attend preschool that used social services later. **Plan:** Use a two-sample *z* test for $p_1 - p_2$ if the conditions are satisfied. *Random:* This was a randomized experiment. *Normal:* The number of successes and the number of failures in both groups are at least 10 (no preschool: 49 successes, 12 failures; preschool: 38 successes, 24 failures). *Independent:* Due to the random assignment, these two groups of children can be viewed as independent. Also, knowing one child's need of social services gives no information about another child's need of social services. **Do:** From the data, $\hat{p}_1 = 0.8033$ and $\hat{p}_2 = 0.6129$. The pooled proportion is $\hat{p}_C = 0.7073$. The test statistic is $z = 2.32$ and the *P*-value is $P(z > 2.32) = 0.0102$. **Conclude:** Since the *P*-value is less than 0.05, we reject H_0 and conclude that children like those in this study who have participated in preschool are less likely to use social services later in life. **2.** We are 95% confident that the interval from 0.033 to 0.347 captures the true difference (no preschool − preschool) in proportion of children who would need social services later in life. This suggests that between 3.3% and 34.7% more children who did not participate in preschool will need social services later in life than those who did participate in preschool. The interval gives us a range of plausible values rather than just making a decision about one specific value.

Answers to Odd-Numbered Section 10.1 Exercises

1. (a) Proportions (categorical data) (b) Observational study
3. (a) Means (quantitative data) (b) Randomized experiment
5. (a) $P(\hat{p}_C \leq \hat{p}_A) = P(\hat{p}_C - \hat{p}_A \leq 0) = P(z \leq -2.03) = 0.0212$ (b) Yes. There is only a 2% chance of getting as few or fewer red jelly beans in the child sample than the adult sample if the company's claim is true.
7. Random, Normal conditions not met.
9. Normal condition is not met.
11. **State:** Our parameters are p_1 = true proportion of young people who use online instant messaging more than email and p_2 = true proportion of older people who use online instant messaging more than email. **Plan:** Two-sample z interval for $p_1 - p_2$. *Random:* Both samples were selected randomly. *Normal:* 73, 85, 26, 117 are all at least 10. *Independent:* There are more than 1580 young people and 1430 older people in the United States. **Do:** (0.1959, 0.3641). **Conclude:** We are 90% confident that the interval from 0.1959 to 0.3641 captures the true difference in the proportions of young people and older people who use online instant messaging more than email.
13. (a) **State:** Our parameters are p_1 = actual proportion of young men who live at home and p_2 = actual proportion of young women who live at home. **Plan:** Two-sample z interval for $p_1 - p_2$. *Random:* Both samples were selected randomly. *Normal:* 986, 1267, 923, 1706 are all at least 10. *Independent:* There are more than 22,530 young men and 26,290 young women in the United States. **Do:** (0.051, 0.123). **Conclude:** We are 99% confident that the interval from 0.051 to 0.123 captures the true difference in the proportions of young men and young women who live at home. (b) Since the interval does not contain 0, there is convincing evidence that the two proportions are not the same.
15. $H_0: p_1 - p_2 = 0$ versus $H_a: p_1 - p_2 \neq 0$, where p_1 is the actual proportion of teens who own an iPod or MP3 player and p_2 is the actual proportion of young adults who own an iPod or MP3 player.
17. (a) **State:** See Exercise 15. **Plan:** Two-sample z test for $p_1 - p_2$. *Random:* Both samples were randomly selected. *Normal:* 632, 168, 268, 132 are all at least 10. *Independent:* There are more than 8000 teens and 4000 young adults who live in the United States. **Do:** z = 4.53, P-value ≈ 0. **Conclude:** Since the P-value < 0.05, we reject H_0 and conclude that the actual proportions of teens and young adults who own iPods and MP3 players are different. (b) **Do:** (0.066, 0.174). **Conclude:** We are 95% confident that the interval from 0.066 to 0.174 captures the true difference in proportions of teens and young adults who own iPods or MP3 players. This is consistent with our answer to part (a). In both cases, we ruled out the difference of proportions being 0 as a plausible value.
19. Here are Holly's mistakes: (1) She fails to define her parameters. (2) Should be a two-sample z test. (3) Switches Normal and Independent conditions and doesn't check the 10% condition. (4) $\hat{p}_2$ should be 0.50. (5) Should use the pooled proportion, $\hat{p}_C$ = 0.54. (6) P-value should be $2P(z > 2.11) = 0.0348$. (7) She should have a two-sided conclusion rather than a one-sided conclusion. And Holly shouldn't have used the word "prove."
21. (a) **State:** We want to perform a test at the $\alpha = 0.05$ significance level of $H_0: p_1 - p_2 = 0$ versus $H_a: p_1 - p_2 \neq 0$, where p_1 and p_2 are the actual proportions of women with a family history of breast cancer that are assigned to the low-fat and normal diet, respectively. **Plan:** Use a two-sample z test for $p_1 - p_2$ if the conditions are satisfied. *Random:* This was a randomized experiment. *Normal:* The number of successes and the number of failures in both groups are at least 10 (low-fat: 3396 successes, 16,145 failures; normal diet: 4929 successes, 24,365 failures). *Independent:* Due to the random assignment, these two groups of women can be viewed as independent. Also, knowing one woman's family history doesn't tell us anything about any other woman's history. **Do:** From the data, $\hat{p}_1 = 0.174$ and $\hat{p}_2 = 0.168$. The pooled proportion is $\hat{p}_C = 0.170$. The test statistic is $z = 1.59$ and the P-value is $2P(z > 1.59) = 0.1118$. **Conclude:** Since the P-value is greater than 0.05, we fail to reject H_0. We do not have enough evidence to conclude that there is a statistically significant difference in the proportions of women assigned to the two groups who have a family history of breast cancer. Note, however, that the null hypothesis would have been rejected at the 10% significance level. (b) A Type I error would be to say that the groups are significantly different when they are not; a Type II error would be to say that the groups are not significantly different when they are. A Type II error would be more serious because the experiment would proceed assuming that the two groups were similar to begin with. Any conclusions made about the difference between the two groups at the end of the study would then be suspect.
23. (a) Two-sample z test for $p_1 - p_2$. *Random:* This was a randomized comparative experiment. *Normal:* 44, 44, 21, 60 are all at least 10. *Independent:* Due to the random assignment, these two groups of women can be viewed as independent. (b) If there is no difference in pregnancy rates of women who are being prayed for and those who are not, there is a 0.07% chance of seeing as many more pregnancies while being prayed for as we did. (c) Since the P-value < 0.05, we reject H_0. We have enough evidence to conclude that the proportion of pregnancies among women like these who are prayed for is higher than that among those who are not prayed for. (d) If the women had known whether they were being prayed for, this might have affected their behavior in some way (even unconsciously) that would have affected whether they became pregnant or not.
25. **Do:** (0.100, 0.382). **Conclude:** We are 95% confident that the interval from 0.100 to 0.382 captures the true difference in proportions of IVF patients who are prayed for and those who are not who get pregnant. The interval gives a range of plausible values for $p_1 - p_2$ that doesn't include 0.
27. **State:** $H_0: p_1 - p_2 = 0$ versus $H_a: p_1 - p_2 \neq 0$, where p_1 is the actual proportion of 4- to 5-year-olds who sorted correctly and p_2 is the actual proportion of 6- to 7-year-olds who sorted correctly. **Plan:** Two-sample z test for $p_1 - p_2$. *Random:* Both samples were selected randomly. *Normal:* 10, 40, 28, 25 are all at least 10. *Independent:* There are more than 500 4- to 5-year-olds and 530 6- to 7-year-olds. **Do:** $z = -3.45$, P-value = 0.001. **Conclude:** Since the P-value < 0.05, we reject H_0. We have enough evidence to conclude that there is a difference in the sorting abilities of 4- to 5-year-olds and 6- to 7-year-olds.
29. b
31. b
33. (a) $\hat{y} = -13{,}832 + 14{,}954x$, where $\hat{y}$ = the predicted mileage and x = the age in years of the cars. (b) For each year older the car is, we predict that the mileage will increase by 14,954 miles. (c) 110,000 − 135,708 = −25,708 miles

Section 10.2

Answers to Check Your Understanding

Page 632 1. Normal because both X_1 and X_2 have Normal distributions. 2. $\mu_{\bar{x}_1 - \bar{x}_2} = 10$ ounces and $\sigma_{\bar{x}_1 - \bar{x}_2} = 0.205$ ounces.

3. $P(\bar{x}_1 - \bar{x}_2 > 12) = P(z > 9.76) \approx 0$. **4.** Since the probability in question 3 is so small, we would be very surprised to find a difference of 12 ounces. ***Page 638*** **State:** Our parameters are μ_1 = the true mean price of wheat in July and μ_2 = the true mean price of wheat in September. We want to estimate $\mu_1 - \mu_2$ at a 99% confidence level. **Plan:** Use a two-sample t interval for $\mu_1 - \mu_2$ if the conditions are satisfied. *Random:* Both samples were selected randomly. *Normal:* Both sample sizes were at least 30. *Independent:* Both samples are less than 10% of their respective populations (there are more than 900 wheat producers in July and 450 wheat producers in September). **Do:** From the data, $n_1 = 90$, $\bar{x}_1 = 2.95$, $s_1 = 0.22$, $n_2 = 45$, $\bar{x}_2 = 3.61$, and $s_2 = 0.19$. Using the conservative df = 44, the 99% confidence interval is $(2.95-3.61) \pm 2.692\sqrt{\frac{(0.22)^2}{90} + \frac{(0.19)^2}{45}} = (-0.759, -0.561)$. **Conclude:** We are 99% confident that the interval from −0.759 to −0.561 captures the true difference in July and September. This suggests that the mean price of wheat in July is between \$0.561 and \$0.759 per bushel less than it is in September. ***Page 644*** **State:** We want to perform a test at the $\alpha = 0.05$ significance level of $H_0: \mu_1 - \mu_2 = 0$ versus $H_a: \mu_1 - \mu_2 > 0$, where μ_1 is the actual mean breaking strength at 2 weeks and μ_2 is the actual mean breaking strength at 16 weeks. **Plan:** Use a two-sample t test if the conditions are satisfied. *Random:* This was a randomized comparative experiment. *Normal:* Since n_1 and n_2 are less than 30, we examine the data. From the dotplot below, neither group displays strong skewness or any outliers. *Independent:* Due to the random assignment, these two groups of pieces of cloth can be viewed as independent. Also, knowing one piece of cloth's breaking strength gives no information about the breaking strength of another piece of cloth.

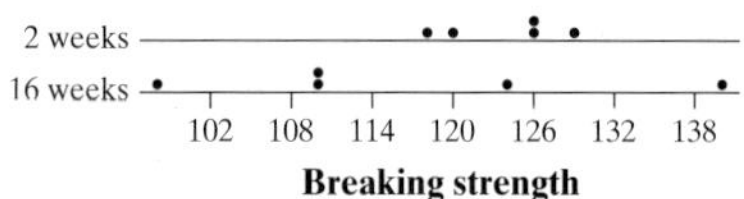

Do: From the data, $n_1 = 5$, $\bar{x}_1 = 123.8$, $s_1 = 4.60$, $n_2 = 5$, $\bar{x}_2 = 116.4$, and $s_2 = 16.09$. Using the conservative df = 4, the test statistic is $t = 0.989$ and the P-value is $P(t > 0.989) = 0.1893$. **Conclude:** Since the P-value is greater than 0.05, we fail to reject H_0. We do not have enough evidence to conclude that there is a difference in the actual mean breaking strength of polyester fabric that is buried for 2 weeks and fabric that is buried for 16 weeks.

Answers to Odd-Numbered Section 10.2 Exercises

35. **(a)** Normal with $\mu_{M-B} = 18$ mg/dl and $\sigma_{M-B} = 50.80$ mg/dl. **(b)** $P(M-B < 0) = P(z < -0.35) = 0.3632$

37. **(a)** Normal with $\mu_{\bar{x}_M - \bar{x}_B} = 18$ mg/dl and $\sigma_{\bar{x}_M - \bar{x}_B} = 9.60$ mg/dl. **(b)** $P(\bar{x}_M - \bar{x}_B < 0) = P(z < -1.88) = 0.0301$ **(c)** Yes. The likelihood that the sample mean of the boys is greater than that of the men is only 3%.

39. No. Normal condition is not met.

41. No. Independent condition is not met.

43. **(a)** The centers of the two groups seem to be quite different, with people drinking red wine generally having more polyphenol in their blood. The spreads, however, are approximately the same. **(b)** **State:** Our parameters are μ_1 and μ_2, the actual mean polyphenol level in the blood of people like those in the study after drinking red wine and white wine, respectively. We want to estimate $\mu_1 - \mu_2$ at a 90% confidence level. **Plan:** Use a two-sample t interval for $\mu_1 - \mu_2$ if the conditions are satisfied. *Random:* This was a randomized experiment. *Normal:* Both sample sizes were less than 30. The dotplots do not indicate serious skewness or outliers. *Independent:* Due to the random assignment, these two groups of men can be viewed as independent. Also, knowing one man's polyphenol level gives no information about another man's polyphenol level. **Do:** From the data, $n_1 = 9$, $\bar{x}_1 = 5.5$, $s_1 = 2.517$, $n_2 = 9$, $\bar{x}_2 = 0.23$, and $s_2 = 3.292$. Using the conservative df = 8, the 90% confidence interval is $(5.5-0.23) \pm 1.860 \sqrt{\frac{(2.517)^2}{9} + \frac{(3.292)^2}{9}} = (2.701, 7.839)$. **Conclude:** We are 90% confident that the interval from 2.701 to 7.839 captures the difference in actual mean polyphenol level in men who drink red wine and men who drink white wine. This interval suggests that men who drink red wine have between a 2.1701 and 7.839 higher polyphenol level than those who drink white wine. **(c)** Since this interval does not contain 0, it does support the researcher's belief that the polyphenol level is different for men who drink red wine than for those who drink white wine.

45. **(a)** The distributions are skewed to the right because the earnings amounts cannot be negative, yet the standard deviation is almost as large as the distance between the mean and 0. The use of the two-sample t procedures is justified because of the large sample sizes. **(b)** **State:** Our parameters are μ_1 = the actual mean summer earnings of male students and μ_2 = the actual mean summer earnings of female students. **Plan:** Two-sample t interval for $\mu_1 - \mu_2$. *Random:* Both samples were randomly selected. *Normal:* Both sample sizes were at least 30. *Independent:* There are more than 6750 males with summer jobs and 6210 females with summer jobs. **Do:** Using df = 620: (413.558, 634.702). **Conclude:** We are 90% confident that the interval from 413.558 to 634.702 captures the actual difference in mean summer earnings of male students and female students. **(c)** If we repeatedly took random samples of 675 males and 621 females from this university, each time constructing a 90% confidence interval in this same way, about 90% of the resulting intervals would capture the actual difference in mean earnings.

47. **(a)** Answers will vary. Assign each day a number from 01 to 60. Move from left to right, looking at pairs of digits. The first 30 distinct pairs between 01 and 60 tell which day Design A will be used. Use Design B on the remaining days. The first three days are 24, 55, and 21. **(b)** Two-sided because we are interested in determining whether there is a difference in either direction. $H_0: \mu_A - \mu_B = 0$ versus $H_a: \mu_A - \mu_B \neq 0$. **(c)** df = 29 **(d)** $0.04 <$ P-value < 0.05. We would reject H_0 and conclude that there is a difference in the actual mean daily sales for the two designs.

49. **State:** $H_0: \mu_1 - \mu_2 = 0$ versus $H_a: \mu_1 - \mu_2 \neq 0$, where μ_1 and μ_2 are the actual mean time to breeding for the birds relying on natural food supply and for the birds with food-supplementation, respectively. **Plan:** Two-sample t test. *Random:* This was a randomized comparative experiment. *Normal:* The comparative dotplot shows no strong skewness or outliers for either group. *Independent:* Due to the random assignment, these two groups of birds can be viewed as independent.

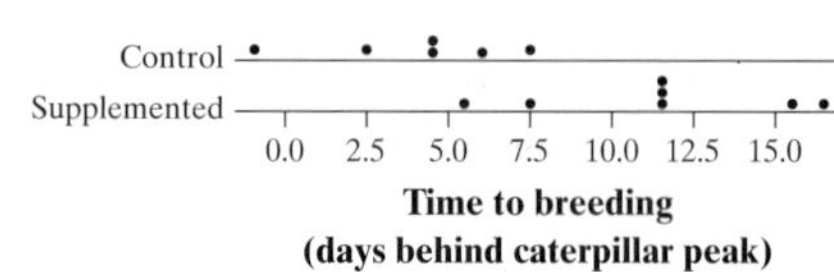

Do: Using df $= 5$, $t = -3.736$, *P*-value $= 0.0134$. **Conclude:** Since the *P*-value is less than 0.05, we reject H_0. We have enough evidence to conclude that there is a difference in the mean time to breeding for birds relying on natural food supply and birds with food supplements.

51. **(a)** The score distribution for the activities group is slightly left-skewed, with a greater mean and smaller standard deviation than the control group, which has a more symmetrical DRP score distribution. **(b) State:** $H_0: \mu_1 - \mu_2 = 0$ versus $H_a: \mu_1 - \mu_2 > 0$, where μ_1 and μ_2 are the actual mean DRP scores for third graders like those who do the activities and who don't do the activities, respectively. **Plan:** Two-sample *t* test. *Random:* This was a randomized comparative experiment. *Normal:* The boxplots show that neither distribution displays strong skewness nor any outliers. *Independent:* Due to the random assignment, these two groups of students can be viewed as independent. **Do:** Using df $= 20$, $t = 2.312$, and *P*-value $=$ 0.0158. **Conclude:** Since the *P*-value is less than 0.05, we reject H_0. We have enough evidence to conclude that there is a difference in the actual mean DRP scores of third graders like those who learn with activities and those who learn without activities. **(c)** Since this was a randomized controlled experiment, we can conclude that the activities caused an increase in the mean DRP score. **(d) Do:** (0.97, 18.95). **Conclude:** We are 95% confident that the interval from 0.97 to 18.95 captures the difference in actual mean DRP scores for students learning with the activities and those learning without the activities. This interval gives a range of plausible values for the difference in the two means that does not include 0.

53. **State:** $H_0: \mu_1 - \mu_2 = 0$ versus $H_a: \mu_1 - \mu_2 \neq 0$, where μ_1 and μ_2 are the actual mean numbers of words spoken per day by female students and by male students at this university, respectively. **Plan:** Two-sample *t* test. *Random:* Both samples were selected randomly. *Normal:* Both n_1 and n_2 are at least 30. *Independent:* There are more than 560 female students and 560 male students at a large university. **Do:** Using df $= 55$, $t = -0.248$, *P*-value $=$ 0.8050. **Conclude:** Since the *P*-value is greater than 0.05, we fail to reject H_0. We do not have enough evidence to conclude that male students and female students speak a different number of words per day on average. **(b)** If males and females at this university speak the same number of words per day on average, then we have about an 80% chance of selecting a sample where the difference between the average number of words spoken per day by males and females is as large as or larger than the difference we actually saw.

55. **State:** $H_0: \mu_1 - \mu_2 = 0$ versus $H_a: \mu_1 - \mu_2 > 0$, where μ_1 and μ_2 are the actual mean knee velocities for skilled rowers and novice rowers, respectively. **Plan:** Two-sample *t* test. *Random:* Both samples were randomly selected. *Normal:* Both n_1 and n_2 are less than 30, but we are told that there were no outliers or strong skewness. *Independent:* There are more than 100 skilled rowers and 80 novice rowers. **Do:** $t = 3.1583$, *P*-value $= 0.0104$ for the two-sided test. We are doing a one-sided test, so $P = 0.0052$. **Conclude:** Since the *P*-value is less than 0.05, we reject H_0. We have enough evidence to conclude that the mean knee velocity is greater for skilled rowers than for novice rowers. **(b) Do:** Using df $= 9.77$, $(0.497, 1.847)$. **Conclude:** We are 90% confident that the interval from 0.497 to 1.847 captures the difference in actual mean knee velocity for skilled rowers and novice rowers. **(c)** Using Table B with df $= 7$ would have led to a larger t^* and therefore a slightly wider interval.

57. **(a)** To help balance out the effects of lurking variables. **(b)** Only about 5 out of the 1000 differences were that big or bigger. We would conclude that the mean rating for those with internal reasons is significantly higher than for those with external reasons. **(c)** Type I error since we rejected $H_0: \mu_A - \mu_D = 0$.

59. **(a)** Two-sample *t* test **(b)** Paired *t* test **(c)** Two-sample *t* test

61. One-sample *z* interval for a proportion

63. Paired *t* interval for the mean difference

65. **(a) State:** We want to perform a test at the $\alpha = 0.05$ significance level of $H_0: \mu_1 - \mu_2 = 10$ versus $H_a: \mu_1 - \mu_2 > 10$, where μ_1 and μ_2 are the actual mean cholesterol reductions for people like the ones in this study when using the new drug and the current drug, respectively. **Plan:** Use a two-sample *t* test if the conditions are satisfied. *Random:* This was a randomized controlled experiment. *Normal:* Both n_1 and n_2 are less than 30, but we are told that no strong skewness or outliers were detected. *Independent:* Due to the random assignment, these two groups of patients can be viewed as independent. Also, knowing one patient's cholesterol reduction gives no information about another patient's cholesterol reduction. **Do:** From the data, $n_1 = 15$, $\bar{x}_1 = 68.7$, $s_1 = 13.3$, $n_2 = 14$, $\bar{x}_2 = 54.1$, and $s_2 = 11.93$. Using the conservative df $= 13$, the test statistic is $t =$ 0.982 and the *P*-value is $P(t > 0.982) = 0.1720$. **Conclude:** Since the *P*-value is greater than 0.05, we fail to reject H_0. We do not have enough evidence to conclude that the mean cholesterol reduction is more than 10 mg/dl more for the new drug than for the current drug. **(b)** Type II error since we failed to reject H_0.

67. d

69. a

71. Jannie treated the data as if she wanted to create a confidence interval for the *difference* between two proportions. In fact, she wants to compute a confidence interval for just one proportion: the proportion of students taking the SAT twice who were coached.

This interval is $0.135 \pm 2.576\sqrt{\dfrac{0.135(0.865)}{3160}} = (0.119, 0.151)$.

73. **(a) State:** We want to perform a test at the $\alpha = 0.05$ significance level of $H_0: \mu_1 - \mu_2 = 0$ versus $H_a: \mu_1 - \mu_2 > 0$, where μ_1 and μ_2 are the actual mean increases in SAT verbal score for coached students and uncoached students, respectively. **Plan:** Use a two-sample *t* test if the conditions are satisfied. *Random:* Both samples were randomly selected. *Normal:* Both n_1 and n_2 are at least 30. *Independent:* Both samples are less than 10% of their respective populations (there are more than 4270 students taking the SAT twice who were coached and 27,330 students taking the SAT twice who were not coached). **Do:** From the data, $n_1 = 427$, $\bar{x}_1 = 29$, $s_1 = 59$, $n_2 = 2733$, $\bar{x}_2 = 21$, and $s_2 = 52$. Using the conservative df $= 426$, the test statistic is $t = 2.646$ and the *P*-value is $P(t > 2.646) = 0.0042$. **Conclude:** Since the *P*-value is less than 0.05, we reject H_0. We have enough evidence to conclude that students who receive coaching have a higher mean increase in SAT verbal score than those who do not receive coaching. **(b) State:** We want to estimate $\mu_1 - \mu_2$ at a 99% confidence level. **Plan:** Use a two-sample *t* interval for $\mu_1 - \mu_2$ since we know from part (a) that the conditions are met. **Do:** Using df $= 426$, the 99% confidence interval is $(29 - 21) \pm 2.587\sqrt{\dfrac{(59)^2}{427} + \dfrac{(52)^2}{2733}} =$ $(0.178, 15.822)$. **Conclude:** We are 99% confident that the interval from 0.178 to 15.822 captures the difference in true mean increase in SAT Verbal scores for students who were coached and students who were not coached. This interval suggests that the

mean increase is between 0.178 and 15.822 points higher for students who are coached than for those who are not coached. **(c)** The number of points gained may not be very large. It does not seem like the money spent on coaching is worth it.

75. **(a)** $1 - (0.95)^2 = 0.0975$ **(b)** 0.0232 **(c)** This is a good criterion. If the process is under control, only 4% of the time would we conclude that it wasn't.

Answers to Chapter 10 Review Exercises

R10.1 **(a)** Paired *t* test for the mean difference **(b)** Two-sample *z* interval for the difference in proportion **(c)** One-sample *t* interval for the mean **(d)** Two-sample *t* interval for the difference between two means

R10.2 **(a)** Observational study; drivers were not assigned to the cities. **(b) State:** Our parameters are p_1 and p_2, the actual proportions of Hispanic female drivers in New York and in Boston, respectively, who wear seat belts. We want to estimate $p_1 - p_2$ at a 95% confidence level. **Plan:** Two-sample *z* interval for $p_1 - p_2$. *Random:* The women were selected randomly. *Normal:* 183, 37, 68, 49 are all at least 10. *Independent:* There are more than 2200 Hispanic women drivers in New York and 1170 Hispanic women drivers in Boston. **Do:** $(0.832 - 0.581) \pm 1.96\sqrt{\frac{0.832(0.168)}{220} + \frac{0.581(0.419)}{117}} = (0.149, 0.353)$. **Conclude:** We are 95% confident that the interval from 0.149 to 0.353 captures the difference in actual proportions of Hispanic women drivers in New York and Boston who wear their seat belts. **(c)** Since 0 is not in the interval, we have good evidence that a smaller proportion of Hispanic women in Boston wear their seat belts.

R10.3 **(a)** The Random and Independent conditions are met because this is a randomized comparative experiment. **(b)** The sample sizes are large enough, $n_1 = n_2 = 45$. **(c)** If there were no difference in the ratings of the product under the two treatments, we would have less than a 1% chance of observing a difference as large as or larger than the one in this experiment.

R10.4 **State:** $H_0: \mu_1 - \mu_2 = 0$ versus $H_a: \mu_1 - \mu_2 \neq 0$, where μ_1 and μ_2 are the actual mean NAEP quantitative skills test scores for men and for women, respectively. **Plan:** Two-sample *t* test. *Random:* Both samples were randomly selected. *Normal:* Both samples had at least 30 observations. *Independent:* There are more than 8400 males who took the NAEP and 10,770 women who took the NAEP. **Do:** From the data, $n_1 = 840$, $\bar{x}_1 = 272.40$, $s_1 = 59.2$, $n_2 = 1077$, $\bar{x}_2 = 274.73$, and $s_2 = 57.5$. Using df = 839, $t = -0.865$ and *P*-value $= 2P(t < -0.865) = 0.3872$. **Conclude:** Since the *P*-value is greater than 0.01, we fail to reject H_0. We do not have enough evidence to conclude that males and females have different mean scores on the NAEP quantitative skills test.

R10.5 **(a) State:** $H_0: p_1 - p_2 = 0$ versus $H_a: p_1 - p_2 < 0$, where p_1 and p_2 are the actual proportions of patients like the ones in this study who develop AIDS after taking AZT and after taking a placebo, respectively. **Plan:** Two-sample *z* test for $p_1 - p_2$. *Random:* This was a randomized comparative experiment. *Normal:* 17, 418, 38, 397 are all at least 10. *Independent:* Due to the random assignment, these two groups of patients can be viewed as independent. Also, knowing whether one patient developed AIDS gives no information about whether another patient developed AIDS. **Do:** From the data, $\hat{p}_1 = 0.039$ and $\hat{p}_2 = 0.087$. $\hat{p}_C = 0.063$, $z = -2.91$, *P*-value $= P(z < -2.91) = 0.0018$. **Conclude:** Since the *P*-value is less than 0.05, we reject H_0. We have enough evidence to conclude that there is a difference in the proportions of patients like these who develop AIDS while taking either AZT or a placebo. **(b)** Neither the subjects nor the researchers who had contact with them (including those determining whether subjects had AIDS) knew which subjects were getting which drug. This is important because if the experiment were not double-blind, the results could not be attributed to the treatments. **(c)** A Type I error means concluding that AZT lowers the risk of developing AIDS when in fact it does not. A Type II error means concluding that we do not have enough evidence that AZT lowers the risk of developing AIDS when in fact it does. Answers may vary as to which one is more serious.

R10.6 **(a)** Normal condition is not met. **(b)** Normal condition is not met.

R10.7 **(a)** The students in the treatment group had generally higher differences in scores and the middle 50% of their differences was much more compact than for the control group. The treatment group differences were also reasonably symmetric, whereas the control group differences were more right-skewed. **(b) State:** Our parameters are μ_1 and μ_2, the actual mean difference in test scores for students like these who get the treatment message and who get the neutral message, respectively. We want to estimate $\mu_1 - \mu_2$ at a 90% confidence level. **Plan:** Two-sample *t* interval for $\mu_1 - \mu_2$. *Random:* This was a randomized controlled experiment. *Normal:* Neither boxplot shows strong skewness or outliers. *Independent:* Due to the random assignment, these two groups of students can be viewed as independent. Also, knowing one student's difference in scores gives no information about another student's difference in scores. **Do:** Using (post − pre) as our difference, $n_1 = 10$, $\bar{x}_1 = 11.4$, $s_1 = 3.169$, $n_2 = 8$, $\bar{x}_2 = 8.25$, and $s_2 = 3.69$. Using df = 7, $(11.4 - 8.25) \pm 1.895\sqrt{\frac{(3.169)^2}{10} + \frac{(3.69)^2}{8}} = (0.03, 6.27)$. **Conclude:** We are 90% confident that the interval from 0.03 to 6.27 captures the true difference in mean scores for those who received the treatment message and those who received the neutral message. **(c)** We cannot generalize to all students who failed the test because our sample was not a random sample of all students who failed the test. It was a group of students who agreed to participate in the experiment.

R10.8 **(a)** About 139 **(b)** Properties of the sampling distribution of the difference $\hat{p}_1 - \hat{p}_2$ can be obtained from properties of the individual sampling distributions used in the individual confidence intervals, but the upper and lower limits of the intervals are not directly related. **(c) State:** Our parameters are p_1 and p_2, the actual proportions of golf clubs returned in 5 days or fewer before the changes and after the changes, respectively. We want to estimate $p_1 - p_2$ at a 95% confidence level. **Plan:** Two-sample *z* interval for $p_1 - p_2$. *Random:* The clubs both before and after changes were selected randomly. *Normal:* 22, 117, 125, 14 are all at least 10. *Independent:* There are more than 1390 golf clubs repaired before the changes and 1390 golf clubs repaired after the changes. **Do:** From the data, $n_1 = 139$, $\hat{p}_1 = 0.16$, $n_2 = 139$, $\hat{p}_2 = 0.90$. $(0.90 - 0.16) \pm 1.96\sqrt{\frac{0.90(0.10)}{139} + \frac{0.16(0.84)}{139}} = (0.661, 0.819)$. **Conclude:** We are 95% confident that the interval from 0.661 to 0.819 captures the difference in the actual proportions of orders sent back to customers in 5 days or fewer before the changes and after the changes.

R10.9 (a) Experiment; a treatment (message or no message) was deliberately imposed on the subjects. **(b) State:** We want to perform a test at the $\alpha = 0.05$ significance level of $H_0: p_1 - p_2 = 0$ versus $H_a: p_1 - p_2 > 0$, where p_1 and p_2 are the actual proportions of people like those who would be contacted when a message is left and when a message is not left, respectively. **Plan:** Use a two-sample z test for $p_1 - p_2$ if the conditions are satisfied. *Random:* This was a randomized comparative experiment. *Normal:* The number of successes and the number of failures in both groups are at least 10 (message: 200 successes, 91 failures; no message: 58 successes, 42 failures). *Independent:* Due to the random assignment, these two groups of people contacted can be viewed as independent. Also, knowing whether one person responded gives no information about whether another person responded. **Do:** From the data, $\hat{p}_1 = 0.687$ and $\hat{p}_2 = 0.58$. The pooled proportion is $\hat{p}_C = 0.66$. The test statistic is $z = 1.95$ and the P-value is $P(z > 1.95) = 0.0256$. **Conclude:** Since the P-value is less than 0.05, we reject H_0. We have enough evidence to conclude that the proportion of people like the ones in this study who would respond when messages are left is higher than the proportion who would respond when no messages are left.

R10.10 Answers will vary, but here is an example. The difference between average female (55.5) and male (57.9) self-concept scores was so small that it can be attributed to chance variation in the samples ($t = -0.83$, df $= 62.8$, P-value $= 0.4110$). In other words, based on this sample, we have insufficient evidence to suggest that mean self-concept scores differ by gender.

Answers to Chapter 10 AP Statistics Practice Test

T10.1 e
T10.2 b
T10.3 a
T10.4 a
T10.5 b
T10.6 e
T10.7 c
T10.8 c
T10.9 b
T10.10 a

T10.11 (a) **State:** Our parameters are μ_1 and μ_2, the actual mean hospital stay for patients like those who get heating blankets during surgery and who have core temperatures reduced during surgery, respectively. We want to estimate $\mu_1 - \mu_2$ at a 95% confidence level. **Plan:** Use a two-sample t interval for $\mu_1 - \mu_2$ if the conditions are satisfied. *Random:* This was a randomized controlled experiment. *Normal:* Both n_1 and n_2 are at least 30. *Independent:* Due to the random assignment, these two groups of patients can be viewed as independent. Also, knowing one patient's length of hospital stay gives no information about another patient's length of hospital stay. **Do:** From the data, $n_1 = 104$, $\bar{x}_1 = 12.1$, $s_1 = 4.4$, $n_2 = 96$, $\bar{x}_2 = 14.7$, and $s_2 = 6.5$. Using the conservative df $= 95$, the 95% confidence interval is

$$(12.1 - 14.7) \pm 1.985\sqrt{\frac{(4.4)^2}{104} + \frac{(6.5)^2}{96}} = (-4.17, -1.03).$$

Conclude: We are 95% confident that the interval from -4.17 to -1.03 captures the difference in the actual mean hospital stay for patients like those who get heating blankets during surgery and those who have their core temperatures reduced during surgery. This interval suggests that the mean hospital stay is between 4.17 and 1.03 days shorter for those receiving the heating blankets in comparison to those who have their core temperatures reduced. **(b)** Yes. Since 0 is not in the interval, the entire interval is negative and we subtracted normothermic − hypothermic. It appears that warming patients during surgery decreases the mean hospital stay.

T10.12 **(a) State:** We want to perform a test at the $\alpha = 0.05$ significance level of $H_0: p_1 - p_2 = 0$ versus $H_a: p_1 - p_2 > 0$, where p_1 and p_2 are the actual proportions of cars with the brake defect in last year's model and in this year's model, respectively. **Plan:** Use a two-sample z test for $p_1 - p_2$ if the conditions are satisfied. *Random:* The samples were randomly selected. *Normal:* The number of successes and the number of failures in both groups are at least 10 (last year: 20 successes, 80 failures; this year: 50 successes, 300 failures). *Independent:* Both samples are less than 10% of their respective populations (there are more than 1000 cars of last year's model and 3500 cars of this year's model). **Do:** From the data, $\hat{p}_1 = 0.2$ and $\hat{p}_2 = 0.143$. The pooled proportion is $\hat{p}_C = 0.156$. The test statistic is $z = 1.39$ and the P-value is $P(z > 1.39) = 0.0823$. **Conclude:** Since the P-value is greater than 0.05, we fail to reject H_0. We do not have enough evidence to conclude that the proportion of brake defects is less on cars of this year's model than on cars of last year's model. **(b)** A Type I error occurs if we reject the null hypothesis and it is really true—in this case, concluding that there are fewer brake defects on this year's car model when in fact there are not fewer. This might result in more accidents because people think that their brakes are safe. A Type II error occurs if we fail to reject the null hypothesis when in fact it is false—in this case, concluding that there are no fewer brake defects this year than last year when the number of brake defects has actually been reduced. This might result in fewer people buying this particular model of car.

T10.13 **(a)** This is a two-sample t test for the difference between two means. *Random:* The samples were selected randomly. *Normal:* Since there are fewer than 30 observations in each sample, we need to examine the data. The dotplot below shows no strong skewness or outliers in either distribution. *Independent:* There are very likely more than 100 one-bedroom and 100 two-bedroom apartments in the area of a college campus.

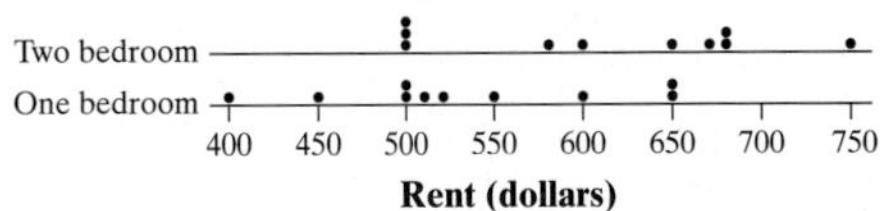

(b) If the mean rent for the two types of apartments is really the same, we have a 5.8% chance of finding a sample where the observed difference in mean rents is as large as or larger than the one in this study. **(c)** Since the P-value is greater than 0.05, Pat should fail to reject H_0. She does not have enough evidence to conclude that the mean rent of two-bedroom apartments is higher than the mean rent of one-bedroom apartments in the area of her college campus.

Answers to Cumulative AP Practice Test 3

AP3.1 e
AP3.2 e
AP3.3 d
AP3.4 c
AP3.5 b
AP3.6 d

AP3.7 c
AP3.8 a
AP3.9 d
AP3.10 c
AP3.11 b
AP3.12 c
AP3.13 c
AP3.14 d
AP3.15 a
AP3.16 e
AP3.17 b
AP3.18 b
AP3.19 e
AP3.20 c
AP3.21 a
AP3.22 d
AP3.23 b
AP3.24 e
AP3.25 a
AP3.26 b
AP3.27 c
AP3.28 d
AP3.29 a
AP3.30 b

AP3.31 Let μ_d = mean difference in weight gain (After – Before)

$$H_0: \mu_d = 0 \text{ versus } H_a: \mu_d < 0 \qquad t = \frac{\bar{x}_d - \mu_d}{\frac{s_d}{\sqrt{n}}}$$

• **Random** A random sample of dieters in the program was taken.
• **Normal** The boxplot of sample differences is relatively symmetric and there are no outliers, so it is plausible that the differences came from a Normal distribution.

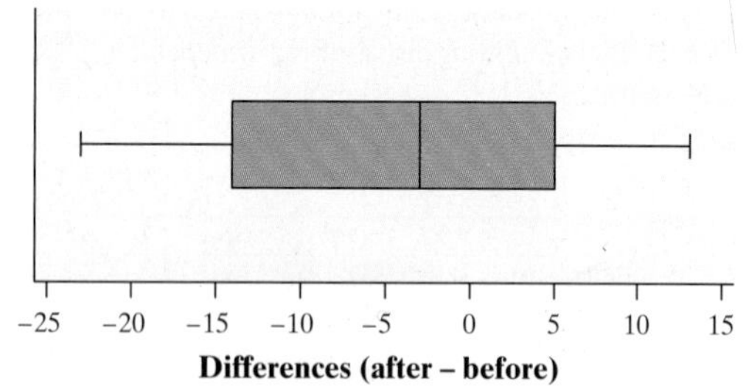

• **Independent** The difference in weights for each dieter can be treated as independent.

$t = -1.21$, P-value $= 0.1232$, df $= 14$. Since the P-value is greater than 0.05, we fail to reject the null hypothesis. We do not have convincing evidence that the diet had a long-term effect, i.e., it does not appear that dieters were able to maintain the initial weight loss.

AP3.32 **(a)** This is an observational study. No treatments were imposed on the two groups. **(b)**

p_1 = actual proportion of all VLBW babies who graduate from high school
p_2 = actual proportion of all normal-birth-weight babies who graduate from high school

$$H_0: p_1 = p_2 \text{ versus } H_a: p_1 < p_2$$

• Both groups of babies were randomly selected and form two independent groups.

$$n_1\hat{p}_1 = (242)\left(\frac{179}{242}\right) = 179 \text{ and } n_1(1-\hat{p}_1) = (242)\left(\frac{63}{242}\right) = 63$$

$$n_2\hat{p}_2 = (233)\left(\frac{193}{233}\right) = 193 \text{ and } n_2(1-\hat{p}_2) = (233)\left(\frac{40}{233}\right) = 40$$

• All values are greater than 10, so the Normal condition is met.

$$\text{Using } \hat{p}_c = \frac{179+193}{242+233} = 0.783, z = \frac{(\hat{p}_1 - \hat{p}_2) - (p_1 - p_2)}{\sqrt{\frac{\hat{p}_c(1-\hat{p}_c)}{n_1} + \frac{\hat{p}_c(1-\hat{p}_c)}{n_2}}} = -2.34$$

The corresponding P-value is 0.0096. Since the P-value is so small, we reject the null hypothesis. There is enough evidence to suggest that the proportion of all VLBW babies who graduate from high school by age 20 is significantly less than the proportion of all normal-birth-weight babies who graduate from high school by age 20.

AP3.33 **(a)** Predicted distance $= -73.64 + 5.7188$ (temperature) **(b)** The slope $= 5.7188$. For every one-degree (Celsius) increase in the water discharge temperature, the predicted increase in the distance of the nearest fish from the outflow pipe is 5.7188 meters. **(c)** Yes. The residual plot shows no apparent pattern and the original scatterplot shows a strong linear relationship between temperature of the discharge water and the distance of the fish from the outflow pipe. **(d)** Predicted distance $= -73.64 + 5.7188(29) = 92.21$ meters. If you examine the residual plot, the point with a fitted value of 92.21 has a negative residual. This means that the model is overpredicting.

AP3.34 Define W = the weight of a randomly selected gift box. We know that

$$\mu_W = 8(2) + 2(4) + 3 = 27$$

$$\sigma_W = \sqrt{8(0.5^2) + 2(1^2) + 0.2^2} = 2.01$$

(b) Since the three distributions are approximately Normal, the weight distribution will be as well. $P(W > 30) = P\left(z > \frac{30-27}{2.01}\right) = P(z > 1.49) = 0.0681$. **(c)** Let X = the number of boxes that weigh more than 30 ounces. This is a binomial setting with $n = 5$ and $p = 0.0681$. $P(X \geq 1) = 1 - P(X = 0) = 1 - (1 - 0.0681)^5 = 0.2972$.

$$\textbf{(d)}\ P(\bar{w} > 30) = P\left(z > \frac{30-27}{\frac{2.01}{\sqrt{5}}}\right) = P(z > 3.34) = 0.000419$$

AP3.35 **(a)** μ_A = mean annual return for stock A; μ_B = mean annual return for stock B.

$$H_0: \mu_A = \mu_B \text{ versus } H_a: \mu_A \neq \mu_B \qquad t = \frac{(\bar{x}_A - \bar{x}_B) - (\mu_A - \mu_B)}{\sqrt{\frac{s_A^2}{n_A} + \frac{s_B^2}{n_B}}}$$

• Both samples are large (50 > 30)
• The two groups of stock prices form independent samples.
• The daily returns for both A and B were randomly selected from the previous five years:

$$t = 2.067; \quad P\text{-value} = 0.0416; \quad \text{df} = 90.53$$

Since $0.0416 < 0.05$, we reject the null hypothesis. There is enough evidence to suggest that there is a significant difference in the mean return on investment for the two stocks. **(b)** σ_A = standard deviation of the return for stock A; σ_B = standard deviation of the return for stock B. $H_0: \sigma_A = \sigma_B$ versus $H_a: \sigma_A > \sigma_B$. **(c)** Values of F that are significantly greater than 1 would indicate that the price volatility for stock A is higher than that for stock B. **(d)** $F = \frac{(12.9)^2}{(9.6)^2} = 1.806$ **(e)** Assuming there is no difference in the standard deviations for the two stocks, we would observe a test statistic of 1.806 or greater only 6 out of 200 times in the simulation. This represents the P-value of the test ($P = 0.03$). Since this value is less than 0.05, we would reject the null hypothesis. There is enough evidence to conclude that the price volatility of stock A is greater than that of stock B.

CHAPTER 11

Section 11.1

Answers to Check Your Understanding

Page 681 **1.** $H_0: p_{blue} = 0.23, p_{orange} = 0.23, p_{green} = 0.15, p_{yellow} = 0.15, p_{red} = 0.12, p_{brown} = 0.12$; H_a: At least one of the p_i's is incorrect **2.** Blue and orange candies: 46(0.23) = 10.58; green and yellow: 46(0.15) = 6.9; red and brown: 46(0.12) = 5.52
3.

$$\chi^2 = \frac{(12-10.58)^2}{10.58} + \frac{(7-10.58)^2}{10.58} + \frac{(13-6.9)^2}{6.9} + \frac{(4-6.9)^2}{6.9} + \frac{(8-5.52)^2}{5.52} + \frac{(2-5.52)^2}{5.52}$$
$$= 0.1906 + 1.2114 + 5.3928 + 1.2188 + 1.1142 + 2.2446 = 11.3724.$$

Page 684 **1.** All expected counts are at least 5. Chi-square distribution with 5 df.
2.

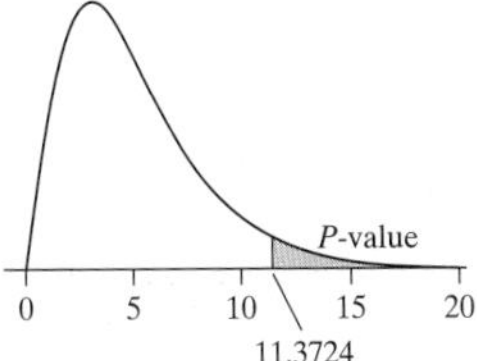

Chi-square distribution with 5 df

3. From Table C: 0.025 < P-value < 0.05. From the calculator: 0.0445. **4.** Since the P-value is less than 0.05, we reject the null hypothesis and conclude that at least one of the proportions of M&M'S® Peanut Chocolate Candies of a particular color is different from what the company reported. *Page 689* **State:** We want to perform a test at the $\alpha = 0.01$ significance level of

$$H_0: p_{red-straight} = \frac{9}{16}, p_{red-curly} = \frac{3}{16}, p_{white-straight} = \frac{3}{16}, p_{white-curly} = \frac{1}{16}$$

Plan: We should use a chi-square goodness-of-fit test if the conditions are satisfied. *Random:* A random sample was used. Large sample size in which the expected counts in each category are: red–straight, 112.5; red–curly and white–straight, 37.5; and white–curly, 12.5—all of which are at least 5. *Independent:* It seems reasonable that there would be more than 2000 pairs of fruit flies. **Do:** The test statistic is $\chi^2 = 6.1867$ and the P-value is 0.1029 using df = 3. **Conclude:** Since the P-value is greater than 0.01, we fail to reject H_0. We do not have enough evidence to reject the biologists' predicted distribution of traits in offspring.

Answers to Odd-Numbered Section 11.1 Exercises

1. (a) $H_0: p_{cashews} = 0.52, p_{almonds} = 0.27, p_{macadamia} = 0.13, p_{brazil} = 0.08$; H_a: At least one of the p_i's is incorrect **(b)** Cashews: 150(0.52) = 78, almonds: 150(0.27) = 40.5, macadamia nuts: 150(0.13) = 19.5, and brazil nuts: 150(0.08) = 12

3.
$$\chi^2 = \frac{(83-78)^2}{78} + \frac{(29-40.5)^2}{40.5} + \frac{(20-19.5)^2}{19.5} + \frac{(18-12)^2}{12}$$
$$= 0.321 + 3.265 + 0.013 + 3.000 = 6.599.$$

5. (a) The expected counts are all at least 5. Chi-square distribution with 3 df.
(b)

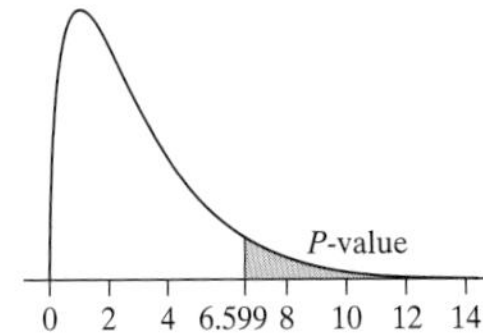

Chi-square distribution with 3 df

(c) Using Table C: 0.05 < P-value < 0.10. Using the calculator: 0.0858. **(d)** Since the P-value is greater than 0.05, we fail to reject H_0. We do not have enough evidence to reject the company's claim about the distribution of nuts.
7. State: We want to perform a test at $\alpha = 0.05$ of H_0: $p_{firs} = 0.54, p_{pines} = 0.40, p_{other} = 0.06$ versus H_a: At least one of the p_i's is incorrect. **Plan:** Use a chi-square goodness-of-fit test if the conditions are satisfied. *Random:* A random sample was used. Large sample size in which the expected counts in each category are: firs, 84.24; pines, 62.4; and other, 9.36; all of which are at least 5. *Independent:* It seems reasonable that there would be more than 1560 red-breasted nuthatches. **Do:** The test statistic is $\chi^2 = 7.418$ and the P-value is 0.0245 using df = 2. **Conclude:** Since the P-value is less than 0.05, we reject H_0 and conclude that nuthatches prefer particular types of trees when they are searching for seeds and insects.
9. The data for each day were not a count.
11. (a) State: We want to perform a test at $\alpha = 0.05$ of H_0: $p_1 = 0.301, p_2 = 0.176, p_3 = 0.125, p_4 = 0.097, p_5 = 0.079, p_6 = 0.067, p_7 = 0.058, p_8 = 0.051, p_9 = 0.046$ versus H_a: At least one of the p_i's is incorrect. **Plan:** Use a chi-square goodness-of-fit test if the conditions are satisfied. *Random:* A random sample was used. Large sample size in which the expected counts are 75.25, 44, 31.25, 24.25, 19.75, 16.75, 14.5, 12.75, 11.5, all of which are at least 5. *Independent:* It seems reasonable that there would be more than 2500 invoices at this company. **Do:** The test statistic is $\chi^2 = 21.563$ and the P-value is 0.0058 using df = 8. **Conclude:** Since the P-value is less than 0.05, we reject H_0 and conclude that the invoices are inconsistent with Benford's law. *Follow-up analysis:* The largest contributors to the chi-square statistic are amounts with first digit 3, 4, and 7. There are too many amounts that start with 3 or 4 and not enough that start with 7. **(b)** Type I error: say that the company's invoices did not follow Benford's law (suggesting fraud) when in fact they were consistent with Benford's law. Type II error: say that the invoices were consistent with Benford's law when in fact they were not. A Type I error would be more serious here—alleging that the company had committed fraud when it had not.
13. (a) $H_0: p_{lemon} = 0.2, p_{lime} = 0.2, p_{orange} = 0.2, p_{strawberry} = 0.2, p_{grape} = 0.2$ versus H_a: At least one of the p_i's is incorrect. **(b)** Each of the 5 flavors: 12. **(c)** 9.49; 13.28 **(d)** Answers will vary. One possibility consists of 6 lemon, 6 lime, 16 orange, 16 strawberry, and 16 grape. The chi-square statistic is $\chi^2 = 10$ with a P-value of 0.04.
15. State: We want to perform a test at $\alpha = 0.05$ of $H_0: p_i = 0.083$ for all astrological signs versus H_a: At least one of the p_i's is incorrect. **Plan:** Use a chi-square goodness-of-fit test if the conditions are satisfied. *Random:* A random sample was used. Large sample size in which the expected count in each category is 362, which is at least 5.

Independent: There are more than 43,440 residents in the United States. **Do:** The test statistic is $\chi^2 = 19.762$ and the *P*-value is 0.0487 using df $= 11$. **Conclude:** Since the *P*-value is less than 0.05, we reject H_0 and conclude that the 12 signs are not equally likely. *Follow-up analysis:* The largest contributors to the statistic are Aries and Virgo. There are fewer Aries and more Virgos than expected.

17. **State:** We want to perform a test at $\alpha = 0.05$ of $H_0: p_{\text{smooth}} = 0.75, p_{\text{wrinkled}} = 0.25$ versus H_a: At least one of the p_i's is incorrect. **Plan:** Use a chi-square goodness-of-fit test if the conditions are satisfied. *Random:* A random sample was used. The expected counts are 417 and 139, both of which are at least 5. *Independent:* It seems reasonable that there would be more than 5560 peas. **Do:** The test statistic is $\chi^2 = 0.3453$ and the *P*-value is 0.5568 using df $= 1$. **Conclude:** Since the *P*-value is greater than 0.05, we fail to reject H_0. We do not have enough evidence to dispute Mendel's beliefs.

19. d

21. c

23. The English grades for the heavy readers are centered on approximately 3.7 with a left-skewed distribution. The *IQR* is approximately 0.4 and there is one low outlier at approximately 2.9. The English grades for light readers have a lower center at approximately 3.35. These grades are more symmetrically distributed with no outliers, and the distribution has an *IQR* of approximately 0.7. In general, the grade distribution for light readers is lower and more spread out than the grade distribution for heavy readers.

25. **(a)** We would predict an English grade of 3.42 for a student who had read no books. **(b)** $2.85 - 3.828 = -0.978$ **(c)** The relationship between English grades and number of books read is not very strong. The value of r^2 is only 0.083, which means that only 8.3% of the variation in English grades is accounted for by the linear relationship with the number of books read.

Section 11.2

Answers to Check Your Understanding

Page 698 **1.** For the main campus: several times a month or less, 6.0%; at least once a week, 23.6%; at least once a day, 70.3%. For the commonwealth campuses: several times a month or less, 12.1%; at least once a week, 25.0%; at least once a day, 62.8%. **2.** Because there was such a big difference in the sample size from the two different types of campuses. **3.** Students on the main campus are more likely to be everyday users of Facebook than are students on the commonwealth campuses, and those on the commonwealth campuses are more likely to use Facebook several times a month or less than are those students on the main campus.

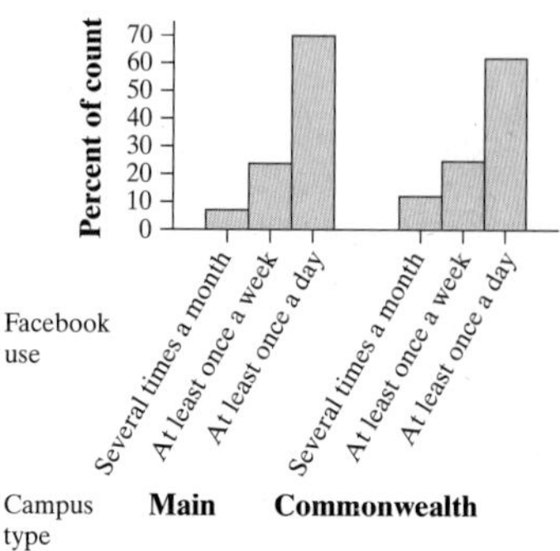

Percent within levels of Campus type.

Page 703 **1.** H_0: There is no difference in the distributions of Facebook use between students at Penn State's main campus and its commonwealth campuses

H_a: There is a difference in the distributions of Facebook use between students at Penn State's main campus and its commonwealth campuses

2. *Main campus:* several times, 77.56; once a week, 220.25; once a day, 612.19. *Commonwealth campus:* several times, 53.44; once a week, 151.75; once a day, 421.81. **3.** $\chi^2 = 19.489$ *Page 705* **1.** Using Table C with df $= (3 - 1)(2 - 1) = 2$: *P*-value < 0.0005. Using the calculator: 0.000059. **2.** Assuming that there is no difference in the distributions of Facebook use between students at Penn State's main campus and students at Penn State's commonwealth campuses, the probability of observing a difference in the distributions of Facebook use for the two samples as large as or larger than the one found in this study is about 6 in 100,000. **3.** Since the *P*-value was so small, reject H_0. It appears that the distribution of Facebook use is different among students at Penn State's main campus and students at Penn State's commonwealth campuses.

Page 708 **1.**

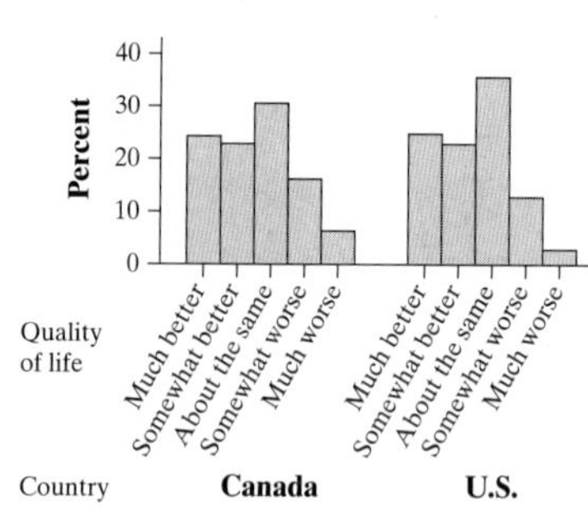

Percent within levels of Country.

2. State: We want to perform a test of

H_0: There is no difference in the distribution of quality of life in Canada and the United States

H_a: There is a difference in the distribution of quality of life in Canada and the United States

at the $\alpha = 0.01$ level. **Plan:** Use a chi-square test for homogeneity if the conditions are satisfied. *Random:* The data came from separate random samples. The expected counts are: *Canada*—much better, 77.37; somewhat better, 71.47; about the same, 109.91; somewhat worse, 41.70; much worse, 10.55; *United States*—much better, 538.63; somewhat better, 497.53; about the same, 765.09; somewhat worse, 290.30; much worse, 73.45. All these counts are at least 5. *Independent:* We clearly have less than 10% of all heart attack victims in the United States and in Canada. **Do:** The test statistic is $\chi^2 = 11.725$ and the *P*-value is 0.0195 using df $= 4$. **Conclude:** Since the *P*-value is greater than 0.01, we fail to reject H_0. There is not enough evidence to conclude that there is a difference in the distribution of quality of life in Canada and the United States. *Page 713* **1.** This was an experiment. Each individual was exposed to a treatment that involved how they were contacted.

2.

Contact method	Answered "Yes"	Answered "No"
Phone	168	632
Personal interview	200	600
Written response	224	576

3. **State:** We want to perform a test of

H_0: There is no difference in the actual proportion who answer yes based on contact method
H_a: There is a difference in the actual proportion who answer yes based on contact method

at the $\alpha = 0.05$ level. **Plan:** Use a chi-square test for homogeneity if the conditions are satisfied. *Random:* The data came from a randomized experiment. Large sample size in which the expected counts are: *Phone*—Yes, 197.33; No, 602.67; *Interview*—Yes, 197.33; No, 602.67; *Written response*—Yes, 197.33; No, 602.67. All these counts are at least 5. *Independent:* Due to the random assignment, these three groups of people can be viewed as independent. Also, knowing one person's response gives no information about another person's response. **Do:** The test statistic is $\chi^2 = 10.619$ and the *P*-value is 0.0049 using df = 2. **Conclude:** Since the *P*-value is less than 0.05, we reject H_0 and conclude that there is convincing evidence of a difference in the proportion of people who answer "Yes" based on how they are contacted. *Page 718* **State:** We want to perform a test of

H_0: There is no association between territory type and success or not in the population of franchises
H_a: There is an association between territory type and success or not in the population of franchises

at the $\alpha = 0.01$ level. **Plan:** Use a chi-square test of association/independence if the conditions are satisfied. *Random:* This was a random sample of franchises. Large sample size in which the expected counts are: *Yes*—success, 102.74; failure, 39.26; *No*—success, 20.26; failure, 7.74. All of these counts are at least 5. *Independent:* There are more than 1700 franchises in the United States. **Do:** The test statistic is $\chi^2 = 5.911$ and the *P*-value is 0.0150 using df = 1. **Conclude:** Since the *P*-value is greater than 0.01, we fail to reject H_0. We do not have enough evidence to conclude that there is an association between whether franchises have an exclusive territory or not and whether they are successful or not.

Answers to Odd-Numbered Section 11.2 Exercises

27. **(a)** *Female:* HSC-HM, 0.209; HSC-LM, 0.104; LSC-HM, 0.313; LSC-LM, 0.373. *Male:* HSC-HM, 0.463; HSC-LM, 0.269; LSC-HM, 0.075; LSC-LM, 0.194.
(b)

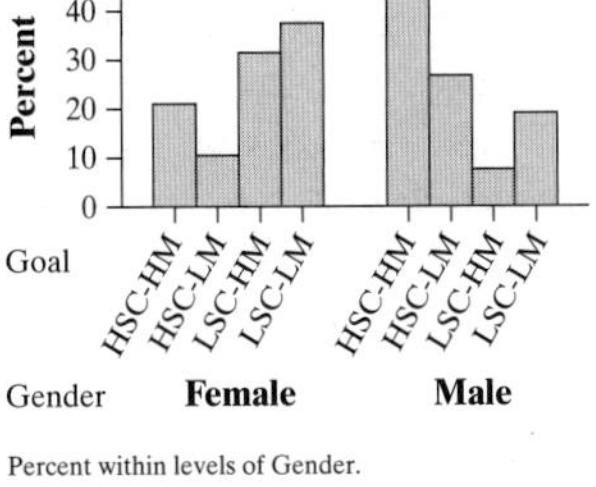

Percent within levels of Gender.

(c) In general, it appears that females were classified mostly as low social comparison, whereas males were classified mostly as high social comparison. Males were most likely to be in the high social comparison/high mastery group, whereas females were most likely to be in the low social comparison/low social mastery group.

29. **(a)**

H_0: There is no difference in the distribution of sports goals for male and female undergraduates
H_a: There is a difference in the distribution of sports goals for male and female undergraduates

(b) *Females:* HSC-HM, 22.5; HSC-LM, 12.5; LSC-HM, 13; LSC-LM, 19. *Males:* HSC-HM, 22.5; HSC-LM, 12.5; LSC-HM, 13; LSC-LM: 19. **(c)** $\chi^2 = 24.898$

31. **(a)** *Random:* The data came from random samples. Large sample size in which the expected counts computed in Exercise 29 were all at least 5. *Independent:* Clearly, we have less than 10% of all male or female students possible. **(b)** Using Table C with df = 3: *P*-value < 0.0005. Using the calculator: 0.00002. **(c)** Assuming that there is no difference in the distributions of goals for playing sports among males and females, the probability of observing a difference in the distributions of goals for the two samples as large or larger than the one found in this study is about 2 in 100,000. **(d)** Since the *P*-value is so small, we reject H_0. There is convincing evidence of a difference in the distributions of goals for playing sports among males and female undergraduates.

33. **(a)**

Environment	Hatched	Did not hatch	Total
Cold	16	11	27
Neutral	38	18	56
Hot	75	29	104
Total	129	58	187

The proportions in each group that hatched are: cold, 0.593; neutral, 0.679; hot, 0.721. These data support the researchers' belief somewhat. A smaller proportion of eggs hatched in the cold water, but a majority of the eggs in cold water did still hatch. **(b) State:** We want to perform a test of

H_0: There is no difference in the proportion of eggs that hatch based on water temperature
H_a: There is a difference in the proportion of eggs that hatch based on water temperature

at the $\alpha = 0.05$ level. **Plan:** Use a chi-square test for homogeneity if the conditions are satisfied. *Random:* The data came from a randomized experiment. Large sample size in which the expected counts are 18.63, 8.37, 38.63, 17.37, 71.74, 32.26. All these counts are at least 5. *Independent:* Due to the random assignment, these three groups of eggs can be viewed as independent. Also, knowing whether one egg hatches does not give information about whether another egg hatches or not. **Do:** The test statistic is $\chi^2 = 1.703$ and the *P*-value is 0.4267 using df = 2. **Conclude:** Since the *P*-value is greater than 0.05, we fail to reject H_0. We do not have enough evidence to say that different proportions of eggs hatch in different water temperatures.

35. We do not have the actual counts of the travelers in each category. We also do not know if the sample was taken randomly,

nor if the same people could have been counted more than once (traveling both for leisure one time and business another).
37. (a)

Treatment	Success	Failure	Total
Nicotine patch	40	204	244
Drug	74	170	244
Patch plus drug	87	158	245
Placebo	25	135	160
Total	226	667	893

(b)

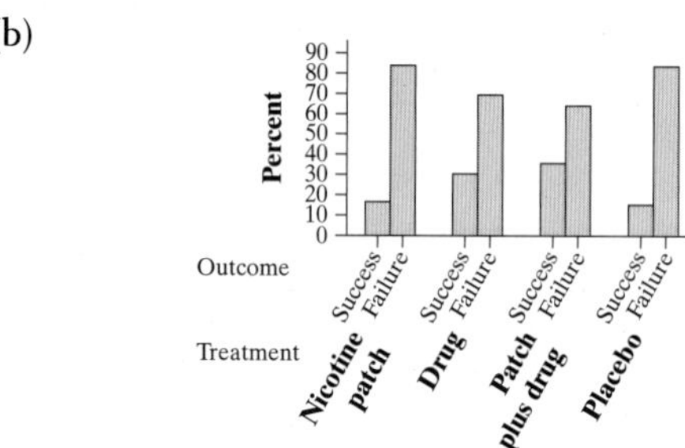

Percent within levels of Treatment.

The best success rate is with both the patch and the drug. The patch alone is not much better than the placebo. (c) The null hypothesis says that each of the four treatments has the same success rate. (d) The expected counts are 61.75, 182.25, 61.75, 182.25, 62.00, 183.00, 40.49, 119.51.
39. **State:** We want to perform a test of

H_0: There is no difference in the smoking cessation rates for the different treatments
H_a: There is a difference in the smoking cessation rates for the different treatments

at the $\alpha = 0.05$ level. **Plan:** Use a chi-square test for homogeneity if the conditions are satisfied. *Random:* The data came from a randomized experiment. Large sample size: from Exercise 37, all expected counts are at least 5. *Independent:* Due to the random assignment, these four groups of subjects can be viewed as independent. Also, knowing whether one person quits smoking gives no information about any other subject's smoking status. **Do:** The test statistic is $\chi^2 = 34.937$ and the *P*-value is approximately 0 using df $= 3$. **Conclude:** Since the *P*-value is less than 0.05, we reject H_0 and conclude that the smoking cessation rate is different for at least one of the four treatments.
41. The largest component of the chi-square statistic comes from those who had success using both the patch and the drug. Far more people fell in this group than would have been expected. The next largest component comes from those who had success using just the patch. Far fewer were in this group than would have been expected.
43. (a) The hypotheses are

H_0: There is no difference in improvement rates for the two treatments
H_a: There is a difference in improvement rates for the two treatments

Assuming that there is no difference in improvement rates between gastric freezing and placebo, the probability of observing a difference in the improvement rates for the two groups as large as or larger than the one found in this study is about 57 in 100. These data do not provide convincing evidence of a difference in improvement rates for gastric freezing and placebo. (b) The *P*-values are the same and $z^2 = \chi^2$.
45. (a)

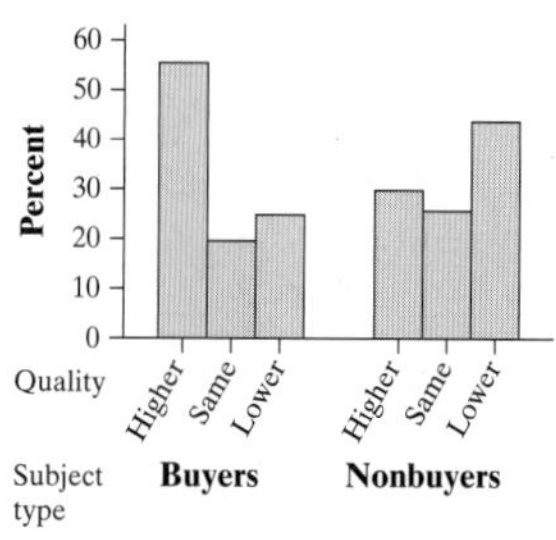

Percent within levels of Subject type.

Buyers are much more likely to think the quality of recycled coffee filters is higher, whereas nonbuyers are more likely to think the quality is lower. (b) **State:** We want to perform a test of

H_0: Buying recycled products is independent of their perceived quality rating
H_a: Buying recycled products is not independent of their perceived quality rating

at the $\alpha = 0.05$ level. **Plan:** Use a chi-square test for independence if the conditions are satisfied. *Random:* The data came from a random sample. Large sample size in which the expected counts on the computer printout are all at least 5. *Independent:* There are more than 1330 adults. **Do:** The test statistic is $\chi^2 = 7.638$ and the *P*-value is 0.022 using df $= 2$. **Conclude:** Since the *P*-value is less than 0.05, we reject H_0 and conclude that buying recycled coffee filters is not independent of their perceived quality.
47. (a) The bar graph below shows the percents. It appears that those who are moderately involved in extracurricular activities (2 to 12 hours per week) earn higher grades than those who are very active and those who are not active.

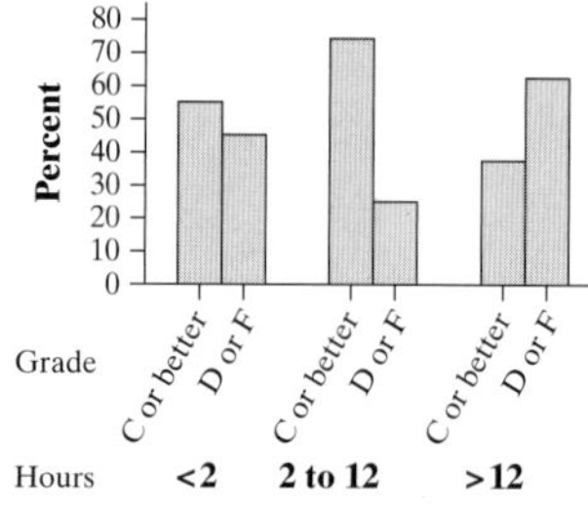

Percent within levels of Hours.

(b) The expected count for >12 hours, D or F is less than 5. We also don't know if the sample was selected at random.
49. (a)

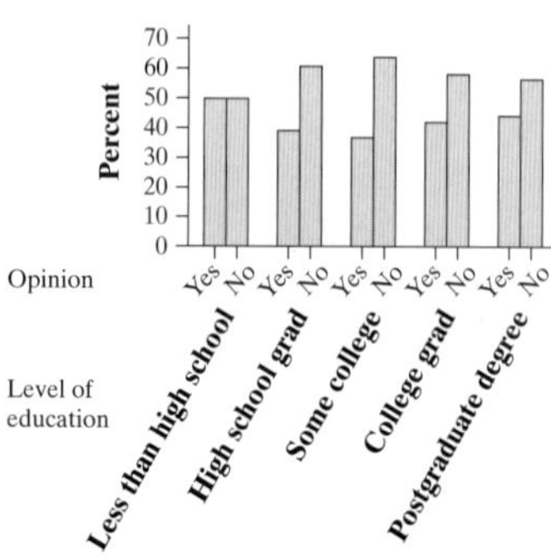

Percent within levels of Level of education.

The majority of people in most education levels seem to oppose such a law. The one education level in which there is an even split is the group that has less than a high school education. (b) **State:** We want to perform a test of

H_0: There is no assocation between education level and support of a gun law in the population of adults
H_a: There is an assocation between education level and support of a gun law in the population of adults

at the $\alpha = 0.05$ level. **Plan:** Use a chi-square test for association/independence if the conditions are satisfied. *Random:* The data came from a random sample. Large sample size in which the expected counts are 46.94, 69.06, 86.19, 126.81, 187.36, 275.64, 94.29, 138.71, 71.22, 104.78. All these counts are at least 5. *Independent:* There are at least 12,010 adults. **Do:** The test statistic is $\chi^2 = 8.525$ and the *P*-value is 0.0741 using df = 4. **Conclude:** Since the *P*-value is greater than 0.05, we fail to reject H_0. We do not have enough evidence to say that there is an association between education level and support of a gun law in the population of adults.

51. (a) Since this represents one sample and the subjects were then classified by their answer and their gender, this would be a chi-square test of association/independence. (b) The hypotheses are

H_0: Gender and where young adults live are independent
H_a: Gender and where young adults live are not independent

(c) *Random:* This was a random sample. Large sample size in which the expected counts given in the output are all at least 5. *Independent:* There are at least 48,540 young adults. (d) If gender and place of living are independent, then we have a 1.2% chance of finding a sample with as much association or more than we found. This is fairly unlikely (less than a 5% chance), so we reject H_0 and conclude that gender and where young adults live are not independent.

53. e

55. a

57. b

59. (a) One-sample *t* interval for the mean. (b) Two-sample *z* test for the difference between two proportions.

61. This was an experiment. The subjects were randomly exposed to a treatment (the type of response they were allowed to make).

63. (a) $\bar{x}_{1-5} = 3.545$; $s_{1-5} = 1.184$ (b) 4.21; 0.568 (c) No. The original measurements are categorical, not quantitative. They are recorded as numbers, but the "distance" between 1 and 2 may not be the same as the distance between 3 and 4 in people's minds.

Answers to Chapter 11 Review Exercises

R11.1 (a)

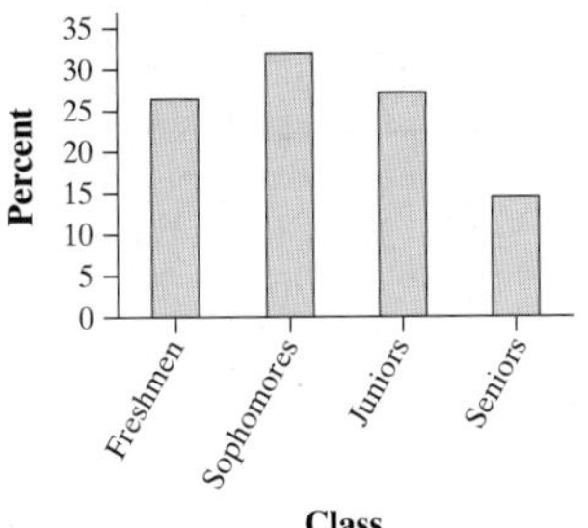

In this school, the largest class consists of the freshmen, and the class sizes decrease each year. But in the sample, the largest groups were the sophomores and the juniors. (b) **State:** We want to perform a test at $\alpha = 0.05$ of $H_0: p_{\text{freshman}} = 0.29$, $p_{\text{sophomore}} = 0.27$, $p_{\text{junior}} = 0.25$, $p_{\text{senior}} = 0.19$ versus H_a: At least one of the p_i's is incorrect. **Plan:** Use a chi-square goodness-of-fit test if the conditions are satisfied. *Random:* A random sample was used. Large sample size in which the expected counts are 59.74, 55.62, 51.5, and 39.14, all of which are at least 5. *Independent:* It seems reasonable that there would be more than 2060 students in a large high school. **Do:** The test statistic is $\chi^2 = 5.016$ and the *P*-value is 0.1706 using df = 3. **Conclude:** Since the *P*-value is greater than 0.05, we fail to reject H_0. We do not have enough evidence to reject the idea that the sample is representative of the population of students.

R11.2 The expected counts are 12.5, 0.5, 4.5, 1.5, 5, 12.5, 0.5, 4.5, 1.5, 5. Note that 6 of the cells have expected counts that are less than 5.

R11.3 (a)

Treatment	Suffered cardiac event	No cardiac event	Total
Stress management	3	30	33
Exercise	7	27	34
Usual care	12	28	40
Total	22	85	107

(b) The success rate for stress management is 0.909; for exercise, 0.794; and for the usual care, 0.70. (c) **State:** We want to perform a test of

H_0: There is no difference in the actual proportion of cardiac events for the three treatments
H_a: There is a difference in the actual proportion of cardiac events for the three treatments

at the $\alpha = 0.05$ level. **Plan:** Use a chi-square test for homogeneity if the conditions are satisfied. *Random:* The data came from a randomized experiment. Large sample size in which the expected counts are 6.79, 26.21, 6.99, 27.01, 8.22, 31.78. All these counts are at least 5. *Independent:* Due to the random assignment, these three groups of patients can be viewed as independent. Also, knowing whether a person has a cardiac event gives no information about whether another person has a cardiac event. **Do:** The test statistic is $\chi^2 = 4.840$ and the *P*-value is 0.0889 using df = 2. **Conclude:** Since the *P*-value is greater than 0.05, we fail to reject H_0. We do not have enough evidence to say that there is a difference in the actual proportion of cardiac events for the three treatments.

R11.4 (a) This should be a chi-square test for association/independence because one random sample was used and people in the ads were classified both by gender of the target audience and by whether the ad was sexual or not. (b) The hypotheses are

H_0: Gender of target audience and whether magazine ads are sexual are independent
H_a: Gender of target audience and whether magazine ads are sexual are not independent

(c) 351/576 = 0.609 or 60.9%. (1113)(576)/1509 = 424.843. $(351 - 424.843)^2/424.843 = 12.835$. (d) Since the *P*-value is so small (essentially 0), we would reject the null hypothesis and

conclude that the gender of the target audience and whether magazine ads are sexual are not independent.
R11.5 (a)

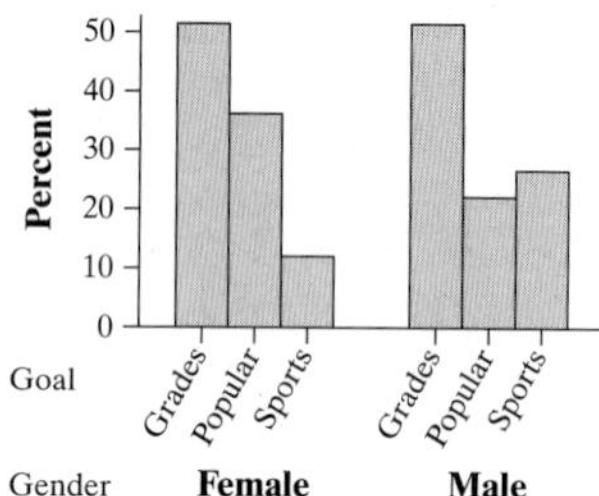

For both groups of children the largest percent report grades as the goal. But after that, boys were more likely to pick sports, whereas girls were more likely to pick being popular. **(b) State:** We want to perform a test of

H_0: There is no association between gender and goal at school for 4th, 5th, and 6th grade students
H_a: There is an association between gender and goal at school for 4th, 5th, and 6th grade students

at the $\alpha = 0.05$ level. **Plan:** Use a chi-square test for association/independence if the conditions are satisfied. *Random:* The data came from a random sample. Large sample size in which the expected counts are 129.70, 74.04, 47.26, 117.30, 66.96, 42.74. All these counts are at least 5. *Independent:* There are at least 4780 students in grades 4, 5, and 6. **Do:** The test statistic is $\chi^2 = 21.455$ and the *P*-value is 0.00002 using df = 2. **Conclude:** Since the *P*-value is less than 0.05, we reject H_0 and conclude that there is an association between gender and goal at school for 4th-, 5th-, and 6th-grade students. **(c)** The largest contributions are from the male/sports cell and the female/sports cell. More males than expected chose sports, and fewer females than expected chose sports.
R11.6 (a)

Type of subject	Answer "Yes"	Answer "No"	Total
Student	22	39	61
Nonstudent	30	29	59
Total	52	68	120

State: We want to perform a test of

H_0: There is no difference in the actual proportion who use unsafe passwords between students and nonstudents
H_a: There is a difference in the actual proportion who use unsafe passwords between students and nonstudents

at the $\alpha = 0.05$ level. **Plan:** Use a chi-square test for homogeneity if the conditions are satisfied. *Random:* The data came from random samples. Large sample size in which the expected counts are 26.43, 34.57, 25.57, 33.43. All these counts are at least 5. *Independent:* We have sampled less than 10% of students and less than 10% of nonstudents. **Do:** The test statistic is $\chi^2 = 2.669$ and the *P*-value is 0.1023 using df = 1. **Conclude:** Since the *P*-value is greater than 0.05, we fail to reject H_0. We do not have enough evidence to say that students and nonstudents have different rates of unsafe password use. **(b) State:** We want to perform a test at $\alpha = 0.05$ of H_0: $p_1 - p_2 = 0$ versus H_a: $p_1 - p_2 \neq 0$, where p_1 is the actual proportion of students who use unsafe passwords and p_2 is the actual proportion of nonstudents who use unsafe passwords. **Plan:** Use a two-sample *z* test for $p_1 - p_2$ if the conditions are satisfied. *Random:* We have random samples from each population. *Normal:* The number of successes and the number of failures in both groups are at least 10. *Independent:* Both samples are less than 10% of their respective populations. **Do:** From the data, $\hat{p}_1 = 0.361$ and $\hat{p}_2 = 0.508$. The pooled proportion is $\hat{p}_C = 0.433$. The test statistic is $z = -1.63$ and the *P*-value is 0.1023. **Conclude:** Since the *P*-value is greater than 0.05, we fail to reject H_0. We do not have enough evidence to conclude that there is a difference in the actual proportions of students and nonstudents who use unsafe passwords. This matches the result from the chi-square test: $(-1.63)^2 \approx 2.669$.

Answers to Chapter 11 AP Statistics Practice Test

T11.1 b
T11.2 c
T11.3 e
T11.4 d
T11.5 c
T11.6 c
T11.7 b
T11.8 a
T11.9 d
T11.10 d
T11.11 **State:** We want to perform a test at $\alpha = 0.05$ of H_0: $p_{\text{regular}} = 0.6$, $p_{\text{premium}} = 0.2$, $p_{\text{supreme}} = 0.2$ versus H_a: At least one of the p_i's is incorrect. **Plan:** Use a chi-square goodness-of-fit test if the conditions are satisfied. *Random:* A random sample was used. Large sample size in which the expected counts are 240, 80, and 80, all of which are at least 5. *Independent:* It seems reasonable that there would be more than 4000 people who stop at this gas station over a long period of time. **Do:** The test statistic is $\chi^2 = 13.15$ and the *P*-value is 0.0014 using df = 2. **Conclude:** Since the *P*-value is less than 0.05, we reject H_0 and conclude that the distribution of type of gas purchased is not what the distributor claims.
T11.12 **(a)** Random assignment was used to create roughly equivalent groups of subjects to receive each of the three possible treatments.
(b)

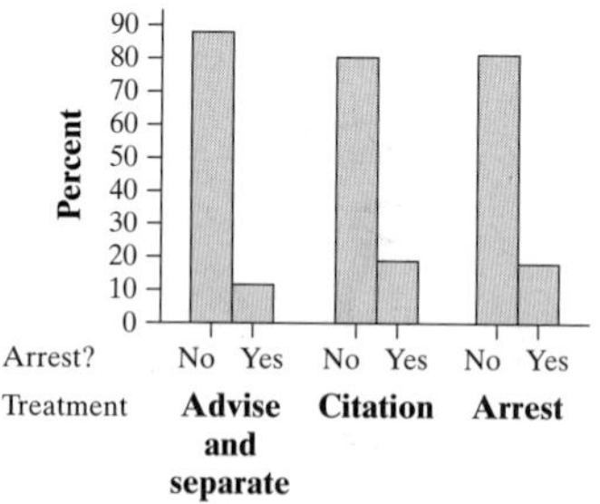

(c) The alternative hypothesis is that at least one p_i is different from the others. **(d)** If the proportion of subsequent arrests is the same for all three techniques, we have a 7.96% chance of getting a difference between the three groups as large as or larger than the one observed ($\chi^2 = 5.063$) in this study due to the random assignment. We fail to reject H_0. We do not have enough evidence to say that the techniques lead to different subsequent arrest rates.
T11.13 **(a) State:** We want to perform a test of

H_0: Smoking status and education are independent among French men aged 20 to 60

H_a: Smoking status and education are not independent among French men aged 20 to 60

at the $\alpha = 0.05$ level. **Plan:** Use a chi-square test for association/independence if the conditions are satisfied. *Random:* The data came from a random sample. The expected counts are 59.48, 50.93, 42.37, 34.22, 44.21, 37.85, 31.49, 25.44, 42.31, 36.22, 30.14, 24.34. All these counts are at least 5. *Independent:* There are at least 4590 French men aged 20 to 60. **Do:** The test statistic is $\chi^2 = 13.305$ and the *P*-value is 0.0384 using df $= 6$. **Conclude:** Since the *P*-value is less than 0.05, we reject H_0 and conclude that education and smoking status among French men aged 20 to 60 are not independent. **(b)** The largest contribution is from the university-educated males who are heavy smokers—there are far fewer of them than expected.

CHAPTER 12

Section 12.1

Answers to Check Your Understanding

Page 750 **1.** The scatterplot must show a linear pattern, the observations must be independent, the residuals need to be approximately Normally distributed, the residuals must show roughly equal scatter for all *x*-values, and the observations must be produced by random sampling or a randomized experiment. **2.** Since there are eight observations, there are 6 degrees of freedom and $t^* = 2.447$. This means that the confidence interval is $0.0908 \pm 2.447(0.02831) = 0.0908 \pm 0.06927 = (0.02153, 0.16007)$. We are 95% confident that the interval from 0.02153 to 0.16007 captures the true slope of the population regression line relating pack weight *y* and body weight *x* among ninth-grade students at the Webb Schools. *Page 755* **1. State:** We want to perform a test of H_0: $\beta = 0$ versus H_a: $\beta \neq 0$, where β is the true slope of the population regression line relating body weight to backpack weight. We will use $\alpha = 0.05$. **Plan:** If the conditions are met, we will do a *t* test for the slope β. We are assuming that the conditions are met here. **Do:** According to the output, the test statistic is $t = 3.21$ and the *P*-value is 0.018 using df $= 6$. **Conclude:** Since the *P*-value is less than 0.05, we reject the null hypothesis and conclude that there is convincing evidence of a linear relationship between body weight and backpack weight for ninth-grade students at the school.
2. Yes. The 99% confidence interval would contain 0 because the *P*-value for the two-sided test is greater than 0.01.

Answers to Odd-Numbered Section 12.1 Exercises

1. No. The conditions for performing inference are not met. The variance of the residuals increases as the laboratory measurement increases.
3. *Linear:* The residual plot can reasonably be considered to be centered on 0, so the scatterplot is approximately linear. *Independent:* Due to the random assignment, the observations can be viewed as independent. *Normal:* The histogram is mound shaped and approximately symmetric, so the residuals could follow a Normal distribution. *Equal variance:* The residual plot shows roughly equal scatter for all *x*-values. *Random:* This was a randomized experiment. The conditions are met.
5. α is the *y* intercept; it would measure the proportion of fish killed if there were 0 fish in the tank to begin with. Obviously, this is extrapolation since the smallest number of fish was 10. The estimate is $a = 0.12$. β is the slope; it tells us how much the proportion of fish killed either increases or decreases, on average, per 1 fish increase in the tank. The estimate is $b = 0.0086$. Finally, σ measures the standard deviation of the proportion killed values about the population regression line. The estimate is $s = 0.1886$.
7. **(a)** If we repeated the experiment many times, the slope of the sample regression line would typically vary by about 0.002456 from the true slope of the population regression line for predicting proportion of fish eaten from the number of fish available. **(b)** $t^* = 1.761$ using df $= 14$. The confidence interval is $0.00857 \pm 1.761(0.002456) = 0.00857 \pm 0.00433 = (0.00424, 0.0129)$. **(c)** We are 90% confident that the interval from 0.00424 to 0.0129 captures the true slope of the population regression line for predicting proportion of fish eaten from the number of fish available. **(d)** If we were to repeat the experiment many times and compute confidence intervals for the regression slope in each case, about 90% of the resulting intervals would contain the slope of the population regression line.
9. **State:** We want to construct a 99% confidence interval for the true slope β of the population regression line relating number of stumps to number of clusters of beetle larvae. **Plan:** Since the conditions are met, we will do a *t* interval for the slope β. **Do:** $t^* = 2.831$ using df $= 21$. The confidence interval is $11.894 \pm 2.831(1.136) = 11.894 \pm 3.216 = (8.678, 15.11)$. **Conclude:** We are 99% confident that the interval from 8.678 to 15.11 captures the true slope of the population regression line for predicting clusters of beetle larvae from the number of stumps.
11. **(a)** $-1.286 + 11.894(5) = 58.184$ clusters **(b)** We would expect our predictions to be off, on average, by about $s = 6.419$ clusters of beetle larvae.
13. **(a)** There is a somewhat weak, negative linear relationship between the number of weeds per meter and the corn yield of the plots. **(b)** Predicted corn yield $= 166.483 - 1.0987$(weeds per meter) **(c)** *Slope:* for each additional weed per meter, we expect the average corn yield to decrease by 1.0987 bushels; *y intercept*: if there are no weeds per meter, we would predict a corn yield of 166.483 bushels. **(d) State:** We want to perform a test at $\alpha = 0.05$ of $H_0: \beta = 0$ versus $H_a: \beta < 0$, where β is the true slope of the population regression line relating weeds per meter to corn yield. **Plan:** If the conditions are met, do a *t* test for β. (Graphs not shown.) *Linear:* The scatterplot is approximately linear. *Independent:* Due to the random assignment, the observations can be viewed as independent. *Normal:* A histogram of the residuals is mound shaped and approximately symmetric, so the residuals could follow a Normal distribution. *Equal variance:* The residual plot shows roughly equal scatter for all *x*-values. *Random:* This was a randomized experiment. **Do:** From the output, $t = -1.92$ and the one-sided *P*-value is 0.0375 using df $= 14$. **Conclude:** Since the *P*-value is less than 0.05, we reject H_0 and conclude that there is convincing evidence of a negative linear relationship between weeds per meter and corn yield.
15. **(a) Do:** The 90% confidence interval for β is $-1.0987 \pm 1.761(0.5712) = -1.0987 \pm 1.0059 = (-2.1046, -0.0928)$. **Conclude:** We are 90% confident that the interval from -2.1046 to -0.0928 captures the true slope of the population regression line for predicting corn yield from weeds per meter. The one-sided test at $\alpha = 0.05$ in Exercise 13 is equivalent to a 90% confidence interval. From the

interval, 0 is not a plausible value for the true slope, which matches our reject H_0 conclusion. **(b)** The typical error when using the regression line to predict corn yield is about 7.98 bushels. About 20.9% of the variation in corn yield can be explained by the linear relationship with the number of weeds per meter. If this experiment were done many times, the estimated slope would differ from the population slope by an average of 0.5712.

17. **(a)** The interval is $0.7902 \pm 2.977(0.0710) = 0.7902 \pm 0.2114 = (0.5788, 1.0016)$. (The slight differences are due to rounding.) **(b)** If one method of measuring wear gives an increase of 1 unit, we hope that the other way of measuring wear will also give an increase of 1 unit. This translates to a slope of 1. **(c)** Since the interval in part (a) includes 1, the slope could plausibly be 1. So we would not reject H_0 that the slope is 1.

19. **(a) State:** We want to test $H_0: \beta = 0$ versus $H_a: \beta < 0$, where β is the true slope of the population regression line relating wine consumption to heart disease death rate at $\alpha = 0.05$. **Plan:** If the conditions are met, do a t test for β. (Graphs not shown.) *Linear:* The scatterplot is approximately linear. *Independent:* There are more than 190 countries in the world. *Normal:* A histogram of the residuals is mound shaped and approximately symmetric, so the residuals could follow a Normal distribution. *Equal variance:* The residual plot shows a roughly equal scatter for all x-values. *Random:* The data come from a random sample. **Do:** From the data, $t = -6.46$ and the P-value is approximately 0 using df = 17. **Conclude:** Since the P-value is less than 0.05, we reject H_0 and conclude that there is convincing evidence of a negative linear relationship between wine consumption and death rate due to heart disease. **(b)** The 95% confidence interval for β is $-22.969 \pm 2.11(3.557) = -22.969 \pm 7.505 = (-30.474, -15.464)$. **Conclude:** We are 95% confident that the interval from -30.474 to -15.464 captures the true slope of the population regression line for predicting heart disease death rate from wine consumption.

21. c

23. c

25. b

27. **(a)** Each student was assigned the two treatments (read the words and say the print color) in random order. **(b)** He used a randomized block design wherein each student was a block (this could also be called a matched pairs design since there were only two treatments per block). He did this to help control for the different abilities of students to read the words (or say the color in which they were printed) with distractions. **(c)** The random assignment was used to help average out the effects of the order in which people did the two treatments.

29. A boxplot of the data shows that there is an outlier, so with only 16 observations, the Normal condition is not satisfied.

31. **(a)** 0.1933; 0.3827; 0.6951 **(b)** No. $P(\text{snowmobile owner}) = 0.1933$, $P(\text{snowmobile owner} \mid \text{environmental club}) = 0.052$. **(c)** $P(\text{both are owners}) = (0.1933)^2 = 0.0374$. $P(\text{at least one belongs to an environmental organization}) = 1 - P(\text{neither belongs}) = 1 - (1 - 0.1999)^2 = 0.3598$.

Section 12.2

Answers to Check Your Understanding

Page 776 **1.** The scatterplot of the original data shows clear curvature. But the scatterplot of years and ln(population) seems quite linear. This suggests that an exponential model would fit the original data. **2.** $\widehat{\ln y} = -1.193 + 0.0287x$, where y is the population and x is years since 1700. **3.** If we want to predict for 1890, that means $x = 1890 - 1700 = 190$. So $\widehat{\ln y} = -1.193 + 0.0287(190) = 4.26$. The predicted population is $\hat{y} = e^{4.26} = 70.81$ million people. **4.** The residual plot suggests that the residual for this point would be negative. This means that our prediction is too high.

Answers to Odd-Numbered Section 12.2 Exercises

33.

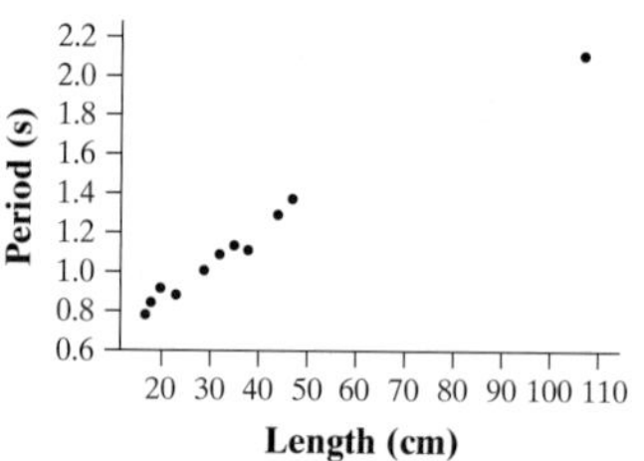

The scatterplot shows a fairly strong, positive, slightly curved relationship between length and period with one very unusual point (106.5, 2.115) in the top-right corner. **(b)** The class used the square root of the length as its explanatory variable and period as its response variable. **(c)** The class used the length as the explanatory variable and the square of the period as its response variable.

35. **(a)** For Transformation 1: $\hat{y} = -0.08594 + 0.21\sqrt{x}$, where y is the period and x is the length. For Transformation 2: $\widehat{y^2} = -0.155 + 0.0428x$, where y is the period and x is the length. **(b)** For Transformation 1: $\hat{y} = -0.08594 + 0.21\sqrt{80} = 1.792$ seconds. For Transformation 2: $\widehat{y^2} = -0.155 + 0.0428(80) = 3.269$, so $\hat{y} = \sqrt{3.269} = 1.808$ seconds. **(c)** For Transformation 1: The typical distance that a predicted value of the period will be from the actual value is about 0.046 seconds. For Transformation 2: The typical distance that the square of a predicted value of the period will be from the square of the actual value is about 0.105 seconds.

37. **(a)**

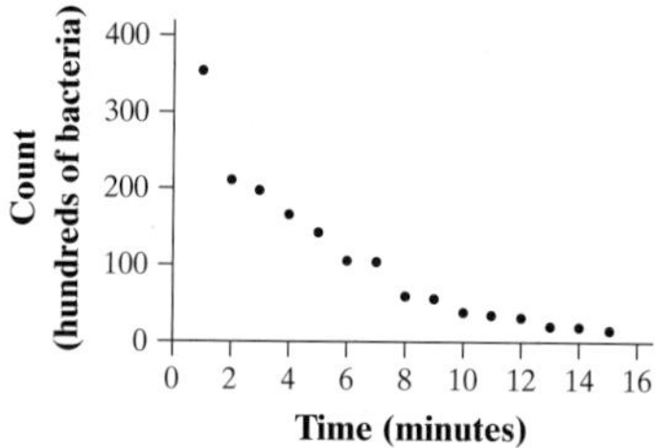

The relationship is strong, negative, and slightly curved with one outlier in the top-left corner. **(b)** Since the graph of the explanatory variable against the natural log of the response is fairly linear, an exponential model would be reasonable. **(c)** $\widehat{\ln y} = 5.973 - 0.218x$, where y is the count of surviving bacteria and x is time in minutes. **(d)** $\widehat{\ln y} = 5.973 - 0.218(17) = 2.267$, so $\hat{y} = e^{2.267} = 9.65$, or 965 bacteria. Since the residual plot shows a random scatter around the value 0, we'd expect this prediction to be about right.

39. If Sasquatch has a body weight of 127 kilograms, we find that $\widehat{\log y} = 1.01 + 0.72\log(127) = 2.525$. This means that $\hat{y} = 10^{2.525} = 334.97$ grams is the estimated brain weight of Sasquatch.

41. **(a)** The power model would work better because the graph of log(abundance) versus log(body mass) is linear, whereas the graph of log(abundance) versus body mass is clearly curved.

(b) $\widehat{\log y} = 1.9503 - 1.0481 \log x$, where y is the abundance (per 10,000 kg of prey) and x is the body mass (kg). (c) $\widehat{\log y} = 1.9503 - 1.0481 \log(92.5) = -0.1104$, so $\hat{y} = 10^{-0.1104} = 0.7755$ per 10,000 kg of prey. (d) The residual plot is randomly scattered about 0. The model looks like it is a good fit.

43. (a)

There is a strong, positive, slightly curved relationship between height and distance. (b) A scatterplot of ln(distance) versus height looks curved, while a scatterplot of ln(distance) versus ln(height) looks linear. So a power model would describe the relationship between distance and height. (c) The equation is $\widehat{\ln y} = 3.7514 + 0.5152 \ln x$, where y is the distance and x is the height. (d) If the ramp height was 700, $\widehat{\ln y} = 3.7514 + 0.5152 \ln(700) = 7.1265$ and $\hat{y} = e^{7.1265} = 1244.53$.

45. c

47. e

49. (a) A point is considered an outlier if it is more than $1.5IQR$ above Q_3. On a standard Normal curve, $Q_3 + 1.5IQR = 0.67 + 1.5(1.34) = 2.68$. Since $\left(\frac{7-4.5}{0.9}\right) = 2.78$, Marcella's shower would be classified as an outlier. (b) The probability that she takes a 7-minute shower in any given day is

$$P(X > 7) = P\left(Z > \frac{7-4.5}{0.9}\right) = P(Z > 2.78) = 0.0027$$

$$\begin{aligned} &P(\text{shower is 7 minutes or more on at least 2 of 10 days}) \\ &= 1 - P(\text{shower is 7 minutes or more on 0 or 1 of 10 days}) \\ &= 1 - (1-0.0027)^{10} - 10(0.0027)(1-0.0027)^9 \\ &= 1 - 0.9733 - 0.0264 = 0.0003 \end{aligned}$$

(c) $$P(\bar{x} > 5) = P\left(z > \frac{5-4.5}{0.9/\sqrt{10}}\right) = P(z > 1.76) = 0.0392$$

51. (a) Answers will vary. A possible response would be that it would be hard to find all the individual teachers in the sample since they were spread out among 40 different sessions. (b) A stratified random sample would allow the statistics teachers to make sure that all subjects were included in appropriate proportions.

Answers to Chapter 12 Review Exercises

R12.1 (a) There is a moderately strong, positive, linear relationship between the thickness and the velocity. (b) $\hat{y} = 70.44 + 274.78x$, where y is the velocity and x is the thickness. (c) The predicted velocity is $\hat{y} = 70.44 + 274.78(0.4) = 180.352$ feet/second. The residual is $104.8 - 180.352 = -75.552$, so the line overpredicts the velocity by 75.552 feet/second. (d) The linear model is appropriate. The scatterplot shows a linear relationship and the residual plot shows a random scatter. (e) *Slope:* For each increase of an inch in thickness, we expect an average increase of 274.78 feet/second for the velocity. s: The typical prediction will vary from the actual value by about 56.36 feet/second. r^2: About 49.3% of the variation in velocity is explained by the linear relationship with thickness. *Standard error of the slope:* If we take many different random samples of 12 pistons and compute the least-squares regression line for each one, the estimated slope will vary from the slope of the population regression line for predicting velocity from thickness by 88.18, on average.

R12.2 **State:** We want to perform a test at $\alpha = 0.05$ of $H_0: \beta = 0$ versus $H_a: \beta \neq 0$, where β is the true slope of the population regression line relating thickness to velocity. **Plan:** If the conditions are met, we will do a t test for β. *Linear:* The scatterplot is approximately linear. *Independent:* There are at least 120 pistons in the population. *Normal:* We can't check this. *Equal variance:* The residual plot shows a roughly equal scatter for all x-values. *Random:* The data come from a random sample. **Do:** The test statistic is $t = \frac{b-0}{s_b} = \frac{274.78}{88.18} = 3.116$ and the P-value is 0.0109 using df = 10. **Conclude:** Since the P-value is less than 0.05, we reject H_0 and conclude that there is convincing evidence of a linear relationship between thickness and velocity.

R12.3 **State:** We want to construct a 95% confidence interval for β. **Plan:** Since the conditions are met, we will construct a t interval for the slope β. **Do:** Using df = 10, $t^* = 2.228$. The confidence interval is $274.78 \pm 2.228(88.18) = 274.78 \pm 196.465 = (78.315, 471.245)$. **Conclude:** We are 95% confident that the interval from 78.315 to 471.245 captures the true slope of the population regression line for predicting velocity from thickness. This interval is consistent with the test in Exercise R12.2 because 0 is not a plausible value for the slope of the population regression line.

R12.4 There is clear curvature to the scatterplot and the residual plot. There may also be a problem with the equal-variance condition.

R12.5 (a) The transformation did achieve linearity because the residuals are scattered randomly above and below the value 0 in the residual plot. (b) The equation is $\hat{y} = -0.000595 + 0.3\frac{1}{x^2}$, so for a bulb at a distance of 2.1 meters, we would predict an intensity of $\hat{y} = -0.000595 + 0.3\frac{1}{(2.1)^2} = 0.0674$ candelas.

R12.6 (a) A possible residual plot is given below.

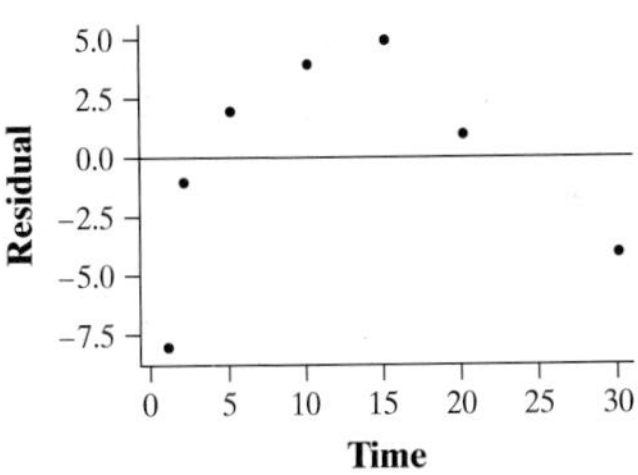

(b) There is clear curvature to these data, so a straight line does not make sense as a model. (c) A power model is more appropriate because the graph of ln(recall) versus ln(time) is linear, while the graph of ln(recall) versus time is curved. (d) The regression equation is $\widehat{\ln y} = 3.48 + 0.293 \ln x$, where y is the recall and x is the time. So for a time of 25 seconds, we predict $\widehat{\ln y} = 3.48 + 0.293 \ln(25) = 4.423$ and $\hat{y} = e^{4.423} = 83.36$ words recalled.

Answers to Chapter 12 AP Statistics Practice Test

T12.1 c. There is no sample size limitation.

T12.2 b. This looks like a power model would make sense.

T12.3 d. The appropriate distribution is t with $n - 2$ degrees of freedom.

T12.4 a. The correlation is the square root of r^2 and has the same sign as the slope.

T12.5 d. The test statistic is the coefficient divided by the standard error of the coefficient.

T12.6 d. The *P*-value refers to the population relationship, not the sample relationship.

T12.7 e. With 67 degrees of freedom, the appropriate t^* is 2.00.

T12.8 d. These are the two outliers at the top of the graph.

T12.9 d. For power models, the graph that is linear is the logarithm of y against the logarithm of x.

T12.10 c. $\widehat{\log y} = -13.5 + 0.01(2020) = 6.7$ so $\hat{y} = 10^{6.7} = 5{,}011{,}872$.

T12.11 **(a)** $\hat{y} = 4.546 + 4.832x$, where y is the weight gain and x is the dose of growth hormone. **(b)** *Slope:* For each 1-milligram increase in growth hormone, we expect 4.832 ounces of weight gain, on average. y intercept: If a chicken is given no growth hormone, we expect 4.54 ounces of weight gain. s: The typical predicted weight gain will vary from the actual weight gain by about 3.13 ounces. *Standard error of the slope:* If we repeated this experiment many times, the estimated slope will vary from the population slope to predict weight gain from amount of growth hormone by an average of 1.0164 ounces/milligram. r^2: About 38.4% of the variability in weight gain is explained by the linear relationship with the amount of growth hormone given. **(c) State:** We want to perform a test at $\alpha = 0.05$ of $H_0: \beta = 0$ versus $H_a: \beta \neq 0$, where β is the true slope of the population regression line relating amount of growth hormone to weight gain. **Plan:** Since the conditions are met, we will do a t test for β. **Do:** From the output, $t = 4.75$ and the *P*-value is 0.0004 using df $= 13$. **Conclude:** Since the *P*-value is less than 0.05, we reject H_0 and conclude that there is convincing evidence of a linear relationship between amount of growth hormone and weight gain. **(d) State:** We want to construct a 95% confidence interval for β. **Plan:** Since the conditions are met, we will construct a t interval for the slope β. **Do:** Using df $= 13$, $t^* = 2.160$ and the confidence interval is $4.8323 \pm 2.160(1.0164) = 4.8323 \pm 2.195 = (2.6373, 7.0273)$. **Conclude:** We are 95% confident that the interval from 2.6373 to 7.0273 captures the true slope of the population regression line for predicting weight gain from amount of growth hormone.

T12.12 **(a)** There is clear curvature evident in both the scatterplot and the residual plot. Also, there may be a problem with the equal-variance condition as seen in the residual plot. **(b)** Option 1: $\hat{y} = 2.078 + 0.0042597x^3$. We would predict the following number of board feet from a tree with a diameter of 30 inches: $\hat{y} = 2.078 + 0.0042597(30)^3 = 117.09$. Option 2: $\widehat{\ln y} = 1.2319 + 0.113417x$. For $x = 30$, we get $\widehat{\ln y} = 1.2319 + 0.113417(30) = 4.63441$ and $\hat{y} = e^{4.63441} = 102.967$ feet. **(c)** The residual plot for Option 2 still shows a large amount of curvature, so the better option is Option 1.

Answers to Cumulative AP Practice Test 4

AP4.1 e
AP4.2 c
AP4.3 e
AP4.4 a
AP4.5 b
AP4.6 e
AP4.7 d
AP4.8 a
AP4.9 e
AP4.10 a
AP4.11 b
AP4.12 d
AP4.13 e
AP4.14 d
AP4.15 b
AP4.16 a
AP4.17 d
AP4.18 c
AP4.19 a
AP4.20 b
AP4.21 e
AP4.22 e
AP4.23 b
AP4.24 c
AP4.25 d
AP4.26 e
AP4.27 c
AP4.28 e
AP4.29 c
AP4.30 b
AP4.31 d
AP4.32 a
AP4.33 e
AP4.34 b
AP4.35 a
AP4.36 b
AP4.37 d
AP4.38 a
AP4.39 b
AP4.40 d

AP4.41 **State:** We want to perform a test of

$$H_0: \mu_1 - \mu_2 = 0$$
$$H_a: \mu_1 - \mu_2 \neq 0$$

where

$\mu_1 =$ mean difference in electrical potential for diabetic mice (right hip − front feet)

$\mu_2 =$ mean difference in electrical potential for normal mice (right hip − front feet)

at the 5% significance level. **Plan:** If conditions are met, we should carry out a two-sample t test.

- **Random** The researchers randomly selected diabetic mice and normal mice for the study.
- **Normal** We are told that graphs of the data show no outliers or strong skewness, so we should be safe using t procedures even if the two population distributions aren't perfectly Normal.
- **Independent** The data came from independent samples of diabetic and normal mice. Knowing one mouse's difference in electrical potential should give no additional information about another mouse's difference in electrical potential, so the individual measurements should be independent. There are more than $10(24) = 240$ diabetic mice and $10(18) = 180$ normal mice.

Do: We will conduct a two-sample t test for $\mu_1 - \mu_2$ on our calculator.

- **Test statistic** $t = 2.549$
- ***P*-value** Using df $= 38.46$, the P-value is 0.0149.

```
2-SampTTest
  μ1≠μ2
  t=2.549830747
  p=.0148807394
  df=38.45983024
  x̄1=13.09
↓ x̄2=10.022
```

Conclude: Since the P-value of 0.015 is less than 0.05, we reject the null hypothesis and conclude that diabetic mice and normal mice differ in the mean electrical potential between their right hip and front feet.

AP4.42 **(a)** We want to test

$$H_0: p_1 - p_2 = 0$$
$$H_a: p_1 - p_2 < 0$$

where

p_1 = proportion of women who were physically active as teens who would suffer a cognitive decline

p_2 = proportion of women who were sedentary as teens who would suffer a cognitive decline

(b) You would perform a two-sample z test for $p_1 - p_2$ (a two-proportion z test). **(c)** No. Because the participants were mostly white women, the findings may not be generalizable to women in other racial and ethnic groups. **(d)** Two variables are confounded when their effects on the response variable (measure of cognitive decline) cannot be distinguished from one another. For example, women who were physically active as teens might have also done other things like engaging in mentally stimulating activities that those who were sedentary during their teen years did not do. We would be unable to determine if it was their physically active youth or the other activities that slowed their level of cognitive decline.

AP4.43 **(a)** Since the first question called it a "fat tax," people may have reacted negatively because they believe this is a tax on those who are overweight. The second question provides extra information that gets people thinking about the obesity problem in the United States and the increased health care that could be provided as a benefit with the tax money, which might make them respond more positively to the proposed tax. The question should be worded in a more straightforward manner. For example, "Would you support or oppose a tax on nondiet sugared soda?" **(b)** This method samples only people at fast-food restaurants. They may go because they like the sugary drinks and thus don't want to pay a tax on their favorite beverages. It is likely that the proportion of those who would oppose such a tax will be overestimated with this method. A random sample of all New York State residents should be taken to provide a better estimate of the level of support for such a tax. **(c)** Use a stratified random sampling method in which each state is a stratum. That is, take a random sample of residents within each state.

AP4.44 Let's start by making a tree diagram that shows the various probabilities. Let W = Mr. Worcester arrives first, L = Mr. Legacy arrives first, C = Dr. Currier arrives first, and S = the coffee is strong.

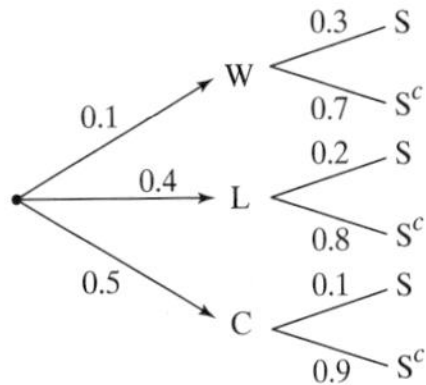

(a) $P(S) = (0.1)(0.3) + (0.4)(0.2) + (0.5)(0.1) = 0.16$

(b) $P(C|S) = \dfrac{P(C \cap S)}{P(S)} = \dfrac{(0.5)(0.1)}{(0.1)(0.3)+(0.4)(0.2)+(0.5)(0.1)} = \dfrac{0.05}{0.16} = 0.3125$

AP4.45 **(a)** A linear model is not appropriate. The scatterplot exhibits a strong curved pattern. **(b)** Model B would be better to use if you wanted to predict seed weight from seed count. The scatterplot of ln(weight) versus ln(count) shows a much more linear pattern, and its residual plot shows a random scatter about the residual = 0 line.

(c) $\ln(\text{weight}) = 15.491 - 1.5222 \ln(3700) \Rightarrow \ln(\text{weight}) = 2.984$

$$\text{weight} = e^{2.984} \approx 19.77 \text{ mg}$$

(d) About 86.3% of the variability in ln(seed weight) is accounted for by the least-squares regression line using ln(seed count) as the explanatory variable.

AP4.46 **(a)** Shape: Normal. Center: $\mu_{\bar{x}} = 4$ inches. Spread: $\sigma_{\bar{x}} = \dfrac{0.02}{\sqrt{25}} = 0.004$ inches. **(b)** $P(\bar{x} < 3.99) + P(\bar{x} > 4.01) = 1 -$ normalcdf (3.99, 4.01, 4.00, .004) $= 0.0124$ **(c)** $P(4.00 < \bar{x} < 4.01) =$ normalcdf (4, 4.1, 4, .004)$= 0.4938$ **(d)** Let X = the number of samples (out of 5) in which the mean is between 4.00 and 4.01. The random variable X is binomially distributed with $n = 5$ and $p = 0.4938$. $P(X \geq 4) = 1 - P(X < 4) = 1 - P(X \leq 3)$ = binomcdf(5, .4938, 3) = 0.1798 **(e)** Since $P(\bar{x} < 3.99) + P(\bar{x} > 4.01) = 0.0124$ from part (b), whereas $P(X \geq 4) = 0.1798$ from part (d), Method 1 would give more convincing evidence that the machine needs to be shut down, since it is much less likely to occur by chance if the machine is working correctly. **(f)** Answers will vary. For example, stop the production process if there are 5 consecutive sample means between 4.00 and 4.01. Using this rule, probability $= (0.4938)^5 = 0.029$. An analogous rule would be to consider 5 consecutive sample means between 3.99 and 4.00. Another possibility is to consider getting 2 consecutive samples means between 4.005 and 4.01. Note that $P(4.005 < \bar{x} < 4.01) = 0.0994$. If 2 consecutive samples are taken, then P(both between 4.005 and 4.01) $= (0.0994)^2 = 0.0099 \approx 0.01$.

Index

Appendix A: AP Exam Tips

CHAPTER 1

• If you learn to distinguish categorical from quantitative variables now, it will pay big rewards later. The type of data determines what kinds of graphs and which numerical summaries are appropriate. You will be expected to analyze categorical and quantitative data effectively on the AP exam. (page 3)

• When comparing distributions of quantitative data, it's not enough just to list values for the center and spread of each distribution. You have to explicitly *compare* these values, using words like "greater than," "less than," or "about the same as." (page 32)

• If you're asked to make a graph on a free-response question, be sure to label and scale your axes. Unless your calculator shows labels and scaling, don't just transfer a calculator screen shot to your paper. (page 39)

• You may be asked to determine whether a quantitative data set has any outliers. Be prepared to state and use the rule for identifying outliers. (page 58)

• Use statistical terms carefully and correctly on the AP exam. Don't say "mean" if you really mean "median." Range is a single number; so are Q_1, Q_3, and *IQR*. Avoid colloquial use of language, like "the outlier *skews* the mean." Skewed is a shape. If you misuse a term, expect to lose some credit. (page 67)

CHAPTER 2

• Don't use "calculator speak" when showing your work on free-response questions. Writing `normalcdf(305,325,304,8)` will *not* earn you credit for a Normal calculation. At the very least, you must indicate what each of those calculator inputs represents. For example, "I used normalcdf on my calculator with lower bound 305, upper bound 325, mean 304, and standard deviation 8." Better yet, sketch and label a Normal curve to show what you're finding. (page 124)

• Normal probability plots are not included on the AP Statistics course outline. However, these graphs are very useful tools for assessing Normality. You may use them on the AP exam if you wish—just be sure that you know what you're looking for (linear pattern). (page 127)

CHAPTER 3

• If you're asked to make a scatterplot on a free-response question, be sure to label and scale both axes. *Don't* copy an unlabeled calculator graph directly onto your paper. (page 150)

• If you're asked to interpret a correlation, start by looking at a scatterplot of the data. Then be sure to address direction, form, strength, and outliers (sound familiar?) and to put your answer in context. (page 154)

• There's no firm rule for how many decimal places to show for answers on the AP exam. *Our advice:* Give your answer correct to two or three nonzero decimal places. *Exception:* If you're using one of the tables in the back of the book, give the value shown in the table. (page 169)

• The formula sheet for the AP exam uses different notation for the equations of the slope and y intercept of the least-squares regression line: $b_1 = r\frac{s_y}{s_x}$ and $b_0 = \bar{y} - b_1\bar{x}$. That's because the least-squares line is written as $\hat{y} = b_0 + b_1x$. We prefer our simpler version without the subscripts: $\hat{y} = a + bx$. (page 172)

• Students often have a hard time interpreting the value of r^2 on AP exam questions. They frequently leave out key words in the definition. Our advice: Treat this as a fill-in-the-blank exercise. Write "____% of the variation in [response variable name] is accounted for by the regression line." (page 182)

• Don't forget to put a "hat" on the response variable when you write a regression equation. Calculator and computer output for regression usually doesn't do this. For the regression in Figure 3.17 (page 182), you should write

$$\widehat{\text{fat gain}} = 3.505 - 0.00344(\text{NEA})$$

CHAPTER 4

• If you're asked to describe how the design of a study leads to bias, you're expected to identify the *direction* of the bias. Suppose you were asked, "Explain how using a convenience sample of students in your statistics class to estimate the proportion of all high school students who own a graphing calculator could result in bias." You might respond, "This sample would probably include a much higher proportion of students with a graphing calculator than in the population at large because a graphing calculator is required for the statistics class. That is, this method would probably lead to an overestimate of the actual population proportion." (page 210)

• If you're asked to identify a possible confounding variable in a given setting, you are expected to explain how the variable you choose (1) is associated with the explanatory variable and (2) affects the response variable. (page 233)

• If you're asked to describe the design of an experiment on the AP exam, you won't get full credit for a diagram like Figure 4.5 (page 239). You are expected to describe how the treatments are assigned to the experimental units and to clearly state what will be measured or compared. Some students prefer to start with a diagram and then add a few sentences. Others choose to skip the diagram and put their entire response in narrative form. (page 239)

• Don't mix the language of experiments and the language of sample surveys or other observational studies. You will lose credit for saying things like "use a randomized block design to select the sample for this survey" or "this experiment suffers from nonresponse since some subjects dropped out during the study." (page 248)

CHAPTER 5

• On the AP exam, you may be asked to describe how you will perform a simulation using rows of random digits. If so, provide a clear enough description of your simulation process for the reader to get the same results you did from *only* your written explanation. (page 291)

• Many probability problems involve simple computations that you can do on your calculator. It may be tempting to just write down your final answer without showing the supporting work. Don't do it! A "naked answer," even if it's correct, will usually earn you no credit on a free-response question. (page 308)

• On probability questions, you may usually choose whether to use words or symbols when showing your work. You can write statements like $P(A \mid B)$ if events A and B are defined clearly, or you can use a verbal equivalent, such as P(reads *New York Times* | reads *USA Today*). Use the approach that makes the most sense to you. (page 325)

CHAPTER 6

• If the mean of a random variable has a non-integer value, but you report it as an integer, your answer will be marked as incorrect. (page 346)

• When you solve problems involving random variables, start by defining the random variable of interest. For example, let X = the Apgar score of a randomly selected baby or let Y = the height of a randomly selected young woman. Then state the probability you're trying to find in terms of the random variable: $P(X \geq 7)$ or $P(68 \leq Y \leq 70)$. (page 352)

• If you have trouble solving problems involving sums and difference of Normal random variables with the algebraic methods of Section 6.2, use the simulation strategy from the Technology Corner on page 376 to earn some (or possibly full) credit. (page 376)

• Don't rely on "calculator speak" when showing your work on free-response questions. Writing `binompdf(5,0.25,3)` = 0.08789 will *not* earn you full credit for a binomial probability calculation. At the very least, you must indicate what each of those calculator inputs represents. For example, "I used `binompdf` on my calculator with $n = 5$, $p = 0.25$, and $k = 3$." Better yet, show the binomial probability formula with these numbers plugged in. (page 389)

CHAPTER 7

• Terminology matters. Don't say "sample distribution" when you mean sampling distribution. You will lose credit on free-response questions for misusing statistical terms. (page 420)

• Notation matters. The symbols $\hat{p}$, $\bar{x}$, p, μ, σ, $\mu_{\hat{p}}$, $\sigma_{\hat{p}}$, $\mu_{\bar{x}}$, and $\sigma_{\bar{x}}$ all have specific and different meanings. Either use notation correctly—or don't use it at all. You can expect to lose credit if you use incorrect notation. (page 444)

CHAPTER 8

• On a given problem, you may be asked to interpret the confidence interval, the confidence level, or both. Be sure you understand the difference: the confidence level describes the long-run capture rate of the method and the confidence interval gives a set of plausible values for the parameter. (page 476)

• If a free-response question asks you to construct and interpret a confidence interval, you are expected to do the entire four-step process. That includes clearly defining the parameter and checking conditions. (page 491)

• You may use your calculator to compute a confidence interval on the AP exam. But there's a risk involved. If you just give the calculator answer with no work, you'll get either full credit for the "Do" step (if the interval is correct) or no credit (if it's wrong). We recommend showing the calculation with the appropriate formula and then checking with your calculator. If you opt for the calculator-only method, be sure to name the procedure (e.g., one-proportion z interval) and to give the interval (e.g., 0.514 to 0.606). (page 492)

CHAPTER 9

• The conclusion to a significance test should always include three components: (1) an explicit comparison of the *P*-value to a stated significance level OR an interpretation of the *P*-value as a conditional probability, (2) a decision about the null hypothesis: reject or fail to reject H_0, and (3) an explanation of what the decision means in context. (page 536)

• When a significance test leads to a fail to reject H_0 decision, as in this example, be sure to interpret the results as "we don't have enough evidence to conclude H_a." Saying anything that sounds like you believe H_0 is (or might be) true will lead to a loss of credit. And don't write text-message-type responses, like "FTR the H_0." (page 554)

• You can use your calculator to carry out the mechanics of a significance test on the AP exam. But there's a risk involved. If you just give the calculator answer with no work, and one or more of your values is incorrect, you will probably get no credit for the "Do" step. We recommend doing the calculation with the appropriate formula and then checking with your calculator. If you opt for the calculator-only method, be sure to name the procedure (one-proportion z test) and to report the test statistic ($z = 1.15$) and *P*-value (0.1243). (page 556)

• Remember: if you just give calculator results with no work, and one or more values are wrong, you probably won't get any credit for the "Do" step. We recommend doing the calculation with the appropriate formula and then checking with your calculator. If you opt for the calculator-only method, name the procedure (t test) and report the test statistic ($t = -0.94$), degrees of freedom (df = 14), and *P*-value (0.1809). (page 573)

CHAPTER 10

• You may use your calculator to compute a confidence interval on the AP exam. But there's a risk involved. If you just give the calculator answer with no work, you'll get either full credit for the "Do" step (if the interval is correct) or no credit (if it's wrong). If you opt for the calculator method, be sure to name the procedure (e.g., two-proportion z interval) and to give the interval (e.g., 0.223 to 0.297). (page 611)

• When checking the Normal condition on an AP exam question involving inference about means, be sure to include a graph.

Don't expect to receive credit for describing a graph that you made on your calculator but didn't put on paper. (page 640)

• When a significance test leads to a fail to reject H_0 decision, as in the previous example, be sure to interpret the results as "We don't have enough evidence to conclude H_a." Saying anything that sounds like you believe H_0 is (or might be) true will lead to a loss of credit. (page 642)

CHAPTER 11

• In the "Do" step, you aren't required to show every term in the chi-square statistic. Writing the first few terms of the sum and the last term, separated by ellipsis, is considered as "showing work." We suggest that you do this and then let your calculator tackle the computations. (page 702)

• If you have trouble distinguishing the two types of chi-square tests for two-way tables, you're better off just saying "chi-square test" than choosing the wrong type. Better yet, learn to tell the difference! (page 718)

CHAPTER 12

• The AP exam formula sheet gives $\hat{y} = b_0 + b_1x$ for the equation of the sample (estimated) regression line. We will stick with our simpler notation, $\hat{y} = a + bx$, which is also used by TI calculators. Just remember: the coefficient of x is always the slope, no matter what symbol is used. (page 744)

• When you see a list of data values on an exam question, don't just start typing the data into your calculator. Read the question first. Often, additional information is provided that makes it unnecessary for you to enter the data at all. This can save you valuable time on the AP exam. (page 752)

Appendix B: TI-Nspire Technology Corners

Texas Instruments released the new TI-Nspire Touchpad in March 2010. The new handheld has a different removable faceplate where the letters are below the numbers and some of the keys have been changed (see below). The "older" version of the faceplate will be referred to as "clickpad" in the instructions, and the new version will be referred to as "touchpad" when there are marked differences in the specific keystrokes. Otherwise, the capabilities of both "plates" are the same with the new 2.0 operating system (OS). If you do not have the new operating system, please update your device's OS to ensure that your handheld has full capabilities. Go to education.ti.com and search under *Downloads* → *Software, Apps, Operating Systems*... to download the new version of the OS. If you have the TI-Nspire computer link software, you should be asked automatically to update your handheld's OS. The new OS is a definite "must have" for use of the Nspire in the classroom.

Clickpad

new Touchpad

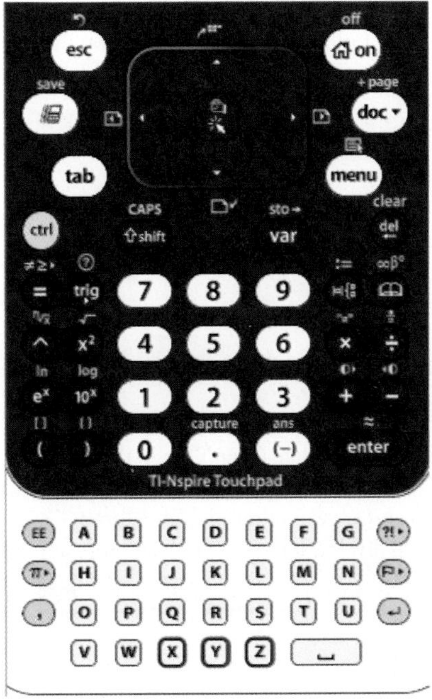

CHAPTER 1 TI-NSPIRE TECHNOLOGY CORNERS

Section 1.2 Histograms on the calculator

1. Insert a *New Document* by pressing (Ctrl) (N).
2. Insert a *Lists & Spreadsheet* page by arrowing down to *Add Lists & Spreadsheet.*
   - Name column A **foreignbrn**.
   - Type the data for the percent of state residents born outside the United States into the list. The data can be found on page 35.

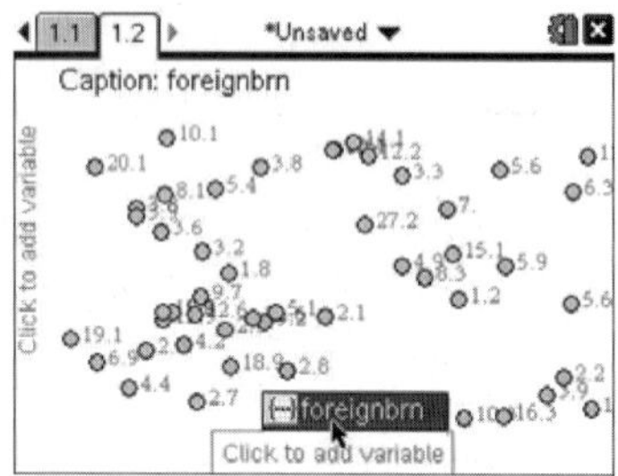

3. Insert a *Data & Statistics* page: press (Ctrl) (I), arrow to *Add Data & Statistics*, and press (enter).
   - Press (tab) and *Click to Add Variable* on the horizontal axis will show the variables available. Select **foreignbrn**.

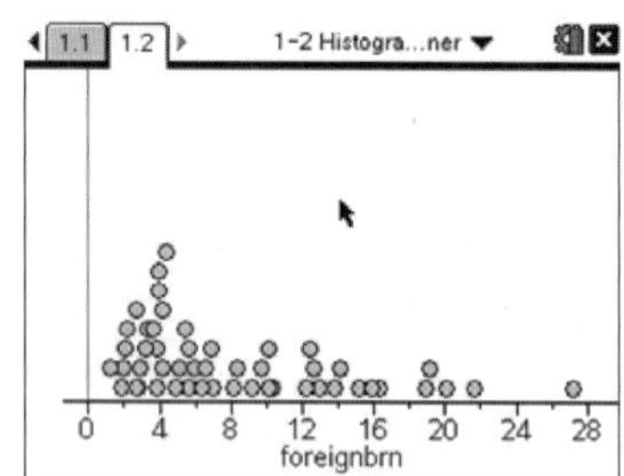

   - The data should now move into a dotplot. Notice the organization of the graph. Even though the data look "lopsided" in some places, you should consider the dots as being directly above each other in each column. To make a better graphical display, let's move the data into a histogram.

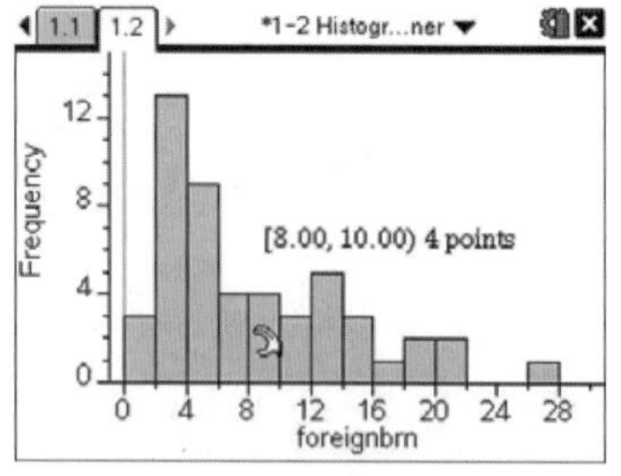

4. Use the Navpad to position the pointer in an empty space within the graph. Press (Ctrl) (menu) and select *Histogram*. You will now see the data move into a histogram.

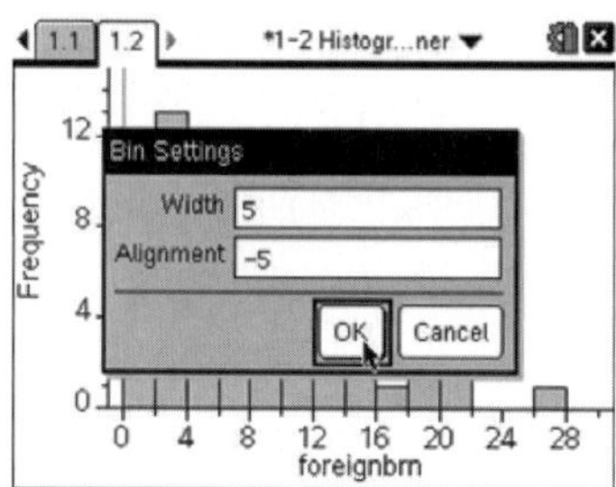

   - You can now position the pointer over each bar to examine the classes. The pointer will become an open hand and the class size will be displayed along with the number of data values in the class.

5. Adjust the classes to match those in Figure 1.16 (page 36).
   - Arrow into an empty space inside the histogram.
   - Press (Ctrl) (menu) and select *Bin Settings*. Enter the values shown. (tab) to (OK) and press (enter). The new histogram should be displayed.

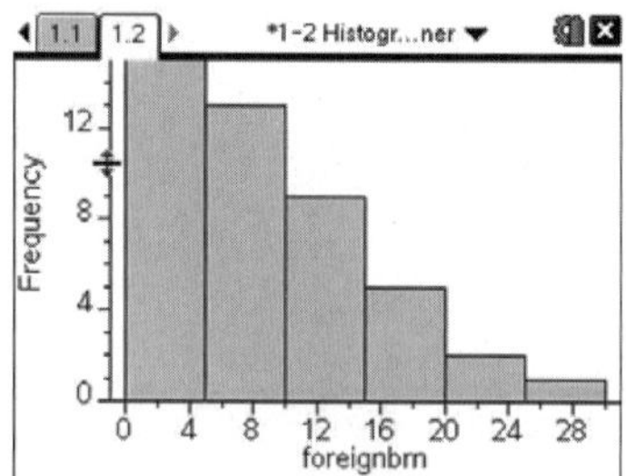

Notice how the first bar is "off the page." To adjust this, arrow over until the pointer becomes ‡.

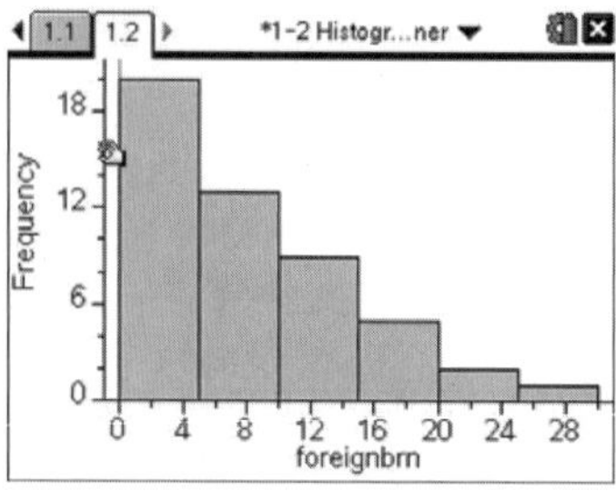

   - Press and hold until ‡ becomes the open hand.
   - Use the Navpad and, with the down-arrow, "pull" the vertical axis down. Keep arrowing down until the top of the tallest histogram class is visible.

6. See if you can match the histogram in Figure 1.17 (page 37).

Section 1.3 Making calculator boxplots

Let's use the calculator to make parallel boxplots of the travel data for the samples from North Carolina and New York.

1. Insert a *Lists & Spreadsheet* page: press (Ctrl) (I), arrow to *Add Lists & Spreadsheet*, and press (enter).
   - Name column A **ncarolina** and column B **newyork**.
   - Enter the travel time data from page 53.

2. Insert a *Data & Statistics* page: press ctrl I, arrow to *Add Data & Statistics*, and press enter.
   - Press tab and *Click to Add Variable* on the horizontal axis will show the variables available. Select **ncarolina**.
   - The data will move into a dotplot. Use the Navpad to position the pointer in an empty space within the graph. Press ctrl menu and select *Boxplot*. You will now see the data move into a boxplot.
   - Using the Navpad, arrow over the plot. You will see the values in the five-number summary display one by one as you move across the boxplot.
3. To add the boxplot of **newyork** travel times, press menu → *Plot Properties* → *Add X Variable*. Select **newyork** and the second boxplot will be added to the page.

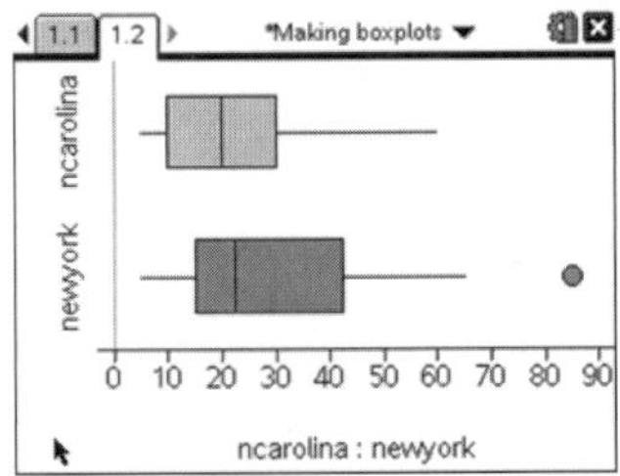

Note: One of the added benefits of the TI-Nspire is its ability to display more than three boxplots at a time in the viewing window.

Section 1.3 Computing numerical summaries with technology

Let's find numerical summaries for the travel times of North Carolina and New York workers from the previous Technology Corner (page 61). If you haven't already done so, enter the North Carolina and New York data.

- Insert a *Lists & Spreadsheet* page: press ctrl I, arrow to *Add Lists & Spreadsheet*, and press enter. Name column A **ncarolina** and column B **newyork**.
- Arrow down to the first empty cell in column A and type in the data. Repeat the process for **newyork** in column B.

1. The Nspire can calculate one-variable statistics for several lists at the same time (unlike the TI-84 or TI-89).
   - Press menu → *Statistics* → *Stat Calculations* → *One Variable Statistics*. A dialogue box should appear asking for the number of lists. Press the up-arrow ▲ to 2 or type "2." tab to OK and press enter.

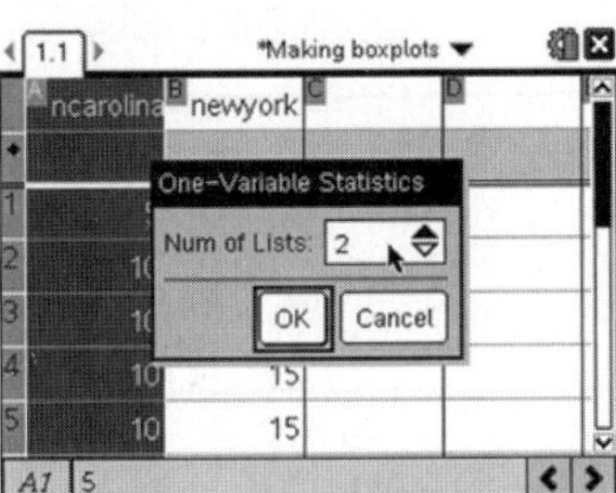

- Another dialogue box should appear. Select the lists in the drop-down boxes: *list1*: **ncarolina** and *list2*: **newyork.** tab between the entry boxes to enter the next list and the column where you want the one-variable stats listed: type "c" and tab to OK and press enter. The numerical summaries for both states should now be displayed.

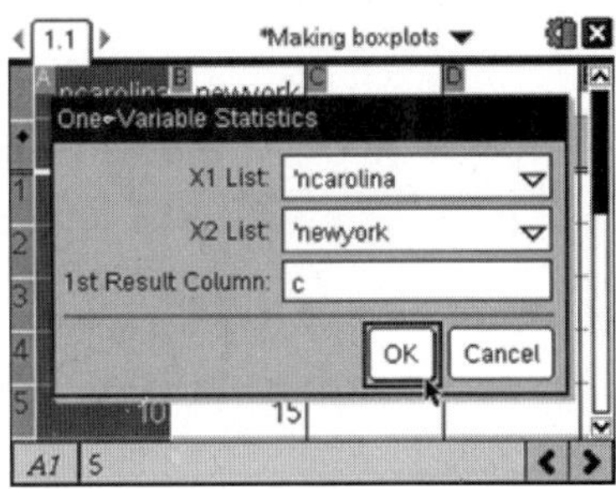

2. You can resize the columns to see which column contains values for which state: D has summary statistics for **ncarolina**, and E has summary statistics for **newyork**.
   - Use the Navpad to place the arrow between the columns. Press and hold [click] until ⟺ appears. Use the arrow keys to increase the column width. Press [click] again to release the column.

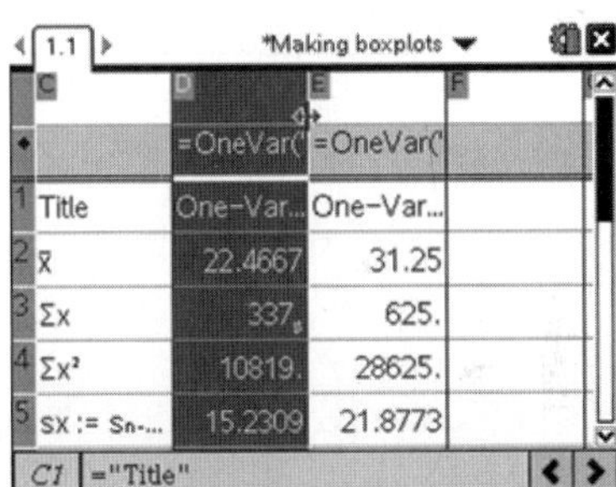

- Repeat the same process to resize the column with one-variable statistics for **newyork**.

CHAPTER 2 TI-NSPIRE TECHNOLOGY CORNERS

Section 2.2 The standard Normal curve

To draw and find areas under a standard Normal curve, you will be using a *Data & Statistics* page. This page requires a variable (a list) to be plotted; therefore, to use a *Data & Statistics* page, you will need to define a "dummy" list: let's call it **z**.

1. Insert a *New Document* by pressing ctrl N.
2. Insert a *Lists & Spreadsheet* page by arrowing down to *Add Lists & Spreadsheet.*
3. Name column A **z** and arrow down to the first cell. Type "100" in the cell.
4. Press ctrl I and use the Navpad to arrow to *Add Data & Statistics*. Press enter.
5. Press tab to select the "Click to add variable" for the horizontal axis. Arrow to **z** and press enter to select it.
6. Arrow up into an open area and press ctrl menu. Select *Histogram*.
7. Press ctrl menu again and select *Scale* → *Density*.
8. Press ctrl menu again and select *Zoom* → *Window Settings*. Enter **Xmin** = −4, **Xmax** = 4, **Ymax** = .5 and press enter.

- Press (menu) and select *Analyze* → *Plot Function* and press (enter).
- Type "normPdf (x,0,1)" and press (enter) to display the curve. You can move **f1(x):=normPdf(x,0,1)** by arrowing over the equation until the hand appears. Click and hold [click]. When the hand closes, you can move the equation using the arrow keys.

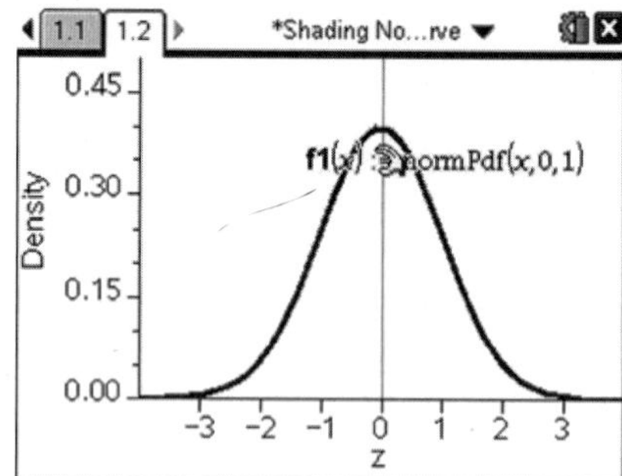

9. To find areas under the standard Normal curve, press (menu) and select *Analyze* → *Shade Under Function* and press (enter). A vertical line should appear; clicking on this line ([click] or (enter)) will set the **lower bound**. *Notice the z-score at the top.*

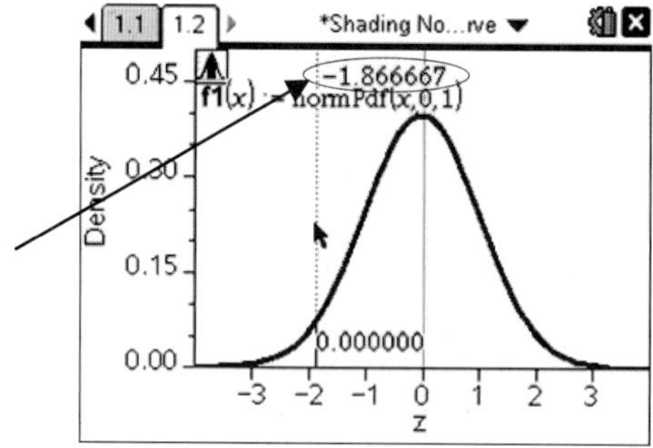

10. Arrow to the right and you should notice a shaded area forming under the Normal curve. The area of the shaded region will be displayed in the middle- to upper-right quadrant of the graph. To stop moving the upper bound, press [click] or (enter).

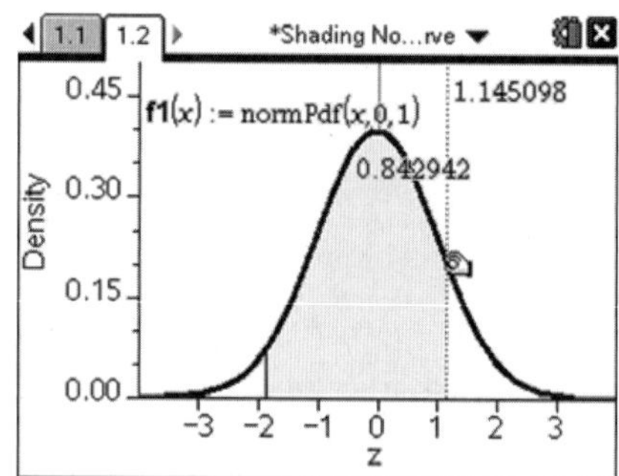

11. You can change the boundaries by using the arrow keys. Arrow over the left boundary line until the arrow becomes a ↔. Press and hold [click] until a closed hand appears.
12. Use the left- and right-arrow keys to move the boundary. Press [click] to deselect the boundary.
13. Follow Steps 11 and 12 to adjust the right boundary.

Note: Keep the left and right boundaries on their respective sides of the graph. Moving one past the other will result in a negative area.

14. For the area to the left of $z = 2.22$, move the left boundary well beyond -3. Then position the right boundary at 2.22.
15. For the area to the right of $z = -1.78$, position the left boundary at -1.78 and move the right boundary well beyond 3.
16. For the area between $z = -1.25$ and $z = 0.81$, position the left boundary at -1.25 and the right boundary at 0.81.

You may not be able to position the boundaries at the exact values above; however, you can get them close.

Note: If you move one of the boundaries too far in one direction and it appears that you have "lost" it, you can move the pointer inside the shaded area and press (Ctrl) (menu) and select *Remove Shaded Region*. Then go to Step 9 to shade under the function again.

Section 2.2 From *z*-Scores to areas, and vice versa

The **normCdf** command on the TI-Nspire can be used to find areas under the Normal curve. This method is quicker than creating a graph of the Normal curve on a *Data & Statistics* page and shading under this graph with *Shade Under Function*. However, you do not get the benefit of seeing a picture associated with the area you are finding.

Let's confirm our earlier answers in the two Tiger Woods examples. Recall that $\mu = 304$ yards and $\sigma = 8$ yards.

1. On the home screen, select the *Calculate* Scratchpad. This will take you to a calculator page that is "outside" your current document. Therefore, you do not have to worry about losing/saving a document you are working on or adding an unneeded page to that document.

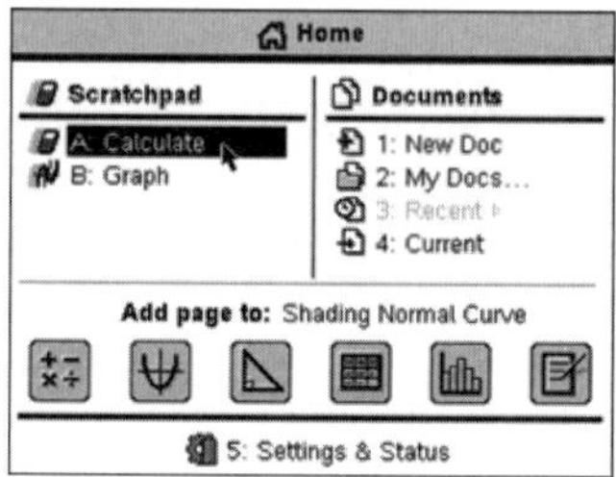

What proportion of Tiger's drives on the range travel at least 290 yards?

2. On the *Calculate* Scratchpad, press (menu) → *Statistics* → *Distributions* → *Normal Cdf*. A dialogue box will appear. Type the numbers in the dialogue box shown. To enter the numbers, (tab) between the entry boxes. When the last number is entered, (tab) to [OK] and press (enter). The proportion should now be displayed on the main screen.

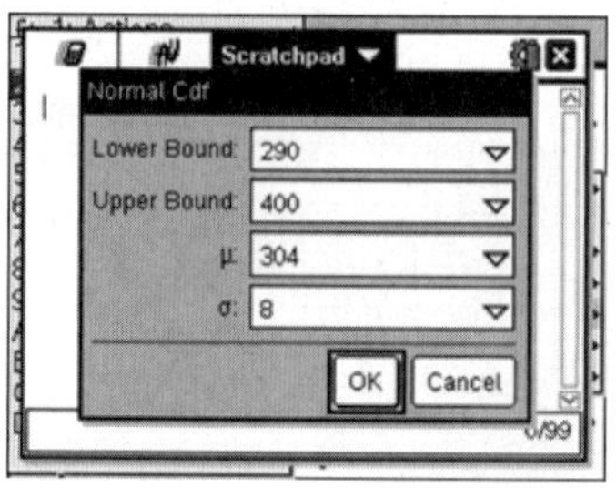

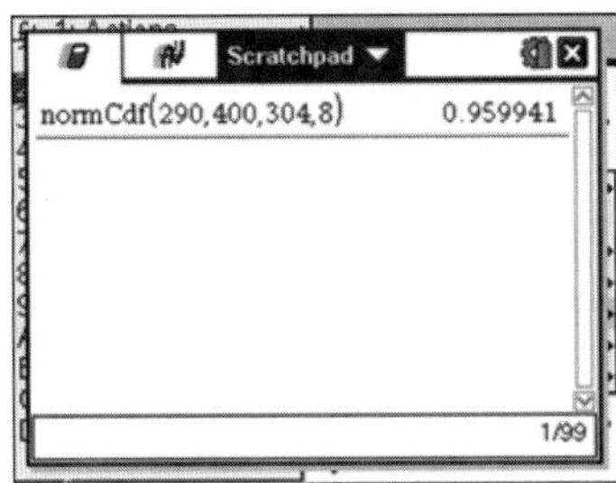

Note: We use 400 because it is *many* standard deviations above the mean.

What proportion of Tiger's drives travel between 305 and 325 yards?

3. The screen shot for this calculated area indicates that our earlier result of 0.4440 using Table A was a little off. This discrepancy was caused by the fact that we rounded our *z*-score to two decimal places in order to use the table.

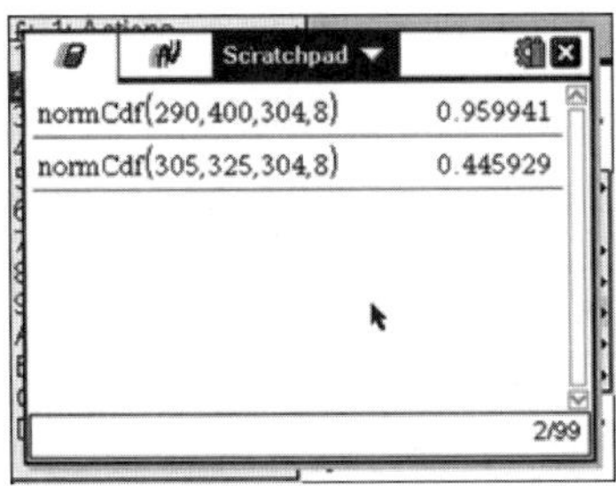

Working backwards: The **invNorm** function calculates the value corresponding to a given percentile in a Normal distribution. For this command, the syntax is **invNorm(percentile, μ, σ)**.

Now let's confirm our earlier answers in the example involving cholesterol levels in 14-year-old boys. Recall that $\mu = 170$ mg/dl and $\sigma = 30$ mg/dl.

1. Let's start with a "clean slate" by clearing the entries on our page. To do this, press (menu) → *Actions* → *Clear History*. Your Scratchpad should now be blank.
2. What is the first quartile of the distribution of blood cholesterol?
   - Press (menu) → *Statistics* → *Distributions* → *Inverse Normal*. A dialogue box will appear. Type the numbers in the dialogue box shown. To enter the numbers, (tab) between the entry boxes. When the last number is entered, (tab) to [OK] and press (enter). Notice in the dialogue box that the first entry line is **area**. This is the same as **percentile**.

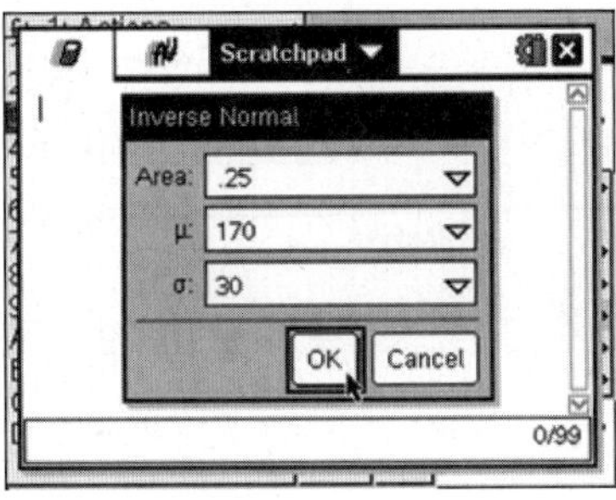

- Compare this result with **invNorm(.25)**. Arrow up to the line: **invNorm(0.25,170,30)** and press (enter). This will put the command on a new calculator line. Press (del) until you have **invNorm(.25)**. Press (enter).

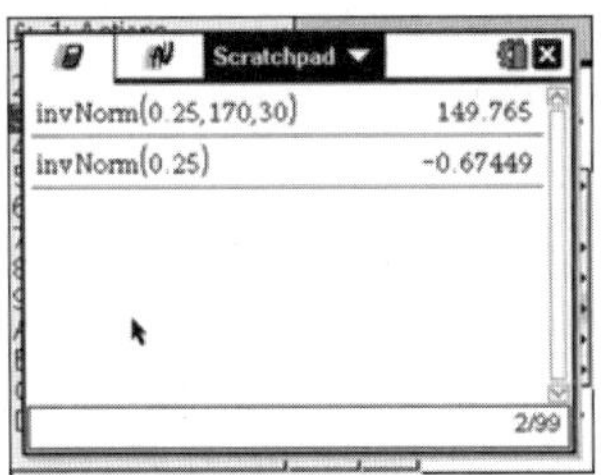

TECHNOLOGY TIP: For **normPdf**, **normCdf**, and **invNorm**, the default values are $\mu = 0$ and $\sigma = 1$.

The first command shows that the first quartile (25th percentile) of the cholesterol distribution for 14-year-old males is 149.765 mg/dl. (Our earlier answer using Table A was 149.9 mg/dl.) The second command shows that in the Normal distribution, 25% of the observations fall below $z = -0.67449$. (We used $z = -0.67$ for our previous calculations, which explains the small discrepancy.)

Section 2.2 Normal probability plots

We will use the state unemployment rates data from page 125 to demonstrate how to make a Normal probability plot for a set of quantitative data.

1. Insert a *New Document* by pressing (Ctrl) [N].
2. Insert a *Lists & Spreadsheet* page by arrowing down to *Add Lists & Spreadsheet.*
   - Name column A **unemploy**.
   - Arrow down to the first cell and type in the 50 data values.
3. Insert a *Data & Statistics* page by pressing (Ctrl) [I] and use the Navpad to arrow to *Add Data & Statistics*. Press (enter).
4. Press (tab) to select the "Click to add variable" for the horizontal axis. Arrow to **unemploy** and press (enter) to select it. The data will now move into a dotplot.
5. Arrow up into an empty region of the dotplot and press (Ctrl) (menu). Select *Normal Probability Plot* and press [☟].

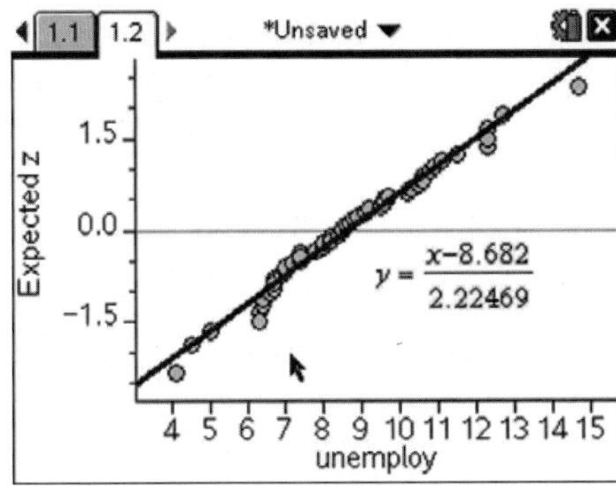

Interpretation: The Normal probability plot is quite linear, so it is reasonable to believe that the data follow a Normal distribution.

CHAPTER 3 TI-NSPIRE TECHNOLOGY CORNERS

Section 3.1 Scatterplots on the calculator

1. Insert a *New Document* by pressing Ctrl N.
2. Insert a *Lists & Spreadsheet* page by arrowing down to *Add Lists & Spreadsheet.*
   - Name column A **body** and column B **backpack**.
   - Type the corresponding values into each column. The data list is given below:

Body weight:	120	187	109	103	131	165	158	116
Backpack weight:	26	30	26	24	29	35	31	28

	A body	B backp...	C	D
◆				
1	120	26		
2	187	30		
3	109	26		
4	103	24		
5	131	29		

1.1 RAD AUTO REAL — B backpack

3. Press Ctrl I and use the Navpad to arrow to *Add Data & Statistics.* Press enter.
4. Press tab to select the "Click to add variable" for the horizontal axis. Arrow to body and press enter to select it.

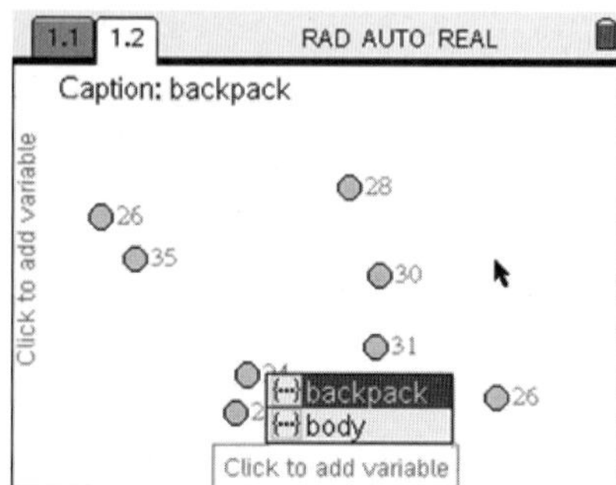

5. Press tab again and the box will move to the vertical axis. Select **backpack**. The data will now move into a scatterplot. TI-Nspire labels the x and y axis with the list names, making a well-labeled graph to insert into documents.

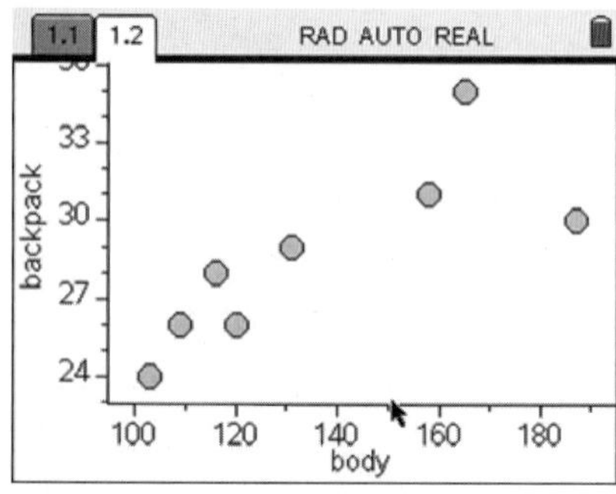

6. You can now trace each data point. Press menu, arrow down to *Analyze → Graph Trace,* enter. By pressing the arrows on the Navpad, you can move between the points, and the ordered pair will be displayed.

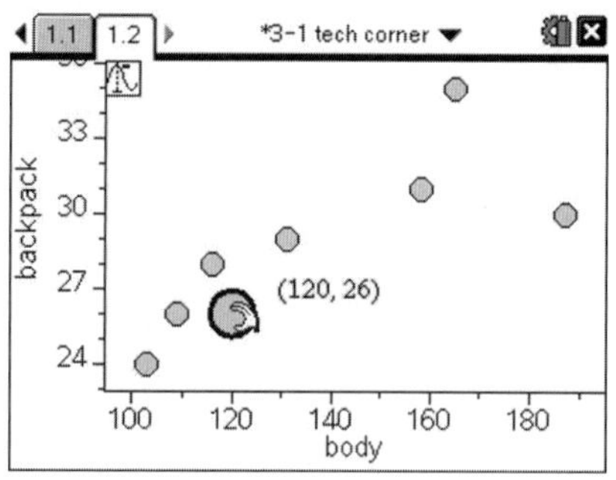

Section 3.2 Least-squares regression lines on the calculator

Use the fat gain and NEA data to find the equation of the least-squares regression line on the TI-Nspire. Here are the data:

NEA change (cal):	−94	−57	−29	135	143	151	245	355
Fat gain (kg):	4.2	3.0	3.7	2.7	3.2	3.6	2.4	1.3
NEA change (cal):	392	473	486	535	571	580	620	690
Fat gain (kg):	3.8	1.7	1.6	2.2	1.0	0.4	2.3	1.1

1. Insert a *New Document* by pressing Ctrl N.
2. Insert a *Lists & Spreadsheet* page by arrowing down to *Add Lists & Spreadsheet.*
   - Name column A **nea** and column B **fatgain**.
   - Type the corresponding values into each column.
3. Graph the data in a scatterplot putting **nea** on the horizontal axis and **fatgain** on the vertical axis. Refer to the TI-Nspire Technology Corner on page 149.
4. To add a least-squares regression line, first Ctrl ◀ back to the *Lists & Spreadsheet* page.
5. Press menu, and arrow to *Statistics → Stat Calculations, Linear Regression* $(a + bx)$, enter. You then see a dialogue box. In the drop-down boxes, arrow down to **nea** for the *X List:*, then press tab and arrow down to **fatgain** for the *Y List:*. tab to OK and press enter.

The linear regression information, a, b, r^2, r, and *resid*, will be displayed in another column within the list.

	tgain	C	D	E
◆				=LinRegBx
1	4.2		Title	Linear Re...
2	3		RegEqn	a+b*x
3	3.7		a	3.50512
4	2.7		b	-0.003441
5	3.2		r²	0.606149
6	3.6		r	-0.778556

E =LinRegBx('nea,'fatgain,1) : CopyVar Sta▸

6. Ctrl ▸ to the *Data & Statistics* page. Press menu; arrow to *Analyze* → *Regression* → *Show Linear* ($a + bx$), and press enter. The least-squares regression line, along with the equation, will appear. If you arrow over the equation, the hand will appear. Click and hold [click]. When the hand closes, you can move the equation using the arrow keys.

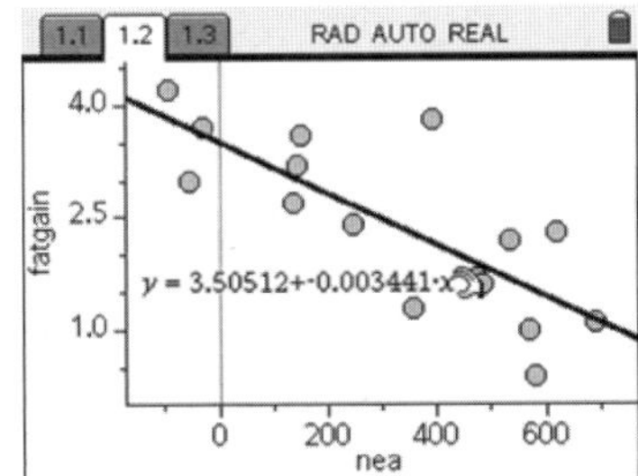

7. To trace points along the regression line, press menu; arrow to *Analyze* → *Graph Trace*, and press enter.

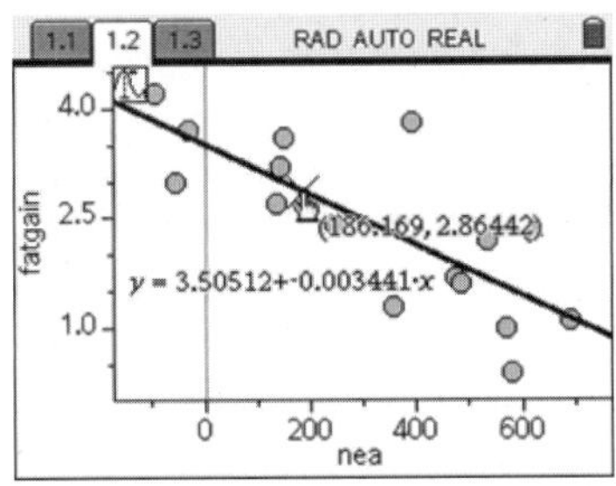

8. Although the equation gives the values of a and b to five or six decimal places, we usually round to fewer decimal places. You would write the equation as $\hat{y} = 3.505 - 0.00344x$.

9. Save the document for later use. Press Ctrl S. Name your document: **NEA Fat gain**.

Section 3.2 Residual plots and *s* on the calculator

Let's continue the analysis of the NEA and fat gain study from the previous Technology Corner (page 170). You should have already made a scatterplot, calculated the equation of the least-squares regression line, and graphed the line on your plot. Now, we want to calculate residuals and make a residual plot. Earlier, we found that $\hat{y} = 3.505 - 0.00344x$.

1. Open the document, **NEA fat gain**. Press Ctrl O, arrow through *My Documents* → **NEA fat gain**, and press enter.
2. Insert a *Lists & Spreadsheet* page.
3. Name column A **fatgain**. The fat gain data should fill the column when you press enter.
4. Name column B **predictedgain**. In the formula bar, press enter and type the formula **3.505−0.00344*(nea)**. Press enter. The predicted values should now fill the list.

1.1 | 1.2 | 1.3 RAD AUTO REAL

	A fatgain	B predictedgain	C residual
◆		=3.505-0.00344*nea	=fatgain-pr
1	4.2	3.82836	0.37164
2	3	3.70108	-0.70108
3	3.7	3.60476	0.09524
4	2.7	3.0406	-0.3406
5	3.2	3.01308	0.18692

C residual:=fatgain-predictedgain

5. Name column C **residual**. In the formula bar, press enter and type the formula **fatgain − predictedgain**. Press enter. The residuals should now fill the list.
6. Insert a *Data & Statistics* page: press Ctrl I, arrow down to *Add Data & Statistics*, and press enter.
7. Press tab to select the "Click to add variable" for the horizontal axis. Arrow to nea and press [click] to select it.
8. Press tab again and the box will move to the vertical axis. Select **residual**. The residuals will now move into the residual plot.

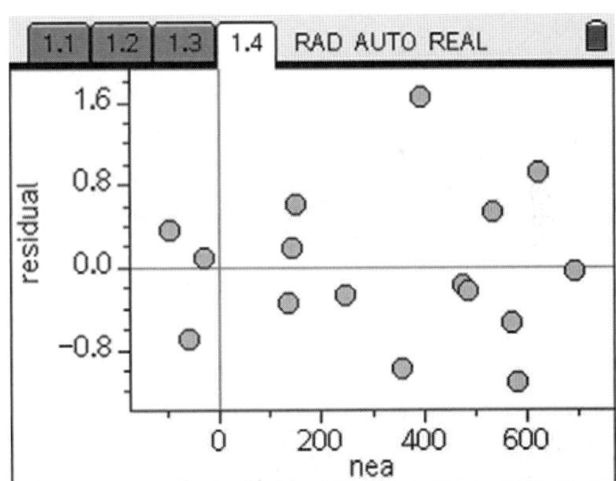

The x axis in the residual plot serves as a reference line: points above this line correspond to positive residuals and points below the line correspond to negative residuals. Using the Navpad, move the arrow to the leftmost point. When the arrow becomes the hand, press [click]. You will now be able to see the ordered pair for this point.

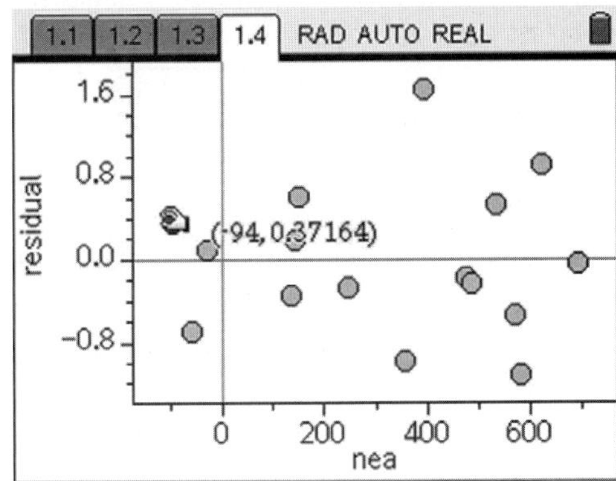

9. Finally, we want to compute the standard deviation s of the residuals.
   - Insert a *Calculator* page. Press (ctrl) (I), arrow down to *Add Calculator,* and press (enter).
   - Press (menu) → *Statistics* → *Stat Calculations* → *One-Variable Statistics.* (tab) to (OK) and in the next dialogue box select **residuals** for the X-List. (tab) to (OK) and press (enter).

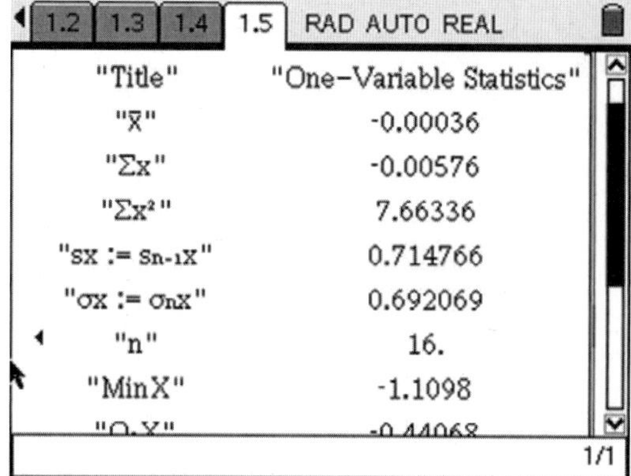

The mean of the residuals is 0 (up to roundoff error). The sum of the squared residuals is $\sum x^2 = 7.6636$. To find s, use the formula

$$s = \sqrt{\frac{\sum \text{residuals}^2}{n-2}} = \sqrt{\frac{7.6636}{14}} = 0.74 \text{ kg}$$

CHAPTER 4 TI-NSPIRE TECHNOLOGY CORNER

Section 4.1 Choosing an SRS

The TI-Nspire has a function called **randSamp** that will randomly select individuals for a sample with or without replacement from a population.

1. Check that your calculator's random number generator is working properly.
   - Open the calculator Scratchpad by pressing (scratchpad) (or (home) (A) on the clickpad).
   - Type **randint (1,28)** and press (enter).
   - Compare results with your classmates. If several students have the same number, you'll need to seed your calculator's random number generator with different numbers before you proceed. Type **randSeed**, press (space) (to insert a space), type <last four digits of your phone number> and press (enter). *Done* should appear. Now your calculator is ready to generate numbers that are different from those of your classmates.

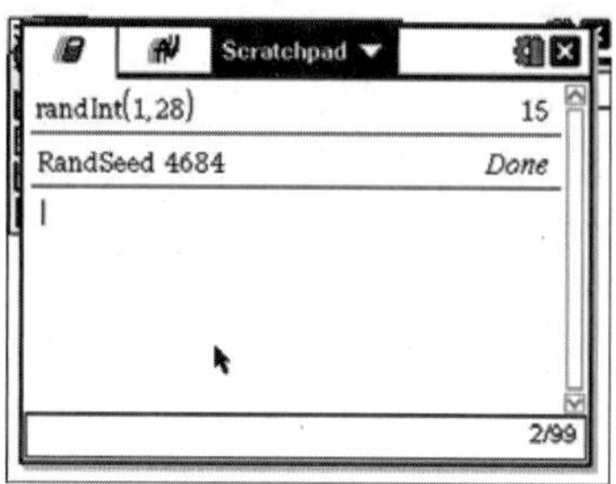

2. Insert a *Lists & Spreadsheet* page. If you already have a document open, press (ctrl) (I) and select *Add Lists & Spreadsheet.* If you do not have a document already open, press (home on) ((home) on the clickpad), then (4). Press (menu) and select *Add Lists & Spreadsheet.*
3. Name column A **hotels**. Arrow down to the formula cell and press (enter). **hotels:=** should appear. Type **seq(x,x,1,28)**. This will put the digits 1 through 28 in this list.
4. Name column B **samphotel**. Arrow down to the formula cell and press (enter). **samphotel:=** should appear. Type **randSamp(hotels, 4, 1)**. This function will take a random sample of four hotels from the list. "1" lets the function know to do the sampling without replacement.

Note: Sampling *with* replacement is the default for this function. You can use **0** as the third input in the **randSamp** command or close the parentheses after the second input.

CHAPTER 6 TI-NSPIRE TECHNOLOGY CORNERS

Section 6.1 Analyzing random variables on the calculator

Let's explore what the calculator can do using the random variable X = Apgar score of a randomly selected newborn.

1. Insert a *Lists & Spreadsheet* page. Press (ctrl) (I), arrow to *Add Lists & Spreadsheet,* and press (enter).
   - Name column A **apgar** and column B **apgrprob**.
   - Enter the values of the random variable (0 – 10) in the **apgar** list and the corresponding probabilities in **apgrprob**.

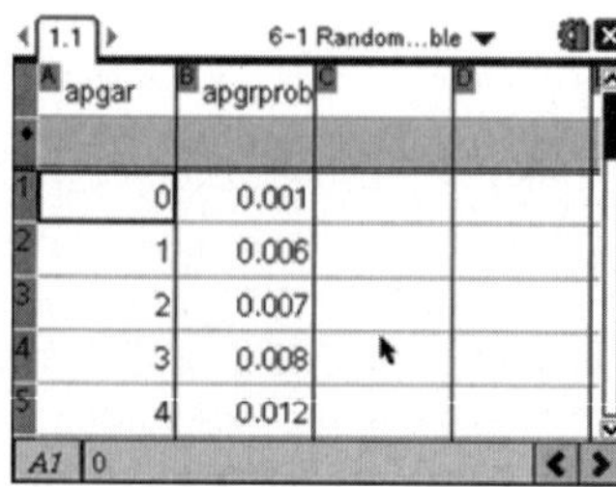

2. Graph a histogram of the probability distribution.
   - Insert a *Data & Statistics* page. Press (ctrl) (I), arrow to *Add Data & Statistics,* and press (enter).
   - Press (ctrl) (menu) and select *Add X Variable with Frequency.* Press (enter) and a dialogue box should appear. **apgar** should be in the *Data List* and **apgrprob** should be in the *Frequency List.* If they are not, use the drop-down boxes to select your variables. When your box looks like the one here, (tab) to (OK) and press (enter).

The probability histogram should now be displayed.

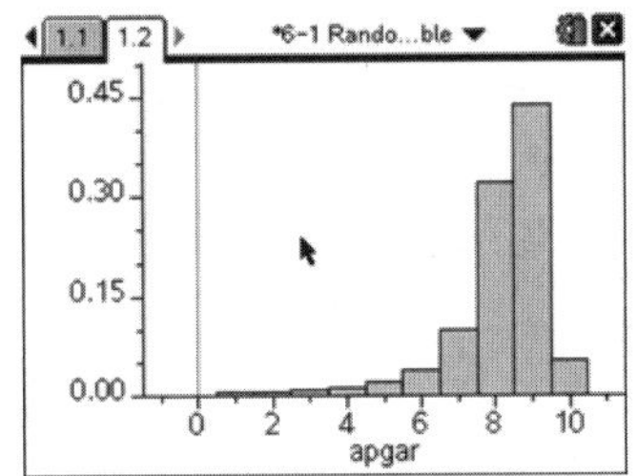

3. To calculate the mean and standard deviation of the random variable, use one-variable statistics with **apgar** as the *Data List* and **apgrprob** as the *Frequency List.*
   - Press (ctrl) ◀ to go back to the *Lists & Spreadsheet* page.
   - Press (menu) → *Statistics* → *Stat Calculations* → *One-Variable Statistics.*
   - Make sure your *X1 List*, *Frequency List*, *1st Result Column* have the variables/values shown (you can press the down-arrow in the drop-down boxes to access the variable names and type [C] for *1st Result Column*). (tab) to [OK] and press (enter).

One-Variable Statistics
X1 List: 'apgar
Frequency List: 'apgrprob
Category List:
Include Categories:
1st Result Column: c
OK Cancel

   - The statistics should now be displayed in your *Lists & Spreadsheet* page.

	B apgrprob	C	D	E
◆			=OneVar(	
2	0.006	x̄	8.128	
3	0.007	Σx	8.128	
4	0.008	Σx²	68.13	
5	0.012	sx := sₙ₋...	#UNDEF...	
6	0.02	σx := σₙx...	1.43723	

D =OneVar('apgar,'apgrprob): CopyV

Section 6.2 Simulating with randNorm

The **randNorm** command on the TI-Nspire allows you to simulate observations from a Normal distribution with a specified mean and standard deviation. For instance, **randNorm(3.98, .02)** will randomly select a value from the Normal distribution with mean 3.98 and standard deviation 0.02. This simulates choosing a large cup lid at random from the fast-food restaurant and measuring its diameter (in inches). To simulate choosing a large drink cup and measuring its width, use **randNorm(3.96, .01)**.

To estimate the probability that a randomly selected lid will fit on a randomly chosen cup:

1. Open a *Lists & Spreadsheet* page: Press (ctrl) [N]. If you already have a document open, you will be given the option to save it. Select *Yes* or *No* and press (enter). Arrow to *Add Lists & Spreadsheet* and press (enter).
2. Name column A **cuplid**. Press (enter) and in the formula cell type **randNorm(3.98, .02, 100)** and press (enter). One hundred random cup lid sizes should now be in this list.
3. Name column B **drinkcup**. Press (enter) and in the formula cell type **randNorm(3.96, .01, 100)** and press (enter). One hundred random drink cup sizes should now be in this list.

	A cuplid	B drinkcup	C	D
◆	=randnorm	=randnorm		
1	3.9798	3.97507		
2	3.95912	3.96525		
3	3.9622	3.97241		
4	3.94152	3.95999		
5	3.95731	3.97362		

B drinkcup:=randnorm(3.96,0.01,100)

4. Now we need to compute the differences in the sizes of the lids and drink cups.
   - Name column C **cupdiff**. Press (enter).
   - In the formula cell press (enter), then (var) (select **cuplid**), press [?], (−), (var) (select **drinkcup**), and then press [?]. Press (enter) and the 100 differences will be in the list.

	A cuplid	B drinkcup	C cupdiff	D
◆	=randnorm	=randnorm	=cuplid-dr	
1	3.9798	3.97507	0.004733	
2	3.95912	3.96525	-0.0061...	
3	3.9622	3.97241	-0.0102...	
4	3.94152	3.95999	-0.0184...	
5	3.95731	3.97362	-0.0163...	

C cupdiff:=cuplid-drinkcup

5. Count the number of differences that are between 0 and 0.06. This number divided by 100 will be your estimate of the probability.
   - On the TI-Nspire there is a function called *Countif*, which will find the values in **cupdiff** that are between 0 and 0.06.
   - Arrow to an empty cell in column D. Press (=) and type *countif*(**cupdiff**, $0 \leq ? \leq 0.06$). Press (enter) and the number of values between 0 and 0.06 will appear. Divide this number by 100.

	A cuplid	B drinkcup	C cupdiff	D
◆	=randnorm	=randnorm	=cuplid-dr	
1	3.9798	3.97507	0.004733	80
2	3.95912	3.96525	-0.0061...	
3	3.9622	3.97241	-0.0102...	
4	3.94152	3.95999	-0.0184...	
5	3.95731	3.97362	-0.0163...	

D1 =countif(cupdiff,0≤?≤0.06)

Note: To type ≤, press (ctrl) (=) and select ≤ (on the clickpad, press (ctrl) (?) and select ≤). To type ?, press (?!▸) and select ? (on the clickpad, press (?)).

Section 6.3 Binomial coefficients on the caculator

To calculate a binomial coefficient like $\binom{5}{2}$ on the TI-Nspire, proceed as follows. Open the calculator Scratchpad by pressing [calculator key] (or [home] [A] on the clickpad). Press [menu] → *Probability* → *Combinations*, and then [enter]. **nCr(** will appear. Complete the command **nCr(5,2)** and press [enter].

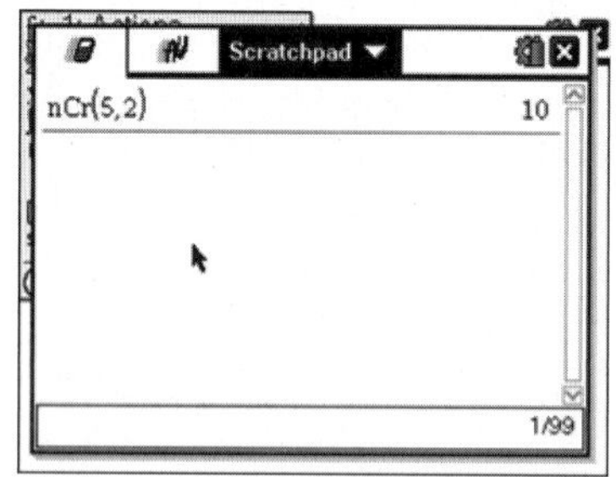

Section 6.3 Binomial probability on the calculator

There are two handy commands on the TI-Nspire for finding binomial probabilities:

`binomPdf(n,p,k)` computes $P(X = k)$
`binomCdf(n,p,k)` computes $P(X \le k)$

You will need to open the *Calculator* Scratchpad (press [calculator key] or [home] [A]). These two commands can be found in the *Distributions* menu within the *Statistics* menu. You can access them by pressing [menu] → *Statistics* → *Distributions*. A dialogue box will appear asking for your input. Type n (the number of observations), p (probability of success), and k (number of successes).

Binomial Pdf **Binomial Cdf**

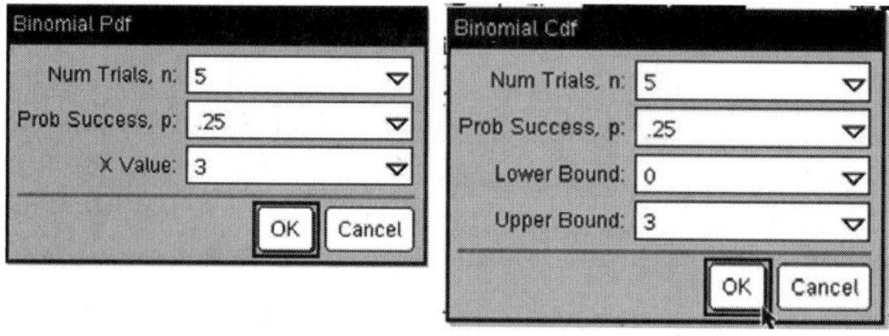

For the parents having $n = 5$ children, each with probability $p = 0.25$ of type O blood:

$P(X = 3) =$ `binomPdf(5,0.25,3)` $= 0.08789$

To find $P(X > 3)$, we used the complement rule:

$P(X > 3) = 1 - P(X \le 3) = 1 -$ `binomCdf(5,0.25,3)`
$= 0.01563$

Of course, we could also have done this as

$P(X > 3) = P(X = 4) + P(X = 5)$

$=$ `binomPdf(5,0.25,4)` $+$
`binomPdf(5,0.25,5)`

$= 0.01465 + 0.00098 = 0.1563$

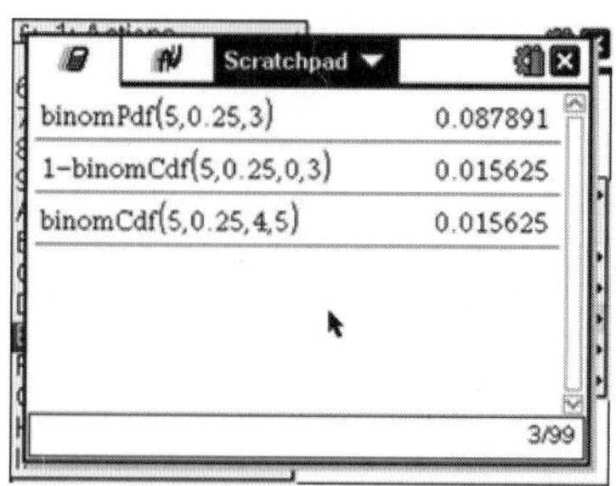

On the TI-Nspire, you can also calculate using

$P(X > 3) = P(X = 4) + P(X = 5)$

$=$ `binomCdf(5,0.25,4,5)`

$= 0.1563$

Section 6.3 Geometric probability on the calculator

There are two handy commands on the TI-Nspire for finding geometric probabilities:

`geomPdf(p,k)` computes $P(X = k)$
`geomCdf(p,k)` computes $P(X \le k)$

You will need to open the *Calculator* Scratchpad (press [calculator key] or [home] [A]). These two commands can be found in the *Distributions* menu within the *Statistics* menu. You can access them by pressing [menu] → *Statistics* → *Distributions*. A dialogue box will appear asking for your input. Type p (probability of success) and k (number of trials to get the first success).

Geometric Pdf **Geometric Cdf**

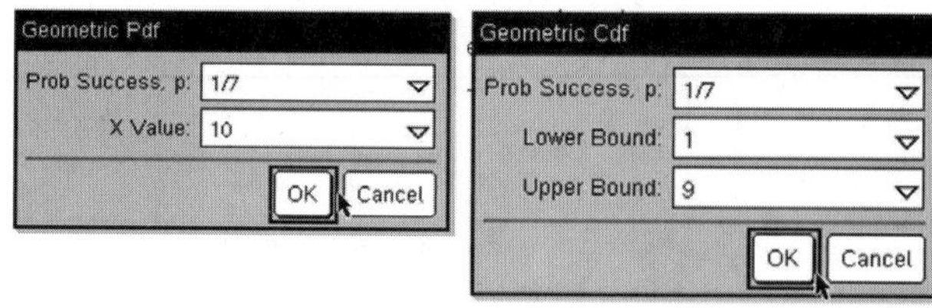

For the Birth Day Game, with probability of success $p = 1/7$ on each trial:

$P(Y = 10) =$ `geomPdf(1/7,10)` $= 0.0357$

To find $P(Y < 10)$, use `geomCdf`:

$P(Y < 10) = P(Y \le 9) =$ `geomCdf(1/7,9)` $= 0.7503$

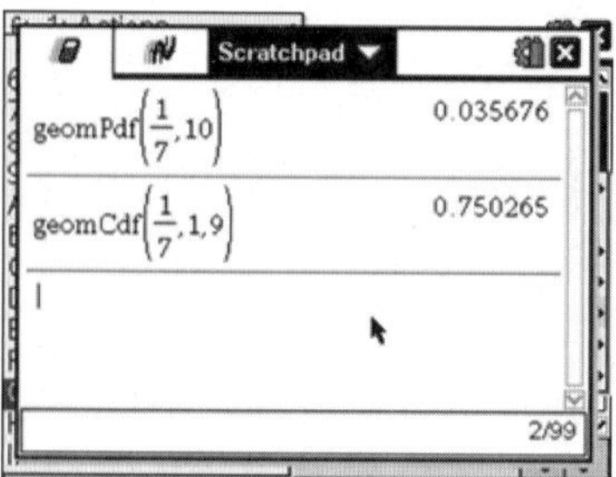

CHAPTER 8 TI-NSPIRE TECHNOLOGY CORNERS

Section 8.2 Confidence interval for a population proportion

The TI-Nspire can be used to construct a confidence interval for an unknown population proportion. We'll demonstrate using the example on page 491. Of $n = 439$ teens surveyed, X = 246 said they thought young people should wait to have sex until after marriage. To construct a confidence interval:

- Press (calculator) ((home) (A)) to insert a *Calculator Scratchpad*.
- Press (menu) → *Statistics* → *Confidence Intervals* → *1 Prop Z Interval*.
- A dialogue box will appear. Enter the values as shown in the box. (tab) to [OK] and press (enter).

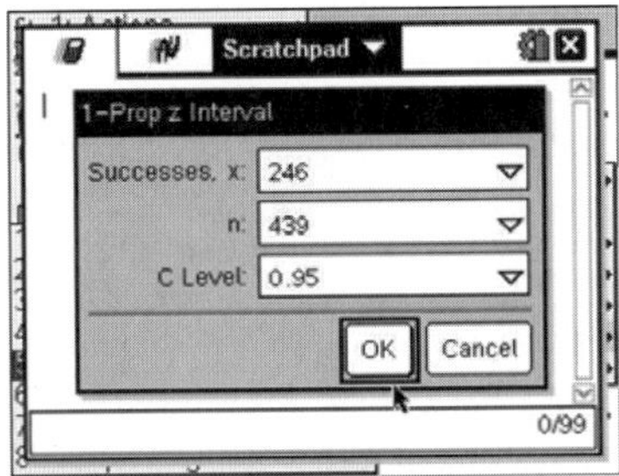

The lower and upper bounds of the confidence interval are reported, along with the sample proportion $\hat{p}$, margin of error (ME), and the sample size.

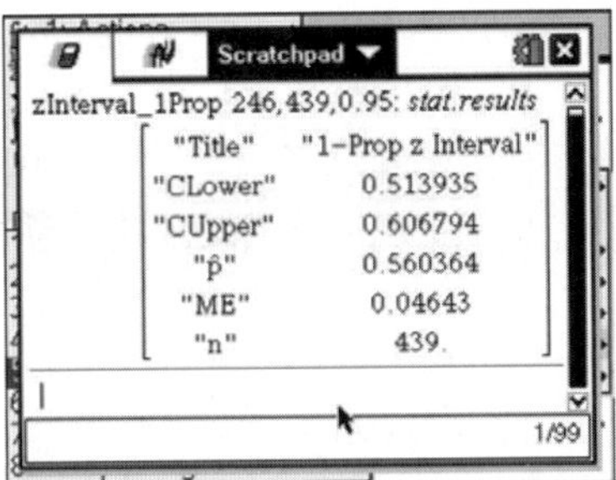

Section 8.3 Inverse t on the calculator

The TI-Nspire allows you to find critical values t^* using the inverse t command. As with the calculator's inverse Normal command, you have to enter the *area to the left* of the desired critical value.

- Press (calculator) ((home) (A)) to insert a *Calculator Scratchpad*.
- Press (menu) → *Statistics* → *Distributions* → *Inverse t*.
- A dialogue box will appear. Enter .975 for the *Area* and 11 for the *Deg of Freedom, df*. (tab) to [OK] and press (enter).

- The critical value t^* should now be displayed.

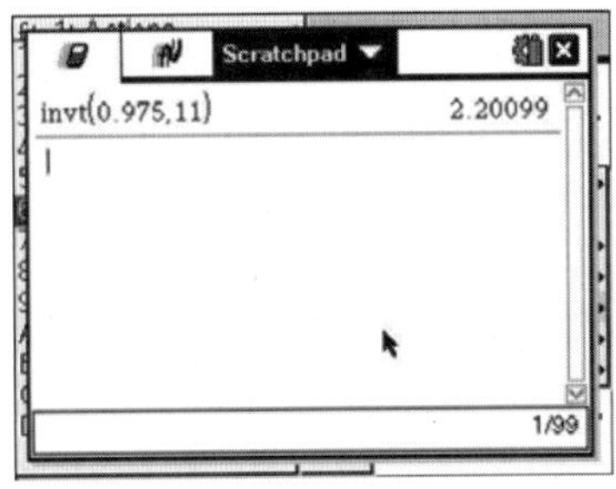

Section 8.3 One-sample t intervals for μ on the calculator

Confidence intervals for a population mean using t procedures can be constructed on the TI-Nspire, thus avoiding the use of Table B. Here is a brief summary of the techniques when you have the actual data values and when you have only numerical summaries.

Using raw data: Video screen tension example, page 508

Enter the 20 video screen tension readings data using the procedure below.

- Insert a *Lists & Spreadsheet*: Press (Ctrl) [I] and select *Add Lists & Spreadsheet*.
- Name the first column **screen**.
- Arrow down to the first cell and enter the 20 values.

	A screen	B	C	D
1	269.5			
2	297.			
3	269.6			
4	283.3			
5	304.8			

To construct the t interval:

- Press (menu) → *Statistics* → *Confidence Intervals* → *t interval*.
- The first dialogue box that appears asks for *Data* or *Stats* in the drop-down box. Select *Data*, (tab) to [OK], and press (enter).
- In the next dialogue box, select the data list, **screen**, and (tab) to [OK]. Press (enter).

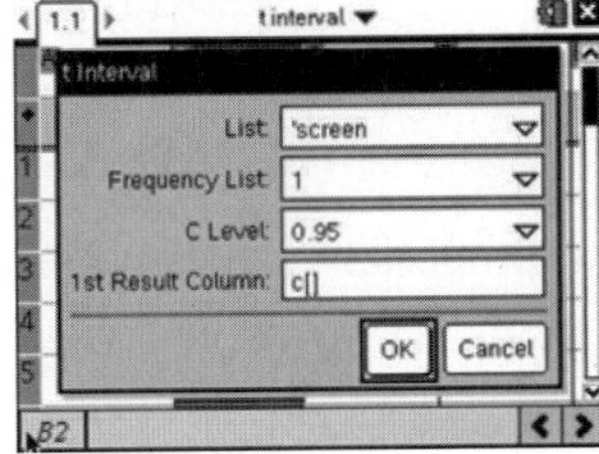

- The results should now appear in the spreadsheet.

	A screen	B	C	D =tInterval(
1	269.5		Title	t Interval
2	297.		CLower	289.374
3	269.6		CUpper	323.266
4	283.3		x̄	306.32
5	304.8		ME	16.9465

Using summary statistics: Auto pollution example, page 509

This time we have no data to enter into a list. Repeat the first step above except this time select *Stats* in the drop-down box.

- In the next dialogue box, enter the values shown.

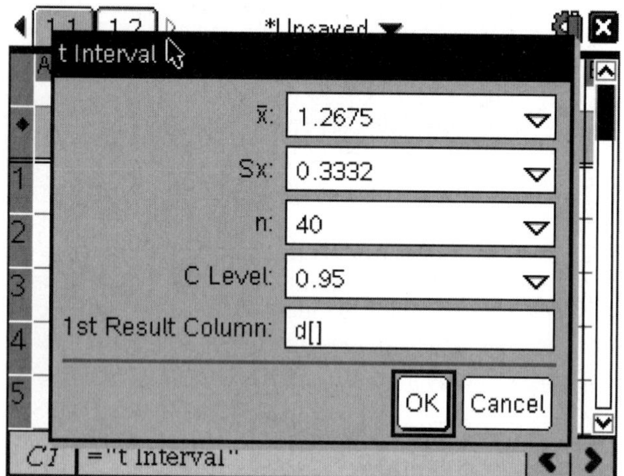

- (tab) to [OK] and press (enter).
- The results should now appear in the spreadsheet.

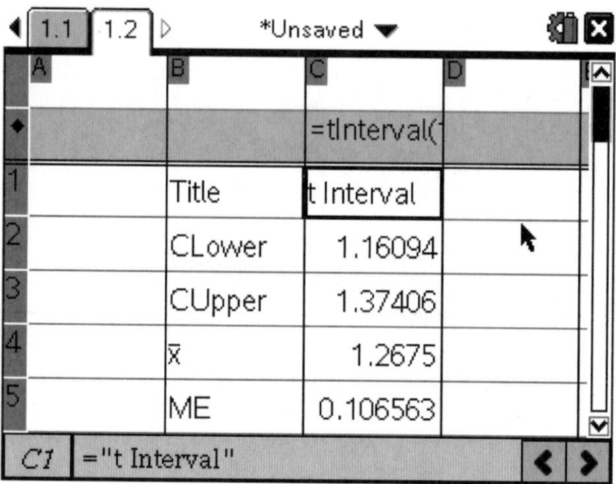

CHAPTER 9 TI-NSPIRE TECHNOLOGY CORNERS

Section 9.2 One-proportion *z* test on the calculator

The TI-Nspire can be used to test a claim about a population proportion. We'll demonstrate using the example on page 553. In a random sample of size $n = 500$, the supervisor found $X = 47$ potatoes with blemishes. To perform a significance test:

- Press (≡▦) (or (⌂) (A)) to insert a *Calculator Scratchpad*.
- Press (menu) → *Statistics* → *Stat Tests* → *1 Prop Z test*.
- A dialogue box will appear. Enter the values shown: $p_0 = 0.08$, $x = 47$, and $n = 500$. Specify the alternative hypothesis as "H_a: prop $> p_0$."

Note: x is the number of successes and n is the number of trials. Both must be whole numbers! Enter the values as shown in the box. (tab) to [OK] and press (enter).

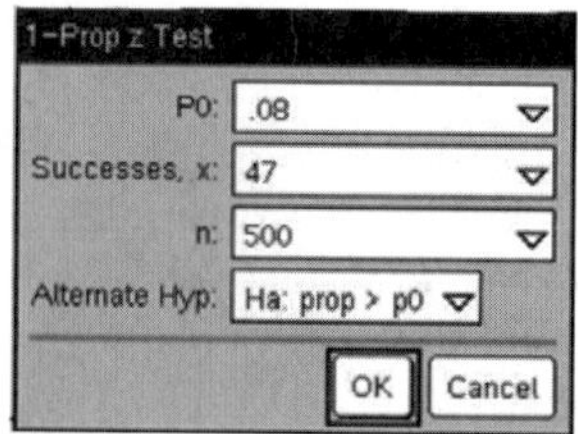

You can see that the test statistic is $z = 1.15392$ and the *P*-value is 0.1243.

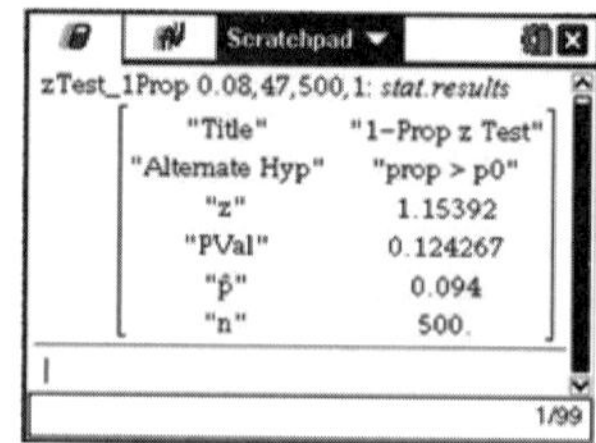

To display the *P*-value as a shaded area under the standard Normal curve:

- Press (⌂ on) and select the *Lists & Spreadsheet* icon.
- Press (menu) → *Statistics* → *Stat Tests* → *1 Prop Z test*.
- A dialogue box will appear. Enter the values as shown in the box. Check the box to *Shade P value*. (tab) to [OK] and press (enter).

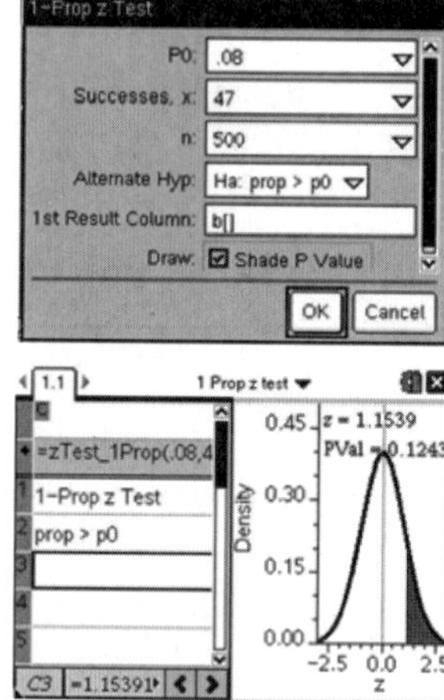

Section 9.3 Computing *P*-values from *t* distributions on the calculator

You can use the *tcdf* command on the TI-Nspire to calculate areas under a *t* distribution curve. The syntax is tcdf(lower bound,upper bound,df). To use this command:

- Press (≡▦) (or (⌂) (A)) to insert a *Calculator Scratchpad.*
- Press (menu) → *Statistics* → *Distributions* → *t Cdf.*
- A dialogue box will appear in which you will enter your lower and upper bound and degrees of freedom.

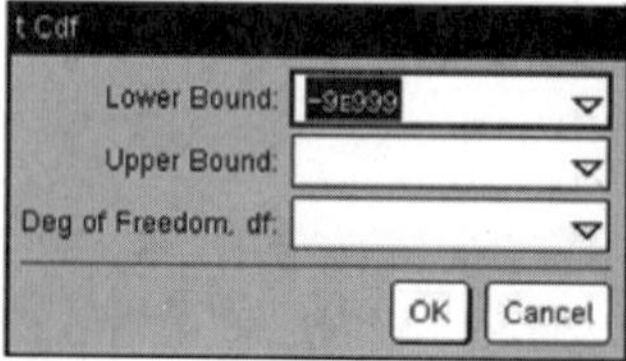

Let's use the *tcdf* command to compute the *P*-values from the examples on pages 566 and 567.

- *Better batteries:* To find $P(t \geq 1.54)$, use Lower Bound: 1.54, Upper Bound: 100, and df: 14.
- *Two-sided test:* To find the *P*-value for the two-sided test with df $= 36$ and $t = -3.17$, execute the command 2* tCdf(−100, −3.17, 36).

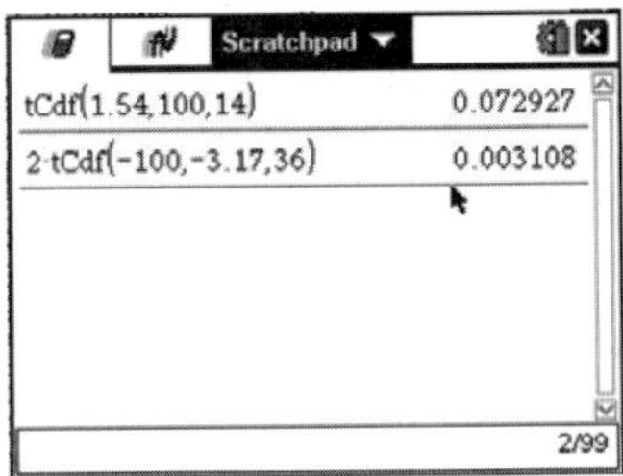

Section 9.3 One-sample *t* test on the calculator

You can perform a one-sample *t* test using either raw data or summary statistics on the TI-Nspire. Let's use the calculator to carry out the test of $H_0: \mu = 5$ versus $H_a: \mu < 5$ from the dissolved oxygen example on page 571.

Start by entering the sample data into a column in a *Lists & Spreadsheet* page. Name the column **oxygen.** Then, to do the test:

- Press (menu) → *Statistics* → *Stats Tests* → *t Test*.
- The first dialogue box that appears asks for *Data* or *Stats* in the drop-down box. Make sure *Data* is selected. (tab) to [OK] and press (enter).
- In the next dialogue box, enter the values shown in the box, and (tab) to [OK]. Press (enter). To just "calculate," leave the *Shade P-value* option unchecked.

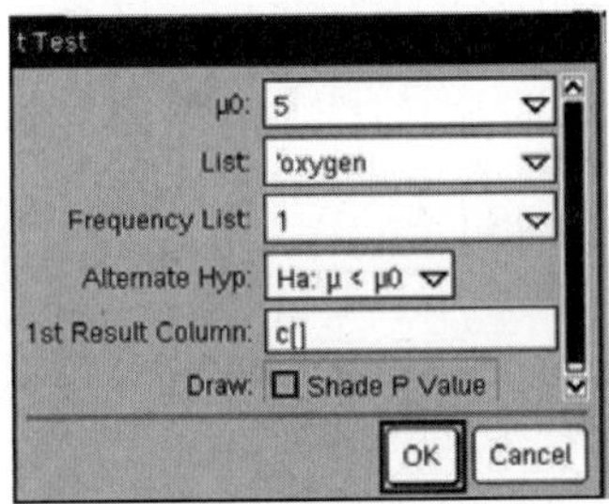

- The results should now appear in the spreadsheet.

	A oxygen	B	C	D
◆				=tTest(5,'o
1	4.53		Title	t Test
2	5.04		Alternate...	μ < μ0
3	3.29		t	-0.9425...
4	5.23		PVal	0.180945
5	4.13		df	14.

The test statistic is $t = -0.94$ and the *P*-value is 0.1809.

If you check *Shade P-Value*, you see a *t* distribution curve (df = 14) with the lower tail shaded.

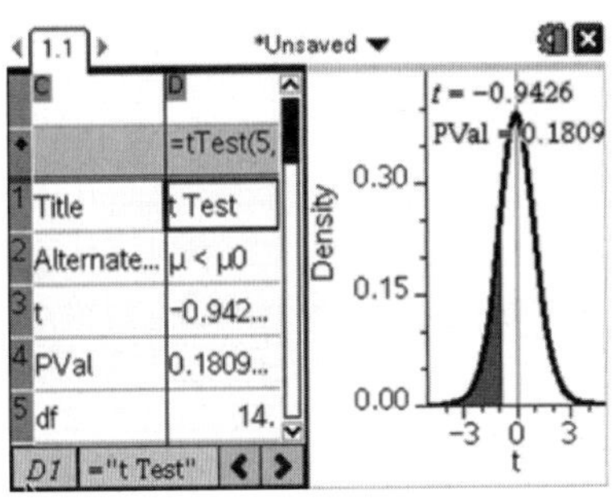

If you are given summary statistics instead of the original data, you would select the "Stats" option in the drop-down box instead of "Data" when the first dialogue box pops up.

CHAPTER 10 TI-NSPIRE TECHNOLOGY CORNERS

Section 10.1 Confidence interval for a difference in proportions

The TI-Nspire can be used to construct a confidence interval for $p_1 - p_2$. We'll demonstrate using the example on page 609. Of $n_1 = 800$ teens surveyed, $X = 584$ said they used social-networking sites. Of $n_2 = 2253$ adults surveyed, $X = 1059$ said they engaged in social networking. To construct a confidence interval:

- Press (home) ((⌂) (A)) to insert a *Calculator Scratchpad*. Press (menu) → *Statistics* → *Confidence Intervals* → *2 Prop Z Interval*.
- A dialogue box will appear. Enter the values as shown in the box. (tab) to [OK] and press (enter).

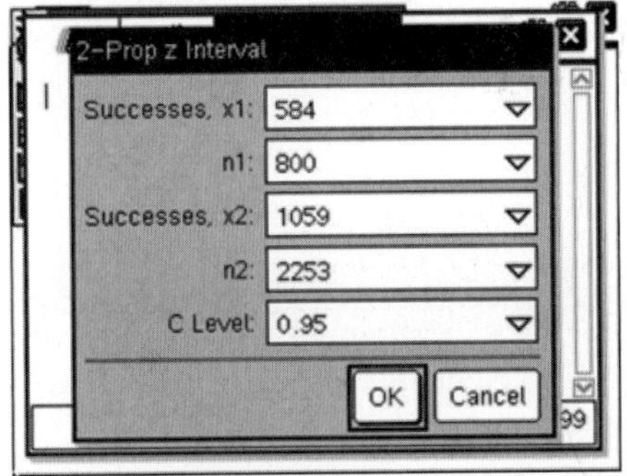

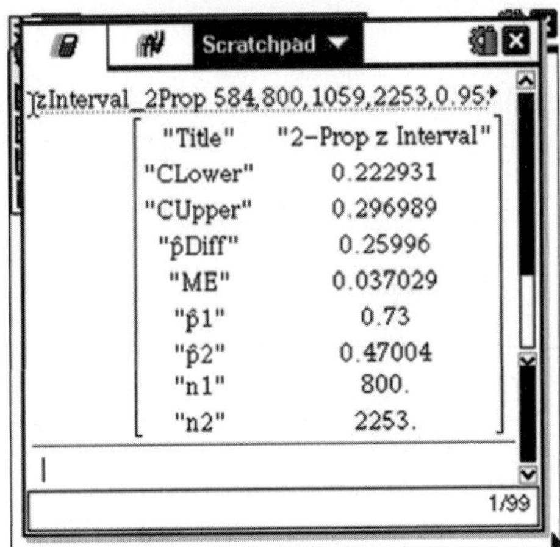

Section 10.1 Significance test for a difference in proportions

The TI-Nspire can be used to perform significance tests for comparing two proportions. Here, we use the data from the Helsinki Heart Study on page 615. To perform a test of $H_0: p_1 - p_2 = 0$:

- Press (or A) to insert a *Calculator Scratchpad.*
- Press menu → *Statistics* → *Stat Tests* → *2 Prop Z test.*
- A dialogue box will appear. Enter the values shown: $x_1 = 56$, $n_1 = 2051$, $x_2 = 84$, $n_2 = 2030$. Specify the alternative hypothesis $H_a: p_1 < p_2$ as shown. tab to OK and press enter.

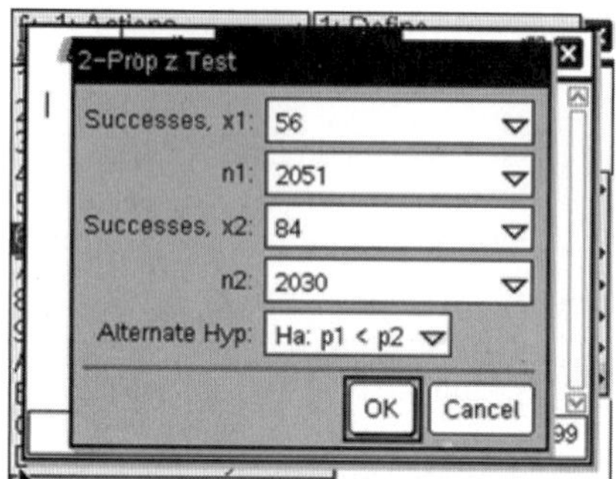

You are told that the z statistic is $z = -2.47$ and the P-value is 0.0068, as shown here. These results agree with those from the previous example. Do you see the combined proportion of heart attacks?

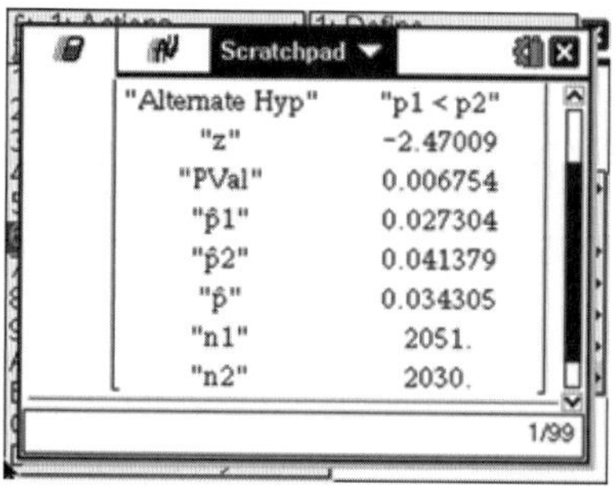

To display the P-value as a shaded area under the standard Normal curve:

- Press on, select the *Lists & Spreadsheet* icon.
- Press menu → *Statistics*→ *Stat Tests* → *2 Prop Z test.*
- A dialogue box will appear. Enter the values as shown in the box. Check the box to *Shade P value*. tab to OK and press enter.

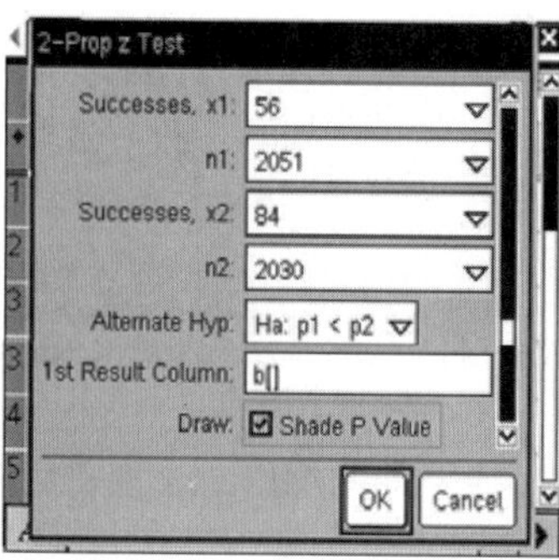

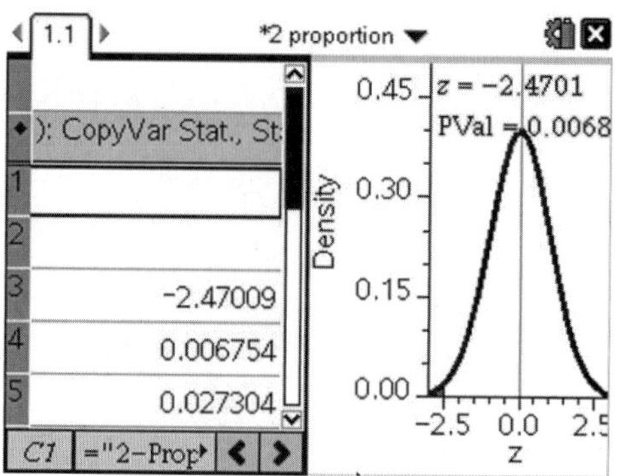

Section 10.2 Two-sample *t* intervals on the calculator

You can use the two-sample t interval command on the TI-Nspire to construct a confidence interval for the difference between two means. We'll show you the steps using the summary statistics from the pine trees example on page 635.

- Press (or A) to insert a *Calculator Scratchpad.*
- Press menu → *Statistics* → *Confidence Intervals* → *2 sample t interval.*
- In the first dialogue box press, select *Stats* in the drop-down box. tab to OK and press enter. Another dialogue box will appear.
  - Enter the summary statistics as shown:

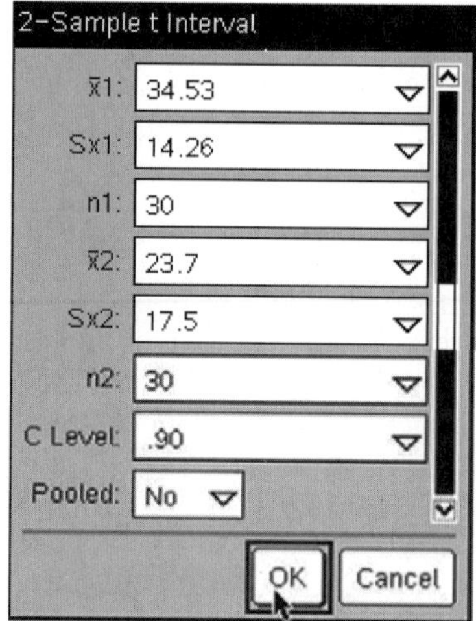

- Enter the confidence level: C-level: .90. For pooled: choose "No." (We'll discuss pooling later.) tab to OK and press enter.

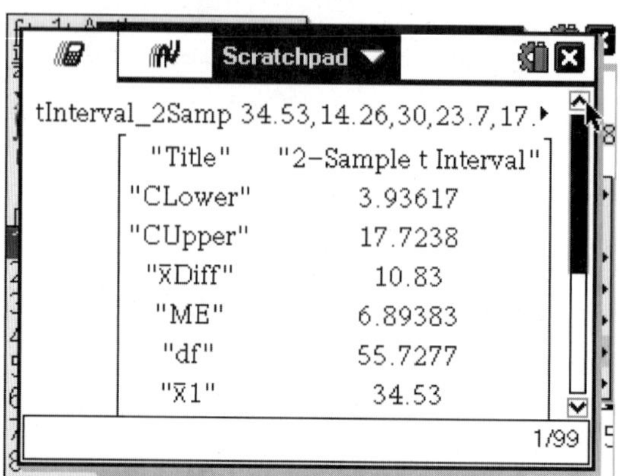

Section 10.2 Two-sample *t* tests with computer software and calculators

Technology gives smaller *P*-values for two-sample *t* tests than the conservative method does. That's because calculators and software use the more complicated formula on page 637 to obtain a larger number of degrees of freedom. Figure 10.13 (reproduced below) gives computer output from Fathom and Minitab for the two-sample *t* test from the calcium experiment. The *P*-values differ slightly because Fathom uses 15.59 degrees of freedom while Minitab truncates to df = 15.

(a)

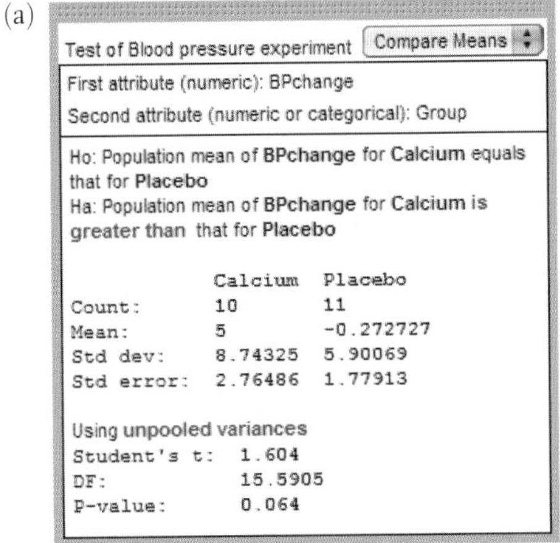

(b)

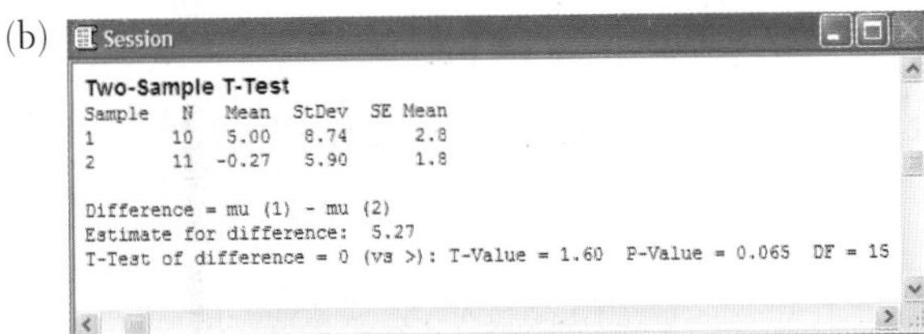

Let's look at what the calculator does. Start by entering the data into a *Lists & Spreadsheet* page. Name column A: **calcium** and enter the Group 1 data. Name column B: **placebo** and enter the Group 2 data. Then, to do the test:

- Press (menu) → *Statistics* → *Stats Tests* → *2 sample t Test*.
- In the first dialogue box select *Data* in the drop-down box. (tab) to [OK] and press (enter).
- In the next dialogue box, enter the values shown and (tab) to [OK]. Press (enter).

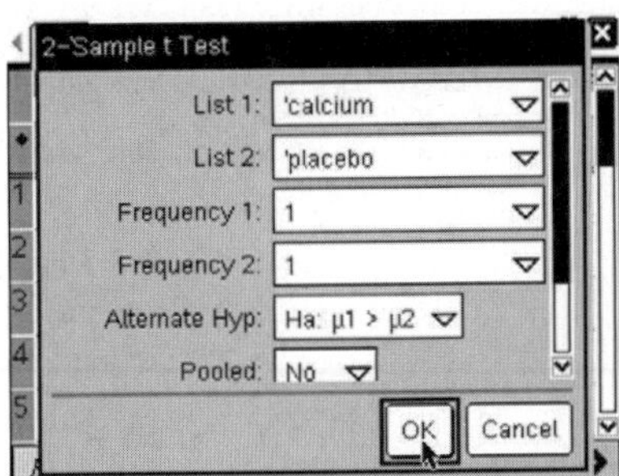

Note: To just "calculate," leave the *Shade P-value* option unchecked.

- The results should now appear in the spreadsheet.

	A calcium	B placebo	C	D
◆				=tTest_2S
1	7	-1	Title	2-Samp...
2	-4	12	Alternate...	μ1 > μ2
3	18	-1	t	1.60372
4	17	-3	PVal	0.06442
5	-3	3	df	15.5905

D1 ="2-Sample t Test"

The results tell us that the two-sample *t* test statistic is $t = 1.60372$, and the *P*-value is 0.0644. There is enough evidence against H_0 to reject it at the 10% significance level but not at the 5% or 1% significance levels.

If you check the *Shade P-value*, the appropriate *t* distribution will also be displayed, showing the same results and the shaded area corresponding to the *P*-value.

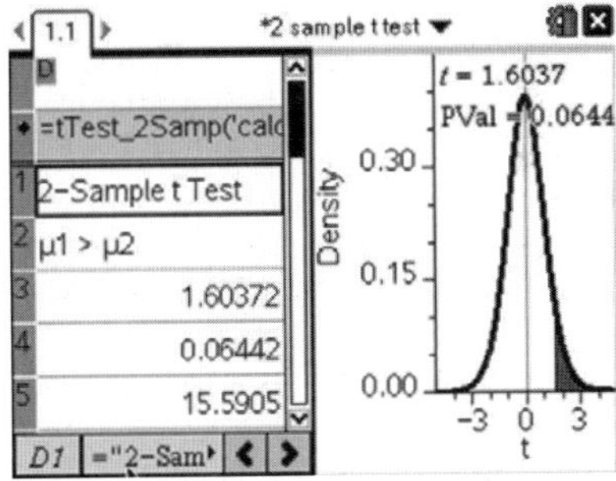

Note: The calculator's 90% confidence interval for the true difference $\mu_1 - \mu_2$ is $(-0.4767, 11.022)$. This is quite a bit narrower than the $(-0.753, 11.299)$ that we obtained earlier.

CHAPTER 11 TI-NSPIRE TECHNOLOGY CORNERS

Section 11.1 Finding *P*-values for chi-square tests on the calculator

To find the *P*-value in the M&M'S® example on page 683 with your calculator, use the χ^2cdf command. We ask for the area between $\chi^2 = 10.180$ and a very large number (we'll use 1000) under the chi-square density curve with 5 degrees of freedom. The command that does this is χ^2Cdf(10.180,1000,5).

- Press (scratchpad) and the *Calculator Scratchpad* should appear.
- Press (menu) → *Statistics* → *Distributions* → *χ^2Cdf*.
- A dialogue box will appear. Enter the values as shown in the box below. (tab) to [OK] and press (enter).

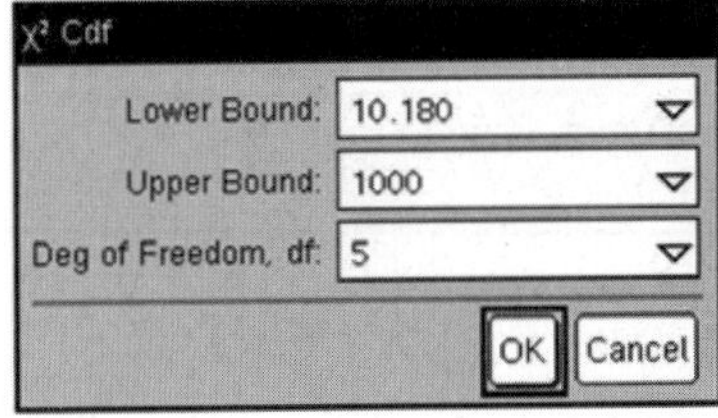

As the calculator screen shot shows, this method gives a more precise *P*-value than Table C.

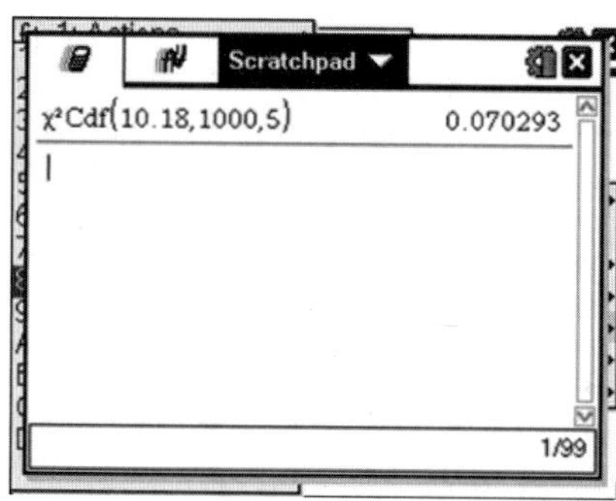

Section 11.1 Chi-square goodness-of-fit test on the calculator

You can use the TI-Nspire to perform the calculations for a chi-square goodness-of-fit test. We'll use Jerome's M&M'S Milk Chocolate Candies data from page 683 to illustrate the steps.

Color	Observed	Expected
Blue	9	14.40
Orange	8	12.00
Green	12	9.60
Yellow	15	8.40
Red	10	7.80
Brown	6	7.80

1. Enter the observed counts and expected counts in two separate columns in a *Lists & Spreadsheet* page. Name the columns **observed** and **expected.**
2. Perform a chi-square goodness-of-fit test.
   - Press (menu) → *Statistics* → *Stat Tests* → χ^2 *GOF*.
   - A dialogue box will appear. Enter the values as shown in the box below. (tab) to [OK] and press (enter).

If the *Shade P value* box remains unchecked, you'll get the test results within the spreadsheet containing the test statistic, *P*-value, and df. If you check the box for *Shade P value,* you'll get a picture of the appropriate chi-square distribution with the test statistic marked and shaded area corresponding to the *P*-value.

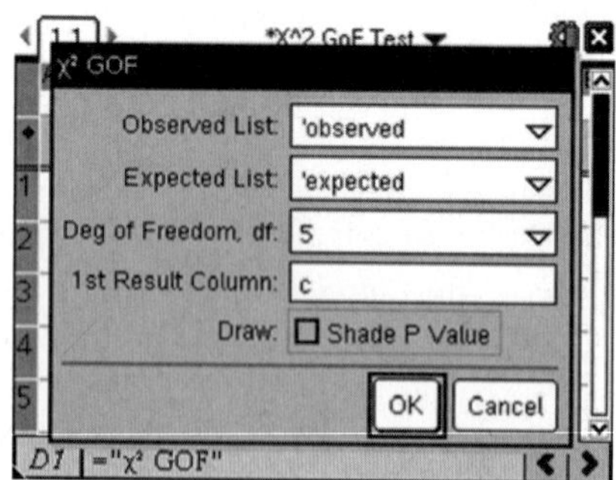

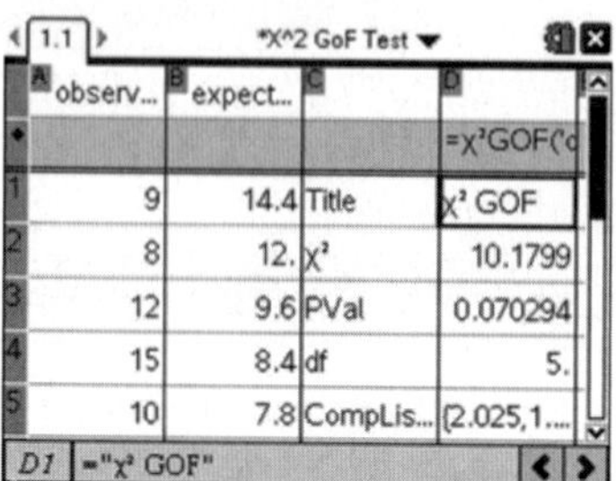

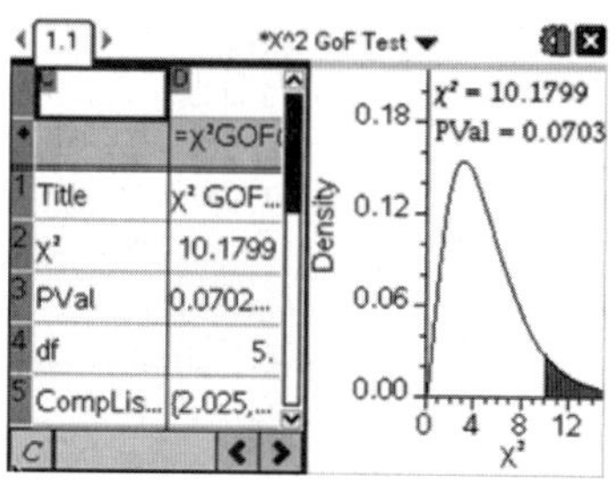

We'll discuss the *Comp List* results later.

Section 11.2 Chi-square tests for two-way tables on the calculator

You can use the TI-Nspire to perform calculations for a chi-square test for homogeneity. We'll use the data from the wine and music study on page 704 to illustrate the process.

1. Press (calc) ((home) (A)) to insert a *Calculator Scratchpad.*
2. Define a matrix by doing the following:
   - Name your matrix by typing **winemusic** (ctrl) [:=] (template). A box will appear with different math type options. Select [matrix] and enter "3" for *Number of rows* and "3" for *Number of columns.*

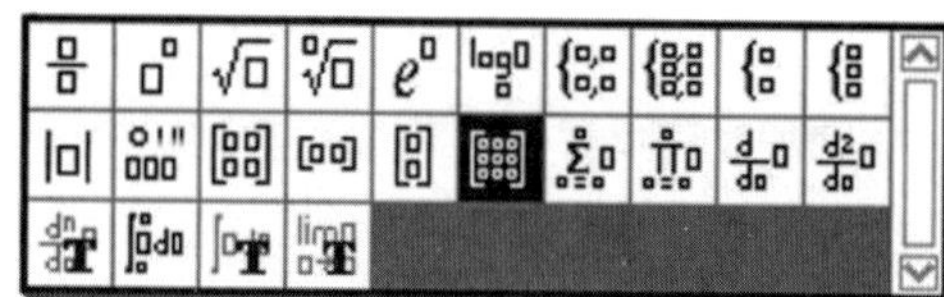

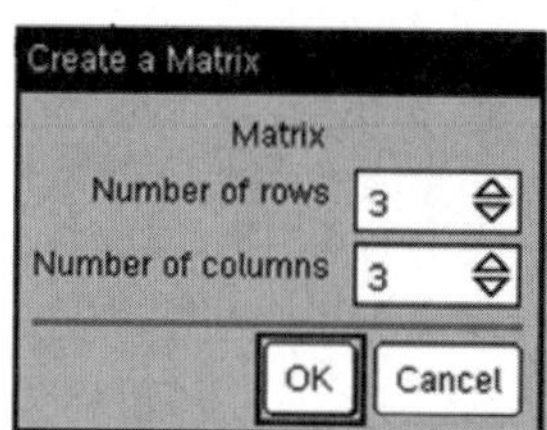

   - Type in the corresponding row data, pressing (tab) between entries. Press (enter) when finished.

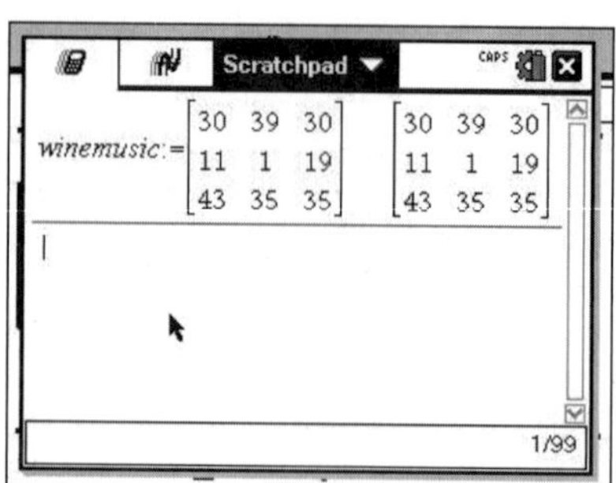

3. To perform the chi-square test, do the following:
   - Press (menu) → *Statistics* → *Stat tests* → χ^2 *2-Way Test …*
   - Specify the observed matrix, (tab) to [OK] and press (enter).

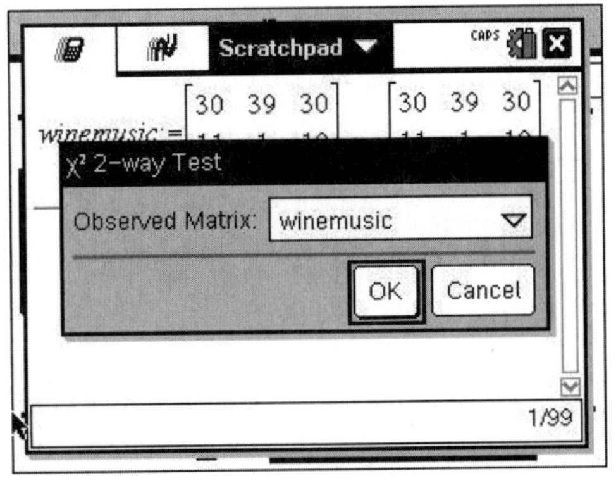

• The results will be displayed, and the expected matrix and component matrix will be calculated.

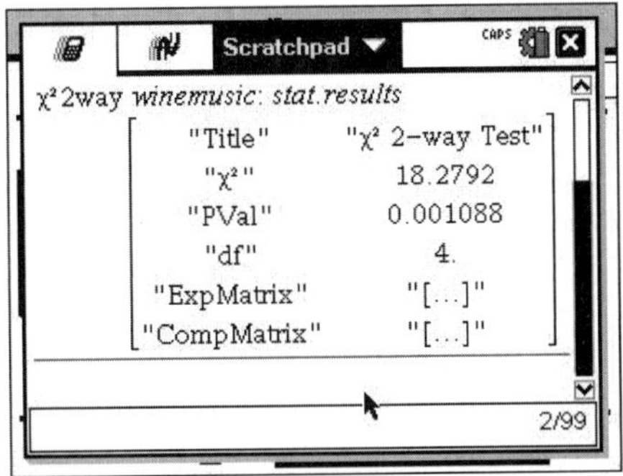

4. To see the expected counts and component matrix, press (var) and select **stat.expmatrix** for the expected matrix or **stat.compmatrix** for the component matrix.

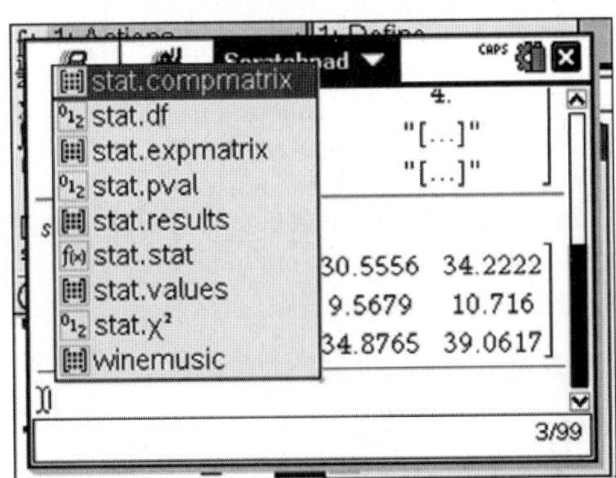

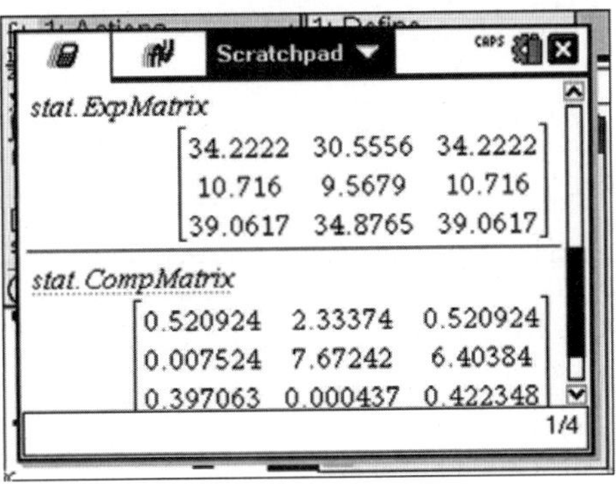

CHAPTER 12 TI-NSPIRE TECHNOLOGY CORNERS

Section 12.1 Regression inference on the calculator

Let's use the data from the nonexercise activity (NEA) and fat gain study on page 749 to illustrate significance tests and confidence intervals on the TI-Nspire.

Here are the data:

NEA change (cal):	−94	−57	−29	135	143	151	245	355
Fat gain (kg):	4.2	3.0	3.7	2.7	3.2	3.6	2.4	1.3

NEA change (cal):	392	473	486	535	571	580	620	690
Fat gain (kg):	3.8	1.7	1.6	2.2	1.0	0.4	2.3	1.1

1. Insert a *Lists & Spreadsheet* page, and name column A: **nea** and column B: **fatgain.** Type the corresponding values into each column.
2. To do a significance test:
   • Press (menu) → *Statistics* → *Stat Tests* → *Linear Reg t test.*
   • Select **nea** for the *X List* and **fatgain** for the *Y List*. Enter the rest of the values as shown. (tab) to [OK] and press (enter).

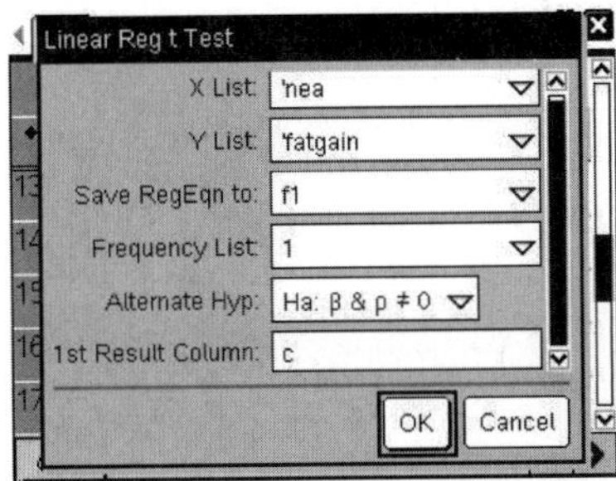

The linear regression *t* test results take three screens to present.

	A nea	B fatgain	C	D
◆				=LinRegtT
1	-94	4.2	Title	Linear Re..
2	-57	3.	Alternate...	β & ρ ≠ 0...
3	-29	3.7	RegEqn	a+b*x
4	135	2.7	t	-4.64182
5	143	3.2	PVal	0.000381
6	3.6	df	14.	
7	2.4	a	3.50512	
8	1.3	b	-0.0034...	
9	3.8	s	0.739853	
10	1.7	SESlope	0.000741	
11	1.6	r²	0.606149	
12	2.2	r	-0.7785...	
13	1	Resid	{0.37137...	
14	0.4			
15	2.3			

D =LinRegtTest('nea,'fatgain,1,0): Co

Compare these results with the Minitab regression output on the next page.

Regression Analysis: Fat gain versus NEA change

```
Predictor      Coef       SE Coef      T      P
Constant     3.5051      0.3036     11.54 0.000
NEA change -0.0034415 0.0007414 -4.64 0.000
S = 0.739853 R-Sq = 60.6% R-Sq(adj) = 57.8%
```

3. To construct a confidence interval:
   - Insert a *Calculator* page: ctrl I → *Add Calculator Page*.
   - Press menu → *Statistics* → *Confidence Intervals* → *Linear Reg t intervals ...*
   - In the first dialogue box, select *Slope*. tab to OK and press enter.

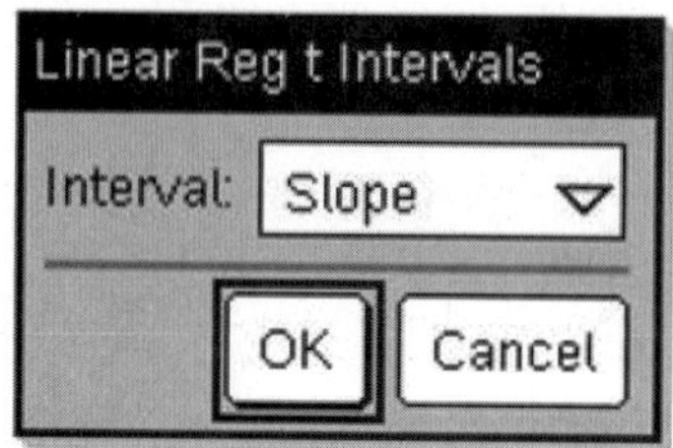

   - In the next dialogue box, select **nea** for the *X List* and **fatgain** for the *Y List* and enter the rest of the values as shown. tab to OK and press enter.

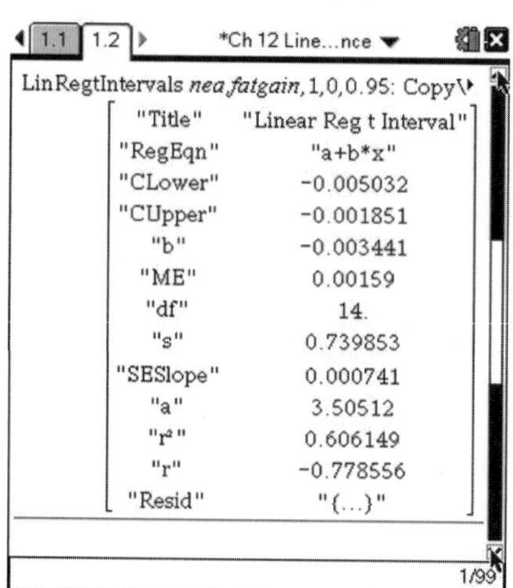

Section 12.2 Transforming to achieve linearity on the calculator

We'll use the planet data on page 778 to illustrate a general strategy for performing transformations with logarithms on the TI-Nspire. A similar approach could be used for transforming data with powers and roots.

1. Insert a *Lists & Spreadsheet* page, and name column A: **distance** and column B: **period.** Type the corresponding values into each column.
2. Make a scatterplot of y versus x and confirm that there is a curved pattern.
   - Insert a *Data & Statistics* page. Press ctrl I and select *Add Data & Statistics.*
   - Press tab and select **distance** for the horizontal axis. Press tab again and select **period** for the vertical axis.

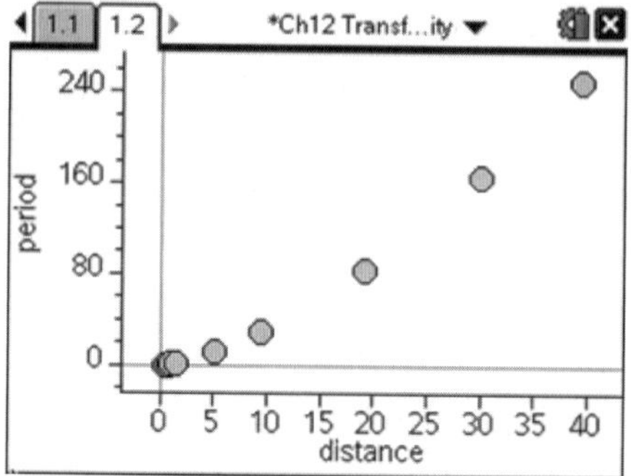

3. To "straighten" the curve (that is, determine the relationship), we can use different models of the explanatory-response data to see which one provides a linear relationship.
   - Press ctrl ◀ to return your spreadsheet. Name column c **lndistance** and column d: **lnperiod.**
   - In the formula cell for **lndistance**, press enter and enter *ln(distance)* to take the natural log of the distance values.
   - Repeat the above step for **lnperiod** using the **period** data.
4. To see if an exponential model fits the data:
   - Insert another *Data & Statistics* page.
   - Put **distance** on the horizontal axis and **lnperiod** on the vertical axis. If the relationship looks linear, then the exponential model is appropriate.

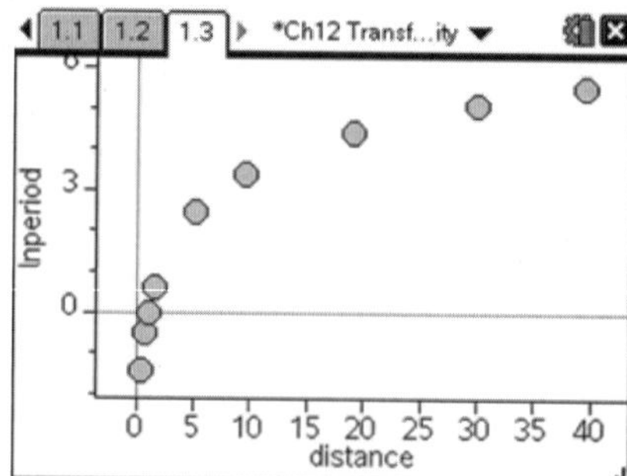

5. To see if a power model fits the data:
   - Using the same *Data & Statistics* page, change the horizontal axis to **lndistance.**

- If this relationship looks linear, than a power model is appropriate.

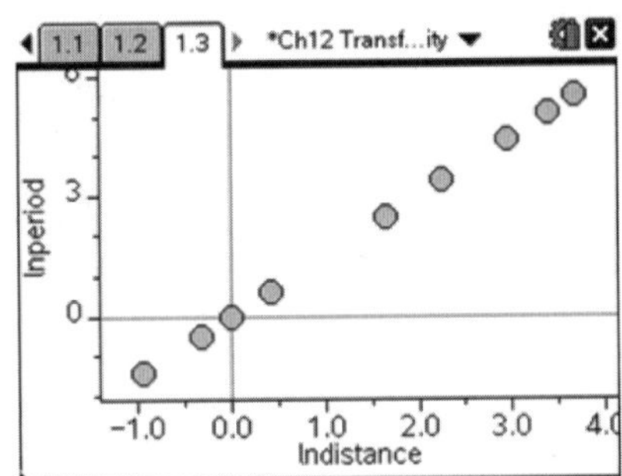

6. If a linear pattern is present, calculate the equation of the least-squares regression line:
   - In the spreadsheet, press (menu) → *Statistics* → *Stat Calculations* → *Linear Regression*$(a + bx)$.
   - In the dialogue box, select **lndistance** for Xlist, **lnperiod** for Ylist, and enter the rest of the values as shown. (tab) to [OK] and press (enter).

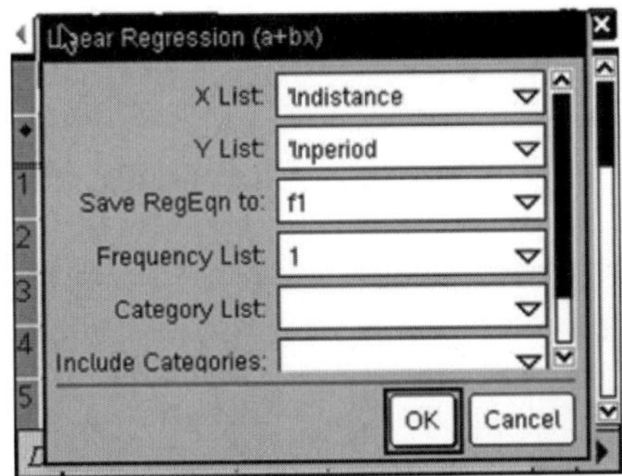

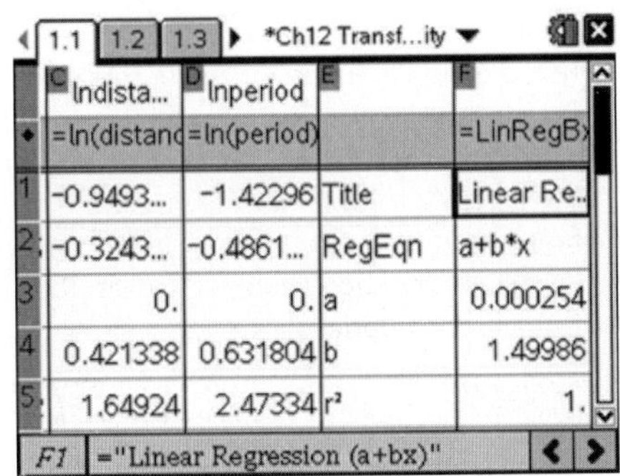

	C lndista...	D lnperiod	E	F
◆	=ln(distanc	=ln(period)		=LinRegB
1	−0.9493...	−1.42296	Title	Linear Re...
2	−0.3243...	−0.4861...	RegEqn	a+b*x
3	0.	0.	a	0.000254
4	0.421338	0.631804	b	1.49986
5	1.64924	2.47334	r²	1.

F1 ="Linear Regression (a+bx)"

7. Construct a residual plot to look for any departures from the linear pattern:
   - Insert another *Data & Statistics* page.
   - For the horizontal axis select **lndistance**. For Ylist, use the **stat.resid** list stored in the calculator.

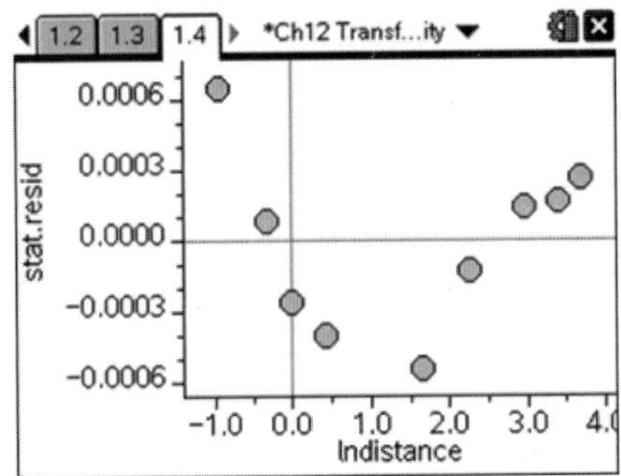

8. To make a prediction for a specific value of the explanatory variable $x = k$, modify the regression equation in **f1** by changing x to $\log(x)$ or $\ln(x)$, if appropriate. Then use **f1**(k) to obtain the predicted value of $\log y$ or $\ln y$. To get the predicted value of y, use 10^Ans or e^Ans to undo the logarithm transformation. Here's our prediction of the period of revolution for Eris, which is at a distance of 102.15 AU from the sun:

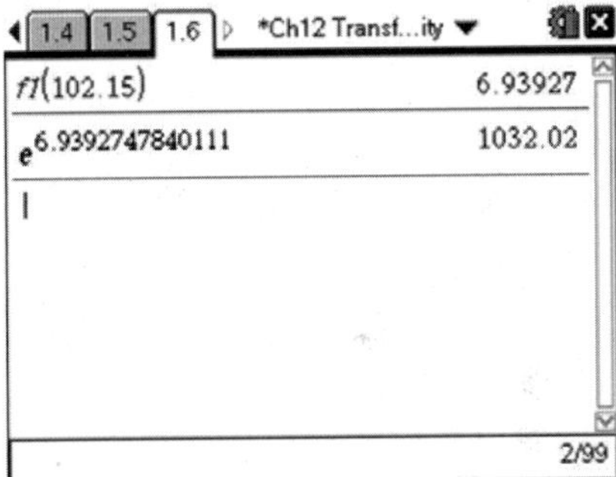

Formulas for AP Statistics Exam

I. Descriptive Statistics

$$\bar{x} = \frac{\sum x_i}{n}$$

$$s_x = \sqrt{\frac{1}{n-1}\sum(x_i - \bar{x})^2}$$

$$s_p = \sqrt{\frac{(n_1-1)s_1^2 + (n_2-1)s_2^2}{(n_1-1)+(n_2-1)}}$$

$$\hat{y} = b_0 + b_1 x$$

$$b_1 = \frac{\sum(x_i - \bar{x})(y_i - \bar{y})}{\sum(x_i - \bar{x})^2}$$

$$b_0 = \bar{y} - b_1\bar{x}$$

$$r = \frac{1}{n-1}\sum\left(\frac{x_i - \bar{x}}{s_x}\right)\left(\frac{y_i - \bar{y}}{s_y}\right)$$

$$b_1 = r\frac{s_y}{s_x}$$

$$s_{b_1} = \frac{\sqrt{\frac{\sum(y_i - \hat{y}_i)^2}{n-2}}}{\sqrt{\sum(x_i - \bar{x})^2}}$$

II. Probability

$$P(A \cup B) = P(A) + P(B) - P(A \cap B)$$

$$P(A|B) = \frac{P(A \cap B)}{P(B)}$$

$$E(X) = \mu_x = \sum x_i p_i$$

$$Var(X) = \sigma_x^2 = \sum (x_i - \mu_x)^2 p_i$$

If X has a binomial distribution with parameters n and p, then:

$$P(X = k) = \binom{n}{k} p^k (1-p)^{n-k}$$

$$\mu_x = np$$

$$\sigma_x = \sqrt{np(1-p)}$$

$$\mu_{\hat{p}} = p$$

$$\sigma_{\hat{p}} = \sqrt{\frac{p(1-p)}{n}}$$

If $\bar{x}$ is the mean of a random sample of size n from an infinite population with mean μ and standard deviation σ, then:

$$\mu_{\bar{x}} = \mu$$

$$\sigma_{\bar{x}} = \frac{\sigma}{\sqrt{n}}$$

III. Inferential Statistics

Standardized test statistic: $\dfrac{\text{statistic} - \text{parameter}}{\text{standard deviation of statistic}}$

Confidence interval: statistic $\pm$ (critical value) $\cdot$ (std. deviation of statistic)

Single-Sample

Statistic	Standard Deviation of Statistic
Sample Mean	$\frac{\sigma}{\sqrt{n}}$
Sample Proportion	$\sqrt{\frac{p(1-p)}{n}}$

Two-Sample

Statistic	Standard Deviation of Statistic
Difference of sample means	$\sqrt{\frac{\sigma_1^2}{n_1} + \frac{\sigma_2^2}{n_2}}$ Special case when $\sigma_1 = \sigma_2$ $\sigma\sqrt{\frac{1}{n_1} + \frac{1}{n_2}}$
Difference of sample proportions	$\sqrt{\frac{p_1(1-p_1)}{n_1} + \frac{p_2(1-p_2)}{n_2}}$ Special case when $p_1 = p_2$ $\sqrt{p(1-p)}\sqrt{\frac{1}{n_1} + \frac{1}{n_2}}$

Chi-square test statistic $= \sum \dfrac{(\text{observed} - \text{expected})^2}{\text{expected}}$

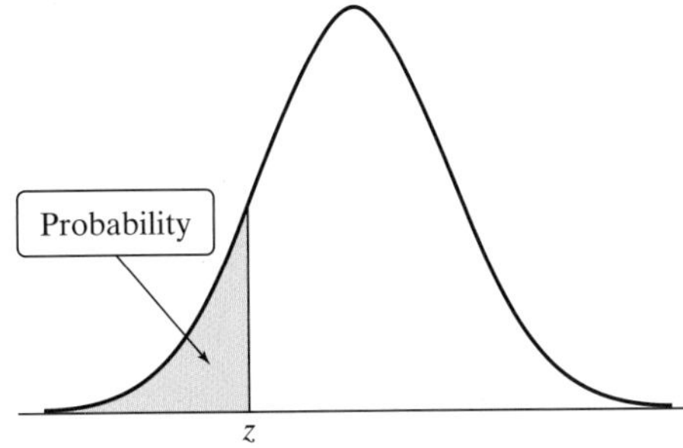

Table entry for z is the area under the standard Normal curve to the left of z.

Table A Standard Normal probabilities

z	.00	.01	.02	.03	.04	.05	.06	.07	.08	.09
−3.4	.0003	.0003	.0003	.0003	.0003	.0003	.0003	.0003	.0003	.0002
−3.3	.0005	.0005	.0005	.0004	.0004	.0004	.0004	.0004	.0004	.0003
−3.2	.0007	.0007	.0006	.0006	.0006	.0006	.0006	.0005	.0005	.0005
−3.1	.0010	.0009	.0009	.0009	.0008	.0008	.0008	.0008	.0007	.0007
−3.0	.0013	.0013	.0013	.0012	.0012	.0011	.0011	.0011	.0010	.0010
−2.9	.0019	.0018	.0018	.0017	.0016	.0016	.0015	.0015	.0014	.0014
−2.8	.0026	.0025	.0024	.0023	.0023	.0022	.0021	.0021	.0020	.0019
−2.7	.0035	.0034	.0033	.0032	.0031	.0030	.0029	.0028	.0027	.0026
−2.6	.0047	.0045	.0044	.0043	.0041	.0040	.0039	.0038	.0037	.0036
−2.5	.0062	.0060	.0059	.0057	.0055	.0054	.0052	.0051	.0049	.0048
−2.4	.0082	.0080	.0078	.0075	.0073	.0071	.0069	.0068	.0066	.0064
−2.3	.0107	.0104	.0102	.0099	.0096	.0094	.0091	.0089	.0087	.0084
−2.2	.0139	.0136	.0132	.0129	.0125	.0122	.0119	.0116	.0113	.0110
−2.1	.0179	.0174	.0170	.0166	.0162	.0158	.0154	.0150	.0146	.0143
−2.0	.0228	.0222	.0217	.0212	.0207	.0202	.0197	.0192	.0188	.0183
−1.9	.0287	.0281	.0274	.0268	.0262	.0256	.0250	.0244	.0239	.0233
−1.8	.0359	.0351	.0344	.0336	.0329	.0322	.0314	.0307	.0301	.0294
−1.7	.0446	.0436	.0427	.0418	.0409	.0401	.0392	.0384	.0375	.0367
−1.6	.0548	.0537	.0526	.0516	.0505	.0495	.0485	.0475	.0465	.0455
−1.5	.0668	.0655	.0643	.0630	.0618	.0606	.0594	.0582	.0571	.0559
−1.4	.0808	.0793	.0778	.0764	.0749	.0735	.0721	.0708	.0694	.0681
−1.3	.0968	.0951	.0934	.0918	.0901	.0885	.0869	.0853	.0838	.0823
−1.2	.1151	.1131	.1112	.1093	.1075	.1056	.1038	.1020	.1003	.0985
−1.1	.1357	.1335	.1314	.1292	.1271	.1251	.1230	.1210	.1190	.1170
−1.0	.1587	.1562	.1539	.1515	.1492	.1469	.1446	.1423	.1401	.1379
−0.9	.1841	.1814	.1788	.1762	.1736	.1711	.1685	.1660	.1635	.1611
−0.8	.2119	.2090	.2061	.2033	.2005	.1977	.1949	.1922	.1894	.1867
−0.7	.2420	.2389	.2358	.2327	.2296	.2266	.2236	.2206	.2177	.2148
−0.6	.2743	.2709	.2676	.2643	.2611	.2578	.2546	.2514	.2483	.2451
−0.5	.3085	.3050	.3015	.2981	.2946	.2912	.2877	.2843	.2810	.2776
−0.4	.3446	.3409	.3372	.3336	.3300	.3264	.3228	.3192	.3156	.3121
−0.3	.3821	.3783	.3745	.3707	.3669	.3632	.3594	.3557	.3520	.3483
−0.2	.4207	.4168	.4129	.4090	.4052	.4013	.3974	.3936	.3897	.3859
−0.1	.4602	.4562	.4522	.4483	.4443	.4404	.4364	.4325	.4286	.4247
−0.0	.5000	.4960	.4920	.4880	.4840	.4801	.4761	.4721	.4681	.4641

(Continued)

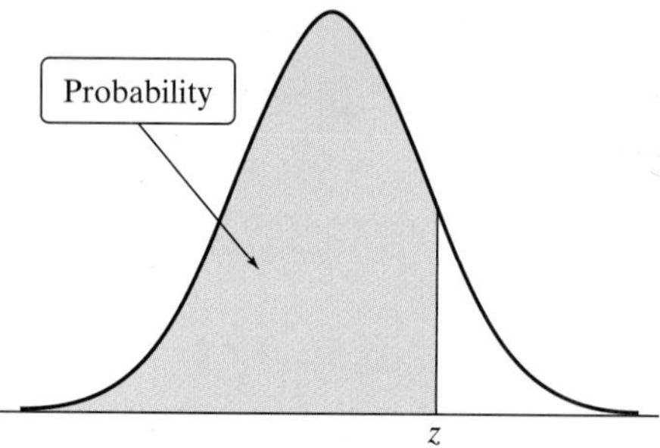

Table entry for z is the area under the standard Normal curve to the left of z.

Table A Standard Normal probabilities (continued)

z	.00	.01	.02	.03	.04	.05	.06	.07	.08	.09
0.0	.5000	.5040	.5080	.5120	.5160	.5199	.5239	.5279	.5319	.5359
0.1	.5398	.5438	.5478	.5517	.5557	.5596	.5636	.5675	.5714	.5753
0.2	.5793	.5832	.5871	.5910	.5948	.5987	.6026	.6064	.6103	.6141
0.3	.6179	.6217	.6255	.6293	.6331	.6368	.6406	.6443	.6480	.6517
0.4	.6554	.6591	.6628	.6664	.6700	.6736	.6772	.6808	.6844	.6879
0.5	.6915	.6950	.6985	.7019	.7054	.7088	.7123	.7157	.7190	.7224
0.6	.7257	.7291	.7324	.7357	.7389	.7422	.7454	.7486	.7517	.7549
0.7	.7580	.7611	.7642	.7673	.7704	.7734	.7764	.7794	.7823	.7852
0.8	.7881	.7910	.7939	.7967	.7995	.8023	.8051	.8078	.8106	.8133
0.9	.8159	.8186	.8212	.8238	.8264	.8289	.8315	.8340	.8365	.8389
1.0	.8413	.8438	.8461	.8485	.8508	.8531	.8554	.8577	.8599	.8621
1.1	.8643	.8665	.8686	.8708	.8729	.8749	.8770	.8790	.8810	.8830
1.2	.8849	.8869	.8888	.8907	.8925	.8944	.8962	.8980	.8997	.9015
1.3	.9032	.9049	.9066	.9082	.9099	.9115	.9131	.9147	.9162	.9177
1.4	.9192	.9207	.9222	.9236	.9251	.9265	.9279	.9292	.9306	.9319
1.5	.9332	.9345	.9357	.9370	.9382	.9394	.9406	.9418	.9429	.9441
1.6	.9452	.9463	.9474	.9484	.9495	.9505	.9515	.9525	.9535	.9545
1.7	.9554	.9564	.9573	.9582	.9591	.9599	.9608	.9616	.9625	.9633
1.8	.9641	.9649	.9656	.9664	.9671	.9678	.9686	.9693	.9699	.9706
1.9	.9713	.9719	.9726	.9732	.9738	.9744	.9750	.9756	.9761	.9767
2.0	.9772	.9778	.9783	.9788	.9793	.9798	.9803	.9808	.9812	.9817
2.1	.9821	.9826	.9830	.9834	.9838	.9842	.9846	.9850	.9854	.9857
2.2	.9861	.9864	.9868	.9871	.9875	.9878	.9881	.9884	.9887	.9890
2.3	.9893	.9896	.9898	.9901	.9904	.9906	.9909	.9911	.9913	.9916
2.4	.9918	.9920	.9922	.9925	.9927	.9929	.9931	.9932	.9934	.9936
2.5	.9938	.9940	.9941	.9943	.9945	.9946	.9948	.9949	.9951	.9952
2.6	.9953	.9955	.9956	.9957	.9959	.9960	.9961	.9962	.9963	.9964
2.7	.9965	.9966	.9967	.9968	.9969	.9970	.9971	.9972	.9973	.9974
2.8	.9974	.9975	.9976	.9977	.9977	.9978	.9979	.9979	.9980	.9981
2.9	.9981	.9982	.9982	.9983	.9984	.9984	.9985	.9985	.9986	.9986
3.0	.9987	.9987	.9987	.9988	.9988	.9989	.9989	.9989	.9990	.9990
3.1	.9990	.9991	.9991	.9991	.9992	.9992	.9992	.9992	.9993	.9993
3.2	.9993	.9993	.9994	.9994	.9994	.9994	.9994	.9995	.9995	.9995
3.3	.9995	.9995	.9995	.9996	.9996	.9996	.9996	.9996	.9996	.9997
3.4	.9997	.9997	.9997	.9997	.9997	.9997	.9997	.9997	.9997	.9998

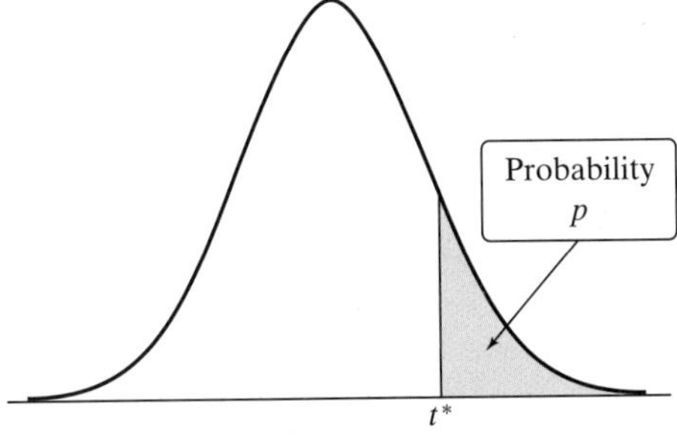

Table entry for p and C is the point t^* with probability p lying to its right and probability C lying between $-t^*$ and t^*.

Table B *t* distribution critical values

	Upper tail probability *p*											
df	**.25**	**.20**	**.15**	**.10**	**.05**	**.025**	**.02**	**.01**	**.005**	**.0025**	**.001**	**.0005**
1	1.000	1.376	1.963	3.078	6.314	12.71	15.89	31.82	63.66	127.3	318.3	636.6
2	0.816	1.061	1.386	1.886	2.920	4.303	4.849	6.965	9.925	14.09	22.33	31.60
3	0.765	0.978	1.250	1.638	2.353	3.182	3.482	4.541	5.841	7.453	10.21	12.92
4	0.741	0.941	1.190	1.533	2.132	2.776	2.999	3.747	4.604	5.598	7.173	8.610
5	0.727	0.920	1.156	1.476	2.015	2.571	2.757	3.365	4.032	4.773	5.893	6.869
6	0.718	0.906	1.134	1.440	1.943	2.447	2.612	3.143	3.707	4.317	5.208	5.959
7	0.711	0.896	1.119	1.415	1.895	2.365	2.517	2.998	3.499	4.029	4.785	5.408
8	0.706	0.889	1.108	1.397	1.860	2.306	2.449	2.896	3.355	3.833	4.501	5.041
9	0.703	0.883	1.100	1.383	1.833	2.262	2.398	2.821	3.250	3.690	4.297	4.781
10	0.700	0.879	1.093	1.372	1.812	2.228	2.359	2.764	3.169	3.581	4.144	4.587
11	0.697	0.876	1.088	1.363	1.796	2.201	2.328	2.718	3.106	3.497	4.025	4.437
12	0.695	0.873	1.083	1.356	1.782	2.179	2.303	2.681	3.055	3.428	3.930	4.318
13	0.694	0.870	1.079	1.350	1.771	2.160	2.282	2.650	3.012	3.372	3.852	4.221
14	0.692	0.868	1.076	1.345	1.761	2.145	2.264	2.624	2.977	3.326	3.787	4.140
15	0.691	0.866	1.074	1.341	1.753	2.131	2.249	2.602	2.947	3.286	3.733	4.073
16	0.690	0.865	1.071	1.337	1.746	2.120	2.235	2.583	2.921	3.252	3.686	4.015
17	0.689	0.863	1.069	1.333	1.740	2.110	2.224	2.567	2.898	3.222	3.646	3.965
18	0.688	0.862	1.067	1.330	1.734	2.101	2.214	2.552	2.878	3.197	3.611	3.922
19	0.688	0.861	1.066	1.328	1.729	2.093	2.205	2.539	2.861	3.174	3.579	3.883
20	0.687	0.860	1.064	1.325	1.725	2.086	2.197	2.528	2.845	3.153	3.552	3.850
21	0.686	0.859	1.063	1.323	1.721	2.080	2.189	2.518	2.831	3.135	3.527	3.819
22	0.686	0.858	1.061	1.321	1.717	2.074	2.183	2.508	2.819	3.119	3.505	3.792
23	0.685	0.858	1.060	1.319	1.714	2.069	2.177	2.500	2.807	3.104	3.485	3.768
24	0.685	0.857	1.059	1.318	1.711	2.064	2.172	2.492	2.797	3.091	3.467	3.745
25	0.684	0.856	1.058	1.316	1.708	2.060	2.167	2.485	2.787	3.078	3.450	3.725
26	0.684	0.856	1.058	1.315	1.706	2.056	2.162	2.479	2.779	3.067	3.435	3.707
27	0.684	0.855	1.057	1.314	1.703	2.052	2.158	2.473	2.771	3.057	3.421	3.690
28	0.683	0.855	1.056	1.313	1.701	2.048	2.154	2.467	2.763	3.047	3.408	3.674
29	0.683	0.854	1.055	1.311	1.699	2.045	2.150	2.462	2.756	3.038	3.396	3.659
30	0.683	0.854	1.055	1.310	1.697	2.042	2.147	2.457	2.750	3.030	3.385	3.646
40	0.681	0.851	1.050	1.303	1.684	2.021	2.123	2.423	2.704	2.971	3.307	3.551
50	0.679	0.849	1.047	1.299	1.676	2.009	2.109	2.403	2.678	2.937	3.261	3.496
60	0.679	0.848	1.045	1.296	1.671	2.000	2.099	2.390	2.660	2.915	3.232	3.460
80	0.678	0.846	1.043	1.292	1.664	1.990	2.088	2.374	2.639	2.887	3.195	3.416
100	0.677	0.845	1.042	1.290	1.660	1.984	2.081	2.364	2.626	2.871	3.174	3.390
1000	0.675	0.842	1.037	1.282	1.646	1.962	2.056	2.330	2.581	2.813	3.098	3.300
z^*	0.674	0.841	1.036	1.282	1.645	1.960	2.054	2.326	2.576	2.807	3.091	3.291
	50%	**60%**	**70%**	**80%**	**90%**	**95%**	**96%**	**98%**	**99%**	**99.5%**	**99.8%**	**99.9%**
	Confidence level *C*											

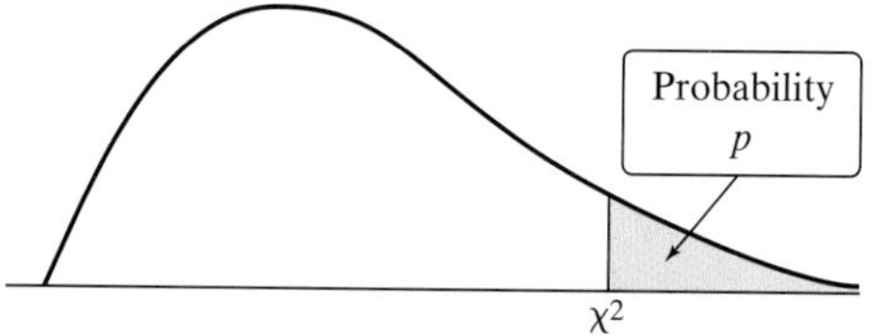

Table entry for p is the point χ^2 with probability p lying to its right.

Table C Chi-square distribution critical values

	p											
df	**.25**	**.20**	**.15**	**.10**	**.05**	**.025**	**.02**	**.01**	**.005**	**.0025**	**.001**	**.0005**
1	1.32	1.64	2.07	2.71	3.84	5.02	5.41	6.63	7.88	9.14	10.83	12.12
2	2.77	3.22	3.79	4.61	5.99	7.38	7.82	9.21	10.60	11.98	13.82	15.20
3	4.11	4.64	5.32	6.25	7.81	9.35	9.84	11.34	12.84	14.32	16.27	17.73
4	5.39	5.99	6.74	7.78	9.49	11.14	11.67	13.28	14.86	16.42	18.47	20.00
5	6.63	7.29	8.12	9.24	11.07	12.83	13.39	15.09	16.75	18.39	20.51	22.11
6	7.84	8.56	9.45	10.64	12.59	14.45	15.03	16.81	18.55	20.25	22.46	24.10
7	9.04	9.80	10.75	12.02	14.07	16.01	16.62	18.48	20.28	22.04	24.32	26.02
8	10.22	11.03	12.03	13.36	15.51	17.53	18.17	20.09	21.95	23.77	26.12	27.87
9	11.39	12.24	13.29	14.68	16.92	19.02	19.68	21.67	23.59	25.46	27.88	29.67
10	12.55	13.44	14.53	15.99	18.31	20.48	21.16	23.21	25.19	27.11	29.59	31.42
11	13.70	14.63	15.77	17.28	19.68	21.92	22.62	24.72	26.76	28.73	31.26	33.14
12	14.85	15.81	16.99	18.55	21.03	23.34	24.05	26.22	28.30	30.32	32.91	34.82
13	15.98	16.98	18.20	19.81	22.36	24.74	25.47	27.69	29.82	31.88	34.53	36.48
14	17.12	18.15	19.41	21.06	23.68	26.12	26.87	29.14	31.32	33.43	36.12	38.11
15	18.25	19.31	20.60	22.31	25.00	27.49	28.26	30.58	32.80	34.95	37.70	39.72
16	19.37	20.47	21.79	23.54	26.30	28.85	29.63	32.00	34.27	36.46	39.25	41.31
17	20.49	21.61	22.98	24.77	27.59	30.19	31.00	33.41	35.72	37.95	40.79	42.88
18	21.60	22.76	24.16	25.99	28.87	31.53	32.35	34.81	37.16	39.42	42.31	44.43
19	22.72	23.90	25.33	27.20	30.14	32.85	33.69	36.19	38.58	40.88	43.82	45.97
20	23.83	25.04	26.50	28.41	31.41	34.17	35.02	37.57	40.00	42.34	45.31	47.50
21	24.93	26.17	27.66	29.62	32.67	35.48	36.34	38.93	41.40	43.78	46.80	49.01
22	26.04	27.30	28.82	30.81	33.92	36.78	37.66	40.29	42.80	45.20	48.27	50.51
23	27.14	28.43	29.98	32.01	35.17	38.08	38.97	41.64	44.18	46.62	49.73	52.00
24	28.24	29.55	31.13	33.20	36.42	39.36	40.27	42.98	45.56	48.03	51.18	53.48
25	29.34	30.68	32.28	34.38	37.65	40.65	41.57	44.31	46.93	49.44	52.62	54.95
26	30.43	31.79	33.43	35.56	38.89	41.92	42.86	45.64	48.29	50.83	54.05	56.41
27	31.53	32.91	34.57	36.74	40.11	43.19	44.14	46.96	49.64	52.22	55.48	57.86
28	32.62	34.03	35.71	37.92	41.34	44.46	45.42	48.28	50.99	53.59	56.89	59.30
29	33.71	35.14	36.85	39.09	42.56	45.72	46.69	49.59	52.34	54.97	58.30	60.73
30	34.80	36.25	37.99	40.26	43.77	46.98	47.96	50.89	53.67	56.33	59.70	62.16
40	45.62	47.27	49.24	51.81	55.76	59.34	60.44	63.69	66.77	69.70	73.40	76.09
50	56.33	58.16	60.35	63.17	67.50	71.42	72.61	76.15	79.49	82.66	86.66	89.56
60	66.98	68.97	71.34	74.40	79.08	83.30	84.58	88.38	91.95	95.34	99.61	102.7
80	88.13	90.41	93.11	96.58	101.9	106.6	108.1	112.3	116.3	120.1	124.8	128.3
100	109.1	111.7	114.7	118.5	124.3	129.6	131.1	135.8	140.2	144.3	149.4	153.2

Table D Random digits

Line								
101	19223	95034	05756	28713	96409	12531	42544	82853
102	73676	47150	99400	01927	27754	42648	82425	36290
103	45467	71709	77558	00095	32863	29485	82226	90056
104	52711	38889	93074	60227	40011	85848	48767	52573
105	95592	94007	69971	91481	60779	53791	17297	59335
106	68417	35013	15529	72765	85089	57067	50211	47487
107	82739	57890	20807	47511	81676	55300	94383	14893
108	60940	72024	17868	24943	61790	90656	87964	18883
109	36009	19365	15412	39638	85453	46816	83485	41979
110	38448	48789	18338	24697	39364	42006	76688	08708
111	81486	69487	60513	09297	00412	71238	27649	39950
112	59636	88804	04634	71197	19352	73089	84898	45785
113	62568	70206	40325	03699	71080	22553	11486	11776
114	45149	32992	75730	66280	03819	56202	02938	70915
115	61041	77684	94322	24709	73698	14526	31893	32592
116	14459	26056	31424	80371	65103	62253	50490	61181
117	38167	98532	62183	70632	23417	26185	41448	75532
118	73190	32533	04470	29669	84407	90785	65956	86382
119	95857	07118	87664	92099	58806	66979	98624	84826
120	35476	55972	39421	65850	04266	35435	43742	11937
121	71487	09984	29077	14863	61683	47052	62224	51025
122	13873	81598	95052	90908	73592	75186	87136	95761
123	54580	81507	27102	56027	55892	33063	41842	81868
124	71035	09001	43367	49497	72719	96758	27611	91596
125	96746	12149	37823	71868	18442	35119	62103	39244
126	96927	19931	36809	74192	77567	88741	48409	41903
127	43909	99477	25330	64359	40085	16925	85117	36071
128	15689	14227	06565	14374	13352	49367	81982	87209
129	36759	58984	68288	22913	18638	54303	00795	08727
130	69051	64817	87174	09517	84534	06489	87201	97245
131	05007	16632	81194	14873	04197	85576	45195	96565
132	68732	55259	84292	08796	43165	93739	31685	97150
133	45740	41807	65561	33302	07051	93623	18132	09547
134	27816	78416	18329	21337	35213	37741	04312	68508
135	66925	55658	39100	78458	11206	19876	87151	31260
136	08421	44753	77377	28744	75592	08563	79140	92454
137	53645	66812	61421	47836	12609	15373	98481	14592
138	66831	68908	40772	21558	47781	33586	79177	06928
139	55588	99404	70708	41098	43563	56934	48394	51719
140	12975	13258	13048	45144	72321	81940	00360	02428
141	96767	35964	23822	96012	94591	65194	50842	53372
142	72829	50232	97892	63408	77919	44575	24870	04178
143	88565	42628	17797	49376	61762	16953	88604	12724
144	62964	88145	83083	69453	46109	59505	69680	00900
145	19687	12633	57857	95806	09931	02150	43163	58636
146	37609	59057	66967	83401	60705	02384	90597	93600
147	54973	86278	88737	74351	47500	84552	19909	67181
148	00694	05977	19664	65441	20903	62371	22725	53340
149	71546	05233	53946	68743	72460	27601	45403	88692
150	07511	88915	41267	16853	84569	79367	32337	03316

TECHNOLOGY CORNERS REFERENCE

Inference Summary

How to Organize a Statistical Problem: A Four-Step Process		
	Confidence intervals (CIs)	**Significance tests**
STATE:	What *parameter* do you want to estimate, and at what confidence level?	What *hypotheses* do you want to test, and at what significance level? Define any *parameters* you use.
PLAN:	Choose the appropriate inference *method.* Check *conditions.*	Choose the appropriate inference *method.* Check *conditions.*
DO:	If the conditions are met, perform *calculations.*	If the conditions are met, perform *calculations.* • Compute the **test statistic.** • Find the ***P*-value.**
CONCLUDE:	*Interpret* your interval in the context of the problem.	*Interpret* the result of your test in the context of the problem.

CI: statistic ± (critical value) · (standard deviation of statistic)

$$\text{Standardized test statistic} = \frac{\text{statistic} - \text{parameter}}{\text{standard deviation of statistic}}$$

Inference about	Number of samples (groups)	Estimate or test	Name of procedure (TI Calculator function) Formula	Conditions
Proportions	1	Estimate	One-sample z interval for p (1-PropZInt) $\hat{p} \pm z^* \sqrt{\frac{\hat{p}(1-\hat{p})}{n}}$	**Random** Data from a random sample or randomized experiment **Normal** At least 10 successes and failures; that is, $n\hat{p} \geq 10$ and $n(1-\hat{p}) \geq 10$ **Independent** observations; *10% condition* if sampling without replacement
		Test	One-sample z test for p (1-PropZTest) $z = \frac{\hat{p} - p_0}{\sqrt{\frac{p_0(1-p_0)}{n}}}$	**Random** Data from a random sample or randomized experiment **Normal** $np_0 \geq 10$ and $n(1-p_0) \geq 10$ **Independent** observations; *10% condition* if sampling without replacement
	2	Estimate	Two-sample z interval for $p_1 - p_2$ (2-PropZInt) $(\hat{p}_1 - \hat{p}_2) \pm z^* \sqrt{\frac{\hat{p}_1(1-\hat{p}_1)}{n_1} + \frac{\hat{p}_2(1-\hat{p}_2)}{n_2}}$	**Random** Data from random samples or randomized experiment **Normal** At least 10 successes and failures in both samples/groups; that is, $n_1\hat{p}_1 \geq 10, n_1(1-\hat{p}_1) \geq 10,$ $n_2\hat{p}_2 \geq 10, n_2(1-\hat{p}_2) \geq 10$ **Independent** observations and independent samples/groups; *10% condition* if sampling without replacement
		Test	Two-sample z test for $p_1 - p_2$ (2-PropZTest) $z = \frac{(\hat{p}_1 - \hat{p}_2) - (p_1 - p_2)}{\sqrt{\frac{\hat{p}_c(1-\hat{p}_c)}{n_1} + \frac{\hat{p}_c(1-\hat{p}_c)}{n_2}}}$ where $\hat{p}_c = \frac{\text{total successes}}{\text{total sample size}} = \frac{X_1 + X_2}{n_1 + n_2}$	**Random** Data from random samples or randomized experiment **Normal** At least 10 successes and failures in both samples/groups; that is, $n_1\hat{p}_1 \geq 10, n_1(1-\hat{p}_1) \geq 10,$ $n_2\hat{p}_2 \geq 10, n_2(1-\hat{p}_2) \geq 10$ **Independent** observations and independent samples/groups; *10% condition* if sampling without replacement